Principles of Tidal Sedimentology

Richard A. Davis, Jr. • Robert W. Dalrymple
Editors

Principles of Tidal Sedimentology

Springer

Editors
Richard A. Davis, Jr.
Harte Research Institute
Texas A&M University
Ocean Drive 6300
Corpus Christi, TX 78412
USA

Coastal Research Laboratory
Department of Geology
University of South Florida
Tampa, FL 33620
rdavis@usf.edu

Robert W. Dalrymple
Department of Geological Sciences and
Geological Engineering
Queen's University
Miller Hall
Kingston, ON K7L 3N6
Canada
dalrymple@geol.queensu.ca

ISBN 978-94-007-0122-9 e-ISBN 978-94-007-0123-6
DOI 10.1007/978-94-007-0123-6
Springer Dordrecht Heidelberg London New York

Library of Congress Control Number: 2011939475

Cover illustration: Fig. 5.13 (upper part) from this book.

Printed on acid-free paper

Springer is part of Springer Science+Business Media (www.springer.com)

Preface

Tides have fascinated humans for millennia. Their regularity and their apparent correlation with lunar behavior intrigued natural philosophers, even the Greeks, who live on an essentially tideless sea although there are strong tidal currents in localized constrictions. Apparently, they learned about tides from areas outside the Straits of Gibralter and from the Arabs who experienced significant tides in the Persian Gulf. From a practical perspective, tidal changes in water elevation and the currents associated with these changes were of great importance for shipping and military purposes. In areas such as the countries surrounding the southern North Sea, such considerations required accurate tidal predictions, which in turn drew the attention of some of the greatest astronomers and mathematicians.

Among the notable individuals who devoted at least part of their careers to the study of tides, and have contributed to our understanding of them are Galileo, Descartes, Bacon, Kepler, Euler, Laplace, and Lord Kelvin (Cartwight 1999). Indeed, many of the widely used mathematical techniques that we now take for granted were developed to help understand the behavior of the tides. More recently, interest in tides and storm surges has been fostered by the need to protect ever-increasing coastal population centers from catastrophic inundation, and by the desire to "reclaim" tidal flats for agricultural and industrial purposes. Foremost in this activity have been The Netherlands, Germany, and adjacent parts of Denmark.

Research on the nature of tidal deposits has been underway for about 50 years. Early studies on the Wadden Sea along the North Sea coast of The Netherlands and Germany were among the original landmark efforts in this area (e.g. van Straaten 1954; Postma 1961; Reineck 1963), and were followed closely by work in England (Evans 1965) and France (Bajard 1966). Such efforts were driven by the dual need to understand the coastal zone for the protection of population centers and to provide an actualistic analog for ancient sedimentary successions. In North America, Klein's work on the Bay of Fundy (Klein 1963) initiated detailed efforts in that part of the world. The early German work in the North Sea had a major biological and ichnological component, a topic that was pursued systematically at the Skidaway Institute of Oceanography in the southeastern United States (e.g. Frey and Howard 1969). Despite having some of the most widespread tidal flats in the world, work along the Chinese coast was relatively slow to develop, although there were notable early studies (e.g. Wang 1963). In the carbonate realm, pioneering studies were conducted on the tidal flats of Andros Island, the Bahamas (e.g. Shinn et al. 1969), and the Persian Gulf (Evans et al. 1969).

In spite of important work on the shallow-marine tidal deposits in the seas of northwestern Europe (e.g. Stride 1963), most of the early work on modern tidal deposits was devoted to study of intertidal environments, mainly because they were readily accessible. This fixation on the intertidal zone is perhaps nowhere more evident in the influential compilation of examples contained in the book *Tidal Deposits: A Casebook of Recent Examples and Fossil Counterparts* (Ginsburg 1975). Indeed, the upward-fining succession developed by the progradation of a tidal flat was among the very first facies models created. Application of these studies to the rock record was widespread in the carbonate literature, with numerous documented examples being published through the 1960s, 1970s and 1980s. By comparison, the extension of the work on the modern tidal deposits to ancient siliciclastic successions was slow. At least one impediment to the widespread application to the ancient was the notion put forward by Irwin (1965), and since largely disproven, at least for siliciclastic sediments, that the expansive epicontinental seas of the past were largely tideless, as a result of frictional damping of the tidal wave. An even greater impediment was the lack of definitive criteria for the recognition of tidal deposits, given that exposure indicators are much less easily preserved in siliciclastic tidal deposits than they are in carbonates. Thus, a milestone in the study of tidal deposits occurred in 1980 with the publication by Visser (1980) of "tidal bundles" in cross beds formed by subaqueous dunes, which provided the first documentation of a definitive indicator of tidal sedimentation, spawned the widespread recognition of ancient tidal deposits in an ever-growing number of localities.

Gradually, the focus of research on modern tidal environments has shifted away from tidal flats, toward a more comprehensive examination of tidal sedimentation in a wide range of settings, including even the deep ocean. Studies have tended to become more holistic in their treatment of entire depositional systems, rather than concentrating on only one part (e.g. tidal flats) of the whole. This more comprehensive approach is evident in many of the papers in this volume.

Because of the increasing attention given to tidal deposits it became important to organize a uniform nomenclature and approach to their study. As a consequence, Robert N. Ginsburg organized and hosted a conference of interested researchers in February of 1973. It included field experiences in both siliciclastic (Sapelo Island, Georgia, USA) and carbonate areas (Florida Keys, USA and the Bahamas), followed by presentations of research on "tidalites" (a term coined by George deVries Klein (1971)) by all in attendance. The next similar conference was held in The Netherlands in 1986, followed in regular succession by a series International Conferences on Tidal Sedimentology that has met in Calgary, Canada (1989), Wilhelmshaven, Germany (1992), Savannah, Georgia USA (1996), Seoul, Korea (2000), Copenhagen, Denmark (2004) and, most recently, in Qingdao, China (2008). The next meeting will be in Caen, France in 2012.

The meeting in 2008 in China was particularly stimulating with an attendance that surpassed any previous meeting. The expansion of interest in tidal deposits appears to be spurred by two factors: the need to understand coastal tidal environments in order to predict how these sensitive environments might respond to sea-level rise and climate change; and providing data and interpretations to help in understanding ancient depositional environments that were influenced by tides. Davis thought it was a good time to assemble a principles-type volume on the topic of tidal sedimentology given that no such synthesis exists, and because there has been so much new research on tidal environments and deposits over the last few years. Dalrymple agreed to be co-editor and the result of their efforts is this volume.

The purpose of this volume is to provide the first-ever, high-level overview of tidal sedimentology. Many of the chapters contain the first-ever synthesis of information on the particular topic! The approach is comprehensive with state-of-the-art reviews of the full spectrum of tidal depositional environments, from supratidal salt marshes, through the full range of coastal environments and continental shelves, to the deep sea. Examples from modern environments and ancient deposits are provided, and both siliciclastic and carbonate environments are discussed. The book is organized in the following four parts. (1) Chapters 1–4 provide overviews of the fundamentals of: the generation of tides, the nature of sediment transport by tidal currents, the criteria by which tidal deposits can be recognized, and the ichnology of tidal deposits. The later chapter represents the first time that the ichnological characteristics of tidal deposits have been reviewed systematically. (2) Chapters 5–14 review the characteristics of the full range of siliciclastic tidal environments, including both tide-dominated estuaries and deltas, as well as the various tidal components of barrier-lagoon systems. These chapters cover all aspects of the sedimentology of these environments, from the details of the physical processes operating in them, through the morphodynamics and facies, and the stratigraphic organization of the deposits. (3) Chapters 15–18 provide syntheses of particular times and places in earth history where tidal deposits are particularly notable. The chapter on the Precambrian reviews tidal sedimentation at a time when the Moon was significantly closer to the Earth and the tide-generating force should have been stronger. The reviews of the tidal deposits in the Illinois Basin (Carboniferous age), Western Interior Seaway (Cretaceous) and Spanish Pyrenean Basin (Eocene) provide unique insights into the large-scale (tectonic and relative sea level) controls on the spatial and temporal distribution of tidal sedimentation. (4) Chapters 19–21 discuss tidal sedimentation in modern and ancient carbonate environments.

Experts from throughout the world have been chosen to be the lead authors on each of the chapters. They and their co-authors build on their considerable personal experience to present insightful syntheses of the latest research in the particular topic. Each chapter has abundant illustrations, many of which are in color to enhance their effectiveness. References are extensive and include historically important ones as well as those on the leading edge of each topic.

Because of the uniquely broad coverage within each of the chapters, and in the volume as a whole, this book should be of value to a wide range of researchers. Workers who study modern sedimentary environments, and especially coastal settings, including environmental managers and coastal engineers, will find much about the dynamics of these environments that will assist them to develop protection strategies that are compatible with the natural behavior of these complex systems, including their response to potentially rising sea level. Geologists who study ancient sedimentary successions, whether for more academic or more applied reasons, will find a wealth of information about the behavior of tidal environments, ranging from the nature of the facies, through small-scale sedimentary successions, to the largest-scale sequence-stratigraphic control on tidal sedimentation.

The editors and authors gratefully acknowledge the financial support of numerous funding agencies that have provided support for their respective research activities. They also thank the people who have provided excellent and constructive reviews (see below). The editors appreciate the cooperation of Dr. Robert Doe and his staff at Springer Publishers.

Chapter Reviewers

Clark Alexander
Serge Berné
Sean Bingham
Ron Boyd
Margie Chan
Kyungsik Choi
Poppe de Boer
Robert Dott
Paul Enos
Jon French
Shu Gao
Murray Gingras
Liviu Giosan
Steven Greb
Gary Hampson
Steve Hasiotis
Christopher Kendall
George Klein
Erik Kvale
Tim Lawton
Don McNeil
Bruce Nocita
Nora Noffke
David Piper
Piret Plink-Bjorklund
Brian Pratt
Denise Reed
Joshiki Saito
Gene Shanmugam
Gene Shinn
Ronald Steel
John Suter
S. Temmerman
Bernadette Tessier
Ad van der Spek
Grant Wach
Ping Wang
Colin Woodruff
Paul Wright

References

Bajard J (1966) Figure et structures sédimentaires dans la partie orientale de la baie de Mont Saint-Michel. Rev Geog Phys Geol Dyn 8:39–112

Cartwright DE (1999) Tides: a scientific history. Cambridge University Press, Cambridge, 292 p

Evans G (1965) Intertidal flat sediments and their environments of deposition in The Wash. J Geol Soc Lond 121:209–245

Evans G, Schmidt V, Bush P, Nelson H (1969) Stratigraphy and geologic history of the Sabkha, Persian Gulf. Sedimentology 12:145–159

Frey RW, Howard JD (1969) A profile of biogenic sedimentary structures in a Holocene barrier island-salt marsh complex, Georgia. Gulf Coast Assoc Geol Soc Trans 19:427–444

Ginsburg RN (1956) Environmental relationships of grain size and constituent particles in some south Florida carbonate sediments. Bull Am Assoc Petrol Geol 40:2384–2427

Ginsburg RN (1975) Tidal deposits: a casebook of recent examples and fossil counterparts. Springer, New York, 426 p

Irwin ML (1965) General theory of epeiric clear water sedimentation. Bull Am Assoc Petrol Geol 49: 445–459

Klein deV G (1971) A sedimentary model for determining paleotidal range. Geol Soc Am Bull 82:2585–2592

Postma H (1961) Transport and accumulation of suspended matter in the Dutch Wadden Sea. Neth J Sea Res 1:148–190

Reineck H-R (1963) Sedimentgefüge im Bereich der südlichen Nordsee. Abhandl Senckenber Naturforsch Ges 505:1–138

Shinn EA, Lloyd RM, Ginsburg RN (1969) Anatomy of a modern carbonate tidal flat, Andros Island, Bahamas. J Sediment Petrol 39:112–123

Stride AH (1963) Current-swept sea floors near the southern half of Great Britain. Q J Geol Soc Lond 119:175–199

van Straaten LMJU (1954) Composition and structure of recent marine sediments in the Netherlands. Leidse Geol Mededel 19:1–110

Visser MJ (1980) Neap-spring cycles reflected in Holocene subtidal large-scale bedform deposits: a preliminary note. Geology 8:543–546

Wang Y (1963) The coastal dynamic geomorphology of the northern Bohai Bay. In: Wang Y (ed) Collected oceanic works of Nanjing University. Nanjing University Press, Nanjing (in Chinese with English abstract)

Corpus Christi, Texas USA
Kingston, Ontario, Canada

Contents

Contributors

Hadi Amin-Rasouli Department of Geosciences, University of Kurdistan, Sanandaj, Iran, H.Aminrasouli@uok.ac.ir

Allen W. Archer Department of Geology, Kansas State University, Manhattan, KS 66506, USA, aarcher@ksu.edu

Jennifer Aschoff Department of Geology and Geologic Engineering, Colorado School of Mines, Golden, CO, USA, jaschoff@mines.edu

Jesper Bartholdy Department of Geography and Geology, University of Copenhagen, 10 Øster Voldgade, Copenhagen DK-3050, Denmark, jb@geogr.ku.dk

Andrew Berkeley Department of Evironmental & Geographical Sciences, Manchester Metropolitan University, John Dalton Extension Building, Chester Street, Manchester M1 5GD, UK

Ilya Buynevich Department of Earth and Environmental Sciences, Temple University, 313 Philadelphia, PA 19122, USA, coast@temple.edu

Kyungsik S. Choi Faculty of Earth Systems and Environmental Sciences, Chonnam National University, Gwangju 500-757, South Korea, tidalchoi@hotmail.com

Robert W. Dalrymple Department of Geological Sciences and Geological Engineering, Queens University, Kingston, ON K7L 3N6, Canada, dalrymple@geol.queensu.ca

Richard A. Davis, Jr. Department of Geology, Coastal Research Laboratory, University of South Florida, Tampa, FL 33620, USA, rdavis@usf.edu

Harte Research Institute for Gulf of Mexico Studies, Texas A&M University – Corpus Christi, TX 78412, USA

Mason Dykstra Department of Geology and Geological Engineering, Colorado School of Mines, Golden, CO 80401, USA, mdykstra@mines.edu

Kenneth A. Eriksson Department of Geosciences, Virginia Tech, Blacksburg, VA 24061, USA, kaeson@vt.edu

Daidu Fan State Key Laboratory of Marine Geology, Tongji University, Shanghai 200092, China, ddfan@tongji.edu.cn

Duncan FitzGerald Department of Earth Sciences, Boston University, Boston, MA 02215, USA, dunc@bu.edu

Burghard W. Flemming Senckenberg Institute, Suedstrand 40, 26382 Wilhelmshaven, Germany, bflemming@senckenberg.de

Murray K. Gingras Department of Earth and Atmospheric Sciences, University of Alberta, Edmonton, AB T6G 2E3, Canada, mgingras@ualberta.ca

Steven L. Goodbred, Jr. Department of Earth and Environmental Sciences, Vanderbilt University, Nashville, TN 37240, USA, steven.goodbred@vanderbilt.edu

Stephen F. Greb Kentucky Geological Survey, University of Kentucky, Lexington, KY 40506, USA, greb@uky.edu

Christopher Hein Department of Earth Sciences, Boston University, Boston, MA 02215, USA, hein@whoi.edu

Zoe J. Hughes Department of Earth Sciences, Boston University, Boston, MA 01778, USA, zoeh@bu.edu

Aitor A. Ichaso Department of Geological Sciences and Geological Engineering, Queens University, Kingston, ON K7L 3N6, Canada, aitorichaso@hotmail.com

Davood Jahani Department of Geology, Faculty of Basic Sciences, North Tehran Branch, Islamic Azad University, Tehran, Iran, d_jahani@iau-tnb.ac.ir

Erik P. Kvale Devon Energy Corporation, 20 North Broadway, Oklahoma City, OK 73102, USA, Erik.Kvale@dvn.com

Yaghoob Lasemi Illinois State Geological Survey, Prairie Research Institute, University of Illinois at Urbana-Champaign, Champaign, IL 61820, USA, ylasemi@illinois.edu

Zakaria Lasemi Illinois State Geological Survey, Prairie Reserarch Institute, University of Illinois at Urbana-Champaign, Champaign, IL 61820, USA, zlasemi@illinois.edu

James A. MacEachern Department of Earth Sciences, Simor Fraser Univeraity, 8888 University Drive, Burnaby, BC V5A 1S6, Canada, jmaceach@suf.ca

Duncan A. MacKay Department of Geological Sciences and Geological Engineering, Queens University, Kingston, ON K7L 3N6, Canada, duncanamackay@yahoo.com

A.W. Martinius Statoil Research and Development, Arkitekt Ebbels vei 10, N-7005 Trondheim, Norway, awma@Statoil.com

Piret Plink-Bjorklund Department of Geology and Geologic Engineering, Colorado School of Mines, Golden, CO, USA, pplink@mines.edu

Eugene C. Rankey Department of Geology, University of Kansas, 1475 Jayhawk Blvd., 120 Lindley Hall, Lawrence, KS 66045, USA, grankey@ku.edu

Stacy Lynn Reeder Schlumberger-Doll Research, One Hampshire Street, Cambridge, MA 02139, USA, sreeder@slb.com

Jean-Yves Reynaud Département Histoire de la Terre – UMR 7193 ISTeP, Muséum National d'Histoire Naturelle, Géologie, CP 48, 43, rue Buffon, F-75005 Paris, France, jyr@mnhn.fr

Yoshiki Saito Geological Survey of Japan, AIST, Central 7, Higashi 1-1-1, Tsukuba 305-8567, Japan, yoshiki.saito@aist.go.jp

Edward Simpson Department of Physical Sciences, Kutztown University, Kutztown, PA 19530, USA, simpson@kutztown.edu

Ronald J. Steel Department of Geological Sciences, University of Texas – Austin, Austin, TX 78712, USA, rsteel@mail.utexas.edu

Bernadette Tessier Morphodynamique Continentale et Côtière, University of Caen, UMR CNRS 6143, 24 Rue des Tilleuls, 14000 Caen, France, bernadette.tessier@unicaen.fr

Ping Wang Coastal Research Laboratory, Department of Geology, University of South Florida, Tampa, FL 33620, USA, pwang@usf.edu

Tidal Constituents of Modern and Ancient Tidal Rhythmites: Criteria for Recognition and Analyses

1

Erik P. Kvale

Abstract

The origin of oceanic tides is a basic concept taught in most introductory college-level sedimentology, geology, oceanography, and astronomy courses. Tides are commonly explained in the context of the equilibrium-tidal theory model. The equilibrium model explains tides in the context of changes in two hemisphere-opposite tidal bulges through which the Earth spins. The position and size of these tidal bulges relative to the Earth's equator is largely controlled by the phases of the Moon and changes in declination and orbital distance of the Moon in its orbit around the Earth. While explaining the driving forces that cause tides, the equilibrium model does not explain most of the tides observed in the Earth's oceans.

A complete explanation of the origin of tides must include a discussion of dynamic tidal theory. In the dynamic tidal model, tides resulting from the motions of the Moon in its orbit around the Earth and the Earth in its orbit around the Sun are modeled as products of the combined effects of a series of phantom satellites. The movement of each of these satellites, relative to the Earth's equator, creates its own tidal wave that moves around an amphidromic point. Each of these waves is referred to as a tidal "constituent" or "species." The geometries of the ocean basins determine which of these constituents are amplified. Thus, the tide-raising potential for any locality on Earth can be conceptualized as the summation of the amplitudes of a series of tidal constituents specific to that region. A better understanding of tidal cycles opens up remarkable opportunities for research on tidal deposits with implications for, among other things, a more complete understanding of the tidal dynamics responsible for sediment transport and deposition, tectonic-induced changes in paleogeographies, and changes in Earth–Moon distance through time.

1.1 Introduction

Tidal rhythmites, small-scale sedimentary structures that include thinly layered, fine grained sediments, record, through the cyclic variations in the thicknesses of successive laminae, changes in current velocities associated with lunar/solar cycles. The thickness of a

E.P. Kvale (✉)
Devon Energy Corporation, 20 North Broadway, Oklahoma City, OK 73102, USA
e-mail: Erik.Kvale@dvn.com

R.A. Davis, Jr. and R.W. Dalrymple (eds.), *Principles of Tidal Sedimentology*,
DOI 10.1007/978-94-007-0123-6_1, © Springer Science+Business Media B.V. 2012

lamina is directly and positively related to tidal current strength, which in turn is directly and positively related to the magnitude of the daily rise and fall of the tide (tidal range). Over periods of days, months, or years, changes in tidal current strengths associated with various lunar/solar cycles are mirrored by the change in thicknesses of the vertically stacked laminae.

Modern and ancient tidal rhythmites have been found on every continent in the world except Antarctica. In modern environments, tidal rhythmites occur in deposits associated with tide-dominated deltas, tidal embayments, and estuaries. Tidal rhythmites can be used for reconstructing ancient paleogeographies and paleoclimates (e.g. this chapter, Hovikoski et al. 2005; Kvale et al. 1994), estimating paleotidal ranges (e.g. Archer 1995; Archer and Johnson 1997), understanding channel migration in the fluvio-estuaring transition (Choi 2010) determining lunar-retreat rates through time (e.g. Williams 1989; Kvale et al. 1999), and most recently, have been used to infer the major tidal constituents associated with the tides that deposited them (e.g. Kvale 2006). In order to understand tidal rhythmites, however, one has to understand how tides are generated and what controls their genesis.

The impact of diurnal, semidiurnal, and semimonthly (neap-spring) tidal cycles on sediment deposition has been well documented since the early 1980s (e.g. Visser 1980; Boersma and Terwindt 1981; Allen 1981). For many geologists these became benchmark papers when they were published because they showed how depositional packages within sedimentary successions can be linked to a tidal origin. However, it was the discovery of modern and ancient tidal rhythmites in the late 1980s and 1990s that showed that a hierarchy of tidal cycles, beyond simple semidaily, daily or fortnightly events, could be preserved in the rock record (e.g. Kvale et al. 1989; Williams 1989; Dalrymple and Makino 1989; Archer et al. 1991; Kvale et al. 1994; Miller and Eriksson 1997). Tidal cycles associated with monthly, semiannual, annual (usually includes a significant seasonal climatic component), and even an approximately 18-year cycle have been identified from ancient tidal rhythmites.

Studies, however, showed that the understanding of one of the most basic of the tidal cycles, the neap-spring or fortnightly tidal cycle, by most geologists, and apparently many oceanographers, and astronomers as well, was over-simplified. Many college-level textbooks today continue to propagate a basic misunderstanding of the neap-spring cycles and the origin of oceanic tides in general (e.g. Duxbury et al. 2002).

The intent of this chapter is neither to outline a history of the study of tides and tidal deposits nor to document the current state of knowledge regarding the history of the Earth-Moon system. These issues are treated in some detail in Klein (1998), Rosenberg (1997), Williams (2000), and Coughenour et al. (2009). Rather, it is to explain some basic tidal theory and show how a more complete knowledge of ancient tides can be extracted from the rock record. Most of the information contained within this chapter is distilled from two summary papers: Kvale et al. (1999) and Kvale (2006).

To truly understand tidal systems and, in particular, the genesis of tidal rhythmites it is useful to understand both an equilibrium tidal model and a dynamic tidal model. The former explains the driving forces behind the formation of tides and is commonly taught to geology, oceanography, and astronomy undergraduates, whereas the later, more accurately explains real-world tides and is more useful in interpreting the rock record. An understanding of both models is essential to anyone who studies tides and tidal deposits, and both will be discussed.

1.2 Equilibrium Tidal Theory

Most geologists understand tidal periodicities in the context of equilibrium tidal theory. Tides are generated by the gravitational forces of the Moon and, to a lesser degree, the Sun on the Earth. The Moon accounts for approximately 70% of the tide-raising force because of its proximity to the Earth. In an equilibrium world, the Earth is covered by an ocean of uniform depth that responds instantaneously to changes in tractive forces (MacMillan 1966). The equilibrium model can be used to explain five of the six tidal periodicities that have been commonly detected in rhythmite successions. These six cycles are illustrated in Figs. 1.1–1.6 (previously illustrated in Kvale et al. 1998). A seventh cycle known as the "nodal cycle", an approximately 18 year-tidal cycle, and very well documented by Miller and Eriksson (1997) within the Pride Shale, a lower Carboniferous succession found in West Virginia, is not illustrated here.

The figures each illustrate (from upper left to lower right): A diagram and explanation of the equilibrium

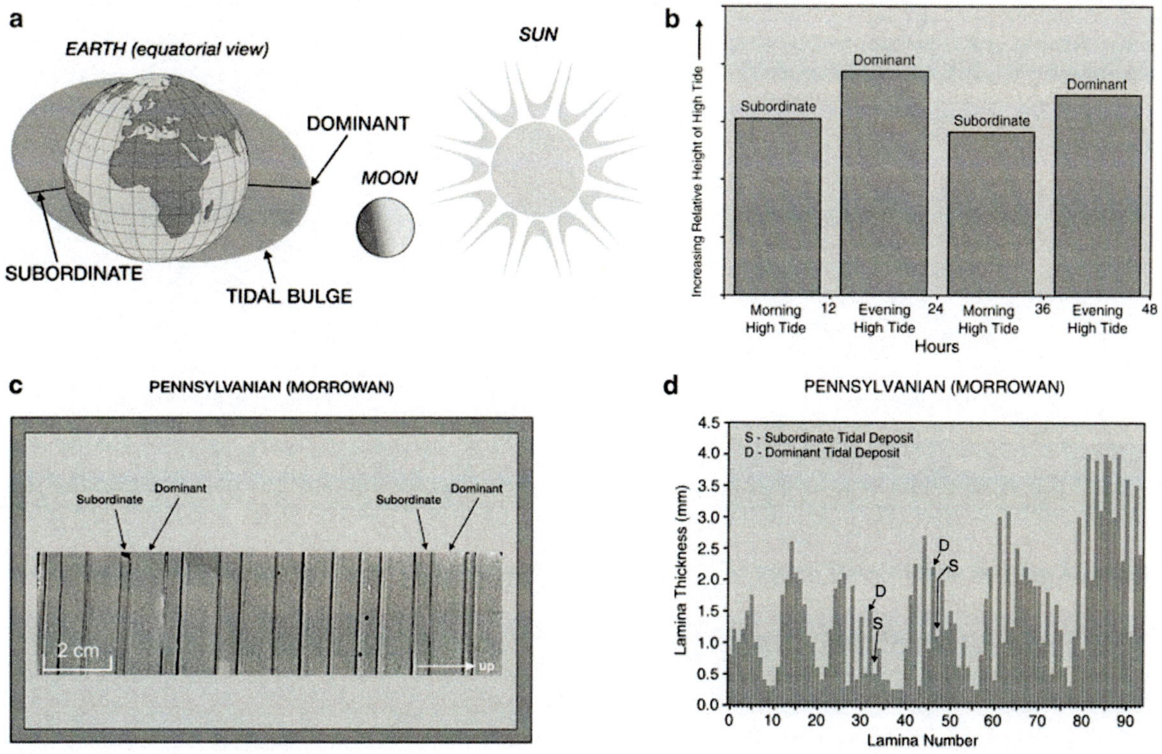

Fig. 1.1 Semidiurnal equilibrium model. (**a**) Two oceanic tidal bulges are produced on opposite sides of the Earth by the gravitational forces of the Sun and the Moon. (**b**) Two tides are produced each day by the spin of the Earth through these bulges. The diurnal inequality is produced when the tidal bulges are not centered above the Earths equator. Semidiurnal tides can be recognized in the rock record by the coupling of thick and thin lamina (**c**) and graphically in the thickness measurements of laminated sequences (**d**) as preserved in the tidal rhythmite succession from the Pennsylvanian Mansfield Formation (Hindostan whetstone beds) from Orange County, Indiana, USA (From Kvale and others (1998) and used by permission from SEPM)

tidal theory of five of the six tidal periods; a bar chart of tidal height data (high tide elevations) from a modern, real-world setting that shows how the astronomical effects are reflected in cyclic changes in daily high tides; a core from an ancient tidal rhythmite succession showing how these cyclic tidal effects might be manifested in a laminated tidal rhythmite; and a bar chart of laminae thicknesses interpreted in the context of the modern tidal cycle.

1.2.1 Semidiurnal (12.42 h)

Within the equilibrium tidal model, the interaction of tidal forces from the Moon and Sun produce two oceanic bulges on opposite sides of the Earth (Fig. 1.1). The rotation of a point on the Earth through these bulges once a day produces two tides (the semidiurnal tide). Typically, these tides are not equal (termed diurnal inequality), as one tide is higher (dominant) than the other (subordinate) because the Moon's orbital plane and the Earth's equatorial plane are not parallel. The angular difference between the two planes is termed lunar declination.

1.2.2 Synodic (29.53 Days)

Daily high tides are higher when the Earth, Moon, and Sun are nearly aligned (full or new moon); this is referred to as "syzygy" (Fig. 1.2). Conversely, lower tides occur when the Sun and Moon are at right angles to the Earth (first or third quarter phase), also known as "quadrature". Tides during full or new moon are referred to as spring tides: "spring" in this context refers to "lively" or "energetic" rather than implying a seasonal connotation. Tides at quarter phases are referred to as neap tides. The neap-spring tidal period

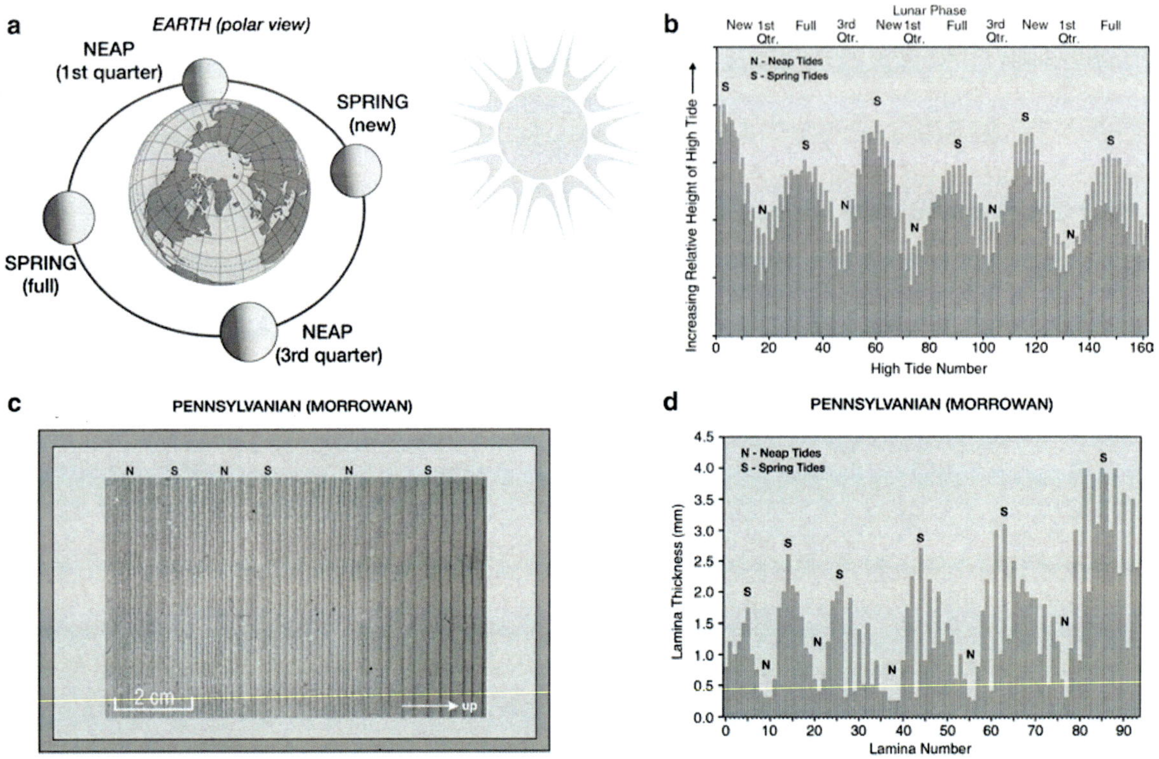

Fig. 1.2 Synodic equilibrium model. (**a**) In an equilibrium tidal model, spring tides occur when the Earth, Moon, and Sun align during full or new moon (also known as "syzygy"). Equlibrium neap tides occur when the Moon-Earth alignment is 90° from an Earth-Sun alignment (also known as "quadrature"). The synodic month (currently 29.53 days) is the time it takes for the Moon to orbit the Earth when measured from a new Moon to the next new Moon. When neap-spring tides can be timed to phases of the Moon they are referred to as "synodic neap-spring tides" (Kvale 2006). (**b**) Graph of tidal heights of a portion of the 1991 predicted high tides for Kwajalein Atoll, Pacific (NOAA 1990) showing the effects of changing lunar phases. (**c**) Portion of a core from the Mansfield Formation (Hindostan whetstone beds), Indiana, USA with neap and spring tidal deposits labeled. (**d**) Measurements of laminae thicknesses from Hindostan whetstone beds with neap and spring tidal deposits labeled (From Kvale et al. (1998) and used by permission from SEPM)

in the equilibrium model is related to the changing phases of the Moon associated with the half-synodic month. The synodic month (new moon to new moon, or full moon to full moon) has a modern period of 29.53 days and encompasses two neap-spring cycles.

1.2.3 Tropical (Semidiurnal, 27.33 Days)

The tidal force also depends on the declination of the Moon (Fig. 1.3). In this usage, "declination" refers to the tilt or angle of the Moon's orbit relative to the Earth's equatorial plane. The period of the variation in declination is called the tropical month – the interval of time it takes the Moon to complete one full orbit from its maximum northern declination to its maximum southern declination and then return. The effect of the tropical month in an equilibrium semidiurnal tidal system is to cause the diurnal inequality of the tides. Ideally, diurnal inequality is greatest when the Moon is at its maximum declination. This inequality is reduced to zero when the Moon is over the equator, producing a crossover in the tidal data (Fig. 1.3). The current length of the tropical month is 27.32 days (2 days shorter than the synodic month – see synodic discussion above). Because of this difference, equatorial passages of the Moon, called crossovers, have a shorter periodicity than the periodicity related to synodic neap-spring tides.

1.2.4 Tropical (Diurnal, 27.32 Days)

In modern, dominantly diurnal systems (primarily one tide per day), the tropical period described above

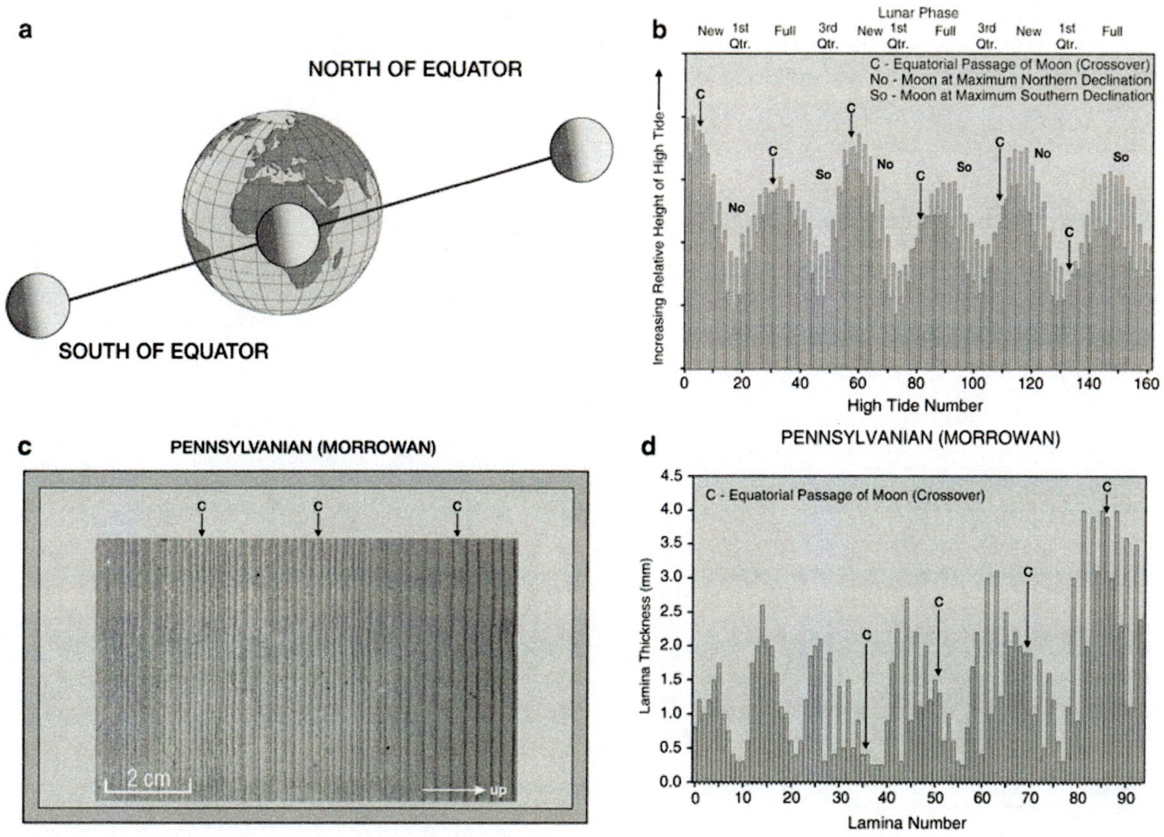

Fig. 1.3 Tropical, semidiurnal equilibrium model. (**a**) Model of the Moon in orbit around the Earth. The lunar declination is exaggerated from its modern range of 18–28°. The tropical month (currently 27.32 days) is the time it takes for the Moon to move from its maximum northern declination to its southernmost declination and back to its northernmost declination in a single orbit. (**b**) Graph of tidal heights of a portion of the same modern tidal record shown in Fig. 1.2b illustrating diurnal inequality of semidiurnal tides. Note diurnal inequality goes to zero when the Moon passes directly over the Earth's equator. (**c**) Image of core shown in Fig. 1.2c showing approximate position (labeled "C") when Moon was above the Earth's equator during deposition. Note the approximate equal thicknesses of the lamina on either side of the *arrow*. (d) *Bar chart* shown in Fig. 1.2d with arrows denoting passages of the Moon above the Earth's equator during deposition (From Kvale et al. (1998) and used by permission from SEPM)

is responsible for generating neap-spring cycles. In contrast to the synodic system, tides in a tropical system behave as though the Sun's gravitational effects are dampened, which is impossible to explain in an equilibrium tidal model (Fig. 1.4). In such cases, the dominant tidal force depends on the declination of the Moon relative to the Earth's equator with the force being greatest when the Moon is most directly over the site in question. In these systems, the predicted and ancient tide data reveal that equatorial passages of the Moon (crossovers) occur in phase with the generation of neap-spring tides, in contrast to the variable relationship exhibited by tropical (semidiurnal) tides.

1.2.5 Anomalistic (27.55 Days)

Another tidal effect arises from the changing distance of the Moon relative to the Earth during the lunar orbit (Fig. 1.5). Because the lunar orbit forms an ellipse, with the Earth slightly offset from the center, the Moon alternates between perigee (closest approach to the Earth) and apogee (the farthest distance from the Earth). During the lunar synodic month there will be two spring tides (see synodic periods described above). These spring tides, however, will be of unequal magnitude producing alternating high-spring and low-spring tides, which correspond to spring tides during or near perigee (high spring) and spring tides during or near apogee (low spring).

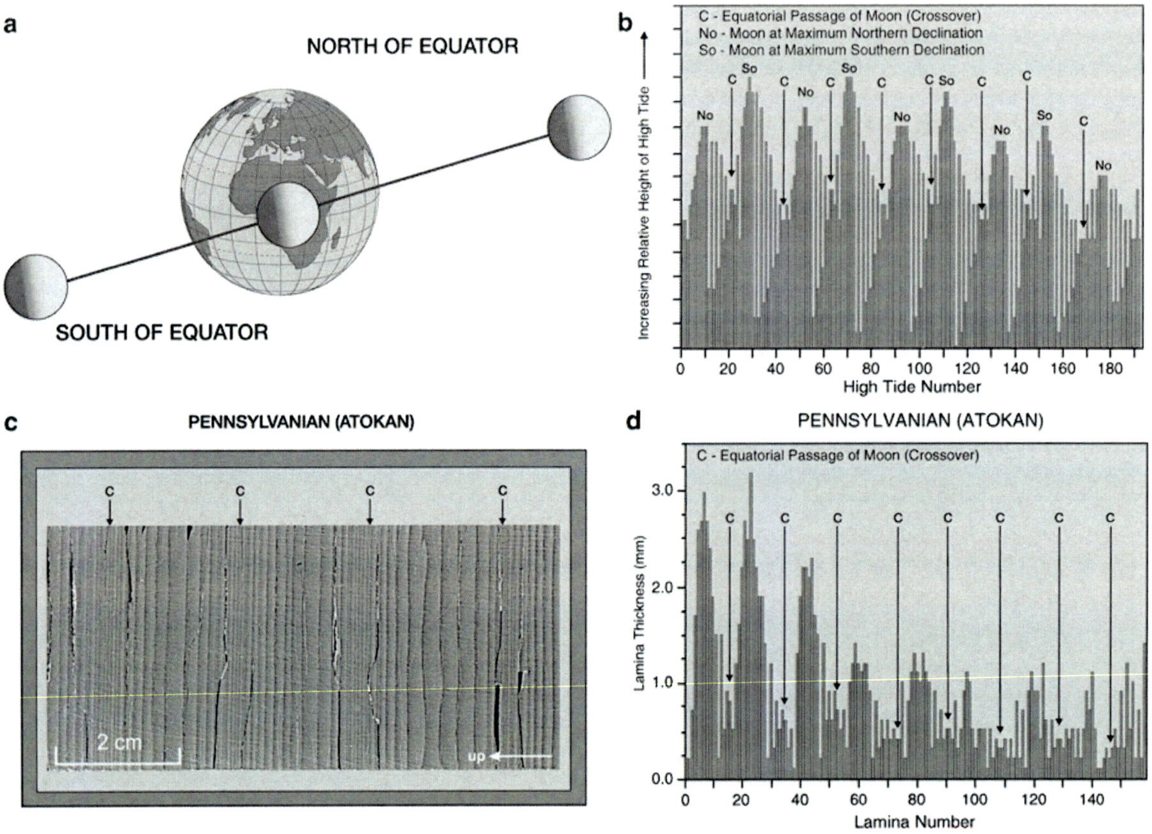

Fig. 1.4 Tropical diurnal model. (**a**) Model of the Moon in its orbit around the Earth (see Fig. 1.3a). (**b**) Graph showing the 1994 predicted relative high tides (mixed, predominantly diurnal) for the Barito River estuary in Borneo (NOAA 1993). Note the passages of the Moon above the Earth's equator perfectly track the neap tides and spring tides to the maximum declinations of the Moon in its orbit around the Earth, a pattern not predicted by equilibrium tidal theory. Such neap-spring tidal cycles are termed "tropical neap-spring tides" (Kvale 2006). (**c**) Photograph of a portion of a core from the Pennsylvanian Brazil Formation, Daviess County, Indiana, USA. *Arrows* indicate lamina deposited with the Moon was above the Earth's equator. (**d**) *Bar chart* of lamina thicknesses measured from core obtained from the Brazil Formation. This unit also is mixed, predominantly diurnal. Note the diurnal inequality of the semidiurnal component goes to zero only in the neap tide deposits. This corresponds to the Moon above the Earth's equator during deposition (From Kvale and others (1998) and used by permission from SEPM)

The semimonthly inequality of the spring tides disappears when the Moon lies along the minor axis of the lunar orbit and the difference in lunar distance is minimized during subsequent spring tides. The time it takes for the Moon to move from perigee to perigee is called the anomalistic month, which is at present 27.55 days.

1.2.6 Semiannual (182.6 Days)

The synodic, tropical, and anomalistic periods have slightly different values. Because of this, these periods will interact constructively twice each year causing tidal forces at these times to reach a maximum (as shown by the dashed line in Fig. 1.6). In the equilibrium tidal model, the date of this tidal maximum is a function of latitude that is related to the declinational effects of the Moon and Sun. An annual inequality has been documented in several ancient tidal rhythmite successions (Kvale et al. 1994). This inequality is interpreted to be climatic (non-tidal) in origin.

1.3 Dynamic Tidal Theory

As noted in the introduction, the equilibrium tidal model explains the driving forces that cause tides but does not explain real-world tides. For instance, the

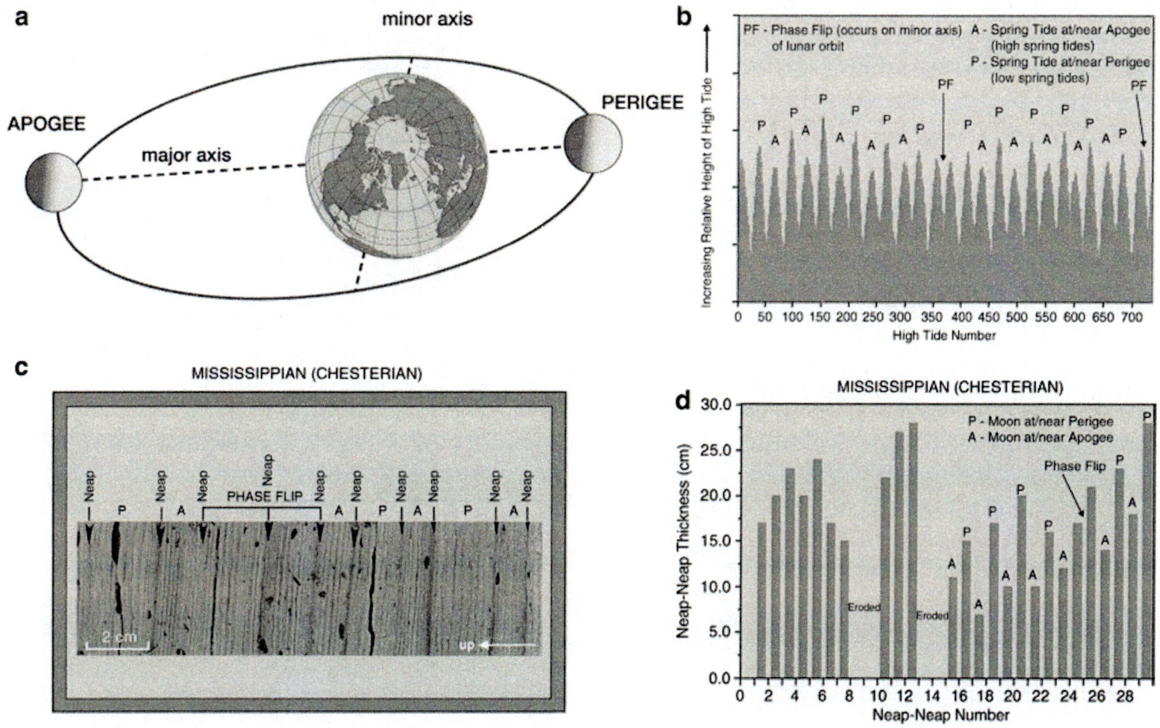

Fig. 1.5 Anomalistic equilibrium model. (**a**) Polar view of the Moon in orbit around the Earth. Note that lunar orbit is not perfectly circular but somewhat elliptical (greatly exaggerated in the diagram) and that the Earth is not position in the direct center of the orbit path. The time it takes for the Moon to go from perigee (closest approach) to apogee (furthest from the Earth) and return is called the "anomalistic month", which is 27.55 days long at present. (**b**) *Graph* showing the 1992 predicted high tides for Saint John, New Brunswick, Canada (NOAA 1991) showing the effects of the anomalistic month on the Saint John tides. Note the semimonthly inequality goes to zero when the Moon and Sun are aligned with the Moon's minor orbital axis (termed "phase flip"). (**c**) Photograph of a core from the Mississippian Tar Springs Formation, Indiana, USA showing the effects of the anomalistic month on neap-spring tidal deposition. (**d**) Graph illustrating thicknesses as measured between neap-to-neap tide deposits from the Tar Springs Formation core, a portion of which is shown in Fig. 1.5c. Note the position of the "phase flip" (From Kvale et al. (1998) and used by permission from SEPM)

world does not spin through two tidal bulges. Instead, oceanic tides rotate as waves around fixed (amphidromic) points within individual ocean basins (Fig. 1.7). Equilibrium tidal theory indicates that diurnal tides should exist only at very high latitudinal positions, which is not the case. For example, the Gulf of Mexico and large tracts in the Indian and western Pacific oceans are dominated by diurnal tides. Tides like those found in Immingham, England, where the semidiurnal tides have minimal diurnal inequality, cannot be explained by equilibrium tidal theory, which requires such tides to exist only in equatorial positions. Finally, equilibrium tidal theory does not explain neap-spring tidal cycles which are synchronous with the 27.32 tropical monthly period such as illustrated in Fig. 1.4.

The difficulties in understanding and explaining real-world tides can be addressed by a dynamic tidal model. This model is built around the concept of a harmonic analysis of the components that compose real-world tides. For instance, the Moon and Sun each generate their own tide within the Earth's oceans. Since the orbits of the Earth around the Sun and the Moon around the Earth are not perfectly circular, the amplitude of the tides generated by each of these bodies, in part, fluctuates depending on the Earth's proximity to the Sun and, much more importantly, the Moon's distance from the Earth. Periodically each of these tides will constructively or destructively interact with each other. The tides associated with changes in Moon-Earth distance or Earth-Sun distance can be considered to be a constituent of the overall tide, which can affect any coastline.

To model these tidal constituents (also known as tidal "species") oceanographers conceptualize each

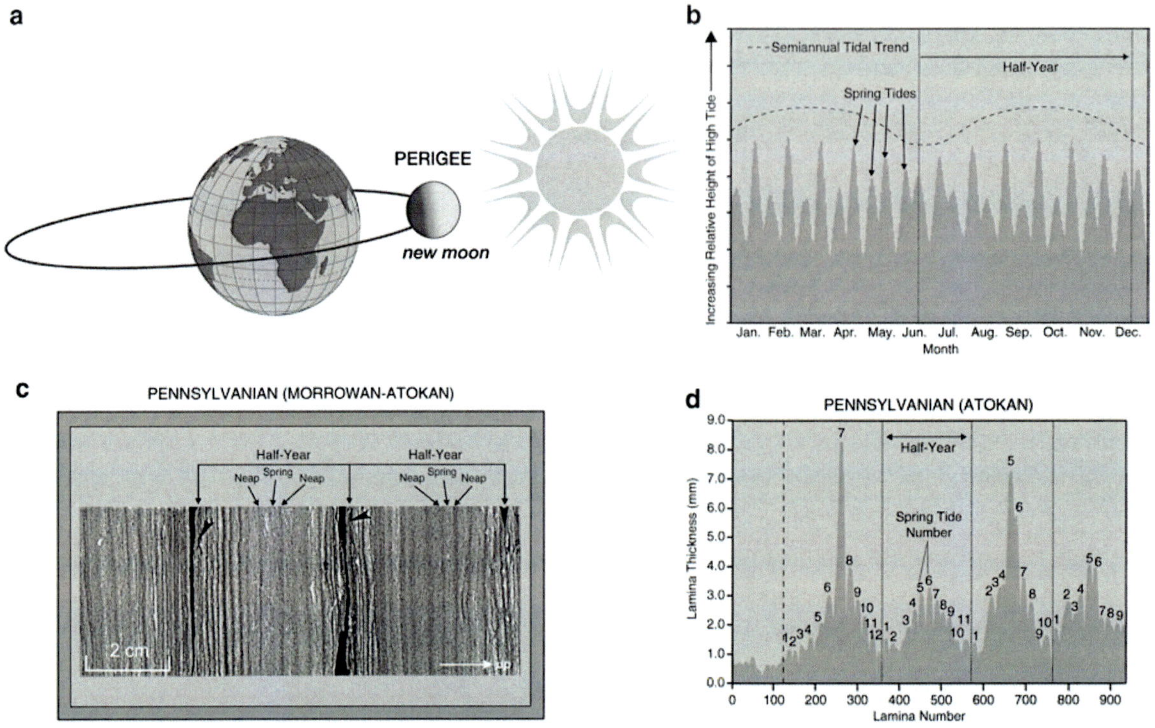

Fig. 1.6 Semiannual equilibrium model. (**a**) View of the configuration of the Earth, Moon, and Sun representing the maximum spring tides formed when the Moon is at perigee, maximum northern declination and new. Such spring tides occur every 182.6 days. (**b**) 1992 predicted high tides from Saint John, New Brunswick, Canada (NOAA 1991) showing the effects of the semiannual convergence of maximum spring tides. (**c**) Photograph of a core from the Pennsylvanian Lead Creek Limestone, Indiana, USA. In this core the neap-spring cycles thicken and thin in a semiannual pattern. (**d**) *Graph* showing the thicknesses of individual lamina from the Brazil Formation, Indiana. These thicknesses are also organized into semiannual tidal cycles. Each number records an individual neap-spring cycle (From Kvale et al. (1998) and used by permission from SEPM)

constituent as a phantom "satellite" that has its own mass (that of the Moon, Sun, or a combination of the two). Each phantom "satellite" has a motion within a plane or is fixed relative to the stars and each generates its own tide with a unique period, response time, and amplitude (Pugh 1987) (Table 1.1). For instance S_2 represents the twice-daily tide generated at a fixed point on the Earth by a "satellite" that has the mass of the Sun in a perfectly circular orbit around the Earth's equator. O_1 represents the daily tide generated at a fixed point on the Earth by a "satellite" with a mass of the Moon and a motion above the Earth's equator. For each of the tidal constituents, the subscript indicates if the tide is diurnal ($_1$) or semidiurnal ($_2$).

The relative intensity for each of these tidal constituents along any oceanic coastline in the world can be determined by a harmonic decoupling of an extended hourly tidal record. These measurements typically are recorded in most major harbors and other tidal stations around the world. More than 100 tidal constituents have been identified from a harmonic extraction of Earth's tides, however, seven of these (Table 1.1) account for more than 80% of any real-world tide (Defant 1961). The resonate amplification or destruction of these tidal constituents determines the resulting tide for a specific area within the Earth's oceans (Fig. 1.8).

As noted above, each of these tidal constituents corresponds to a unique tidal wave. These waves do not travel around the world as predicted by equilibrium tidal theory, but rather rotate around a point (referred to as an "amphidromic point") within a region of the ocean at a speed determined by their constituent's orbital periodicity or the periodicity of the Earth's spin (Fig. 1.7). The location of these points is determined by basin geometries and the Coriolis force.

Ideally, amphidromic circulation should be counterclockwise in the Northern Hemisphere and clockwise in the Southern Hemisphere and never on the equator but, as shown above, real-world tides don't always follow convention and exceptions are known (Open University Course Team 1999).

The major tidal cycles discussed under the equilibrium model can be understood in the context of the dynamic model and tidal constituents. Specifically, the synodic neap-spring cycle is generated through the interaction of the S_2 and M_2 constituents. In the modern world, these two tides come into phase and amplify the resulting tide every 14.77 days. The result is a synodic spring tide. Conversely, every 13.66 days K_1 and O_1 converge and generate a tropical spring tide. Whether a spring tide along a specific coastline is dominated by the synodic spring tide or the tropical spring tide is determined by the basin geometry. For instance, the Gulf of Mexico is dominated by the K_1 and O_1 tides, therefore neap-spring tides cycle with the tropical month (Fig. 1.9). The east coast of the USA, however, is dominated by S_2 and M_2 tides resulting in neap-spring tides that cycle with the synodic month (Fig. 1.9). The semimonthly inequality of spring tides occurs because of the convergence of M_2 and N_2 every 27.55 days. A diurnal inequality is driven by the interaction of O_1 and M_2 (in phase once a day) and is noted in coastal tides when these constituents are of sufficient amplitude.

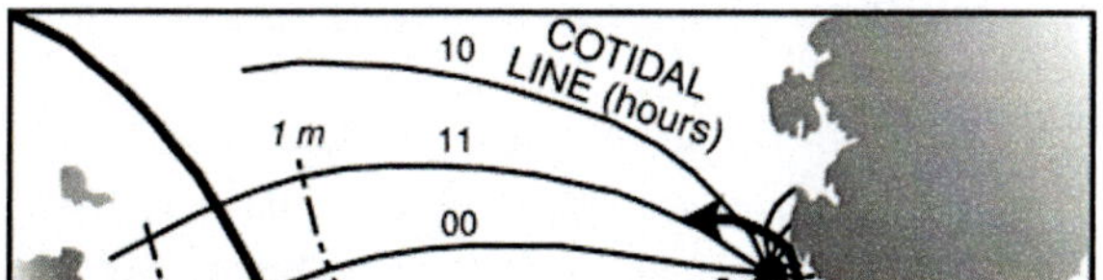

Fig. 1.7 Diagram showing the amphidromic circulation for the M_2 tide in the North Sea. Co-tidal lines indicate times of high water. And co-range lines indicate lines of equal tidal range. Figure is modified from Dalrymple (1992) which was based on a map first drawn by J. Proudman and A. T. Doodson (From information found in Cartwright 1999) (From Kvale (2006) and used by permission from Marine Geology)

One can look at the progressive change in relative intensity of particular tidal constituent along a coast and see how that affects the resulting tides. For example, Figs. 1.10 and 1.11 shows the amplitudes for the seven dominant tidal constituents for the Gulf of Carpentaria, Australia and the tidal patterns that result from changes in the relative amplitudes of the various constituents (from Kvale 2006). At the mouth of the gulf at Booby Island, the tides are dominated by M_2, K_1 and O_1. Given the dominance of O_1 and K_1, the neap-spring cycle occurs every 27.32 days and corresponds to the tropical monthly period. However, unlike many regions whose neap-spring cycles are tropically driven, there is a relatively strong M_2 tide (but relatively weak S_2 tide) at the mouth of the gulf. The resultant

Table 1.1 List of the seven most common tidal constituents, their rotational speed (number of degrees a tidal wave generated by the constituent can travel around its amphidromic point in 1 h), description, and period in solar hours (Defant 1961)

Tidal constituent	Speed (degrees/hour)	Origin	Period in solar hours
M_2	28.9841	Principal lunar	12.42
S_2	30	Principal solar	12
N_2	28.4397	Larger elliptical lunar	12.66
K_2	30.0821	Combined declinational lunar and declinational solar	11.97
K_1	15.0411	Combined declinational lunar and declinational solar	23.93
O_1	13.943	Principal lunar	25.82
P_1	14.9589	Principal solar	24.07

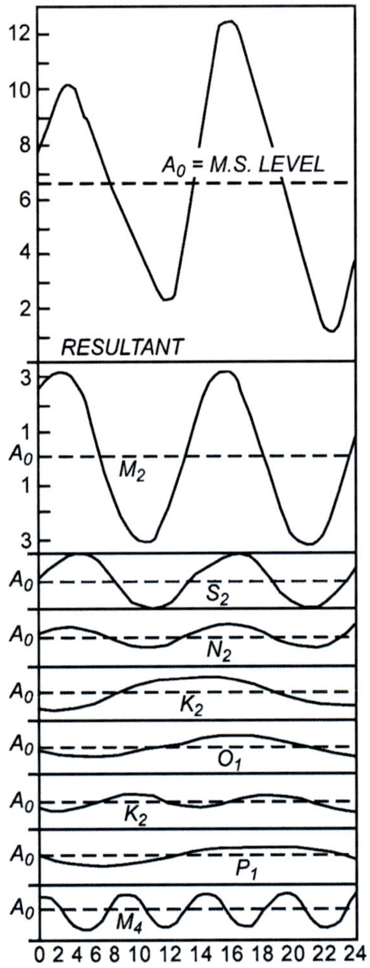

Fig. 1.8 Resulting tide predicted from the stacking of 9 different tidal constituents. Horizontal units are in hours (Modified from MacMillan, 1966 in Kvale, (2006) and used by permission from Marine Geology)

tide at Booby Island exhibits a tropically driven neap-spring cyclicity comparable to the tide depicted in Fig. 1.4 except that it also exhibits a strong semidiurnal component that is driven by M_2. Progressing further south into the Gulf of Carpentaria, the strengths of K_1 and O_1 increase relative to M_2 creating a tide that is dominantly diurnal.

1.4 Ancient Tides

Some tidal rhythmites in the rock record preserve long (several months worth), relatively complete successions of daily or semidaily tidal deposition. Particularly complete records can be interpreted in the context of the dynamic tidal model and several examples are noted below.

1.4.1 Hindostan Whetstone Beds (Pennsylvanian, Indiana)

Figures 1.2 and 1.3 show both a segment of core and a bar chart of the laminae thicknesses from the Hindostan Whetstone beds found in Orange County, Indiana (Kvale et al. 1989). Neap-spring cycles in this chart occur more frequently than crossovers indicating that these tides were synodically driven and hence related to the dominance of the M_2 and S_2 over the O_1 and K_1 constituents. Some caution is needed, however, in interpreting crossover patterns because the absence of a single half-day event could cause an apparent crossover. Ways to infer completeness of a tidal pattern are discussed by Kvale et al. (1999). Suffice it to state that with suitably long tidal rhythmite records, such as presented here, it is possible to interpret crossover patterns with some confidence.

This example clearly shows a diurnal inequality, and, as such, O_1 must be significant. There appears to be a lack of a pronounced semimonthly inequality (anomalistic cycle) suggesting that N_2 was relatively weak. Therefore, tides that deposited the Hindostan Whetstone beds were dominated by the constituents M_2, S_2, and O_1 followed by K_1 and N_2.

1.4.2 Brazil Formation (Pennsylvanian, Indiana)

Figure 1.4 show a segment of core and a bar chart of laminae thicknesses from the Brazil Formation of Daviess County, Indiana (Kvale and Archer 1990; Kvale and Mastalerz 1998). The neap-spring cycles in this example occur at the same frequency as the crossovers indicating that these tides were driven by the tropical period and hence reflect a dominance of O_1 and K_1 over S_2 and M_2. A weak semidiurnal signal occurs during the neap tides and indicates that M_2 had some amplitude and importance in the resulting tide. The Brazil Formation rhythmites, like the whetstone beds discussed above, lack a prominent semimonthly inequality suggesting a weak N_2 tidal constituent. It can be inferred from this data base that the Brazil

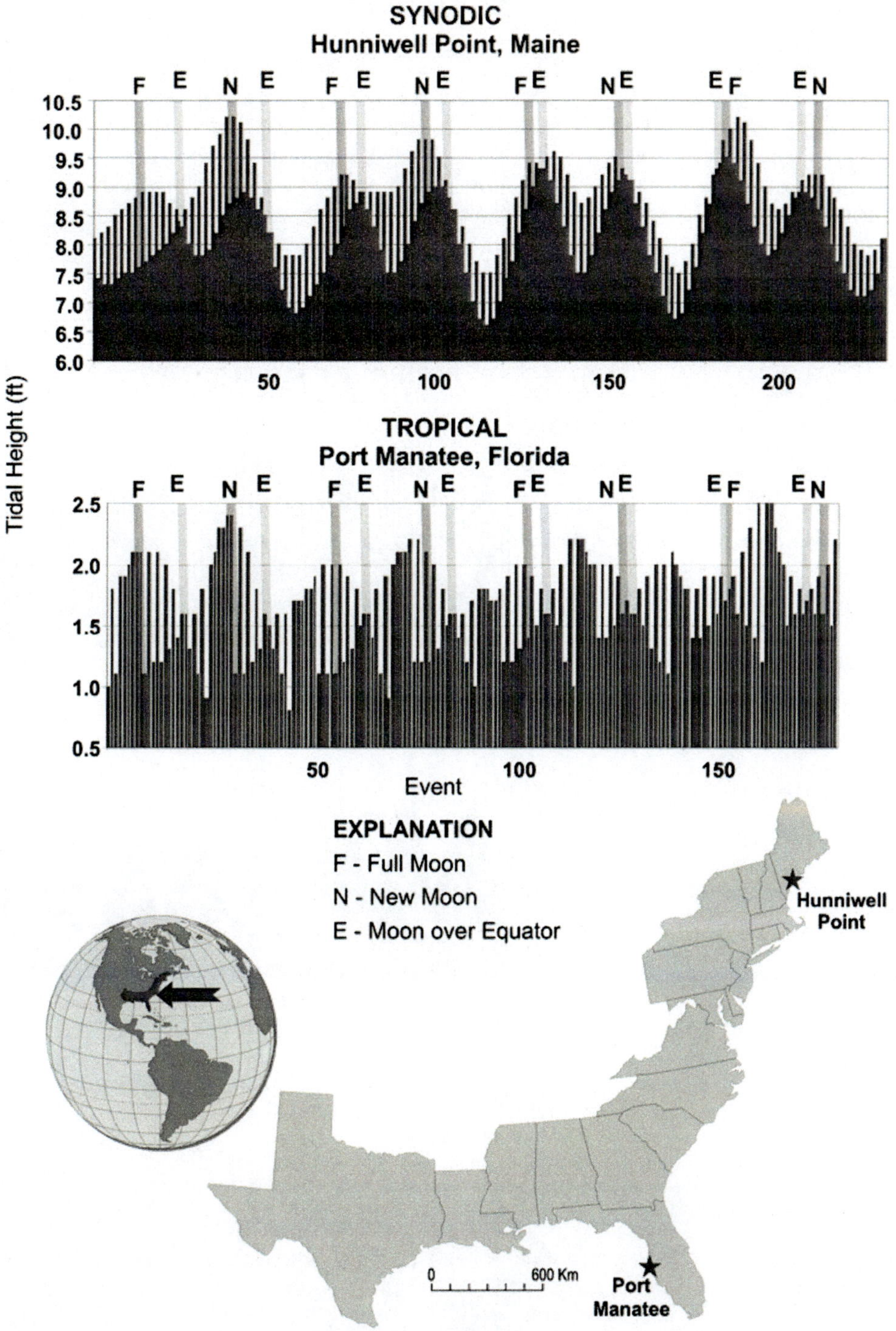

Fig. 1.9 Graphs showing predicted high-data for two tidal references stations from the east coast and Gulf coast USA. The Port Manatee example is typical of the tides in the Gulf coast and the Hunniwell graph typifies east coast tides. Both tidal records cover the same interval of time from January through early May, 2005 (National Oceanographic and Atmospheric Administration Web site 2004). Note that the equatorial passages of the Moon are fixed with the neap tides in the Gulf coast station but move through the graph in the east coast example. As such, Gulf coast neap-spring tides are driven by the tropical month but the east coast neap-spring tides are controlled by the phase changes of the Moon associated with the synodic month (From Kvale (2006) and used by permission from Marine Geology)

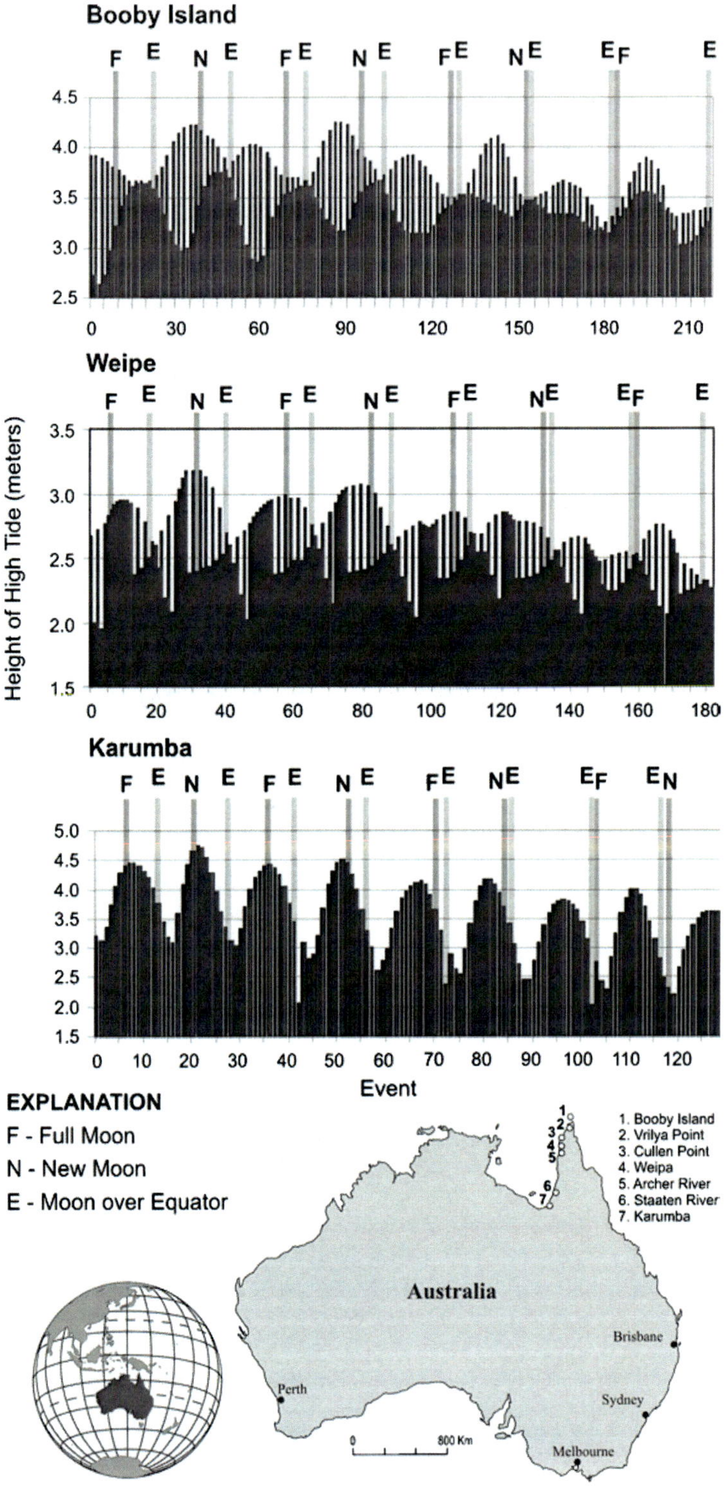

Fig. 1.10 Graphs and location map for predicted high-tide data from three tidal reference station in the Gulf of Carpentaria, Australia. The time interval for each graph spans January through early June, 2004 (Australian National Tidal Centre, Bureau of Meteorology Web site, 2004) (From Kvale (2006) and used by permission from Marine Geology)

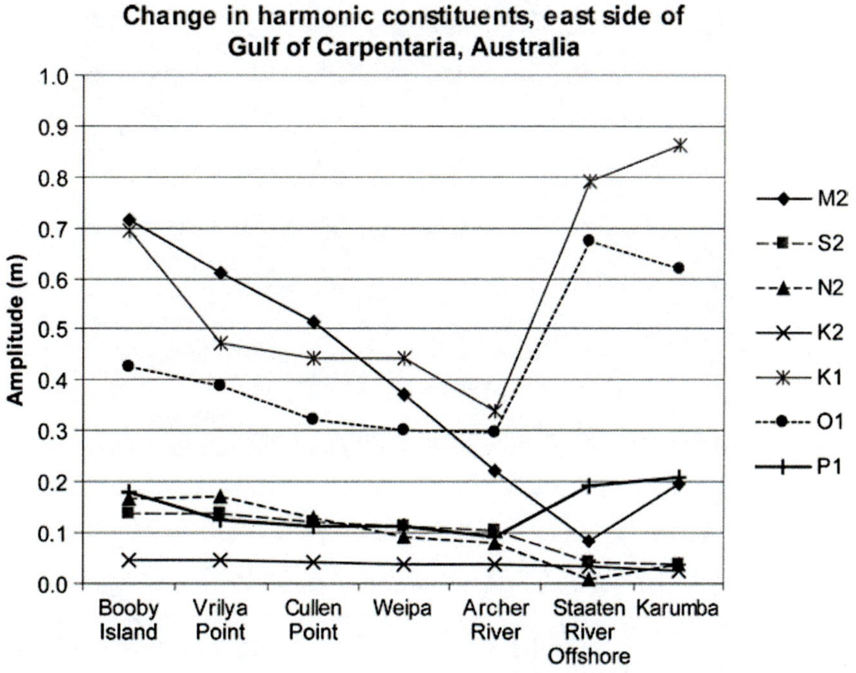

Fig. 1.11 Line graph showing the changes in tidal amplitude for the seven most dominant tidal constituents for several tidal reference stations located along the eastern side of the Gulf of Carpentaria (locations noted in Fig. 1.10. Constituent data was extracted using the Seafarer Tides software package by the Australian National Tidal Centre, Bureau of Meteorology and provided to Kvale (2006) (From Kvale (2006) and used by permission from Marine Geology)

Formation tides were dominated by O_1, K_1, followed by M_2 with very weak contributions from S_2 and N_2.

1.4.3 Abbott Sandstone (Tradewater Formation, Pennsylvanian, Illinois)

Figure 1.12 shows an outcrop and bundle thicknesses from some flaggy, large-scale tidal bundles along Interstate 57 in Johnson County, Illinois (Kvale and Archer 1991). A histogram of bundle thicknesses indicates a strong semidiurnal signal throughout the record. While not as clean a tidal record as the two previous examples, the Abbott sandstone example appears to exhibit minimal diurnal inequality during the neap tides. When the diurnal inequality tracks neap tides, it indicates that neap-spring cyclicity is driven by the tropical period (e.g. Fig. 1.4). As such, the Abbott Sandstone tidal record resembles that of Booby Island, Australia (Fig. 1.10), in which M_2, O_1 and K_1 dominate the resultant tide over S_2. There is a suggestion of a semimonthly inequality to the Abbott sandstone record indicating that N_2 was stronger than S_2 and sufficiently strong to influence the tidal record.

These examples illustrate that tidal constituents can be extracted from the rock record in well-preserved tidal rhythmites. While it is not always possible to draw conclusions regarding so many tidal constituents, deposits can generally be determined to be either diurnal or semidiurnal in nature based on the absence or occurrence of alternating thick-thin laminae. Most, but not all, semidiurnal tidal deposits can be related to the synodic period and the convergence of M_2 and S_2 constituents. Exceptions of semidiurnal, tropically driven neap-spring tides or tidal deposits, such as Booby Island and the Abbott Sandstone, are known and can be discerned if the tidal record is long and clean enough. All diurnal deposits should have been deposited in tropically driven neap-spring cycles. Semidiurnal depositional systems that lack strong K_1 or O_1 constituents (like Effingham, England), and in which tidal sediments were deposited only on high intertidal zones might mimic a diurnal tidal deposit (Archer and Johnson 1997). In such a case, additional outcrop work might result in the discovery of lower intertidal or subtidal facies that would resolve the issue.

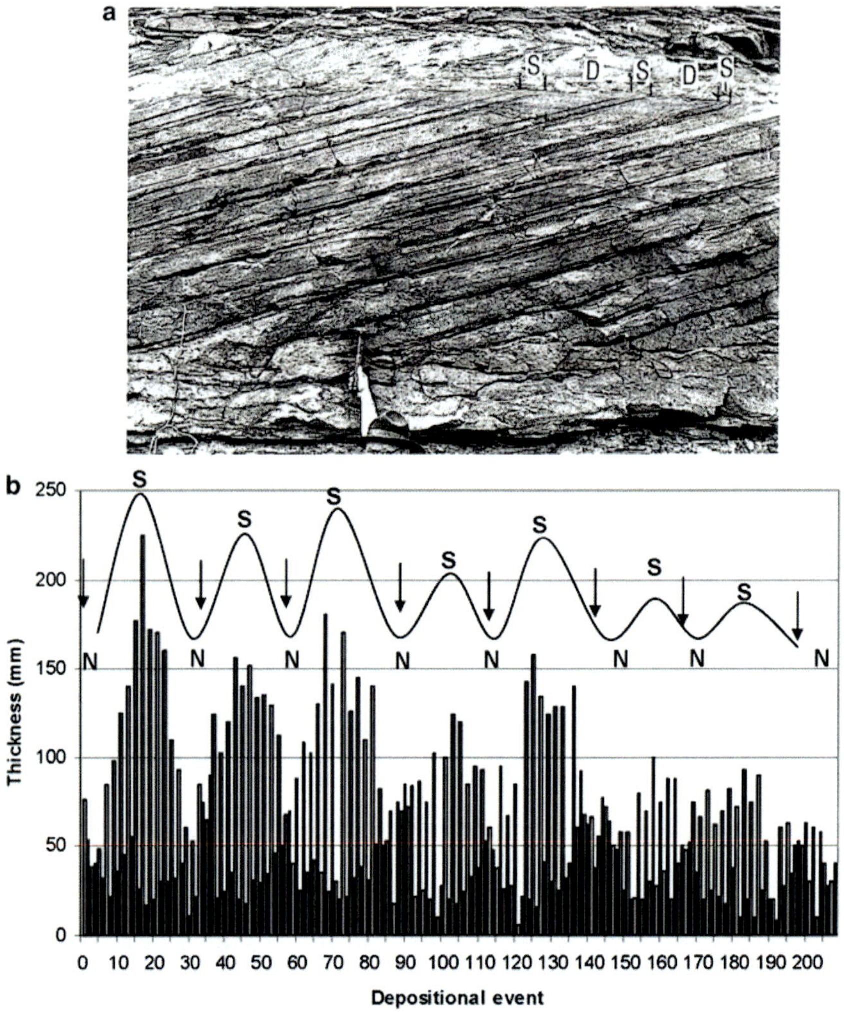

Fig. 1.12 Tradewater Formation, (**a**) Photo of the Abbott sandstone outcrop. This is part of a much more extensive dune mesoform. Examples of dominant (*D*) and subordinate (*S*) semidiurnal foresets are labeled. Rock hammer for scale (*lower part* of photo) (**b**) *Bar chart* showing foreset (depositional event) thickness variability with spring tides (*S*), neap tides (*N*) and lunar crossover (*arrows*) events labeled. Notice the semimonthly inequality of the spring tides related to perigee and apogee effects. Also note that the lunar passages of the equator (*arrows*) track the neap tide deposits fairly closely suggesting that the neap-spring cycles are in phase with the tropical month

1.5 Summary and Implications

The equilibrium tidal model is very useful for explaining the gravitational forces that generate tides on the Earth. However, it is an over-simplification and does not explain the tides in most of the oceans of the world. To explain real-world tides requires a basic understanding of the dynamic tidal model. The dynamic tidal model has been used to estimate changes in the Earth-Moon distance through time (Williams 1989; Kvale et al. 1999) and has even been suggested as a way to better understand the impact that tides have on biological systems (Kvale 2006). It has also been used to model tidal basin dynamics for determining the importance of tidal facies within a basin or region (e.g. Ericksen and Slingerland 1990; Wells et al. 2007). In the Abbott example, an interpretation of neap-spring cyclicity could be done with both the equilibrium and dynamic model, but interpretation of the relative importance of the M_2,

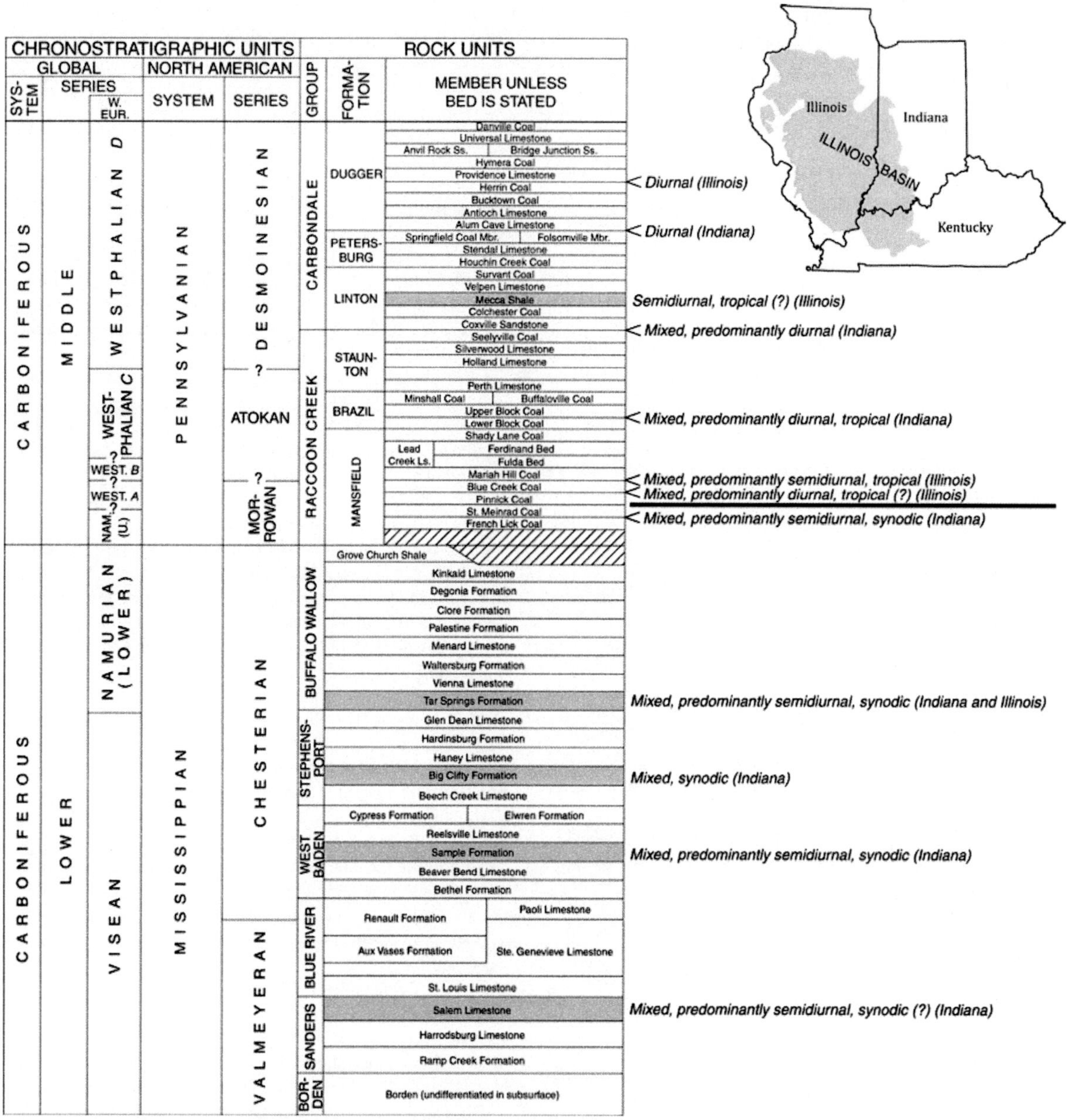

Fig. 1.13 Stratigraphic chart for the Indiana portion of the Illinois Basin showing stratigraphic intervals where good tidal rhythmite records have been identified by the author. The *solid grey line* marks the boundary below which tidal rhythmites seem to be controlled primarily by the synodic monthly cycle and above which the tidal rhythmites appear to reflect the tropical monthly cycle

S_2, O_1, K_1, and N_2 constituents using the dynamic model allows much more specific comparisons to be made to real-world analogues (in this case Booby Island tides) than would otherwise be possible. In fact, utilizing this approach within the Illinois Basin one can interpret the dominance of diurnal (O_1 and K_1) tides versus semidiurnal (M_2 and S_2) tides for various tidal rhythmite packages that span the Mississippian-Pennsylvanian systems (Fig. 1.13). As Fig. 1.13 shows, tidal rhythmites older than the upper Morrowan Blue Creek Coal appear to have been deposited within synodically driven systems dominated by M_2 and S_2. Younger tidal rhythmites appear to have been deposited within tropically driven systems. This change from synodically driven to tropically driven tidal systems may reflect the closure of the Iapetus Ocean during the early Pennsylvanian and a major change in tidal dynamics within the midcontinent Carboniferous sea of North America.

While teaching and understanding the dynamic tidal system represents a bit of a paradigm shift to most geologists, it creates possible research venues not accessible through an understanding of equilibrium tidal theory alone.

References

Allen JRL (1981) Lower Cretaceous tides revealed by cross-bedding with mud drapes. Nature 289:579–581

Archer AW (1995) Modeling of tidal rhythmites based on a range of diurnal to semidiurnal tidal-station data. Mar Geol 123:1–10

Archer AW, Johnson TW (1997) Modeling of cyclic tidal rhythmites (Carboniferous of Indiana and Kansas, Precambrian of Utah, USA) as a basis for reconstruction of intertidal positioning and paleotidal regimes. Sedimentology 44:991–1010

Archer AW, Kvale EP, Johnson HR (1991) Analysis of modern equatorial tidal periodicities as a test of information encoded in ancient tidal rhythmites. In: Smith DG, Reinson GE, Zaitlin BA, Rahmani RA (eds) Clastic tidal sedimentology. Canadian Soc Petrol Geol Mem 16:189–196

Boersma JR, Terwindt JHJ (1981) Neap-spring tide sequences of intertidal shoal deposits in a mesotidal estuary. Sedimentology 28:151–170

Cartwright DE (1999) Tides: a scientific history. Cambridge University Press, Cambridge, UK, 292 pp

Choi K (2010) Rhythmic climbing cross-lamination in inclined heterolithic stratification (IHS) of a macrotidal estuarine channel, Gomso Bay, west coast of Korea. J Sediment Res 80:550–561

Coughenour CL, Archer AW, Lacovera KJ (2009) Tides, tidalites, and secular changes in the Earth-Moon system. Earth Sci Rev 97:59–79

Dalrymple RW (1992) Tidal depositional systems. In: Walker RG, James NP (eds) Facies models response to sea level changes. Geological Association of Canada, St. Johns, pp 195–218

Dalrymple RW, Makino Y (1989) Description and genesis of tidal bedding in the Cobequid Bay-Salmon River estuary, Bay of Fundy, Canada. In: Taira A, Masuda F (eds) Sedimentary facies in the active plate margin. Terra Science Publication Co., Tokyo

Defant A (1961) Physical oceanography, vol 11. Pergamon, New York, 598 pp

Duxbury AB, Duxbury AC, Sverdrup KA (2002) Fundamentals of oceanography, 4th edn. McGraw Hill, Boston, 344 pp

Ericksen MC, Slingerland R (1990) Numerical simulations of tidal and wind-driven circulation in the Cretaceous Interior Seaway of North America. Geol Soc Am Bull 102:1499–1516

Hovikoski J, Rāsānen M, Gingras M, Roddaz M, Brusset S, Hermosa W, Romero-Pittman L, Lertola K (2005) Miocene semidiurnal tidal rhythmites in Madra de Dios, Peru. Geology 33:177–180

Klein GD (1998) Clastic tidalites-a partial retrospective view. In: Alexander C, Davis RA, Henry VJ (eds) Tidalites: processes and products, vol 61, Special publication (SEPM (Society for Sedimentary Geology)). Society of Sedimentary Geology, Tulsa, pp 5–14

Kvale EP (2006) The origin of neap-spring tidal cycles. Mar Geol 235:5–18

Kvale EP, Archer AW (1990) Tidal deposits associated with low-sulfur coals, Brazil formation (lower Pennsylvanian), Indiana. J Sediment Petrol 60:563–574

Kvale EP, Archer AW (1991) Characteristics of two Pennsylvanian-age semidiurnal tidal deposits in the Illinois Basin, U.S.A. In: Smith DG Reinson GE Zaitlin BA Rahmani RA (eds), Clastic tidal sedimentology. Canada Soc Petrol Geol Mem 16:179–188

Kvale EP, Mastalerz M (1998) Evidence of ancient freshwater tidal deposits. In: Alexander C, Davis RA, Henry VJ (eds) Tidalites: processes and products, vol 61, Special publication (SEPM (Society for Sedimentary Geology)). Society of Sedimentary Geology, Tulsa, pp 95–107

Kvale EP, Archer AW, Johnson HR (1989) Daily, monthly, and yearly tidal cycles within laminated siltstones of the Mansfield formation (Pennsylvanian) of Indiana. Geology 17:365–368

Kvale EP, Fraser GS, Archer AW, Zawistoski A, Kemp N, McGough P (1994) Evidence of seasonal precipitation in Pennsylvanian sediments in the Illinois Basin. Geology 22:331–334

Kvale EP, Sowder KH, Hill BT (1998) Modern and ancient tides. Poster and explanatory notes, SEPM, Tulsa, OK, and Indiana Geological Survey, Bloomington, IN

Kvale EP, Johnson HW, Sonett CP, Archer AW, Zawistoski A (1999) Calculating lunar retreat rates using tidal rhythmites. J Sediment Res 69:1154–1168

MacMillan DH (1966) Tides. American Elsevier Publishing Company, New York, 240 pp

Miller DJ, Eriksson KA (1997) Late Mississippian prodeltaic rhythmites in the Appalachian Basin: a hierarchical record of tidal and climatic periodicities. J Sediment Res 67:653–660

National Oceanographic and Atmospheric Administration (2004) http://www.co-ops.nos.noaa.gov/tides04/ 2004 date accessed, Sept

NOAA (1990) Tide tables 1991, high and low water predictions, Central and Western Pacific Ocean and Indian Ocean, U.S. Department of Commerce, National Oceanic and Atmospheric Administration, Riverdale, Maryland

NOAA (1991) Tide tables, 1992 high and low water predictions, Central and Western Pacific Ocean and Indian Ocean, U.S. Department of Commerce, National Oceanic and Atmospheric Administration, Riverdale, Maryland

NOAA (1993) Tide tables, 1994 high and low water predictions, Central and Western Pacific Ocean and Indian Ocean, U.S. Department of Commerce, National Oceanic and Atmospheric Administration, Riverdale, Maryland

Open University Course Team (1999) Waves, tides and shallow-water processes, 2nd edn. Open University, Butterworth–Heinemann, Oxford, 227 p

Pugh DT (1987) Tides, surges and mean sea level. Wiley, New York, 472 p

Rosenberg GD (1997) How long was the day of the dinosaur? And why does it matter? In: Wolberg DL, Stump E, Rosenberg GD (eds) Dinofest international: proceeding symposium sponsored by Arizona State University, The Academy of Sciences, Philadelphia, pp 493–512

Visser MJ (1980) Neap-spring cycles reflected in Holocene subtidal large-scale bedform deposits; a preliminary note. Geology 8:543–546

Wells MR, Allison PA, Piggott MD, Gorman GJ, Hampson GJ, Pain CC, Fang F (2007) Numerical modeling of tides in the late Pennsylvanian midcontinent seaway of North America with implications for hydrography and sedimentation. J Sediment Res 77:843–865

Williams GE (1989) Late Precambrian tidal rhythmites in South Australia and the history of the Earth's rotation. J Geol Soc Lond 146:97–111

Williams GE (2000) Geological constraints on the Precambrian history of Earth's rotation and the Moon's orbit. Rev Geophys 38:37–59

Principles of Sediment Transport Applicable in Tidal Environments

2

Ping Wang

Abstract

Physical processes of sediment transport in tidal environments are extremely complicated and are influenced by numerous hydrodynamic and sedimentological factors over a wide range of temporal and spatial scales. Both tide and wave forcing play significant roles in the entrainment and transport of both cohesive and non-cohesive particles. Present understanding of sediment transport is largely empirical and based heavily on field and laboratory measurements. Sediment transport is composed of three phases: (1) initiation of motion (erosion), (2) transport, and (3) deposition. In tidal environments, the coarser non-cohesive sediments are typically transported as bedload, forming various types of bedforms. The finer cohesive sediments tend to be transported as suspended load, with their deposition occurring mostly during slack tides under calm conditions. Rate of sediment transport is generally proportional to flow velocity to the 3rd to 5th power. This non-linear relationship leads to a net transport in the direction of the faster velocity in tidal environments with a time-velocity asymmetry. Due to the slow settling velocity of fine cohesive sediment and a difference between the critical shear stress for erosion and deposition, scour and settling lags exist in many tidal environments resulting in a landward-fining trend of sediment grain size. The periodic reversing of tidal flow directions results in characteristic bi-directional sedimentary structures. The relatively tranquil slack tides allow the deposition of muddy layers in between the sandy layers deposited during flood and ebb tides, forming the commonly observed lenticular, wavy, and flaser bedding.

Notations and Conventional Units

a: a reference level (typically defined at the top level of the bedload layer) for suspended sediment concentration. (m)

c: suspended sediment concentration (dimensionless for volume concentration, kg/m^3 for mass concentration)

c_a: reference concentration (dimensionless for volume concentration, kg/m^3 for mass concentration)

$c(z)$: suspended sediment concentration profile (dimensionless for volume concentration, kg/m^3 for mass concentration)

P. Wang (✉)
Coastal Research Laboratory, Department of Geology, University of South Florida, Tampa, FL 33620, USA
e-mail: pwang@usf.edu

R.A. Davis, Jr. and R.W. Dalrymple (eds.), *Principles of Tidal Sedimentology*,
DOI 10.1007/978-94-007-0123-6_2, © Springer Science+Business Media B.V. 2012

$\overline{c}$: depth averaged concentration (dimensionless for volume concentration, kg/m^3 for mass concentration)
D: sediment grain size (m)
D_*: dimensionless sediment grain size (dimensionless)
D_w: wave-energy dissipation due to breaking (kg/s^3)
d_m: mean sediment grain size (m)
d_{50}: 50th percentile sediment grain size (m)
E: wave energy per unit water volume (kg/s^2)
f_c: bottom friction coefficient (dimensionless)
H: wave height (m)
h: water depth (m)
k_d: empirical coefficients used in suspended sediment concentration profile modeling (dimensionless)
k_x: dispersion coefficient in x direction (dimensionless)
k_y: dispersion coefficient in y direction (dimensionless)
L: wave length (m)
L_s: turbulent mixing length (m)
Q_b: volumetric bed-load transport rate ($m^3/m/s$)
q_s: volume rate of suspended sediment transport ($m^3/m/s$)
S = source and sink terms
s: sediment specific density = ρ_s/ρ_w (dimensionless)
T: wave period (s)
U_δ: near bottom wave orbital velocity (m/s)
$u(z)$: current velocity with respect to depth z (m/s)
$\overline{u}$: depth-averaged current velocity (m/s)
u_*: current related bed-shear velocity (m/s)
u_{*_c}: critical bed shear velocity (m/s)
u_{*_crs}: critical shear velocity for sediment suspension (m/s)
$\overline{u_{cr}}$: depth-averaged critical velocity (m/s)
$\overline{v}$: depth average velocity in y direction (m/s)
w_s: settling velocity (m/s)
w_{s_s}: settling velocity of single suspended particle in clear water used in the calculation of the settling velocity of flocs (m/s)
z: vertical coordinate representing water depth (m)
z_o: vertical level with zero velocity, also often referred to as bed roughness (m)
α_1: empirical coefficients used in suspended sediment concentration profile modeling (dimensionless)
α_2: empirical coefficients used in suspended sediment concentration profile modeling (dimensionless)
β: empirical coefficients used in suspended sediment concentration profile modeling (dimensionless)
ε_s: sediment mixing coefficient
θ: Shields parameter (dimensionless)
θ_c: critical Shields parameter (dimensionless)
θ_{crs}: critical Shields parameter for sediment suspension (dimensionless)
κ: Von Karman's constant, typically taken as 0.4 (dimensionless)
μ: an efficiency factor to incorporate the influence of bedforms on bedload transport used in the Meyer-Peter and Mueller (1948) bedload transport formula (dimensionless)
ν: kinematic viscosity (m^2/s)
ρ_s: sediment density (kg/m^3)
ρ_w: density of water (seawater in the case of tidal environment) (kg/m^3)
τ_b: bed shear stress (N/m^2)
τ_c: critical bed shear stress (N/m^2)
ϕ_{floc}: flocculation factor (dimensionless)
ϕ_{hs}: hindered settling factor (dimensionless)

2.1 Introduction

Coastal sedimentology and morphodynamics are controlled by a variety of interactive factors, including forces from ocean tides and waves, trends and rates of sea-level changes, sediment supply, climatic and oceanographic settings, and antecedent geology. Depending on the relative dominance of wave and tide forcing, coastal environments can be classified as tide-dominated and wave-dominated (Davis and Hayes 1984). This chapter focuses on general physical processes of sediment transport that are applicable to the tide-dominated environments. In this chapter, tidal environments are defined generally as shallow marine environments that are significantly influenced by tides.

The rise and fall of tides provide the main mechanism for sediment transport and morphology changes in tidal environments. In addition to generating tidal current which constitutes the dominant forcing in tidal environments, this regulated water-level fluctuation can also modulate wave action. For example, higher waves were often measured at a fixed location on a tidal flat during higher tides due to less friction related wave dissipation (Lee et al. 2004; Talke and Stacey 2008). Sediment transport by wave forcing can be significant locally, as well as during storm conditions. Bottom shear stress, and therefore initiation of sediment motion and transport, is also strongly influenced by water depth, which varies substantially in tidal environments. When the tidal water-level fluctuations are confined by channels, e.g., tidal inlets and creeks, strong tidal-driven flows can be generated. As compared to other types of channelized flow, tidal flow reverses direction periodically with a slack water period in between, which may create unique bi-directional sedimentary structures. In the case of tidal inlets between barrier islands, large flood and ebb tidal deltas can be deposited through the interaction of tide and wave forcing. The cyclical rising, slacking, and falling tide and the associated flow variation leave signature sedimentary records through geological history, providing valuable information for understanding earth history (e.g. Kvale et al. 1989).

Sediment grain size in tidal environment typically ranges from non-cohesive medium sand to cohesive clay. Compositionally, tidal sediments can be siliciclastic, carbonate, and organic materials. A variety of sedimentary structures, ranging from millimeter-scale sand-mud laminations on tidal flats to subaqueous dunes of tens of meters in tidal channels, exists in tidal environments, indicating a wide range of sediment transport and deposition processes. Transport and deposition of mixed cohesive and non-cohesive sediments are poorly understood and provide cutting edge research topics (Van Rijn 2007a, b, c)

Given the wide range of both cohesive and non-cohesive sediments, and the energetic and highly variable hydrodynamic processes driven by both tides and waves, sediment transport processes in tidal environments are extremely complicated. This chapter aims at providing a basic review of the principals of sediment transport applicable in tidal environments. Various transport formulas and their general applications in tidal environments are discussed. It is worth emphasizing that methods of computing the rates of sediment transport are largely empirical, based heavily on field and laboratory experiments. Calibration and verification based on site-specific data are essential to accurate applications of the formulas. The transport principles and formulae can also be applied qualitatively to interpret the sedimentary processes observed in the field, and to design field experiments. More detailed and further in-depth mechanics of sediment transport can be found in several dedicated texts, e.g., Mehta (1986b), Fredsoe and Deigaard (1992), Nielsen (1992),Van Rijn (1993), Pye (1994), Allen (1997), and Soulsby (1997).

2.2 Principles of Sediment Transport

Transport of sediment in coastal environments results from the interaction between moving fluid (seawater in this case) and sediment. Present knowledge on sediment transport processes is largely empirical, based on numerous field and laboratory experiments. Insightful parameterization is crucial in describing the complicated fluid-sediment interaction. In the following section, key parameters describing fluid motion, sediment, and fluid-sediment interaction are discussed, followed by the presentation of the commonly-used methods for the calculation of non-cohesive and cohesive sediment transport, respectively.

2.2.1 Fundamental Parameters

Fluid motion over a sediment bed exerts a horizontal drag force and a vertical lift force. Generally, when these forces overcome the gravity of a sediment grain, transport is initiated. A theoretical analysis of the initiation of motion of an individual grain typically starts with a force balancing between the drag-lift forces and the gravitational force on the grain. The sediment grain is put in motion if the moments of the fluid drag (F_D) and lift (F_L) forces exceed the moments of the submerged gravitational force (F_G) on the grain (Fig. 2.1). However, due to our limited understanding of the very complicated fluid-sediment interaction, sediment transport in the natural environments cannot be quantified from the force analysis of each grain. Instead, it is quantified empirically through insightful parameterization of sediment-fluid interaction, as discussed in the following.

When viscous fluid, e.g., seawater, flows over a surface, a shear stress is generated by the fluid flow. This shear stress is responsible for entraining and transporting sediment. On the other hand, the friction at the fluid-sediment interface exerts a drag on the fluid flow,

Fig. 2.1 Schematic force balancing of individual grains on a horizontal bed

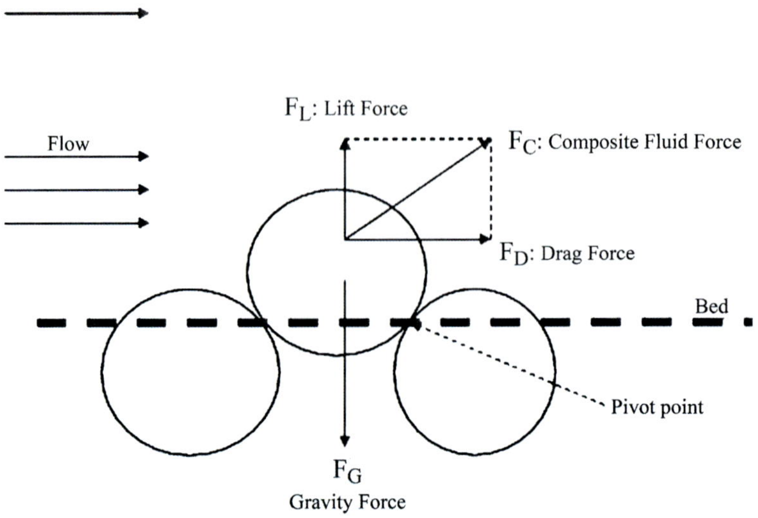

yielding the commonly observed logarithmic velocity profile over depth, i.e., the law of the wall:

$$u(z) = \frac{u_*}{\kappa}\ln\left(\frac{z}{z_o}\right) \quad (2.1)$$

Where $u(z)$ = current velocity with respect to depth, z = vertical coordinate representing water depth, u_* = current related bed-shear velocity, κ = Von Karman's constant, typically taken as 0.4, and z_o = vertical level with zero velocity, also often referred to as bed roughness. A list of notation and conventional units are provided at the beginning of this chapter. Figure 2.2 illustrates an example of a logarithmic profile. The dynamics of the bottom boundary layer where the current velocity decreases rapidly with respect of depth is crucial to sediment entrainment and transport. For plane bed, the bed roughness (Fig. 2.2) is a function of sediment grain size. When bedforms exist, the bed roughness is related to the geometry of the bedform. The bed shear velocity is directly related to bed shear stress (τ_b) as:

$$\tau_b = \rho_w u_*^2 \quad (2.2)$$

where ρ_w = density of water (seawater in the case of tidal environments). The bed shear velocity and bed shear stress are two of the key parameters describing the fluid-sediment interaction and are commonly used in computing sediment transport. Determining bed shear velocity and bed shear stress can be difficult and often comprises an essential part of a sediment transport study. By measuring a velocity profile through the water column,

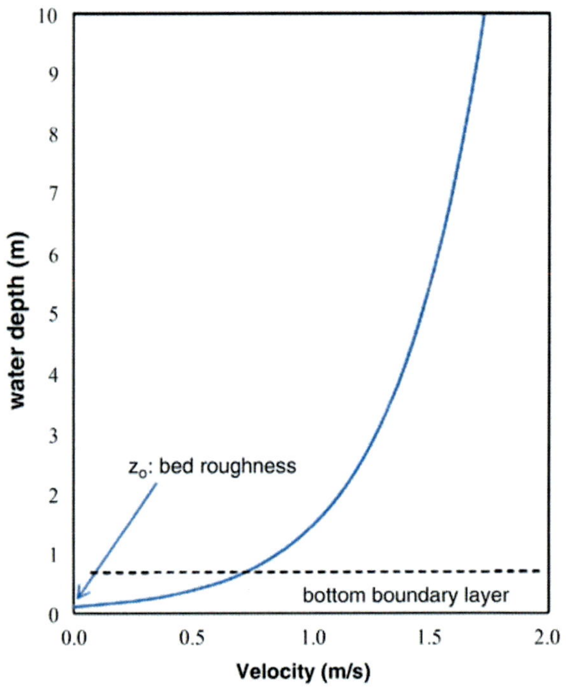

Fig. 2.2 An example of a logarithmic current profile, showing the bed roughness (z_o) and the schematic bottom boundary layer

Eq. 2.1 can be used to determine bed shear velocity and bed shear stress, as well as the bottom roughness. Another commonly used approach to determine the bottom shear stress, especially for depth-averaged models, is to relate bottom stress to velocity squared as:

$$\tau_b = \frac{1}{2}\rho_w f_c \bar{u}^2 \quad (2.3)$$

where f_c is a bottom friction coefficient, determined experimentally, and $\overline{u}$ = depth-averaged current velocity. Equation 2.3 describes the so-called "quadratic friction law", i.e., the friction exerted by a fluid flow is proportional to its velocity squared. Equations 2.1–2.3 suggest that the faster the flow velocity and the rougher the bed, the greater the shear stress, and therefore the greater potential of sediment transport.

Although wave forcing is not the dominant factor in determining the overall morphology and sedimentation pattern in tidal environments, it is important in local sediment entrainment and transport. For example, numerous studies have shown that wave forcing can have significant influence on the sedimentology and morphodynamics of tidal flats (Christie et al. 1999; Dyer 1998; Dyer et al. 2000; Li et al. 2000; Talke and Stacey 2003, 2008; Lee et al. 2004). Wave motion can be visualized as a circular motion of an imaginary water particle. This wave orbital velocity, especially near the bottom, can induce considerable shear stress to entrain and transport sediment. Based on linear wave theory, the maximum value of near bottom orbital velocity (U_δ) is:

$$U_\delta = \frac{\pi H}{T \sinh\left(\frac{2\pi h}{L}\right)} \tag{2.4}$$

where h = water depth, L = wave length, T = wave period, and H = wave height.

In a more simplified larger scale approach, wave-induced sediment transport is often evaluated based on the amount of energy that is carried by the wave. Higher wave energy typically results in more active sediment transport. Wave energy per unit water volume (E) is determined as:

$$E = \frac{1}{8}\rho_w g H^2 \tag{2.5}$$

Equation 2.5 shows that wave energy is proportional to wave height squared, e.g., a 2 m wave will carry four times the energy than a 1 m wave.

Waves break as they propagate from deep water into shallow water. The energy that is carried by the wave motion is dissipated rapidly through wave breaking. A large portion of this energy is expended to transport sediment. Due to the intense turbulence generated by wave breaking, sediment transport associated with wave breaking tends to be much greater than that under non-breaking waves and a typical current. Various empirical formulas were developed to evaluate "when" waves break (Kaminsky and Kraus 1994), one of the simplest and also reasonably accurate formulas is:

$$H_b = 0.78 h_b \tag{2.6}$$

where H_b = breaking wave height, h_b = water depth at which waves break. In other words, waves break when their height is about 80% of the water depth. Wells and Kemp (1986) found that muddy bottoms, typical of some tidal environments, can dissipate wave energy to such an extent that the above breaking criterion is never reached. Although wave forcing is secondary in tidal environments, it can contribute significantly to local sediment transport, especially in the nearshore region and during storm conditions, and should not be neglected.

In addition to the basic hydrodynamic parameters described in Eqs. 2.1–2.6, sediment grain size (D) and density (ρ_s) also play a crucial part in understanding and estimating sediment entrainment and transport. Mean grain size (d_m) and the 50th percentile size (d_{50}) are typically used to represent natural sediment that is composed of grains with a range of sizes. The ratio of the fluid force and the submerged particle weight yields probably the most commonly used dimensionless parameter, the Shields parameter (θ), in quantifying sediment entrainment and transport:

$$\theta = \frac{\tau_b}{(\rho_s - \rho_w) g D} = \frac{u_*^2}{(s-1) g D} \tag{2.7}$$

where D = grain diameter, and s = sediment specific density = ρ_s/ρ_w.

Settling velocity (w_s) is another key parameter, especially for suspended sediment transport. Under most circumstances, the settling velocity of sediment particles is defined as the terminal velocity through tranquil water. Therefore, it is regarded as one of the physical properties of sediment particle and is not related to the flow regime, although actual settling velocity through turbulent water can be very different from that through tranquil water. Numerous studies have been conducted on particle settling resulting in the development of a variety of empirical formulas. For particles that are smaller than fine sand, Stoke's law of viscous drag can be applied to derive the settling velocity as:

$$w_s = \frac{1}{18}\frac{(s-1)gD^2}{\nu} \quad (2.8)$$

where ν=kinematic viscosity. For larger grains that have faster settling velocities, the drag force is determined based on the quadratic friction law (e.g., Eq. 2.3). Soulsby (1997) examined a large amount of existing data and developed an empirical formula as:

$$w_s = \frac{\nu}{D}\left[(10.36^2 + 1.049D_*^3)^{\frac{1}{2}} - 10.36\right] \quad (2.9)$$

where the dimensionless grain size, another commonly used parameter for sediment transport, is:

$$D_* = \left[\frac{(s-1)g}{\nu^2}\right]^{\frac{1}{3}} D \quad (2.10)$$

Based on the above Eqs. 2.8 and 2.9, the settling velocity for coarse silt (5.0 phi or 0.031 mm) to medium sand (1.0 phi or 0.5 mm) ranges approximately from 0.1 to 8 cm/s.

Sediments which are finer than medium silt (6 phi or 0.016 mm) are often referred to as cohesive sediments. They tend to form aggregates which are larger than the individual grains but with lighter "overall" density than the mineral grains. The settling of cohesive grains is complicated and comprises a significant part of the processes of cohesive sediment transport, and is discussed in the following section on cohesive sediment transport.

2.2.2 Transport of Non-cohesive Sediments in Tidal Environments

Transport of non-cohesive sediment has been studied extensively and is much better understood than the transport of cohesive sediment and of mixed sediment. The following discussion on non-cohesive sediment transport serves two purposes. Firstly, some tidal environments or parts of them are composed of non-cohesive sediments and the subsequent transport relationships are directly applicable. Secondly, theories on non-cohesive sediment transport provide the fundamentals for the present understanding of cohesive and mixed sediment transport.

2.2.2.1 Initiation of Motion

Generally speaking, sediment motion is initiated when the fluid force exceeds the submerged gravitational force (Fig. 2.1). In natural environments, initiation of motion can be very complicated and influenced by numerous factors including the characteristics of the flow (laminar or turbulent), sediment size and shape, sediment sorting, and by presence and characteristics of bedforms. One of the most commonly used tools is the Shields parameter (Eq. 2.7) and the Shields diagram (Fig. 2.3). A critical Shields parameter (θ_c), above which sediment motion is initiated, is defined in the same form as Eq. 2.7:

$$\theta_c = \frac{\tau_c}{(\rho_s - \rho)gD} = \frac{u_{*\_c}^2}{(s-1)gD} \quad (2.11)$$

where the bed shear stress in the original Shields parameter is replaced by the critical bed shear stress (τ_c) and u_{*_c}=critical bed shear velocity. The original Shields diagram has shear velocity u_* on both the horizontal and vertical axes and is quite difficult to use. Soulsby (1997) provided a direct relationship (Fig. 2.4) between the critical Shields parameter (θ_c: Eq. 2.11) and the dimensionless grain diameter (D_*: Eq. 2.10):

$$\theta_{cr} = \frac{0.24}{D_*} + 0.055(1 - e^{-0.020D_*}) \text{ for } D_* > 5 \quad (2.12)$$

$$\theta_{cr} = \frac{0.30}{1 + 1.2D_*} + 0.055(1 - e^{-0.020D_*}) \text{ for } D_* \le 5 \quad (2.13)$$

Equations 2.12 and 2.13 yield the critical shear stress conveniently from sediment grain size. Intuitively, the larger the grain size, the more fluid power (i.e., a higher critical shear stress) it needs for the initiation of motion. However, the relationship is not linear. Soulsby (1997) suggested that the above simple and straightforward equations are also valid for wave motion and combined wave and current.

2.2.2.2 Bedload Transport

After the sediment motion is initiated, it can be transported in three modes, i.e., bedload, suspended load, and washload. Washload has little to no significance in

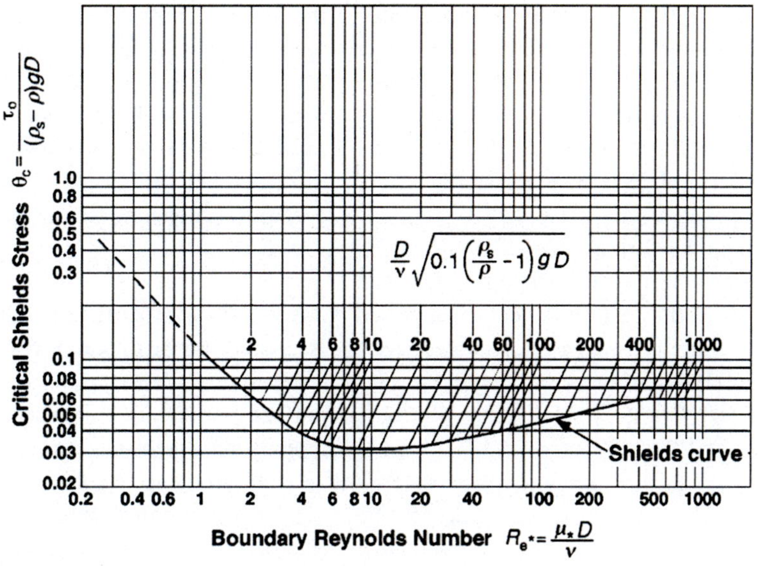

Fig. 2.3 The Shields diagram for initiation of motion under steady currents. In order to calculate the critical shear stress, it is first necessary to determine the value of $\frac{D}{\nu}\sqrt{0.1\left(\frac{\rho_s}{\rho}-1\right)gD}$, the critical Shields parameter is the intersection between this value and the Shields curve

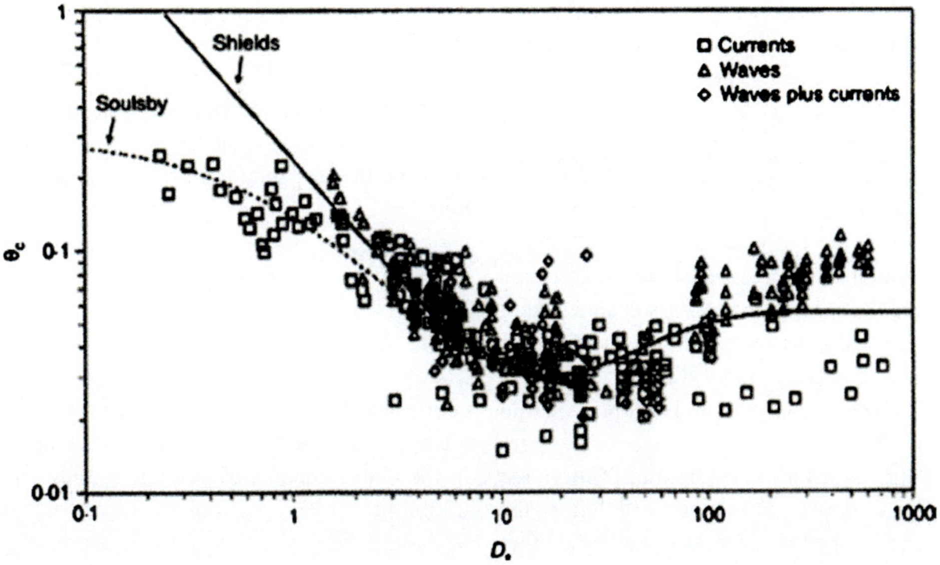

Fig. 2.4 The revised Shields diagram relating the critical Shields stress directly with dimensionless grain diameter (Modified from Soulsby 1997)

sediment deposition and morphology change. In most tidal environments and excluding few local areas (e.g., nearshore breaker zone), most of the sediments that are coarser than fine sand tend to be transported as bedload. The definition of bedload is not always clear. Theoretically, the Bagnold (1956) definition, defined as the part of the total load that is supported by intergranular forcing, is commonly used. This definition is convenient for mathematical modeling. However, from a measurement point of view, especially during field measurements when both bedload and suspended load exist, the Bagnold definition is difficult, if not impossible, to apply. Experimentally, the bedload is sometimes defined as the part of the total load that travels below a certain level (Nielsen 1992). Several modes of bedload motion have been described, including rolling, sliding, and saltating.

Numerous empirical formulas predicting bedload transport by currents have been developed Van Rijn (1984a). One of the earliest and still a commonly used bedload transport formulas was developed by Meyer-Peter and Mueller (1948) as

$$Q_b = 8(\mu\theta - 0.047)^{1.5}(s-1)^{0.5} g^{0.5} d_m^{1.5} \tag{2.14}$$

where Q_b = volumetric bed-load transport rate, θ = the Shields parameter (Eq. 2.7), and μ = an efficiency factor to incorporate the influence of bedforms on bedload transport. Meyer-Peter and Mueller (1948) used mean grain size (d_m) in both Q_b and θ instead of the d_{50} used by many other formulas. Since the Shields parameter θ is proportional to velocity squared, the 1.5 power of θ implies that bedload transport rate is proportional to velocity cubed. This strong non-linear relationship yields much greater transport rates at larger velocities, e.g., during peak ebbing or flooding.

Another commonly used bedload transport formula and approach were developed by Bagnold (1966), via balancing the work needed to be done by the grain-shear stress in moving the bedload particles and the fluid energy per unit area. The Bagnold (1966) formula can account for the gravity forcing associated with a sloping bed. Similar to the Meyer-Peter and Mueller (1948) formula, the Bagnold (1966) formula also suggests that bedload transport rate is proportional to velocity cubed.

Due to nonlinear distortion by bottom friction, the tidal wave may become asymmetrical, with half of the tidal cycle shorter but with faster flow, while the other half lasts longer with slower flow. Generally, the non-linear friction (e.g., Eq. 2.3) in shallow water may result in greater resistance during low tide than during high tide. Therefore, the time delay between low water in the inlet and low water in the inner tidal basin is longer than the time delay at high water. Due to mass conservation, this leads to a shorter duration of the flood and higher flood velocity, as compared to the ebbing tide. Because transport rate is proportional to the velocity cubed, much greater rate of transport occurs during period of greater velocity (Fig. 2.5). Therefore, this time-velocity asymmetry will result in a net transport in the direction of the faster (often flood) flow. Time-velocity asymmetry may have significant influence on sedimentation and morphology in a certain tidal environment or a certain part of a tidal environment.

2.2.2.3 Suspended-Load Transport

Bagnold (1966) suggested that when turbulent eddies have dominant vertical velocity components exceeding the particle fall velocity (w_s), the particle may remain in suspension. Experiments indicate that the vertical turbulent intensity (w') has a maximum value of the same order as the bed-shear velocity (u_*). Therefore, assuming the vertical turbulent velocity roughly equals bed-shear velocity and modifying the Shields parameter, the initiation of sediment suspension (not to be confused with the initiation of motion discussed above) can be described as

$$\theta_{crs} = \frac{u_{*\_crs}^2}{(s-1)gD} = \frac{w_s^2}{(s-1)gD} \tag{2.15}$$

where the subscription "*crs*" denotes critical value for sediment suspension. Generally, suspended particles are assumed to move at the same velocity (u) as the fluid and the suspended sediment transport (q_s) is computed as

$$q_s = \int_a^h u(z)c(z)dz \tag{2.16}$$

where c = suspended sediment concentration, and a = the top level of the bedload layer. The sediment concentration at the a level, c_a, is often referred to as the reference concentration, which is a key parameter in the determination of suspended sediment concentration profile $c(z)$. Equation 2.16 can also be used in designing field or laboratory measurements of suspended sediment transport rate. In other words, both $c(z)$ and $u(z)$ should be measured simultaneously to obtain the transport rate. As discussed above, the current velocity profile typically follows a logarithmic curve (Eq. 2.1). Numerous studies have been conducted on sediment suspension resulting in various models quantifying suspended sediment concentration profiles. Given that many tidal environments tend to have a large amount of fine-grain sediments,

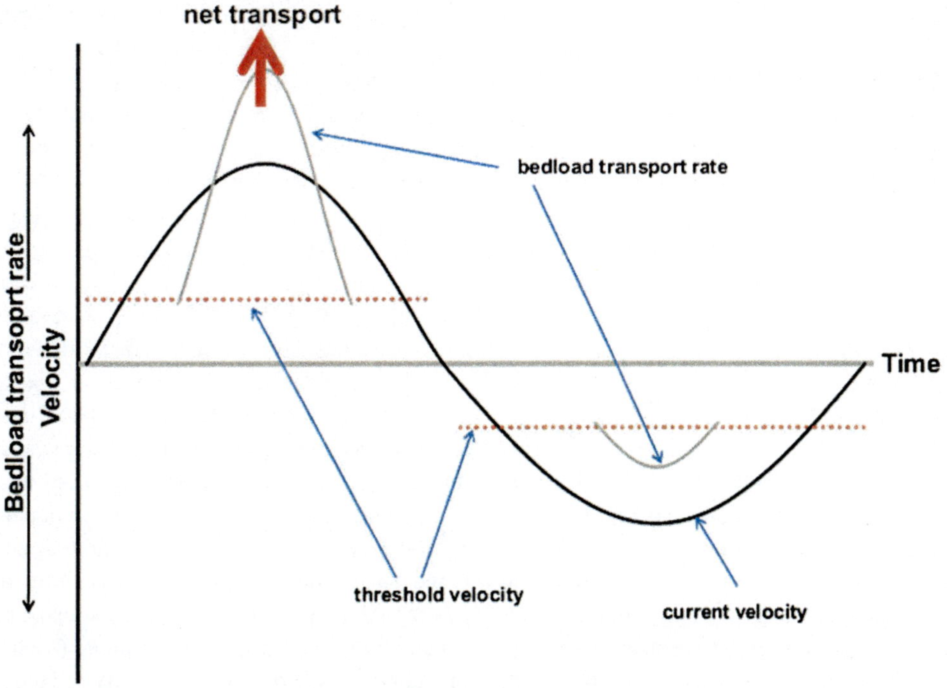

Fig. 2.5 Schematic illustration of time-velocity asymmetry. Because the bedload transport rate is proportional to velocity cubed, much more sediment is transported in the direction of the greater velocity, which results in a net transport toward that direction

suspended sediment transport is an important mode of transport.

From Eq. 2.16, suspended-load transport is strongly influenced by the shapes of the current and sediment-concentration profiles, which are controlled by the intensity of fluid and sediment mixing. Active mixing by highly turbulent flow results in a more homogeneous concentration profile throughout the water column. Weak mixing results in a profile with rapidly decreasing concentration upward. A general understanding is that the turbulence that is responsible for the mixing of sediment through the water column is generated at the sediment-fluid interface. A mixing coefficient (ε_s) is developed to parameterize the sediment mixing (summarized by Van Rijn 1993). Sediment concentration profiles can be obtained by solving the sediment convection and diffusion equation:

$$cw_s + \varepsilon_s \frac{dc}{dz} = 0 \tag{2.17}$$

Equation 2.17 is valid where sediment concentration is low and fall velocity is largely constant. It can be solved analytically with a known mixing coefficient which is determined based on field and laboratory experiments. Two of the commonly used sediment concentration profiles solved from Eq. 2.17 are:

$$\frac{c(z)}{c_a} = e^{-\alpha_1\left(\frac{w_s}{\beta\kappa u_*}\right)\left(\frac{z-a}{h}\right)} \tag{2.18}$$

$$\frac{c(z)}{c_a} = \left(\frac{a}{z}\right)^{-\alpha_2\left(\frac{w_s}{\beta\kappa u_*}\right)} \tag{2.19}$$

where $c(z)$ = suspended sediment concentration profile, c_a = reference concentration, and α_1, α_2, and β are empirical coefficients. Equation 2.18 describes a logarithmic decrease of sediment concentration upward through the water column, solved assuming the mixing coefficient is constant throughout the water column. Equation 2.19 describes a power-function decrease of sediment concentration upward, solved assuming the mixing coefficient is a linear function of depth. Both Eqs. 2.18 and 2.19 show that suspended sediment concentration decrease rapidly upward through the water column. The reference concentration, (a maximum concentration near bed) is determined largely based on field and laboratory data and is the subject of active research.

Once the current and suspended sediment concentration profiles are determined (Eqs. 2.1, 2.18, and 2.19), the suspended-load transport rate can be calculated (Eq. 2.16). However, accurately determining current and suspended sediment concentration profiles is difficult, especially for complicated flow regime. Various simplified formulas have been developed to estimate a total rate of suspended load transport (q_s). A commonly used formula was developed by Van Rijn (1984b):

$$\frac{q_s}{\overline{u}h} = 0.012\left(\frac{\overline{u}-\overline{u_{cr}}}{((s-1)gd_{50})^{0.5}}\right)^{2.4}\left(\frac{d_{50}}{h}\right)\left(\frac{1}{D_*}\right)^{0.6} \quad (2.20)$$

where $\overline{u}$ and $\overline{u_{cr}}$ = depth-averaged velocity and critical velocity, respectively. According to Van Rijn (1984b), Eq. 2.20 is valid for water depth from 1 to 20 m, velocity from 0.5 to 2.5 m/s, and grain size from 0.1 to 2.0 mm, which is applicable to many tidal environments. Equation 2.20 suggests that suspended load transport is proportional to velocity to the power of 3.4. Therefore, similar to the bedload transport, the time-velocity asymmetry will also induce a net suspended sediment transport (Fig. 2.5).

Based on a series of experiments in the Large-scale Sediment Transport Facility at the US Army Engineer Research and Development Center, Wang et al. (2002a, b, 2003) combined a suspended sediment concentration model of Nielsen (1984, 1986) for non-breaking waves and that of Kraus and Larson (2001) for breaking waves and proposed a model predicting the sediment concentration profile under waves as:

$$c(z) = c_a \exp\left(-z\left(\frac{w_s}{k_d\left(\frac{D_w}{\rho}\right)^{\frac{1}{3}}}\frac{1}{h}+\frac{1}{L_s}\right)\right) \quad (2.21)$$

where L_s = turbulent mixing length (Nielsen 1984, 1986), k_d = empirical coefficient, D_w = wave-energy dissipation due to breaking. Due to their oscillatory nature of motion, waves may play a significant role in suspending sediment. The direction of the net suspended-load transport is controlled by tidal flow (the u term in Eq. 2.16); wave forcing may contribute significantly to the c term, and therefore to the magnitude of the transport.

2.2.3 Transport of Cohesive Sediments in Tidal Environments

When sediment grain size is very small, from fine silt to clay, the electrostatic forces between individual particles become comparable to the gravitational forces. The sediments do not behave as individual particles but tend to cohere together forming aggregates, or flocs (Mehta and Patheniades 1975). Sedimentologically, these aggregates behave differently than the individual small particles and non-cohesive particles of similar size due to their lower density and weak strength (Krone 1986). The entrainment and settling of cohesive particle aggregates are complicated and controlled not only by physical properties but also by chemical and biological conditions. Present understanding of cohesive sediment transport is limited and largely based on laboratory experiments. A limitation of the laboratory studies is that natural chemical and biological conditions are difficult to simulate (Mehta 1986a). In addition to the poor compatibility of laboratory and field measurements, compatibility among field measurements is also influenced by data collection methods (Dye et al. 1996; Eisma et al. 1996). Basic processes of flocculation, settling, erosion, and transport are discussed below. As emphasized by all the studies, calibration and verification using *in-situ* field data are crucial for quantifying cohesive sediment transport.

In salt water, the positively charged sodium ions tend to form a cloud of cations around the negatively charged clay particles promoting the formation of flocs via the process of flocculation. Flocculation is caused by particle collisions due to Brownian motion, turbulent mixing, and differential settling, with turbulent mixing identified as the dominant process for most natural systems. Flocculation is influenced by many factors including particle size, sediment concentration, salinity, temperature, and organic content. The size of flocs typically ranges from 0.01 mm to over 1.0 mm. However, the density of flocs is much lower than that of the clay minerals, or that of a quartz particle of similar size. In addition, the floc density decreases with increasing size. When the fluid shearing forces exceed the strength of the flocs, they will break into smaller flocs or particles (Winterwerp 2002).

Settling of fine-grain particles comprises a substantial part in the understanding of cohesive sediment transport, in which flocculation plays an essential role.

In addition to particle size, the settling velocity of flocs is influenced by many factors including salinity and organic content, temperature, sediment concentration, water depth, as well as flow velocity. All the factors that influence flocculation processes, as discussed above, also influence the settling velocity of flocs. Results from numerous laboratory studies are summarized in Mehta (1986b). Methods and results from *in situ* field measurements are summarized in Dyer et al. (1996). Van Rijn (2007b) proposed a general formula to estimate the settling velocity of cohesive particles, or flocs (w_{s_f}):

$$w_{s\_f} = w_{s\_s}\phi_{floc}\phi_{hs} \tag{2.22}$$

where w_{s_s} = the settling velocity of single suspended particle in clear water, ϕ_{floc} = flocculation factor, and ϕ_{hs} = hindered settling factor. For cohesive fine sediment, the individual particles are small. The w_{s_s} can be determined based on the Stokes law (Eq. 2.8). The flocculation factor (ϕ_{floc}) is determined empirically to account for the influence of flocculation on the fine grain settling. Another important consideration for fine-grain deposition is the hindered settling, which occurs at a concentration that is greater than 10 kg/m^3. Hindered settling is the effect that the settling velocity of the flocs is reduced due to an upward flow of fluid displaced by the large amount of flocs. The hindered settling factor (ϕ_{hs}) is less than one and is determined empirically. Significant hindered settling may occur near the bed during slack tides (Van Rijn 2007b).

Based on a relatively large amount of field measurements in several northern European estuaries using an *in-situ* video method (INSSEV) developed by Fennessy et al. (1994), Manning and Dyer (2007) developed a series of empirical formulas predicting the settling velocity of flocs. Different from most other approaches, Manning and Dyer (2007) separated the flocs into two populations based on the size, the macrofloc (>160 μm) and the microfloc (<160 μm), arguing that their settling behavior is significantly different. The settling velocity of macroflocs is a function of suspended particle matter concentration (in mg/l) and turbulent shear stress, while the settling velocity of microflocs is only controlled by the turbulent shear stress. The advantage of the Manning and Dyer (2007) model is that it is based on a large set of in-situ field measurement under a variety of conditions. A weakness of the Manning and Dyer (2007) formulas is that they are purely empirical and the dimensions of the parameters were neglected.

Unique to cohesive sediment, the settling velocity of flocs is also influenced by water depth. This is due to differential settling, because larger flocs settle faster than smaller flocs and therefore may collide with them during the settling. The collisions may result in even larger flocs. Therefore, the size of the flocs increases with water depth in tranquil water. This results in an increasing trend of settling velocity of flocs with water depth. On the other hand, the floc size and corresponding settling velocity are controlled by the shear stress from fluid motion and settling, which tends to break the weakly bounded flocs (Hunt 1986; Winterwerp 2002). This balance tends to create an equilibrium floc size under a particular set of conditions.

Cohesive sediments are transported mostly as suspended load and the rate of transport can be estimated by integrating the product of concentration and velocity with respect to depth (Eq. 2.16). Equation 2.16 provides transport by convection only and may not be applicable for many cases. A numerical model of cohesive sediment transport is typically developed by solving the convection-diffusion equation. A simpler depth-averaged convection-diffusion equation neglecting settling can be written as

$$\begin{aligned}\frac{\partial \overline{c}}{\partial t} + \overline{u}\frac{\partial \overline{c}}{\partial x} + \overline{v}\frac{\partial \overline{c}}{\partial y} - \frac{1}{h}\frac{\partial}{\partial x}\left[hk_x\frac{\partial \overline{c}}{\partial x}\right] \\ -\frac{1}{h}\frac{\partial}{\partial y}\left[hk_y\frac{\partial \overline{c}}{\partial y}\right] - \frac{S}{h} = 0\end{aligned} \tag{2.23}$$

where $\overline{c}$ = depth averaged concentration, $\overline{u}$ = depth average velocity in x direction, $\overline{v}$ = depth average velocity in y direction, k_x and k_y = dispersion coefficient in x, y direction, respectively, h = water depth, and S = source and sink terms. Equation 2.23 can be solved numerically with boundary conditions describing a particular tidal environment.

Deposition of cohesive sediment is a complicated process due to the concentration- and depth-dependent flocculation. Based on a series of laboratory experiments, Krone (1962) found that deposition occurs when the bed shear stress falls below a critical value for deposition, e.g., during slack tide. Krone (1962) further proposed that deposition can be quantified based on a critical shear stress for deposition. Adopting and expanding the concept of critical shear stress for deposition, Mehta and Partheniades (1975) conducted an extensive laboratory investigation on the deposition of cohesive sediment, with an initial sediment concentration ranging from 1 to 10 kg/m^3, which were much greater than the concentrations used in the Krone (1962) experiments. After a short period of rapid deposition, an equilibrium concentration (c_{eq}) was typically observed for a specific set of conditions (Fig. 2.6). Furthermore, the ratio between the equilibrium and initial concentrations remains largely constant and is independent of the initial concentration (c_o). This leads to an insightful conclusion that a given flow can maintain a constant fraction of sediment in suspension regardless of the absolute value of the concentration. It was found that the c_{eq}/c_o ratio depends solely on the bed shear stress, leading to the development of several empirical formulas relating the c_{eq}/c_o to the bed shear stress (Mehta and Partheniades 1975). Three deposition regimes were distinguished based on the bed shear stress, or the ratio of the equilibrium concentration and the initial concentration. They are full deposition, hindered or partial deposition, and no deposition. It is worth noting that wave motion, which generates additional shear stress, may prevent full deposition during slack tides.

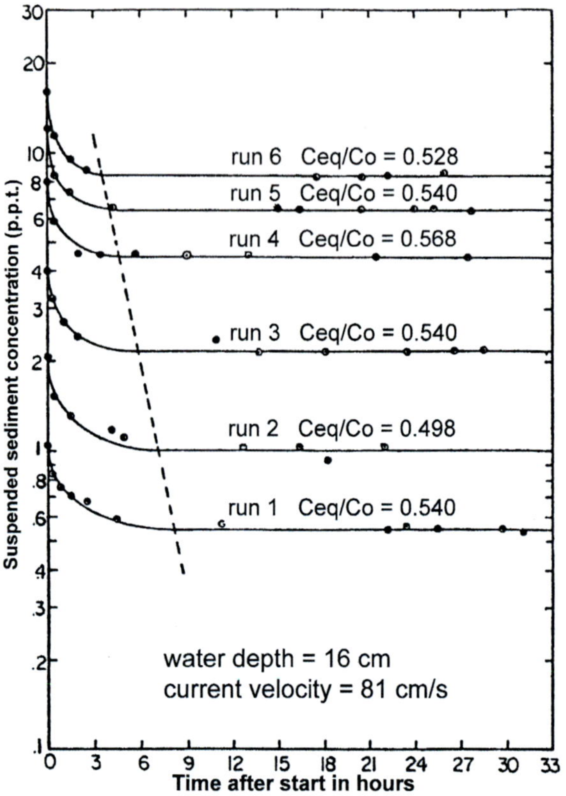

Fig. 2.6 Variation of suspended sediment concentration with time. After an initial rapid decrease, an equilibrium sediment concentration is reached. For a given flow condition, the ratio between the equilibrium concentration and the initial concentration remains constant and is dependent of the initial concentration (Modified from Partheniades 1986)

The complicated flocculation processes and the settling/depositional behavior of flocs, as discussed briefly above, can be applied to understand the commonly observed turbidity maximum in tide-dominated estuaries. Turbidity maximum is one of the most distinctive regional scale sediment transport phenomena in meso- and macro-tidal estuaries with abundant fine-grain sediment (Nichols and Biggs 1985; Dyer 1986). It is a zone with suspended sediment concentrations that is higher than those in the input river as well as in further seaward in the estuary. The turbidity maximum typically occurs near the head of the salt water intrusion with its formation controlled by erosion due to the tidal flow, interaction between fluvial and tidal flows, salinity (typically 1–5‰), and mixing patterns of freshwater and seawater (partially or fully mixed). The influences of these factors on fluctuation and the settling of flocs lead to the formation and maintenance of turbidity maximum. The location and suspended sediment concentration of turbidity maximum vary with variations of fluvial discharge and tidal fluctuations.

Because of fluctuation, settling of fine-grain particles towards the bed can be quite rapid around slack tide. If the sediment flux is high, the water trapped between the flocs may not escape, resulting in the formation of a high concentration layer, i.e., fluid mud, above the bed. Fluid mud can also be formed at locations with overall low energy. Fluid mud can be easily eroded by the tidal currents during the subsequent tide. Extensively developed fluid mud can significantly dissipate the incident wave energy (Wells and Kemp 1986). Fluid mud may behave like Bingham plastics (Sills and Elder 1986). During the flow-accelerating phase of the next tide, the increasing shear stress at the top of the viscous fluid mud layer may re-suspend some of the sediment. Alternatively, the shear stress may also induce failure at the bottom of the layer, leading to the "flow" of the entire fluid mud layer.

Laboratory studies on erosion of cohesive sediments, or the initiation of motion for non-cohesive sediment, are conducted from a similar approach as

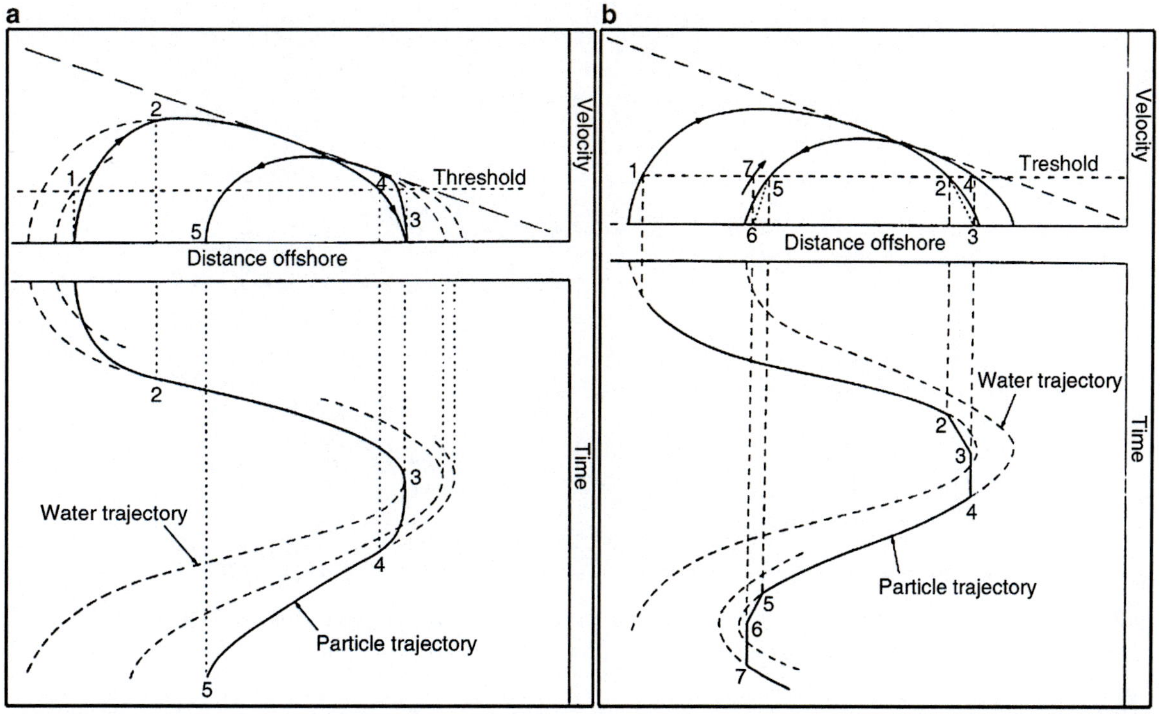

Fig. 2.7 Schematics of scour lag (**a**) and settling lag (**b**) for fine-grain sediments. (**a**) Scour lag: a particle on the bed is suspended into the water column when the threshold velocity is exceeded at point 1. It does not, however, achieve the depth-averaged velocity till point 2, a relatively seaward position. It then travels with the water trajectory to point 3, where we assume it is instantaneously deposited. On the following ebb tide, the particle is suspended, but again lags the flow till point 4 is reached. It is eventually re-deposited at point 5. Considerable landward movement has occurred during the tidal cycle because of the scour lag. (**b**) Settling lag: at position 1, the particle is entrained from the bed and travels with water till point 2, where it starts to settle. Because of the settling lag, it reaches the bed at point 3. On the following ebb tide, it is not entrained till later in the tide cycle when the threshold velocity (greater than the velocity for settling) is reached. The deposition at low water is at position 6. Consequently, the particle has moved shoreward due the settling lag (Modified from Dyer 1994)

their deposition discussed above. Erosion of cohesive sediment from the bed occurs when the bed shear stress from the fluid (current, wave, or combined) exceeds a critical value for erosion. The critical value for erosion is greater than the critical value for full deposition (Mehta 1986a). In other words, more fluid power is needed to erode the cohesive sediments than that to deposit them. This difference in the shear stresses for erosion and deposition, in addition to the time needed for the fine grains to settle, is responsible for the so-called settling lag and scouring lag, which is an important sedimentary process on many mud flats (Postma 1961; Dyer 1998; Bartholdy 2000). The schematics of scour lag and settling lag are well illustrated (Fig. 2.7) and explained by Dyer (1994). Theoretically, the scour and settling lag should result in a net deposition of fine grain sediment landward in the area where maximum velocity during the tidal cycle equals the grain's threshold velocity. At many tidal flats, the net sedimentation on the upper intertidal zone due to settling lag during calm season is often eroded by storm waves during storm season, result in a seasonal cycle (Allen and Duffy 1998; Dyer et al. 2000; Talke and Stacey 2008).

Erosion of cohesive sediment is strongly influenced by the degree of consolidation, as well as the strength of the newly deposited flocs. Therefore, time history of the sediment, which is highly variable temporally and spatially, plays a significant part in cohesive sediment erosion. Consolidation history, as examined in laboratory studies, is usually evaluated using the bulk density of the bed sediment at different depths. The bulk density typically increases with depth below the sediment surface (Partheniades 1986). Under many cases, the erosional process is stopped when it reaches a well-consolidated layer, although the fluid power may remain the same. Bioturbation can play a significant,

Fig. 2.8 Mud clasts formed by bank failure in a small drainage creek in a tidal flat, Changjiang River delta, China

but extremely difficult to quantify, role in the initiation of motion of cohesive sediment (Montague 1986). The influence of bioturbation is illustrated by seasonal variations of erosion and deposition. Allen and Duffy (1998) found that the active algal mat growth during the summer contributed to reduced erosion, when compared to the erosion during the winter season. It has been estimated that the total volume of water in an estuary can be cycled in a few weeks by bottom filter feeders (Dyer 1994).

Overall, two types of erosion can be distinguished: (1) initiation of motion of individual particles and/or aggregates as discussed above; and (2) mass erosion in the form of mud clasts. Freshly deposited mud with little to no consolidation favors type (1) erosion, while consolidated mud may erode in the form of mud clasts, which is often observed on the tidal flats along the banks of tidal creeks of all sizes (Fig. 2.8). Significance of the mass erosion is controlled by the areal distribution of the ephemeral creeks. The physics of mass erosion of a consolidated bed is more of a geotechnical issue than a transport topic, and is beyond the scope of this chapter.

In summary, a sequence of cohesive bed erosion starts with the erosion of the top layer with the lowest critical erosion shear stress. As the erosion continues, the critical erosion stress for the bed increases as the previously buried, more consolidated and erosion-resistant layers becomes exposed. The erosion stops at the level where the critical erosion stress equals the bed shear stress from the fluid. Many other sedimentological (e.g., fluid mud), physical, chemical, and biological factors influence the critical erosion shear stress. Generally, the stress decreases with increasing temperature, pH values, and sand content, while τ_e increases with increases of clay content, organic content, and salinity (Mehta 1986a).

2.2.4 Tidal Currents as Non-steady Flows

Due to the relatively fast fall velocity of non-cohesive particles, (on the order of cm/s), the settling time of non-cohesive sediment through typical water depths of tidal environments is much shorter than the period of the tide. Furthermore, most of the non-cohesive sediments are transported as bedload. Therefore, for non-cohesive sediment transport, tidal flow can be largely regarded as steady flow.

For cohesive sediment, the fall velocity is typically on the order of mm/s (or less) and the settling time through the water column in tidal environments may range from tens of minutes to hours, which is comparable to the tidal period. Therefore, tidal flow should be regarded as non-steady for cohesive sediment transport. Because the critical bed shear stress for deposition requires small velocity, settling of cohesive sediment occurs mostly during slack tides. The regulated acceleration and deceleration of tidal current, in addition to the different bed shear stress for erosion and deposition, is also the cause of the scour and settling lags. Under most circumstances in tidal environments, fine-grain sediments (except rip-up mud clasts) are transported as suspended load and the vertical component of the turbulent fluctuation is typically greater than the settling velocity of the cohesive particles. The movement of cohesive particles can be approximated with the movement of water particles.

When the flow intensity increases, more sediment is suspended into the water column resulting in an increasing sediment concentration. As the flow intensity decreases toward slack tide, deposition occurs resulting in a decreasing sediment concentration throughout the water column. Therefore, the fluctuating intensity of the tidal flow is crucial to the erosion and deposition of fine-grain particles. Due to the large temporal and spatial variations of sediment concentration and the slow settling velocity of the fine-grain particle, sediment transport by dispersion can be significant (Eq. 2.23). Accurately quantifying the critical shear stress for erosion and deposition is essential to the quantification of cohesive sediment transport and deposition. Unfortunately, these two parameters are influenced by many site specific factors including physical, chemical, as well as biological ones. *In situ* data are crucial to the quantification of cohesive sediment transport and deposition.

2.3 Summary

Physical processes of sediment transport in tidal environments are extremely complicated and influenced by numerous hydrodynamic and sedimentological factors over a wide range of temporal and spatial scales. Both tides and waves play significant roles in the entrainment and transport of both cohesive and non-cohesive particles. Sediment transport is composed of three phases, initiation of motion (erosion), transport, and deposition. Various commonly used empirical formulas are provided in this chapter for the quantification of the three phases.

In tidal environments, the coarser, non-cohesive sediments are typically transported as bedload, forming various types of bedforms. The bedforms in turn influence the fluid motion and sediment transport. The finer cohesive sediments tend to be transported as suspended load; their deposition occurs mostly during slack tides. Rate of sediment transport is generally proportional to flow velocity to the third to fifth power. This non-linear relationship leads to a net transport in the direction of the faster velocity in the tidal environments with a time-velocity asymmetry. Due to the slow settling velocity of the fine cohesive sediment and a difference between the critical shear stress for erosion and deposition, a scour lag and a settling lag exists in many tidal environments resulting in a fining depositional trend landward. The periodic reversing of tidal flow directions results in the commonly observed bi-directional sedimentary structures (e.g., the herringbone cross-stratification). The relatively tranquil slack tides between the flood and ebb tides allow the deposition of muddy layers in between sandy layers deposited during the flood and ebb tides, forming lenticular, wavy, and flaser bedding.

References

Allen PA (1997) Earth surface processes. Blackwell Science, London, 404 p

Allen JRL, Duffy MJ (1998) Temporal and spatial depositional patterns in the Severn Estuary, southwestern Britain: intertidal studies at spring-neap and seasonal scales, 1991–1993. Mar Geol 146:147–171

Bagnold RA (1956) Flow of cohesionless grains in fluid. R Phil Soc Lond Trans 249:235–297

Bagnold RA (1966) An approach to the sediment transport problem from general physics. U.S. geological survey professional paper 422-I, Washington, DC

Bartholdy J (2000) Process controlling import of fine-grained sediment to tidal areas: a simulation model. In: Pye K, Allen JRL (eds) Coastal and estuarine environments: sedimentology, geomorphology, and geoarchaeology. Geological Society Lond, Spec Publ 175:13–29

Christie MC, Dyer KR, Turner P (1999) Sediment flux and bed level measurements from a macro tidal mudflat. Estuar Coast Shelf Sci 49:667–688

Davis RA, Hayes MO (1984) What is a wave dominated coast? Mar Geol 60:313–329

Dyer KR, Cornelisse M, Dearnaley MP, Fenness MJ, Jones SE, Kappenberg J, McCave IN, Pejrup M, Puls W, Van Leussen W, Wolfstein KA (1996) A comparison of *in situ* techniques for estuarine floc settling velocity measurements. J Sea Res 36:15–29

Dyer KR (1986) Coastal and estuarine sediment dynamics. Wiley, Chichester, New York, 342 p

Dyer KR (1994) Estuarine sediment transport and deposition. In: Pye K (ed) Sediment transport and depositional processes. Blackwell Science Publication, Oxford, pp 193–216

Dyer KR (1998) The typology of intertidal mudflats. In: Black KS, Paterson DM, Cramp A (eds) Sedimentary processes in the intertidal zone. Geological Society London, Spec Publ 139:11–24

Dyer KR, Christie MC, Wright EW (2000) The classification of intertidal mudflats. Cont Shelf Res 69:1039–1060

Eisma D, Bale AJ, Dearnaley MP, Fennessy MJ, Van Leussen W, Maldiney M-A, Pfeiffer A, Wells JT (1996) Intercomparison of *in situ* suspended matter (floc) size measurements. J Sea Res 36:3–14

Fennessy MJ, Dyer KR, Huntley DA (1994) INSSEV: an instrument to measure the size and settling velocity of flocs in-situ. Mar Geol 117:107–117

Fredsoe J, Deigaard R (1992) Mechanics of coastal sediment transport. World Scientific, Singapore, 369 p

Hunt JR (1986) Particle aggregate breakup by fluid shear. In: Mehta AJ (ed) Estuarine cohesive sediment dynamics. Springer, New York, pp 85–109

Kaminsky G, Kraus NC (1994) Evaluation of depth limited breaking wave criteria. In: Proceedings of ocean wave measurement and analysis, Waves'94, ASCE Press, New York, pp 180–193

Kraus NC, Larson M (2001) Mathematical model for rapid estimation of coastal inlet entrance channel infilling. Coastal engineering technical note CETN-IV-35, U.S. Army Engineering Research and Development Center, Vicksburg, Mississippi

Krone RB (1962) Flume studies of the transport of sediment in estuarial shoaling processes. Final report, Hydraulic Engineering and Sanitary Engineering Research Laboratory, University of California, Berkeley

Krone RB (1986) The significance of aggregate properties to transport processes. In: Mehta AJ (ed) Estuarine cohesive sediment dynamics. Springer, New York, pp 66–84

Kvale EP, Archer AW, Johnson HR (1989) Daily, monthly, and yearly tidal cycles within laminated siltstones of the Mansfield Formation of Indiana. Geology 17:365–368

Lee HJ, Jo HR, Chu YS, Bahk KS (2004) Sediment transport on macrotidal flats in Garolim Bay, west coast of Korea: significance of wind waves and asymmetry of tidal currents. Cont Shelf Res 24:821–832

Li C, Wang P, Fan D, Dang B, Li T (2000) Open coast intertidal deposits and the preservation potential of individual lamina: a case study from east-central China. Sedimentology 47:1039–1051

Manning AJ, Dyer KR (2007) Mass settling flux of fine sediments in Northern European estuaries: measurements and predictions. Mar Geol 245:107–122

Mehta AJ (1986a) Characterization of cohesive sediment properties and transport processes in estuaries. In: Mehta AJ (ed) Estuarine cohesive sediment dynamics. Springer, New York, pp 290–325

Mehta AJ (1986b) Estuarine cohesive sediment dynamics. Springer, New York, 473 p

Mehta AJ, Patheniades E (1975) An investigation of the deposition properties of flocculated fine sediments. J Hydraul Res 12:361–381

Meyer-Peter E, Mueller R (1948) Formulas for bed-load transport. In: Proceedings of international association hydraulic research, 2nd Congress, Stockholm, Sweden, pp 39–64

Montague CL (1986) Influence of biota on erodibility of sediments. In: Mehta AJ (ed) Estuarine cohesive sediment dynamics. Springer, New York, pp 251–269

Nichols MM, Biggs RB (1985) Estuaries. In: Davis RA Jr (ed) Coastal sedimentary environments. Springer, Berlin, pp 77–186

Nielsen P (1984) Field measurements of suspended sediment concentrations under waves. Coast Eng 8:51–73

Nielsen P (1986) Suspended sediment concentrations under waves. Coast Eng 10:23–31

Nielsen P (1992) Coastal bottom boundary layers and sediment transport. World Scientific, Singapore, 324 p

Partheniades E (1986) A fundamental frame for cohesive sediment dynamics. In: Mehta AJ (ed) Estuarine cohesive sediment dynamics. Springer, New York, pp 219–250

Postma H (1961) Transport and accumulation of suspended matter in the Dutch Wadden Sea. Neth J Sea Res 1:148–190

Pye K (1994) Sediment transport and depositional processes. Blackwell Scientific, London, p 397

Sill GC, Elder DM (1986) The transition from sediment suspension to settling bed. In: Mehta AJ (ed) Estuarine cohesive sediment dynamics. Springer, New York, pp 192–205

Soulsby R (1997) Dynamics of marine sands. Thomas Telford, London, 249 p

Talke SA, Stacey MT (2003) The influence of oceanic swell on flows over an estuarine intertidal mudflat in San Francisco Bay. Estuar Coast Shelf Sci 58:541–554

Talke SA, Stacey MT (2008) Suspended sediment fluxes at an intertidal flat: the shifting influence of wave, wind, tidal, and freshwater forcing. Cont Shelf Res 28:710–725

Van Rijn LC (1984a) Sediment transport, Part I: Bed load transport. J Hydraul Eng 110:1431–1456

Van Rijn LC (1984b) Sediment transport, Part II: Suspended load transport. J Hydraul Eng 110:1613–1641

Van Rijn LC (1993) Principles of sediment transport in rivers, estuaries and coastal seas. Aqua Publications, The Netherlands

Van Rijn LC (2007a) Unified view of sediment transport by currents and waves. I: Initiation of motion, bed roughness, and bed-load transport. J Hydraul Eng 133:649–667

Van Rijn LC (2007b) Unified view of sediment transport by currents and waves. II: Suspended transport. J Hydraul Eng 13:668–689

Van Rijn LC (2007c) Unified view of sediment transport by currents and waves. III: Graded beds. J Hydraul Eng 133:761–775

Wang P, Smith ER, Ebersole BA (2002a) Large-scale laboratory measurements of longshore sediment transport under spilling and plunging breakers. J Coast Res 18:118–135

Wang P, Ebersole BA, Smith ER, Johnson BD (2002b) Temporal and spatial variations of surf-zone currents and suspended-sedment concentration. Coast Eng 46:175–211

Wang P, Ebersole BA, Smith ER (2003) Beach profile evolution under plunging and spilling breakers. J Waterway Port Coastal Ocean Eng 129:41–46

Wells JT, Kemp GP (1986) Interaction of surface waves and cohesive sediments: field observations and geological significance. In: Mehta AJ (ed) Estuarine cohesive sediment dynamics. Springer, New York, pp 43–65

Winterwerp JC (2002) On the flocculation and settling velocity of estuarine mud. Cont Shelf Res 22:1339–1360

Tidal Signatures and Their Preservation Potential in Stratigraphic Sequences

3

Richard A. Davis, Jr.

Abstract

Several indicators of tidal influence are preserved in modern and ancient stratigraphic sequences. These tidal signatures are dominated by cyclic deposition. The cycles may represent as short as semi-diurnal tides or as long as multiple years. Most commonly they represent daily or lunar cycles. These cycles are most commonly represented by some alternation of sand and mud; so-called heterolithic deposits. Some are monolithic. The scale of these rhythmic packages ranges from a few millimeters to several decimeters.

The depositional environments in which these tidal sequences accumulate include intertidal and subtidal positions to depths of at least hundreds of meters. The most common are intertidal flats and their contained channels in estuaries and deltas as well as in coastal bays and open coasts. Preservation potential ranges from poor to very good. Tidal channels tend to be among the best preserved whereas the upper intertidal zone is the most poorly preserved.

3.1 Introduction

Tides are significant process factors in coastal environments as discussed in the previous chapter. They produce currents that move sediment and eventually deposit it, and in so doing, they create signatures that may be preserved in the stratigraphic record. These tidal signatures are important in reconstructing the ancient environment of deposition in which the stratigraphic sequences of interest were deposited. This chapter is the third of four "generic" chapters in this volume in that they are not tied to a specific environment. The aim here is to acquaint the reader with various tidal signatures and the possible environmental settings in which they occur. Details of their environmental and stratigraphic positions are found in the subsequent chapters where individual tidally-influenced depositional environments are discussed in detail.

The movement of sediments by tidal currents was recognized in the times of the ancient Greeks and Romans. Those people noticed that the regularity of tidal fluctuations was related to lunar cycles. The detailed investigation of how tidal currents move sediment and how this sediment accumulates, however. has been a subject of sedimentological research for less than a century. Much of the initial work was on tidal flats on the North Sea coast of Germany and The Netherlands. Much of the efforts of these investigators

R.A. Davis, Jr. (✉)
Department of Geology, Coastal Research Laboratory, University of South Florida, Tampa, FL 33620, USA

Harte Research Institute for Gulf of Mexico Studies, Texas A&M University – Corpus Christi, TX 78412, USA
e-mail: rdavis@usf.edu

R.A. Davis, Jr. and R.W. Dalrymple (eds.), *Principles of Tidal Sedimentology*,
DOI 10.1007/978-94-007-0123-6_3,

focused on muddy environments and on ichnology (see Chap. 4). Many would credit George deVries Klein with the first detailed work on tidal sedimentology in North America during his doctoral research on the Bay of Fundy while at Yale University under the direction of John Sanders (Klein 1963).

Although rudimentary, he showed how basic tidal processes produced recognizable sedimentary structures that could be present in the stratigraphic record. The term *tidalite* had is origin as the result of this work (Klein 1971, 1972). It is now applied to all sediments and sedimentary structures that have accumulated under the influence of tides. The following discussion will address how tidal signatures can be recognized in stratigraphic successions and how they can be distinguished from similar features that are deposited under non-tidal conditions. Some structures and sequences can form in multiple depositional environments, e.g. fluvial, deltaic, shallow marine or lacustrine. Some situations exist where sediments also accumulate under tidal conditions but there are no tidal signatures preserved, e.g. estuaries, tidal flats, or tidal inlet systems. The lack of tidal signatures may be due to energy conditions with little or no sediment transport, or may be due to the influence of waves that mask or rework any evidence of tidal activity. Because of these circumstances, it is pretty certain that the strata that are interpreted as containing tidalites represent only a portion of those sediments that have actually accumulated under tidal conditions.

As discussed in the Chap. 1, tides occur in cycles of various durations. These cycles are an important key to the production of tidal signatures. The most important and most easily recognizable of these cycles in the stratigraphic record are those of daily and monthly duration, but longer cycles can also be recognized. The daily or diurnal cycles may represent one or two tidal cycles depending upon the location; diurnal or semi-diurnal tides. The lunar monthly cycles represent 2-week or fortnightly periods associated with phases of the moon. A gradual change occurs from relatively high energy to relatively low and back to high energy again in both cycles. These energy cycles are reflected in a combination of grain size and bed thickness changes. In the absence of erosion and the presence of sufficient sediment supply, the packages of sediment that accumulate will display easily recognizable tidal signatures. Longer cycles also exist that can be recognized; monthly, seasonal and longer (see Chap. 1).

Although tides and related processes produce various signatures that permit us to recognize tidal influence in various environments, there are other features that are associated with various tidal environments, especially tidal flats. Some are biogenic, some are chemical and others are physical.

3.2 Sedimentary Structures

Various types of features are known that form on the sediment surface that are typical products of tidally influenced phenomena; a few that are indicative of only tidal conditions. These features include biogenic structures and physically produced structures. Because of their development on the sediment-fluid interface, they may be destroyed during burial and even if preserved, they may not be visible because most of them require exposure of the bed surface to be recognized.

3.2.1 Biogenic Structures

3.2.1.1 Surface Structures

Benthic organisms occupy several tidally-influenced environments; on both intertidal and subtidal surfaces. Generally these are somewhat low physical energy environments because regular and intense sediment mobility precludes colonization by epifaunal or infaunal organisms. Less dynamic substrates will permit organisms to be present on a regular basis. Those animals that are exclusively epifaunal leave tracks, trails, resting places, and surface expressions of burrows that have a low preservation potential (Fig. 3.1). The only circumstances whereby such features might become preserved in the stratigraphic record would be during a fairly deep and rapid burial. Recognition generally requires exposure of a bedding surface.

The one biogenic surface feature that is likely to be preserved depending on sediment composition, and displays environmental significance is microbial mats that are composed of blue-green algae, now more commonly termed cyanobacteria. These mats typically are associated with the spring high tide to supratidal portion of tidal flats. They are especially common along wind tidal flats (Fig. 3.2), such as along the coast of Texas (Miller 1975). This environment typically does not experience regular lunar tides but is subjected to water level change produced by wind. The south

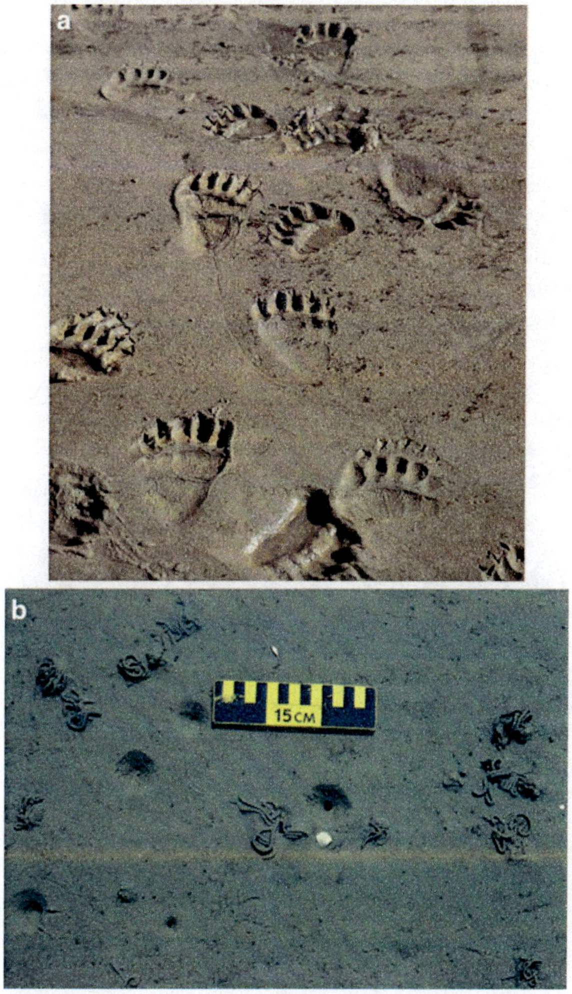

Fig. 3.1 Examples of intertidal tracks (bear) (**a**) and burrow spoil (**b**) that are common on intertidal flats but that have very little preservation potential

Texas coast has extensive such flats. These microbial mats are useful to interpret the stratigraphic record because they are extensive, easily recognized, and have a fairly high preservation potential because they provide cohesion to the substrate, especially in carbonate successions (e.g. sabkhas) where cementation facilitates preservation. Additionally, they are they are restricted to the intertidal and supratidal zone where erosion is relatively uncommon permitting them to be buried by washover sediment that is carried across barrier islands.

3.2.1.2 Three-Dimensional Structures

Three-dimensional cyanobacterial structures, typically called stromatolites or algal stromatolites, are also important surface structures in the intertidal zone. For many years these features were exclusively associated with the intertidal zone with the most spectacular examples being in Shark Bay in Western Australia (Fig. 3.3). The combination of the intertidal environment with carbonate sediment and tropical climate provides for rapid lithification and high preservation potential. These structures are present in the rock record from the Precambrian to the Holocene (Fig. 3.4). They are now known to develop subtidally also, but in locations where tidal currents are swift (Dill et al. 1986) so that it is appropriate to consider them as tidalites. Stromatolites are typically carbonate features but have been reported from siliciclastic sediments as well (Davis 1966).

Many organisms are infaunal and produce distinct structures, typically in the form of burrows. These

Fig. 3.2 Microbial mats (*arrows*) on the surface and in the stratigraphy of a wind-tidal flat on Padre Island, Texas, USA

Fig. 3.3 Intertidal algal stromatolites in Shark Bay, Western Australia (Photo by M Marsden)

Fig. 3.4 Well-preserved algal stromatolites in the Shakopee Formation (L. Ord.) in southwestern Wisconsin, USA

develop in intertidal and subtidal deposits and some are restricted to specific environments. Most of these burrows are near vertical in orientation cutting across bedding structures. In actuality, these burrows are typically found in sediments that are extensively bioturbated and have little or no stratification preserved (Fig. 3.5). They do tend to be preserved easily because of their occurrence in the stratigraphy beneath the surface.

Because some taxa live only in certain environments influenced by tidal processes, the trace fossils that they produce are used to recognize specific depositional environments (see Chap. 4). These may be intertidal or subtidal. The burrows or escape structures incorporated in these sediments cannot be considered as uniquely tidalites. They may be present, however, in tidalites. In other words, they do not have a distinctly tidal signature but they may occur in tidal environments that may or may not have tidal signatures.

3.2.2 Physical Structures

These structures form as the result of sediment transport and accumulation. Most are three-dimensional with both surface and stratigraphic expression.

Fig. 3.5 Example of a lug worm, *Arenicola*, burrow from an intertidal flat on the Wadden Sea of Germany

3.2.2.1 Surface Structures

The sediment surface may show a wide range of configurations produced by physical processes. Bedforms are by far the most common and diverse of the surface structures, but desiccation features can also be common.

Those sediments that have even a modest amount of cohesiveness contain enough mud so that exposure to the atmosphere leads to desiccation. The result is the formation of mud cracks (Fig. 3.6). These features are easily recognized on bedding planes and may also be recognized from the stratigraphic perspective (Fig. 3.7). Typically the surface of them is slightly concave upward, especially at the margins of each "crack". Desiccation features can develop on intertidal surfaces but may also form in various other environments that have no tidal influence such as fluvial flood plains. They are not the result of tidal processes but of alternating wet and dry conditions. Such alternations may coincide with tidal cycles, e.g. the spring high-tide position on a tidal flat. Here the sediment might be deposited during spring high tide and is then followed by several days of exposure when desiccation can take place. The alternations of wet and dry can be non-cyclic also, especially on wind-tidal flats. In summary, desiccation features can be associated with tidal conditions and with tidalites but they do not themselves, serve as distinctive tidalite signatures.

Bedforms

Interaction of fluid motion and the sediment substrate develops regularly undulating surfaces called bedforms. Tidal flux produces currents of varying velocities over time and space that produce a spectrum of scales and geometries of these bedforms. Scale, that is wave length and height, is dependent on current strength, water depth and grain size. Morphology is also influenced by current strength. Some bedforms are linear and others are so-called three-dimensional bedforms (Fig. 3.8).

Bedforms are produced in many environments and in subtidal, intertidal and non-tidal environments. They are commonly preserved in the ancient record and are displayed on bedding planes (Fig. 3.9). None of this necessarily means tidal processes are required to produce such features; they are not. All conditions where fluid flow moves over the sediment substrate at speeds in excess of the threshold of sediment movement produce bedforms and many depositional environments experience such conditions.

3.2.2.2 Three-Dimensional Structures

A special aspect of bedform development and preservation is commonly associated with tidal conditions. Those bedforms that develop on intertidal flats and subtidally in estuaries and deltas may accumulate fine sediment in their troughs (Fig. 3.10) as the result of

Fig. 3.6 Good examples of desiccation and the formation of mud cracks in (**a**) modern sediments, and (**b**) in Ordovician strata of western Maryland, USA

Fig. 3.7 Stratigraphic expression of desiccation features in an ancient carbonate sequence

Fig. 3.8 Photograph showing (**a**) linear bedforms on the Wadden Sea of Germany and (**b**) three-dimensional bedforms in Louisa Creek, Queensland, Australia

sediment settling during high-tide (or low tide), slack water conditions. Here a combination of relatively high-energy conditions exist as tides flood and ebb producing currents that move sediment, and low-energy conditions during slack tide when fine sediment can settle into bedform troughs. Sediment accumulation under these circumstances incorporate the mud and produce what is called *flaser bedding* (Reineck and Wunderlich 1968) after the German word for "flame". Depending on the relative amount of mud versus sand in these successions, wavy bedding or lenticular bedding (Fig. 3.11) may occur.

Because of the alternation of sand and mud on a regular basis, tidal processes are interpreted to control sedimentation. These types of bedding are called heterolithic because they contain both relatively coarse sediment and fine sediment. They are very common on intertidal environments and are present in some subtidal environments such as estuaries, and the interdistributaries of deltas. They are easily preserved in the stratigraphic record (Fig. 3.12) and are an excellent tidal signature. However, they may also form under non-tidal conditions, especially in fluvial environments. The primary requirement is a combination of

Fig. 3.9 Example of bedforms (ripples) preserved in the ancient stratigraphic record in the Cretaceous of the San Juan Basin, New Mexico, USA. Note also that there are trails on the surface, probably made by a mobile invertebrate

Fig. 3.10 Ripples with mud in the troughs; the beginnings of the formation of flaser bedding: (**a**) a modern example from the Bay of Fundy, Canada and (**b**) from the Precambrian Baraboo Quartzite of Wisconsin, USA

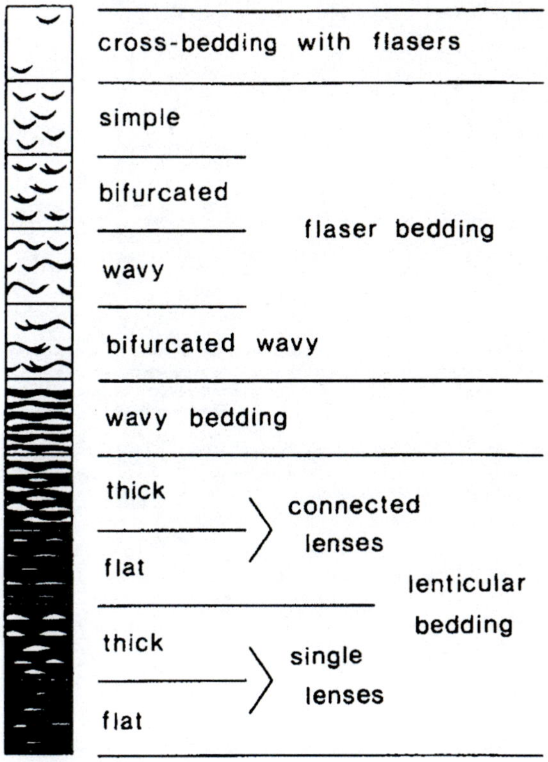

Fig. 3.11 Schematic diagram showing flaser bedding, wavy bedding, and lenticular bedding (After Reineck and Wunderlich 1968)

relatively high and relatively low energy conditions with accompanying sediment. These need not be tidal or cyclic in nature; the same products can be produced. Stacked sequences of such bedding types do, however, strongly suggest tidal phenomena as modes of accumulation because of their cyclicity.

Cross-Strata

Migrating bedforms produce cross stratification with a scale that is related to the wave height of the bedform that produced it and is in proportion to the original bedform that produced it. These features develop wherever currents move over non-cohesive sediments of sand or fine gravel particle size. Tidal-influenced environments produce special types of cross-strata that leave tidal signatures. These types are (1) bidirectional or herringbone cross-strata, (2) reactivation surfaces and (3) tidal bundles. Production of each of these structures is discussed in the following paragraphs.

The flood and ebb oriented currents in tidal environments are commonly in opposite directions or at least in directions with opposing currents at greater than 90° angles. As bedforms migrate with flood and ebb currents, the direction of cross-strata changes to reflect these currents. With the addition of sediment, the result is a stacking of cross-sets dipping in opposite

Fig. 3.12 Photo showing preserved flaser bedding in the lower half of this modern stratigraphic sequence from the Bay of St. Malo, France

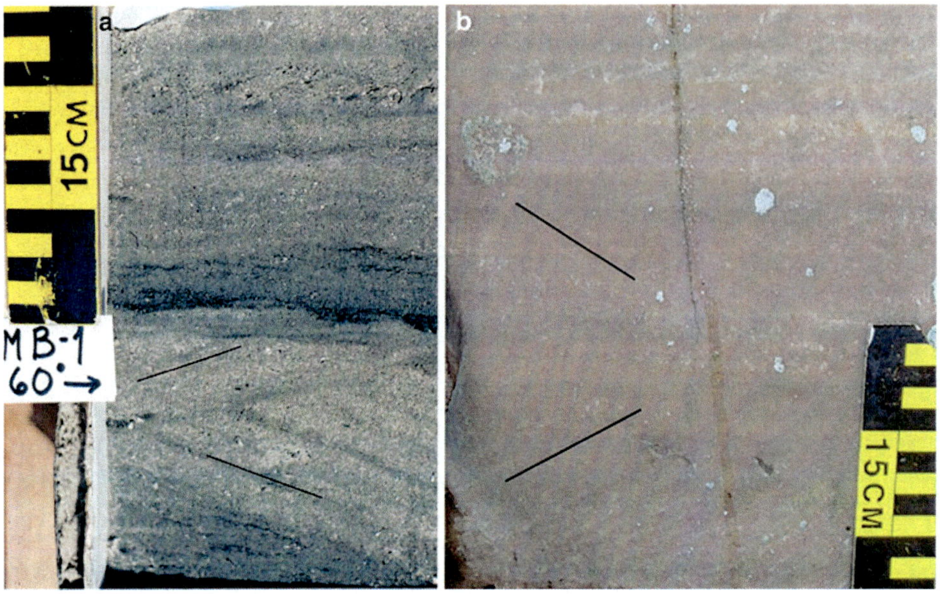

Fig. 3.13 Examples of (**a**) modern bidirectional cross-strata from Martens Plate, German Wadden Sea and (**b**) and example of bidirectional cross-strata from 1.7 billion year old Baraboo Quartzite in Wisconsin, USA

directions (Fig. 3.13). Depending on sediment availability and the balance between flood and ebb currents, herringbone cross-strata will form. It is not common, even in known tidal environments because herringbone cross-strata require near equal flood and ebb tidal current conditions and this condition is atypical. Generally tidal currents display a distinct asymmetry during flood and ebb conditions at a given location with the development of mutually exclusive channels (see Chap. 2). Misinterpretation of bidirectional processes can occur when observing nested sets of trough cross-beds where apparent herringbone may be present.

Because of time-velocity asymmetry, it is unusual that the velocity and amount of tidal energy is the same at any given location for flood and ebb currents. These factors contribute to the other special type of bedding features that can be attributed to tidal activity. In a tidal cycle at a specific location there will be one current, it can be either flood or ebb, that is dominant, and the other is subordinate. The dominant current will move the bedforms and produce cross-stratification. The subordinate current, moving in the opposite direction, is not strong enough to reverse the direction of the migrating bedforms, but it does scour the upper portion of the bedforms, removing some sediment of the upper part of the bedform and producing an erosional, sometimes undulating surface. When the next cycle takes place, the dominant current again forms migrating bedforms. The cycle repeats itself during each tidal cycle. The contact that forms between successive migrating bedforms is called a *reactivation surface* (Klein 1970). A reactivation surface can be recognized as one that interrupts the cross-strata that has the same direction and inclination above and below it (Fig. 3.14). Reactivation surfaces are quite prominent in many ancient stratigraphic sequences (Fig. 3.15). They are representative of subtidal, tidal-influenced environments such as channels. Although not the intent of Klein (1970) the parallel or near-parallel surfaces that separate stacked cross-sets are also produced in the same fashion and are technically a surface that separates two periods of active sedimentation. Excellent reactivation surfaces can also be produced in unidirectional environments such as streams (Collinson 1968; Nio and Yang 1991). These occur when one megaripple (dune) overtakes the other or when wave action or water level changes lead to the erosion of the peak of the bedform.

Horizontal Laminations

Some diagnostic sedimentary structures are based on thin, flat beds. Like other structures formed under tidal influence, these also reflect the flood, slack, ebb, slack conditions of a tidal cycle. The currents during flood and/or ebb tide move and deposit relatively coarse sediment, typically sand. During high slack and low slack, fine suspended sediment settles to the bottom. The result is an alternation of sand and mud forming what is called tidal bedding (Reineck 1967). This is one of several types of *rhythmites*; sequences of sediments that are produced by cyclic conditions. Another common type of planar rhythmites is varves, cyclic annual sediment layers in mid-latitude lakes.

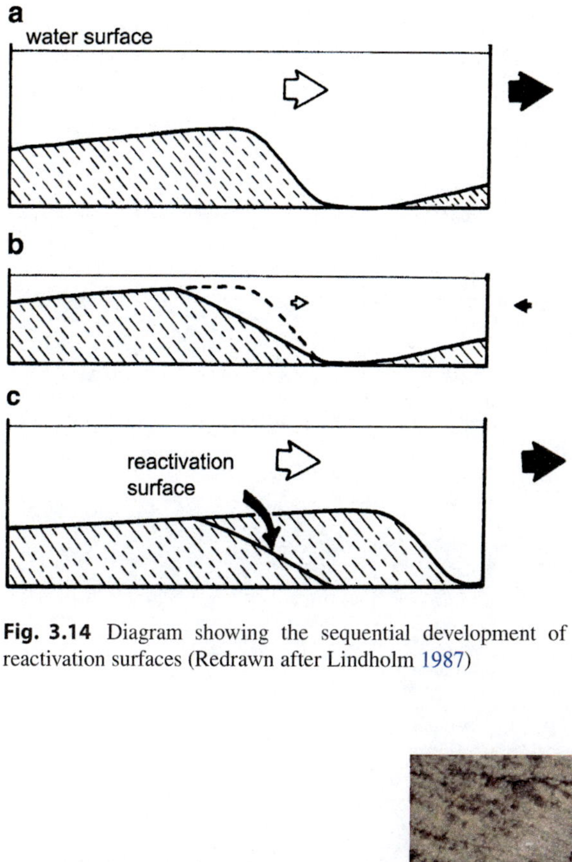

Fig. 3.14 Diagram showing the sequential development of reactivation surfaces (Redrawn after Lindholm 1987)

Tidal bedding units tend to be quite thin; only millimeters to a centimeter or so in thickness (Fig. 3.16). They represent individual tidal cycles and may accumulate in very thick successions that represent many tidal cycles (Fig. 3.17). Their environment of deposition is typically intertidal, but they may also form in very shallow subtidal locations. Most are *heterolithic*, that is, they display more than one distinct sediment type. They may, however, also be monolithic (Fig. 3.18).

The rhythmites that comprise tidal bedding may include two to four individual layers in a semi-diurnal setting (Archer 1998). The number depends on the time-velocity asymmetry. The dominant current, ebb or flood, deposits the coarser material, typically sand. The strength of the subordinate current that determine the presense/absence of relatively fine layers and the number of elements in the individual tidal record (Fig. 3.19). Such conditions can produce four laminations; two relatively thin and the other two relatively coarse. The reduction in sedimentation by the subordinate current will reduce the number from four to three and eventually to two layers; both of the coarser type (Fig. 3.19). The nature of the tidal bedding rhythmites changes within the spring and neap lunar cycle (Fig. 3.20). Measurement of individual layers and plotting them over time shows this spring-neap cyclicity well. Detailed analysis of these cycles permits establishing the moon-earth relationship over geologic time (Kvale et al. 1999). Additionally, outcrops and preservation permitting, a transition exists from a complete spring-neap sequence of tidal bedding to one that

Fig. 3.15 Photo of a stratigraphic sequence showing examples of reactivation surfaces in the Precambrian Baraboo Quartzite of Wisconsin, USA

Fig. 3.16 Diagram showing the nature and development of tidal bedding (From Dalrymple 1992)

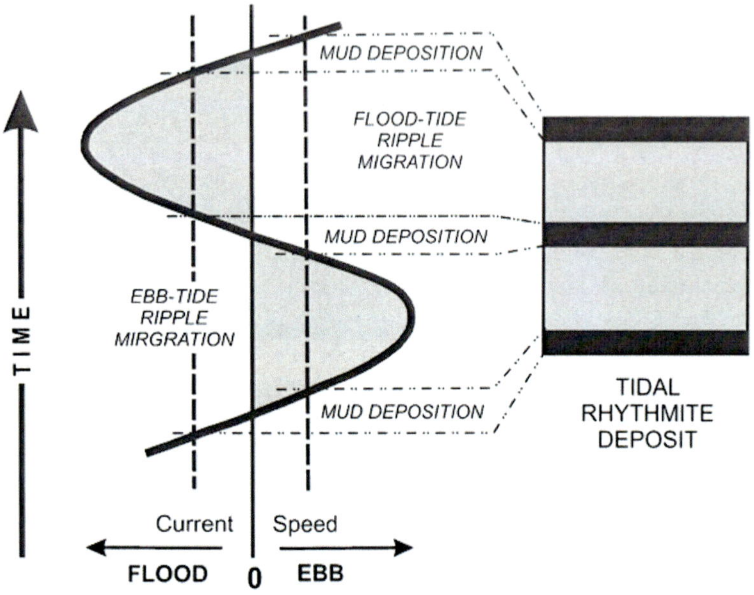

Fig. 3.17 Photos showing examples of (**a**) modern tidal bedding from the central coast of China and (**b**) tidal bedding from the Miocene of Florida, USA

Fig. 3.18 Photo of tidal bedding that is monolithic from the Baraboo Quartzite (Precambrian) of Wisconsin, USA

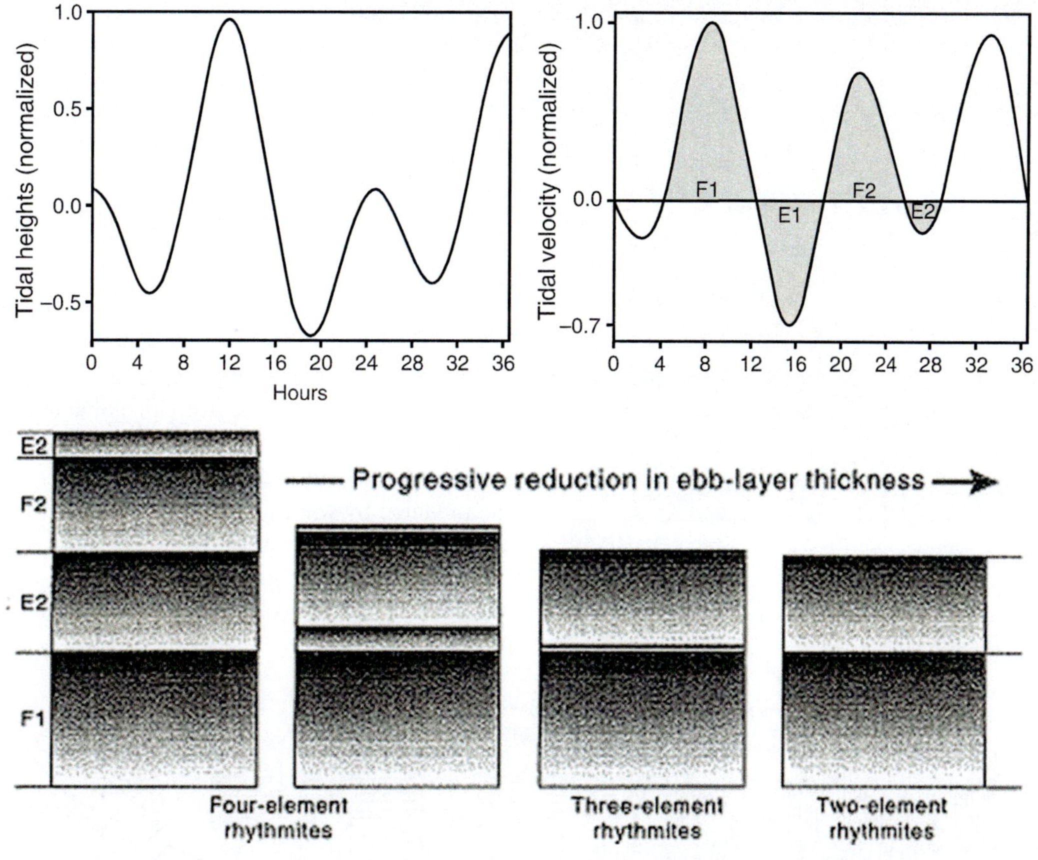

Fig. 3.19 Diagram showing time-velocity curves and the reduction in individual tidal layers. The dominant current here is the flood and the subordinate current is the ebb (From Archer 1998)

records only a portion of this sequence because of the slope of the intertidal surface (Fig. 3.21).

Films of biological material such as bacteria and diatoms may also play a role in tidal rhythmites. These biofilms cause some stabilization of the sediment surface. Accretion by mineral sediment produces rhythmites of light and dark material in the muddy coast of French Guiana (Debenay et al. 2007). Cyanobacteria also may produce similar rhythmic bedding (e.g. Krumbein et al. 2003). An excellent summary of the influence of biofilms on tidal sedimentation is presented by Decho (2000). These organic materials may assist in preservation and can be recognized in ancient stratigraphic sequences.

Fig. 3.20 Photo of tidal bedding sequences from the Baraboo Quartzite (1.7 BY) in Wisconsin (scale is in centimeters). These rhythmites are monolithic except for some iron staining that gives them their definition

Tidal rhythmites are typically associated with intertidal flats that border marine or estuarine water bodies. Similar sequences that accumulate under tidal conditions may occur in a fluvial, fresh-water environment. These are present in both modern systems and ancient sequences (Kvale and Mastalerz 1998). It is common for tidal environments and tidal sediment transport to extend beyond the salt-fresh water boundary. Examples include the St. Lawrence Seaway in Canada, Hudson River in New York and the Gironde River in France. Recognition in the ancient record involves the preserved biogenic skeletal material and the placement of the sequence in the context of the entire depositional system.

Heterolithic tidal beds are also produced in fluvial deltaic environments with a strong tidal influence. They have been reported and described from the Fly River delta of Papua New Guinea (Harris et al. 1993; Dalrymple et al. 2003) and the Mahakam River delta of Indonesia (Gastaldo et al. 1995). Similar ancient tidalite sequences have been reported from delta front deposits in the Mississippian of the Appalachians (Adkins and Eriksson 1998; Miller and Eriksson 1997).

The preservation potential of tidal bedding is high where tidal flats are aggrading and prograding. As is typical for intertidal flats, the sediment is delivered by tidal currents from the adjacent subtidal environment(s). Such intertidal deposits are still vulnerable to erosion by wind tides and storms in general. Raising the water level above its normal condition would also be accompanied by wave action. This combination will rework the tidal rhythmites. Although there are preserved sequences that have several hundred continuous tidal cycles, most show scour areas and local unconformities.

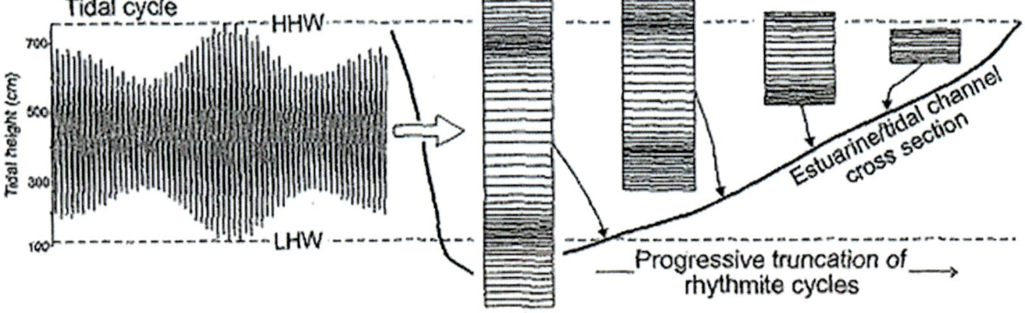

Fig. 3.21 Diagram showing spring-neap cycle and the change in the accumulated sequence as you move across and up the tidal flat (From Archer 1998)

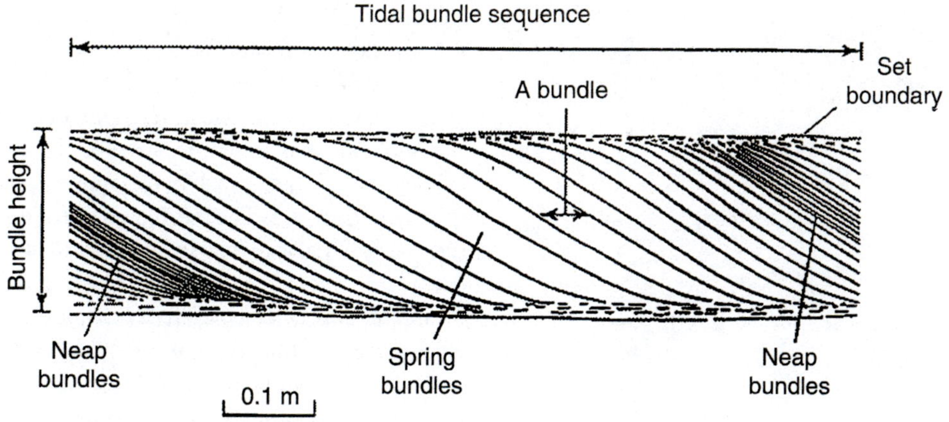

Fig. 3.22 Diagram showing the spring-neap cycles in tidal bundles (From Visser 1980)

Cross-Strata Structures

The cross-stratification produced by migrating bedforms occurs in scales that range from a few centimeters to tens of centimeters. Migrating ripples that develop when small bedforms develop may display climbing ripple cross-strata when an abundant sediment supply is present (i.e. sediment fallout is much greater than bedform migration). These distinctive structures are typically present in multiple depositional environments; especially fluvial and those with tidal influence. Unless a cyclic reversal of the directional orientation of the cross-strata occurs, there is no reason to interpret them as tidalites. Such a bimodal organization of migrating ripples is rare.

A more important cross-strata structure that qualifies as a tidalite is *tidal bundles* (Fig. 3.22). Tidal bundles are special cross sets that are generally at least 10 cm thick and are commonly near 50 cm. Each bundle is a couplet of typically heterolithic cross-strata that develop from the migration of small dunes (megaripples) but they may also be monolithic. As these bedforms migrate with the tidal currents there is variation in the velocity over the spring-neap cycle that produces differences in bundle thickness due to changes in distances of bedform migration. Under the influence of a strong predominant current, either flood or ebb, and a very weak subordinate current there is a succession of sandy cross-strata separated by thin mud drapes from slack tide conditions (Fig. 3.22).

The tidal bundles accumulate in a sequence of trough cross-strata that shows rhythmic changes in individual bed thickness from spring to neap and back as each tidal cycle takes place (Visser 1980). Individual bundles may only be a centimeter or two thick during spring conditions and a few millimeters during neap conditions (Fig. 3.23). These bundles are excellent examples of tidalites and show a high preservation potential because they develop on channel margins and floors where burial can be rapid. They are almost always oriented in a single direction in a sequence of several stacked cross-sets (Fig. 3.24).

Because tidal bundles occur in medium scale cross-strata that were formed by similar size bedforms, their development is the result of relatively strong currents. As a result there is a high potential for reactivation surfaces to be present within the bundle sequence, especially near spring tide conditions. This limits their environment of formation to tidal channels, channel margins or intertidal flats where mesoscale bedforms (< 2 m wavelength) are present.

Miscellaneous Tidal Influence Indicators

There are other sedimentary phenomena that may be considered as tidalites. These may be located within the intertidal zone but do not show any evidence of tidal processes. Examples include the plants that grow only in this specific environment. Salt marsh grasses are excellent indicators of the position of sea level and are associated with tidal processes. In general, marsh grasses are restricted to the upper part of the intertidal zone; typically between neap and spring high tide (Frey and Basan 1985). *Spartina* is typically lower in elevation than *Juncus*. These two taxa are the most common marsh species at present and can be recognized easily in peats. The surfaces upon which these plants are located receive sediment by tidal processes

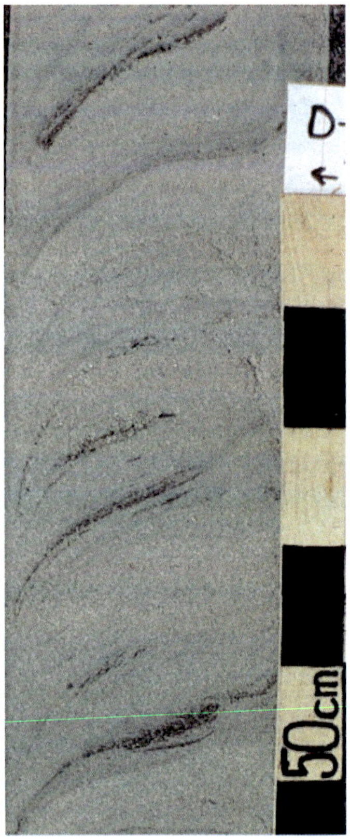

Fig. 3.23 Photo of a tidal bundle sequence showing spring and neap packages from the margin of a tidal channel in Martens Plate, German Wadden Sea

and also as the result of suspended sediment delivered during times of storm surge.

In low latitude areas, mangroves occupy this same general environment. These trees extend from the shallow subtidal environment to the supratidal level. They occupy water bodies that range from normal marine salinities to fresh water but they do require tidal flux (Dawes 1998). Like marsh grasses, mangroves produce easily recognizable peat deposits that represent the intertidal zone and therefore provide a good indicator of sea level at the time of accumulations.

Caution should be exercised with the interpretation of peats in the stratigraphic record. Some peats are deposited *in situ* and some are not. Transported plant material that produces peats is most commonly composed of sea grass debris. Recognition of any *in situ* root material, especially in marsh and mangrove peat is the best indicator of the intertidal zone and sea level position.

Another type of tidal cyclicity has been recognized in the shells of estuarine mollusks (Murakoshi et al. 1995). Detailed examination of the growth lines on the shells of the bivalve, *Potamucorbula*, show cyclic patterns of thick and thin layers representing spring-neap cycles. These patterns appear as miniature tidal bundles (Fig. 3.25). Such a bivalve can be transported but if found *in situ* it would be considered as a tidalite.

Fig. 3.24 Photo of stacked sequences of tidal bundles, all oriented in the same direction from the Cretaceous in the San Juan Basin of New Mexico

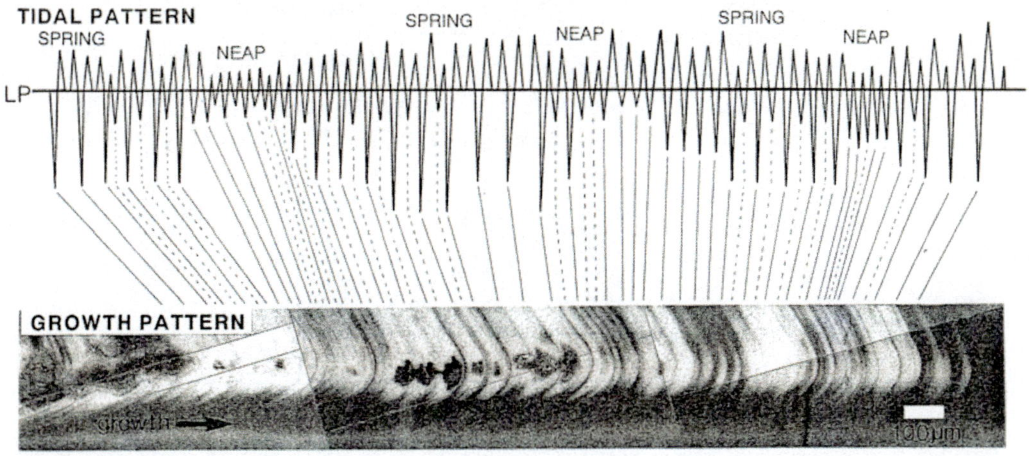

Fig. 3.25 Photo and diagram showing the tidal cyclicity recorded in the growth lines in a bivalve shell from the Pleistocene of Tokyo Bay, Japan (From Murakoshi et al. 1995)

3.3 Paleotidal Range

Another aspect of tidalites concerns the tidal range in which they were deposited. This information is important in reconstructing the environments of deposition of the respective tidalites. Because tidalites have been recognized in sediments that were billions of years old it is likely that tidal range has changed overall. We know that the earth/moon relationship has changed over that time as the distance between them has increased (Kvale et al. 1999; Chap. 1 this book). This undoubtedly also caused changes in tidal range, most likely a decrease on a global scale. A generalization by Shaw (1964) stated that ancient epeiric seas had small tidal ranges. Additionally, Irwin (1965) and later Boggs (2001) were of the opinion that tidal waves could not progress over the shallow environments of these seas. This has been shown to be untrue by Klein and Ryer (1978) who gave both modern and ancient examples of epeiric seas and broad shallow shelves where tide-dominated conditions were common (see Chap. 13) and their respective tidal ranges were greater than microtidal (<2 m). There have been efforts in the past to determine the tidal range under which specific tidalite sequences were deposited (e.g. Klein 1971, 1975).

In modern tidalite sequences there is a general tendency to equate a complete sequence from sediments at the base of the intertidal environment to those at the upper portion with the tidal range (Evans 1975; Knight and Dalrymple 1975). The stratigraphic models are hypothetical and include the entire potential sequence (Fig. 3.26). In both of the conceptual models the thickness of the complete stratigraphic sequence is equal to the tidal range. Other similar models have been proposed, however a complete sequence such as included in these models is almost never preserved as such in the stratigraphic record. Most commonly the top portion is missing due to erosion, but other components might also be missing just because of the specific circumstances at the site of accumulation.

Both transgressive and progradational tidalite sequences occur in the modern and ancient stratigraphic records. These also are partial sequences and therefore do not permit accurate paleotidal range for the environment of deposition. A detailed study of the Wood Canyon Formation in Nevada permitted Klein (1972) to construct a paleotidal model for the stratigraphy from the base of the intertidal zone to the supratidal (Fig. 3.27). Such theoretical models provide a decent answer to the question but problems exist. The base is sometimes difficult to determine and, as mentioned above, the upper part of the sequence is commonly removed by erosion prior to the overlying accumulation. The base can be difficult to recognize because wave influence can destroy any tidal signatures that might accumulate there. The top is easier to identify if marsh deposits and/or desiccation features are present.

To date good methods do not exist to determine paleotidal range other than the presence of a complete stratigraphic sequence from the base to the top of the intertidal zone and these are quite rare.

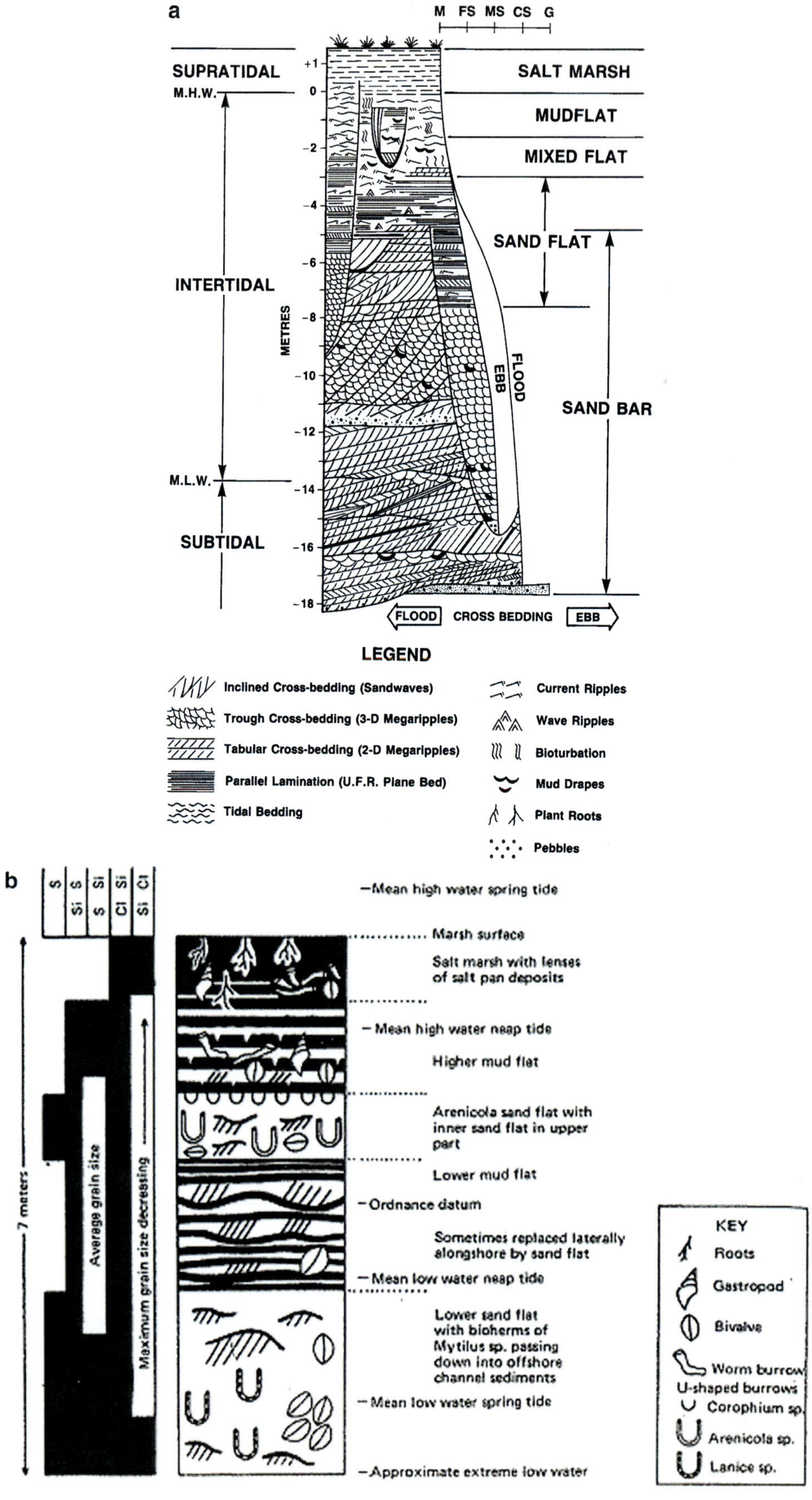

Fig. 3.26 Schematic stratigraphic sections showing complete tidalite sequences for (**a**) the Bay of Fundy, Canada and (**b**) the Wash in England (Courtesy of R. Dalrymple and from Evans (1975) respectively)

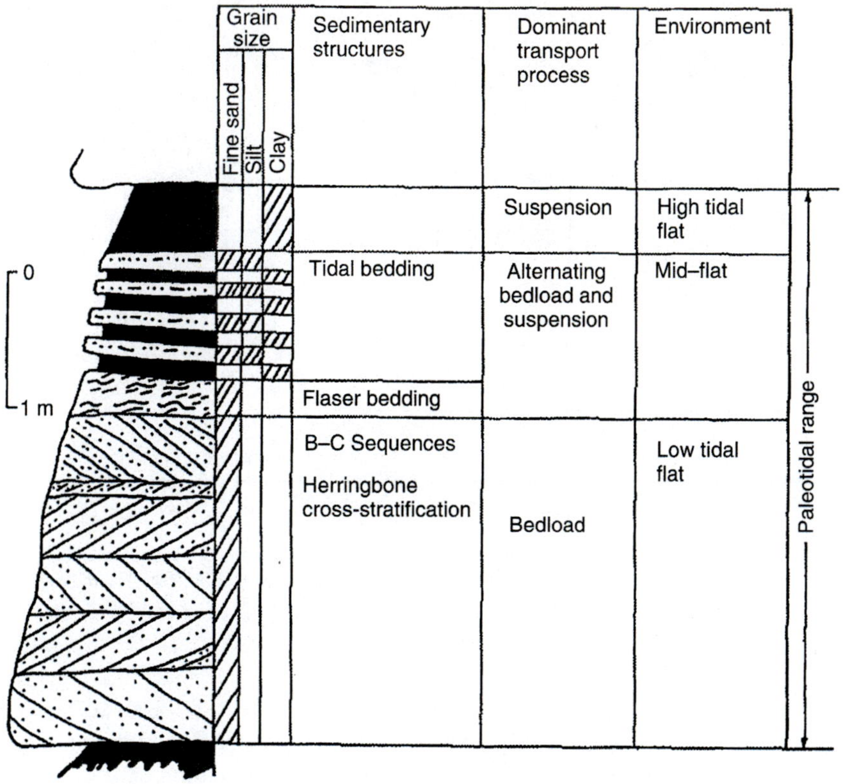

Fig. 3.27 Theoretical sequence based on studies of the Wood Canyon Formation in the Precambrian of Nevada (USA) that would equal the tidal range (From Klein 1975)

3.4 Lack of Tidalite Production in Tidal Environments

Not all tidally-influenced environments produce tidalites. What is preserved in the stratigraphic record as tidalites represents the minimum of tidal environments that existed in the geologic past. Production of tidalites requires sediment transport. Conditions exist whereby sediment is not moved or is actually removed even though tidal phenomena are present. That is, too little or too much energy may be present.

One of the most common environments where this situation prevails is the transition from fluvial to tidal domination in estuaries (see Chap. 5). A wide range of conditions may influence this transition including the rate of discharge, tidal stage and lunar tidal stage. Storm surge can also influence this transition. An excellent schematic model shows the range of tide- and fluvial-domination (Dalrymple and Choi 2007). Where the tidal currents actually reverse direction during flood and ebb conditions the possibility exists for tidalite deposition (Fig. 3.28). Landward of this position the tides cause modulation of fluvial currents so that during ebb tide the fluvial currents are a maximum and during flood tide they are a minimum. Good examples of such estuaries are the Hudson River in New York (USA) where tidal influence extends more than 150 km up the valley but where tidalite accumulation is limited to only a few kilometers. The Gironde River in France shows some tidal influence all the way to Bourdeaux, a distance of 100 km, but tidal sedimentation stops only a few kilometers from its mouth (Allen and Posamentier 1994).

Other environments where significant tidal flux is present but where tidalites are absent include many tidal flats. Waves in intertidal environments typically rework sediment. In addition, burrowing organisms are typically common in these environments. This combination of wave action and bioturbation reworks deposits that originally accumulated under tide-dominated

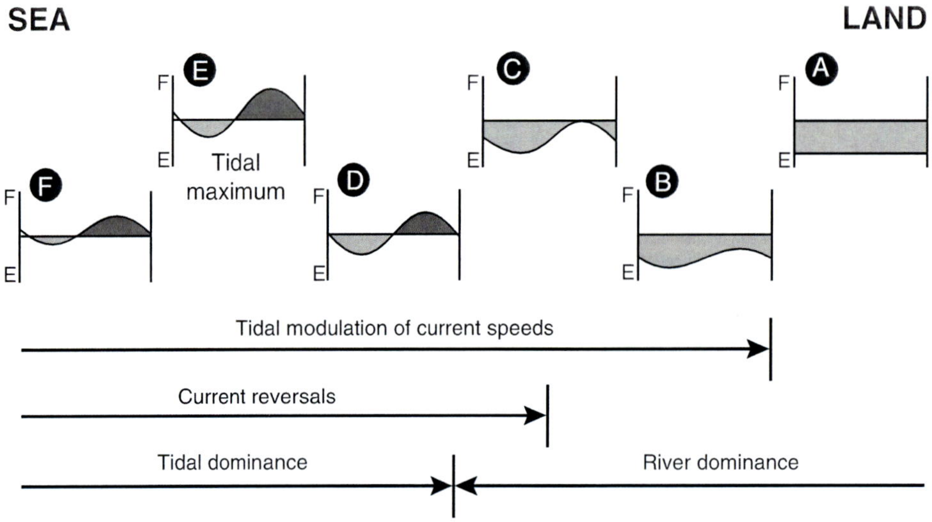

Fig. 3.28 Schematic diagram showing various relationships between fluvial discharge and tidal flux in an estuary as tidal influence is modulated. This diagram may cover only a kilometer or so, or it may extend for many tens of kilometers (From Dalrymple and Choi 2007)

conditions. An excellent example of this is present in tidal flats of the German Wadden Sea (Davis and Flemming 1995; see Chap. 10). Here extensive tidal flat accumulations lack tidal signatures but the sediments in and adjacent to the tidal channels show good tidalite signatures.

3.5 Summary

Tidalites accumulate in a wide range of depositional environments. In nearly all cases they represent conditions where tidal flux transported sediment that accumulated incorporating the typical tidal signature of heterolithic bedding of some type, usually in the form of tidal bedding or tidal bundles. Scale may range widely, especially in tidal bundles. Monolithic tidalites occur but are not common. It is likely that tidal deposits in monolithic depositional environments commonly do not leave tidal signatures and are therefore, not recognized in the stratigraphic record. Exceptions include the Precambrian Baraboo and the Cambrian Jordan sandstones of Wisconsin (Pape et al. 2003) where tidalites comprised entirely of well-sorted sand occur. Waves and bioturbation can destroy tidal signatures and commonly do so. Extensive environments occur where tidal flux is present but without enough energy to transport sediments. What we see as tidalites in the stratigraphic record may represent the minimum of preserved tidally-influenced environments.

References

Adkins RM, Eriksson KA (1998) Rhythmic sedimentation in a mid-Pennsylvanian delta-front succession, Magoffin Member (Four Corners Formation, Breathitt Group), eastern Kentucky: a near-complete record of daily, semi-monthly, and monthly tidal periodicities. In: Alexander CR, Davis RA, Henry VJ (eds.) Tidalites: processes and products, SEPM special publication 61. SEPM Society for Sedimentary Geology, Tulsa, pp 85–94

Allen GP, Posamentier HW (1994) Transgressive facies and sequences architecture in mixed tie- and wave-dominated incised valleys: examples from the Gironde estuary, France. In: Dalrymple RW, Zaitlin BA, Boyd R (eds.) Incised valley systems: origin and sedimentary sequences. SEPM special publication 51. SEPM Society for Sedimentary Geology, Tulsa, pp 225–240

Archer AW (1998) Hierarchy of controls on cyclic rhythmite deposition, Carboniferous basins of eastern and mid-continental USA. In: Alexander CR, Davis RA, Henry VJ (eds.) Tidalites: processes and products, SEPM special publication 61. SEPM Society for Sedimentary Geology, Tulsa, pp 59–68

Boggs SN (2001) Principles of sedimentology and stratigraphy, 3rd edn. Prentice-Hall, Englewood Cliffs

Collinson JD (1968) Deltaic sedimentation in the Upper Carboniferous of northern England. Sedimentology 10:223–254

Dalrymple RW (1992) Tidal depositional systems. In: Walker RG, James NP (eds.) Facies models. Geological Association of Canada, Toronto

Dalrymple RW, Choi K (2007) Morphologic and facies trends through the fluvial – marine transition in tide-dominated depositional systems: a schematic framework for environmental and sequence-stratigraphic interpretation. Earth Sci Rev 81:135–174

Dalrymple RW, Baker EK, Harris PT, Hughes MG (2003) Sedimentology and stratigraphy of a tide-dominated foreland-basin delta (Fly River, Papua New Guinea. In: Sidi H, Nummedal D, Imbert P, Darman H, Posamentier HW (eds.) Tropical deltas of Southeast Asia – sedimentology, stratigraphy, and petroleum geology, SEPM special publication 76. SEPM (Society for Sedimentary Geology), Tulsa, pp 147–173

Davis RA (1966) Algal stromatolites composed of quartz sandstone. J Sediment Petrol 38:953–955

Davis RA, Flemming BW (1995) Stratigraphy of a wave- and tide-dominated intertidal sand body: Martens Plate, east Frisian Wadden Sea, Germany. Int Assoc Sediment Spec Publ 24:121–132

Dawes CJ (1998) Marine botany, 2nd edn. Wiley, New York

Debenay J-P, Jouanneau J-M, Sylvestre F, Weber O, Guiral D (2007) Biological origin of rhythmites in sediments of French Guiana. J Coastal Res 26:1431–1442

Decho AW (2000) Exopolymer microdomains as a structuring agent for heterogeneity within microbial biofilms. In: Ridey R, Awramik S (eds.) Microbial sediments. Springer, Berlin, pp 9–15

Dill RF, Shinn EA, Jones AT, Kelly K, Steinen RP (1986) Giant subtidal stromatolites forming in normal salinity waters. Nature 62:116–121

Evans G (1975) Intertidal flat deposits of the Wash, western margin of the North Sea. In: Ginsburg RN (ed.) Tidal deposits. Springer, New York, pp 13–20

Frey RW, Basan P (1985) Salt marshes. In: Davis RA (ed.) Coastal sedimentary environments, 2nd edn. Springer, Heidelberg, pp 225–301

Gastaldo RA, Allen GP, Huc AY (1995) The tidal character of flluvial sediments of the Recent Mahakam River delta, Kalimantan, Indonesia. Int Assoc Sedinent Spec Publ 24:171–181

Harris PT, Baker EK, Cjole AR, Short SA (1993) A preliminary study of sedimentation in the tidally dominated Fly River Delta, Gulf of Papua. Cont Shelf Res 13:441–472

Irwin ML (1965) General theory of epeiric clear water sedimentation. Am Assoc Petrol Geol 49:445–459

Klein GD (1963) Bay of Fundy intertidal zone sediments. J Sediment Petrol 33:844–854

Klein GD (1970) Depositional and dispersal dynamics of tidal sand bars. J Sediment Petrol 40:1095–1127

Klein GD (1971) A sedimentary model for determining paleotidal range. Geol Soc Am Bull 82:2585–2592

Klein GD (1972) Determination of paleotidal range in clastic sedimentary rocks. In: 24th international geological Congress, Montreal Compte Rendus, Section 6, Montreal, 1972, pp 397–405

Klein GD (1975) Paleotidal range sequences, Middle Member, Wood Canyou Formation (late Precambrian), east California and western Nevada. In: Ginsburg RN (ed.) Tidal deposits. Springer, Heidelberg, pp 171–177

Klein GD, Ryer TA (1978) Tidal circulation patterns in Precambrian, Paleozoic and Cretaceous epeiric and mioclinal shelf seas. Geol Soc Am Bull 89:1050–1058

Knight RJ, Dalrymple RW (1975) Intertidal sediments from the south shore of Cobequid Bay, Bay of Fundy, Nova Scotia, Canada. In: Ginsburg RN (ed.) Tidal deposits. Springer, New York, pp 47–55

Krumbein WE, Brehm U, Gerdes G, Gorbushina AA, Levot G, Lalomsla K (2003) Bio. biodictyon amd biomat – biolaminates, oolites, stromatolites – geophysiology, global mechanisms and parahistology. Kluwer, Dordrecht, pp 1–28

Kvale EP, Mastalerz M (1998) Evidence of ancient freshwater tidal deposits. In: Alexander CR, Davis RA, Henry VJ (eds.) Tidalites: processes and products, SEPM special publication 61. SEMP Society for Sedimentary Geology, Tulsa, pp 95–108

Kvale EP, Johnson HW, Sonnett CP, Archer AW, Zawistocki A (1999) Calculating lunar retreat ratio using tidal rhythmites. J Sediment Res 69:1154–1168

Lindholm RC (1987) A practical approach to sedimentology. Springer, New York, 276 p

Miller JA (1975) Facies characteristics of Laguna Madre wind-tidal flats. In: Ginsburg RN (ed.) Tidal deposits. Springer, New York, pp 67–73

Miller DJ, Eriksson KA (1997) Late Mississippian prodelta rhythmites in the Appalachian Basin: a hierarchical record of tidal and climatic periodicities. J Sediment Res 67:653–660

Murakoshi N, Nakayama N, Masuda F (1995) Diurnal inequality pattern of the tide in the upper Pleistocene Palaeo-Tokyo Bay: reconstruction from tidal deposits and growth-lines of fossil bivalves. Int Assoc Sediment Spec Publ 24:289–300

Nio SD, Yang C (1991) Diagnostic attributes of clastic tidal deposits: a review. In: Smith DG, Reinson GE, Zaitlin BA, Rahmani RA (eds.) Clastic tidal sedimentology, Canadian Society of Petroleum Geologists Memoir 16. Canadian Society of Petroleum Geologists, Calgary, pp 3–28

Pape CH, Cowan CA, Runkel AC (2003) Tidal bundle sequences in the Jordan Sandstone (Upper Cambrian), southeastern Minnesota, USA: evidence for tides along inboard shorelines of the Sauk epicontinental sea. J Sediment Res 71:354–366

Reineck H-E (1967) Layered sediments of tidal flats, beaches and shelfbottom of the North Sea. In: Lauff GH (ed.) Estuaries, American Association for the advancement of science publication 83. American Association for the Advancement of Science, Washington, DC, pp 191–206

Reineck H-E, Wunderlich F (1968) Classification and origin of flaser and lenticular bedding. Sedimentology 11:99–104

Shaw AB (1964) Time in stratigraphy. McGraw-Hill, New York, 365 p

Visser MJ (1980) Neap-spring cycles reflected in Holocene subtidal large-scale bedform deposits: preliminary note. Geology 8:543–546

Tidal Ichnology of Shallow-Water Clastic Settings

4

Murray K. Gingras and James A. MacEachern

Abstract

This chapter explores using ichnological data as an indicator of tidal influence on sedimentation. Presently, the most commonly used method for establishing tidal influence with trace-fossil datasets is the presence of a brackish-water suite of trace fossils. Brackish-water trace-fossil suites generally comprise low-diversity, comparably diminutive trace fossils including forms such as *Cylindrichnus, Planolites, Thalassinoides, Teichichnus, Arenicolites, Skolithos* and cryptobioturbation. A problem with this method is that tidal currents and brackish-water do not always accompany one another.

We attempt to build on previous brackish-water studies by examining the influence of tides on trace-fossil assemblages by considering 5 major impacts of tidal sedimentation. (1) The impact of rhythmic changes in current velocity on (passive) burrow infill, which can generate rhythmically infilled, large-diameter trace fossils (e.g. *Thalassinoides* and *Psilonichnus*) that correspond to tubular tidalites. (2) The influence of longer periodicity tidal rhythms on bed colonization, which produce regular waxing and waning of bioturbation intensity at the bed and/or bed-set scale. (3) The effect of tidal currents on burrow distributions, which produces a characteristic bioturbation-increasing-landwards trend into the middle part of bays and estuaries. (4) The impact of differing sedimentation rates between subtidal and intertidal settings, which contributes to a characteristic bioturbation-increasing-upwards profile on tidal bars. (5) And, the influence of variable resource distribution due to variable current energy, which results in a preponderance of interface deposit-feeding structures, stellate feeding traces and the presence of systematic deposit feeding trace fossils.

Notably, the potential of using trace fossils to identify tidal influence on sedimentation and faunal colonization is at its inception. Substantial data is still needed to refine the ichnological model proposed herein.

4.1 Introduction

Biogenic sedimentary structures (i.e. trace fossils or ichnofossils) are well known for helping to refine interpretations of depositional conditions (Seilacher 1964; Frey and Howard 1970; Frey and Howard 1990;

M.K. Gingras (✉)
Department of Earth and Atmospheric Sciences, University of Alberta, Edmonton, AB T6G 2E3, Canada
e-mail: mgingras@ualberta.ca

J.A. MacEachern
Department of Earth Sciences, Simor Fraser University, 8888 University Drive, Burnaby, BC V5A 1S6, Canada
e-mail: jmaceach@suf.ca

R.A. Davis, Jr. and R.W. Dalrymple (eds.), *Principles of Tidal Sedimentology*,
DOI 10.1007/978-94-007-0123-6_4, © Springer Science+Business Media B.V. 2012

Pemberton et al. 1990; Pemberton et al. 1992b; Taylor et al. 2003; MacEachern and Bann 2008; McIlroy 2008). This is due to their origins as preserved vestiges of animal behavior, dictated by the trace-makers' responses to environmental factors such as sedimentation rate, sea-bottom consistency, sediment caliber, water turbidity, dissolved-oxygen content of the water, and salinity. In concert with other sedimentological data, ichnological data can reliably distinguish between inner shelf/proximal offshore, shoreface, bay and estuary, and delta deposits in the shallow water clastic regime. A useful characteristic of trace-fossil suites is that they improve interpretations that otherwise are based solely on physical sedimentary structures and the vertical sedimentary succession, such that depositional subenvironments and detailed process details can be inferred.

This chapter focuses on the ichnology of tidally influenced environments in "shallow-water" (i.e., proximal) clastic settings: dominantly embayments, bays and estuaries (henceforth marginal-marine settings). Although tides may occur in oceanic (i.e., distal) locales, the amplification of tidal currents in marginal-marine settings leads to elevated sedimentation, and may be accompanied by continentally derived mud and/or salinities lower than those of marine settings. Nevertheless, the distributions of physico-chemical parameters in tidal settings are complex and dynamic. For example, the magnitude of tidal and fluvial fluxes shift cause the position of salinity variation to shift markedly. Thus, burrowing animals may cope with reduced salinity in bays, but that stress is compounded in estuaries in that salinity varies so much spatially and temporally. Fluctuations in energy may also be significantly different in a bay setting, wherein wave agitation may be more persistent, which may be more effective in inhibiting settling of mud. In contrast, tidal currents are more effective at mobilizing fluid mud and transporting it as bedload. At the system scale, most physico-chemical factors (i.e. sedimentation rate, sediment caliber, substrate consistency, turbidity, salinity and oxygenation) are variable due to the shift of tidal waters along the longitudinal profile of the system, leading to greater spatial variability (summarized in Hubbard et al. 2004).

The behavioral responses of animals to tidally imposed physico-chemical stresses suggest that an ichnological signal of tidally influenced sedimentation should be discernible, but beyond animal response to brackish-water (discussed below), this has not been strenuously explored. This ultimately limits workers, because brackish-water conditions may be present in non-tidal settings (e.g., wave-dominated restricted bays and estuaries), and strong tidal currents may be present in mesohaline through to marine waters. Additionally, although no studies have been directed toward establishing the influence of tides on freshwater fauna, tidal processes can be detected well landward of salt-water invasion, so there is the potential for continental suites to be affected by tidal processes as well.

This chapter uses observations from modern depositional environments and from (high certainty) occurrences in the rock record to establish a general framework for identifying ichnological suites indicative of tidal processes. Towards this goal, we summarize key aspects of the brackish-water ichnological model, and broaden our view to consider the influence of: (1) rhythmic changes in current velocity on (passive) burrow infill; (2) longer periodicity tidal rhythms on bed colonization; (3) tidal currents on burrow distributions; (4) differing sedimentation rates between subtidal and intertidal settings; (5) variable resource distribution resulting from variable current energy; and (6) individual behavioral reactions of infauna to tides.

4.2 Background

The interpretation of tidal sedimentary features has improved substantially in the last three decades. The establishment of widely recognized physical sedimentary structures best explained by tidal processes has provided workers with a surprising range of interpretive tools that can be applied to tidally influenced sedimentary deposits (for a summary, see Chap. 10 of this volume). Trace-fossil workers have supported this effort by improving the recognition of brackish-water settings (summarized in MacEachern and Gingras 2007) and by progressively improving interpretations of nearshore depositional subenvironments (McIlroy 2007; MacEachern and Bann 2008; Carmona et al. 2009; MacvEachern et al. 2010). To date, contributions to tidal ichnology have focused on the ichnological characterization of depositional environments deemed tide influenced, determined on the basis of careful scrutiny of primary physical sedimentary features (Pemberton et al. 1982; McIlroy 2004, 2007; Mangano and Buatois 2004; Rebatah et al. 2006; Hovikoski et al. 2007, 2008; Carmona et al. 2009). However, the inverse question, "can tides be inferred *from* ichnological data" has yet to be tested.

By far, the largest volume of work has depended on the association of brackish-water ichnofossil assemblages with tidal settings. Through the establishment of a brackish-water trace-fossil model, and in coordination with the documentation of physical features ascribed to tidal sedimentation, several studies have established that a passive relationship exists between the occurrence of brackish-water ichnofossils and the presence of tides (Beynon et al. 1988; Pemberton et al. 1982; MacEachern et al. 1999; Gingras et al. 2002a, b; McIlroy 2004; Bann et al. 2004; Mangano and Buatois 2004; Buatois et al. 2005; Rebatah et al. 2006; MacEachern and Gingras 2007; Hovikoski et al. 2007; Carmona et al. 2009). However, it is fair to say that the identification of brackish-water settings has been conflated with the identification of tidally influenced settings and that the presence of a brackish-water trace-fossil assemblage, in and of itself, should not be taken as an indicator of tidal conditions.

Some effort has been directed toward establishing an animal response to variable, but not necessarily rhythmic sedimentation rates. In particular, the ichnological response and recovery to event sedimentation has been carefully documented from numerous systems, most notably with respect to tempestite and turbidite deposition (Crimes et al. 1974, 1981; Seilacher 1982; Wanless et al. 1988; Pemberton and Frey 1984; Pemberton et al. 1992a; Pemberton and MacEachern 1997; Savrda and Nanson 2003). Storm- and turbidity-current events, and perhaps also river floods, however, are characterized by recurrence intervals of months to centuries, and are not analogous to tidal sedimentation. Perhaps more germane to this paper are studies of bioturbation under conditions of persistent but irregular sedimentation in colonized subaqueous dunes (Bromley 1996; Taylor and Goldring 1993; Savrda 2002; Dashtgard et al. 2008). However, the potential of identifying *rhythmic* distributions of bioturbation levels in association with short-period, rhythmic sedimentation remains essentially unexplored (Gingras et al. 2011).

Several workers have identified trace-fossil morphologies that recur in some tidal environments. Interface-feeding behaviors that consist of foraging for food from a central burrow, for example, produce stellate traces that radiate outwards from the burrow aperture. Such burrows are commonly, but not exclusively found on tidal flats (Aitken et al. 1988; Dashtgard et al. 2008): the importance of such trace fossils is discussed below. Detailed analyses of burrow-infill and burrow-lining characteristics have recently been investigated for potential tidal associations (Gingras et al. 2011); this work is also discussed below.

4.3 Process Sedimentological Importance of Some Selected Ichnological Characteristics

Tidal processes and the sedimentary structures they generate are understood to a far greater degree than are the ichnological responses to them. Rhythmic variations in current strength, direction and duration generate characteristic and predictable sedimentary structures that are most easily interpreted in the context of tidally influenced sedimentation. In contrast, with the exception of rhythmic infilling of trace fossils and perhaps regular variation in bioturbation intensity, both of which are rare (see below), the morphologies of ichnofossils are not directly influenced by the presence of tidal currents. On the other hand, some aspects of trace-fossil occurrence and distribution are directly impacted by the tidal depositional processes. These include parameters that can be related to sedimentation rates (e.g. trace-fossil distribution and diversity), altered water chemistry due to tidal mixing (e.g. trace-fossil size and diversity), variations in sediment calibre at the bed (e.g. burrow linings and infills), and the details of food allocation (e.g., characteristic feeding behaviors).

In marine and marginal-marine deposits, trace-fossil size and diversity are generally taken to reflect the degree of physico-chemical stress in the depositional environment. Highly diverse and robust suites correspond to optimal environmental conditions, whereas low-diversity suites and those composed of diminutive structures are regarded to be indicative of environmental duress. As such, it is useful to understand that, for the deposit under scrutiny, the dataset ideally should be compared to an ichnological baseline (cf. MacvEachern et al. 2010). This helps to identify intervals that have been influenced by elevated sedimentation, brackish water, overall low oxygenation concentrations, biased larval recruitment, and/or various substrate stresses (such as anomalously firm or soft/soupy consistencies). If possible, an ichnological baseline should be drawn from the marine part of the depositional system that was characterized by uniform salinity, fully marine waters, variable substrate consistencies, generally reduced deposition rates, and variable but abundant food resources such as suites of the *Cruziana* Ichnofacies occurring in proximal offshore / inner shelf locales (MacEachern and Bann 2008; MacvEachern et al. 2010). If any of these parameters are compromised—as is common in tidal settings—a downward shift in the size and diversity of the ichnogenera present will likely be observed.

4.3.1 Vertical Spatial Distribution of Trace Fossils in Tidal Settings

At the core and outcrop scale, the vertical spatial distribution of trace fossils primarily reflects the degree of stability and temporal persistence of physico-chemical conditions in a sedimentary environment. Trace fossils can be distributed: (1) homogeneously; (2) regularly heterogeneously; or (3) sporadically heterogeneously. Thorough, homogeneously distributed bioturbation is generally associated with readily available food and oxygen, coupled with slow sedimentation. In such settings, tidal effects mainly revolve around the regular resupply of food materials to the setting and replenishment of marine to brackish waters. Regularly heterogeneous trace-fossil distributions result from recurrent (rhythmic) variability in local physico-chemical parameters. Depositional environments characterized by such regular but uneven colonization show a response to tidal as well as seasonal to annual rhythms. Regular heterogeneous distributions associated with tidal settings most commonly are expressed as similarly burrowed horizons of approximately recurring composition and thicknesses, interbedded with unburrowed or sparsely burrowed media, also of regular thicknesses. Such distributions are exemplified by migrating tidal dunes and within sets of Inclined Heterolithic Stratification (IHS).

Sporadically heterogeneous distributions, on the other hand, are the result of persistent spatio-temporal variability in physico-chemical conditions. Sedimentary environments that are characterized by episodic erosion and/or sediment deposition, such as estuaries and deltas, are particularly susceptible to such sporadically heterogeneous distributions of bioturbation.

As a result of tidal influence on epifaunal and infaunal organisms, three styles of trace-fossil distribution are expected: (1) owing to low overall sedimentation rates and commonly abundant food, homogeneous distributions probably characterize most microtidal and mesotidal intertidal flats; (2) regular heterogeneous distributions probably result from neap-spring tidal processes; and (3) seasonal / annual regular heterogeneous distributions can result from seasonal interactions between the tidal and fluvial waters. Sporadically heterogeneous distributions can also occur in tidal settings, but in other settings as well. Being less diagnostic, such fabrics are not discussed here further.

Perhaps counter to intuition, regular heterogeneous ichnofossil distributions are not typically observed in most (neap-spring) tidal rhythmites. This is due to the high sedimentation rates that characteristize deposition (several milimeters to a few centimeters per week) of (diurnal/semidiurnal) tidalites. Neap-spring bundles may contain laterally regular, heterogeneous trace-fossil distributions in the distal toe- and bottom-sets of migrating tidal dunes (Fig. 4.1d, e); however, this relationship is difficult to observe in cored vertical successions, and is most readily observed with good lateral exposure (Savrda 2002; Gingras et al. 2002a, b, Pearson and Gingras 2006).

Seasonal variations in sedimentary conditions are strongly associated with regular heterogeneous distributions of burrowing, especially in marginal-marine environments. Several examples of such distributions are associated with inclined heterolithic stratification (IHS) common to tidal-bar deposits (Gingras et al. 1999; Pearson and Gingras 2006; Hovikoski et al. 2007; Lettley et al. 2007). Bioturbation within IHS dominantly points to seasonal colonization (Fig. 4.1c). Such successions consist of beds that accumulated under conditions of lower energy, intercalated with beds deposited during high fluvial-discharge events (Gingras et al. 2002a; Pearson and Gingras 2006), which, during times of lower sedimentation and higher salinity, favors predictable periods of recolonization. Regularly but non-uniformly distributed burrowing in IHS has been closely associated with deposition in bays, estuaries and, to a lesser degree, deltas. Additionally, Dalrymple et al. (1991) reported seasonal bioturbation (summer layers bioturbated / winter layers unburrowed) recorded in macrotidal intertidal deposits of the Bay of Fundy: this is dominantly a reflection of strong seasonality and its effect on the metabolism of bioturbating organisms (Van den Berg 1981).

In micro- and mesotidal settings, tidal-flat deposits characteristically display high bioturbation intensities and homogeneous burrow distributions (Fig. 4.1b, f). This is the consequence of high infaunal biomass and a markedly low sedimentation rate (Gingras et al. 1999; Dashtgard 2011). Under increasingly macrotidal conditions, rapidly shifting sediment and high velocity currents may counteract the efforts of bioturbators, such that lamination and bedding are dominantly preserved (Klein 1970; Dalrymple 1984; Dalrymple et al. 1990); nevertheless, there are zones in macrotidal settings where sedimentation rates are sufficiently low and conditions more or less stable, leading locally to intense bioturbation (e.g. upper intertidal flats). Notably, it is under these somewhat more energetic conditions that regular heterogeneous trace-fossil distributions may develop in dune-associated neap-spring bundles (Fig. 4.1d).

The ichnological characteristics of supratidal, intertidal and subtidal bars are summarized in Table 4.1 and Fig. 4.2. The ichnological identification of intertidal-flat deposits should not be under-estimated as an interpretive tool for the rock record. The identification of such deposits infers the presence of tidal processes and reveals the depositional base level. Further, these deposits are sufficiently distinctive that they provide local correlation levels, provided the tidal flats were broad and/or prograded long distances. Moreover, the preserved thickness of tidal-flat deposits, especially where subtidal to supratidal aspects of the deposits are discernible, can act as a proxy for tidal range (discussed in Mangano et al. 1998; Gingras et al. 1999). Tidal-flat deposits are generally situated above subtidal bars (Fig. 4.1a), resulting in a distinctive bioturbation-increasing upwards signal that is readily observed in vertical succession.

The lowermost channel/bar deposits display minor bioturbation, whereas the medial parts of the bars contain seasonally induced regularly heterogeneous trace-fossil distributions. Trace fossil assemblages are dominated by either horizontal or vertical trace-fossil forms in monospecific or bispecific suites (e.g. Fig. 4.2c–h). The commonest vertical traces observed are *Skolithos* and *Cylindrichnus* and more rarely *Arenicolites*, *Ophiomorpha* and *Polykladichnus* (Fig. 4.2b, d, f). The most common horizontal ichnofossils are *Planolites* and *Teichichnus* with subordinate *Thalassinoides* (Fig. 4.2c, e). Overall, the degree of bioturbation attenuated compared to overlying intertidal strata; this is owing to higher sedimentation rates on the subtidal bar. Most commonly, unburrowed beds are intercalated with burrowed beds (BI 2–3), and in environmentally hostile conditions (very low salinity, higher depositional energies, very turbid water), bioturbation is rare throughout and sporadically distributed.

The top of the bars grade into the intertidal zone, which displays highly bioturbated (up to BI 6), homogeneous to sporadic heterogeneous trace-fossil distributions (Figs. 4.1b, f and 4.2g–j). The intertidal zone is most commonly characterized by a mixed assemblage of co-occurring vertical and horizontal trace fossils (*Planolites, Teichichnus, Thalassinoides, Siphonichnus, Skolithos*, and *Arenicolites* with subordinate *Cylindrichnus*, *Polykladichnus*, *Ophiomorpha*, *Lockeia*, and stellate interface trace fossils) (Fig. 4.2g–j). Mud-dominated intertidal strata are normally highly bioturbated (BI 4–6) (Fig. 4.2g, i) with bioturbation increasing in intensity upwards. Primary bedding, if preserved, is associated with tidal run-off creeks. Sand-dominated intertidal flats typically contain more abundant primary sedimentary structures. This is due to wave-reworking of the tidal flat, tidal-dune migration or the presence of variably scaled tidal run-off creeks (Fig. 4.2h). Within sandy intertidal flat deposits, BI ranges between 0 and 3, with rare beds displaying higher intensities of bioturbation.

If the entire vertical succession is preserved, supratidal deposits overlie the intertidal units. The supratidal zone is commonly eluviated, locally massive appearing, and contains rhizoliths (Fig. 4.2k–o) and displays rare trace fossils, including *Psilonichnus*, *Scoyenia* and/or *Naktodemasis*.

As a result of variability in the position of the high and low tide levels, the transition from subtidal through intertidal to supratidal sedimentation is represented by gradational sedimentological and ichnological changes. Correspondingly, a precise level demarcating subtidal from intertidal levels may be difficult to discern—this is exacerbated with increasing tidal range.

It is notable that the vertical succession outlined above (i.e. bioturbation intensity increasing markedly upwards) occurs repeatedly in tidal settings such as the Lower Cretaceous McMurray Formation, Pleistocene and modern Willapa Bay, and the present-day Shepody

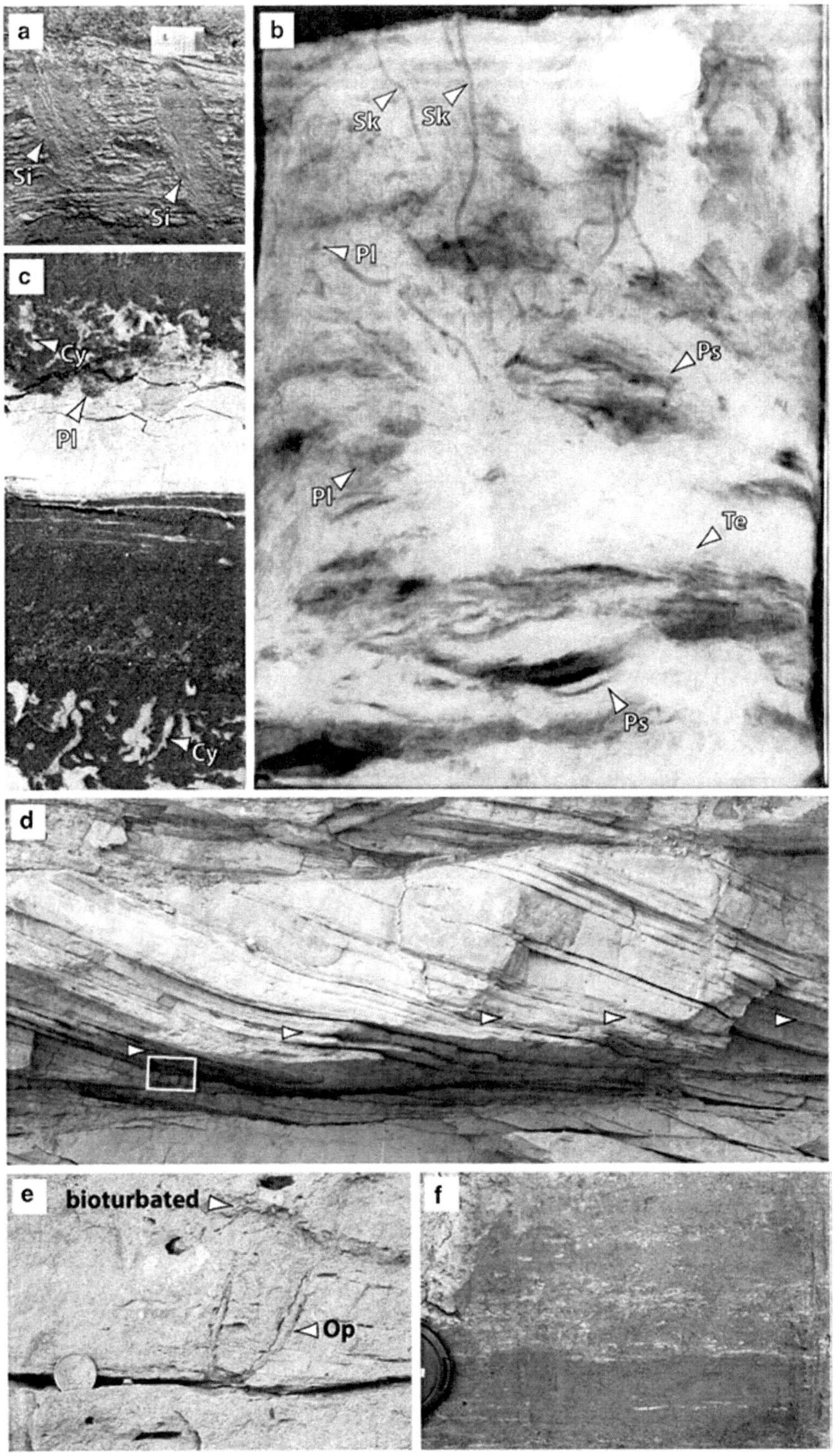
a
Si
Si
b
Sk
Sk
Pl
Ps
Pl
Te
Ps
c
Cy
Pl
Cy
d
e
bioturbated
Op
f

River (Gingras et al. 1999; Pearson and Gingras 2006), and appears to be diagnostic of tidal-bar to tidal-flat deposits.

4.3.2 Trace-Fossil Size and Diversity

Diminution is a first-order response to chemical stress and occurs in salinity-stressed environments and settings with low oxygen levels (Pemberton et al. 1982; Savrda and Bottjer 1988; Wignal 1991; Gingras et al. 1999; Hauck et al. 2009). Discerning brackish-water-induced diminution from low-oxygen diminution can be challenging. Salinity-stressed trace-fossil suites generally comprise simple, facies-crossing forms such as *Cylindrichnus, Planolites, Thalassinoides, Teichichnus, Arenicolites, Skolithos,* and cryptobioturbation. Oxygen-stressed environments have been identified by the domination of a very low-diversity of exceedingly small trace fossils such as *Planolites* and *Chondrites* (Ekdale and Mason 1988; Savrda and Bottjer 1988; Wignal 1991), although less oxygen-reduced facies can show zones that contain euryhaline-related ichnogenera such as *Zoophycos*, *Helminthorhaphe*, and *Cosmorhaphe*. More recent work by Martin (2004) has questioned the recurrence of suites dominated by *Chondrites* as indicative of oxygen reduction, however. Savrda (2007) has re-evaluated distinctive expressions of ORI (oxygen related ichnocoenosis) and piped zones for evaluating reduced-oxygen regimes. More problematic, of course, are salinity-reduced settings that are also oxygen stratified, such as some euxinic basins and some restricted bays. Strong tidal mixing tends to preclude this scenario.

Focusing on marginal-marine settings, size and diversity trends are demonstrably influenced by progressively brackish water conditions (i.e., as salinity goes from marine levels towards fresh). The brackish-water ichnological model is well established (Pemberton et al. 1982; Beynon et al. 1988; MacEachern and Pemberton 1994; Buatois et al. 2005, see summary in MacEachern and Gingras 2007). In short, Pemberton et al. (1982) showed that brackish-water trace-fossil suites are characterized by: (1) a low diversity of trace-fossil forms (Fig. 4.3a–m); (2) the presence of an impoverished marine assemblage (Fig. 4.3a–m); (3) a preponderance of morphologically simple, vertical and horizontal structures (Fig. 4.3c, h, l); (4) the presence of suites dominated by a single ichnospecies (Fig. 4.3i, m); (5) diminutive trace fossils (Fig. 4.3a–m); (6) locally high densities of a particular ichnospecies (Fig. 4.3a, b, g, h. i, m); and (7) the predominance of trophic-generalist behaviors represented by trace fossils such as *Gyrolithes*, *Palaeophycus*, *Cylindrichnus*, *Arenicolites*, *Skolithos*, *Planolites*, *Thalassinoides*, and *Teichichnus* (Fig. 4.3a–m). The veracity and utility of these criteria have been confirmed in many modern studies (Howard et al. 1975; Gingras et al. 1999; Hauck et al. 2009).

More broadly speaking, trace-fossil size is most indicative of chemical stress. Size reduction is a population-level response that is necessitated by the need of animals that live under chemical duress to maintain a small mass-to-surface-area ratio for the purpose of efficient osmoregulation. Additionally, the physiologically taxing energy demand imposed by such settings leads to opportunistic colonization, a higher mortality rate, and a higher proportion of juvenile-generated structures (Slobdkin and Sanders 1969; Levinton 1970; Pianka 1970). Ichnofossil diversity appears to be strongly affected by both physical (e.g. high sedimentation, energetic depositional conditions) and chemical

Fig. 4.1 Photographs of typical burrow distributions in tidally influenced deposits. (**a**) Heterogeneous distribution of *Siphonichnus* (probably made by bivalves) at the subtidal-intertidal transition. Scale card is 9 cm wide with eight 1-cm divisions located at *top*. Pleistocene, Willapa Bay, Washington, USA. (**b**) Bioturbated homogeneous intertidal–flat deposits (x-ray negative: *white* is sand rich, *dark* is mud rich). Indicated traces include *Skolithos* (Sk), *Planolites* (Pl), *Psilonichnus* (Ps), and *Teichichnus* (Te). Box core is 20 cm wide. Modern tidal flat, Willapa Bay, Washington, USA. (**c**) Core showing regularly sporadic distribution of *Cylindrichnus* (Cy) and *Planolites* (Pl) in IHS. Slabbed core is 10 cm wide. Cretaceous, McMurray Formation, Alberta, Canada. (**d**) Tidal dune showing neap-spring variation as thin and thick foresets. *White arrows* show zones of intense bioturbation in the lower foreset to toeset (regular sporadic distribution) associated with neap-tide deposition. Interpretation after Yang and Nio (1989). *White rectangle* indicates location of closer view provided in (**e**). Field of view is 2.5 m across. Eocene, Roda Sandstone, Spain. (**e**) Close-up view from (**d**), showing thin bioturbated zone and *Ophiomorpha* (Op) near toeset. Coin 24 mm diameter. Eocene, Roda Sandstone, Spain. (**f**) Homogeneous burrow fabric from an intertidal-flat deposit. Bioturbated sand laminae make bioturbation appear to be localized, but that is not the case. Field of view is approximately 30 cm across. Pleistocene, Willapa Bay, Washington, USA

Table 4.1 Ichnological criteria for the differentiation of subtidal, intertidal and supratidal settings

	Trace-fossil Content	Bioturbation Index	Distribution of Trace Fossils	Character of Lower Contact	Other Key Sedimentological Observations
Supratidal Deposits	(1) Commonly contain rhizoliths and otherwise generally devoid of burrows (2) Where present, trace fossils may include care *Psilonichnus* and upwards, rare *Scoyeria* and / or *Noktodermois*	(1) Degree of animal-generated bioturbation is generally obscured by rhizoturbation and eluvination, lending the sediment a massive appearance	(1) Notwithstanding the rooted fabric, trace fossils are sporadically and heterogeneously distributed	(1) Generally abrupt transition from burrowed intertidal to rhizoturbated, massive appearing supratidal media (2) Roots and eluviation commonly descend into intertidal unit	(1) Common coalified plant fragments (2) Locally common Fe-minealization (3) Deposits of steep-banked, small scale run-off creeks may be present
Intertidal Deposits	(1) A mixed assemblage of co-occurring vertical and horizontal trace fossils, most commonly dominated by *Planolives, Teichichrus, Thalassinoides. Siphonichnus, Skolithos*, and *Arenicolies* (2) Subordinate numbers of *Cylindrichnus, Polyklodichnus, Lockeia*, and <$$$> tellate interface trace fossils (the latter two observed on bedding planes) may be preserved	(1) Mud-dominated flats are commonly highly bioturbated (B1 4–5) with bioturbation increasing in intensity upwards. Primary bedding is locally preserved in association with tidal run-off creeks (B1 0–2) 0–2 (2) Sand-dominated flats commonly display more primary sedimentary structures due to wave-reworking, tidal-dune migration or the presence of variably scaled tidal run-off creeks. In these cases all ranges between 0 to 3, potentially with rare beds displaying B1 6	(1) Lower energy tidal flats characterized by slow sedimentation rates display homogeneous bioturbation at the core / small outcrop scale (2) Increased sediment shifting due to wave exposure and increased tidal run-off leads to sporadic heterogeneous distributions (3) More rarely, seasonally influenced tidal flats may display nearly horizontal, planiform regular heterogeneous distributions	(1) Gradational with underlying subtidal deposits, but is commonly discenible by an abrupt increase in bioturbation intensity and a switch to comparably homogeneous burrow distributions. In modern settings, this switch in burrowing intensity and distribution approximately corresponds to the outer part of the middle intertidal flat (2) With increasing tidal range, intertidal flat deposits and the transition zone become thicker, leading to an increasingly gradational transition	(1) Most commonly, sand and mud are admixed by bioturbation (2) Where present, preserved sedimentary features include a range of primary structures, such as lenticular through wavy to flaser bedding (3) Local scour and fill associated with tidal run off and the presence of larger scale tidal run-off creeks creeles

Subtidal Deposits	(1) A mixed assemblage of trace fossils may be present, but are generally dominated by either horizontal or vertical trace-fossil forms in (nearly) monospecific suites. Most common vertical traces include *Slolithos, Cylindrichus*, as well as a subordinate *Arenicolites, Cphiomorpha* and *Polyklodichnus*. Most common horizontal forms include *Planolites, Teichichnus* with subordinate *Thalassinoides*	(1) Bioturbation is commonly attenuated, owing to higher sedimentation rates. Most commonly, unburrowed beds are intercalated with burrowed beds (B12-3) (2) In environmentally hostile conditions, bioturbation is rare throughout, with B10 to 1 representing the dominant modes of bioturbation intensity	(1) Under conditions of effective larval recruitment and in the presence of brackish water, regular heterogeneous distributions dominate (2) Where living conditions are poor (e.g., very low salinities, persistently elevated sedimentation rates, very turbid water, common fluid muds), sporadic heterogeneous distributions are most common	(1) Sharp-based and erosive (2) Rarely, suites of the *Glassifungites* <$$$> chnofacies may be observed at the basal contact (3) Lags may be present, but are not requisite. Thicker mud-beds associated with fluid mud accumulation may be present near lower contact	(1) Dominantly expressed as inclined heterolithics tratification (IHS) comprising cm-scale through to <$$$>–scale sand / mud couplets (2) In the case of cm-scale IHS, 3+ to <$$$> dipping lenticular through wavy to flaser bedding are common

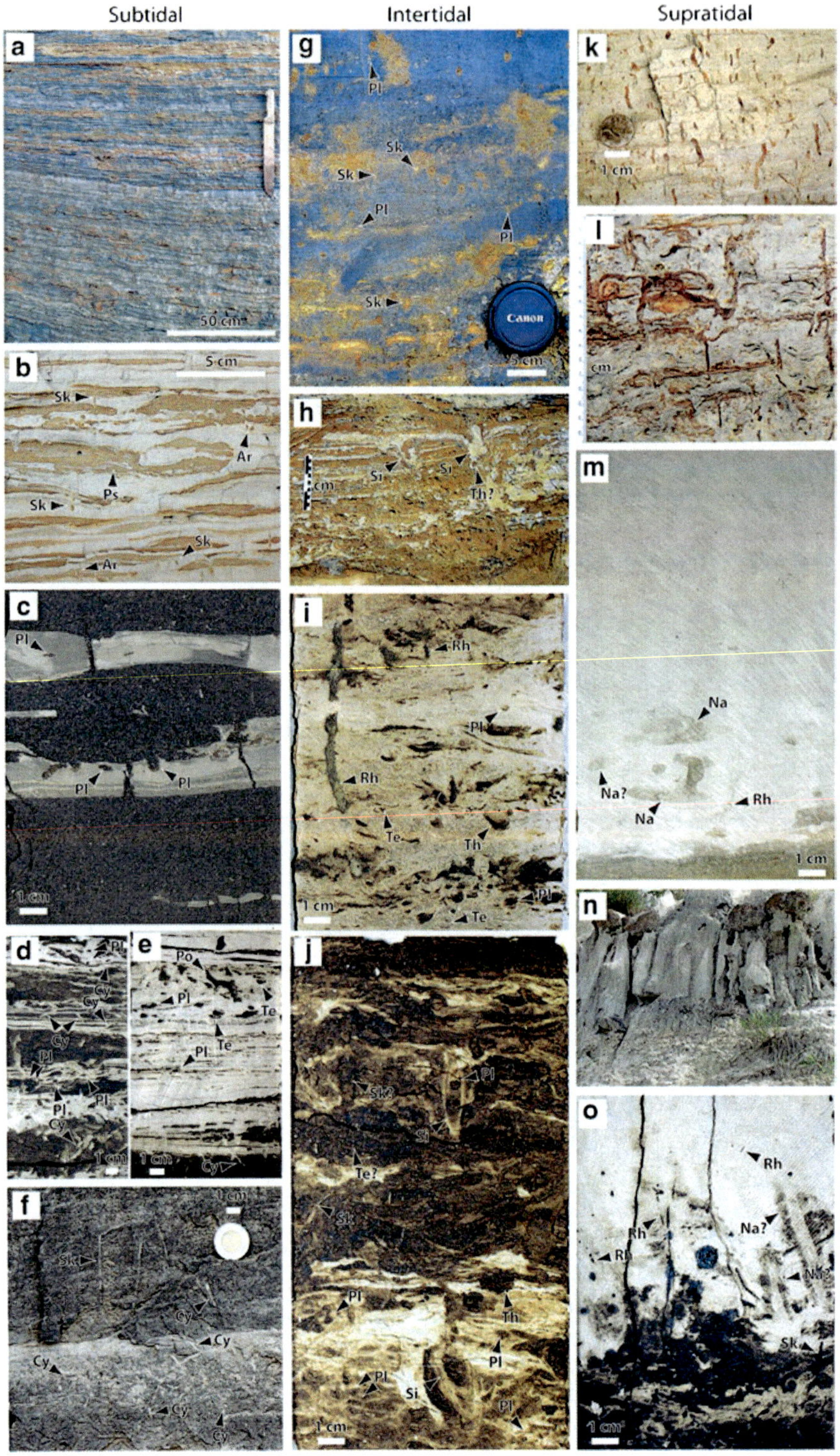
Subtidal
Intertidal
Supratidal
a
50 cm
b
5 cm
Sk
Ar
Ps
Sk
Sk
Ar
c
Pl
Pl
Pl
1 cm
d
Pl
Cy
Cy
Cy
Pl
Pl
Pl
Cy
1 cm
e
Po
Pl
Te
Pl
Te
1 cm
Cy
f
1 cm
Sk
Cy
Cy
Cy
Cy
Cy
g
Pl
Sk
Sk
Pl
Pl
Sk
5 cm
h
cm
Si
Si
Th?
i
Rh
Pl
Rh
Te
Th
Pl
Te
1 cm
j
Pl
Sk?
Si
Te?
Sk
Pl
Th
Pl
Pl
Si
Pl
1 cm
k
1 cm
l
cm
m
Na
Na?
Na
Rh
1 cm
n
o
Rh
Rh
Na?
Rh
Na
Sk
1 cm

(salinity reduction, salinity variation) factors. Thus, tidally influenced trace-fossil suites typically display low-diversities with more normally sized burrows in tidal marine settings, and low diversities coupled with diminutive trace fossils in tidal settings prone to brackish waters.

Size and diversity trends proffer a mappable ichnological function that can be conceptually linked to tidal range. Gunn et al. (2008) and Hauck et al. (2009) mapped burrow sizes and their distributions in two modern estuaries; the lower mesotidal Ogeechee River, Georgia, USA and the microtidal Kouchibouguac, New Brunswick, Canada, respectively (Fig. 4.4). They used a simple function: the diameter of the largest burrow observed at a site (measured in millimeters) was multiplied by the number of burrow types observed, to create the Size-Diversity Index (SDI). Results show a geometric increase in the SDI in an oceanward direction. For the lower mesotidal Ogeechee, the SDI is very high near the mouth of the bay (although the sound itself is approximately SDI 500), but dissipates on a log-linear basis into the proximal tidal reach where salinities approach zero (Gunn et al. 2008). At Kouchibouguac, which is wave-dominated and microtidal, the central basin has a high SDI, but this decreases abruptly in the landward direction (Hauck et al. 2009). Conceptually, SDI trends *should be* distinctive for bays and estuaries of various tidal ranges. Although this has not yet been tested extensively in modern environments, an interpretive framework for SDI based on our observations in modern settings is provided in Fig. 4.5.

4.3.3 The Significance of Burrow Linings and Infills

Animals in tidal settings commonly concentrate fine-grained sediment in or adjacent to their burrows (Zorn et al. 2010). This is a result of selective ingestion of the small-caliber sediment, which is generally comparably rich in refractory organic carbon (Konhauser and Gingras 2007). Thus, some burrows associated with tidal settings display thickened linings composed of mud. Due to the tidally driven settling of food at the sediment-water interface, several animals (e.g. Nereid

Fig. 4.2 Examples of subtidal, intertidal and supratidal deposits with ichnological content indicated. (**a**) Mud-dominated tidally influenced subtidal pointbar. At this scale of view, trace fossils are difficult to observe, however, evident are lenticular sand beds defining characteristic inclined heterolithic stratification (IHS). Pleistocene, Willapa Bay, Washington, USA. (**b**) Closer view of same outcrop as in (**a**). Ichnofossil assemblage dominated by invertebrate domiciles including *Skolithos* (Sk), *Arenicolites* (Ar), and *Psilonichnus* (Ps). The trace fossil *Psilonichnus* is more common to intertidal and supratidal deposits and may be indicative of relatively low energy conditions in the subtidal channel. Pleistocene, Willapa Bay, Washington, USA. (**c**) Sand-dominated IHS with sparse bioturbation consisting of *Planolites* (Pl). Sand (*black* due to heavy-oil content) may be burrowed as well, but in the absence of lithological definition, this is difficult to assess. Cretaceous, Bluesky Formation, Peace River area of Alberta, Canada. (**d**) Sand-mud IHS attributed to deposition on a tidally influenced subtidal pointbar. *Cylindrichnus* (Cy) and *Planolites* (Pl) are indicated. Cretaceous, McMurray Formation, Athabasca area of Alberta, Canada. (**e**) Mud-dominated IHS attributed to deposition on a tidally influenced subtidal pointbar. *Cylindrichnus* (Cy), *Teichichnus* (Te), *Planolites* (Pl) and *Polykladichnus* (Po) are indicated. Cretaceous, McMurray Formation, Athabasca area of Alberta, Canada. (**f**) Bioturbated bottom sets and toe sets attributed to deposition in tidally influenced subtidal compound dunes. *Cylindrichnus* (Cy), and *Skolithos* (Sk) are indicated. Cretaceous, McMurray Formation, Athabasca area of Alberta, Canada. (**g**) Bioturbate texture with vestigial sand laminae locally preserved. The degree of bioturbation is high, but discrete trace fossils are observable where sand contributes lithological definition. *Planolites* (Pl), and *Skolithos* (Sk) are indicated. (**h**) Pleistocene, Willapa Bay, Washington, USA. Intertidal sand flat deposit with tidal runoff channels crosscut by *Siphonichnus* (Si) and *Thalassinoides* (Th). Pleistocene, Willapa Bay, Washington, USA. (**i**) Intensely burrowed strata common to intertidal deposits. This example contains *Planolites* (Pl), *Teichichnus* (Te) and *Thalassinoides* (Th). Cross-cutting the bioturbate texture are pyrite-replaced *rhizoliths* (Rh), evidencing vertical proximity to the supratidal setting. Cretaceous, Bluesky Formation, Peace River area of Alberta, Canada. (**j**) Similar to (**i**) with the addition of the characteristic intertidal trace-fossil *Siphonichnus* (Si). Cretaceous, McMurray Formation, Athabasca area of Alberta, Canada. (**k**) Intensely rooted supratidal strata (located in outcrop 60 cm above bioturbated intertidal media). Pleistocene, Willapa Bay, Washington, USA. (**l**) Similar to (**k**), but at the transition zone from intertidal to supratidal strata. Pleistocene, Willapa Bay, Washington, USA. (**m**) Eluviated supratidal strata with rare *rhizoliths* (Rh) and putative insect-larval traces *Nactodemasis* (Na). Cretaceous, Bluesky Formation, Peace River area of Alberta, Canada. (**n**) Outcrop view of eluviated supratidal media. Note the ironstone nodules. Cretaceous, McMurray Formation, Athabasca area of Alberta, Canada. (**o**) Eluviated supratidal strata with rare *rhizoliths* (Rh), putative insect-larval traces *Nactodemasis* (Na) and *Skolithos* (Sk). Cretaceous, McMurray Formation, Athabasca area of Alberta, Canada

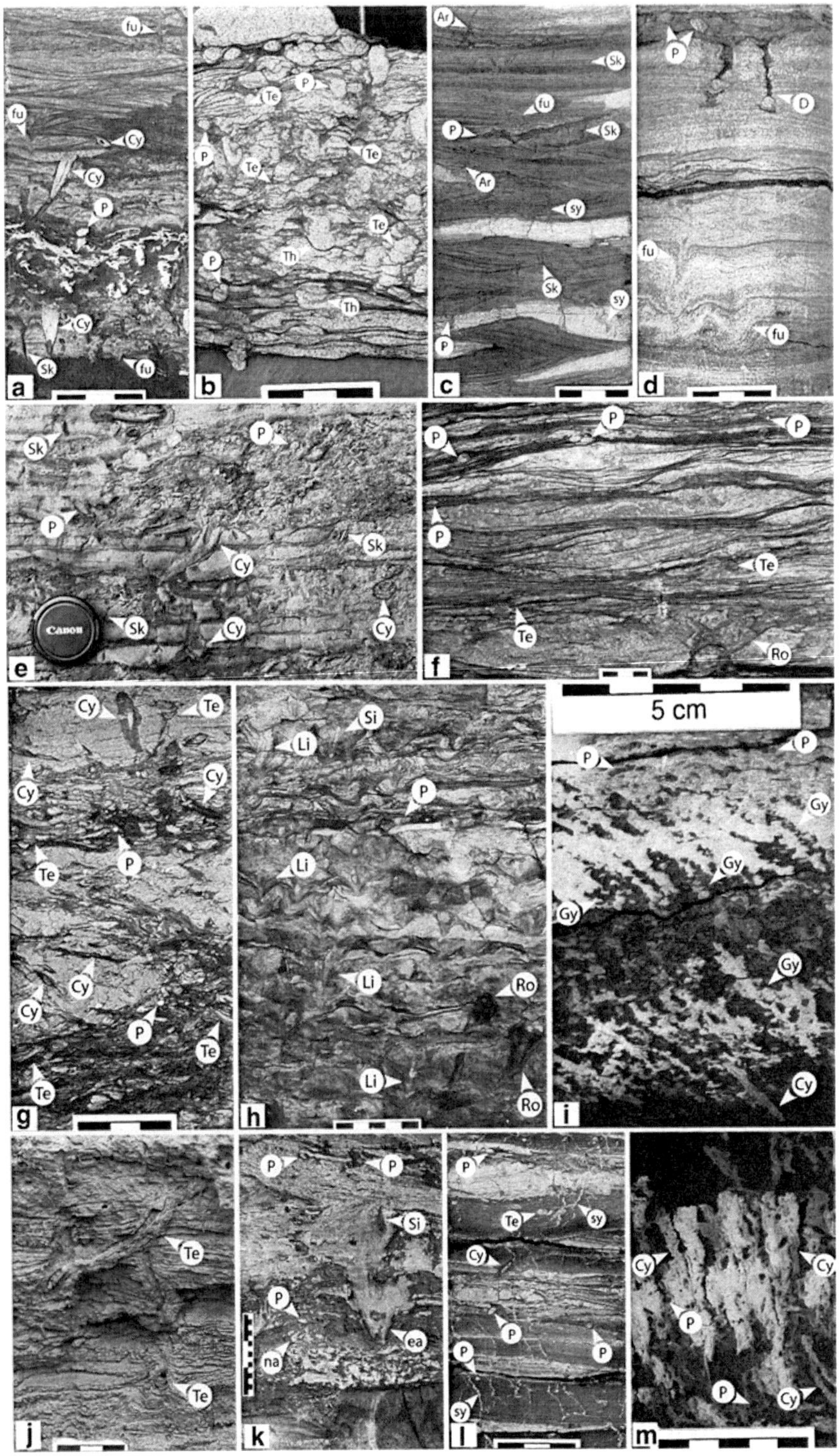
fu
fu
Cy
Cy
P
Cy
Sk
fu
a
P
Te
P
Te
Te
Te
P
Th
Th
b
Ar
Sk
fu
P
Sk
Ar
sy
Sk
sy
P
c
P
D
fu
fu
d
Sk
P
P
Sk
Cy
Sk
Cy
Cy
e
P
P
P
P
Te
Te
Ro
f
5 cm
Cy
Te
Cy
Cy
Te
P
Cy
Cy
P
Te
Te
g
Si
Li
P
Li
Li
Ro
Li
Ro
h
P
P
Gy
Gy
Gy
Gy
Cy
i
Te
Te
j
P
P
Si
P
na
ea
k
P
Te
sy
Cy
P
P
P
sy
l
Cy
Cy
P
P
Cy
m

polychaetes, the bivalve *Macoma balthica*, and the amphipod *Corophium volutator*; Gingras et al. 1999; Dashtgard et al. 2008) selectively feed at this interface, thereby allowing interment of mud (as burrow linings), even if the mud tends to hydraulically bypass the depositional locale. In marginal-marine settings, thickened linings are exemplified in the ichnogenera *Ophiomorpha* and *Cylindrichnus* (Fig. 4.6). Of course, similar mud stowage may also be observed in other depositional settings (e.g. lower shoreface, prodelta, inner shelf), and of itself is not diagnostic. However, low-diversity assemblages that can be ascribed to brackish-water settings and which *also* display biogenic stowage of fine-grained sediment are best interpreted as ichnological features related to rhythmic availability of mud.

Burrow infill may also be affected by the presence of tidal currents. Specifically, sediment that is moving at or near the seafloor can be trapped in burrows that possess a large, open aperture, or in burrows where a narrowed aperture has been breached by erosion. In tidal settings, open burrows may be rhythmically infilled by tidal processes forming "tubular tidalites" (Gingras et al. 2007) (Fig. 4.7). The most common feature observed in tubular tidalites is physically generated sedimentary couplets. However, apparent neap-spring bundles also have been observed in large examples of *Thalassinoides* and *Arenicolites* (e.g., Miocene-aged deposits of Amazonia; Gingras et al. 2002b). Tubular tidalites are dominantly observed in large diameter (>1 cm), vertical to inclined shafts, and in J- through U-shaped burrows. As such, the trace

Fig. 4.3 Examples of trace fossil suites characteristic of brackish-water settings. (**a**) Current-rippled sandstones showing BI 2. Ichnological suite includes diminutive and sporadically distributed *Cylindrichnus* (Cy), *Planolites* (P), *Skolithos* (Sk), and *fugichnia* (fu). Scale is 3 cm. Lower Cretaceous McMurray Fm, Alberta, Canada. (**b**) Bioturbated muddy sandstone of a sandy bay, showing BI 5. Suite is very low diversity, and dominated by *Teichichnus* (Te), *Planolites* (P), and *Thalassinoides* (Th). Scale is 3 cm. Lower Cretaceous Basal Colorado Sandstone, Alberta, Canada. (**c**) Sandy bay deposit showing BI 1–2, with alternating horizontal structures such as *Planolites* (P), and vertical structures such as *Arenicolites* (Ar), and *Skolithos* (Sk). Oscillation and combined-flow ripples show *fugichnia* (fu). Diminutive syneresis cracks (sy) might suggest salinity fluctuations. Scale is 3 cm. Lower Cretaceous Grand Rapids Fm, Alberta, Canada. (**d**) Estuary-mouth deposit in an incised valley, consisting of low-angle undulatory parallel-laminated sandstone and oscillation ripples with dark, carbonaceous mudstone interlaminae. Unit shows sporadically distributed bioturbation (BI 0–3). Stacked event beds contain *fugichnia* (fu), and display recolonization suites of *Diplocraterion* (D), and *Planolites* (P). Scale is 3 cm. Lower Cretaceous Viking Formation, Alberta, Canada. (**e**) Muddy IHS showing sporadically distributed bioturbation (BI 1–3), with a low-diversity suite consisting of diminutive, facies-crossing ichnogenera such as *Cylindrichnus* (Cy), *Skolithos* (Sk), and *Planolites* (P). Lens cap is ~8 cm in diameter. Pleistocene, Willapa Bay, Washington, USA. (**f**) Heterolithic current-rippled sandstone with mudstone drapes forms wavy-bedded composite bedsets, interpreted as a tidal-estuarine bay deposit. Bioturbation intensity is low (BI 0–2), with a low-diversity suite of diminutive *Planolites* (P), *Teichichnus* (Te), and uncommon *Rosselia* (Ro). Permian Pebbley Beach Formation, south Sydney Basin, Australia. (**g**) Sandy central-basin deposits of a riverine (tidally influenced) estuary. Facies consists of wavy-bedded sandstone and sandy mudstone containing remnant, wavy parallel lamination toward the top of the photo. The unit shows BI 3–5, with diminutive *Planolites* (P), *Teichichnus* (Te), and *Cylindrichnus* (Cy). Scale is 3 cm. Lower Cretaceous Glauconite Formation, Alberta, Canada. (**h**) Sandy estuarine-bay deposit showing BI 4–5, and a low-diversity suite dominated by re-equilibration structures and vertical dwelling/deposit-feeding structures. Dominant ichnogenera are *Lingulichnus* (Li) and *Rosselia* (Ro), with subordinate *Siphonichnus* (Si), and *Planolites* (P). Scale is 5 cm. Permian Pebbley Beach Formation, south Sydney Basin, Australia. (**i**) Thoroughly bioturbated (BI 4) IHS, with a monospecific suite of *Gyrolithes* (Gy) in the sand layers, and *Planolites* (P) in the mud beds. An isolated *Cylindrichnus* (Cy) occurs near the base of the photo. Down-slope creep of the sediment has resulted in deformation of the burrows. Lower Cretaceous McMurray Fm, Alberta, Canada. (**j**) Laminated sand and mud near the top of a tidal-creek point bar, showing Nereid-generated horizontal dwelling burrows that have been shifted upward during sedimentation. This structure is attributable to the ichnogenus *Teichichnus* (Te). Scale is 3 cm. Modern, Bay of Fundy, New Brunswick, Canada. (**k**) Sandy heterolithic estuarine-bay interval, containing bivalve-generated equilibrium adjustment structures (ea) and *Siphonichnus* (Si). Sediment-swimming structure or navichnia (na) is associated with water-rich muds. *Planolites* (P) are common to the mud layers. Scale is 15 cm. Pleistocene, Willapa Bay, Washington, USA. (**l**) Muddy bay deposits with heterolithic lenticular-bedding containing remnant oscillation ripples, wavy parallel lamination, dark unburrowed fissile mudstone drapes, and abundant synaeresis cracks (sy). The facies shows BI 2–3, with a low-diversity suite of diminutive *Planolites* (P), *Teichichnus* (Te), *Cylindrichnus* (Cy). Scale is 3 cm. Lower Cretaceous McMurray Formation, Alberta, Canada. (**m**) Mud bed capping a sand layer in tidally influenced IHS. The mud layer shows BI 4, with abundant *Cylindrichnus* (Cy). Rare *Planolites* (P) occur locally. Scale is 5 cm. Lower Cretaceous McMurray Formation, Alberta, Canada

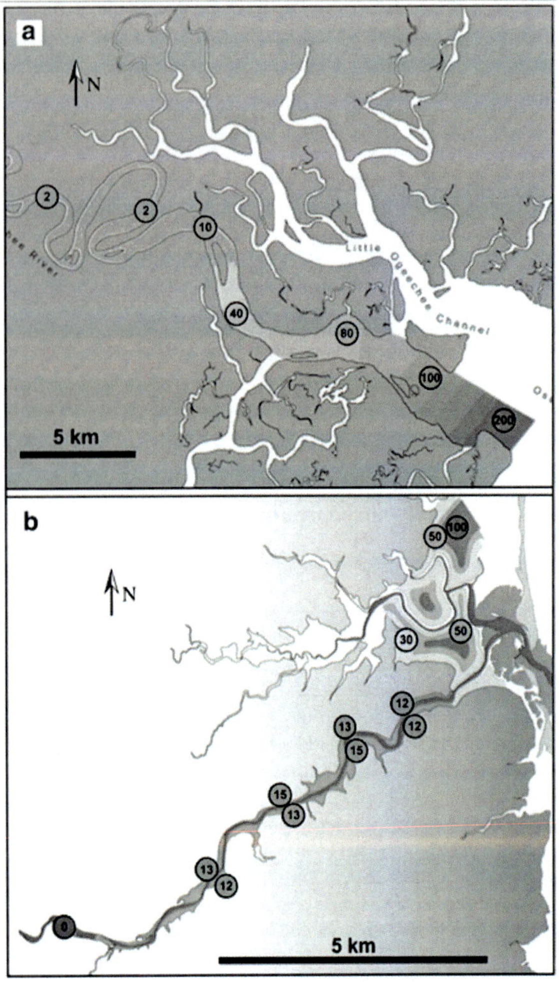

Fig. 4.4 Size-diversity index (SDI = maximum burrow diameter X diversity of trace types) based on field studies in two modern locales (**a**) Ogeechee River, Georgia, USA and (**b**) Kouchibouguac River, New Brunswick, Canada (Data after Gunn et al. 2008 and Hauck et al. 2009)

fossils *Skolithos*, *Thalassinoides*, *Psilonichnus* and *Arenicolites* are most commonly associated with tubular tidalites.

Although tubular tidalites are distinctive, their appearance may bear some similarities to meniscate backfill (*Scolicia*, *Taenidium*) or to spreite (*Teichichnus*, *Zoophycos*). Key differentiating features from menisicae and spreite include: burrow-infill laminae that are consistently oriented with a dip of less than 20° from (depositional) horizontal; the presence of laminae that abut the burrow wall (i.e. are non-tangential to the burrow margin); and the presence of sedimentary couplets.

4.3.4 Characteristic Feeding Behaviors in Tidal Settings

Food distribution is temporally and spatially heterogeneous in tidal settings. Due to attenuated tidal-current energies and the settling of water-borne organics with falling tide and during slack water, intertidal-flat deposits are generally food-rich compared to subtidal settings.

Subtidal locales tend to suffer higher energy tidal currents, which inhibit the deposition of food; this is exacerbated with increasing tidal-current strength. In subtidal settings, food is deposited most commonly with fine-sediment IHS or with slack-water mud drapes that characterize wavy through to flaser bedded composite bedsets. This localized resource is either exploited at the sediment-water interface in the form of stellate feeding traces (Figs. 4.2m and 4.7a), or is targeted in the substratum using simple (e.g., *Planolites*; Fig. 4.3e, g, m) through to more complex (e.g., *Teichichnus* Fig. 4.3b; "*Phycodes*" Fig. 4.7e; possibly *Gyrolithes* Fig. 4.3i) deposit-feeding strategies.

With more thickly interbedded media (IHS and wavy bedding), both surface feeding and substratal deposit feeding are commonly employed (Fig. 4.8). Surface feeding is observed as abundant *Cylindrichnus* or *Skithos* generated by organisms that colonized either sand or mud beds and gathered tidally delivered food from the surface (Figs. 4.1c and 4.2i, m). Intrastratal deposit-feeding ethologies are most commonly observed as *Planolites* and *Teichichnus*, and reflect organisms that bioturbated the deposit with varying degrees of thoroughness (Figs. 4.3b, e, g, m and 4.7c). These are behaviors that are well suited to the relatively rapid exploitation of intastratal food. Unlike the interface-feeding behaviors, it is likely that the intrastratal activities (especially within IHS) are aimed at exploiting seasonally deposited beds as opposed to directly benefitting from tidally delivered resources (Gingras et al. 2002b; Pearson and Gingras 2006). As such, the presence of vertically oriented trace fossils in seasonally generated IHS is likely a better indicator of the presence of tidal currents. Inclined heterolithic stratification that consist predominantly of tidally deposited strata with little or no seasonal cycle will likely be unburrowed, due to high sedimentation rates.

In contrast, the abundance of resources in tidal-flat deposits leads to a preponderance of characteristic ichnogenera. Although not limited to these settings, feeding behaviors that facilitate the rapid exploitation of

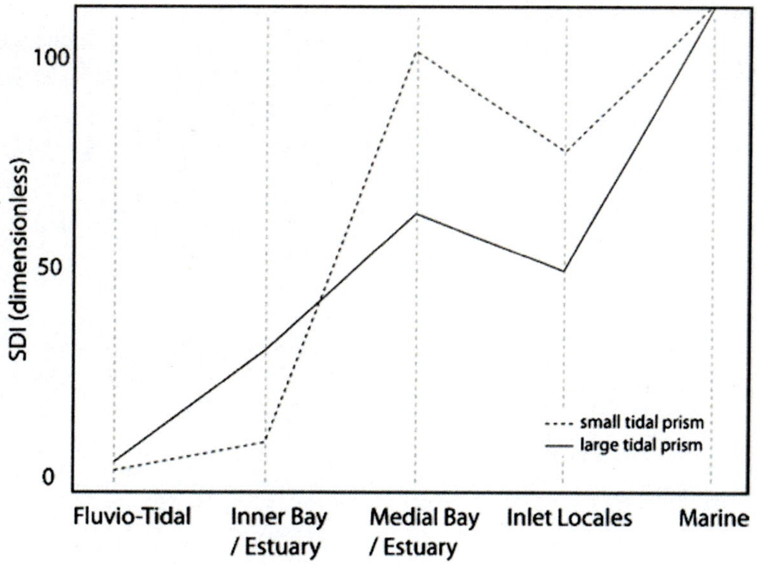

Fig. 4.5 Conceptual graph of SDI distribution in embayed depositional settings that are influenced by small versus large tidal prisms. Key factors in the construction of this graph include the greater landward propagation of marine waters with large tidal prisms, increased residence time of fresh water in settings with small tidal prisms, and the inhospitable nature of outer-bay settings in locales where diurnal/semidiurnal tidal currents are exceedingly strong

intrastratal food (e.g., *Planolites, Teichichnus* and *Thalassinoides*) or resources that have settled on the sediment-water interface (e.g., *Skolithos, Arenicolites,* bivalve-generated *Siphonichnus*, and surficial stellate probing marks) provide the dominant elements of the preserved ichnocoenose (Gingras et al. 1999; Hertweck et al. 2007). Stellate interface-feeding traces radiate from a central location (Fig. 4.8a, b), and are the surface expression of vertical traces that include *Skolithos* (e.g., Fig. 4.3c), *Arenicolites* (ibid.), *Siphonichnus* (Figs. 4.1a and 4.7d) and *Cylindrichnus* (e.g., Fig. 4.3m). The stellate expression can only be seen on uncommon bedding planes, however, and it is far more common to observe abundant vertical traces that most probably were associated with stellate surface/interface feeding (Figs. 4.1c and 4.2m). An important distinction here is that although elements of the *Skolithos* Ichnofacies are abundant in tidal settings, many of the vertically oriented ichnofossils represent an engagement in deposit feeding rather than filter feeding (Gingras et al. 1999; Dashtgard et al. 2008). This is, again, an ethological strategy made possible by the large amounts of resource that commonly accumulate during slack-water periods.

In flaser bedded sediment, comparably systematic deposit feeding may be used to harvest the localized resource. Within large-scale flasers, this may be seen as regular perforation of the flasers with *Planolites* or even *Phycodes*-like trace fossils (Fig. 4.7e). Small and thin flasers are not typically exploited through such selective efforts. These are more likely to be targeted by large-scale convection of the sediment, as conducted by "sand shrimp" (e.g., *Neotrypea, Callianassa*) producing *Thalassinoides,* or "lug worms" (e.g., *Arenicola*) excavating large *Arenicolites* or *Polykladichnus*. Additionally, these animals are also mining organics associated with the sand deposit as well (Swinbanks and Luternauer 1987).

4.4 Ichnologic Recognition of Tidally Influenced Deposits

As with the primary physical sedimentary structures, there are very few ichnological features that are produced solely in response to the influence of tides. As such, the success of an ichnological interpretation of tidal sedimentation depends upon accruing a preponderance of supporting observations. In practice, with our current level of understanding, it is unlikely that ichnological data would be put forth as the primary evidence for tidal reworking in the absence of more diagnostic physical sedimentary structures. However, in the interest of exploring the potential of trace fossils as tidal indicators, this paper deliberately avoids such dependence upon other corroborating datasets.

To establish an ichnological response to tidal processes effectively, the trace-fossil dataset should be

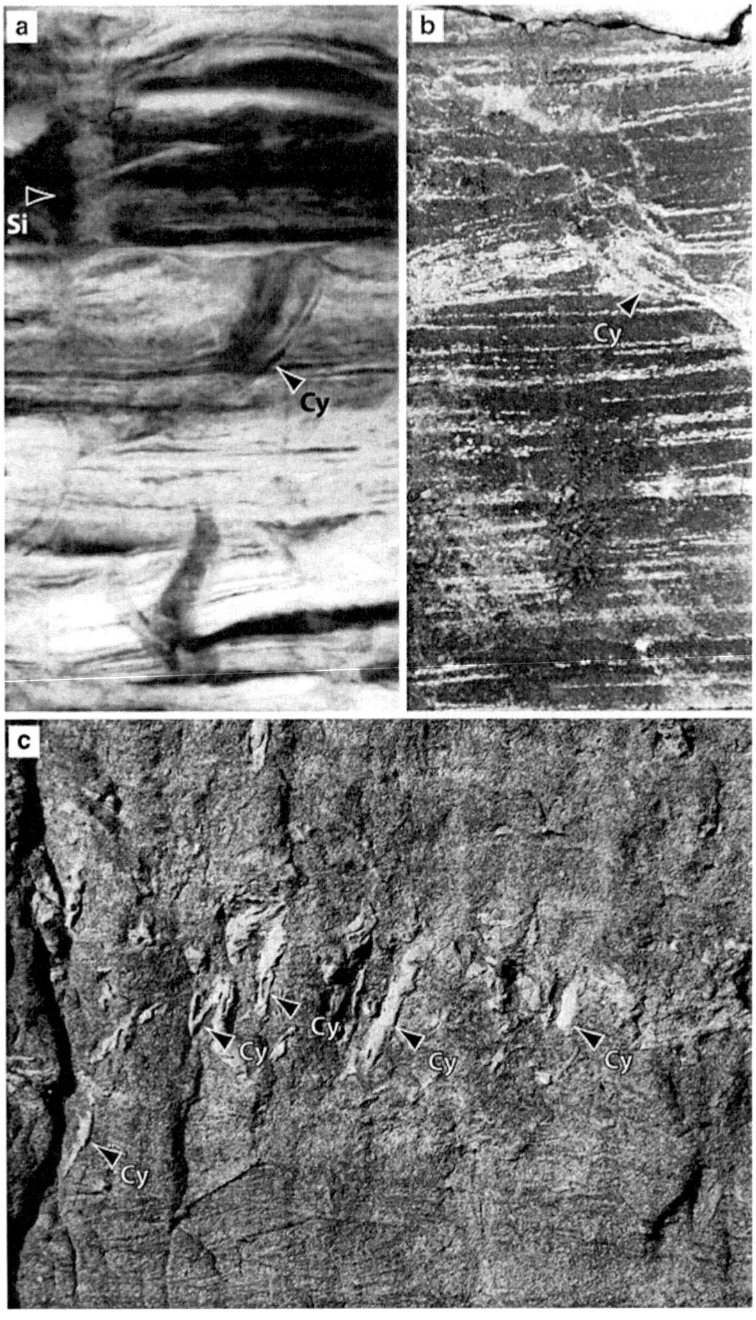

Fig. 4.6 Examples of mud stowage by animals in tidal settings. (**a**) *Cylindrichnus* (Cy) and *Siphonichnus* (Si) where mud has been reamed/packed into the burrow lining (x-ray negative: *white* is sand rich, *dark* is mud rich). Modern subtidal point-bar, 1-m depth Willapa Bay, Washington, USA. Box core is 20 cm wide. (**b**) *Cylindrichnus* (Cy) with concentric, iterative mud linings, indicating temporal fluctuation of mud availability. Slabbed core is 10 cm wide. Cretaceous Centenario Fm, Argentina. (**c**) Thickly lined *Cylindrichnus* (Cy). Slabbed core is 10 cm wide. Cretaceous, McMurray Formation, Alberta, Canada

viewed from as many scales as possible. Hand-sample, core-, and outcrop-scale observations include the types of ichnogenera present, the nature of their linings and infill (if present), and the distribution of trace fossils in beds and bed-sets. Mappable data, such as the distribution of key ichnogenera and their sizes, and diversity trends within genetically linked strata also stand to enhance a tidal interpretation: it should be emphasized that very few studies of this nature presently exist, and that a full understanding of ichnofossil distributions in various tidal settings is still forthcoming.

The general framework for identifying tidal influence from ichnological data depends upon the ability

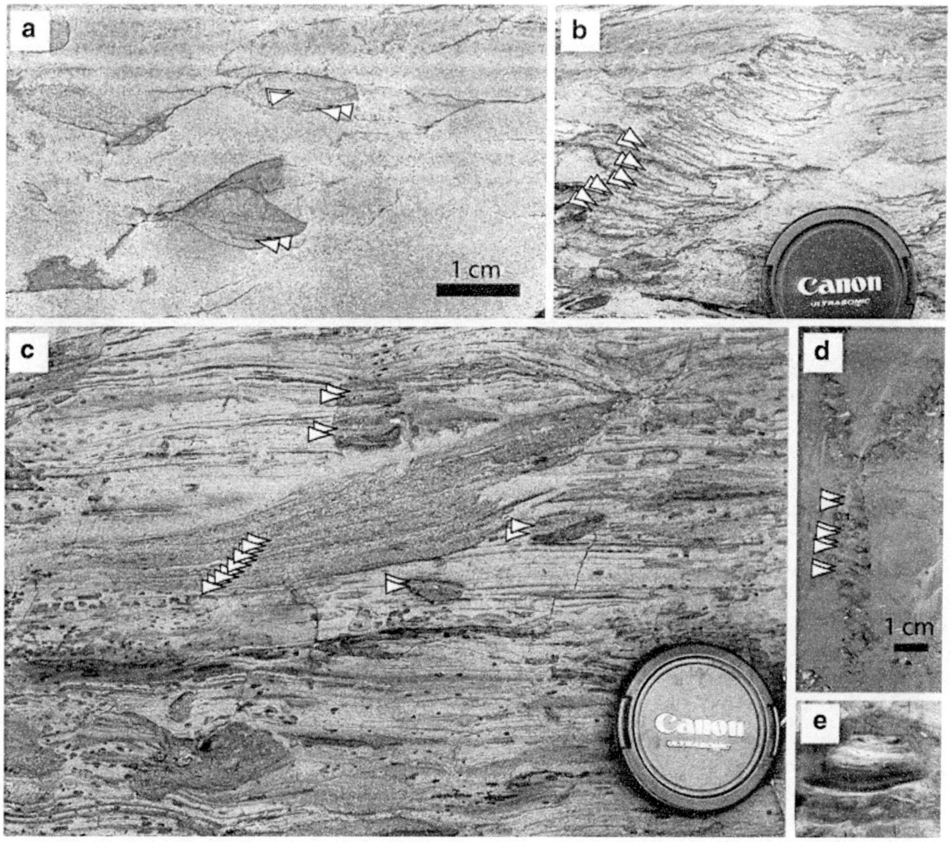

Fig. 4.7 Examples of tubular tidalites. In all photographs, the *white arrows* indicate the position of discernible sedimentary couplets. (**a**) ?*Thalassinoides* with poorly developed couplets. Jurassic, Upper Monteith Member, Nikanassin Formation, British Columbia, Canada. (**b**) Undiagnosed domicile (similar to *Psilonichnus*) displaying several sedimentary couplets. Lens cover is 6.2 cm diameter. Intertidal flat deposit, Pleistocene, Willapa Bay, Washington, USA. (**c**) Undiagnosed domicile from the same outcrop as (**b**) showing a side view through an analogous domicile. This trace fossil also displays several sedimentary couplets. Lens cover is 6.2 cm diameter. Intertidal flat deposit, Pleistocene, Willapa Bay, Washington, USA. (**d**) Vertical tube with well developed couplets. Miocene, Pebas Formation, Peru. (**e**) Cropped x-ray image (x-ray negative: *white* is sand rich, *dark* is mud rich) from a modern intertidal-flat deposit showing a recently abandoned crab burrow similar to the Pleistocene examples shown in (**b**) and (**c**). Image is 4 cm wide. Modern tidal flat, Willapa Bay, Washington, USA

of workers to establish as many of the following criteria as possible:

1. The presence of thickly lined ichnofossils, with the linings composed of sediment not otherwise preserved in associated beds. If this is so, it should be established that the trace-fossil assemblage is also of lower *diversity* than the baseline assemblage.
2. Rhythmically (but passively) infilled, larger-diameter trace fossils such as *Thalassinoides* and *Psilonichnus* are present, corresponding to tubular tidalites.
3. Regular waxing and waning of bioturbation intensity is observable at the bed and/or bed-set scale.
4. Ichnofossil suites in highly bioturbated sediment comprise both horizontally and vertically oriented trace fossils (i.e., concurrent substratal- and interface-deposit feeding can be established), and highly bioturbated zones lie gradationally above dominantly bedded strata.
5. In heterolithic bedding, zones dominated by (presumed interface deposit-feeding) vertically oriented trace fossils are present. In outcrop studies, surficial stellate feeding traces are commonly observed as well.
6. The presence of systematic deposit feeding within flaser-bedded intervals.

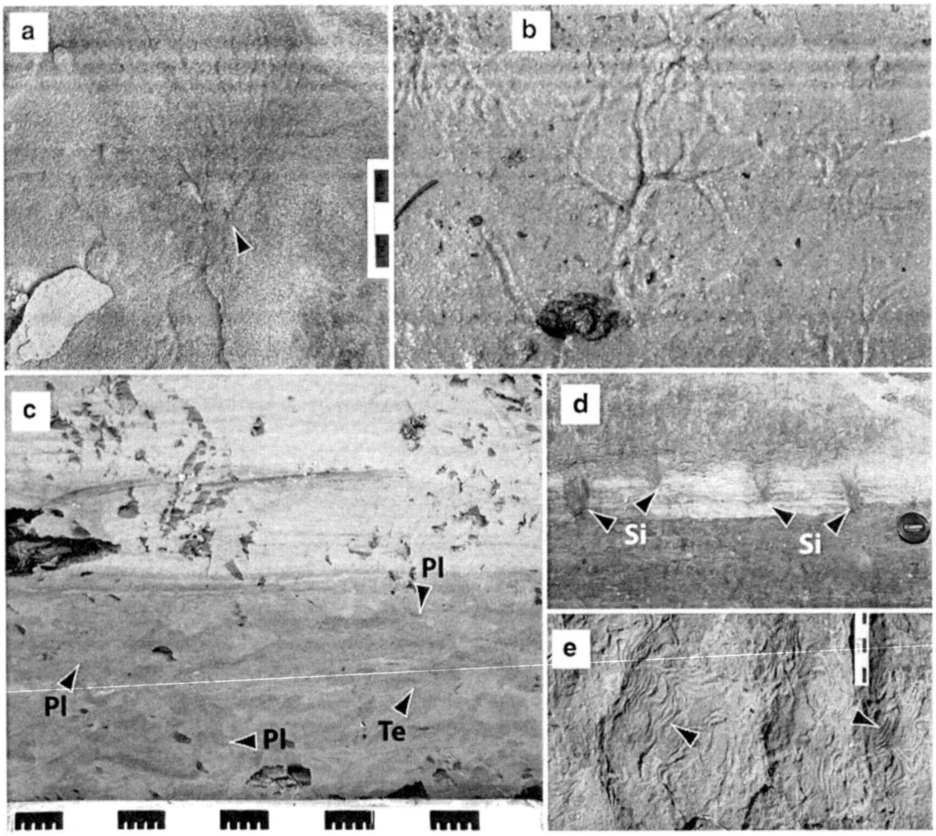

Fig. 4.8 Distinctive trace fossils of tidal settings. (**a**) Stellate, iteratively branched interface-feeding trace (presumably made by an annelid). Eocene Baronia Formation, Spain. (**b**) Modern feeding traces analogous to that shown in (**a**) made by Nereid polychaetes. Field of view 14 cm across. Modern tidal flat, Kouchibouguac, New Brunswick, Canada. (**c**) Highly bioturbated zone dominated by *Planolites* (Pl) and *Teichichnus* (Te). Miocene Pebas Formation, Peru. (**d**) *Siphonichnus* (Si) at transition to intertidal flat. Field of view is approximately 110 cm across. Pleistocene, Willapa Bay, Washington, USA. (**e**) Undiagnosed meandering, dominantly horizontal mud-filled burrows showing high level of resource optimization. Tidal dune toeset, Eocene, Baronia Formation, Spain

7. Demonstrably mappable ichnological data (particularly size and diversity trends) that can be related to the landward decrease of salinity in the tidal-fluvial transition zone.
8. The common presence of a brackish-water fauna (as expressed by the dominance of simple, facies-crossing ichnogenera).

4.5 Summary

The potential of trace fossils in helping to identify tidal influence on sedimentation and faunal colonization is at its inception. Although relatively sophisticated analyses of tidally generated sedimentary structures have been established and refined, the same cannot be said for ichnological datasets. Limitations of identifying tidal influence from ichnological datasets are, in part, intrinsic: due to the seasonal nature of larval recruitment and the longer time scales represented by biogenic sedimentary structures, trace fossils generally do not represent organism responses to rhythmic changes in energy associated with diurnal or semidiurnal tidal cycles. On the other hand, careful documentation of specific trace-fossil characteristics (lining and infill), documentation of the episodic character of mud/food deposition and their distributions in the sediment, interpreted in the context of the detailed analysis of the composition of the trace-fossil assemblage, can be linked strongly to tidal transport of sediment and food. Characterization of vertical successions and the identification of intertidal-flat deposits, which are typically ichnologically distinctive, provide a reliable means of identifying tidal influence on sedimentation. At the map scale, ichnological size and diversity trends also can be related to tidal processes operating at the scale of entire environments.

A substantial amount of work is still required before an ichnological model that is fundamentally related to tidal processes *per se* can be developed. First, analysis of (tidal) trace-fossil compositions must continue, but this work should be conducted with a focus on characterizing the ethological response(s) (i.e. principal feeding behaviors represented) as opposed to simply cataloging the ichnogenera present. Decoupling the observed "tidal" behaviors from the recurring (marine) behavioral paradigms of the *Skolithos* Ichnofacies and the *Cruziana* Ichnofacies will be a necessary aspect of tidal trace-fossil interpretations. Continued analysis of tidally modulated and tidally influenced shoreface settings (Dashtgard et al. 2009; in press; Frey and Dashtgard in press) hold the promise of discerning the tidal signal on animal-sediment responses. Second, more examples of high-certainty tidal deposits in the rock record need to be examined, in order to establish better the prevailing behaviors used to exploit spatially and temporally heterogeneous resources. In other words, what behaviors are employed to exploit anisotropically distributed resources that are present in mud drapes? Third, the size and diversity trends in various marginal-marine settings must be established better; this includes documentation of the spatial distribution of SDI in microtidal through to megatidal settings, and testing these patterns in a range of different depositional environments. Although establishing SDI for a range of sedimentary environments is forthcoming, the resultant knowledge will be readily applicable to the rock record.

References

Aitken AE, Risk MJ, Howard JD (1988) Animal-sediment relationships on a subarctic intertidal flat, Pangnirtung Fiord, Baffin Island, Canada. J Sediment Petrol 58:969–978

Bann KL, Fielding CR, MacEachern JA, Tye SC (2004) Differentiation of estuarine and offshore marine deposits using integrated ichnology and sedimentology; Permian Pebbley Beach Formation, Sydney Basin, Australia; the application of ichnology to palaeoenvironmental and stratigraphic analysis. Geol Soc Spec Publ 228:179–211

Beynon BM, Pemberton SG, Bell DD, Logan CA (1988) Environmental implications of ichnofossils from the Lower Cretaceous Grand Rapids Formation, Cold Lake oil sands deposit; sequences, stratigraphy, sedimentology; surface and subsurface. Can Soc Petrol Geol Mem 15:275–289

Bromley RG (1996) Trace fossils; biology, taphonomy and applications. Chapman & Hall, London, UK (GBR)

Buatois L, Gingras M, MacEachern J, Mangano M, Zonneveld J, Pemberton S, Netto R, Martin A (2005) Colonization of brackish-water systems through time: evidence from the trace-fossil record. Palaios 20:321–347

Carmona N, Buatois L, Ponce J, Mangano M (2009) Ichnology and sedimentology of a tide-influenced delta, Lower Miocene Chenque Formation, Patagonia, Argentina: trace-fossil distribution and response to environmental stresses. Palaeogeog Palaeoclim Palaeoecol 273:75–86

Crimes TP, Marcos A, Perez-Estaun A (1974) Upper Ordovician turbidites in western Asturias: a facies analysis with particular reference to vertical and lateral variations. Palaeogeog Palaeoclim Palaeoecol 15:169–184

Crimes TP, Goldring R, Homewood P, Van Stuijvenberg J, Winkler W (1981) Trace fossil assemblages of deep-sea fan deposits, Gurnigel and Schlieren flysch (Cretaceous-Eocene), Switzerland. Ecol Geol Helv 74:953–995

Dalrymple RW (1984) Morphology and internal structure of sandwaves in the Bay of Fundy. Sedimentology 31:365–382

Dalrymple RW, Knight RJ, Zaitlin BA, Middleton GV (1990) Dynamics and facies model of a macrotidal sandbar complex, Cobequid Bay – Salmon River estuary (Bay of Fundy). Sedimentology 37:577–612

Dalrymple RW, Makino Y, Zaitlin BA (1991) Temporal and spatial patterns of rhythmite deposition on mud flats in the macrotidal Cobequid Bay-Salmon River estuary, Bay of Fundy, Canada; clastic tidal sedimentology. Can Soc Petrol Geol 16:137

Dashtgard SE (2011) Neoichnology of the lower delta plain: Fraser River Delta, British Columbia, Canada: Implications for the ichnology of deltas. Palaeogeography, Palaeoclimatology, Palaeoecology. doi:10.1016/j.palaeo.2011.05.001

Dashtgard S, Gingras M, Pemberton S (2008) Grain-size controls on the occurrence of bioturbation. Palaeogeog Palaeoclim Palaeoecol 257:224–243

Dashtgard SE, Gingras MK, MacEachern JA (2009) Tidally modulated shorefaces. J Sediment Res 79:793–807

Dashtgard SE, MacEachern JA, Frey SE, Gingras MK (in press) Tidal effects on the shoreface: towards a conceptual framework. Sed Geol. doi: 10.1016/j.sedgeo.2010.09.006, 21 p

Ekdale AA, Mason TR (1988) Characteristic trace-fossil associations in oxygen-poor sedimentary environments. Geology 16:720–723

Frey SE, Dashtgard SE, (in press) Sedimentology, ichnology and hydrodynamics of strait-margin, sand and gravel beaches and shorefaces: Juan de Fuca Strait, British Columbia, Canada. Sedimentology. doi: 10.1111/j.1365-3091.2010.01211.x

Frey RW, Howard JD (1970) Comparison of upper Cretaceous ichnofaunas from siliceous sandstones and chalk, Western Interior region, U.S.A; trace fossils. Geol J Spec Issue 3:141–166

Frey RW, Howard JD (1990) Trace fossils and depositional sequences in a clastic shelf setting, Upper Cretaceous of Utah. J Paleo 64:803–820

Gingras MK, Pemberton SG, Saunders T, Clifton HE (1999) The ichnology of modern and Pleistocene brackish-water

deposits at Willapa Bay, Washington; variability in estuarine settings. Palaios 14:352–374
Gingras MK, Rasanen M, Ranzil A (2002a) The significance of bioturbated inclined heterolithic stratification in the southern part of the Miocene Solimoes Formation, Rio Acre, Amazonia Brazil. Palaios 17:591–601
Gingras MK, Rasanen ME, Pemberton SG, Romero LP (2002b) Ichnology and sedimentology reveal depositional characteristics of bay-margin parasequences in the Miocene Amazonian foreland basin. J Sediment Res 72:871–888
Gingras MK, Bann KL, MacEachern JA, Waldron W, Pemberton SG (2007) A conceptual framework for the application of trace fossils. In: MacEachern JA, Bann KL, Gingras MK, Pemberton SG (eds) Applied ichnology. SEPM Short Course Notes 52, pp 1–25
Gingras MK, Dashtgard SE, MacEachern JA (2011) The potential of trace fossils as tidal indicators. Sediment Geol. doi:10.1016/j.sedgeo.2011.05.007
Gunn SC, Gingras MK, Dalrymple RW, Pemberton SG (2008) Ichnological gradation of subtidal deposits, Ogeechee Estuary, Georgia, USA. In: 2008 AAPG annual convention & exhibition; abstracts volume. Abstracts of annual meeting, American Association of Petroleum Geology
Hauck TE, Dashtgard SE, Pemberton SG, Gingras MK (2009) Brackish-water ichnological trends in a microtidal barrier island-embayment system, Kouchibouguac National Park, New Brunswick, Canada. Palaios 24:478–496
Hertweck G, Wehrmann A, Liebezeit G (2007) Bioturbation structures of polychaetes in modern shallow marine environments and their analogues to *Chondrites* group traces. Palaeogeog Palaeoclim Palaeoecol 245:382–389
Hovikoski J, Gingras M, Rasanen M, Rebata LA, Gjuerrero J, Ranzi A, Melo J, Romero L, Nunez Del Prado H, Jaimes F, Lopez S (2007) The nature of Miocene Amazonian epicontinental embayment; high-frequency shifts of the low-gradient coastline. Bull Geol Soc Am 119:1506–1520
Hovikoski J, Rasanen M, Gingras M, Ranzi A, Melo J (2008) Tidal and seasonal controls in the formation of late Miocene inclined heterolithic stratification deposits, western Amazonian foreland basin. Sedimentology 55:499–530
Howard JD, Elders CA, Heinbokel JF (1975) Animal-sediment relationships in estuarine point bar deposits, Ogeechee River-Ossabaw Sound, Georgia. Senckenbergiana Maritima, 7:181–203
Hubbard SM, Gingras MK, Pemberton SG (2004) Palaeoenvironmental implications of trace fossils in estuarine deposits of the Cretaceous Bluesky formation, Cadotte region, Alberta, Canada. Foss Strat 51:1–20
Klein GD (1970) Depositional and dispersal dynamics of intertidal sand bars. J Sediment Petrol 40:1095–1127
Konhauser KO, Gingras MK (2007) Linking geomicrobiology with ichnology in marine sediments. Palaios 22:339–342
Lettley CD, Pemberton SG, Gingras MK, Ranger MJ, Blakney BJ (2007) Integrating sedimentology and ichnology to shed light on the system dynamics and paleogeography of an ancient riverine estuary. In: MacEachern JA, Bann KL, Gingras MK, Pemberton SG (eds) Applied ichnology. SEPM Short Course Notes 52:142–160
Levinton JS (1970) The paleoecological significance of opportunistic species. Lethaia 3:69–78
MacEachern JA, Bann KL (2008) The role of ichnology in refining shallow marine facies models; recent advances in models of siliciclastic shallow-marine stratigraphy. SEPM Spec Publ 90:73–116
MacEachern JA, Gingras MK (2007) Recognition of brackish-water trace-fossil suites in the Cretaceous Western Interior Seaway of Alberta, Canada; sediment-organism interactions; a multifaceted ichnology. SEPM Spec Publ 88: 149–193
MacEachern JA, Pemberton SG (1994) Ichnological aspects of incised-valley fill systems from the Viking Formation of the Western Canada sedimentary basin, Alberta, Canada; incised-valley systems; origin and sedimentary sequences. SEPM Spec Publ 51:129–157
MacEachern JA, Stelck CR, Pemberton SG (1999) Marine and marginal marine mudstone deposition; paleoenvironmental interpretations based on the integration of ichnology, palynology and foraminiferal paleoecology; isolated shallow marine sand bodies; sequence stratigraphic analysis and sedimentologic interpretation. SEPM Spec Publ 64: 205–225
MacvEachern JA, Pemberton SG, Gingras MK, Bann KL (2010) Ichnology and facies models. In: Dalrymple RW, James NP (eds) Facies models, 3rd edn. Geological Association of Canada, St. John's, NL, pp 19–58
Martin KD (2004) A re-evaluation of the relationship between trace fossils and dysoxia; the application of ichnology to palaeoenvironmental and stratigraphic analysis. Geo Soc Spec Publ 228:141–156
Mangano MG, Buatois LA (2004) Ichnology of Carboniferous tide-influenced environments and tidal flat variability in the North American Midcontinent; the application of ichnology to palaeoenvironmental and stratigraphic analysis. Geol Soc Spec Publ 228:157–178
Mangano MG, Buatois LA, West RR, Maples CG (1998) Contrasting behavioral and feeding strategies recorded by tidal-flat bivalve trace fossils from the Upper Carboniferous of eastern Kansas. Palaios 13:335–351
McIlroy D (2004) Ichnofabrics and sedimentary facies of a tide-dominated delta; Jurassic Ile Formation of Kristin Field, Haltenbanken, offshore mid-Norway; the application of ichnology to palaeoenvironmental and stratigraphic analysis. Geol Soc Spec Publ 228:237–272
McIlroy D (2007) Ichnology of a macrotidal tide-dominated deltaic depositional system; Lajas Formation, Neuquen Province, Argentina; sediment-organism interactions; a multifaceted ichnology. SEPM Spec Publ 88:195–211
McIlroy D (2008) Ichnological analysis: the common ground between ichnofacies workers and ichnofabric analysts. Palaeogeog Palaeoclim Palaeoecol 270:332–338
Pearson NJ, Gingras MK (2006) An ichnological and sedimentological facies model for muddy point-bar deposits. J Sediment Res 76:771–782
Pemberton SG, Frey RW (1984) Ichnology of storm-influenced shallow marine sequence; Cardium Formation (Upper Cretaceous) at Seebe, Alberta; The Mesozoic of middle North America. Can Soc Petrol Geol Mem 9:281–304
Pemberton SG, MacEachern JA (1997) Ichnological signature of storm deposits; the use of trace fossils in event stratigraphy. In: Brett CC, Baird GC (eds) Paleontological events;

stratigraphic, ecological, and evolutionary implications. Columbia University Press, New York
Pemberton SG, Flach PD, Mossop GD (1982) Trace fossils from the Athabasca Oil Sands, Alberta, Canada. Science 217:825–827
Pemberton SG, Frey RW, Saunders TDA (1990) Palaeoecology; trace fossils [modified]. In: Briggs DEG, Crowther PR (eds) Palaeobiology; a synthesis. Blackwell Science, Oxford
Pemberton SG, MacEachern JA, Ranger MJ (1992a) Ichnology and event stratigraphy; the use of trace fossils in recognizing tempestites; applications of ichnology to petroleum exploration. A core workshop. SEPM core workshop note 17, pp 85–117
Pemberton SG, Reinson GE, MacEachern JA (1992b) Comparative ichnological analysis of late Albian estuarine valley-fill and shelf-shoreface deposits, Crystal Viking Field, Alberta; applications of ichnology to petroleum exploration. A core workshop. SEPM core workshop notes 17, pp 291–317
Pianka ER (1970) On r and k selection. Am Nat 104:592–597
Rebatah LA, Gingras MK, Rasanen ME, Barberi M (2006) Tidal channel deposits on a delta plain from the upper Miocene Nauta formation, Maranon foreland subbasin, Peru. Sedimentology 53(5):971–1013
Savrda C (2002) Equilibrium responses reflected in a large *Conichnus* (Upper Cretaceous Eutaw formation, Alabama, USA). Ichnos 9:33–40
Savrda CE (2007) Trace fossils and marine benthic oxygenation. In: William Miller, III (ed) Trace fossils: concepts, problems, prospects, pp 149–158
Savrda CE, Bottjer DJ (1988) Recognition of paleo-oxygenation fluctuations in California Neogene basins; abs. Bull Am Assoc Petrol Geo 72:394
Savrda CE, Nanson LL (2003) Ichnology of fair-weather and storm deposits in an Upper Cretaceous estuary (Eutaw Formation, western Georgia, USA). Palaeogeog Palaeoclim Palaeoecol 202:67–83
Seilacher A (1964) Biogenic sedimentary structures. In: Imbrie J, Newell ND (eds) Approaches to paleoecology. Wiley, New York, pp 296–316
Seilacher A (1982) General remarks about event deposits. In: Einsele G, Seilacher A (eds), Cyclic and event stratification. Springer-Verlag, New York, pp 161–174
Slobdkin LB, Sanders HL (1969) On the contribution of environmental predictability to species diversity. Brookhaven Symp Biol 22:82–93
Swinbanks DD, Luternauer JL (1987) Burrow distribution of thalassinidean shrimp on a Fraser Delta tidal flat, British Columbia. J Paleontol 61:315–332
Taylor AM, Goldring R (1993) Description and analysis of bioturbation and ichnofabric; organisms and sediments; relationships and applications. J Geol Soc Lond 150: 141–148
Taylor A, Goldring R, Gowland S (2003) Analysis and application of ichnofabrics. Earth Sci Rev 60:227–259
Van den Berg JH (1981) Rhythmic seasonal layering in a mesotidal channel fill sequence, Oosterchelde Mouth, the Netherland. In: Nio S-D, Shuttenhelm RTE, van Weering TjCE (eds) Holocene marine sedimentation in the North Sea Basin. International Association of Sedimentology Special Publication 5, pp 147–159
Wanless HR, Tedesco LP, Tyrrell KM (1988) Production of subtidal tubular and surficial tempestites by Hurricane Kate, Caicos Platform, British West Indies. J Sediment Petrol 58:739–750
Wignal PB (1991) Dysaerobic trace fossils and ichnofabrics in the Upper Jurassic Kimmeridge Clay of southern England. In: 13th international sedimentological congress, Ichnologic symposium. Palaios 6:264–270
Yang C, Nio S (1989) An ebb-tide delta depositional model; a comparison between the modern Eastern Scheldt tidal basin (Southwest Netherlands) and the lower Eocene Roda Sandstone in the southern Pyrenees (Spain). Sediment Geol 64:175–196
Zorn ME, Gingras MK, Pemberton SG (2010) Variation in burrow-wall micromorphologies of select intertidal invertebrates along the Pacific Northwest coast, USA; behavioral and diagenetic implications. Palaios 25:59–72

Processes, Morphodynamics, and Facies of Tide-Dominated Estuaries

5

Robert W. Dalrymple, Duncan A. Mackay, Aitor A. Ichaso, and Kyungsik S. Choi

Abstract

As defined in this chapter, an estuary forms during a shoreline transgression and then fills during a progradational phase that is transitional to a delta. The spatial distribution of processes, grain sizes and facies within tide-dominated estuaries is predictable in general terms. Tidal currents dominate sedimentation along the axis, with wave-dominated sedimentation occurring along the flanks of the estuary in its outer part. Tidal energy increases into the estuary but then decreases toward the tidal limit, with a gradual transition to river-dominated sedimentation at its head. The interaction of the tidal wave with the morphology of the estuary, and with river currents, causes the outer estuary to be flood-dominant, with a net landward movement of sand. By contrast, the inner estuary is ebb-dominant, creating a bedload convergence within the estuary. The axial sandy deposits are typically finest at this location. In transgressive-phase estuaries, the main channel shows a low—high—low pattern of sinuosity, with the tightest bends (sinuosity ≥ 2.5) occurring at the bedload convergence. These bends experience neck cutoff in the transition to the progradational phase of estuary filling. The estuary-mouth region is characterized by cross-bedded sands deposited on elongate sand bars, although wave-generated structures can be important in some cases. Estuaries that are down-drift of major rivers have anomalously muddy outer estuarine deposits. Further landward, upper-flow-regime parallel lamination can be widespread. The margins of the inner estuary are flanked by muddy salt-marsh and tidal-flat deposits that can contain well-developed tidal rhythmites and evidence of seasonal variations in river discharge.

5.1 Introduction

The term 'estuary' is fraught with confusion, with two overlapping but distinct definitions. The broadest definition is that of Pritchard (1967) that states that an *estuary* is "*a semi-enclosed coastal body of water in which the salinity is measurably diluted by fresh water derived from land drainage*". In this definition, the key element is the presence of brackish water; the specific geographic, geologic or stratigraphic context is

R.W. Dalrymple (✉) • D.A. Mackay • A.A. Ichaso
Department of Geological Sciences and Geological Engineering, Queens University, Kingston, ON K7L 3N6, Canada
e-mail: dalrymple@geol.queensu.ca; duncanamackay@yahoo.com; aitorichaso@hotmail.com

K.S. Choi
Faculty of Earth Systems and Environmental Sciences, Chonnam National University, Gwangju 500-757, South Korea
e-mail: tidalchoi@hotmail.com

R.A. Davis, Jr. and R.W. Dalrymple (eds.), *Principles of Tidal Sedimentology*,
DOI 10.1007/978-94-007-0123-6_5, © Springer Science+Business Media B.V. 2012

immaterial, except for the criterion of partial enclosure. Thus, the presence of a salt wedge that is over-ridden by fresh water supplied by a river is referred to as *estuarine circulation*, regardless of whether it occurs in the distributary channels of the Changjiang River delta, which is actively creating new land as a result of sediment deposition (Hori et al. 2001), or the mouth of the Severn River, which is migrating landward by means of coastal erosion (Allen 1990). Of course, in a geological context, these two situations (progradational and transgressive, respectively) are polar opposites because they generate stratigraphic successions that are upside down relative to each other. This distinction is particularly important in a sequence-stratigraphic context, which aims to reconstruct shoreline behavior in response to changes in eustatic sea level, tectonic movement and sediment supply.

As a result, Dalrymple et al. (1992) (see also Dalrymple 2006) proposed a 'geological' definition that states that *"an estuary is a transgressive coastal environment at the mouth of a river, that receives sediment from both fluvial and marine sources, and that contains facies influenced by tide, wave and fluvial processes. The estuary is considered to extend from the landward limit of tidal facies at its head to the seaward limit of coastal facies at its mouth"* (Dalrymple 2006, p. 11). This definition represents a subset of the environments covered by the Pritchard (1967) definition because it is restricted to transgressive settings. This is the definition used in this chapter. It is noteworthy, however, that this definition indicates that estuaries as defined here *import* sediment from the sea (i.e. there is a strong element of flood dominance), whereas deltas *export* sediment to the sea (i.e. they are ebb dominated). This is an important process distinction that has featured prominently in process-oriented literature on coastal environments (e.g. Friedrichs and Aubrey 1988; Friedrichs et al. 1990) and which is discussed further below. Estuaries are, therefore, ephemeral features, in that they are formed by relative sea-level rise that creates *accommodation* (i.e. the space available for sediment accumulation; Catuneanu 2006) in the river-mouth area, which is then filled by sediment input by both river and marine processes. Estuaries are abundant today because of the recent post-glacial transgression. Depending on the local circumstances, some of them are still actively transgressing, whereas others are in various stages of transition to deltas. Therefore, the nature of this transition is considered in this chapter. Systems that have made the full transition to deltas are discussed in Chap. 7 of this volume.

The focus in this chapter is on estuaries in which tidal currents are the *dominant* agent of sediment transport. Tidal dominance is produced either by the presence of a large tidal range and/or by the presence of weak wave action in the coastal zone (Davis and Hayes 1984). There has been a tendency in the literature to associate tidal dominance with macrotidal conditions (i.e. tidal range >4 m), but tidal dominance can also occur in microtidal and mesotidal areas, provided wave energy is low enough. Well-studied examples of tide-dominated estuaries include the Cobequid Bay-Salmon River estuary, Bay of Fundy (Dalrymple et al. 1990, 1991; Dalrymple and Zaitlin 1994), the Severn River estuary, Great Britain (Harris and Collins 1985; Allen 1990; McLaren et al. 1993), Mont-Saint-Michel Bay, France (Tessier et al. 2006, 2010; Billeaud et al. 2007) and the Fitzroy River estuary, Australia (Bostock et al. 2007; Ryan et al. 2007). Such estuaries show an exponential seaward widening that is referred to as a 'funnel-shaped' mouth (Fig. 5.1). Strong tidal currents flowing into and out of the river mouth create a series of channels that are approximately perpendicular to the main shoreline trend. At their mouth, these channels are separated by elongate tidal bars that are typically, but not everywhere, composed of sand. Broad tidal flats are widespread. Further landward, these channels become more sinuous and are flanked by tidal point bars. Tidal flats are narrower here as are the channels themselves. In the following sections we first describe the processes that operate in these systems, and then examine how the morphology and facies respond to these processes. The stratigraphy of tide-dominated estuaries is considered in Chap. 6.

5.2 Process Framework

5.2.1 Waves, River, Tidal Currents, and Bed-Material Movement

Although tidal currents are the most important process responsible for sediment erosion and deposition in tide-dominated estuaries, waves and river currents also play an important role locally (Figs. 5.2 and 5.3) at certain times. Waves control sedimentation on the seaward flanks of the estuary because the *tidal prism* (i.e. the volume of water moving past a location during each half tidal cycle) is small. Thus, the open coast adjacent to a tide-dominated estuary is typically wave dominated (Fig. 5.2; Yang et al. 2005, 2007). However, near the mouth of the estuary, the tidal prism and the resulting tidal currents become larger, generating

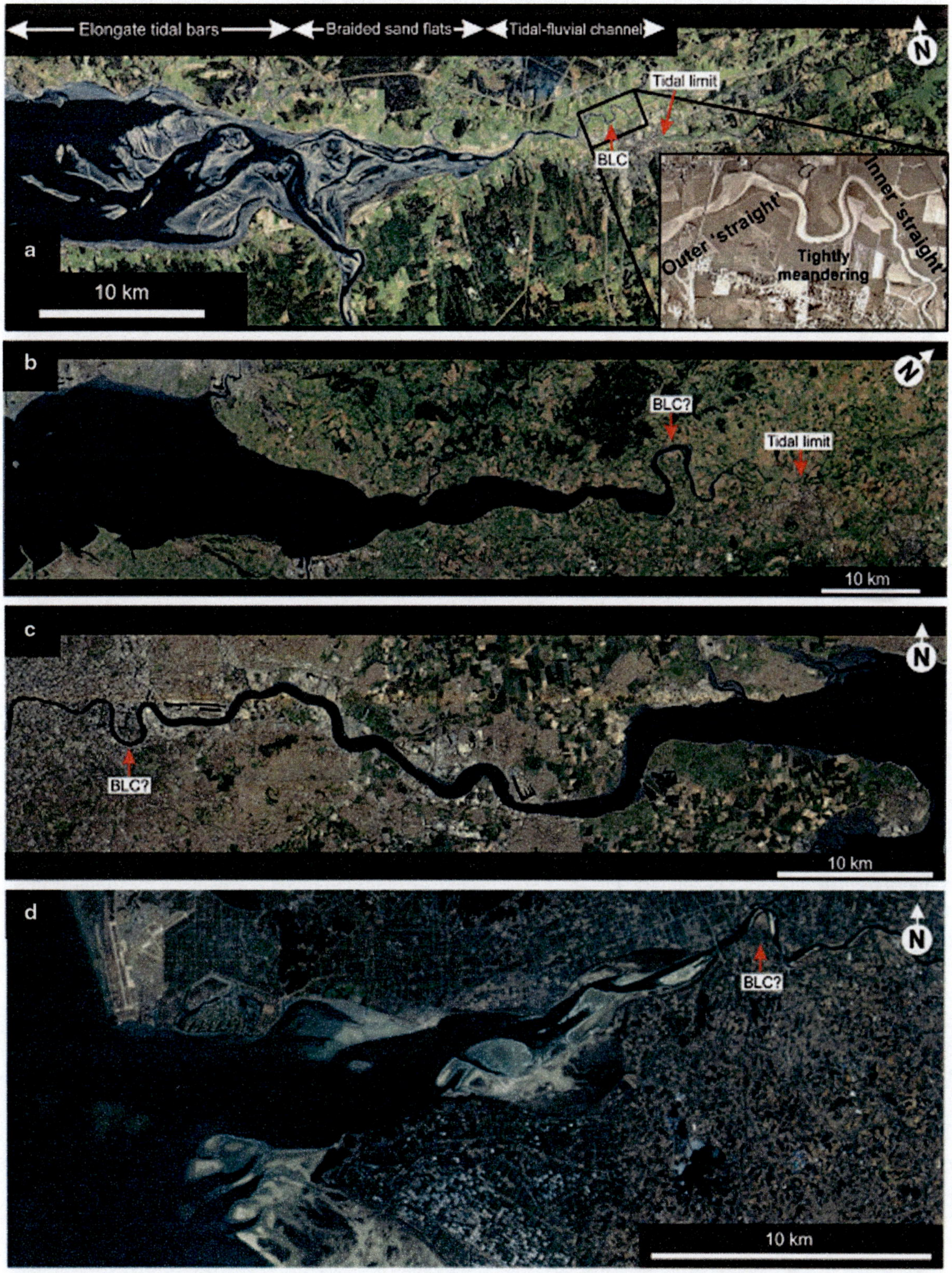

Fig. 5.1 Composite satellite images of tide-dominated estuaries: (**a**) the Cobequid Bay—Salmon River (CB—SR) estuary, Bay of Fundy; (**b**) the Severn estuary, England; (**c**) the Thames estuary, England; and (**d**) the Mangyeong estuary, Korea. Note the exponential seaward widening in the mouth region and the presence of a very tightly meandering zone in the inner estuary where the bedload convergence (BLC) is known to occur in the CB—SR estuary, and is presumed to occur in the other systems. The morphological zones discussed in the text are shown for the CB—SR estuary (Images courtesy of Flash Earth)

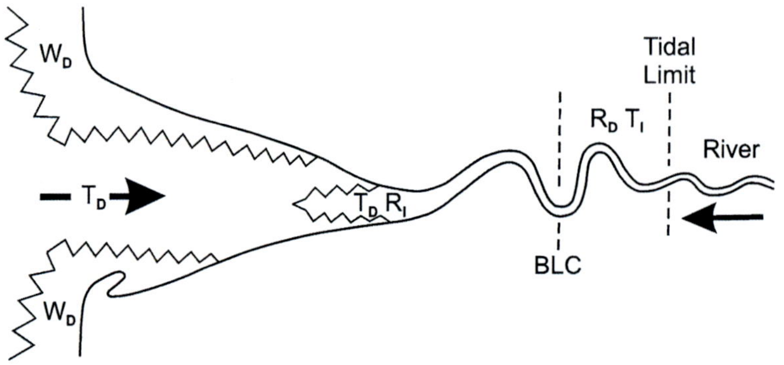

Fig. 5.2 Simplified map view of a tide-dominated estuary showing the spatial distribution of processes: W_D = wave dominated; T_D = tide dominated; $T_D R_I$ = tide dominated, river influenced; and $R_D T_I$ = river dominated, tide influenced. Large *black arrows* indicate the directions of predominant sediment transport: note the presence of two sediment sources and of a bedload convergence (*BLC*) within the estuary. As the relative importance of waves increases, the seaward extent of tidal dominance decreases until the entire front and mouth of the estuary becomes wave dominated, with the production of a barrier island—tidal inlet system (see Chap. 12). Many estuaries close to the tide-dominated end of the spectrum have one or two small spits that extend a short distance into the estuary

tide-dominated but wave-influenced conditions. Even here, however, intense wave action during storms can exert a strong influence on sediment movement, and might promote rapid morphological change. As one moves into the estuary, wave action is attenuated by friction (Pethick 1996) and sedimentation becomes tide dominated, except along the high-tide margins of the outer estuary where wave-dominated conditions exist because the tidal currents are weak and the fetch is large (e.g. Pye 1996; Tessier et al. 2006).

Tidal domination persists inland along the axis of the estuary but with a progressively larger influence of river currents (Fig. 5.3b). Moving landward, one encounters first tide-dominated, river-influenced, and then river-dominated, tide-influenced conditions (Fig. 5.2). The landward limit of the estuary is taken where tidal influence is no longer evident, a position that can be many tens to hundreds of kilometers inland from the main coast (cf. Van den Berg et al. 2007). This *tidal limit* can be defined easily over a short time, but is a diffuse zone over longer time periods for two reasons.

1. The gradual weakening of the tides in a landward direction causes flow patterns to evolve gradually from reversing flow with a seaward residual movement because of the river current, to seaward-directed flow that stops periodically, and then to continuous seaward flow that slows down and speeds up periodically in response to the tidal backwater effect (cf. Dalrymple and Choi 2007, Fig. 14).
2. All of these zones can migrate up and down river over long distances as a result of variations in the intensity of river flow. Thus, during periods of low flow, tidal influence penetrates further up the river than it does during river floods (Fig. 5.4; Allen et al. 1980; Uncles et al. 2006; Kravatsova et al. 2009). Changes in the intensity of the tides, because of neap-spring and longer-term astronomic cyclicity, have a similar but smaller effect, with the tidal influence penetrating further into the estuary during spring tides, for example.

Because of the funnel shape of tide-dominated estuaries (Fig. 5.1), the energy of the incoming tidal wave is concentrated into an ever-decreasing cross-sectional area as it propagates up the estuary. This tendency is not initially offset fully by friction, so the tidal range increases into the estuary, reaching a maximum value some distance landward of the coast (cf. Dalrymple and Choi 2007, their Fig. 5; Li et al. 2006, their Fig. 4). Beyond a certain point in the estuary, however, the decreasing water depth causes friction to become more important than convergence, and the tidal range decreases toward the tidal limit. Such a hydrodynamic pattern (i.e. a landward increase in the intensity of the tides) has been termed *hypersynchronous* (Salomon and Allen 1983; Nichols and Biggs 1985; Dyer 1997).

Within tide-dominated estuaries, the tidal wave adopts the characteristics of a standing wave (cf. Dyer 1997) with the fastest currents occurring approximately at mid-tide, and little or no water movement at both high and low water, creating two *slack-water* periods (Fig. 5.5). Because of the lateral constraint provided by the estuary margins, the currents are

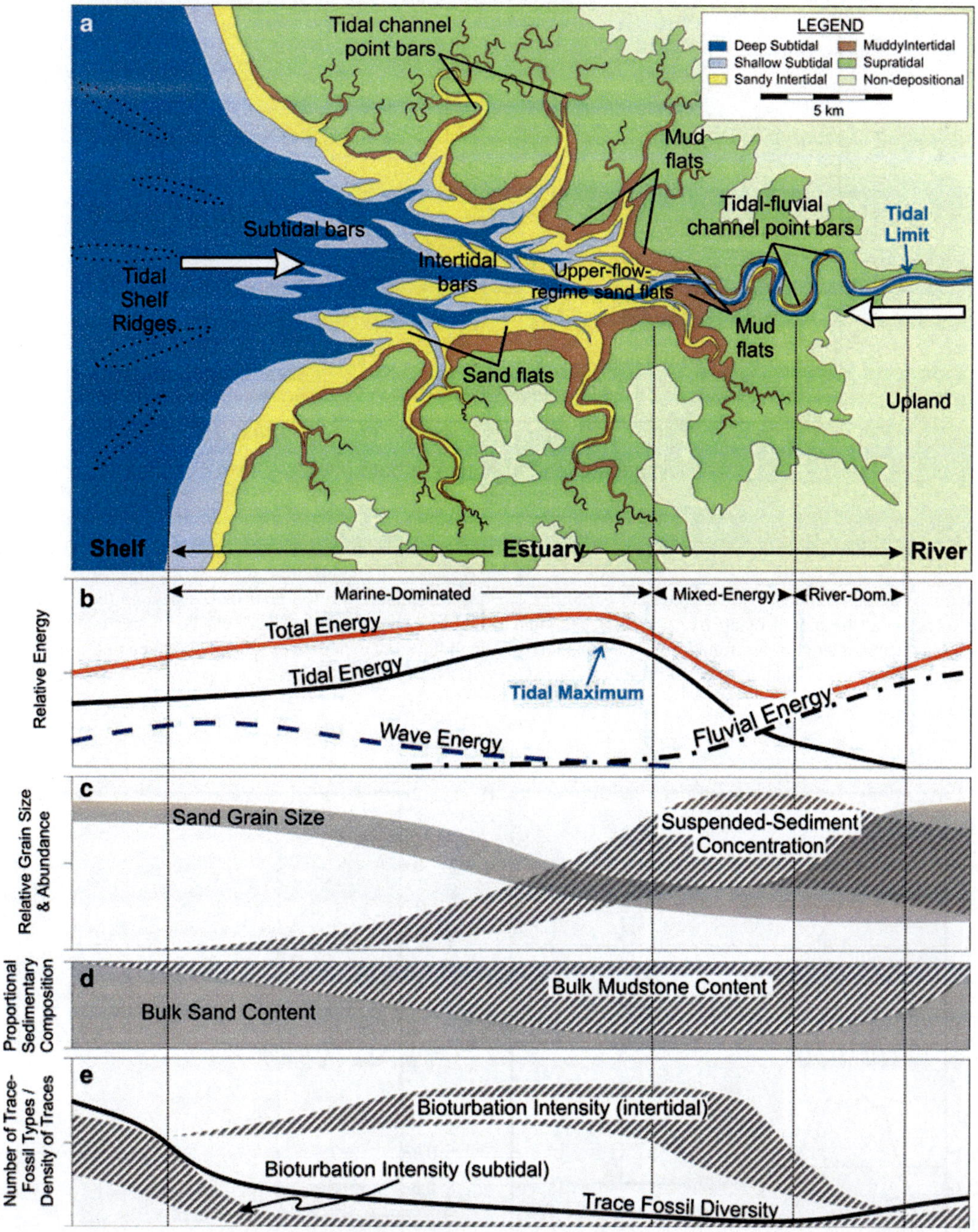

Fig. 5.3 (**a**) Schematic map showing the typical distribution of channel forms and subenvironments in a sandy macrotidal estuary, based on systems such as the Cobequid Bay-Salmon River and Bristol Channel-Severn River estuaries. The large *white arrows* indicate sediment movement into the estuary from both the landward (fluvial) and seaward directions. (**b**) Longitudinal distribution of wave, tidal and river energy (Modified after Dalrymple et al. 1992 and Dalrymple and Choi 2007). The 'tidal maximum' is the location where the tidal-current speeds are greatest. (**c**) Longitudinal distribution of bed-material (sand) grain size, showing the presence of a grain-size minimum near the location where flood-tidal and river currents are equal (i.e., the bedload convergence), and of suspended-sediment concentrations, showing the turbidity maximum. (**d**) Longitudinal distribution of the relative proportion of sand- and mud-sized sediment in the deposits. (**e**) Longitudinal distribution of trace-fossil characteristics, based on Lettley et al. (2005) and MacEachern et al. (2005)

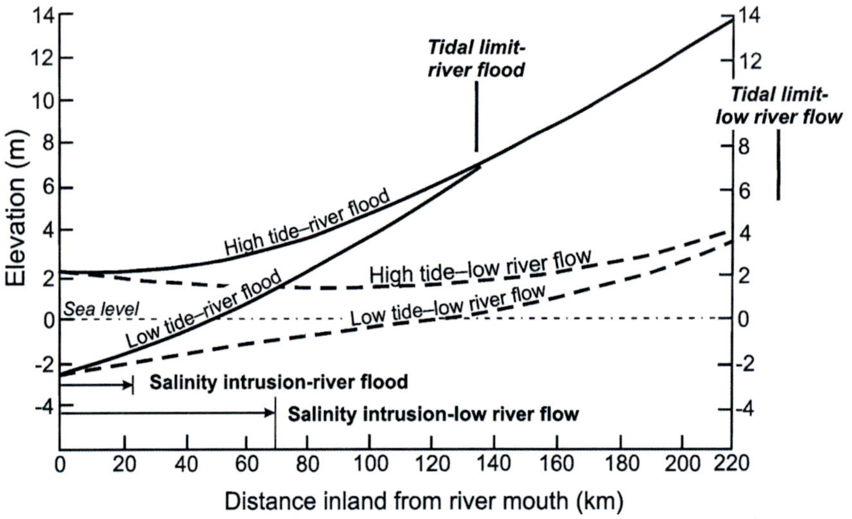

Fig. 5.4 Variation in the upstream penetration of tidal influence and salt water as a function of river discharge in the Irrawaddy River, Myanmar (after Kravatsova et al. 2009, their Fig. 5). Although this system is deltaic, a similar pattern of variations is expected to occur at the mouth of all river systems, although with different excursion lengths as a function of the variation in river discharge and slope. Smaller rivers will generally have shorter distances and smaller changes in the distance of marine influence. In rivers with a greater variability of discharge between high and low flow, the area of saline water can penetrate further inland, into the area that is beyond the high-flow tidal limit. In such situations, there can be an area that is non-tidal at high flow, but experiences brackish-water conditions during low river flow

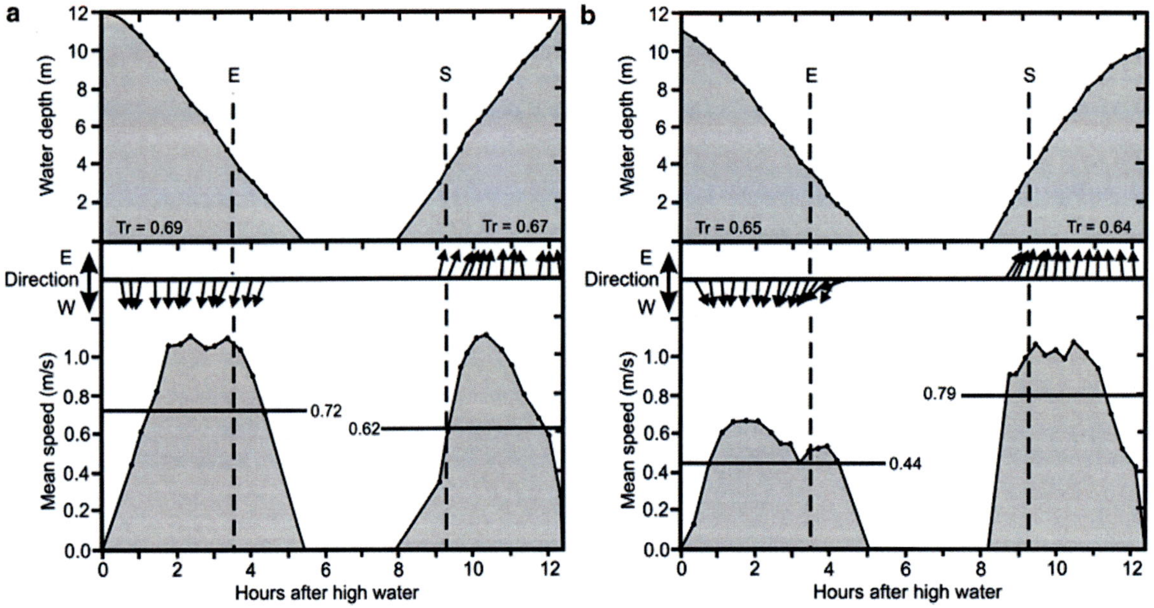

Fig. 5.5 Plots of water-depth, current direction and mean (depth-averaged) current speed over complete tidal cycles for ebb-dominated (**a**) and flood-dominated (**b**) locations on Diamond Bar, Cobequid Bay, Bay of Fundy. See Dalrymple et al. (1990) for more information about this bar. 'E' and 'S' refer to the time of emergence and submergence of the adjacent bar crest. *Tr*=tidal coefficient, which is the tidal range for the half cycle divided by the mean range for large spring tides (16.1 m). (The mean tidal range has a Tr value of 0.73). The *horizontal lines* in the current-speed panels indicate the average mean speed over the half tidal cycle. The differences in the peak speeds have a more important influence on the direction of movement of bed material than the differences in the average speeds

essentially rectilinear, and reverse by 180° between the flood and ebb tides (Fig. 5.5). The longitudinal variation in the peak tidal-current speeds mimics the distribution of tidal range, increasing landward to some maximum value (Dalrymple et al. 1991), termed the *tidal maximum* by Dalrymple and Choi (2007) (Fig. 5.3b), before decreasing to zero at the tidal limit.

In general terms, the incoming tidal wave is typically asymmetric because the crest migrates onshore more quickly that the trough, a feature that is analogous to the behavior of wind waves as they approach the beach (Dyer 1995, 1997). The shorter duration of the flood tide causes the flood currents to be faster than the ebb currents (e.g. Li and O'Donnell 1997; Moore et al. 2009), which, in turn, creates a flood dominance and a net onshore movement of bed material (i.e. sand and/or gravel), at least in the seaward part of estuaries (Dalrymple et al. 1990). This occurs because the amount of bed material that can be moved is a power function of the current speed, so that the direction of net sediment movement is determined more by an inequality in the peak speeds than by differences in the durations of the flood and ebb currents (Chap. 2; Dalrymple and Choi 2003). The inner part of estuaries, by contrast, experiences an ebb dominance as a result of the superposition of river currents on the tides. As a result of these opposing directions of net bedload movement, tide-dominated estuaries contain a *bedload convergence* (Johnson et al. 1982; Dalrymple and Choi 2007), a location toward which bedload migrates from both directions when averaged over a period of years. This process, supplemented by the trapping of suspended sediment (see more below), is responsible for filling the accommodation (i.e. unfilled space) that is created by flooding and transgression of the river mouth. In general, filling of an estuary is most rapid in the inner part, and progresses in a seaward direction. Thus, as the space fills, the bedload convergence migrates seaward until river-dominated seaward transport of bed material extends all the way to the main coast. At this point, the estuary has been filled, river-supplied sediment is exported to the ocean, and the system is considered to be a delta. Here, this transitional phase is referred to as the *progradational phase* of estuary evolution, as opposed to the *transgressive phase* when the estuary is created.

The time-velocity asymmetry between the flood and ebb currents, and the resulting patterns of net sediment transport described above, are accentuated by the longitudinal variation in the cross-sectional shape of the channels (Friedrichs and Aubrey 1988; Friedrichs et al. 1990; Pethick 1996). In situations with relatively small intertidal areas, the average water depth (across the entire channel) is less at low tide than at high tide (Fig. 5.6a). However, in situations with broad intertidal areas, the water depth averaged across the entire width of the channel and flats is actually less at high tide (Fig. 5.6b) because of the inundation of the wide, shallow tidal flats. In the first case, the crest of the tidal wave moves more quickly than the trough, because of the greater water depth at high water, causing the flood tide to be shorter than the ebb, which then creates flood dominance. By contrast, in the second case, the tidal-wave crest moves into the estuary more slowly than the trough, generating a shorter ebb tide and ebb dominance. In most estuaries, the latter situation tends to occur in the inner part because this is where infilling occurs first. Consequently, there is a tendency for the inner part to be ebb dominated, independent of the river current, whereas the outer part tends to be flood dominated. As the estuary fills, more and more of the system has the cross-channel morphology (Fig. 5.6b) that promotes ebb dominance, and, eventually, the system becomes a sediment-exporting delta. (For a discussion of the factors controlling tidal-flat morphology see Chaps. 9 and 10, and Roberts et al. 2000).

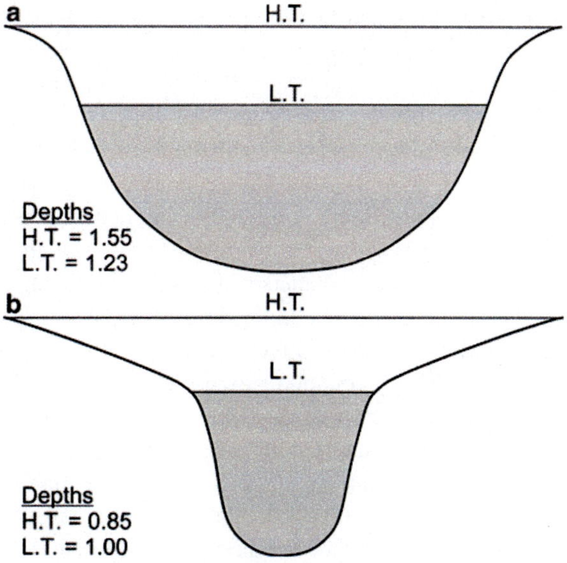

Fig. 5.6 Contrasting channel cross-sectional shapes for (**a**) an unfilled part of the estuary near the mouth, and (**b**) a more completely filled part of the estuary near the head. The shape in (**a**) promotes flood dominance because the tidal-wave crest (i.e., high water) migrates faster than the trough (i.e., low water), whereas the shape in (**b**) promotes ebb dominance because the progression of the tidal-wave crest is retarded because of the broad shallow tidal flats

It should be noted that the patterns of dominance referred to above represent generalities that average out a great deal of local variability, both temporally and spatially. For instance, it is widely observed that the channel thalweg tends to be ebb dominant, whereas the flanking tidal flats are flood dominant (Li and O'Donnell 1997; Moore et al. 2009). In addition, the morphological irregularities that exist because of the presence of channel meanders and elongate tidal bars, which are slightly oblique to the flow, create localized areas of ebb- and flood-directed residual movement of sediment. This is commonly expressed as a series of mutually evasive channels. Typically, the two sides of an elongate tidal bar, or the upstream and downstream flanks of a tidal point bar, experience opposing directions of net sediment transport (Dalrymple et al. 1990; Choi 2010), because they are alternately exposed and sheltered from the reversing current. In addition, temporal variability in the strength of the tidal and river currents can cause temporary reversals in the direction of net sediment transport. As a result of these complexities, spot measurements of currents and sediment transport have the potential to be misleading. The geomorphic setting and temporal context of a measurement station must be documented with care before the significance of a data set can be assessed.

5.2.2 Salinity, Residual Circulation and Suspended-Sediment Behavior

The interaction of marine and fresh water generates longitudinal and vertical salinity gradients within an estuary (Haas 1977; Uncles and Stephens 2010). The location of the longitudinal gradient is highly sensitive to both the phase of the tide, moving up and down the estuary with the flood and ebb tides, respectively, and also to variations in river discharge, potentially moving down river a considerable distance when the river is in flood (Uncles et al. 2006). Turbulence associated with the strong tidal currents minimizes the tendency for density stratification, producing partially mixed or well-mixed conditions (Dyer 1997). Stratification is least pronounced during times of weak river flow and at spring tides, but can become better developed when the fresh-water input is greater (Allen et al. 1980; Castaing and Allen 1981). Such density stratification generates so-called estuarine circulation, which has a net landward-directed residual flow in the bottom-hugging *salt wedge*, and a residual seaward flow in the lighter overriding fresher water. The currents associated with this circulation are extremely weak and have little or no influence on the transport of bed material, but they do control the longer-term movement of the suspended sediment (Dalrymple and Choi 2003).

Flocculation of the river-born suspended sediment as it moves into the area with measureable salinity, coupled with the density-driven residual circulation (termed *baroclinic flow*; Dyer 1997), tends to trap suspended sediment within the estuary, generating a *turbidity maximum* (Fig. 5.3c), within which suspended-sediment concentrations (SSC) can be elevated to very high levels (Dyer 1995). The peak of this turbidity maximum typically lies near the tip of the salt wedge (Allen et al. 1980), although the broader zone of elevated turbidity can stretch from the fresh-water tidal zone near the tidal limit, out beyond the mouth of the estuary (e.g. Guan et al. 1998; Uncles et al. 2006). Suspended-sediment concentrations in the water column generally decrease upward from the bed, and vary in phase with, but commonly with some lag relative to, the speed of the tidal currents (Fig. 5.7) because of erosion and resuspension of material from the bed (Allen et al. 1980; Castaing and Allen 1981; Wolanski et al. 1995; Ganju et al. 2004). During slack-water periods, however, the suspended particles settle to the bed and can generate a thin near-bed layer of very high concentrations. If these concentrations exceed 10 *g/l*, then this dense suspension is termed a *fluid mud* (Faas 1991; Mehta 1991). They are being found in a growing number of strongly tide-influenced or tide-dominated estuaries (Thames Estuary: Inglis and Allen 1957; Gironde estuary: Allen 1973; Castaing and Allen 1981; Bristol Channel—Severn River: Kirby and Parker 1983; James River: Nichols and Biggs 1985; Jiaojiang River: Guan et al. 1998) and deltas (Fly River delta: Wolanski et al. 1995; Dalrymple et al. 2003; the Amazon delta: Kuehl et al. 1996; Seine River: Lesourd et al. 2003; Weser River: Schrottke et al. 2006), apparently because the strong tidal currents resuspend large amounts of mud; it is possible that such high-concentration suspensions are present in most tide-dominated estuaries.

The intensity of the turbidity maximum is highly sensitive to the strength of the tidal currents, with the highest turbidity generally associated with spring tides (Allen et al. 1980; Kirby and Parker 1983; Wolanski et al. 1995), because of their ability to resuspended more sediment. Its location is strongly influenced by

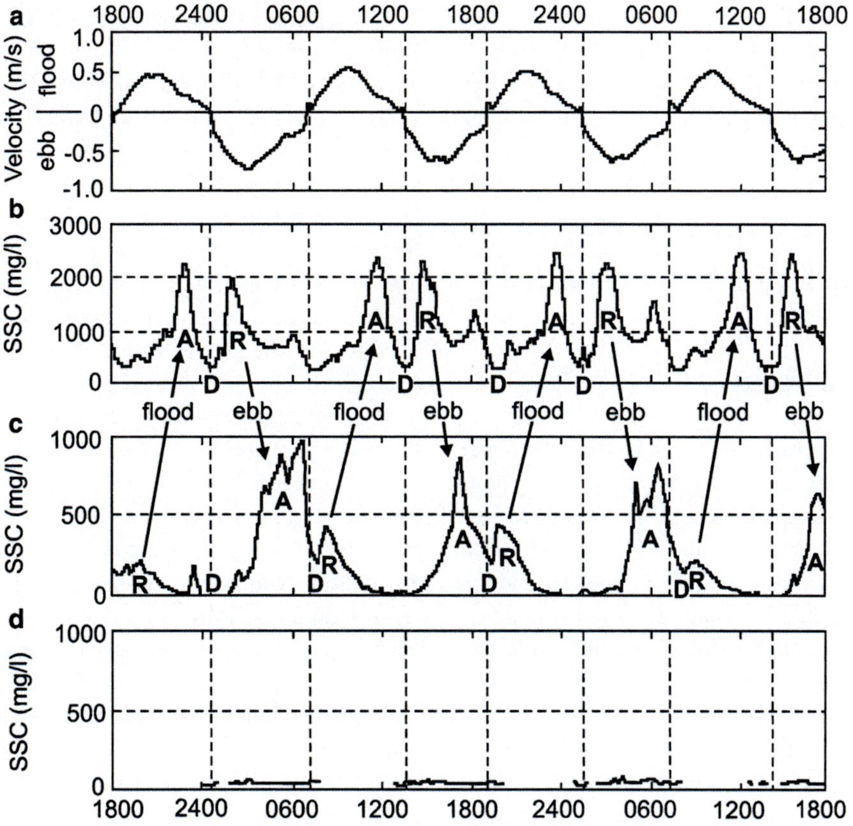

Fig. 5.7 Plots of current speed (**a**) and suspended-sediment concentration (*SSC*; **b–d**) for three locations in a tributary of the San Francisco Bay estuary, showing the lateral movement (advection—**a**) of the turbidity maximum in response to the tides, coupled with deposition (*D*) of the suspended sediment during slack-water periods and resuspension (*R*) of material from the bed as the current accelerates after slack water. Location (**b**) lies at the position of the turbidity maximum at high tide; location (**c**) lies near the low-tide location of the turbidity maximum; and location (**d**) lies seaward of the influence of the turbidity maximum even at low tide. Note the overall decrease in SSC values from (**b**) to (**d**). The *arrows* between panels (**b**) and (**c**) reflect the advection of the turbidity maximum: landward during the flooding tide, and seaward during the ebbing tide. The excursion distance between the high-tide and low-tide positions of the turbidity maximum is of the order of 15 km in this micro-mesotidal system (Modified after Ganju et al. 2004, Fig. 3)

tidal water motions and the river discharge (Lesourd et al. 2003; Ganju et al. 2004). The distance that the water moves during a half tidal cycle is termed the *tidal excursion* (Uncles et al. 2006) and varies from a few to many kilometers (Fig. 5.7). As a result of this movement, any property of the water that varies longitudinally (e.g. salinity, temperature, SSC, and the concentration of any pollutants) will show a variation at any one location because of the back-and-forth movement of the longitudinal gradient. Thus, salinity is least at low tide and greatest at high tide. The SSC value will be greatest at low tide at locations that lie seaward of the average position of the turbidity maximum, but will be greatest at high tide in areas landward of the average turbidity-maximum position. At times of low river flow, the turbidity maximum is located relatively far up the river, whereas the turbidity maximum shifts down river as the discharge increases (Doxaran et al. 2009), perhaps even being expelled from the estuary at times of highest discharge (Castaing and Allen 1981; Lesourd et al. 2003). A useful parameter for studies of both the deposition of fine-grained sediment and the fate of pollutants is the *trapping efficiency* of an estuary, which is related to the 'flushing rate' (Dyer 1995, 1997; Wolanski et al. 2006) and estuarine 'capacity' (O'Connor 1987), and which is the ratio of the amount of sediment input by the river to that which accumulates in the estuary. In estuaries with a large water volume and large, aggrading intertidal areas, the trapping efficiency is high and can even exceed 100% if sediment is input from the ocean, whereas small estuaries and deltas will have a low efficiency. The trapping efficiency is also a function of grain size, with estuaries exporting fine-grained suspended sediment to the ocean earlier than sand during their transition to a delta.

5.3 Morphology of Tide-Dominated Estuaries

5.3.1 General Aspects

Tide-dominated estuaries show the typical funnel-shaped geometry that characterizes all coastal systems in which there is appreciable tidal influence (Myrick and Leopold 1963; Wright et al. 1973; Fagherazzi and Furbish 2001; Rinaldo et al. 2004). This exponential decrease in width in a landward direction (Figs. 5.1–5.3) is a result of the landward decrease in the tidal flux (Myrick and Leopold 1963; Wang et al. 2002), which reaches zero at the tidal limit. By comparison, river channels are nearly parallel sided and show only a very slow seaward increase in width in the coastal zone, because there is only a small increase in fresh-water discharge, derived from small tributaries, direct precipitation and groundwater discharge. In the end-member case of strongly tide-dominated estuaries (Fig. 5.1), the tidally created funnel extends right to the open coast. However, as the wave influence increases, longshore drift becomes capable of building a spit into one or both sides of the estuary mouth, producing a constriction. Gomso Bay, which has an 'incipient barrier' (Yang et al. 2007), represents a situation that is close to the tide-dominated end-member of the wave-tide spectrum of estuary types. The Gironde estuary, France (Allen 1991), with its tide-dominated bayhead delta and muddy 'central basin' that is enclosed by a wave-built spit, and the Westerschelde estuary, the Netherlands, are more mixed-energy settings because of the presence of a wave-built barrier-inlet complex at their mouth (Dalrymple et al. 1992). For more on such barrier-inlet systems, see Chap. 12.

Every river entering an estuary possesses a main channel that continues seaward through the estuary as an ebb-dominated channel. Main channels issuing from tributaries join the main ebb channel, but seaward branching of this channel in a distributary-like pattern is not obvious, although the *swatchways* that dissect the elongate tidal bars in the estuary mouth serve a similar hydraulic function. The main ebb channel generally becomes more sinuous in a landward direction. Near the mouth of the estuary, it can be essentially straight, but the radius of curvature of the meander bends decreases (i.e. the bends become tighter) and the sinuosity increases in a landward direction (Dalrymple et al. 1992; Billeaud et al. 2007; Burningham 2008) (Figs. 5.1 and 5.8). Qualitative observations and quantitative measurements indicate that the main channel reaches a peak sinuosity that exceeds a value of about 2.5 (and may be greater than 3) some distance inland, after which it becomes less sinuous again near the limit of tidal influence (Ichaso and Dalrymple 2006). The sinuosity of the river above the limit of tides varies widely between examples, and can be quite sinuous, but rarely reaches a value as high as 2.5. Dalrymple et al. (1992) was the first study to note the presence of this pattern, which they termed 'straight'—meandering—'straight' (SMS; Fig. 5.1a), where 'straight' refers to a channel of relatively low sinuosity and not to a truly straight channel. Subsequent quantitative studies reveal that the SMS pattern even exists in small tidal creeks (Fagherazzi and Furbish 2001; Solari et al. 2002; see also Chap. 11), provided there is little or no fluvial influence. Systems that are known to be prograding and, thus, are 'deltas' in the sense used here, do not show this pattern (Ichaso and Dalrymple 2006; see also Chap. 7). Instead, there is a progressive straightening of the channel from the river to the mouth of the estuary (Dalrymple et al. 2003, their Fig. 6). As a result, the presence or absence of a short zone (typically only one or two meander-bends long) with very tight and generally symmetrical meanders appears to be an easy way to distinguish between estuaries and deltas. The reason for this SMS pattern is not known with certainty, but observations in the Cobequid Bay–Salmon River estuary (Zaitlin 1987; Dalrymple et al. 1991) show that the tightly meandering zone lies approximately at the location of the long-term (i.e. multi-year) bedload convergence, a suggestion supported by observations reported by Ayles and Lapointe (1996). As the estuary fills and the bedload convergence migrates seaward, the zone of tight meanders should migrate with it, but gradual migration of the meandering zone is apparently not possible. In the Fitzroy estuary (Bostock et al. 2007; Ryan et al. 2007), for example, the point of bedload convergence, as indicated by the facing directions of large subaqueous dunes in the main channel, lies approximately 10 km seaward of the very tight meander bend. The predicted

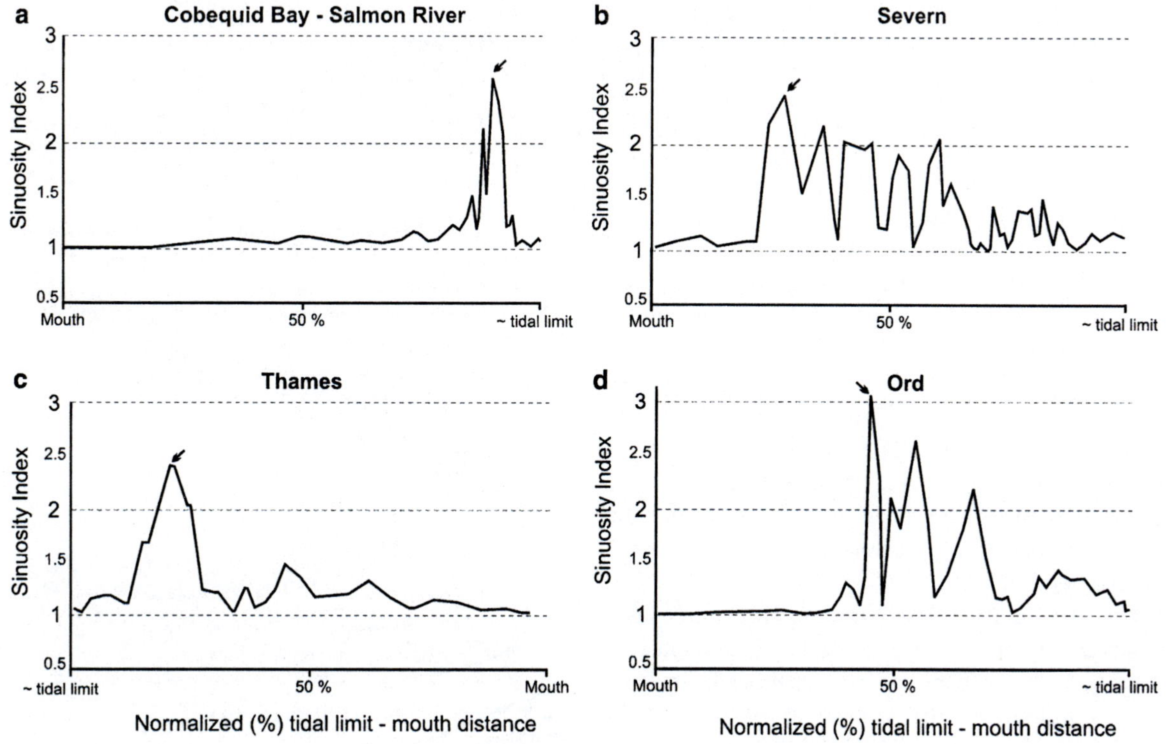

Fig. 5.8 Plots of sinuosity as a function of position within each of four tide-dominated estuaries. See Fig. 5.1 for satellite images of the Cobequid Bay—Salmon River, Severn and Thames estuaries; note that the plots shown here are oriented in the same way as the satellite images in Fig. 5.1. The 'sinuosity index' is the ratio of the along-channel length divided by the straight-line distance between the tidal limit and estuary mouth. In all four cases, the sinuosity increases inland from the mouth, commonly quite abruptly, reaching a maximum (indicated by *arrows*) where the sinuosity is greater than about 2.5, before decreasing to lower values further inland. This zone of maximum sinuosity is the tightly meandering zone of the 'straight'—meandering—'straight' channel pattern. Note the much greater variability of channel form in the area landward of the sinuosity maximum. Systems that export sediment to the sea (i.e., deltas) do not show this peak. Instead, the sinuosity increases inward

straightening of this bend occurred suddenly by means of a neck cutoff in 1991 during a particularly large river flood, and the river shows no sign of reoccupying the tight bend, which is passively filling with sediment (Bostock et al. 2007). The South Alligator River in Northern Australia also shows morphological evidence that it was once more highly sinuous in the inner part of the coastal plain and is now exporting sediment to its mouth (Woodroffe et al. 1989). The Ord River in Northern Australia, which is commonly cited as a tide-dominated delta, possesses the tightly meandering zone, so it is either an estuary or has evolved into a sediment-exporting deltaic system so recently that it has not yet lost its estuarine channel pattern (Fig. 5.8d).

Flood-dominant channels flank the main ebb channel. Unlike the main ebb channel, these channels are invariably discontinuous, terminating headward into tidal flats or sand bars. They are separated from the main ebb channel by an elongate tidal bar that attaches to the shoreline or to another, commonly larger, tidal bar. The morphology of the 'blind' flood channel and its flanking bar looks like a fish hook, and the short, flood-dominant channel has been termed a *flood barb* (Robinson 1960). Overall, these channels become shorter in a landward direction and are absent beyond the inner end of the tide-dominated portion of the estuary (Fig. 5.2).

In general terms, tide-dominated estuaries can be subdivided into two main morphological zones based on the nature of the channel network:

1. A broader outer estuary with several ebb- and flood-dominated channels that separate elongate tidal bars and/or sand flats (zones 1 and 2 of Dalrymple et al. 1990) that are commonly flanked by wave-generated beaches and shorefaces (Fig. 5.2); and

2. A narrower inner estuary that is characterized by a single main ebb channel with, or without, flanking flood channels (zone 3 of Dalrymple et al. 1990) that are bordered by muddy tidal flats and salt marshes.

5.3.2 Outer Estuary

In the broad, outer part of tide-dominated estuaries, the ebb- and flood-dominant channels form a *mutually evasive* system of channels that are separated by elongate tidal bars (Figs. 5.1 and 5.3). The morphology and size of these elongate tidal bars has been reviewed by Dalrymple and Rhodes (1995). These bars and channels form seemingly complex patterns (Fig. 5.1a), the morphology of which follows a few general rules. In general, the bars lie approximately parallel to the main ebb and flood currents, but with a deviation of approximately 20° from the peak currents. The largest bars commonly occupy one or both flanks of the main ebb channel, with the opposite side of these large bars being bordered by the largest of the headward-terminating flood channels (Fig. 5.9a). These large bars, therefore, form a linear or very gently curved 'bar chain' (Dalrymple et al. 1990) that attaches to the side of the estuary at its landward end. It is composed of an *en echelon* series of bars or 'bar elements' (Dalrymple et al. 1990) that are separated by oblique channels, called *swatchways* (Robinson 1960), that dissect the bar chain and connect the ebb and flood channels. These swatchways diverge from the ebb channel in a seaward direction (Fig. 5.9a) because this orientation allows the flood currents to pass across the bar from the flood-dominant channel into the main channel, and the ebb currents to exit the main channel in the same way that distributary channels accommodate part of the river's discharge. The tidal bars can also occur as essentially free-standing, seaward-opening U-shaped bars that contain a flood-dominant channel between their arms.

Individual elongate bars range in length from 1 to 15 km, although bar chains can reach 40 km long. Bar widths range from only a few hundred meters to about 4 km. The relief from the bottom of the adjacent channels to the bar crest can be as much as 20 m, but relief as low as only a few meters is possible, especially toward the outer end of the bar complex and particularly in cases where wave action acts to flatten the topography. The slope of the channel-bar flanks can be as little as a fraction of a degree to nearly vertical,

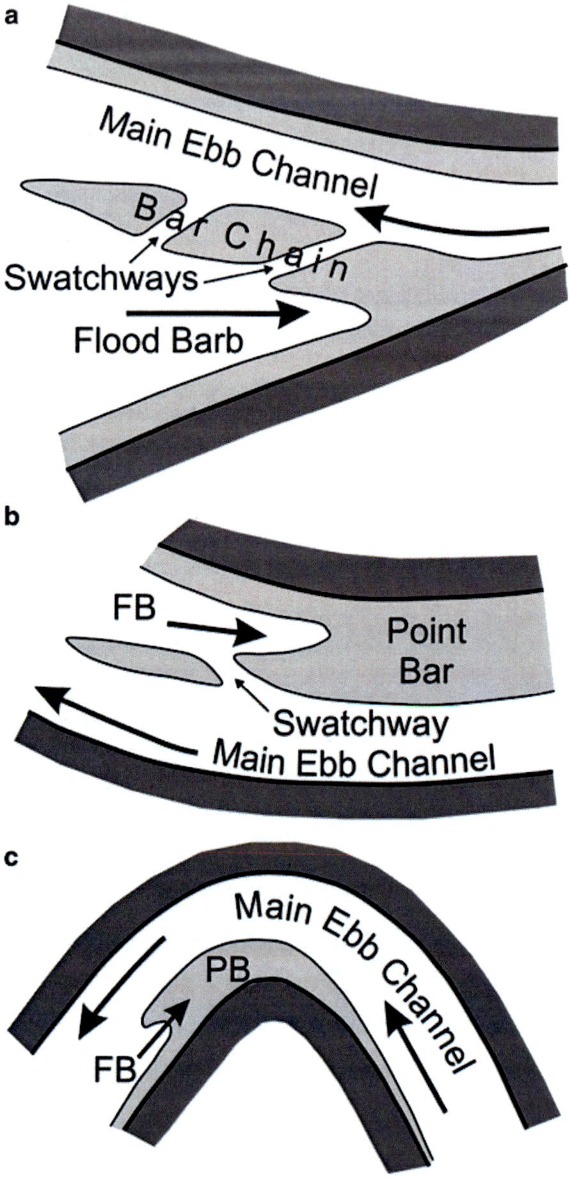

Fig. 5.9 Schematic diagrams showing the morphology of channel-bar systems in (**a**) the broad outer part of an estuary, (**b**) the relatively straight outer part of the fluvial-marine transition, and (**c**) the more tightly meandering reach. *PB* = point bar; *FB* = flood barb. The three parts are not to the same scale; (**a**) is several kilometers to several tens of kilometers wide; (**b**) is a few hundred to about 10 km wide; and (**c**) is less than about 2–3 km wide. See text for more discussion

depending on the sediment that comprises the bars. If the sediment is sandy, slopes are typically in the range of 1–3° (cf. Fig 5.10a); steeper slopes occur if the elongate bars are composed of muddy material, as is the case, for example, in the Mangyeong estuary, Korea

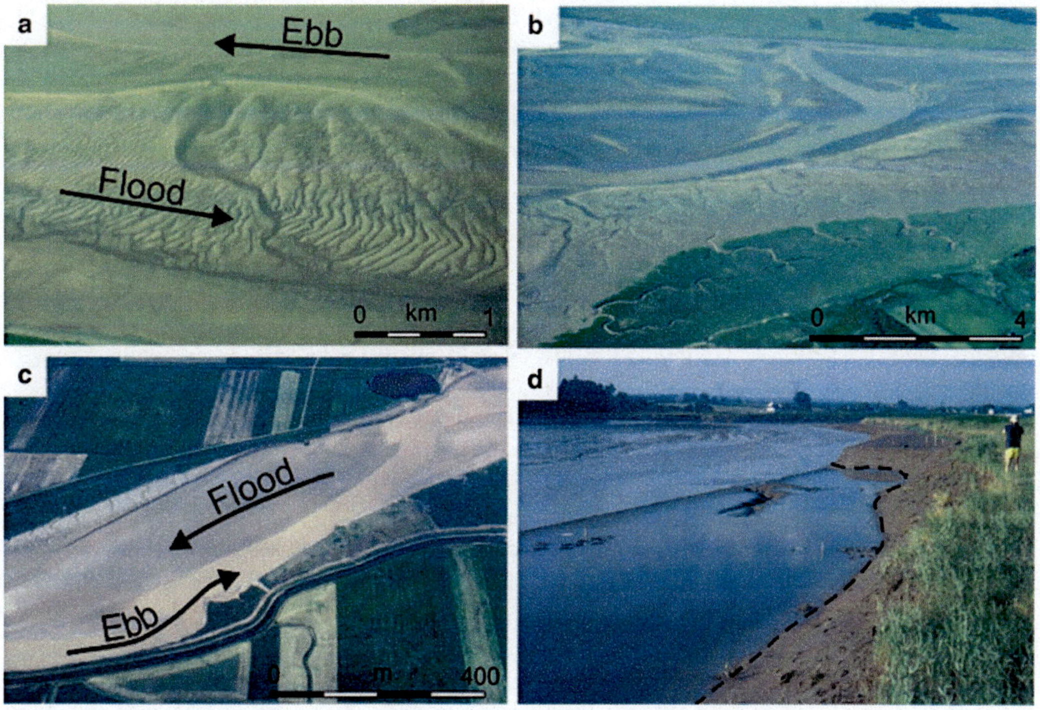

Fig. 5.10 Morphology and facies zonation in the Cobequid Bay—Salmon River estuary, Bay of Fundy, Nova Scotia. (**a**) Elongate sand bar in the outer part of the estuary, covered by large compound and simple dunes. The featureless area to the south of the bar (*at bottom*) is an erosional, wave-dominated foreshore/shoreface. (**b**) Upper-flow-regime sand flats that lie landward of the elongate sand bars, flanked on the south (*foreground*) by mudflats and salt-marsh. Note the dendritic tidal-gully networks that dissect the muddy deposits. Until the 1950s, the main ebb channel lay along this south shore. It then abruptly switched to its present course along the north shore, allowing 7–8 m of mudflat and salt-marsh deposits to fill the old channel. (**c**) Subtle elongate bar and flood barb (Fig. 5.9b) on the seaward side of a gentle point bar (to the *left* of the image) in the outer 'straight' portion of the Salmon River. The surface sediment in the channel is fine sand. A narrow band of mudflat separates the channel—bar sands from the salt-marsh, most of which has been reclaimed for agriculture. (**d**) Mudflat terraces, separated by former cutbank cliffs, near the transition from the outer 'straight' to the tightly meandering zone in the Salmon River (Fig. 5.1a, *inset*). The dashed line is the former cutbank location of the channel

(Fig. 5.1d). Bars are commonly asymmetric, with the steeper side facing in the direction of the stronger of the ebb and flood currents; because of the overall flood dominance that characterizes the outer estuary, this is generally the flood current. Bar crests vary from relatively narrow and sharp-crested to broad and flat. As described first by Harris (1988), and noted subsequently by other workers (Dalrymple et al. 1990; Ryan et al. 2007), the sharp-crested bar form represents situations that are underfilled, whereas the flat-topped form occurs in situations where the bar has aggraded as high as it can, and has expanded laterally, through deposition on one or both flanks. It is invariably the case that the broad, flat-topped bars occur in the inner part of sand-bar complexes, whereas the narrow, sharp-crested forms occur at the seaward end (unless wave action prevents this). For this reason, the crest of individual bars, and of the bar complex as a whole, rises in a landward direction.

The rate of morphologic change of the channels that separate the elongate tidal bars is not known with confidence. The most dramatic and frequent changes occur as a result of 'tidal avulsions' whereby a swatchway becomes large enough that it captures the main ebb flow, causing an abrupt change in the path of the main channel. This appears to have occurred repeatedly in the outer part of the Ribble Estuary, Great Britain (Van der Wal et al. 2002), and has been documented in the Cobequid Bay (Bay of Fundy) estuary (Dalrymple et al. 1990). Major storms might play an important role in triggering such channel switches. Sediment then fills the abandoned channel (Van der Wal et al. 2002), provided there is not enough tidal flux to maintain the channel. Slow, progressive shifting of the gentle meanders in the main channels is to be expected, but detailed documentation of such changes are rare, so it is not known whether there is a systematic behavior of the meander bends. The swatchways also migrate, apparently preferentially in a headward direction because of the flood-dominated sediment transport that prevails. In the Cobequid Bay estuary, one large swatchway (relief ca. 5 m) has been documented from sequential air photos to have migrated 2.1 km over a 35-year period (average rate 61 *m/a*), with a maximum rate of slightly more than 80 *m/a* (Dalrymple et al. 1990). Smaller swatchways with a relief of only about 1 m migrated more than 150 *m/a*.

In most tide-dominated estuaries, the zone of elongate tidal bars passes gradationally into the narrower inner part of the estuary. This transition involves the gradual simplification of the channel—bar morphology through the loss of channels, until there is only a single, main ebb channel (Fig. 5.9). The Cobequid Bay—Salmon River estuary appears to be unusual, if not unique, in having a 'braided' sand-flat area (i.e. 'zone 2' of Dalrymple et al. 1990) (Fig. 5.10b) between the zone of high-relief elongate tidal bars and the 'single-channel' inner estuary. In this area, which owes its existence to the shallowness of the estuary, the very strong tidal currents that exist here, and the fine sand that characterizes this area (see below), cause the widespread development of upper-flow-regime conditions. The resulting morphology consists of an apparently disorganized 'braided' network of subtle, only slightly elongate bars, most of which show a headward (flood-dominant) asymmetry. The relief of these bars is typically less than a meter, but can reach as much as 2 m, and slopes are rarely more than 0.5°.

The areas along the margins of the outer part of tide-dominated estuaries tend to be wave dominated (Fig. 5.2) because waves can penetrate into the estuary at high tide, and because tidal-current speeds are minimal in the upper intertidal zone at that time. As a result, the margins have a concave-up shoreface profile, with a beach at the high-water level if coarse sediment is available (Dalrymple et al. 1990; Pye 1996; Tessier et al. 2006). If the estuary mouth is transgressing, this shoreface is erosional (Fig. 5.10a): this erosional transgression can continue even though the margins of the inner part of the estuary are prograding (Allen 1990; Dalrymple et al. 1990; Dalrymple and Zaitlin 1994; Allen and Duffy 1998; Pye 1996; Tessier et al. 2006). At some point in the estuary, the beaches end abruptly and are replaced by tidal flats and salt marshes; a good example of this has been documented in the Dee estuary, England (Pye 1996, his Figs. 2.11–2.13). The location of this beach-marsh boundary commonly lies near the headward end of the elongate sand-bar complex, but presumably depends in part on the evolutionary stage of the estuary, migrating further into the estuary as the estuary transgresses.

5.3.3 Inner Estuary

The axial channel system in the inner part of tidal-dominated estuaries consists of a single ebb channel that connects to the river(s) that feed into the estuary, and displays the 'straight'—meandering—'straight' channel pattern discussed above (Figs. 5.1 and 5.8). The depth of the ebb channel is deepest on the outside of each bend and is shallowest in the cross-over areas (Jeuken 2000). In those portions of the channel where there is appreciable tidal influence (i.e. in the outer 'straight' reach [zone 3A of Dalrymple et al. 1990]), the channel shows a repetitive pattern of channel bends, flood barbs and elongate tidal bars (Fig. 5.1; Jeuken 2000; Schuttelaars and de Swart 2000). Each 'estuary section' or 'estuary compartment' comprises a single channel bend between two successive inflection points and consists of a point bar or alternate bar that is cut by a flood barb. The flood and ebb channels are separated by an elongate tidal bar that can be either simple and continuous (Barwis 1978), or a complex series of bars separated from each other by one or more swatchways (Jeuken 2000; Schuttelaars and de Swart 2000). These flood barbs and adjacent tidal bars become progressively shorter in a landward direction because of the decreasing wavelength of the meanders (Fig. 5.9b, c); the number of swatchways also decreases inward as the bars become shorter (Fig. 5.11; Jeuken 2000). On occasion, the flood channel and a swatchway can become large enough that they assume the role of the main channel for a period of time. This can lead to the alternation of channel location between two discrete locations (van Proosdij and Baker 2007; Burningham 2008), and the episodic creation of channel-center bars.

The meander bends tend to be asymmetric, or skewed, with a tendency for the asymmetry to alternate between landward-directed and seaward-directed in successive bends (Burningham 2008). Overall, there might be a tendency for the meanders to be skewed

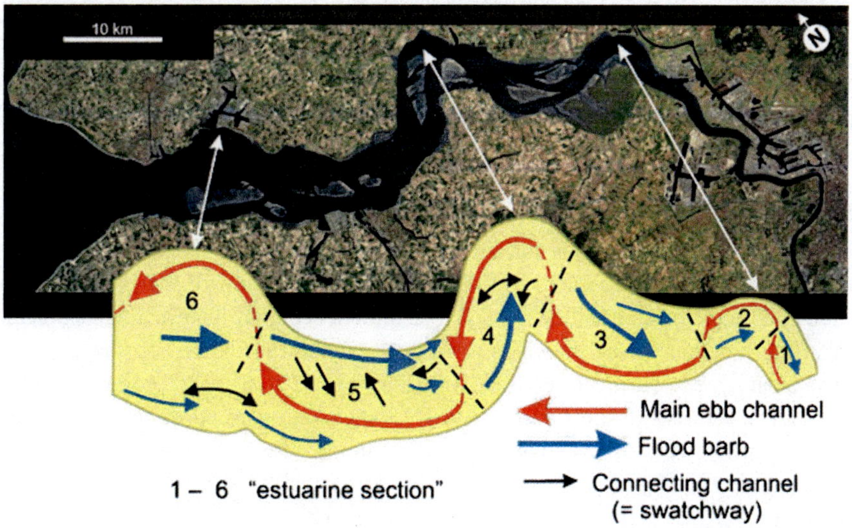

Fig. 5.11 Composite satellite image of the Westerschelde estuary, The Netherlands (Image courtesy of Flash Earth), and a schematic representation of the directions of net sediment transport (Modified after Schuttelaars and de Swart 2000 and Jeuken 2000). Note that the main ebb channel is continuous along the length of the estuary, whereas there is a series of discrete flood-dominant channels, each successive one being on the opposite side of the channel relative to the adjacent ones. Each ebb-flood channel pair comprises an 'estuarine section' (Jeuken 2000), with a major tidal bar situated between these channels (i.e. at the location of the numbers indicating the estuarine sections) These bars are dissected by 'connecting channels', which are here termed 'swatchways'

downstream in situations where there is flood dominance (Fagherazzi et al. 2004; Burningham 2008). The direction and rate of propagation of the bends is not known in most cases, but, in general, it is likely that the rate of change is less than that seen in meandering fluvial channels because of the partial counterbalancing effects of the reversing tidal currents. In the Westerschelde estuary (Fig. 5.11), the bends tended to migrate outward at a rate of 20–80 m per year before significant human intervention in the early 1800s, but they then became essentially stable after they encountered the muddy sediments of the flanking marshes and the training walls along the estuary margin. Channel stability has characterized the inner part of the Cobequid Bay—Salmon River estuary over the period of airphoto coverage, perhaps because of the confinement by muddy deposits. A very detailed study of the Avon River estuary also shows that the channel system has remained essentially the same over the approximately 150 years of map and airphoto coverage (van Proosdij and Baker 2007). Small-scale changes in the path of the channel thalweg do occur, causing local erosion of the channel bank, but the channel typically returns to the original location after only a few years.

In the more tightly meandering reach of the channel (i.e. zone 3B of Dalrymple et al. 1990), where flood-tidal currents and river currents are essentially equal when averaged over the span of years to decades, the meander bends are typically more or less symmetrical (Fig. 5.1, Dalrymple et al. 1992). Two meander shapes are common: *cuspate*, in which the apex of the point bar is pointed with concave flanks (e.g. the meander in the centre of Fig. 5.1c), and *box* in which the meander is square with channel bends that are nearly 90° (see the tightest meander bends in Fig. 5.1a–c, cf. Galay et al. 1973). Meander cutoffs and oxbow lakes are rare and appear to occur only in those cases where the tightly meandering zone has been lost as a result of channel straightening during the transition from an estuary to a delta as discussed above (Woodroffe et al. 1989; Bostock et al. 2007).

In the inner estuary, the channel belt is flanked by mudflats (see Chap. 10) and salt marshes (see Chap. 8) or mangrove swamps that occupy the area between the channel and the valley walls. In the early stage of valley filling, the intertidal flats tend to be broad, but the tidal flats generally become narrower, and the vegetated upper-intertidal zones increase in width, as the unfilled volume (i.e. the accommodation) within the estuary decreases. This happens because the area around the high-tide elevation accumulates sediment faster than the subtidal and lower intertidal areas

(Van der Wal et al. 2002). However, when the estuary becomes nearly filled and broad tidal flats and salt marshes occupy most of the area, the locus of maximum deposition shifts to the channel margins as has been noted in Arcachon Bay (Allard et al. 2009). Overall, the width of the intertidal flats increases seaward. In some cases, the mudflats slope gently into the main channels, producing smooth point-bar surfaces. In other situations, cliffed margins are created by episodic erosion of the outer edge of the mudflats, either because of shifts in the location of the channels, or because of channel enlargement during river floods. Aggradation of the area at the foot of the cliff occurs when the channel migrates away, or the river-flow decreases, leading to the development of a terraced channel-margin morphology (Fig. 5.10d).

The tidal flats and salt marshes are dissected by networks of smaller channels (see Chap. 11) that are oriented approximately at right angles to the larger channels (Fig. 5.10b, c). Some of these small channels connect to terrestrial drainage, but many have no freshwater input, except for local rainfall. They have a meandering pattern and appear to show the 'straight'—meandering—'straight' pattern described above (Fagherazzi et al. 2004). The larger pattern is typically dendritic, with the first-order tributaries consisting of small rills only a few decimeters wide. Higher-order channels become progressively wider. The banks of these runoff channels are gentle in sandy sediments, but may be steeper than 20° in muddy sediments.

5.4 Sediment Facies

As described above, the axial portion of tide-dominated estuaries is occupied by a network of channels that contain sandy and, locally, gravelly sediment, whereas the fringing tidal flats and salt marshes consist of muddy deposits. The spatial organization of sediment caliber and sedimentary facies is relatively predictable because of the process organization discussed above.

5.4.1 Axial Grain-Size Trends

The grain size and its spatial distribution within tide-dominated estuaries is a function of two factors: the nature of the sediment supplied by the terrestrial and marine sources (cf. Figs. 5.2 and 5.3), and the sediment-sorting process that occurs within the estuary.

The sediment supplied by the river can range from gravel-dominated, as is the case in the Cobequid Bay—Salmon River estuary (Figs. 5.1a and 5.12), to quite fine grained and predominantly mud, as a result of differences in the nature of the river's catchment area. Because there is deposition in the river-dominated inner portion of the estuary, the river-supplied sediment becomes finer in a downstream direction (see the general discussion of the causes of fining in Dalrymple 2010a). The sediment supplied by marine processes can also be quite variable in caliber. Most commonly, the sediment entering the mouth of the estuary consists of sandy material that can be quite coarse. This occurs because transgressive erosion (i.e. ravinement) of coastal and shallow-marine areas commonly reworks older fluvial deposits that are characteristically relatively coarse grained. This marine-sourced sediment also becomes finer as it moves into the estuary, again because of deposition. Consequently, the sediment in tide-dominated estuaries is typically coarsest at its mouth and head, and finest in the vicinity of the bedload convergence (Fig. 5.12; Lambiase 1980; Dalrymple et al. 1990).

Superimposed on this general trend, there can be an abrupt decrease in grain size at the inner end of the complex of elongate sand bars that occupies the outer part of the estuary (Fig. 5.12). As explained by Dalrymple et al. (1990), this is attributable to the differential transport speeds of the sediment fractions moving as traction load (generally medium sand and coarser) and in intermittent suspension (mainly fine and very fine sand). Sediment entering the estuary by way of the headward-terminating flood channels must pass through or over an ebb-dominated region before continuing its migration into the estuary. The slow-moving traction material cannot do this and is recycled back out of the estuary, and remains 'trapped' in the zone of elongate sand bars. By contrast, the fast-moving grains that travel by intermittent suspension are capable of reaching the inner parts of the estuary. Thus, sediment in the outer estuary, and in the flood-dominant areas in particular, tends to be composed of medium to coarse, or even very coarse, sand, whereas the middle and inner estuary are characterized by fine and very fine sand. The ebb-dominant channels in the outer estuary that pass through the inner estuary first also tend to be finer grained than the adjacent flood channels. This pattern

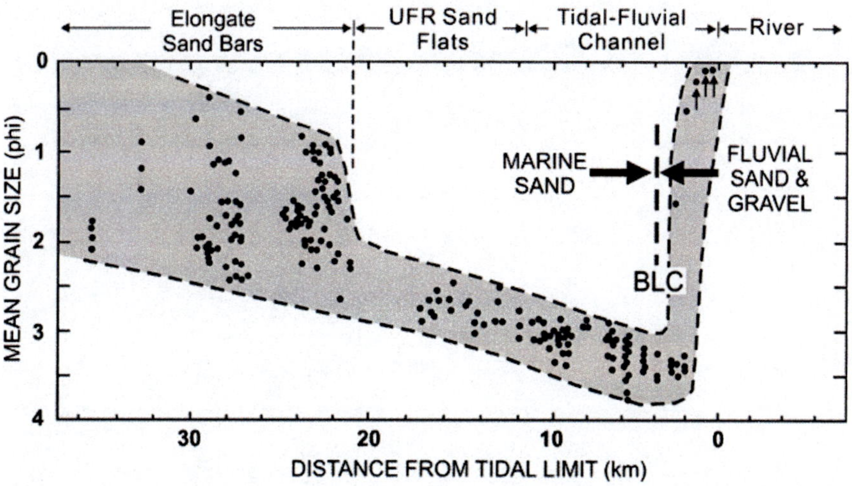

Fig. 5.12 Distribution of mean grain size (each *dot* is an individual sample mean) in the axial channels as a function of position within the Cobequid Bay—Salmon River estuary, Bay of Fundy (Fig. 5.1a). Note that the sediment is coarsest at the mouth and head of the estuary, and finest at the bedload convergence (cf. Fig. 5.10). The abrupt decrease in the size of the coarsest sediment at 21 km is coincident with the inner end of the complex of elongate tidal sand bars, and, more specifically, with the termination of the large flood barb that lies to the north of the main bar chain. See text for further discussion

has been documented in greatest detail in the Cobequid Bay—Salmon River estuary, but is also evident in the Bristol Channel—Severn River estuary (Hamilton 1979; Harris and Collins 1985).

The above pattern of grain-size variation is conspicuously absent in a small number of tide-dominated estuaries, the best documented example being the Hangzhou Bay-Qiantangjiang estuary, China (Zhang and Li 1996; Li et al. 2006). In this system, the outer estuary is muddy rather than sandy, and sediment becomes sandier into the estuary. The cause of this anomalous trend lies in the fact that the local seafloor beyond the mouth of the estuary is mantled with mud that escapes from a nearby, updrift river, namely the Changjiang River to the north, and is carried into the Qiantangjiang estuary because of the flood-tide dominance of the outer estuary (Xie et al. 2009). The landward coarsening trend is caused by the inward increase in tidal-current speeds, coupled with the addition of coarse sediment by the river at the head of the estuary. The Charente estuary, on the western coast of France, shows some similarity to this trend, because of the input of mud from the Gironde estuary to the south (Chaumillon and Weber 2006). It has been discovered in recent years that the suspended sediment issuing from major rivers tends to be advected in one direction along the coast, as a result of the Coriolis affect, oceanic circulation and/or coastal winds. Thus, down-drift estuaries are likely to have muddy rather than sandy mouths, whereas estuaries up-drift of major rivers are more prone to being sandy in their outer part.

5.4.2 Facies Characteristics

5.4.2.1 Outer Estuary: Axial Deposits

In the majority of tide-dominated estuaries, three facies zones can be distinguished in the outer part of the estuary: an erosional lag seaward of the area of sand accumulation, elongate tidal sand bars, and an area of upper-flow-regime sedimentation.

The sea floor beyond the tip of the elongate tidal sand bars is generally erosional and is the marine source area for the estuary. Stratigraphically, it represents a tidal ravinement surface. Older sediments can be exposed here, and the surface is mantled by a lag of coarser sediment if such coarse sediment is available; erosional scours, sand ribbons, and isolated dunes or dune fields can occur (Harris and Collins 1985; see also discussion of *bedload-parting zones* in Chap. 13).

The elongate tidal bars at the mouth of the estuary are typically composed of medium to coarse sand (Fig. 5.12); consequently, they are generally covered by various types of subaqueous dunes (Figs. 5.10a, 5.13a and 5.14a; cf. Ashley 1990). The morphology and dynamics of these bedforms have been reviewed

Fig. 5.13 (**a**) Field of ebb-oriented 3D dunes on the surface of an elongate sand bar, Cobequid Bay. (**b**) Trench through a flood-asymmetric dune, with an ebb cap and two internal reactivation surfaces that define a tidal bundle; the dune migrated a distance of approximately 1 m during one tidal cycle. The surface at the right side of the dune will be buried when the flood current resumes and the ebb cap is eroded

in detail by Dalrymple and Rhodes (1995) and only the main points are summarized here (see also Chap. 13). In estuaries, tidal dunes commonly scale with water depth (height approximately 20% of the depth; wavelength approximately five times the depth, where the depth is that which corresponds with the maximum current speed, and not the depth at high tide; Dalrymple et al. 1978), such that the largest dunes occur in the bottom of channels. In these channels, dunes can reach several meters in height. However, dune size is influenced by factors other than water depth, including current speed, grain size and sediment availability; consequently, there can be deviations from this generalization. Bedforms that are less than about 10 m in wavelength tend to be simple dunes (*sensu* Ashley 1990), whereas larger dunes are generally 'compound', with smaller, simple dunes covering all or part of their stoss and lee sides. The smaller, simple dunes can be either 2D or 3D, whereas the larger *compound dunes* are typically 2D and lack scour pits. Dunes tend to be approximately perpendicular to the main flow, but an oblique orientation is possible in cases where the flood and ebb currents are not 180° apart, or because of lateral gradients in the dune migration rate. As a result, caution is required when using the crestline orientation to deduce sediment-transport directions in detail. Almost all dunes are asymmetric, but the significance of a given asymmetry is strongly dependent on the size of the dune, because the *lag time* (the time required for the bedform to equilibrate with the flow) increases

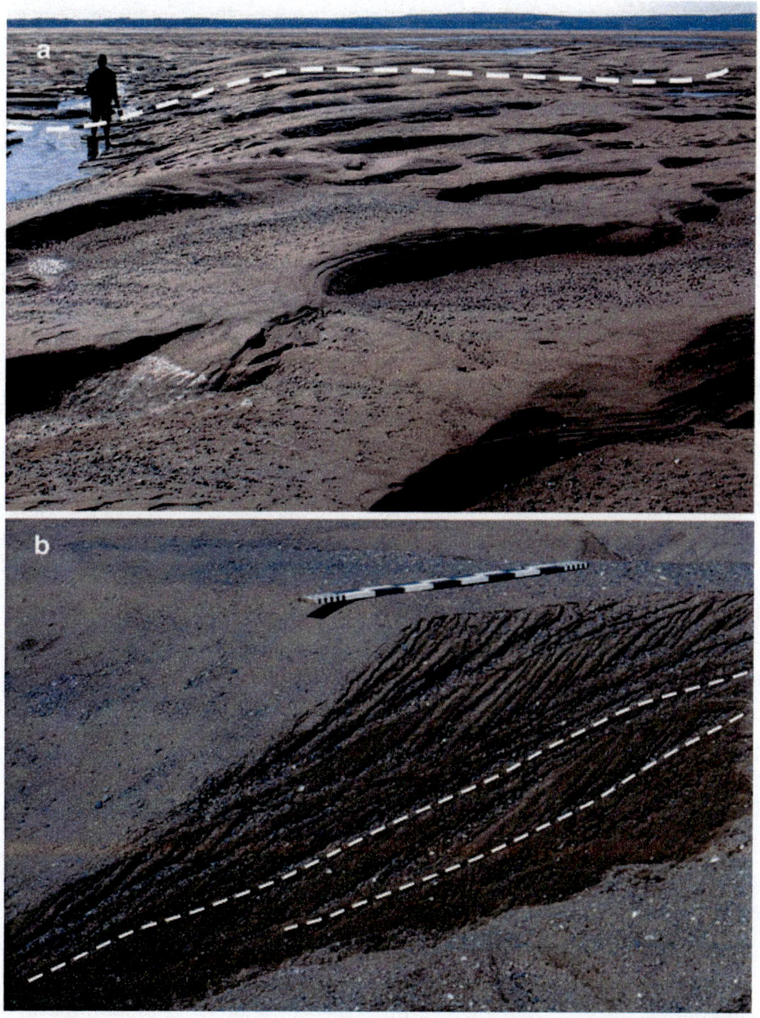

Fig. 5.14 Surface morphology (**a**) and cross section (**b**) through a compound dune in Cobequid Bay. In (**a**), the compound dune, whose profile is outlined by the *dashed white line*, is flood asymmetric, whereas the superimposed simple dunes are ebb oriented at an oblique angle to the crest of the compound dune. In (**b**), the cross beds formed by the superimposed simple dunes have internal cross bedding that dips in the same direction as the master bedding planes (*white dashed lines*) that were formed as the troughs of the simple dunes migrated over the brink of the compound dune

approximately as the square of dune size. Small simple dunes can reverse partially or completely during each half tidal cycle; thus, their facing direction records only the most recent flow. By contrast, large to very large compound dunes have lag times of months to years and are a good indicator of the residual-transport direction over such periods. In this case, seasonal changes in river discharge can play a role in dune reversal (Berné et al. 1993).

The deposits of the elongate sand bars consist predominantly of cross beds (Figs. 5.10a, 5.13b and 5.14b). Within simple dunes, reactivation surfaces and tidal bundles (Visser 1980; see also Chap. 3) are variably developed. In areas with relatively slow currents, such as where 2D dunes occur, the reactivation surfaces are closely spaced (i.e. a few centimeters to decimeters apart; Fig. 5.13b), but they can be as much as a 1–2 m apart in areas with strong currents; such is the case with 3D dunes that migrate rapidly. In all dunes, erosional removal of the dune crest during the passage of a subsequent dune can make recognition of the reactivation surfaces difficult. Compound dunes generate *compound cross bedding* (Dalrymple 1984, 2010b), in which gently dipping (typically $<10°$) master bedding planes separate smaller cross beds generated by the superimposed simple dunes as they migrate down the master surfaces (Fig. 5.14b); see Dalrymple (1984, 2010b) and Dalrymple and Rhodes (1995) for more detail. In general, the deposits of a compound dune coarsen upward because the trough experiences lower currents speeds than the dune's crest. Mud drapes are not abundant in the deposits of the elongate sand bars because the suspended-sediment concentration is low (Fig. 5.3c), but they are most common in relatively sheltered areas, and especially in the troughs of the compound dunes. Mud drapes, including those formed by fluid mud, might also be common in the subtidal part of the main ebb channel because the turbidity maximum can come to rest here during slack water at low tide, at the seaward end of its tidal excursion. At any one location, the cross bedding is likely to have a unidirectional paleocurrent direction because of the local dominance of the flood or ebb current (Dalrymple et al. 1990). Throughout the entire sand body, however, there should be a bimodal paleocurrent pattern, perhaps with an overall flood dominance. Wave-generated structures, such as wave ripples and hummocky cross stratification (HCS), are most likely to occur at the seaward end of the sand-bar complex, because this is the area with the greatest exposure to open-ocean waves (Fig. 5.3b).

Very few benthic organisms are capable of inhabiting these sand bars because of the rapidly shifting nature of the bedforms and the great thickness of the surface mobile layer (equal to the bedform height). As a result, shelled organisms are scarce, and are typically limited to mesohaline bivalves. They occur most commonly as a comminuted shell hash that can be leached in ancient sediments. Trace fossils are also generally scarce in subtidal areas (Fig. 5.3e), and consist mainly of a low-diversity suite of deep vertical burrows of the *Skolithos* Ichnofacies (see Chap. 4 for a more detailed examination of the ichnology of tidal deposits).

The large-scale internal architecture of the elongate sand bars is not well known. The limited seismic data that have been published (e.g. Dalrymple and Zaitlin 1994) suggest that deposition on the bar flanks generates large-scale 'master bedding' that generally dips at only 2–3°, although values as high as 10° are possible. The cross bedding is oriented approximately along the strike of this bedding, forming lateral-accretion deposits. These bar-flank deposits can reach 10–15 m in thickness, but complete preservation is unlikely because of truncation by later channels. The grain-size trend in these deposits generally fines upward because the fastest currents occur in the channels, and the slowest currents on the bar crests. The swatchways, which migrate toward the head of the estuary, generate smaller, upward-fining successions in which lateral-accretion bedding is also present; the dip of these beds should fan obliquely outward relative to the axis of the estuary because of the skewed orientation of the swatchways.

In estuaries that are exposed to large ocean waves, the sands at the mouth can be subjected to significant wave reworking (Fig. 5.3b). Ridge-and-runnel systems, which are typical of beach-like settings, have been reported from the outer part of The Wash, eastern England (McCave and Geiser 1978; Ke et al. 1996), and wave-formed swash bars are present in Mont-Saint-Michel Bay, France (Billeaud et al. 2007) and Gomso Bay, Korea (Yang et al. 2007), and hummocky cross stratification can be present, if the sediment is fine or very fine sand (Yang et al. 2007).

The area that lies landward of the elongate sand bars consists of fine to very fine sand (Fig. 5.12) that occupies the zone of strongest tidal currents (Fig. 5.3b). In this area, tidal-current speeds that can exceed 2 *m/s* generate extensive upper-flow-regime sand flats in shallow water. At low tide, most surfaces are covered by current (Fig. 5.15a) and/or combined-flow ripples, but the internal structures consist predominantly of parallel lamination, with scattered ripple cross-lamination (Fig. 5.15b). The ripples can show bipolar dips, but ebb-oriented sets outnumber flood ripples, even though this area is flood-dominant, overall. The parallel lamination is typically flat-lying, but gently dipping stratification can be formed on the flanks and lee side of the subtle braid bars that occupy this zone in shallow estuaries such as the Cobequid Bay, Bay of Fundy (Figs. 5.1a and 5.10a). Ripple-laminated sand becomes more common along the margins of the estuary, in the transition to the flanking mudflats. Dune cross bedding is uncommon, and is most common in the transition to the elongate tidal sand bars because this is the area where grain size is coarse enough to support dunes. In deeper systems such as the Severn River estuary (Fig. 3.1b), this braided sand-flat zone appears to be absent, although upper-flow-regime conditions do occur on the point bars (Hamilton 1979) that occur in the outer part of the tidal-fluvial channel zone (see below). Biologically, very few organisms can live in these high-energy sand flats (Fig. 5.3e), because of the rapid movement of sand, the reduced salinity (typically in the range of 5–15‰), and the generally high suspended-sediment concentrations. Because of the absence of dunes, the depth of frequent reworking is, however, less than it is on the elongate tidal sand bars, which allows a small number of deeply burrowing, opportunistic organisms to colonize the substrate. Mud drapes are not abundant (Fig. 5.15b), despite the high suspended-sediment concentration, because of erosion

Fig. 5.15 (**a**) Surface of upper-flow-regime sand flat at low tide, covered with current ripples. Beneath the surface, the predominant structure is parallel lamination. (**b**) Epoxy peel of a core from the upper-flow-regime sand flats, showing abundant parallel lamination, with scattered sets of current ripples, here showing bipolar paleocurrent directions. Although the suspended-sediment concentration is high in this area, there are few mud drapes (one is present at 23–24 cm depth) because of subsequent erosion (Both images from the Cobequid Bay—Salmon River estuary)

by subsequent currents. They are most prominent in situations where one of the channels that occur in this area gets cut off and fills with heterolithic strata that might include fluid-mud layers, and in the transition to the flanking mudflats. Comminuted organic detritus, which is commonly referred to as *coffee grounds* or *tea leaves* because of its granular appearance, can also form drapes.

In estuaries that lie immediately down-drift (with respect to mud dispersal) of a major river, the erosional area at the mouth is replaced by muddy deposits (e.g., the Hangzhou Bay-Qiantangjiang estuary, Zhang and Li 1996; Li et al. 2006). Descriptions of this facies lack detail, but indicate the presence of sandy laminae, 1–2 mm thick, interbedded with mud layers several centimeters thick. It is likely that this stratification reflects the action of storm waves (cf. Fig. 5.2). Based on observations in tide-dominated deltas (Kuehl et al. 1996; Dalrymple et al. 2003), it is possible that these muddy layers could be rapidly deposited from high-density, wave-generated suspensions, rather than having accumulated by slow settling. Vertical burrows and shell debris are also reported from this facies. Terrestrial organic material is also present and probably increases in abundance in the landward transition into fine sand and/or silty sand. The nature of the structures in this transition zone is not reported; more detailed studies are needed.

5.4.2.2 Inner Estuary: Tidal-Fluvial Transition

This zone (zone 3 of Dalrymple et al. 1991) stretches from the limit of tidal action to the location where significant widening occurs, allowing the development of several ebb and flood channels. Note that this is defined more broadly than the 'tidal-fluvial transition' subdivision in Dalrymple and Choi (2007), and encompasses the entire 'straight'—meandering—'straight' channel pattern discussed above (Figs. 5.1 and 5.8). In this zone as distinguished here, there is a single main ebb channel that is only locally flanked by flood barbs on the seaward side of the point bars that occur along the channel (Fig. 5.10c). The nature of the deposits in this zone, which is transitional between purely fluvial deposition beyond the tidal limit and almost purely tidal sedimentation at the seaward end, is not known in detail and more work is needed. Based largely on theoretical considerations, supplemented by the limited available information (Billeaud et al. 2007; Van den Berg et al. 2007), Dalrymple and Choi (2007) have speculated on the deposit characteristics. In at least some systems with a large tidal range, upper-flow-regime conditions prevail in the outer, tide-dominated part of the transition, occupying the thalweg and/or lower part of the point bars (Hamilton 1979; Lambiase 1980; Dalrymple et al. 1990; Billeaud et al. 2007), producing deposits that are similar to those in the braided sand-flat zone that lies immediately seaward (i.e.

Fig. 5.16 Photo of the channel in the tightly meandering reach of the Salmon River, Bay of Fundy (Fig. 5.1a, *inset*). The gravel in the channel thalweg was deposited by river floods, whereas the horizontally bedded sediment on the bank, which consists of very fine sand, silt and clay with tidal rhythmites, was deposited by tidal processes

parallel-laminated fine to very fine sand with scarce mud drapes and limited bioturbation). In deeper channels that contain coarser sediment, dunes will be present, and the deposits there will be cross bedded. In the outer part of the tidal-fluvial transition, fluid-mud deposits can be an important component of the channel-bottom facies (cf. Schrottke et al. 2006). These fluid-mud layers can be recognized by the presence of anomalously thick (i.e. >1 cm before compaction), structureless to faintly-laminated mud layers that lack contemporaneous bioturbation (Ichaso and Dalrymple 2009). The sediment interbedded with the fluid-mud layers is likely to be the coarsest material that occurs in that part of the system, producing a markedly bimodal association of river-flood deposits and tidally deposited fluid muds. This bimodality is likely to be most pronounced near the bedload convergence area, where depositional conditions alternate seasonally (Fig. 5.16). If dunes are present on the channel floor, the fluid muds are preferentially preserved in their troughs (Fig. 5.17; cf. Schrottke et al. 2006), generating muddy bottomset and toeset deposits. The sands in these channel deposits will fine upward, whereas the amount of mud and mud-layer thickness will decrease upward, producing an upward-cleaning, but upward fining succession (Dalrymple 2010b). In channels that lack significant river input of coarse material, such as the smaller tributary channels that drain low-lying coastal areas (Fig. 5.3a), the channel-bottom deposits can consist almost entirely of thick fluid-mud layers, with channel-bank slump deposits and patchy development of mud-clast breccias.

5.4.2.3 Fringing Facies

The axial deposits described in the two preceding sections are flanked by a suite of generally fine-grained deposits, that accumulate in the space been the active funnel-shaped network of channels and any valley walls that border the estuary. In narrow, rock-walled estuaries, the channels can occupy the entire width of the valley (e.g. Cobequid Bay, Bay of Fundy; Dalrymple et al. 1990), whereas broad valleys in soft, coastal-plain sediments can have wide muddy tidal flats and marshes (e.g. the South Alligator River, Northern Australia; Woodroffe et al. 1989). The nature of these fringing facies varies with position along the length of the estuary, and with distance away from the channels (Dalrymple et al. 1991).

The margins of the outer part of most estuaries are erosional, and older material, including mudflat and salt-marsh deposits that accumulated earlier in the transgression, can be exposed on the intertidal foreshore (cf. Allen 1990; Cooper et al. 2001). This erosional surface can be covered by a blanket of mud during periods of low wave activity (e.g. the summer), but it is typically removed by winter waves. Bioturbation

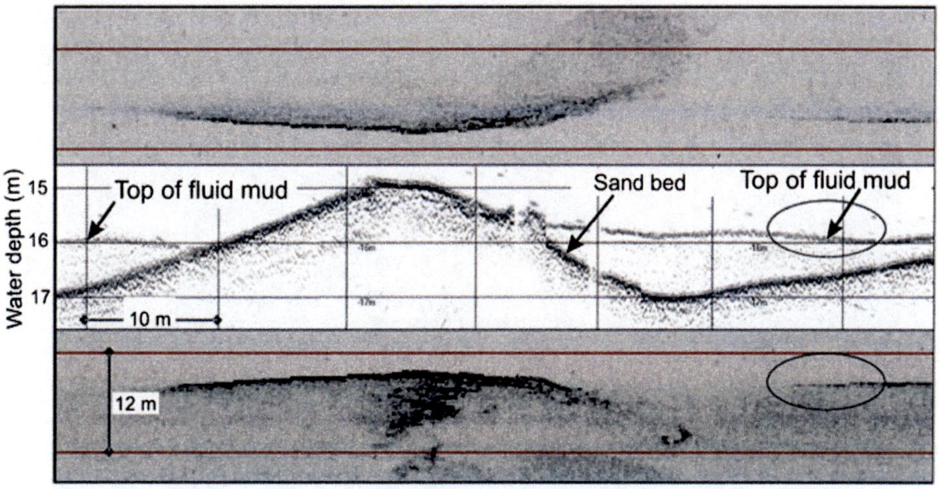

Fig. 5.17 Cross section and sidescan sonar images (*top and bottom*) of a dune on the bed of the Weser River, showing the presence of fluid mud in the troughs between the dunes. The ellipses show locations where the fluid mud becomes so soft that no acoustic reflection is detected in the sidescan sonar record. The firm sand on the dune crest that is not buried by fluid mud appears dark on the sidescan sonar record (Modified after Schrottke et al. 2006, Fig. 5.9b)

can be intense in this mud layer, and consists of a relatively diverse assemblage (Fig. 5.3e). At their inner end, the high-tide beaches interfinger with mudflat and salt-marsh deposits, and form coarse-grained cheniers encased in muddy deposits (Fig. 5.18b) (Lee et al. 1994; Pye 1996; Tessier et al. 2006).

The mudflats that flank the channels in the inner estuary become broader in a seaward direction, ranging from only a few meters wide in the largely filled innermost part of the estuary (Fig. 5.10c, d), to several 10s to 100 s of meters wide near the seaward end of active mudflat sedimentation, which typically occurs in the middle estuary (Fig. 5.10b). At any given location, the width of the mudflats decreases through time as the estuary fills. In the inner estuary where the mudflats lie closest to the fast currents in the channels, and where, consequently, the delivery of sediment to the mudflats is rapid, the sedimentation rate can reach several meters per year, generating well-developed tidal rhythmites (Fig. 5.19a; Dalrymple et al. 1991; Tessier 1993; Choi 2010). Further seaward where the mudflats are, on average, a greater distance from the strong currents in the channel, the sedimentation rate is lower (several millimeters to several decimeters per year), allowing the development of annual cyclicity as a result of seasonal changes in temperature and/or the intensity of wave action (Van den Berg 1981; Dalrymple et al. 1991; Allen and Duffy 1998). These cycles typically consist of alternations of layers with physical lamination, in which tidal rhythmites might be present, and intensely bioturbated sediment (Fig. 5.19b). Although this bioturbation can be intense, the diversity of traces is usually lower than in areas further seaward (Fig. 5.3e) because of the lower salinity. Overall, there is considerable diversity in the intensity of bioturbation spatially, with a much lower level of bioturbation in areas of higher sedimentation rate near channels, and a higher level in the more slowly aggrading tidal flats further from the channels. Deformation structures produced by grounding ice are present in mudflats in temperate to polar settings (Dionne 1985; Dalrymple et al. 1991). Seasonal cyclicity can also occur in the innermost, fluvially dominated portion of the estuary, but here the primary seasonal signal appears to be variations in river discharge. The diversity and intensity of bioturbation in these inner-estuarine mudflats are low because of the stress imposed by the low salinity.

A salt-marsh (see Chap. 8), or mangrove swamp in tropical areas, lies at a greater distance from the channel, typically in the elevation range between about neap and spring high tide. The deposits here are intensely rhizoturbated (Fig. 5.19b), and contain a variable amount of organic material. The development of a levee along the margin of the channel can lead to the development of boggy conditions at greater distances from the channel, commonly in the area adjacent to the valley walls (Woodroffe et al. 1989). Organic-rich sediments, including potentially peat, accumulate in such areas.

Fig. 5.18 (**a**) Erosional foreshore along the margin of Cobequid Bay, Bay of Fundy, with cliffs composed of Triassic sandstone, with a beach at the high-tide level. (**b**) Gravel beach in Cobequid Bay that has migrated in front of and is encroaching on salt-marsh deposits. The gravel is sourced from coastal erosion of Pleistocene till and glaciofluvial outwash

The nature of the contacts between the sand flats, mudflats and salt-marsh can be either gradational (Fig. 5.10b) or erosional (Fig. 5.10d). Lateral migration of a channel, or enlargement of a channel because of increased fluvial discharge, causes frequent erosion of the outer edge of the mudflat and/or salt-marsh (Fig. 5.10c, d). The cliffs created by these processes generate steeply inclined or even vertical erosion surfaces that can be mantled by a mud-pebble conglomerate. Once the channel migrates away, or the river flow returns to a lower value, the previously erosional area becomes depositional, and rapid vertical aggradation occurs, producing a terraced margin to the channel (Fig. 5.10d). Such situations generate upward-fining vertical successions with a thickness (before compaction) that is equal to the channel depth, in which the tidal deposits are essentially horizontal. In other cases, the banks of the channel are more gently sloping, with gradational facies contacts, and produce inclined heterolithic stratification (IHS; Thomas et al. 1987) that dips toward the channel with inclinations typically of 5–15°. The conditions under which each of these two channel-bank morphologies exist are not known.

Smaller tidal channels, or the channels of tributary streams, dissect the mudflats and salt marshes (Fig. 5.10b; Chap. 11). These channels become wider in a seaward direction, and their banks become less steep as they pass from the mudflats out into the sandflats. The floor of these channels will consist of a patchy lag of mud pebbles derived from erosion of the bank. Shell debris can be present locally, but is typically monospecific in character because of the reduced salinity. Sand is rarely present in the channels that do not have terrestrial drainage, but can be present in channels that have their

Fig. 5.19 (**a**) Tidal rhythmites from a location just seaward of the tightly meandering reach in the Salmon River. The section is located at the site of Fig. 5.10d. *Sp* = spring-tide layers; *N* = neap-tide layers. Each sand layer was deposited by a single flood tide. In general, the ebb tide does not deposit a recognizable layer. In some of the mud drapes during spring tides, however, a separate silt stringer is present in the middle of the mud layer (highlighted by the inscribed line in the mud layer just below layer '16'). This was deposited by the ebb tide. (**b**) Mudflat deposits from the middle of the Cobequid Bay—Salmon River estuary, with well-developed annual cycles. *W* = fall, winter and spring deposits that are weakly bioturbated and laminated. *S* = summer deposits that are completely homogenized by bioturbation. Note how the annual layers become thinner upward as the surface rises higher in the tidal frame. The top of the section is partially turbated by roots of salt-marsh plants

headwaters on land. Deposition on the point bars of these channels generates IHS (De Mowbray 1983; Pearson and Gingras 2006; Choi 2010). Because the position of these channels is relatively stable, the 'channel belt' that they produce is narrow, and the bulk of the mudflat and salt-marsh deposits is horizontally stratified.

5.5 Summary

Tide-dominated estuaries are dynamic environments, because of the strong and widespread action of tidal currents, with lesser influence from waves and river currents. The spatial organization of processes, morphology and facies within these estuaries is predictable in general terms, if not in detail, because of the regular way in which the intensity of these three processes varies along the length and across the width of the estuary. A large amount of information exists on these processes, because of the great amount of research that has been done in order to understand the dynamics of sediment transport, a topic of considerable interest with regard to human utilization of these estuaries. There is a growing body of research that has examined the morphodynamics of tide-dominated estuaries, and the broad patterns are understood reasonably well, but more needs to be done to document the rates and patterns of morphological change. In general terms, tide-dominated estuaries can be in one of two evolutionary states: active transgression, during which all shorelines within the estuary experience net erosion as a result of wave action in the outer part, and channel-bank scour in the inner reaches, as the estuarine funnel translates landward; and progradational filling when the rate of sediment input from fluvial and marine sources exceeds the rate of creation of accommodation as a result of sea-level rise. The transition between these two states begins in the inner part of the estuary and migrates seaward as filling progresses; many modern estuaries are part way through this transition, and show continued erosion in their outer part, while their inner margins prograde. Any human activity that alters the sediment supply (e.g. the building of dams in inland areas, or breakwaters and training walls at the estuary mouth), the propagation of the tidal wave (e.g. dredging, the construction of impermeable causeways), or the space available for sediment accumulation (e.g. marsh reclamation) has predictable consequences when viewed in this general context.

Although much has been learned in recent years about the stratigraphy of the deposits of tide-dominated estuaries (see Chap. 6), much less is known about the detailed nature of the facies within them. The discovery that fluid mud is a common occurrence within the channels beneath the turbidity maximum has been a significant addition to the criteria for interpreting estuarine (and deltaic) deposits, but much remains to be done to refine our ability to determine where in the fluvial-marine transition a given deposit in an ancient succession might have formed.

References

Allard J, Chaumillon E, Féniès H (2009) A synthesis of morphological evolutions and Holocene stratigraphy of a wave-dominated estuary: the Arcachon lagoon, SW France. Cont Shelf Res 29:957–969

Allen GP (1973) Suspended sediment transport and deposition in the Gironde estuary and adjacent shelf. Memoire, Institute Geologique du Bassin d'Aquitaine 7:27–36

Allen GP (1991) Sedimentary processes and facies in the Gironde estuary: a recent model for macrotidal estuarine systems. In: Smith DG, Reinson GE, Zaitlin BA, Rahmani RA (eds) Clastic tidal sedimentology. Can Soc Petrol Geol Mem 16:29–40

Allen GP, Salomon JC, Bassoulet P, Du Penhoat Y, De Grandpré C (1980) Effects of tides on mixing and suspended sediment transport in macrotidal estuaries. Sediment Geol 26:69–90

Allen JRL (1990) The Severn estuary in southwest Britain: its retreat under marine transgression and fine-sediment regime. Sediment Geol 66:13–28

Allen JRL, Duffy MJ (1998) Temporal and spatial depositional patterns in the Severn estuary, southwestern Britain: intertidal studies at spring-neap and seasonal scales, 1991–1993. Mar Geol 141:147–171

Ashley GM (1990) Classification of large-scale subaqueous bedforms: a new look at an old problem. J Sediment Petol 60:160–172

Ayles CP, Lapointe DMF (1996) Downvalley gradients in flow patterns, sediment transport and channel morphology in a small macrotidal estuary: Dipper Harbour Creek, New Brunswick, Canada. Earth Surf Proc Land 21:829–842

Barwis JH (1978) Sedimentology of some South Carolina tidal-creek point bars, and a comparison with their fluvial counterparts. In: Miall AD (ed) Fluvial sedimentology. Can Soc Petrol Geol Mem 5:129–160

Berné S, Castaing P, Le Drezen E, Lericolais G (1993) Morphology, internal structure, and reversal of asymmetry of large subtidal dunes in the entrance to the Gironde estuary (France). J Sediment Petrol 63:780–793

Billeaud I, Tessier B, Lesueur P, Caline B (2007) Preservation potential of highstand coastal sedimentary bodies in a macrotidal basin: example from the Bay of Mont-Saint-Michel, NW France. Sediment Geol 202:754–775

Bostock HC, Brooke BP, Ryan DA, Hancock G, Pietsch T, Packett R, Harle K (2007) Holocene and modern sediment storage in the subtropical macrotidal Fitzroy River estuary, Southeast Queensland, Australia. Sediment Geol 201:321–340

Burningham H (2008) Contrasting geomorphic response to structural control: the Loughros Estuaries, northwest Ireland. Geomorphology 97:300–320

Castaing P, Allen GP (1981) Mechanisms controlling seaward escape of suspended sediment from the Gironde: a macrotidal estuary in France. Mar Geol 40:101–118

Catuneanu O (2006) Principles of sequence stratigraphy. Elsevier, Amsterdam, 375 p

Chaumillon E, Weber N (2006) Spatial variability of modern incised valleys on the French Atlantic coast: comparison between the Charente and the Lay-Sèvre incised valleys. In: Dalrymple RW, Leckie DA, Tillman RW (eds) Incised valleys in time and space. SEPM special publications 85. Society for Sedimentary Geology, Tulsa, pp 57–85

Choi KS (2010) Rhythmic climbing-ripple cross-lamination in inclined heterolithic stratification (IHS) of a macrotidal estuarine channel, Gomso Bay, west coast of Korea. Jour Sed Res 80:550–561

Cooper NJ, Cooper T, Burd F (2001) 25 years of salt marsh erosion in Essex: Implications for coastal defence and nature conservation. J Coast Conserv 7:31–40

Dalrymple RW (1984) Morphology and internal structure of sandwaves in the Bay of Fundy. Sedimentology 3:365–382

Dalrymple RW (2006) Incised valleys in space and time: an introduction to the volume and an examination of the controls on valley formation and filling. In: Dalrymple RW, Leckie DA, Tillman RW (eds) Incised valleys in space and time. SEPM special publications 85. Society for Sedimentary Geology, Tulsa, pp 5–12

Dalrymple RW (2010a) Introduction to siliciclastic facies models. In: James NP, Dalrymple RW (eds) Facies models 4. Geological Association of Canada, St. John's, pp 59–72

Dalrymple RW (2010b) Tidal depositional systems. In: James NP, Dalrymple RW (eds) Facies models 4. Geological Association of Canada, St. John's, pp 199–208

Dalrymple RW, Choi KS (2003) Sediment transport by tides. In: Middleton GV (ed) Encyclopedia of sediments and sedimentary rocks. Springer, Dordrecht, pp 606–609

Dalrymple RW, Choi KS (2007) Morphologic and facies trends through the fluvial-marine transition in tide-dominated depositional systems: a systematic framework for environmental and sequence-stratigraphic interpretation. Ear Sci Rev 81: 135–174

Dalrymple RW, Zaitlin BA (1994) High-resolution sequence stratigraphy of a complex, incised valley succession, the Cobequid Bay Salmon River estuary, Bay of Fundy, Canada. Sedimentology 41:1069–1091

Dalrymple RW, Knight RJ, Lambiase JJ (1978) Bedforms and their hydraulic stability relationships in a tidal environment, Bay of Fundy. Nature 275:100–104

Dalrymple RW, Knight RJ, Zaitlin BA, Middleton GV (1990) Dynamics and facies model of a macrotidal sand bar complex. Sedimentology 35:577–612

Dalrymple RW, Makino Y, Zaitlin BA (1991) Temporal and spatial patterns of rhythmite deposition on mud flats in the macrotidal, Cobequid Bay-Salmon River. In: Reinson GE, Zaitlin BA, Rahmani RA (eds) Clastic tidal sedimentology. Can Soc Petrol Geol Mem 16:137–160

Dalrymple RW, Rhodes RN (1995) Estuarine dunes and barforms. In: Perillo GM (ed) Geomorphology and sedimentology of estuaries. Development in sedimentology 53. Elsevier, Amsterdam, pp 359–422

Dalrymple RW, Zaitlin BA, Boyd R (1992) Estuarine facies models: conceptual basis and stratigraphic implications. J Sediment Petrol 62:1130–1146

Dalrymple RW, Baker EK, Harris PT, Hughes M (2003) Sedimentology and stratigraphy of a tide-dominated, foreland-basin delta (Fly River, Papua New Guinea). In: Sidi FHD, Nummedal D, Imbert D, Darman H, Posamentier HW (eds) Tropical deltas of Southeast Asia. Sedimentology, stratigraphy, and petroleum geology. SEPM Spec Publ 77: 147–173

De Mowbray T (1983) The genesis of lateral accretion deposits in recent intertidal mudflat channels, Solway Firth, Scotland. Sedimentology 30:425–435

Dionne JC (1985) Forms, figures et facies sedimentaires glaciels dans des estrans vaseux des regions froides. Palaeogeogr Palaeoclim Palaeoecol 54:415–451

Davis RA, Hayes MO (1984) What is a wave-dominated coast? Mar Geol 60:313–329

Doxaran D, Froidefond JM, Castaing P, Babin M (2009) Dynamics of the turbidity maximum zone in a macrotidal estuary (the Gironde, France): observations from field and MODIS satellite data. Estuarine Coast Shelf Sci 83:321–332

Dyer KR (1995) Sediment transport processes in estuaries. In: Perillo GME (ed) Geomorphology and sedimentology of estuaries. Development in sedimentology 53. Elsevier, Amsterdam, pp 423–449

Dyer KR (1997) Estuaries—a physical introduction, 2nd edn. Wiley, New York, 195 p

Faas RW (1991) Rheological boundaries of mud. Where are the limits? Geo-Mar Lett 11:143–146

Fagherazzi S, Furbish D (2001) On the shape and widening of salt marsh creeks. J Geophys Res Oceans 106:991–1005

Fagherazzi S, Gabet EJ, Furbish DJ (2004) The effect of bidirectional flow on tidal channel planforms. Earth Surf Proc Land 29:295–309

Friedrichs CT, Aubrey GD (1988) Non-linear tidal distortion in shallow well-mixed estuaries. Estuar Coast Shelf Sci 27:521–545

Friedrichs CT, Aubrey DG, Speer PE (1990) Impacts of relative sea-level rise on evolution of shallow estuaries. In: Cheng RT (ed) Residual currents and long-term transport. Coastal estuarine studies 38. Springer, New York, pp 105–122

Galay VJ, Kellerhals R, Bray DI (1973) Diversity of river types in Canada. In: Fluvial process and sedimentation. Proceedings of hydrology symposium. National Research Council of Canada, Edmonton, pp 217–250

Ganju NK, Schoellhamer DH, Warner JC, Barad MF, Schladow SG (2004) Tidal oscillation of sediment between a river and a bay: a conceptual model. Estuar Coast Shelf Sci 60:81–90

Guan WB, Wolanski E, Dong LX (1998) Cohesive sediment transport in the Jiaojiang River estuary, China. Estuar Coast Shelf Sci 46:861–871

Haas LW (1977) The effect of the spring-neap tidal cycle on the vertical salinity structure of the James, York and Rappahannock Rivers, Virginia, USA. Estuar Coast Mar Sci 5:485–496

Hamilton D (1979) The high-energy sand and mud regime of the Severn estuary, S.W. Britain. In: Severn RT, Dinely D, Hawker LE (eds) Tidal power and estuary management. Albuquerque, Transatlantic Arts Incorporated, Colston paper 30. Scientechnica, Bristol, pp 62–172

Harris PT (1988) Large scale bedforms as indicators of mutually evasive sand transport and the sequential infilling of wide-mouthed estuaries. Sediment Geol 57:273–298

Harris PT, Collins MB (1985) Bedform distribution and sediment transport paths in the Bristol Channel and Severn estuary, UK. Mar Geol 62:153–166

Hori K, Saito Y, Zhoa Q, Cheng X, Wang PY, Li C (2001) Sedimentary facies and Holocene progradation rates of the Changjiang (Yangtze) delta, China. Geomorphology 41: 233–248

Ichaso AA, Dalrymple RW (2006) On the geometry of tidal channels. American Association of Petroleum Geologists annual meeting, Houston, 9–12 April, Search and Discovery Article #90052

Ichaso AA, Dalrymple RW (2009) Tidal and wave-generated fluid-mud deposits in the Tilje Formation (Jurassic), offshore Norway. Geology 37:539–542

Inlgis CC, Allen FH (1957) The regimen of the Thames estuary as affected by currents, salinities, and river flow. Proc Inst Civ Eng London 7:827–868

Jeuken MCJL (2000) On the morphologic behaviour of tidal channels in the Westerschelde estuary. Netherlands Geographical Studies 79, 378 p

Johnson MA, Kenyon NH, Belderson RH, Stride AH (1982) Sand transport. In: Stride AH (ed) Offshore tidal sands, processes and deposits. Chapman and Hall, London, pp 58–94

Ke X, Evans G, Collins MB (1996) Hydrodynamics and sediment dynamics of The Wash embayment, eastern England. Sedimentology 43:157–174

Kirby R, Parker WR (1983) Distribution and behavior of fine sediment in the Severn estuary and inner Bristol Channel, U.K. Can J Fish Aquat Sci 40:83–95

Kravatsova VI, Mikhailov VN, Kidyaeva VM (2009) Hydrologic regime, morphological features and natural territorial features of the Irrawaddy River delta (Myanmar). Water Res 36:259–276

Kuehl SA, Nittrouer CA, Allison MA, Faria LEC, Dakut DA, Maeger JM, Pacioni TD, Figueiredo AG, Underkoffler EC (1996) Sediment deposition, accumulation, and seabed dynamics in an energetic fine-grained coastal environment. Cont Shelf Res 16:787–815

Lambiase J (1980) Sediment dynamics in the macrotidal Avon River estuary, Bay of Fundy, Nova Scotia. Can J Earth Sci 17:1628–1641

Lee HJ, Chun SS, Chang JH, Han SJ (1994) Landward migration of isolated shelly sand ridge (chenier) on the macrotidal flat of Gomso Bay, west coast of Korea: controls of storms and typhoon. J Sediment Res 64:886–893

Lesourd S, Lesueur P, Brun-Cottan JC, Garnaud S, Poupinet N (2003) Seasonal variations in the characteristics of superficial sediments in a macrotidal estuary (the Seine inlet, France). Estuar Coast Shelf Sci 58:3–16

Lettley CD, Pemberton SG, Gingras MK, Ranger MJ, Blakney BJ (2005) Integrating sedimentology and ichnology to shed light on the system dynamics and paleogeogrpahy of an ancient riverine estuary. In: MacEachern JA, Bann KL, Gingras MK, Pemberton SG (eds) Applied ichnology. SEPM Short Course Notes 52:144–162

Li C, O'Donnell J (1997) Tidally driven residual circulation in shallow estuaries with lateral depth variation. J Geophys Res 102:27, 915–927, 929

Li C, Wang P, Fan D, Yang S (2006) Characteristics and formation of late Quaternary incised-valley-fill sequences in sediment-rich deltas and estuaries: case studies from China. In: Dalrymple RW, Leckie DA, Tillman RW (eds) Incised valleys in time and space. SEPM special publications 85. Society for Sedimentary Geology, Tulsa, pp 141–160

MacEachern JA, Pemberton SG, Bann KL, Gingras MK (2005) Departures from the archetypal ichnofacies: effective recognition of physico-chemical stresses in the rock record. In: MacEachern JA, Bann KL, Gingras MK, Pemberton SG (eds) Applied ichnology. SEPM short course notes 52. SEPM, Tulsa, pp 65–93

McCave IN, Geiser AC (1978) Megaripples, ridges and runnels on intertidal flats of The Wash, England. Sedimentology 26:353–369

McLaren P, Collins MB, Gao S, Powys RIL (1993) Sediment dynamics of the Severn estuary and inner Bristol Channel. J Geol Soc Lond 150:589–603

Mehta AJ (1991) Understanding fluid mud in a dynamic environment. Geo-Mar Lett 11:113–118

Moore RD, Wolf J, Souza AJ, Flint SS (2009) Morphological evolution of the Dee estuary, Eastern Irish Sea, UK: a tidal asymmetry approach. Geomorphology 103:588–596

Myrick RM, Leopold LB (1963) Hydraulic geometry of a small tidal estuary. U.S. Geological Survey Professional Paper 422-B, 18 p

Nichols MM, Biggs RB (1985) Estuaries. In: Davis RA (ed) Coastal sedimentary environments, 2nd edn. Springer, New York, pp 77–186

O'Connor BA (1987) Short and long term changes in estuary capacity. J Geol Soc Lond 144:187–195

Pearson NJ, Gingras MK (2006) An ichnological and sedimentological facies model for muddy point-bar deposits. J Sediment Res 76:771–782

Pethick JS (1996) The geomorphology of mudflats. In: Nordstrom KF, Roman CT (eds) Estuarine shores: evolution, environments and human alterations. Wiley, New York, pp 185–211

Pritchard DW (1967) What is an estuary? Physical viewpoint. In: Lauff GH (ed) Estuaries. Am Assoc Adv Sci Publ 83:3–5

Pye K (1996) Evolution of the shoreline of the Dee estuary, United Kingdom. In: Nordstrom KF, Roman CT (eds) Estuarine shores: evolution, environments and human alterations. Wiley, New York, pp 15–37

Rinaldo A, Belluco E, D'Alpaos AF, Lanzoni S, Marani M (2004) Tidal Networks: form and function. In: Fagherazzi S, Blum L, Marani M (eds) Ecogeomorphology of tidal marshes. American Geophysical Union, Coastal and estuarine monograph series 59. American Geophysical Union, Washington, DC, pp 75–91

Roberts W, Le Hir P, Whitehouse RJS (2000) Investigation using simple mathematical models of the effect of tidal currents and waves on the profile shape of intertidal mudflats. Cont Shelf Res 20:1079–1097

Robinson AHW (1960) Ebb-flood channel systems in sandy bays and estuaries. Geography 45:183–199

Ryan DA, Brooke BP, Bostock HC, Radke LC, Siwabessy PJW, Margvelashvili N, Skene D (2007) Bedload sediment transport dynamics in a macrotidal embayment, and implications for export to the southern Great Barrier Reef shelf. Mar Geol 240:197–215

Salomon JC, Allen GP (1983) Role sedimentologique de la mare dans les estuaires a fort marnage. Compagnie Francais des Pétroles, Notes et Memoires 18:35–44

Schrottke K, Becker M, Batholomä A, Flemming BW, Hebbeln D (2006) Fluid mud dynamics in the Weser estuary turbidity zone tracked by high-resolution side-scan sonar and parametric sub-bottom profiler. Geo-Mar Lett 26:185–198

Schuttelaars HM, de Swart HE (2000) Multiple morphodynamic equilibria in tidal embayments. J Geophys Res 105:24, 105–124, 118

Solari L, Seminara G, Lanzoni S, Marani M, Rinaldo A (2002) Sand bars in tidal channels; Part II: Tidal meanders. J Fluid Mech 451:203–238

Tessier B (1993) Upper intertidal rhythmites in the Mont-Saint-Michel Bay (NW France): perspectives for paleoreconstruction. Mar Geol 110:355–367

Tessier B, Billeaud I, Lesueur P (2006) The Bay of Mont-Saint-Michel northeastern littoral: an illustrative case of coastal sedimentary body evolution and stratigraphic organization in a transgressive/highstand context. Bull Soc géol Fr 177:71–78

Tessier B, Billeaud I, Lesueur P (2010) Stratigraphic organization of a composite macrotidal wedge: the Holocene sedimentary infilling of the Mont-Saint-Michel Bay (NW France). Bull Soc géol Fr 181:99–113

Thomas RG, Smith DG, Wood JM, Visser J, Calverley-Range EA, Koster EH (1987) Inclined heterolithic stratification—terminology, description, interpretation and significance. Sediment Geol 53:123–179

Uncles RJ, Stephens JA (2010) Turbidity and sediment transport in a muddy sub-estuary. Estuar Coast Shelf Sci 87:213–224

Uncles RJ, Stephens JA, Harris C (2006) Runoff and tidal influences on the estuarine turbidity maximum of a turbid system: the upper Humber and Ouse estuary, UK. Mar Geol 235: 213–228

Van den Berg JH (1981) Rhythmic seasonal layering in a mesotidal channel fill sequence, Oosterschelde Mouth, the

Netherland. In: Nio S-D, Shuttenhelm RTE, van Weering TjCE (eds) Holocene marine sedimentation in the North Sea Basin. International Association of Sedimentologists special publications 5. Blackwell, Oxford, pp 147–159

Van den Berg JH, Boersma JR, Van Gelder A (2007) Diagnostic sedimentary structures of the fluvial–tidal transition zone. Evidence from deposits of the Rhine Delta. Neth J Geosci 86:253–272

Van der Wal D, Pye K, Neal A (2002) Long-term morphological change in the Ribble estuary, northwest England. Mar Geol 189:249–266

van Proosdij D, Baker G (2007) Intertidal morphodynamics of the Avon River estuary. Final report submitted to Nova Scotia Department of Transportation and Public Works, 186 p. Available at http://www.gov.ns.ca/tran/highways/Hwy101twinningWindsor.asp

Visser MJ (1980) Neap-spring cycles reflected in Holocene subtidal large-scale bedform deposits: a preliminary note. Geology 8:543–546

Wang ZB, Jeuken MCJL, Gerritsen H, de Vriend HJ, Kornman BA (2002) Morphology and asymmetry of the vertical tide in the Westerschelde estuary. Cont Shelf Res 22: 2599–2609

Wolanski E, King B, Galloway D (1995) Dynamics of the turbidity maximum in the Fly River estuary, Papua New Guinea. Estuar Coast Shelf Sci 40:321–337

Wolanski E, Williams D, Hanert E (2006) The sediment trapping efficiency of the macro-tidal Daly estuary, tropical Australia. Estuar Coast Shelf Sci 69:291–298

Woodroffe CD, Chappell JMA, Thom BG, Wallensky E (1989) Depositional model of a macrotidal estuary and flood plain, South Alligator River, Northern Australia. Sedimentology 36:737–756

Wright LD, Coleman JM, Thom BG (1973) Processes of channel development in a high-tide-range environment: Cambridge Gulf-Ord River delta, western Australia. J Geol 81:15–41

Xie D, Wang Z, DeVriend HJ (2009) Modeling the tidal channel morphodynamics in a macro-tidal embayment, Hangzhou Bay, China. Cont Shelf Res 29:1757–1767

Yang BC, Dalrymple RW, Chun SS (2005) Sedimentation on a wave-dominated, open-coast tidal flat, southwestern Korea: summer tidal flat – winter shoreface. Sedimentology 52:235–252

Yang BC, Dalrymple RW, Gingras MK, Chun SS, Lee HJ (2007) Up-estuary variation of sedimentary facies and ichnocoenoses in an open-mouthed, macrotidal, mixed-energy estuary, Gomso Bay, Korea. J Sediment Res 77:757–771

Zaitlin BA (1987) Sedimentology of the Cobequid Bay–Salmon River estuary, Bay of Fundy, Canada. Unpublished Ph.D. thesis, Queen's University, Kingston, Ontario, 391 p

Zhang G, Li C (1996) The fills and stratigraphic sequences in the Qiantangjiang incised paleo-valley, China. J Sed Res 66:406–414

Stratigraphy of Tide-Dominated Estuaries

6

Bernadette Tessier

Abstract

Tide-dominated estuaries have received less attention than wave-dominated estuaries due mainly to the fact that they are less common coastal systems. Consequently, the data available on the sedimentary infill of tide-dominated estuaries are limited. The present chapter describes several modern (Holocene) examples for which seismic, sediment core and 14C age data are available, allowing reconstruction of sediment fills. Some ancient examples are also given. The distribution and preservation of some key features such as systems tracts and ravinement surfaces are discussed in light of the different examples, as well as the controlling factors of infilling. Only a few features and factors can finally be assigned specifically to tide-dominated estuary infills. However, two points must be emphasized: (1) wave-built bodies are common features preserved within tide-dominated estuary infills; and (2) the potential for preservation of estuarine sedimentary bodies is primarily controlled by tidal accommodation, defined by the depth of the main tidal channel belt.

6.1 Introduction

Estuaries are usually defined as the seaward portion of a drowned incised valley (Dalrymple et al. 1992). Interest in the study of incised-valley systems increased tremendously in the 1990s both because they represent key objects for a better understanding of sequence stratigraphy of marine-to-continental successions, and they potentially constitute good clastic oil reservoirs (Dalrymple et al. 1994; Zaitlin et al. 1994). Additionally, since estuaries are known as ephemeral coastal systems, sensitive to sea-level and climate fluctuations (Masselink and Hughes 2003), many projects have been devoted over the last decades to study their sedimentary infilling for defining the forcing factors of their evolution. As a consequence, a huge amount of data has been published on the stratigraphy of estuarine fills, both from modern and ancient examples. Literature on tide-dominated or tide-influenced sedimentary coastal systems, and more generally on estuaries, is particularly vast. The most common classification of estuaries used by sedimentologists is that defined by Dalrymple et al. (1992), and slightly revised recently (Dalrymple 2006). According to the prevailing hydrodynamics at the mouth of the estuary, waves or tidal currents, two

B. Tessier (✉)
Morphodynamique Continentale et Côtière, University of Caen, UMR CNRS 6143, 24 rue des Tilleuls, 14000 Caen, France
e-mail: bernadette.tessier@unicaen.fr

R.A. Davis, Jr. and R.W. Dalrymple (eds.), *Principles of Tidal Sedimentology*,
DOI 10.1007/978-94-007-0123-6_6, © Springer Science+Business Media B.V. 2012

end-members are distinguished, wave-dominated estuaries and tide-dominated estuaries. Surprisingly, few papers have been published on the stratigraphy of tide-dominated estuaries in spite of the huge literature dealing both with incised-valley infilling and tidal environments.

The aim of the present chapter is undertake a synthetic overview of the stratigraphy of the sedimentary infilling of tide-dominated estuaries. After setting out the main elements that typify the morphosedimentary organization of tide-dominated estuaries, the main body of the chapter is based on the description of several modern examples where available data on the sedimentary infill are published. These descriptions allow comparison of tide-dominated estuaries located in various contexts, especially in terms of sediment supply. This includes recently published works on tide-dominated estuaries located along the French coasts of the English Channel and Atlantic (Seine, Mont St Michel and Vilaine estuaries). The probably best-known example of a tide-dominated estuary, the Cobequid Bay–Salmon River estuary, is also described, as well as the South Alligator River estuary. All these estuaries are characterized by low sediment supply. The paper also includes the Gironde estuary, although the latter is a mixed wave- and tide-dominated estuary according to the classification of Dalrymple et al. (1992). Descriptions of the Holocene Yangtze estuary and delta, and of the Qiantang River estuary are provided as examples of systems located in contexts of high sediment discharges. Finally, four ancient examples (Pleistocene, Eocene, Cretaceous) are also described. At the light of these different examples, the factors that control the infill of tide-dominated estuaries, such as sea-level fluctuations, sediment supply, bedrock morphology, and climate changes, are discussed, and some criteria for recognition of such estuaries in the rock record are proposed, especially regarding distinction with wave-dominated systems.

Other chapters in the present book provide definitions and descriptions that should be used as additional information for this chapter. In particular, for information on sedimentary dynamics, morphological evolution and facies, refer to Chap. 5, and on tidal shelf bodies (tidal banks, tidal bars) comparable to those present in the outermost entrance of tide-dominated estuaries, refer to Chap. 13.

6.2 Tide- vs. Wave-Dominated Estuaries: A Few Reminders

According to the definition of Dalrymple et al. (1992), tide-dominated estuaries refer to estuaries, the sediment dynamics of which are dominated by tidal currents at the mouth. By contrast, sediment transport and deposition at the mouth of wave-dominated estuaries is predominantly due to wave action. As a result, the main morphosedimentary component of wave-dominated estuaries consists in wave-built coarse-grained coastal barrier scoured by a tidal inlet of variable width and depth. Sheltered from high-energy marine dynamics by the sand-dominated mouth body, fine-grained deposits, mostly originating from fluvial sources, aggrade in a central basin, while coarser fluvial sediments concentrate at the head to the estuary, forming a prograding bay-head delta (cf. The Gironde estuary example in Fig. 6.3). Many examples of wave-dominated estuaries, both in modern and from ancient deposits, have already been described around the world, with pioneering work along the eastern coast of the USA and in Australia (cf. Chaps. 10 and 12). The morphosedimentary organization in tide-dominated estuaries differs quite significantly from the typical tripartite sandy mouth/clayey central basin/sandy bay-head delta distribution that typifies wave-dominated estuaries. In the ideal case of a tide-dominated estuary with a well-defined funnel shape and a hypersynchronous mode of tidal wave propagation, the morphosedimentary distribution consists in longitudinal tidal bars at the mouth, followed landward by a sandy tidal channel-and-bars complex. This braided system that corresponds to the area of highest tidal energy evolves to a single tidal channel that is transitional with the fluvial one (Fig. 6.1, cf. Fig. 6.3). An important feature of this single channel is the sinuous to meandering shape that it develops in the bedload convergence zone (BLZ on Fig. 6.3) between landward flood-dominated and seaward fluvial-dominated net transports (for more details, refer to Dalrymple et al. in this volume).

According to these sea-to-land distributions of sedimentary bodies and facies, conceptual stratigraphic models for sedimentary infilling of wave-dominated and tide-dominated estuaries have been proposed (Dalrymple et al. 1992, cf. Boyd (2010) and Dalrymple 2010 for slightly modified models). The two end-member models are not drastically different as both

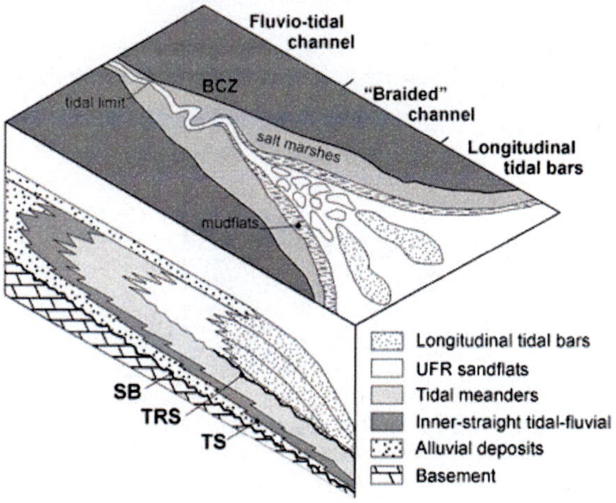

Fig. 6.1 A tide-dominated estuary: ideal distribution of sedimentary bodies and facies both in plan view and section. *BCZ* bedload convergence zone, *UFR* upper flow regime, *SB* sequence boundary, *TS* transgressive surface, *TRS* tidal ravinement surface (After Zaitlin et al. 1994, Emery and Myers 1999)

illustrate a single transgressive–regressive infilling cycle, with landward and then seaward shift of facies and sedimentary bodies (see Fig. 6.1 for the tide-dominated estuary model). The model for wave-dominated estuaries, because of its distinct tripartite character, appears to be more easily applicable. Independent to the fact that wave-dominated estuaries are more abundant around the world, this explains why many examples of estuarine infillings have been described using the wave-dominated estuaries model.

6.3 Stratigraphy of Tide-Dominated Estuary Infill: Case Studies

As previously introduced, only a few examples of stratigraphic studies describing the sedimentary infill of tide-dominated estuaries are available compared with wave-dominated estuaries. Surprisingly, one of the most commonly cited examples is the Gironde estuary, although the latter is defined as a mixed wave- and tide-dominated estuary. The Gironde estuary is macrotidal, but according to Dalrymple et al. (1992), it is not a tide-dominated estuary, as powerful oceanic swells largely control the morphodynamic behaviour of the mouth. Published cases of tide-dominated estuaries, as defined by Dalrymple et al. (1992), and for which data including sedimentology, architecture and dating of the infilling are available, remain rare (cf. next section). The Cobequid Bay–Salmon River estuary (Bay of Fundy, Canada, cf. Fig. 6.3) is undoubtedly the best-known example, as this tide-dominated estuary is the basis of the morphosedimentary model of Dalrymple et al. (1992). Almost no data are available for instance on the infill stratigraphy of the Thames or Severn tide-dominated estuaries (Harris 1988). The aim of this present section is thus to briefly describe the sedimentary infilling of the few main examples of tide-dominated estuaries that have been published. The list includes mostly modern (Holocene) estuaries (the Cobequid Bay–Salmon River; the Seine and Mont St Michel, the Vilaine, France; the South Alligator, Australia; the Yangtze and Qiantang, China) but also some ancient cases (Pleistocene, Eocene, Cretaceous). The tide-dominated estuary examples are compared to the mixed-energy Gironde estuary.

6.3.1 Progress in the Assessment of Estuary Stratigraphy: The Use of Very High-Resolution Seismic Data

The lack of data and reconstruction studies on the sedimentary infill of tide-dominated estuaries partly results from the difficulties to investigate such shallow water, and sometimes dangerous (because of powerful tidal currents) coastal settings. Most studies performed in tide-dominated estuaries as well as wave-dominated estuaries are based on sediment vibracores that are relatively easy to collect, at least on the estuary rims at low tide, but which provide only a partial knowledge of the whole infill. In the 1990s and 2000s, the development of very high-resolution seismic devices, more adapted for coastal studies, allows the collection of new data on coastal sediment wedge architecture. In particular, boomer sources, the vertical resolution of which (<0.5 m) is convenient to image modern sedimentary bodies, were designed to be used on small boats. As an example, along the coasts of France, characterized by numerous estuaries and lagoons, a huge amount of very high-resolution seismic data have been collected since the beginning of the 2000s, providing new advances in our understanding of incised-valley infill in different geomorphological and hydrodynamical contexts (Chaumillon et al. 2010). Over the last 10 years, a considerable effort has been made in carrying out integrative studies combining seismic, core and

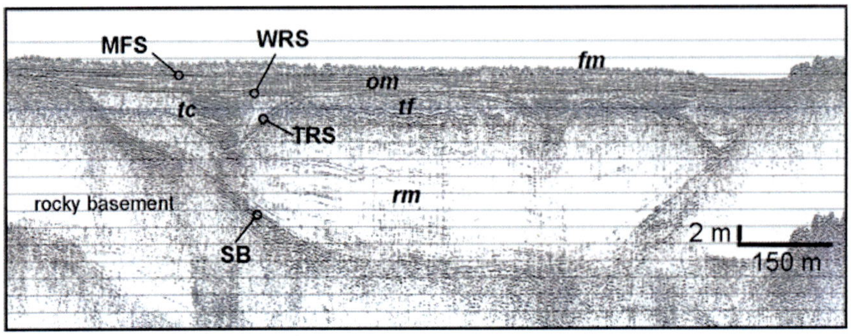

Fig. 6.2 Very high-resolution seismic profile (boomer IKB-Seistec, UMR CNRS M2C/University of Caen) shot in the outer tide-dominated rocky coast estuary of the Vilaine (Southern Brittany, NW France). VHR seismic data provide detailed images of incised-valley infill and, combined with core data and radiocarbon dating, have contributed to significantly improve our knowledge of estuary stratigraphy. *SB* sequence boundary, *TRS* tidal ravinement surface, *WRS* wave ravinement surface, *MFS* maximum flooding surface, *rm* ria mud, *tc* and *tf* tidal channels and tidal flats, *om* offshore muds, *fm* fluvial muds (Modified after Menier et al. 2010)

radiocarbon data, increasing significantly our knowledge of coastal stratigraphy. The seismic profile shown in Fig. 6.2 illustrates how very high-resolution seismic data allow detailed imaging of the different depositional units and surfaces that partly characterize an estuary infill. However, such accurate very high-resolution seismic data are not yet available in many tide-dominated estuaries, not to mention that in many cases, biogenic gas that is produced in the infilling sediment, frequently composed of organic-rich deposits, prevents the acquisition of good-quality seismic images. At last, seismic data should be ground-truthed by core data, and collecting good-quality long cores in soft sediments in subtidal zones still remains a challenging technical objective.

6.3.2 Modern Estuaries with Low River Sediment Supply

6.3.2.1 Cobequid Bay–Salmon River Estuary, Bay of Fundy

The infill of the Cobequid Bay–Salmon River estuary was investigated by the end of the 1980s on the basis of high-resolution seismic data (sparker source) and cores. Tidal range is up to 16 m, and sediment supply is mainly derived from a marine source, at least since the mid- to late Holocene. As in most other modern estuaries around the world, sediment infill is related to the last post-glacial sea-level evolution. Due to the glacio-isostatic rebound, the post-glacial sea level reached a maximum (+15 m/present-day zero) 14,000 years ago, followed by a minimum (−10–15 m/present-day zero) about 7,500 years ago, before rising to the present-day level. As a result, the Cobequid Bay–Salmon River estuary comprises a compound infill composed of a Pleistocene unit and a Holocene unit (Fig. 6.4).

The Pleistocene unit is principally composed of glacio-fluvial to glacio-marine deposits that are not described here. The Holocene succession is divided into two stages that coincide with the early to mid-Holocene lowstand to early transgression (9,000–5,000 years BP) and the subsequent mid- to late Holocene transgression (5,000 years BP – present). Based on numerical modeling and the geometrical aspects of sedimentary bodies, the succession that fills the Cobequid Bay–Salmon River valley is interpreted to have accumulated in a micro- to mesotidal wave-dominated estuary during the lowstand – early transgression stage, comprising spit barrier, tidal inlet, washover, flood delta, central basin deposits and bay-head delta deposits. A drastic increase in tidal range occurred during the second stage, probably in relation with the destruction of the barrier at the wave-dominated estuary mouth (Shaw et al. 2010). The Cobequid Bay–Salmon River estuary then evolved as a tide-dominated estuary. The associated infilling is marked at the base by an extensive erosional surface underlying tidal facies made up of axial sands and fringing mudflats and marshes. Tidal sand is the thickest above the present-day longitudinal bars and contains superimposed sets of gently inclined stratification. This tide-dominated estuary unit, overlying the wave-dominated estuary lowstand succession, is interpreted as the transgressive systems tract, bounded at the base by the transgressive surface amalgamated with the tidal ravinement surface. Data indicate that throughout most of the estuary, this second stage of infilling is an aggradational unit. It is suggested that the estuary has entered in a progradational phase within the last 200 years only. Rapidly accreting tidal flat successions containing tidal rhythmites that were emplaced along the margin of the estuary are assigned to this highstand systems tract. Prograding tidal rhythmites lie above vertically aggrading marsh deposits through a tidal ravinement surface that is therefore responsible for removing landward facies during the transgression (Dalrymple and Zaitlin 1994).

6.3.2.2 South Alligator Estuary, van Diemen Gulf, North Australia

The work of Woodroffe et al. (1989) on the South Alligator estuary is certainly the pioneer study regarding the infilling stratigraphy of a tide-dominated estuary. Although tidal range is 'only' 6 m at the entrance, the South Alligator estuary comprises all characteristics of a tide-dominated estuary (Fig. 6.3). Like in the Cobequid Bay–Salmon River estuary, the fluvial sediment supply is low. The main sediment source is of marine origin, and dominated by fine-grained sediment. The reconstruction of the infill is based on drilling data. Related to the Holocene sea level that rose quickly until about 6,000 years BP and then remained stable until present, three main phases of infilling are distinguished (Fig. 6.4). The transgressive phase (8,000–6,800 years BP) corresponds to the marine incursion into the funnel-shaped valley. Mangroves developed on the margins, whereas marine and estuarine sands and muds are found along the valley axis. The second phase (6,800–5,300 years BP) is marked by the expansion of mangrove forests throughout the estuary, followed by the third phase (after 5,300 years BP) during which a sinuous-to-cuspate tidal channel floodplain developed. As inferred by radiocarbon dating, shoreline almost attained its present-day position some 3,000 years ago, indicating that floodplain aggradation has been very slow during the late Holocene. Although the architectural reconstruction for this tide-dominated estuary is only partial, one can expect that the transgressive surface and tidal ravinement surface are amalgamated together and located below the marine-to-estuarine sands that compose the transgressive systems tract.

More recently, Bostock et al. (2007) documented the Holocene infill of the tide-dominated Fitzroy River estuary in Southeast Queensland, Australia. The general infilling stratigraphy is very comparable with that of the South Alligator estuary, three main phases being distinguished as well during the Holocene evolution: (1) post-glacial inundation, (2) mangrove development, (3) floodplain aggradation and estuarine mouth progradation. As in the South Alligator estuary, sediments in the Fitzroy River estuary are dominantly fine-grained, but contrary to the South Alligator estuary, they are mainly of fluvial origin. No sedimentary unit is assigned to the transgressive systems tract, the whole infill succession being related to the post-7,000 years BP sea-level highstand. According to these data, the Fitzroy River estuary should rather be defined as a tide-dominated delta.

6.3.2.3 Gironde Estuary, Central Bay of Biscay, SW France

Most works on tide-dominated estuaries refer to the Gironde estuary case at least as a comparative example. Moreover, the stratigraphic model established by Allen (1991) and Allen and Posamentier (1993) provided elements used by Zaitlin et al. (1994) when constructing their model. Spring tidal range in the Gironde is macrotidal (5.5 m), but the mouth is affected by very high-energy Atlantic waves, reaching up 8 m during winter storms. As a consequence, the Gironde entrance is under the influence of both strong tidal currents that incise a deep tidal inlet and construct well-developed tidal deltas, and powerful wave-induced dynamics that construct spit barriers and rework the ebb delta sand.

A large amount of sand and mud is delivered to the estuary by the Garonne and Dordogne Rivers, but only 25% of this fluvial sediment is exported seaward, the rest remaining in the estuary, feeding the bay-head delta and central basin (Fig. 6.3). Hence, most sands at the mouth are marine sourced. The stratigraphic reconstruction of the Gironde estuary infill is based on a compilation of core and well log data. The infill above the bedrock is related to the last post-glacial transgression. Holocene sea-level rise was quick until 4,000 years BP, and present-day sea level is inferred to have been almost attained at that time. Above late Pleistocene fluvial deposits concentrated into the talweg of the incised valley, the bulk of the infill is made of a

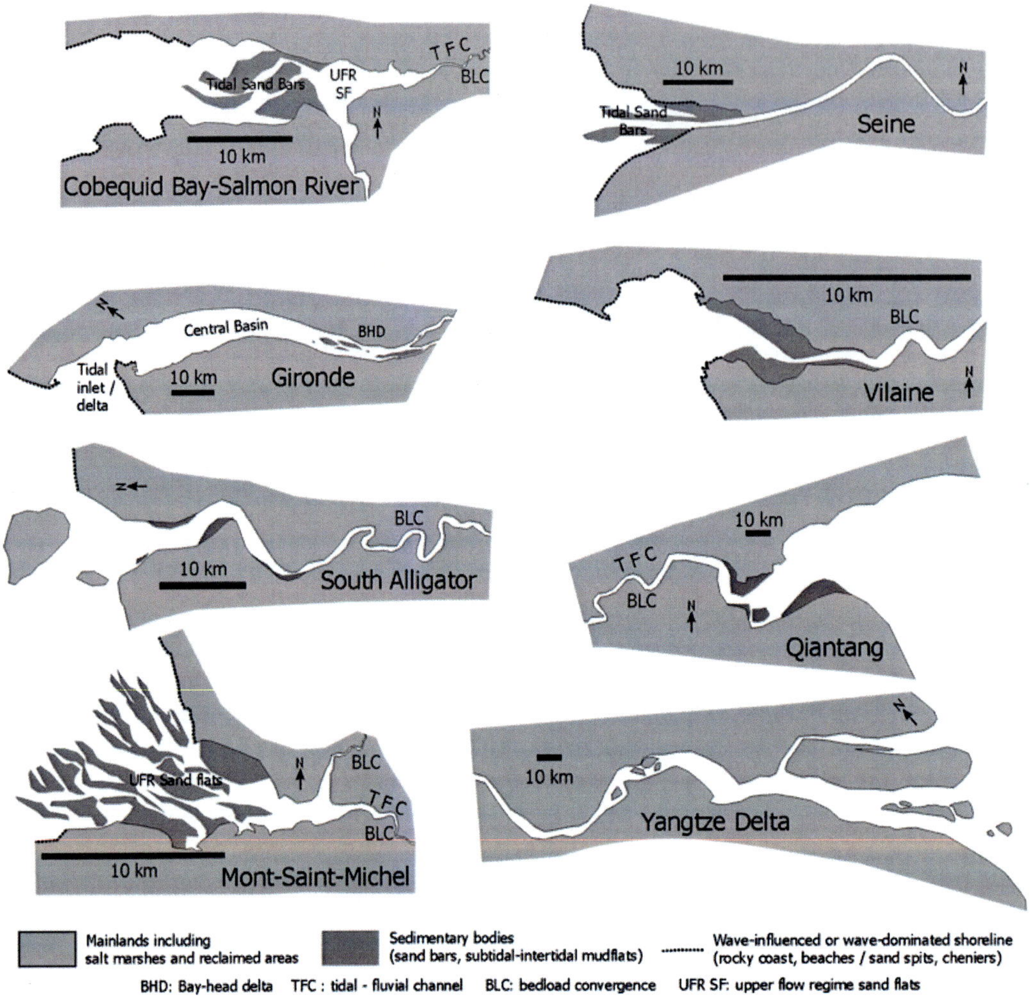

Fig. 6.3 Schematic plan view of the modern tide-dominated estuaries described in this chapter (drawn after Google Earth and Landsat images). All sedimentary bodies have not been represented. The Gironde estuary and the Yangtze delta are drawn to show the main morphological differences respectively with wave-dominated estuaries and tide-dominated deltas. Be careful with scales (*all bar* scales represent 10 km)

landward thinning wedge of aggrading tidal estuarine sands and muds. At the seaward end of the estuary, the tidal–inlet complex sands sharply overlay the aggrading estuarine facies through the tidal ravinement surface. Seaward, in response to high-energy wave reworking, the shoreline retreat creates a wave ravinement surface, eroding the estuarine mouth sand.

Since 4,000 years BP, the tide-dominated bay-head delta has begun to prograde as sea level ceased to rise. This results in the development of huge tidal bars fed by fluvial sands and that migrate seaward. Simultaneously, at the seaward end, the wave-induced coastal retreat continues, meaning that at present, the highstand systems tract develops landward, whereas the transgressive systems tract is still in construction seaward (Fig. 6.4).

The Holocene sedimentary infill of the Charente estuary–Marennes-Oléron Bay, classified like the Gironde as a mixed-energy wave- and tide-dominated system, and located about 50 km north the Gironde, has been investigated recently using very high-resolution seismic and core data (Chaumillon and Weber 2006; Weber et al. 2004; Allard et al. 2010, cf. review in Chaumillon et al. 2008). This example of a mixed energy system differs quite significantly from the Gironde model because it is characterized by very low sediment supply of fluvial origin. As a consequence, no bay-head delta is developed. In spite of this context

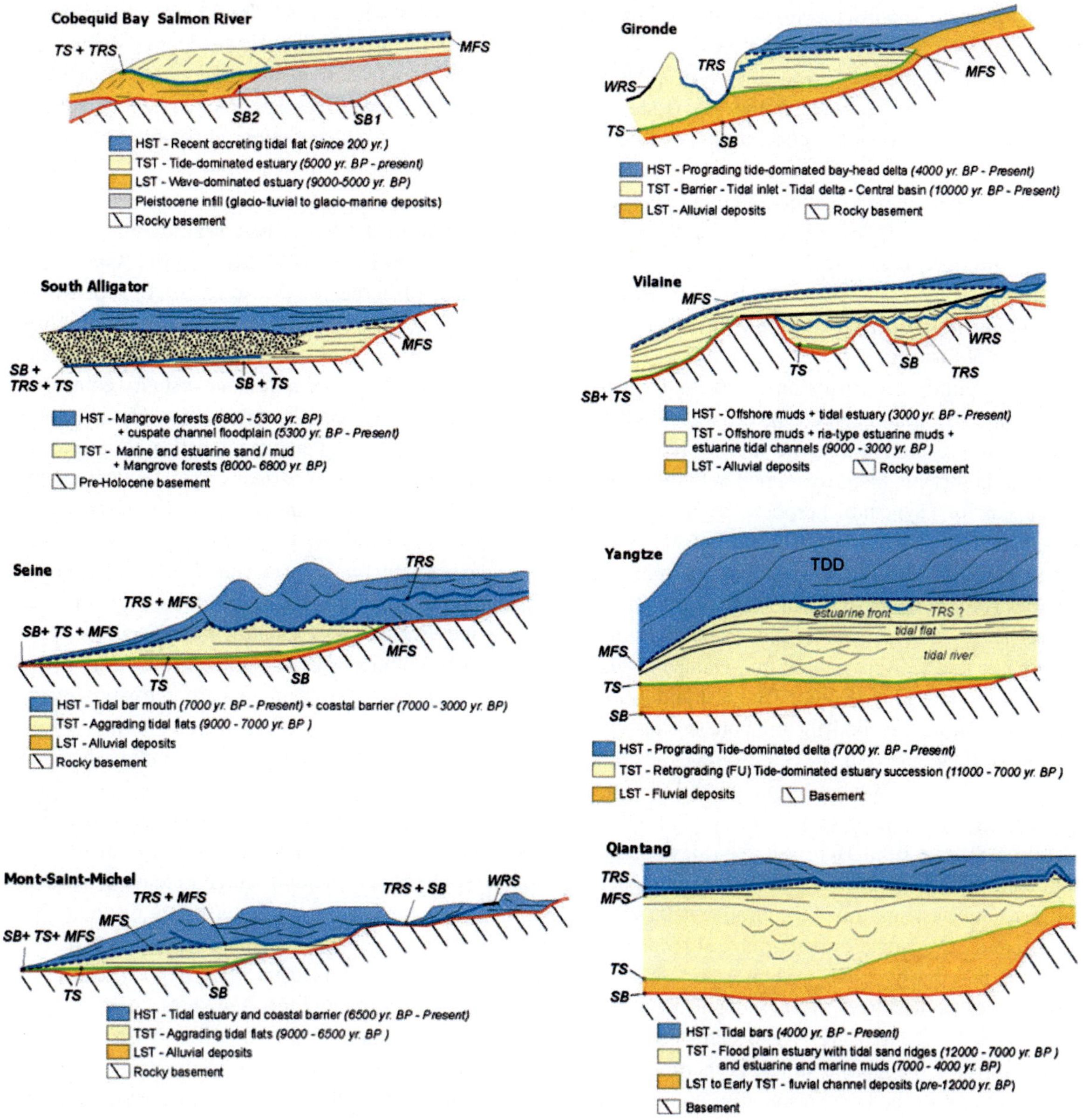

Fig. 6.4 Stratigraphic organization of the sedimentary infill of the main modern (Holocene) tide-dominated estuaries, including as well the mixed-energy estuary of the Gironde, and the tide-dominated delta of the Yangtze (a tide-dominated estuary during the transgressive systems tract deposition). The Cobequid Bay–Salmon River estuary (Bay of Fundy, Canada) after Dalrymple and Zaitlin (1994); the South Alligator estuary (van Diemen Gulf, North Australia) after Woodroffe et al. (1989, 1993); the Gironde estuary (Central Bay of Biscay, SW France) after Allen (1991), Allen and Posamentier (1993); the Seine estuary (Bay of Seine, NW France) after Tessier et al. (2010a); the Mont-Saint-Michel estuary (Norman-Breton Gulf, NW France) after Tessier et al. (2010b); the Vilaine estuary (Northern Bay of Biscay, NW France) after Menier et al. (2010); the early Holocene Yangtze estuary (East China Sea, China) after Hori et al. (2001); the Qiantang River estuary (Hangzhou Bay, China) after Lin et al. (2005). *SB* sequence boundary, *TS* transgressive surface, *TRS* Tidal ravinement surface, *WRS* wave ravinement surface, *MFS* maximum flodding surface, *LST*, *TST*, *HST*-lowstand, transgressive, highstand systems tracts, *TDD* tide-dominated delta

of low sediment supply, the highstand systems tract started to develop in the Marennes-Oléron Bay as soon as the rate of Holocene sea-level rise dropped, i.e. around 6,000 years BP, because the bay is sheltered from marine erosion by structural highs. In the outer segment of the system, the transgressive systems tract constitutes most of the Holocene infill, as in the Gironde, and is locally deeply scoured due to the action of tidal currents and waves.

6.3.2.4 Seine Estuary, Bay of Seine, NW France

The stratigraphic reconstruction of the outer estuary of the Seine River has been conducted recently thanks to very high-resolution seismic and vibracore data. Although highly modified by human activities mainly carried out for navigation purpose, the mouth of the Seine estuary displays a well-shaped funnel with two prominent longitudinal bars, typical of tide-dominated estuaries (Fig. 6.3). Spring tidal range at the entrance approaches 8 m. Sediment originates both from the Seine River that delivers mainly fine-grained sediment and from the sea. Fluvial muds feed a turbidity maximum that can be expelled far offshore during severe river floods (Garnaud et al. 2003; Lesourd et al. 2003).

The sedimentary infill is simple and related to the last post-glacial transgression that was rapid until 7,000–6,500 years BP (−10 m) and then slow until now (Fig. 6.4). Above late Pleistocene terraces assigned to the lowstand systems tract, the infill comprises two main units. Aggrading organic-rich estuarine facies compose the transgressive systems tract. No high-energy tidal bodies have been recognized. Around 7,000 years BP, the transgressive systems tract, above the main axis of the incised valley, is sharply overlain by a massive sand body corresponding to the estuary mouth tidal bar complex. The erosive basal limit of this sand unit amalgamates the transgressive surface and the tidal ravinement surface.

On the shallower edges of the valley, where tidal currents are not so powerful as compared with the estuary axis, wave-built sandy barrier and back-barrier tidal flats were emplaced. The latter together with the tidal sands belong to the highstand systems tract. At about 3,000 years BP, the barrier/back-barrier unit is in turn eroded by a tidal channel belt, indicating the expanding of the tidal system throughout the estuarine funnel. The destruction of the barrier 3,000 years ago is believed to be related to a period of enhanced storminess (cf. discussion in next section). The combination between the natural highstand infilling processes and the heavy human activities in the estuarine system, including the Seine River catchment, results in a rapid seaward shift, i.e. progradation, of the tidal bars since at least the two last centuries.

6.3.2.5 The Mont-Saint-Michel Estuary, Norman-Breton Gulf, NW France

In the eastern corner of the so-called Bay of Mont-Saint-Michel, where the tidal range reaches up 15 m during high spring tides, an estuary forms at the mouth of three rivers. This tide-dominated estuary comprises an extensive tidal channel-and-bar complex evolving landward to a single tidal channel that passes gradually into a meandering bedload convergence zone occurs here (Fig. 6.3). No longitudinal tidal bars are present at the seaward end of the estuary, probably because of the restricted length of the estuarine funnel that widens rapidly on the open sea. Hence, sediment dynamics in this wide entrance is probably influenced by wave action. The latter is very significant on the northern margin of the estuary where a locally retreating wave-built barrier is developed (Figs. 6.3 and 6.4).

The Bay of Mont-Saint-Michel including the estuary is a mixed carbonate/siliciclastic environment, and sediment is almost exclusively of marine origin. The sedimentary infill, reconstructed on the basis of very high-resolution seismic and vibracore data, is related, as for the Seine estuary, to the last post-glacial transgression that slowed down at about 6,500 years BP. The Holocene infill is divided into two depositional units resembling the Seine infill (Fig. 6.4). Above some probable remnants of Pleistocene fluvial terraces assumed to form the lowstand systems tract, an aggradational unit made of organic-rich fine-grained facies fills the bottom of the valley. Near the estuarine mouth, this unit passes to high energy-migrating tidal dunes and banks, or to shoreface sands. These deposits compose the transgressive systems tract.

In the estuarine axis, the transgressive systems tract is overlain by a sandy tidal channel belt through an erosive surface corresponding to the tidal ravinement surface, and dated at 6,500 years BP. This high-energy estuarine unit, constituting the highstand systems tract, pinches out progressively landward and seaward. In the single tidal channel zone, the bottom of the present-day channel reaches bedrock, which is higher. Hence, the highstand systems tract in this area is composed of a single tidal channel succession, the upper part of which is composed of tidal rhythmites (Tessier 1993; Billeaud et al. 2007). On the sides of the estuarine mouth, the highstand systems tract corresponds either to an aggradational unit of tidal flat deposits or to retrograding/aggrading back-barrier facies. A wave ravinement surface signals the retrogradation of the present-day barrier (Fig. 6.4). Seismic data clearly show bottomset geometry at the seaward end of the tidal estuarine sands that prograde therefore over the transgressive systems tract.

6.3.2.6 Vilaine Estuary, Northern Bay of Biscay, NW France

The Vilaine estuary, situated along a rocky coast, is a ria-type estuary, characterized by a very low sand supply of marine origin. As a consequence, it does not develop sand bodies such as longitudinal tidal bars and sand flats at the mouth (Fig. 6.3). The Vilaine estuary belongs to the category of rocky-coast estuaries defined in Chaumillon et al. (2008). Maximum tidal range is about 4.5 m, and wave and fluvial dynamics are of low energy. Hence, tidal currents control sediment processes. The estuary is mud-dominated and composed of one main tidal channel with fringing tidal mudflats and salt marshes.

The sediment infill, reconstructed with core and very high-resolution seismic data, is simple and related to the Holocene transgression. It consists of four main depositional units above the sequence boundary (cf. Figs. 6.2 and 6.4). The basal unit has the highest volume and is an aggradational unit made of fine-grained organic-rich facies interpreted as muds deposited in a well-sheltered estuarine environment such as a ria. This unit is sharply eroded by a channelized surface overlain by sandy to muddy facies interpreted as estuarine tidal channels and tidal flats. The channelized surface is the tidal ravinement surface and is dated at about 6,500 years BP.

This unit is in turn sharply eroded by a flat surface covered by offshore muds. This flat surface is the wave ravinement surface and is dated around 5,500–4,500 years BP. Ages of 3,000 years BP are found at the top of the offshore mud unit, indicating that sedimentation rate has been extremely low during the last stages of the Holocene. Hence, in this tide-dominated estuary, characterized by very low sediment supply, the sedimentary infill records a continuously retrograding bayline. The maximum flooding is placed in the uppermost part of the infill. In the most internal zones only, a last unit, composed of muds supplied by the fluvial source, can be assigned to the highstand systems tract.

6.3.3 Modern Estuaries with High River Sediment Supply

All of the aforementioned modern case studies described estuaries with a low to insignificant river sediment supply. Systems characterized by high to very high fluvial sediment input are mentioned here. Most of these systems are tide-dominated deltas, such as the Fly River delta (Dalrymple et al. 2003), but some of them are described as tide-dominated estuaries during the deposition of the transgressive systems tract. The Yangtze delta illustrates this case. The Qiantang River estuary located immediately south the Yangtze is described for comparison.

6.3.3.1 Early Holocene Yangtze Estuary, East China Sea, China

The Yangtze River is among the largest rivers in the world, ranking fourth in terms of sediment discharge. Present-day maximum tidal range reaches 4.5 m, and maximum wave height at the mouth is 6.5 m. The present-day Yangtze River mouth is thus defined as a mixed-energy environment. The stratigraphy of the Yangtze incised-valley infill, reconstructed using enormous numbers of cores and well logs, comprises two main depositional stages related to the last post-glacial sea-level rise that almost stopped at about 7,000 years BP (Fig. 6.4). The transgressive systems tract is assumed to comprise a succession of units that typify a transgressed tide-dominated estuary. The main difference with tide-dominated estuaries as defined by Dalrymple et al. (1992) is that this succession is fining upward because sediment is mainly supplied by the Yangtze River. In this transgressive tide-dominated estuary succession, longitudinal tidal bars and tidal channel and shoal complexes do not exist. The whole estuary is occupied by a tidal distributary channel complex passing seaward to muddy flats and a mud-dominated estuarine front. Tidal rhythmites of different types are very well preserved in almost all facies, indicating the tide-dominated character of the setting. Finally, this succession resembles that of a mixed-energy estuary such as the present-day Gironde estuary, with a tide-dominated bay-head delta and central basin fed by river supply, but without the high-energy sand-dominated seaward end-member. As soon as the transgression stopped at 7,000 years BP, the system evolved into a tide-dominated delta whose progradational succession forms the highstand systems tract.

6.3.3.2 Qiantang River Estuary, Hangzhou Bay, China

This last modern case study is similar in many aspects to the Yangtze case but the present-day Qiantang estuary is defined as a tide-dominated estuary, not a tide-dominated delta. Tidal range reaches up 9 m, twice the range in the Yangtze. Sediment is supplied by the river but in much less quantity than in the Yangtze delta. Moreover it is suggested that a considerable part of fine-grained sediment is also sourced by longshore drift from the Yangtze River located about 100 km to the north.

The sediment infill is simple and related to the last post-glacial transgression. Sea-level rise was very rapid until 12,000 years BP (35 mm/year), and slower until 7,500 years BP (10 mm/year). Till 4,000 years BP, sea level rose with a rate of 3 mm/year and became stable after this. Four depositional units have been distinguished in the infill thanks to numerous cores (Fig. 6.4). Above a fluvial channel unit deposited during base level rise until 12,000 years BP, a retrograding/aggrading clay- to silt-dominated estuarine unit accumulated. This unit contains tidal sand ridges. Tidal facies are common, including tidal rhythmites. At 7,500 years BP, the maximum rate of sea-level rise was reached and estuarine to marine muds blanketed the whole estuary. Finally, at 4,000 years BP, when sea level stabilized, tidal bars made of marine silts to fine sands developed, prograding over the estuarine muds. The tidal ravinement surface is believed to be located below the upper unit.

6.3.4 Ancient Estuaries

Examples of successions assigned to tide-dominated estuaries in the rock record remain relatively rare. Many sedimentological and sequence stratigraphic analyses refer to tidal facies and deposition in estuarine environments, but, as previously mentioned, most interpretations finally refer to the wave-dominated estuary model or to the mixed-energy wave- and tide-dominated estuary model based on the Gironde. Few examples of ancient sediment successions are described and interpreted explicitly to be the result of tide-dominated estuary infill. Some of them are briefly reported below.

6.3.4.1 Pleistocene Dong Nai River Succession, Vietnam

A very similar case to the Yangtze system (China) is reported by Kitazawa (2007) who described a Pleistocene succession outcropping along the Dong Nai River in Vietnam within the Quaternary Mekong Basin. The succession comprises two superimposed transgression–regression cycles related to Pleistocene sea-level fluctuations, from MIS8 (270 kyr. BP) and MIS5 (about 100 kyr. BP). The transgressive systems tract of each cycle is interpreted as a tide-dominated estuary succession, stacking fluvio-tidal aggradational deposits that are partially eroded laterally at the seaward edge by marine sand bodies lying on a tidal ravinement surface. The transgressive systems tract is overlain by a progradational deltaic succession forming the highstand systems tract.

6.3.4.2 Aspelintoppen Formation, Eocene Central Basin, Spitsbergen

This sedimentological and high-resolution sequence stratigraphy work conducted by Plink-Björklund (2005) is undoubtedly the most detailed study that has been published to date on tide-dominated estuary facies and successions recognized in the rock record. The Aspelintoppen Formation is a mud-prone, aggradational coastal plain formation that contains 18 stacked depositional sequences. Each sequence consists in a lowstand systems tract/transgressive systems tract/highstand systems tract succession interpreted to be the result of a tide-dominated estuary infill during a fourth-order cycle (400 kyr) of sea-level fluctuation. Outcrops permit each individual tide-dominated estuary in successive sequences to be followed from their upstream to their downstream end, and to assign each facies recognized in the field to a typical morphological component of a tide-dominated estuary, from the fluvio-tidal channel with the low- to high-sinuosity zones, through the upper-flow-regime tidal flats, into the tidal sand bars.

In an ideal sequence, the lowstand systems tract consists of fluvial deposits, overlying an erosional sequence boundary, that grade upward to aggradational fluvial facies influenced by tidal dynamics. This upper part of the lowstand systems tract could be assigned to the early transgressive systems tract. The early transgressive systems tract is capped by a transgressive surface, overlain by tide-dominated estuarine facies that form the transgressive systems tract.

The transgressive character of the succession is evidenced by facies superimposition that demonstrates a landward shift of the bay-line. At the seaward end of the profile, tidal sand bar facies, overlying a tidal ravinement surface, compose the upper part of the transgressive systems tract. The maximum flooding surface is usually located at the top of the bars. The highstand systems tract displays similar facies as in the transgressive systems tract, but is characterized by a seaward shift of the inner- and central-estuarine facies, and preferential preservation of root horizons, coal layers and marsh deposits. During the highstand systems tract, the environment is assumed to remain a tide-dominated estuary, and not to become a tide-dominated delta, since highly sinuous tidal channel facies are still present and demonstrate the existence of the bedload convergence zone in the inner estuary.

6.3.4.3 Chimney Rock Tongue, Upper Cretaceous, Campanian, the Flaming Gorge Area, Utah–Wyoming, USA

Plink-Björklund (2008) provided another very accurate description of tide-dominated estuarine successions in a formation from the Western Interior basin. The Chimney Rock Tongue comprises three distinct stratigraphic intervals: wave-dominated delta deposits, mixed-energy estuary deposits as an incised-valley fill and tide-dominated estuary deposits. The tide-dominated estuary succession, about 60-m thick, consists of three transgressive–regressive units. Each unit is composed of tide-influenced fluvial deposits, inner-estuary tidal-flat and marsh deposits, and outer-estuary upper-flow-regime tidal-flat and tidal-sand-bar deposits. Tidal ravinement surfaces, located at the base of tidal-sand-bar deposits, mark the base of each transgressive–regressive unit.

6.3.4.4 Cujupe Formation, Upper Cretaceous – Lower Tertiary São Luis Basin, N Brazil

This study reported by Rossetti (1998) documents an interesting case resembling the infill evolution of the Cobequid Bay–Salmon River estuary. Facies analyses and architectural reconstruction demonstrate that the Cujupe Formation is made of tide-influenced facies that infill an incised valley. The infill is divided into five stratigraphic intervals. Above the sequence boundary, the two basal-most intervals are interpreted as a lowstand systems tract to early transgressive systems tract. The facies succession is indicative of a wave-dominated estuary environment. Above these wave-dominated estuary deposits, a flooding surface is overlain by a third interval assigned to the transgressive systems tract and interpreted to be a tide-dominated estuary succession on the basis of the occurrence of facies related to tidal channels, tidal flat and sand shoal and estuary-mouth tidal bars. At the top, the highstand systems tract consists in two intervals similar to the tide-dominated estuary interval but with increasing fluvial influence, indicative of the progradational pattern of the infill. This ancient example thus reports an evolution from a wave-dominated estuary to a tide-dominated estuary, similar to what happened during the Holocene lowstand and subsequent transgression of the Cobequid Bay–Salmon River estuary.

Other studies, not detailed herein, document analyses of facies successions interpreted as tide-dominated estuary infill deposits such as for instance those of Khin and Myitta (1999, Miocene, Central Myanmar), Shanmugam et al. (2000, Cretaceous, Ecuador), Mellere (1994, Cretaceous, USA), Johnson and Levell (1995, Cretaceous, UK), Archer et al. (1994, Carboniferous, USA) and Pontén and Plink-Björklund (2009, Devonian, Baltic Basin).

6.4 Key Features of Tide-Dominated Estuary Successions

These brief descriptions of modern (Holocene) and of some ancient case allow the key points that characterize the stratigraphy of sediment fills of tide-dominated estuaries to be highlighted.

From a general point of view, most studies related to tide-dominated estuary sediment infill refer to basic sequence stratigraphy concepts allowing the distinction of systems tracts (lowstand, transgressive and highstand systems tracts) and key surfaces such as the sequence boundary, the transgressive surface, the tidal and wave ravinement surfaces and the maximum flooding surface (Figs. 6.4 and 6.5). Due mainly to the diversity of contexts, interpretations differ from one place to another. This clearly applies to the recognition and placement of the maximum flooding surface and the related distinction between the transgressive and highstand systems tracts. The differences in the type of data (cores and/or seismic), in data quality (coring or drilling, seismic resolution) and in location of the studied area with respect to the estuary system (outer to inner) account as well for the difference of interpretation.

The sequence boundary in most cases is defined as the bottom of a fluvial valley incised during the previous relative sea-level drop. Recently, Dalrymple (2006) proposed a revision of his original definition of an estuary in order to take into account that estuaries are transgressive coastal environments that do not form necessarily as the result of a river valley drowning. Plink-Björklund (2008) suggested that the successive tide-dominated estuarine intervals that compose the upper stratigraphic unit of the Chimney Rock Tongue occupied a drowned river mouth rather than an incised valley. There is no evidence for fluvial incision in this unit, and each tide-dominated depositional interval reflects an episode of tidal ravinement and reshaping of the river mouth. High rates of subsidence explain the total aggradation of the 60-m-thick tide-dominated estuarine succession. Shanmugam et al. (2000, Cretaceous, Ecuador) suggested as well that the tide-dominated estuary succession they studied does not fill an incised valley. However, the authors admitted that their available data and cross-sections may be insufficient to image an extra large incised valley.

The lowstand systems tract is usually assigned to alluvial deposits preserved in the bottom of the valley. In most cases, the lowstand systems tract is very reduced in volume, represented only by remnants of fluvial terraces reworked during the subsequent transgression by powerful tidal currents and/or waves. In only few cases does part of the lowstand systems tract consist of tide-influenced fluvial deposits (Plink-Björklund 2005), of to marine (coastal) facies such as in the Cobequid Bay–Salmon River estuary (Dalrymple and Zaitlin 1994) due to the early Holocene 'regional' sea-level lowstand.

The transgressive systems tract, as predicted by the model of Zaitlin et al. (1994), is usually described as the bulk of tide-dominated estuary infill. In many cases, the transgressive surface that corresponds to the basal limit of the transgressive systems tract is amalgamated with the tidal ravinement surface at least in the seaward zone. The transgressive systems tract normally contains all facies successions and sedimentary bodies that typify the different morphosedimentary components of a tide-dominated estuary (e.g. Plink-Björklund 2005). In some cases such as the Seine and Mont St Michel estuaries, the transgressive systems tract is of reduced volume and described as an aggrading depositional unit made of fine-grained organic-rich facies that accumulates prior to the 'active' tide-dominated estuary representing the highstand systems tract.

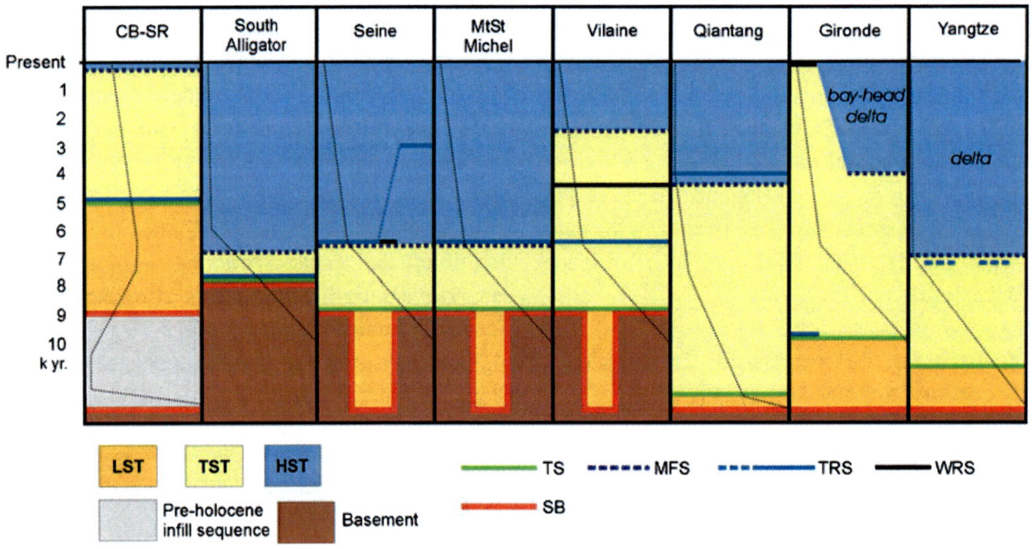

Fig. 6.5 Synthetic overview of the infill stratigraphy of the main modern (Holocene) tide-dominated estuaries as well as the mixed-energy estuary of the Gironde, and the tide-dominated delta of the Yangtze (a tide-dominated estuary during the transgressive systems tract deposition). Vertical scale in time (k/year). The *thin dotted line* in each example indicates the regional relative sea-level curve (low to the right). All infill cases are simple, except that of the Cobequid Bay–Salmon River estuary described as compound (cf. Fig. 6.4 for captions and references)

The highstand systems tract, separated from the underlying transgressive systems tract by the maximum flooding surface, differs greatly from one place to another both in terms of volume and facies successions. In the case of the Yangtze, the highstand stystems tract is a tide-dominated delta. In tide-dominated estuaries, the highstand systems tract is assigned either to the most recent stages of infill, represented by mud-dominated facies (e.g. Cobequid Bay–Salmon River and Vilaine estuaries), or to the bulk of the infill, containing the whole tide-dominated estuary succession, such as in the Seine and Mont St Michel estuaries.

In those cases, the tidal ravinement surface that lies below the tidal channel-and-bar body is amalgamated with the maximum flooding surface. The distinction between the transgressive systems tract and the highstand systems tract is based on the recognition of a seaward shift of the successive estuarine environments such as is precisely described, for example, in the Eocene Central Basin of Spitsbergen (Plink-Björklund 2005), and by progradational facies stacking and downlap configurations above the maximum flooding surface. This last criterion is fulfilled in the case of the Seine and Mont St Michel estuaries. This is as well the case in the mixed-energy estuary of the Gironde, where the highstand systems tract is represented by the prograding tide-dominated bay-head delta.

6.5 Factors Controlling Tide-Dominated Estuary Infilling

As demonstrated through the different examples previously described, both in modern settings and from the rock record, sediment infills of tide-dominated estuaries show a large diversity in terms of geometry and relative proportion of facies within the preserved systems tracts. This variability is related first of all to the diversity of the sites. Hence, the different factors that govern the sediment infill of tide-dominated estuaries can be discussed in the light of this diversity.

6.5.1 Tidal Dynamics and Inherited Bedrock Morphology

The first factor is related to tidal dynamics since tide-dominated estuaries need tides to develop. This does not mean that very large tidal ranges are necessary (Davis and Hayes 1984) although all modern examples discussed are macrotidal. As a consequence, most tide-dominated estuaries are associated with tide-dominated shelves such as the English Channel or the China Sea, and more generally to shelves that are large enough to amplify the oceanic tidal wave. Elongated bays or gulfs are then favourable coastal configurations for extreme amplification of tidal waves that propagate on shelves (Bay of Fundy, Canada; Bristol Channel, UK; Norman-Breton Gulf, France; Hangzhou Bay, China).

It thus appears that coastal configuration by controlling tidal dynamics is a critical factor. This underlies the major role of bedrock morphology on tide-dominated estuary infill. Most studies of incised valleys highlight the importance of bedrock inheritance on sediment infilling (cf. review in Dalrymple 2006 or in Chaumillon et al. 2010). Bedrock inheritance on tide-dominated estuary infill should be considered from different aspects. As previously mentioned, the valley shape determines the possibility for tides to be or not amplified as the valley is transgressed. Funnel-shaped valleys, and more generally, valleys with a high length/width ratio, primarily favour hypersynchronous behaviour of the tidal wave and thus tide-dominated estuary occurrence.

Nordfjord et al. (2006) have drawn such a conclusion from seismic data collected on Pleistocene to Holocene incised-valley fills on the New Jersey shelf: narrow valleys are assumed to promote tide-dominated estuary development rather than wave-dominated estuaries, whereas broad valleys might not provide enough constriction to create strong tidal currents, causing them to be wave dominated. This major role of bedrock (valley) morphology probably implies that the 'unfilled spaces where tide-dominated estuaries can form as transgressive coastal environments' (Dalrymple 2006) are necessarily (incised) valleys. Consequently, the initial definition of estuaries by Dalrymple et al. (1992) stating, 'an estuary is the seaward portion of a drowned valley system…' should probably be considered as still valuable for most tide-dominated estuaries. The cross-sectional shape of the valley is also important to consider as demonstrated for instance by the Seine Estuary case. Irregular valley walls shaped by plateaus (probably wave-cut platforms originating from previous Pleistocene sea-level stillstands) constitute geomorphological features promoting the construction and preservation of wave-dominated coastal barriers on the margin of a tide-dominated estuary.

Similar morphological features of the valley are evoked by Araújo da Silva et al. (2009) to explain the occurrence of thick wave-dominated sandy units at the mouth of the Marapanim tide-dominated estuary (Amazon, Brazil). However, wave-dominated coasts and wave-built sedimentary bodies (beaches, sand-spits, cheniers) are very common on the seaward flanks of tide-dominated estuaries (cf. Fig. 6.3), and do not necessarily require the presence of bedrock plateaus.

In the rocky coast estuary of Vilaine, the highly irregular morphology of the bedrock (Menier et al. 2006) has played a major role on the timing of the valley inundation and consequently on the timing of tidal amplification (Menier et al. 2010; Sorrel et al. 2010). The deepness of the incision is another critical aspect to consider regarding bedrock inheritance. Since it controls directly the accommodation space, the incision deepness determines the stage of infill of an estuary with respect to sediment supply, i.e. its degree of maturity from unfilled to completely filled (Dalrymple et al. 1992). However, this is important both for tide-dominated estuaries and wave-dominated estuaries. Regarding more specifically tide-dominated estuaries, the incision deepness governs the potential for preservation of the infill.

Tide-dominated estuaries are associated with powerful tidal currents and therefore to potentially deep tidal scouring. As a consequence of a deep tidal ravinement surface, preservation potential of underlying deposits is low. This is particularly noticeable in the Mont St Michel estuary that is characterized by a shallow bedrock incision; consequently, the tidal ravinement surface reaches the bedrock throughout the whole internal estuary and reworks almost all older depositional units, in particular the transgressive systems tract (Billeaud et al. 2007; Tessier et al. 2010b). This partly explains why the latter is poorly developed compared to the highstand systems tract in this tide-dominated estuary. The same process occurred in the Seine estuary since the longitudinal tidal bar body has remained active with deeply scoured tidal channels throughout the highstand infill above the main axis of the incised valley.

6.5.2 Sea-Level Fluctuations

Sea-level change is evidently another important factor to take into account regarding sediment infill of estuaries, in the case of both tide-dominated estuaries or wave-dominated estuaries. In all Holocene examples, the main changes in the infill architecture and facies are the result of the change from rapid to slow sea-level rise. Concerning the control of sea-level changes on tide-dominated estuary infill specifically, it controls mainly the possibility for the valley to enter into a tidal amplification window as it is transgressed. With respect to a fifth order relative sea-level cycle, such as the last late Pleistocene–Holocene cycle, this depends clearly on the location of the valley on the shelf, between the lowstand shoreline and the highstand shoreline.

Tidal resonance or at least tidal amplification has a low potential to occur in a valley that is inundated too rapidly, or, if tidal amplification occurs, it does not last enough time for a tide-dominated estuary to develop. Such a concept of a tidal resonance window during a sea-level cycle has been already applied to interpret tide-dominated sedimentary body occurrence and architecture in the rock record and in a Quaternary shelf succession (e.g. Sztano and De Boer 1995; Reynaud et al. 1999), and has been recently re-considered to be integrated in sequence stratigraphic analyses (Yoshida et al. 2007). The entrance of the Cobequid Bay–Salmon River estuary into tidal resonance during the middle Holocene transgression could be partly responsible for the rapid shift from a wave-dominated to a tide-dominated estuary (Dalrymple and Zaitlin 1994).

6.5.3 Sediment Supply

The infill of all types of estuaries depends as well on sediment supply or, more precisely, on the balance between rate of sea level change and sediment supply, i.e. the very common A/S parameter used in sequence stratigraphic analyses. In turn, sediment supply should be considered as a complex factor that integrates the availability of sediment (of both marine and and fluvial origins) and the potential of hydrodynamics, i.e. waves, tidal currents, and river flows, to rework this sediment. Typically, high to moderate wave energy associated with meso- to macrotidal range but high volume of available marine sediment promotes the construction of coastal barriers and thus the development of wave-dominated or mixed-energy estuaries (e.g. Gironde), and of tide-dominated estuaries if tidal range is very large.

This explains probably why the transgressive systems tract in the infill of the Cobequid Bay–Salmon River estuary is assigned to a wave-dominated estuary; a large amount of sediment was already available during that period due to the Pleistocene glacio-marine sand stock that was emplaced previously at the seaward edge of the estuary. When considering the cases of the Seine or Mont St Michel, prior to becoming tide-dominated after 6,500 years BP, these estuaries were probably characterized by moderate- to high-energy wave dynamics, meso- to macrotidal range, but overall by a low volume of available marine sediment. This fundamentally explains why the transgressive systems tract in both estuaries is made of fine-grained organic-rich facies that was deposited as aggrading tidal flats along an open coast, rather than as tide-dominated estuarine or either wave-dominated sediment successions.

Relative sediment supply increased as soon as sea-level rise slowed down, allowing tidal currents to stock sediment and construct a tide-dominated estuary. In the case of the rocky coast tide-dominated estuary of Vilaine, the very low sediment supply is interpreted as the main factor responsible for the late position of the maximum flooding surface, compared for instance with the Seine or Mont St Michel tide-dominated estuaries.

Two additional aspects should be considered about sediment supply. Firstly, in most tide-dominated estuary case studies, sediments are dominantly siliciclastic. The Mont St Michel estuary is an exception with mixed siliciclastic/carbonate sediments. Carbonate production, as represented by bivalves and red algae, is high in this siliciclastic-sediment-starved rocky area of the English Channel (the so-called 'Norman-Breton Gulf'). This additional carbonate sediment source could explain why the highstand systems tract in the Mont St Michel tide-dominated estuary is particularly well developed. Secondly, major estuaries and deltas can deliver during river floods a significant amount of fine-grained sediment to the proximal shelf.

This is the case for example with the Yangtze delta that delivers large amounts of fine-grained sediment to the adjacent Qiantang estuary. Along the French Atlantic coasts (Bay of Biscay), fine-grained sediments are delivered to the shelf by the Gironde and Loire estuaries (Chaumillon et al. 2008). These 'estuary'-borne sources are known to contribute significantly in supplying sediment to adjacent smaller estuaries such as the Charente estuary–Marenns-Oléron Bay located north the Gironde (Chaumillon and Weber 2006) or the Vilaine tide-dominated estuary located north the Loire estuary. These additional sources partly explain the mud-dominated character of the infill of these systems (Menier et al. 2010; Allard et al. 2010).

6.5.4 Climate Changes and Human Influences

Lastly, climate change also should be considered as an important controlling factor of estuary infill since it directly governs river-borne sediment supply. More importantly, with respect to tide-dominated estuary infill is that climate change has an impact on wave dynamics at the mouth. It is suggested for instance that periodic enhanced storminess episodes resulted in successive wave-dominated coarse-grained facies in the tide-dominated infill of the Seine estuary (Sorrel et al. 2009) and Vilaine estuary (Sorrel et al. 2010). Such an enhanced storminess episode is believed to be responsible for the destruction around 3,000 years BP of the coastal barrier that formed on the margin of the Seine estuary (Tessier et al. 2010a). The 3,500–2,500 years BP period is indeed recognized all along the coasts of northern Europe as a period of climatic deterioration (cf. review in Sorrel et al. 2009).

After the barrier destruction, the resulting smoothed shape of the Seine estuarine entrance would have triggered the passage to full tidal conditions throughout the estuary. It is proposed that a similar event occurred at about 3,400 years BP in the Bay of Fundy (Shaw et al. 2010). In relation probably with enhanced storm action, the coastal barrier located at the mouth of the Minas Basin was destroyed, provoking a sudden tidal expansion in the Bay of Fundy and consequently the passage, as it is recorded in the sedimentary infill, from lagoonal–mesotidal to macrotidal environment. The environmental change was so fast that it gave rise to an old aboriginal legend. Middle to late Holocene millennial climate changes have probably influenced significantly the Mont St Michel estuary behaviour since they have also caused periodic destruction of marginal barriers (Billeaud et al. 2009).

Lastly, it is worth noting that many tide-dominated and mixed-energy estuaries along the French coasts experienced since about 1,000 years a significant increase in fine-grained sedimentation. This is the case in particular of the Vilaine estuary (Menier et al. 2010; Sorrel et al. 2010) and the Charente estuary–Marennes-Oléron Bay system (Billeaud et al. 2005; Poirier et al. 2009; Allard et al. 2010). This mud supply is believed to originate mainly from the catchments where land use changes, in relation with deforestation and agriculture development, have dramatically enhanced soil erosion. Such a phenomenon has probably been amplified, if not triggered, during episodes of climate deterioration marked by heavy rain seasons. Such a general increase in sediment supply of fluvial origin led to the deposition of a progradational mud unit. This implies that these estuaries tend to evolve since recent times towards deltaic environments.

6.6 Tide-Dominated vs. Wave-Dominated Estuaries

The objective of this last section is to provide some keys to help the recognition of tide-dominated estuaries successions in ancient coastal sediment wedges. Recently, Dalrymple and Choi (2007) provided a synthetic and helpful overview of all sedimentological and morphological criteria allowing the recognition of tide-dominated fluvio-marine transitional environments (i.e. tide-dominated estuaries and tide-dominated deltas). The purpose herein is not to recall these criteria, but rather to highlight some features that are assumed to typify tide-dominated estuary infill stratigraphy and could be useful in particular for distinguishing them from wave-dominated estuaries.

As demonstrated by long-lasting debates about the interpretation of the depositional environment of some well-known tide-dominated successions (e.g. Cretaceous of the Interior Seaway, North America; Eocene Roda sandstone, Spain), accurate facies models are still missing to allow the distinction between different tide-dominated environments (estuary vs. delta vs. shelf) as preserved in the rock record.

With respect to estuaries, deciphering tide-dominated estuaries and wave-dominated estuaries is not a task as easy as it seems, mainly because outcrop or subsurface data are not sufficient and/or their quality is not good enough to reconstruct precisely the depositional palaeoenvironment. The very detailed and accurate facies analyses and architectural reconstructions made by Plink-Björklund (2005, Eocene, Spitsbergen, 2008, Cretaceous, U.S.A.) benefited from exceptional outcrop conditions, allowing continuous observations throughout marine-to-fluvial transitions.

Reconstruction analyses probably tend to apply too strictly one of the two end-member models, the tide-dominated or the wave-dominated estuary model. Dalrymple (2006) noted that mixed-energy estuaries such as the Gironde (Allen and Posamentier 1993) or the Charente estuary (Chaumillon and Weber 2006) are probably much more common in stratigraphic records than what has been previously interpreted. As discussed in the previous sections, many factors play a major role in the infilling processes of estuaries, and their interaction is consequently very complex.

When considering wave-dominated infilling, three stratigraphic components (depositional units) can be identified: (1) a transgressive sand-dominated wave-built barrier, incised by a tidal inlet (transgressive systems tract); (2) a prograding sand-dominated bay-head delta (highstand systems tract); and (3) an aggrading mud-dominated central basin body (transgressive and highstand systems tracts). According to the degree of tidal influence, the tidal inlet is more or less deep and wide, the morphodynamics of the bay-head delta is more or less influenced by tidal currents, and tidal signatures are more or less pronounced into the central mud facies.

The descriptions of the Holocene case studies have pointed out that in the infill of almost all tide-dominated estuaries, these wave-dominated estuary components can be preserved. Wave-built coastal barriers are preserved on the margin of the Seine and Mont St Michel estuaries. A bay-head delta unit with tidal bars is preserved in the Yangtze tide-dominated estuary transgressive systems tract. A central mud-like depositional unit constitutes the bulk of the transgressive systems tract in the Mont St Michel and Seine estuaries. The identification of such units in the rock record would have probably led to proposals that the depositional environment was a wave-dominated estuary, or at least a mixed-energy estuary. Hence, one can suspect that the abundance of tide-dominated estuary successions has been underestimated in stratigraphic records.

The tidal ravinement surface is certainly the most striking stratigraphic feature for differentiating tide-dominated estuaries and wave-dominated estuaries. As predicted by Zaitlin et al. (1994), the tidal ravinement surface is restricted to the mouth (tidal inlet) in wave-dominated estuaries, whereas it extends throughout tide-dominated estuaries (Fig. 6.6). Moreover, in mixed-energy estuaries, the tidal ravinement surface is usually very deep due to the constriction in the inlet of relatively powerful tidal flows. As soon as the tidal range increases and/or wave power decreases, tidal action expands, laterally and upstream, and the tidal ravinement surface becomes relatively shallower but extends throughout the whole estuary. As a result, sediment fills of wave-dominated estuaries and tide-dominated estuaries are differently preserved (Fig. 6.6). At the seaward end of wave-dominated estuaries and mixed-energy

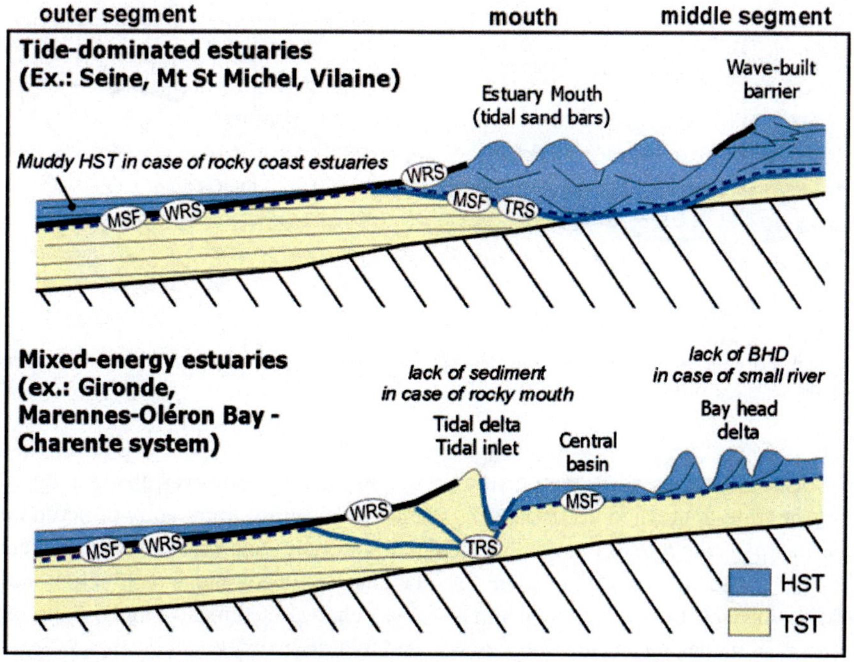

Fig. 6.6 Schematic longitudinal cross sections showing the variability of the sedimentary fill of tide-dominated and mixed wave- and tide-dominated estuaries along French coasts (after Chaumillon et al. 2010). Except in the outer segment of some rocky coast estuaries (such as the Vilaine), the highstand systems tract (HST) is much more developed than the transgressive systems tract (*TST*) in tide-dominated estuaries. This is due to the combination of shallow bedrock incision, deep tidal ravinement and the general sediment-starved context of the French Atlantic and English Channel shelves. For simplification, the lowstand systems tract and transgressive surface have not been represented. *MSF*, *TRS*, *WRS* maximum flooding, tidal ravinement, wave ravinement surfaces

estuaries, highstand systems tract sand-dominated tidal units should be poorly to very poorly preserved as the tidal inlet area is a zone of sediment bypass.

Moreover, transgressive systems tract deposits can be deeply eroded. This is predicted by Allen and Posamentier (1993). At tide-dominated estuary mouths, by contrast, highstand systems tract tidal sand bodies are better preserved. But the major difference arises from the upstream extension of the tidal ravinement surface in tide-dominated estuaries. Indeed, reworking processes by the tidal ravinement surface can occur all along a tide-dominated estuary, leading in some places to a complete erosion of underlying units (Fig. 6.6). The most recent studies carried out on estuary fills along the French coasts (cf. review in Chaumillon et al. 2010) point out clearly this main difference: the highstand systems tract constitutes the bulk of tide-dominated estuary infills, whereas the reverse configuration typifies wave-dominated and mixed-energy estuaries. However, as it has been stated previously, French inner shelves and coasts are sediment-starved systems, this partly explaining such a pronounced difference.

The finest-grained depositional unit of tide-dominated estuaries infill corresponds normally to the sinuous tidal channel area (bedload convergence zone). In the absence of data indicating tidal point-bar deposits, this unit could be misinterpreted as a wave-dominated or mixed-energy estuary central basin (although in terms of facies strictly the occurrence of well-developed and preserved tidal rhythmites definitely indicates a tide-dominated setting). Recent studies on inner mudflat basins (Allard et al. 2010; Billeaud et al. 2009) demonstrate that wide mudflats in mixed energy systems are incised by tidal creeks during maximum flooding and during the highstand. It is suggested that these secondary tidal ravinement surfaces preserved inside the highstand systems tract can be developed when mudflat surfaces are sufficiently wide to allow powerful drainage processes during falling tides in macrotidal settings. Such tidal creeks are probably more common in wave-dominated and mixed energy estuaries inasmuch as the cross-section of central basins and consequently of fringing mudflat surfaces are generally wider compared with mudflats in tide-dominated estuaries.

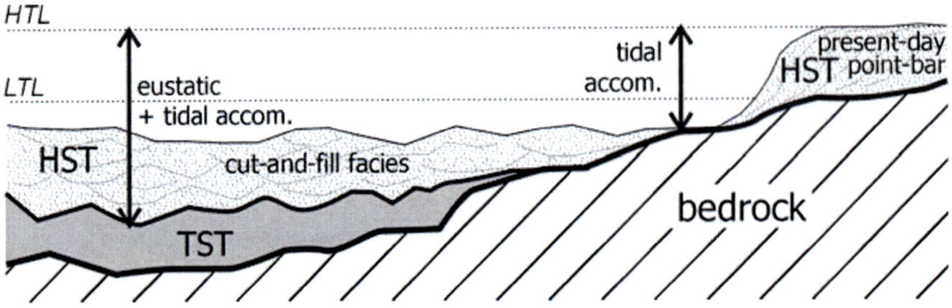

Fig. 6.7 Downstream–upstream cross section in a tide-dominated estuary along the axis of the main tidal channel. This illustrates how tidal accommodation should be considered as a major factor of preservation of systems tracts, especially in the internal domain where the bottom of the channel reaches potentially the bedrock. *TST*, *HST* transgressive, highstand systems tracts, *HTL* high tide level, *LTL* low tide level

The final question to be addressed in the perspective of distinguishing tide-dominated estuaries and wave-dominated estuaries is the role of the tidal range in infill stratigraphy. Estimation of tidal range in ancient environments through the thickness of intertidal–supratidal successions has received some attention (Terwindt 1988), but considering tidal range as a forcing parameter of infill stratigraphy is not common. Tidal range plays necessarily a significant role regarding the volume of preserved systems tracts, particularly in macrotidal settings with extreme tidal ranges such as the Cobequid Bay–Salmon River estuary or the Mont-Saint-Michel estuary. This 'tidal accommodation' should be added to the initial accommodation, especially if tidal range is supposed to have changed significantly during the transgression and thus during the infill of the estuary, such as in the Cobequid Bay–Salmon River estuary. Anyway, tidal accommodation that can be defined as the depth of the active channel belt (Billeaud et al. 2007; Tessier et al. 2011) controls the preservation of the entire estuarine channel body in the area where the tidal ravinement surface reaches the basement (Fig. 6.7).

6.7 Summary

This chapter is an attempt to synthesize available data on the stratigraphy of the deposits filling tide-dominated estuaries. Compared with those published for wave-dominated estuaries, these data remain relatively rare, primarily because tide-dominated estuaries are not as common. Moreover, data quality (the resolution of seismic and age data, and the portion of the estuary studied) is extremely variable and unequal from one site to another. However, this synthesis demonstrates that the infilling stratigraphy of tide-dominated estuaries is closely controlled, as it is for all other coastal deposits, by a complex combination of the rate of sea-level change, sediment supply, bedrock morphology, and hydrodynamics.

Only a few features can be assigned specifically to tide-dominated estuaries in terms of infill stratigraphy. Tidal accommodation appears as the most critical factor in as much as the tidal ravinement surface can potentially rework part of, if not all of, the underlying deposit throughout the estuary. Another important point is that in almost all tide-dominated estuary-infill successions, wave-built sedimentary bodies can be preserved, especially along the seaward flanks of the valley. The recognition of such facies in outcrops or subsurface data could lead to misinterpretations. It should be noted also that, in spite of the predominant action of powerful tidal currents, climate changes can exert a critical control on tide-dominated estuary infilling, especially in terms of the nature of the sediment. This control is recorded through fluctuations in fluvial discharges, but also in the morphodynamics of marginal wave-built barriers that induce changes in tidal-channel behavior.

A great deal of new data have been collected during the last decade on modern tide-dominated estuaries. However, further studies, based on seismic, core, and high-resolution age data, need to be done in order to improve our understanding with respect to two aspects: (1) the longitudinal variability of the infill (i.e., the downstream-upstream evolution) remains poorly illustrated in most modern cases since data are usually not available along the entire length of the estuarine system; and (2) the influence of human activities on the infill stratigraphy. Most tide-dominated estuaries have experienced an accelerated rate of infilling during the last millennium. Human activities combined with climate changes are assumed to be responsible for this process, but the respective role of each one of these factors remains to be more precisely quantified.

References

Allard J, Chaumillon E, Bertin X, Ganthy F, Poirier C (2010) Secular and millenary sedimentary record of morphological changes and human interferences in a macrotidal bay: the Marennes-Oléron Bay (SW France). Bull Soc Géol France 181:151–169

Allen G (1991) Sedimentary processes and facies in the Gironde estuary: a recent model for macrotidal estuarine system. In: Smith G, Reinson GE, Zaitlin BA, Rahmani RA (eds) Clastic tidal sedimentology. Society for Sedimentary Geology special publication 51, pp 3–10

Allen GP, Posamentier HW (1993) Sequence stratigraphy and facies model of an incised valley fill: the Gironde Estuary, France. J Sediment Petrol 63:378–391

Araújo da Silva C, Souza-Filho PWM, Rodrigues SWP (2009) Morphology and modern sedimentary deposits of the macrotidal Marapanim Estuary (Amazon, Brazil). Cont Shelf Res 29:619–631

Archer AW, Feldman HR, Kvale EP, Lanier WP (1994) Comparison of drier- to wetter-interval estuarine roof facies in the Eastern and Western Interior coal basins, USA. Palaeogeog Palaeoclimat Palaeoecol 106:171–185

Billeaud I, Chaumillon E, Weber O (2005) Correlation between VHR seismic profiles and cores evidences a major environmental change recorded in a macrotidal bay. Geo Mar Lett 25:1–10

Billeaud I, Tessier B, Lesueur P, Caline B (2007) Preservation potential of highstand coastal sedimentary bodies in a macrotidal basin: example from the Bay of Mont-Saint-Michel, France. Sediment Geol 202:754–775

Billeaud I, Tessier B, Lesueur P (2009) Impacts of late Holocene rapid climate changes as recorded in a macrotidal coastal setting (Mont-Saint-Michel Bay, France). Geology 37:1031–1034

Bostock HC, Brooke BP, Ryan DA, Hancock G, Pietsch T, Packett R, Harle K (2007) Holocene and modern sediment storage in the subtropical macrotidal Fitzroy River estuary, Southeast Queensland, Australia. Sediment Geol 201:321–340

Boyd R (2010) Transgressive wave-dominated coasts. In: James NP, Dalrymple RW (eds) Facies models 4. Geological Association of Canada, St. John's, pp 265–294

Chaumillon E, Weber N (2006) Spatial variability of modern incised valleys on the French Atlantic coast: comparison between the Charente and the Lay-Sèvre incised valleys. In: Dalrymple RW, Leckie DA, Tillman RW (eds) Incised valleys in time and space. Society for Sedimentary Geology special publication 85, pp 57–85

Chaumillon E, Proust J-N, Menier D, Weber N (2008) Incised-valley morphologies and sedimentary-fills within the inner shelf of the Bay of Biscay (France): a synthesis. J Mar Syst 72:383–396

Chaumillon E, Tessier B, Reynaud JY (2010) Stratigraphic records and variability of incised valleys and estuaries along French coasts. Bull Soc Géol France 181:75–86

Dalrymple RW (2006) Incised valleys in time and space: an introduction to the volume and an examination of the controls on valley formation and filling. In: Dalrymple RW, Leckie DA, Tillman RW (eds) Incised valleys in time and space, vol 85, Special publication (SEPM (Society for Sedimentary Geology)). Society for Sedimentary Geology, Tulsa, pp 5–12

Dalrymple RW (2010) Introduction to siliciclastic facies models. In: James NP, Dalrymple RW (eds) Facies models 4. Geological Association of Canada, St. John's, pp 59–72

Dalrymple RW, Baker EK, Harris PT, Hughes M (2003) Sedimentology and stratigraphy of a tide-dominated, foreland–basin delta (Fly River, Papua New Guinea). In: Sidi FH, Nummedal D, Imbert P, Darman H, Posamentier HW (eds) Tropical deltas of southeast Asia – sedimentology, stratigraphy, and petroleum geology, vol 76, Special publication (SEPM (Society for Sedimentary Geology)). Society for Sedimentary Geology, Tulsa, pp 147–173

Dalrymple RW, Boyd R, Zaitlin BA (1994) History of research, types and internal organization of incised-valley systems: introduction to the volume. In: Dalrymple RW, Boyd RJ, Zaitlin BA (eds) Incised-valley systems: origin and sedimentary sequences. Society for Sedimentary Geology special publication 51, pp 5–12

Dalrymple RW, Choi K (2007) Morphologic and facies trends through the fluvial–marine transition in tide-dominated depositional systems: a schematic framework for environmental and sequence-stratigraphic interpretation. Earth Sci Rev 81:135–174

Dalrymple RW, Zaitlin BA (1994) High-resolution sequence stratigraphy of a complex, incised valley succession, the Cobequid Bay Salmon River estuary, Bay of Fundy, Canada. Sedimentology 41:1069–1091

Dalrymple RW, Zaitlin BA, Boyd R (1992) Estuarine facies models: conceptual basis and stratigraphic implications. J Sediment Petrol 62:1130–1146

Davis RA, Hayes MO (1984) What is a wave-dominated coast? Mar Geol 60:313–329

Emery D, Myers KJ (1999) Sequence stratigraphy. Blackwell Science, Oxford, 297 p

Garnaud S, Lesueur P, Clet M, Lesourd S, Garlan T, Lafite R, Brun-Cottan J-C (2003) Holocene to modern fine-grained sedimentation on a macrotidal shoreface to inner-shelf setting (Eastern Bay of the Seine, France). Mar Geol 202:33–54

Harris PT (1988) Large-scale bedforms as indicators of mutually evasive sand transport and the sequential infilling of wide-mouthed estuaries. Sediment Geol 125:31–49

Hori K, Saito Y, Zhao Q, Cheng X, Wang P, Sato Y, Li C (2001) Sedimentary facies of tide-dominated paleo-Changjiang (Yangtze) estuary during the last transgression. Mar Geol 177:331–351

Johnson HD, Levell BK (1995) Sedimentology of a transgressive, estuarine sand complex: the Lower Cretaceous Woburn Sands (Lower Greensand), Southern England. In: Plint AG (ed) Sedimentary facies analysis, vol 22, International

Association of Sedimentologists Special publications. Blackwell Scientific, Oxford, pp 17–46
Khin K, Myitta (1999) Marine transgression and regression in the Miocene sequences of northern Pegu (Bago) Yoma, Central Myanmar. J Asian Earth Sci 17:369–393
Kitazawa T (2007) Pleistocene macrotidal tide-dominated estuary-delta succession, along the Dong Nai River, Southern Vietnam. Sediment Geol 194:115–140
Lesourd S, Lesueur P, Brun-Cottan J-C, Garnaud S, Poupinet N (2003) Seasonal evolution in the characteristics of superficial sediments in a macrotidal estuary (the Seine inlet, France). Estuar Coastal Shelf Sci 58:3–16
Lin C-M, Zhuo H, Gao S (2005) Sedimentary facies and evolution in the Qiantang River incised valley, Eastern China. Mar Geol 219:253–259
Masselink G, Hughes MG (2003) Introduction to coastal processes and geomorphology. Arnold, London, 354 p
Mellere D (1994) Sequential development of an estuarine valley fill: the Twowells tongue of the Dakota Sandstone, Acoma Basin, New Mexico. J Sediment Res 4:500–515
Menier D, Reynaud J-Y, Proust JN, Guillocheau F, Guennoc P, Bonnet S, Tessier B, Goubert E (2006) Basement control on shaping and infilling of valleys incised at the southern coast of Brittany, France. In: Dalrymple RW, Leckie DA, Tillman RW (eds) Incised valleys in time and space, vol 85, Special publication (SEPM (Society for Sedimentary Geology)). Society for Sedimentary Geology, Tulsa, pp 37–55
Menier D, Tessier B, Proust JN, Baltzer A, Sorrel P, Traini C (2010) The Holocene transgression as recorded by incised-valley infilling in a rocky coast context with low sediment supply (Southern Brittany, Western France). Bull Soc Géol France 181:115–128
Nordfjord S, Goff JA, Austin JA, Gulick SPS (2006) Seismic facies of incised-valley fills, New Jersey continental shelf; implications for erosion and preservation processes acting during latest Pleistocene-Holocene transgression. J Sediment Res 12:1284–1303
Plink-Björklund P (2005) Stacked fluvial to tide-dominated estuarine deposits in high-frequency (fourth-order) sequences of the Eocene Central Basin, Spitsbergen. Sedimentology 52:391–428
Plink-Björklund P (2008) Wave-to-tide facies change in a Campanian shoreline complex, Chimney Rock Tongue, Wyoming-Utah, U.S.A. In: Hampson GJ, Steel RJ, Burgess PM, Dalrymple RW (eds) Recent advances in models of siliciclastic shallow-marine stratigraphy. Society for Sedimentary Geology special publication 90, pp 265–291
Poirier C, Sauriau PG, Chaumillon E, Allard J (2009) Can molluscan assemblages give insights into Holocene environmental changes other than sea level rise? A case study from a macrotidal bay (Marennes-Oléron, France). Palaeogeog Palaeoclimat Palaeoecol 280:105–118
Pontén A, Plink-Björklund P (2009) Regressive to transgressive transits reflected in tidal bars, Middle Devonian Baltic Basin. Sediment Geol 218:48–60
Reynaud J-Y, Tessier B, Proust J-N, Dalrymple R, Marsset T, De Batist M, Bourillet J-F, Lericolais G (1999) Eustatic and hydrodynamic controls on the architecture of a deep shelf sand bank (Celtic Sea). Sedimentology 46:703–721
Rossetti DdeF (1998) Facies architecture and sequential evolution of an incised-valley estuarine fill: the Cujupe Formation (Upper Cretaceous to ?Lower Tertiary), São Luis Basin, Northern Brazil. J Sediment Res 68:299–310
Shanmugam G, Poffenberger M, Toro Alava J (2000) Tide-dominated estuarine facies in the Hollin and Napo ('T' and 'U') formations (Cretaceous), Sacha Field, Oriente Basin, Ecuador. Am Assoc Petrol Geol Bull 84:652–682
Shaw J, Amos CL, Greenberg DA, O'Reilly CT, Parrot DR, Patton E (2010) Catastrophic tidal expansion in the Bay of Fundy, Canada. Can J Earth Sci 47:1079–1091
Sorrel P, Tessier B, Demory F, Delsinne N, Mouazé D (2009) Evidence for millennial-scale climatic events in the sedimentary infilling of a macrotidal estuarine system, the Seine estuary (NW France). Quat Sci Rev 28:499–516
Sorrel P, Tessier B, Demory F, Baltzer A, Bouaouina F, Proust JN, Menier D, Traini C (2010) Sedimentary archives of the French Atlantic coast (inner Bay of Vilaine, South Brittany): depositional history and late Holocene climatic signals. Cont Shelf Res 30:1250–1266
Sztano O, De Boer PL (1995) Basin dimensions and morphology as controls on amplification of tidal motions (the early Miocene North Hungarian Bay). Sedimentology 42:665–682
Terwindt JHJ (1988) Palaeo-tidal reconstructions of inshore tidal depostional environments. In: de Boer PL, van Gelder A, Nio SD (eds) Tide-influenced sedimentary environments and facies, sedimentology and petroleum geology. D. Reidel Publishing, Dordrecht, pp 233–263
Tessier B (1993) Upper intertidal rhythmites in the Mont-Saint-Michel Bay (NW France): perspectives for paleoreconstruction. Mar Geol 110:355–367
Tessier B, Delsinne N, Sorrel P (2010a) Holocene infilling of a tide-dominated estuarine mouth. The example of the macrotidal Seine estuary (NW France). Bull Soc Géol France 181:87–98
Tessier B, Billeaud I, Lesueur P (2010b) Stratigraphic organization of a composite macrotidal wedge: the Holocene infill of the Mont-Saint-Michel Bay. Bull Soc Géol France 181:99–113
Tessier B, Billeaud I, Sorrel P, Delsinne N, Lesueur P (2011) Infilling stratigraphy of macrotidal tide-dominated estuaries. Controlling mechanisms: sea-level fluctuations, bedrock morphology and climatic changes (The examples of the Seine estuary and the Mont-Saint-Michel Bay, English Channel, NW France). Sediment Geol doi:10.1016/j.sedgeo.2011.02.003
Weber N, Chaumillon E, Tesson M, Garlan T (2004) Architecture and morphology of the outer segment of a mixed tide and wave-dominated-incised valley, revealed by HR seismic reflection profiling: the paleo-Charente River, France. Mar Geol 207:17–38
Woodroffe CD, Chappell J, Thom BG, Wallensky E (1989) Depositional model of a macrotidal estuary and floodplain, South Alligator River, Northern Australia. Sedimentology 36:737–756
Woodroffe CD, Mulrennan ME, Chappell J (1993) Estuarine infill and coastal progradation, southern van Diemen Gulf, Northern Australia. Sediment Geol 83:257–275
Yoshida S, Steel RJ, Dalrymple RW (2007) Changes in depositional processes – an ingredient in a new generation of sequence-stratigraphic models. J Sed Res 77:447–460
Zaitlin BA, Dalrymple RW, Boyd R (1994) The stratigraphic organization of incised-valley systems. In: Dalrymple RW, Boyd R, Zaitlin BA (eds) Incised-valley systems: origin and sedimentary sequences. Society for Sedimentary Geology special publication 51, pp 45–60

Tide-Dominated Deltas

7

Steven L. Goodbred, Jr. and Yoshiki Saito

Abstract

Among tidally influenced sedimentary environments, tide-dominated deltas are perhaps the most variable and difficult to characterize. This variability is due in part to the major role that fluvial systems play in defining their delta, with rivers differing widely in discharge, sediment load, seasonality, and grain size. Tide-dominated deltas also tend to be large systems that can extend hundreds of kilometers across and along the continental margin. The associated sediment transport regimes are typically high energy, but they vary considerably at the scale of tidal cycles and seasonal river discharge. As a consequence of varying transport energy, the sedimentary successions formed in tide-dominated deltaic settings tend to be heterolithic, with interbedded sands, silts, and clays and both fining- and coarsening-upward facies associations. The deltaic nature of tide-dominated deltas that distinguishes them from other tidally influenced settings is defined by the cross- or along-shelf progradation of a clinoform, or 'S' shaped, sedimentary deposit. Under the influence of strong bed shear in tidally dominated margins, this prograding clinoform is often separated into two distinct units, one associated with the subaerial deltaplain and one with an offshore subaqueous delta. Onshore, the large, fertile deltaplains built by many modern tide-dominated deltas, especially in South and East Asia, are heavily populated and sustain large economies, making them global important settings. However, the reduction of fluvial inputs by damming and water extraction, as well as intense agricultural, urban, and industrial land uses, threaten the stability and sustainability of these environments.

S.L. Goodbred, Jr. (✉)
Department of Earth and Environmental Sciences,
Vanderbilt University, Nashville, TN 37240, USA
e-mail: steven.goodbred@vanderbilt.edu

Y. Saito
Geological Survey of Japan, AIST, Central 7, Higashi 1-1-1,
Tsukuba 305-8567, Japan
e-mail: yoshiki.saito@aist.go.jp

7.1 Introduction

River deltas are variably defined by their geography, morphology, or stratigraphy, but are most generally considered to be a sedimentary deposit formed by a river at its mouth. Here, to distinguish deltas from rivermouth estuaries that also receive fluvial sediment

R.A. Davis, Jr. and R.W. Dalrymple (eds.), *Principles of Tidal Sedimentology*,
DOI 10.1007/978-94-007-0123-6_7,

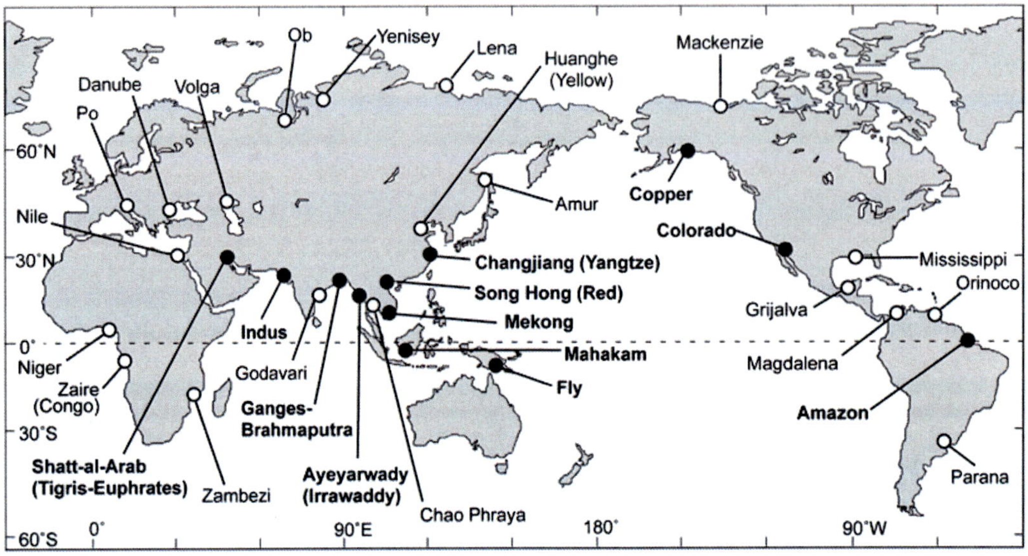

Fig. 7.1 Map of the world's major river delta systems, with those forming tide-dominated deltas indicated (*bold type; filled circle*) (Modified after Hori and Saito 2007)

input (see tide-dominated estuaries chapter), river deltas must receive adequate sediment from the river to build a clinothem, which is a sedimentary deposit having characteristic topset-foreset-bottomset morphology, often in a sigmoidal or 'S' shape. In this way river-fed coastal systems may be depositional, but they are not deltaic if lacking a definable clinoform morphology and progradational features. The surfaces defining many deltaic clinothems are very low-gradient (<3°) for fine-grained deltas and may be difficult to recognize in core or outcrop, so other criteria discussed in this chapter may be important in recognizing deltaic settings from such data. In simplest terms, it is expected that large volumes of heterolithic mud will be found offshore of deltaic rivermouths, which should be a distinguishing character from most other river-influenced settings. Inherent in this definition, deltaic systems will be controlled at a first order by river discharge and fluvial sediment load and secondarily to the rate of reworking by marine processes, primarily waves, tides, and coastal currents.

Although modern and ancient deltas may share a general clinoform morphology, examples from around the world show considerable variability in their surface geomorphology, lithology, process, and response to external forcing. To account for some of this variability, deltas are commonly classified by the dominant process controlling sediment dispersal, and hence surface geomorphology (Galloway 1975). The end-members in this ternary classification scheme are river-, wave- and tide-dominated delta systems, with many examples exhibiting intermediate characteristics that can be classified as mixed-energy (Figs. 7.1 and 7.2). Large deltas may also comprise a composite system, where different portions of the delta are morphologically distinct and controlled differently by fluvial, wave, or tidal processes (Bhattacharya and Giosan 2003). More recent variations of this scheme have in addition considered grain size (Orton and Reading 1993), sediment supply, and sea level (Boyd et al. 1992), although the original Galloway classification arguably remains the most useful for large river deltas.

7.2 Background

7.2.1 Past Research

Although the study of river deltas was active during the first half of the twentieth century (e.g. Russell and Russell 1939), comparatively little research was done on tidally dominated systems, due in part perhaps to their large size, remote locations, and challenging navigation. In the 1970s when delta classification models were first emerging (e.g. Wright and Coleman 1971; Galloway 1975), the only "tide-dominated" end-members that had been studied in any detail were the very small Klang-Langat delta of Malaysia

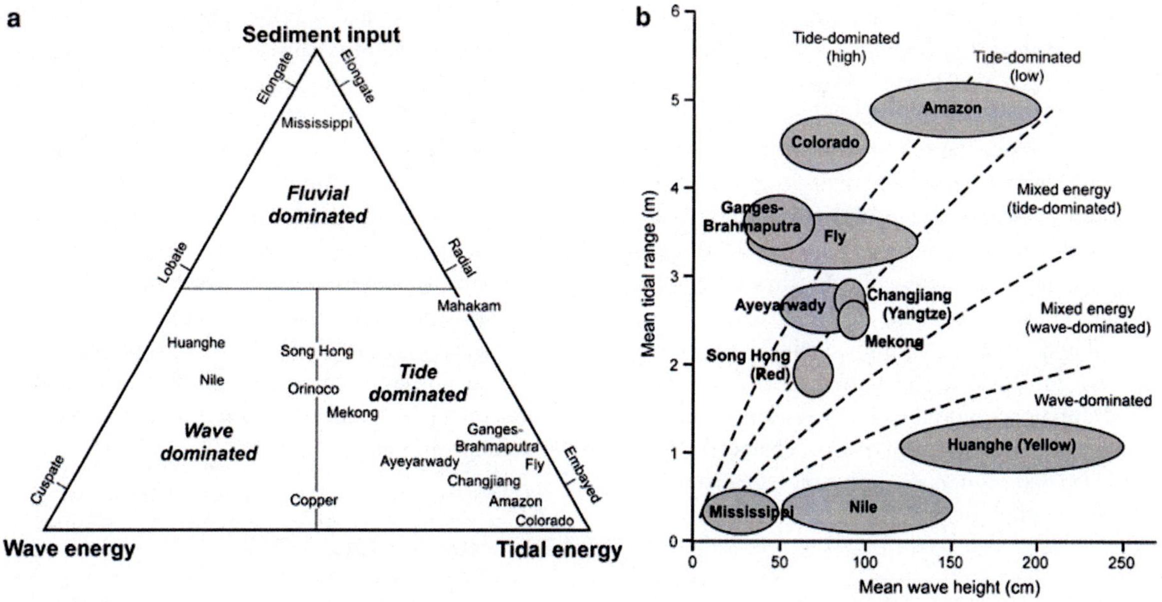

Fig. 7.2 (**a**) Major river deltas classified by the relative influence of river, wave, and tidal processes (After Galloway 1975). (**b**) Mean wave height versus mean tidal range for major large river deltas. The areas are grouped into five morphological classes after the classification of Davis and Hayes (1984) (Modified after Hori et al. 2002a)

(Coleman et al. 1970) and the Yalu and Ord rivers of Korea and Australia, respectively (Coleman and Wright 1978). None of these systems are discussed at length in this chapter as they are best reclassified as tide-influenced deltas (Klang-Langat) or as tidal estuaries (Yalu and Ord; see Chap. 5). In the 1980s the Amazon and Changjiang (i.e. Yangtze) were the first large tide-influenced deltas to be studied in detail through large, comprehensive, and multidisciplinary investigations. The Amazon project, called AMASEDS, collected observational data simultaneously at the seabed and water column over different phases of the river hydrograph and tidal conditions, demonstrating the tremendous benefits of such an integrated approach (Nittrouer and DeMaster 1986). Combined with sediment coring and seismic-reflection surveys, AMASEDS defined the modern approach for studying complex, river-fed continental margin systems. A similar comprehensive study was done for the Changjiang in Asia (Milliman and Jin 1985). However research of tide-dominated deltas remained limited as most studies were of river- or wave-dominated examples (e.g. Mississippi, Nile, Ebro, Rhine).

Middleton (1991) pointed out that a majority of very large rivers in terms of sediment load discharge along meso- to macrotidal coasts, forming tide-dominated or tide-influenced deltas (Fig. 7.1). In response, research was initiated in several tidally affected deltas, with the Fly river being among the first major tide-dominated deltas to be studied in detail (Harris et al. 1996; Wolanski et al. 1995). Since that time the rate of investigation has accelerated and today most major tide-dominated delta systems have received some formal investigation. Most studies have employed stratigraphic or seismic-reflection approaches, but observational and hydrodynamic data remain rare for many systems. Among several coordinated research programs, recent efforts have focused on the Changjiang, Mekong, and other nearby Asian deltas (e.g. Hori et al. 2001; Ta et al. 2005), and the Gulf of Papua 'continuum' that includes the tide-dominated Fly and Kikori deltas (e.g., Ogston et al. 2008; Walsh et al. 2004). The Ganges-Brahmaputra has been reasonably well studied by individual working groups (Goodbred and Kuehl 2000; Kuehl et al. 2005; Michels et al. 1998), and to a lesser extent the Indus (Giosan et al. 2006) and Colorado (Carriquiry and Sanchez 1999; Thompson 1968) deltas. The Ayeyarwady (i.e., Irrawaddy) and Tigris-Euphrates deltas, however, remain notable exceptions with very little published research.

Other more general studies have advanced our understanding of continental margin systems with great implications for tide-dominated deltas, including developments in shelf hydrodynamics and sediment transport (Wright and Friedrichs 2006), and the quantitative modeling of delta evolution, stratigraphy (Fagherazzi and Overeem 2007), and clinothem development (Swenson et al. 2005; Slingerland et al. 2008). One continuing challenge, though, is the difficulty in numerically modeling tidal sediment transport due to complications of the bidirectional flow, thus limiting our ability to assess impacts of environmental changes such as discharge variations, sediment loading, and sea-level change. Although effective modeling of tidal sediment transport remains elusive, progress is being made in understanding hydrodynamics of the complex network of tidal channels (Fagherazzi 2008) and compound clinoform morphology (Swenson et al. 2005; Wright and Friedrichs 2006) that characterize tide-dominated delta systems. These topics are discussed in detail later in this chapter.

7.2.2 Modern Examples

In this chapter we focus primarily on tide-dominated deltas, including examples of the Colorado, Fly, Ganges-Brahmaputra, Indus, Irrawaddy, and Changjiang, with some discussion of tide-influenced deltas such as the Amazon, Mahakam, and Mekong. Overall these systems are best characterized by their wide river mouths that have a pronounced upstream taper and well-developed channel bars and islands. All examples are subject to mesotidal to macrotidal conditions with spring tidal ranges typically ≥3 m. Because of this continual exposure to tidal exchange and sediment transport, tide-dominated deltas along open shorelines are typically fed by large rivers that discharge high sediment loads, although smaller rivers may form deltas in more embayed settings (e.g. Gironde River, France). Indeed, 10 of the river deltas listed above (excluding the Mahakam) rank among the world's top 25 rivers in terms of their fluvial sediment discharge (Milliman and Meade 1983; Milliman and Syvitski 1992). Rankings for the Colorado, Tigris-Euphrates, and Indus rivers are based on historical estimates prior to major damming and sediment trapping.

Most tide-dominated deltas today are located in tectonically active, low-latitude regions, including South Asia, East Asia, and Oceania (Fig. 7.1). Many factors relevant to the development of tide-dominated delta systems are common to these areas. First, amplification of the M2 tidal component in high tidal-range areas is supported by broad, relatively shallow continental shelves and seas that are well connected to the open ocean, and in many instances taper in width toward their apex. Prominent examples include the Arabian Sea (Indus), Bay of Bengal (Ganges-Brahmaputra), Andaman Sea (Ayeyarwady), Gulf of Papua (Fly), and East China Sea (Changjiang). A second factor common to most tide-dominated deltas, and many deltas in general, is that they drain high-standing, tectonically active mountains. Such active orogens yield the abundant sediment required for deltas to form in high-energy coastal basins. In particular the Himalayan-Tibetan uplift and Indonesian archipelago sustain among the world's highest sediment yields (Milliman and Syvitski 1992).

7.2.3 Humans and Deltas

Many tide-dominated deltas are among the world's largest in areal extent (Woodroffe et al. 2006), and the immense, agriculturally rich, lowland delta plains that have formed at the mouths of the Ganges-Brahmaputra, Indus, Ayeyarwady, Mekong, and Changjiang rivers support nearly 200 million people. These populations, like those in all deltas, are at risk from flooding, tropical cyclones, sea-level rise and related environmental hazards. Unfortunately, our current understanding of the process-response (morphodynamic) relationships in tide-dominated deltas is inadequate to assess the likely outcome of various environmental-change scenarios. Much may be learned by further investigation of the several tide-dominated deltas that have already been severely degraded due to river damming, water extraction, and reduced sediment discharge, notably the Indus, Colorado, and Tigris-Euphrates (Syvitski et al. 2009). Despite risk and uncertainty, major dams continue to be constructed on rivers that feed high-energy, tide-dominated delta systems, such as the Three Gorges Dam on the Changjiang and the Xiaowan Dam on the Mekong (Yang et al. 2006; Kummu et al. 2010). Not all tide-dominated deltas are strongly human-impacted, however, with the Amazon, Copper, and Fly river systems draining relatively natural catchments and having sparsely populated delta plains. Similarly, the Ganges-Brahmaputra and Ayeyarwady rivers remain undammed despite their heavily populated catchments, and so their large water discharge and sediment loads sustain stable, if still locally dynamic, delta systems.

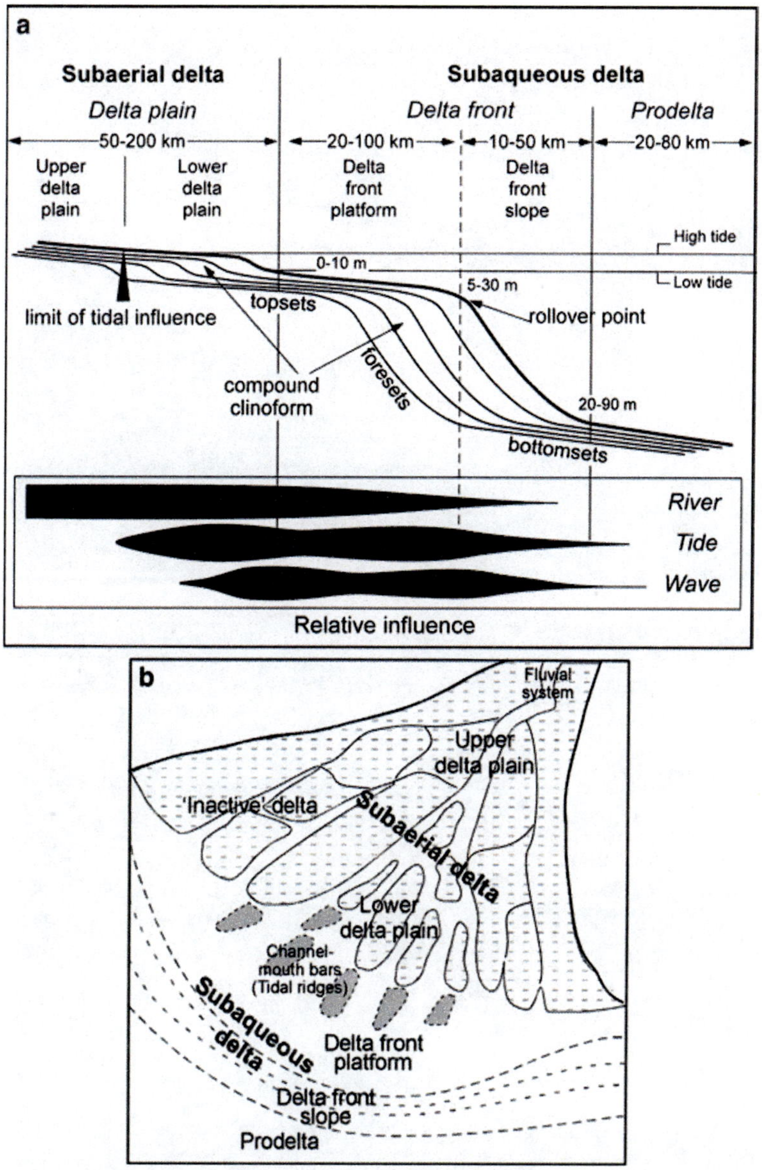

Fig. 7.3 Major physiographic and morphologic features of tide-dominated delta systems shown in (**a**) cross-section and (**b**) planform. Note the well developed subaerial and subaqueous portions of the delta, each represented by a prograding clinoform. The rivermouth is also characterized by channel-mouth bars that build just seaward of the shoreline, and in many cases become emergent and amalgamate into large channel-mouth islands (Modified from Hori and Saito 2007)

7.3 Hydrodynamics

Tide-dominated deltas have complex hydrodynamics that are strongly influenced by river discharge, tidal exchange, and other marine processes such as waves and storms (Fig. 7.3). Each of these controls varies considerably with time (e.g., fortnightly, seasonal, episodic) and location (e.g. active rivermouth, 'inactive' delta plain, subaqueous delta). By definition, tides are perhaps the overarching control on tide-dominated delta systems, but the fact that these are prograding deltas and not transgressing tidal estuaries also reflects the tremendous influence of large fluvial systems feeding them. In addition, large riverine

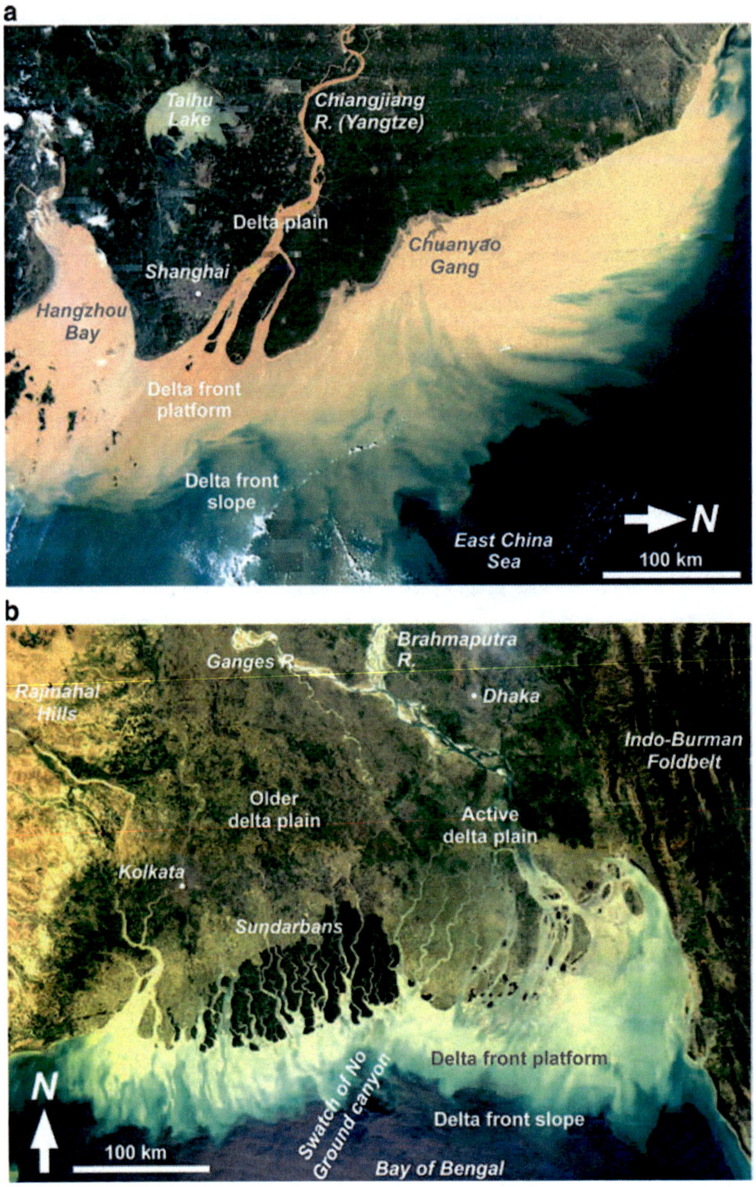

Fig. 7.4 MODIS satellite images of two major tide-dominated delta systems, (**a**) the Changjiang river delta taken near the end of the flood season on 25 October 2000 and (**b**) the Ganges-Brahmaputra river delta taken late in the dry season on 19 March 2002. Major geographic and physiographic features of the delta and surrounding areas are labeled. Both images show high suspended sediment concentrations that extend 50–100 km offshore and hundreds of kilometers alongshore, largely due to suspension by tidal currents (Images from NASA MODIS, http://modis.gsfc.nasa.gov/)

sediment fluxes and persistent tidal energy sustain high suspended-sediment concentrations offshore, where these particulates are subject to widespread dispersal by coastal and ocean currents that are normally too slow for entraining sediments without the addition of a tidal-velocity component (Fig. 7.4). Overall, though, tide-dominated deltas bear the mark of not only strong tidal influence, but also fluvial and marine processes that play critical roles in defining the character and behavior of these complex margin systems.

7.3.1 Tidal Processes

7.3.1.1 Tidal Amplification

At the offshore limits of the delta system, the incoming ocean tide first interacts with the clinoform delta-front, where water depths shoal from 20 to 90 m at the bottomsets to 5–30 m at the topset-foreset roll-over point, a distance typically of a few tens of kilometers for megadeltas to a few kilometers for smaller deltas (Fig. 7.3; Storms et al. 2005). Tidal currents accelerate across this zone from <20 cm/s on the open shelf to 30–80 cm/s on the outer delta-front platform (ie., topsets), still tens of kilometers offshore. This acceleration across the prograding delta-front represents an important morphodynamic feedback that in large part is responsible for forming the compound clinoform that is typical of most tide-dominated delta systems. In this case strong bed shear on the inner shelf (i.e., delta platform) defines a zone of limited deposition that separates the prograding subaqueous and subaerial clinoforms (Fig. 7.3a; see also Sect. 7.3.3.2).

After crossing the delta-front platform (i.e. topsets) the progressive tide wave becomes channelized as it propagates upstream of the shoreline, inducing a second phase of energy focusing that accelerates tidal currents to velocities of 50 to >100 cm/s. This acceleration continues for a significant distance upstream (10s of km) due to tidal amplification. Although tidal energy is lost to friction, the local tidal power is actually amplified by the decreasing cross-sectional area of the narrowing channels. This is called a hypersynchronous channel system, whereby tidal height and current velocities increase steadily upstream before declining to zero as tidal energy becomes increasingly attenuated by frictional forces.

Due to this positive feedback of tidal amplification across the shallow prograding delta-front and tapering delta-plain channels, tides actually influence a much larger reach of the continental margin than they would in the absence of the delta. In larger tide-dominated deltas, this enhanced tidal influence may extend 100–200 km across the margin (Fig. 7.5). In general tidal-bed shear in this broad reach is sufficient to impart a strong influence on sediment transport and deposition, although preservation of tidal signatures in the sedimentary record is less certain (see 7.4.2).

7.3.1.2 Tidal Asymmetry

An important consequence of hypersynchronous tidal amplification is the development of an asymmetry in the ebb and flood limbs of the tidal wave. In this case the wave crest (high tide) propagates faster than the wave trough (low tide), causing the flood period (low to high tide) to shorten and ebb period (high to low tide) to lengthen. This time asymmetry requires higher current velocities for the flooding tide to accommodate the tidal prism, and is described as being a flood-dominant tidal system.

Given that the rate of sediment transport (y) increases as a power function (b) of current velocity (x), where $y = ax^b$ with $b = 1.6$–2.0, most flood-dominant tidal systems result in a net onshore-directed transport of sediment, an effect called "tidal pumping" (Postma 1967). This effect may have fundamental implications for the morphology and behavior of tide-dominated delta systems (see Sect. 7.4.1), but its nfluence likely varies spatially and temporally with such factors as river discharge. For example, where river discharge is high the net flow and sediment transport patterns may be significantly altered or even reversed from the tidal signature alone. In general low river discharge allows a net upstream (landward) transport of sediment (e.g., during the dry season), whereas high discharge weakens this tidal-pumping effect and forces net offshore transport. These natural patterns in tidal pumping and sediment transport may be considerably altered on rivers with large dams used to artificially control water discharge (Wolanski and Spagnol 2000).

7.3.2 Fluvial and 'Estuarine' Processes

The evolution of tidal hydrodynamics at the coast is not only influenced by seabed and shoreline morphology but also by interactions with freshwater discharge. In the case of most tide-dominated deltas, the interaction of large river-water fluxes and meso- to macro-tidal regimes tend to generate strong horizontal shear, turbulent eddies, and vigorous vertical mixing. Such dynamic flows are generally adequate to preclude density stratification and result in a well-mixed estuary at the delta rivermouth. Therefore buoyancy-driven gravitational circulation is not as significant in

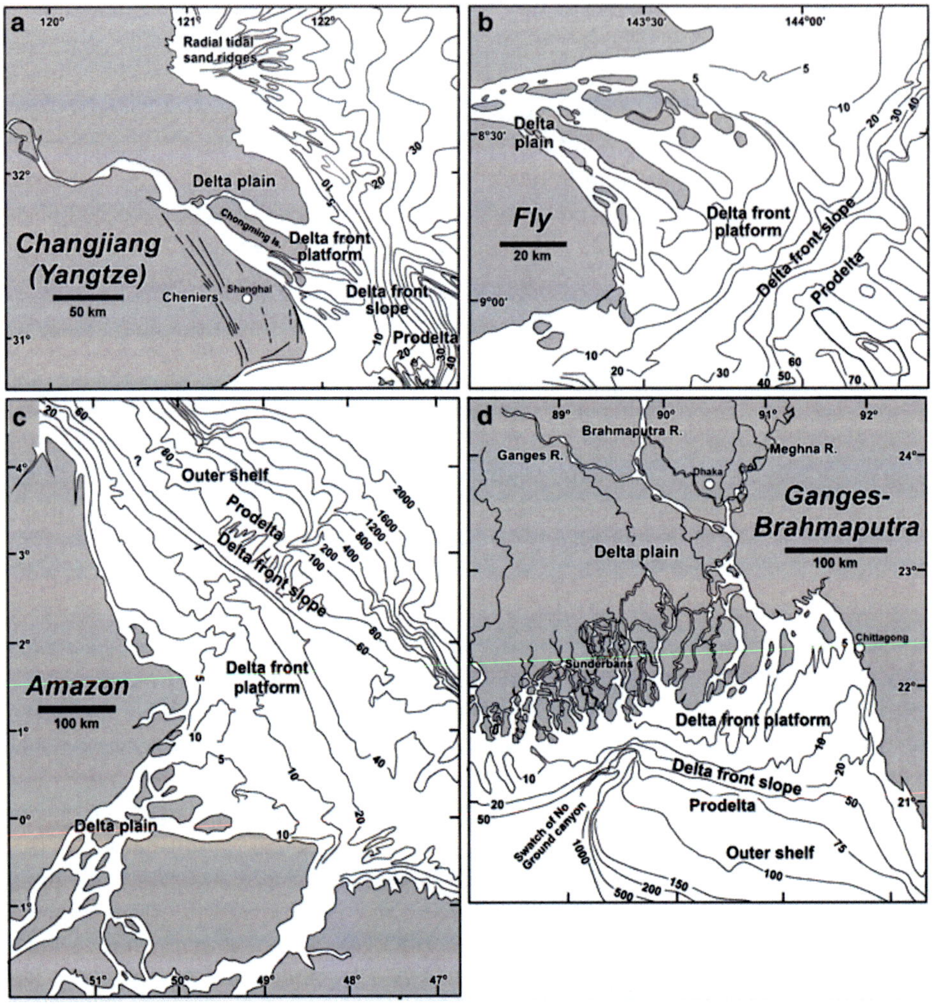

Fig. 7.5 Physiographic maps of four major tide-dominated delta systems, including the (**a**) Changjiang, (**b**) Fly, (**c**) Amazon, and (**d**) Ganges-Brahmaputra. Note the variable scale but similar funnel-shaped morphology of the rivermouths, each with characteristic channel-margin bars that are many tens of kilometers long. Each delta system is also characterized by a large muddy clinothem deposit that is forming off the rivermouth, at a similar length-scale of many tens of kilometers offshore (Compiled after Hori et al. 2002a; Harris et al. 2004; Nittrouer et al. 1986; Goodbred and Kuehl 2000)

tide-dominated deltas as it is at many less energetic river mouths. As with any complex natural system, though, partially mixed stratification and weak estuarine circulation may develop locally within tide-dominated deltas given spatiotemporal differences in tidal energy (spring vs. neap) and river discharge (seasonality and flow splitting amongst the distributary channels).

7.3.2.1 Sediment Transport Convergence

Although stratification is not generally important in tide-dominated delta systems, river discharge plays at least two other key roles in defining system-scale hydrodynamics. First, the flux of freshwater from the river relative to the incoming tidal prism determines the position of sediment transport convergence within the rivermouth or on the shelf. Sediment convergence occurs where sediments are trapped by outflowing river discharge and onshore tidal transport, causing a high concentration of suspended sediment, often referred to as the turbidity maximum, and high deposition rates on the underlying seabed. In general, high river discharge relative to the tidal prism forces the location of this sediment convergence further seaward and defines an important location of dynamic-scale sediment accretion (i.e., where sediments may be stored for short time periods, less than a year, and subject to later reworking). In contrast to tide-dominated estuaries where the flux convergence of suspended sediment tends to be located near the apex of the rivermouth embayment, the convergence in tide-dominated deltas is typically near the mouth of the embayment or slightly seaward (e.g. Fly, Ganges-Brahmaputra, Changjiang; Dalrymple and Choi 2007). The convergence may shift many kilometers upstream during the low-discharge dry season (Wolanski et al. 1996), but, with the majority of sediment delivered during high river flow, the wet-season transport regime is more important to delta evolution. In an extreme case the Amazon River, with more discharge than any other river on Earth, actually forces its tide-river flux convergence 60–90 km offshore onto the middle continental shelf where most sediment accumulates in the subaqueous clinothem (Kuehl et al. 1986; Nittrouer et al. 1986). Although depositional patterns here are strongly tide influenced, no saltwater enters the Amazon rivermouth at any time of the year despite a spring tidal range of ~7 m. Finally, it is important to note that the location of flux convergence for coarser-grained bedload may lie considerably landward of that for suspended load (Montaño and Carbajal 2008).

7.3.2.2 Residual Flow

The second important interaction of river discharge with tidal hydrodynamics is that river flow, at least seasonally, dominates the residual flow in tide-dominated deltas. Residual flow is the resultant current vector (i.e., 'net drift') that emerges from averaging all flow components (tidal, fluvial, and marine) over a period of weeks to a year. Residual flow can be difficult to determine from short-term instrumental deployments because of the dominance of non-steady synoptic-scale forces (e.g., waves, storms, flood discharge), and thus results may differ depending on the time-scale over which observations or calculations are made. Ultimately, though, it is the asymmetry in tidal currents and the unidirectional flow of river discharge that tend to generate residual flows and dominate the net *fluid* transport in tide-dominated delta systems (Barua et al. 1994).

Because residual flow is a purely fluid transport phenomenon, its role in sediment transport will vary depending on the timing, magnitude, and duration that sediments are suspended in the water column. Thus, along lower-energy margins where suspended sediment concentrations are comparatively low and much of the sediment is relatively coarse (i.e., sand-sized) bedload, time-averaged residual flows may not be important to overall morphologic development. However, on high-energy, tide-dominated deltaic margins where suspended-sediment concentrations are consistently high, the weak but persistent residual flows may account for much of the long-term net sediment transport and resulting morphological evolution of the rivermouth delta and adjacent tidal delta plain.

7.3.3 Marine Processes

The large rivers that feed most modern tide-dominated deltas export much of their sediment load to the shelf, where it is subject to a suite of marine processes – tides, waves, storms, geostrophic currents – that ultimately define the morphology and development of the subaqueous portion of the delta (Walsh and Nittrouer 2009). Often the greatest effect of these processes on sediment dispersal and development of the subaqueous delta occurs when they are coincident with high river discharge. Complex, non-linear interactions that emerge during high-energy stochastic events (e.g., storms, floods) may account for large-scale transport and redistribution of fine-grained sediment to all portions of the delta, but has been demonstrated to be especially important to offshore transport (e.g. Ogston et al. 2000). In this case the importance of such offshore mud transport has long been recognized (Swift et al. 1972) from the widespread occurrence of accreting mud wedges on the shelf, but the mechanisms of such transport remained uncertain and controversial until recently (Hill et al. 2007). In the past two decades direct instrumental observations have revealed the regular occurrence of gravity-driven cross-shelf transport occurring off the mouths of most of the world's major rivers (Wright and Friedrichs 2006). This transport phenomenon, which is generated by the interaction of fluvial and marine processes, shares many of the same conditions shown to be necessary for the development of a subaqueous muddy clinothem (Swenson et al. 2005), and probably defines much of the shelf morphology found offshore of large rivers in high-energy settings.

7.3.3.1 Gravity-Driven Sediment Transport

Widespread occurrence of mud deposits and active mud accretion on the middle of continental shelves has long drawn speculation as to the mechanisms responsible for their emplacement (Swift et al. 1972). General observations of focused, rapid accumulation imply an association of these deposits with sediment-laden density currents, which are near-bottom fluid flows that are denser than the overlying water column because of a high concentration of suspended sediment. Turbidity currents are an example of gravity-driven transport, but the gradient of the shelf is typically too low to sustain the high flow velocities needed to maintain continuous sediment suspension and the downslope propagation of such gravity flows. Not only are shelf gradients low, but very few rivers discharge sediment plumes that are hyperpycnal (i.e., denser than the ambient coastal seawater), and this is especially true of the larger, relatively dilute rivers offshore of which shelf mud deposits are most prevalent.

In the past two decades, though, repeated synoptic-scale observations of seabed and water column dynamics during storms and high-discharge flood events have demonstrated that gravity-driven near-bed density flows are a common mode of cross-shelf mud transport (Wright and Friedrichs 2006). The controlling processes and boundary conditions can vary widely, but the fundamental requirements are hyperpycnal near-bed sediment concentrations and a mechanism for maintaining sediment suspension on the low-gradient shelf, typically accomplished by waves and/or tidal currents. These specialized requirements are most typically met when rivers are discharging peak sediment loads onto an energetic shelf, which arguably occurs with the greatest regularity along tide-dominated deltaic margins (Harris et al. 2004). It is uncertain whether this assertion is true because gravity-driven transport is recognized in many margin systems, but it can be said that gravity-driven transport has been documented in all tide-dominated deltas with adequate observations (Wright and Friedrichs 2006).

7.3.3.2 Compound Clinoform Development

As gravity-driven transport is generally associated with high-discharge and high-energy conditions, so too is the development of a compound-clinoform morphology in delta systems (Fig. 7.3; Swenson et al. 2005). The concept of compound clinoforms emerged from investigations of tide-dominated and tide-influenced deltas in the 1980s (e.g. Amazon, Huanghe), when it became clear that these systems supported actively accreting subaqueous deltas that are located substantial distances offshore of, and separate from, their better recognized subaerial landforms (Fig. 7.5; Nittrouer et al. 1986; Prior et al. 1986). The presence of well-developed subaqueous deltas has also been documented for the tide-dominated Ganges-Brahmaputra, Indus, and Changjiang river deltas (Chen et al. 2000; Kuehl et al. 1997; Giosan et al. 2006). In these systems the subaerial clinoform includes primarily the lower delta plain and advancing shoreline that form at the convergence of onshore-directed marine processes and river discharge, whereas the subaqueous clinoform develops at the boundary between shallow-water and deep-water processes (i.e., wave-tide-current transport vs. gravity-driven transport; Swenson et al. 2005).

7.3.4 Sediment Budgets

Tide-dominated deltas are commonly large sediment dispersal systems controlled both by high-energy coastal processes and high-discharge rivers. Their sediment load is widely dispersed with active deltaic sedimentation occurring tens to hundreds of kilometers across and along the continental margin. Therefore, developing sediment budgets for these systems is inherently useful in understanding how they respond to external forcings (e.g., climate, sea level) and how their fluvial, coastal, and marine reaches interact.

One of the first budgets developed for a tide-dominated delta was in the Fly River system, where Harris et al. (1993) could only account for about half ($55 \pm 20\%$) of the annual sediment load of $\sim 85 \times 10^6$ metric tons within the tide-dominated portion of the delta (note: load estimate prior to construction of the Ok Tedi mine). Of the sediment that could be located, roughly equal volumes were apportioned to the lower delta plain (i.e., subaerial clinothem) and deltafront/prodelta system (i.e., subaqueous clinothem). Subsequent work has shown that most of the ‘missing’ fraction is split between deposition on the Fly's vast lowland river floodplain (Swanson et al. 2008) and the actively growing alongshelf clinothem (Slingerland et al. 2008). A similar distribution of sediment was determined for the Ganges-Brahmaputra river delta, where both modern and Holocene budgets show that ~40% of the annual load is trapped within the prograding subaerial and subaqueous clinothems of the tide-dominated portion of the delta. The remaining 60% is distributed about evenly to the fluvial deltaplain through overbank sedimentation and to the Swatch of No Ground canyon that feeds the deep-sea Bengal Fan (Goodbred and Kuehl 1999).

Liu et al. (2009) recently developed budget approximations for several tide-dominated or tide-influenced deltas, showing that 30–40% of the sediment load for the Huanghe (Yellow), Mekong, and Changjiang rivers escape the deltaic depocenters located in the vicinity of the river mouth, similar to the portion observed for the Fly and Ganges-Brahmaputra dispersal systems. In the case of these East Asian examples, though, sediments are advected distances of up to 500–800 km before being deposited as an alongshelf clinothem at inner- to mid-shelf water depths. Prior to these recent studies, it was thought that only the Amazon dispersal system supported such long-distance alongshelf-export of sediment from its river delta (Allison et al. 2000). Aside from their distance, though, these remote clinothems share nearly all characteristics of a prodelta mud wedge, raising the question of whether they should be considered part of the delta system. Regardless of their classification, these findings emphasize that tide-dominated deltas are only part of a larger source-to-sink continuum of interacting continental-margin components (e.g., Goodbred 2003).

7.4 Sedimentary Environments

The sedimentary environments of tide-dominated delta systems can be largely divided into those associated with the 'subaerial' and 'subaqueous' portions of the compound clinoform (Figs. 7.3 and 7.5). The subaerial delta can be further subdivided into a 'lower delta plain' that is influenced by tides and other marine processes and an 'upper delta plain' that is above the tidal influence and dominated by fluvial processes. Offshore the subaqueous delta has often been subdivided into the 'delta front' and 'prodelta', but here we subdivide the clinothem into the 'delta-front platform' (or subtidal delta plain), the 'delta-front slope', and 'prodelta' based on both morphology and sediment facies (Fig. 7.3a). In river-dominated delta systems the subaerial delta, together with the delta-front platform, comprises the topsets of a single deltaic clinoform, with wave-dominated systems often having a definable but closely spaced double clinoform. In the case of most tide-dominated deltas though, these environments are separated by a broad high-shear zone of limited sediment accumulation that separates the prograding subaerial and subaqueous clinoforms of the compound delta system (Nittrouer et al. 1986; Swenson et al. 2005). Beyond the rollover point (i.e. topset-foreset transition) the 'foreset' and 'bottomset' regions of the clinoform correspond to the delta-front slope and prodelta, respectively. Another feature of tide-dominated deltas is that this zonation is irregular along the coast with multiple, wide distributary channels and islands occurring within a funnel-shaped embayment (Fig. 7.5), as compared with wave-dominated deltas where environmental zonation is roughly parallel to the shoreline.

7.4.1 Subaerial Delta

As noted by Middleton (1991) many of the largest rivers discharging to tide-dominated coasts have a principally fine-grained sediment load that forms a mud-dominated delta system. The shoreline of such deltas is often fringed by expansive tidal flats, marshes, and/or mangroves threaded by tidal channels (see Chaps. 8–10). These tidally-dominated environments are characteristic of the intertidal to shallow subtidal zone, particularly at the rivermouth and along adjacent coasts, and may include salt marshes, mangroves, muddy tidal flats, tidal channels, and channel-mouth bars. In tropical to subtropical tide-dominated deltas the subaerial deltaplain comprises broad mangrove-colonized plains that extend from the limits of salt intrusion downward to the upper half of the intertidal zone, where they merge with wide intertidal mud and sand flats in the lower intertidal zone.

This transition between subtidal and supratidal environments is the principal zone of subaerial delta progradation and is largely defined by the development of channel-mouth bars within and just seaward of the active river mouth (Allison 1998). These bars are generally large (10^2–10^4 m) elongate features that extend from shallow subtidal to supratidal elevations, forming within or along the active distributaries of the rivermouth estuary and comprising muddy, sandy, to heterolithic sediments (Fig. 7.5; Chen et al. 1982; Dalrymple 2010). For rivers discharging large sediment loads, such tidal ridges accrete vertically and horizontally, and ultimately merge to form shallow, intertidal flats. These flats eventually become emergent and vegetated to form new delta-plain environments. In this way the growth of tidal ridges marks the incipient stage of delta-plain progradation and is a defining process in tide-dominated deltas (Allison et al. 2003).

The sedimentary facies that characterize the tide-influenced distributaries comprise laminated to thinly bedded sand-mud alternations with tidal signatures, although these are not always well preserved or statistically definable (Dalrymple et al. 2003). Due to the saltwater intrusion into distributary and tidal channels, marine to brackish fauna (e.g. molluscs, foraminifera and ostracods) can be found >100 km upstream of the shoreline. Foraminifera transported by flood tides are recognized even further upstream, presumably transported during low river discharge and high astronomical tidal conditions. However, such patterns are expected to be temporally and spatially variable in complex delta systems, where differences in discharge among active and abandoned distributary may strongly affect onshore transport distances for marine-derived particles.

Along the distributary channel margins, inclined sand-mud alternations are reported from channel slope to tidal flats, which are termed inclined heterolithic stratification (IHS) (Choi et al. 2004). Rhythmic climbing-ripple cross-lamination and neap-spring cycles may also be associated with IHS (Choi 2009). These distributary-channel deposits contain well-sorted fine silt to clay, often derived from near-bed fluid muds (e.g. Fly River; Ichaso and Dalrymple 2009). These sediments with high accumulation rate and large sediment supply can provide indirect evidence of river deltas in the rock record, although they do not necessarily distinguish them from tide-dominated estuaries unless other indicators, such as a progradational stacking of facies, can also be recognized.

Muddy tidal flats are one of the most important components of tide-dominated deltas. The typical sediment facies of this environment comprises sand-mud alternations with flaser, lenticular and wavy laminations or bedding, especially close to the river mouth where sedimentation rates are high and bedding is well preserved (Reineck and Singh 1980). Bidirectional features of sand-layer stacking and cross-laminations, and mud-drapes or double mud-drapes, indicate tidally influenced deposition. These sand-mud layers are basically controlled by cycles of flood-slack-ebb-slack tidal currents, where slack periods produce the draping muds and flood and ebb currents form planar to ripple-laminated sand layers. However, neap-spring tidal cycles are not often recorded in the laminations (Dalrymple and Makino 1989), as much of the record is destroyed by bioturbation, waves, storms, and other events (Fan and Li 2002; Fan et al. 2004, 2006). From the subtidal to intertidal zones, these sediment facies typically show an upward-fining and thinning succession. The thicker and coarser layers in the lower intertidal zone result from more mud settling from the water column at slack tide and stronger currents during flood and ebb for sand transport. The migration of tidal channels and creeks across tidal flats may also generate a typically fining-upward and thinning-upward succession (e.g., Gulf of Papua, Walsh and Nittrouer 2004). Toward the top of the succession in the upper intertidal zone, plant rootlets and peat/peaty sediments become common and reflect transition to a vegetated delta-plain facies with subaerial soil formation (Allison et al. 2003). In tropical to subtropical areas woody mangroves dominate these environments, with tree roots, leaves, and other plant fragments forming peats and organic-rich sediments.

Alternating sand-mud layers also commonly occur within subtidal shoals that form on the delta-front platform and likely represent the incipient phase of channel-mouth bar formation. In the Amazon and Ganges-Brahmaputra deltas these deposits are interbedded or interlaminated sand and mud that are formed under the strong influence of tides, especially the neap–spring cycle (Jaeger and Nittrouer 1995; Michels et al. 1998). The daily tidal exchange is not typically recorded, though, either not being formed or not preserved. The sand layers within the delta-front platform develop through erosion and bedload transport during spring tides, whereas muddy layers are produced under relatively low-energy conditions during neap tides. In case of the Gulf of Papua shelf of the Fly and Kikori deltas, the delta-front platform (topset) shows massive mud with laminated sandy mud, interbedded mud and sand, and bioturbated sandy mud (Dalrymple et al. 2003; Walsh et al. 2004). Some of these thick mud sets

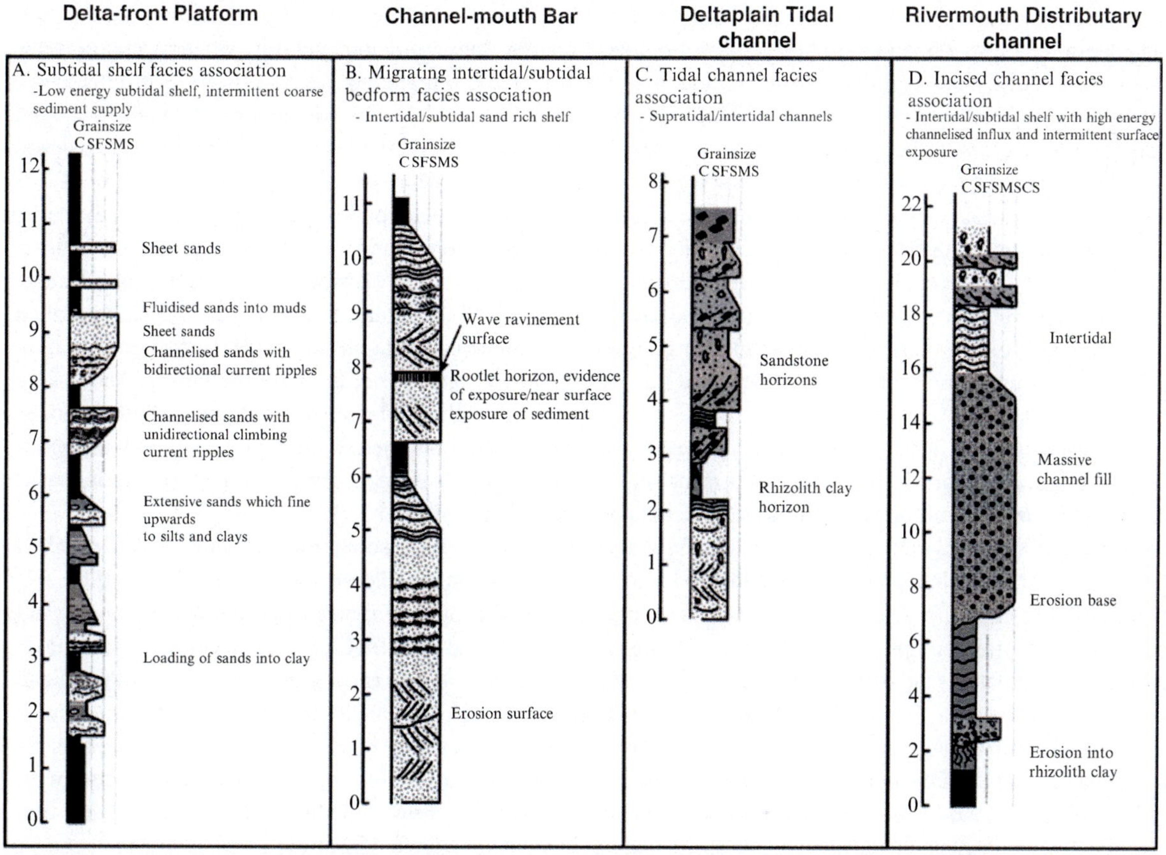

Fig. 7.6 Sketch logs of major facies associations identified from a 500-m thick Miocene-age sequence of the tide-dominated Ganges-Brahmaputra river delta. These facies associations comprise juxtaposed deltaic environments (see Fig. 7.3) that can be found within 50 km of one another in the modern Ganges-Brahmaputra delta system (see Figs. 7.4b and 7.5d). Note that neither the fluvially dominated upper delta plain nor the marine-dominated delta-front slope or prodelta are represented in this thick deltaic section, suggesting limited transgression/regression during this time (After Davies et al. 2003)

on the delta-front slope are likely formed by wave-supported hyperpycnal flows during storm events (Kudrass et al. 1998) and may be correlative with local wave-scoured erosion surfaces on the delta-front platform.

Where wave influence is high at the shoreline, sediment facies in the intertidal zone change significantly with the development of sandy beaches and longshore bars. The Mekong and Red river deltas of Vietnam both have beach ridges with aeolian dunes and foreshore with longshore bars in an intertidal zone in parts of the delta (Thompson 1968; Ta et al. 2005; Tanabe et al. 2006; Tamura et al. 2010). Portions of these deltas are also tide-dominated and characterized by mangroves and tidal channels. Where changes in river, wave, and tidal influence vary through time, reductions in sediment supply to muddy tidal flats can induce erosion and the downdrift formation of sand/shell-mound along the shoreline, called 'cheniers'. Such episodic changes locally form a series of cheniers on the prograding delta plain (Fig. 7.5a; e.g., Changjiang, Mekong).

7.4.2 Subaqueous Delta

Seaward of the muddy subaerial delta and inner delta-front platform, sediments typically coarsen again on the outer delta-front platform toward the rollover point (e.g., Changjiang, Gulf of Papua, Mekong; Hori et al. 2001; Ta et al. 2005). This situation is common for deltas with a relatively shallow rollover where abrupt shoaling across the delta-front slope exposes the outer platform to high wave energy and tidal-current acceleration (Figs. 7.5 and 7.6). Structures on this outer portion of the delta-front platform include fine to medium-scale bedding with wave ripples, hummocky and trough cross-stratification and frequent sharp-based erosional contacts formed by storm-wave scour. Subaqueous dunes are also occasionally reported from this zone of the delta-front platform (Gagliano and McIntire 1968; Kuehl et al. 1997). Overall tidal signatures are not well developed in these deposits despite the strong cross-shelf tidal currents, because of generally lower sedimentation rates and frequent bed resuspension by waves.

At water depths below fair-weather wave base (~5–30 m), sedimentary facies of the delta-front slope are characterized by a coarsening-upward succession of alternating sand and mud deposits (e.g., Changjiang, Mekong, Ganges-Brahmaputra) or laminated to bioturbated muds (e.g., Gulf of Papua, Amazon). Individual bedding units often comprise graded (upward fining) and finely laminated sand–silt layers with sharp basal contacts, such as in the Ganges-Brahmaputra (Michels et al. 1998) and Changjiang deltas. Ripples are also found on the seabed of the delta front of the Changjiang (Chen and Yang 1993). However, clear tidal signatures are not always present in the delta-front slope sediments of tide-dominated deltas, because tidal currents are not usually well-developed this far offshore. Similarly, prodelta sediments even further offshore are often highly bioturbated and intercalated with silt stringers and thin shell beds. The shell beds result primarily from storms, which may also transport coarser-grained sediments to the prodelta. In contrast to the prevalent tide-dominated facies formed in the delta-plain distributaries and the adjacent intertidal to subtidal delta-front platform, the delta-front slope to prodelta environments are mostly influenced by waves, ocean currents, and storms.

7.4.3 Facies Associations

Because many factors can influence the formation of stratigraphic sequences over 10^3–10^5 years, it is also useful to consider mesoscale facies associations that characterize the various subenvironments of tide-dominated deltas (Fig. 7.6; Gani and Bhattacharya 2007; Heap et al. 2004). A facies association is a group of sedimentary facies that are typically found together and define a particular environment, but also allow for local variability in lithology, structure, and stratal relationships. In deltaic settings where accretion rates are relatively high, facies associations record delta progradation and lobe development that typically occurs at timescales of 10^1–10^3 years . For tide-dominated deltas the most frequently described facies association is that of the lower delta plain, which captures the advancing deltaic shoreline and subtidal to supratidal transition (Allison et al. 2003; Harris et al. 1993; Hori et al. 2002a, b; Ta et al. 2002; Dalrymple et al. 2003). As described from numerous delta-plain systems, the facies association comprises an 8–10 m thick, fining upward succession starting with sandy, cross-stratified subtidal shoals, which grade into heterolithic intertidal mud-sand couplets and are capped by a rooted mud-dominated supratidal soil (Fig. 7.7).

Other facies associations that have been described for tide-dominated deltas include tidal bars, tidal gullies and channels, incised distributary channels, and the subtidal shelf (Fig. 7.7; Davies et al. 2003; McCrimmon and Arnott 2009; Tänavsuu-Milkeviciene and Plink-Björklund 2009). The tidal-bar facies association is variably described as a fining-up or coarsening-up succession of cross-stratified sand with bidirectional flow indicators and inclined planes that is very similar to, if not the same as, the portion of the delta-plain facies association (Fig. 7.6b). The difference between the upward-fining and upward-coarsening descriptions is likely related to their proximity to the active distributary mouth, the fining-up example being more proximal to the rivermouth and receiving abundant sediment to make a rapid transition from subtidal to vegetated intertidal setting, whereas the coarsening-up succession may be a more wave-tide dominated downdrift littoral deposit. The tidal gullies and distributary channels are regularly described as fining-up, current-rippled to planar-bedded deposits with a sharp, often incised, lower contact. However, the most characteristic features of these facies associations is the regular occurrence of mud clasts that reflect the local reworking of shallow intertidal and supratidal delta-plain deposits as channels migrate, avulse, and incise (Fig. 7.6c; Dalrymple et al. 2003; Davies et al. 2003; Tänavsuu-Milkeviciene and Plink-Björklund 2009). On average, though, tidal channels are relatively laterally stable (e.g. Fagherazzi 2008) and so the muddy delta-plain deposits that cap tidal-channel sands are commonly preserved in the upper stratigraphy of the subaerial delta clinothem.

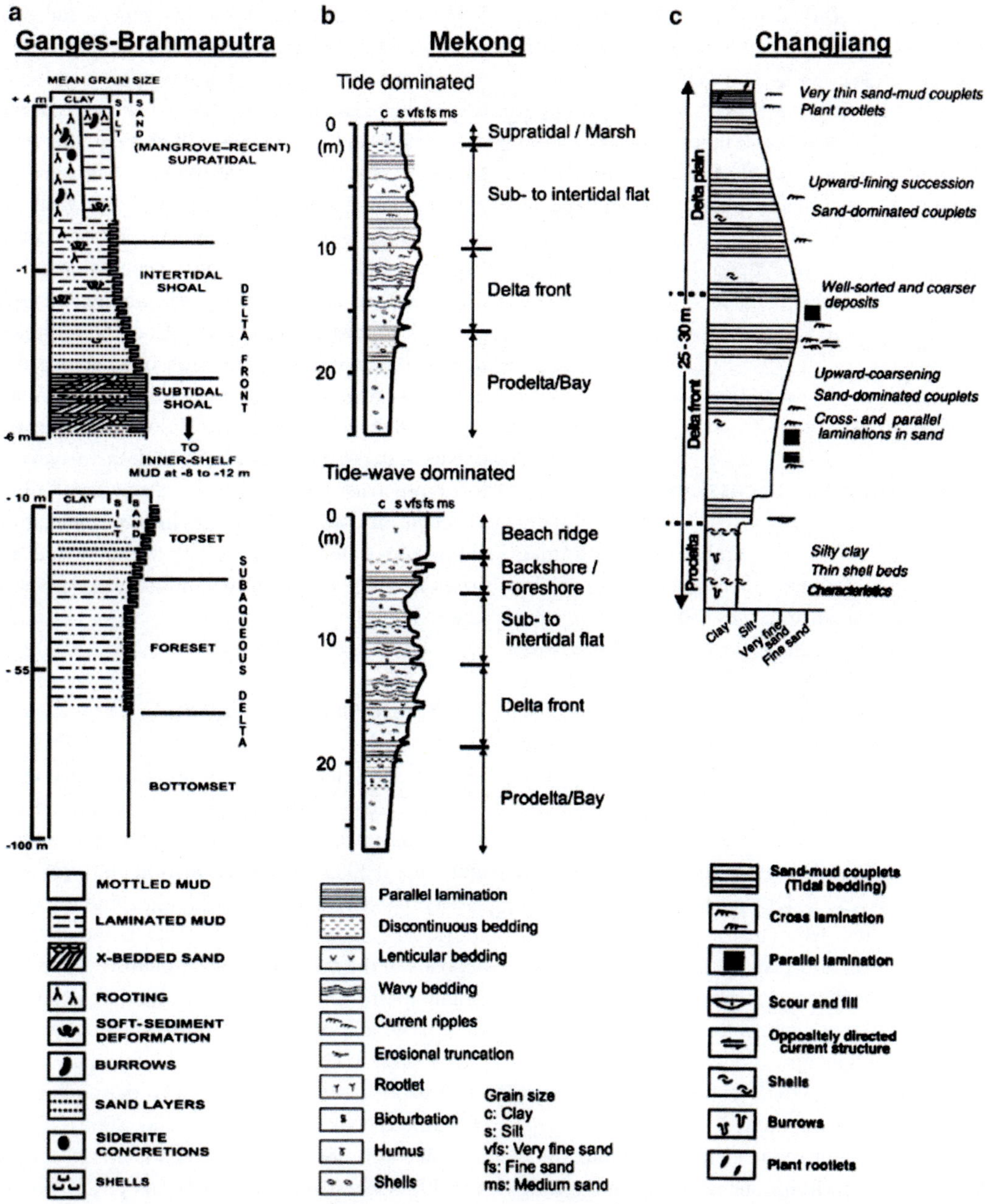

Fig. 7.7 Stratigraphic succession models for three major tide-dominated delta systems, (**a**) the Ganges-Brahmaputra, (**b**) the Mekong, and (**c**) the Changjiang. Each model includes the lower coarsening-up subaqueous clinothem overlain by the upper, generally fining-up, subaerial clinothem. The Mekong example also shows an alternate coarsening-up model that is characteristic of more wave-influenced portions of the delta where beach ridges are well developed at the shoreface (Modified after Kuehl et al. 2005; Ta et al. 2002; Hori et al. 2002a, respectively)

Offshore facies associations are less frequently described for tide-dominated deltas, in part because of sampling constraints in modern examples, but also because tidal signatures become increasingly weak offshore and may not be recognized in the rock record. This potential bias may explain early confusion with interpreting the Sego sandstones (Book Cliffs, USA), which are incised into marine shales and thus described as various types of forced regression deposits in a tidally influenced setting (Van Wagoner et al. 1991; Yoshida et al. 1996). Willis and Gabel (2001, 2003) have since argued that the Sego Sandstone actually represent the tidal channels and inner shelf sand sheet of a tide-dominated delta system, which incised into its own muddy delta-front platform and prodelta deposits during progradation. Such a mud-incised succession of prograding tidal channel deposits has also been described from the Miocene-age record of the Ganges-Brahmaputra delta (Fig. 7.6d; Davies et al. 2003).

7.5 Stratigraphy

7.5.1 Stratigraphic Successions

Deltas are defined as discrete shoreline deposits formed where rivers supply sediment more rapidly than can be redistributed by basinal processes (Elliott 1986); thus shoreline advance is essential for distinguishing them from estuaries, which also occur at river mouths but are transgressive depositional systems. As defined, deltas are regressive prograding to aggrading systems (Boyd et al. 1992; Dalrymple et al. 1992). Therefore deltaic successions will overall shallow upward, ideally including facies associations from prodelta, delta-front slope, delta-front platform, and delta-plain environments, in ascending order (Fig. 7.7; Dreyer et al. 2005).

In tide-dominated deltas that support a compound clinothem with prograding subaerial and subaqueous deltaic units, the idealized stratigraphic succession can be subdivided into two major intervals (Fig. 7.7). The lower portion shows an upward-coarsening facies succession from the prodelta to delta-front slope and outer platform deposits that is marked at its top by sharp-based wave and current scours. This lower interval is overlain by an upward-fining succession of prograding deposits from the inner delta-front platform and shoaling to subaerial delta-plain facies. The upper interval is most typically represented by the delta-plain facies association (see Sect. 7.4.1), but may also include local sub-environments such as tidal channel bars or estuarine distributary associations. Within the overall deltaic succession, the coarsest and most well-sorted deposits typically occur in the boundary zone between the delta-front platform and slope, and secondarily in the prograding, distributary-mouth channel bars (Coleman 1981; Hori et al. 2001, 2002b; Dalrymple et al. 2003; Tänavsuu-Milkeviciene and Plink-Björklund 2009). With only modest variation this general succession of an upward-coarsening subaqueous-delta unit overlain by an upward fining subaerial-delta unit has been documented in many of the world's modern tide-dominated delta systems, including the Ganges-Brahmaputra (Allison et al. 2003), Mekong (Ta et al. 2002), Changjiang (Hori et al. 2001), and Fly (Harris et al. 1993; Dalrymple et al. 2003). Such similarity suggests that this stratigraphic succession may be a useful tool in distinguishing tide-dominated deltas in the rock record (Willis 2005). Local variation in the tide-dominated delta succession has been recognized in the Mekong system, which has become increasingly wave influenced in the late Holocene and shows an upward-coarsening succession ending in wave-swept foreshore to aeolian beach-ridge deposits (cf. Fig. 7.6b, lower profile; Ta et al. 2002). In the Mahakam delta, alongshore heterogeneity in stratigraphic successions arises from the greater fluvial influence relative to tidal reworking (Gastaldo et al. 1995).

7.5.2 Delta Progradation

The rate of delta progradation can strongly influence the delta facies succession. As the subaerial delta progrades basinward, the tidal distributary channels can incise up to 20 m into the delta-front platform deposits, and a relative rise of sea level (e.g., commonly through subsidence) is important in order to preserve topset deposits of the outer delta-front platform. The Ganges-Brahmaputra and Mahakam deltas are examples of such progradational and aggradational deltas that display a largely continuous and conformable Holocene succession from prodelta to delta-plain facies (Goodbred et al. 2003; Storms et al. 2005). If distributary channels are stable relative to delta progradation, a delta succession will form as described above. However, if the lateral migration of distributaries is fast relative to delta progradation, then much of the delta-front facies will be replaced by distributary-channel fill, which is thought to occur in the Fly river delta (Dalrymple et al. 2003).

7.5.3 Role of Sea-Level Change

Sea-level change can also force environmental changes that may appear similar to delta progradation in the stratigraphic record. During periods of sea-level fall there is a forced regression of the shoreline that drives delta progradation and potentially downward incision. If the drop in sea level is relatively fast compared to the rate of delta progradation, then the succession should shift toward a more fluvially dominated stratigraphy with decreasing marine and tidal influence (Bhattacharya 2006). However, with further sea-level fall and a narrowing of the shelf, tidal range will ultimately drop and tidal energy will decrease considerably relative to a growing wave influence. It might therefore be inferred that tide-dominated deltas are more generally highstand features, as adequate tidal energy is less well developed during lowstands due to narrow shelf widths. Indeed meso- to macrotidal conditions in the modern are associated exclusively with broad shelves or large drowned valleys and embayments. Regional morphology of the continental margin (e.g. rift settings, epicontinental seas) could maintain tidal amplification even during lowstand, though, in such settings as the Cretaceous Western Interior Seaway (Bhattacharya and Willis 2001) and the Gulf of California.

Sea-level rise following a lowstand leads to the transgression and marine inundation of incised valleys formed during the previous fall of sea level. Riverine sediments are effectively trapped in these valleys to form fluvial and coastal plains, resulting in sediment starvation on the adjacent shelf and the formation of a ravinement surface and condensed section (Hori et al. 2004; Goodbred and Kuehl 2000). Continued sea-level rise and transgression of the shelf and valleys will tend to favor tidal amplification and the development of tide-influenced or tide-dominated environments (Uehara et al. 2002; Uehara and Saito 2003), although such responses are also dependent on shelf and shoreline physiography. If sediment supply is sufficient relative to the rate of sea-level rise, though, then these transgressive estuarine settings will evolve into deltas with an associated change in shoreline trajectory from landward to seaward. When constrained within the incised valleys, such highstand deltaic successions typically overlie transgressive estuarine sediments along the maximum flooding surface (Hori et al. 2002a, b; Tanabe et al. 2006). Where deltas have infilled their lowstand valley, channel avulsion and migration to interfluve areas will lead to delta-lobe formation directly on the lowstand exposure surface and sequence boundary (Goodbred and Kuehl 2000; Ta et al. 2005). In some cases, such as the Mekong and Red river deltas, tidal dominance may wane as the delta progrades into the estuarine embayment and coastal morphology shifts from concave to convex, making the system more wave-dominated as the delta lobe faces more open ocean (Ta et al. 2005; Tanabe et al. 2006).

7.6 Summary

Tide-dominated deltas are an end member of the river-wave-tide ternary delta classification and have been studied in earnest only since the 1970s. Several comprehensive research programs during the 1980s and 1990s developed a sound knowledgebase on the hydrodynamics, sediment transport and marine processes, and strata formation in tide-dominated deltaic settings. More recent research on modern deltas, particularly studies involving the drilling of cores and the collection of observational data, have accelerated our understanding of the specific sedimentary environments, processes, and stratigraphic successions found within and around tide-dominated deltaic settings.

Today most modern tide-dominated deltas are building seaward through modestly prograding deltaplains and more rapidly prograding muddy subaqueous clinothems. The sedimentary facies within these settings are typically, perhaps characteristically, heterolithic and often mud-dominated (e.g. Changjiang, Fly), although some systems may have an appreciable sand component (e.g. Ganges-Brahmaputra). In contrast, most sections of the rock record that have been interpreted as tide-dominated deltas comprise sand-dominated, or alternating sand-mud, sedimentary facies. This apparent bias toward coarse-grained ancient examples may arise from the difficulty of distinguishing deltaic successions from other mud-dominated sedimentary facies, many of which may lack clear indicators of fluvial origin due to the strong overprint of tidal processes. The broad distances across which many modern tide-dominated deltas develop also present a challenge at the outcrop scale, and differences in fluvial sediment input (e.g., coarse vs. fine) may further limit the recognition of unique facies characteristics.

In terms of human impacts, more than 200 million people live in tide-dominated delta systems today, ranking them among the world's most economically and culturally important environments. In many systems the mangroves, salt marshes, and tidal flats typical of tide-dominated delta systems are threatened by human activities. Several modern deltas are already severely degraded due to decreases in sediment and freshwater delivery caused by damming and water extraction, respectively (e.g. Colorado, Indus). Similar modifications and activities have been implemented along the Yangtze river system, with anticipated negative impacts; damming and water consumption remain likely threats to the heavily populated Ganges-Brahmaputra and Ayeyarwady basins as well. Regardless, sustainable ways to conserve and use these environments will be a continuing challenge.

References

Allison MA (1998) Historical changes in the Ganges–Brahmaputra delta front. J Coast Res 14:1269–1275

Allison MA, Lee MT, Ogston AS, Aller RC (2000) Origin of mudbanks along the northeast coast of South America. Mar Geol 163:241–256

Allison MA, Khan SR, Goodbred SL, Kuehl SA (2003) Stratigraphic evolution of the late Holocene Ganges–Brahmaputra lower delta plain. Sediment Geol 155(317–342):1594–1597

Barua DK, Kuehl SA, Miller RL, Moore WS (1994) Suspended sediment distribution and residual transport in the coastal ocean off the Ganges-Brahmaputra river mouth. Mar Geol 120:41–61

Bhattacharya JP (2006) Deltas. In: Posamentier HW, Walker RG (eds) Facies Models Revisited, SEPM, Spec Publ 84. SEPM, Tulsa, pp 237–292

Bhattacharya JP, Giosan L (2003) Wave-influenced deltas: geomorphological implications for facies reconstruction. Sedimentology 50:187–210

Bhattacharya JP, Willis BJ (2001) Lowstand deltas in the Frontier Formation, Powder River basin, Wyoming: Implications for sequence stratigraphic models. Am Assoc Petol Geol Bull 85:261–294

Boyd R, Dalrymple RW, Zaitlin BA (1992) Classification of clastic coastal depositional environments. Sediment Geol 80:139–150

Carriquiry JD, Sanchez A (1999) Sedimentation in the Colorado River delta and upper Gulf of California after nearly a century of discharge loss. Mar Geol 158:125–145

Chen W, Yang Z (1993) Study of subaqueous slope instability of the modern Changjiang River delta. In: Hopley D, Wang Y (eds) Proceedings of the 1993 PACON China symposium, estuarine coastal processes. Estuarine Coastal Processes, Resource Exploration and Management, pp 133–142

Chen J, Zhu S, Lu Q, Zhou Y, He S (1982) Descriptions of the morphology and sedimentary structures of the river mouth bar in the Changjiang estuary. In: Kennedy VS (ed) Estuarine comparisons. Academic, New York, pp 667–676

Chen ZY, Song BP, Wang ZH, Cai YL (2000) Late Quaternary evolution of the sub-aqueous Yangtze Delta, China: sedimentation, stratigraphy, palynology, and deformation. Mar Geol 162:243–441

Choi K (2009) Rhythmic climing-ripple cross-lamination in inclined heterolithic stratification. (HIS) of a macrotidal estuarie channel, Gomso Bay, west coast of Korea. J Sediment Res 80:550–561

Choi KS, Dalrymple RW, Chun SS, Kim S-P (2004) Sedimentology of modern, inclined heterolithic stratification (IHS) in the macrotidal Han River delta. Korea J Sed Res 74:677–689

Coleman JM, Gagliano SM, Smith WG (1970) Sedimentation in a Malaysian high tide tropical delta. In: Morgan JP (ed) Deltaic sedimentation: Modern and Ancient, SEPM Spec Publ 15. SEPM, Tulsa, pp 185–197

Coleman JM (1981) Deltas: processes of deposition and models for exploration. Burgess Publishing, Minneapolis

Coleman JM, Wright LD (1978) Sedimentation in an arid macrotidal alluvial river system: Ord River, Western Australia. J Geol 86:621–642

Dalrymple RW (2010) Tidal depositional systems. In: James NP, Dalrymple RW (eds) Facies models 4. Geological Association Canada, St. John's, pp 199–208

Dalrymple RW, Choi KS (2007) Morphologic and facies trends through the fluvial–marine transition in tide-dominated depositional systems: a schematic framework for environmental and sequence-stratigraphic interpretation. Earth Sci Rev 81:135–174

Dalrymple RW, Makino Y (1989) Description and genesis of tidal bedding in the Cobequid Bay-Salmon River estuary, Bay of Fundy, Canada. In: Taira A, Masuda F (eds) Sedimentary facies in the active plate margin. Terra Science, Tokyo, pp 151–177

Dalrymple RW, Zaitlin BA, Boyd RA (1992) A conceptual model of estuarine sedimentation. J Sediment Petrol 62:1130–1146

Dalrymple RW, Baker EK, Harris PT, Hughes M (2003) Sedimentology and stratigraphy of a tide-dominated, foreland–basin delta (Fly River, Papua New Guinea). SEPM Spec Publ 76:147–173

Davies C, Best J, Collier R (2003) Sedimentology of the Bengal shelf Bangladesh: comparison of late Miocene sediments, Sitakund anticline, with the modem, tidally dominated shelf. Sediment Geol 155:271–300

Davis RA, Hayes MO (1984) What is a wave-dominated coast? Mar Geol 60:313–329

Dreyer T, Whitaker M, Dexter J, Flesche H, Larsen E (2005) From spit system to tide-dominated delta: integrated reservoir model of the Upper Jurassic Sognefjord Formation on the Troll West Field. Geol Soc Lond, Petrol Geol Conf Ser 6:423–448. doi:10.1144/0060423

Elliott T (1986) Deltas. In: Reading HG (ed) Sedimentary environments and facies, 2nd edn. Blackwell Scientific, Oxford, pp 113–154

Fagherazzi S (2008) Self-organization of tidal deltas. Proc Natl Acad Sci USA 105:18692–18695

Fagherazzi S, Overeem I (2007) Models of deltaic and inner continental shelf evolution. Ann Rev Earth Planet Sci Rev 35:685–715

Fan D, Li C (2002) Rhythmic deposition on mudflats in the mesotidal Changjiang estuary, China. J Sediment Res 72:543–551

Fan D, Li CX, Wang DJ, Wang P, Archer AW, Greb SF (2004) Morphology and sedimentation on open-coast intertidal flats of the Changjiang delta, China. J Coast Res SI 43:23–35

Fan D, Guo Y, Wang P, Shi JZ (2006) Cross-shore variations in morphodynamic processes of an open-coast mudflat in the

Changjiang Delta, China: with an emphasis on storm impacts. Cont Shelf Res 26:517–538

Gagliano SM, McIntire WG (1968) Reports on the Mekong Delta. Coastal Studies Institute, Louisiana State University Technical Report 57, 144 p

Galloway WE (1975) Process framework for describing the morphologic and stratigraphic evolution of deltaiv depositional systems. In: Broussard ML (ed) Deltas: models for exploration. Houston Geological Society, Houston, pp 87–98

Gani MR, Bhattacharya JP (2007) Basic building blocks and process variability of a Cretaceous delta: internal facies architecture releals a more dynamic interaction of river, wave, and tidal processes. J Sediment Res 77:284–302

Gastaldo RA, Allen GP, Huc A-Y (1995) The tidal character of fluvial sediments of the modern Mahakam River delta, Kalimantan, Indonesia. Int Assoc Sediment Spec Publ 24:171–181

Giosan L, Constantinescu S, Clift PD, Tabrez AR, Danish M, Inam A (2006) Recent morphodynamics of the Indus delta shore and shelf. Cont Shelf Res 26:1668–1684

Goodbred SL Jr (2003) Response of the Ganges dispersal system to climate change: a source-to-sink view since the last interstade. Sediment Geol 162:83–104

Goodbred SL Jr, Kuehl SA (1999) Holocene and modern sediment budgets for the Ganges–Brahmaputra River: evidence for highstand dispersal to floodplain, shelf, and deep-sea depocenters. Geology 27:559–562

Goodbred SL Jr, Kuehl SA (2000) The significance of large sediment supply, active tectonism and eustasy on margin sequence development: Late Quaternary stratigraphy and evolution of the Ganges–Brahmaputra delta. Sediment Geol 133:227–248

Goodbred SL Jr, Kuehl SA, Steckler MS, Sarker MH (2003) Controls on facies distribution and stratigraphic preservation in the Ganges–Brahmaputra delta sequence. Sediment Geol 155:301–316

Harris PT, Baker EK, Cole AR, Short SA (1993) A preliminary study of sedimentation in the tidally dominated Fly River Delta, Gulf of Papua. Cont Shelf Res 13:441–472

Harris PT, Pattiaratchi CB, Keene JB, Dalrymple RW, Gardner JV, Baker EK, Cole AR, Mitchell D, Gibbs P, Schroeder WW (1996) Late Quaternary deltaic and carbonate sedimentation in the Gulf of Papua foreland basin: response to sea-level change. J Sediment Res 66:801–819

Harris PT, Hughes MG, Baker EK, Dalrymple RW, Keene JB (2004) Sediment transport in distributary channels and its export to the pro-deltaic environment in a tidally dominated delta: Fly River, Papua New Guinea. Cont Shelf Res 24:2431–2454

Heap AD, Bryce S, Ryan DA (2004) Facies evolution of Holocene estuaries and deltas: a large-sample statistical study from Australia. Sediment Geol 168:1–17

Hill PS, Fox JS, Crockett JS, Curran KJ, Friedrichs CT, Geyer WR, Milligan TG, Ogston AS, Puig P, Scully ME, Traykovski PA, Wheatcroft RA (2007) Sediment delivery to the seabed on continental margins. In: Nittrouer CA, Austin JA, Field MA, Kravitz JH, Syvitski JPM, Wiberg PL (eds) Continental margin sedimentation: from sediment transport to sequence stratigraphy. Blackwell Publishing, Oxford, pp 49–99

Hori K, Saito Y (2007) Classification, architecture, and evolution of large-river deltas. In: Gupta A (ed) Large rivers: geomorphology and management. Wiley, Chichester, pp 75–96

Hori K, Saito Y, Zhao Q, Cheng X, Wang P, Sato Y, Li C (2001) Sedimentary facies and Holocene progradation rates of the Changjiang (Yangtze) delta, China. Geomorphology 41:233–248

Hori K, Saito Y, Zhao Q, Wang P (2002a) Architecture and evolution of the tide-dominated Changjiang (Yangtze) River delta, China. Sediment Geol 146:249–264

Hori K, Saito Y, Zhao Q, Wang P (2002b) Evolution of the coastal depositional systems of the Changjiang (Yangtze) river in response to late Pleistocene– Holocene sea-level changes. J Sediment Res 72:884–897

Hori K, Tanabe S, Saito Y, Haruyama S, Nguyen V, KitamuraI A (2004) Delta initiation and Holocene sea-level change: example from the Song Hong (Red River) delta, Vietnam. Sediment Geol 164:237–249

Ichaso AA, Dalrymple RW (2009) Tide- and wave-generated fluid mud deposits in the Tilje Formation (Jurassic), offshore Norway. Geology 37:539–542

Jaeger JN, Nittrouer CA (1995) Tidal controls on the formation of fine-scale sedimentary strata near the Amazon river mouth. Mar Geol 125:259–281

Kudrass HR, Michels KH, Wiedicke M, Suckow A (1998) Cyclones and tides as feeders of a submarine canyon off Bangladesh. Geology 26:715–718

Kuehl SA, DeMaster DJ, Nittrouer CA (1986) Nature of sediment accumulation on the Amazon continental shelf. Cont Shelf Res 6:209–225

Kuehl SA, Levy BM, Moore WS, Allison MA (1997) Subaqueous delta of the Ganges–Brahmaputra river system. Mar Geol 144:81–96

Kuehl SA, Allison MA, Goodbred SL, Kudrass HR (2005) The Ganges-Brahmaputa Delta. In: Giosan L, Bhattacharya JP (eds) River Deltas–Concepts, Models, and Examples, SEPM Special Publication, 83. SEPM, Tulsa, pp 87–129

Kummu M, Lu XX, Wang JJ, Varis O (2010) Basin-wide sediment trapping efficiency of emerging reservoirs along the Mekong. Geomorphology 119:181–197

Liu JP, Xue Z, Ross K, Wang HJ, Yang ZS, Li AC, Gao S (2009) Fate of sediments delivered to the sea by Asian large rivers: long-distance transport and formation of remote alongshore clinothems. Sediment Rec 7:4–9

McCrimmon GG, Arnott RWC (2009) The Clearwater Formation, Cold Lake, Alberta: a worldclass hydrocarbon reservoir hosted in a complex succession of tide-dominated deltaic deposits. Bull Can Petrol Geol 50:370–392

Michels KH, Kudrass HR, Hubscher C, Suckow A, Wiedicke M (1998) The submarine delta of the Ganges–Brahmaputra: cyclone-dominated sedimentation patterns. Mar Geol 149:133–154

Middleton GV (1991) A short historical review of clastic tidal sedimentology. In: Smith DG, Reinson GE (eds) clastic tidal sedimentology, Can. Soc. Petrol. Geol. Memoir 16. Canadian Society of Petroleum Geologists, Calgary, pp ix–xv

Milliman JD, Jin QM ed (1985) Sediment dynamics of the Changjiang Estuary and the Adjacent East China Sea. Cont Shelf Res 4:1–254

Milliman JD, Meade RH (1983) World-wide delivery of river sediment to the oceans. J Geol 91:1–21

Milliman JD, Syvitski JPM (1992) Geomorphic/tectonic control of sediment discharge to the ocean: the importance of small mountainous rivers. J Geol 100:525–544

Montaño Y, Carbajal N (2008) Numerical experiments on the long-term morphodynamics of the Colorado River Delta. Ocean Dyn 58:19–29

Nittrouer CA, DeMaster DJ (eds) (1986) Sedimentary processes on the Amazon continental shelf. Cont Shelf Res 6:1–361

Nittrouer CA, Kuehl SA, DeMaster DJ, Kowssmann RO (1986) The deltaic nature of Amazon shelf sedimentation. Geol Soc Am Bull 97:444–458

Ogston AS, Cacchione DA, Sternberg RW, Kineke GC (2000) Observations of storm and river flood-driven sediment transport on the northern California continental shelf. Cont Shelf Res 20:2141–2162

Ogston AS, Sternberg RW, Nittrouer CA, Martin DP, Goñi MA, Crockett JS (2008) Sediment delivery from the Fly River tidally dominated delta to the nearshore marine environment and the impact of El Niño. J Geophys Res 113:F01S11. doi:10.1029/2006JF000669

Orton GJ, Reading HG (1993) Variability of deltaic processes in terms of sediment supply, with particular emphasis on grain size. Sedimentology 40:475–512

Postma H (1967) Sediment transport and sedimentation in the marine environment. In: Lauff GH (ed) Estuaries, American Association for the Advancement of Science, Publication 83. American Association for the Advancement of Science, Washington, DC, pp 158–186

Prior DB, Yang Z-S, Bornhoid BD, Keller GH, Lin ZH, Wiseman WJ Jr, Wright LD, Lin TC (1986) The subaqueous delta of the modern Huanghe (Yellow River). Geo-Mar Lett 6:67–75

Reineck HE, Singh IB (1980) Depositional Sedimentary Environments. Springer, Berlin, 551 pp

Russell RJ, Russell RD (1939) Mississippi River delta sedimentation. In: Trask PD (ed) Recent Marine Sediments. Amer Assoc Petrol Geol, Tulsa, pp 151–177

Slingerland R, Driscoll NW, Milliman JD, Miller SR, Johnstone EA (2008) Anatomy and growth of a Holocene clinothem in the Gulf of Papua. J Geophys Res 113:F01S13. doi:10.1029/2006JF000628

Storms JEA, Hoogendoorn RM, Dam RAC, Hoitink AJF, Kroonenberg SB (2005) Late-Holocene evolution of the Mahakam delta, east Kalimantan, Indonesia. Sediment Geol 180:149–166

Swanson KM, Watson E, Aalto R, Lauer JW, Bera MT, Marshall A, Taylor MP, Apte SC, Dietrich WE (2008) Sediment load and floodplain deposition rates: comparison of the Fly and Strickland rivers, Papua New Guinea. J Geophys Res 113:F01S03. doi:10.1029/2006JF000623

Swenson JB, Paola C, Pratson L, Voller VR, Murray AB (2005) Fluvial and marine controls on combined subaerial and subaqueous delta progradation: Morphodynamic modeling of compound-clinoform development. J Geophys Res 110(F2):1–16

Swift DJP, Duane D, Pilkey OH (eds) (1972) Shelf sediment transport: process and pattern. Douden Hutchinson & Ross, Stroudsburg

Syvitski JPM, Kettner AJ, Overeem I, Hutton EWH, Hannon MT, Brakenridge GR, Day J, Vörösmarty CJ, Saito Y, Giosan L, Nicholls RJ (2009) Sinking deltas due to human activities. Nat Geosci 2:681–689

Ta TKO, Nguyen VL, Tateishi M, Kobayashi I, Saito Y, Nakamura T (2002) Sediment facies and late Holocene progradation of the Mekong River Delta in Bentre Province, southern Vietnam: an example of evolution from a tide-dominated to a tide- and wave-dominated delta. Sediment Geol 152:313–325

Ta TKO, Nguyen VL, Tateishi M, Kobayashi I, Saito Y (2005) Holocene delta evolution and depositional models of the Mekong River Delta, southern Vietnam. In: Giosan L, Bhattacharya JP (eds) River Deltas – Concepts, Models and Examples, SEPM Special Publication 83. SEPM, Tulsa, pp 453–466

Tamura T, Horaguchi K, Saito Y, Nguyen VL, Tateishi M, Ta TKO, Nanayama F, Watanabe K (2010) Monsoon-influenced variations in morphology and sediment of a mesotidal beach on the Mekong River delta coast. Geomorphology 116:11–23

Tanabe S, Saito Y, Vu QL, Hanebuth TJJ, Ngo QL (2006) Holocene evolution of the Song Hong (Red River) delta system, northern Vietnam. Sediment Geol 187:29–61

Tänavsuu-Milkeviciene K, Plink-Björklund P (2009) Recognition of tide-dominated versus tide-influenced deltas: Middle Devonian strata of the Baltic Basin. J Sediment Res 79:887–905

Thompson RW (1968) Tidal flat sedimentation on the Colorado River delta, northwestern Gulf of California. Geological Society of America Memoir, 107, 133 p

Uehara K, Saito Y (2003) Late Quaternary evolution of the Yellow/East China Sea tidal regime and its impacts on sediments dispersal and seafloor morphology. Sediment Geol 162:25–38

Uehara K, Saito Y, Hori K (2002) Paleotidal regime in the Changjiang (Yangtze) Estuary, the East China Sea, and the Yellow Sea at 6 ka and 10 ka estimated from a numerical model. Mar Geol 183:179–192

Van Wagoner JC, Nummedal D, Jones CR, Taylor DR, Jennette DC, Riley GW (eds) (1991) Sequence stratigraphy applications to shelf sandstone reservoirs. American Association of Petroleum Geologists, Field Conference Guidebook

Walsh JP, Nittrouer CA (2004) Mangrove sedimentation in the Gulf of Papua, Papua New Guinea. Mar Geol 208:225–248

Walsh JP, Nittrouer CA (2009) Understanding fine-grained river-sediment dispersal on continental margins. Mar Geol 263:34–45

Walsh JP, Nittrouer CA, Palinkas CM, Ogston AS, Sternberg RW, Brunskill GJ (2004) Clinoform mechanics in the Gulf of Papua, New Guinea. Cont Shelf Res 24:2487–2510

Willis BJ (2005) Deposits of tide-influenced river deltas. In: Giosan L, Bhattacharya JP (eds) River Deltas–Concepts, Models, and Examples, SEPM Special Publication, 83. SEPM, Tulsa, pp 87–129

Willis BJ, Gabel S (2001) Sharp-based, tide-dominated deltas of the SegoSandstone, Book Cliffs, Utah, USA. Sedimentology 48:479–506

Willis BJ, Gabel SL (2003) Formation of deep incisions into tide-dominated river deltas: implication for the stratigraphy

of the Sego sandstone, Book Cliffs, Utah, U.S.A. J Sediment Res 73:246–263
Wolanski E, Spagnol S (2000) Environmental degradation by mud in tropical estuaries. Reg Environ Chang 1:152–162
Wolanski E, King B, Galloway D (1995) Dynamics of the turbidity maximum in the Fly River estuary, Papua New Guinea. Estuarine, Coast Shelf Sci 40:321–337
Wolanski E, Nguyen NH, Le TD, Nguyen HN, Nguyen NT (1996) Fine-sediment dynamics in the Mekong River estuary, Vietnam. Estuarine, Coast Shelf Sci 43:565–582
Woodroffe CD, Nicholls RJ, Saito Y, Chen Z, Goodbred SL (2006) Landscape variability and the response of Asian megadeltas to environmental change. In: Harvey N (ed) Global change and integrated coastal management: the Asia-Pacific region. Springer, Berlin, pp 277–314
Wright LD, Coleman JM (1971) The discharge/wave power climate and the morphology of delta coasts. Assoc Am Geograph Proc 3:186–189
Wright LD, Friedrichs CT (2006) Gravity-driven sediment transport on continental shelves: a status report. Continental Shelf Research 26:2092–2107
Yang Z, Wang H, Saito Y, Milliman JD, Xu K, Qiao S, Shi G (2006) Dam impacts on the Changjiang (Yangtze River) sediment discharge to the sea: the past 55 years and after the Three Gorges Dam. Water Resources Research 42:W04407. doi:10.1029/2005WR003970
Yoshida S, Willis A, Miall AD (1996) Tectonic control of nested sequence architecture in the Castlegate Sandstone (Upper Cretaceous), Book Cliffs, Utah. J Sediment Res 66:737–748

Salt Marsh Sedimentation

8

Jesper Bartholdy

Abstract

This chapter deals with salt marsh sedimentation with emphasis on depositional processes and resulting products. Salt marsh sedimentation and related dynamic conditions are evaluated and described with examples from a wide range of locations. General mechanisms and depositional conditions are primarily illustrated by examples from the Danish Wadden Sea based on the author's own experience. The chapter opens with an overview over measurements of salt marsh sedimentation through time and a general description of salt marsh morphodynamics, including an assessment of the effects of vegetation. Salt marsh sediments and autocompaction are discussed prior to a description of salt marsh accretion models. The latter is used to give examples of salt marsh stability in relation to different tidal conditions and sea-level-rise scenarios. The chapter concludes with a description of salt marshes in the geological record.

8.1 Introduction

As part of the coastal zone, salt marshes are defined here as: 'vegetated areas located between coastal hinterlands and daily (or permanently) flooded coastal areas'. Thus, salt marshes form a buffer zone between areas where coastal and estuarine processes act on a daily basis and areas that are either never flooded or only flooded during infrequent events such as severe storms. Most salt marsh areas are flooded for longer or shorter periods in relation to high water during at least part a normal spring–neap tidal cycle. Some, however, can maintain a level above the highest astronomical high tide as a result of wind tide effects. Salt marshes exist in all climate zones from the tropics to high-arctic coastal environments. In the tropics, mangroves represent a special vegetated coastal environment which departs from typical coastal marshes in many ways. Part of the processes related to sedimentation in mangroves are for obvious reasons similar to those related to regular marshes, and no separate discussion of the tropical zone's special conditions is included in this chapter. See Augustinus (1995) for a comprehensive description of the geomorphology and sedimentology of mangroves. Along the banks of tidally influenced rivers, there exists a gradual transition between salt marshes in the outer part and fresh water tidal marshes further inland. Likewise, freshwater marshes can develop along the banks of lakes subject to wind tides. This chapter deals with salt marshes. No separate distinction will be made in relation to these freshwater marsh environments in which the sedimentary

J. Bartholdy (✉)
Department of Geography and Geology,
University of Copenhagen, 10 Øster Voldgade,
Copenhagen DK-3050, Denmark
e-mail: jb@geogr.ku.dk

R.A. Davis, Jr. and R.W. Dalrymple (eds.), *Principles of Tidal Sedimentology*,
DOI 10.1007/978-94-007-0123-6_8, © Springer Science+Business Media B.V. 2012

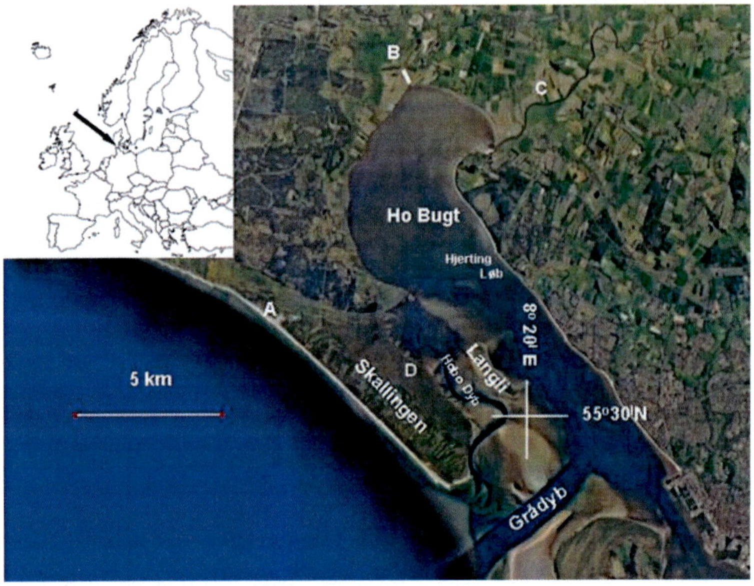

Fig. 8.1 Orthophoto of the northernmost part of the Wadden Sea located at the eastern part of the North Sea (*arrow* at the schematic map of Europe, *upper left*). Two generations of barriers (Skallingen and Langli) divide the northern part of Grådyb tidal area, Ho Bugt, into two parallel tidal areas, Hobo Dyb and Hjerting Løb. The mainland to the east consists of glacial deposits facing the tidal area with an active coastal cliff. The mean tidal range is 1.6 m but wind tide can raise the water level up to about 4 m above the mean water level. The locations referred to in the text includes: (**A**) the location of ^{14}C-dated samples from an old salt marsh platform emerging on the exposed west coast, (**B**) the location of an auger cored profile at Kjelst, (**C**) the Varde Å Estuary, and (**D**) location of a measuring station in the salt marsh creek Store Lo at the barrier spit Skallingen

processes in general are very similar to those of salt marshes. Salt marsh sedimentation and related dynamic conditions will be evaluated and described with examples from a wide range of locations. General mechanisms and depositional conditions, however, are primarily illustrated by examples from the Danish Wadden Sea based on the author's own experience. These examples are concentrated in an area belonging to the northern part of the Grådyb tidal area shown in Fig. 8.1. Key locations referred to in the text are marked with capital letters from A to D.

8.2 Measurements of Salt Marsh Sedimentation Through Time

The first to publish on relations between salt marsh and tidal levels was Mudge (1858). Later, Shaler (1886) followed with studies on salt marsh formation. Both recognized the interplay between plants and salt marsh–forming processes and related peat formation to vertical zones relative to sea level in the peaty New England marshes on the north/east coast of USA. Davis (1910) also reported on sea-level change in relation to salt marsh accretion. Later, Yap et al. (1916; 1917) formulated the perhaps first conceptual description of salt marsh morphodynamics from work in the Dovey Estuary, UK. They concentrated on interactions between plants, tidal inundations and creek formation. Studies on salt marsh formation in North Norfolk, UK, were initiated in the 1910s–1920s by Oliver (1913) and followed up by Steers (1936, 1938). Similar studies were carried out parallel with these in the USA (Chapman 1938). The first time series of salt marsh accretion based on direct measured accretion rates was published by Richards (1934), based on studies from the Dovey Estuary, followed by Nielsen (1935) in the backbarrier marsh at Skallingen, Denmark. After these pioneers in salt marsh accretion research, a great many studies using marker horizons have been published (e.g. Stevenson et al. 1986).

Direct measurements of salt marsh levels represent another frequently used technique for analysing salt marsh sedimentation. These types of analysis are either related to comparisons of maps, surveyed lines of different age, or to point measurements carried out at more precise time intervals. The latter is similar to the use of marker horizons, but differs with its relation to autocompaction. Esserlink et al. (1998) used detailed surveyed lines in the Dollard Estuary, the Netherlands, to determine vertical salt marsh accretion. Pethick (1980, 1981) used maps and historical data to determine the time of marsh inception at spots where he, through standard field levelling techniques, also determined the actual salt marsh level. Based on data from 14 salt marsh areas in North Norfolk (UK), he found an asymptotic age-surface elevation relationship. Lately, the incorporation of LIDAR data in such studies (e.g. Van der Wall et al. 2002) represents a new and powerful tool to detect and examine detailed morphology and morphological changes through time. Direct measurements of elevation change at carefully selected points have also been made by many researchers. Boumans and Day (1993) presented a so-called sedimentation-erosion table (SET) which was used by Cahoon et al. (2000) together with measurements based on marker horizons to evaluate the role of autocompaction. Lately, Charman et al. (2007) have come up with an automatic devise, primarily developed for measuring downwearing rates (TEB) that can be seen as a modified modern version of the SET.

Modern dating methods represent a vital tool for measurements of salt marsh accretion over longer timescales. ^{14}C dating is widely used in salt marsh environments (e.g. Gehrelds et al. 2006). In many instances, however, it fails to give useful results concerning accurate salt marsh accretion rates, partly because of the upper time limitations and accuracy of the method, and partly because of problems with a comprehensive way of treating autocompaction. Following Goldberg (1963) who developed the ^{210}Pb dating technique for snow accumulation studies in Greenland, this method was used to date sediments (Krishnaswami et al. 1971). A combination of ^{210}Pb dating and the use of the fallout product of nuclear testing, ^{137}Cs, became widely used since the 1970s as a tool for measuring sedimentation rates in lakes (Robbins and Edington 1975; Oldfield et al. 1978). Koide et al. (1972) were the first to use ^{210}Pb dating of sediments in the marine environment. Two different assumptions can be used in the process of ^{210}Pb dating of sediments: (1) the constant net rate of supply (c.r.s.) and (2) the constant initial concentration (c.i.c.). The use of one or the other was strongly debated in the late 1970s (e.g. Megumi 1978; Oldfield et al. 1978; Appleby et al. 1979). Appleby and Oldfield (1978) suggested dating by means of the c.r.s. method. In salt marsh environments, the c.i.c. method has been preferred (Kirchner and Ehlers 1998; Bartholdy et al. 2004; Pedersen and Bartholdy 2006) and verified by means of marker horizons (Madsen 1981; Pedersen et al. 2007). ^{210}Pb dating was quickly implemented in sediment budget studies of tidal areas (Bartholdy and Madsen 1985) and represents today an important tool for evaluating a vast majority of aspects related to salt marsh accretion such as heavy metal accumulation (Christiansen et al. 2001), and accumulation of other pollutants (French et al. 1994). An important aspect of ^{210}Pb dating is that not all sediment types have the same ability to retain this radionuclide (Ackermann 1980). Studying heavy metal concentrations in sediment, Ackermann et al. (1983) advocate the use of grain sizes <20 μm. Other authors (e.g. Olsen et al. 1982) have argued that also organic matter is able to bind ^{210}Pb. So, if a sediment core is not homogeneous in terms of organic matter and/or grain size, it is important to normalize the unsupported ^{210}Pb, in order to avoid unwanted effects of sediment-type variation. Kirchner and Ehlers (1998) suggested the unsupported ^{210}Pb to be normalized by the sum of loss-on-ignition and grain-size content of <20 μm. Lately, also the OSL method (optical stimulated luminescence) has been used to assess salt marsh accretion rates (Madsen et al. 2007). This type of sediment dating no doubt will be of great importance in future salt marsh research, as it can be used to date sediments over small (decades) as well as large (millennia) timescales and is found to work well in the estuarine environment (Madsen et al. 2010).

8.3 Morphodynamics of Salt Marshes

Salt marsh sedimentation can be separated into three main types: (1) sedimentation associated with channel flow in the vicinity of salt marsh creeks, (2) sedimentation associated with sheet flow over vegetated salt marsh surfaces, and (3) sedimentation associated with exposed salt marsh edges. The two former has large resemblance with sedimentation in fluvial systems. There are, however, some obvious differences between the fluvial and the tidal dominated environment. First of all, the bidirectional flow in tidal areas can alter the channel-related morphology compared to the unidirectional flow in fluvial systems. Furthermore, and as a consequence of this, the flow velocity is at a minimum in the tide-dominated environment at the time of both

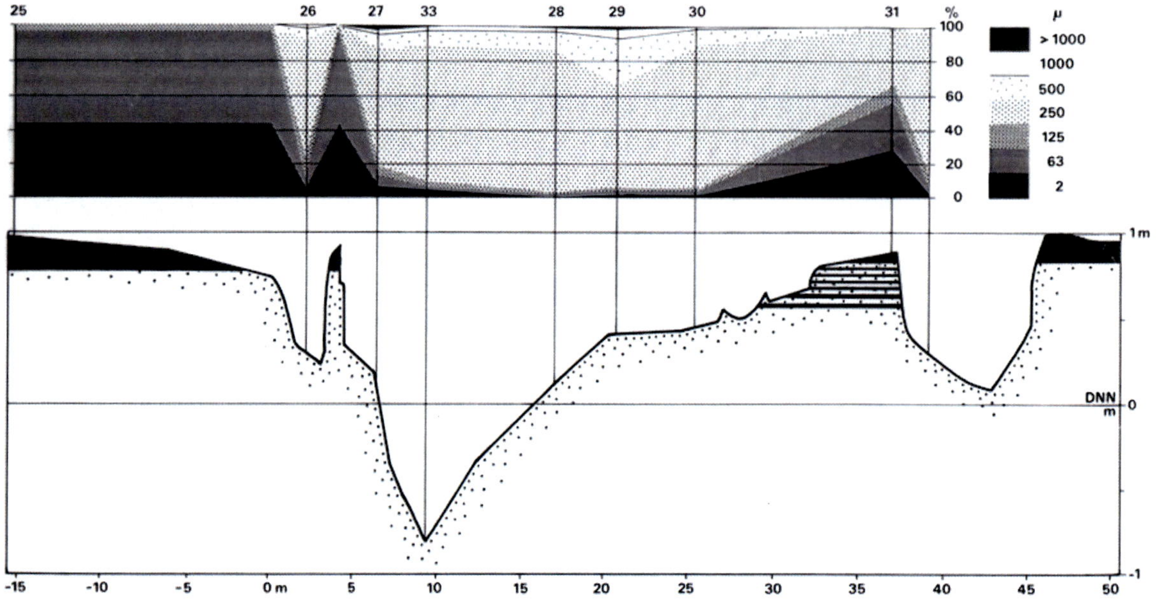

Fig. 8.2 Profile across the salt marsh creek Store Lo in the backbarrier salt marsh on Skallingen. For location, see Fig. 8.1 (Modified from Bartholdy 1983)

low and high waters; the latter being in direct contrast to fluvial systems. A comparison between key factors dominating the fluvial and those prevailing in the tidal salt marsh system is presented below in order to highlight the special conditions of the salt marsh sedimentary environment.

8.3.1 Relationship to Salt Marsh Creeks

Sedimentation associated with channel flow in salt marsh creeks includes all the typical sedimentary units from the fluvial system, of which point bars and natural levees are directly related to salt marsh deposits. Barwis (1978) attempted to distinguish between tidal creek point-bar deposits and their fluvial counterparts and found several differences related to the vertical sequence. Apart from the origin of the organic material capping the point bar, however, none of these relates directly to the salt marsh deposits. Bioturbation was regarded by Barwis as far more intense in the surface layers of tidal compared to fluvial point bars. This, however, is only true for warm climates like the subtropical/tropical zone. Temperate salt marshes are generally only slightly affected by bioturbation. If affected, it is primarily as a result of plants with vertical root structures that rarely destroy the lamination to the point of complete obliteration (Bartholdy and Madsen 1985; Reineck and Gerdes 1996; Chang et al. 2006a). As illustrated in Fig. 8.2, the gradual transition between point-bar and salt marsh deposition on top of point bars is associated with horizontal lamination with interlayered sand/mud bedding similar to the classical example described by Reineck and Singh (1975). It contains sandy deposits related to deposition during periods of relative high water levels and current velocities primarily associated with wind setup during storms, alternating with mud deposits also related to rough weather conditions but deposited around high water slack. During storms, fine-grained sediment is typically mobilized by wave action in the surrounding tidal area resulting in increased concentrations of mud in the water entering from the adjacent tidal area during flood. This type of 'layer cake' deposition is also found in natural levees along salt marsh creeks. In environments with relatively little mud and a potentially high organic production, such beddings can consist of layers of sand alternating with layers of organic fibrous material. This type of bedding was described by Redfield (1972) as a kind of varve in New England (USA) marshes. The fibrous material is associated with algal mats with subsequent concentrations of root fibres formed during the summer. The sand horizons (usually with added silt) that separate these

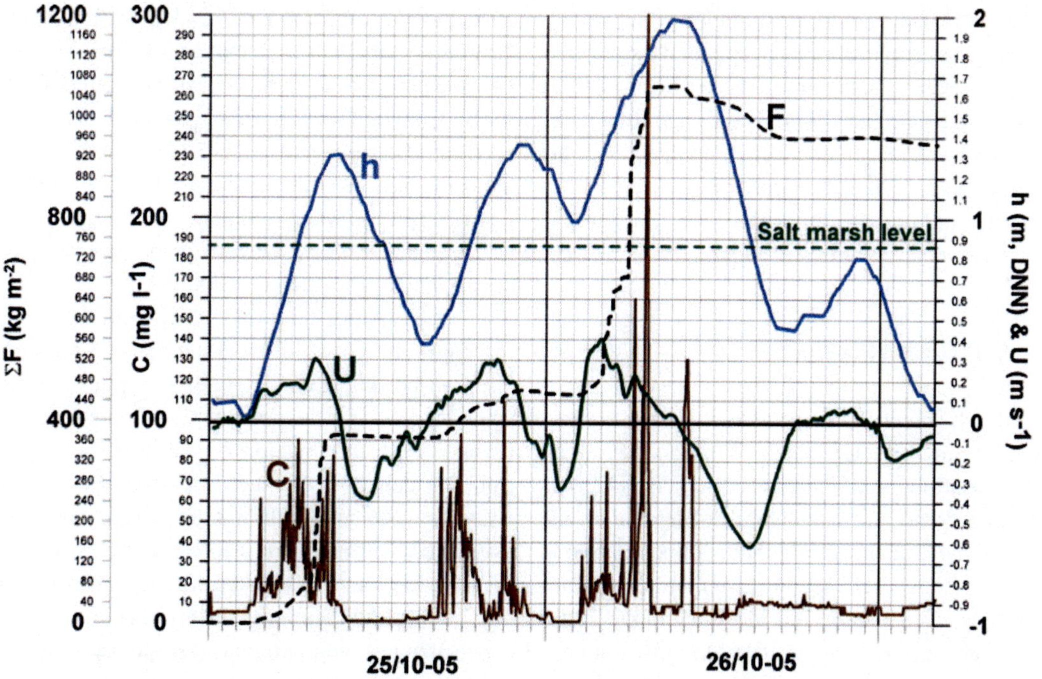

Fig. 8.3 Concentration of suspended sediment (*C*), current velocity (*U*, +/– equals flood/ebb), water level (*h*) and cumulated sediment flux (*F*, calculated as the sum of $C \cdot U \cdot \Delta t$ based on measurements every 5 min.) as function of time in the salt marsh creek Store Lo at Skallingen on October 25 and 26, 2005. *C* and *U* are measured 0.5 m above the bed in the deepest part of the channel. The cross section in which the measurements were carried out is that of Fig. 8.2

fibrous horizons are derived from deposition of suspended sediment during winter gales.

Wave activity is an important factor for mobilizing fine-grained material in tidal areas. This is illustrated in Fig. 8.3, which shows data obtained from the same location as that of the cross section of the salt marsh creek shown in Fig. 8.2 (a sheltered backbarrier, representative for such). The data set is typical for weather types leading to import of fine-grained sediment to salt marsh areas affected by wind tides. The suspended sediment concentration and current velocity shown in Fig. 8.3 were measured 0.5 m above the bed in the deepest part of the creek. During the observed period, a southwesterly gale caused a setup that flooded the salt marsh. Waves in the adjacent tidal area resuspended sediment and resulted in raised concentrations in the water entering the creek on the flood tide. The highest high water inundated the salt marsh with a water depth of approximately 1 m. The efficiency with which the salt marsh traps sediment is emphasized by the almost clean water running back during ebb, and the resulting flood-directed flux of suspended sediment.

A number of general characteristics can be demonstrated from this data set. First of all, the presence of two plumes of fine-grained suspended sediment: (1) a local plume, associated with sediment mobilized in the immediate vicinity of the salt marsh creek associated with low water; and (2) a subsequent 'regional' plume, associated with wave suspended sediment from further out in the adjacent tidal area associated with high water. The first represents relatively turbid water entering the creek system at the beginning of the gale, where it reached a maximum concentration of about 90 mg l^{-1}. During the three overmarsh tides associated with the gale, this plume passed the measuring station with successively smaller concentrations as the local sediment sources 'dried out'. The second plume did build up during the storm and reached a maximum of about 300 mg l^{-1} in the descending flood current during the peak of the gale. This reflects the general concentration level in water picking up suspended matter from muddy parts in the tidal area with a large storage capacity. More or less all sediment associated with these two plumes was deposited on the salt marsh during high water slack. Only the last and largest plume was able to cause enough deposition in the channel region itself to form a resuspension peak of any noticeable size at the beginning of the succeeding ebb tide. There are numerous studies of tidal creek hydrodynamics and related suspended sediment transport in salt marsh areas (e.g. Boon 1975; Settlemyre and Gardner 1977; Bayliss-Smith et al. 1979; Nixon 1980; Healy et al. 1981; Dankers et al. 1984; Green et al. 1986; French and Stoddart 1992; Reed et al. 1999) just to mention a few. This topic will be treated elsewhere in the book.

8.3.2 Salt Marsh Platform

A flooded salt marsh is, in principle, without water movement during high tide slack. Wind effects, topographical variations, and hydrodynamic delays can all create local currents, but these will in general be small compared to those acting in relation to a flooded floodplain where water is exchanged between turbulent flow in the channel-near region and the floodplain. Thus, floodplain deposits are largely related to diffusion processes (Pizutto 1987), whereas the distribution of sediments in salt marsh areas is primarily related to advection (Woolnough et al. 1995). The relative importance of diffusion and advection, however, depends on a number of local conditions such as topography and tidal conditions. Even if it seems logical that diffusion is more active in fluvial than in tidal environments, both processes are present in both places. The diffusion model of Pizutto (1987) has actually been demonstrated to be applicable in salt marsh areas (Bartholdy et al. 2004). A number of empirical studies (e.g. Letzch and Frey 1980; Carling 1982; Reed 1988; Stoddart et al. 1989; French and Spencer 1993; Bartholdy 1997; Bartholdy et al. 2010a) have shown that deposition of salt marsh sediments decreases away from the source primarily in form of the salt marsh edge, and followed in significance by the higher-order salt marsh creeks. The small-order creeks have little influence (Stoddart et al. 1989; Bartholdy et al. 2010a). This agrees well with a general picture of tides having an increasing part of the inundating water derived from the salt marsh edge as they get higher. From an initial concentration of fine-grained material that is present when the water passes the salt marsh boundary, sedimentation commences and continues as long as enough material is left in the water column or the water leaves the marsh area. The major deposition takes place close to the marsh edge or creek margin, where the coarsest particles settle out. This suggests that at some distances from the source, there will be a deposition minimum. Ultimately, this will lead to a concave profile between two primary salt marsh creeks and between the salt marsh front and its hinterland (Allen 2000).

As is the case with river banks, levees are formed along the banks of salt marsh creeks as a result of fallout from suspension when sediment-laden water flows from the turbulent channel region to the calmer overbank environment. In general, however, salt marsh creek levees are smaller than their fluvial counterparts. The reason for this is twofold: (1) salt marsh creeks are usually transporting relatively fine grain sizes with only a small tendency for developing pronounced levee deposits and (2) a combination of the astronomically controlled water level and the development of salt marsh drainage systems. As the period around high water is associated with moderate to no currents, overmarsh tides are in general associated with less energetic conditions than is the case for their fluvial counterparts. This means that a smaller amount of suspended sandy material is carried over the banks of salt marsh creeks than over fluvial banks during otherwise similar conditions. The less elevated salt marsh creek levees are in general also less exposed to the formation of crevasse splays in weak spots than is the case for their fluvial counterparts where the most vigorous flow coincides with the highest water level. Breached locations in salt marsh creek levees, however, attract water draining the salt marsh during subsequent and usually relatively quick tidal-controlled water level descends after overmarsh tides. As a consequence, these locations easily become focus points for small backward eroding first-order creeks. Eventually, these creeks will dominate the local area behind breached levees as erosional features (Fig. 8.4) where, in the fluvial counterpart, crevasse splays dominate as small depositional systems.

Headward-eroding creek systems, from first to higher orders, represent the way in which a drainage system is established in salt marsh areas. Some creeks will develop faster than others and thereby 'steal' drainage area from less fortunate creeks and leave them behind to degenerate. An example of this can be seen in the central lower right of Fig. 8.4. This example is from a creek system in Georgia, USA, with little chances for small-order tributaries to develop into larger higher-order creeks, as the primary drainage structure has already matured. In more juvenile salt marsh formations, headward-eroding first-order creeks will mature as higher-order creeks as the drainage system develops.

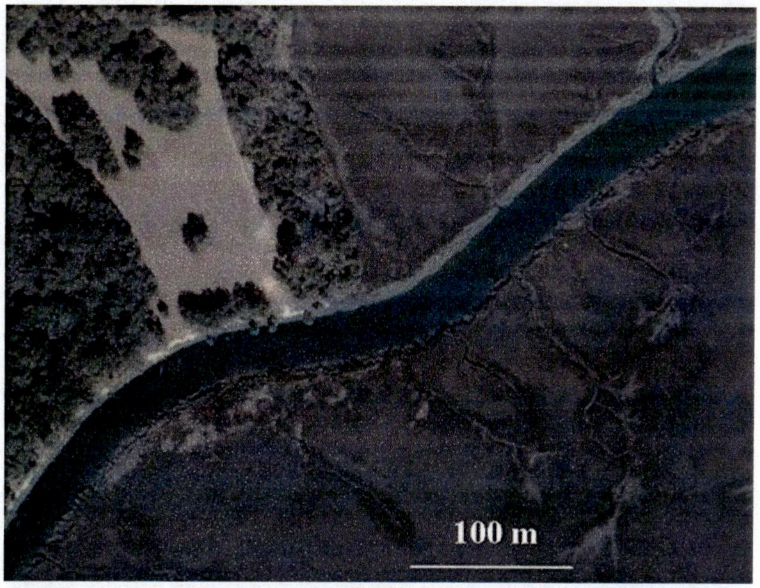

Fig. 8.4 Breached levees as focus points for small salt marsh creek systems in Groves Creek, Georgia, USA

The hydraulic conditions related to the contact between almost impermeable salt marsh clay in relatively young salt marsh areas and permeable sand beneath it with a high hydraulic conductivity enable a phenomenon called piping. Piping consists of small drainage tunnels formed in the sand beneath the clay. This type of tunnels or pipes results from an uneven pressure distribution in the clay-sealed sand after overmarsh tides. Small fountains of water up to decimetres high at the upper end of small first-order creeks have been observed in the Skallingen backbarrier salt marsh. As the submerged channels collapse, they create elongated depressions in the salt marsh surface (Fig. 8.5, upper left) that dictate the direction of the creek's further headward erosion. This kind of piping in salt marsh environments was first observed by Kesel and Smith (1978) working in the Nigg Bay salt marsh in Scotland, UK. They discovered a relation between the disappearance of salt pans and the formation of such subsurface channels.

Salt pans consist of isolated nonvegetated small depressions in the marsh surface that need to be drained before vegetation can invade them. Before this happens, salt will precipitate on the surface due to evaporation of trapped water after overmarsh tides. The first to publish on the formation of salt pans was Warming (1904 in Danish with a summary in French). He described the reason for these depressions in salt marshes as a result of either: (1) an uneven colonizing of marsh plants resulting in unvegetated spots surrounded by growing vegetated areas, (2) depressions formed on the exposed coast by beach ridges prior to the salt marsh formation, (3) uneven coastal erosion resulting in small bays, later closed by beach ridges prior to the salt marsh formation, or (4) direct wave attack on weak spots on the salt marsh surface during storms. Warming (1904) stressed, from observations in the Wadden Sea, that such depressions will be prone to collect seaweed and algae which will putrefy and obstruct the growth of salt marsh plants. He also suggested that the treading of cattle could initiate weak spots for the waves to excavate. Unfortunately, the hypothesis of weak spots in interplay with putrefying organic matter seems to be the only thing remembered in literature from this paper with many original observations and interpretations (Redfield 1972; Pethick 1974; Kesel and Smith 1978; Boston 1983). The paper was cited by Yapp et al. (1917) who stated that Warming was of the opinion that 'pans may originate in a variety of ways'. They tested the weak spot hypothesis by establishing an artificial pan as a weak spot, which they found was not being further excavated. This kind of test of cause is strongly dependent on the location. Pethick (1974) carried out a statistical analysis of pan distribution in salt marshes between Scolt Head Island and Blakeney Point, UK. He used aerial photographs for the position and levelling for the altitude, and ended up supporting Warming's weak spot hypothesis. No doubt salt pans in marsh environments can be formed in a variety of ways, either (1) by chance because of an

Fig. 8.5 Pictures from the backbarrier salt marsh of Skallingen (Denmark). *Upper left*: A collapsed 'pipe' at the upper end of a first-order creek The depression is invaded by *Spartina townsendii*, while the surrounding salt marsh is dominated by *Halimione portulacoides*, about 30 cm high. *Upper right*: Oxbow lake as a result of a meander neck cutoff in the central part of the salt marsh area. Most of it is still connected to the creek system, and will be drained at low water. Eventually, however, such cutoffs will all be disconnected by sedimentation along the banks of the active channels and hereafter transformed into channel salt pans. *Lower left*: Salt pan created by an exploded Second World War mine. Note the desiccation cracks due to drying in the now inundated depression. *Lower right*: Nest of the yellow meadow ant. The depression around the nest is caused by collapsed galleries and sand removed to build up the mound. For scale, the diameter of the mound is approximately 1 m

unevenly spread vegetation; (2) as a result of inherit depressions at the marsh base; (3) because of wave attacks on weak spots; and, (4) as pointed out by Yapp et al. (1917), as a result of blocking of shallow channels. Small oxbow lakes belong to this last category (Fig. 8.5, upper right). Salt pans can also be man-made as the one shown in Fig. 8.5 (lower left) where a World War II mine has created a circular depression of about 2 m in diameter. The desiccation cracks, due to dewatering through drying, seen at the bottom are covered by about 5 cm of water. When dry, this pan is covered by small visible salt crystals giving evidence of a salt suffering environment not suitable for plants. This depression is located in the central part of the Skallingen backbarrier salt marsh and experiences only restricted wave action even during storms (Bartholdy and Aagaard 2001). As a result, it is not enlarging even if it very clearly represents a 'weak spot'. It is interesting to notice that it has kept its circular shape for over 60 years and significantly lags behind its vegetated surroundings in terms of accretion. Isolated salt pans are by some called 'rotten spots' (Redfield 1972), especially when they are related to more organic-rich salt marsh areas as those of New England, USA. In general, most topographic variations in vegetated salt marsh areas are dampened out as the salt marsh grows and matures. Local exceptions for this will often be associated with the fauna. An example of this is represented by nests of the Yellow Meadow Ant (*Lasius Flavus*). This ant, which is common in Central Europe and also can be found in Northern Europe, North America, Northern Africa and Asia, excavates the ground with galleries and feeds on the honeydew from root aphids, which they breed here. The nest can be up to

Fig. 8.6 *Upper left*: Border between a bare mudflat and Spartina marsh at Klægfledningen in the north/eastern part of Ho Bugt, Denmark. *Upper right*: Border zone between a sandy tidal flat and the backbarrier salt marsh of Skallingen, Denmark. *Below left*: Wave exposed salt marsh edge from the northern part of Ho Bugt, Denmark. *Below right*: An expanding exposed salt marsh area northwest of Ribe, Denmark

about half a metre high and is in salt marsh areas located in upper primary sandy parts (Fig. 8.5 lower right).

Also, ice rafting can contribute to salt marsh sedimentation and topographic variations on salt marsh surfaces. It takes place when ice flakes containing sediment from the intertidal zone are moved to rest on the salt marsh surface and subsequently release melt out pockets of sediment. This type of sedimentation creates an uneven salt marsh surface and is documented to be present even in relatively mild temperate climates (Bartholdy 1997; Pejrup and Andersen 2000). This process is responsible for moving large quantities of sediment in areas characterized by cold winters (e.g. Dionne 1984; van Proosdij et al. 2006a).

8.3.3 Exposed Salt Marsh Zones

Processes related to the zone close to the salt marsh edge are primarily connected to wave action and can affect the general picture described above significantly. If the salt marsh edge is heavily exposed to waves, wave-induced turbulence can obstruct sedimentation such that the deposition maximum is moved inland. This is documented for a salt marsh area in the Bay of Fundy (e.g. Davidson-Arnott et al. 2002; van Proosdij et al. 2006b). In contrast, for sheltered backbarrier salt marshes, waves may be of little importance, even under storm conditions (Bartholdy and Aagaard 2001).

If expanding, lee-type salt marshes generally develop on top of the adjacent tidal flat either as a vegetation front (Fig. 8.6, upper left, where *Spartina* grows directly out over a mudflat) or in patches of vegetation that eventually merge to form a coherent salt marsh (Fig. 8.6, upper right). Wave activity forms salt marsh cliffs at the edge of exposed salt marshes (Fig. 8.6, below left), and if expanding, the salt marsh in the active cliff area will usually be uneven and dominated by vegetated hummocks separated by unvegetated gaps (Fig. 8.6, lower right). This is a result of both depositional and erosional activity. Hummocks of salt marsh can grow with lumps of clay liberated from the salt marsh cliff as their base, or develop from colonies of pioneer vegetation. Between the hummocks and in

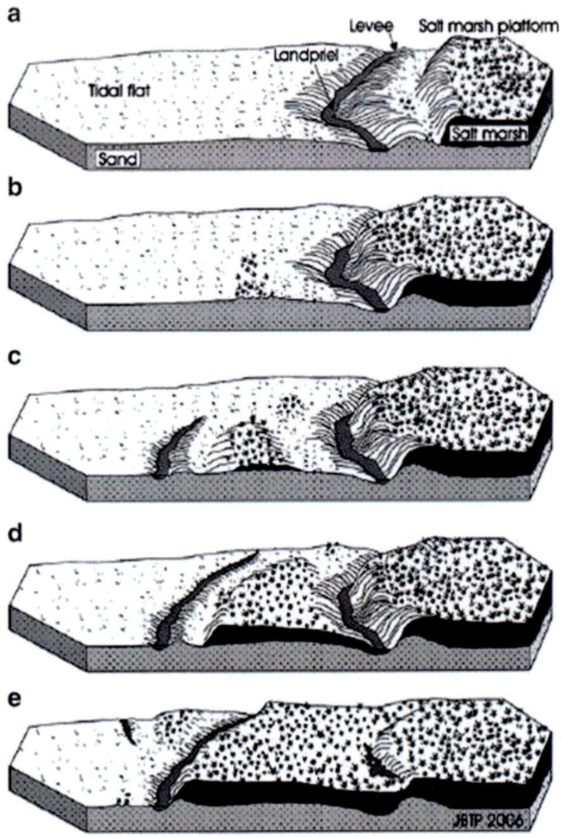

Fig. 8.7 Schematic model of the cyclic growth of an exposed salt marsh area. For scale, the length of the horizontal front is in the order of 1 km (From Pedersen and Bartholdy 2007)

weak spots along the salt marsh frontage, wave erosion keeps patches free of vegetation. This creates an uneven relief that only slowly transforms into a coherent salt marsh surface if the salt marsh front advances and the area becomes more sheltered against direct wave attack. In areas with an expanding salt marsh, such cliff areas (sometimes called microfalaise after the French word for cliff) can be found as elongated steps in the marsh, inland of the active cliff. In exposed salt marsh areas, marsh expansion has been found to take place in sequences (Jakobsen 1954; Pedersen and Bartholdy 2006) as illustrated in Fig. 8.7.

As a result of differences in hydraulic roughness between the relatively smooth tidal flat and the relatively rough plant-covered salt marsh surface (Tsihrintzis and Madiedo 2000), the transition between tidal flats and salt marshes will often function as a zone where oblique tidal or wind-induced currents are captured and guided to run parallel to the edge. The resulting shallow channel running parallel to the salt marsh edge was identified by Jakobsen (1954) as a natural part of the intertidal landscape and called *landpriel* (Fig. 8.7a). Because of natural levee formation along the banks of the *landpriel*, the level along its banks will eventually become elevated and salt marsh plants will start to invade. This is most pronounced on the marsh side facing the small salt marsh cliff. On the tidal flat side, the surplus material will be reworked by waves and forms small ridges in a broader belt of unevenly spread microtopographic highs which eventually also form a somewhat wider platform for salt marsh growth (Fig. 8.7b). This creates the necessary conditions for the formation of a new *landpriel* and a new small cliff on the exposed side of the fillet of newly formed salt marsh which continue to grow (Fig. 8.7c, d). Eventually, the old *landpriel* silts up, and a new generation of this cyclic salt marsh growth starts to form. This leaves the old salt marsh cliff as a small elongated ridge in the marsh hinterland (Fig. 8.7e). Through time, sedimentation in salt marsh areas has been strongly influenced by human activity for agricultural purposes. The normal pattern of this activity consists of an artificial draining of tidal flat areas by means of a system of small parallel channels open for tidal action by means of a system of larger channels (Fig. 8.8). During the normal evolution of such areas, dredged material is spread between the channels raising the level between them. This enables salt marsh plants to invade and further increase import of fine-grained material from the adjacent tidal area. When such an area in time has reached a level high enough for an adequate agricultural production, it will usually be closed by a dike. As a role, a new part of the tidal flat area in front of the dike will hereafter be treated in the same way and so on. This type of land reclamation has been used in many suitable coastal and estuarine areas all over the world during the last millennium. Because of the sea-level rise and as a result of sediment compaction, this has produced coastal zones with a landscape consisting of a succession of reclaimed areas (polders) where the first, the furthest from the sea, has the lowest level. Along with the rising environmental awareness during the last century, this type of land reclamation has been abandoned in many parts of the world. In the Wadden Sea, there are even examples of attempts to restore reclaimed areas as shown in Fig. 8.8 that illustrates a natural restoration project on the German Wadden Sea island Langeoog. Here, the dike has been bridged in order to let the normal tidal action get access to the former reclaimed area.

Fig. 8.8 A reclaimed salt marsh area at the backbarrier of the German Wadden Sea island Langeoog has been 'opened' by bridging the dike as part of a nature restoration project

8.3.4 Effects of Vegetation

Salt marsh sedimentation starts to form at a level close to the mean high water level. This seems to be a general border reported from a wide range of climates and tidal conditions. In subtropical salt marshes, Edwards and Frey (1977) and Letzsch and Frey (1980) found the level of salt marsh initiation to be close to the mean neap high water level at Sapelo Island, USA. For salt marshes in California, USA, Phleger (1970) found the first plants (*Spartina*) to grow in a zone between what he called 'mean lower high water level' and 'mean higher high water level' above which also other plants became abundant. In temperate areas, Coldewey and Erchinger (1992), Jakobsen (1954) and Bartholdy (1997) found the border to be close to the mean high water level which is also close to the level reported from an arctic salt marsh by Nielsen (1969). Below this level, in the pioneer zone (Fig. 8.9), the most salt-tolerant plants can exist in clusters of scattered vegetation as far down as to where the average tidal flooding last for approximately 3 h (Coldewey and Erchinger 1992). Even if the dominant species change with climate, this overall pattern seems to be relatively general.

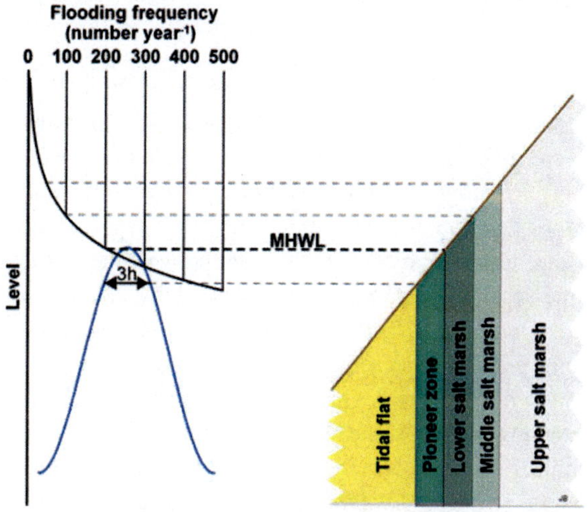

Fig. 8.9 Definition diagram relating zones in the intertidal area to flooding duration and frequency. The *lower zone borders* like the concept are adopted from Coldewey and Erchinger (1992). The flood frequency curve is from the Skallingen peninsula, Denmark

The vegetation zones in the temperate salt marsh of Skallingen have been analysed by Kim et al. (2009a, b, 2010). Here, the pioneer zone is dominated by *Salicornia herbacea* and *Spartina townsendii*. Colonies of the latter (Fig. 8.6, upper right) are capable of accumulating sediment that, because of the correspondingly higher level, enables other plants to invade. The ungrassed part of the salt marsh area of the Skallingen peninsula (Fig. 8.1) can be subdivided into mainly three vegetation zones that roughly follow the topography divided into low, middle and upper salt marsh zones (Figs. 8.9 and 8.10). The dominating plant in the ungrazed part of this salt marsh is *Halimione portulacoides*. The fence between the grazed inner part and the ungrazed outer part is clearly visible on the aerial

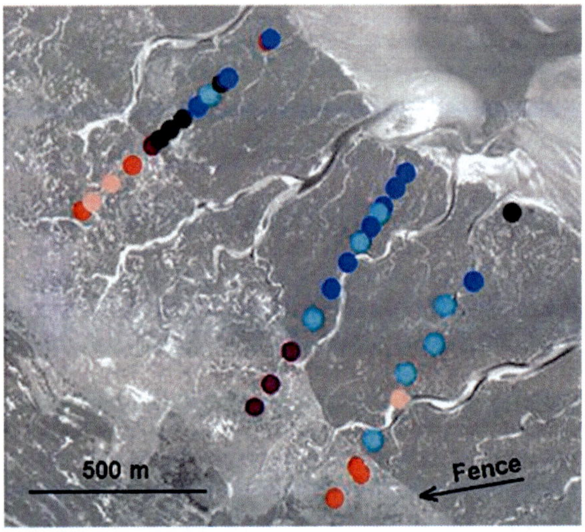

Fig. 8.10 Aerial photo of the central part of the Skallingen backbarrier. There is a clear difference in the vegetation cover between grazed (*lower left side*) and ungrazed (*upper right side*) parts. The coloured bullets correspond to analysed plant associations (see the text) (Modified from Kim et al. 2010)

photo (Fig. 8.10). In the low marsh zone (light blue bullets), this species is followed by a number of other species of which *Puccinellia maritima*, *Limonium vulgare*, *Plantago maritima*, *Triglochin maritima* and *Aster tripolium* are most abundant. This is in contrast to the middle marsh (dark blue bullets) whereby the same standard *Halimione* is followed by *Artemisia maritima*, *Aster tripolium* and *Limonium vulgare*. Other plants are present but these are the most abundant with an overwhelming abundance of *Halimione*. In the lowest part of the high marsh (black bullets), the dominating plant is *Artemisia maritima* followed by *Juncus gerardii*, *Halimione portulacoides* and *Festuca rubra*. Further up (landward) in this zone, the plant community gradually changes into less salt-tolerant species. The first of the four mentioned salt-tolerant plants to disappear during this transformation is *Halimione*. Notice that with some scatter, the vegetation zones follow the concave topographic profile from a relative high outer part close to the sediment source (the salt marsh rim) through the lowest central part in some distance from this that further inland merges with the higher central parts of the peninsular. In this case, this part is only reached in the northern line in the form of an isolated ridge.

It is not the purpose of this chapter to go into further details about plant community variations in various types of salt marsh environments, and interested readers are referred to botanical literature for further information. The general effects of plants on salt marsh deposition, however, is of great importance, and details in these mechanisms need to be understood in order to be able to assess the complexity of salt marsh formation. The ability of plants to form the necessary dynamic conditions for deposition and their role in counteracting resuspension, is of vital importance for salt marsh deposition and represents an important example in sedimentology, where the interplay between physical and organic processes is of fundamental importance for the final product. This was recognized early in the history of salt marsh studies, and attempts to quantify the influence of vegetation were made by pioneers of salt marsh accretion studies like Richards (1934) and Nielsen (1935). Stumpf (1983) is most likely the first to attempt to measure the flow through salt marsh vegetation. He injected dyed water with needle and syringe and timed the movement of the dye in a densely vegetated salt marsh in Delaware, USA. By means of the measurements and traditional turbulent boundary layer theory, he concluded that both the turbulence and flow velocities are one order of magnitude less in the salt marsh canopy than in the adjacent creek environment. Furthermore, by injecting dye close to the bottom, he was able to show that the flow here was smooth with the presence of a laminar sublayer which kept the dye at the bottom with only little vertical exchange.

Technological progress allows still more detailed measurements of water flow even in environments like inside a salt marsh canopy. Wang et al. (1993) measured current velocity in a salt marsh on the Mississippi River deltaic plain with an electromagnetic current meter and concluded that tidal currents in the marshes were about 10 to 20% of the current in the adjacent bayou (creek). Christensen et al. (2000) used an acoustic doppler velocimeter (ADV) in a creek/marsh environment on the eastern shore of Virginia, USA, and found that the velocity in the salt marsh area even at spring tide never exceeded 0.01 $m\,s^{-1}$. Shi et al. (2000) found by means of electromagnetic current meter measurements on a mudflat/salt marsh transition in the Changjiang Estuary, China, an average reduction in current velocity between the bare flat and the salt marsh canopy by 16%. Likewise by means of electromagnetic current meters, van Proosdij et al. (2000, 2006a) and Davidson-Arnott et al. (2002) studied the near-bed velocity conditions in a macrotidal and wave-exposed salt marsh in the Bay of Fundy. Their results about the significance of vegetation in the sedimentation process are less conclusive, most likely because of the extreme tidal and wave conditions of their study site. Studying wave dynamics in vegetation canopies, Neumeier and Amos (2006) found that in submerged *Spartina*, a low-turbulence zone of near-constant velocity is often separated from faster and more turbulent flow above it. This result supports the general conclusion from investigations of the impact of vegetation on tidal current and wave dynamics in the canopy of salt marsh vegetation, and underlines its ability to increase sedimentation and counteract resuspension of already deposited sediment. In addition, waves are dampened by vegetation. Möller et al. (1999) found that salt marsh vegetation induced an attenuation of waves which was approximately 50% higher than that of sand flats with the same water depth.

Apart from studying the dynamics of flow in salt marshes, Stumpf (1983) also addressed another important interplaying factor between salt marsh plants and sedimentation. With reference to the works of Ginsberg and Lowenstan (1958) and Schubel (1973), he concluded that suspended material can adhere to plants and subsequently be deposited at the salt marsh surface, either as faecal pellets from grazing gastropods, washed-down material by rainfall, or by the death and collapse of the plants.

Even if vegetation can be regarded as a factor that enhances sedimentation, vegetation in its outset also causes channel erosion in tidal landscapes. This is documented by Temmerman et al. (2007), who pointed out that vegetation can be locating erosion if patches of vegetation obstruct the flow. Such conditions can lead to flow concentrations that again will lead to channel erosion. In this way, a patchy vegetation cover enhances the ability of flow to concentrate and be able to erode. It is also a well-known fact that vegetation and its roots tend to bind together the substrate in which they are formed. Therefore, a channel developed without riparian vegetation (e.g. on tidal flats) will have a much more gently changing and shallower/wider cross section than channels developed on a vegetated surface where riparian plants will enhance bank resistance and promote a deeper and narrower cross section. On the other hand, in many areas suited for salt marsh growth, channels form the basis for a necessary drainage in order for the plants to invade. In that way, plant cover and channel formation to some extent represent a chicken and egg situation.

8.4 Physical Properties of Salt Marshes

8.4.1 Sediments

Salt marsh material consists of fine-grained sediments with a smaller or larger amount of coarser material; primarily sand (Fig. 8.11) and organic matter. In silty salt marsh sediments, most of the organic material will, as a rule, be part of suspended material settling onto the salt marsh surface. A smaller part is produced as underground organic production due to salt marsh plants. In peaty salt marshes, plants play the major role as supplier of material. These marsh types are primarily controlled by biogenic rather than sedimentological processes.

As salt marsh sediments fall out of suspension, the size and content of the coarsest fractions in general reflect the distance to the source. This is demonstrated in Fig. 8.12 which shows a profile across the central part of the Skallingen backbarrier (just north of 'D' in Fig. 8.1) with information on grain-size content of the surface sediments. Sand fractions become gradually less abundant inward from the salt marsh edge until aeolian

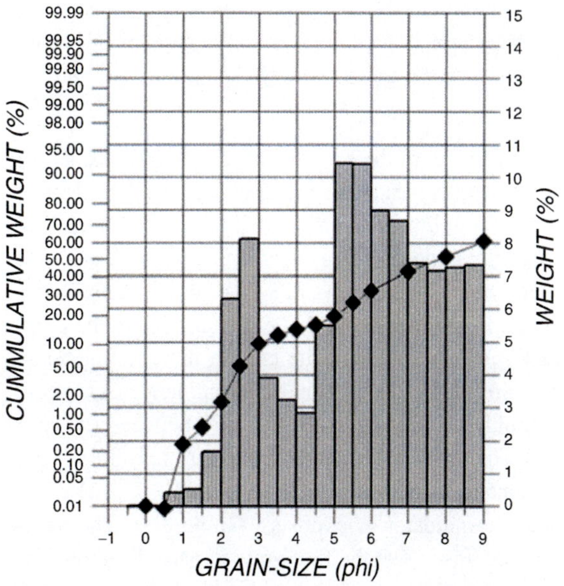

Fig. 8.11 Average grain-size distribution of the deposited salt marsh sediment on the backbarrier of Skallingen (west Denmark). Only grain sizes above 2 μm are included. The histogram describes the distribution of the 60.5% sand and silt recalculated to 100% (From Bartholdy et al. 2010b)

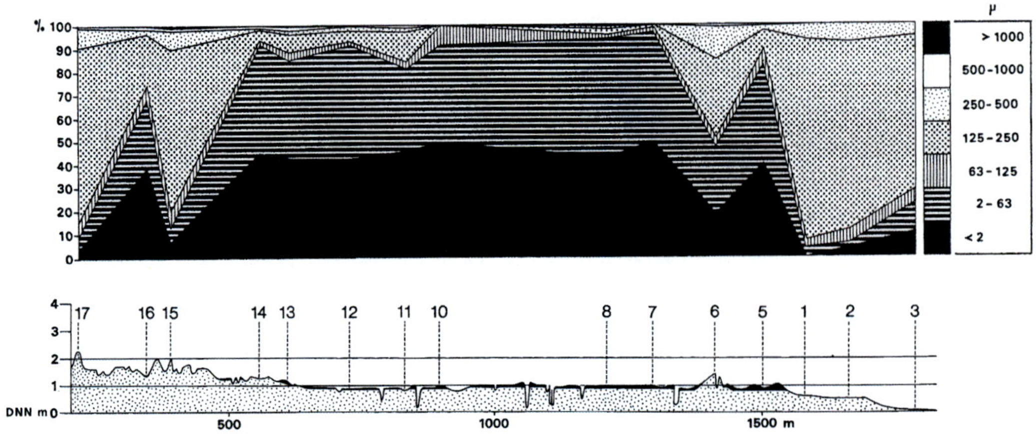

Fig. 8.12 Beneath: The topography (in relation to Danish Ordinance Level) in a cross section across the central part back-barrier of Skallingen. From *right* to *left*, the tidal flat area meets the salt marsh at about 1,600 m, the outer marsh – with a ridge at about 1,400 m – extends to about 1,100 m. Inland of here, the lower inner marsh is located between the higher laying outer marsh and the aeolian reinforced beach ridges at about 700 m. The profile crosses the peninsular from south/west towards north/east. *Above*: The grain-size distribution across the profile divided into clay (*black*), silt (horizontally striped) and sand in various size classes (Modified from Bartholdy 1983). The topographical profile is adopted from Nielsen and Nielsen (1973)

Table 8.1 The ratio between the content of 5 Φ–7 Φ and 5 Φ–9 Φ particles in salt marsh clay in relation to the exposure of the sedimentary environment of the salt marsh in question. Based on Bartholdy (1985)

Content of 5 Φ–7 Φ/content of 5 Φ–9 Φ	Exposure
Above 0.65	High
Between 0.55 and 0.65	Medium
Below 0.55	Low

blown sand from the exposed beach area (to the left of the shown profile) again increases the sand content of the marsh sediment. Thus, even if the sand content will reflect the exposure of a given salt marsh, it can also reflect other conditions, and will often be source controlled. In order to avoid this effect and relate the composition of salt marsh sediments to the environmental exposure of the depositional environment in question, Bartholdy (1985) suggested the ratio between the content of material in the range 5 Φ to 7 Φ (31–8 μm) and the content in the range 5 Φ to 9 Φ (31–2 μm) as an adequate indicator of exposure (Table 8.1). The argument for this is that the fine-grained part of the investigated salt marsh sediments can be divided into two log/normal distributions: one, primarily associated with deposition of single grains with the mean grain size varying between 4.5 and 6 Φ (63 μm to 16 μm) and a sorting coefficient (standard deviation) of about 1 Φ; and one primarily associated with flocculated grains with a mean grain size of 9 Φ to 10 Φ (2–1 μm) and a sorting coefficient of about 2.5 Φ. These two fine-grained populations can easily be recognized to the right of the sand population in Fig. 8.11. Thus, the proposed ratio describes a measure of the amount of fine-grained material deposited as single particles in relation to the combined amount of silt.

In energetic environments, flocculated material suffers from mechanical dispersion (e.g. Dyer 1989; Pejrup and Mikkelsen 2010) and is therefore less abundant in the resulting deposit than coarser grains that can be deposited as single particles. It needs here to be remembered that particles close to the size of clay have a very low settling velocity in relation to just slightly coarser silt particles. A particle of 6 Φ (16 μm) in water of 15°C and 30‰ salinity needs approximately 1.5 h to settle 1 m, whereas a particle of 8 Φ (4 μm) under the same conditions needs 23 h to settle the same distance. Thus, it is practically impossible for the latter to settle out, if it is not part of a flocculated larger particle.

The 5 Φ limit was chosen to avoid any influence from sand populations, and the 9 Φ limit was chosen both for reasons of symmetry and because it represents the lower limit of a pipette grain-size analysis. The 7 Φ limit has later been recognized by others as the lower limit for the so-called sortable silt particles with arguments similar to those stated above (McCave et al. 1995; Chang et al. 2006b). It is important to stress that these limits are based on settling diameters. The increasing use of grain-size measuring devices based on lacer diffraction opens for possible misinterpretations as these devices tend to overestimate the size of especially the finest silt fractions (e.g. Konert and Vandenberghe 1997; McCave et al. 2006; Ramaswamy and Rao 2006). The primary reason for this is that the laser diffraction technique has a tendency of measuring platey particles as their large projected grain area, whereas the same platey particle's settling diameter is much smaller. McCave et al. (2006) found that settling diameters of 2 and 16 μm are equivalent to ~8 and 22 μm sizes, respectively, when measured by a laser particle sizer. In Φ units, this corresponds to a change for a clay particle of 9 Φ to be seen as a silt particle of 7 Φ and in the coarse end to a more modest but similar change from 6 to 5.5 Φ.

8.4.2 Autocompaction

Any interpretation of salt marsh sedimentation in relation to dynamics is restricted to deal with the amount of sediment in form of concentration (mass per unit volume) in the water. When evaluating sedimentation as a result of salt marsh dynamics, the most appropriate measure is therefore weight per unit area and time. Such data can be obtained by means of sediment traps as in French and Spencer (1993) and van Proosdij et al. (2006b). The most common and easy measure of deposition, however, is the accretion rate measured in length (level or thickness) per unit time. This is also the measure that is relevant in order to compare salt marsh growth to sea-level rise. In order to evaluate accretion rates in relation to dynamically controlled deposition rates, the bulk dry density and its variation in the uppermost layers, therefore, become of vital importance. The bulk dry density depends on grain-size parameters, organic content, and level of compaction. For the same location in a salt marsh environment, the sediment type might be regarded as constant with depth. Autocompaction is then usually related to the sediment deposited above the observed layer. Models relating this overburden to autocompaction are described by Skempton (1970). According to Allen (2000), Skempton's empirical findings suggest that:

$$T = (T_0 - T_{min})e^{-kH} + T_{min} \tag{8.1}$$

where T is the actual thickness of a layer which originally right after deposition had a thickness of T_0 and which limiting thickness ('zero' porosity) is T_{min}. H is the depth below the surface and k (m^{-1}) is an empirical coefficient describing the compressibility of the layer. However, this attempt to describe the natural compacting behaviour of shallow silty salt marsh sediments fails to describe autocompaction in the uppermost layers as pointed out by Bartholdy et al. (2010b). They designed a method by which the down core bulk dry density BDD_z (kg m^{-3}) at level z (m) beneath the surface can be directly related to the bulk dry density of the uppermost 5 cm ($BDD_{0-0.05}$) in uniform silty salt marsh clay. The basis for this method is the finding that bulk dry density varies down core as a logarithmic function of depth (Fig. 8.13):

$$BDD_z = A\ln(z) + B \tag{8.2}$$

and that the two empirical constants A and B can be directly related to the bulk dry density of the uppermost 5 cm by the following two empirically derived equations:

$$A = 0.16BDD_{0-0.05} + 20.63 \tag{8.3}$$

$$B = 1.64BDD_{0-0.05} + 82.44 \tag{8.4}$$

In case $BDD_{0-0.05}$ is unknown, Bartholdy et al. (2010b) suggested the following relationship between $BDD_{0-0.05}$ and the loss on ignition (*LOI*, %) and sand content (*S*, %):

$$BDD_{0-0.05} = 4478 \cdot LOI^{-0.73} \cdot 0.98S^{0.036} \tag{8.5}$$

Integration of Eq. 8.2 from the surface and down to the level, z under the surface gives the mass depth of *z*:

$$MSD_z = A \cdot z\ln(z) + z(B - A) \tag{8.2'}$$

By means of the above equations, it is possible to calculate the most likely variation of BDD_z with depth under the salt marsh surface based on either $BDD_{0-0.05}$

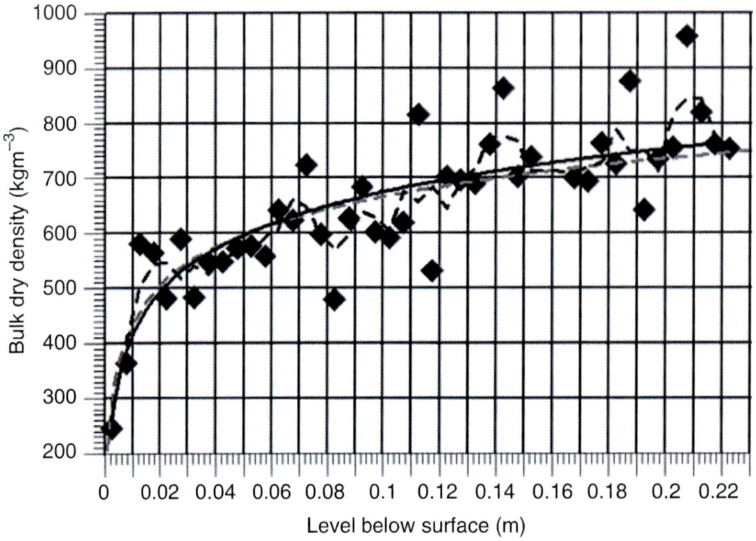

Fig. 8.13 Bulk dry density as a function of depth beneath the salt marsh surface at Skallingen. The diamonds represent measurements of ½ cm slices from the *top down* to *just above* the sand flat beneath the clay. The *broken black line* represents the running mean of 3 measurements (1½ cm); the *full black line* is the logarithmic best fit curve: $BDD = 113 \ln(z) + 935$; $R^2 = 0.67$; $P < 0.01$; the *stippled grey line* ($BDD = 100 \ln(z) + 897$) represents the suggested method (see the text) (From Bartholdy et al. 2010b)

or LOI and S in Eq. 8.5. The stippled grey line in Fig. 8.13, which in praxis is identical with the regression line, is constructed on the basis of this method. The method is based on empirical relations derived from silty salt marsh types like those present in the Wadden Sea.

The importance of adjusting accretion measurements to autocompaction is evident from Fig. 8.14. Here, a constant rate of deposition of 1.7 kg m^{-2} year^{-1} of the same sediment type as that from Fig. 8.13 is modelled to show the position of any given yearly surface relative to an incompressible base each year over a 75-year period. If a marker horizon is established on the surface, it will, with the given conditions after a 5-year period, be located exactly 2.19 cm below the surface no matter when it is established. Because of autocompaction, however, the correct accretion in relation to the fixed level of the base will decrease to half its start value after approximately 75 years (~20 cm). It is therefore crucial for any studies of accretion with marker horizons to be aware of both position of the marker below the surface and the degree of autocompaction. Accretion rates based on markers spread on the surface will, if these matters are not addressed, overestimate the correct accretion rates in relation to, for example, a measured sea-level rise.

8.5 Salt Marsh Accretion Models

8.5.1 Model Formulation

Vertical salt marsh accretion consists of elevation change as a function of net accumulation due to sediment supply plus the net accumulation of internally derived plant detritus. Furthermore, it is necessary to take into consideration the rate of deposit thinning due to autocompaction and possible isostatic changes in order to keep track with the absolute level variations. If the level is to be evaluated in relation to the relative sea level at a given location, eustatic changes also need to be considered (Fig. 8.15).

The continuity equation for salt marsh formation can be expressed as:

$$\Delta e = \Delta s_{\mathrm{sed}} + \Delta s_{\mathrm{org}} \tag{8.6}$$

where Δe is the change in salt marsh mass per unit area and unit time, Δs_{sed} is the sediment supply per unit area and unit time and Δs_{org} is the change in organic matter due to biological production per unit area and unit time. In order to translate this equation into levels, it is necessary to incorporate autocompaction and isostatic movements:

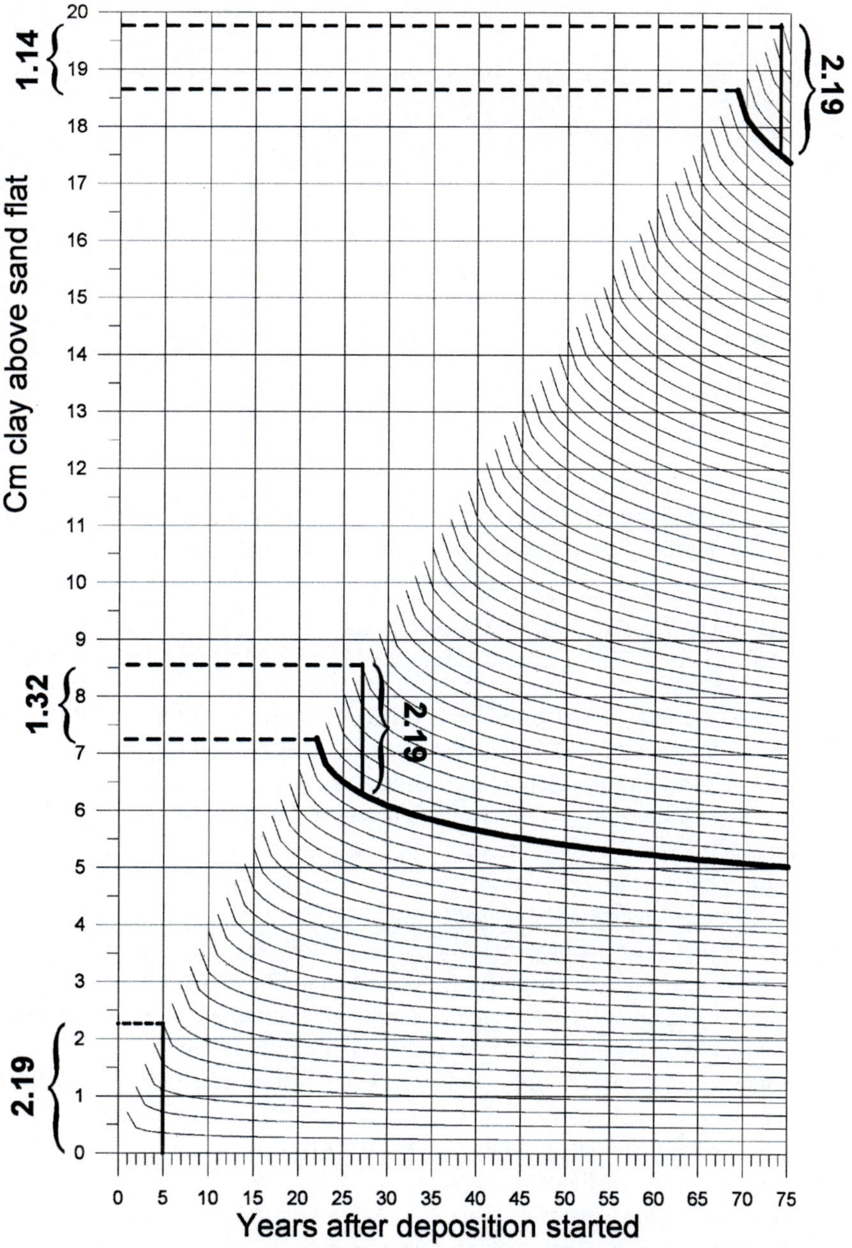

Fig. 8.14 Yearly isochrons of sediment surfaces buried during a salt marsh accretion equivalent to 1.7 kg m^{-2} year^{-1}. The physical parameters of the salt marsh in question are equal to those of the marsh type illustrated in Fig. 8.13. The deposition starts on an incompressible base here thought of as the sand flat beneath the salt marsh deposits. The absolute level of the salt marsh surface relative to the sand flat surface is indicated by the *left* end of the isochrons (From Bartholdy et al. 2010b)

$$\Delta E = \Delta S_{sed} + \Delta S_{org} - \Delta P - \Delta I \tag{8.7}$$

ΔE is the change in absolute level over the considered time interval which is usually taken as 1 year, ΔS_{sed} is the accretion rate of sediment supplied to the salt marsh, ΔS_{org} is the accretion rate due to biomass production, ΔP is the rate of autocompaction and ΔI is the rate of isostatic movement in the area.

Sometimes it may be preferable to evaluate the salt marsh level change in relation to the relative sea level.

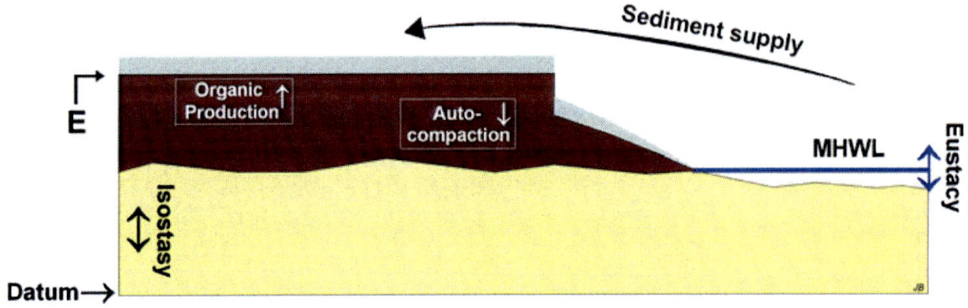

Fig. 8.15 Schematic description of factors influencing the surface level, E, of a salt marsh located above an incompressible base of consolidated sand

In this case, the salt marsh level change rate, ΔE, has to be corrected for the eustatic level change rate, ΔEu, in order to give the rate of salt marsh level change in the moving tidal frame ΔE_{rsl}:

$$\Delta E_{rsl} = \Delta S_{sed} + \Delta S_{org} - \Delta P - \Delta I - \Delta EU = \Delta S_{sed} + \Delta S_{org} - \Delta P - \Delta M \tag{8.8}$$

It is here convenient to merge the isostatic and the eustatic changes into one term describing the relative sea-level change rate, ΔM. Eq. 8.8 is equivalent to the equation suggested by Allan (Allan 1990, p. 79) with the exception that ΔS_{sed} in his version is exchanged with ΔS_{min} being 'the time-rate of build up by mineral sediment (m a^{-1})'. The choice of using ΔS_{sed} instead of ΔS_{min} addresses the fact that most of the organic material in silty salt marsh deposits is actually derived from biogenic processes in the adjacent tidal area. Here, more or less decomposed organic material combines with mineral particles to form the resulting sediment source which feeds the salt marsh areas. Separating organic and mineral material, therefore, creates a problem as the measurable sediment supply consists of both mineral and organic material. This was also recognized by French (1993, p. 69) naming this contribution 'externally derived (predominantly clastic) material' and the organic one as 'that due to internally derived plant detritus'. The latter, which represents the organic below ground production, is the same as ΔS_{org} and was estimated by French (1993) to be on the order of 0.2×10^{-3} m year^{-1} for temperate silty salt marshes.

In the following example, the salt marsh sediment described in Figs. 8.13 and 8.14 is used as background for practical solutions of the above stated variations of the continuity equation for salt marsh sedimentation:

Given parameters: $\Delta S_{sed} = 4.18 \times 10^{-3}$ m year^{-1}, $\Delta S_{org} = 0.2 \times 10^{-3}$ m year^{-1}, $\Delta P = 2.10 \times 10^{-3}$ m year^{-1}, $\Delta I = 0.5 \times 10^{-3}$ m year^{-1}, $\Delta Eu = 1.78 \times 10^{-3}$ m year^{-1} (8.8)

According to Eq. 8.7, the absolute salt marsh level is raised by:

$$\Delta E = 4.18 \times 10^{-3} + 0.2 \times 10^{-3} - 2.10 \times 10^{-3} - 0.5 \times 10^{-3} \text{m year}^{-1} = 1.78 \times 10^{-3} \text{m year}^{-1}$$

Notice that if the isostatic term is neglected, we get:

$$\Delta E_{(\text{without } \Delta I)} = 4.18 \times 10^{-3} + 0.2 \times 10^{-3} - 2.10 \times 10^{-3} = 2.28 \times 10^{-3} \text{m year}^{-1}$$

This is equivalent to the level rise that would be measured down to a marker horizon if it was located under the autocompacting salt marsh, for example in or at the sandy base (Fig. 8.15). Over 5 years, it adds up to 11.4×10^{-3} m or 1.14 cm as indicated at the top of the salt marsh modelled in Fig. 8.14.

Likewise, if we furthermore neglect the autocompaction term, we get:

$$\Delta E_{(\text{without } \Delta I \text{ and } \Delta P)} = 4.18 \times 10 + 0.2 \times 10^{-3} = 4.38 \times 10^{-3} \text{m year}^{-1}.$$

This is the level rise which would be measured down to a marker horizon established on the surface the year before the measurement. Over 5 years, it adds up to 21.9×10^{-3} m or 2.19 cm as indicated in Fig. 8.14.

According to Eq. 8.8, the salt marsh level change relative to the rising tidal frame is:

$$\Delta E_{\text{rsl}} = 4.18 \times 10^{-3} + 0.2 \times 10^{-3} - 2.10 \times 10^{-3} - 0.5 \times 10^{-3} \text{m year}^{-1} - 1.78 \times 10^{-3} = 0 \text{ m year}^{-1}.$$

Thus, the example here illustrates a salt marsh which is in equilibrium with the stated relative sea-level rise.

As ΔP, ΔI and ΔEu are given properties and as ΔS_{org} in many cases is so small that it can be neglected, the challenge, when trying to evaluate a salt marsh in relation to its accretion rate, mainly consists in finding a way of simulating ΔS_{sed} (or ΔS_{min} as used by Allan, 1990 and Temmerman et al. 2003).

French (1993) and Temmerman et al. (2003) adopted an approach to the determination of ΔS_{sed} based on that of Krone (1987). Following Temmerman et al. (2003), the temporal variation of the depth average suspended concentration is described by the following mass balance equation:

$$\frac{d\left[\left(h\langle t\rangle - E\right)C\langle t\rangle\right]}{dt} = -w_s C\langle t\rangle + C_0 \frac{dh}{dt} \quad (8.9)$$

where h(t) is the time-dependent water level, E is the salt marsh level, C(t) is the time-dependent depth average suspended sediment concentration in the water, w_s is the settling velocity of the suspended sediment and C_0 is the sediment concentration in the flooding water.

The left-hand side of Eq. 8.9 describes the change in the content of suspended sediment in the water with time as h(t)-E is the water depth which multiplied with the depth average suspended sediment concentration quantifies the mass of suspended sediment over a unit area. The first term on the right-hand side describes the result of the vertical settling of suspended sediment and the second term on the right is supposed to describe the lateral flux of sediment in water with a suspended sediment concentration of C_0. C_0 is given a specific value during flood tide, while during ebb tide it is set equal to $C(t)$.

Carrying out the differentiation of the left-hand side in Eq. 8.9 gives:

$$\left(h\langle t\rangle - E\right)\frac{dC\langle t\rangle}{dt} + C\langle t\rangle \frac{d\left[h\langle t\rangle - E\right]}{dt} = \left(h\langle t\rangle - E\right)\frac{dC\langle t\rangle}{dt} + C\langle t\rangle \frac{d\left[h\langle t\rangle\right]}{dt} = -w_s C\langle t\rangle + C_0 \frac{dh}{dt}$$

$$\Rightarrow$$

$$\left(h\langle t\rangle - E\right)\frac{dC\langle t\rangle}{dt} = -w_s C\langle t\rangle + \left[C_0 - C\langle t\rangle\right]\frac{dh\langle t\rangle}{dt} \quad (8.10)$$

This is the original equation suggested by Krone (1987) by which the temporal variation of the depth average suspended concentration, $C(t)$, can be found and multiplied by the settling velocity to give the time-dependent sedimentation on a unit area of salt marsh. Integrating this product over a tidal period and adding up similar results from every tidal period in a year, then give the combined deposition in mass per unit

Fig. 8.16 Excavation of one of the test sides on the salt marsh of the Skallingen backbarrier. The location at the time of excavation in 2003 is shown to the *left*. To the right is seen the excavated trench with about 20 cm salt marsh clay (*brown*) above the sand base (*light grey*). A few cm *above* the old sand surface there is a *red-collared* horizon. This is the marker horizon consisting of *red-collared* sand spread on the salt marsh surface by Nielsen in the 1930s (Nielsen 1935) (Photos courtesy of Jørn Bjarke Torp Pedersen)

area per year which, by division with the bulk dry density of newly deposited material, gives ΔS_{sed}. This procedure has been programmed with various editions of Eqs. 8.7 and 8.8 by Temmerman et al. (1993) who also incorporated a mathematical description of the tidal curve and by French (2006), who incorporated the compaction term described in Eq. 8.1.

The basic assumptions by means of which this model is implemented provide limitations. Perhaps the most important of these is the assumption of a constant value of C_0, as it is clear from observations (Fig. 8.3) that C_0 is not constant over the tidal period. Temmerman et al. (2003) allow C_0 to vary as a linear function of the high tide level (but keep it constant over the tidal period), whilst French (2006) considers it to be constant for and during all modelled tidal periods. Both regard the settling velocity, w_s, to be constant, even if it undoubtedly decreases with time during a tidal period, and most likely in general follows an expected variation in C_0. Being in the open, these limitations do not disqualify the two models, but contribute to classify them as first approximations based on which more work should be done to improve their performance.

Bartholdy et al. (2004, 2010a) used another approach. Their database allowed them to evaluate accurate accretion measurements in three lines across the Skallingen backbarrier (Fig. 8.1) over a period of more than 60 years. This database was founded by Nielsen (1935) who spread out red-coloured sand on the surface of what then was a juvenile backbarrier salt marsh. The marker horizon can be seen in an excavated trench dug in 2003 (Fig. 8.16).

This model assumes that in time, the average deposition (kg m^{-2} tidal period^{-1}) at a specific site, related to a specific high water level (*HWL*), is equivalent to:

$$\Delta s_{sed} = \Delta C_{(HWL)}\left[HWL - E(t)\right] \tag{8.11}$$

where $E(t)$ is the salt marsh elevation at the specific location and $\Delta C_{(HWL)}$ is the characteristic concentration difference available for deposition. $\Delta C_{(HWL)}$ is regarded as place specific but dependent on the high water level relative to the mean high water level (*MHWL*):

$$\Delta C_{(HWL)} = \alpha \ln\left(HWL - MHWL\right) + \beta; \text{ valid for HWL} > MHWL \tag{8.12}$$

The mean high water level is supposed to be the level of salt marsh initiation. The model was formulated in a Fortran program where all high water levels were grouped in classes of 0.2 m steps above the mean high water level and assigned the class midpoint as its level. Observations in the salt marsh creeks (e.g. time series as the one shown in Fig. 8.3) suggested that no deposition takes place for over-marsh tides in the lowest water level class. Therefore, ΔC in the lowest class of HWL-$MHWL = 0.1$ m was set as 0. According to Eq. 8.12, this means that:

$$\alpha = \beta / 2.303 \Rightarrow \Delta C_{(HWL)} = (\beta / 2.303)\ln(HWL - MHWL) + \beta;\ \text{valid for } HWL > MHWL \qquad (8.12')$$

where the physical meaning of β is ΔC for a high water level of 1 m above MHWL.

Substituting Eq. 8.12′ into Eq. 8.11 and adding the result for all tidal periods in a year give Δs_{sed} for the year in question.

In order to correct this for autocompaction, it is necessary to know the mass depth (kg m^{-2}) of the salt marsh from the surface to the basement under the salt marsh deposits at the specific location. If Δs_{sed} for the calculated year is added to this and introduced in Eq. 8.2′, this can be solved for z, giving the salt marsh level on top of the basement after the modelled year.

Using the procedure described above in parallel with the model of Temmerman et al. (2003) gave a linear relationship between ΔC and the overmarsh high tide level, which indicates correspondence between this very simple, purely empirical model and the more complicated model based on the theoretical considerations of Krone (1987). Even if the conditions are radically changed when going from a semitheoretical model as that of Temmerman et al. (2003) to a purely empirical model like this one, it can be criticized for the same drawbacks as stated above. On the other hand, if we assign the suspended sediment a settling diameter of 25 μm in accordance with 'normal' suspended fine-grained sediment in the area around Skallingen (Bartholdy and Anthony 1998), this corresponds to a settling velocity of about $ws = 0.4\ 10^{-3}$ m s^{-1} and a settling time for 1 m of about 1.5 h, which is less than the expected period of very small velocities around high water. In this environment, the model's ΔC-value can therefore be thought of as the 'full' delivery of available sediment imported from the tidal area during a given over-marsh tide (i.e. all the sediment in suspension will settle out).

8.5.2 Examples of the Use of Accretion Models

8.5.2.1 Salt Marsh Stability in Relation to Sea-Level Rise

By means of accretion measurements over the approximately 60-year period in three lines across the Skallingen backbarrier (Nielsen 1935; Jakobsen 1953; Bartholdy et al. 2004, 2010a), former and present measurements of clay thickness corrected for autocompaction by means of Eq. 8.2′ and sea-level data from a nearby tide gage were used to calibrate 32 points scattered over the backbarrier for their β-value (the deposition potential). β was found to correlate with two variables: (1) distance to marsh edge ($X1$) and (2) distance to creeks of second or higher order ($X2$). The distance to salt marsh edge was able to explain 61% of the variation, and the distance to second or higher-order creeks was able to add 10% to this. Thus, the combined correlation explained 71% of the variation in the deposition potential. The best correlation in both cases was achieved by means of a logarithmic relation giving the final empirical equation the following appearance:

$$\beta = 4.095\ln(X1)\ln(X2) - 36.402\ln(X1) - 32.421\ln(X2) + 288.224 \qquad (8.13)$$

By means of Eq. 8.13, a map of the characteristic concentration difference available for deposition for a high water level of 1.3 m DNN was constructed as shown in Fig. 8.17. This high water level was chosen as it represents the most 'efficient' level in terms of salt marsh deposition when frequency is also considered (Bartholdy et al. 2004). It is clear from Fig. 8.13 how the depositional environment reflects the general pattern discussed above giving rise to a salt marsh surface level that accretes most rapidly in the outer part of

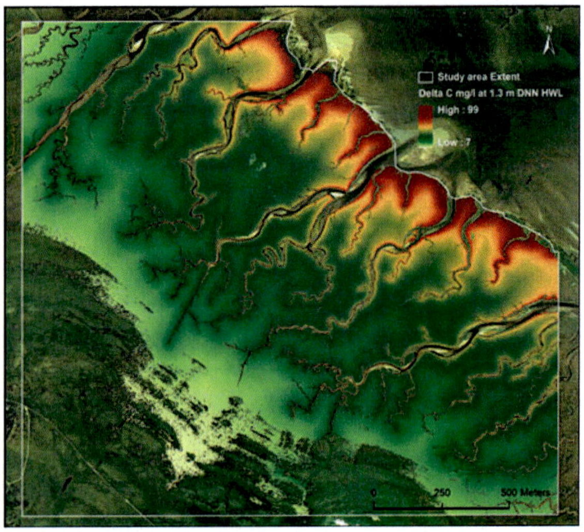

Fig. 8.17 Map of the distribution of the characteristic concentration difference available for deposition, ΔC in mg l^{-1}, on the central part of the Skallingen backbarrier in a tidal period with a high water level of 1.3 m. Areas above 1.3 m DNN and areas associated with creeks have been cut out leaving the underlying orthophoto visible (After Bartholdy et al. 2010a)

the backbarrier as well as in the areas along the major creek systems. This pattern, as will be discussed beneath, complicates the concept of salt marsh equilibrium.

Based on high water statistics for the period 1949–2007, the model by Bartholdy et al. (2010a) was run for longer periods with a stable sea level simply by using multiples of this distribution. An inner and an outer position relative to the salt marsh edge with typical β-values and a sand base level of 0.80 m DNN (close to the actual conditions on the Skallingen backbarrier) were chosen to illustrate developments in the salt marsh level from the time where deposition started on top of the underlying sand flat.

The difference between the outer and inner salt marsh (Fig. 8.18a, b) continues to grow during the marsh development for a long period. After about 1,700 years, it reaches a maximum of a little less than 21 cm. Both the outer and the inner area continue to grow with a few cm per 100 year after this where the level above the highest astronomical tide (*HAT* = 1.3 m DNN) has reached 44 cm and 63 cm, respectively. A stable sea level for such a long period is unrealistic, and the result is therefore in every respect highly hypothetical. It shows, nevertheless, that the idea of equilibrium salt marsh topography in this type of environment is highly problematic. Theoretically, the sedimentation should ultimately decline to almost zero. This would happen first in the outer part and later the inner part where after the two levels should even out. But even after 5,000 years (not shown), the modelled difference between the two locations has only declined to a little under 20 cm with a growth rate of approximately 0.5 cm per 100 year and a level of the outer part of about 1.9 m above *HAT*.

The concept of salt marshes being able to reach dynamic equilibrium in a rising tidal frame under a relative sea-level rise has been suggested by Allan (1990) and discussed by others such as French (2006). Bartholdy et al. (2010a) tested this idea by letting MHWL in Eq. 8.12′ follow a rising tidal frame of different sea-level rise scenarios using conservatively raised high water statistics similar to the one used above. The results are presented in Fig. 8.18c (inner marsh) and Fig. 8.18d (outer marsh). The starting point is the same as that above, with the marsh at its present elevation after about 100 years of simulated deposition on top of the bare sand flat. Simulated levels are plotted as the difference between the modelled salt marsh level and the rising *HAT*. When this difference becomes less than ÷0.5 m (corresponding to the actual *MHWL* of 0.8 m DNN with the present *HAT* of 1.3 m DNN), the salt marsh is assumed to degrade back to unvegetated tidal flat, as this is the level of salt marsh initiation. There might be a hysteresis effect keeping already established salt marshes 'alive' below this level, but to what extent this is the case is unknown. The constant rising salt marsh level in relation to *HAT* in the stable case (a sea-level rise of 0.0 mm year^{-1}) is similar to the results given for the two locations in Fig. 8.18a. The overall impression of the different sea-level rise scenarios is that there is a significant difference between the topographical reaction of the inner and outer marsh, which in principle rules out any possibility of topographical equilibrium within a realistic timescale. For each site, there is a specific sea-level rise where the equilibrium concept of Allan (1990) actually exists. This is for the inner marsh close to 0.5 mm year^{-1} and for the outer marsh close to 1.0 mm year^{-1}. With a sea-level rise of 0.5 mm year^{-1}, the inner marsh would 'mature' to a constant level in the rising tidal frame about 1,000 years after salt marsh initiation, and with a sea-level rise of 1.0 mm year^{-1}, the outer marsh would reach the same stability in about 500 years. It is interesting that both marsh areas reach stability at the same

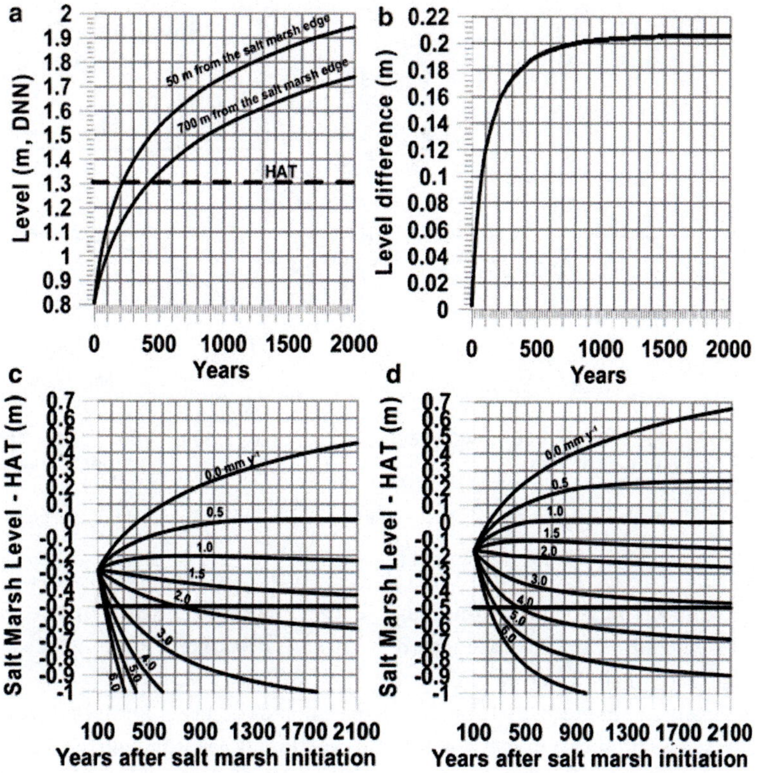

Fig. 8.18 (**a**) Simulated growth of the outer and inner salt marsh at Skallingen based on a constant sea level and tidal conditions corresponding to those present in the period 1949–2007. (**b**) Simulated level difference between the outer and inner salt marsh at Skallingen under similar conditions as those in **a**. (**c**) Simulated difference between highest astronomical tide (*HAT*) and salt marsh level under different sea-level rise scenarios at the inner part of the Skallingen salt marsh. The tidal conditions are similar to those present in the period 1949–2007 conservatively raised parallel to the rising sea level. The simulation starts at present, about 100 years after salt marsh initiation. (**d**) Simulated difference between highest astronomical tide (*HAT*) and salt marsh level under different sea-level scenarios at the outer part of the Skallingen salt marsh. The tidal conditions are similar to those in **c**. The simulation starts at present, about 100 years after salt marsh initiation. The fat horizontal line in **c** and **d** represents the present level difference between *HAT* and the level where salt marsh starts to form (From Bartholdy et al. 2010a)

level relative to HAT – namely 0. The two salt marsh locations would therefore both stabilize at a level very close to *HAT* but not with the same sea-level rise. A stable inner part would coincide with a development where the outer part would grow towards the top of the rising tidal frame, whereas a stable outer part would coincide with a development where the inner part would eventually drown. For any other sea-level rise scenarios, both the inner and the outer salt marsh would be in disequilibrium, either in a state of downing or growing towards the top of the tidal frame. The reason for the apparent paradox that a salt marsh area can establish and grow during a sea-level rise which eventually will be too high for it to keep pace with the rising tidal frame (e.g. 1.5 mm year^{-1} in Fig. 8.18d) is that along with the salt marsh growth, the level increase of the salt marsh surface slows down because of auto-compaction (Fig. 8.14).

8.5.2.2 Relationship to Different Tidal Conditions

The model can also be used to describe and quantify variations in the development of different salt marsh types. In Fig. 8.19, the accretion of a wind-tide-influenced salt marsh in the Wadden Sea (Skallingen) is compared to that of a predominantly astronomical controlled salt marsh on the east coast of USA (Georgia). The model is ran for a constant sea level

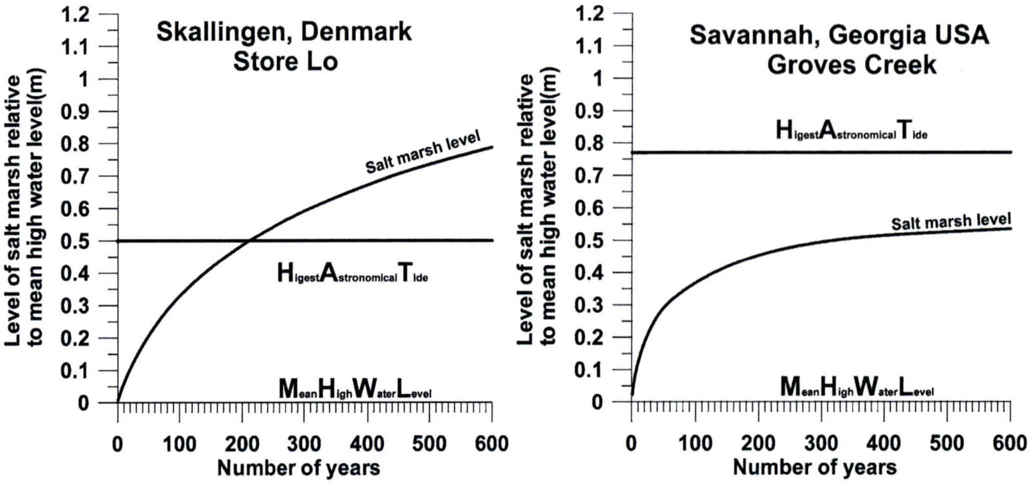

Fig. 8.19 Comparison between modelled salt marsh accretion in the Wadden Sea (*left*) and at the east coast of USA (*right*)

and the mean tidal conditions (repeated) at the harbour of Esbjerg (Denmark) and at Fort Pulaski located at the entrance of the Savannah River in Georgia (USA). The mean tidal range in Esbjerg is ~1.5 m and at Fort Pulaski ~2.1 m. The initial salt marsh level in both places is taken as the mean high water level and the used β-value is at Skallingen put to 90 mg/l (a typical value from Bartholdy et al. 2010a) and in Georgia to 230 mg/l (judged as a typical value at Sapolo Island a little south of Savannah, from Howard and Frey (1985)). It is apparent that the wind-tide-affected salt marsh (Fig. 8.19 *left*) relatively quickly (~200 years) accretes up to a level above the highest astronomical tide (HAT) and continues to accrete beyond this level, whereas the almost solely astronomical controlled salt marsh (Fig. 8.19 *right*) most likely will never reach that level. Even after 600 years of deposition, the salt marsh level here is still as much as 20 cm lower than *HAT*. These model results reflect and quantify general observed differences between the relatively dry salt marsh areas along the North Sea coast (capable of being grassed by cattle in the summer time) and the constant soft, muddy and wet salt marsh areas along the east coast of USA. Frequent wind-tide setup events above the level of *HAT* enable the Wadden Sea salt marsh to grow above this level, whereas the growth of the Georgian salt marsh will, even if the β-value is more than twice that of the Wadden Sea example, accrete asymptotically towards the highest astronomical level which is not exceeded under stable sea-level conditions.

8.6 Salt Marshes in the Geological Record

Salt marsh deposits are found in a number of coastal depositional environments. Following the classification of Boyd et al. (1992, Fig. 8.20), salt marshes can exist on open tidal dominated coasts under transgression as well as regression/progradation. During transgression, salt marsh deposits typically form in the inner part of wave-dominated estuaries and they are abundant along the entire coast of tidal dominated estuaries. This is also the case along coastal lagoons sheltered from the open coast by barriers. Deltas form a special case of prograding coastal areas where salt marsh deposition is widespread. Viewed separately, salt marsh formation is independent of coastal type, and will in general look the same and be associated with the same primarily morphological features no matter where it develops. This, as already discussed, is a result of the fact that salt marsh deposition is dependent on a special combination of vegetation, water-level variation, underlain topography, sediment supply and dynamical relations. Regardless of where this combination appears, it will result in basically the same type of sedimentation. Variations like those related to, for example, degree of wave exposure, tidal range and the occurrence of ice rafting will, of cause, give rise to variations in morphology. However, such variations can occur more or less, regardless of the coastal types defined by Boyd et al. (1992). It is possible to have both large and small tidal ranges on a

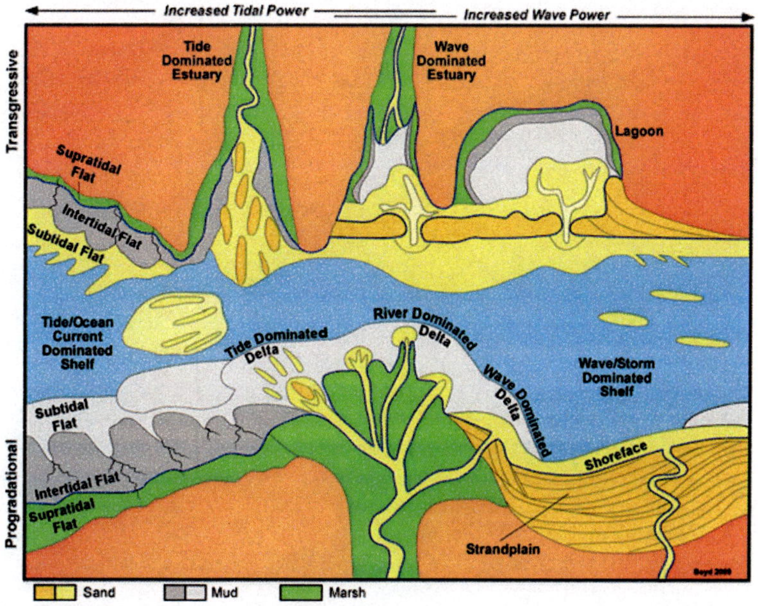

Fig. 8.20 The distribution of major coastal depositional features (After Boyd et al. 1992)

wave-dominated transgressive barrier coast as well as local wave-exposed areas in tidal dominated estuaries. Even sheltered areas in wave-dominated estuaries are possible. The reason for mentioning salt marsh deposition in relation to these coastal types is solely to point out that the facies associations related to the base of salt marsh deposits can vary significantly and should be evaluated in a broader sequence stratigraphic perspective. Deposition and stratigraphy in estuaries, tide-dominated deltas and open coast tidal flats are treated elsewhere in this book and will therefore not be further discussed in the present chapter. Tide-dominated estuaries will briefly be mentioned at the end of this section.

8.6.1 Mainland and Backbarrier Salt Marsh Deposits

On a transgressive barrier coast, there are two focus points for salt marsh formation in a profile perpendicular to the coast. One is located on the mainland coast where the salt marsh in general forms on top of basal peat, and another is located on the back of barrier islands/spits, where salt marsh formation as a rule is formed on top of washover or tidal flat deposits. The reason for this general pattern in the stratigraphic relationship is twofold. As a rule, the groundwater supply (and thus, peat formation) on a mainland coast is substantial compared to that of barrier islands. Furthermore, the substrate on a mainland coast can consist of a variety of surface types, while the substrate of transgressive barrier islands is either washover sand or tidal flat deposits formed in the rim of the lagoon. There are of course a number of variations over this theme, for example missing peat in some subtropical barrier sequences like those of Georgia, USA (Howard and Frey 1985), and peat underlying also the barriers in New Jersey, USA (Pusty 1980). See Davis (1994) for an overview of different types of barrier island systems. The following generalized stratigraphic model of a transgressive barrier system, with regressive episodes (Fig. 8.21), is based on personal observations in the Danish Wadden Sea and results published in Davis et al. (2001), Bartholdy et al. (2004), Gehrelds et al. (2006) and Pedersen et al. (2009).

The evolution illustrated in Fig. 8.21 shows two important anomalies in relation to the general schematic model of Galloway and Hobday (1983). First of all, most transgressions start with peat formation at the mainland coast (often just regarded as part of the salt marsh formation) and secondly, even if the preservation potential in some cases is small, salt marshes form

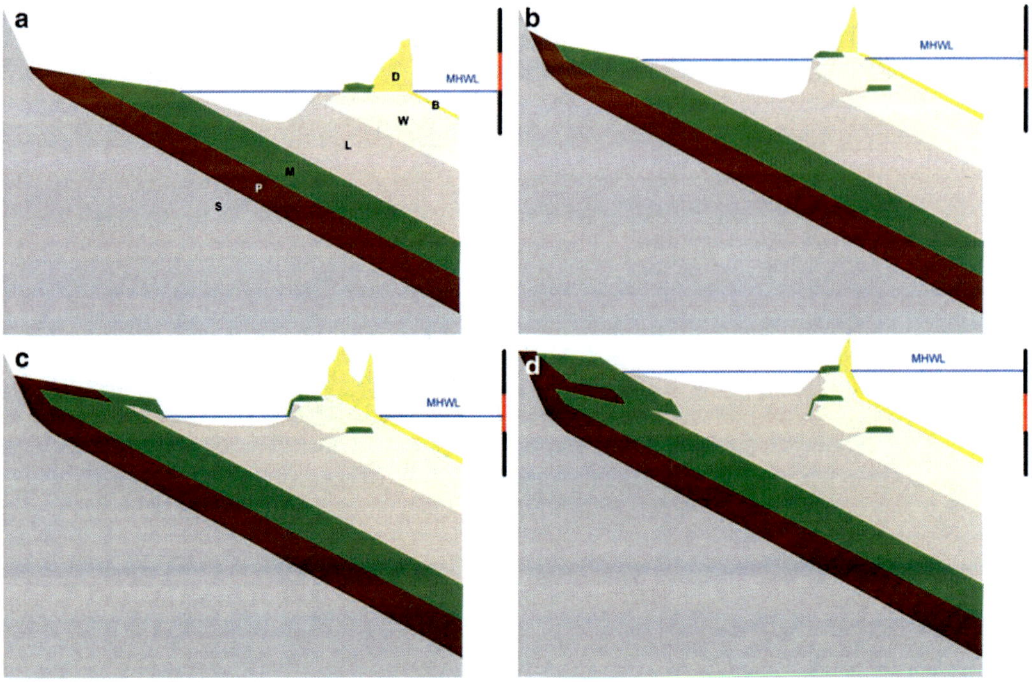

Fig. 8.21 Generalized diagram illustrating the stratigraphic relationship in a transgressive barrier system typical for the Danish Wadden Sea. The mean high water level (MHWL) is indicated by the position of the *horizontal blue lines* scaled on the *vertical black* and *red* range to the *right*. The typical behaviour of this type of barrier system is illustrated from a traditional overall model (e.g. Galloway and Hobday 1983) in **a**. In **b**, the water level is raised under transgression. In **c**, the water level is lowered. **d** illustrates the final stage after another transgressive sea-level rise. The six facies illustrated in the diagram are: **b** beach and shore face, **d** aeolian dunes, *L* lagoon, *M* salt marsh, *W* washover, *P* peat, *S* substratum

on the backbarrier of the barrier islands/spits. Both should be considered as part of the possible resulting deposition of fully developed salt marsh environments during deposition in a transgressive barrier system. In cases of sparse sediment supply to a retreating coastline, the shore face front can be steep and leave nothing else but lagoonal deposits behind during a transgression. If this is not the case, however, these backbarrier salt marshes will be preserved in the geological record as horizontal slabs of salt marsh deposits buried in washover sand overlaying either lagoonal or older washover deposits as illustrated in Fig. 8.21. It is important here to note that even if salt marsh deposition in such an environment is small (a few tens of centimetres), the very fact that it exists actually has an influence on a several metre thick depositional units formed by associated salt marsh creeks.

Observation of a clay bench emerging close to the mean spring low water level on the exposed west coast of the barrier spit Skallingen (point A in Fig. 8.1) confirms this and suggests that these benches of salt marsh clay form as episodic isolated depositional events, most likely as a result of longer periods of a relatively steady sea level. Under such conditions and if the supply of fine-grained sediment to the backbarrier is sufficient, salt marsh is prone to form and grow on top of the washover sand in the lee of the foredune area. The mentioned clay bench on the exposed coast of Skallingen is located about 2 m below the actual MHWL and dates back to a period of approximately 400 years centred around 1350 AD (based on three calibrated ^{14}C dates of shell material: AAR-8203, AAR-8204 and AAR-8205 from AMS 14C Dating Laboratory, University of Aarhus). This coincides with a cooler global temperature following the Mediaeval warm period which culminated about 1150 AD and was followed by the Little Ice Age with a temperature minimum at about 1650 AD (IPCC 1990). In this period – most likely – the postglacial sea-level rise was dampened by the cooling climate conditions. From old maps (i.e. Johannes Meyer 1654), we know that Skallingen in the mid-seventeenth century existed in

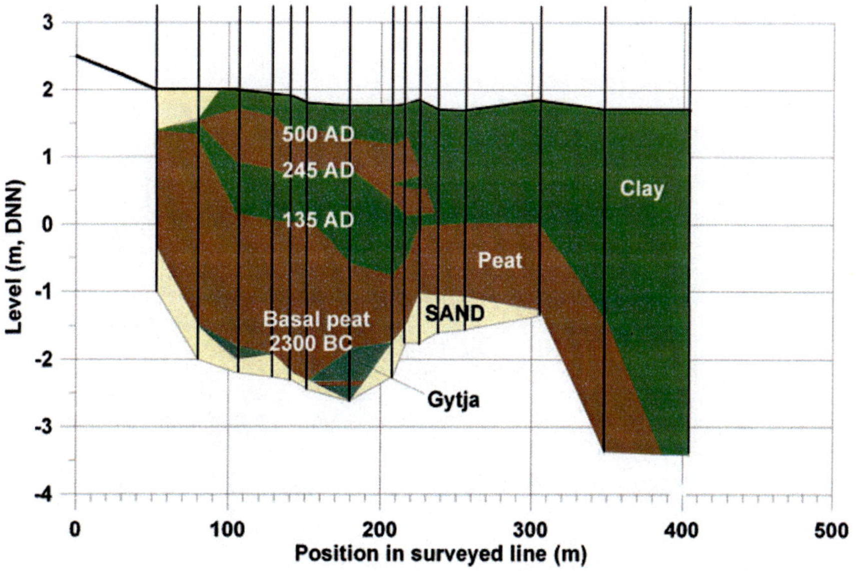

Fig. 8.22 Stratigraphic profile based on corings in a line directed approximately north/south from the salt marsh at Kjelst (location **b** in Fig. 8.1). The profile starts at the Pleistocene northern rim and extends about 400 m towards Ho Bugt. The four ^{14}C datings represent *top* and *bottom* of the two peat horizons. The *vertical lines* indicate the locations of corings based on which the profile was constructed

the form of an elongated sand flat. A foredune chain was beginning to form in the beginning of the nineteenth century (Society of Natural Science 1804). Thereafter, modern maps (from 1870) document the new salt marsh formation on the Skallingen backbarrier as discussed above starting in the small global cooling period from 1880 to 1910 (HadCRUT3; http://www.cru.uea.ac.uk/cru/data/temperature) and getting a foothold close to the beginning of the twentieth century (Nielsen 1935). Thus, this documented evolution with two sequences of salt marsh formation over a period of 700–1,000 years corresponds to the modelled transgressive evolution between Fig. 8.21a, b, where the salt marsh formation is assumed to coincide with pauses in the relative sea-level rise.

During sea-level fall, the level where peat can start to form is lowered which causes peat to grow out over the salt marsh formed on the mainland coast. This is illustrated in Fig. 8.22 showing the results of auger corings (hand-operated Eijkelkamp type) in a profile from the salt marsh at Kjelst in the northern part of Ho Bugt (location B in Fig. 8.1).

At the base of the profile, sand is interpreted as the Pleistocene substrate on which the Holocene sediments are deposited. The surface sand at the start of the profile is most likely part of more recent material moved by either soil erosion or human interference – the beginning of the profile is located in a farm area. The outer part of the profile shows a typical transgressive sequence with peat below fine-grained deposits overlaying the substrate (this is not reached by the auger at the end of the profile). When the postglacial Holocene sea level reached about −4 m DNN about 4,000 years ago (Fig. 8.25), peat formation started and continued to build up during the following 2,000 years. The sea level reached a maximum of about −1 m DNN in the first centuries after AD, and the whole area represented by the profile was flooded and covered by mudflats under deposition of marine clay, most likely followed by salt marsh deposits hereafter. After this sea-level summit, a minor sea-level fall caused peat to spread out over the fine-grained sediment surface, before the whole area again was flooded and eventually turned into a salt marsh where accretion today exceeds that of the relative sea-level rise in the area. Thus, this last part of the stratigraphic development in the profile corresponds to the modelled regressive evolution between Fig. 8.21b, c. The last development between Fig. 8.21c, d schematizes another sea-level rise in order to complete the overall transgressive development characterized by both transgressions and regressional periods.

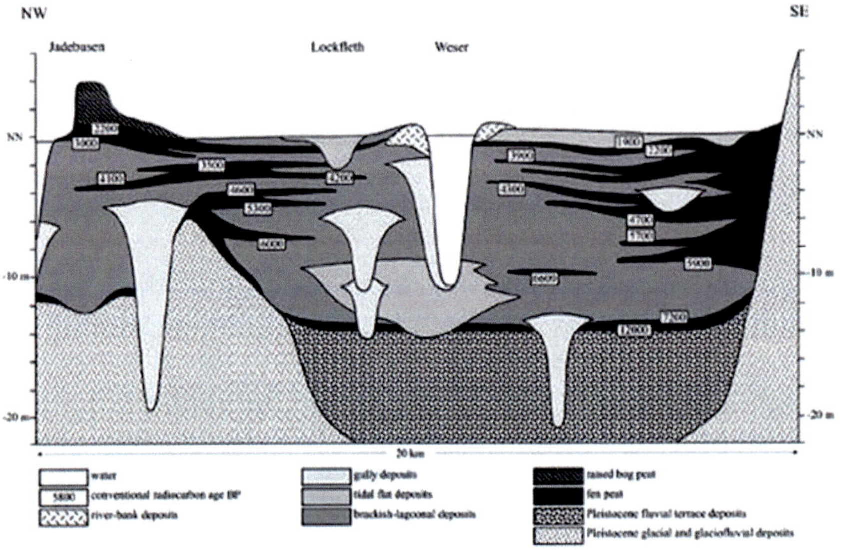

Fig. 8.23 Cross section of the Weser Estuary about 15 km upstream of Bremerhaven (From Streif 2004)

The profile at Kjelst (Fig. 8.22) represents a mainland salt marsh type which is located relatively close to the higher Pleistocene hinterland and forms the inland border for marine deposits. Along a transgressive barrier coast, such areas on the mainland will typically exists between smaller or larger low-laying areas belonging to valleys of the hinterland's drainage system. The large valleys will also accommodate rivers and form tide-dominated estuaries before entering the sea. In such environments, salt marsh deposits gradually transform into freshwater marshes in an inland direction, and, like in the example from a section across the Weser Estuary (Fig. 8.23, Streif 2004), become interrupted by a number of channel-fill deposits (by Streif called 'gully deposits'). Apart from this, the pattern is the same as that of the Kjelst profile. The sedimentation has kept phase with the Holocene sea-level rise, and basal peat covers the substrate followed by fine-grained deposits and intercalated peat horizons.

In Figs. 8.22 and 8.23, no distinction has been made between marsh and mudflat deposits. They are both part of the clay and brackish-lagoonal deposits stated in the legends. It is difficult to distinguish between these sediment types in the geological record, and both will typically replace each other in turns as a result of different degrees of inundation during the infill. This is in Fig. 8.24 illustrated in a reconstruction of the Holocene evolution of the Varde Å Estuary (for location, see Fig. 8.1 point c). The resulting relative sea-level curve is shown in Fig. 8.25. Both figures are from Pedersen et al. (2009), who suggested a method to distinguish between salt marsh clay and tidal flat clay in the geological record based on presence or absence of small (63–355 μm) red iron concretions. When these concretions were present in the analysed core material, there were no foraminifers and vice versa. The concretions were therefore interpreted as reminiscence of gleying (mobilization and subsequent precipitation of iron compounds), which is typical for salt marsh deposits.

At the start of its evolution during the last part of the Pleistocene (Fig. 8.24a), the central area constituted a melt water valley with (presumably) a braided river system forming the substrate with valley walls of glacial deposits from the second last glaciation (Saale) on both sides. At that time, the water level was far below today's sea level. At about 8,000 years BP (Fig. 8.24b), the rising sea reached a level of about −12 m DNN (Fig. 8.25) and peat started to form in the valley (the top of the braided river deposits is located at about −10 m DNN). With a pause close to 3,000 BP (a local sea-level fall), the depositional evolution continued with salt marsh forming on top of the peat until a little before 2,000 years BP (Fig. 8.24c, d). Hereafter, the valley was flooded in its outer part (Fig. 8.24e), resulting in deposition of lagoonal mud as far inland as approximately 3 km from the actual coastline. During

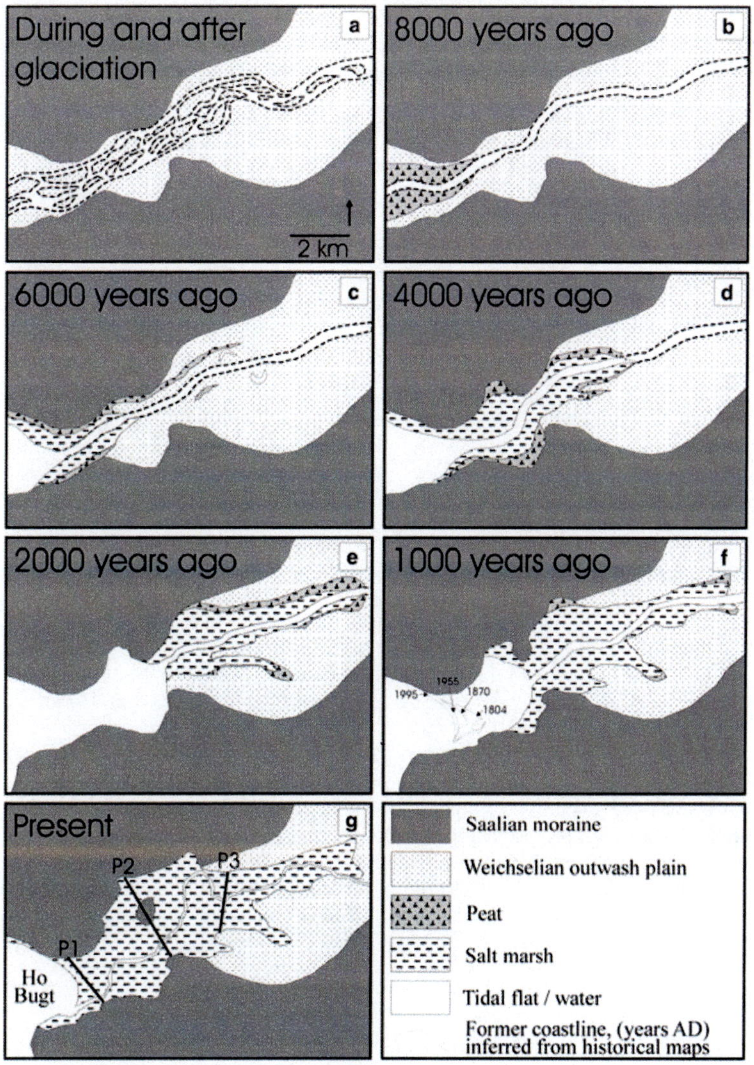

Fig. 8.24 The Holocene evolution of the Varde Å Estuary. The relatively recent coastlines in **f** are drawn from maps. The three lines in **g** indicate the position of the coring lines after which the reconstruction was carried out. The base of each sketch is about 10 km long (After Pedersen et al. 2009)

the subsequent small sea-level fall and moderate sea-level rise (Fig. 8.25), the salt marsh grew outward on top of the lagoonal mud (Fig. 8.24f, g). This last part of the evolution shows that if sediment supply is large enough, a salt marsh is capable of maintaining a regressive coastline, even if the sea level is rising (see Fig. 8.6 upper left which shows this transition as it looks today). The reason is that the area is importing large amounts of fine-grained sediment from the North Sea (Bartholdy and Madsen 1985; Pedersen and Bartholdy 2006) which builds up a huge mudflat area in front of the mouth of the estuary. From here, fine-grained sediment is imported during storms and deposited inside the estuary (Bartholdy 1984). The accretion rate in the salt marsh area of the estuary facing the mudflats is presently between 5 and 10 mm year^{-1}.

The preservation potential of salt marsh is highest for the mainland type, where deposits are also potentially thickest. In the geological record, this type of salt marsh should be found as elongated enclaves between high laying substrates and consists of a basal peat overlain by fine-grained sediment interbedded with peat and frequently interrupted by channel-fill deposits. The backbarrier type should be found as interbedded slaps of salt marsh deposits in washover sand, signalizing a slowing down of the relative sea-level rise. The characteristic salt marsh sediment is associated with a hierarchy of channels from very small (less than

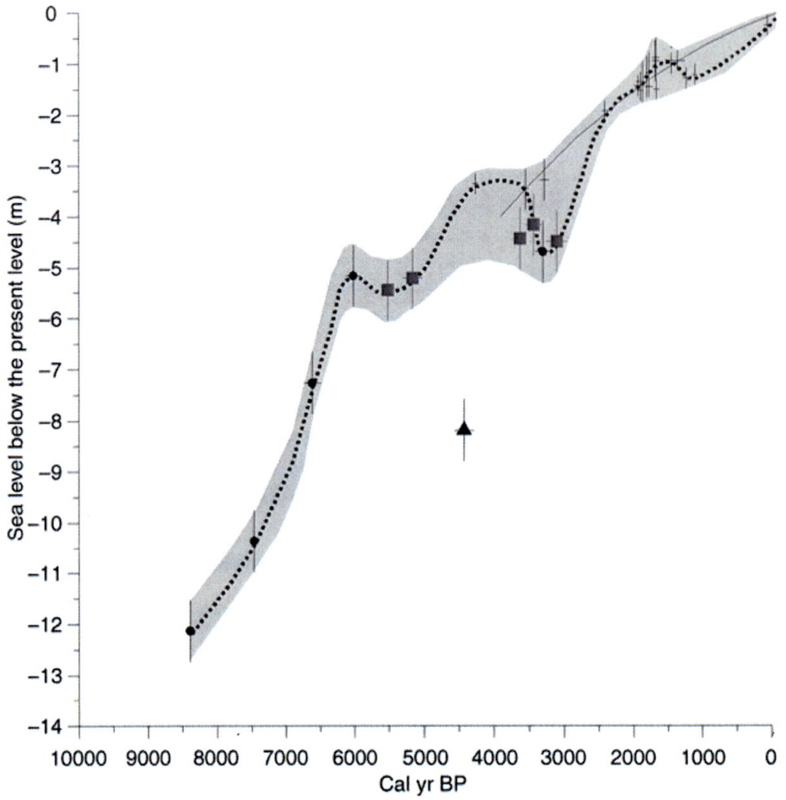

Fig. 8.25 Sea-level reconstruction (m below the present level) based on the data from Gehrelds et al. (2006) (*crosses*) and ^{14}C datings from Pedersen et al. (2009) (*dots*: ^{14}C datings of basal peat, *squares*: ^{14}C datings of nonbasal peat, *triangle*: Rejected ^{14}C dating of peat). *Solid line*: Glacio-isostatic rebound model of Gehrelds et al. (2006). *Punctuated line*: New sea-level reconstruction. The grey envelope encompasses the inaccuracy of all samples (After Pedersen et al. 2009)

a metre wide) to, according to size of the tidal area and tidal range, large channels (up to on the order of 100 m wide and 5 m deep) with cut banks, point bars, channel lags and natural levees.

8.6.2 Facies Associations

Sedimentological and ichnologically centred facies analysis for identifying salt marsh deposits are hard to find in literature. Dashtgard and Gingras (2005) claim novelty in their facies description of an open-coast to open-embayment mature (high) marsh environment with a spring tidal range of 12 m in the Bay of Fundy. They identified 9 facies and 3 facies associations (Fig. 8.26). The 9 facies are: (F1) horizontal to undulatory, rhythmically laminated silt (greenish-grey); (F2) weakly laminated clayey silt (grey); (F3) horizontal to undulatory, rhythmically laminated silt (reddish-brown and greenish-grey) more bioturbated than F1; (F4) low-angle, interbedded clayey silt and silty gravel (brown); (F5) massive-bedded to chaotically bedded clayey silt (brown); (F6) low-angle, laminated silt (reddish-brown); (F7) bioturbated clayey silt (reddish-brown to dark-grey); (F8) steeply dipping sand and gravel beds; (F9) nearly horizontal, parallel-laminated sand and gravel.

The 3 facies associations consists of: (1) salt marsh deposits including F(1), F(2) and F(3); (2) tidal-creek deposits including F(4), F(5), F(6) and F(7); and (3) beach-related deposits including F(8) and F(9). These facies associations are in good agreement with the above stated discussion of the general morphodynamic conditions for salt marshes deposition. Even if the results of Dashtgard and Gingras (2005) are strongly related to the relatively extreme tidal and wave conditions for salt

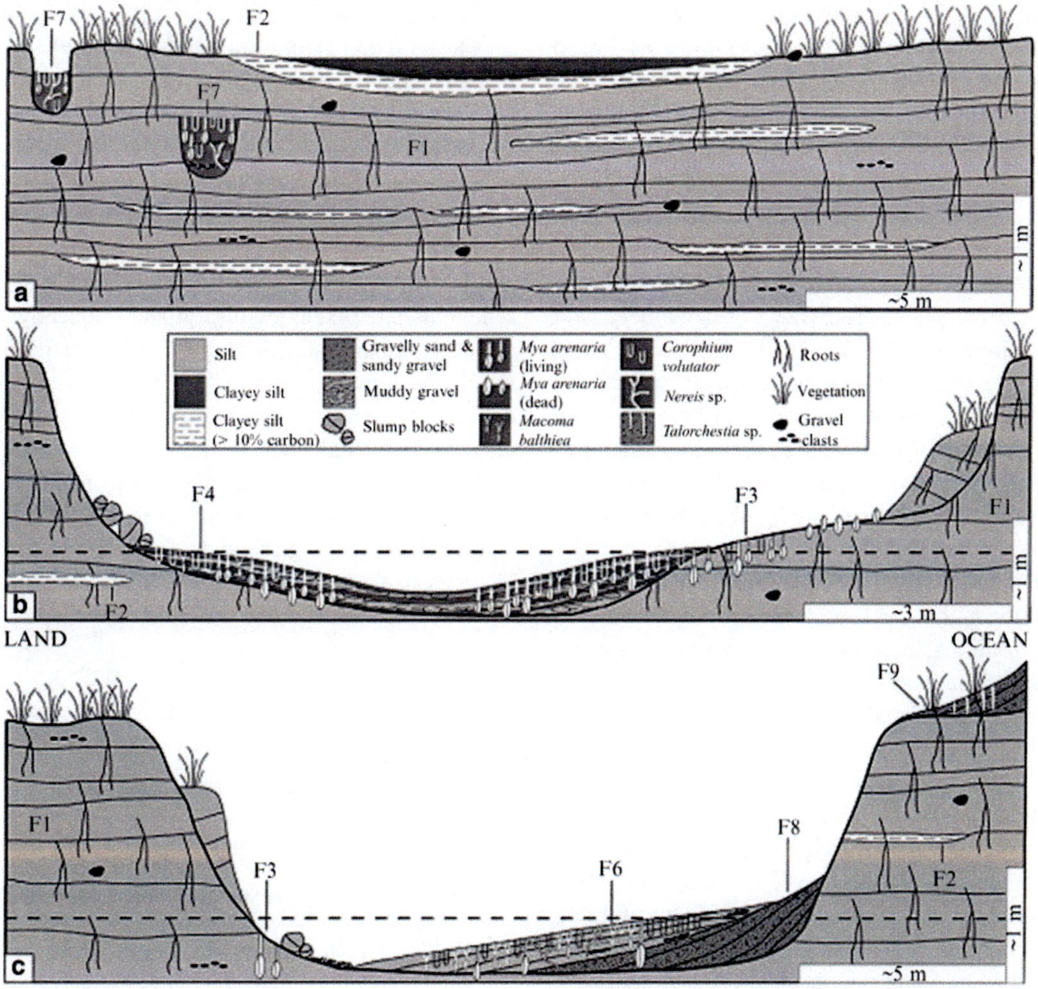

Fig. 8.26 Examples of three facies associations from an open-coast to open-embayment mature (*high*) salt marsh environment with a spring tidal range of 12 m in the Bay of Fundy (From Dashtgard and Gingras 2005). (**a**) Nonorientated profile through the salt marsh centring on a panne (*salt pan*). (**b**) Creek-normal section illustrating sediment deposition via channel aggradation. (**c**) Shore-normal section through a tidal-creek point bar. The nine facies F1–F9 are described in the text. The *dashed line* in **b** and **c** indicates the level of the *upper* erosional boundary exposed in the foreshore

marshes in the Bay of Fundy, their results have a degree of general significance and should inspire similar studies of salt marsh deposits in other climate zones and under different dynamic conditions.

8.7 Summary

Salt marshes defined as 'vegetated areas located between coastal hinterlands and daily (or permanently) flooded coastal areas' are analysed with emphasis on depositional processes and resulting products. Salt marsh sedimentation and related dynamic conditions are evaluated and described with examples from a wide range of locations. General mechanisms and depositional conditions, however, are primarily illustrated by examples from the Danish Wadden Sea based on the author's own experience.

The chapter opens with an overview over measurements of salt marsh sedimentation through time, including direct measurements by means of marker horizons and salt marsh levels and indirect measurements by means of different types of dating methods.

In a description of salt marsh morphodynamics, salt marsh sedimentation is separated into three main types: (1) sedimentation associated with channel flow in the vicinity of salt marsh creeks, (2) sedimentation associated with sheet flow over vegetated salt marsh surfaces, and (3) sedimentation associated with exposed salt marsh edges. A comparison between key factors dominating the fluvial and those prevailing in the tidal salt marsh system is presented in order to highlight the special conditions related to salt marsh sedimentary environments. Import of fine-grained material to salt marsh areas are analysed in relation to salt marsh creek dynamics, including a discussion of shared morphodynamic features between the creek and the salt marsh system like natural levees and crevasse splays. The nature of fine-grained sediment transport over salt marsh platforms is discussed and related to the headward migration of salt marsh creeks and morphological features like salt pans, piping, ice rafting and salt marsh edge morphodynamics.

Salt marsh formation is reported to start at a level close to the high water level from a variety of different climate zones. An example of plant zones across a salt marsh platform is given from the Skallingen backbarrier, Denmark. Effects of vegetation are illustrated by means of recent examples of direct measurements of dynamics in the salt marsh vegetation canopy. Even if vegetation can be regarded as a factor that enhances sedimentation, it is also discussed how vegetation in its outset can cause channel erosion in tidal landscapes when a patchy vegetation cover enhances the ability of flow to concentrate and erode.

The textural composition of salt marsh sediments are described and related to a measure of the environmental exposure of the depositional environment via a ratio between the content of material in the range 5 Φ to 7 Φ (31–8 μm) and that of 5 Φ to 9 Φ (31–2 μm). Autocompaction is discussed in relation to a procedure enabling a description of the bulk dry density variation with level under the salt marsh surface by means of the bulk dry density of the upper 5 cm. Furthermore, the role of autocompaction is discussed and shown to be of crucial importance for a correct interpretation of different types of salt marsh accretion measurements.

Existing dynamic salt marsh accretion models are discussed, explained and related to the salt marsh accretion continuity equation. One of the discussed models is used to give examples of salt marsh stability in relation to different tidal conditions and sea-level rise scenarios, and to discuss the concept of salt marshes being able to reach dynamic equilibrium in a rising tidal frame.

The chapter is concluded by a description of salt marshes in the geological record with distinctions made between mainland coast salt marshes and backbarrier salt marsh formations. The preservation potential of salt marsh is highest for the mainland type, where deposits are also potentially thickest. In the geological record, this type of salt marsh should be found as elongated enclaves between high laying substrates and consists of a basal peat overlain by fine-grained sediments interbedded with peat and frequently interrupted by channel-fill deposits. The backbarrier type should be found as interbedded slaps of salt marsh deposits in washover sand, signalizing a slowing down of the relative sea-level rise. The characteristic salt marsh sediment is associated with salt marsh creeks with cut banks, point bars, channel lags and natural levees. Sedimentological and ichnologically centred facies analysis for identifying salt marsh deposits are described and discussed at the end of the chapter.

References

Ackermann F (1980) A procedure for correcting the grain size effect in heavy metal analyses of estuarine and coastal sediments. Environ Technol Lett 1:518–527

Ackermann F, Bergmann H, Schleichert U (1983) Monitoring of heavy metals in coastal and estuarine sediments – a question of grain-size: <20 μm versus <60 μm. Environ Technol Lett 4:317–328

Allan JRL (1990) Salt-marsh growth and stratification: a numerical model with special reference to the Severn estuary, southwest Britain. Mar Geol 95:77–96

Allen JRL (2000) Morphodynamics of Holocene salt marshes: a review sketch from the Atlantic and southern North Sea coasts of Europe. Quat Sci Rev 19:1155–1231

Appleby PG, Oldfield F (1978) The calculation of ^{210}Pb dates assuming a constant rate of supply of unsupported ^{210}Pb to the sediment. Catena 5:1–8

Appleby PG, Oldfield F, Thompson R, Huttunen P, Tolonen K (1979) ^{210}Pb dating of annually laminated lake sediments from Finland. Nature 280:53–55

Augustinus PGEF (1995) Geomorphology and sedimentology of mangroves. In: Perillo GME (ed) Developments in sedimentology, vol 53. Elsevier Science B.V, Amsterdam, pp 333–357

Bartholdy J (1983) Recent sedimentologi i Hobe Dybs tidevandsområde. PhD thesis, Institute of Geography, University of Copenhagen, Copenhagen, 217 pp

Bartholdy J (1984) Transport of suspended matter in a Bar-Built Danish estuary. Estuar Coast Shelf Sci 18:527–541

Bartholdy J (1985) Dynamic interpretation of grain size relations in silt. Geo-Mar Lett 5:67–70

Bartholdy J (1997) The backbarrier sediments of the Skallingen Peninsula, Denmark, Geografisk Tidskrift. Danish J Geogr 97:11–32

Bartholdy J, Aagaard T (2001) Storm surge effects on a back-barrier tidal flat of the Danish Wadden Sea. Geo-Mar Lett 20:133–141

Bartholdy J, Anthony D (1998) Tidal dynamics and seasonal dependent import and export of finegrained sediment through a back-barrier tidal channel of the Danish Wadden Sea. Tidalites: Processes and Products. SEPM Spec Publ 61:43–52

Bartholdy J, Madsen PP (1985) Accumulation of fine grained material in a Danish tidal area. Mar Geol 67:121–137

Bartholdy J, Christiansen C, Kunzen H (2004) Long term variations in backbarrier salt marsh deposition on the Skallingen peninsula – the Danish Wadden Sea. Mar Geol 203:1–21

Bartholdy AT, Bartholdy J, Kroon A (2010a) Salt marsh stability and patterns of sedimentation across a backbarrier platform. Mar Geol 278:31–42

Bartholdy J, Pedersen JBT, Bartholdy AT (2010b) Autocompaction in shallow silty salt marsh clay. Sediment Geol 223:310–319

Barwis JH (1978) Sedimentology of some South Carolina tidal-creek point bars and a comparison with their fluvial counterparts. In Miall AD (ed) Fluvial sedimentology. Can Soc Pet Geol, 129–160

Bayliss-Smith TP, Healey R, Lailey R, Spencer T, Stoddart DR (1979) Tidal flows in salt marsh creeks. Estuar Coast Mar Sci 9:235–255

Boon JD (1975) Tidal discharge asymmetry in a salt marsh drainage system. Limnol Oceanogr 20:71–80

Boston KG (1983) The development of salt pans on tidal marshes, with particular reference to south-eastern Australia. J Biogeogr 10:1–10

Boumans RMJ, Day JW (1993) High precision measurements of sediment elevation in shallow coastal areas using a sedimentation-erosion table. Estuaries 16:375–380

Boyd R, Dalrymple R, Zaitlin BA (1992) Classification of clastic coastal depositional environments. Sediment Geol 80:139–150

Cahoon DR, Marin PE, Black BK, Lynch JC (2000) A method for measuring vertical accretion, elevation, and compaction of soft, shallow-water sediments. J Sediment Res 70:1250–1253

Carling PA (1982) Temporal and spatial variation in intertidal sedimentation rates. Sedimentology 29:17–23

Chang TS, Flemming BW, Tilch E, Bartholomä A, Wöstmann R (2006a) Late Holocene stratigraphic evolution of a back-barrier tidal basin in the East Frisian Wadden Sea, southern North Sea: transgressive deposition and its preservation potential. Facies 52:329–340

Chang TS, Joerdel O, Flemming BW, Bartholomä A (2006b) The role of particle aggregation/disaggregation in muddy sediment dynamics and seasonal sediment turnover in a back-barrier tidal basin, East Frisian Wadden Sea, southern North Sea. Mar Geol 235:49–61

Chapman VJ (1938) Coastal movement and the development of some New England salt marshes. Proc Geol Assoc 49:373–384

Charman R, Cane T, Cherith M, Williams R (2007) A device for measuring downwearing rates on cohesive shore platforms. Earth Surf Process Landf 32:2212–2221

Christiansen T, Wiberg PL, Milligan TG (2000) Flow and sediment transport on a tidal salt marsh surface. Estuar CoastShelf Sci 50:315–331

Christiansen C, Bartholdy J, Kunzendorf H (2001) Effects of morphological changes on metal accumulation in a salt marsh sediment of the Skallingen peninsula, Denmark. Wetl Ecol Manag 10:11–23

Coldewey HG, Erchinger HF (1992) Diechvorland: Seine Entwicklung zwischen Ems und Jade und die Untersuchungen in Forschungsvorhaben 'Erosionsfestigkeit von Hellern'. Die Küste 54:169–187

Dankers N, Binsbergen M, Zegers K, Laane R, van der Loeff MR (1984) Transportation of water, particulate and dissolved organic and inorganic matter between a salt marsh and the Ems-Dollard estuary, The Netherlands. Estuar Coast Shelf Sci 19:143–165

Dashtgard E, Gingras MK (2005) Facies architecture and ichnology of recent salt-marsh deposits: Waterside Marsh, New Brunswick, Canada. J Sediment Res 75:596–607

Davidson-Arnott RGD, van Proosdij D, Ollerhead J, Schostak L (2002) Hydrodynamics and sedimentation in salt marshes: examples from a macrotidal marsh, Bay of Fundy. Geomorphology 48:209–231

Davis CA (1910) Salt marsh formation near Boston and its geological significance. Econ Geol 5:623–639

Davis RA (1994) Geology of Holocene barrier island systems. Springer, Berlin, 464 pp

Davis RA, Bartholdy J, Lykke-Andersen H (2001) Sedimentary depositional environments and Holocene geologic evolution of the northern Danish Wadden Sea. Korean Society of Oceanography, Special Publication 2001, Proceedings of Tidalites 2000 in Seoul, South Korea: 63–75

Dionne J-C (1984) An estimate of ice-drifted sediments based on the mud content of the ice cover at Montmagny, middle St. Lawrence estuary. Mar Geol 57:149–166

Dyer KR (1989) Sediment processes in estuaries: future research requirements. J Geophys Res 94:14327–14399

Edwards JM, Frey RW (1977) Substrate characteristics within a Holocene salt marsh, Sapelo Island, Georgia. Senckenberg Mar 9:215–259

Esselink P, Dijkema KS, Reents S, Hageman G (1998) Vertical accretion and profile changes in abandoned man-made tidal marshes in the Dollard Estury, the Netherlands. J Coast Res 14:570–582

French J (2006) Tidal marsh sedimentation and resilience to environmental change: exploratory modelling of tidal, sea-level and sediment supply forcing in predominantly allochthonous systems. Mar Geol 235:119–136

French JR (1993) Numerical simulation of vertical marsh growth and adjustment to accelerated sea-level rise, North Norfolk, U.K. Earth Surf Process Landf 18:63–81

French JR, Spencer T (1993) Dynamics of sedimentation in a tide-dominated backbarrier salt marsh, Norfolk, UK. Mar Geol 110:315–331

French JR, Stoddart DR (1992) Hydrodynamics of salt marsh creek systems: implications for marsh morphological development and material exchange. Earth Surf Process Landf 17:235–252

French P, Allen JRL, Appleby PG (1994) 210-Lead dating of a modern period salt marsh deposit from the Severn estuary (Southwest Britain), and its implications. Mar Geol 118:327–334

Galloway WE, Hobday DK (1983) Terrigenous clastic depositional systems. Springer, New York, 423 pp

Gehrelds RW, Szkornik K, Bartholdy J, Kirby JR, Bradley SL, Marshall WA, Heinemeier J, Pedersen JBT (2006) Late Holocene sea-level changes and isostasy in Western Denmark. Quat Res 66:288–302

Ginsberg RN, Lowenstan HA (1958) The influence of marine bottom communities on the depositional environments of sediments. J Geol 66:310–318

Goldberg ED (1963) Geochronology with Lead-210. In: Radioactive dating. International Atomic Energy Agency, Vienna, pp 121–131

Green HM, Stoddart DR, Reed DR, Bayliss-Smith TP (1986) Saltmarsh tidal creek hydrodynamics, Scolt Head Island, Northfolk, England. In: Sigbjarnarson G (ed) Iceland coastal and river symposium proceedings, Reykjavik:93–103

Healy RG, Pye K, Stoddart DR, Bayliss-Smith TP (1981) Velocity variations in salt marsh creeks, Northfolk, England. Estuar Coast Shelf Sci 13:535–545

Howard JD, Frey RW (1985) Physical and biogenic aspects of backbarrier sedimentary sequences, Georgia coast, USA. Mar Geol 63:77–127

IPCC (1990) Climate Change. The IPCC scientific assessment. Report prepared for IPCC by working group 1. Cambridge, UK

Jakobsen B (1953) Landskabsudviklingen i Skallingmarsken. Geografisk Tidsskrift. Danish J Geogr 52:147–158

Jakobsen B (1954) The tidal area in south-west Jutland and the process of salt marsh formation. Geografisk Tidsskrift. Danish J Geogr 53:49–61

Johannes Meyer (1654) Historic map of Denmark. Kort over det danske rige. N.E. Nørlund, Copenhagen 1942.

Kesel RH, Smith JS (1978) Tidal creek and pan formation in intertidal salt marshes, Nigg bay Scotland. Scottish Geogr Mag 94:159–168

Kim D, Cairns DM, Bartholdy J (2009a) Spatial heterogeneity and domain of scale on the Skallingen salt marsh, Denmark. Danish J Geogr 109:95–104

Kim D, Cairns DM, Bartholdy J (2009b) Scale-dependent interactions and community structure along environmental gradients on a coastal salt marsh. J Coast Res SI56:429–433

Kim D, Cairns D, Bartholdy J (2010) Wind-driven sea-level variation influences dynamics of Salt Marsh vegetation, Ann Assoc Am Geogr 101:231–348

Kirchner G, Ehlers H (1998) Sediment coastal geochronology in changing coastal environments. J Coast Res 14:483–492

Koide M, Soutar A, Goldberg ED (1972) Marine geochronology with ^{210}Pb. Earth Planet Sci Lett 14:442–446

Konert M, Vandenberghe J (1997) Comparison of laser grain size analysis with pipette and sieve analysis: a solution for the underestimation of the clay fraction. Sedimentology 44:523–535

Krishnaswami S, Lal D, Martin M, Meybeck M (1971) Geochronology of lake sediments. Earth Planet Sci Lett 11:407–414

Krone RB (1987) A method for simulating historic marsh elevation. In: Kraus NC (ed) Coastal sediments '87. ASCE, New York, pp 316–323

Letzch WS, Frey RW (1980) Deposition and erosion in a Holocene salt marsh, Sapelo Island, Georgia. J Sediment Petrol 50:529–542

Madsen PP (1981) Accumulation rates of heavy metals determined by ^{210}Pb dating: experience from a marsh area in Ho Bay. Rapp. P.-v.Réun. Cons Int Explor Mer 181:59–63

Madsen AT, Murray AS, Andersen TJ, Pejrup M (2007) Temporal changes of accretion rates on an estuarine salt marsh during the late Holocene – reflections of local sea level changes? The Wadden Sea, Denmark. Mar Geol 242:221–233

Madsen AT, Murray AS, Andersen TJ, Pejrup M (2010) Spatial and temporal variability of sediment accumulation rates on two tidal flats in Lister Dyb tidal basin, Wadden Sea, Denmark. Earth Surf Process Landf. doi:10.1002/esp. 1999

McCave IN, Manighetti B, Robinson SG (1995) Sortable silt and fine sediment size/composition slicing: parameters for paleocurrent speed and paleoceanography. Paleoceanography 10:593–610

McCave IN, Hall IR, Bianchi GG (2006) Laser vs. settling velocity differences in silt grainsize measurements: estimation of paleocurrent vigour. Sedimentology 53:919–928

Megumi K (1978) A problem in 210Pb geochronologies of sediments. Nature 274:885–887

Möller I, Spencer T, French JR, Leggett DJ, Dixon M (1999) Wave transformation over salt marshes: a field and numerical modelling study from North Norfolk, England. Estuar Coastal Shelf Sci 49:411–426

Mudge BF (1858) The salt marsh formation of Lynn. Essex Inst. Proc II (1856–1860):117–119

Neumeier U, Amos CL (2006) The influence of vegetation on turbulence and velocities in European salt-marshes. Sedimentology 53:259–277

Nielsen N (1935) Eine Methode zur Exakten Sedimentationsmessung. Det Kgl. Danske Videnskabernes Selskab. Biologiske Meddelelser XII 4:1–97

Nielsen N (1969) Morphological studies on the eastern coast of Disko, West Greenland. Geografisk Tidsskrift. Danish J Geogr 68:1–35

Nielsen N, Nielsen J (1973) Skallingen Niveauforholdene på det marine forland. Geonoter 2:72–77

Nixon SW (1980) Between coastal marshes and coastal waters – a review of 20 years of speculation and research in the role of salt marshes in estuarine productivity and water chemistry. In: Hamilton R, McDonald KB (eds) Estuarine and wetland processes. Plenum, New York, pp 437–525

Oldfield F, Appleby PG, Battarbee RW (1978) Alternative ^{210}Pb dating: results from the New Guinea Highlands and Lough Erne. Nature 271:339–342

Oliver F (1913) Report of the Blakeney Point Research Station. Trans Norf Norway Nat Soc 9:485–542

Olsen CR, Cutshall NH, Larsen I (1982) Pollutant – particle associations and dynamics in coastal marine environments: a review. Mar Chem 11:501–533

Pedersen JBT, Bartholdy J (2006) Budgets for fine-grained sediment in the Danish Wadden Sea. Mar Geol 235:101–117

Pedersen JBT, Bartholdy J (2007) Exposed salt marsh morphodynamics: an example from the Danish Wadden Sea. Geomorphology 90:115–125

Pedersen JBT, Bartholdy J, Christiansen C (2007) ^{137}Cs in the Danish Wadden Sea: contrast between tidal flats and salt marshes. J Environ Radioact 97:42–56

Pedersen JBT, Svinth S, Bartholdy J (2009) Holocene evolution of a drowned melt-water valley in the Danish Wadden Sea. Quat Res 72:68–79

Pejrup M, Andersen TJ (2000) The influence of ice on sediment transport, deposition and reworking in a temperate mudflat area, the Danish Wadden Sea. Cont Shelf Res 20:1621–1634

Pejrup M, Mikkelsen OA (2010) Factors controlling the field settling velocity of cohesive sediments in estuaries. Estuar Coast Shelf Sci 87:177–185

Pethick JS (1974) The distributions of salt pans on tidal salt marshes. J Biogeogr 1:57–62

Pethick JS (1980) Salt-marsh initiation during the Holocene transgression: the example of North Norfolk marshes, England. J Biogeogr 7:1–9

Pethick JS (1981) Long-term accretion rates on tidal salt marshes. J Sediment Petrol 51:571–572

Phleger FB (1970) Foraminiferal populations and marine marsh processes. Limnol Oceanogr 15:522–534

Pizutto J (1987) Sediment diffusion during overbank flows. Sedimentology 34:301–317

Pusty NP (1980) The forces that shape the islands. In: Brown PM, Renwick H (eds) New Jersey's barrier islands: an ever-changing public resource. DNJ-DEP 80 5681 barrrier islands Publications, Center for Coastal and Environmental Studies/State University of New JerseyRutgers/New Brunswick, pp 2–10

Ramaswamy V, Rao PS (2006) Grain size analysis of sediments from the Northern Andaman Sea: comparison of laser diffraction and sieve-pipette techniques. J Coast Res 22:1000–1009

Redfield AC (1972) Development of a New England Salt Marsh. Ecol Monogr 42:201–237

Reed DJ (1988) Sediment dynamics and deposition in a retreating coastal salt marsh. Estuar Coast Shelf Sci 26:67–79

Reed DJ, Spencer T, Murray A, French JR, Leonard L (1999) Marsh surface sediment deposition and the role of tidal creeks: implications for created and managed coastal marshes. J Coast Conserv 5:81–90

Reineck H-E, Gerdes G (1996) A seaward progressive siliciclastic sequence from upper tidal flats to salt marsh facies (Southern North Sea). Facies 34:209–218

Reineck H-E, Singh IB (1975) Depositional sedimentary environments. Springer, Berlin, 439 pp

Richards FJ (1934) The salt marsh of the Dovey estuary. Ann Bot 2:225–259

Robbins JA, Edington DN (1975) Determination of recent sedimentation rates in Lake Michigan using 210-Pb and Cs-137. Geochim Cosmochim Acta 39:285–304

Schubel JR (1973) Some comments on sea grasses and sedimentary processes. Chesapeake Bay Institute Special Report 33, John Hopkins University, Baltimore, 33 pp

Settlemyre JL, Gardner LR (1977) Suspended sediment flux through a salt marsh drainage basin. Estuar Coast Mar Sci 5:653–663

Shaler NS (1886) Sea-coast swamps of the Eastern United States. U.S. Geological Survey, 6th Annual Report, pp 359–368

Shi Z, Hamilton LJ, Wolanski E (2000) Near-bed currents and suspended sediment transport in salt marsh canopies. J Coast Res 16:909–914

Skempton AW (1970) The consolidation of clay by gravitational compaction. Quat J Geol Soc Lond 125:373–411

Society of Natural Science (1804) Historic map of Denmark. Det Kongelige Danske Videnskabelige Selskab, Copenhagen.

Steers JA (1936) Some notes on the North Norfolk coast from Hunstanton to Brancaster. Geogr J 87:35–46

Steers JA (1938) The rate of sedimentation on salt marshes on Scolt Head Island. Geol Mag 75:26–29

Stevenson JC, Ward LG, Kearney MS (1986) Vertical accretion in marshes with varying rates of sea level rise. In: Wolfe DA (ed) Estuarine variability. Academic Press Inc, San Diego, pp 241–259

Stoddart DR, Reed DJ, French JR (1989) Understanding salt marsh accretion, Scott Head Island, Norfolk, England. Estuaries 12:228–236

Streif H (2004) Sedimentary record of Pleistocene and Holocene marine innundations along the North Sea coast of Lower Saxony, Germany. Quat Int 112:3–28

Stumpf RP (1983) The process of sedimentation in the surface of a salt marsh. Estuar Coast Shelf Sci 17:495–508

Temmerman S, Bouma TJ, Van de Koppel J, Van de Wal D, De Vries MB, Herman PMJ (2007) Vegetation causes channel erosion in a tidal landscape. Geology 35:631–634

Temmermann S, Govers G, Meire P, Wartel S (2003) Modelling long-term tidal marsh growth under changing tidal conditions and suspended sediment concentrations, Scheldt estuary, Belgium. Mar Geol 193:151–169

Tsihrintzis VA, Madieo EE (2000) Hydraulic resistance determination in marsh wetlands. Water Resour Manag 14:285–309

Van der Wall D, Pye K, Neal A (2002) Long-term morphological changes in the Ribble Estuary, Northwest England. Mar Geol 189:249–266

van Proosdij D, Ollerhead J, Davidson-Arnott RGD (2000) Controls on suspended sediment deposition over single tidal cycles in a macrotidal saltmarsh, Bay of Fundy, Canada. In: Pye K, Allen JRL (eds) Coastal and Estuarine Environments: sedimentology, geomorphology and geoarchaeology, vol 175, Geol Society Special publication. Geological Society, London, pp 43–57

van Proosdij D, Ollerhead J, Davidson-Arnott RGD (2006a) Seasonal and annual variations in the volumetric sediment balance of a macro-tidal salt marsh. Mar Geol 225:103–127

van Proosdij D, Davidson-Arnott RGD, Ollerhead J (2006b) Controls on spatial patterns of sediment deposition across a macro-tidal salt marsh surface over single tidal cycles. Estuar Coast Shelf Sci 69:64–86

Wang FC, Tiesong L, Sikora WB (1993) Intertidal marsh suspended sediment transport processes, Terrebonne Bay, Louisiana, USA. J Coast Res 9:209–220

Warming E (1904) Bidrag til vadernes, sandenes og marskens naturhistorie. Det Kgl. Danske Videnskabernes Selskab Skrifter 7. række Naturvidenskaber og Matematik Afd. II. 1:1–56

Woolnough SJ, Allen JRL, Wood WL (1995) An exploratory numerical model of sediment deposition over tidal salt marshes. Estuar Coast Shelf Sci 41:515–543

Yapp RH, Johns D, Jones OT (1916) The salt marshes of the Dovey estuary, Part I. Intro J Ecol 4:27–42

Yapp RH, Johns D, Jones OT (1917) The salt marshes of the Dovey estuary. Part II. The salt marshes. J Ecol 5:65–103

Open-Coast Tidal Flats

9

Daidu Fan

Abstract

Recent research advances highlight the importance of open-coast tidal-flat depositional system in both modern and ancient coastal environments. The system is unique in its wave- and tide-dominated physical setting, notably distinct from the tide-dominated barred tidal flats and the wave-dominated shorefaces. Interactions of waves and tides over different time scales produce not only cyclic morphologic variations in terms of erosion and deposition, but also rhythmic depositional units consisting of storm-generated sand-dominated layers (SDLs) and post-storm mud-dominated layers (MDLs). Ancient deposits of the open-coast tidal flats can be distinguished by abundance of storm-generated structures and scarcity of tidal-channel deposits from those of barred tidal flats, and by abundance of the structures created by combined flows or the interactions of waves and tides from those of (tidal) shorefaces. Difference is also remarkable between muddy and sandy open-coast tidal flats. Muddy open-coast tidal flats tend to develop along mega-river deltas and the adjacent chenier plains, have a general accretional convex-up profile with clear zonation, and produce aggradational fining-upward intertidal successions. Sandy open-coast tidal flats are common in the open-mouth estuaries of small rivers and the adjacent strand plains, usually develop an erosional concave-up profile with common presence of inner swash bars having the coarsest sediment near the high water, and produce coarsening-upward retrogradational successions. The vertical successions of sandy open-coast tidal flats generally contain more storm-generated beds volumetrically than those of muddy open-coast tidal flats. Notably, there are some accretional sandy open-coast tidal flats, lying in between the above two settings. A new spectrum of coastal morphodynamic settings is therefore proposed to change from the tide-dominated barred tidal flats, the wave-influenced and tide-dominated muddy open-coast tidal flats, the wave- and tide-dominated accretional sandy open-coast tidal flats, the wave-dominated erosional sandy open-coast tidal flats, to the wave-dominated tidal shorefaces.

9.1 Introduction

Tidal flats can be developed in numerous environments, like lagoons, embayments, estuaries, deltas, and coastal plains, ranging from sheltered environments to those

D. Fan (✉)
State Key Laboratory of Marine Geology, Tongji University, Shanghai 200092, China
e-mail: ddfan@tongji.edu.cn

R.A. Davis, Jr. and R.W. Dalrymple (eds.), *Principles of Tidal Sedimentology*,
DOI 10.1007/978-94-007-0123-6_9, © Springer Science+Business Media B.V. 2012

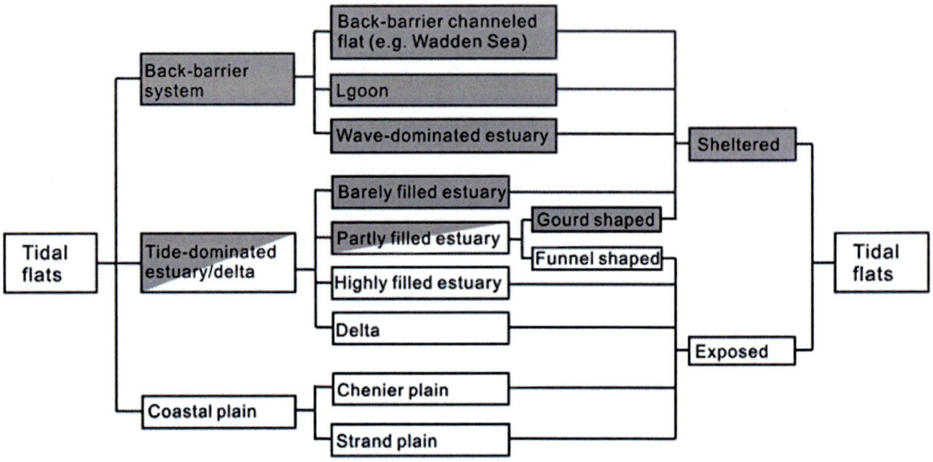

Fig. 9.1 Classifications of tidal flats and their relationships

completely exposed to the coastal sea or the open ocean (Fig. 9.1). Rudolf Richter, a German paleontologist, began the first systematic geologic studies of tidal flats along the German North Sea coast in the early 1920s. These studies were not only well carried on by his successors at the Senckenberg Institute but also rapidly spread out to the Netherlands, UK, and North America after the World War II (Ginsburg 1975; Middleton 1991; Klein 1998). Increased interest in fossil tidal deposits ignited comparative studies, and criteria for recognizing tidal-flat deposits were summarized: (1) cross-bedding with evidence for current reversals, like herringbone cross-stratification, reactivation surfaces, and mud couplets/drapes; (2) tidal bundles with evidence for tidal rhythms, e.g., diurnal, semilunar, and lunar cycles; (3) flaser and lenticular bedding in fining-upward successions (Middleton 1991; Nio and Yang 1991).

The well-known criteria for tidal-flat deposition obviously bear their regional background of the Wadden Sea and the Bay of Fundy (e.g., Klein 1985; Dalrymple 1992; Boggs 2005), that are classified as sheltered tidal flats (Fig. 9.1). The Wadden Sea is the tidal-channel-flat complex separating the barrier island system from the mainland, which in turn connects with the North Sea through the tidal inlets between each two barrier islands (Ginsburg 1975; Davis et al. 1998). In the Bay of Fundy, tidal flats occupy the innermost part of the estuary, separating the outer erosion zone with elongate sand bars (Dalrymple et al. 1991). The sheltered tidal flats are characterized by the presence of tidal sand bodies (linear shoals or bars) at the lower intertidal zone or the subtidal zone, and delicate tidal-channel systems cutting into the flats. Lateral migration of channel bars and/or high sedimentation rate in the sheltered settings are potential to produce rhythmic tidal-bundle successions or cyclic tidal rhythmites (Boersma and Terwindt 1981; Nio and Yang 1991; Dalrymple et al. 1991; Dalrymple 2010).

Over the past three decades, a deep passion has been intrigued toward finding tidalites with cyclic variations in bundle/lamina thickness, encoding tidal periodicities of neap-spring cycles, diurnal, fortnightly, and other longer inequalities (e.g., Boersma and Terwindt 1981; Kvale et al. 1989; Dalrymple et al. 1991; Tessier 1993; Williams 1997; Coughenour et al. 2009). The quantitative features (tidal periodicities) of the strata are significantly important in ascertaining their tidal origin, considering that most of qualitative features like flaser and lenticular bedding are not exclusive from non-tidal environments. The neap-spring periodicity of cyclic tidal rhythmites is also highly valued for reconstruction of the history of tides and lunar orbit throughout the geologic time from the Archean to the present (Coughenour et al. 2009). It is, however, noteworthy that the preservation of cyclic tidal rhythmites requires special conditions like sheltered areas with high sediment and accommodation-space availability to achieve the continuous rapid deposition (Tessier 1993; Greb and Archer 1995; Fan and Li 2002; Fan et al. 2004a, b; Dalrymple 2010). Cyclic tidal rhythmites are therefore quite limited in their spatial and temporal distributions throughout the geological time (Davis et al. 1998).

Noncyclic tidalites are actually much more widely distributed than the cyclic tidal rhythmites, considering that most of tidal flats are not exclusive from wave impacts. The exposure to wave increases from tidal flats fringing open-mouth estuaries, deltas, to coastal plains, which are all referred as open-coast tidal flats in this context (Fig. 9.1). The recent studies highlight that the interaction of tides and waves is major mechanism for sediment transport and morphological and stratigraphic formation on open-coast tidal flats (Allen and Duffy 1998; Li et al. 2000; Fan et al. 2004a, b, 2006; Lee et al. 2004). Moreover, some sandy open-coast tidal flats can be wave dominated (Yang et al. 2005; Dalrymple et al. 2006). The strata of open-coast tidal flats are characterized by containing a mixture of tide- and wave-generated sedimentary structures with a general increase in the thickness ratio of storm deposition and tidal deposition as the wave exposure increases. It should therefore undoubtedly lead to misidentification of the open-coast tidal-flat deposition, if the pure tide-depositional criteria were employed for facies interpretation (Dalrymple 2010).

Open-coast tidal flats remained less studied until the current century, although they are much more widely distributed than the sheltered tidal flats along the world coast, and their importance for both modern and ancient tidal sedimentology has been highlighted since the mid-1970s (Klein 1975; Wang 1983; Ren 1985; Wells et al. 1990; Fan et al. 2004a; Yang et al. 2005). Several reasons account for the situation. Open-coast tidal flats are majorly composed of fine-grained sediment with mud domination, which is usually considered less economically important for resource exploitation. Muddy open-coast tidal flats are principally distributed along the coasts of South and East Asia, Oceania, and South America, receiving gigantic volume of fluvial fine sediment (Figs. 9.2–9.4), which are less accessible to Western European and North American. Although Chinese scientists began pilot studies of tidal flats in the early 1960s (e.g., Wang 1963; Li et al. 1965), most of their research results were published in Chinese (see a review given by Shi and Chen 1996), also unreadable to the western scientific community. Sandy exposed tidal flats are common along the open coast neighboring small river mouth, nourished by riverine sediment principally composed of coarse grains (coarse silt and sand), like the central west coast of Taiwan (Reineck and Cheng 1978), the southwest coast of Korea (Yang et al. 2005), the east and northwest coast of India (Mukherjee et al. 1987) and the northwest coast of Australia (Semeniuk 1981). They are also common in parts of Western Europe and North America, like the coast of the southwestern English Channel, the Irish Sea, and the Strait of Georgia in Canada (Hale and McCann 1982; Deloffre et al. 2007; Yang et al. 2008a). The sandy open-coast tidal flats lie at a transitional ground between the tide-dominated muddy open-coast tidal flats and wave-dominated shorefaces, and their affinity to wave dynamics and shoreface morphology makes them easy to be misinterpreted as tidal beach. All of these may account for their having been much less studied until the present decade (Yang et al. 2005, 2008a; Dalrymple 2010).

There has been a rapid increase in publications on open-coast tidal flats since the beginning of the new century. The topics cover all fields including hydrodynamics, sediment- and morphodynamics, sedimentology, and stratigraphy, especially those in China and Korea (Li et al. 2000, 2005a; Chang and Choi 2001; Fan 2001; Fan and Li 2002; Fan et al. 2001, 2002, 2004a, b, 2006; Kim 2003; Yang et al. 2005, 2006a, b, 2008a, b; Dalrymple et al. 2006; Gao 2009; Wang et al. 2009). In these studies, open-coast tidal flats have been highlighted as a major coastal setting, significantly different from the well-known sheltered tidal flats and tidal beaches. This chapter attempts to summarize these research progresses on open-coast tidal flats, cultivating a general sedimentary model for future comparison studies of the systems.

9.2 Depositional Systems

Tidal flats are low-relief environments typically flanking the coast of broad shelf with marked tidal rhythms (Fig. 9.2). Macrotidal conditions undoubtedly favor the development of extensive tidal flats, but they are also common in mesotidal to microtidal coasts (Eisma 1998). Sediment supply (source, flux, and size) and the magnitude of exposure to waves are two other key factors controlling tidal-flat morphology and sedimentology.

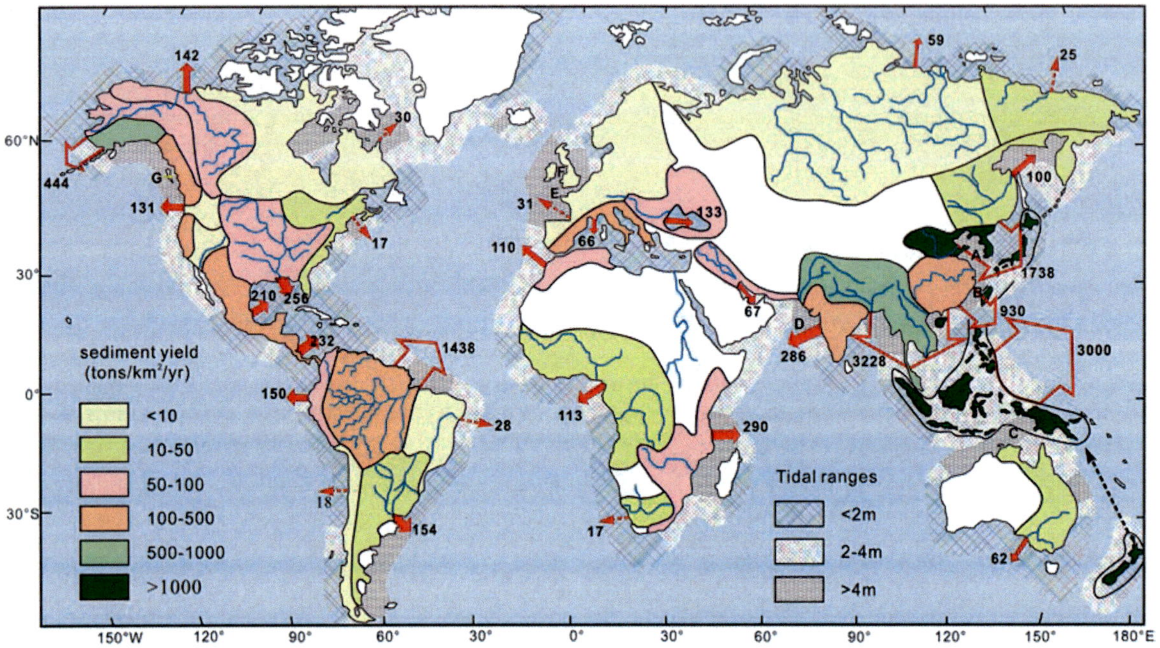

Fig. 9.2 Distribution of annual sediment discharge into the sea (millions of tons per year, modified after Milliman and Meade 1983; Milliman and Syvitski 1992) and spring tidal range (m) along the world's coasts (after Davies 1972). *A*: Yellow Sea, *B*: Taiwan Strait, *C*: Torres Strait, *D*: Arabian Sea, *E*: English Channel, *F*: Irish Sea, *G*: Strait of Georgia

Tidal flats are therefore classified into three major types: (1) back-barrier coast, (2) tide-dominated estuary/delta, and (3) coastal plain, with increasing wave exposure from low to high (Figs. 9.1 and 9.5).

Back-barrier tidal flats are referred to those occupying the landward side of barrier islands, spits, and bars, which serve as wave breakers (Fig. 9.5a, b). The back-barrier area can be partly occupied by tidal flats like those in lagoons and wave-dominated estuaries, or entirely filled with the tidal channel and flat system like the Wadden Sea. All of them receive limited sediment input from land. Back-barrier tidal flats typically develop extensively along the North Sea coast, and the Atlantic coast of Europe and North America (Flemming, Chap. 10 in this volume), which are both trailing edge coast receiving limited fluvial sediment input. They can also develop along the leading edge coast receiving abundant sediment like the Pacific US coast (e.g., Alaska and North California).

Tidal flats fringing the tide-dominated estuaries/deltas have varied magnitudes of wave exposure. Those are generally exempted from wave influence in the innermost part of the estuarine-deltaic system or in the estuaries having a guard (bottle-neck) shape (Fig. 9.5c). The outer part of the open-mouth estuaries and deltas are subject to wave impacts, especially during storm seasons (Fig. 9.5d–f). For example, wave influence is generally absent for the tidal flats in the inner part of the Xiangshan Bay deeply cutting into the land and Sanmeng Bay, while it gradually increases toward the baymouth. The straight coast of the highly filled Jiaojiang Estuary is completely open to the sea in the Zhejiang Province, central east of China (Fig. 9.6). In general, small river estuaries of tide domination usually foster sandy tidal flats because of limited fine sediment input from the rivers, except those receiving large volume of fine-grained sediment from the adjacent mega-river deltas by alongshore transport, like the estuaries along the Zhejiang coast nourished by Changjiang-sourced sediment (Figs. 9.3 and 9.6). Therefore, sandy estuarine and strand plain tidal flats are commonly distributed along the coast of islands or peninsulas enclosing a broad and shallow strait or epeiric sea, where fine-grained sediment is generally exempted and macrotide is highly favorable owing to the morphological amplification, like the Yellow Sea, the Taiwan Strait, the Torres Strait, the Arabian Sea, the English Channel, the Irish Sea, the Strait of Georgia, and so on (Fig. 9.2).

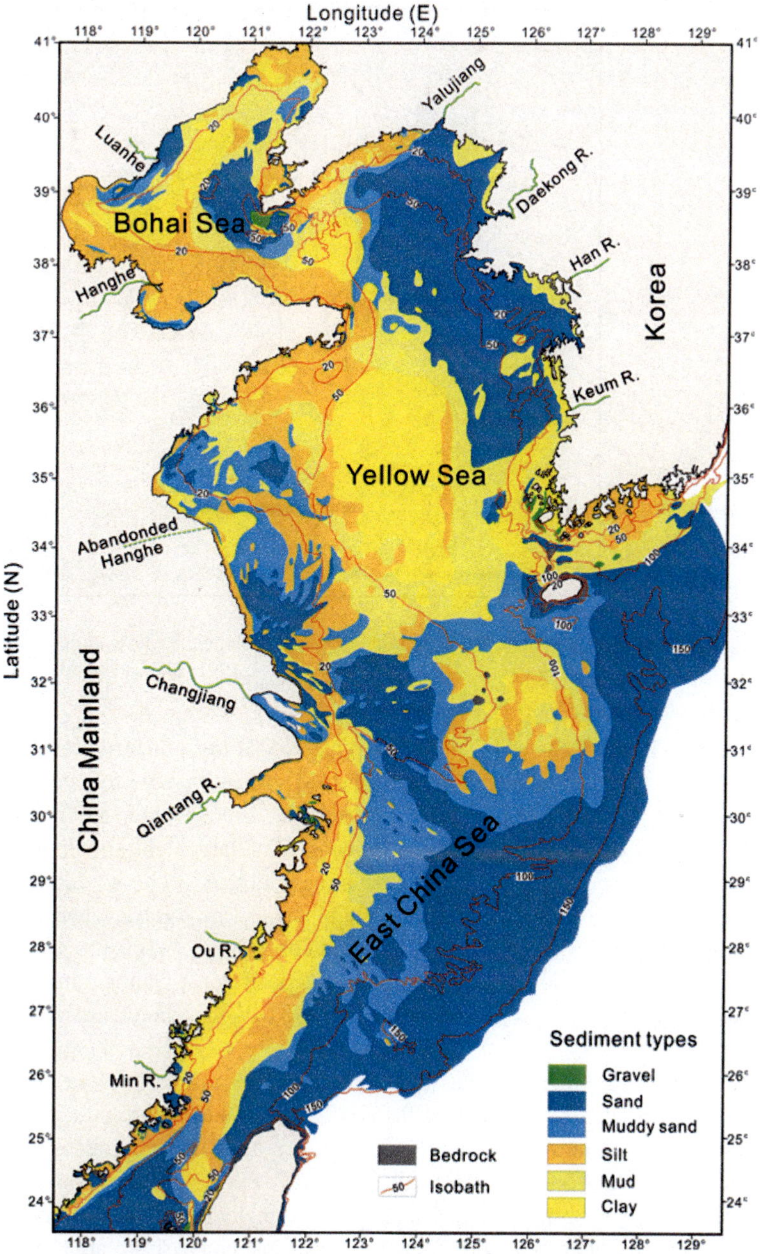

Fig. 9.3 Distribution of bottom sediment types and tidal flats that the coasts are surrounded directly by fine-grained deposits in light to dark yellow in the Bohai Sea, the Yellow Sea, and the East China Sea (After Li et al. 2005b)

Tidal flats flanking the coastal plains are directly open to the sea, thereby named as truly open-coast tidal flats (Dalrymple, pers comm 2010). Muddy and sandy open-coast tidal flats are generally linked to tide-dominated deltas/estuaries of large and small rivers, respectively. Larger river deltas tend to foster wider (cross-shore) and longer (alongshore) muddy tidal flats bordering the extensive chenier plains. Muddy open-coast tidal flats are globally widely extensive, considering that most of mega-river deltas are tide dominated or under significant tidal influence, building up extensive Holocene chenier plains in the world (Fig. 9.2, Table 9.1). The two longest stretches of muddy open-coast tidal flats are linked to the river deltas of the Changjiang and the Amazon, respectively. North Jiangsu tidal flats stretch over 600 km long between the abandoned Huanghe Delta and the Changjiang Delta at a macrotidal setting (Figs. 9.2 and 9.3, Wang et al. 2002). Guiana tidal flats extend over 1,600 km long from the Amazon River Delta to the

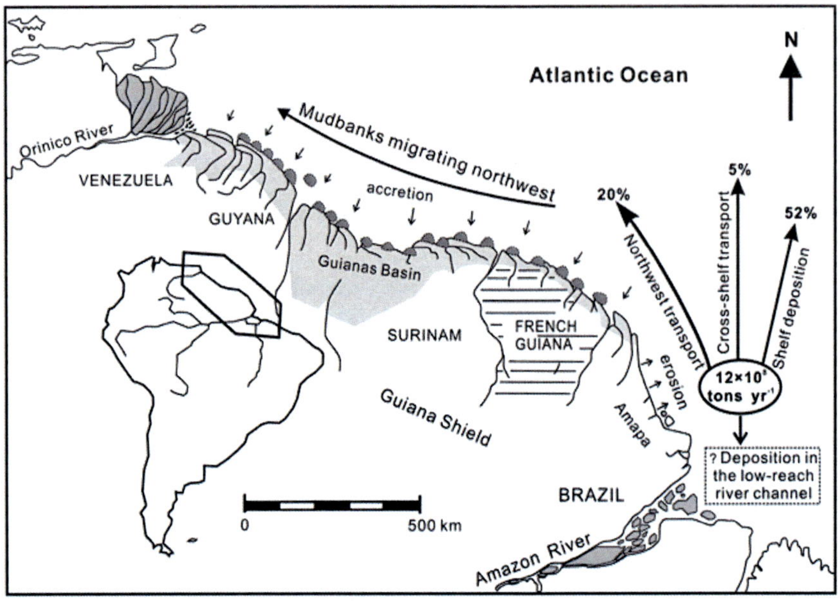

Fig. 9.4 Distribution of tidal flats (migrating mudbanks) along the coasts of northwest South America (Guiana) from the Amazon River mouth to the Orinoco River Delta (After Eisma et al. 1991; Plaziat and Augustinus 2004)

Orinoco River Delta (Fig. 9.4) at a macrotidal-to-mesotidal setting (Wells and Coleman 1981; Meade et al. 1985; Froidefond et al. 1988; Kineke et al. 1996; Baltzer et al. 2004). Sandy (truly) open-coast tidal flats are generally coexisting with sandy estuarine tidal flats (Fig. 9.2), bordering the Holocene strand plains of varied width (Reineck and Cheng 1978; Semeniuk 1981), or against the erosional Late Quaternary deposits or the rocky cliff (Thompson 1968; Semeniuk 1981; Yang et al. 2005).

Tidal flats are therefore classified into nine types in terms of coastal morphology (Figs. 9.1 and 9.5). They are in turn grouped into sheltered or exposed (open-coast) tidal flats on the basis of the magnitude of wave exposure. So the open-coast tidal flats in this context include both the truly open-coast tidal flats (i.e., coastal-plain tidal flats) and the partially sheltered to highly exposed tidal flats fringing the outer part of open-mouth estuaries and deltas (Fig. 9.1).

9.3 Physiography and Morphology

9.3.1 Tide, Wave, and Wind Climate

Strong tidal amplification occurs when tidal waves propagate into the broad and shallow shelf sea, consequently raising tidal range shoreward. The extreme case to induce tidal bores takes place when tidal waves enter a funnel-shaped embayment rapidly narrowing and shoaling landward (Archer and Hubbard 2003). The world's largest tidal bore occurs in the Hangzhou Bay, where the width of the bay decreases from ~100 km across the baymouth to <20 km along the Ganpu transection, while the mean tidal range increases landward from <2 m at the baymouth to over 5.5 m near Ganpu (Fig. 9.7). Epeiric seas and funnel-shaped bays were considered to be more prevalent in the geologic past, especially during the time of the supercontinent Pangea (Archer 1998), so macrotidal coastal settings should be common at that time.

Tidal currents are usually rotary on the open offshore area, gradually changing into linear when tides propagate into the distributary/estuarine channels or toward the shore. Main flows of flood and ebb tides in the estuaries and deltaic distributaries are steered toward different directions because of the Coriolis effect, resulting in different current strength and depositional patterns along the two different banks. In the Hangzhou Bay, flooding flows are steered north, and the convergent flows strengthen the current toward the north bank, producing the erosional and narrow tidal flats, while ebbing flows are steered south, and divergent and weakening flows favor developing the depo-

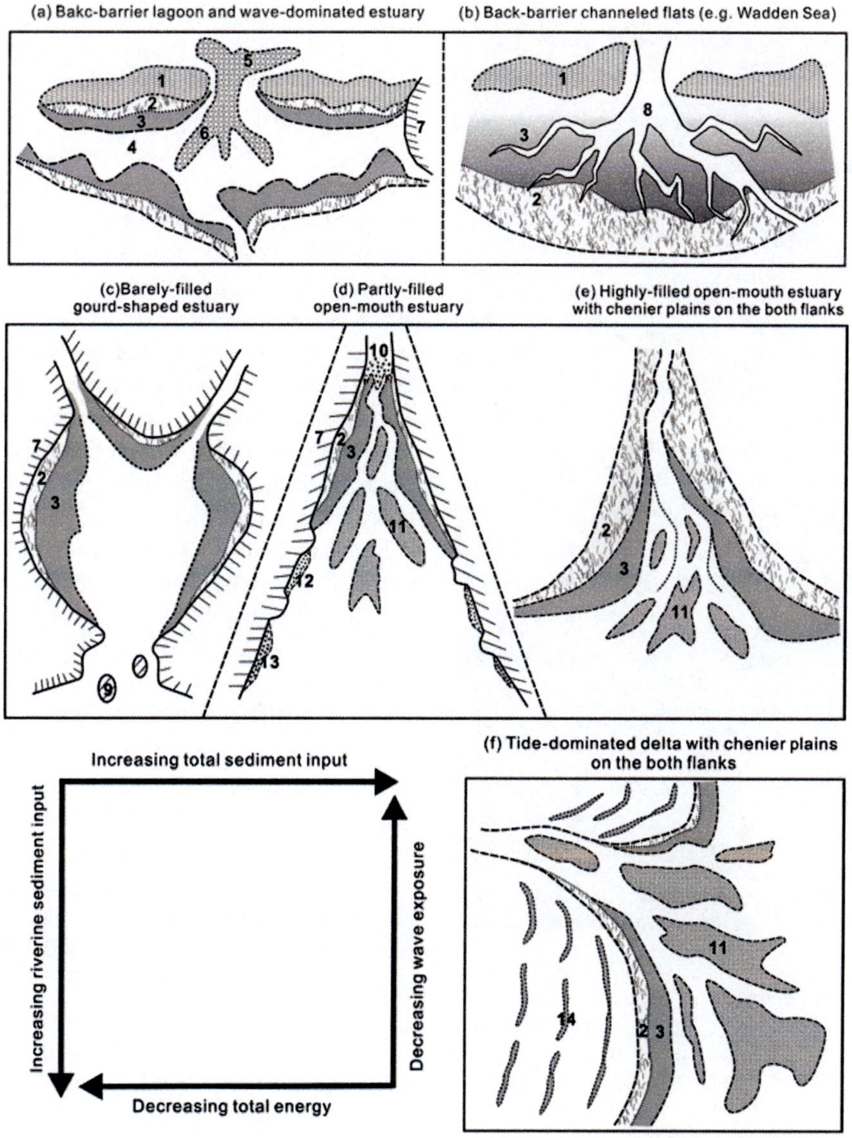

Fig. 9.5 Thematic plots of tidal flats in the different coastal systems with varying sediment input and wave exposure (**a–f**). *1* – barrier island, *2* – salt marsh, *3* – bare intertidal flat, *4* – lagoon, *5* – ebb tidal delta, *6* – flood tidal delta, *7* – rocky coast, *8* – tidal channel/creek, *9* – rocky island, *10* – bay-head delta, *11* – tidal sand ridge/bar, *12* – sand beach, *13* – gravel beach, *14* – chenier ridge

sitional and wide tidal flats along the south bank (Fig. 9.7; ECCE 1992; Fan et al. 2005). Constrained flows in the main channels of estuaries and deltaic distributaries are usually shore-parallel and ebb dominated, gradually changing into shore-normal direction and flood domination as they flow over the intertidal flats. Shore-normal tidal flow is generally quite weak due to limited tidal prism over the intertidal flats. The velocities of flood flows decrease from the subtidal zone to the upper intertidal flat owing to the shoaling effect. For example, the spring flood flow velocities decreased from 0.86 cm/s at a subtidal station (−2 m, elevation referred to Wusong Datum), 0.49 cm/s at 0 m, to 0.35 cm/s at an intertidal station (+2 m) on the Nanhui Mudbank in the Changjiang Delta (Li 1990).

The open-coast tidal flats are highly exposed to wave impact, and they are potential to change from the tide-dominated setting temporally (a few hours to days) or seasonally into the wave-dominated setting, especially where they are highly affected by the

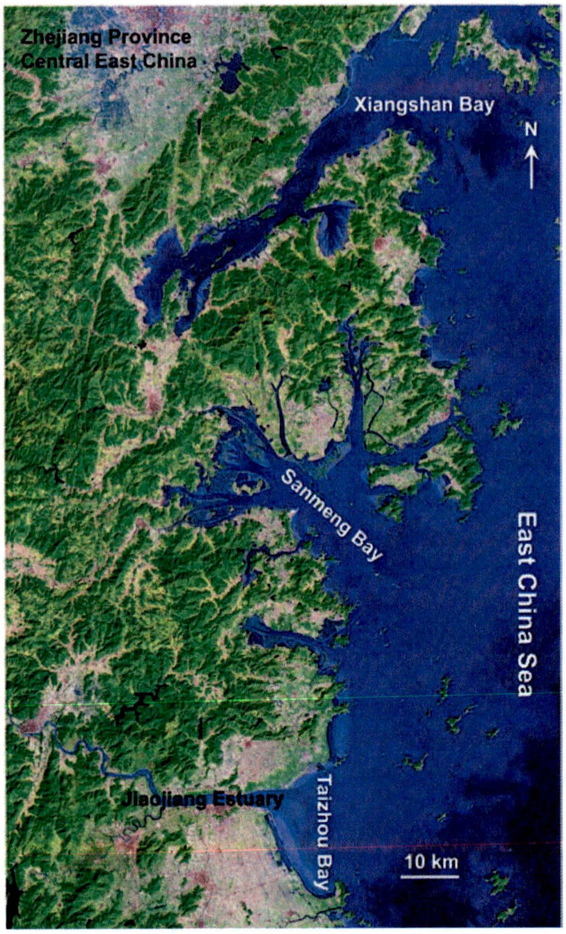

Fig. 9.6 Satellite photo of central-east Zhejiang coast showing the development of different tidal flats in the estuaries or embayments: (**a**) Xiangshan Bay, partly filled type with narrow baymouth and "¬" shape; (**b**) Sanmeng Bay, partly filled type with relatively narrow baymouth and gourd shape; (**c**) Jiaojiang Estuary-Taizhou Bay, highly filled coastal plain type with a wide baymouth and straight coastline Đ They are typical representative for the three major estuarine tidal flats in the classification lists of the Fig. 9.1

monsoon climate (Harris et al. 1993; Li et al. 2000; Yang et al. 2005; Fan et al. 2006). During the longer calm-weather season, shallow-water waves can be greatly attenuated when propagating over the gentle and broad subtidal zone, and the attenuated waves finally die out without breaking over the intertidal flats, especially where present with fluid mud (Wells and Coleman 1981; Lee et al. 1997; Kim 2003; Fan et al. 2006). Open-coast tidal flats can therefore be assigned as a tide-dominated or mix-energy tide-dominated coastal setting on the basis of Davis and Hayes' (1984) model (Fig. 9.8). While entering the storm season, rough seas generated by single events of tropical storms or winter cold fronts commonly sustain a few hours to days. They may also continue for a few weeks when more than two storms occur one after another. The open-coast tidal flats are consequently shifted into a mixed-energy wave-dominated or wave-dominated coastal setting for a few hours to weeks during the storms (Fig. 9.8). For example, the mean wave height on the Baeksu coast (SW Korea) is 0.5–1.0 m in summer, drastically elevated to 2–3 m in winter (Kim 2003; Yang et al. 2005), accounting for the development of typical depositional pattern "summer tidal flat – winter shoreface" (in the paper title of Yang et al. 2005).

In short, the open-coast tidal flats are generally characterized by large tidal range, weak shore-normal tidal flow, and high exposure to waves. These typical physical dynamics are conceived to develop the unique sedimentology and morphology of the open-coast tidal flats.

9.3.2 Landforms and Zonation

Tidal flats in this context are not limited to the intertidal zone, but also include supratidal and subtidal zones. The zonation can be generally defined: (1) by grain size into muddy, mixed, and sandy flats; (2) by vegetation into bare and vegetated flats; and (3) by tidal level into supratidal, intertidal, and subtidal flats. The intertidal flats can be further divided into upper, middle, and lower subdivisions (Fig. 9.9). However, the zonation boundary is quite varied among the different criteria of grain size, vegetation, and tidal level, and a worldwide zonation can only be distinguished by tidal levels (Amos 1995; Eisma 1998). Here we describe the zonation of tidal flats in a synthetical way instead of using single criteria, also considering the difference between muddy and sandy open-coast tidal flats.

The supratidal and the upper parts of intertidal flats are usually covered with vegetation, characterized by extraordinarily gentle relief, fine-grained deposits principally consisting of clay and fine silt, and higher concentration of organic matter. The canopies can be salt-marsh plants in the temperate zone like the Jiangsu and Zhejiang coast in China (Ren 1985; Wang and Eisma 1988, 1990; Fan et al. 2004a), and mangroves in the tropical zone like the Guiana coast (Fig. 9.10; Plaziat and Augustinus 2004), the northwest Australia coast (Semeniuk 1981), and the coast of the Gulf of Papua (Allison and Lee 2004; Walsh and Nittrouer 2004).

Table 9.1 Delta class for the 16 largest rivers with annual sediment load larger than 100 million tons (based on data of Milliman and Syvitski 1992)

Rank	River	Country	Sediment load (million tons/year)	Spring tidal range (m)	Delta class
1	Amazon	Brazil	1,200	4.9	T
2	Huanghe (Yellow)	China	1,100	1.13	R to R/W
3	Ganges-Brahmaputra	Bangladesh	1,060	3.63	T
4	Changjiang (Yangtze)	China	480	3.66	T
5	Mississippi	USA	400(210)[a]	0.43 (M)	R to R/W
6	Irrawaddy	Burma	260	2.71	T
7	Indus	Pakistan	250(59)[a]	2.62	T/W
8	Magdalena	Colombia	220	1.1	R/W
9	Godavari	India	170	1.2	W
10	Mekong	Vietnam	160	3.2	T/W
11	Orinoco	Venezuela	150	1.77	W/R/T
12	Song Hong (Red)	Vietnam	130	3.2	T-W
13	Narmada	India	125	9.0 (Max)	T
14	Colorado	USA	120(0.1)[a]	8.0 (Max)	T
15	Nile	Egypt	120(0)[a]	0.43	W
16	Fly	PNG	115	3.8	T

[a]400(210): sediment load before and after river damming
M mean tidal range, *Max* maximum tidal range, *R* river-dominated, *T* tide-dominated, *W* wave-dominated, *T-W* subequal wave and tidal domination, *R/W* river-dominated and wave-modified, *PNG* Papua New Guinea

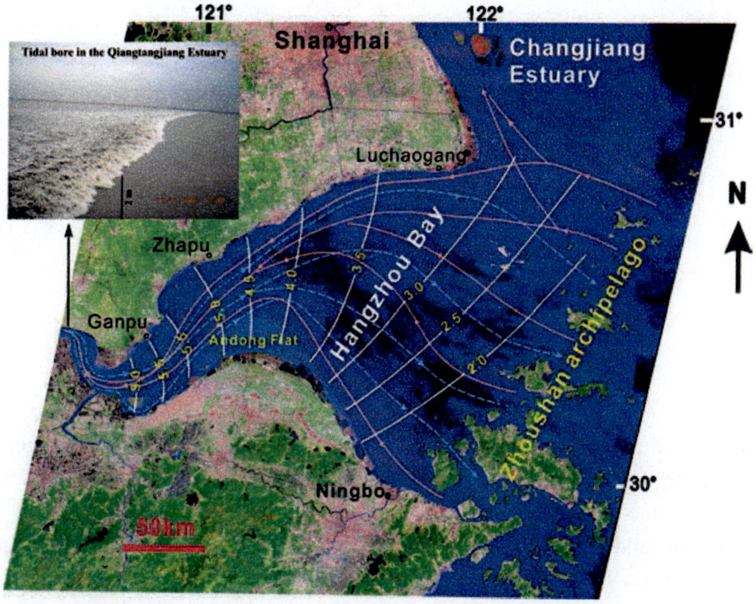

Fig. 9.7 Distribution patterns of mean tidal range, major tidal flow tracts in the Qiangtangjiang Estuary-Hangzhou Bay

In some dry regions, the higher part of tidal flats support only scant vegetation where salt pans can be well developed with sediment having higher concentration of evaporate minerals, like the Colorado Delta in the Northern Gulf of California (Thompson 1968, 1975; Meckel 1975). The vegetated flats usually grade landward into deltaic/chenier plains with little relief, which range from a few 100 m to over tens of kilometers in width, depending on the sediment supply. For example, the Holocene North Jiangsu coastal plain is more than

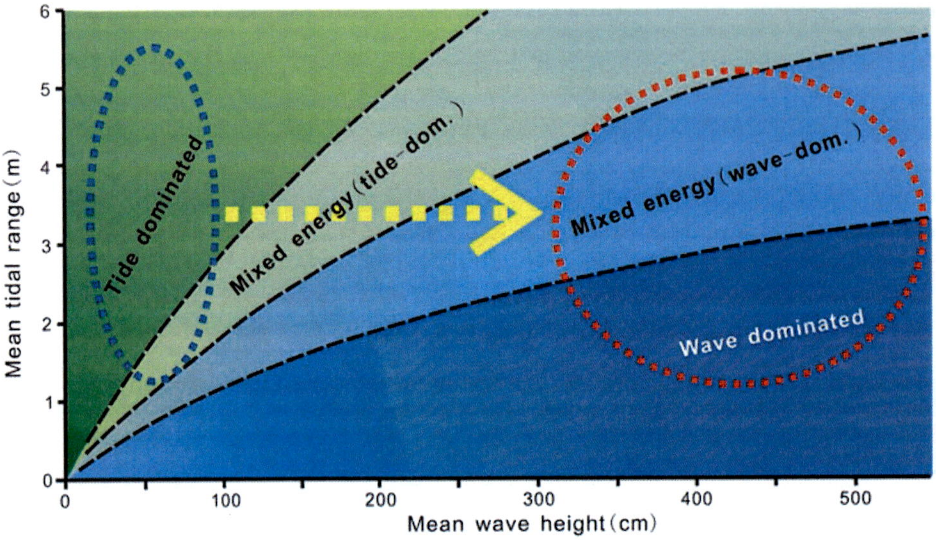

Fig. 9.8 Open-coast tidal flats shifting temporally from tide-dominated or mix-energy (tide-dom.) to wave-dominated regimes for a few days to weeks during summer/winter storms (Adapted from Davis and Hayes 1984)

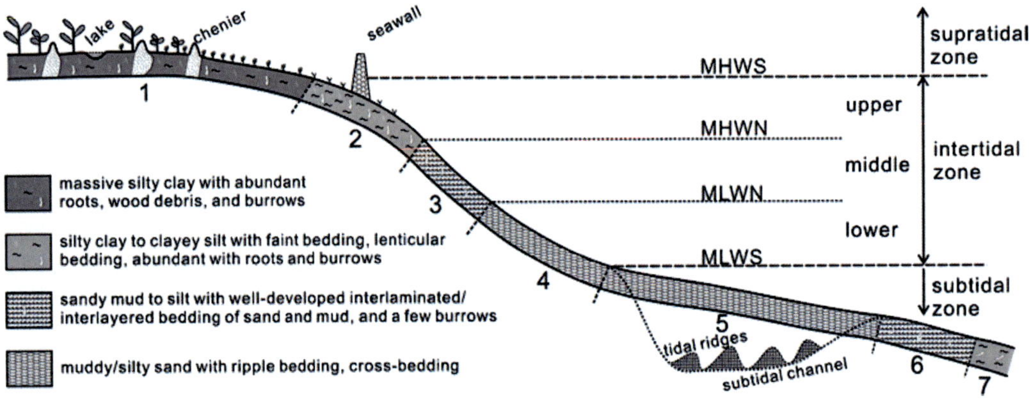

Fig. 9.9 Schematic map showing the cross-profile variations of tidal-flat morphology and sedimentology from the supratidal flats to the subtidal zones

60 km wide on average (Figs. 9.11 and 9.13), and the Holocene Guiana coastal plain ranges from 10 to 100 km wide with an average of ~30 km (Figs. 9.4 and 9.14, Rine and Ginsburg 1985; Allison et al. 1995a). In some sediment-starved coasts, the supratidal flats are narrow or even absent, bordering directly on the rocky hills like those along the Korean coast (Alexander et al. 1991; Yang et al. 2005).

The transition from the vegetated to the bare flats can be smooth or drastic with erosional cliff or channels. The mangrove land is usually surrounded by deeply cut tidal channels or cliffs (Semeniuk 1981; Plaziat and Augustinus 2004). A smooth transition generally occurs ahead of the accretional salt marshes, while recessional salt marshes tend to have an erosional escarpment at the front. Swash bars can be developed at the salt-marsh front on both accretional and recessional flats, resulting from wave breaking when storm waves ride on the high tides. On the Chongming Eastern Flat (Changjiang Delta), an accretional and smoothly transitional profile is favorable during weak-wave conditions (Fig. 9.12a), while it is replaced by the presence of erosional escarpment, scouring ponds, and swash sand sheets/bars consisting of coarse

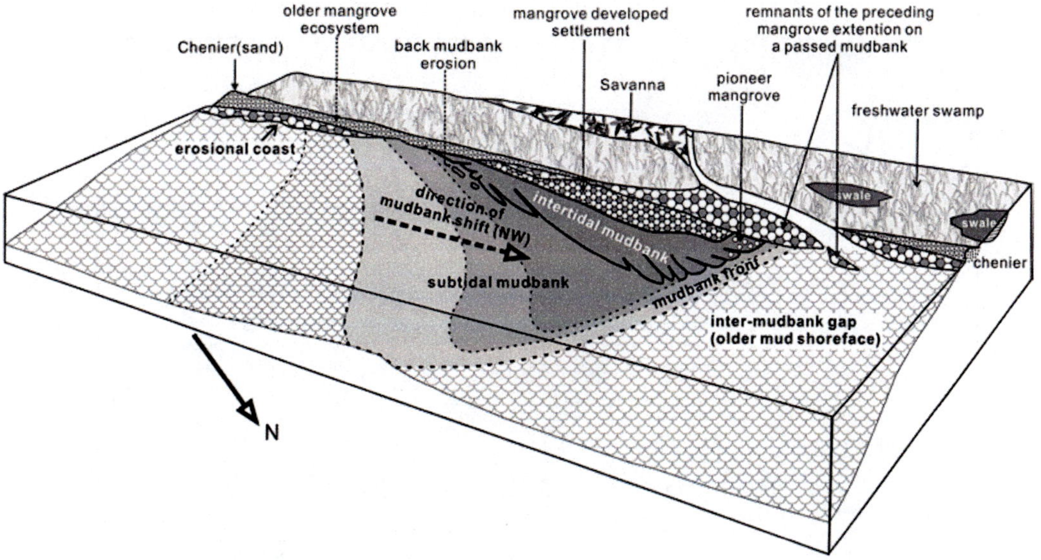

Fig. 9.10 Block diagram showing the zonations (including mangrove) and the shape of a shifting mudbank along the coast of Guianas (After Plaziat and Augustinus 2004). The intertidal area of the largest mudbanks is about 20–40 km long (alongshore) and 2–5 km wide

sediment and abundant shell debris during storm climates (Fig. 9.12b–f). Swash bars are conceivably the embryo of chenier ridges, which potentially develop into full chenier ridges if the processes continue for a longer term (over tens of to hundreds of years). Older chenier ridges are typically common on the extensive Holocene coastal plains abutting mega-river deltas, conceivably resulting from the long-term alternations of erosion (producing chenier) and accretion (depositing mud) linked to the secular fluctuation of river sediment discharge or shifting of the river main channel (Figs. 9.13 and 9.14, Rine and Ginsburg 1985; Liu and Walker 1989; Wang and Ke 1989; Eisma 1998).

Across the bare intertidal and subtidal flats, muddy and sandy open-coast tidal flats tend to have different cross-shore profiles and sediment distribution patterns. Muddy open-coast tidal flats usually have an accretional, convex-up cross-shore profile. The transition between the intertidal and the subtidal flats is generally smooth without a discernible relief, except those drop into the estuarine or deltaic-distributary channels with a higher slope (e.g., Ren 1985; Wang and Eisma 1988, 1990; Frey et al. 1989; Wang and Ke 1997; Fan et al. 2004a; Plaziat and Augustinus 2004, Figs. 9.9–9.11). Sediment on the accretional flats tends to be coarsest near the mean lower water springs, gradually fining landward and seaward from there. The intertidal zonations are roughly displayed with the lower sandy flat, the middle mixed (sand-mud) flat, and the upper muddy flat (Table 9.3). The zonation pattern was clearly shown by the surficial sediment distribution on the North Jiangsu tidal flats, typically along the rapid depositional section from Sheyang to Dafeng (Fig. 9.11; Wang and Ke 1997). On the central west coast of Korea where the estuarine/embayment tidal flats are well developed with mediate to high wave exposure, surficial sediment generally coarsens from fine silt/clay at the upper intertidal flats to coarse silt/fine sand at the lower intertidal flats (Frey et al. 1989; Alexander et al. 1991). It is noteworthy that the Guiana muddy tidal flats have the finest deposition mainly composed of silty clay (Table 9.3), showing little trend variations in grain size across the entire mudbank from the intertidal to the subtidal zones (Fig. 9.10, Rine and Ginsburg 1985; Allison et al. 1995a, b; Lefebvre et al. 2004).

Sandy open-coast tidal flats tend to have a concave-up cross-shore profile, commonly with very low (<1–2 mm year^{-1}) or even negative sedimentation rates (Semeniuk 1981; Hale and McCann 1982; Yang et al. 2005). The sandy lower intertidal flat is ordinarily very gentle and broad, but the seaward end is commonly bordered by ridge and runnel systems with significant undulations. The ridges are commonly shore-parallel with the crest-to-trough height of several decimeters to

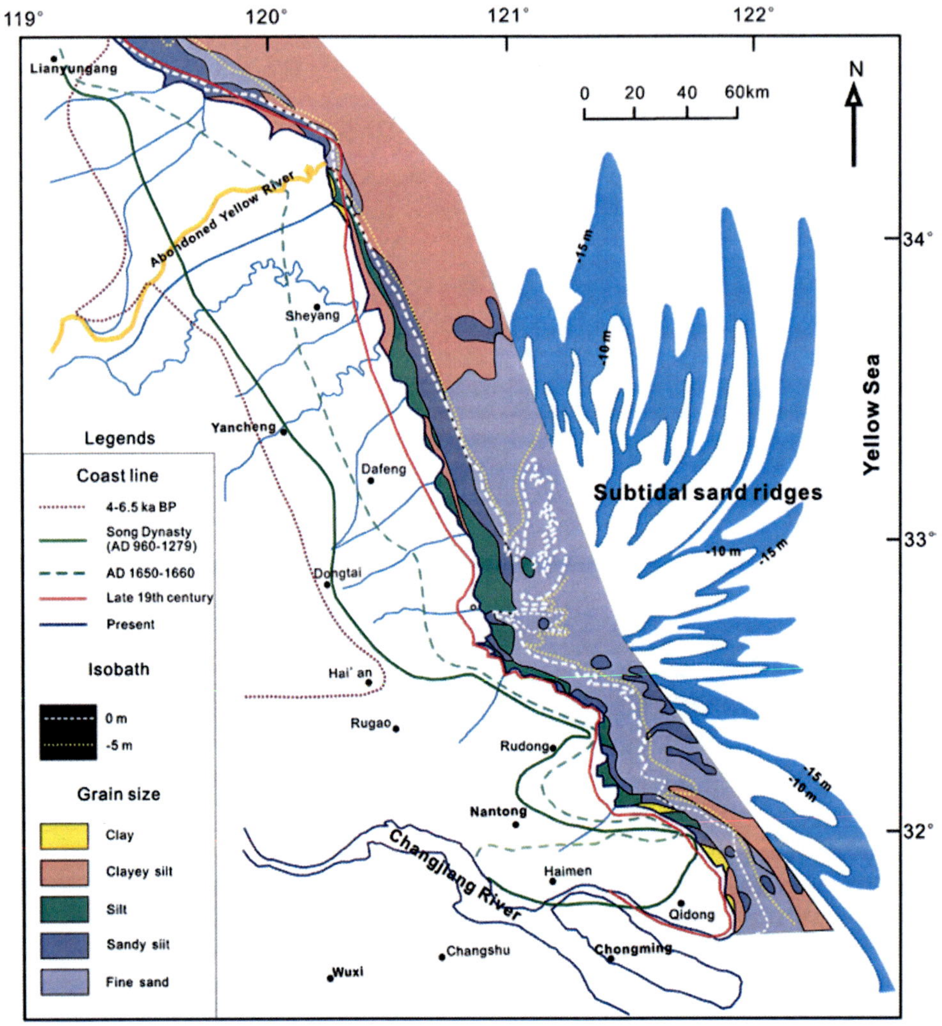

Fig. 9.11 Sediment distribution pattern for the extant tidal flats rimming the extensive Holocene coastal (chenier) plain of Jiangsu Province, China (After Wang and Ke 1997)

2–3 m, which are conceived to be generated like swash bars/ridges on the beach (Reineck and Cheng 1978; Yang et al. 2005). The sandy lower intertidal flat is landward continuous either with a gentle sand-mud middle intertidal flat and a mud upper intertidal flat for the accretional open coast like that on the central west coast of Taiwan (Reineck and Cheng 1978), or with a narrow inner slope like that on the coast of the Parksville Bay in the Strait of Georgia, Canada (Hale and McCann 1982), or with the inner swash bar/ridge and the inner muddy flat behind the bar/ridge like that on the southwest coast of Korea. The inner swash bar/ridge can be initially formed at the middle intertidal ground, migrates landward intermittently driven by seasonal storm events, and is finally welded to the coast as a part of the strand plain (Lee et al. 1994; Yang et al. 2005, 2008a; Ryu et al. 2008). The intertidal surficial sediment distribution has generally a landward-fining trend for the sandy open-coast intertidal flats without inner swash bar, similar to that of muddy open-coast tidal flats. However, a reverse trend, i.e., a landward-coarsening tendency except the inner muddy flat, is founded for those developing the inner swash bar composed of the coarsest sediment, which is highly mimic to the shoreface (Yang et al. 2005).

Tidal-channel networks are generally not much developed or event absent on the open-coast tidal flats in comparison with their delicate development on the

Fig. 9.12 Photos showing different morphologic and sedimentary features at the transitional zone between the salt marsh and the bare flats: (**a**) a gradual transition profile with soft muddy deposits on the flats, denoting accretion; (**b**) the erosion cliff of several centimeters high at the front of the salt marsh; (**c**) small erosion ponds on the salt-marsh land; (**d**) erosion remnant patches of partly consolidated salt-marsh deposits; (**e**) abundance of mud pebbles and shell debris on the erosion remnant patches; (**f**) swash sand sheet over the salt-marsh land

barred tidal flats (Frey et al. 1989; Wells et al. 1990; Alexander et al. 1991; Fan et al. 2004a; Yang et al. 2005). This is typically true for non-estuarine open-coast tidal flats, where the flats are exposed directly to the open sea. In the open-mouth estuaries, the main channels usually run parallel or oblique to local shoreline, but their landward branches are highly potential to cut deeply into the tidal flats, typically for those in the inner, less-exposed part of the estuaries. Although tidal channels are less developed on the bare flats, the broad vegetated flats tend to have tidal-creek networks as intricate as those on the sheltered tidal flats (Ren 1985;

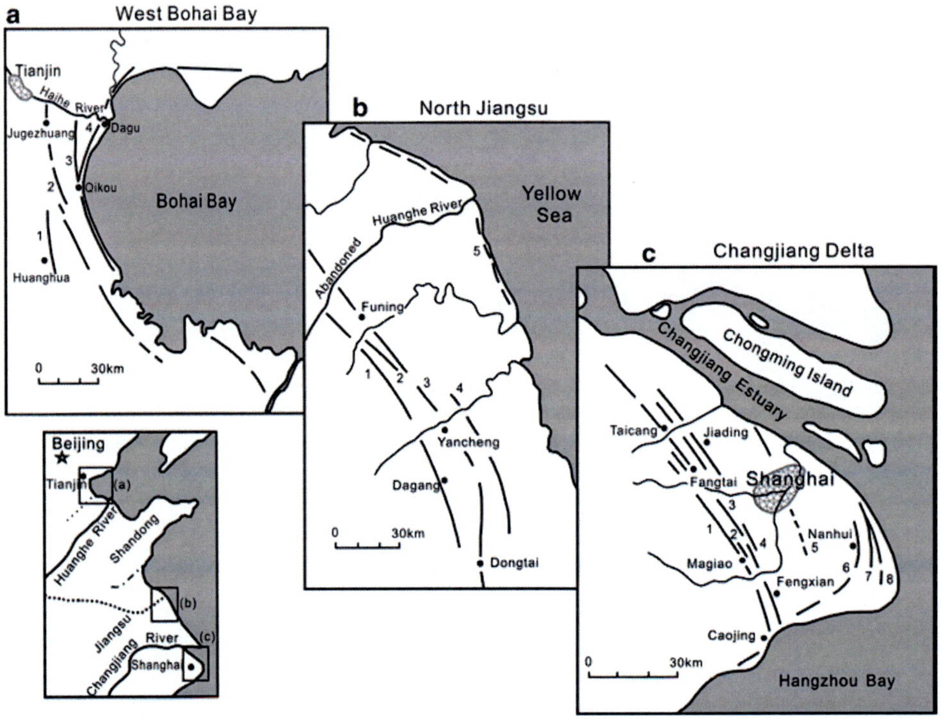

Fig. 9.13 The distribution of the chenier-ridge series on the coastal plain in East China (After Liu and Walker 1989)

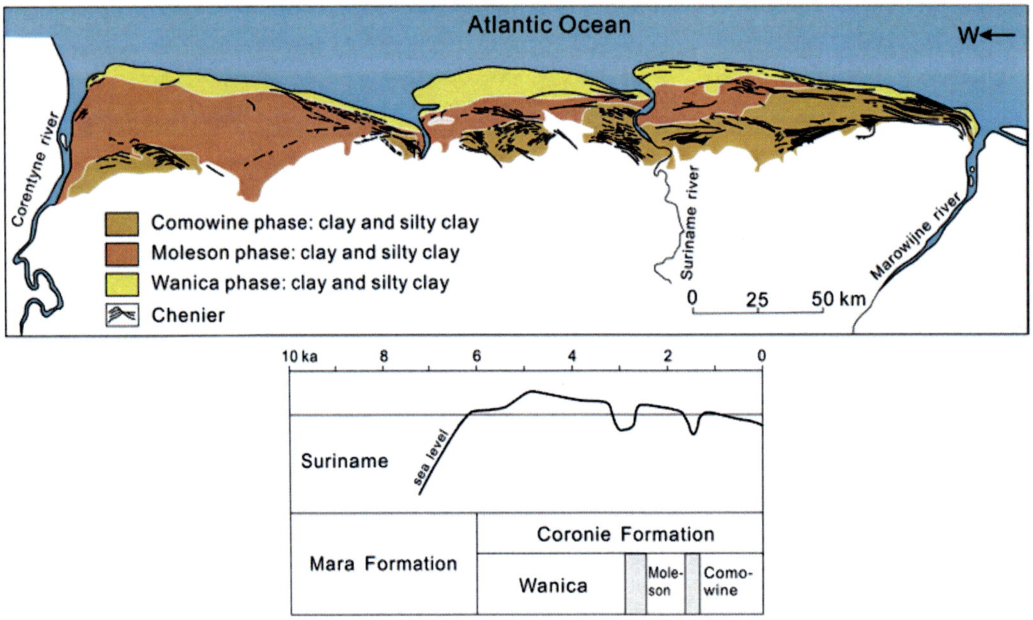

Fig. 9.14 Sketch map of the Surinam coastal plain showing the three-phase sedimentation units, separated by extensive chenier-bundle series denoting erosion phases (*upper panel*), and the relationship of regional sea-level fluctuations with alternations of accretion and non-deposition (erosion) phases during the Coronie Formation (*lower panel*, after Augustinus 2004)

Froidefond et al. 1988; Wang and Eisma 1988; Frey et al. 1989; Wells et al. 1990; Alexander et al. 1991; Zhang and Wang 1991; Eisma 1998; Fan et al. 2004a). On the Chongming Eastern Flats, the extensive salt-marsh land accommodates more than 20 tidal-creek dendritic networks (Fig. 9.15a). The larger the drainage basin, the more complicate the creek network. Small creeks usually terminate near the salt-marsh front, because the stability of creek banks decreases without protection by the vegetation and the slowing weak flows lose power to encroach the flats as the channel width increases toward the bare flats (Fig. 9.15b). Only a few large creeks continue their course on the bare flats from the marshland to the sea (Fig. 9.15a). The bare-flat channels are commonly shallow, muddy, and intermittent. They are potentially widened by big waves typically during storm or/and increasing flows during spring tides or heavy raining days (Fig. 9.15c–e), while they can be silted during small tides and waves. The tidal creek/channel system is therefore mainly functioned as a drainage network collecting and funneling water back to the sea during ebb tide, and flood current is less constrained by the shallow channels. Fine sediment dominates the creek/channel floor with frequent presence of fluid mud draining from the adjacent flats after exposure (Fan et al. 2004a).

In general, muddy and sandy open-coast tidal flats tend to have an accretional, convex-up and recessional, concave-up profile, respectively, which has been considered to be common morphodynamic features of tidal flats (Kirby 2000). The smoothed open-coast tidal flats are less dissected by tidal channels, conceivably linked to rotary current pattern, weak shore-normal component, and strong wave action. The most obvious relief over the entire tidal flats is the presence of inner and outer swash bars/ridges near the mean higher and lower water level, produced by storm waves during high tides and low tides, respectively. They are permanent or mobile, migrating landward intermittently driven by storm waves. The zonation pattern of intertidal sediment distribution is marked on the muddy open-coast tidal flats with a fining-landward trend, while it becomes blurry toward the sandy open-coast tidal flats. The latter potentially resembles the tidal beach in terms of a general coarsening-landward tendency and well-developed swash bars/ridges, occupying a transitional position between muddy open-coast tidal flats and tidal beach (Yang et al. 2008a; Dalrymple 2010).

9.4 Morphodynamics and Sediment Dynamics

9.4.1 Erosion and Deposition Cycles

The primary forces shaping intertidal flats are tidal currents and wind-induced waves. Waves are considered to be important for sediment suspension and currents for sediment advection. The interactions of waves and tides determine the magnitude and direction of sediment flux, and the resulting erosion and deposition are significantly site-specific over a small temporal scale ranging from a few minutes (wave frequency) to a few days (periods for intense waves/swells by meteorological events like tropical and subtropical storms). Intermediate erosion/deposition erosion cycles are mainly related to neap-spring cycles and seasonal alternations of wind and wave climate. Longer-term (decades to millennia) cycles of erosion and deposition are envisioned to take place over a larger spatial scale like the entire coastline of single deltas/estuaries or coastal basins. The super cycles are conceivably linked to longer-term variations in sediment supply (basin climate controlling), delta-lobe switching or the main channel shifting from one distributary to another, secular variations in sea-level change, and long-term sea wave climate change.

The short-term natural behavior of tidal flats can be monitored directly through the bed-level measurements or deduced indirectly through high-frequency measurements of waves, currents, and suspended sediment concentrations (SSC) during tidal inundation (Green et al. 1997; O'Brien et al. 2000; Lee et al. 2004; Thomas and Ridd 2004; Talke and Stacey 2008). Bed-level measurements are generally made during tidal exposure through using graduated stakes/poles, sediment erosion table (SET), and buried accretion plates, and sampling intervals usually vary from a single tidal cycle to 1 month (O'Brien et al. 2000; Thomas and Ridd 2004; Fan et al. 2006). Recent technical advances make it possible to monitor bed-level changes continuously during tidal inundation by employing acoustic transducer or submarine video camera (Christie et al. 1999; O'Brien et al. 2000; Thomas and Ridd 2004; Deloffre et al. 2007). Longer-term erosion/deposition cycles can be studied by the collection and comparison of the different-age satellite/aero photos or historical maps/charts. Dated sedimentary cores and chenier-ridge series are useful evidence for centennial to millennial cycles.

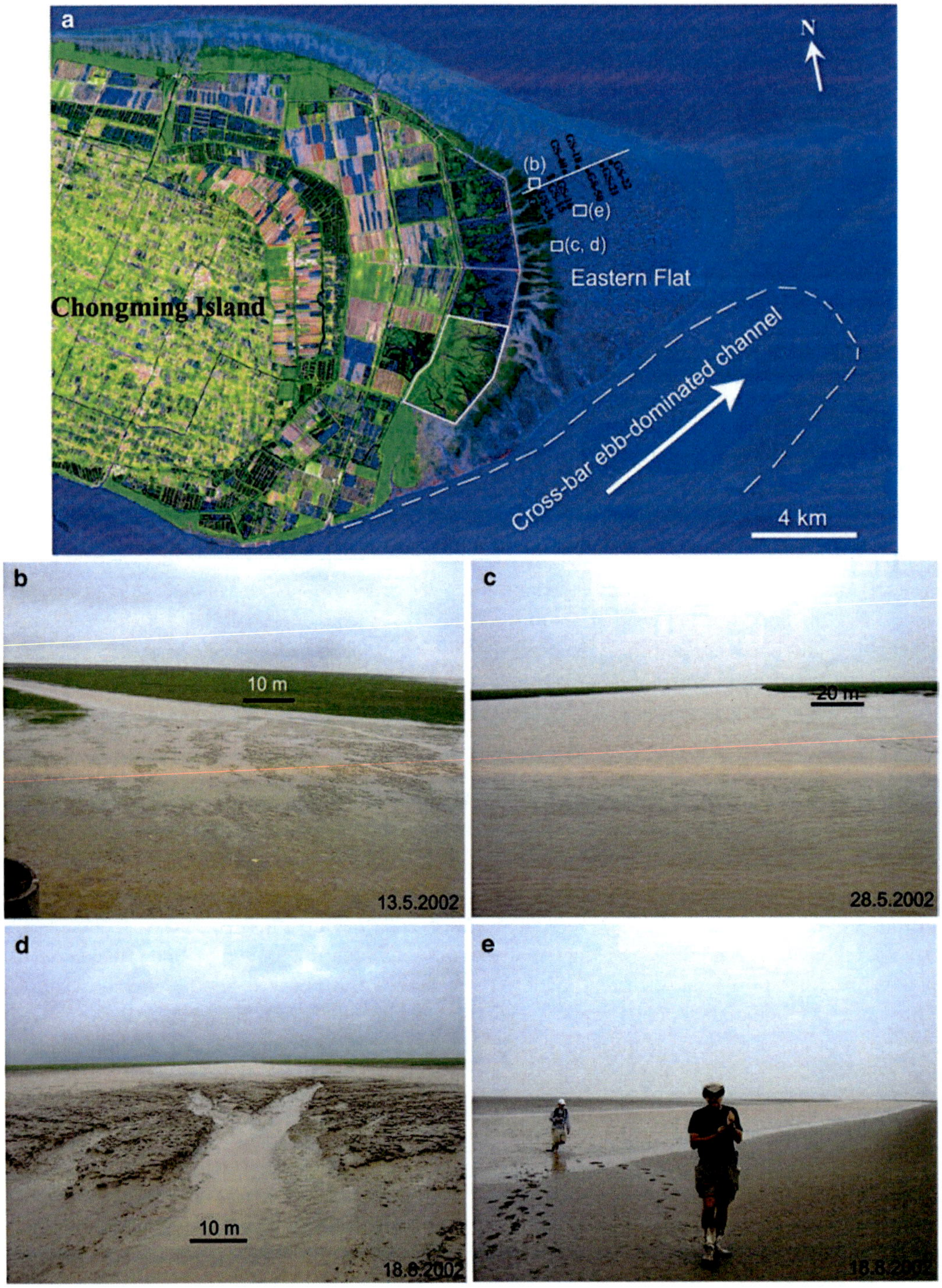

Fig. 9.15 (**a**) A satellite photo showing tidal creek/channel networks on the Chongming Eastern Flat in the Changjiang Delta, and the locations of the photos (**b–e**) taken and the elevation-monitoring stakes. (**b**) A tidal creek terminated near the boundary of the vegetated and bare flats. (**c**) A tidal creek continuing its course on the bare flat to the sea during spring tides or heavy raining days; (**d**) the shoal blocking the marsh creek dissected by head erosion due to increased creek discharge during spring tides with heavy raining. (**e**) A channel on the bare flat being widened by storm waves

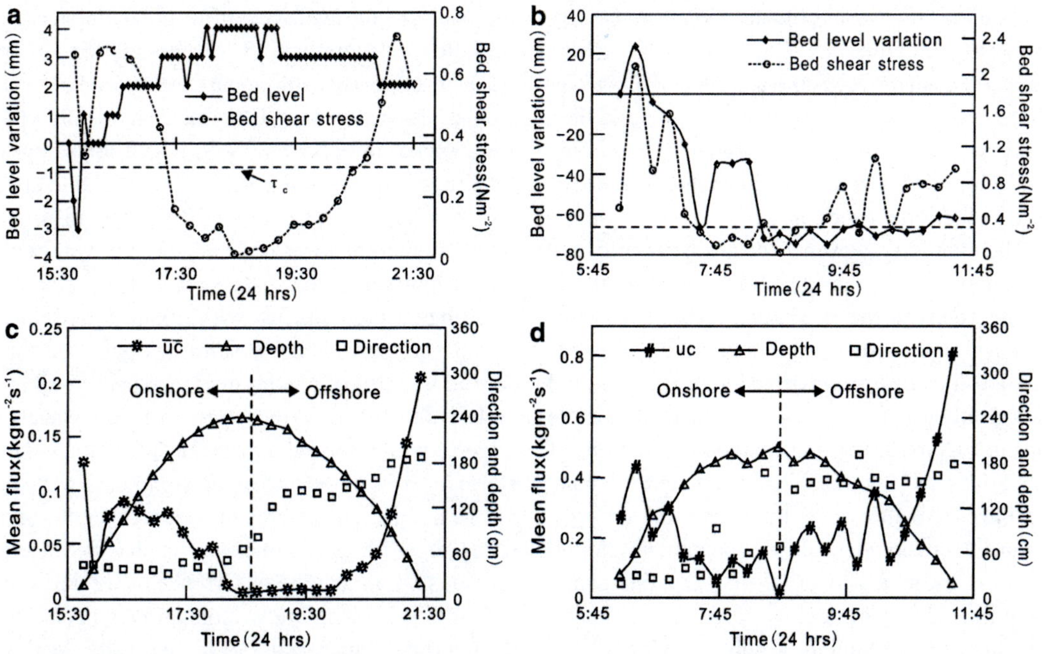

Fig. 9.16 Variation in hydrodynamic and morphodynamic processes within a tidal cycle on the Skeffling intertidal flat (Humber Estuary, UK) during the calm (**a** and **c**) and stormy (**b** and **d**) conditions. τ_c, the critical shear stress of sediment, was set to 0.3 N m^{-2}. Erosion occurs when $\tau_b > \tau_c$ (After Christie et al. 1999)

9.4.1.1 Short-Term Cycles (A Few Minutes to Days)

The interactions of tides and waves over a tidal cycle are fundamental processes and mechanisms to induce deposition and erosion on the open-coast tidal flats. Bottom shear stress is the important parameter to assess the deposition and erosion of sediments (e.g., Christie et al. 1999; Le Hir et al. 2000). Over the intertidal flats, purely tide-induced stresses are generally low to trigger significant erosion, while wave-induced stresses are much higher due to their orbital movement character. It has been widely acknowledged that even presence of small waves potentially enhances erosion of the surficial sediment significantly (Anderson et al. 1981; Christie et al. 1999; Le Hir et al. 2000; Lee et al. 2004). Enhanced erosion by large waves was nicely described by a comparison study of Christie et al. (1999) on a semi-closed and mega-tidal mudflat in the Humber Estuary (UK) with the average tidal range of ~6 m. The bed shear stresses (τ_b) of combined flows and waves were highly elevated toward the large-wave condition, producing several centimeters of erosion, which is strongly contrastive with a few millimeters of accretion during the small-wave condition (Fig. 9.16a, b).

The difference is highly outstanding between wave and tide processes over a tidal cycle. Tidal currents and tide-induced bottom stresses decrease landward from the lower intertidal flats, and drop to nearly zero at the high, slack tide. However, waves and wave-induced bottom stresses are significantly strengthened by the increased water depth during the rising tide owing to the relationship between wave height and water depth (Green et al. 1997; Le Hir et al. 2000).

So peak waves and the related processes hypothetically attain at high tide for any given location at the intertidal flats, and the slack (quiet) condition is consequently detained at high tide by the presence of large waves (Christie et al. 1999). The phenomena were clearly shown by the variations in $\overline{uc}$ value (mean suspended sediment flux, the product of mean current velocity and mean concentration of suspended sediment). The curve was flat with roughly zero value over approximately 1.5 h at high tide during small-wave condition (Fig. 9.16c), but was drastically replaced by a seesaw curve section during large-wave condition (Fig. 9.16d, Christie et al. 1999).

The shallow wave-break/swash zone should shift several 100–1,000 m landward and seaward over a tidal cycle under the storm climate, violently disturbing the intertidal morphology over a wider area (Mao 1987; Shi and Chen 1996; Fan et al. 2006). The wave-break zone tends to stall respectively at low and high water line for longer time, hypothetically accounting for the development of inner and outer swash ridges as discussed in the former section.

Episodic high-energy events occur infrequently, but are the fiercest force to produce marked erosion and deposition cycles on the tidal flats. A large wave or storm event usually lasts several hours to days, so the resulted tidal-flat erosion and deposition pattern should be revealed only by a finer time scale than the life cycle of the events. Considering the complex interactions of waves and tides and a philosophy that a finer time scale tends to have a smaller spatial attribute, the episodically induced erosion/deposition phenomena should therefore be examined over a finer spatial scale. Following this, an experiment was carried out along a cross-profile on the Nanhui Mudbank (Changjiang Delta) in 1999, through using graduated-stake elevation-monitoring technology (Fan 2001). Sixty-two stakes were fixed on the intertidal ground with a distance of 30–50 m between two neighboring stakes, and they were regularly monitored every single or 2 days. The result shows that net erosion switches on when waves exceed 1.5 m high during non-typhoon conditions (Fig. 9.17). Serious erosion occurs during peak storm periods with a maximum of >15 cm deflations over two to four tidal cycles at some locations (Fan 2001). There are generally existent two erosion zones separated by the accretional zone. It was hypothesized that the erosion and the deposition zones were respectively produced by series of wave breaking and reforming processes over the gentle and broad mudflats (3.4 km wide across entire intertidal zone with a mean tidal range of 2.6 m) before the wave was dying out (Fig. 9.18, Fan et al. 2006). The sites of erosion or deposition change alternatively into deposition or erosion over a next few tidal cycles (Fig. 9.17), denoting the same mechanism that waves tend to break over the previous deposition zones because of shoaling and produce new erosion zones, and vice versa for the previous erosion zones which turn into wave-reforming area to promote deposition.

9.4.1.2 Intermediate Cycles (Neap-Spring Tidal Cycle to Season)

The modulation effect of large-wave processes by tides is significantly different over a single tidal cycle and the neap-spring cycle, exerting great impacts on the tidal-flat development. Tide modulation of storm waves was clearly exhibited by the hydrodynamic data from a field experiment on the Baeksu tidal flats (southwest Korea) in February 1999, where the semidiurnal tidal range varies from 2.3 m at neap to 5.5 m at spring with a mean of ~3.9 m (Kim 2003). The instrumentation system to monitor wave and current was deployed on the lower intertidal flats for 2 weeks, catching two invading winter storms with main wind direction toward the shore. The first storm was stronger than the second in terms of maximum wind speed (~18 m/s vs. ~14 m/s). However, the storm-induced waves were larger for the second (weaker) storm than the first storm in terms of the maximum significant wave height (3 m vs. 2 m). The discrepancy of weaker storm generating larger waves is actually ascribed to the wave-depth relationship, in that the occurrence of weaker storm at spring tide tends to allow larger waves penetrate into the intertidal flats because of higher water depth rising by spring tide, whereas neap tide is potential to damp storm energy more seaward (Kim 2003).

The fact that the intensity of storm-related processes is greatly modulated by the neap-spring cycle was also addressed by Fan et al. (2006) in terms of the erosion magnitude and the distribution of erosion zones across the intertidal flats at the Nanhui Mudbank. They discussed that a weaker storm at spring tides was potential to induce more intense erosion than a stronger storm at neap tides (Fig. 9.18). It is not only due to the higher wave energy and the related mightier vertical scouring capability at spring tides than at neap tides which are determined by water depth as discussed above, but also linked with the higher spring current speed and the related mightier horizontal advection capability than neap tides. Consequently, more sediment is suspended and carried out of the erosion zone by the combined action of storm waves and currents at spring tides than at neap tides, producing higher magnitude of the erosion.

Seasonal alternations of accretion and erosion are the most significant and widely addressed features of the tidal flats (Ren 1985; Shi and Chen 1996; O'Brien et al. 2000; Yang et al. 2005; Fan et al. 2006; Yang et al. 2008b). The processes leading to this periodic development are principally related to seasonal wind climate change, and less to other factors like changes in tidal level, floral and faunal distributions, solar intensity, and estuarine turbidity maximum location. The Baeksu tidal flat in southwest Korea has more

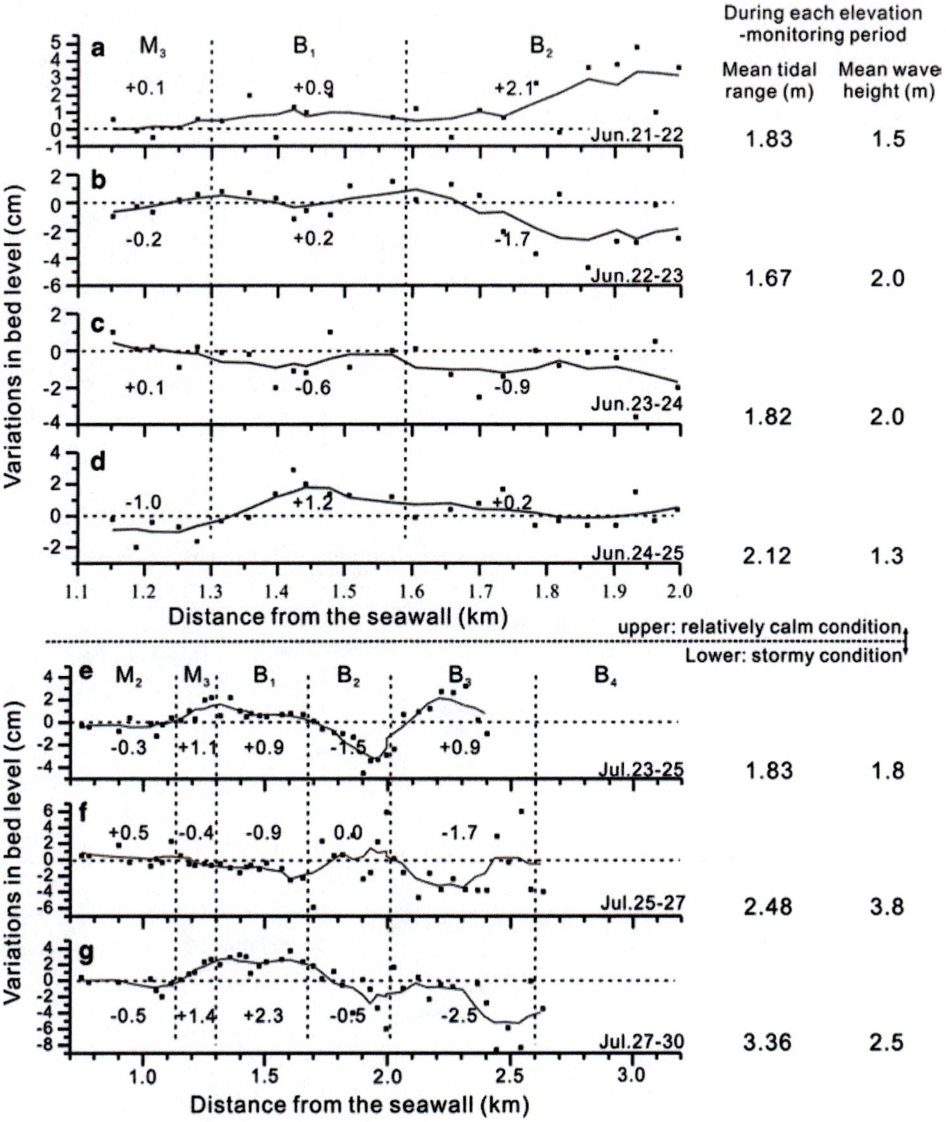

Fig. 9.17 Short-term (1–2 days) variations in bed level of the intertidal flat in the Changjiang Delta (China) over a relatively calm period June 21–25 in 1999 (**a–d**) and a stormy period July 23–30 in 1999 (**e–g**), respectively. Individual stake measurement data were analyzed by three point adjacent average to generate a smoothed curve, representing a general trend of erosion and deposition across the profile. Different sub-zones of the intertidal flat had different erosion/accretion trends in response to waves. Mean changes in bed level over the different sub-zones were calculated and marked in the figures (After Fan et al. 2006)

frequent and intense storms in winter than in summer, and the resulting seasonal cycles of erosion and deposition are highly contrasting with summer accretional and muddy profile of a normal tidal flat and winter erosional and sandy profile resembling a shoreface (Yang et al. 2005).

Most of the subtropical coast is actually subjected to both summer and winter storms. On the Chongming Eastern Flat (Fig. 9.15), the tidal-flat elevation surveying data showed that the bare flats underwent erosion both in summer and winter (Fig. 9.19). The bare flats were lowered by a few decimeters during sporadic typhoon strikes, and the same magnitude of deposition usually took place soon after the erosion events in summer. By contrast, the winter erosion season began in mid-October with sharp erosion of a few decimeters, and followed with light erosion throughout late March. It lasted for more than 5 months before entering a gradual

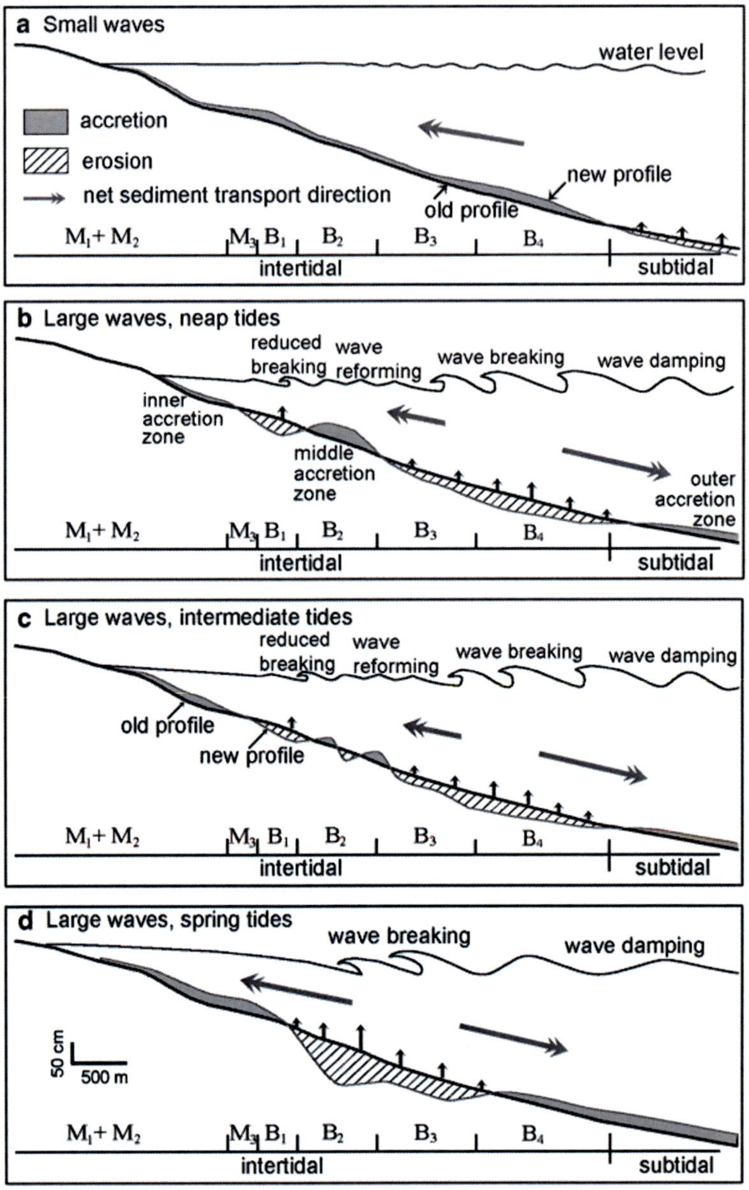

Fig. 9.18 Schematic models showing intertidal morphodynamic processes and sediment transport patterns under different wave and tidal regimes: (**a**) small waves, (**b**) large waves and neap tides, (**c**) large waves and intermediate tides, (**d**) large waves and spring tides (After Fan et al. 2006)

accretion season. It is noteworthy that the salt-marsh land underwent continuous slight accretion throughout the monitoring period, reasonably linked to the protection by the salt-marsh canopy (Fig. 9.19).

The shoreline strike can play important role in the seasonal cycle development owing to the selecting effect of onshore and offshore wind. Along the north bank of the Hangzhou Bay (Fig. 9.7), the east-west orientation shoreline makes the tidal flats sensitive to summer tropical storms with onshore wind domination, whereas sluggish to winter storms with offshore wind domination. This was clearly exhibited by the alternations of the summer sandy erosional flats and the winter muddy accretional flats (Fig. 9.20, Yang et al. 2008c).

Seasonal cycles are also the most outstanding morphodynamic action on the Guiana open-coast tidal flats. The Guiana tidal flats differ from others by their

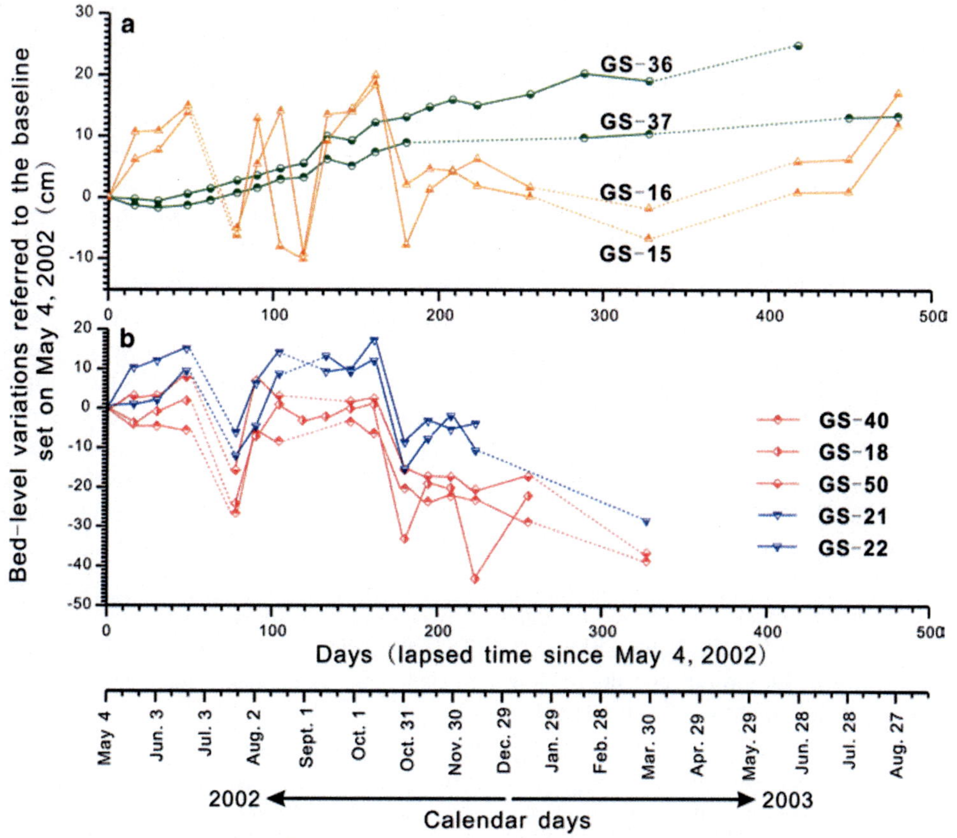

Fig. 9.19 Seasonal variations in bed level of the salt-marsh land and the bare intertidal flats along a northern transect of the Chongming Eastern Flat in the Changjiang Delta. Four groups of elevation-monitoring stakes (GS-36, 37), (GS-15, 16), (GS-18, 40, 50), and (GS-21, 22) are positioned at the lower salt marsh and the upper, middle, and lower bare intertidal flats, respectively. See their detailed locations in Fig. 9.15a

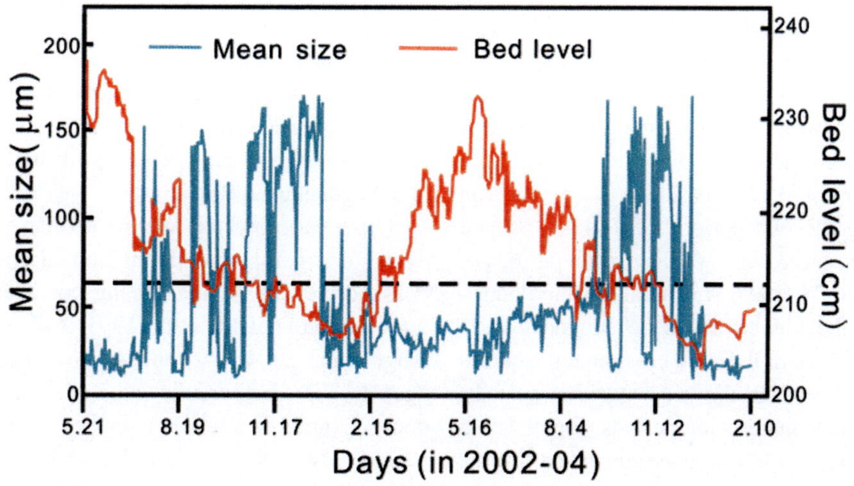

Fig. 9.20 Daily variations in bed level and grain size of surface sediment at a fixed station on the middle intertidal flat along the north bank of Hanzhou Bay. The dash line denotes the division between sand and mud at 63 μm (After Yang et al. 2008c)

typical longshore migration instead of cross-shore evolution (Wells and Coleman 1981; Froidefond et al. 1988; Eisma et al. 1991; Augustinus 2004). There are totally 20–25 mudbanks along the coast of the Guianas (Fig. 9.4), and each is generally 10–40 km in length and several tens of kilometers in width extending gently from the shore to nearly 20-m isobaths (Fig. 9.10). The mudbanks migrate westward, driven by the alongshore current and northwest wind waves generated by trade winds. The mudbank migration is completed by deposition at the leading (western) edge and erosion at the trailing (eastern) edge, which is highly related to the attenuation of the incoming waves significantly by presence of fluid mud at the leading edge or less by presence of the compacted clay deposits at the trailing edge (Fig. 9.10, Wells and Coleman 1981; Allison and Lee 2004). The mudbanks migrate downdrift at mean rates of 0.9–1.5 km year^{-1} (Froidefond et al. 1988; Eisma et al. 1991). Over an annual interval, the windy season from January to April accounts for major sediment exchange between the trailing and the leading edges, producing the most rapid mudbank migration. The annual sediment exchange from the trailing to leading edges along the entire 1,400-km-long Guiana coast adds up to an amazing figure, approximately equal to the average input of new sediment from the Amazon (Allison and Lee 2004).

9.4.1.3 Long-Term Cycles (A Few Years to Decades)

Long-term cycles of tidal-flat developments are commonly envisioned along the deltaic plains/coastal plains, where the shoreline can change rapidly and vastly with a few kilometers or more in a few decades, driven by different mechanisms. Coastal development of the Nanhui Mudbank in the Changjiang Delta was analyzed by using time series of nautical charts from 1842 to 2004 (Fig. 9.21). It was shown that net erosion occurred during the two periods of 1842–1864 and 1911–1958, alternating with two deposition periods of 1864–1911 and 1958–2004. A general erosion/deposition pattern of the mudbank can be summarized as "accretion at the eastern flat with erosion at the southern flat (smaller area)" in the net-deposition periods, and vice versa in the net-erosion periods (Fig. 9.21). It is presumed to mainly result from the coincident shifting of the Changjiang's main channel away (erosion) or toward (deposition) the mudbank. Another pattern was also noted as "accretion at the higher flat (above −2-m) and erosion at the lower flat (below −2-m)" in the net-deposition periods, especially during 1864–1911, whereas a reversed pattern of "erosion at the higher flat and accretion at the lower flat" occurred in the net-erosion periods, especially during 1842–1864 (Fig. 9.21). The alternations of erosion and deposition phases between the higher and lower flats were presumably linked to multi-decadal alternations of the frequency of intense tropical storms. In other words, the coastal morphology tends to preserve the storm-induced profile in terms of "erosion at the higher flat and accretion at the lower flat" in the stormy decades, whereas the normal-wave morphology with "the higher flat accretion vs. the lower flat erosion" is cumulatively present in the relative calm decades (Huo et al. 2010).

The multi-decadal variations in the Guiana coastline are produced by its typical alongshore migration of series of the mudbank and the interbank units, which continuously pass through a given location (Fig. 9.22). The mudbanks migrate with an average rate of 1.5 km year^{-1} along the Surinam coast, and the average alongshore width is 45 km for each geomorphologic unit including a mudbank and the neighboring interbank area (Fig. 9.10). It is therefore estimated a roughly 30-year cycle of erosion and deposition along the Surinam coast (Augustinus 2004). Due to huge sediment input from the Amazon, a net coastal-plain growth is generally produced after each mudbank-interbank cycle (Fig. 9.22, Allison and Lee 2004). It was noteworthy that the comparison studies using historical maps and Holocene core data demonstrated the difference between the present and Holocene sedimentary processes of the mudbanks. The present, roughly 30-year cycle of high rates of accretion and erosion associated with the mudbank migration extended at least as far back as the last two and a half centuries, but the beginning of this cycle is still uncertain (Plaziat and Augustinus 2004).

There was reported to exist another multi-decadal cycle of erosion and deposition along the Guiana coast, driven by secular change of ocean wind and wave climate (Eisma et al. 1991; Allison et al. 2000; Augustinus 2004). A comparison study of different ages of air photographs showed that the Surinam coast changed as a whole from net erosion during the period 1947–1966 to net deposition in the period 1966–1981 (Fig. 9.23, Table 9.2). The change was found coincidently with an increase in mean wind velocity and a general shift of wind direction from

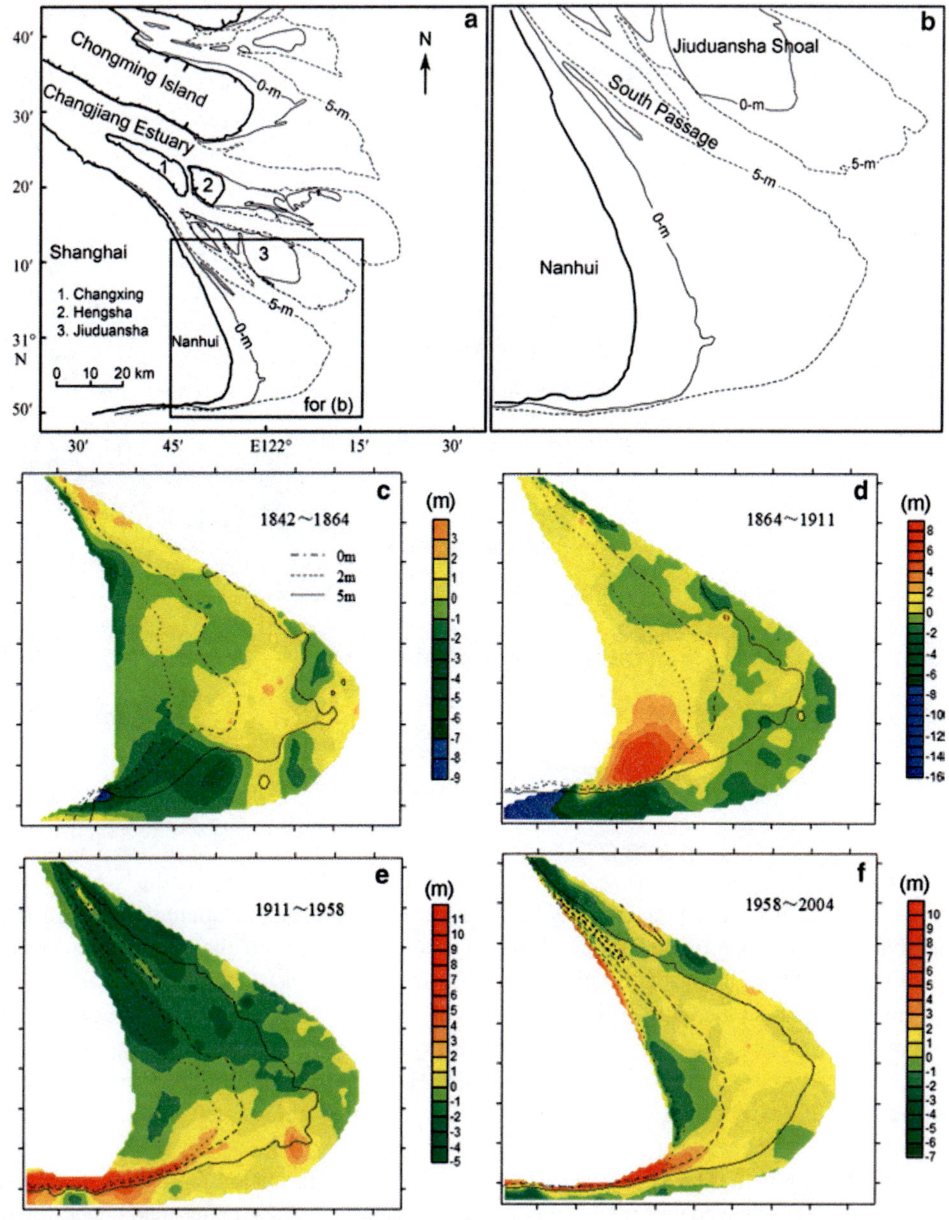

Fig. 9.21 Multi-decadal coastal development of the Nanhui Mudbank in the Changjiang Delta (After Huo et al. 2010). Color bars denoting the magnitude of accretion (positive) and erosion (negative)

more NE to more ENE from 1959 onward. The stronger winds and the change in direction toward a smaller angle with the coastline resulted in an enhanced alongshore transport and a reduction of the onshore wave energy component. This led to net deposition in the Surinam coast (Eisma et al. 1991). Enhanced alongshore transport also favored development of longer mudbanks (Fig. 9.23, Augustinus 2004). Multi-decadal changes in the Guiana coastal development were therefore presumed to be determined by the variations in the strength and the direction of the trade winds instead of the flux of sediment supply from the Amazon (Eisma et al. 1991). Note that the coast of Guiana and French Guiana has an SE-NW orientation, different from the nearly east-west orientation of the Surinam coast. The difference in angle

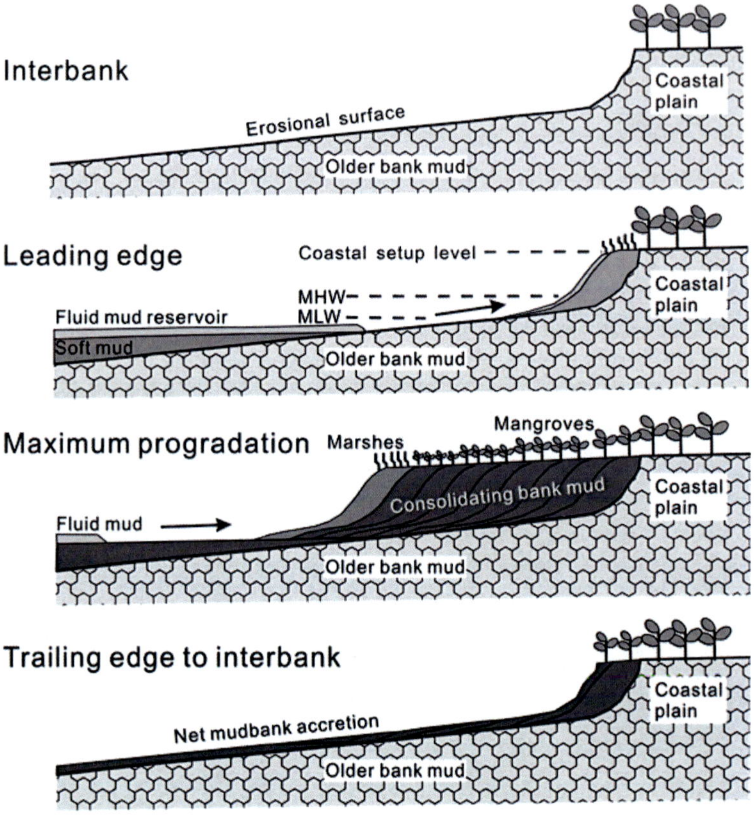

Fig. 9.22 Schematic model for shoreline evolution cycles in the Guiana coast. Serious erosion takes place at the interbank phase (*top panel*). It is succeeded by leading edge mudbank deposition (*second panel*). The accretion continues till the passage of the leading edge to reach the maximum progradation (*third panel*). And erosion takes place again by the swing of the trailing edge (*bottom panel*). Note that there is a net coastal plain growth with each mudbank-interbank cycle (After Allison and Lee 2004)

between the coastline and the direction of wave propagation accounts for the different behaviors of the respective mudbanks. Consequently, the mudbanks in Guiana are shorter than those in Surinam, and their behaviors are more erratic than the latter (Eisma et al. 1991; Augustinus 2004).

9.4.1.4 Megacycles (Hundreds to Thousands of Years)

Mega-scale coastal development is referred to the geological evolution of the Holocene coastal plains over a time scale of centuries to millennia. It is generally addressed through using cores or large-scale coastal morphological features. Chenier ridges are typical and extensive features on coastal plains near the mouths of rivers (typically large rivers), excellent indicator for the ancient coastline location and evolution (Meckel 1975; Liu and Walker 1989; Wang and Ke 1989; Lees 1992). They demarcate the flat coastal plains with coastline-parallel ridge series, mainly composed of sand and shell debris, a few decimeters to meters higher than the surround mudflat/marsh deposits. There are generally four chenier ridges along the western coastal plain of the Bohai Bay, five in the North Jiangsu coastal plain, and eight in the Changjiang Delta (Fig. 9.13). The oldest chenier in the Changjiang Delta and the North Jiangsu coastal plain was dated 6,460 year BP and 6,160 year BP, respectively. The two chenier plains were therefore assumed to begin to develop since the mid-Holocene maximum flooding period, but have afterward been interrupted several times to produce other cheniers.

The chenier formation represents an episode of erosion, whereas the intervening mud deposition represents a period of coastal progradation. The development of chenier plains is generally presumed to mainly result from sediment starvation by the distributary channel switching (Lees 1992). The lower reach of the Huanghe switched several times between debouching into the Bohai Bay and the Yellow Sea, accounting for the formation of the cheniers on the western coastal plain of the Bohai Bay and the North Jiangsu coastal plain (Liu and Walker 1989; Wang and Ke 1989). Each chenier on the southern coast of the Changjiang Delta was formed when the Changjiang main channel switched north and away from the south bank. The same mechanism was also employed to interpret the chenier development in the Mississippi Delta (Byrne et al. 1959) and the Colorado Delta in the Northern Gulf of California (Meckel 1975).

Sediment starvation for a coast can also result from a reduction in fluvial input from the drainage basin instead of the delta channel switching. The chenier ridges on the coastal plains of North Australia were presumably formed during low mud influx accompanying the long-term dry periods in the drainage basins, while mudflat deposition occurred during the wetter periods (Rhodes 1982; Lees 1992).

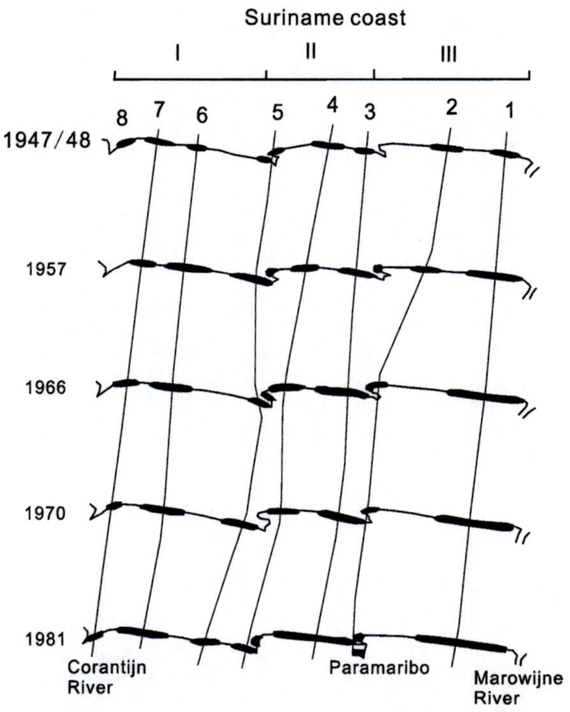

Fig. 9.23 Changes in position and length of the mudbanks (*heavy lines*) along the Surinam coast between 1947 and 1981 (After Eisma et al. 1991)

The formation of cheniers could have resulted from not only the reduced sediment availability but also the increased energy of marine processes. There are two series of well-developed chenier bundles on the Surinam coastal plain, denoting that the coastal development was interrupted at least twice by longer intervals of erosion during the Coronie Formation (<6,000 year BP). These two hiatuses among the three sedimentation phases of Wanica, Moleson, and Comowine coincided with a slight drop in sea level (Fig. 9.14) and a systematic increase of more northerly wind frequencies. The chenier formation is therefore simply considered to link with the more northerly wind frequencies and a fall in sea level (Eisma et al. 1991; Augustinus 2004).

9.5 Sedimentary Structures and Bedding

Bedforms and sedimentary structures are highly related to sediment size and hydrodynamics (Boguchwal and Southard 1990). Open-coast tidal flats vary greatly in major grain-size composition from fine silt to sand (Table 9.3). Wave energy can be dissipated higher or less when propagating over the muddy or sandy flats due to presence of fluid mud or not, and the wave is also greatly modulated by tidal fluctuations. The difference

Table 9.2 Total amounts of mud yearly eroded (–) or deposited (+) along the Surinam coast over different periods (After Eisma et al. 1991)

Period	Section	Total amount ($\times 10^8$ tons)	Amount per year ($\times 10^6$ tons year^{-1})	Net amount over the entire coast ($\times 10^6$ tons year^{-1})
1947–1957	I	–15.90	–1.59	
	II	+9.85	+0.99	–0.82
	III	–2.09	–0.21	
1957–1966	I	–14.19	–1.58	
	II	+0.45	+0.05	–1.00
	III	+4.71	+0.52	
1966–1970	I	+6.55	+1.64	
	II	+1.28	+0.32	+2.18
	III	+0.89	+0.22	
1970–1981	I	+30.85	+2.81	
	II	+44.57	+4.05	+7.61
	III	+8.29	+0.75	

Table 9.3 Typical sedimentary structures on the open-coast flats

Types of tidal flats			Muddy flats					Sandy flats
Examples			The coast of the Guianas	Northwestern Gulf of California	The Fly Delta, Papua New Guinea	Kyonggi Bay and Namyang Bay	Wanggang tidal flats, Jaingsu Province, China	Baeksu, Doowoori, and Dongho tidal flats, Korea
Grain size			Silty clay	Clayey silt to silty clay	Mean percentiles of sand, silt, and clay are 29.4%, 53.2%, and 17.4%	4.3–5.8 ϕ in Kyonggi Bay; 4.6–8.8 ϕ in Namyang Bay	4.38–6.25 ϕ	Seasonally changing from 4–5.5 ϕ in summer to <3–4 ϕ in winter
Zonations	Supratidal flat		Mangrove occupied; massive beds with abundant roots and benthic traces	Not distinct laminations or beddings; prevalent mud desiccation and salt crystallization; rare shell fragments	Mangrove and salt marsh occupied; prevalent homogenous mud; infrequent thin interlayered laminations	Absent	Salt marsh occupied; intense bioturbation of original parallel to slight-wavy laminations	Absent
	Intertidal flats	Upper		Common laminae; some small-scale cross-beddings in sand; common burrows; fine shell hash in sandy laminae and complete mollusk shells in mud		Intense bioturbation of original wavy beds; infrequent flaser beds, clasts, and convolutions	Interlayered beddings of thick mud layers and thin sand layers; common parallel laminations in the mud beds and small ripple laminations in the thick sand beds; intense bioturbation	Interlayered bedding of winter sand deposits and summer mud deposits and the features preserved in the strata
		Middle	Common massive beds and parallel to subparallel laminations; some wavy laminations; rare lenticular laminations, micro-cross-laminations, scour and fill structures, and biogenetic traces		Prevalent planar to wavy laminations; common soft sediment deformation; sand layer thickening and coarsening as decreasing elevation; limited bioturbation	Common parallel to wavy beddings; infrequent ripple and parallel laminations; slight bioturbation	Interlayered beddings of thick sand layers and thin mud layers; common wave beddings in the thick sand layers; coarser sediment in the tidal creeks containing bipolar cross-beddings	The summer muddy flat with interbedded sand and mud, changing into the winter sandy flat prevalent sandy parallel laminations and hummocky cross-stratification; the strata preserved majorly the winter storm coarse deposits with little bioturbation
		Lower				Prevalent small ripple laminations, intense bioturbation		
	Subtidal flat/ channel			Thin laminae, typically lenticular and irregular	Prevalent interbedded sand and mud; present ripple cross-laminations and slump features	Common thick cross-stratification, slightly mottled to better stratified	Thin parallel to wavy beddings; presence of small cross-beddings and herringbone cross-beddings	Sediment finer than the lower intertidal flat
Sedimentation rates over a 100-year scale			Several millimeters to centimeters	A few centimeters	A few centimeters to decimeters	A few millimeters	A few centimeters	≤1 mm
Major references			Rine and Ginsburg, (1985); Allison et al. (1995a, b)	Meckle (1975); Thompson (1975)	Baker et al. (1995), Walsh and Nittrouer (2004)	Frey et al. (1989); Alexander et al. (1991), Park et al. (1995)	Ren (1985)	Yang et al. (2005, 2008b)

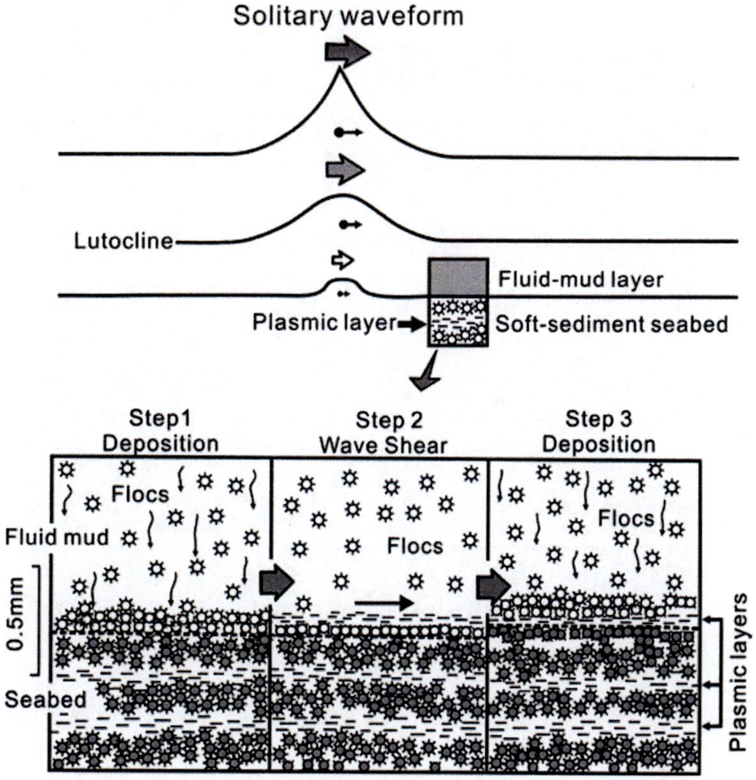

Fig. 9.24 Schematic model for the genesis of plasmic fabric layers (~0.01–0.1 mm thick) on the muddy banks along the coast of Guianas. Because of the viscosity difference, shear is produced at the boundary of fluid mud and sediment with passage of solitary wave crests (*upper panel*). The shear is considered to break the surface flocs and orient platy mineral grains, creating a plasmic layer. Between weaves, floc deposition is possible, producing unoriented interlaminations (After Allison et al. 1995a)

in grain size and combined wave-current energy consequently determine the sedimentary character changing in a spectrum from the wave-influenced, tide-dominated on the muddy (fine silt domination) open-coast tidal flats, through the wave- and tide-dominated on the muddy (coarse silt domination) open-coast tidal flats, to wave-dominated on the sandy open-coast tidal flats. The spectrum change is also evident in the strata with increasing volume of storm deposits from the muddy to sandy open-coast tidal flats.

The mudbanks along the Guiana coast are muddy open-coast tidal flats, predominantly consisting of fine silt and clay. The prevalent bedding is massive beds, or parallel to subparallel laminations of a few micrometers to millimeters thick (Table 9.3; Rine and Ginsburg 1985; Allison et al. 1995b; Allison and Lee 2004). The millimeter-scale laminations of silt enrichment are commonly layered structures in X-radiographs. The micrometer scale of the structures should be examined in thin sections under microscopes, named plasmic or unistrial fabric (Kuehl et al. 1988; Allison et al. 1995a, b). The plasmic fabric is composed of alternating layers (about 0.01 mm thick) of oriented and unoriented clay and mica with the oriented layers showing uniform extinction under polarized light. The formation of plasmic fabric was presumed to result from in situ shearing by surface gravity waves in sediments being rapidly deposited from a fluid-mud suspension (Fig. 9.24, Rine and Ginsburg 1985; Allison et al. 1995b). Solitary waves have almost unidirectional flow approaching the shore (Wells and Coleman 1978). Because of the viscosity difference, passage of solitary waves induces shear along the fluid mud/sediment boundary. This shear is postulated to break flocs and orient platy particles (micas and clays) in the direction of shear. Floc deposition takes place between waves, producing unoriented interlaminations (Allison et al. 1995a). The plasmic fabric is the finest scale of sedimentary structures having wave imprints as known so far, but it is difficult to be distinguished in fossil rocks and linked to wave generation.

Most of the muddy open-coast tidal flats are predominantly composed of silt, with small fractions of clay and fine sand (Table 9.3). The preferred surface structures are small ripples on the muddy flats if no presence of storm waves. On the Chongming Eastern Flat in the Changjiang Delta, small symmetrical wave ripples and combined flow ripples are the commonest

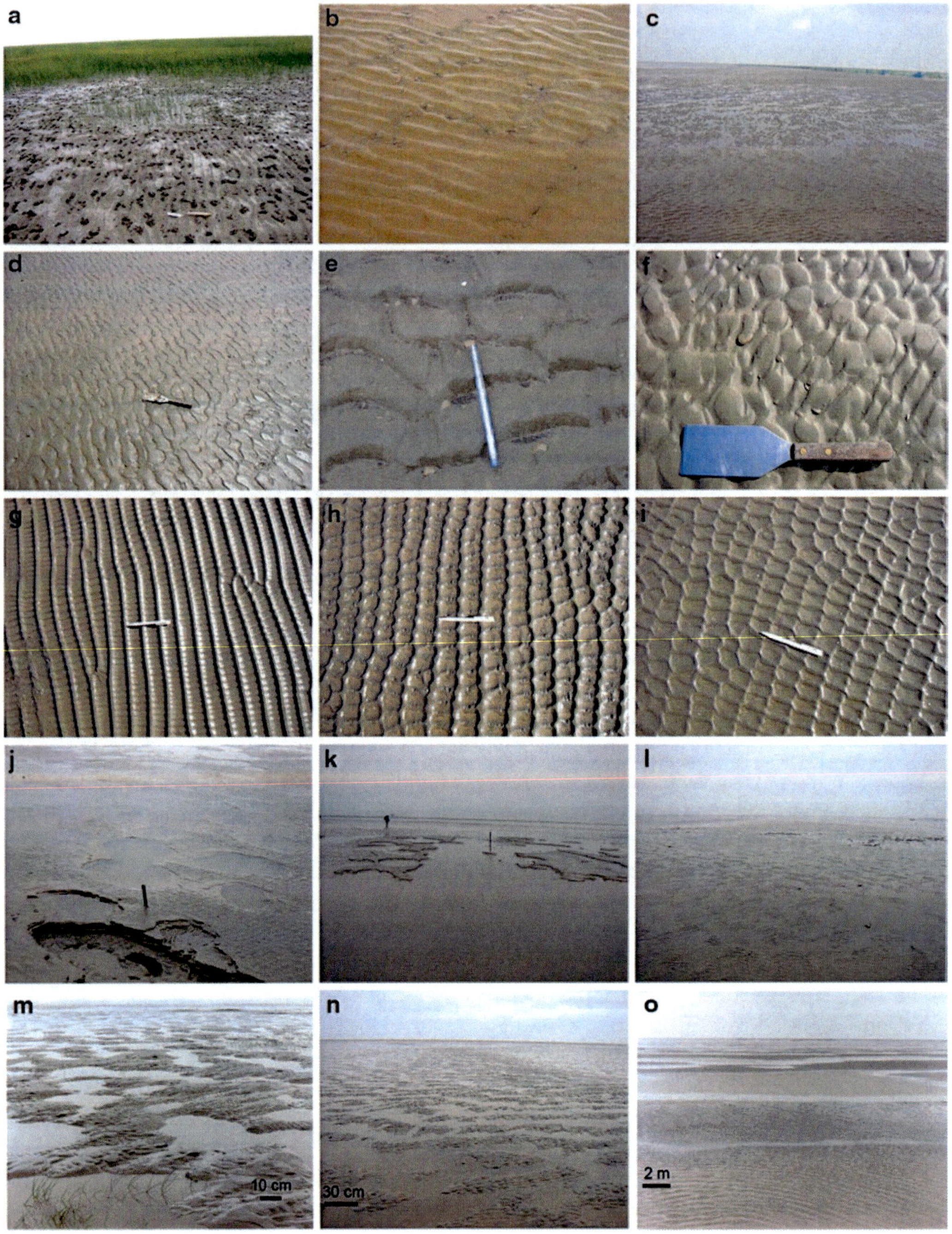

Fig. 9.25 Surface structures on the Chongming Eastern Flat of the Changjiang Delta. Small symmetrical wave ripples on the marshland (**a**) and on the upper bare flat (**b**). Small combined-flow ripples on the bare flat (**c–f**) including the flatten-crested ripples (**d**) and undulatory to lingoid current ripples (**e**, **f**). Interfering ripples with the coast-parallel straight-crested ripples increasingly modified by orthogonal waves seaward from (**g**) to (**i**); ballpoint pen is 14 cm long and pointing toward the sea. Increasing wave reworking on the muddy flat to produce small separate erosion holes (**j**), broaden and connect the holes, turn muddy deposited layer into patches (**k**), and develop the embryonic dunes (**i**) and well-developed dunes topped by small ripples on the marsh-frontal zone (**m**) and the middle and lower intertidal flats (**n**, **o**)

bedforms (Fig. 9.25a–f). Interfering ripples are also frequently present, with the modification magnitude of one group of wave ripples by another increasing seaward from the upper to the lower bare flats (Fig. 9.25g–i). Accumulation of these small rippled beds tends to produce lenticular and wavy bedding, which is commonly seen on the tile sedimentation and along the erosion cliff of remnant muddy patches (Fig. 9.26). On the fortnightly (one neap-spring-neap cycle) tiles, 3–6-cm-thick deposits were not found to have direct link with neap-spring cycles in terms of lamina number and thickness variation (Fig. 9.26e–h), even the extrapolated sediment rate reaching up to 72–144 cm year^{-1}. It is reasonably linked with the formation of rippled laminae, lenticular and wavy bedding by waves or combined wave and tide flows instead of purely tides (usually known as tidal bedding, Reineck and Singh 1980), and thicker and sandier laminae represent higher energy events of larger waves or the combined flows of larger waves and higher tides instead of purely higher tides.

The muddy open-coast tidal flats can temporarily shift into sandy flats during the storm conditions, developing erosion features, dunes, and storm-generated bedding. On the Chongming Eastern Flat, erosion by rising storm waves starts at discrete points, and the erosion holes gradually expand to unite each other until there are only a few isolated muddy patches on the sandier deflated flats, following with the bedforms growing from small ripples into large dunes (Fig. 9.25j–o). Storm decaying initiates to deposit first a sandy lag layer with abundant shell debris and mud pebbles over the erosion surfaces, and follows with a thinning- and fining-upward succession, that both grain size and thickness of sandy laminae decrease upward, gradually returning the normal tidal-flat thinly interlayered deposition (Fig. 9.27). During a storm season, previous storm deposits tend to be reformed by the following storms, producing a single amalgamated storm-deposited succession. For example, units b, c, and d were deposited by typhoons Neil, Olga, and Paul, respectively, over the typhoon season in 1999; the former two units were the remnants by subsequent storm reworking (Fig. 9.27; Fan et al. 2004a). A thinning- and fining-upward succession is therefore a storm-related small succession consisting of a lower half of sand-dominated layers (SDLs, storm deposition) and an upper half of mud-dominated layers (MDLs, after storm, normal tidal-flat deposition). The small storm-related succession usually has approximately half-and-half ratios of SDLs and MDLs, commonly seen in the Changjiang Delta where a high sedimentation rate is generally achievable with several centimeters per year (Li et al. 2000; Fan and Li 2002).

The deposits of sandy open-coast tidal flats may consist predominantly of high-energy storm deposits with volumetrically minor amounts of tidally induced lamination (Yang et al. 2005, 2008a). The Baeksu sandy flats in southwest Korea were finely explored by Yang et al. (2005, 2006, 2008a, b). In summer lower energy season, the flats are commonly veneered by mud layer of several centimeters thick, consisting of thinly interbedded to interlaminated sand and mud. The mud layer can be partitioned into two to three smaller-scale upward-fining successions, interpreted as weak summer storm deposits (Fig. 9.28). In winter higher energy season, the flats turn into sandy substrate topped by dune field. The deposits contain extensive wave-generated parallel lamination and short-wavelength (0.3–2 m) hummocky cross-stratification (HCS, Table 9.3, Fig. 9.28), highly similar with those of shoreface facies. Yang et al. (2008a) suggested that the storm deposits on the sandy open-coast tidal flats contained evidence of tidal modulation of storm processes, in which single storm layer is composed of three distinctive rippled intervals: (1) landward-dipping, ripple cross-lamination at the base, produced by combined flows during rising tide; (2) symmetrical buildup of wave-ripple cross-lamination in the middle, formed by oscillatory wave motion at high tide when currents are weak; and (3) seaward-dipping, ripple cross-lamination at the top, deposited by combined flows again during falling tide. Because of limited input of fine sediments and lower sedimentation rate, the summer muddy laminated successions were less preserved, leading to the strata mainly composed of winter sandy deposits with typical HCS (Yang et al. 2005).

It is generally concluded that deposits of open-coast tidal flats are characterized by abundant sedimentary features generated by waves or combined flows, making them distinctly different from the sheltered tidal flats. The features of tidal modulation of wave action distinguish them from the shoreface facies. The most extensive cyclic successions are strongly asymmetric with only the upper half of a fining-upward cycle (Figs. 9.27 and 9.28), denoting the annual depositions of seasonal prevalent large waves alternating with small waves (Baker et al. 1995; Li et al. 2000; Dalrymple et al. 2003; Fan et al. 2004a; Yang et al. 2005).

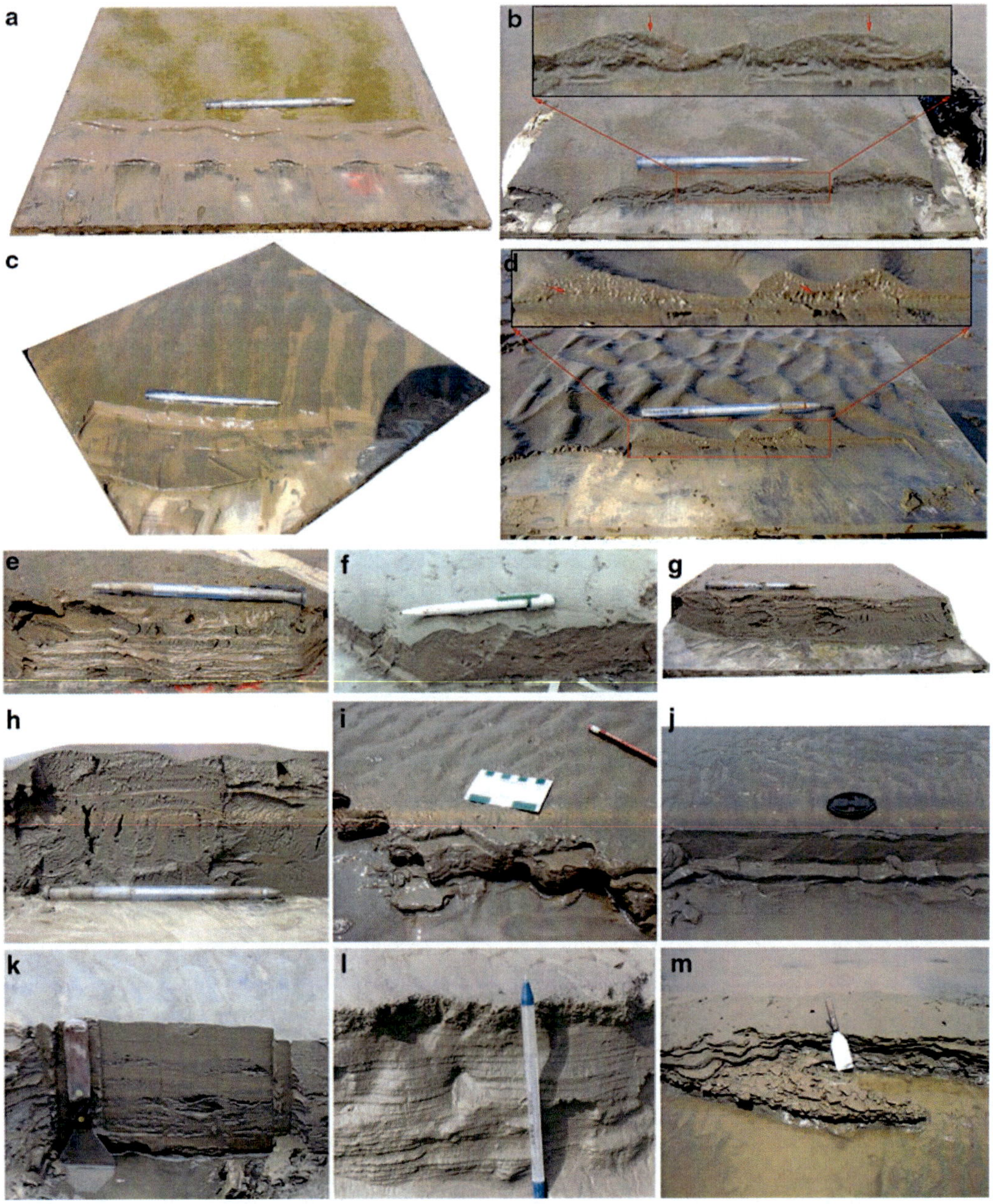

Fig. 9.26 Sedimentary beddings and internal structures of the tidal-flat deposits in the Changjiang Delta. Semidiurnal tile observations showing (**a**) one sand-mud couplets pair with small symmetrical wave ripples capped by a thick frozen fluid-mud layer and a thin algal mat on the surface, and (**b**) two sand-mud couplets with the ebb-current ripples laid on the stoss-side of the underlying flood-current ripples. Diurnal tile observations exhibiting two thin sand laminae separated by a thick mud layer (**c**), or by a hiatus surface with different color and composition (**d**). On the fortnightly observation tiles, 3–6-cm-thick deposits consisting of ≤7 sand-mud couplets, exhibiting thinly interlayered bedding (**e**); two thin sand layers sandwiched by a thick mud layer with a few thin sand lenses (**f**); fining-upward successions with massive sand layer at the bottom and parallel to wavy beddings on the *top* (**g**), or developing load structures with the lower heterolithic bedding (**h**). The erosion cliff of the mud patches showing finely laminated bedding (**i**), fining-upward succession with parallel to wavy bedding (**k**, similar with (**g**) and (**h**) on the fortnightly ties), and thick massive mud layer capped by thin sand layer (**j**, like those of (**f**)). The underlying sandy deposits exposed by deep erosion containing parallel bedding (**l**) and mud-pebble concentration layer (**m**). Ballpoint pen always pointed to the sea

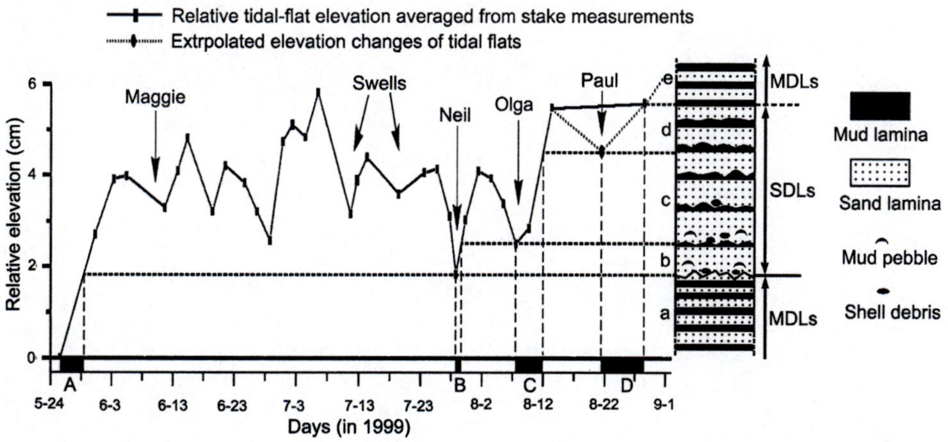

Fig. 9.27 Genesis interpretation of a small succession consisting of sand-dominated layers (*SDLs*) and mud-dominated layers (*MDLs*) using elevation-monitoring data at the Nanhui Mudbank, the Changjiang Delta. The net sediment increments (*a–d*) in the right column were deposited at the time intervals (*A–D*) in the bottom column. Maggie, Neil, Olga, and Paul are names for four typhoons having exerted great impact on the study area in 1999 (After Fan et al. 2002, 2004b)

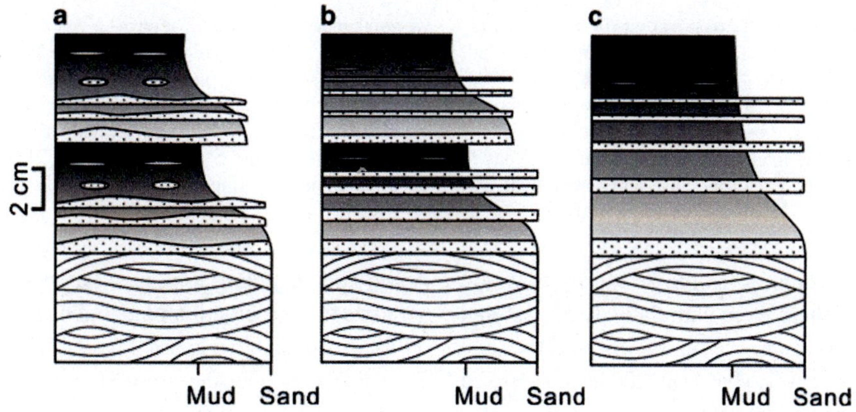

Fig. 9.28 Sketch drawing of the superposition of the winter large-wave coarse deposits and the summer weak-wave fine deposits. The coarse-bedding package characteristic of hummocky cross-stratification (*HCS*) was formed during the waning stage of bigger storm. The fine-bedding package was composed of one or two smaller upward-fining successions, each succession presumably being formed during the waning stage of smaller storm and during the immediate post-storm period in summer (After Yang et al. 2005)

9.6 Preservation Potential

Preservation potential of the deposits is general very low on the open-coast tidal flats due to frequent reformation by waves, especially storms. It has be recently studied by the preservation potential of individual couplets, ratios of gross and net sedimentation rates, and numerical modeling (Li et al. 2000; Fan 2001; Fan and Li 2002; Gao 2009).

Sand-mud couplets are basic sedimentary units of tidal-flat deposits, and their formation and preservation potential have been extensively discussed by a series of field observations in the Changjiang Delta (Li et al. 2000; Fan 2001; Fan et al. 2001, 2002, 2004b; Fan and Li 2002). The method is simply based on the comparison study of number and thickness of couplets using the tile observation. A group of tiles were fixed closely on the tidal-flat surface and regularly visited by different time intervals spanning from a semidiurnal tide to a month or longer (Fig. 9.29). Two couplets were sometimes observed to accumulate on the semidiurnal tile, denoting both flooding and ebbing flows potential

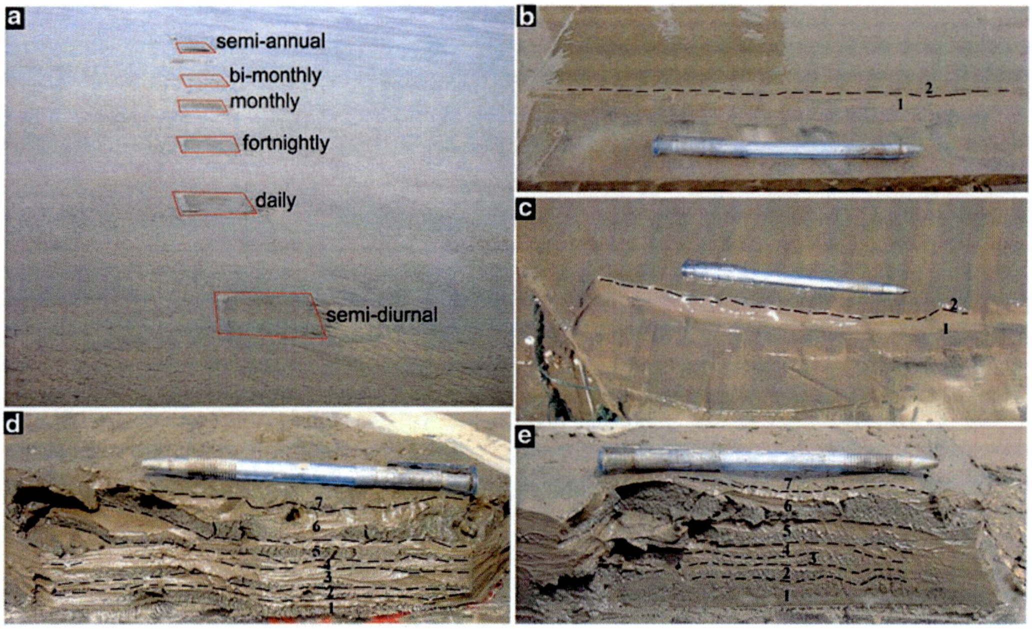

Fig. 9.29 Sedimentation on the observation tiles (40-cm wide and 40-cm long) visited by regular intervals spanning from a half day to a month or longer. The field monitoring experiment was carried in 2002 with six tiles fixed on the middle intertidal-flat surface at the same time, and each tile was scheduled to visit and redeploy at different time intervals spanning from a half day to six months (**a**). However, the bi-monthly and semiannual tiles were taken away by waves after 6–7-week deployment, so no record was available from these two tiles. Photos (**b–e**) showed examples of sedimentation on the tiles. There are two sand-mud couplets on the semidiurnal and daily tiles (**b** and **c**) and seven couplets on the fortnightly and monthly tiles (**d** and **e**). The dashed line marks the couplet boundaries. The ballpoint pen is 14 cm long, and pointing toward the sea (After Fan et al. 2004a)

to form their own couplets. In the period from May 4 to June 4 in 2002, there were cumulatively 55, 13, and 7 couplets observed on the daily, fortnightly, and monthly tiles, respectively (Fan et al. 2004a). Compared to the maximum of 120 couplets potentially deposited by 60 semidiurnal tides in the period, the preservation rates of couplets were 45.8%, 10.8%, and 5.8%, respectively, for the daily, fortnightly, and monthly intervals. It is indicated that the preservation rate of couplets decreases rapidly as time intervals increase (Fan et al. 2004a). The same conclusion has been achieved from the tile observation experiment in 1999 at the Nanhui Mudbank (Table 9.4, Fan and Li 2002). Based on the core study, Li et al. (2000) further extrapolated that the preservation rate of couplets could be lowered to 0.2% over a 100-year scale. Also, the sedimentation rates calculated over different time intervals decrease exponentially as the time intervals increase (Table 9.4, Fan et al. 2001).

The problem of the low spatial resolution of the preservation potential studies by in situ measurements can be effectively solved by numeric modeling. A forward modeling approach has been employed to simulate the preservation potential of tidal-flat deposits on the North Jiangsu coast (Fig. 9.11, Gao 2009). The results showed that the preservation potential was the highest over the upper part of the intertidal flat and the lower part of the subtidal flat, and the lowest near the mean sea level and the mean low water springs (Fig. 9.30). Also, the preservation potential decreased as the tidal flats prograded seaward. The slope of tidal flats has significant influence on the preservation potential, in that the minimum value is approximately four times greater for the slope scenario of $\tan \beta = 0.5 \times 10^{-3}$ than that of $\tan \beta = 1.0 \times 10^{-3}$ (Gao 2009). The simulation result of the preservation potential is comparable with those from field experiments in the Changjiang Delta (Li et al. 2000; Fan 2001; Fan and Li 2002).

The lower preservation rates indicate that the intertidal-flat deposition is riddled with various diastems. The incompleteness of tidal-flat deposition has been explored in detail along the 4-m-thick intertidal-flat deposition in the Changjiang Delta (Fan et al. 2002). The 4-m-thick strata were extrapolated to deposit in

Table 9.4 Comparison of couplet number and thickness on the daily and fortnightly tiles for the field observation on the Nanhui Mudbank in the Changjiang Delta during the period May 24 to July 8 in 1999 (After Fan and Li 2002)

	Daily tiles		Fortnightly tiles		Preservation rates (%)	
	1	3	2	4	2 vs. 1	4 vs. 3
Cumulative thickness of couplets (mm)	303.2	335.4	65.0	92.2	21.4	27.5
Cumulative number of couplets	81	77	16	16	19.8	20.8
Average couplet thickness (mm)	3.7	4.4	4.1	5.8		
Sedimentation rate (cm year^{-1})	245.9	272.0	52.7	74.8		

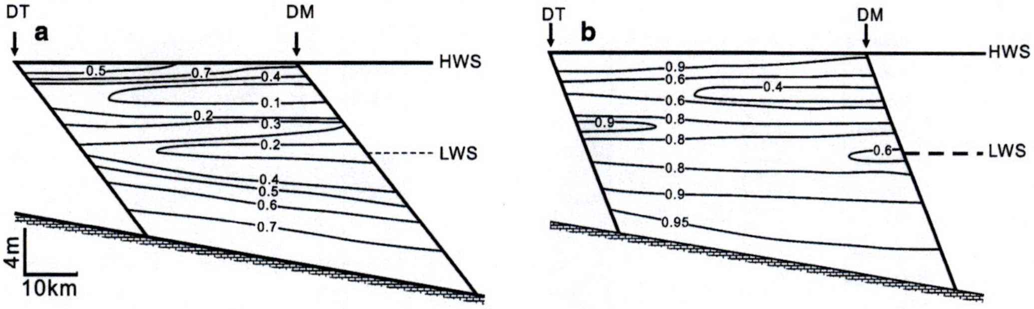

Fig. 9.30 Modeling output of the distribution pattern of the preservation potential over the transect DT-DM in Northern Jiangsu coast for different bed slopes: (**a**) tan $\beta = 0.5 \times 10^{-3}$; (**b**) tan $\beta = 1.0 \times 10^{-3}$ (After Gao 2009)

roughly 96 years, and were on average composed of 32 small successions (Fig. 9.31). The small succession is a storm-related thinning- and fining-upward succession, consisting of a couple of SDLs and MDLs, which are a group of sand- or mud-dominated layers, respectively, generated by storms or after the storms. Single small succession generally represents annual cycle of storm season and non-storm season deposition (Fig. 9.27), so only one third of 96-year depositional intervals contain their own deposits, and the diastems occupy the other two-third time intervals. Assuming the even distribution of the diastems, their temporal distribution is shown in Fig. 9.34a. Over 1-year interval, the temporal distribution of diastems looks like that of Fig. 9.34b, where the typhoon season and the following 1–2 months are marked in black with deposition of SDLs and MDLs. The latter can be deposited in a few weeks after the typhoon season on the basis of modern sedimentation rates, leaving the other several months blank or diastems during the non-typhoon season. Single SDLs can be formed by amalgamation of several storm deposits over a typhoon season as that shown in Fig. 9.27, and only the time intervals B, C, and D had the corresponding deposits preserved in the strata, leaving other time intervals blank (Fig. 9.27). The temporal distributions of diastems over the fortnightly and daily intervals were extrapolated from the tile observations. Meanly five to six sand-mud couplets on the fortnightly tiles and two sand-mud couplets on the daily tiles denote that 11 out of 14 days and 2 out of 4 semidiurnal tides were blank or without deposits over the fortnightly and daily intervals, respectively (Fig. 9.31c, d). It is therefore concluded that a sedimentary unit complete on a longer time scale actually contains many diastems on a shorter time scale. Diastems in the tidal-flat deposits can be generated by erosion of storm waves, and also by small waves and tides. They are discernible or non-discernible, representing the missing sediment intervals from a few hours to several years or longer (Fan et al. 2002).

9.7 Sedimentary Facies and Successions

9.7.1 Holocene Examples

9.7.1.1 Progradational Open-Coast Tidal-Flat Successions

The progradational open-coast tidal flats generally have a convex-up profile with marked fining-landward intertidal zonations. The regressive vertical stratigraphic

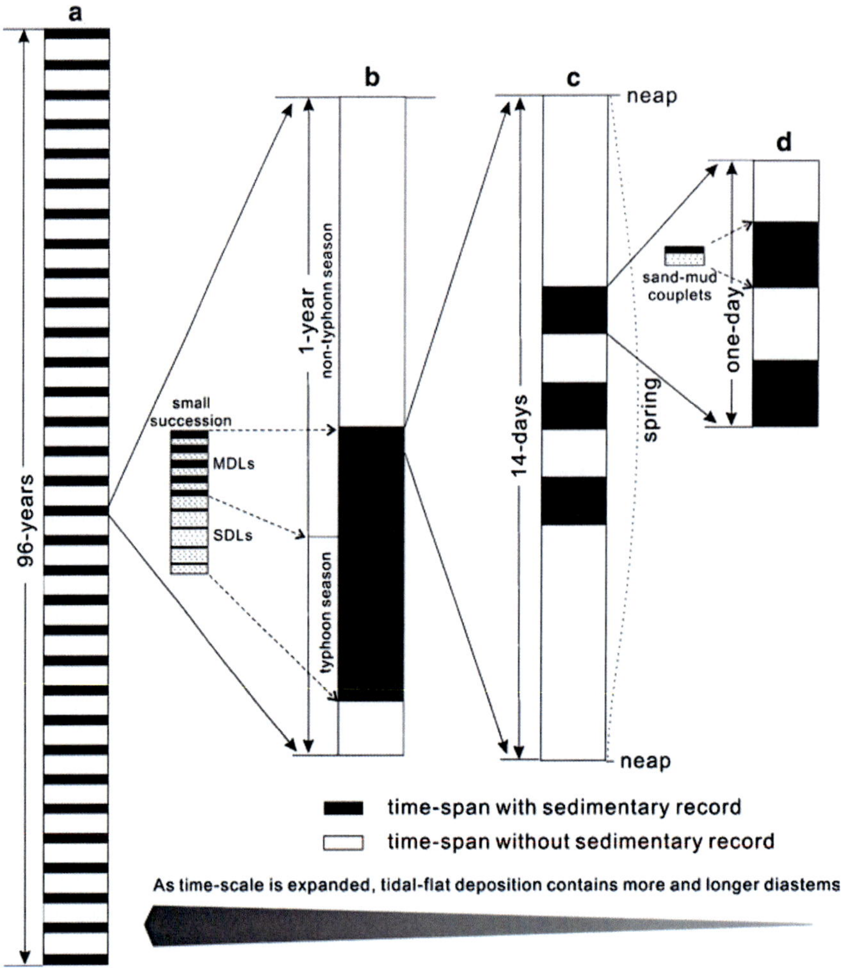

Fig. 9.31 The stratigraphic completeness of an open-coast intertidal-flat succession examined at different time scales, showing that a complete stratigraphic unit at a longer term was examined to fill with diastems at a shorter time scale. The intertidal-flat succession is roughly 4 m thick, deposited in ~96 years in the Changjiang Delta. The thickness of individual small successions and sand-mud couplets is generally a few centimeters to decimeters, and a few millimeters to centimeters (After Fan et al. 2002)

succession reflects this trend of variations, grading conformably upward from the lower intertidal sand, through the middle intertidal mud-sand mixture, to the upper intertidal mud (Fig. 9.32, Semeniuk 1981; Li and Li 1982; Li et al. 1992). The (shelly) sand facies generally consists of thick wave-rippled or massive sand layers with clay lenses or seams, and thick sand-dominated layers (SDLs, Fig. 9.26) of storm generation, developing cross-stratified bedding of slight bioturbation. The heterolithic mud and sand facies is characterized by abundant wavy bedding with moderate bioturbation and the small fining-upward successions consisting of alternative sand- and mud-dominated layers in roughly equal thickness. The mud facies grades upward from the upper bare to vegetated intertidal flats and usually continuing toward the supratidal flats, lithologically from finely laminated silt and clay with silty parallel to wavy laminae of a few grains to millimeters to massive mud with rippled sand lenses, scattered pigmentation mottles, and abundant in situ salt-marsh-plant rootlets or mangrove stumps. The laminated to bioturbated mottled mud may be interbedded with lensed beds of muddy and shelly sand of several decimeters thick or more, which are swash bar/chenier ridge deposits produced by storm waves (Semeniuk 1981; Li et al. 1992).

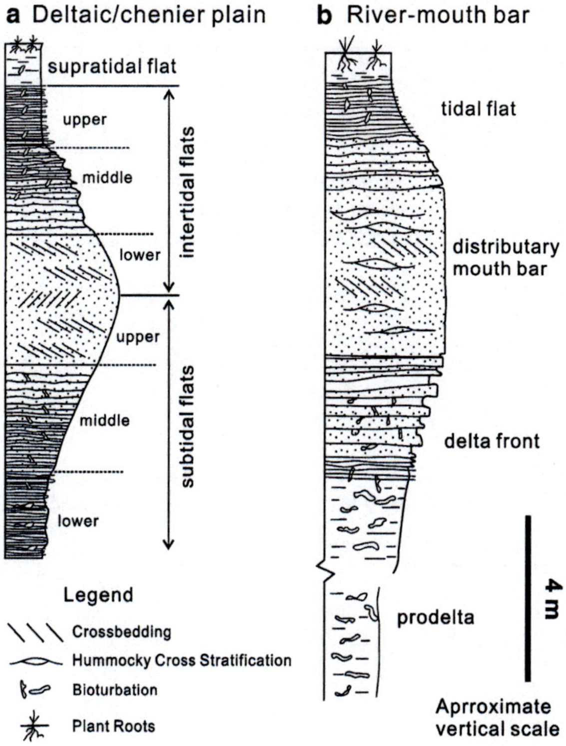

Fig. 9.32 Schematic models showing two most common progradational tidal-flat successions on the open-coast environment (After Li and Li 1982; Li et al. 1992; Dalrymple et al. 2003)

The progradational fining-upward intertidal-flat succession commonly continues with a coarsening-upward succession toward the subtidal flats for the muddy open-coast tidal flats with a gently smooth intertidal-subtidal profile (Fig. 9.32a, Li et al. 1992). It may also be underlain by the thick sand deposits when the muddy intertidal flats prograde over the subtidal sand-ridge systems like those on the central North Jiangsu coast (Ren 1985) and the distributary-mouth bars in the deltas (Fig. 9.32b; Dalrymple et al. 2003), or the sandy intertidal flats prograde over the subtidal ridge-runnel complex (Reineck and Cheng 1978; Semeniuk 1981).

9.7.1.2 Retrogradational Open-Coast Tidal-Flat Successions

The retrogradational sandy open-coast tidal flats commonly develop along the coast receiving slight sediment input, having a concave-up profile with general coarsening-landward trend of intertidal sediment distribution except the inner parts behind the swash bars/ridges (Yang et al. 2005, 2008a). They produce a transgressive coarsening-upward succession in response to Holocene sea-level rise (Fig. 9.33, Kim et al. 1999; Chang and Choi 2001; Lim et al. 2004; Yang et al. 2006b). The transgressive succession, usually unconformally underlain by either bedrock or pre-Holocene deposits which are commonly stiff mud (Kim et al. 1999), consists of three or four depositional units including the facies of salt-marsh, mud-, mixed-, and sand flats in an ascending order (units B_1–B_4 in Fig. 9.33). The lowermost unit (B_1), early Holocene salt-marsh deposit, is composed of intensively bioturbated dark-gray mud with minor sand-rippled lamination, rich in organic matter, and small plant roots. Unit B_2, mudflat deposit, is characterized by intensely bioturbated dark-gray mud with sporadic rhythmic lamination, grading conformably both downward and upward into units B_1 and B_3. Unit B_3, mixed-flat deposit, is composed of dark-gray, moderately bioturbated to laminated sandy silt or silty sand with relative abundance of shell fragments. The uppermost unit B_4, sand-flat deposit and unconformably overlying unit B_3, is characterized by greenish to olive gray, very fine to fine sand with slight bioturbation and relatively well-preserved lamination, typically developing storm-generated small fining-upward successions with hummocky cross-stratification (Fig. 9.28).

The whole Holocene coarsening-upward succession was previously interpreted as the result of the continual retrogradation of the non-barred tidal flats (Kim et al. 1999; Chang and Choi 2001; Lim et al. 2004), whereas the interpretation was questioned by Yang et al. (2006b). They noted that there was generally relative abundance of tidal creek/channel deposits (Fig. 9.33c) and absence of storm- or wave-generated structures in units B_1–B_3. These three units were consequently interpreted to be deposited in a back-barrier tidal-flat setting, because tidal channels are commonly rare on the modern open-coast tidal flats of the study area. As the transgression continued, the former barriers migrated landward over the back-barrier tidal flats, which thereafter changed into open-coast tidal flats (Fig. 9.33c). They become a wave-dominated setting with a volumetric majority of storm-generated beds (Yang et al. 2006b).

9.7.1.3 Estuarine-Deltaic Channel Filling Successions with Tidal Rhythms

A few examples have been reported to have the neap-spring cycles in recent and Holocene estuarine-deltaic channels, which lie between the truly open-coast and

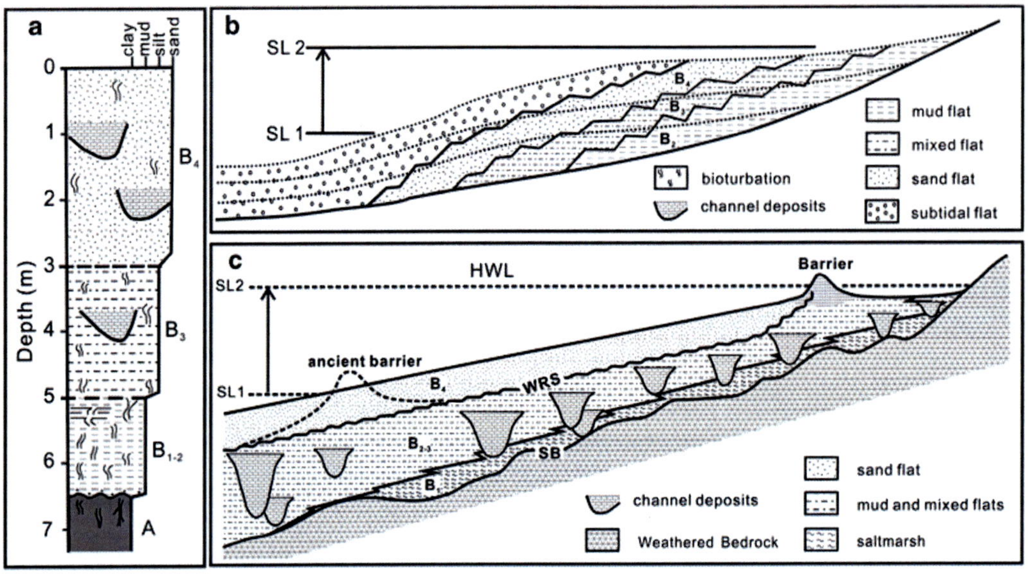

Fig. 9.33 (**a**) Schematic drawing of a retrogradational coarsening-upward tidal-flat succession, and (**b**, **c**) cross-shore profiles of vertical stacked tidal-flat sub-facies in response to sea-level rise on an open-coast setting with limited sediment input (After Kim et al. 1999; Yang et al. 2006b)

highly sheltered settings (Fan 2001; Hori et al. 2001; Dalrymple et al. 2003). In the Fly Delta, the recent vertical succession of tidal bar facies was found at a few core sections to exhibit cyclic changes in lithology. Each single cycle contains a lower half coarsening-upward succession and an upper half fining-upward succession, counting 26 sand laminae which are close to 28 tides for a semidiurnal tidal setting (Fig. 7a in Dalrymple et al. 2003). These features, together with the thick-thin alternation of adjacent sand laminae for the spring tidal deposits, have been undoubtedly interpreted to be the tide-generated neap-spring cycles and the diurnal inequality, respectively (Dalrymple et al. 2003).

Sand-mud couplets with cyclical changes in lamina thickness are common features in the early Holocene estuarine facies of the Changjiang Delta (Fan 2001; Hori et al. 2001). The core section of 47.12–48.49-m depth in the borehole CM-97 contains four complete neap-spring cycles (Fig. 9.34). The cycles are more clearly shown by variations in sand-laminae thickness (not including the very thin sand laminae within the mud couplets) through using 5-point adjacent averaging smoothing method. Discrete Fourier analysis of the smoothed data shows an average peak period at 28.4 laminae (25.2–31.5), matching well with 28 semi-diurnal tides within a neap-spring cycle. Mud couplets and the superposition of current ripples with opposite foreset dipping directions are other good indicators of tidal origin. It is therefore hypothesized that the estuarine and deltaic-distributary channels potentially accommodate some rapid filling successions containing the neap-spring cycles.

9.7.2 Ancient Examples

Ancient tidalites have been extensively studied in the last three decades, typically those registering tidal cycle signals. Due to an excessive passion for periodic cycles, the dramatic growing publications are biased toward the cyclic tidal rhythmites with the neap-spring cycles. Noncyclic rhythmites have been greatly neglected even though they may contain the seasonal wave-climate cycles. Only a few ancient tidal rhythmites have been undoubtedly interpreted as the open-coast tidal-flat environments (Klein 1970; Fan et al. 2004b).

9.7.2.1 Tonglu (Late Ordovician)

Tonglu tidalites of the Late Ordovician age are well outcropped along a roadcut near Tonglu County, Zhejiang Province, east-central China (Fan et al. 2004b). They are the uppermost member subdivision

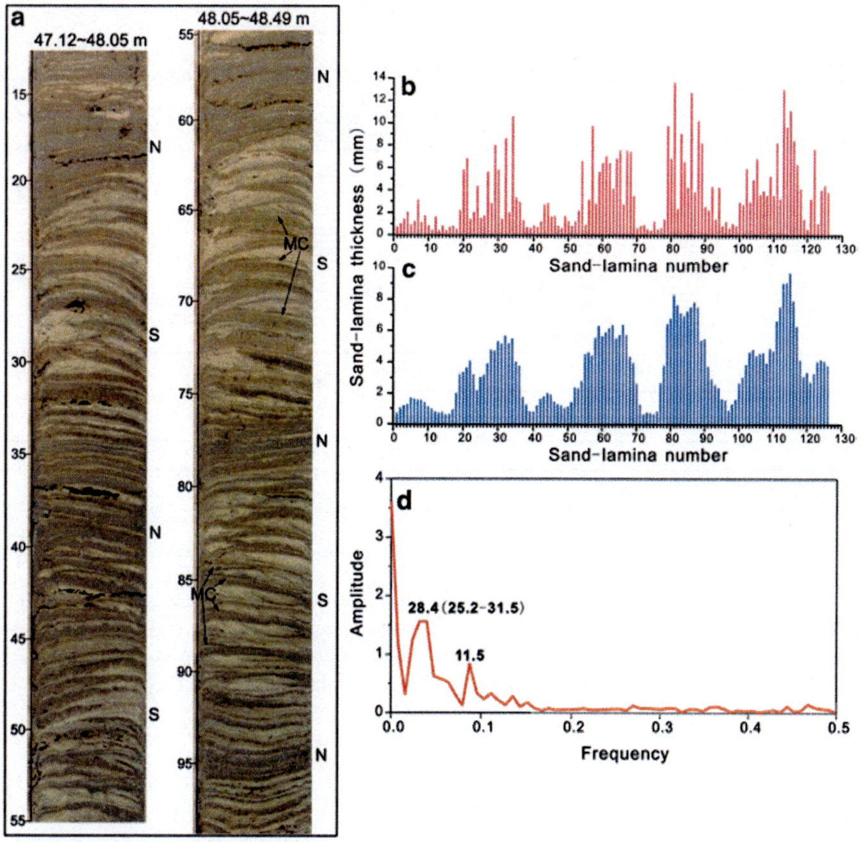

Fig. 9.34 Cyclic variations in sand-lamina thickness of the estuarine facies in the Changjiang Delta (depth of 47.12–48.49 cm along the borehole CM-97) deciphering neap-spring cycles. (**a**) Core photos; (**b** and **c**) plots of original data and smoothed data (using 5-point adjacent averaging method) of lamina thickness over lamina number; (**d**) FFT amplitude-frequency plot of the lamina-thickness data showing two major peak periods at 28.4 and 11.5 laminae. *N*, *S*, and *MC* in (**a**) are shortened for neap tide, spring tide, and mud couplets, respectively (After Fan 2001, photos courtesy of Yoshiki Saito)

of the Wenchang Formation, a shallowing-upward progradational succession from shallow marine to open coastal settings (Fig. 9.35). Tonglu tidalites exhibit three orders of periodicities in terms of sandstone and mudstone layer thickness. Millimeter- to centimeter-thick alternations of sandstone and mudstone laminae were ascribed to be deposited by single tidal cycles. Centimeter- to decimeter-thick alternations of sand-dominated layers (SDLs) and mud-dominated layers (MDLs) were interpreted to be formed by seasonal alternations of storm- and calm-wave climates. The storm-genesis interpretation of each single SDLs was convincingly based on the abundance of wave and storm action products, like intraformational mud pebbles, symmetrical wave ripples, and the asymmetrical small successions of thinning-upward trends which began with an erosion surface and overlain thick sandstone bed with abundant shell debris and mud pebbles, similar to modern storm-generated SDLs in the Changjiang Delta. The megacycle of several meters thick, composed of a lower half coarsening-upward succession and an upper half fining-upward succession, was interpreted as a vertical regressive succession produced by gradual shoaling from the lower subtidal zone to the upper intertidal zone with the coarsest and thickest sand layers at the middle, similar with that of modern open-coast tidal-flat depositional succession in Fig. 9.32a. Other evidence like general lack of tidal-channel filling deposits and abundance of wave-generated structures and small depositional successions also supports the open-coast tidal-flat environmental interpretation (Fan et al. 2004b).

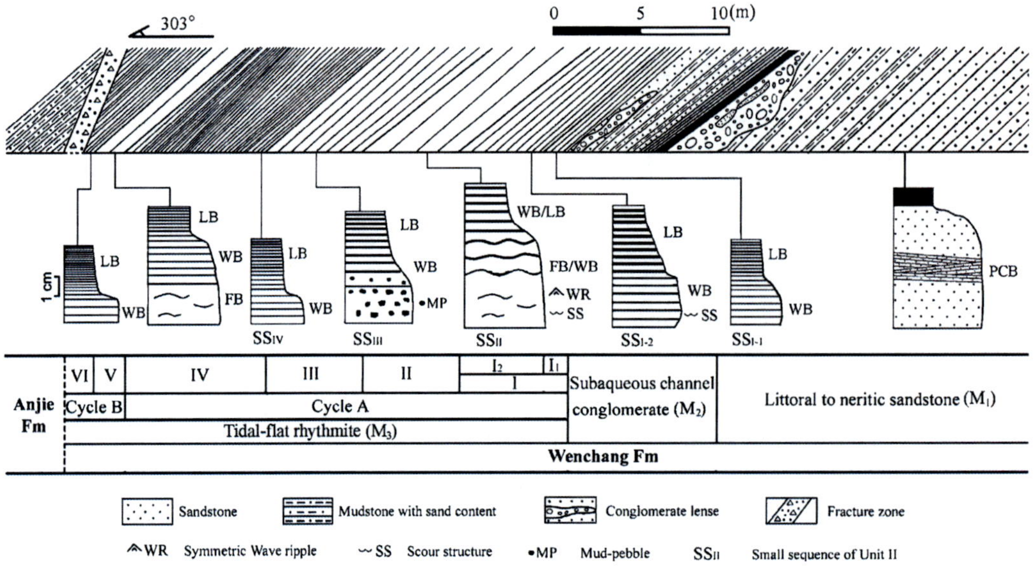

Fig. 9.35 Features of typical stratigraphic units compromising Tonglu tidal-flat successions, upper Wenchang Formation of Late Ordovician. The Tonglu tidal-flat successions consist of a complete cycle (A) from the subtidal facies to the upper intertidal facies and a half cycle (B) with upper part of intertidal facies associations. *PCB*, low-angle planar cross-bedding; *LB*, lenticular bedding; *WB*, wavy bedding; *FB*, flaser bedding; Anjie Fm, Anjie Formation of Early Silurian (After Fan et al. 2004b)

9.7.2.2 Islay (Late Proterozoic)

The Lower Fine-grained Quartzite of Middle Dalradian (Late Proterozoic) age in Islay, Scotland, consists of massive-bedded, cross-stratified, and rippled orthoquartzites (Facies 1), and siltstone and mudstone (Facies 2). The two facies are organized into a sharp-based, fining-upward succession with basal shallow subtidal sandstones, intertidal sand-flat or sand-bar sandstone, and high tidal-flat mudstones (Klein 1970). The succession is highly identical to that of back-barrier tidal-flat deposition (Klein 1985; Dalrymple 1992), except the general devoid of tidal-channel deposits. Comparison study prefers the modern analogue in the Wash of Eastern England with few presence of intertidal channels (Klein 1970), falling into the divisions of open-coast tidal flats.

9.7.2.3 Ramgundam (Middle Proterozoic)

The Ramgundam Sandstone of Middle Proterozoic age is well exposed along the Godavari Valley of south-central India (Chaudhuri and Howard 1985). It is composed of lens-shaped bodies of arkosic sandstone and interbedded sandstone and shale. The fining-upward succession devoid of tidal-channel deposits presumably represents a transgressive intertidal depositional succession building over linear-bar shoals frequently subjected to sufficient storm-induced wave activity (Chaudhuri and Howard 1985). Such facies interpretation sought no acceptable modern analogue at that time (Chaudhuri and Howard 1985), but is now comparable with the facies model of open-coast intertidal deposits building over the distributary-mouth or estuarine-channel bars (Fig. 9.32b).

9.7.2.4 Hazel Patch (Late Carboniferous)

The Hazel Patch sandstone of the Pennsylvanian age (Late Carboniferous) in eastern Kentucky (USA) has been highlighted for developing cyclic tidal rhythmites registering three orders of tidal cyclicities, including diurnal inequality, neap-spring cycle, and monthly tidal cycle (Greb and Archer 1995). Actually, cyclic tidal rhythmites are spatially limited in the lower part of a single major channel-filled succession. Within the channel fill, cyclic rhythmites grade upward into amalgamated rhythmites. The broad flats outside of the major channel are exclusively dominated by noncyclic rhythmites, containing small rhythmic successions similar to those of storm-generated successions on modern open-coast tidal flats (Figs. 9.27 and 9.28). The Hazel Patch sandstone is generally a fining-upward succession, consisting of the subtidal sand-bar deposits and the intertidal broad sand-flat deposits (both containing extensive structures of combined-flow and storm genesis) and high intertidal-flat deposits containing numerous small tidal-channel fills (Greb and Archer 1995). The succession is also considered to be deposited on an open-coast intertidal flat over the distributary-mouth or estuarine-channel bars.

9.8 Summary

Tidal flats occupy a large section of the world's unsheltered shoreline, especially along the coast receiving large volumes of terrigenous fine sediments from rivers that build up broad and gentle shelf deposits. Open-coast tidal flats have received increasing interest because of their importance in global environmental issues posed by rising sea level, decreased sediment fluxes linking to river damming, and increasing human usage, also providing a modern analogue for fossil facies interpretation.

Large tidal range favors but is not a prerequisite for tidal-flat development. Sediment supply and the magnitude of wave exposure are two key controlling factors of tidal-flat morphology and sedimentology. Open-coast tidal flats develop in wide environments, ranging from partly exposed embayments and estuaries to highly exposed deltas and coastal plains. The commonest open-coast tidal flats are principally composed of mud, extensively distributing along the tide-dominated mega-deltas and their adjacent chenier plains. Most of the world's largest river deltas are tide dominated or under significant tidal influence. Littoral currents carry the resuspended sediment from the deltas, downdrift for tens to hundreds of kilometers along the coast to nourish tidal flats. The longest stretches of open-coast tidal flats are of this type, including the muddy coast along the East China and the Guianas. Sandy open-coast tidal flats majorly develop in the open-mouth estuaries and the adjacent strand plains, where large tidal ranges usually occur owing to tide amplification by typical coastal morphology, like narrow and shallow straits or funnel-shaped estuaries.

Open-coast tidal flats bare some common features to distinguish them from other coastal environments. These features include: (1) developing broad and gentle flats without significant morphological break along the shore-normal profile, (2) fronting an open sea or ocean without barriers, (3) exposing to different magnitudes of wave action with clear seasonal variations from tide-dominated into wave-dominated morphodynamic conditions, (4) developing few tidal channels on the bare flats except complex tidal-creek systems in the adjacent salt-marsh land, (5) exhibiting clear intertidal zonations of a coarsening seaward trend, and (6) containing abundant combined-flow and wave-induced structures and bedding.

It is noteworthy that there is a significant difference between the muddy and sandy open-coast tidal flats. The muddy classification tends to have an accretional convex-up profile with the coarsest sediment near the mean lower water springs, and develop a cyclic progradational succession consisting of a lower half of subtidal coarsening-upward succession and an upper half of a fining-upward intertidal-to-supratidal succession. The sandy type usually has an erosional concave-up profile with the coarsest sediment near the mean high water, and develops a coarsening-upward retrogradational succession. Typical hummocky cross-stratification (HCS) of short wavelength can be common in the sandy open-coast tidal-flat deposits, whereas not present in muddy open-coast tidal-flat deposits. The vertical succession of sandy open-coast tidal flats generally has higher abundance of storm-generated beds volumetrically than that of muddy open-coast tidal flats. A spectrum of coastal morphodynamic settings is therefore reorganized to change from the wave-influenced, tide-dominated muddy open-coast tidal flats, the wave- and tide-dominated accretional sandy open-coast tidal flats, wave-dominated erosional sandy open-coast tidal flats, to wave-dominated tidal beaches.

The open-coast tidal-flat deposition is far from complete (or continuous) due to reworking by complex physical processes. The well-developed sand-mud couplets do not represent the continuous deposition of tidal cycles in the vertical stacking succession. The preservation potential of strata is very low in terms of preservation rates of couplets and a general decreasing trend of sedimentation rates over different time scales. So the basic stratigraphic tenet is highlighted that deposition by shorter-time cyclic processes (e.g., tide) is highly reworked by successive longer-time cyclic processes (e.g., seasonal alternations of wave climate). Rapid continuous deposition with the neap-spring cycles is generally exempted from the open-coast tidal flats, except the relatively protected setting like the estuarine or distributary channels.

Following the recent research advances in open-coast tidal flats, the classification of clastic coastal environments should be changed to account for this new knowledge. New efforts should undoubtedly be steered to build new facies models for newly proposed subdivisions and to clarify the inter-relationships among any two transitional facies. Open-coast tidal flats are shaped by the interactions of tides and waves instead of their separate action, so the modulation of waves by tide should be stressed in the roles of sediment dynamics and morphodynamics. A renaissance is highly expected using integrative available data on both descriptive and quantitative features for ancient tidal facies interpretation after the three decades of research on tidal cycles. Muddy coasts adjacent to river deltas are undergoing great impacts (e.g., coastal erosion, wetland degradation) from human activities and global change, so modern environmental issues should be included and studied in the geological and sedimentologic aspects.

References

Alexander CR, Nittrouer CA, DeMaster DJ et al (1991) Macrotidal mudflats of west Korea: a model for interpretation of intertidal deposits. J Sediment Pet 61:805–824

Allen JRL, Duffy MJ (1998) Temporal and spatial depositional patterns in the Severn Estuary, SW Britain: intertidal studies at spring-neap and seasonal scales, 1991–1993. Mar Geol 146:147–171

Allison MA, Lee MT (2004) Sediment exchange between Amazon mudbanks and shore-fringing mangroves in French Guiana. Mar Geol 208:169–190

Allison MA, Nittrouer CA, Faria LEC (1995a) Shoreline morphology downdrift of the Amazon river mouth. Mar Geol 125:373–392

Allison MA, Nittrouer CA, Kineke GC (1995b) Seasonal sediment storage on mudflats adjacent to the Amazon River. Mar Geol 125:303–328

Allison MA, Lee MT, Ogston AS (2000) Origin of mud banks along the northeast coast of South America. Mar Geol 163:241–256

Amos C (1995) Siliclastic tidal flats. In: Perillo GME (ed) Geomorphology and sedimentology of estuaries, Advancement in Sedimentology 53. Elsevier, Amsterdam, pp 273–306

Anderson FE, Black L, Watling LE et al (1981) A temporal and spatial study of mudflat erosion and deposition. J Sediment Pet 51:729–736

Archer AW (1998) Hierarchy of controls on cyclic rhythmite deposition: Carboniferous basins of eastern and mid-continental, U.S.A. In: Alexander CR, Davis RA Jr, Henry VJ (eds) Tidalites: processes and products, SEPM Special Publications 61. SEPM, Tulsa, pp 59–68

Archer AW, Hubbard MS (2003) Highest tides of the world. In: Chan MA, Archer AW (eds) Extreme depositional environments: mega and members in geologic time, Geological Society of America Special Publication 370. Geological Society of America, Boulder, pp 151–173

Augustinus PGEF (2004) The influence of trade winds on the coastal development of the Guianas at various scale levels. Mar Geol 209:145–151

Baker EK, Harris PT, Keene JB et al (1995) Patterns of sedimentation in the macrotidal Fly River delta, Papua New Guinea. In: Flemming BW, Bartholoma A (eds) Tidal signatures in modern and ancient sediments, International Association of Sedimentology Special Publication 24. Elsevier, New York, pp 193–211

Baltzer F, Allison M, Fromard F (2004) Material exchange between the continental shelf and mangrove-fringed coasts with special reference to the Amazon-Guianas coast. Mar Geol 208:115–126

Boersma JR, Terwindt JHJ (1981) Neap-spring tide sequences of intertidal shoal deposits in a mesotidal estuary. Sedimentology 28:51–170

Boggs S Jr (2005) Principle of sedimentology and stratigraphy, 4th edn. Prentice Hall, New Jersey

Boguchwal LA, Southard JB (1990) Bed configurations in steady unidirectional water flows. Part 2. Synthesis of flume data. J Sediment Pet 60:658–679

Byrne JV, Jeroy DO, Riley CM (1959) The chenier plain and its stratigraphy, southwestern Louisiana. Trans Gulf Coast Assoc Geol Soc 9:237–260

Chang JH, Choi JY (2001) Tidal-flat sequence controlled by Holocene sea-level rise in Gomso Bay, west coast of Korea. Estuar Coast Shelf Sci 52:391–399

Chaudhuri A, Howard JD (1985) Ramgundam Sandstone: a Middle Proterozoic shoal-bar sequence. J Sediment Pet 55:392–397

Christie MC, Dyer KR, Turner P (1999) Sediment flux and bed level measurements from a macro tidal mudflat. Estuar Coast Shelf Sci 49:667–688

Coughenour CL, Archer AW, Lacovara KJ (2009) Tides, tidalites, and secular changes in the Earth-Moon system. Earth Sci Rev 97:59–79

Dalrymple RW (1992) Tidal depositional systems. In: Walker RG, James NP (eds) Facies models: response to sea level change, 2nd edn. Geological Association of Canada, St. John's, pp 195–218

Dalrymple RW (2010) Tidal depositional systems. In: James NP, Dalrymple RW (eds) Facies models 4, 2nd edn. Geological Association of Canada, St. John's, pp 195–218

Dalrymple RW, Makino Y, Zaitlin BA (1991) Temporal and spatial patterns of rhythmite deposition on mud flats in the macrotidal Cobequid Bay-Salmon River estuary, Bay of Fundy, Canada. In: Smith DG, Reinson GE, Zaitlin BA et al (eds) Clastic tidal sedimentology, Canadian Society of Petroleum Geologists Memoir 16. Canadian Society of Petroleum Geologists, Calgary, pp 137–160

Dalrymple RW, Baker EK, Harris PT et al (2003) Sedimentology and stratigraphy of a tide-dominated, foreland-basin delta (Fly River, Papua New Guinea). In: Sidi FH, Nummedal D, Imbert P et al (eds) Tropical deltas of southeast Asia-sedimentology, stratigraphy, and petroleum geology, SEPM Special Publication 76. SEPM, Tulsa, pp 147–173

Dalrymple RW, Yang BC, Chun SS (2006) Sedimentation on a wave-dominated, open-coast tidal flat, south-western Korea: summer tidal flat-winter shoreface – reply. Sedimentology 53:693–696
Davies JL (1972) Geographical variation in coastal development. Hafner Publishing, New York
Davis RA Jr, Hayes MO (1984) What is a wave-dominated coast? Mar Geol 60:313–329
Davis RA Jr, Alexander CR, Henry VJ (1998) Tidal sedimentology: historical background and current contributions. In: Alexander CR, Davis RA Jr, Henry VJ (eds) Tidalites: processes and products, SEPM Special Publication 61. SEPM, Tulsa, pp 1–4
Deloffre J, Verney R, Lafite R et al (2007) Sedimentation on intertidal mudflats in the lower part of macrotidal estuaries: sedimentation rhythms and their preservation. Mar Geol 241:19–32
ECCE (Editorial Committee for Chinese Embayments) (1992) Chinese embayments (Part V): Shanghai and North Zhejiang. China Ocean Press, Beijing (in Chinese)
Eisma D (1998) Intertidal deposits: river mouths, tidal flats, and coastal lagoons. CRC Press, New York
Eisma D, Augustinus PGEF, Alexander CA (1991) Recent and subrecent changes in the dispersal of Amazon mud. Neth J Sea Res 28:181–192
Fan DD (2001) Formation and preservation of rhythmic Deposition on the mudflats and quantitative analyses on diastems. PhD thesis, Tongji University, Shanghai (in Chinese with an English abstract)
Fan DD, Li CX (2002) Rhythmic deposition on mudflats in the mesotidal Changjiang estuary, China. J Sediment Res 72:543–551
Fan DD, Li CX, Chen MF et al (2001) Preservation potential of individual couplet and deposition rates on mudflats in the Changjiang Estuary. Sci China Ser B 44(supp):33–3
Fan DD, Li CX, Archer AW et al (2002) Temporal distribution of diastems in deposits of an open-coast intertidal flat with high suspended sediment concentrations. Sediment Geol 186:211–228
Fan DD, Li CX, Wang DJ et al (2004a) Morphology and sedimentation on open-coast intertidal flats of the Changjiang Delta, China. J Coast Res 43(Spec Issue):23–35
Fan DD, Li CX, Wang P (2004b) Influences of storm erosion and deposition on rhythmites of the Upper Wenchang Formation (Upper Ordovician) around Tonglu, Zhejiang Province, China. J Sediment Res 74:527–53
Fan DD, Guo YX, Li CX et al (2005) Grain-size distributions and their applications on Andong intertidal facies analyses in Hangzhou Bay. J Tongji Univ Nat Sci 33:687–691 (in Chinese with an English abstract)
Fan DD, Guo YX, Wang P (2006) Cross-shore variations in morphodynamic processes of an open-coast mudflat in the Changjiang Delta: with an emphasis on storm impacts. Cont Shelf Res 26:517–538
Frey RW, Howard JD, Han SJ et al (1989) Sediments and sedimentary sequences on a modern macrotidal flat, Inchon, Korea. J Sediment Petrol 59:28–44
Froidefond JM, Pujos M, Andre X (1988) Migration of mudbanks and changing coastline in French Guiana. Mar Geol 84:19–30
Gao S (2009) Modeling the preservation potential of tidal flat sedimentary records, Jiangsu coast, eastern China. Cont Shelf Res 29:1927–1936
Ginsburg RN (1975) Tidal deposits, a casebook of recent examples and fossil counterparts. Springer, Berlin
Greb SF, Archer AW (1995) Rhythmic sedimentation in a mixed tide and wave deposit, Hazel Patch sandstone (Pennsylvanian), eastern Kentucky coal field. J Sediment Res 65:96–106
Green MO, Black KP, Amos CL (1997) Control of estuarine sediment dynamics by interactions between currents and waves at several scales. Mar Geol 114:97–116
Hale PB, McCann SB (1982) Rhythmic topography in a mesotidal, low-wave-energy environment. J Sediment Pet 52:415–429
Harris PT, Baker EK, Cole AR et al (1993) A preliminary study of sedimentation in the tidally dominated Fly River Delta, Gulf of Papua. Cont Shelf Res 13:441–472
Hori K, Saito Y, Zhao QH et al (2001) Sedimentary facies of the tide-dominated paleo-Changjiang (Yangtze) estuary during the last transgression. Mar Geol 177:331–351
Huo M, Fan DD, Lu Q, et al (2010) Decadal variations in the erosion/deposition pattern of the Nanhui Mudbank and their mechanism in the Changjiang Delta. Acta Oceanologica Sinica 32(5):41–51
Kim BO (2003) Tidal modulation of storm waves on a macrotidal flat in the Yellow Sea. Estuar Coast Shelf Sci 57:411–420
Kim YH, Lee HJ, Chun SS et al (1999) Holocene transgressive stratigraphy of a macrotidal flat in the southeastern Yellow Sea: Gomso bay, Korea. J Sediment Res 69:328–337
Kineke JC, Sternberg RW, Trowbridge JH et al (1996) Fluid-mud processes on the Amazon continental shelf. Cont Shelf Res 16:667–696
Kirby R (2000) Practical implications of tidal flat shape. Cont Shelf Res 20:1061–1077
Klein GD (1970) Tidal origin of a Precambrian quartzite – the Lower Fine-grained Quartzite (Middle Dalradian) of Islay, Scotland. J Sediment Pet 40:973–985
Klein GD (1975) Epilogue-tidal sedimentation: some remaining problems. In: Ginsburg RN (ed) Tidal deposits. Springer, Heidelberg, pp 407–410
Klein GD (1985) Intertidal flats and intertidal sand bodies. In: Davis RA Jr (ed) Coastal sedimentary environments, 2nd edn. Springer, New York, pp 187–224
Klein GD (1998) Clastic tidalites – a partial retrospective view. In: Alexander CR, Davis RA Jr, Henry VJ (eds) Tidalites: processes & products, SEPM Special Publication 61. SEPM, Tulsa, pp 5–14
Kuehl SA, Nitrouer CA, DEMaster DJ (1988) Microfabric study of fine-grained sediments: observations from the Amazon subaqueous delta. J Sediment Pet 58:12–23
Kvale EP, Archer AW, Johnson HR (1989) Daily, monthly, and yearly tidal cycles within laminated siltstones of the Mansfield Formation (Pennsylvanian) of Indiana. Geology 17:365–368
Le Hir P, Roberts W, Cazaillet O et al (2000) Characterization of intertidal flat hydrodynamics. Cont Shelf Res 20:1433–1459
Lee HJ, Chun SS, Chang JH et al (1994) Landward migration of isolated shelly sand ridge (chenier) on the macrotidal flat of

Gomso Bay, west coast of Korea: controls of storms and typhoon. J Sediment Res 64:886–893
Lee SC, Mehta AJ, Members of ASCE (1997) Problems in characterizing dynamics of mud shore profiles. J Hydraul Eng 123:351–361
Lee HJ, Jo HR, Chu YS et al (2004) Sediment transport on macrotidal flats in Garolim Bay, west coast of Korea: significance of wind waves and asymmetry of tidal currents. Cont Shelf Res 24:821–832
Lees BG (1992) The development of a chenier sequence on the Victoria Delta, Joseph Bonaparte Gulf, northern Australia. Mar Geol 103:215–224
Lefebvre JP, Dolique F, Gratiot N (2004) Geomorphic evolution of a coastal mudflat under oceanic influences: an example from the dynamic shoreline of French Guiana. Mar Geol 209:191–205
Li J (1990) Sediment transport on the Nanhui mudflat in the Changjiang Estuary. Acta Oceanologica Sinica 12:74–82 (in Chinese with English abstract)
Li CX, Li P (1982) Sediments and sand bodies on the tidal flats. Oceanol Limnol Sinica 13(1):48–59 (in Chinese with English abstract)
Li CX, Yang X, Zhuang Z et al (1965) Formation and evolution of the intertidal mudflat. J Shangdong Oceanogr Coll 2:21–31 (in Chinese with English abstract)
Li CX, Han C, Wang P (1992) Depositional sequences and storm deposition on low-energy coast of China. Acta Sedimentol Sinica 10:119–127 (in Chinese with English abstract)
Li CX, Wang P, Fan DD et al (2000) Open-coast intertidal deposits and the preservation potential of individual lamina: a case study from East-central China. Sedimentology 47:1039–1051
Li CX, Wang P, Fan DD (2005a) Tidal flat, open ocean coasts. In: Schwartz ML (ed) Encyclopedia of coastal science. Springer, Heidelberg, pp 975–978
Li GX, Yang ZG, Liu Y et al (2005b) Studies on the genetic types of sub-sottom sedimentary facies in the east China seas. Chinese Science Press, Beijing
Lim DI, Choi J, Shin IH (2004) Late Quaternary sedimentation on a macrotidal mudflat deposit in Namyang Bay, west coast of Korea. J Coast Res 20:478–488
Liu CZ, Walker HJ (1989) Sedimentary characteristics of cheniers and the formation of the chenier plains of East China. J Coast Res 5:353–368
Mao ZC (1987) The role of wave action in scouring and siltation processes of Nanhui Eastern Flats. Trans Oceanol Limnol 4:21–29
Meade RH, Dunne T, Richey JE et al (1985) Storage and remobilization of suspended sediment in the lower Amazon River of Brazil. Science 228:488–490
Meckel LD (1975) Holocene sand bodies in the Colorado Delta area, Northern Gulf of California. In: Broussard ML (ed) Deltas-models for exploration. Houston, Houston Geological Society, pp 239–265
Middleton GV (1991) A short historical review of clastic tidal sedimentology. In: Smith DG, Reinson GE, Zaitlin BA et al (eds) Clastic tidal sedimentology, Canadian Society of Petroleum Geologists Memoir 16. Canadian Society of Petroleum Geologists, Calgary, pp ix–xv
Milliman JD, Meade RH (1983) World-wide delivery of river sediment to the oceans. J Geol 91:1–21
Milliman JD, Syvitski JPM (1992) Geomorphic/tectonic control of sediment discharge to the ocean: the importance of small mountainous rivers. J Geol 100:525–544
Mukherjee KK, Das S, Chakrabarti AA (1987) Common physical sedimentary structure in a beach-related open-sea siliciclastic tropical tidal flat at Chandipur, Orissa, India and evaluation of the weather conditions through discriminant analysis. Senckenberg Marit 19:261–293
Nio SD, Yang CS (1991) Diagnostic attributes of clastic tidal deposits: a review. In: Smith DG, Reinson GE, Zaitlin BA, Rahmani RA (eds) Clastic tidal sedimentology, Canadian Society of Petroleum Geologists Memoir 16. Canadian Society of Petroleum Geologists, Calgary, pp 3–28
O'Brien DJ, Whitehouse RJS, Cramp A (2000) The cyclic development of a macrotidal mudflat on varying timescales. Cont Shelf Res 20:1593–1619
Park YA, Wells JT, Kim BW et al (1995) Tidal lamination and facies development in the macrotidal flats of Namyang Bay, west coast of Korea. In: Flemming BW, Bartholoma A (eds) Tidal signatures in modern and ancient sediments, International Association of Sedimentation Special Publication 24. Blackwell, Berlin, pp 183–191
Plaziat C, Augustinus PGEF (2004) Evolution of progradation/erosion along the French Guiana mangrove coast: a comparison of mapped shorelines since the 18th century with Holocene data. Mar Geol 209:127–143
Reineck HE, Cheng YM (1978) Sedimentology and faunistics of tidal flats in Taiwan. I. Marine geology. Senckenberg Marit 10:85–115
Reineck HE, Singh IB (1980) Depositional sedimentary environments, 2nd edn. Springer, Berlin
Ren ME (1985) Modern sedimentation in coastal and nearshore zone of China. Springer, Berlin
Rhodes EG (1982) Depositional model for a chenier plain, Gulf of Carpentaria, Australia. Sedimentology 29:201–222
Rine JM, Ginsburg RN (1985) Depositional facies of a mud shoreface in Suriname, South America – a mud analogue to study, shallow-marine deposits. J Sediment Pet 55:633–652
Ryu JH, Kim CH, Lee YK et al (2008) Detecting the intertidal morphologic change using satellite data. Estuar Coast Shelf Sci 78:623–632
Semeniuk V (1981) Long-term erosion of the tidal flats King Sound, northwestern Australia. Mar Geol 43:21–48
Shi Z, Chen JY (1996) Morphodynamics and sediment dynamics on intertidal mudflats in China (1961–1994). Cont Shelf Res 16:1909–1926
Talke SA, Stacey MT (2008) Suspended sediment fluxes at an intertidal flat: the shifting influence of wave, wind, tidal and freshwater forcing. Cont Shelf Res 28:710–725
Tessier B (1993) Upper intertidal rhythmites in the Mont-Saint-Michel Bay (NW France): perspectives for paleoreconstruction. Mar Geol 110:355–367
Thomas S, Ridd PV (2004) Review of methods to measure short time scale sediment accumulation. Mar Geol 207:95–114
Thompson RW (1968) Tidal flat sedimentation on the Colorado River delta, northwestern Gulf of California. Geol Soc Am Mem 107:413 p
Thompson RW (1975) Tidal-flat sediment of the Colorado River delta, northwestern Gulf of California. In: Ginsburg RN (ed) Tidal deposits: a casebook of recent examples and fossil counterparts. Springer, Heidelberg, pp 57–65

Walsh JP, Nittrouer CA (2004) Mangrove-bank sedimentation in a mesotidal environment with large sediment supply, Gulf of Papua. Mar Geol 208:225–248

Wang Y (1963) The coastal dynamic geomorphology of the northern Bohai Bay. In: Wang Y (ed) Collected oceanic works of Nanjing University. Nanjing University Press, Nanjing, pp 25–35 (in Chinese with English abstract)

Wang Y (1983) The mudflat system of China. In: Gordon DC Jr, Hourston AS (eds) Proceedings of the symposium on the dynamics of turbid coastal environments, Canadian Journal of Fish Aquatic Science 40. Government of Canada, Fisheries and Oceans, Ottawa, pp 160–171

Wang BC, Eisma D (1988) Mudflat deposition along the Wenzhou coastal plain in southern Zhejiang, China. In: DE Boer PL, VAN Gelder A, Nio SD (eds) Tide-influenced sedimentary environments and facies. D. Reidel Publishing Company, Dordrecht, pp 265–274

Wang BC, Eisma D (1990) Supply and deposition of sediment along the north bank of Hangzhou Bay, China. Neth J Sea Res 25:377–390

Wang Y, Ke SK (1989) Cheniers on the east coastal plain of China. Mar Geol 90:321–335

Wang XY, Ke XK (1997) Grain-size characteristics of the extant tidal flat sediments along the Jiangsu coast, China. Sediment Geol 112:105–122

Wang Y, Zhu DK, Wu XG (2002) Tidal flats and associated muddy coast of China. In: Healy T, Wang Y, Healy JA (eds) Muddy coasts of the world: processes, deposits and function. Elsevier, Amsterdam, pp 319–346

Wang AJ, Gao S, Chen J et al (2009) Sediment dynamic response of coastal salt marsh to typhoon "Kaemi" in Quanzhou Bay, Fujian Province, China. China Sci Bull 53:120–130

Wells JT, Coleman JM (1978) Longshore transport of mud by waves: northeastern coast of South America. Geol Mijn 57:353–359

Wells JT, Coleman JM (1981) Physical processes and fine-grained sediment dynamics, coast of Surinam, South America. J Sediment Pet 51:1053–1063

Wells JT, Adams CE Jr, Park YA et al (1990) Morphology, sedimentology and tidal channel processes on a high-tide-range mudflat, west coast of South Korea. Mar Geol 95:111–130

Williams GE (1997) Precambrian length of day and the validity of tidal rhythmite paleotidal values. Geophys Res Lett 24:421–424

Yang BC, Dalrymple RW, Chun SS (2005) Sedimentation on a wave-dominated, open-coast tidal flat, southwestern Korea: summer tidal flat – winter shoreface. Sedimentology 52:235–252

Yang BC, Dalrymple RW, Chun SS (2006a) The significance of hummocky cross- stratification (HCS) wavelengths: evidence from an open-coast tidal flat, south Korea. J Sediment Res 76:2–8

Yang BC, Dalrymple RW, Chun SS et al (2006b) Transgressive sedimentation and stratigraphic evolution of a wave-dominated macrotidal coast, western Korea. Mar Geol 235:35–48

Yang BC, Dalrymple RW, Chun SS et al (2008a) Tidally modulated storm sedimentation on open-coast tidal flats, southwestern coast of Korea: distinguishing tidal-flat from shoreface storm deposits. In: Hampson GJ, Steel RJ, Burgess PM, Dalrymple RW (eds) Recent advances in models of Siliciclastic shallow-marine stratigraphy, SEPM Special Publication 90. SEPM, Tulsa, pp 161–176

Yang BC, Gingras MK, Pemberton SG et al (2008b) Wave-generated tidal bundles as indicator of wave-dominated tidal flats. Geology 36:39–42

Yang SL, Li H, Ysebaert T et al (2008c) Spatial and temporal variations in sediment grain size in tidal wetlands, Yangtze Delta: on the role of physical and biotic controls. Estuar Coast Shelf Sci 77:656–671

Zhang RS, Wang XY (1991) Tidal creek system on tidal mud flat of Jiangsu Province. Acta Geog Sinica 46:195–206 (in Chinese with English abstract)

Siliciclastic Back-Barrier Tidal Flats

10

Burghard W. Flemming

Abstract

Back-barrier tidal flats occur along micro- to mesotidal coasts landward of barrier islands and in the shelter of coastal sand spits and bars. Tidal flats are generally flood dominated, the grain size progressively decreasing shoreward. The sediment can be divided into sand, slightly muddy sand, muddy sand, sandy mud, slightly sandy mud, and mud. The mud fraction consists of non-cohesive sortable silt and cohesive flocs and aggregates. Important physical and biological surface structures include wave- and current-generated ripples, ladderback ripples, washed out ripples and other late-stage emergence runoff features, shell pavements, fluid mud sheets, tool marks, crawling, feeding and resting traces of intertidal organisms, as well as the feeding traces and tracks of birds. Internal sedimentary structures range from rare dune cross-bedding to ubiquitous ripple cross-bedding in sand, through flaser, wavy and lenticular bedding in mixed sediment, and homogenous or laminated mud toward the high-water line. Bioturbation may be intense, but the preservation potential depends on the frequency and depth of reworking. The transition from land to sea is typically marked by laminated versicolored microbial mats. The interaction between sea-level rise and sediment supply defines the sediment budget and hence the stratigraphy. Prograding, aggrading or transgressive systems are easily distinguished by their stratigraphic architecture.

10.1 Introduction

This chapter deals with non-vegetated or bare (inter) tidal flat depositional systems that occur in the shelter of coastal barriers and which are predominantly composed of siliciclastic sediments (sand and mud) of terrigenous origin. In these systems, bioclastic material derived from shell-bearing organisms, especially molluscs, forms an overall subordinate sedimentary component, although it may locally be enriched in the form of shell beds and channel lag deposits. Back-barrier tidal flats commonly occur along micro- to mesotidal coasts (tidal ranges of ~0.3–3.5 m) in the rear of barrier islands, and in the shelter of coastal sand spits. Specifically excluded are carbonate tidal deposits (Pratt et al. 1992), intertidal sand bodies occurring along lower courses of many barred estuaries (Dalrymple et al. 1992), episodically flooded back-barrier wind flats (Miller 1975; Schneider 1975), tidal lagoons without substantial intertidal flats (Ashley 1988;

B.W. Flemming (✉)
Senckenberg Institute, Suedstrand 40, 26382 Wihelmshaven, Germany
e-mail: bflemming@senckenberg.de

R.A. Davis, Jr. and R.W. Dalrymple (eds.), *Principles of Tidal Sedimentology*,
DOI 10.1007/978-94-007-0123-6_10, © Springer Science+Business Media B.V. 2012

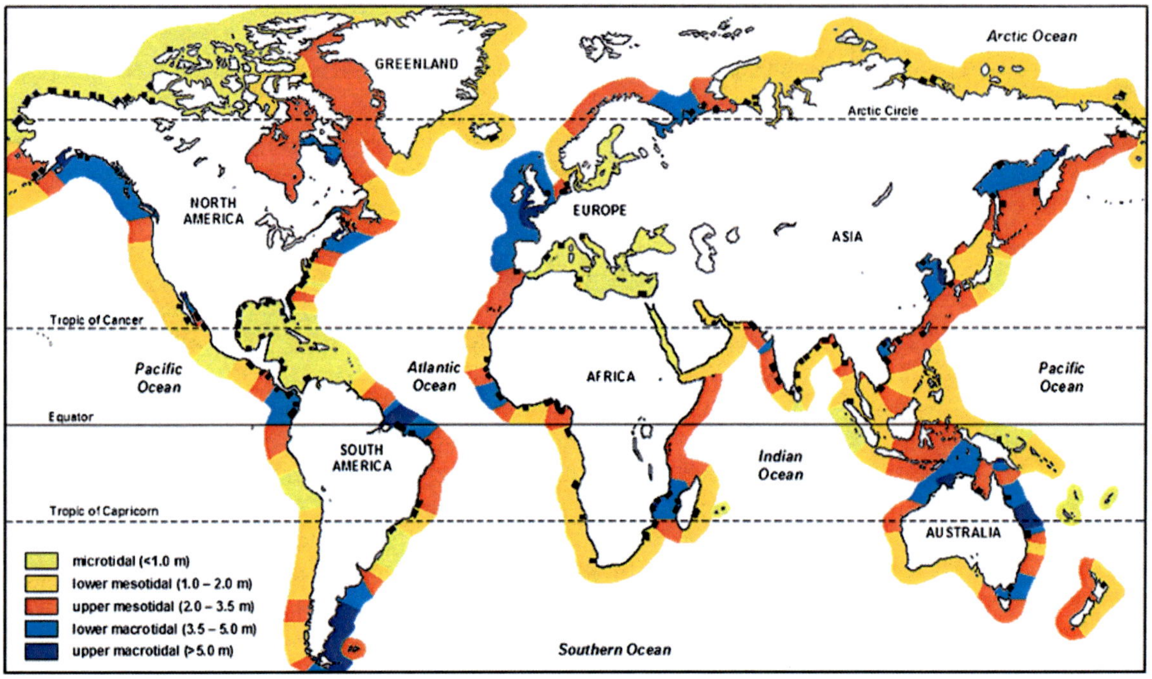

Fig. 10.1 Global distribution of coastal barriers backed by tidal flats and/or lagoons (Amended after Pilkey 2003) in relation to tidal regime (Modified after Flemming 2005)

Boothroyd et al. 1985), and extensively vegetated intertidal flats (Pestrong 1972; Frey and Basan 1985), in particular comprising cordgrass (*Spartina sp.*) marshes or mangrove forests. The stratigraphy and facies successions of these latter systems are distinctly different from those of typical non-vegetated, back-barrier tidal flat systems (Kraft et al. 1979), of which the Ria Formosa along the Algarve coast of Portugal (Pilkey et al. 1989) is a partial and the Wadden Sea lining the coasts of The Netherlands, Germany and Denmark a prime example (Bartholdy and Pejrup 1994; Flemming and Davis 1994; Oost and de Boer 1994).

A comprehensive global inventory and classification of barred tidal flat systems is currently still lacking, but as most are associated with coastal barrier systems, the amended global map of the latter (Fig. 10.1) provides a reasonable, if incomplete, picture of their geographic distribution. From Fig. 10.1 it is clearly evident that barrier islands and other types of barred coasts are not evenly distributed along the shores of the world, the vast majority being associated with low-lying coastal plains (~72%) and river deltas (~28%) (Pilkey 2003). In the context of global tectonics, 49% of barrier islands are located along trailing-edge coasts, 24% along collision coasts, and 27% along marginal sea coasts (Glaeser 1978). Furthermore, of those located along trailing-edge coasts, 75% occur along amero-trailing-edge, 19% along afro-trailing-edge, and only 6% along neo-trailing-edge coasts. These barrier systems occupy 12–13% of the world's shoreline, the greater part being represented by the lagoonal type where fringing intertidal flats are heavily vegetated by cordgrass (from subtropical to boreal climates) or mangrove forests (from subtropical to tropical climates), bare intertidal flats being reduced to narrow belts along tidal channels.

The unique nature of bare tidal flat landscapes had already been recognized by the Roman geographer Pliny the Elder (ca. AD 45) who, after having personally visited the Wadden Sea coast, describes it in his epochal geographic compendium "Historia Naturalis" as an immeasurable expanse of land inundated twice a day by sea water and of which it was uncertain whether it formed part of the land or the sea. Proper tidal flat research, however, merely dates back to the first part of the twentieth century (Kindle 1917), especially when, in 1928, Rudolf Richter founded the Senckenberg Research Station for Marine Geology and Palaeontology in Wilhelmshaven on the North Sea coast of Germany. It was the first institution worldwide specifically dedicated to tidal flat research (Ginsburg 1975). Earlier studies either focused on regional physiographic descriptions (Arends 1833), coastal barrier formation (de Bèaumont 1845), or shore processes in general (Johnson 1919). Selections of historical benchmark papers on barrier islands and tidal flats can be found in Schwartz (1973) and Klein (1976). Recent summaries of the main characteristics of tidal flats and tidal environments can be found in Flemming (2003a, b, 2005).

10.2 Hydrological Constraints

Barrier island systems, and hence back-barrier tidal flats, are typically restricted to tidal ranges of up to about 3.5 m (Hayes 1979). Above this limit, the tidal prism or water masses moving toward and away from the coast during each tidal cycle are generally so large that there is literally no room left for barrier islands to exist, wave action being unable to counteract the strong tidal currents. The tidal prism is a function of tidal range, basin surface area and filling efficiency, the latter depending on the inlet cross-section (Van Veen 1950). As a consequence, barrier islands progressively decrease in size the larger the tidal prism gets with increasing tidal range (Oost and de Boer 1994; Davis and Flemming 1995) before degenerating into scattered ephemeral sand bank islands when a certain limit is exceeded (Reineck 1987). Because of this, Hayes (1979) proposed a new tidal classification in which five subdivisions are distinguished (<1 m: microtidal; 1–2 m: lower mesotidal; 2–3.5 m: upper mesotidal; 3.5–5.0 m: lower macrotidal; >5.0 m: upper macrotidal) (Fig. 10.2). It represents a refinement of the more commonly used classification of Davies (1964) that only distinguishes three categories (<2 m: microtidal; 2–4 m: mesotidal; >4 m: macrotidal). The geographic distribution of tidal ranges according to the more detailed classification of Hayes (1979) has been included in Fig. 10.1 and can also be found in Flemming (2005). In addition to being morphogenetically more meaningful, it also provides a much better spatial resolution of tidal regimes around the world than the older one.

As shown by Davis and Hayes (1984), a second important hydrodynamic factor limiting barrier stability

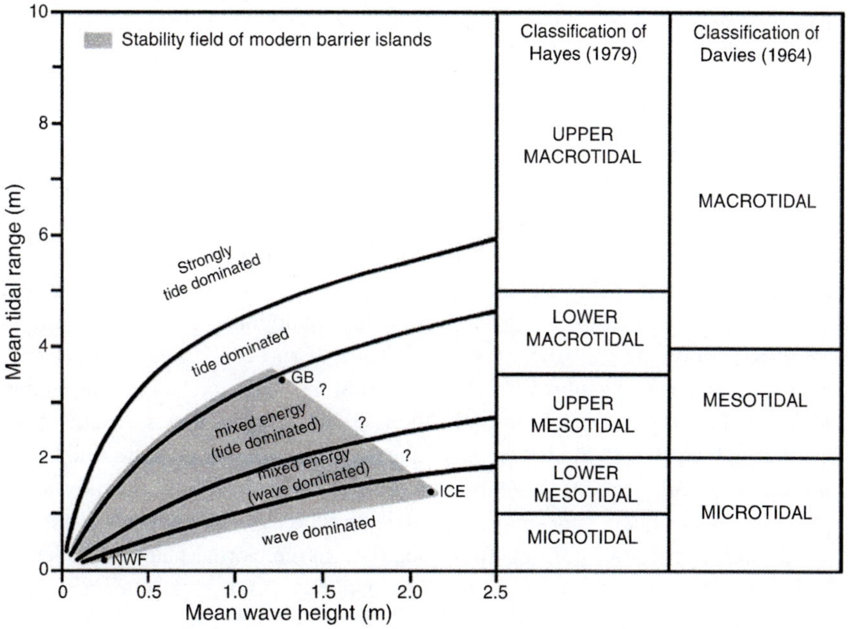

Fig. 10.2 Barrier-island stability as a function of wave climate and tidal range relative to the classification schemes of Davies (1964) and Hayes (1979)

is the wave climate. Thus, sandy barriers are today restricted to coasts exposed to mean wave heights of less than about 2.2 m (Hayes 1979). In terms of wave climate and tidal range, the stability field of modern sand barriers essentially occupies the mixed wave- and tide-dominated energy regimes (Fig. 10.2). In this interacting, relative energy constellation the Gulf coast of north-west Florida (NWF), for example, represents a low wave/low tidal energy endmember, the German Bight (GB) an intermediate wave/high tidal energy endmember, and the barrier coast of south-eastern Iceland (ICE) a high wave/intermediate tidal energy endmember. To date it is not clear whether the boundaries of this stability field, especially the one between the GB and ICE endmembers, are definitive or purely fortuitous, the occurrence of gravelly barriers in macrotidal environments suggesting that grain size may play an important additional role (Jennings and Coventry 1973; Hayes 1994).

In contrast to the processes along the open coast, the tidal basins on the landward side of barriers are predominantly controlled by tidal energy fluxes, although wave action is an important secondary factor, as emphasised by the ubiquitous occurrence of wave-generated sedimentary structures. The high correlations between physical parameters such as the surface area of a tidal basin, tidal prism, tidal discharge, inlet width, inlet cross-section, inlet depth, channel depth, and ebb-delta area and volume document the overriding control by the tides (Walther 1972; Jarrett 1976; Walton and Adams 1976; Hume and Herdendorf 1992; Flemming and Davis 1994; van Dongeren and de Vriend 1994; Biegel and Hoekstra 1995; van der Spek 1995; Williams et al. 2002). With respect to back-barrier tidal flats, important hydrological factors are the time/distance velocity asymmetries between flood and ebb currents, tidal flats being generally flood dominated, whereas deeper channels are ebb dominated (Groen 1967; Boon and Byrne 1981; Aubrey and Speer 1985; Speer and Aubrey 1985; Dronkers 1986; Ridderinkhof 1988; Friedrichs and Aubrey 1988; Friedrichs et al. 1992; Stanev et al. 2007). This has two important implications. First, the residual current over intertidal shoals (tide-induced drift) results in a net shoreward transport of suspended sediment, a process that may be enhanced or retarded by wind stress and wave action. An additional factor may be the development of horizontal density gradients over tidal flats, as recently proposed by Burchard et al. (2008). Suspended particulate matter (SPM) eventually settles out in places where the settling velocity exceeds the erosion velocity. This process acts in conjunction with the settling lag/scour lag mechanism (van Straaten and Kuenen 1957; Postma 1961) which is responsible for an overall stepwise net displacement of resuspended particles in the direction of the flood current. By this mechanism suspended particles settle out at high water slack tide before being resuspended in the course of the subsequent ebbing tide. As the particles require higher velocities to be resuspended than to settle out, the time-velocity asymmetry between the ebb and flood phase produces a net landward transport. This mechanism proceeds until a balance between settling velocity and erosion velocity is reached. The resulting shoreward decrease in grain size is one of the main consequences and hence a fundamental diagnostic criteria for intertidal deposits (e.g., van Straaten and Kuenen 1957, 1958; Nyandwi and Flemming 1995; Chang and Flemming 2006).

Both mechanisms outlined above may be strongly enhanced by seasonal changes in water temperature which, at higher latitudes, may differ by >20°C. The higher kinematic viscosities of the seawater in winter result in lower settling velocities of equivalent particles, i.e. the same particles behave as coarser sediment in summer and finer sediment in winter (Anderson 1983; Krögel and Flemming 1998; Chang et al. 2006a). That this effect is significant is demonstrated by the fact that, for example in the Wadden Sea (>55°N), particles with equivalent settling velocities in winter (T <5°C) and summer (T >20°C) are spatially separated by as much as 3 km (Fig. 10.3).

A second implication is that the channel systems of back-barrier tidal basins are not landward-facing flooding systems, but rather seaward-facing drainage systems analogous to terrestrial drainage networks (Flemming and Davis 1994). However, as the flow is bidirectional, there are some morphological modifications associated with flow separation between the dominant ebb and the subordinate flood current (Jakobsen 1962; van Straaten 1964). This flow separation is modulated by the Coriolis effect, which deflects the flow to the right in the Northern and to the left in the Southern Hemisphere. As a consequence, tidal channels are frequently split longitudinally into ebb- and flood-dominated sections that can, for example, be identified by the corresponding orientation of larger bedforms.

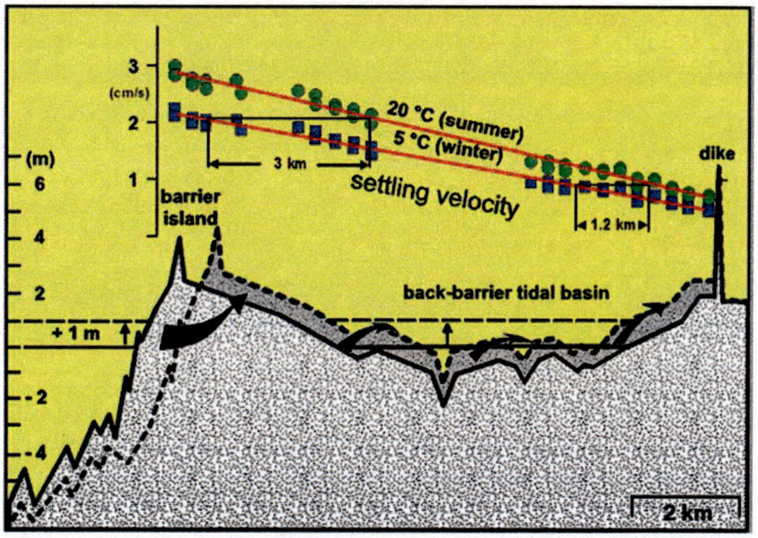

Fig. 10.3 The back-barrier energy gradient, as reflected in the progressively shoreward decreasing mean settling velocity of the sediment (Modified after Flemming 2002). Note the pronounced spatial displacement between the summer (high water temperature, low kinematic viscosity) and winter (low water temperature, high kinematic viscosity) gradient

Tidal flow patterns and suspended matter transport in back-barrier tidal basins have to date been successfully simulated over a number of tidal cycles (Stanev et al. 2007, 2009; Lettman et al. 2009). In addition, promising advances in morphodynamic modeling on decadal to millennial time scales have been made in recent years (Fortunato and Oliveira 2004; Dastgheib et al. 2008; van der Wegen and Roelvink 2008; Dissanayake et al. 2009; Ganju et al. 2009; van der Wegen et al. 2010).

As mentioned above, wave action is an important secondary hydrodynamic factor on tidal flats. Indeed, it is doubtful whether barrier islands, supratidal flats and a number of other coastal environments located above the spring high-tide level would exist at all without the influence of waves. In the case of back-barrier tidal flats, longer-period open ocean swells (T>8 s) and wind waves (T=4–7 s) penetrating tidal inlets loose as much as 95% of their energy through friction and breaking when crossing the ebb-deltas (Lettmann et al. 2009). As a consequence, back-barrier sedimentary processes are more strongly influenced by locally generated short-period wind waves (T=2–3 s). This is reflected by the preponderance of small-scale wave ripples in both intertidal sand and mud deposits (Davis and Flemming 1995).

10.3 Morphology, Sedimentology and Mass Physical Properties

10.3.1 Morphological Characteristics

Back-barrier tidal basins are typically bounded by barrier islands on the seaward side and the mainland coast on the landward side. Laterally they are separated from neighbouring tidal basins by slightly elevated watersheds (tidal divides). The location of the watersheds depends on the angle of approach of the tidal wave relative to the orientation of the coast. If the tide approaches normal to the coast, the watersheds are located midway between the two heads of the islands. However, the more acute the angle of approach, the stronger the displacement of the watersheds in the direction of tidal wave propagation. In the case of Fig. 10.4, for example, the tidal wave approaches from the left (west), as a consequence of which the watersheds are displaced toward the right (east). As a rule of thumb, the watersheds are located where two separating flow paths at the head of the tidal wave meet at high tide behind the islands after having travelled roughly equal distances through adjacent inlets.

Individual back-barrier tidal basins are composed of a number of characteristic morphological elements.

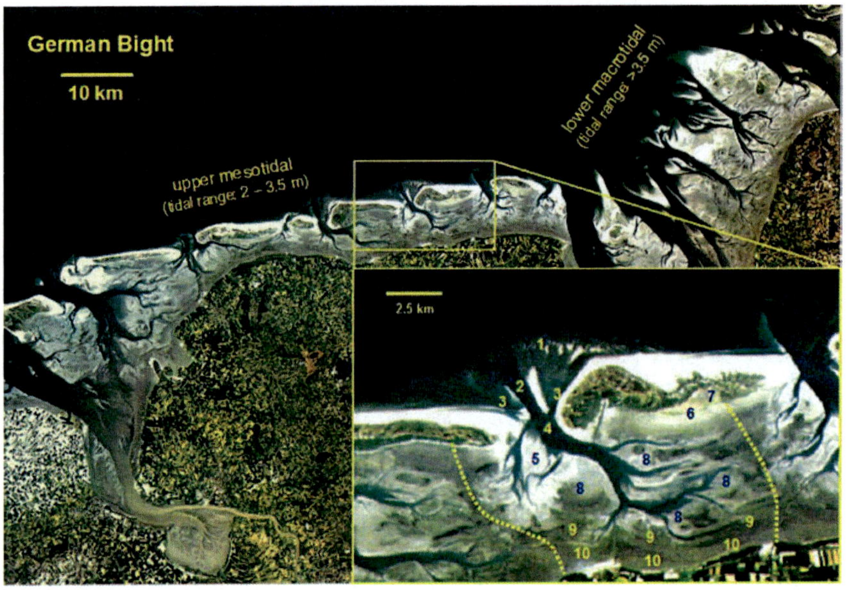

Fig. 10.4 Typical morphological elements of a barrier island depositional system, here illustrated by an example from the German Wadden Sea. *1* ebb-delta, *2* main ebb channel, *3* marginal flood channels, *4* inlet with back-barrier channel system, *5* flood ramp, *6* overwash fans, *7* back-barrier salt marsh and microbial mats, *8* sand flats, *9* mixed flats, *10* mud flats, *yellow dotted line*: tidal watersheds

This is illustrated by an example from the Wadden Sea (Fig. 10.4, inset). The depositional system begins on the seaward side of an inlet with an ebb-delta shoal (marked 1 in the inset of Fig. 10.4), which is subdivided by a central ebb channel (2) and marginal flood channels (3) that converge on the inlet (4). Inlets typically reach depths of 15–30 m, in exceptional circumstances up to 50 m, the depth being highly correlated with the tidal prism (Oost and de Boer 1994; van der Spek 1995). The same relationship applies to any location of the intra-basin channels and the fractional tidal prism discharging through that location (van der Spek 1995). In many respects the tidal channel systems comply with the morphometric rules known from fluvial drainage systems (Hack 1957; Leopold et al. 1964; Flemming and Davis 1994; Rinaldo et al. 2004).

The tidal drainage systems are cut into what are known as the back-barrier tidal flats. These can be subdivided into a number of morpho-sedimentological units that reflect particular hydrodynamic processes, in particular the shoreward decreasing energy gradient. As evident from Fig. 10.4, the intertidal flats at low tide occupy a much larger area than the tidal channels, a feature that distinguishes these systems from back-barrier lagoonal systems where the water-covered area at low tide is much larger than that of the fringing intertidal flats. It is possibly due to this fact that the type of back-barrier tidal flat system discussed here (cf. Fig. 10.4) does not display morphologically distinct flood deltas *sensu* Hayes (1979). They are instead replaced by ‘flood ramps’ (5) located along the outer margins of tidal-flat sand bodies facing the inlet. These ramps are barely visible on the ground but can be clearly identified on aerial photographs or satellite images by their lobate or crescentic shapes and the lighter colour of the sediment that is typical for highly mobile, drained sand that is almost devoid of biological activity. In contrast to classical flood deltas, flood ramps represent the current- and wave-reshaped margins of tidal flat sand bodies opposite the inlet.

The tidal flats and salt marshes in the rear of the islands are shaped by overwash fans (6) composed of beach sand transported across the islands during storms. On the aerial photograph in Fig. 10.5, two generations of such fans can be seen. The larger ones stem from storm events at a time when the facing part of the island was occupied by a bare supratidal flat without a protective eolian dune belt (pre-1962). Overwash activity during storms was thus unimpeded, resulting in large fans. After the establishment of a dune belt,

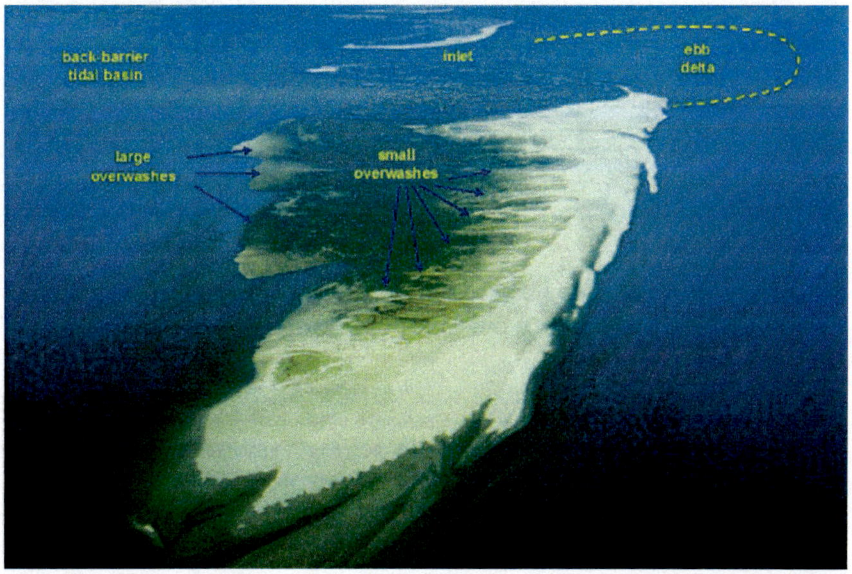

Fig. 10.5 Aerial view of a barrier island showing the location of overwash fans (Spiekeroog island, East Frisian Wadden Sea). Note the large fans that formed without obstruction of an eolian dune belt (here pre-1962) in comparison to the much smaller fans that formed after the establishment of the dune belt (post-1962)

salt marshes (7 in inset of Fig. 10.4) rapidly spread eastward and overwash activity thus became channelized through breaches in the dune belt, resulting in a larger number of smaller fans terminating in the salt marshes. Of particular interest here is the remarkably regular spacing of the breaches, which suggest a genetic link with resonant processes in the surf zone (Flemming and Davis 1994). The lobate overwash splays on the upper tidal flats are not very thick and hence difficult to identify on the ground. However, besides clear evidence on aerial photographs (e.g., Fig. 10.5), they can also be recognised by their grain-size composition, which is essentially identical to that of the adjacent beach sand (Flemming and Ziegler 1995). Within the back-barrier salt marsh, recent overwash activity is highlighted by distinctly laminated, decimetre thick sand layers penetrated by the stems and roots of salt marsh plants.

Toward the mainland shore follow the tidal flats proper. These comprise sand flats (8), mixed flats (9) and mud flats (10) (cf. Fig. 10.4). While the margins of sand flats slope more steeply toward the channels, they are almost level some distance away at elevations just below mean sea level. Along many channel margins, sand flats display slightly elevated levees formed by the interaction of tidal currents and wave action. Mixed flats and mud flats, by contrast, gradually rise toward the mean high water mark, convex profiles indicating accretion, concave ones erosion (Kirby 2000). As illustrated in Fig. 10.3, this sedimentary facies progression is hydrodynamically finely tuned, the energy gradient being associated with a progressive reduction in particle settling velocity.

10.3.2 Sedimentological Characteristics

In addition to indicators for emergence, the most striking feature of tidal flats is the pronounced shoreward fining in grain size. In effect this means that grain-size distributions gradually shift from coarser to finer mean diameters (Bartholomä and Flemming 2007), settling velocity data being generally more sensitive than sieve data because of the hydraulic sorting process (Flemming 2007). This shift in mean grain size implies a gradual change in the textural composition of the sediments. In order to describe such sediments in a consistent way, a variety of classification schemes have been devised, ternary sand/silt/clay diagrams having been the most commonly used ones (Shepard 1954; Folk 1954). Because the determination of silt and clay content is technically demanding, a simpler two-component classification based on routinely determined mud content has recently been proposed by Flemming (2000).

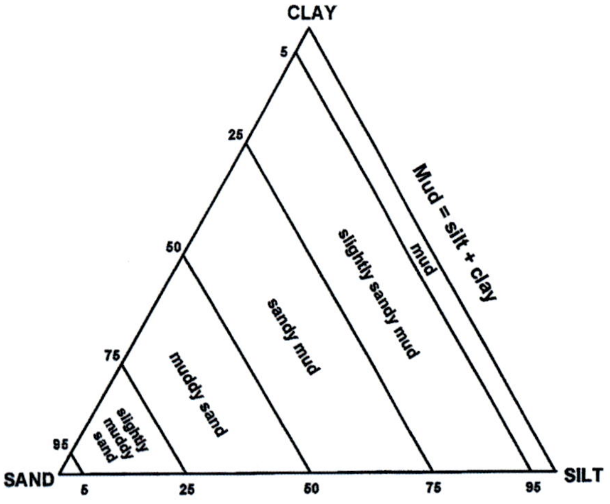

Fig. 10.6 Sediment classification based on sand/mud ratios (After Flemming 2000)

It allows the distinction of six sediment types (Fig. 10.6). These are: sand (<5% mud), slightly muddy sand (5–25% mud), muddy sand (25–50% mud), sandy mud (50–75% mud), slightly sandy mud (75–95% mud), and mud (>95% mud). The scheme provides a good spatial resolution of textural sediment composition, the textural classes also forming good descriptors of sedimentary environments or facies. For example, an intertidal area consisting of muddy sand would be called a 'muddy sand flat' or a 'muddy sand facies', etc. A more detailed scheme based on sand/silt/clay ratios, constructed by adding lines to the diagram of Fig. 10.6 fanning out from the sand endmember toward the silt-clay baseline, can be found in Flemming (2000).

In the past, it was generally thought that mixed sediments were immature, being principally more poorly sorted than sand. In recent years, however, it has been recognised that mud is composed of two major particle groups, one comprising non-cohesive 'sortable' silt (McCave et al. 1995) consisting of particles coarser than about 8 μm (medium, coarse and very coarse silt), the other comprising flocs and aggregates consisting of particles finer than about 8 μm (fine silt, very fine silt, and clay) (Chang et al. 2007). The aggregated nature of suspended sediment is illustrated in Fig. 10.7, in which a laser-based *in situ* size distribution (a) is compared with that of a dispersed sample (b) collected at the same location.

As the aggregates also get size-sorted according to the principle of hydraulic equivalence, they are deposited together with mineral grains (sand, sortable silt) having similar settling velocities. Field evidence suggests that the largest aggregates have equivalent 'grain sizes' corresponding to sand grains about 180 μm in diameter (fine sand) (Chang et al. 2007). As a consequence, the mud content of the sediment gradually increases toward finer-grained sediments in accordance with the rapidly increasing number of smaller aggregates having lower settling velocities than the larger ones. Once deposited, the aggregates are mixed into the ambient sediment, which will become increasingly more cohesive once the clay content of the total sediment exceeds 5–10% (van Ledden et al. 2004). Laboratory analyses of dispersed mud thus introduce mechanical artefacts into grain-size distributions that suggest poor sorting. In hydraulic terms, such sediments are actually very well sorted, the standard deviation of the sand fraction being a good approximation of the 'true' sorting of the total sediment. The deposition of mud on tidal flats is thus controlled by the settling velocities of the differently sized flocs and aggregates and not by those of the constituent particles. At a water temperature of 18°C the critical lower size limit for individual 'sortable' silt particles (8 μm) corresponds to a settling velocity of ~0.01 cm s^{-1}, smaller particles being rapidly scavenged to be incorporated into aggregates ranging from floccules to fecal pellets. Conceptually this is in excellent agreement with earlier findings about the settling velocity of suspended matter in a variety of environments (Nichols and Biggs 1985, based on data of Migniot 1968; Haven and Morales-Alamo 1968; Owen 1971, and Krone 1972) (Fig. 10.8).

The diagram in Fig. 10.8 shows the settling velocity range of dispersed clay particles at 18°C relative to that of composite particles (flocs and aggregates) in quiet and turbulent water, as well as that of fecal pellets. The corresponding equivalent grain size of quartz spheres shows that the bulk of aggregated material generally exceeds the critical size, fecal pellets and some of the flocs and aggregates being hydraulically equivalent to grain sizes as large as fine sand. While sortable silt particles, as in the case of sand, respond in a predictable way to changing hydrodynamic conditions, flocs and aggregates constantly change their size and composition in the course of transport, deposition and resuspension due to continually changing shear forces in the course of a tidal cycle (Chang et al. 2006b). Because the number of aggregates increases

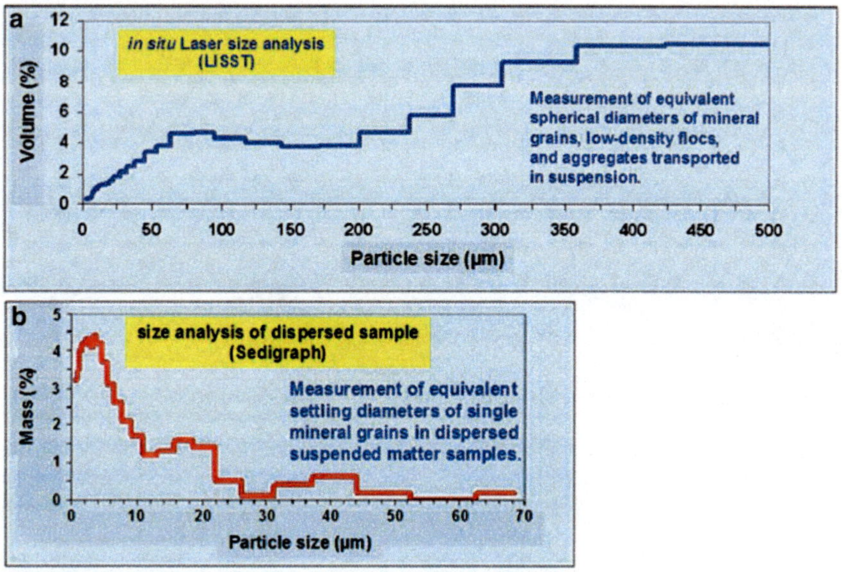

Fig. 10.7 Comparison of particle-size distributions carried out *in situ* on suspended matter by means of a laser particle sizer (**a**) and in the laboratory by means of a Sedigraph™ (**b**). In the latter case, the analysis was carried out on a disaggregated (dispersed) sample collected at the same site (Based on Chang et al. 2007). It is clearly evident that the bulk of the suspended material consists of a wide range of differently sized aggregates which, when disaggregated, is seen to be primarily composed of constituent particles <10 µm in size

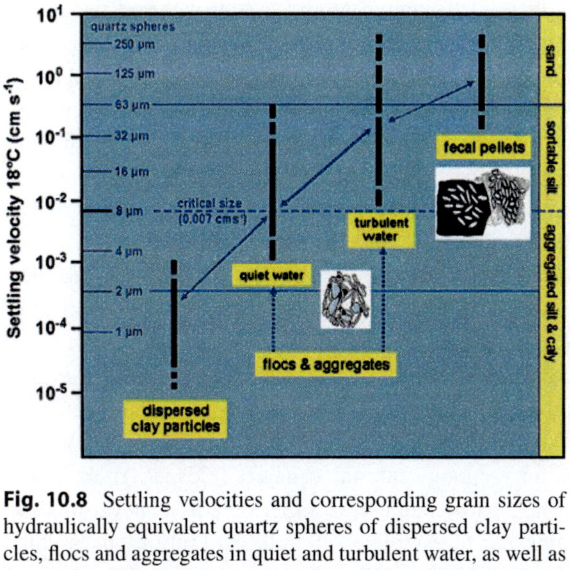

Fig. 10.8 Settling velocities and corresponding grain sizes of hydraulically equivalent quartz spheres of dispersed clay particles, flocs and aggregates in quiet and turbulent water, as well as fecal pellets (Modified after Nichols and Biggs 1985; based on data of Migniot 1968; Haven and Morales-Alamo 1968; Owen 1971; and Krone 1972). Note that the bulk of aggregated material exceeds the critical size of 8 µm. The *arrows* indicate continuous exchange in the course of aggregation and disaggregation in response to changes in current shear and bioactivity

rapidly with decreasing size, the gradual increase in mud content in the direction of the energy gradient is plausibly explained.

The partitioning of mud into two fundamentally different particle groups challenges the conventional wisdom of plotting sand/silt/clay ratios in ternary diagrams and suggests that important information may also be gained from plotting the ratios of sand, sortable silt (8–63 µm fraction), and flocs & aggregates (<8 µm fraction). This is illustrated by the two comparative plots in Fig. 10.9 that were generated from the same Wadden Sea dataset. While the conventional plot (a) shows a silt-clay partitioning at proportions of about 37–63%, the modified plot (b) reveals that, in this particular example, sortable silt and aggregated material contribute about equal amounts to the mud fraction of the back-barrier tidal basin.

The ternary diagrams in Fig. 10.9 show that sediment composition in back-barrier tidal basins is represented by narrow bands extending across the entire spectrum of sedimentary facies from sand to mud as defined in Fig. 10.6. Such trends are typical of many

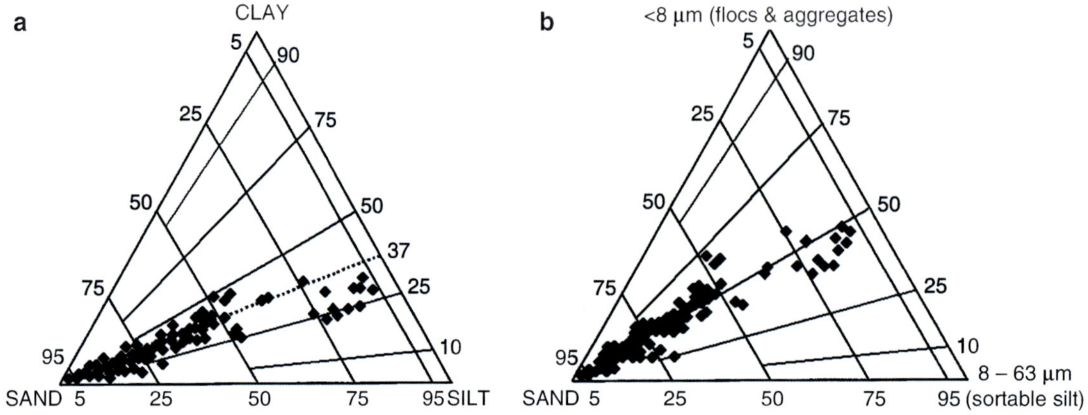

Fig. 10.9 Ternary diagram of sand/silt/clay ratios (**a**) and sand/sortable silt/flocs & aggregates (**b**) observed in a back-barrier tidal basin of the German Wadden Sea. Note the 50/50 partitioning of sortable silt and aggregates in the latter case, as opposed to a 63/37 partitioning of silt and clay in the former case (Based on Chang et al. 2007; subdivisions after Flemming 2000)

mixed sedimentary environments (Flemming 2000). At the same time the progression reveals energy gradients from sand to mud, on one hand, and between silt and clay, on the other (cf. Pejrup 1988; Molinaroli et al. 2009). In diagram a, the position of the data band between the silt and clay endmembers suggests a relatively exposed depositional environment, whereas in diagram b it occupies a more intermediate energy position. Which of the diagrams is hydraulically more relevant in this context requires further investigation as comparative data for the case b are currently lacking. Nevertheless, in both cases the grain-size composition allows a relative energy classification of the environment (Flemming 2000; Molinaroli et al. 2009).

10.3.3 Mass Physical Sediment Properties

Sedimentary environments such as back-barrier tidal flats are highly dynamic systems that constantly change their outward appearance in response to energy fluctuations, on a regular basis in the course of the spring-neap tidal cycle and episodically by sediment reworking during storms. To quantify such changes, the import and/or export of material to or from a tidal flat area is commonly achieved by repeated elevation surveys with subsequent calculation of volume changes between surveys. The volume changes then need to be converted into material masses. This is achieved by determining critical mass physical sediment properties. Important parameters in this context are wet and dry bulk densities, porosity, water content, and organic matter content. Mass balancing exercises are particularly important in disciplines such as sedimentology, geochemistry, biology, microbiology, and biochemistry. Good examples can be found in Bartholomä et al. (2000) for the import and export of sand and mud, and in Delafontaine et al. (2000) for organic matter.

In tidal flat environments, both wet and dry bulk density have been found to be highly correlated with mud content and average values of the former can thus be calculated from the latter on the basis of regression analyses. Examples from the Wadden Sea are shown in Fig. 10.10. From the calibration curves it can be seen that pure sand has an average wet bulk density (BD_w) of ~2.0 g cm^{-3} and a corresponding dry bulk density (BD_d) of ~1.6 g cm^{-3}. At the other end, the wet and dry bulk densities of pure mud are ~1.2 and ~0.3 g cm^{-3}, respectively. More precise average values can be calculated on the basis of the regression equations. Although the Wadden Sea trends should be generally valid for many other tidal flat systems composed of terrigenous material (quartz, feldspar, rock fragments, carbonate, clay minerals), it is nevertheless advisable to establish separate calibration curves for other areas, especially if organic matter contents are high (Delafontaine et al. 2004).

Wet and dry bulk densities can also be determined from the water content (Wc) of intertidal sediments. An example for dry bulk density is illustrated in Fig. 10.11.

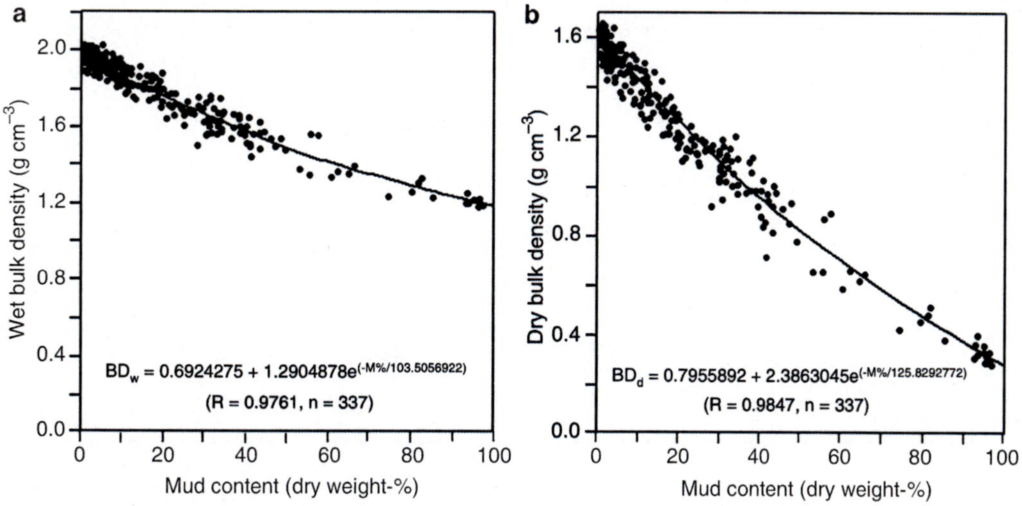

Fig. 10.10 Relationship between mud content and wet (**a**) and dry (**b**) bulk densities in Wadden Sea sediments (Based on Flemming and Delafontaine 2000)

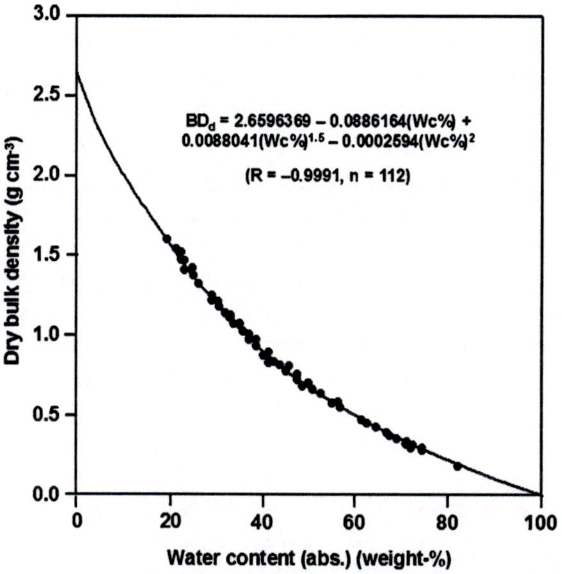

Fig. 10.11 Relationship between dry bulk density and absolute water content in intertidal sediments of the Wadden Sea (Based on Flemming and Delafontaine 2000). Note the very high correlation. The relationship has universal character for average terrigenous material

As can be seen, the correlation is superior to that for mud content (Fig. 10.10b) and, for average terrigenous material ($\delta = 2.65$ g cm^{-3}), this relationship has universal character (cf. regression equation in Fig. 10.11). It is important, however, to carefully distinguish between absolute and relative water contents, the former being defined as the ratio between the mass of pore water and the mass of the total water-saturated sample, the latter as the ratio between the mass of pore water and the mass of the dry solids. Relative water content can reach several hundred percent, i.e. the mass of the water can greatly exceed the mass of the dry solids, whereas the absolute water content is always a fraction of one-hundred. Relative water content <100% can therefore be confused with absolute content if not identified as such. Excellent treatments of these and other mass physical properties can be found in Lambe and Whitman (1969), Carver (1971), Inderbitzen (1974), Dunn et al. (1980), Hillel (1998), and Warrick (2002).

Mass concentrations of sand and mud relative to the total sediment (sand + mud) are illustrated in Fig. 10.12a. Of particular interest here is the counter-intuitive trend described by the dry mass concentration of the mud component (Fig. 10.12b). Thus, with increasing mud content, the mass concentration of mud initially increases as would intuitively be expected. At higher mud content, however, the trend changes in an unexpected manner, i.e. it flattens off, peaks (in this case at a mud content of about 60%), and thereafter decreases again. This counter-intuitive trend is caused by a progressive change in the network structure or fabric of the sediment as the water content increases with increasing mud content. Beyond the apex of the

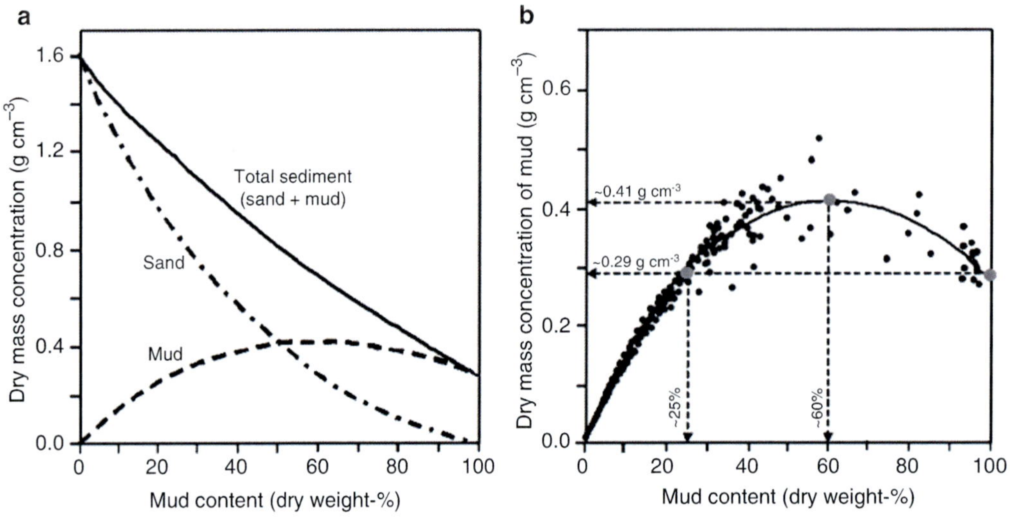

Fig. 10.12 Dry mass concentrations of sand and mud relative to the total sediment (**a**). Note the counter-intuitive trend in the progression of the mud component (**b**) which reverses after peaking around 60% in this dataset from the Wadden Sea (Based on Flemming and Delafontaine 2000)

regression curve, the network structure is very loose, being merely supported by flocs and aggregates. As sand is added, the grains initially fill the voids by expelling water without breaking down the network structure that results in a proportional increase in the mass concentration of mud. This continues up to the apex point where the sand content in this case is about 40%. As this limit is approached, the network structure begins to break down and the sediment is increasingly grain supported, water and mud now filling the voids between the sand grains. In effect this means that, in a unit volume of intertidal sediment, the mass concentration of mud in sediment consisting of pure mud (>95% mud content) is equal to that at mud contents as low as 25%, while the highest mass concentration of mud is registered at the apex of the regression curve. As in the case of bulk density, other tidal flat environments may have slightly different trends to the Wadden Sea example shown here. If required, corresponding calibration curves should therefore be established for other tidal flat environments.

The unexpected trend observed in mud mass concentration has far-reaching implications because any other parameter linked to the mud fraction (e.g., organic matter, trace elements, pollutants) will by necessity follow a similar trend. Contrary to common perception, highest mass concentrations of mud, and hence of any substances linked to the mud fraction, are found in mixed sediments (muddy sand and sandy mud) and not in pure mud. This potentially confusing issue and its pitfalls are discussed in detail by Flemming and Delafontaine (2000). A particularly common mistake is to relate measures of concentration, i.e. masses per unit volume or area, e.g., animal density per m^2, to measures of content, i.e. masses per unit mass, e.g. weight-% organic matter. By ignoring the dimensional incompatibility between contents and concentrations, it goes unnoticed that corresponding masses occupy increasingly larger volumes as the water content and the mud content increases. Thus, the volume occupied by a unit mass of pure mud with a dry bulk density ~0.3 $g\,cm^{-3}$ is more than five times larger than that occupied by the same mass of pure sand having a dry bulk density ~1.6 $g\,cm^{-3}$. For organic carbon, which is a measure commonly associated with the amount of food available to organisms, this disparity is illustrated in Fig. 10.13. The positive correlation between POC content and mud content (Fig. 10.13a) is commonly assumed to indicate that the largest amount of food is contained in pure mud as reflected by the highest POC content. However, as in the case of mud mass concentration (Fig. 10.12b), the relationship between POC mass concentration and mud content (Fig. 10.13b) clearly demonstrates that this assumption is wrong, the amount of POC per unit volume of sediment that corresponds to the dimensional measure for animal

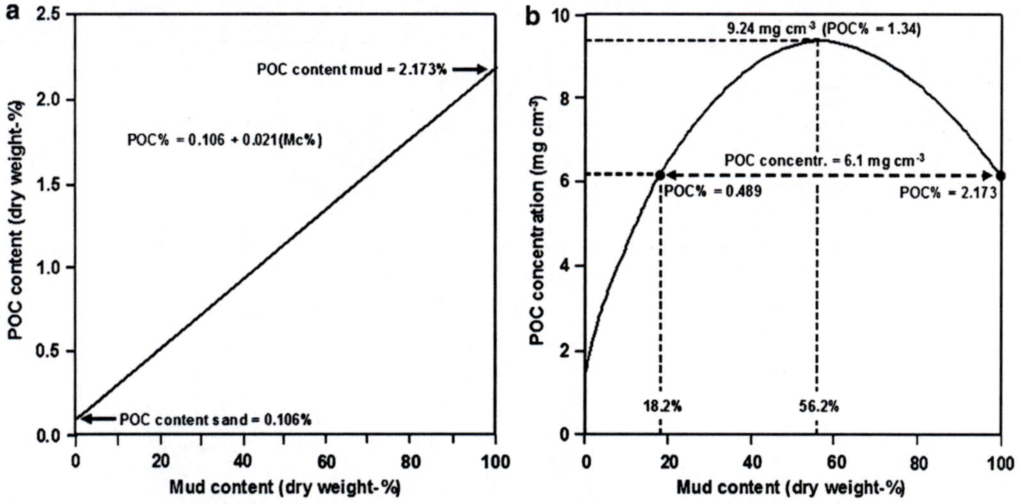

Fig. 10.13 Comparison of the trends in organic carbon content (**a**) and organic carbon concentration (**b**) as a function of mud content (data points omitted for clarity) (Based on Flemming and Delafontaine 2000). Note the similarity of the POC concentration curve to that of mud mass concentration in Fig. 10.12b

density being identical at 18% mud content, and up to 50% larger at intermediate mud content, in comparison to that at 100%. As pointed out earlier, this would also apply to any other sediment component linked to the mud fraction, e.g. heavy metals, trace elements, organic pollutants, toxic substances.

A mass physical property of intertidal sediment that plays an important role in the mobility or stability, i.e. the erosion resistance, of sediment, is the shear strength of the substrate. This parameter is conveniently determined in the field by a so-called vane shear apparatus where cross-vanes of different dimensions are calibrated such as to provide the shear strength upon yield after being inserted into the sediment and twisted against the resistance of a spring. Shear strength of intertidal sediment in relation to mud content, porosity, wet bulk density, and dry bulk density is illustrated in Fig. 10.14. Overall, the shear strength of intertidal sediment decreases with increasing mud content and porosity, and consequently increases with increasing wet and dry bulk density. Two features in the illustrated trends are of particular interest here. First, in all four cases, the highest and lowest shear strength for any value of the other parameter is well defined in what could be called an upper and lower boundary criterion. Both have a similar positive or negative trend as the mean trend line that would be defined by a regression analysis. In the case of mud content this means that for each of the criteria (i.e. upper boundary, mean, lower boundary) the shear strength progressively decreases as mud content increases, the reverse being true for bulk density. This applies in corresponding manner to any other correlating parameter.

The other interesting feature is the increasing range in shear strength (increasing standard deviation) toward lower mud content and porosity, and higher bulk density. The increasing scatter of the data points reflects an increasing variability in the degree of compaction (grain packing density) toward more sandy sediments. This is not unexpected as the hydrodynamic energy also increases toward higher sand content. The trends therefore trace the shoreward energy gradient together with its local variability, which is highest in sand. Thus, wave-compacted sands will have relatively high shear strength, whereas waterlogged sand will display a shear strength that may be as low as that at intermediate to high mud content. Excluded from these examples is dewatered mud commonly found in the subsurface of mixed flats, in channel-fill sequences, and between the neap and spring high-tide level where desiccation over the neap-tide period results in compaction and corresponding higher bulk density associated with lower porosity and water content.

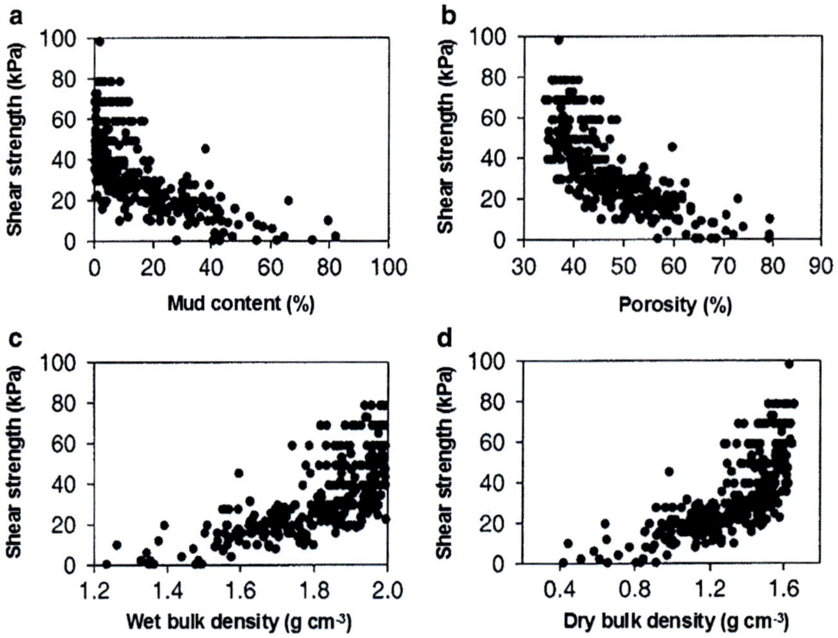

Fig. 10.14 Shear strength as a function of mud content (**a**), porosity (**b**), wet bulk density (**c**), and dry bulk density (**d**). Note the clearly defined maximum and minimum shear strengths for any value of the given parameters, and the increasing range (standard deviation) in shear strength with decreasing mud content and porosity, but increasing bulk density (Based on Xu 2000)

10.4 Depositional Facies and Sedimentary Structures

Barrier island depositional systems include a variety of facies that are intimately related to the morphological elements illustrated in Fig. 10.4. In the course of vertical accretion, the individual facies take on the form of three-dimensional, interfingering sedimentary units that define the internal architecture of the depositional system. In Fig. 10.15, the most important sedimentary units of back-barrier tidal flats are illustrated in a schematic geological cross-section of a Holocene barrier system (Fig. 10.15) that is aligned perpendicular to the shore and cuts through the middle section of a typical barrier island system of the Wadden Sea. Similar examples can be found in van Straaten (1964), Reineck and Singh (1980), Beets et al. (1996), and Vos and van Kesteren (2000).

The base of the depositional system is commonly formed by an erosional unconformity (here above Pleistocene deposits). In some places, pre-existing and partly eroded brackish-water deposits and basal peat can still be found. On the seaward side, the barrier system commences with shoreface/beach/eolian dune or shoreface/ebb-delta/inlet deposits not shown here, depending on where the cross-section is located relative to the barrier shoreline. The bulk of the back-barrier depositional system consists of channel fills and tidal-flat deposits, the latter getting progressively finer-grained toward the mainland shore. The land-sea transition is commonly marked by extensive salt marsh deposits. In the immediate rear of the coastal barriers, salt marshes and overwash deposits complete the depositional sequence. Intercalated brackish-water deposits and peat horizons may occur up to variable distances from the mainland shore, indicating temporary sea-level still-stands or short-lived regressions. Superimposed on this depositional system is a variety of biofacies comprising particular invertebrate animal communities adapted to exist in particular parts of the system (Frey and Howard 1969; Schäfer 1972; Howard and Frey 1975; Hertweck 1994). In addition to depending on the geographic (climatic) location, the community structure also depends on the energy gradient (current and wave exposure), on the tidal gradient (exposure or immersion period), and on sediment composition, including organic matter (food resources). Many of

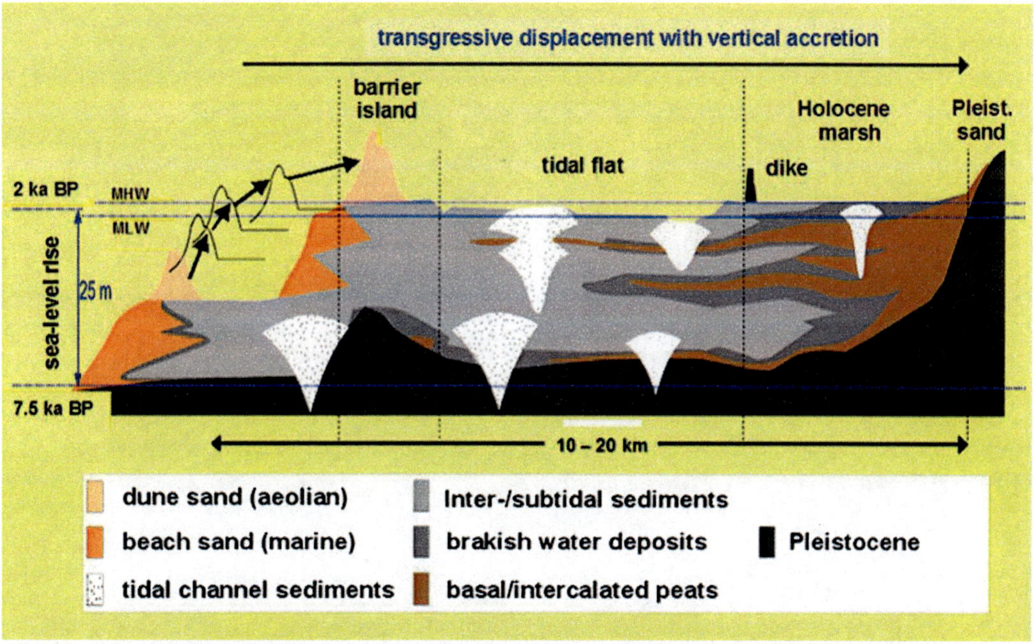

Fig. 10.15 Schematic geological cross-section through a transgressive barrier island depositional system as exemplified by the Wadden Sea and showing typically stacked sedimentary facies relating to particular coastal environments and facies (Modified after Streif 1990)

these organisms are shell-producing and hence contribute to the total and bioclastic material budgets of the back-barrier depositional system. In addition, most of the organisms are responsible for a partial or complete destruction of primary sedimentary structures up to the depth of their burrowing and feeding activity. This 'bioturbation' process also includes the activity of some higher order animals such as birds and fish.

In a geological context, bioturbation and the production of bioclastic material are tidal flat attributes that have only evolved in the course of the Phanerozoic (600 Ma BP – Present), older (Precambrian) deposits being generally devoid of such features. Finally, any list of biological influences would be incomplete if algal and bacterial activity were omitted. In this context, true algal mats, which consist predominantly of green algae and mostly occur on muddy tidal flats and in salt marsh pools, must be distinguished from mats produced around the mean high-water level by so-called 'blue-green algae' (cyanobacteria). These latter mats should correctly be referred to as 'microbial' mats (e.g., Gerdes and Krumbein 1987; Noffke 2010).

At smaller spatial scales, the basic depositional building blocks of back-barrier tidal flat systems outlined above are characterized by a large variety of physical and biological surface structures as well as internal sedimentary structures which, if comprehensively illustrated, would fill a whole book. Good summaries of typical clastic tidal facies and their sedimentary structures can be found in de Raaf and Boersma (1971), Klein (1977), and Reineck and Singh (1980), while biogenic structures and ichnofacies of temperate tidal environments are comprehensively dealt with in Schäfer (1972). For the purpose of this contribution, a selection of features is presented that, alone or in combination, have some degree of diagnostic power in identifying tidal flat deposits in the rock record.

10.4.1 Biological Surface Structures

The evolution of organisms in the course of the Phanerozoic, and their frequent adaptation to specific environmental conditions, has greatly facilitated the identification of particular depositional environments in the rock record. Intertidal flats are no exception in this context.

In Fig. 10.16a, a well preserved and still rooted tree stump in the middle of an intertidal flat suggests transgressive inundation in the East Frisian Wadden

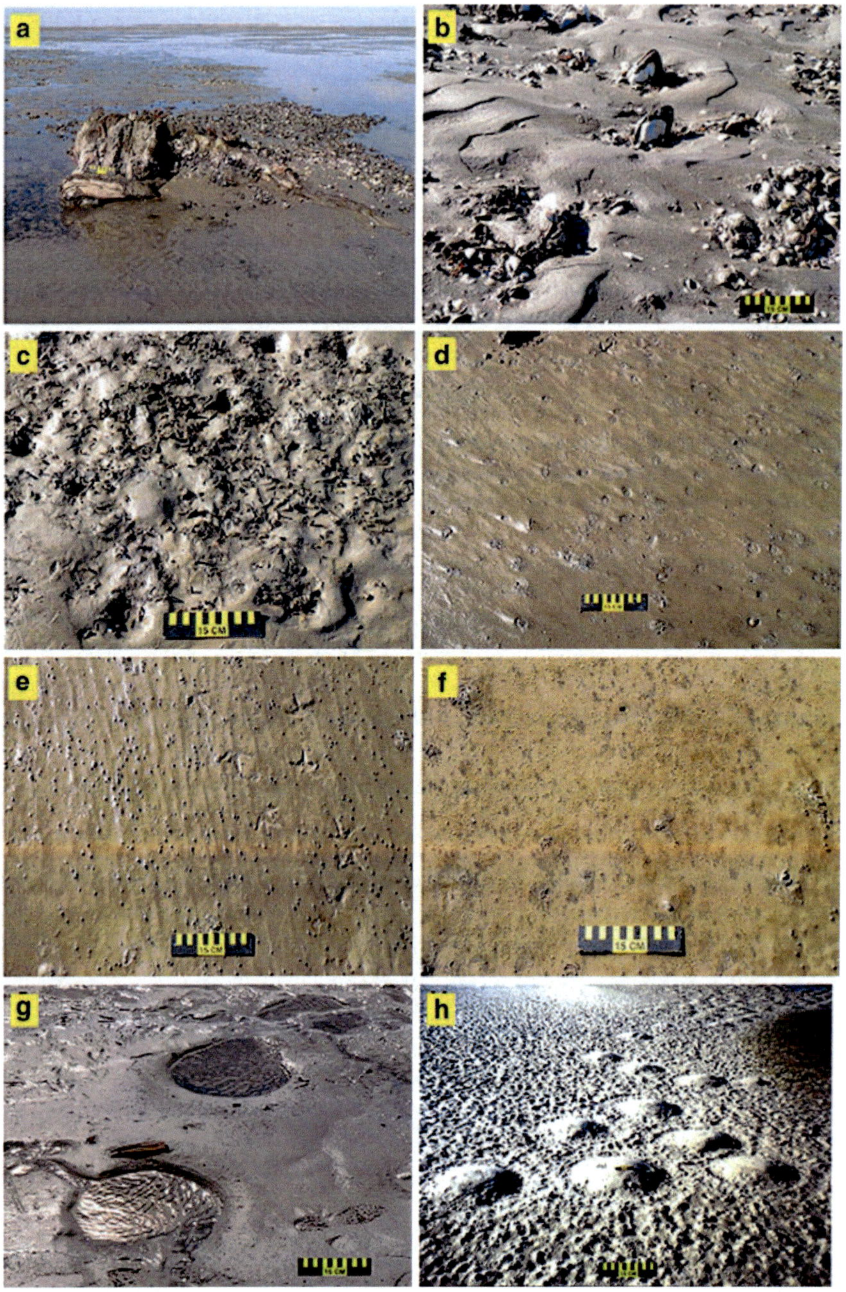

Fig. 10.16 Evidence for biological activity on tidal flats. (**a**) Rooted tree stump; (**b**) Shell lag together with articulated bivalve shells in live position (*Mya arenaria*); (**c**) Protruding polychaete tubes (*Lanice conchilega*) together with bird tracks and crawling traces of intertidal snails; (**d**) Sandy mud flat colonised by juvenile bivalves (*Cerastoderma edule*) living just beneath the sediment surface. Note bird track at the top and the uniformly aligned mounds and streaks indicating current flow from lower right to upper left; (**e**) Muddy sand flat with small polychaete fecal mounds (*Heteromastus filiformis*). Note the bird tracks and the current-aligned streaks emanating from the fecal heaps (current from bottom to top); (**f**) Slightly muddy sand flat colonised by *Arenicola marina* (large stringy fecal heaps) and *Heteromastus filiformis* (small *gray* fecal patches). Note patches of diatoms (*brownish discoloration*) producing gas bubbles (oxygen); (**g**) Feeding hollows created by trampling seagulls; similar hollows are made by rays; (**h**) Feeding hummocks created by flamingos (here in Langebaan Lagoon, South Africa)

Sea, Germany. Of similar diagnostic potential are organisms that only occur in intertidal environments, especially when preserved in live position such as the bivalve *Mya arenaria* in Fig. 10.16b. More difficult to assign to a tidal flat setting are organisms that also live subtidally. In such cases, additional evidence is required to diagnose an intertidal setting. The polychaete *Lanice conchilega* in Fig. 10.16c is a case in point. It is only in conjunction with other diagnostic criteria such as bird tracks that an intertidal setting can be allocated with some degree of confidence. The same applies to the cockle or heart mussel (here *Cerastoderma edule*) that lives just below the sediment surface and that can be recognised by the bumpy surface or the scars produced by slightly protruding shells (Fig. 10.16d). Similarly, while the presence of the polychaete *Heteromastus filiformis* is betrayed by the occurrence of small (black or gray) fecal heaps on the surface of modern tidal flats (Fig. 10.16e), it would be the bird tracks on the same bedding plane that would identify the depositional environment as being intertidal in the rock record. In principle this also applies to the lugworm *Arenicola marina* (Fig. 10.16f), here in community with *Heteromastus*. Circular resting hollows of rays or feeding hollows of wading birds (Fig. 10.16g, here seagulls) and feeding hummocks created by flamingos (Fig. 10.16h, here in Langebaan Lagoon, South Africa) complete the picture.

Considering that large-scale exposures of fossil bedding planes are relatively rare in comparison to vertical sections, it is inherently difficult to identify intertidal settings with confidence in the rock record on the basis of surface structures alone unless additional unequivocal diagnostic evidence is available. Such evidence includes late-stage emergence runoff features and traces of organisms restricted to the intertidal, including the tracks and feeding structures of wading birds. In cold climates, such evidence would in addition encompass tool marks induced by moving ice floes. In all other cases, it would be the association of a multitude of features which, by application of the exclusion principle, could eventually justify a decision in favour of a particular environment.

10.4.2 Physical Surface Structures

Prominent physical surface structures frequently observed on tidal flats include wave and current ripples that display features characteristic of very shallow water and late-stage emergence. Prominent among these are extensive sheets of symmetrical and asymmetrical wave ripples (Fig. 10.17a). Because ripples also occur in subtidal environments, one should in addition look for evidence of late-stage emergence. Such features include ladderback ripples (Fig. 10.17b), especially where smaller trough-bound wave ripples aligned perpendicular to the larger ripple crests are associated with water-level marks (Fig. 10.17c). The small wave ripples in the troughs in Fig. 10.17c formed when the water level had dropped below the crest level of the larger ripples but before the water-level marks formed which later dissected the crests of the small ripples where they merge with the steep slopes of the larger ones.

Features formed during late-stage run-off are particularly diagnostic for emergence at low tide. Among these are narrow streams of linguoid current ripples dissecting wave-rippled surfaces, the crests of the latter having in this case been flattened just before emergence (Fig. 10.17d). Shallow, laterally migrating intertidal creeks are commonly paved by shell beds (Fig. 10.17e), in this case overlain by narrow sand ribbons formed during upper-plane-bed flow shortly before emergence. It should be noted here, however, that shell concentrations can also result from the burrowing activity of intertidal organisms, in particular *Arenicola marina* (van Straaten 1952). Near steeper channel margins, such creeks display a multitude of late-stage runoff features such as scour pits around shells, grooves, rill marks, microbars and small fan structures (Fig. 10.17f).

Wave- and current-generated ripples are frequently observed in muddy sediments upon exposure at low tide (Fig. 10.17g). This contradicts the common perception that such bedforms do not form in fine-grained sediments. Their occurrence has been explained by the aggregated nature of the mud during transport and deposition, the aggregates and also fecal pellets initially responding to waves and currents as non-cohesive particles would, similar to the fine or very fine sand to which they are hydraulically equivalent (Schieber and Southard 2009). In contrast to rippled mud, tidal flat surfaces may locally become draped by thin blankets of fluid mud (Fig. 10.17h) that often display erosional windows revealing the underlying sediment together with any surface structures on them (in this case wave ripples). An analogous feature can be observed in

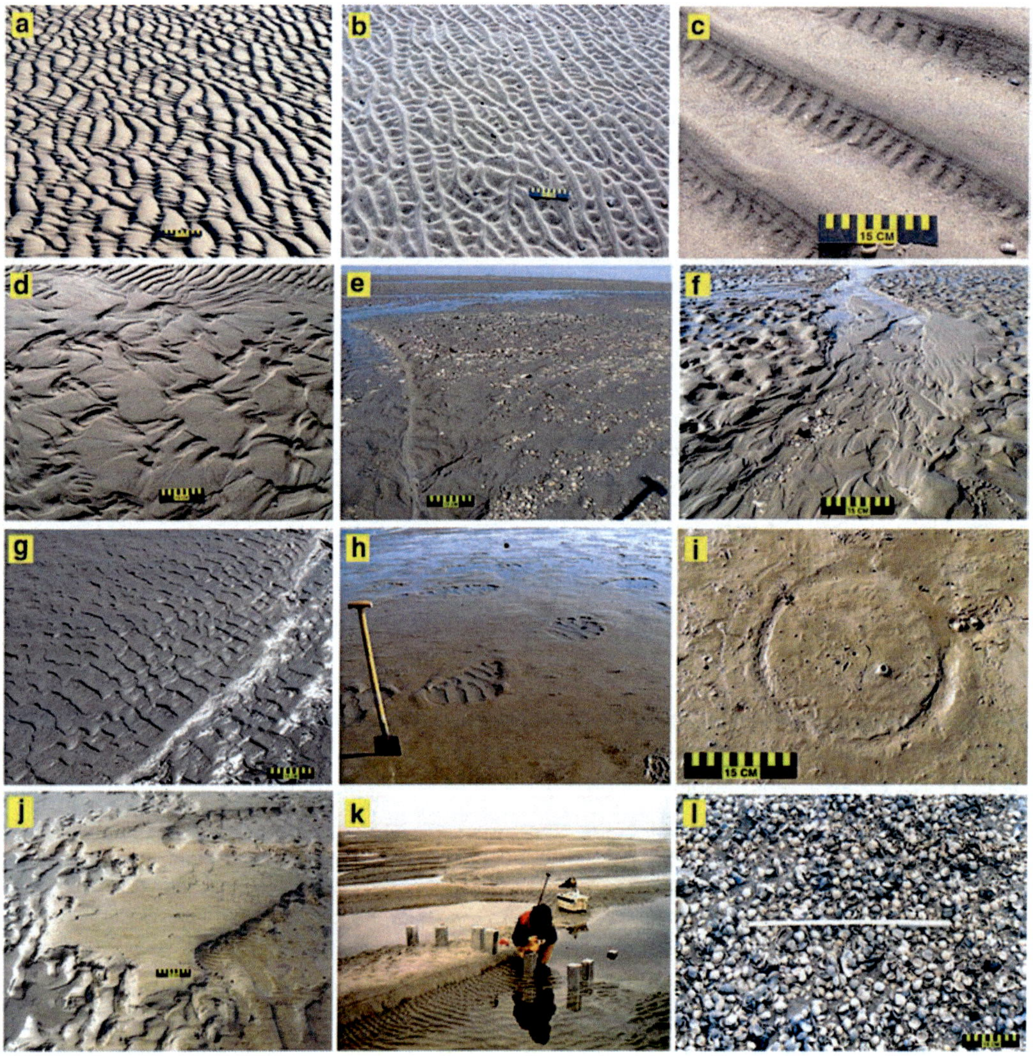

Fig. 10.17 Physical surface structures frequently observed on tidal flats. (**a**) Asymmetrical wave ripples; (**b**) Ladderback ripples; (**c**) Small wave ripples in the troughs of larger ripples and water-level marks; (**d**) Late-stage runoff with linguoid current ripples dissecting a field of flat-crested wave ripples; (**e**) Shallow intertidal creek with small sand ribbons over shell pavement; (**f**) Late-stage runoff features; (**g**) Current ripples in mud; (**h**) Thin fluid mud sheet with scour windows displaying ripples on the surface of underlying sand. (**i**) Circular tool mark formed by the rotation of a protruding polychaete tube. Note the bird tracks surrounding the structure; (**j**) Rippled sand bed with patchy wash-outs formed shortly before emergence; (**k**) Intertidal dunes; (**l**) Shell pavement

sandy sediment where a rippled surface is overlain by slightly elevated, smooth-topped sand patches (Fig. 10.17j). These patches are the remnants of an eroded sand sheet that was locally stabilized by diatoms. The lower surface was subsequently covered by wave ripples, whereas the surfaces of the elevated patches were smoothed by wind-induced washover shortly before emergence.

Tool marks are less frequent than other surface structures on tidal flats but may on occasion be encountered where driftwood or dislodged algae have scraped or rolled across the sediment surface in shallow water. In contrast to this, a large variety of scour, prod and roll marks induced by drifting ice floes are ubiquitous on tidal flats in cold regions (Dionne 1974; Reineck 1976; Dionne 1988; Pejrup and Andersen 2000). A rather unusual concentric tool mark is illustrated in Fig. 10.17i where a protruding polychaete tube has excavated a circular groove around its holdfast (diameter ca. 20 cm). Larger-scale current-generated features on tidal flats are represented by 2D and 3D dunes (Fig. 10.17k) that, in back-barrier tidal basins, are usually best developed at spring tide. Upon emergence, the dunes commonly display well-developed water-level marks along their steep slipfaces. Finally, extensive and often very selective shell beds swept together by wave action, in contrast to current-generated lag deposits, can be found locally on more exposed parts of intertidal flats (Fig. 10.17l).

10.4.3 Internal Sedimentary Structures

Reconstruction of ancient depositional environments is commonly based on the interpretation of internal sedimentary structures, bedding types, and stratification sequences observed in rock outcrops or cores. In modern environments, internal structures are either visualized by trenching and preparation of lacquer peels or, where conditions prevent this, by coring and preparation of relief casts using suitable resins (e.g., Bouma 1969). Due to the water-saturated nature of the sediments, coring is the only feasible procedure in tidal flat research. While sedimentary structures are well preserved in cores, they have the disadvantage of only revealing narrow sections of laterally more extensive structures.

In spite of this, the systematic preparation of both short box-cores since the 1950s and longer vibro-cores since the late 1970s has revolutionized our understanding of tidal flat deposits (Reineck and Singh 1980). This is illustrated in Fig. 10.18 by a small selection of relief casts ranging from exposed sand flats to protected salt marshes. As mentioned earlier, small subaqueous dunes are best developed along the ebb-dominated, outer tips of flood ramps on sand bodies facing the inlet. This is exemplified by the cross-bedding in Fig. 10.18a where the flow was dominated by the current flowing from right to left. The tangential cross-beds are indicative of 3D dunes and hence relatively strong flow as opposed to planar cross-beds indicative of 2D dunes and weaker flow. Considerably weaker and more evenly distributed bidirectional currents of uniform strength are reflected in the sequences of vertically-stacked herringbone cross-stratified units in Fig. 10.18b. The term 'herringbone' is strictly reserved for sets of ripple cross-stratified beds displaying opposing dip directions formed in the course of individual ebb-flood or flood-ebb cycles. Not all opposing cross-beds comply with this definition because individual units may be separated by hiatuses of varying duration. Furthermore, misinterpretations can result where bidirectional currents are wrongly inferred from trough cross-beds cutting each other at odd angles (Reineck and Singh 1980).

A typical stratification sequence found along shallow, migrating intertidal creeks is illustrated in Fig. 10.18c where the partly excavated and still articulated shells of *Mya arenaria* in live position protrude through a shell lag deposit that accrued as a tidal creek migrated across it. The core reflects in a remarkable way the subsurface conditions of a surface situation as illustrated in Fig. 10.16b. A second shell layer near the bottom of the core indicates the depth of the tidal creek during a previous crossing. The horizontally stratified channel fill above this layer is partly obliterated by bioturbation. Quite different are the deposits found along the margins of larger and deeper channels. Such tidal flat margins are frequently composed of horizontally laminated, partly waterlogged beds that, at depth, may be deformed into convolute beds by sudden liquefaction events (Fig. 10.18d; cf. Wunderlich 1967). In other cases, channel-margin deposits comprise small current- and/or wave-rippled cross-bedded sets (Fig. 10.18e). Convolute lamination has also been observed to form as a result of entrapped air (de Boer 1979). Processes and concepts of convolute bed formation, also including tidal flats, have been comprehensively described by Williams (1960) and Einsele (1963).

Proceeding from sand flats to mixed flats, the degree of bioturbation gradually increases (Fig. 10.18f, i). At low mud content (slightly sandy mud), ripple troughs initially get draped by thin mud layers that, in cross-section, produce the well-known flaser structures (Fig. 10.18g). As the mud content increases with decreasing energy, the mud drapes get thicker and eventually form interconnected wavy layers alternating with rippled sand layers to produce the characteristic wavy bedding around the transition between intertidal muddy sand and sandy mud facies. Finally, as the sand content decreases (sandy mud to slightly sandy mud facies), the internal sedimentary structures are now dominated by thick mud drapes interrupted by connected or disconnected sand lenses (starved ripples)

Fig. 10.18 Internal sedimentary structures typical for tidal flat deposits. (**a**) Bidirectional dune cross-bedding with current dominance from right to left; (**b**) Bidirectional ripple cross-bedding with well-developed herringbone structures; (**c**) Partly bioturbated sand with shell layer at depth and shell lag at the surface. Note partly excavated shell of the bivalve *Mya arenaria* in live position; (**d**) Horizontally bedded sands above several convoluted bedsets; (**e**) Multidirectional wave and current ripples in sand. Note the absence of clear herringbone structures; (**f**) Partly bioturbated, horizontally bedded sand in lower part of core, grading into partly bioturbated rippled sand in upper part; (**g**) Flaser bedding typical for muddy sand facies; (**h**) Lenticular bedding typical for sandy mud facies; (**i**) Weakly laminated sand in lower core, followed by well preserved lamination in upper core, both penetrated by a large worm tube, possibly of *Arenicola marina* (u-part hidden); (**j**) Rooted salt marsh deposit

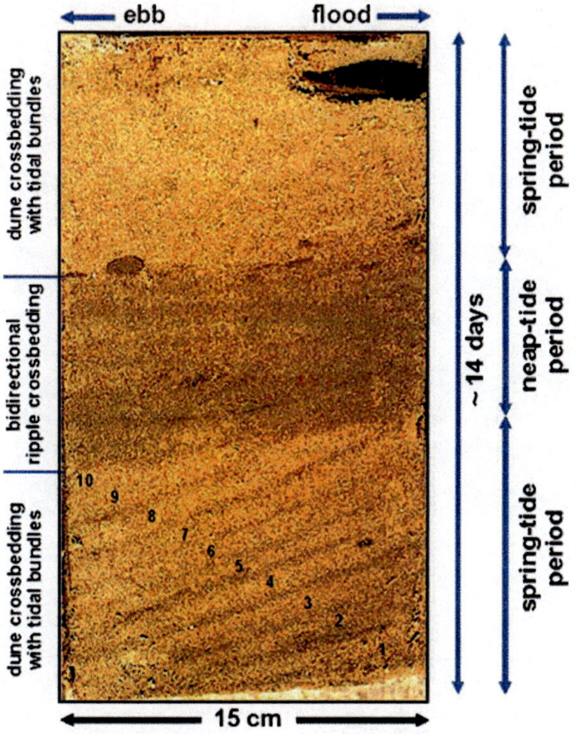

Fig. 10.19 A vertically stacked spring-neap-spring cycle with tidal bundles preserved in intertidal dune cross-beds formed over spring tide. The bundles are clearly visible in the *bottom sequence*, but only faintly so in the *upper* one. Note that the bundles are not separated by mud drapes, but instead by finer-grained sand

to form so-called lenticular bedding (Fig. 10.18h, cf. Reineck and Wunderlich 1968. Flemming 2003a). Due to the high water content of the mud, overburden pressure can result in the formation of convolute bedding in this environment (Fig. 10.18h, bottom). Pure mud is either completely homogenized or thinly laminated, depending on the degree of local bioturbation. These eventually grade into salt marshes, the laminated deposits of which are usually intensely bioturbated by root structures (Fig. 10.18j).

The cores illustrated in Fig. 10.18 do not include any evidence of tidal bedding such as sand-mud couplets or tidal bundles associated with deposition in the course of neap-spring cycles. Such rhythmic sedimentary structures are well documented from subtidal channels (Visser 1980; Allen and Homewood 1984) and from estuarine mudflats in macrotidal settings (Dalrymple et al. 1991), but have rarely been reported from back-barrier tidal flats. A cross-bedded example from the Wadden Sea is presented in Fig. 10.19.

The box-core was recovered from a dune of the type shown in Fig. 10.17k. It is interpreted to show a vertically-stacked spring-neap-spring sequence with clearly visible tidal bundles in the crossbeds of the lower unit and more faintly preserved ones in the upper unit. The two units were formed during the ebb tide over successive spring-tide periods and are separated by a bipolar, current-rippled sequence formed over the intervening neap-tide period. Due to the small width of the core, it is not precisely clear how many bundles were actually formed in each case, at least ten (representing 5 days) having been identified in the lower unit.

In contrast to the rather rare occurrence of tidal bundles in cross-bedded sand of back-barrier tidal flats, tidal bedding represented by sand-mud couplets is more frequently encountered. These are preferentially formed along mobile intertidal creeks as long as sufficient suspended matter is available to settle out at high tide. However, as intertidal creeks are rather shallow, one rarely finds more than just a few sand-mud couplets stacked above each other (Fig. 10.20). Each cycle begins on the rising tide as the tidal flat is inundated and the flood current begins to move sand across the sediment surface formed during the previous falling tide. Mud then settles out during the slack-water period over high tide and is subsequently covered by a sand layer in the course of the ebb tide. Each sand layer may be composed of two opposing current-generated ripple cross-stratified units, the thickness of each depending on the relative dominance of one current component over the other. In contrast to subtidal (de Boer et al. 1989) or intertidal estuarine rhythmites (Dalrymple et al. 1991), one would not expect large numbers of stacked couplets or any clear evidence of the daily inequality of the tide. At a larger spatial scale, a characteristic depositional facies is the so-called 'inclined heterolithic stratification' (Thomas et al. 1987). These form in the process of lateral channel migration or meandering, and are identified on the ground by what has also been called 'longitudinal' or 'lateral-accretion' bedding (Reineck 1958; Bridges and Leeder 1976).

As sand content decreases and mud content increases toward the mainland coast, the tidal flat gradually transforms into an almost featureless muddy plain, tidal channels or creeks being now restricted to locations where freshwater streams drain the hinterland. This is in stark contrast to non-barred macrotidal mud flats that are commonly sculptured into meandering

Fig. 10.20 Tidal bedding in form of vertically stacked mud-sand couplets along laterally migrating intertidal creeks. The mud settles out at high tide and is subsequently covered by a sand layer during the following ebb tide

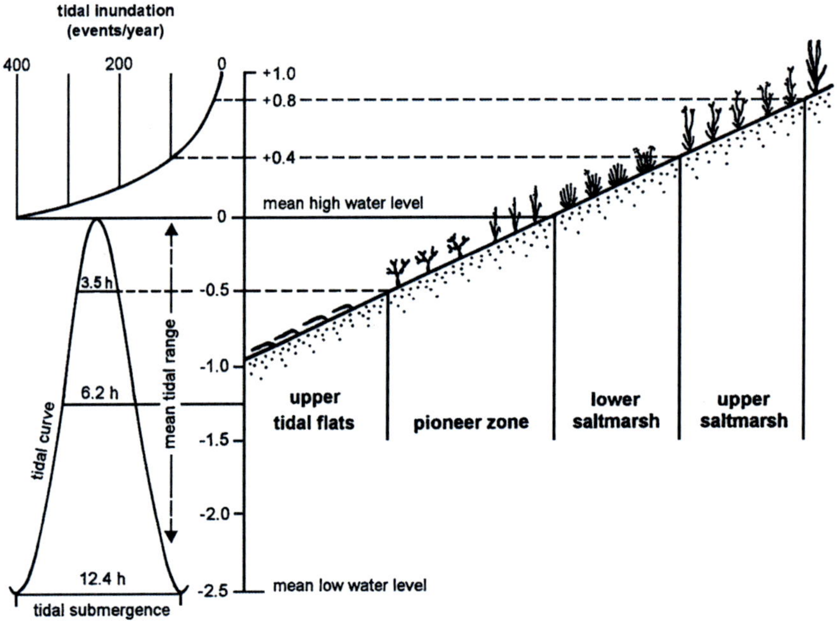

Fig. 10.21 Plant zonation pattern marking the transition from tidal flat to salt marsh (Modified after Streif 1990). In the Wadden Sea, the pioneer zone, which is occupied by *Salicornia* sp. and *Spartina* sp., begins at the elevation where the duration of tidal submergence is <3.5 h and terminates at the mean high-tide level where the salt marsh proper begins. Note that salt marsh zonation is primarily controlled by the annual frequency of inundation by seawater

tidal creeks and/or longitudinal mud ridges separated by erosional trenches (e.g., Gouleau et al. 2000; O'Brien et al. 2000). In back-barrier tidal basins, the monotonous muddy landscape only changes with the onset of vegetation, the uppermost tidal flats being colonized by so-called pioneer plants comprising halophytes (e.g., *Salicornia* and *Spartina*) which occupy a zone between the elevation where tidal submergence is <3.5 h (approx. 0.5 m below MHT in upper mesotidal settings) and the mean high-water line (Fig. 10.21).

The salt marsh proper begins at the mean high-water line, the transition between the halophytes of the pioneer zone and the salt resistant plants of the salt marsh being exceptionally sharp. The salt marsh itself

Fig. 10.22 Characteristic sedimentary features around the mean high-tide level. (**a**) Final meander of a salt marsh creek at low tide before draining onto the upper intertidal flat. Note that the water in the creek is not completely drained because it is dammed by a small 'ebb-delta' lobe at the transition to the open tidal flat; (**b**) Well developed mud cracks in sparsely vegetated *Salicornia* marsh; (**c**) Cut meander bank in *Spartina* marsh revealing preserved laminae near the surface and complete obliteration of physical structures at depth

can be subdivided further by specific plant associations comprising different species in different climates and geographic locations. In contrast to the smooth intertidal mud flats, the salt marshes are drained by an intricate network of meandering marsh creeks (Fig. 10.22a). Although the creek beds are generally excavated to depths below the elevation of the adjacent mud flats, they rarely extend into the latter as they commonly terminate in small 'ebb-delta' lobes at the salt marsh boundary where the channel-confined flow spreads out onto the open tidal flat. These depositional lobes often prevent the marsh creeks from draining completely at low tide.

Mud-cracked surfaces are frequently regarded as good indicators of emergence in tidal environments (Klein 1977). This applies in particular to tidal flat depositional systems with large differences between the elevations of neap high tide and spring high tide. The larger this difference, the longer the period of emergence over neap tide and the more extensive the mud-cracked surfaces. In regions where this difference in elevation is small, for example in the Wadden Sea, such surfaces are narrow and patchy (Fig. 10.22b). Mud-cracks and the roots of salt marsh plants tend to destroy any lamination in the course of time, as can be seen in Fig. 10.22c where the lamination is still preserved in the upper few centimetres but completely obliterated below (cf. also Fig. 10.18j).

In addition to the characteristic salt marsh zones associated with specific plants, the transition from upper intertidal to lower supratidal flats is locally characterized by laminated sediments (mats) produced by microbial activity, especially in places where this transition is more sandy (Fig. 10.23). As pointed out earlier, these microbial mats have to be carefully distinguished from algal mats produced by green algae (Gerdes and Krumbein 1987). Microbial mats commence at the sediment surface with a thin filamentous

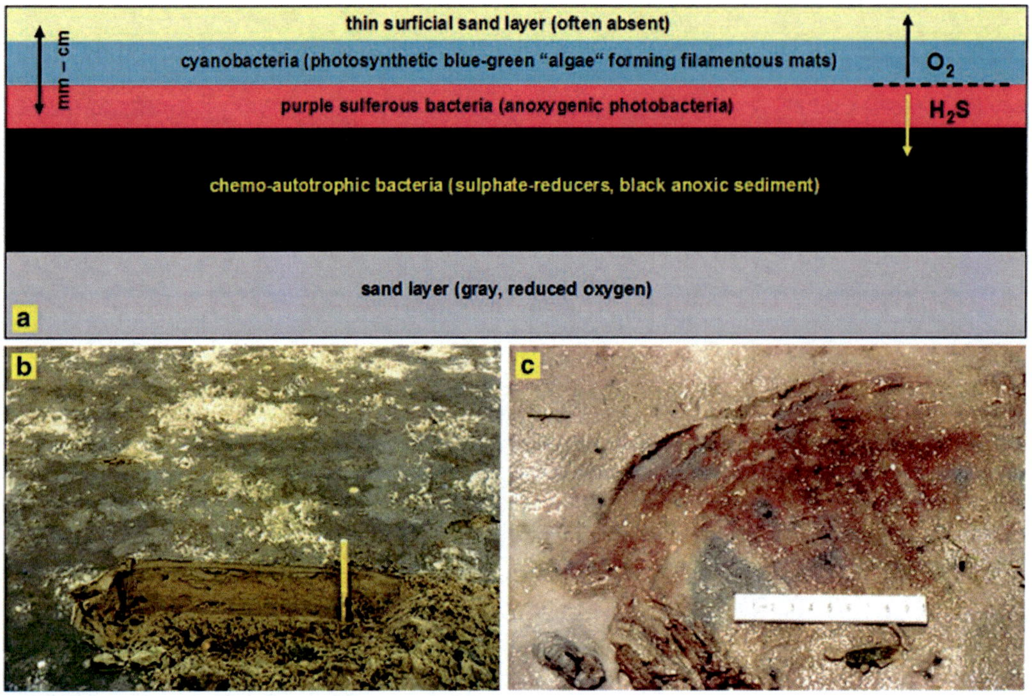

Fig. 10.23 Structure of a typical microbial mat occurring at the land-sea transition around the mean high-tide level (upper intertidal to lower supratidal). (**a**) Schematic model illustrating the vertical succession of bacteria in a typical microbial mat; (**b**) Photograph of a back-barrier microbial mat on the supratidal flat of a Wadden Sea island; (**c**) Intensely purple colored sulferous bacteria beneath a thin layer of sand in the rear of a Gulf Coast barrier island of Florida

carpet composed of oxygen-producing cyanobacteria (blue-green algae) (Fig. 10.23b). If covered by sand, the cyanobacteria migrate to the surface where a new mat is constructed, leaving behind the organic material of the old mat below the sand layer. The organic matter of the abandoned mat is then decomposed by so-called chemo-autotrophic sulphate-reducing bacteria in the course of which oxygen is depleted to produce a black anoxic layer. Just beneath the cyanobacterial mat anoxygenic sulferous photobacteria, identifiable by their intense purple color, are frequently observed. Being photobacteria, their activity increases markedly from the darker high latitudes towards the brighter low latitudes. Because of this, they are sometimes hard to spot in places like the Wadden Sea (55°N), whereas they occur in profusion in places like Florida (28°N) (Fig. 10.23c, cf. Davis 1994b). Because sand covering and subsequent upward migration of cyanobacteria occurs relatively frequently, several black horizons may be stacked above each other, the depletion of oxygen also affecting the sand below the mats which takes on a dark gray color. Due to the varied color scheme of the mats, this laminated microbial facies straddling the land-sea boundary has been given the apt name 'versicolored' tidal flat (Gerdes et al. 1985). When preserved, the characteristic lamination associated with specific bacteria makes it an excellent diagnostic tool for the identification of the land-sea boundary in the rock record (Schieber 2004; Noffke et al. 2006).

The trends of major parameters characterizing intertidal flats along the energy gradient between the mean low-tide and mean high-tide levels are summarized in Fig. 10.24. Parameters that decrease toward mean high tide include hydrodynamic energy, duration of water cover, submergence time, sand content of the sediment, and physical sediment reworking. The opposite trend is observed for mud content, exposure time, organic matter content, phytobenthos (diatoms), and plant fragments. Notable exceptions are seaweeds which preferentially occur on mixed flats, microbial mats that are restricted to a narrow zone around mean high tide, and last but not least the mass concentration

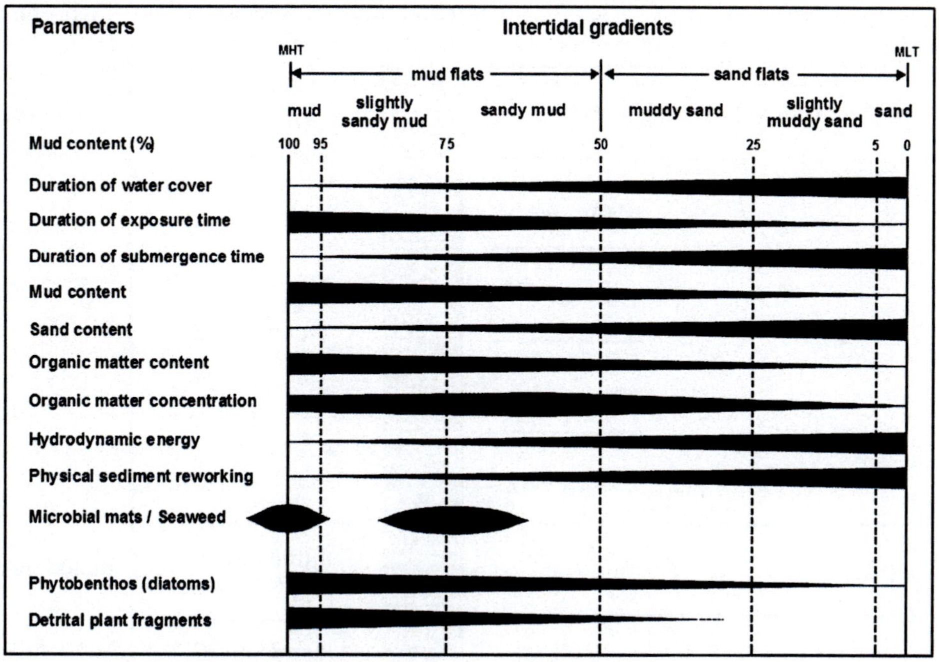

Fig. 10.24 Diagram summarizing the trends of major parameters characterizing intertidal flats. While most parameters either increase or decrease along the gradient between mean low tide (*MLT*) and mean high tide (*MHT*), the concentration (mass per unit volume) of organic matter peaks in the sandy mud facies, as would any other substance linked to the mud fraction (Modified after Hertweck 1994)

(mass per unit volume) of mud and any substances linked to the mud fraction. As outlined earlier, this would, for example, not only apply to the concentration of mud itself, but also to organic matter, heavy metals, trace elements, organic pollutants and toxic substances.

10.5 Stratigraphic Relationships

Having discussed and illustrated a variety of typical intertidal sedimentary structures and bedding types, the question arises of how these might be preserved in the rock record. Applying Walther's Law by vertically stacking typical sedimentary facies characterizing the intertidal gradient, a synthetic upward-coarsening transgressive facies model has been constructed (Fig. 10.25). The model is based on Reineck and Wunderlich (1968) and includes the most characteristic sedimentary structures generally observed along the intertidal gradient from sand flats near MLT to mud flats near MHT. While highlighting the most important primary sedimentary structures (flaser, wavy and lenticular bedding), the model is inherently incomplete because bioturbation, shell accumulations and sedimentary structures typical of intertidal creeks have been excluded. As such it would be more applicable to Precambrian than Phanerozoic tidal flats. However, as shown below, it serves a very useful purpose in that it simplifies the interpretation of generally much more complicated real-world situations.

Such a real-world situation, in this case representing a particular location in the modern Wadden Sea, is illustrated in Fig. 10.26 (after Chang et al. 2006c). The upward-coarsening sedimentary sequence recorded in the core clearly documents a transgressive setting commencing with mudflat deposits at the bottom and ending with sand flat deposits at the top. A closer look at the sequence reveals a number of features that complicate the interpretation. Thus, while the mudflat deposit in the lower core section is interspersed with thin lenticular beds as one would expect, the sequence does not progressively grade upward into wavy and lenticular bedding as required by the idealized model, but is

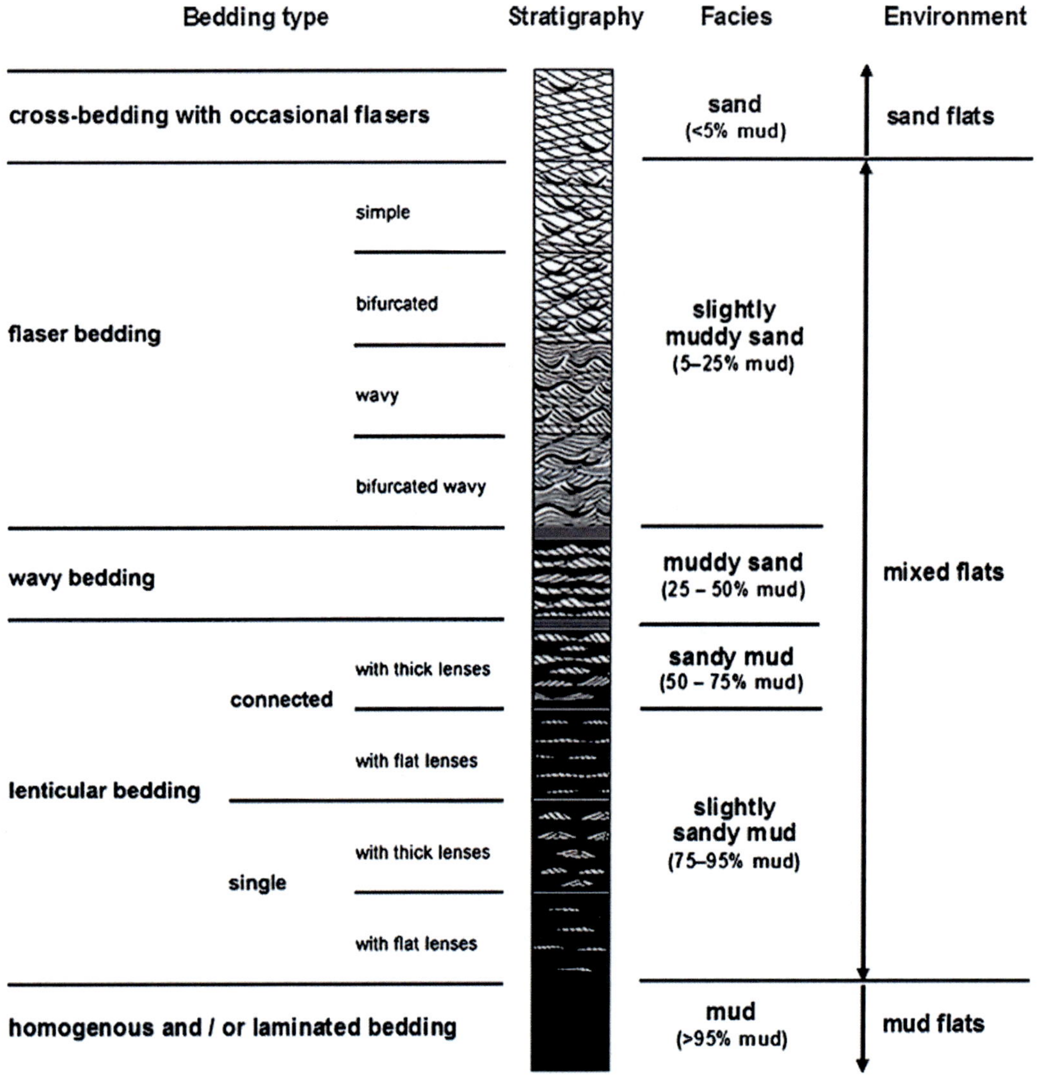

Fig. 10.25 Synthetic section in which successive sedimentary facies occurring between the low-tide and the high-tide level of intertidal flats (without tidal creeks and bioturbation) have been vertically stacked in an idealized transgressive facies model (Modified after Reineck and Wunderlich 1968; cf. also Flemming 2003a)

instead truncated by an erosional surface that is followed by an 80-cm thick sequence of alternating sand and shell beds, the latter being partly graded inversely. In the course of transgression and accretion, the former mudflat at the core site was evidently crossed a number of times by a migrating intertidal creek that reworked and destroyed any lenticular and wavy bedding that may have existed, leaving behind the reworked sand and shells in the form of stacked lag deposits. Above the channel deposits, the sequence continues with sparsely interspersed flaser beds and more prominently displayed cross-bedded sand of mostly wave-generated origin, alternating with thin mud layers typical of muddy and slightly muddy sand flats. Toward the top, the sequence grades into exposed sand flats. Significantly, bioturbation is only preserved in the uppermost layer.

As outlined above, the depositional sequence preserved in the core can be rationally explained on the basis of the idealized model, part of the expected

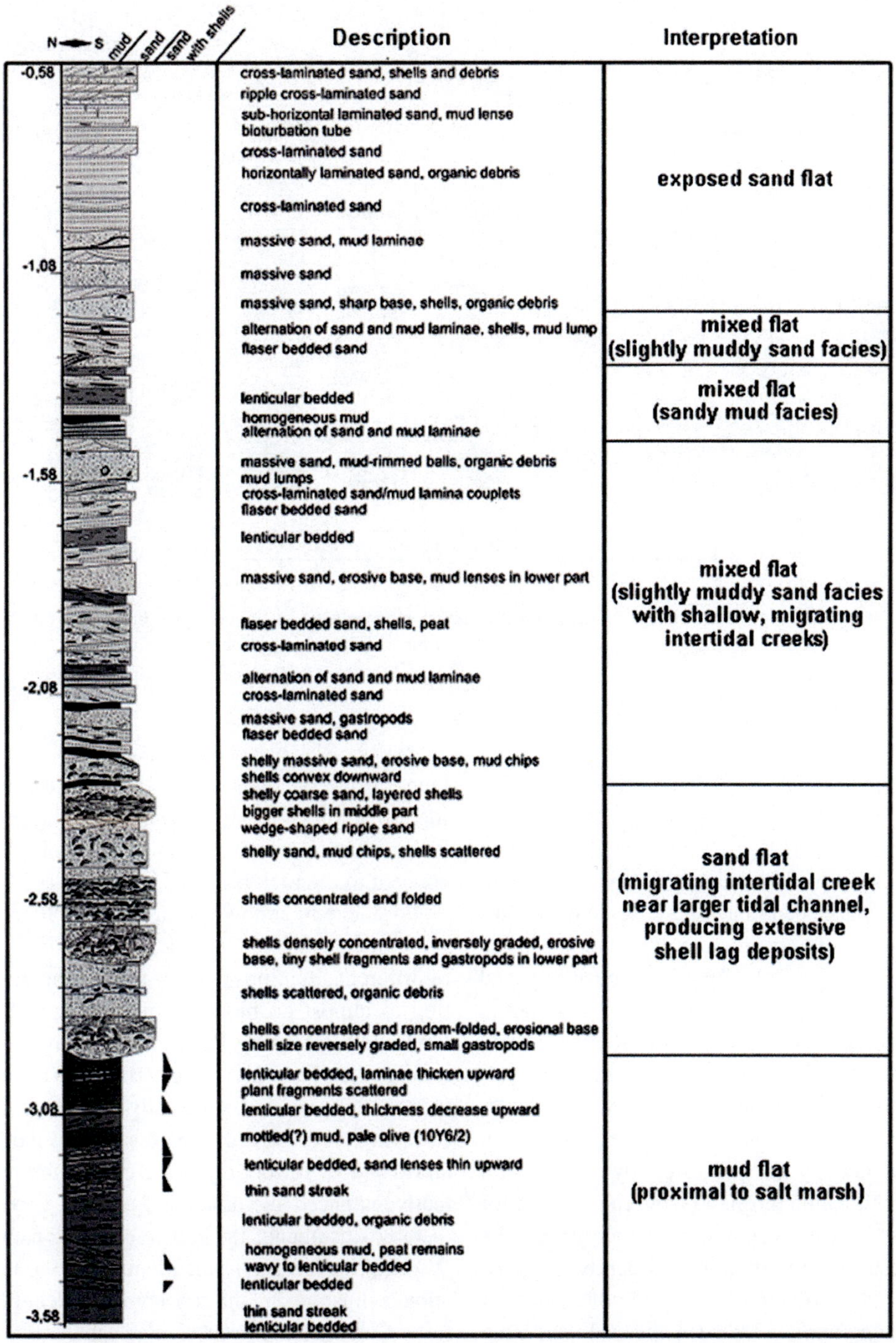

Fig. 10.26 Real-world transgressive section as observed in a vibro-core from the Wadden Sea (Modified after Chang et al. 2006c). Note the discontinuous nature of the succession and the multiple erosion surfaces

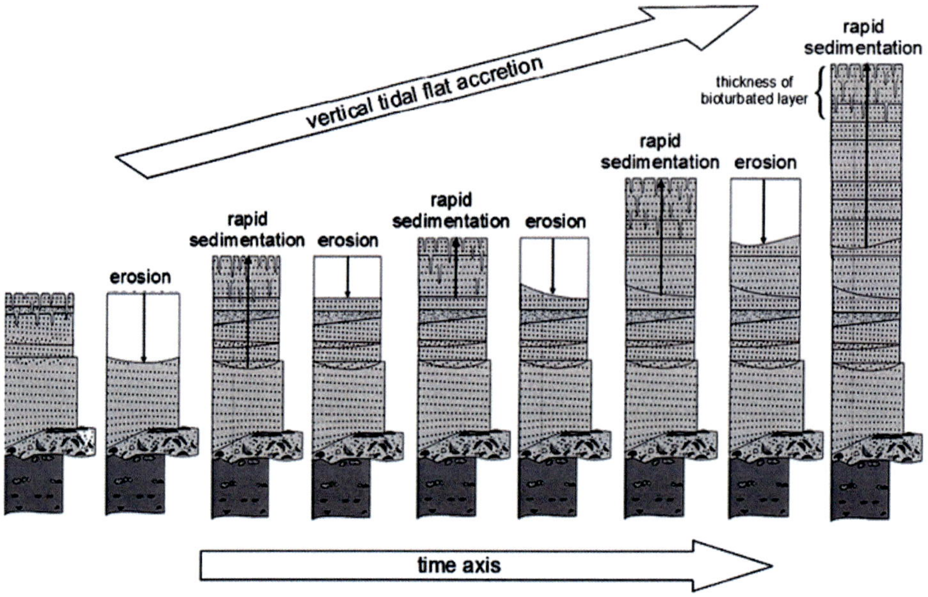

Fig. 10.27 Conceptual model illustrating the situation where physical reworking of the sediment outpaces bioturbation in the course of vertical accretion as observed in many parts of the Wadden Sea (Modified after Chang et al. 2006c)

sequence having simply been removed by migrating channels. Tidal-flat deposition above the channel sequence commences at higher energy conditions than required for the formation and preservation of lenticular, wavy and flaser bedding. The core demonstrates that, in the course of transgression, local conditions on a tidal flat can vary strongly, frequent depositional and erosional events ultimately resulting in only partial preservation of the potential record (Reineck 1960; van der Spek 1996). This not only concerns the preservation of particular sedimentary sequences as illustrated in the idealized model, but also the preservation of biological activity as suggested by the general absence of bioturbation throughout the core, except for the uppermost active layer. Within certain limits, the degree of preserved bioturbation is evidently an excellent criterion for the depth of reworking of a tidal flat by waves and/or currents. Because physical reworking of intertidal flats is generally restricted to the depth of intertidal creeks (~50 cm), this concept can only be applied to regions where the bioturbated layer is relatively thin (20–30 cm). This is generally the case in temperate climates. In subtropical and tropical climates, by contrast, callianassid shrimps will completely bioturbate the sediment to depths exceeding 1 m and hence obliterate the effects of physical reworking. Identification of the type of bioturbating organism(s) is thus a crucial prerequisite for the application of this concept to the rock record (Dott 1983, 1988).

For the temperate Wadden Sea, van Straaten (1954) presented a conceptual model that considers a number of different situations ranging from complete bioturbation to almost no bioturbation. The model was later modified slightly by Reineck and Singh (1980). As remarked earlier, only the uppermost 15 cm of the core in Fig. 10.26, i.e. the biologically active layer at the time of coring, is preserved. This means that the frequency and depth of reworking at the coring site has consistently outpaced bioturbation from the very start and thereby documents the rather exposed nature of the Wadden Sea (cf. Davis and Flemming 1995). This situation is illustrated in the conceptual model of Fig. 10.27 (after Chang et al. 2006c; cf. also van der Spek 1996).

As shown above, the succession of depositional facies along the intertidal gradient and their characteristic sedimentary structures will produce typical and easily recognized stratigraphic sequences in the course of vertical accretion. The main driving forces on geological time scales are changes in relative sea level and sediment supply, both being primarily controlled by

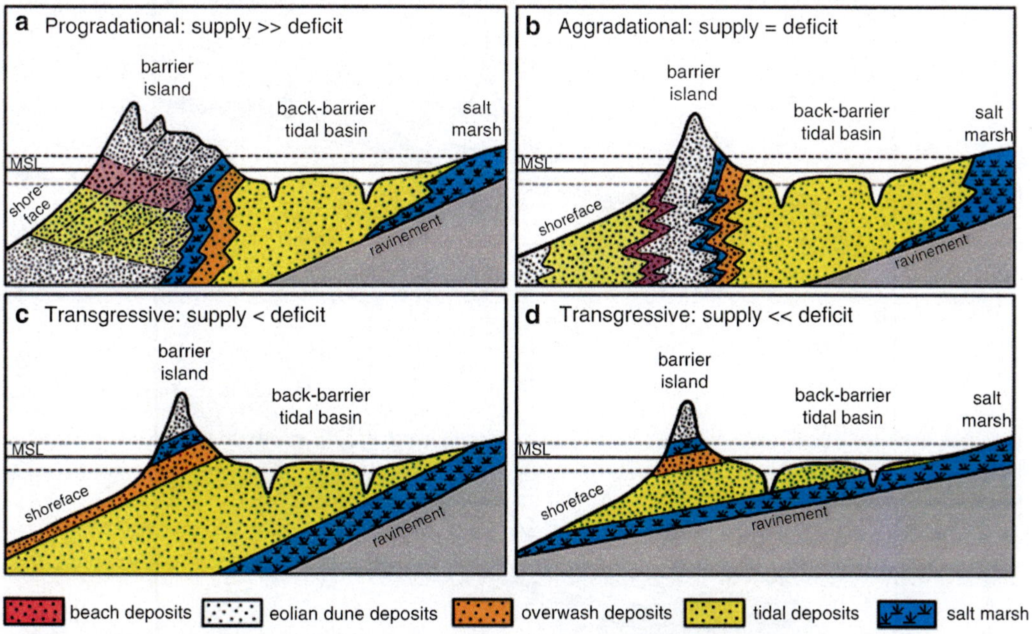

Fig. 10.28 Four geological cross-sections of barrier island depositional systems reflecting the stratigraphy produced by particular sediment budget situations controlled by the interaction between relative sea-level rise and sediment supply (Modified after Flemming 2002). Note that the shoreward fining sedimentary facies succession characterising the back-barrier tidal flat deposits (*yellow*) are not shown. Legend to color scheme includes beach, eolian dune, overwash, tidal flat (with channels), is salt marsh facies, but excludes the upper and lower shoreface

tectonic, isostatic and climatic processes acting either together or independently of each other. The interaction between the rate of relative sea-level rise and the rate of sediment supply defines the stratigraphic response of the system. Because sea-level fall leaves barrier islands and back-barrier tidal flats stranded, sustained accretion is usually associated with sea-level rise. In this context three basic stratigraphic response types can be distinguished, namely progradational, aggradational and retrogradational ones. Each has a characteristic and hence diagnostic stratigraphic expression (Galloway and Hobday 1975). The former two occasionally act in conjunction to produce aggrading progradational systems. In the case of both prograding and aggrading tidal flats, the entire depositional sequence is conserved (total retention), while in the retrogradational (transgressive) case varying parts of the sequence are lost, depending on the overall sediment deficit. For this reason it is useful to distinguish between partly conserved systems in which the depositional sequence is partially retained, and totally reworked systems in which only the final high-stand sequence is retained (Kraft 1971).

The main four types of stratigraphic responses outlined above are illustrated in the schematic cross-sections of Fig. 10.28. Each type represents a particular sediment budget situation that reflects the stratigraphic response resulting from the rate of sediment supply from external sources relative to the deficit created by sea-level rise over the same time interval. When keeping one of the control parameters constant, a change in the other will automatically affect the sediment budget, as a consequence of which the stratigraphic response changes from one state to another. It should be noted, however, that the four examples presented here represent time slices of particular sediment budget situations and that, in nature, one may find transitional systems reflecting budgets intermediate between any two of these. Furthermore, in the course of time, the sediment budget of a particular locality may change and an existing depositional type will then grade into another.

Cases a and b in Fig. 10.28 reflect the prograding and aggrading stratigraphies that result from a positive sediment budget where as much or more sediment is imported from external sources than required for the

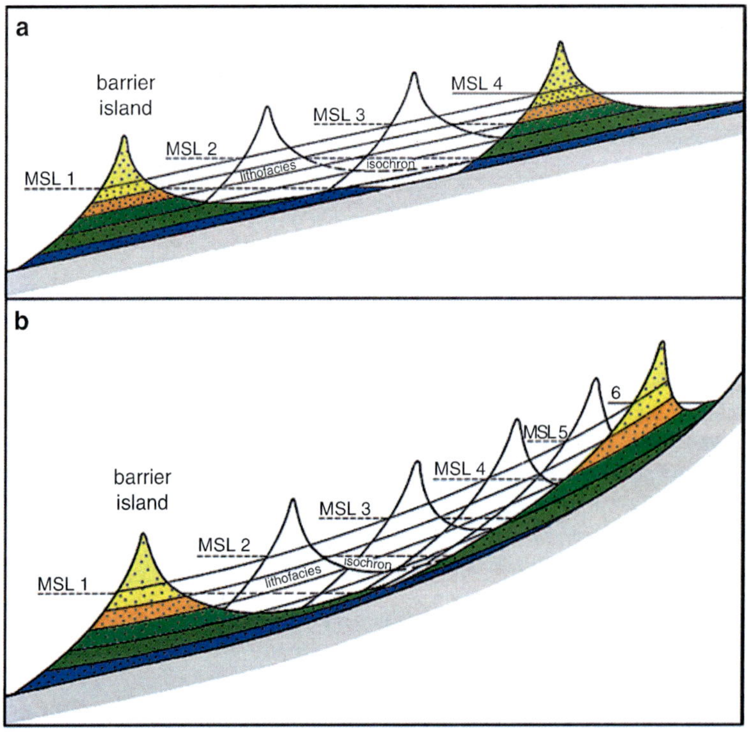

Fig. 10.29 Schematic conceptual models illustrating the stratigraphic situation in the case of a uniformly sloping (**a**) and a progressively steepening (**b**) shore. Note the squeezing out of individual sedimentary facies on the landward side in the latter case (Modified after Flemming and Bartholomä 1997)

compensation of the deficit created by sea-level rise. The difference between Figs. 10.28a, b mainly concerns the stratigraphy below the barrier island and the shoreface, the vertically aggrading back-barrier tidal flat systems (yellow colour) being more or less identical. As outlined earlier, the back-barrier deposits can also be identified by their progressive shoreward fining grain-size gradient which is not shown in Fig. 10.28. If less sediment is supplied from external sources than required to compensate sea-level rise, then the remaining deficit must be replaced by sediment from the existing reservoir. This involves moving sediment from the beach and upper shoreface toward the back-barrier basin. As a consequence, the barrier island is forced to migrate landward across its own back-barrier tidal flat. In the case where the remaining deficit is small, most of the depositional sequence is retained, the loss in the stratigraphic section being restricted to upper shoreface, beach, and barrier sands (Fig. 10.28c). The less sediment is available from external sources, the larger the remaining deficit and the fewer the depositional sequences retained in the stratigraphic section. At the same time the speed of barrier island migration increases. This may evolve to the point where no sediment is available from external sources and the deficit now has to be replaced entirely from the existing sediment reservoir (Fig. 10.28d). In this case the whole depositional sequence is progressively lost and only the final highstand deposit is retained for potential preservation in the rock record. The marked difference in depositional architecture produced by the variable interplay between sediment supply and sea-level rise thus turns out to be a powerful diagnostic tool for the interpretation of the stratigraphic record in terms of the temporal sediment budget evolution and the rate of sea-level change.

Thus far, the retention or loss of back-barrier deposits in the course of sea-level rise has been considered only in terms of changes in the sediment budget. In such cases, the progressive loss of sedimentary facies occurs on the seaward side, the capping barrier sand being eliminated first, the basal salt marsh last (Fig. 10.29a). This model applies to situations where a transgressive barrier-island depositional system encroaches upon a low-lying coastal plain along a uniformly sloping surface. A reverse situation, or rather the simultaneous loss of sedimentary facies at both ends, appears to occur where a barrier system migrates up against a coastal cliff or a progressively steepening shore. Such a loss of finer-grained sedimentary facies along the mainland coast of the Wadden Sea was first recognised by Flemming and Nyandwi (1994). It was suggested to reflect the response of the back-barrier basin to land reclamation, the hydrodynamic energy – especially wave action – having increased as the water depth at high tide increased along the foot of the dike. This interpretation subsequently received strong support from a study in which the loss of accommodation space and the grain-size composition of tidal flats lost to land reclamation was numerically reconstructed (Mai and Bartholomä 2000). Thus, contrary to intuition, the widths of individual sedimentary facies belts do not simply adjust to fit into the reduced space, but the finer-grained ones are instead progressively eliminated.

A similar effect is postulated to occur along cliffed coasts or where a steepening slope obstructs normal barrier evolution, the latter case being illustrated in Fig. 10.29b. The process can be conceived to continue until the entire back-barrier depositional system has been removed and the former barrier sand has evolved into a perched coastal dune (Roy et al. 1994). This aspect in the stratigraphic evolution of barrier-island depositional systems has received little attention thus far. Nevertheless, some evidence favouring such an interpretation, even though not entirely conclusive, can be found in the literature (Curray et al. 1969; Belknap and Kraft 1977; Kraft et al. 1979. Vos and van Kesteren 2000).

10.6 Modern Examples and Ancient Analogues

10.6.1 Modern Examples

As pointed out earlier, most larger-scale back-barrier depositional systems documented in the literature lack substantial bare tidal flats, being instead dominated by salt marshes or mangroves and estuarine or lagoonal water bodies. As a consequence, there are few documented examples of modern bare siliciclastic back-barrier tidal flats that can match the Wadden Sea in scale and studied detail (Klein 1976). Excellent regional summaries covering various aspects of Wadden Sea research, including comprehensive literature citations up to the time of publication, can be found in Reineck and Singh (1980), Dijkema et al. (1980), Postma (1982), Ehlers (1988), Oost and de Boer (1994), Flemming and Davis (1994), Flemming and Hertweck (1994), and Bartholdy and Pejrup (1994).

Among back-barrier tidal flat systems that merit being mentioned here are the ones off the Copper River delta located along the Pacific coast of Alaska (Reimnitz 1966; Galloway 1976; Hayes and Ruby 1994). However, little detail on intertidal sediment distribution and sedimentary structures has been published. The same, in principle, holds for the enigmatic Ria Formosa (Algarve) barrier island system off southern Portugal, which is unique in the sense that it is a non-coastal plain system backed by a steeply rising coastal cliff. The accessible literature (Pilkey et al. 1989; Davis 1994a) mostly concentrates on the barrier islands and provides little information on the back-barrier tidal flats. Some geomorphological, sedimentological and geochemical data exist but are difficult to access (Granja 1984; Granja et al. 1984; Monteiro et al. 1984; Dias 1986). Because the barrier island chain gradually approaches the coastal cliff towards the border of Spain, it could represent an ideal case to validate the concept of progressive loss of finer-grained sediment facies along the foot of the cliff as the back-barrier tidal flats get narrower.

The back-barrier systems along the east and south coasts of the North American continent are mostly of the lagoonal and/or estuarine type with extensive salt marshes covering the intertidally exposed parts with no or only very narrow bare intertidal flats. These locally display stratigraphies that are similar to more extensive tidal flat systems such as the Wadden Sea. Examples can be found in Leatherman (1979), FitzGerald et al. (1994), Hayes (1994) and Oertel and Kraft (1994).

A variety of back-barrier tidal flat system that does not quite correspond to the classic type discussed here, are coastal lagoons in which intertidal flats occupy a large part of the area. Well documented examples of this type are Willapa Bay along the Pacific coast of the USA (Clifton et al. 1989; Dingler and Clifton 1994) and Langebaan Lagoon along the west coast of South Africa (Flemming 1977). The main difference to classic back-barrier tidal flats is the fact that the grain-size gradient (and hence the energy gradient) is not perpendicular to the coast, but instead more or less aligned along the main axis of the lagoon between the mouth and the head. Other than this, the intertidal flats display many of the features discussed in this contribution.

10.6.2 Ancient Analogues

Examples of tidal deposits span the time period from the Early Archaean to the present (Noffke et al. 2006; Noffke 2010) and it would go beyond the scope of this contribution to cite them all. This section therefore restricts itself to a listing of important collective works that deal with clastic tidal deposits and which also include ancient examples. A first symposium proceedings specifically dedicated to tidal research was edited by Ginsburg (1975). Hobday and Errikson (1977) present the results of a tidal conference with special reference to South African case studies. An exhaustive summary of both modern and ancient examples up to the year 1977 can be found in Klein (1977). Other conference proceedings containing case studies from modern and ancient tidal environments include de Boer et al. (1988), Smith et al. (1991), Flemming and Bartholomä (1995), Alexander et al. (1998), Park and Davis (2001), and Bartholdy and Kvale (2006).

The cited works demonstrate that tidal-flat deposits can generally be recognized in the rock record with some degree of confidence. However, it is much more difficult, and in many cases impossible, to differentiate between micro-mesotidal back-barrier tidal flats and macrotidal open coast counterparts. Both display essentially the same range of surface features and internal sedimentary structures. Being more energetic, macrotidal flats should display a greater proportion of larger-scale dune cross-bedding relative to smaller-scale ripple cross-bedding on sand flats than would be the case on mesotidal flats where dune cross-bedding is the exception rather than the rule. Another feature that is more prominent on macrotidal flats is the occurrence of relatively high (>1 m) intertidal bars entirely composed of bioclastic material (mollusc shells) that may be found up to and even within salt marsh deposits (e.g., Larsonneur 1994; Schneider-Storz et al. 2008).

As the features addressed above are not unequivocally diagnostic, in addition to requiring excellent outcrop conditions, some researchers have looked for ways to estimate paleotidal ranges in order to solve this problem (Klein 1971; Allen 1981; Terwindt 1988). In each case, a different approach was used. Thus, Klein (1971) suggested that the thickness of upward-fining sequences approximated the paleotidal range, whereas Allen (1981) used the thickness of cross-bedding sets displaying mud drapes to derive at such estimates. Terwindt (1988), in turn, used a complex combination of stratigraphic criteria to reconstruct paleotidal ranges, but submits that this was very difficult because reliable criteria to identify the low-water line were lacking. Although very persuasive, this issue has not really been resolved to this day.

An elegant way to decide whether a tidal deposit was formed in a back-barrier setting would be the identification of the ancient barrier itself, or at least remnants thereof, in the rock record. Criteria for this have been summarized by Dickinson et al. (1972). This would side-step the issue of having to estimate paleotidal range, but would again require good exposures in the field. Finally, tidal euphoria can also lead to astonishing misinterpretations. Thus, the supposedly 'transgressive-barrier and shallow shelf interpretation of the lower Paleozoic Peninsula Formation, South Africa' (Hobday and Tankard 1978) is more likely a large alluvial fan delta that incorporates a few thin marine transgressions (Brian Turner, personal communication 1993). The interpretation can be shown to have been based on non-conclusive evidence and that other features such as the occurrence of massive pebble beds, ubiquitous floating pebbles, stacked linguoid bars, sand-draped mud cracks, and especially the total absence of any tidal rhythmites in any of the excellent exposures collectively favor an alluvial origin (Flemming, unpublished).

References

Alexander CR, Davis RA, Henry VJ (eds) (1998) Tidalites: processes and products. SEPM Spec Publ 61

Allen JRL (1981) Palaeotidal speeds and ranges estimated from cross-bedding sets with mud drapes. Nature 293:394–396

Allen PA, Homewood P (1984) Evolution and mechanics of a Miocene tidal sandwave. Sedimentology 31:63–81

Anderson FE (1983) The northern muddy intertidal: seasonal factors controlling erosion and deposition – a review. Can J Fish Aquat Sci 40:143–159

Arends F (1833) Physische Geschichte der Nordsee-Küste und deren Veränderungen durch Sturmfluthen seit der Cymbrischen Fluth bis jetzt. Woortmann, Emden, vol 1,

384 pp; vol 2, 355 pp. Facsimile reprinting in one volume 1974. Verlag Schuster, Leer

Ashley GM (1988) The hydrodynamics and sedimentology of a back-barrier lagoon-salt marsh system, Great Sound, New Jersey. Mar Geol 82:1–132

Aubrey DG, Speer PE (1985) A study of non-linear tidal propagation in shallow inlet/estuarine systems, Part I: observations. Estuar Coast Shelf Sci 21:185–205

Bartholdy J, Kvale EP (eds) (2006) Proceedings of the 6th international congress on tidal sedimentology (Tidalites 2004), 02–05 Aug 2004, Copenhagen. Mar Geol 235, 272 pp

Bartholdy J, Pejrup M (1994) Holocene evolution of the Danish Wadden Sea. In: Flemming BW, Hertweck G (eds) Tidal flats and barrier systems of continental Europe: a selected overview. Senckenberg marit 24:187–209

Bartholomä A, Flemming BW (2007) Progressive grain-size sorting along an intertidal energy gradient. Sed Geol 202:464–472

Bartholomä A, Flemming BW, Delafontaine MT (2000) Mass balancing the seasonal turnover of mud and sand in the vicinity of an intertidal mussel bank in the Wadden Sea (southern North Sea). In: Flemming BW, Delafontaine MT, Liebezeit G (eds) Muddy coast dynamics and resource management. Elsevier Science, Amsterdam, pp 85–106

Beets DJ, Fischer MM, de Gans W (eds) (1996) Coastal studies on the Holocene of the Netherlands. Mededel Rijks Geologische Dienst 57

Belknap DF, Kraft JC (1977) Holocene relative sea-level changes and coastal stratigraphic units on the northwest flank of the Baltimore Canyon trough geosyncline. J Sed Petrol 47:610–629

Biegel E, Hoekstra P (1995) Morphological response characteristics of the Zoutkamperlaag, Frisian inlet (The Netherlands), to a sudden reduction in basin area. In: Flemming BW, Bartholomä A (eds) Tidal signatures in modern and ancient sediments. Int Assoc Sediment Spec Publ 24:85–99

Boon JD III, Byrne RJ (1981) On basin hypsometry and the morphodynamic response of coastal inlet systems. Mar Geol 40:27–48

Boothroyd JC, Friedrich NE, McGinn SR (1985) Geology of microtidal coastal lagoons, Rhode Island. Mar Geol 63:35–76

Bouma AH (1969) Methods for the study of sedimentary structures. Wiley-Interscience, New York

Bridges PH, Leeder MR (1976) Sedimentary model for intertidal mudflat channels, with examples from the Solway Firth, Scotland. Sedimentology 23:533–552

Burchard H, Flöser G, Staneva JV, Badewien T, Riethmüller R (2008) Impact of density gradients on net sediment transport into the Wadden Sea. J Phys Oceanog 38:566–587

Carver RE (ed) (1971) Procedures in sedimentary petrology. Wiley-Interscience, New York

Chang TS, Flemming BW (2006) Sedimentation on a wave-dominated, open-coast tidal flat, southwestern Korea: summer tidal flat–winter shoreface – discussion. Sedimentology 53:687–691

Chang TS, Bartholomä A, Flemming BW (2006a) Seasonal dynamics of fine-grained sediments in a back-barrier tidal basin of the German Wadden Sea (southern North Sea). J Coast Res 22:328–338

Chang TS, Joerdel O, Flemming BW, Bartholomä A (2006b) Importance of flocs and aggregates in muddy sediment dynamics and seasonal sediment turnover in a back-barrier tidal basin of the East Frisian Wadden Sea (southern North Sea). Mar Geol 235:49–61

Chang TS, Flemming BW, Tilch E, Bartholomä A, Wöstmann R (2006c) Late Holocene stratigraphic evolution of a back-barrier tidal basin in the East Frisian Wadden Sea, southern North Sea: transgressive deposition and its preservation potential. Facies 52:329–340

Chang TS, Flemming BW, Bartholomä A (2007) Distinction between sortable silts and aggregated particles in muddy intertidal sediments of the southern North Sea. In: Flemming BW, Hartmann D (eds) From particle size to sediment dynamics. Proceeding of a workshop, Hanse Institiute for Advanced Study, Delmenhorst (Germany), 15–18 Apr 2004. Sed Geol 202:453–463

Clifton HE, Phillips RL, Anima RJ (1989) Sedimentary facies of Willapa Bay, Washington; a field guide. Can Soc Petrol Geol Mem 16

Curray JR, Emmel FJ, Crampton PJS (1969) Holocene history of a strand plain, lagoonal coast, Nayarit, Mexico. In: Castanares AA, Phleger FB (eds) Coastal lagoons – a symposium. Universidad Autonoma, Mexico City, pp 63–100

Dalrymple RW, Makino Y, Zaitlin BA (1991) Temporal and spatial patterns of rhythmite deposition on mudflats in the macrotidal, Cobequid Bay-Salmon River estuary, Bay of Fundy, Canada. In: Smith DG, Reinson GE, Zaitlin BA, Rahmani RA (eds) Clastic tidal sedimentology. Can Soc Petrol Geol Mem 16:137–160

Dalrymple RW, Zaitlin BA, Boyd R (1992) Estuarine facies models: conceptual basis and stratigraphic implications. J Sed Petrol 62:1130–1146

Dastgheib A, Roelvink JA, Wang ZB (2008) Long-term process-based morphological modeling of the Marsdiep tidal basin. Mar Geol 256:90–100

Davies JL (1964) A morphogenetic approach to the world shorelines. Zeitschr f Geomorph 8:127–142

Davis RA Jr (ed) (1994a) Geology of Holocene barrier island systems. Springer, Berlin

Davis RA Jr (1994b) Barriers of the Florida Peninsula. In: Davis RA Jr (ed) Geology of Holocene barrier island systems. Springer, Berlin, pp 167–205

Davis RA Jr, Flemming BW (1995) Stratigraphy of a combined wave- and tide-dominated intertidal sand body: Martens Plate, German Wadden Sea. In: Flemming BW, Bartholomä A (eds) Tidal signatures in modern and ancient sediments. Spec Publ Int Assoc Sediment 24:121–132

Davis RA Jr, Hayes MO (1984) What is a wave-dominated coast? Mar Geol 60:313–329

de Bèaumont LE (1845) Septième Leçon: Léeves de sable et de galet. In: Bertrand P (ed) Leçons de géologie practique. Masson, Paris, pp 221–252

de Boer PL (1979) Convolute lamination in modern sands of the estuary of the Oosterschelde, the Netherlands, formed as a result of entrapped air. Sedimentology 26:283–294

de Boer PL, van Gelder A, Nio SD (eds) (1988) Tide-influenced sedimentary environments and facies. D Reidel Publishing Company, Dordrecht

de Boer PL, Oost AP, Visser MJ (1989) The diurnal inequality of the tide as a parameter for recognizing tidal influences. J Sed Petrol 59:912–921
de Raaf JFM, Boersma JR (1971) Tidal deposits and their sedimentary structures (seven examples from Western Europe). Geol Mijn 50:479–504
Delafontaine MT, Flemming BW, Bartholomä A (2000) Mass balancing the seasonal turnover of POC in mud and sand on a back-barrier tidal flat (southern North Sea). In: Flemming BW, Delafontaine MT, Liebezeit G (eds) Muddy coast dynamics and resource management. Elsevier Science, Amsterdam, pp 107–124
Delafontaine MT, Flemming BW, Thimm M (2004) Large-scale trends in some mass physical properties of Danish Wadden Sea sediments, and implications for organism-sediment interactions. Danish J Geog 104:27–36
Dias JMA (1986) Observacoes sobre a origem das areias das ilhas barreira da Ria Formosa. 4th Congr Algarve. Textos das Comunicaçoes 1:579–587
Dickinson KA, Berryhill HL Jr, Holmes CW (1972) Criteria for recognizing ancient barrier coastlines. In: Rigby JK, Hamblin WK (eds) Recognition of ancient sedimentary environments. SEPM Spec Publ 16:192–204
Dijkema KS, Reineck H-E, Wolff WJ (eds) (1980) Geomorphology of the Wadden Sea area. In: Wadden Sea Working Group (eds) Final report of the section 'Geomorphology'. AA Balkema, Rotterdam
Dingler JR, Clifton HE (1994) Barrier systems of California, Oregon, and Washington. In: Davis RA Jr (ed) Geology of Holocene barrier island systems. Springer, Berlin, pp 115–165
Dionne JC (1974) How drift ice shapes the St. Lawrence. Can Geogr J 88:4–9
Dionne JC (1988) Characteristic features of modern tidal flats in cold regions. In: de Boer PL, van Gelder A, Nio SD (eds) Tide-influenced sedimentary environments and facies. D Reidel Publishing Company, Dordrecht, pp 301–332
Dissanayake DMPK, van der Roelvink JA, Wegen M (2009) Modelled channel patterns in a schematized tidal inlet. Coast Eng 56:1069–1083
Dott RH Jr (1983) Episodic sedimentation – how normal is average? How rare is rare? Does it matter? J Sed Petrol 53:5–23
Dott RH Jr (1988) An episodic view of shallow marine clastic sedimentation. In: de Boer PL, van Gelder A, Nio SD (eds) Tide-influenced sedimentary environments and facies. D Reidel Publishing Company, Dordrecht, pp 3–12
Dronkers J (1986) Tidal asymmetry and estuarine morphology. Neth J Sea Res 20:107–131
Dunn IS, Anderson LR, Kiefer FW (1980) Fundamentals of geotechnical analysis. Wiley, New York
Ehlers J (1988) Morphodynamics of the Wadden Sea. AA Balkema, Rotterdam, 397 pp
Einsele G (1963) "Convolute bedding" und ähnliche Sedimentstrukturen im rheinischen Oberdevon und anderen Ablagerungen. N Jb Paläont, Abh 116:162–189
FitzGerald DM, Rosen PS, van Heteren S (1994) New England barriers. In: Davis RA Jr (ed) Geology of Holocene barrier island systems. Springer, Berlin, pp 305–394
Flemming BW (1977) Langebaan Lagoon: a carbonate–siliciclastic tidal environment in a semi-arid climate. Sed Geol 18:61–95
Flemming BW (2000) A revised textural classification of gravel-free muddy sediments on the basis of ternary diagrams. Cont Shelf Res 20:1125–1137
Flemming BW (2002) Effects of climate and human interventions on the evolution of the Wadden Sea depositional system (southern North Sea). In: Wefer G, Berger W, Behre KE, Jansen E (eds) Climate development and history of the North Atlantic Realm. Springer, Berlin, pp 399–413
Flemming BW (2003a) Flaser. In: Middleton GV (ed) Encyclopedia of sediments and sedimentary rocks. Kluwer, Dordrecht, pp 282–283
Flemming BW (2003b) Tidal flats. In: Middleton GV (ed) Encyclopedia of sediments and sedimentary rocks. Kluwer, Dordrecht, pp 734–737
Flemming BW (2005) Tidal environments. In: Schwartz M (ed) Encyclopedia of coastal science. Springer, Berlin, pp 1180–1185
Flemming BW (2007) The influence of grain-size analysis methods and sediment mixing on curve shapes and textural parameters: implications for sediment trend analysis. Sed Geol 202:425–435
Flemming BW, Bartholomä A (eds) (1995) Tidal signatures in modern and ancient sediments. Int Assoc Sed Spec Publ 24
Flemming BW, Bartholomä A (1997) Response of the Wadden Sea to a rising sea level: a predictive empirical model. Ger J Hydrogr 49:343–353
Flemming BW, Davis RA Jr (1994) Holocene evolution, morphodynamics and sedimentology of the Spiekeroog barrier island system (southern North Sea). In: Flemming BW, Hertweck G (eds) Tidal flats and barrier systems of continental Europe: a selected overview. Senckenberg marit 24:117–155
Flemming BW, Delafontaine MT (2000) Mass physical properties of intertidal muddy sediments: some applications, misapplications and non-applications. Cont Shelf Res 20:1179–1197
Flemming BW, Hertweck G (eds) (1994) Tidal flats and barrier systems of continental Europe: a selected overview. Senckenberg marit 24
Flemming BW, Nyandwi N (1994) Land reclamation as a cause of fine-grained sediment depletion in backbarrier tidal flats (southern North Sea). Neth J Aquat Ecol 28:299–307
Flemming BW, Ziegler K (1995) High-resolution grain size distribution patterns and textural trends in the backbarrier tidal flats of Spiekeroog Island (southern North Sea). Senckenberg marit 26:1–24
Folk RL (1954) The distinction between grain size and mineral composition in sedimentary-rock nomenclature. J Geol 62:344–359
Fortunato AB, Oliveira A (2004) A modelling system for tidally driven long-term morphodynamics. J Hydraul Res 42:426–434
Frey RW, Basan PB (1985) Coastal salt marshes. In: Davis RA Jr (ed) Coastal sedimentary environments, 2nd edn. Springer, New York, pp 225–301
Frey RW, Howard JD (1969) A profile of biogenic sedimentary structures in a Holocene barrier island salt marsh complex, Georgia. Trans Gulf Coast Assoc Geol Soc 19:427–444
Friedrichs CT, Aubrey DG (1988) Non-linear tidal distortion in shallow well-mixed estuaries: a synthesis. Estuar Coast Shelf Sci 27:521–545

Friedrichs CT, Lynch DR, Aubrey DG (1992) Velocity asymmetries in frictionally-dominated tidal embayments: longitudinal and lateral variability. In: Prandle D (ed) Dynamics and exchanges in estuaries and the coastal zone. American Geophysical Union, Washington, DC, pp 277–312
Galloway WE (1976) Sediments and stratigraphic framework of the Copper River fan delta. J Sed Petrol 46:726–737
Galloway WE, Hobday DK (1975) Terrigenous clastic depositional systems. Springer, Berlin, pp 234
Ganju NK, Schoellhamer DH, Jaffe BE (2009) Hindcasting of decadal-timescale estuarine bathymetric change with a tidal-timescale model. J Geophys Res 114:F04019. doi:10.1029/2008JF001191
Gerdes G, Krumbein WE (1987) Biolaminated deposits. Lecture notes in earth sciences 9. Springer, Berlin
Gerdes G, Krumbein WE, Reineck H-E (1985) The depositional record of sandy, versicolored tidal flats (Mellum Island, southern North Sea). J Sed Petrol 55:265–278
Ginsburg RN (ed) (1975) Tidal deposits – a casebook of recent examples and fossil counterparts. Springer, Berlin/Heidelberg/New York
Glaeser JD (1978) Global distribution of barrier islands in terms of tectonic setting. J Geol 86:283–297
Gouleau D, Jouanneau JM, Weber O, Sauriau PG (2000) Short- and long-term sedimentation on Montportail–Brouage intertidal mudflat, Marennes–Oléron Bay (France). Cont Shelf Res 20:1513–1530
Granja H (1984) Etude géomorphologique, sédimentologique et géochimique de la "Ria Formose" (Algarve-Portugal). Thése Zeme cycle, n 1944, Université de Bordeaux I, Bordeaux, France
Granja H, Froidefond J-M, Pera T (1984) Processus d'évolution morpho-sédimentaire de la Ria Formosa (Portugal). Bull Inst Géol Bassin d'Aquitaine 36:37–50
Groen P (1967) On the residual transport of suspended matter by an alternating tidal current. Neth J Sea Res 3:564574
Hack JT (1957) Studies of longitudinal stream profiles in Virginia and Maryland. US Geol Surv Prof Paper 219B
Haven DS, Morales-Alamo R (1968) Occurrence and transport of fecal pellets in suspension in a tidal estuary. Sed Geol 2:141–151
Hayes MO (1979) Barrier island morphology as a function of tidal and wave regime. In: Leatherman SP (ed) Barrier islands. Academic, New York, pp 1–27
Hayes MO (1994) The Georgia Bight barrier systems. In: Davis RA Jr (ed) Geology of Holocene barrier island systems. Springer, Berlin, pp 233–304
Hayes MO, Ruby CH (1994) Barriers of Pacific Alaska. In: Davis RA Jr (ed) Geology of Holocene barrier island systems. Springer, Berlin, pp 395–433
Hertweck G (1994) Zonation of benthos and lebensspuren in the tidal flats of the Jade Bay, southern North Sea. In: Flemming BW, Hertweck G (eds) Tidal flats and barrier systems of continental Europe: a selected overview. Senckenberg marit 24:157–170
Hillel D (1998) Environmental soil physics. Academic, San Diego
Hobday DK, Errikson KA (eds) (1977) Tidal sedimentation with special reference to South African examples. Sed Geol 18
Hobday DK, Tankard AJ (1978) Transgressive-barrier and shallow-shelf interpretation of the lower Paleozoic Peninsula Formation, South Africa. Geol Soc Am Bull 89:1733–1744
Howard JD, Frey RW (1975) Estuaries of the Georgia coast, U.S.A.: Sedimentology and Biology. II. Regional animal-sediment characteristics of Georgia estuaries. Senckenberg marit 7:33–103
Hume TM, Herdendorf CE (1992) Factors controlling tidal inlet characteristics on low drift coasts. J Coast Res 8:355–375
Inderbitzen AL (ed) (1974) Deep-sea sediments: physical and mechanical properties. Plenum, New York
Jakobsen B (1962) The formation of ebb and flood channels in tidal channels described on basis of morphological and hydrological observations. Geogr Tijdskrift 61:119–149
Jarrett JT (1976) Tidal prism-inlet area relationships. US Army Corps of Engineers, Coastal Engineering Research Center and Waterways Experimental Station, Vicksburg. GITI report 3
Jennings JN, Coventry RJ (1973) Structure and texture of a gravely barrier island in the Fitzroy estuary, Western Australia, and the role of mangroves in the shore dynamics. Mar Geol 15:145–167
Johnson DW (1919) Shore processes and shore line development. Wiley, New York, 584 pp
Kindle EM (1917) Recent and fossil ripple marks. Geol Surv Can Museum Bull 25
Kirby R (2000) Practical implications of tidal flat shape. Cont Shelf Res 20:1061–1077
Klein G (1971) A sedimentary model determining paleotidal range. Geol Soc Am Bull 82:2585–2592
Klein G (1977) Clastic tidal facies. CEPCO, Champaign
Klein G (ed) (1976) Holocene tidal sedimentation. Benchmark papers in geology 30. Dowden, Hutchinson & Ross, Stroudsburg
Kraft JC (1971) Sedimentary facies patterns and geologic history of a Holocene marine transgression. Geol Soc Am Bull 82:2131–2151
Kraft JC, Allen EA, Belknap DF, John CJ, Maurmeyer EM (1979) Processes and morphologic evolution of an estuarine and coastal barrier system. In: Leatherman SP (ed) Barrier islands – from the Gulf of St. Lawrence to the Gulf of Mexico. Academic, New York, pp 149–183
Krögel F, Flemming BW (1998) Evidence for temperature-adjusted sediment distributions in the backbarrier tidal flats of the East Frisian Wadden Sea (southern North Sea). In: Alexander CR, Davis RA, Henry VJ (eds) Tidalites: processes & products. SEPM, Tulsa, Spec Publ 61:31–41
Krone RB (1972) A field study of flocculation as a factor in estuarine shoaling processes. Committee on Tidal Hydraulics, US Army Corps of Engineers. Technical Bulletin 19
Lambe WT, Whitman RV (1969) Soil mechanics. Wiley, New York
Larsonneur C (1994) The bay of Mont-Saint-Michel: a sedimentation model in a temperate macrotidal environment. In: Flemming BW, Hertweck G (eds) Tidal flats and barrier systems of continental Europe: a selected overview. Senckenberg marit 24:3–63
Leatherman SP (ed) (1979) Barrier islands – from the Gulf of St. Lawrence to the Gulf of Mexico. Academic, New York
Leopold LB, Wolman MG, Miller JP (1964) Fluvial processes in geomorphology. Freeman, San Francisco
Lettmann KA, Wolff J-O, Badewien TH (2009) Modelling the impact of wind and waves on suspended particulate matter fluxes in the East Frisian Wadden Sea (southern North Sea). Ocean Dyn 59:239–262

Mai S, Bartholomä A (2000) The missing mud flats of the Wadden Sea: a reconstruction of sediments and accommodation space lost in the wake of land reclamation. In: Flemming BW, Delafontaine MT, Liebezeit G (eds) Muddy coast dynamics and resource management. Elsevier Science, Amsterdam, pp 257–272

McCave IN, Manighetti B, Robinson SG (1995) Sortable silt and fine sediment size/composition size slicing: parameters for palaeocurrent speed and palaeoceanography. Paleoceanography 10:593–610

Migniot C (1968) A study of the physical properties of various forms of very fine sediments and their behaviour under hydrodynamic action. La Houille Blanche 7:591–620

Miller JA (1975) Facies characteristics of Laguna Madre wind-tidal flats. In: Ginsburg RN (ed) Tidal deposits. Springer, New York, pp 67–72

Molinaroli E, Guerzoni S, De Falco G, Saretta A, Cucco A, Como S, Simeone S, Perilli A, Magni P (2009) Relationships between hydrodynamic parameters and grain size in two contrasting transitional environments: the Lagoons of Venice and Cabras, Italy. Sed Geol 219:196–207

Monteiro JH, Pilkey O, Dias JA, Gaspar LC, Paixo G (1984) Origem, evoluçao e processos geologicos das ilhas barreira a sua importancia para o desenvolvimento destas ilhas. 3 rd Congresso Sobre o Algarve. Textos das Comunicaçoes 2:713–719

Nichols MM, Biggs RB (1985) Estuaries. In: Davis RA Jr (ed) Coastal sedimentary environments. Springer, New York, pp 77–186

Noffke N (2010) Geobiology – microbial mats in sandy deposits from the Archaean Era to today. Springer, Heidelberg

Noffke N, Eriksson KA, Hazen RM, Simpson EL (2006) A new window into Early Archean life: microbial mats in Earth's oldest siliciclastic tidal deposits (3.2 Ga Moodies Group, South Africa). Geology 34:253–256

Nyandwi N, Flemming BW (1995) A hydraulic model for the shore-normal energy gradient in the East Frisian Wadden Sea (southern North Sea). Senckenberg marit 25:163–171

O'Brien DJ, Whitehouse RJS, Cramp A (2000) The cyclic development of a macrotidal mudflat on varying timescales. Cont Shelf Res 20:1593–1619

Oertel GF, Kraft JC (1994) New Jersey and Delmarva barrier islands. In: Davis RA Jr (ed) Geology of Holocene barrier island systems. Springer, Berlin, pp 207–232

Oost AP, de Boer PL (1994) Sedimentology and development of barrier islands, ebb-tidal deltas, inlets and backbarrier areas of the Dutch Wadden Sea. In: Flemming BW, Hertweck G (eds) Tidal flats and barrier systems of continental Europe: a selected overview. Senckenberg marit 24:65–115

Owen MW (1971) The effect of turbulence on the settling velocities of silt flocs. In: 14th congress of the international association of hydraulic research, Paris, Proceedings, pp 27–32

Park YA, Davis RA Jr (eds) (2001) Proceedings of Tidalites 2000. Special publication, Korean Society of Oceanography, Seoul, 103 pp

Pejrup M (1988) The triangular diagram used for classification of estuarine sediments: a new approach. In: de Boer PL, van Gelder A, Nio SD (eds) Tide-influenced sedimentary environments and facies. D Reidel Publishing Company, Dordrecht, pp 289–300

Pejrup M, Andersen TJ (2000) The influence of ice on sediment transport, deposition and reworking in a temperate mudflat area, the Danish Wadden Sea. Cont Shelf Res 20:1621–1634

Pestrong R (1972) Tidal-flat sedmentation at Colley Landing, southwest San Francisco Bay. Sed Geol 8:251–288

Pilkey OH (2003) A celebration of the world's barrier islands. Columbia University Press, New York

Pilkey OH, Neal WJ, Monteiro JH, Dias JMA (1989) Algarve barriers islands: a noncoastal-plain system in Portugal. J Coast Res 5:239–261

Postma H (1961) Transport and accumulation of suspended matter in the Dutch Wadden Sea. Neth J Sea Res 1:148–190

Postma H (1982) Hydrography of the Wadden Sea: movements and properties of water and particulate matter. In: Wadden Sea Working Group (eds) Final report on 'hydrography'. AA Balkema, Rotterdam

Pratt BR, James NP, Cowan CA (1992) Peritidal carbonates. In: Walker RG, James NP (eds) Facies models – response to sea level change, 2nd edn. Geological Association of Canada, pp 303–322

Reimnitz E (1966) Late Quaternary history and sedimentation of the Copper River Delta and vicinity, Alaska. Ph.D. thesis, University of California, San Diego

Reineck H-E (1958) Longitudinale Schrägschichtung im Watt. Geol Rundschau 37:73–82

Reineck H-E (1960) Über Zeitlücken in rezenten Flachseesedimenten. Geol Rundschau 49:149–161

Reineck H-E (1976) Drift ice action on tidal flats, North Sea. Rev Geogr Montr 30:197–200

Reineck H-E (1987) Morphologische Entwicklung der Insel Mellum. In: Gerdes G, Krumbein WE, Reineck H-E (eds) Mellum – Portrait einer Insel. Kramer, Frankfurt, pp 87–99

Reineck H-E, Singh IB (1980) Depositional sedimentary environments, 2nd edn. Springer, Berlin

Reineck H-E, Wunderlich F (1968) Classification and origin of flaser and lenticular bedding. Sedimentology 11:99–104

Ridderinkhof H (1988) Tidal and residual flows in the western Dutch Wadden Sea: I. Numerical model results. Neth J Sea Res 22:1–26

Rinaldo A, Belluco E, D'Alpeos A, Feola A, Lanzoni S, Marani A (2004) Tidal networks: form and function. In: Fagherazzi S, Marani A, Blum LK (eds) The ecogeomorphology of tidal marshes. American Geophysical Union, Washington, DC, pp 75–91

Roy PS, Cowell PJ, Ferland MA, Thom BG (1994) Wave-dominated coasts. In: Carter RWG, Woodroffe CD (eds) Coastal evolution: Late Quaternary shoreline evolution. Cambridge University Press, Cambridge, pp 121–185

Schäfer W (1972) Ecology and palaeoecology of marine environments. Oliver & Boyd, Edinburgh, 538 pp

Schieber J (2004) Microbial mats in the siliciclastic rock record. In: Eriksson PK, Altermannn DR, Nelson DR, Mueller WE, Catuneanu O (eds) The Precambrian earth: tempos and events. Elsevier, Amsterdam, pp 663–673

Schieber J, Southard JB (2009) Bedload transport of mud by floccules ripples – Direct observation of ripple migration processes and their implications. Geology 37:483–486

Schneider JF (1975) Recent tidal deposits, Abu Dhabi, UAE, Arabian Gulf. In: Ginsburg RN (ed) Tidal deposits. Springer, New York, pp 209–214

Schneider-Storz B, Nebelsick JH, Wehrmann A, Federolf C (2008) Comparative taphonomy of three bivalves from mass shell accumulation in the macrotidal regime of North Sea tidal flats. Facies 54:461–478

Schwartz ML (ed) (1973) Barrier islands. Benchmark papers in geology 9. Dowden, Hutchinson & Ross, Stroudsburg

Shepard FP (1954) Nomenclature based on sand-silt-clay ratios. J Sed Petrol 24:151–158

Smith DG, Reinson GE, Zaitlin BA, Rahmani RA (eds) (1991) Clastic tidal sedimentology. Can Soc Petrol Geol Mem 16

Speer PE, Aubrey DG (1985) A study of non-linear tidal propagation in shallow inlet/estuarine systems, Part II: Theory. Estuar Coast Shelf Sci 21:207–224

Stanev EV, Flemming BW, Bartholomä A, Staneva JV, Wolff J-O (2007) Vertical circulation in shallow tidal inlets and back-barrier basins. Cont Shelf Res 27:798–831

Stanev EV, Grayek S, Staneva J (2009) Temporal and spatial circulation patterns in the East Frisian Wadden Sea. Ocean Dyn 59:167–181

Streif HJ (1990) Das ostfriesische Küstengebiet – Nordsee, Inseln, Watten und Marschen. Sammlung geologischer Führer 57. Gebr Borntraeger, Berlin

Terwindt JHJ (1988) Palaeo-tidal reconstructions of inshore tidal depositional environments. In: de Boer PL, van Gelder A, Nio SD (eds) Tide-influenced sedimentary environments and facies. D Reidel Publishing Company, Dordrecht, pp 233–263

Thomas RG, Smith DG, Wood JM, Visser J, Calverley-Range EA, Koster EH (1987) Inclined heterolithic stratification – terminology, description, interpretation and significance. Sed Geol 53:123–179

van der Spek AJF (1995) Reconstruction of tidal inlet and channel dimensions in the Frisian Middelzee, a former tidal basin in the Dutch Wadden Sea. In: Flemming BW, Bartholomä A (eds) Tidal signatures in modern and ancient sediments. Int Assoc Sediment Spec Publ 24:239–258

van der Spek AJF (1996) Holocene depositional sequences in the Dutch Wadden Sea south of Ameland. In: Beets DJ, Fischer MM, de Gans W (eds) Coastal studies on the Holocene of the Netherlands. Mededel Rijks Geol Dienst 57:41–69

van der Wegen M, Roelvink JA (2008) Long-term morphodynamic evolution of a tidal embayment using a two-dimensional, process-based model. J Geophys Res C03016. doi:10.1029/2006JC003983

van der Wegen M, Dastgheib A, Roelvink JA (2010) Morphodynamic modeling of tidal channel evolution in comparison to empirical PA relationship. Coastal Eng 57:827–837

van Dongeren AR, de Vriend HJ (1994) A model of morphological behaviour of tidal basins. Coastal Eng 22:287–310

van Ledden M, van Kesteren WGM, Winterwerp JC (2004) A conceptual framework for the erosion behaviour of sand-mud mixtures. Cont Shelf Res 24:1–11

van Straaten LMJU (1952) Biogene textures and the formation of shell beds in the Dutch Wadden Sea. Koninkl Ned Akad Wetenschap, Amsterdam, Ser B 55:500–516

van Straaten LMJU (1954) Composition and structure of recent marine sediments in the Netherlands. Leidse Geol Mededel 19:1–110

van Straaten LMJU (1964) De bodem der Waddenzee. In: Anderson WF, Abrahamse J, Buwalda JD, van Straaten LMJU (eds) Het Waddenboek. Nederl geol Vereeniging, pp 75–151

van Straaten LMJU, Kuenen PH (1957) Accumulation of fine-grained sediments in the Wadden Sea. Geol Mijn 19:329–354

van Straaten LMJU, Kuenen PH (1958) Tidal action as a cause of clay accumulation. J Sed Petrol 28:406–413

van Veen J (1950) Eb- en vloedschaar systemen in de Nederlandse getijwateren. Tijdschrift van het Koninklijk Nederlandsch Aardrijkskundig Genootschap, Tweede Reeks LXVII:303–325

Visser MJ (1980) Neap-spring cycles reflected in Holocene subtidal large-scale bedform deposits: a preliminary note. Geology 8:543–546

Vos PC, van Kesteren WP (2000) The long-term evolution of intertidal mudflats in the northern Netherlands during the Holocene; natural and anthropogenic processes. Cont Shelf Res 20:1687–1710

Walther F (1972) Zusammenhänge zwischen der Größe der Ostfriesischen Seegaten mit ihren Wattgebieten sowie den Gezeiten und Strömungen. Jahrber Forschungstelle Norderney 23:7–32

Walton TL Jr, Adams WD (1976) Capacity of inlet outer bars to store sand. In: Proceedings, 15th international conference on coastal engineering. ASCE, Honolulu, Hawaii, pp 1919–1937

Warrick AW (2002) Soil physics companion. CRC Press, Boca Raton, 389 pp

Williams E (1960) Intra-stratal flow and convolute folding. Geol Mag 97:208–214

Williams PB, Orr MK, Garrity NJ (2002) Hydraulic geometry: a geomorphic design tool for tidal marsh channel evolution in wetland restoration projects. Restoration Ecol 10:577–590

Wunderlich F (1967) Die Entstehung von "convolute bedding" an Platenrändern. Senckenberg lethaea 48:345–349

Xu W (2000) Mass physical sediment properties and trends in a Wadden Sea tidalbasin. Berichte, Fachbereich Geowissenschaften, Universität Bremen, No. 157

Tidal Channels on Tidal Flats and Marshes

11

Zoe J. Hughes

Abstract

In shallow coastal settings channels provide a pathway for the tide to propagate and are, thus, a primary control on the sedimentation and ecology of these environments. Being shaped by bidirectional flows, tidal channels exhibit morphologies, which, despite apparent similarities, bear significant and fundamental differences to fluvial channels, specifically their scaling with size. This chapter considers the classification of tidal channels and the networks they form. We examine the hydrodynamics of shallow tidal channels, including asymmetry in period or velocity between the ebb and flood tides, and the hysteresis seen in stage-velocity curves in regions with large intertidal areas. Channel initiation may occur either through incision or by variations in rates of deposition. Tidal channels evolve over time and a number of relationships are presented that have been derived to describe the geometry of tidal channels. Mutually-evasive pathways of flood and ebb flows may produce cuspate meanders; a morphology unique to tidal channels. Of particular importance, in terms of preservation potential, is the development of meanders in channels and the resulting pointbars. Pointbars in tidal environments are often fully or partially detached from the bank by a channel formed by the subordinate tidal current, however their exact morphology varies being dependent on channel sinuosity and tidal asymmetry. Channels are preserved through infilling (as tidal prism is reduced) and through lateral accretion, particularly at meanders. Pointbars in tidal regions are generally heavily bioturbated in the upper tidal range, and mid-tidal zones will exhibit inclined stratigraphy, often with intercalated beds of muddier and sandier deposits.

11.1 Introduction

Within tidally dominated coastal landscapes, channels provide the conduit through which the tidal wave propagates, driving the exchange of water and sediment between the outer and inner regions of a coastal water body. The nature of the channel network will influence local tidal conditions, specifically tidal range, and tidal flow velocity. Within tidal flats and marshes, which are in the intertidal zone, this translates to the period and depth of inundation and potential for erosion and deposition. These conditions in turn determine the flux of sediment, nutrients and biota across an environment, ultimately impacting the long-term morphological evolution of the region.

Z.J. Hughes (✉)
Department of Earth Sciences, Boston University,
Boston, MA 01778, USA
e-mail: zoeh@bu.edu

R.A. Davis, Jr. and R.W. Dalrymple (eds.), *Principles of Tidal Sedimentology*,
DOI 10.1007/978-94-007-0123-6_11, © Springer Science+Business Media B.V. 2012

Channels are, therefore, a primary control on coastal environments.

Tidal channels are generally defined by bidirectional tidal flow. The term *tidal channel* can describe features across a range of scales, from large distributaries or cuts between tidal sand bars to small marsh creeks and shallow runnels across tidal flats. Networks formed by connected tidal channels are dynamic in nature, experiencing changes on timescales shorter than that of the evolution of the tidal landscape as a whole (D'Alpaos et al. 2005) and at times may appear comparatively transient. Active channel systems may reflect present conditions, or exhibit inheritance from paleo- or preexisting networks. For example, marsh systems often develop over tidal flats or bars with the channels preserved as creeks (Pethick 1969; Perillo and Iribarne 2003; Temmerman et al. 2007). Alternatively, rapid changes in sediment supply, sea-level, or freshwater inputs can change the hydrodynamics of a system, and the resulting morphological adaptation may rework deposits, obliterating the record of past environments. The migration and evolution of channels in response to changing physical conditions can lead, therefore, to complicated architecture in the resulting sedimentary deposits, including the presence of multiple erosive surfaces. The transgressive nature of many modern shorelines adds to the difficulty of interpreting tidal channel deposits (Dalrymple and Choi 2007). Yet, understanding the evolution of modern systems, explaining changing morphology and quantifying rates of network expansion or reduction, can provide improved insights into coastal response to sea-level change, both past and present.

Previous chapters have described the channels, and the associated facies, in a number of different tidal settings. This chapter aims to give an overview of the evolution and common characteristics of channels within coastal systems, drawing comparisons with fluvial channels. We will start with a general overview of the nature of tidal channels and then compare several classifications of tidal channel network, according to planform, with a focus on shallow intertidal settings such marshes and tidal flats. The remainder of the chapter will examine the defining physical processes and the resulting geomorphologic relationships that have been observed for channels in these environments. Deposits created by tidal channels and the potential for their preservation within the stratigraphic record in specific settings have been explored in previous chapters. However, in the last Section we will provide a description of certain tidal facies that can are particular to channels.

11.2 General Characteristics of Tidal Channel Systems

The response of a region to the repetitive flooding of tidal water is to self-organize into shallow areas that are periodically flooded, and channels that drain them. As a consequence, in shallow, intertidal landscapes there tend to be three major morphological components: (1) unvegetated tidal flats or bars; (2) vegetated marsh platforms or mangroves; and (3) channels, which dissect and interconnect the other two zones (D'Alpaos et al. 2005; Fagherazzi et al. 2006). These channels may be intertidal (drying out or having standing water in only the very deepest parts during low water) or peri-subtidal (in which the wetted perimeter of the channel is large in comparison to the tidal range). In systems that exhibit extensive subtidal regions, channel-shoal morphology is often seen, in which very deep (compared to the tidal range) channels run between bank-attached bars or mid-channel linear sand bars, parts of which may be exposed at low tide (for example the Wash, the Gironde Estuary, or the mouth of the Fly River).

Although, large-scale features, such as an estuary (a flooded river mouth; Dalrymple et al. 1992), are undoubtedly to be considered a tidal channel, features on this scale are complicated by extreme variations along their length. For simplicity, here we will focus on meso-scale features; channels that fit within macro-scale features, such as flood-tidal deltas, or mega-scale features, such as estuaries or back-barrier basins. These tidal channels contain micro-scale morphological forms such as bedforms or bar-forms (de Vriend 1996; Hibma et al. 2004a, b). Their evolution occurs over medium timescales (days to centuries), as they equilibrate to forcing such as storm events, sea level rise or gradual infilling where there is an adequate supply of sediment. Tidal inlets and high-order tidal channels have a relatively high preservation potential (Belknap and Kraft 1985), while shallower tidal features are vulnerable to erosion during shoreline transgression.

Meso-scale tidal channels share a number of characteristics: (1) some sinuosity; (2) depositional bed morphology such as ripples and bars; (3) low channel-bed

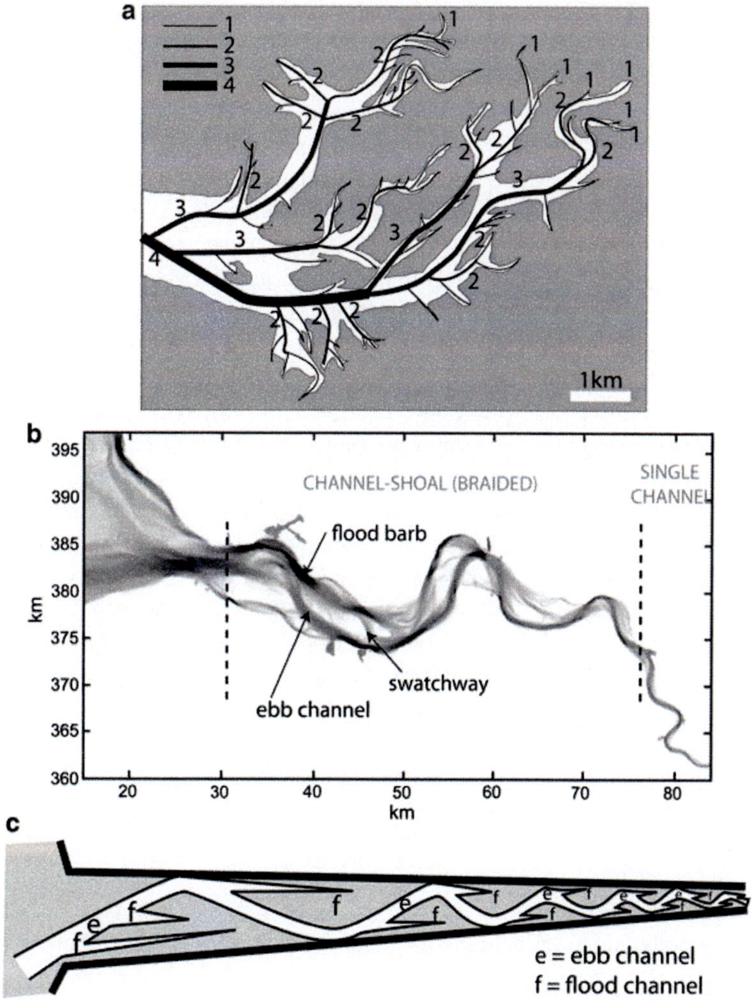

Fig. 11.1 Types of tidal channels: (**a**) a dendritic or fractal network in the Dutch Wadden Sea, stream orders are numbered (Cleveringa and Oost 1999); (**b**) braided/interconnected channeling in the Western Scheldt Estuary, Netherlands; and (**c**) a sketch of ebb and flood channels in a braided network (van Veen (1950); adapted from Hibma et al. (2004a))

gradients; and (4) width to depth ratios greater than 5 (Steers 1969; D'Alpaos et al. 2005). In general, the channels tend to narrow inland; seen from above, coastal waterways often appear funnel-shaped. This relates to a reduction in tidal prism upstream (the rate of this reduction is sometimes explained by tidal resonance; for further discussion see Wright et al. (1973) and Van der Wegen et al. (2008)). This strong spatial gradient of channel width, which occurs in shallow tidal channels of all orders, is arguably one of the key characteristics that distinguishes tidal from fluvial systems, along with the notably higher width of tidal channels with respect to the inter-channel region that they 'drain' (Fagherazzi et al. 1999). This difference in channel width to drainage area means that tidal channels would seem more closely spaced when compared to rivers of a similar width.

Tidal channels are ubiquitous, occurring across macro-, meso- and microtidal environments. They often form dendritic networks (i.e. branching and blind ended; Ashley and Zeff 1988), commonly of low-order. The smallest creeks at the edge of a network are the lowest order, these meet to form a channel of the next order (Fig. 11.1a, Horton 1945). Tidal channel networks have been described by some studies as fractal (Perillo et al. 1996; Fagherazzi et al. 1999; Schwimmer 2008). Pestrong (1965) observed that the dendritic tidal networks in San Francisco Bay resembled fluvial systems. However, despite their apparent similarity, he determined that the tidal channels did not follow Hortonian laws of drainage networks. Tidal networks, unlike their fluvial counterparts, are not true scaling structures (Fagherazzi et al. 1999). Marani et al. (2002) concluded that "in any real case of fluvial versus tidal patterns, differences are the norm rather than the exception once carefully examined".

Variability may, in fact, be the primary characteristic of channel systems in tidal environments. In tidal marshes, multiple sub-basins may exist with quantifiably different channel distributions (Fagherazzi et al. 1999; Marani et al. 2002) as a result of highly localized changes in sediment type or vegetation or broader changes in hydrodynamics. Neighboring drainage basins may have entirely different planform morphology and exhibit different relationships between drainage area and channel dimensions (Marani et al. 2002, 2003; Rinaldo et al. 1999, 2004, c.f. Eisma 1998). Some will exhibit values closer to fluvial systems than others.

The explanation for such variation lies in the large number of factors that influence the evolution of tidal channels. These can be broken down into either physical-environmental constraints or hydrodynamic factors. Physical attributes that are important in channel development include antecedent geology, sediment deposition patterns and grain size, and the presence and type of vegetation. These will all impact the erodibility of the substrate and consequently the stability of the channel features. Stability controls persistence, and therefore evolutionary complexity, but it is also a factor in channel shape (both planform and cross-sectional profile).

Hydrodynamic influences on channel evolution encompass the balance of exposure to tidal and wave forces. The tidal flows in a channel may either result from external or remote forcing (i.e. the offshore tide) or be a response to the local morphology, but it is not always easy to separate these. For example, channel size and shape respond to the portion of the tidal prism that passes through it. This depends not only on the regional tidal range and the size of the basin being flooded, but also on the local morphology of the surrounding channels, which modify the advancing tidal wave (Marani et al. 2003). Other factors influencing the hydrodynamics are: the gradient over which the drainage occurs (ranging from very abrupt, local effects relating to a change in underlying stratigraphy or local vegetation, to regional variations in tidal range), the dominance of the ebb (seaward) or flood (landward) tidal velocities, the curvature of the channel, and, lastly, the hydraulic radius of the channel (a function of the width to depth ratio). Assessing these relationships is complicated by interdependent feedbacks between factors, particularly channel curvature and hydraulic radius. Note that here, we will not consider the impact of meteorological tides and waves in any detail.

Processes controlling the initiation and evolution of channel systems operate within both the vertical and the horizontal plane. Vertical processes include: deepening through erosion and suspension of sediment, through compaction, or due to sea-level rise; shallowing through inorganic sediment deposition; or relative change due to the erosion or accretion of the surrounding platform or tidal flat. Laterally, processes include channel widening through bank erosion; 'elaboration' i.e., a change in the intensity of meandering or channel migration; and headward erosion (D'Alpaos et al. 2005).

Channels within the same system may not only result from different processes, they may also function differently depending upon their origin. The observations of Zeff (1988) and Ashley and Zeff (1988) illustrate this. These studies identify two types of tidal channel within the salt marshes of New Jersey. The first type are 'through-flowing' channels that connect channel to channel or to lagoons; the second type are 'dead-end' channels which end within the marsh, and often start at a through-flowing channel.

As well as notable differences in channel size, width to depth ratio, sediment properties, sedimentary type and structure, and the variation of width inland, there is also a significant difference in hydraulics between these two types of channel. Peak currents in the dead-end channels occur close to bank-full conditions, whereas, in the through-flowing channels, they occur at mid to low tide. The maximum currents are generally an order of magnitude larger in the through-flowing channels than the dead-end channels. Zeff (1988) proposes that through-flowing channels are formed during the infilling of the back barrier as the flood-tidal delta was stabilized by vegetation, (i.e. they are flood-formed channels that are now essentially relict). In contrast, the dead-end channels have eroded headward into the marsh platform, post vegetation, and as such are formed by ebb flows and are still likely to be actively evolving. These two channel forms are therefore fundamentally different, yet proximal, with very different sediments and resulting facies. As this example illustrates, it would be easy to assume that smaller tidal channels are a scaled version of the larger channels in a system, but this is often not the case.

11.3 Classification of Channels and Channel Network Morophologies

Several authors have classified tidal channel network morphologies according to their planform. Hibma et al. (2004a) broadly view an entire estuarine system, looking at the large channel forms and making a general classification into two morphologies: fractal (i.e. dendritic systems) and braided (meandering, interconnected channels separated by shoals; Fig. 11.1). A similar classification is made by van Veen (1950) who describes them as 'apple tree' and 'poplar tree' morphologies, respectively. These two morphologies loosely relate to the shallow, intertidal and peri-tidal environments of tidal flats and marshes, and the deeper subtidal environments, respectively, as described above.

Eisma (1998) examines intertidal channels on a variety of scales in a range of coastal settings, including estuaries, back-barrier systems and open coast tidal flats and marshes. His classification is more detailed, recognizing ten types of channels within three categories: (1) single channels: straight, sinuous, and meandering (sinuosity ratio > 1.5; see Sect. 11.5.3); (2) channel systems: parallel channels, dendritic and elongate dendritic, distributary, braided, and interconnecting; and (3) few or no channels.

Further narrowing the environment considered, Pye and French (1993) identify seven categories of network within marsh systems. These overlap or expand on those of Eisma (1998): linear single, dendritic and linear dendritic, meandering dendritic; reticulate, complex, and superimposed (Fig. 11.2). Several types of channels or channel network may occur concurrently within a tidal system. The Wash (UK) is a classic example of variability within a single estuary: in an area approximately 25 km^2, one may find extensive salt marshes, tidal flats and channel-shoal morphologies.

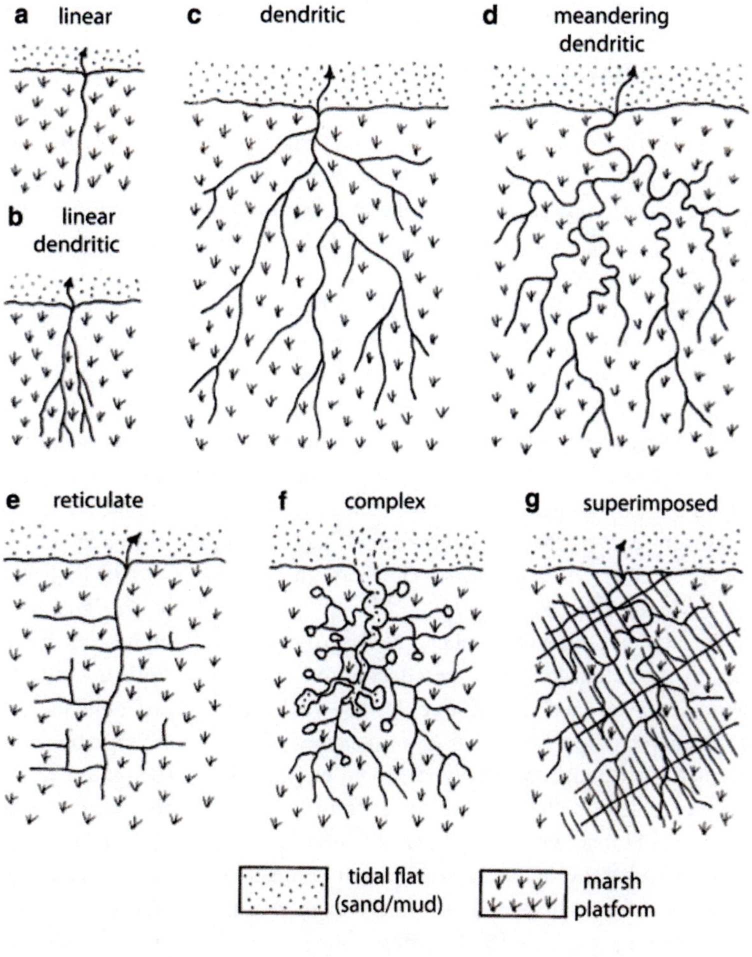

Fig. 11.2 A classification for salt marsh creek networks (After Pye and French 1993)

Within these systems there are single channels alongside complex dendritic networks; parallel, straight channels alongside sinuous, elongate, dendritic channels; channels that meander strongly inland but gradually straighten as they extend seaward; in short, there is a diverse array of channel forms, created by local variations in tides, sediment and vegetation.

The overlaps and differences between the classifications occur because of the scale of the area that is considered by each. Hibma et al. (2004a) provide a large-scale view, whereas Pye and French (1993) concentrate on marsh systems only and present essentially a detailed classification of the possible variation in dendritic systems. The classification of Eisma (1998) falls somewhere between these scales, overlapping with each. However, the classifications of both Eisma (1998) and Pye and French (1993) incorporate two key morphological observations: they make differentiations based on the level of channel complexity (*elaboration*), and whether or not the system consists of a single channel or developed networks.

11.3.1 Elaboration

The morphology of an individual channel may range from straight, its simplest form, to meandering, and further to convolutions involving the incorporation of ponding or man-made drainage ditches (as are common, for example, in the marshes of New England, USA; Figs. 11.2f and 11.3e). Straight or linear channels, despite their name, will have some natural irregularities. This may make the boundary between a channel that is straight and one that has some gentle curving less clear. However, as curving increases, a channel is described as sinuous or weakly meandering (Fig. 11.3a, b). In general, larger channels tend to be straighter (e.g. in the Wash or Zaire River estuary; Fig. 11.3a, d, f, g; Eisma 1998; Ginsberg and Perillo 2004; Marani et al. 2002, 2004).

The sinuosity of a channel can be described by the ratio of the actual length of the channel to the downstream distance (in a straight line) of the 'wavelength' of the curve. When this *sinuosity ratio* exceeds 1.5 the channel is termed *meandering* (Leopold et al. 1964). Many authors note that easily eroded, non-cohesive or unvegetated substrates are more likely exhibit straighter channels, whereas channels extending into vegetated regions, such as salt marsh, are likely to increase in sinuosity (e.g. Fig. 11.3h, Pestrong 1965; Garofalo 1980; Eisma 1998). Some marsh channel networks develop beyond meanders to highly complex morphologies incorporating ponding (e.g., Tollesbury marsh, Essex, UK; Figs. 11.2e and 11.3d). The processes leading to the development of meanders and the resulting channel-bed morphology is discussed below (Sect. 11.5.3).

On a macro-scale, Dalrymple et al. (1992) describe a pattern of straight-meandering-straight. This configuration is observed in channels within the inner reaches of estuaries but not seen within deltas (i.e., on regressive shorelines). The outer straight relates to deeper subtidal environments where flow and sediment transport is generally directed landward because of asymmetry in tidal flows, the upper straight occurs in a region where the sediment transport is directed seaward because of river dominance and the central, meandering section exhibits a region of fine sediment (grain size decreasing towards it from both directions). A similar pattern is also suggested in the data presented for a single salt marsh creek by Solari et al. (2002) (their Fig. 2). A physical explanation for this pattern has yet to be identified.

11.3.2 Dendritic Networks

Dendritic channel networks are the most commonly observed form on tidal flats and salt marshes (Figs. 11.2 and 11.3). The smallest, or *first-order*, channels, end abruptly on the marsh platform or tidal flat, fed by sheet flow over the inter-channel areas. In a classic dendritic system two of these smaller channels join to form a larger (second-order) channel, and so on, until the highest-order channel in the system is reached (Fig. 11.1a). In tidal channels, third- or higher-order channels are relatively rare (Eisma 1998).

In a fluvial system, the low-order streams feed water into the higher-order streams. Within a tidal system, all of the channels experience bidirectional flow, with high-order streams both feeding and receiving flow to/from lower-order creeks. The ratio of low to higher-order channels in low gradient fluvial systems is 2; this *bifurcation ratio* is higher in tidal channels, closer to 4 (Knighton et al. 1992; Novakowski et al. 2004). However, the data presented by Novakowski et al. (2004) for North Inlet, South Carolina, USA, suggest that for low-order channels the ratio falls nearer to 2, increasing with stream order to 7.25 for the highest orders observed (forth- to fifth-order).

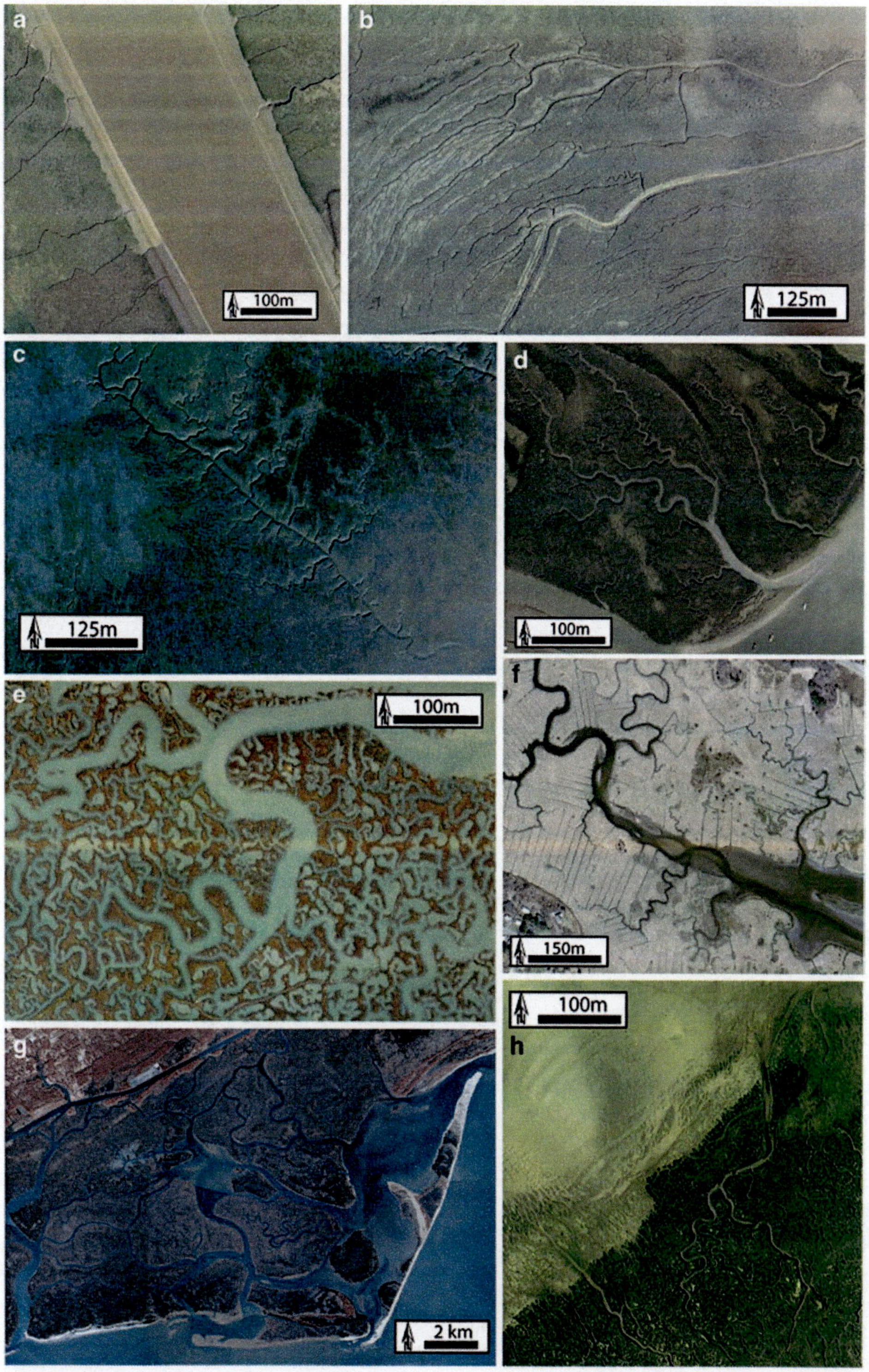

Fig. 11.3 Examples of tidal channel morphology: (**a**) straight, parallel creeks meeting a larger straight tidal channel in the Wash (UK); (**b**) an elongate dendritic network reaching from the tidal flats onto the vegetated marsh, Wash (UK); (**c**) an example of a reticulate network, although the smaller channels exhibit high sinuosity, West coast of Korea; (**d**) a highly meandering dendritic network, Norfolk (UK); (**e**) a complex morphology, Tollesbury Marsh (UK); (**f**) superimposed man-made drainage ditches and natural channels, Essex Marsh, Massachusetts (USA); (**g**) Cape Romain, South Carolina (USA), formed as part of the Santee River delta, exhibiting both interconnected (through-flowing) and dead-end channels, all channels have a level of sinuosity, however meandering is more extreme in the smaller creeks; (**h**) a meandering, dendritic network in the Dyfi Estuary (UK), where the channels extend across the boundary between the sandy tidal flats and vegetated salt marsh

Reticulate channels (Fig. 11.3e) can be considered as a form of dendritic channel; however, they are notable for the 90° angle at which the low-order channels meet higher-order channels. First-order tidal creeks commonly end at 90° to the higher-order channel (Zeff 1988; Eisma 1998; Ginsberg and Perillo 2004), whereas higher-order channels commonly meet at a lower angle. In fluvial systems a 90° attachment of a low-order stream is usually associated with a high bed gradient in the low-order channel compared to the higher-order one. Pestrong (1965) observed that in San Franscico Bay, low-order tidal channels often had steeper bed gradients than the higher-order channels. However, 90° attachment angles in tidal systems have also been attributed to the nature of the tidal flow, when small channels experience high tidal asymmetry relative to the larger channels that have more equally balanced ebb and flood flows (Zeff 1988; Eisma 1998).

11.3.3 Braided, Distributary and Interconnected Channels

The term 'braided' is used by Hibma et al. (2004a) to describe the channel systems in the deeper subtidal regions of an inner estuary. Here, a complex system of ebb- and flood-dominated tidal channels occurs within a relatively straighter section of the estuary. The mutually evasive channels meander, slightly out of phase, the ebb channel is generally well formed and the flood channels may be continuous or form *flood barbs* across the shoals amongst which the ebb channel weaves (Fig. 11.1c). Periodic overlapping of the flood and ebb channel and small swatchways connect the channels. Shoals may become vegetated and eventually form islands (Fig. 11.1b, c).

In fluvial networks, the term braided is applied to channel complexes, which form in regions of higher gradient and where sediment supply overwhelms hydraulic transport potential. In contrast, in tidal environments this channel morphology occurs in the middle parts of estuaries, where peak ebb currents and peak flood currents occur at a similar stage of the tide. In plan view this morphology is similar to that of terrestrial braided channel systems. The process of formation in tidal environments is not well understood, although it is likely different from fluvial setting as tidal flow is bi-directional and water surface slope is normally more influential than bed gradient in driving the flow.

As noted, spatial scale seems to be an important control on the expression of tidal channels within a given system. Eisma (1998) examines smaller intertidal systems, and neither distributary channels nor braided channels are common within the collected observations upon which he based his classification. The term distributary channels is used to describe ebb dominated channels which form on small deltas building out of the entrance of larger tidal channels. On this smaller scale, braided channel morphologies tend to occur in macrotidal environments such as King Sound (Australia) or the Bay of Fundy (Canada). These channels have a low gradient and a low topography, suggesting they are active during lower water levels. They form in the region of maximum tidal energy. On a very small scale, braided channels have also been observed on tidal flats of loose sediment where the gradient of the flat is steep, forming either near a river mouth or over loose debris at the base of cliffs (Eisma 1998). Likewise, Dalrymple et al. (1992) describe similar patterns of channelization on sand flats in macro-tidal regions with very large tidal range.

Interconnected channels begin and end at another channel, or link a lagoon to the ocean (Ashley and Zeff 1988, Fig. 11.3g). These often occur in conjunction with dendritic channels; in fact, it is common to see many of the different categories of channel morphology or network existing concomitantly. Interconnected channels are not exclusive to any tidal range and are likely to meander, although sinuous and straight forms are also observed (Eisma 1998) and may purely be inherited as by marshes as flood tidal deltas are stabilized by vegetation (Zeff 1988). Based on observations in the Niger Delta, Allen (1965) suggests that interconnected channels form as tidal flats grow vertically and horizontally (due to the sediment supplied by the river), or as blind channels join together. Both of these studies describe the evolution of a delta (the first tidal, the second riverine) with stabilization and increased accretion on the higher flats, while the channels are maintained by the tidal flows. The term 'interconnected' channel is, thus, fairly broad.

11.3.4 Parallel Channels or No Channels

Systems displaying parallel channels or no channels at all are relatively rare, and most are found in regions with large tidal ranges (macrotidal). Parallel channels frequently develop where the sediment is erodible, such as unvegetated, fine silts and sands on tidal flats. Often a sign of an immature drainage pattern, they commonly occur on flats that are regularly impacted by storms, thus 'resetting' them either partially or totally. Such behavior is observed in open regions of Kyenoggi Bay (Korea) and along the Jiangsu coast (China; Lee et al. 1992; Ren 1986; described in Eisma 1998), where the wave energy is high and tidal currents are weaker in the open-coast environment. The more sheltered regions of Kyenoggi Bay exhibit dendritic channel networks (Lee et al. 1992). Gullies in sandy sediment are generally shallower and wider than those in finer sediment, and are more likely to be ephemeral or even absent (van Straaten 1954). The implication is that parallel channels, often also straight, are transient, potentially being removed and recreated with every storm.

Areas with few or no channels may also occur in regions that experience only infrequent tidal inundation, or freezing or arid conditions for long periods of time, stabilizing the sediment (e.g. James Bay in southern Hudson Bay, which is covered with ice for up to 6 months of the year; Eisma 1998).

There are of course exceptions: in New South Wales, Australia, there is a distinct lack of channels in the marshes (Adams 1997). These systems are of limited size, sitting landward of mangrove forests. Where drainage does exist it is often inherited from river systems. The region is microtidal and doesn't fit most patterns as described above. It is not clear why these regions lack channels, perhaps it is purely that the strength of the vegetated soils are sufficiently high, and the size of the areas sufficiently small, that the sheet flow across the marsh surface is unable to initiate channels, but further research is undoubtedly needed. Likewise, Hughes et al. (2009) observe a system of parallel channels forming and actively incising into vegetated salt marsh platforms across the Santee Delta (SC, USA). The authors propose that burrowing and herbivory by crabs weakens the soils in the region surrounding the head, allowing the creek to erode headward more easily than on other vegetated marsh platforms.

11.4 Hydrodynamics

Along the continuum from marine to terrestrial settings, tidal environments experience variations in tidal, fluvial and wave energy (Dalrymple and Choi 2007). Tidal currents are the dominant hydrodynamic forcing in the generation and maintenance of tidal channels. Fluvial currents (if present) decrease in influence with distance seaward of the tidal limit (i.e. the landward extent of the tidal wave). The intensity of fluvial flow depends upon river stage and precipitation, but can be considered constant over the timescale of a tidal cycle. Wave energy decreases swiftly with distance from the ocean. Locally generated wind waves may occur within very large channels and bays, producing local erosion of the marsh edge and channel banks. This may create gullies in tidal flats or a "cleft and neck" morphology on salt marshes (Pethick 1992; Watzke 2004; Schwimmer 2001, 2008). Clefts, are narrow channel-like indents in the edge of the marsh platform and necks are the tracts of marsh remaining between the clefts. The influence of waves in smaller channels tends to be low because of sheltering.

Tidal areas experience two peak velocities during a full tidal cycle, which occurs once or, more commonly, twice a day (tidal period = 25.8 and 12.4 h respectively). The flood velocity is directed landward and the ebb is directed seaward. Depending on the position within a tidal system, these velocities will vary both in absolute magnitude and in comparison to each other (*tidal asymmetry*). The bidirectionality of tidal flows makes them distinctly different from fluvial systems and has a significant impact on channel morphology. In general, flows within tidal channels are often driven by gradients in water slope rather than bed slope (Rinaldo et al. 1999). In many areas channel bed slopes are low, yet fast currents are generated by the variation in water depth related to the tide. In small, first-order creeks and across a tidal flat or marsh platform, however, bed slope may have more influence becoming a significant force driving flows at low stages of the tide. Unlike rivers, maximum current velocities within tidal channels do not necessarily coincide with maximum stage (water depth), but instead occur at some midpoint during the tidal cycle.

Spatial variations occur both in the magnitude of tidal flows and in the asymmetry of the ebb and flood periods or velocities. These variations result from: (1) variation of tidal range (prism) across the system, (2) water depth and its effects in terms of modifying the tidal wave, and (3) the morphology of the surrounding intertidal area (e.g., vegetation will retard flows during over-bank events; low versus high gradients on the regions between channels will produce different flow rates).

11.4.1 Tidal Range

Tidal range is proscribed by the offshore tidal wave, which varies according to latitude, the shape of the ocean basin and the width of continental shelf (Davis and FitzGerald 2004). Within a geographically extensive tidal system (mega-scale), tidal range may vary in both timing and magnitude. Given the forcing of the offshore tidal range, the magnitude of a tidal signal in a region is primarily a response to bed morphology. In wide-open basins and back-barrier areas, the signal will experience a gradual reduction in amplitude inland (*hyposynchronous*). However, a funnel shaped estuary may experience amplification of the tidal wave inland, before reducing to zero at the tidal limit (*hypersynchronous*, Dyer 1997). This results in two zones of similarly weak tidal influence occurring seaward and landward of a strongly tide-dominated zone (Dalrymple and Choi 2007), the Bay of Fundy (Canada) being the classic example.

11.4.2 Asymmetry of Tidal Currents

Essentially there are two reasons for inequalities between the magnitude of flood and ebb velocities or the respective periods over which they flow. The first is the finite amplitude effect (also called the shallow-water effect). In shallow water, the difference in depth between the crest and trough of the tidal wave is significant; therefore water under the crest (i.e., high water) will move faster than water under the trough as the celerity of a wave is proportional to the water depth ($c \propto \sqrt{gh}$ where g is gravity and h is the water depth) (Dronkers 1986; Parker 1977, 1991; French and Stoddart 1992).

The second cause of tidal asymmetry is morphological. The presence of extensive intertidal regions has an impact on the timing of the flood and ebb (particularly in the presence of vegetation). The slower propagation of the flood and the ebb over the platform leads to both a delay in the turn to ebb and a slower returning flow to first-order channels. The delay in the turn of the tide shortens the ebb, and continuity requires that the velocities need to be faster to move the same tidal prism during this shorter period of time. Physically the flows in the channel can move more easily than flows over the platform, so during the ebb tide the relative water surface slope between the platform and the channel is steeper, creating faster flows. As water on the platform surface becomes very shallow, flows returning to the channel may be driven by bed slope. As a consequence, the magnitude and timing of peak velocity during the ebb tide are altered (Friedrichs and Aubrey 1988; Fagherazzi et al. 2008).

While these two factors are the principle controls on asymmetry in most tidal environments, in these complex systems there are often other factors. Location can be of great importance to the tidal asymmetry and very local variations may be seen across a channel or either side of a shoal. This is particularly notable in meandering channels or in the deeper subtidal regions of the inner estuary. Li and O'Donnell (2005) examine the behavior of flows is subtidal channels, comparing long and short channels. This study neatly demonstrates that in estuaries that are long in comparison to the tidal wave, the seaward regions are likely to experience ebb dominance in deeper regions, with flood dominance on shoals. In contrast, short estuaries and the upper reaches of long estuaries will exhibit flood dominance in deep channels and ebb dominance in shallower subtidal regions. This is the result of the nature of the tidal wave, whether it behaves as a standing wave (in short channels) or a progressive wave (in the outer part of long channels). Residual sediment transport within an estuary will be integrated across these local variations and, thus, it will be influenced by the tidal asymmetry throughout the entire system and calculations of this parameter should not be based purely on measurements in the main channel.

In regions with diurnal tides (e.g. the Louisiana coastal plain), where the K1 and O1 tidal constituents are very significant in comparison to the semi diurnal M2 tide, tidal asymmetry (in the ebb direction) is directly related to the ocean tidal wave rather than to shallow water effects (known as overtides) or the hypsometry of the drainage network (i.e. the relative extent of the marsh platform or tidal flat to the channel; Howes 2009). This asymmetry of the flow at the tidal inlet may propagate throughout the system, underlying further modulations upbasin.

Finally, there is a potential influence of fluvial discharge, which, if significant, can produce apparent ebb dominance towards the tidal limit as the flows are superimposed (Wolanski et al. 2006; Dalrymple and Choi 2007).

11.4.3 Overtopping and Velocity-Stage Relationships

Where water depth is deep relative to the tidal range, the tidal wave is progressive. Just like a wind wave, velocities are highest under the crest and the trough. Thus, peak flood currents occur at high water and peak ebb currents at low water. Under a standing wave, by contrast, peak flows occur at mid tide, which is the common model for coastal tidal flows. The latter occurs in regions where the water depth is shallow compared to the tidal range. Where the tidal wave is progressive, little energy, and thus tidal amplitude, is lost over distance. In shallow regions friction acts to reduce the amplitude of the tidal wave and so tidal range is reduced up channel. Large systems will experience some combination of these two stage-velocity models (progressive and standing wave conditions) as the tidal wave moves up estuary (Wright et al. 1973; Hibma et al. 2004a; Howes 2009).

In regions of extensive intertidal areas, the stage-velocity is complicated further. Hydrodynamically, it is possible to distinguish two types of tidal channel, which represent the two end members along a continuum. Low-order channels, in which the tidal range is significant in terms of the channel depth, derive the majority of the water flux that passes through them from sheet flow leaving tidal flats or the marsh platform. Higher-order channels, for which the change in volume experienced over a tidal cycle is small in comparison to their size, in contrast, receive a significant volume of water from other channels, rather than from overbank flow which has been strongly effected by shallow water and frictional effects. It is possible that these two end members could be compared to the dead-end and through-flowing channels of Ashley and Zeff (1988); however, any high-order channel within a dendritic system may fall into the larger subtidal category. The two types of channel will experience different flows. Low-order creeks experience velocity transients (surges) at close to bankfull conditions (Fig. 11.4, Bayliss-Smith et al. 1979; French and Stoddart 1992; Fagherazzi et al. 2008). The higher-order creeks are more likely to have their highest velocities near mid-tide (if the tidal wave is a standing wave), have a lower tidal asymmetry, and experience significantly higher velocities (~1 m/s at compared to ~0.1–0.6 m/s in low-order salt marsh creeks; Ashley and Zeff 1988; Hughes et al. 2009).

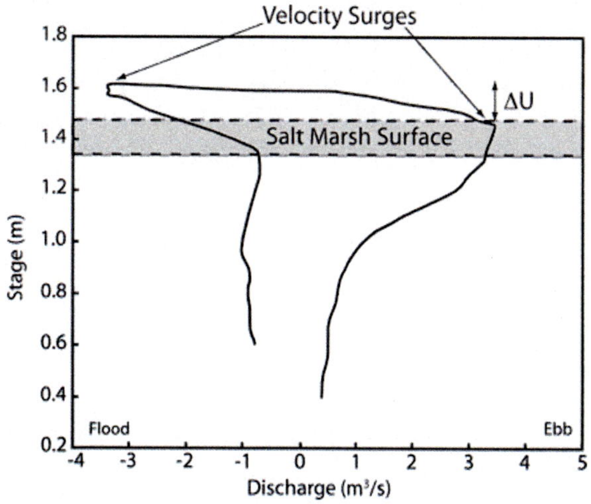

Fig. 11.4 The hysteresis observed in tidal velocity versus water depth (stage). Velocity is highly variable, but two distinct peaks are seen, one during the flood just above bankfull conditions when the water level is at the level of the marsh surface, and one during the ebb. In terms of symmetry around either high tide or the timing of bankfull conditions, the peak ebb velocities lag the flood transients, occurring later, at a lower stage of the tide, just below bankfull. ΔU indicates the difference in the height at which the peak velocity occurs (From the observations of Bayliss-Smith et al. (1979), adapted from Fagherazzi et al. (2008))

This has significant implications for the net transport and erosion patterns in each type of channel; it may also help to explain why tidal channels are not scale invariant in the way of fluvial systems (Fagherazzi et al. 1999; Rinaldo et al. 1999; Marani et al. 2003).

The frequency of bankfull and overtopping tides varies; it occurs with every tide on unvegetated tidal flats, but may occur as few as 6–8 times a month on the high marsh. When overbank flow does occur, a distinct hysteresis is seen in the discharge of low-order channels (Fagherazzi et al. 2008 Fig. 11.4). During the flood, a surge is seen when the platform is inundated (as an increased volume of water is drawn through the channel in order to fill the platform area). During the ebb, flow peaks when the water level is at or just below the marsh surface. As water drains from the marsh platform, a steep hydraulic gradient between the water on the platform and the water level in the channel creates fast flows and focuses the flow into the creeks, particularly at the head (which serves a greater area of unchanneled platform). The discharge within the channel will be a function of the inundated surface area (S) and the water depth (h) (Boon 1975):

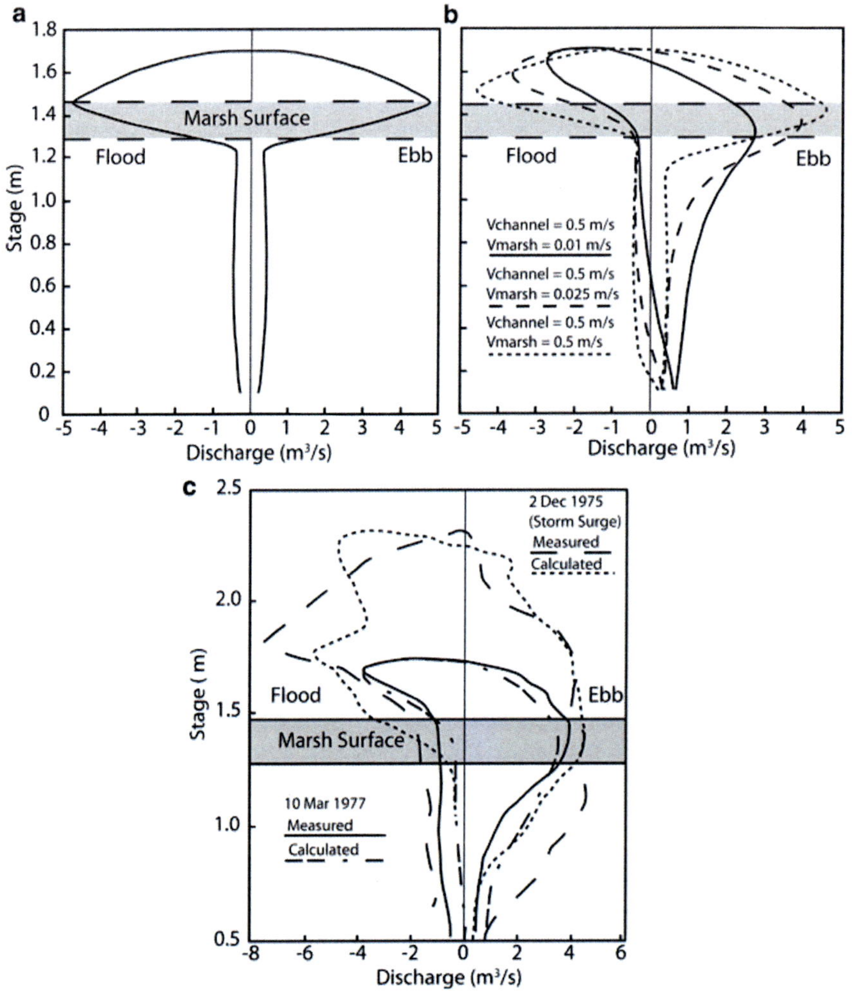

Fig. 11.5 (**a**) The stage-discharge relationship based on the simple continuity model of Boon (1975); (**b**) the impact on the asymmetry of the velocity peaks (as seen in the channel during the flood and ebb) of reducing the velocity (and thus apparent friction) across the tidal platform relative to that in the channels; and (**c**) the application of the TIGER model to a real system in Norfolk (UK) using a channel flow of 0.5 m/s and an overmarsh velocity of 0.05 m/s to reproduce the observed stage-discharge relationship

$$Q = S\, dh / dt \tag{11.1}$$

However, this relationship does not fully capture the asymmetry of the hysteresis loop (Fig. 11.5a, Fagherazzi et al. 2008). Pethick (1980) added an influence of asymmetry from the tidal inlet to this model in order to address this inconsistency, yet the result still does not reproduce the relative delay in the peak ebb flows. Fagherazzi et al. (2008) demonstrate that the delay in travel time of water moving across the flats contributes significantly to this behavior (Fig. 11.5b). Taking this into account, they successfully use their model ("Tidal Instantaneous Geomorphologic Elementary Response" or TIGER) to predict the delay in velocity surge during the ebb (Fig. 11.5c). Using this observation in reverse, a hydrograph from a tidal channel can provide information about the travel distance and thus, the residence time of water on the marsh surface (Fagherazzi et al. 2008).

11.4.4 Shear Stress and Erosion Potential

Numerical models of flow variation across a marsh surface demonstrate that shear stress reaches maximum a value at the tip of channels and near bends

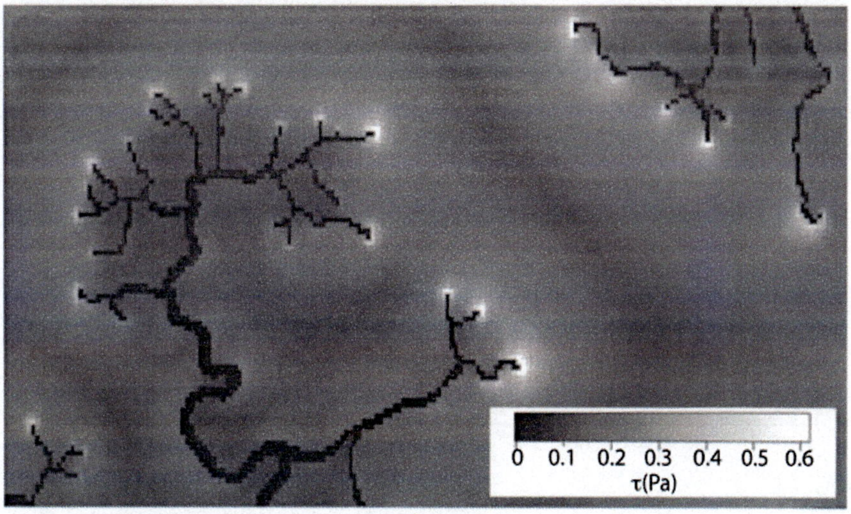

Fig. 11.6 Distribution of shear stress within a tidal channel – tidal flat system (Adapted from D'Alpaos et al. 2005)

where flow from the platform is focused into the creek (D'Alpaos et al. 2005). The shear stress is calculated based on the gradient of the water surface across the marsh surface using a Poisson model (Rinaldo et al. 1999), which assumes that the microtopography on the marsh surface and the water surface slope are both much smaller than the absolute water depth, and that friction is only applied to the marsh surface rather than to the channel. The reported shear stresses near the channel head are sufficient to cause erosion (Fig. 11.6). This supports the idea of headward erosion and lateral erosion as mechanisms for channel growth and elaboration, respectively.

11.4.5 Implication for Sediment Transport

A large number of researchers have investigated the sediment flux within tidal channels (Settlemyre and Gardner 1977; French and Stoddart 1992; Mudd et al. 2010). In general, sediment transport within intertidal systems is highly complex; this is a direct result of the equally complicated hydrodynamics. Tidal range and asymmetry vary throughout systems; thus, as mentioned previously, the measurement of flood dominance in one creek does not mean the entire system is experiencing the same net flux of sediment. Furthermore, the occurrence of overbank flow creates convoluted transport pathways and residence times. Conservation of mass or momentum within an individual channel may not hold due to overbank flows to adjacent channels, or because of the loss of integrity of defined 'creek-sheds' (i.e., in situations where the watershed is not defined by a topographic high and is overtopped during a spring tide, water on the marsh surface or tidal flat may flood and ebb through different creeks; French and Stoddart 1992). Sediment transport is also complicated by bioturbation and biostabilization (either by biofilm or vegetation). Vegetation may also influence sediment transport though baffling of flow or by inducing scour (Temmerman et al. 2007).

Figure 11.7 depicts the typical behavior within a tidal gully in the Wadden Sea. During the period when the banks are overtopped, ebb velocities are low (the opposite of fluvial systems). As discussed above peak velocities occur just after water depth in the channel falls below bankfull. A peak in sediment transport is associated with this velocity maximum, as fast flows erode the channel and tidal flats. Net flux out of the small channel may still not necessarily be indicative of the behavior of other channels in the system. In many tidal systems high suspended sediment loads are advected around the system, either coming from nearby rivers or from offshore.

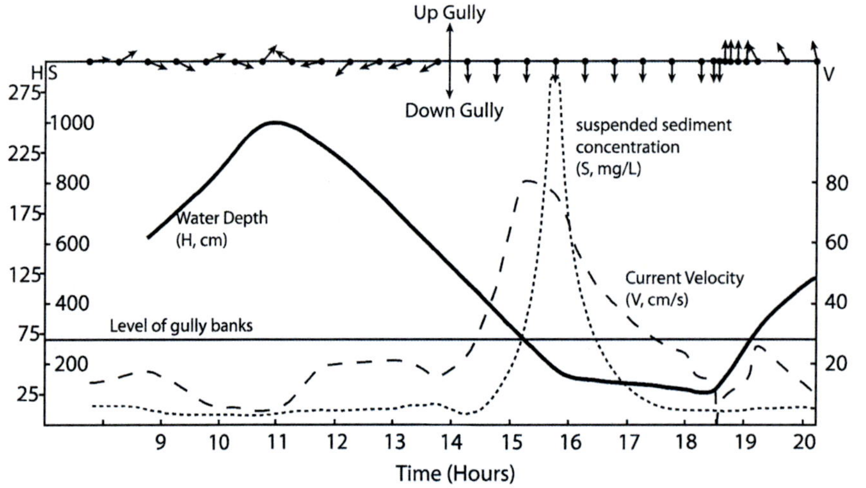

Fig. 11.7 An example of the temporal variation in water depth, velocity and suspended sediment concentration in a small tidal gully in the Wadden Sea (Adapted from van Straaten 1954; in Eisma 1998). Water depth (H) in cm is given on the *outer left* y axis and sediment concentration (*S*) in mg/L is displayed on the inside of this axis, velocity (*V*) in cm/s is given on the *right* y axis. The direction of the velocity is indicated by the *arrows* running along the *top* of the diagram

11.5 Tidal Channel Morphology

11.5.1 Initiation

Observational evidence suggests that there are two ways in which a channel may develop: incision into a surface or deposition, i.e., accumulation of sediment around a channel. In the first of these, initial formation is followed by a slower elaboration (deepening or increase in sinuosity; D'Alpaos et al. 2005; Symonds and Collins 2007; Knighton et al. 1992). Conceptual models describing this process have been put forward by a number of authors (Pethick 1969; French and Stoddart 1992; Steel and Pye 1997; Allen 1997). High shear stress at creek heads and the behavior of first-order channels suggests that headward erosion is the major process in the development of a network of channels. Thus the formation of a network is decoupled from any subsequent evolution (meander development and ecogeomorphological development of intertidal areas), which happens gradually over longer time-scales.

In general, very shallow flows over a flat surface will occur as sheet flow. However, after a certain distance of flow the converging volume and velocity of the flow will reach a sufficient magnitude to erode the surface of the flats. This is known as the *critical length* of a flow and depends upon surface slope, the intensity of the run off (which in tidal environments will vary with spring–neap cycles, meteorological tides and precipitation), the infiltration capacity of the sediment, and the resistance of the sediment on the flats to erosion. In intertidal environments, sufficiently high velocities are most likely to occur on an ebb tide because of the stronger hydraulic gradients that can be generated between platform and channel, however Pethick (1992) suggested that some channels form as the result of flood inundation. Given the relative erodibility of non-cohesive versus cohesive sediment, and unvegetated versus vegetated soils, channel initiation will occur more easily on sandy tidal flats (Eisma 1998). The initiation of channel formation on a previously bare surface could be related to a number of potential perturbation to the system, it may be as little as a small change in the height of the tidal flat as sediment is deposited, but the resulting ebb flow may be increased just enough to exceed the critical value for erosion. Once channels start to form, cross-grading (the slope tangential to the main channel gradient) and micropiracy (the capture of flow by a slightly deeper channel) lead to the combination of channels and to the formation of dendritic networks (Leopold et al. 1964). Depending on how easily the substrate can be eroded, this development may take a few tidal cycles or many years (Knighton et al. 1992; Symonds and Collins 2007; D'Alpaos et al. 2007b; Hughes et al. 2009).

Once a channel system has formed, flow convergence, and thus the erosive forces, will be focused at the head of the channels (Fig. 11.6), which receive water from the broad area of the platform beyond the channel as well as from the sides. If shear stress is sufficient, channels may erode headward. Rates of headward erosion reported in the literature vary, the highest rates being reported by Knighton et al. (1992) in Northern Australia, where tidal inundation of a flat coastal plane and the reoccupation of paleochannels led to channel growth of up to 500 m/year. Symonds and Collins (2007) monitored the development of channels over a tidal flat in the Wash (UK), finding 'natural condition' extension of 15 m/year. After the managed breaching of a seawall, channel extensions of 400 m/year were measured because of increased sheet flow across the flats due to insufficient capacity of the existing channels given the enlarged tidal prism. Shi et al. (1995), report that five channels in the sandy salt marshes of the Dyfi Estuary (UK) extended at an average rate of 2.5 m/year. Newly formed channels in the muddy salt marshes of South Carolina are extending by 2 m/year (Hughes et al. 2009).

Channels on tidal flats and marshes are not always formed through erosional processes (Eisma 1998). Depositional models for channel formation in marshes have been put forward by Hood (2006, 2010) and Temmerman et al. (2007). Vegetation is seen to colonize tidal flats, creating raised 'islands', and ultimately extending the marsh edge seaward. Accumulation rates on the marsh platform are enhanced in comparison to those on the tidal flats or in the channels, by the contribution of organic material by vegetation (primarily root development) and increased baffling of tidal waters, enhancing inorganic deposition. Both scouring at the edge of vegetation patches and inheritance of pre-existing tidal flat channels produce conduits where flow is focused, prohibiting accumulation of sediment, while the marsh islands grow up around them. This process is central to the formation of channels within the numerical models of Kirwan and Murray (2007). While inheritance from an antecedent network is not a necessary part of this paradigm, it is likely the most common underlying cause of this phenomenon in nature. Salt marshes have been observed to inherit their channels from both tidal flat systems (as they prograde seaward; Pethick 1969) and fluvial systems and streams (as they expand inland; Adams 1997).

11.5.2 Secondary Processes of Initiation or Evolution

Secondary processes operate to alter existing networks, playing a part in their elaboration. These processes include the connection of existing channel sections and the extension or blocking of channels by the collapse of blocks from the channel bank (Allen 1965; Pestrong 1972; Collins et al. 1987; Eisma 1998). In the high marshes of New England, first-order channels are seen to fluctuate in length in conjunction with ponding and drainage on the marsh surface (Wilson et al. 2009). These processes operate over moderate time scales, changes being seen over a number of years, sometimes decades. It is possible that marsh channels envolve through geochemical as well as physical processes. Ponded water on the marsh surface can lead to increased salinity, and thus changes in vegetation (Perillo and Iribarne 2003), and may also alter the rate of decomposition of organic matter. These changes can change both the topography of the marsh surface, influencing flow patterns, and the erodibility of the sediment through reduction in rooting.

A similar phenomenon is observed by Perillo and Iribarne (2003) and modeled by Minkoff et al. (2006) in salt marshes in Argentina, where the interaction of crabs and vegetation cause bare patches on the marsh surface. These de-vegetated regions coalesce to create creeks. Analogous behavior is seen in the marshes of South Carolina, whereby straight creeks erode headward into a mature marsh platform as a result of low soil strengths within transient de-vegetated regions that move with the head of the creeks, again as a result of crab herbivory and burrowing (Hughes et al. 2009).

The continued existence of a channel is a balance between erosion and deposition. If the tidal prism changes (due to sea-level rise or fall, anthropogenic basin modification, or changes in sedimentation) such that velocities in the channels are reduced, then the channel will infill (Symonds and Collins 2007). Likewise, events such as heavy precipitation, storm surges and increases in tidal prism may also lead to erosion of sediment due to increased flows across the tidal flat or marsh (Murphy and Voulgaris 2006; Hughes et al. 2009).

The impact of changing salinity and ecology within tidal channels is an additional consideration. Recent research into channel elaboration has focused on the importance of the interaction of biological, biogeochemical and physical processes in the geomorphological evolution of creek systems in tidal flats and marshes. Processes such as scouring around vegetation (Temmerman et al. 2007), reduction of current and wave energy through baffling by vegetation causing deposition of sediment (Leonard and Croft 2006; Neumeier 2007) and bioturbation (Perillo and Iribarne 2003; Minkoff et al. 2006; Hughes et al. 2009) demonstrate the complex eco-geomorphic feedbacks that exist in tidal environments. Changes in tidal range will influence the vegetation and biota, thus influencing the geomorphology. A recent study in Louisiana showed that fresh-water tidal soils were notably weaker that saltwater marsh soils as a result of rooting (Howes et al. 2010). This has the potential to influence the development of channel networks (Garofalo 1980).

11.5.3 Meander Evolution (Elaboration)

Although elaboration by meandering is a secondary process of channel evolution, this process warrants a detailed examination. This is primarily because of the great influence that meandering has on the stratigraphy of intertidal regions through lateral channel migration and point-bar deposition.

There is a natural tendency for a stream to undulate, very few channels being truly straight, and over time complex meanders and channel-forms may evolve (Dury 1971; Lanzoni and Seminara 2002; Hibma et al. 2003, 2004a, b; Seminara 2006). Bejan (1982) demonstrated that the equilibrium shape of a river is a sinusoid where the wavelength is proportional to the width. This is supported in tidal channels by observations that the narrow, inland portions of creeks have a higher curvature than those farther seaward, which are wider (Marani et al. 2002, 2004).

Eisma (1998) describes three theories of meander formation derived for fluvial systems. The first is mechanical, where secondary currents develop from a slight irregularity along the channel. This instability creates a deviation in the main streamline of the flow, creating a build up of water on one side of the channel. The cross-channel gradient of the water surface creates a secondary circulation, the result of which is the erosion of the outside bank and deposition on the inner, ultimately forming a pointbar (Seminara 2006). The second theory of meander formation is the stochastic or bend instability model, where a small perturbation disrupts the flow in a straight channel. The initial disturbance creates a response in the bed topography at a certain spatial frequency that encourages meanders to develop and, ultimately, a stable condition is reached. The last theory is the hydraulic theory stating that a stream that is not 'at grade' is lengthened by meanders, thereby lowering the along-channel gradient, until equilibrium is reached (i.e., the meanders widen lowering the velocity until erosion of the banks ceases). Bagnold (1960) suggested that this occurs when the meander radius is two to three times the channel width. Both of the latter theories of formation require a physical process such as that described by the mechanical theory in order to reach their equilibrium curvature or length. The latter theory draws upon the channel bed gradient, which is perhaps an unlikely driving force within tidal channels where the bed slope is often very low. Likewise, we need to ask how well would bend instability theory hold in a marsh creek with cohesive substrate on the channel bed, preventing topographic response to the initial disturbance of the channel planform? Seminara (2006) admits that it is hard to substantiate any of these theories with field observations or in the laboratory, where creating a scaled model of a meandering system has proven difficult. Recent studies, modeling meanders in fluvial systems have seen a break though in the lab. By using alfalfa seedlings to stabilize the sediment, increasing the erosion threshold of the banks relative to the channel bed, scientists were able to emulate fluvial meander formation and migration (c.f., Seminara 2006; Braudrick et al. 2009). The problem of meander formation in tidal channels, however, seems open for further research.

The evolution of a channel from straight to meandering takes time (Hibma et al. 2004a, b). In rivers, the ratio of meander wavelength to channel width is 2–3 for young channels and 6.5–11 for very mature systems (Leopold and Wolman 1960). Thus, it is a reasonable assumption that short-lived or new tidal channels are less likely to exhibit sinuosity. In non-cohesive or unconsolidated sediment, meanders may be washed out by overbank flow, bank collapse or wave action. Meanders are more likely to be stable in vegetated areas or areas with cohesive sediment, such as muddy tidal flats and marsh platforms. Garofalo (1980) concluded that channels in tidal freshwater marshes have a lower sinuosity than channels in salt marsh. Although the study documented little migration in either types of marsh channel, the rates that were observed were higher in the muddier freshwater tidal channels than in the heavily rooted salt marsh channels. This is consistent with the observations made in Sect. 11.3.

In general, large channels are more stable than smaller channels in a similar setting due to the relative volume of sediment transport that is required to make any change (Eisma 1998). Large channels tend to be less sinuous, and flow speeds are often lower in straighter sections of a given channel (Eisma 1998). Elaboration or migration of large-scale (tens of meters wide) or macro-scale (hundreds of meters wide) channels would be likely to occur on the scale of decades to centuries (Eisma 1998, c.f. van Proosdij and Baker 2007).

The formation of meanders is very likely to be related to the periods of strongest flow as their evolution depends on erosion (Ashley 1980). The timing of peak currents varies throughout a tidal environment (Sect. 11.4), but in large creeks this condition is likely to occur at mid tide, when water is lower in the channel. In smaller creeks this peak current velocity may occur closer to high tide (just after bankfull conditions; Figs. 11.5 and 11.7). It is unclear if this has any effect on meander evolution. The key observation to make when considering meanders in tidal channels is that tidal flows are not steady, but reverse on a relatively short time scale (compared to meander evolution), and high velocities are not maintained for long periods. This may limit the time during which erosion thresholds are exceeded and prevent the development of full meanders (as proposed by bend-instability theories for rivers).

Meanders in fluvial systems may be skewed, a geometry that is sometimes termed *goose-necking* (Fig. 11.8b, Fagherazzi et al. 2004; Seminara 2006). It occurs because the streamline of highest velocity does not necessarily coincide with the channel axis. Thus, the peak erosion on the outside edge of a meander may not coincide with the apex of the meander curve. If the erosion is sufficient, the feature may migrate in the direction of the skewness (Seminara 2006). In tidal channels the flow is bidirectional, but the streamline of the highest velocities during an ebb tide may not take the same path as the streamline during the flood (Figs. 11.8 and 11.9). As a result, peak erosion occurs at different points of the meander during the flood and the ebb. Depending on the relative strength of the ebb and flood (tidal asymmetry) the meanders may be skewed or symmetrical (Fig. 11.8, Fagherazzi et al. 2004).

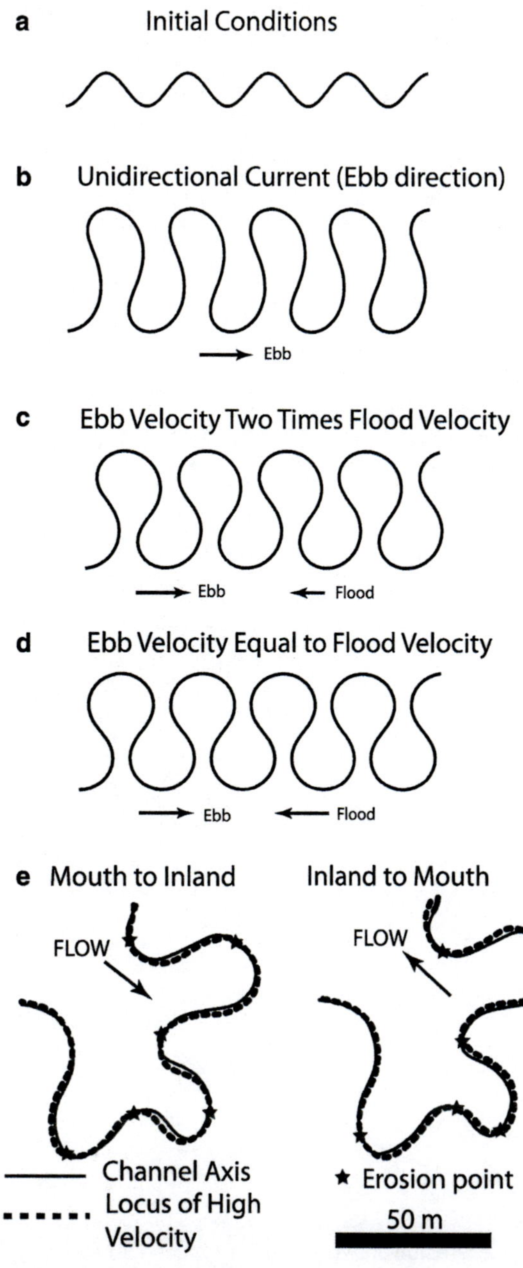

Fig. 11.8 Meander morphology evolving from (**a**) an initially sinuous channel, under conditions of: (**b**) unidirectional flow; (**c**) bidirectional flow with a notable ebb-dominance; and (**d**) bidirectional flow with equal flood and ebb currents. (**e**) Shows the position of the streamline of highest velocity flow in comparison to the central channel axis, for flood and ebb conditions. The position of peak erosion is indicated for each case on each meander by a *star*, here the tidal streamline is closest to the bank and this highest velocities experienced along the bank will occur at these point. These positions vary notably between the flood and ebb flows (Adapted from Fagherazzi et al. 2004)

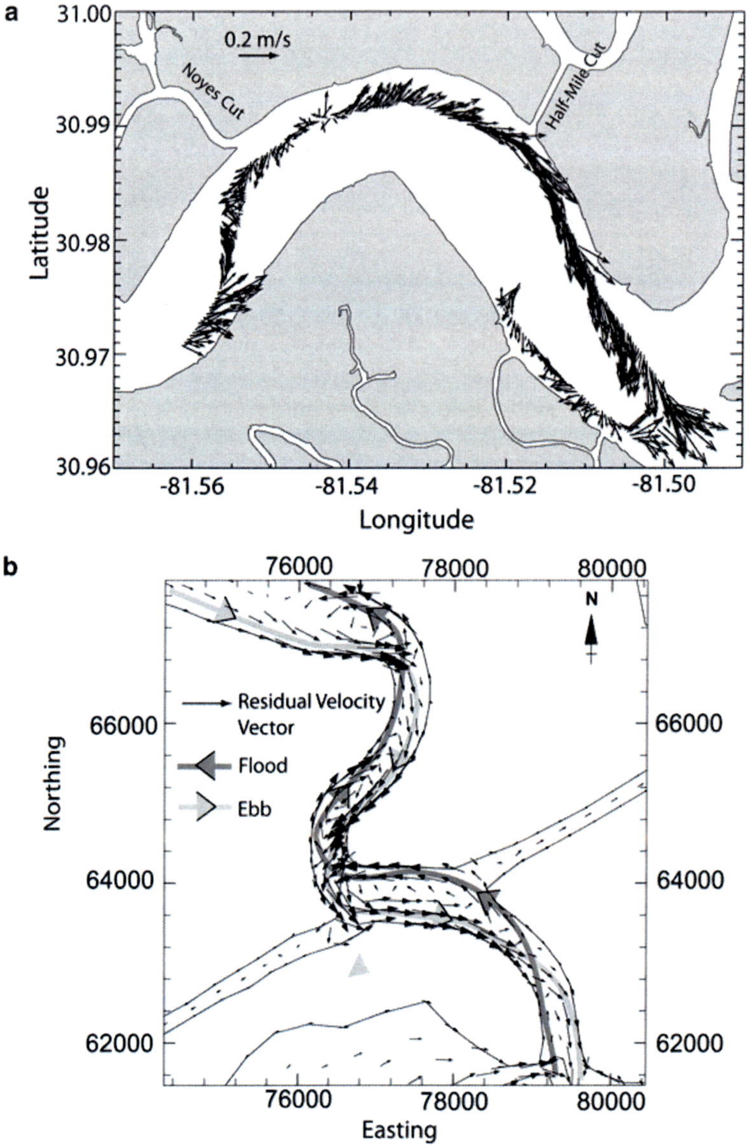

Fig. 11.9 Residual circulation over tidal pointbars. Residual velocity vectors calculated from observations in: (**a**) in the Satilla River Estuary, GA, USA (conducted on 17–18 November 2004); and (**b**) modeled using FVCOM (Finite Volume Coastal Ocean model) for a meander in southeast Louisiana, USA (Chef Menteur Pass) (Adapted from Li et al. 2008). The position of the high velocity stream lines during the ebb and flood, respectively are illustrated using *grey lines*

Likewise, erosion at two points on a meander bend may lead to the formation of *cuspate* (or box) meanders, also described as 'pinch and swale', seeing this planform morphology on meanders is a clear indication of a tidal influence (Figs. 11.9 and 11.10)

The bidirectionality of flow also impacts the resulting cross-sectional morphology of meandering tidal channel. As flow moves around a curve, momentum draws the streamline of high velocity towards the inside bank of the channel, before forcing it to the outside of the curve where it erodes the bank. When this high flux of water is forced to the outside of the curve, it creates a sufficient gradient in the water surface that a secondary circulation is set up, moving water and sediment towards the inside of the curve, building up a *pointbar*. The hydraulics and morphology of fluvial pointbars are well documented (Abad and Garcia 2009a, b; Parker et al. 2010), however, studies concerning bars in tidal channels are scarce. Under unidirectional flow, the growth of the pointbar creates a shallow zone close to the inner bank. This reduction in depth also acts to direct the streamline of high velocity toward the outside of the bank further

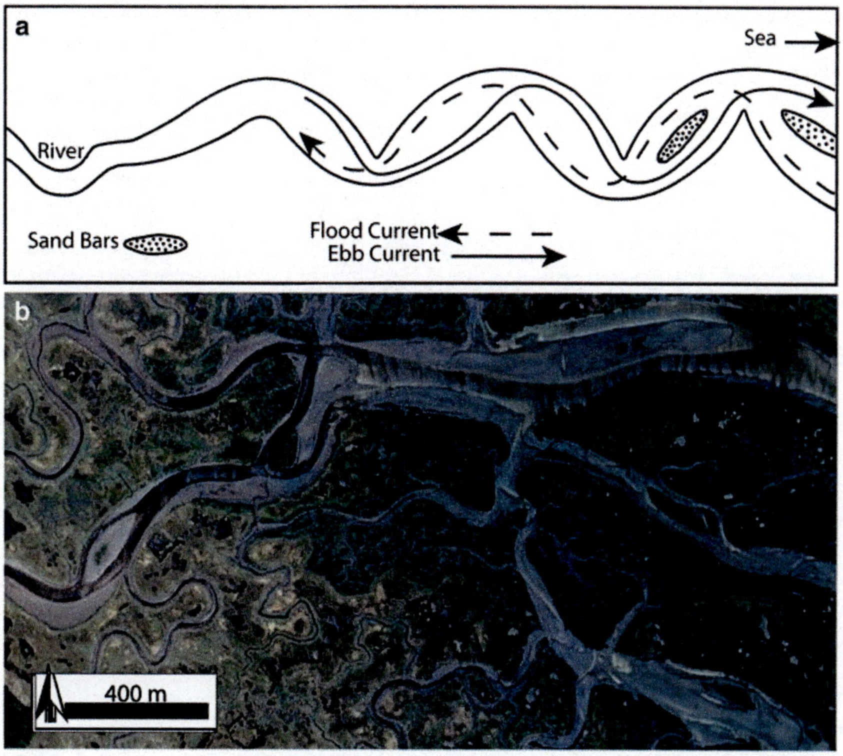

Fig. 11.10 (**a**) Estuarine meanders showing the mutually evasive flow within channels and the development of mid-channel islands (After Bird 1984), this also illustrates the cuspate nature of tidal meanders (**b**) Examples of meanders, pointbars and mid-channel islands in the Rowley River, MA, USA. The Rowley River has very little freshwater input and a tidal range of almost 4 m during springs tides

enhancing the secondary circulation and pointbar formation (Seminara 2006).

In a tidal system, the secondary circulations set up during the ebb or flood are likely to be offset, acting in different directions, and of different magnitudes. The reversing flow causes deposition or erosion on the upstream and downstream bank of a meander alternately. Figure 11.9 shows measured and modeled, depth-averaged, residual currents, in planform, over point-bars in a meandering tidal channel. The bars all show clear rotational residual circulations resulting from the interaction of the differential paths of the high-velocity streamlines during the flood and the ebb (Li et al. 2008). The inner bank of the meander experiencing reversed flow compared to the direction of the stronger flows on the outer bank. Evidence for these opposing, but offset flows can also be visualized by observing the bedforms that occur on each side of the pointbar (Fenies and Faugères 1998). The inner bank of a tidal pointbar often exhibits bedforms of the opposite symmetry and orientation to the dominant flow (Barwis and Hayes 1979; Barwis 1978; Dalrymple and Choi 2007). This hydrodynamic regime leads to complex pointbar formations (Barwis 1978).

In a fluvial system small erosional channels may form across the inside of the meanders when the river stage is high; these are known as *chutes* (Van Straaten 1954; Eisma 1998). These may form blind channels or cut entirely across part of the pointbar or meander, shortening and straightening the channel (Seminara 2006). In meso-scale tidal creeks and channels, similar morphology can be observed, but will be compounded as each side of the inner meander bend is periodically exposed to bank-normal velocities (Fig. 11.9) This may lead to the creation of a tidal barb in direction of the subordinate tidal flow, the dominate tidal flow occupying the outer region of the meander. If the meander is cut off completely, a secondary channel may form carrying the subsidiary tide. Mid-channel islands are common in large meandering tidal channels and pointbars often exhibit some level of detachment from the bank (Barwis 1978). Figure 11.10b shows

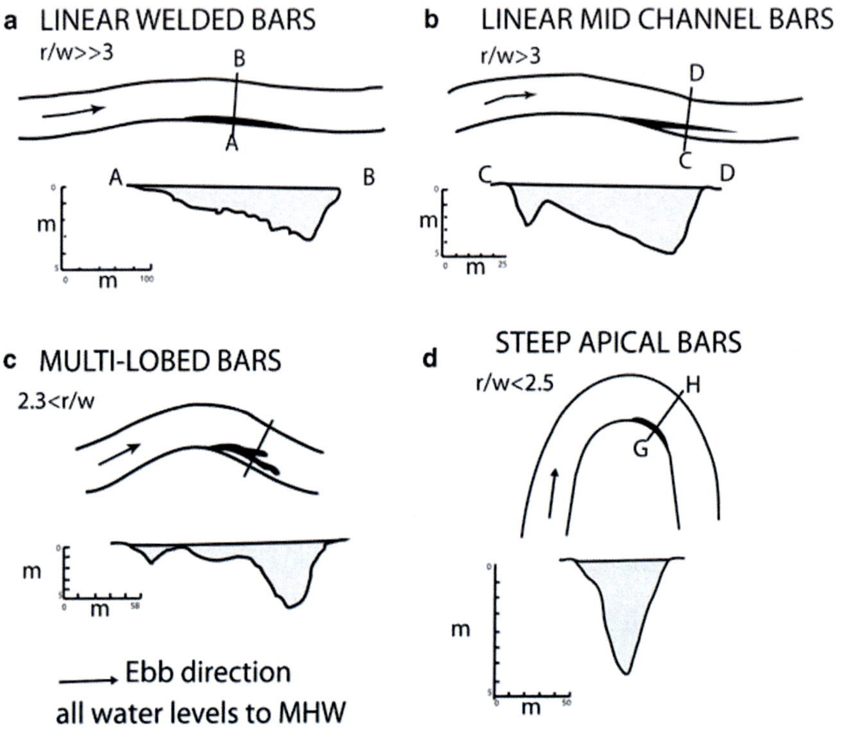

Fig. 11.11 Four planform pointbar morphologies observed by Barwis (1978) in tidal channels in South Carolina (USA)

several such islands in meanders in the Rowley River, a tidal river in Plum Island Sound (Massachusetts, USA).

Barwis (1978) undertook detailed investigations of the morphology and resulting vertical succession of deposits within tidal-creek pointbars in South Carolina. Figure 11.11 illustrates the four common pointbar planforms, which the study identified within the back-barrier area of an ebb-dominated, meso-tidal barrier system. Pointbars are categorized according to morphology and the ratio of the radius of the channel curvature (r) to the channel width (w): (a) linear welded bars ($r/w >> 3$); (b) linear mid-channel bars ($r/w > 3$); (c) multi-lobed bars ($2.5 < r/w < \sim 3$); and (d) steep apical bars ($r/w < 2.5$). As r/w decreases sinuosity increases.

Unless forming on a very straight or a very tight meander, pointbars in tidal systems tend to be elongate, stretching out in the direction of the dominant tidal current. From this, one could surmise that the system shown in Fig. 11.10b is ebb-dominated as the pointbars that are visible extends seaward from the apex of the meanders. When forming on a meander of intermediate sinuosity ($2.5 < r/w < \sim 3$), pointbars are more complex. The bars are detach from the inner bank at all but the tip closest to the meander apex because of the presence of a barb, which carries the subordinate current while the main channel carries the dominant current, in this case flood and ebb respectively (Fig. 11.11b, Barwis 1978). A value of r/w closer to 2.3 produces a pointbar with multiple lobes. Multi-lobed pointbars also display segregation of currents. This is caused by topographic shielding as the high-velocity streamlines occur in different positions during the flood and ebb.

It is interesting to note the similarities in pointbar and barb morphology of large tidal channels which occur in varied tidal settings, and the 'braided' channel-shoal networks seen deeper subtidal regions in the middle of estuaries (Hibma et al. 2004a, b; Dalrymple and Choi 2007). Comparisons can be also be drawn between the mutually-evasive channels observed mid-estuary and the mutually evasive streamlines in meandering tidal channels (Figs. 11.9c and 11.10b). This suggests a continuum where similar processes act under slightly different forcing conditions.

The evolution of estuarine morphology has been modeled (using a 2-D depth-averaged model of flow and non-cohesive sediment transport) by Hibma et al.

(2003, 2004a, b). The study discussed but did not determine the process of this evolution. Non-linear interactions in the model lead to a stable regular pattern developing from an initial perturbation, producing realistic estuarine morphologies that change progressively up estuary from alternating bars in the outer estuary, to channel-shoal mid-estuary and meandering channels with bars in the inner reaches. The decrease in meander wavelength (and thus shoal size) inland seems to be a response to changing depth and width to depth ratio. The braided channeling mid-estuary occurs where the ebb and flood currents both reach high velocities at approximately the same water depths, whereas the inner estuary is more likely to exhibit peak flows closer to high tide. This variation in velocity-stage relationship along the length of the channel is a result of the gradual change from a progressive to a standing tidal wave within a long estuary. This could perhaps explain the resulting morphology, however, questions remain. The use of differing sediment-transport formula in the model produces different scales of morphology and the actual processes causing these morphological responses to the tidal wave are still not understood fully. The model is also yet to include cohesive sediments or vegetation (Hibma et al. 2004a).

Seminara (2006) questions the similarity of the processes forming meandering channels in cohesive and/or vegetated soils, to those in more easily eroded sediment. He conjectures that in small, dead-end, salt marsh channels, meandering may occur purely by erosion. Often in the smallest first-order creeks no depositional features, such as pointbars, are seen. A symmetrical cross-section might limit morphological feedback with flow and thus the position of the erosional maxima, potentially creating a slightly different shape of meander. These questions warrant further investigation.

11.5.4 Channel Migration

Migration of channels has the potential to produce significant depositional features through lateral accretion. In fluvial systems, migrating meander bends may produce a series of asymmetrical ridges, parallel to the meander described as *scroll-bars*, however, these features are less common in tidal environments (Howard 1996; Seminara 2006; Hood 2006). Migration of a channel and creation of lateral deposition requires a sufficient supply of sediment and flows capable of eroding sediment from channel margins (Braudrick et al. 2009). The latter condition will be a function of both the flow conditions and the erodibility of the sediment. As a consequence the migration of tidal channels is related to setting as well as the size of the channel, which will control the rate at which it can migrate (larger channels are more stable).

Tidal channels in salt marshes are considered highly stable and lateral movement ranges from a few centimeters a year to imperceptible depending on the vegetation and the channel size (Redfield 1972; Garofalo 1980; Gabet 1998). On the contrary, Hood (2010) observes active development of meanders in tidal channels in a deltaic setting, with lateral channel migration varying with channel width but on the order of meters per year. In the mid and outer reaches of estuaries, where sediment is more likely to be non-cohesive, channels may be more dynamic. Likewise, in channels that periodically experience a strong fluvial influence may also experience periodic migration or channel bank erosion (Allen and Duffy 1998).

In general the rates of migration decrease toward the tidally-influenced sections of river systems (French and Stoddart 1992; Gabet 1998; Fagherazzi et al. 2004). Reworking of the sediments by tidal channels is significantly lower than that in river systems in comparison to vertical accretion (Howard 1996). This explains why the morphology of meander bends in tidal systems is unlikely to display the typical scroll bar deposits observed in fluvial systems.

11.6 Geomorphic Relationships

A number of relationships has been determined to quantify the morphology of tidal channels in tidal flats and salt marshes using a combination of aerial photography and field surveys (Rinaldo et al. 1999; Fagherazzi et al. 1999; Marani et al. 2002, 2003). The relationships reported here describe channel dimensions and network distributions in shallow intertidal settings. Where stated, they may also apply to subtidal environments, but will not necessarily scale up to deeper coastal zones, such as the outer reaches of an estuary (Rinaldo et al. 1999).

11.6.1 Channel Width

In salt marsh networks, channel width is consistently seen to reduce towards the head of a channel (Fagherazzi and Furbish 2001). Marani et al. (2002) compare the reduction in width with distance along-channel for seven meandering tidal channels in three locations globally (Venice Lagoon, Barnstaple Marsh MA, USA and Petaluma CA USA). They find a tendency toward an exponential relationship, but this *e-folding* relationship (i.e. the length of channel over which the width decreases by a factor of *e*), is not consistent amongst the channels. The ratio of e-folding length to total channel length is larger for shorter channels, indicating that they widen at a faster rate than longer channels.

On much larger scales, estuaries also demonstrate a similar exponential decrease in width, or funneling, towards the inner estuary. Macrotidal estuaries exhibit longer, relatively narrower funnels, while in mesotidal estuaries the shape is broader and shorter (Wright et al. 1973; Pethick 1984; Eisma 1998). A channel with a purely progressive wave is likely to exhibit parallel banks (Wright et al. 1973).

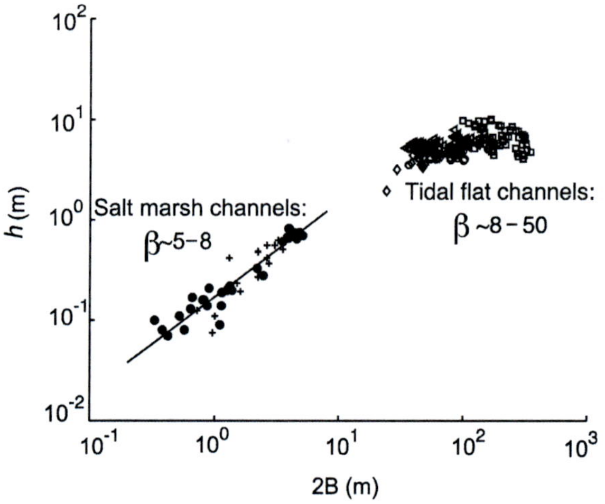

Fig. 11.12 Plot of width versus depth showing the two discrete populations of tidal channels. Channels on vegetated salt marshes show a distinctly different width-depth ratio ($\beta = 2B/h$) than channels over tidal flats, which tend to behave more like their fluvial counterparts

11.6.2 Width-to-Depth Ratio

While there is great variability throughout tide-dominated systems, the channel width-to-depth ratio ($\beta = 2B/h$) can be split into two populations: marsh creeks ($5 < \beta < 8$) and tidal flat channels ($8 < \beta < 50$) (Fig. 11.12, Zeff 1999; D'Alpaos et al. 2005).

This bi-modality of channel type has implications in terms of hydraulics and implies that vegetated creeks and channels in bare flats respond differently to erosional and depositional processes. Factors contributing to this distribution of width to depth ratios include the different processes and rates of bank erosion, e.g. the tendency for undercutting and slumping when channel banks are heavily rooted near the marsh surface where the live root biomass is most dense (van Eerdt 1985; Huat et al. 2009; Howes et al. 2010). Vegetative baffling of flow will also retard currents once the water level overtops the channel bank, leading to increased deposition close to channel edges and the potential for enhanced accretion close to the bank (Leonard and Luther 1995; Brown 1998), thus increasing channel depth. Within lower tidal flats, sediments are coarser, potentially non-cohesive and are more easily eroded (Feagin et al. 2009), these are conditions that support development of wider, shallower channels.

11.6.3 Channel Cross-Sectional Area

The existence of a relationship between cross-sectional area (Ω) and tidal prism within tidal inlet channels is widely accepted, such that

$$\Omega \propto aP^b \qquad (11.2)$$

where *P* is the volume of the spring tidal prism and a and b are empirically derived constants (Escoffier 1940; O'Brien 1969; Jarrett 1976, see Chap. 12 discussion). This relationship suggests that there exists a dynamic equilibrium whereby cross-sectional area will adjust in response to discharge given that a set volume, *V*, of water must pass through the area during the fixed period of half a tidal cycle. This produces erosion or deposition within the channel. Friedrichs (1995) noted that, although this relationship is complicated at tidal inlets by exposure to wave energy and littoral drift, in more sheltered regions in the interior of a tidal embayment, the cross-sectional area of the channel is more closely related to shear stresses resulting from tidal currents alone. As the nature of this equilibrium would suggest, tidal prism may be substituted with peak discharge (Q), a value more easily derived or measured given the indeterminate division of the total tidal prism between channels in a network:

$$\Omega \alpha Q^{c} \quad (11.3)$$

where, based on observations in 242 cross sections, c, the exponent of Q, falls within the range 0.73–1.34, with an average of 0.96 (i.e. ~1; Friedrichs 1995).

The equilibrium theory requires that the peak discharge (Q), and thus the peak velocity ($U = Q/A$) produces a 'stability' shear stress, τ_s, which controls the sediment transport within the channel. The τ_s will be just greater than the critical shear stress, τ_c, required for initiation of sediment movement; and based on laboratory experiments $\tau_c < \tau_s < 0.15\tau_c$. (Diplas 1990, from Friedrichs 1995). However, this theory is complicated by the variation of sediment type through tidal systems and it would stand to reason that τ_c in marsh channels will differ to that in channels on tidal flats because of the difference in grain size, organic content and level of vegetative stabilization. Lastly, application of this theory is complicated further by lateral friction, which will vary with hydraulic radius, itself a function of channel width and shape; with sediment type, organic content or biomass; and with the variation of the hydraulic radius over the tidal cycle (i.e., the ratio of mean water depth to tidal range).

Further studies have explored the idea that the area, A, which a certain channel drains (sometimes called the *creekshed*) is representative of the volume of water which flows through it. Based on this, an alternative relationship may be used (Fagherazzi et al. 1999; Rinaldo et al. 1999):

$$\Omega \alpha A^{d} \quad (11.4)$$

where d is of the order ~1.

This relationship has recently been explored even further using numerical models of hydro- and morphodynamics and successfully used to represent the evolution of tidal networks (D'Alpaos et al. 2005, 2010).

The validity of the assumption of dynamic equilibrium is supported by early observations by Steers (1969) that headward-eroding marsh creeks exhibit a gradient (albeit low) along the channel bed. This gradient becomes zero once the creek has stopped extending and reached equilibrium with its local tidal prism or drainage area. This suggests further that headward erosion of previously stable creeks is indicative of an increased tidal prism (Hughes et al. 2009).

11.6.4 Sinuosity

A relationship exists between the length of meanders and the channel width (Fig. 13, Marani et al. 2002, 2004; Dalrymple and Choi 2007). This relationship holds for all meandering channels from fluvial to tidal, including salt marsh and tidal flats channels, and channels within estuaries and deltas (Marani et al. 2002; Seminara 2006; Hood 2010). Salt marsh channels do not form a distinct population in terms of meander-to-width geometry as they do for width to depth ratio (Fig. 11.13, D'Alpaos et al. 2005). This is consistent with the observations that marsh creeks, which tend to be narrower, exhibit tighter meanders than channels over tidal flats (Figs. 11.2 and 11.13) and implies that depth does not significantly influence meander width.

11.6.5 Stream Order and Drainage Density

Pestrong (1965) observed that, unlike fluvial systems, neither drainage basin area, nor channel lengths and widths, scaled with stream order. Knighton et al. (1992) found closer agreement to fluvial behavior in channels in the Van Diemen Gulf (Australia), and Novakowski et al. (2004) concluded that tidal networks in South Carolina, USA were similar but more elongate than fluvial networks. The disagreement in these observations may be explained by variation in scaling from basin to basin that can be observed in tidal flats and salt marshes (Rinaldo et al. 1999; Fagherazzi et al. 1999).

Within networks, the drainage density is defined as the ratio of total channel length (Σl) divided by the watershed area (A). This parameter, which provides a measure of channelization, was examined for tidal channel networks within salt marshes by Marani et al. (2003). The study considers 136 creeksheds within the Venice Lagoon, Italy, and makes several poignant observations; firstly the probability distribution of length of pathways across the unchanneled surface follow an exponential decay, similar to that seen in fluvial networks. As with the variation of width along channel (e-folding lengths), different decay rates were seen within individual basins. Secondly, a linear relationship exists between total channel length (in any creekshed) and tidal prism (for that basin). A similar correlation, although not specified as linear, was found by Allen (1997). The exact relationship varies between

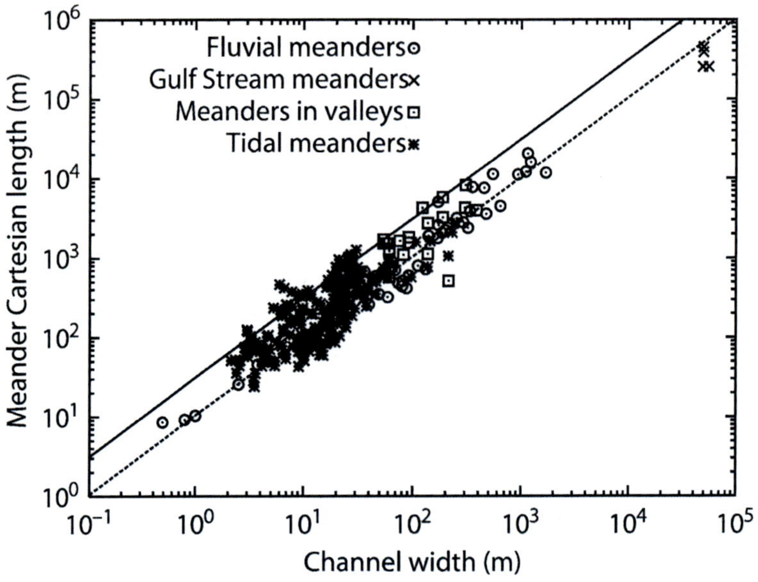

Fig. 11.13 Plot showing the relationship between meander wavelength and channel width (Adapted from Marani et al. 2002 in Seminara 2006)

basins, however, when total channel length is compared to creekshed area as a proxy to prism, the relationship is more consistent and is of the order $\Sigma l = 0.02A$. This implies a constant Hortonian drainage density. Novakowski et al. (2004) find that $\Sigma l = 0.03A^{0.88}$, based on analysis of 725 creeksheds in South Carolina, US, with drainage densities for ranging from 0.0008 to 0.069 m/m^2 (a wider range than is seen in fluvial systems 0.0023–0.0137 m/m^2). Steel and Pye (1997) also see a similar relationship in British salt marshes. The implication of this is that there may be a common network geometry within marsh systems, potentially an underlying similarity in branching.

Marani et al. (2003) go further to confirm this hypothesis by examining the mean length of unchanneled pathways (L) for a given basin with respect to creekshed area and the Hortonian characteristic path length (the inverse of the drainage density; $l_H = A/\Sigma l$). Hortonian length l_H provides a measure of how the catchment is dissected by the channel network, whereas L is essentially the mean distance that flow must travel from a point on the flats to reach a channel and indicates how efficiently the network drains (ebb) or feeds (flood) the creekshed. A direct comparison of these two parameters does not provide any clear relationship. Further, when the ratio l_H/L is use as a proxy for drainage efficiency (a high value indicating relatively short unchanneled paths) and is compared to branching frequency (i.e., the ratio of lower and higher-order streams), the relationship is also poor. The conclusion must be that traditional Hortonian drainage density does not provide a good measure of the variability of network patterns seen on salt marshes, because unlike rivers, these systems are not scale invariant.

11.7 Preservation Potential

Preservation of sedimentary deposits formed in tidal channels may occur vertically and horizontally, through infilling and lateral accretion. Reduced tidal prism because of changing tidal range or modification of the surrounding tidal system will naturally lead to a reduction in cross-sectional area and infilling of the channel with fine-grained sediment (Rieu et al. 2005). An upward fining in sediment and change from sandy to heterolithic or muddy bedding indicates reduction in flow strength and can be observed in both infilling channels and where lateral movement of the channel alters the tidal conditions at a particular point. Dalrymple et al. (1992) suggest that estuarine tidal channels are continuously infilling during rising sea level, where sediment supply is adequate.

Lateral migration of channels produces both lateral and vertical sedimentation; cut and fill facies, which exhibit upward fining sediment over a sharp erosional base (van Straaten 1954; Terwindt 1988). Figure 11.14 provides a conceptual sketch of such a succession; the scale of the channel and bedding would vary according to hydrodynamic and sedimentary setting.

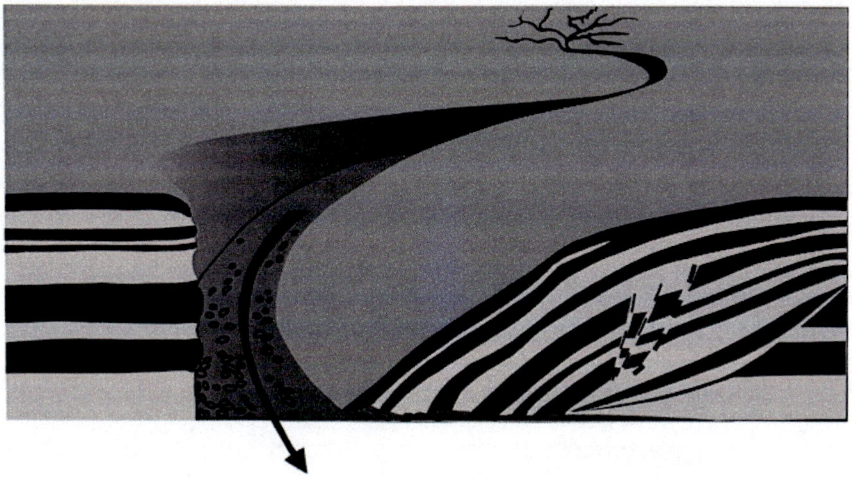

Fig. 11.14 A simplified sketch of the cut-and-fill succession produced on a tidal flat by lateral migration of a channel meander/pointbar. Note the healing of a slump mid-succession (Adapted from Reineck (1967) in Eisma (1998))

The base of a channel can be recognized by a concave upward erosional bounding surface, which indicates confined flows (Santos and Rossetti 2006). In intertidal channels, the base is often identifiable by a lag of coarse sediment or shell, although in very muddy systems this may be more difficult to distinguish (Klein 1977; Barwis and Hayes 1979; Terwindt 1988; Rieu et al. 2005; Pearson and Gingras 2006). The thalweg of salt marsh creeks may present only as increase in sand content. In larger channels, lag deposits range in thickness from a few decimeters to a few meters and in intertidal channels shell lags of a few centimeters in thickness are expected (Barwis 1978; Terwindt 1988). Similarly, mud blocks (breccia) from bank slumping and channel edge erosion may form part of a channel lag, creating lithologies such as mud chip conglomerates (Klein 1977; Terwindt 1988; Santos and Rossetti 2006). In mesotidal back-barrier environments, bank-margin slump blocks up to a meter in diameter and containing preserved rhizomes and burrows can occur (Barwis 1978). Large-scale slumping has also been observed on the meter scale in regions of the Bay of Fundy (Pearson and Gingras 2006) in areas on pointbars that are dissected by tributaries.

In regions experiencing seasonal variation in temperature, where ice periodically forms in channel beds (such as the north east coast of the USA and the east coast of Canada), ice rafts may also produce patchy granule and pebble lags and deposits of marsh peats across pointbars (Pearson and Gingras 2006). In the Bay of Fundy these were observed to form part of a repetitive set of bedding associated with seasonal variability, rather than occurring over a clear erosional contact as would be expected in a channel base.

Using the known relationship between cross-sectional area and peak discharge, reasonable estimates of maximum paleo-current velocity and historical variation in tidal prism may be estimated from preserved channel cross-sections. Rieu et al. (2005) examine a preserved tidal channel, offshore of the western Netherlands (Fig. 11.15a). The channel fill is characterized by alternating sub-parallel high- and low-amplitude seismic reflectors. A clear lateral accretion unit can be seen proximal to the channel, indicating channel migration. The thickening of these lateral units toward the final channel position is interpreted to indicate an increasing tidal prism, followed by channel fill related to a decrease in tidal prism (Fig. 11.15b). In salt marshes where meandering channels are stable and lateral migration is close to zero, no such lateral accretion would be expected, accretion would occur purely in the vertical (Redfield 1972; Gabet 1998).

Common indicators of tidal influence in a channel include: reactivation surfaces (formed as the tidal flow changes direction) and mud drapes in cross-sets, and low angle dipping cross-sets with alternating thicker and thinner packages of sands and mud, or muds and silts (Santos and Rossetti 2006). Tidal deposits typically

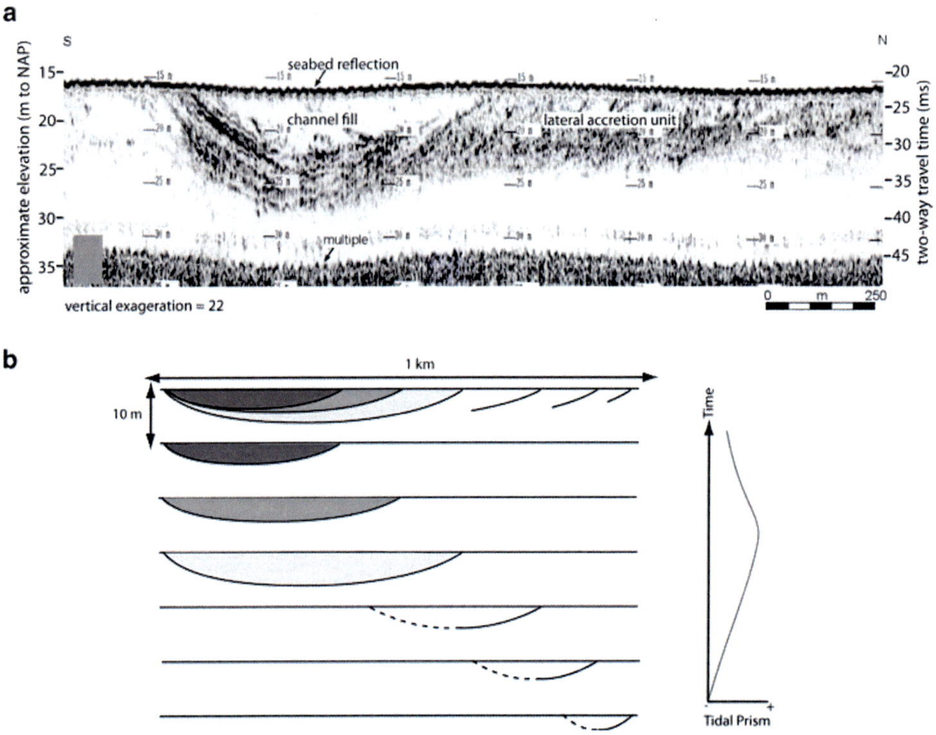

Fig. 11.15 (**a**) Shallow seismic records of cut-and-fill deposits of a tidal channel preserved offshore, a region of lateral accretion is clearly visible to the north of a channel, determined to be a main channel close to the tidal inlet; (**b**) Schematic evolution of channel with the inferred change in tidal prism responsible for the growth, lateral accretion and eventual infilling (Adapted from Rieu et al. 2005)

contain high proportions of heterolithics (intercalated sand and mud). *Tidal bundl*es associated with differences in flow are formed as a result of periodic variations in tidal energy. Sandier deposits relate to higher-energy periods such as spring tides and muddier to lower-energy neap tides.

Close to the marine or fluvial sediment sources the channel deposits will consist of coarser sediment with a lower mud content, and heterolithic deposits will take the form of flaser bedding. Moving away from high-energy environments and sediment sources, towards the mid regions of an estuary or the back of a lagoon system, deposits become increasingly muddy (wavy or lenticular bedding). In the lowest energy reaches of an intertidal system, channel deposits are entirely muddy making it difficult to distinguish between deposits from lateral movement and channel-fills that result from abandonment (Barwis and Hayes 1979). However, in a predominantly muddy system Pearson and Gingras (2006) were able to discern rhythmic silt-mud and sandy-mud couplets within tidal pointbars, interpreted as neap-spring bundles.

Inclined Heterolithic Stratification (IHS) is commonly associated with tidal pointbars, representing lateral accretion. These dipping, interbedded mud, silt, and slightly sandy beds are formed as sediment accumulates across a sloping face (either through suspended or bedload deposition) and would be expected in meandering tidal channels (Dalrymple et al. 1992; Santos and Rossetti 2006; Pearson and Gingras 2006; Dalrymple and Choi 2007). These deposits can lie between 1° and 30° (angle of repose for sands) and, may exhibit cross-stratification if bedforms were present during formation. In contrast to fluvial settings, stratification in tidal pointbars is inclined towards the thalweg of the channel (Barwis 1978; Pearson and Gingras 2006), as opposed to dipping predominantly downstream. IHS is indicative of the high frequency variability in hydrodynamics, which occur in tidal systems. In pointbars in the Bay of Fundy, Pearson and Gingras (2006) observe laterally continuous IHS over a horizontal distance of 26 m.

In sandy environments, the migration of 2D- and 3D-bedforms in tidal settings typically results in cross-sets that display low angle dipping foresets and may be used as evidence of tidal influence (Santos and Rossetti 2006). Bidirectional tidal flow can create distinct cross-lamination (ripples) or cross-bedding (dunes). Sets of ebb-oriented cross-laminae, bounded by flood-oriented cross-laminae (or vice versa) are known as *herringbone cross-stratification* and are a good indicator of tidal deposition and may be seen in deep subtidal portions of a channel. Degree of symmetry in the herringbone structures provides insight into tidal asymmetry at the point of deposition in time and space. If one tidal current is weaker than the other, the subordinate current may create a 'cap' of smaller oppositely directed foresets at the crest of the bedform created by the dominant current (Mowbray and Visser 1984). However, the complex recirculation and flow-segregation, which occur in most braided or meandering channels, can create sets of exclusively flood- or ebb-oriented cross-stratification in shallower regions of the channel. In low-energy settings, such as small channels with slower flows (~0.3 m/s), bedforms are unlikely to form but parallel laminations may be seen where mud settles out of suspension during low-flow periods and sand is moved as bedload during times of faster flow.

The crests of tidal pointbars are often heavily populated with worms, mollusks and burrowing crustaceans. A high level of bioturbation is a notable feature of intertidal regions, providing differentiation between tidal and fluvial systems, where infauna are scarce. Species diversity increases inland from saline to brackish environments (Barwis and Hayes 1979). Using this information, in hypersynchronous systems, where similar tidal ranges can exist at two or more sites, ichnology can help to differentiate between regimes based on species tolerance to salinity and diversity.

Bioturbation differs with position in the tidal range; below mean low water, bioturbation is relatively sparse, decreasing into the channel thalweg. Likewise, in regions of recent slumping, bioturbation may be less frequent. In the upper regions of a tidal pointbar, however, faunal activity can be intense. Pearson and Gingras (2006) observed burrow densities of up to 60,000 burrows/m^2 in the upper-intertidal zone of a muddy pointbar in the Bay of Fundy. Ichnological investigation showed different assemblages across the bar (in the upper-subtidal and lower-intertidal zones of *Polykladichnus*- and *Skolithos*-like traces characterized the pointbars; Arenicolites-, Diplocraterion-, Polykladichnus-, *Palaeophycus*-, and *Planolites*-like forms were found in the middle-intertidal portions of the pointbars; and in the upper-intertidal, *Siphonichnus*- and *Polykladichnus*-like burrows were found). These assemblages are consistent with brackish water conditions.

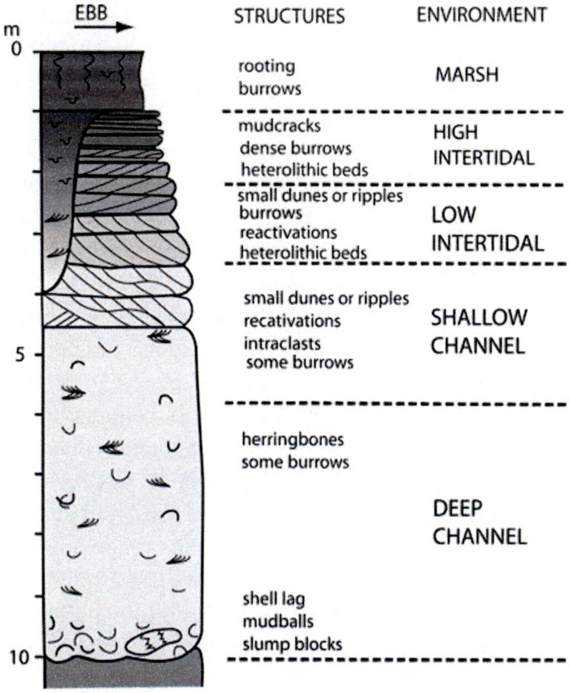

Fig. 11.16 (**a**) A general pointbar facies model (Barwis and Hayes 1979)

Tidal facies are more likely to be preserved when bioturbation is low. This would be the case in channels where the thalweg and pointbars have a higher sand content, as muddy sediment supports more active infauna. Likewise regions of moderately high velocities also discourage faunal activity and stratigraphy is more likely to be preserved (Ashley and Zeff 1988). In regions with low deposition rates, the activity of burrowers may completely obscure bedding (Barwis and Hayes 1979; Pearson and Gingras 2006). However, if rates of deposition are sufficiently high, then both bedding and burrows may be distinct (Barwis 1978). Variation in seasonal bioturbation may be reflected in deposits as intercalated, laminated and burrowed beds. The laminated beds characterize early winter when bioturbation is low, whereas the bioturbated beds are formed during summer when faunal activity is high (in response to temperature and salinity variations, which are commonly a response to fluvial inputs).

A general model for tidal pointbar facies is illustrated in Fig. 11.16 (Barwis and Hayes 1979). The deep channel is typified by a shell lag underlying a thick subtidal unit. In sandy channels herringbone cross-stratification may occur with some low level of bioturbation. The main channel may exhibit unidirectional dunes and ripples oriented with the dominant tide, due to segregation of the flood and ebb flows either side of the pointbar. In a muddier regime, the sub-tidal and low-intertidal regions are typified by planar, horizontal bedding (Pearson and Gingras 2006). Moving up the tidal range, the middle of the intertidal zone exhibits predominantly low-angle, planar-bedded and potentially IHS. In sandy environments the presence of small dunes and ripples may result in cross-bedding with increasing mud content as the pointbar emerges into the intertidal zone. The high intertidal zone reverts back to planar, horizontal bedding, highly bioturbated and the deposits may exhibits desiccation marks (Pearson and Gingras 2006). The unit is finally topped by marsh sediment as the channel moves laterally and ultimately, the marsh follows on. Where the bar is detached from the channel bank, mud deposits are seen in the blind-ended, subordinate barb channel that crosses the surface of the bar. If enough sand is present for bedforms, ripples and dune in this region will be oriented in the opposite direction to the main channel.

Barwis (1978) identified distinct vertical succession associated with each of the four tidal pointbar morphologies that he observed. Each form has subtle differences in the distribution of flows, sedimentation and biota. Steep apical bars are the only morphology that would create a continuous, unbroken succession with a thickness equal to the channel depth. This is because these features are fully welded to the inner channel bank up to the elevation of the marsh itself. Additionally, this type of bar is steeper, with less suitable habitat for infauna and has no sheltered tidal barb behind the bar crest. Thus, bioturbation is comparatively low compared to sedimentation. Detailed descriptions of each are presented in Barwis (1978).

While both the planform morphology and vertical facies in tidal pointbars have been described in the literature, a full three-dimensional description is still missing to fully document the internal structure and horizontal variations, which result from the heterogeneity of the physical (and biological) processes both across and along the forms.

In salt marsh systems, deposits in lower-order creeks are steep-sided and narrow. They consist of predominantly massive mud units, which lack the high level of organics that would be seen in the surrounding marsh platform deposits. In marsh sediments (specifically on high marshes which sit at the high-water elevation), standing pools of water (pannes or ponds) may produce similar muddy facies, devoid of rhizomes. They can be distinguished from creek deposits by the presence of *Ruppia maritime*, a submerged vegetation, which is commonly seen in ponds but not in channels (Wilson et al. 2009). A notable levee of coarser sediment may be present along a channel edge due to the baffling of flow speeds by vegetation (Allen 2000).

There is no clear relationship between network planform and the sedimentary structures observed in tidal channels (Eisma 1998), beyond the obvious influence of meanders on pointbar geometry. Terwindt (1988) suggests that the number and the dimensions of drainage channels could be used give an indication of the tidal regime: a low tidal range producing a low number of small channels, indicating micro- or mesotidal conditions; a large number of deep channels indicating macrotidal conditions. This seems unlikely based on the wide variability seen between sub-basins within intertidal systems such as Venice Lagoon (Marani et al. 2002). Shallow channels are observed over exposed flats in macrotidal environments (Eisma 1998). Likewise, deep channels can be found in microtidal regions, such as the back-barrier areas in New Jersey, where the through-flowing channels of Ashley and Zeff (1988) are 5–100 m wide and 2–5 m deep. In general, there are few differences between the faces generated in macro- and mesotidal environments (Terwindt 1988) with the exceptions, however, where current velocities are exceptionally high and parallel laminated sand-rich facies may be deposited across bars during upper sheet flow (e.g., Cobequid Bay-Salmon River estuary, Bay of Fundy; Dalrymple et al. 1991, 1992).

11.8 Summary

Channels provide the pathway for the tidal wave to propagate and are a primary control on the sedimentation and ecology of coastal environments. Defined by the alternating flow of ebb and flood currents, tidal channels occur across a range of scales within macro-, meso- and microtidal environments. They often form dendritic networks, which, despite being described by some studies as fractal, exhibit a great deal of variation and do not truly scale with size. The complexity and variability of tidal channels is ultimately a function of the heterogeneity of the flows, sediments and ecosystems within the intertidal and subtidal systems. Perhaps it was this that led Rinaldo et al. (2004) to describe tidal channel networks as: "arguably a consequence of a frustrated tendency towards critical self-organization", because so many factors act to inhibit this self-organization and thwart scaling within the system. Despite the scale invariance seen within intertidal channel networks, all tidal channels consistently maintain an equilibrium between channel cross-section and tidal prism. Likewise, there seems to be a continuum of bar-channel morphology within estuaries, although further research is needed to explore this hypothesis.

Classifications of network morphology differentiate between channels based on the level of elaboration, and the shape of a network (if a network is indeed formed). Initiation of channels and the formation of networks occur through both erosive and depositional processes. Active channel systems may reflect present conditions, or exhibit inheritance from paleo-channels. Residual circulation patterns and presence of bidirectional flow create high spatial variability in hydrodynamics and there is, as a consequence, a great deal of potential for overlap in both the processes and the resulting morphology and stratigraphy that are observed in tidal channels.

Sinuosity is common and is exhibited by channels in all tidal environments, however, channels tend to be more sinuous where the substrate is vegetated or more difficult to erode (cohesive sediments). The process by which meanders form is still not well understood. It is also unclear how mobile the meanders in channels in salt marsh or in very cohesive sediments may be. Salt marshes have the same width to meander-length relationships as other channels but have an individual population when width is compared to depth. This perhaps supports the hypothesis that they inherit their form from tidal flats or fluvial systems, vegetation stabilizing their original meander geometry while the channel bed deepens and the banks accrete vertically (Marani et al. 2003). This observation, however, does not seem consistent with the higher sinuosity associated with meanders in vegetated environments, which implies increased elaboration after colonization by vegetation. Furthermore, it is unknown whether meanders in small salt marsh creeks experience a different evolution because of the lack of morphological feedback through the formation of pointbars. Additional research is necessary to address these questions.

In planform, tidal channels can often be identified by cuspate meanders, associated with the mutually evasive flood and ebb flow paths. Tidal point bars are often skewed in direction of dominant flow, and detached from the channel bank, with a subordinate current barb forming at the inner meander bend.

Preservation of tidal channels occurs through infilling as tidal prism changes over time and or lateral accretion as a channel migrates. Deposition occurs in particular at tidal pointbars, making our understanding of meandering in these channels all the more important. The range of facies expected within a pointbar varies with morphology and with setting (according to mud content) but the presence of IHS bedding and moderate to high levels of bioturbation are two key indicators of deposition in a tidal channel environment. The three-dimensional internal architecture of the tidal pointbars has not yet been extensively examined and is another topic that warrants further research.

References

Abad JD, Garcia MH (2009a) Experiments in a high-amplitude Kinoshita-generated meandering channel. 2: implications of bend orientation on bed morphodynamics. Water Resour Res. doi:10.1029/2008WR007017

Abad JD, Garcia MH (2009b) Experiments in a high-amplitude Kinoshita-generated meandering channel. 1: implications of bend orientation on mean and turbulence flow structure. Water Resour Res. doi:10.1029/2008WR007016

Adams P (1997) Absence of creeks and pans in temperate Australian salt marshes. Mangroves Salt Marshes 1:239–241

Allen JRL (1965) Coastal geomorphology of eastern Nigeria: Beachridge barrier islands and vegetated tidal flats. Geol Mijn 44:1–21

Allen JRL (1997) Simulation models of salt marsh morphodynamics: some implications for high-intertidal sediment couplets related to sea-level change. Sed Geol 113:211–223

Allen JRL (2000) Morphodynamics of Holocene salt marshes: a review sketch from the Atlantic and Southern North Sea coasts of Europe. Q Sci Rev 19:1155–1231

Allen JRL, Duffy MJ (1998) Temporal and spatial depositional patterns in the Severn Estuary, Southwestern Britain: intertidal studies at spring-neap and seasonal scales, 1991–1993. Mar Geol 146:147–171

Ashley G (1980) Channel morphology and sediment movement in a tidal river Pitt River British Columbia. Ear Surf Process 5:347–368

Ashley GM, Zeff ML (1988) Tidal channel classification for a low-mesotidal salt marsh. Mar Geol 82:17–32

Bagnold RA (1960) Some aspects of the shape of river meanders. US Geol Surv Prof Pap 282(E):135–144

Barwis JH (1978) Sedimentology of some South Carolina tidal-creek pointbars and a comparison with their fluvial counterparts. In: Miall AD (ed) Fluvial sedimentology, Canadian Society of Petroleum Geologists Memoir 5. Canadian Society Petroleum Geologists, Calgary, pp 129–160

Barwis JH, Hayes MO (1979) Regional patterns of modern barrier island and tidal inlet deposits as applied to paleoenvironmental studies. In: Ferm JC, Horne JC, Weisenfluh GA, Staub JR (eds) Carboniferous depositional environments in the Appalachian region. Univ So Carol, Dep Geol, Carol Coal Group, Columbia, pp 472–498

Bayliss-Smith TP, Healey R, Lailey R, Spencer T, Stoddart DR (1979) Tidal flow in salt marsh creeks. Estuar Coast Mar Sci 9:235–255

Bejan A (1982) Theoretical explanation for the incipient formation of meanders in straight rivers. Geophys Res Lett 9:831–834

Belknap DF, Kraft JC (1985) Influence of antecedent geology on stratigraphic preservation potential and evolution of Delaware's barrier system. Mar Geol 63:235–262

Bird ECF (1984) Coasts. Blackwell, Oxford

Boon JD (1975) Tidal discharge asymmetry in a salt marsh drainage system. Limnol Oceanogr 20:71–80

Braudrick CA, Dietrich WE, Leverich GT, Sklar LS (2009) Experimental evidence for the conditions necessary to sustain meandering in coarse-bedded rivers. Proc Natl Acad Sci. doi:10.107/pnas.0909417106

Brown SL (1998) Sedimentation on Humber salt marsh sedimentary process. In: Black KS, Paterson DM, Cramp A (eds) Sedimentary processes in the intertidal zone, Geological Society Special Publication 139. The Geological Society, London, pp 69–84

Cleveringa J, Oost AP (1999) The fractal geometry of tidal-channeltidal channel systems in the Dutch Wadden sea. Geol Mijn 78:21–30

Collins LM, Collins JN, Leopold LB (1987) Geomorphic processes of an estuarine marsh. Prelim Hypotheses Int Geomorph 1:1049–1072

D'Alpaos A, Lanzoni S, Marani M, Fagherazzi S, Rinaldo A (2005) Tidal network ontogeny: channel initiation and early development. J Geophys Res. doi:101029/2004JF000182

D'Alpaos A, Lanzoni S, Marani M, Rinaldo A (2007a) Landscape evolution in tidal embayments: modeling the interplay of erosion sedimentation and vegetation dynamics. J Geophys Res. doi:101029/2006JF000537

D'Alpaos A, Lanzoni S, Marani M, Bonometto A, Cecconi G, Rinaldo (2007b) Spontaneous tidal network formation within a constructed salt marsh: observations and morphodynamic modeling. Geomorphology 91:186–197

D'Alpaos A, Lanzoni S, Marani M, Rinaldo A (2010) On the tidal prism-channel area relations. J Geophys Res. doi:101029/2008JF001243

Dalrymple RW, Choi KS (2007) Morphologic and facies trends through the fluvial-marine transition in tide-dominated depositional systems: a systematic framework for environmental and sequence-stratigraphic interpretation. Earth Sci Rev 81:135–174

Dalrymple RW, Makino Y, Zaitlin BA (1991) Temporal and spatial patterns of rhythmite deposition on mud flats in the macrotidal, Cobequid Bay Salmon River estuary, Bay of Fundy, Canada. In: Smith DG, Reinson GE, Zaitlin BA, Rahmani RA (eds) Clastic tidal sedimentology, Canadian Society of Petroleum Geologists Memoir 16. Canadian Society of Petroleum Geologists, Calgary, pp 137–160

Dalrymple RW, Zaitlin BA, Boyd R (1992) Estuarine facies models: conceptual basis and stratigraphic implications. J Sed Petrol 62:1130–1146

Davis RA, FitzGerald DM (2004) Beaches and Coasts, Blackwell Science, Oxford, England. 419 p

De Vriend HJ (1996) Mathematical modelling of meso–tidal barrier island coasts Part I Empirical and semi-empirical models. In: Liu PL, Liu PL-F (eds) Advances in coastal and ocean engineering, Fluid Mechanism 386. World Scientific, Singapore, pp 15–42

Diplas P (1990) Characteristics of self-formed straight channels. J Hydraul Eng ASCE 116:707–728

Dronkers J (1986) Tidal asymmetry and estuarine morphology. Neth J Sea Res 20:117–31

Dury GH (1971) Channel characteristics in a meandering tidal channel: crooked river Florida. Geog Ann Ser A Phys Geogr 53:188–197

Dyer KR (1997) Estuaries—physical introduction, 2nd edn. Wiley, Chichester

Eisma D (1998) Intertidal deposits: river mouths tidal flats & coastal lagoons. CRC Press, New York

Escoffier FF (1940) The stability of tidal inlets. Shore Beach 8:114–115

Fagherazzi S, Furbish DJ (2001) On the shape and widening of salt marsh creeks. J Geophys Res 106(C1):991–1003

Fagherazzi S, Bortoluzzi A, Dietrich WE, Adami A, Marani M, Lanzoni S, Rinaldo A (1999) Tidal networks: 1 Automatic network extraction and preliminary scaling features from digital terrain maps. Water Resour Res 35:3891–3904

Fagherazzi S, Gabet EJ, Furbish DJ (2004) The effect of bidirectional flow on tidal channel planforms. Earth Surf Proc Land 29:295–309

Fagherazzi S, Carniello L, D'Alpaos A, Defina A (2006) Critical bifurcation of shallow microtidal landforms in tidal flats and salt marshes. Proc Natl Acad 103:8337–8341

Fagherazzi S, Hannion M, D'Odorico P (2008) Geomorphic structure of tidal hydrodynamics in salt marsh creeks. Water Resour Res. doi:101029/2007WR006289

Feagin RA et al (2009) Does vegetation prevent wave erosion of salt marsh edges? Proc Natl Acad Sci 106:10109–10113

Fenies H, Faugères JC (1998) Facies and geometry of tidal channel-fill deposits (Arcachon lagoon, SW France). Mar Geol 150:131–148

French JR, Stoddart DR (1992) Hydrodynamics of salt marsh creek systems: implications for marsh morphological development and material exchange. Earth Surf Process Land 17:23–252

Friedrichs CT (1995) Stability shear stress and equilibrium cross-sectional geometry of sheltered tidal channels. J Coast Res 11:1062–1074

Friedrichs CT, Aubrey DG (1988) Non-linear tidal distortion in shallow well-mixed estuaries: a synthesis. Estuar Coast Shelf Sci 27:521–545

Gabet EJ (1998) Lateral migration and bank erosion in a salt marsh tidal channel in San Francisco Bay, California. Estuaries 4B:745–753

Garofalo D (1980) The influence of wetland vegetation on tidal stream migration and morphology. Estuaries 3:258–270

Ginsberg SS, Perillo GME (2004) Characteristics of tidal channels in a mesotidal estuary of Argentina. J Coast Res 20:489–497

Hibma A, de Vriend HJ, Stive MJF (2003) Numerical modelling of shoal pattern formation in well-mixed elongated estuaries. Estuar Coast Shelf Sci 57:98–191

Hibma A, Stive MJF, Wang ZB (2004a) Estuarine morphodynamics. Coast Eng 51:765–778

Hibma A, Schuttelaars HM, de Vriend HJ (2004b) Initial formation and long-term evolution of channel-shoal patterns. Cont Shelf Res 24:1637–1650

Hood WG (2006) A conceptual model of depositional rather than erosional tidal channel development in the rapidly prograding Skagit River Delta (Washington USA). Earth Surf Proc Land. doi:101002/esp1381

Hood WG (2010) Tidal channel meander formation by depositional rather than erosional processes: examples from the prograding Skagit River Delta (Washington, USA). Earth Surf Proc Land 35:319–330

Horton RE (1945) Erosional development of streams and their drainage basins: hydrophysical approach to quantitative morphology. Bull Geol Soc Am Bull 56:275–370

Howard AD (1996) Modelling channel evolution and floodplain morphology. In: Anderson MG, Walling DE, Bates PD (eds) Floodplain processes. Wiley, Chichester

Howes NC (2009) The impact of wetland loss on inlet morphology and tidal range within Barataria Bay Lousiana. MSc Thesis Boston University, Boston

Howes NC et al (2010) Wetland loss during hurricanes: failure of low salinity marshes. Proc Natl Acad Sci. doi:101073/pnas0914582107

Huat BBK, Asadi A, Kazemian S (2009) Experimental investigation on geomechanical properties of tropical organic soils and peat. Am J Eng Appl Sci 2:184–1888

Hughes ZJ, FitzGerald DM, Wilson CA, Pennings SC, Wieiski K, Mahadevan A (2009) Rapid headward erosion of marsh creeks in response to relative sea-level rise. Geophys Res Lett. doi:101029/2008GL036000

Jarrett JT (1976) Tidal prism-inlet area relationships. Gen Invest Tidal Inlets Rep 332 US Army Coastal Engineering Research Center, Fort Belvoir

Kirwan ML, Murray AB (2007) A coupled geomorphic and ecological model of tidal marsh evolution. Proc Natl Acad Sci 104:6118–6122

Klein GdeVries (1977) Clastic tidal facies. CEPCO, Champaign, p 148

Knighton AD, Woodroffe CD, Mills K (1992) The evolution of tidal creek networks Mary River Northern Australia. Earth Surf Proc Land 17:67–190

Lanzoni S, Seminara G (2002) Long-term evolution and morphodynamic equilibrium of tidal channels. J Geophys Res. doi:101029/2000JC000468

Lee CB, Yoo HR, Park KS (1992) Distribution and properties of intertidal surface sediments of Kyeonggi Bay west coast of Korea. J Oceanol Soc Korea 27:277–289

Leonard LA, Croft AL (2006) The effect of standing biomass on flow velocity and turbulence in *Spartina alterniflora* canopies. Estuar Coast Shelf Sci 69:325–336

Leonard LA, Luther ME (1995) Flow hydrodynamics in tidal marsh canopies. Am Soc Limnol Ocean 40:1474–1484

Leopold LB, Wolman MG (1960) River meanders 1. Geol Soc Am Bull 71:769–793

Leopold LB, Wolman MG, Miller JP (1964) Fluvial processes. In: Geomorphology. W H Freeman, San Francisco

Li C, O'Donnell J (2005) The effect of channel length on the residual circulation in tidally dominated channels. J Phys Ocean 35:1826–1840

Li C, Chen C, Guadagnoli D, Georgiou I (2008) Geometry induced residual eddies in estuaries with curved channels – observations and modeling studies. J Geophys Res. doi:101029/2006JC004031

Marani M, Lanzoni S, Zandolin D, Seminara G, Rinaldo A (2002) Tidal meanders. Water Resour Res. doi:101029/2001WR000404

Marani M, Belluco E, D'Alpaos A, Defina A, Lanzoni S, Rinaldo A (2003) On the drainage density of tidal networks. Water Resour Res. doi:101029/2001WR001051

Marani M, Lanzoni S, Silvestri S, Rinaldo A (2004) Tidal landforms patterns of halophytic vegetation and the fate of the lagoon of Venice. J Mar Syst 51:191–210

Minkoff DR, Escapa M, Ferramola FE, Maraschin SD, Pierini JO, Perillo GME, Delrieux C (2006) Effects of crab-halophytic plant interactions on creek growth in a SW Atlantic salt marsh: a Cellular Automata model. Estuar Coast Shelf Sci 69:403–413

Mowbray T, Visser MJ (1984) Reactivation surfaces in subtidal channel deposits, Oosterschede, Southwest Netherlands. J Sed Pet 54:811–824

Mudd SM, D'Alpaos A, Morris JT (2010) How does vegetation affect sedimentation on tidal marshes? Investigating particle capture and hydrodynamic controls on biologically mediated sedimentation. J Geophys Res. doi:101029/2009JF001566

Murphy S, Voulgaris G (2006) Identifying the role of tides, rainfall and seasonality in marsh sedimentation using long-term suspended sediment concentration data. Mar Geol 227:31–50

Neumeier U (2007) Velocity and turbulence variations at the edge of salt marshes. Cont Shelf Res 27:1046–1059

Novakowski KI, Torres R, Gardner LR, Voulgaris G (2004) Geomorphic analysis of tidal creek networks. Water Resour Res. doi:10.1029/2003WR002722

O'Brien MP (1969) Equilibrium flow areas of inlets in sandy coasts. J Water Harbors Coastal Eng Div Am Soc Civ Eng 95:43–52

Parker RS (1977) Experimental study of drainage basin evolution and its hydrologic implications. In: Hydrol Papers 90. Colorado State University, Fort Collins

Parker BB (1991) The relative importance of the various nonlinear mechanisms in a wide range of tidal interactions (review). In: Parker BP (ed) Tidal hydrodynamics. Wiley, New York

Parker et al (2010) A new framework for modeling the migration of meandering rivers. Earth Surf Proc Land. doi:10.1002/esp. 2113

Pearson NJ, Gingras MK (2006) An ichnological and sedimentological facies model for muddy point-bar deposits. J Sed Res 76:771–782

Perillo GME, Iribarne OO (2003) Processes of tidal channel development in salt and freshwater marshes. Earth Surf Proc Land 28:1473–1482

Perillo GME, Garcia Martinez MB, Piccolo MC (1996) Geomorfología de canales de marea: Aná lisis de fractales y espectral. In: Actas VI Reunión Argentina de Sedimentología, Bahía Blanca, pp 155–160

Pestrong R (1965) The development of drainage patterns on tidal marshes. Stanford Univ Publ Geol Sci Tech Rep 10:1–87

Pestrong R (1972) Tidal flat sedimentation at Cooley Landing southwest San Francisco Bay. Sed Geol 8:251–288

Pethick JS (1969) Drainage in tidal marshes. In: Steers JA (ed) The coastline of England and Wales, 2nd edn. Cambridge University Press, Cambridge

Pethick JS (1980) Velocity surges and asymmetry in tidal channels. Estuar Coast Mar Sci 11:331–345

Pethick JS (1984) An introduction to coastal geomorphology. Edward Arnold, London

Pethick JS (1992) Salt marsh geomorphology. In: Allen JRL, Pye K (eds) Salt marshes: morphodynamics conservation and engineering significance. Cambridge University Press, Cambridge

Pye K, French PW (1993) Erosion and accretion processes on British Salt Marshes Vol 1 Introduction: salt marsh Processes and Morphology. Cambridge Environmental Research Consultants, Cambridge

Redfield AC (1972) Development of a New England salt marsh. Ecol Mono 42:201–237

Reineck HE (1967) Layered sediments of tidal flats beaches and shelf bottoms of the North Sea. In: Lauff GH (ed) Estuaries: American Association for the Advancement of Science, Spec Publ 83

Ren Mei-e (1986) Modern sedimentation in the coastal and nearshore zones of China. China Ocean Press, Bejing and Springer, Berlin/Heidelberg

Rieu R, Van Heteren S, Van der Spek AJF, De Boer PL (2005) Development and preservation of a mid-Holocene tidal channel network offshore the western Netherlands. J Sed Res 75:409–419

Rinaldo A, Fagherazzi S, Lanzoni S, Marani M, Dietrich WE (1999) Tidal networks: 3 Landscape-forming discharges and studies in empirical geomorphic relationships. Water Resour Res 35(12):3919–3929

Rinaldo A, Belluco E, D'Alpaos A, Feola A, Lanzoni S, Marani A (2004) Tidal networks: form and function. In: Fagherazzi S, Marani M, Blum LK (eds) The ecogeomorphology of tidal marshes. Amer Geophys Union, Washington, DC

Santos A, Rossetti D (2006) Depositional model of the Ipixuna Formation (Late Cretaceous-? Early Tertiary), Rio Capim area, northern Brazil. Latin Am J Sed Basin Anal 13:101–117

Schwimmer RA (2001) Rate and processes of marsh shoreline erosion in Rehoboth Bay Delaware USA. J Coast Res 17:672–683

Schwimmer RA (2008) A temporal geometric analysis of eroding marsh shorelines: can fractal dimensions be related to process? J Coast Res 24:152–158

Seminara G (2006) Meanders. J Fluid Mech 554:271–297

Settlemyre JL, Gardner RL (1977) Suspended sediment flux though a salt marsh drainage basin. Estuar Coast Mar Sci 5:653–663

Shi Z, Lamb HF, Collins RL (1995) Geomorphic change of salt marsh tidal creek network in the Dyfi Estuary Wales. Marine Geol 128:73–83

Solari L, Seminara G, Lanzoni S, Marani M, Rinaldo A (2002) Sand bars in tidal channels Part 2. Tidal meanders. J Fluid Mech 451:203–238

Steel TJ, Pye K (1997) The development of salt marsh tidal creek networks: evidence from the UK. Proc Can Coast Conf 1:267–280

Steers JA (1969) The coastline of England and Wales, 2nd edn. Cambridge University Press, Cambridge

Symonds AM, Collins MB (2007) The establishment and degeneration of a temporary creek system in response to managed coastal realignment: the Wash UK. Earth Surf Proc Land 32:1783–1796

Temmerman S, Bouma TJ, Van de Koppel J, Van der Wal D, De Vries MB, Herman PMJ (2007) Vegetation causes channel erosion in a tidal landscape. Geology 35:631–634

Terwindt JHJ (1988) Palaeo-tidal reconstructions of inshore tidal depositional environments. In: de Boer PL, van Gelder A, Nio S-D (eds) Tide-influenced Sedimentary environments and facies. D Reidel Publishing Company, Boston

van der Wegen M, Wang ZB, Savenije HHG, Roelvink JA (2008) Long-term morphodynamic evolution and energy dissipation in a coastal plain tidal embayment. J Geophys Res. doi:101029/2007JF000898

van Eerdt MM (1985) The influence of vegetation on erosion and accretion in salt marshes of the Oosterschelde The Netherlands. Vegetation 62:367–373

van Proosdij D, Baker G (2007) Intertidal morphodynamics of the Avon River estuary. Final Report submitted to Nova Scotia Department of Transportation and Public Works. http://www.gov.ns.ca/tran/highways/Hwy101twinning Windsor.asp

van Straaten LMJU (1954) Composition and structure of recent marine sediments in the Netherlands. Leid geol Meded 19:1–110

van Veen J (1950) Ebb and flood-channel systems in the Netherlands tidal waters (in Dutch English summary) KNAG 2ed Series Part 67. Republished translated and annotated by Delft University of Technology 2001 ISBN 9040723389

Watzke DA (2004) Short-term evolution of a marsh island system and the importance of cold front forcing, Terrebone Bay, Louisiana. Masters Thesis, Louisiana State University, Baton Rouge

Wilson KR, Kelley JT, Croitoru A, Dionne M, Belknap DF, Steneck R (2009) Stratigraphic and ecophysical characterizations of salt pools: dynamic landforms of the webhannet salt marsh, Wells ME, USA. Estuaries Coasts 32:855–870

Wolanski E, Williams D, Hanert E (2006) The sediment trapping efficiency of the macro-tidal Daly Estuary, tropical Australia. Estuar Coast Shelf Sci 69:291–29

Wright LD, Coleman JM, Thom BG (1973) Processes of channel development in a high-tide range environment: Cambridge Gulf-Ord River delta Western Australia. J Geol 81:15–41

Zeff ML (1988) Sedimentation in a salt marsh–tidal channel system. South N J Mar Geol 82:33–48

Zeff ML (1999) Salt marsh tidal channel morphometry: applications for wetland creation and restoration. Restor Ecol 7:205–211

Morphodynamics and Facies Architecture of Tidal Inlets and Tidal Deltas

12

Duncan FitzGerald, Ilya Buynevich, and Christopher Hein

Abstract

Tidal inlets are highly dynamic systems marking positions along barrier coasts where dominant wave and longshore sand transport processes are juxtaposed with a tide-dominated regime in which onshore-offshore sand movement is manifested in the formation of flood- and ebb- tidal deltas. The morphodynamics of tidal inlets and distribution of their associated sand shoals are governed by the tidal prism, wave versus tidal energy, and the regional geological framework. Sand that is delivered to the inlet channel via longshore transport can be sequestered in the backbarrier, moved onto the ebb-tidal delta, or can bypass the inlet. Such bypassing is accomplished through wave and tidal processes and ultimately results in the landward migration and welding of large sand bar complexes to the downdrift shoreline. Tidal inlet-fill deposits typically exhibit a sharp basal contact with underlying units and consist of a fining-upward sequence in contrast to the generally coarsening-upward barrier lithosome. The preservation potential of inlet and associated tidal-delta deposits is high in regressive sequences, but relatively poor in transgressive systems due to the shallow nature of inlet-fill deposits compared to the base of the erosional wave- or tidal- ravinement surfaces. Exceptions occur in paleotidal inlet regions having large bay tidal prisms and deep inlet channels. Although tidal-inlet deposits have been reported in the rock record and may serve as important petroleum reservoirs, to date they are not readily recognized.

High-resolution geophysical and sedimentological research of both active and relict inlets is providing a wealth of information necessary to improve the inlet facies models for ancient sedimentary sequences.

D. FitzGerald (✉) • C. Hein
Department of Earth Sciences, Boston University, Boston, MA 02215, USA
e-mail: dunc@bu.edu; hein@whoi.edu

I. Buynevich
Department of Earth and Environmental Sciences, Temple University, 313 Philadelphia, PA 19122, USA
e-mail: coast@temple.edu

12.1 Introduction

Tidal inlets are openings along barrier coasts through which tidal waters penetrate the land, thereby providing a hydraulic connection between the ocean and bays, lagoons, and marsh and tidal creek systems. Inlets are societally important because they provide a pathway

R.A. Davis, Jr. and R.W. Dalrymple (eds.), *Principles of Tidal Sedimentology*,
DOI 10.1007/978-94-007-0123-6_12,

to major ports and serve as conduits through which nutrients are exchanged between lagoons and estuaries and the coastal ocean. Numerous species of finfish and shellfish rely on inlet channels for access to backbarrier regions for feeding, breeding, and nursery grounds of their young. Sedimentologically, inlet-fill sequences may comprise a substantial portion of the barrier lithosome (Moslow and Tye 1985) and their presence may be the only means of identifying the existence of former barrier island chains on the continental shelf (Hine and Snyder 1985; Rieu et al. 2005) or in the rock record (Tye and Moslow 1993). Tidal inlets not only serve as important interruptions in the longshore transport system, but their fills represent subsurface anomalies that influence groundwater exchange and reservoir permeability when preserved in the rock record.

Most tidal inlets are associated with coastal plain settings along passive margins where abundant sediment has formed extensive barrier chains, such as those along the East and Gulf Coasts of the United States, the Friesian Islands along the North Sea, and the North Slope of Alaska. Barrier island coasts and tidal inlets are most prevalent in the Northern Hemisphere where sea-level rise during the Holocene flooded deltaic and other coastal plain settings, forming lagoons and bays behind developing barrier systems. It is the filling and drainage of these open-water backbarrier areas that produce tidal currents and maintain the inlet channel. The formation of tidal inlets has been a time-dependent process. In the early evolution of barrier systems, tidal inlets developed preferentially in the low areas along developing barrier systems, coinciding with existing or former river valleys. In other settings, abundant sand supplies led to the construction of spit systems across embayments. Tidal inlets also became stabilized at the downdrift ends of spits, if tidal prisms were large enough to keep them open. Subsequent to these initial phases of inlet development and extending to the present time, new tidal inlets have been a product of barrier breaching during large magnitude storms. If the hypsometry and hydrodynamics of the backbarrier are conducive to the capture of a sufficiently large tidal prism, the new inlet will remain open. Tidal inlets are much less common in the Southern Hemisphere because falling sea level (1–4 m; Angulo et al. 2006) since the mid-Holocene has resulted in strandplain, rather than barrier island, development. Exceptions to this trend occur where abundant sand resources have built spits and barriers at the mouths of estuaries and in front of large embayments, such as those that occur along the coasts of Australia and Brazil.

Tidal inlets are normally accompanied by an ebb-tidal delta on the seaward side and a flood-tidal delta in the bays. Inlets and their associated tidal deltas exhibit a wide diversity in morphology, hydrodynamic character, and sediment transport patterns, which are a function of the variability in oceanographic, meteorologic, and geologic settings, including parameters such as tidal range, wave energy, sediment supply, storm magnitude and frequency, fresh water influx, and antecedent geologic controls.

This chapter examines the general morphodynamics and hydrodynamics of tidal inlets for the purpose of illustrating how inlet processes, such as barrier breaching, inlet evolution, channel shifting, and bar migrations, produce the facies architecture and large-scale bedding surfaces that define tidal inlet and associated tidal delta sedimentary deposits. An excellent treatment of textural characteristics, bedform patterns, and sedimentary sequences at tidal inlets exists in Boothroyd (1985). In addition to tidal-delta morphology, this discussion covers the development of bedforms, their migrational trends, and resulting sedimentary structures in the various tidal inlet sub-environments. Smith (1991), Boothroyd (1985), Reinson (1984), and Hubbard et al. (1979) have also provided stratigraphic models for inlet-fill sequences and the tidal deltas; however, these are hypothetical models based primarily on surficial data. During the past 25 years, extensive coring efforts and geophysical studies carried out at numerous tidal inlets offer new perspectives on inlet dynamics and geological legacy, and provide empirical data for more detailed stratigraphic models of inlet-fill sequences and tidal deltas. The second half of the chapter reviews many of these field studies and discusses the variability and complexity of inlet systems. Finally, the preservation potential and an overview of tidal inlet deposits as recognized in the rock record are presented.

12.2 Morphology and Stability: General Concepts

A tidal inlet is the narrowed region of water between two barrier islands or between a single barrier and an adjacent bedrock or glacial headland (Fig. 12.1).

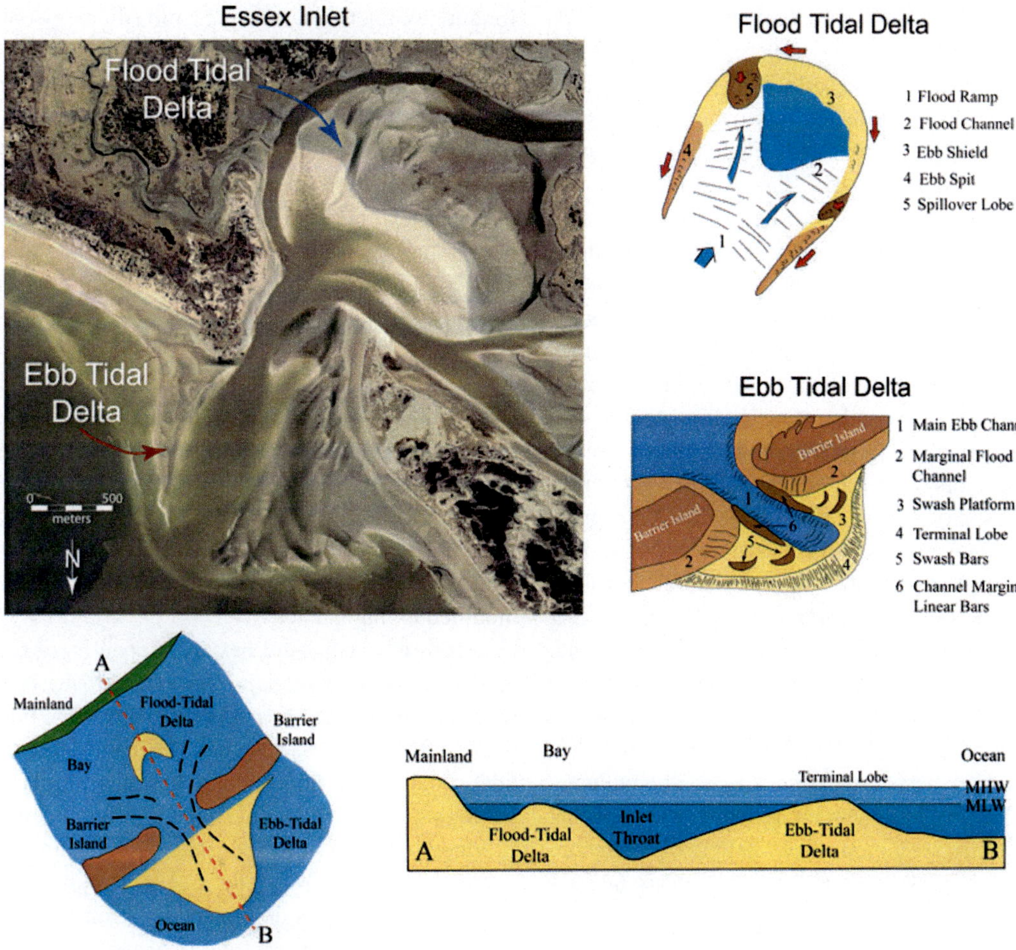

Fig. 12.1 Ebb- and flood-tidal delta models (From Hayes 1979). Aerial photograph of Essex Inlet, MA. Inlet cross-section model from FitzGerald (1996)

Commonly, the sides of the inlet are formed by the recurved ridges of spits, consisting of sand that was transported toward the backbarrier by refracted waves and flood-tidal currents. The inlet throat is the constricted, deepest part of an inlet normally coinciding with the minimum channel width and maximum tidal current velocities. The strength of the tidal currents in the throat section is generally sufficient to remove sand from the channel floor leaving behind a lag deposit consisting of gravel or shells, or in some locations revealing bedrock or consolidated mud.

The stability of an inlet is related to the framework geology of the region and littoral sediment transport trends. Migrating inlets occur along coasts having a pronounced net longshore transport system and where the channel erodes through unconsolidated sediment (e.g. Kiawah River Inlet, SC, Tye and Moslow 1993). Conversely, natural stable inlets are situated in channels that resist lateral erosion such as those scoured into former river valleys (Virginia; Halsey 1979), cut into limestone (Florida; Zarillo et al. 2003), or abutting bedrock or till outcrops (New England; FitzGerald 1993).

Several classifications of tidal inlets have been published based on the distribution of intertidal and subtidal sand bodies and the general morphology of the ebb-tidal delta and inlet shoreline including Oertel (1975), Nummedal and Fischer (1978), Hubbard et al. (1979), Hayes (1979), Davis and Gibeaut (1990), and Galvin (1994). A discussion of these schemes can be found in FitzGerald (1996).

12.2.1 Tidal Delta Morphodynamics

Tidal deltas were first described in detail by Hayes (1975) and then incorporated into his coastal classification based on the relative magnitude of wave versus tidal energy (Hayes 1979). Wave-dominated coasts have long barriers with few inlets, whereas mixed-energy coasts have short stubby barriers with numerous tidal inlets. Later, it was shown that this scheme could also be influenced by tidal prism, which is the volume of water flowing through a tidal inlet during a half tidal cycle (Davis and Hayes 1984). They showed that regardless of coastal setting, barrier coasts with large tidal prisms tend to have larger and/or greater number of inlets to accommodate this tidal exchange. The formation of ebb- and flood-tidal deltas is a product of sand deposition by the ebb- and flood-tidal jets, respectively. As tidal waters flow beyond the constriction of the barriers, the currents expand laterally, losing their velocity and their capacity to transport sand. Flood deltas are commonly built or enlarged during storm events. Their size, geometry, and facies relationships are a function of tidal range, tidal prism, storm processes, and accommodation space. The volume, morphology, and sedimentary sequences comprising ebb-tidal deltas are a product of tidal prism, nearshore slope, sand bypassing processes, as well as interactions between wave and tidal energy.

12.2.1.1 Flood-Tidal Deltas

These shoals exhibit a characteristic horseshoe-shaped morphology worldwide (Figs. 12.1 and 12.2). Their presence or absence is related to the availability of sediment and extent of open water in the backbarrier. Along mixed-energy coasts, tidal inlets are connected to a broad marsh and tidal creek system containing a single, relatively large flood-tidal delta, if space permits (i.e., Essex Inlet, Massachusetts, Fig. 12.1; Hayes 1979). Contrastingly, inlets such as Drum Inlet along the Outer Banks of North Carolina (wave-dominated coast), that are backed by large shallow bays, may contain multiple flood-tidal deltas. Along some microtidal coasts, such as Rhode Island, flood deltas form at the end of narrow inlet channels cut through the barrier. Changes in the locus of deposition at these deltas produce a multi-lobate morphology resembling a lobate river delta (Boothroyd et al. 1985).

To some extent, delta size is related to the amount of open water area in the backbarrier, which is a function of inlet size. Along the mixed-energy coast of Maine, where tidal inlets are comparatively small (width < 100 m), flood-tidal deltas are correspondingly small and stacked in an alternating pattern along the main tidal creek (FitzGerald et al. 1990). Tidal inlets along the barrier coasts of central South Carolina and Georgia have no flood-tidal deltas because the backbarrier has been almost completely filled with fine-grained sediment and marsh deposits, resulting in tidal channels that are too narrow and deep for delta development. Along some mixed-energy coasts (Hayes 1979), flood-tidal deltas are indistinguishable from the surrounding marsh due to sedimentation and infilling of backbarrier followed by the colonization of sand shoals by saltwater vegetation. This may be the case for some inlets in central South Carolina. At other sites, portions of flood-tidal deltas are dredged to provide navigable waterways and thus they become highly modified by human activity.

Flood-tidal deltas are best developed in areas with moderate to large tidal ranges (1.5–3.0 m) because in these regions they are well exposed at low tide. As tidal range decreases, flood deltas become largely subtidal shoals. Most flood-tidal deltas have similar morphologies consisting of the following components:

1. Flood ramp- Landward shallowing channel that slopes upward toward the intertidal portion of the delta. The ramp is dominated by strong flood-tidal currents and landward sand transport.
2. Flood channels- The flood ramp splits into two shallow flood channels. Like the flood ramp, these channels are dominated by flood-tidal currents. Sand is delivered through these channels onto the flood delta.
3. Ebb shield- This defines the highest and landward-most part of the flood delta and may be partly covered by marsh vegetation. It shields the rest of the delta from the effects of the ebb-tidal currents.
4. Ebb spits- These spits extend from the ebb shield toward the inlet. They form from sand that is eroded from the ebb shield and transported back toward the inlet by ebb-tidal currents.
5. Spillover lobes- These are lobes of sand that form where the ebb currents have breached through the ebb spits or ebb shield, depositing sand in the interior of the delta.

In wave-dominated settings, flood-tidal deltas are commonly present at small to large inlets exhibiting a variety of forms from sub-tidal shoals to multi-lobate

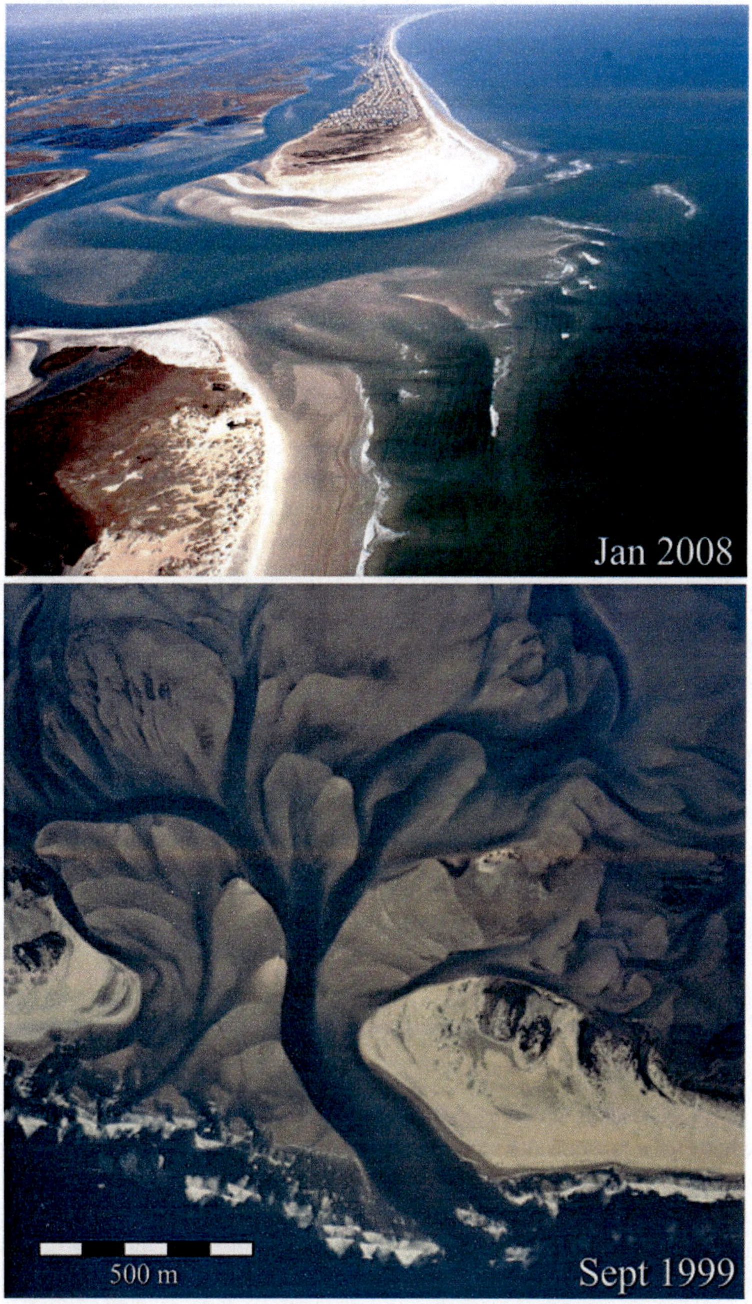

Fig. 12.2 New Topsail Inlet (*top*) and Drum Inlet (*bottom*), North Carolina. (Courtesy of William Cleary)

deltas (i.e. Rhode Island, North Carolina, Texas). Their multi-lobate morphology is a product of different periods of deposition. The small tidal range of these coasts prevents their reworking by ebb-tidal currents as occurs on mesotidal coasts, except in regions having large tidal prisms (i.e., west coast of Florida).

Through time, some flood-tidal deltas accrete vertically and/or grow in size. This is evidenced by an increase in areal extent of marsh grass, which requires a minimum elevation above mean low water to exist. At many inlets sand contribution to flood deltas is event driven. During storms, high wave energy delivers large quantities of sand to the inlet via longshore transport, while at the same time the storm surges augments the hydraulic slope of the flooding tide. Thus, the increased supply of sand to the inlet channel is directed into the backbarrier and to the flood-tidal delta due to the increased capacity of the tidal currents. At migrating inlets, new flood-tidal deltas are sequentially formed as the inlet moves along the coast and encounters new open water areas in the backbarrier. An almost continuous, mostly sub-tidal shoal system has been produced by this process along the bayside of Katama Spit on the southern coast of Martha's Vineyard, Massachusetts (FitzGerald and Pendleton 2002). At most stable inlets, however, sand comprising the flood delta is mostly recirculated between a relatively fixed number of depocenters. The transport of sand to and from flood deltas is controlled by the time-velocity asymmetry of the tidal currents (Boothroyd and Hubbard 1975). During the rising tide, flood currents reach their strongest velocities near high tide when the entire flood-tidal delta is covered by water. Hence, there is a net transport of sand up the flood ramp, through the flood channels, and onto the ebb shield. Some of the sand is moved across the ebb shield and into the surrounding tidal channel. During the falling tide, the strongest ebb currents occur near mid to low water. At this time, the emerging ebb shield diverts the currents around the delta. The ebb currents erode sand from the landward face of the ebb shield and transport it along the ebb spits and eventually into the inlet channel where once again it will be moved onto the flood ramp thus completing the sand gyre.

12.2.1.2 Ebb-Tidal Deltas

An ebb-tidal delta is a complex of shoals and channels on the seaward side of the inlet. It consists of sand that is intercepted from the longshore transport system and is carried seaward and deposited by ebb-tidal currents, where it is subsequently modified by incident waves and ambient tidal currents. Ebb deltas exhibit a variety of forms dependent on the relative magnitude of wave and tidal energy of the region as well as geological controls. Despite this variability, most ebb-tidal deltas contain the same general features including:

1. Main ebb channel- This is a seaward shallowing channel that is scoured in the ebb-tidal delta sand and is dominated by ebb-tidal currents.
2. Terminal lobe- Sediment transported out the main ebb channel is deposited in a lobe of sand forming the terminal lobe. The deposit slopes relatively steeply on its seaward side. The outline of the terminal lobe is well defined by breaking waves during storms or periods of large wave swell at low tide.
3. Swash platform – This is a broad shallow sand platform located on both sides of the main ebb channel, defining the general extent of the ebb delta and dominated by wave action.
4. Channel margin linear bars – These are bars that border the main ebb channel and sit atop the swash platform. These bars tend to confine the ebb flow and are exposed at low tide.
5. Swash bars – Waves breaking over the terminal lobe and across the swash platform form arcuate-shaped swash bars that migrate onshore. The bars are usually 50–150 m long, 50 m wide, and 0.5–2.0 m in height.
6. Marginal-flood channels. These are shallow channels (up to 2 m deep at mean low water) located between the channel margin linear bars and the onshore adjacent beaches. The channels are dominated by flood-tidal currents.

The general shape of an ebb-tidal delta and the distribution of its sand bodies reveal the relative magnitude of different sand transport processes operating at a tidal inlet. Ebb-tidal deltas that have elongated main ebb channel and channel margin linear bars that extend far offshore are common for tide-dominated inlets, including those having strong tidal-versus-wave energy and/or large tidal prisms (e.g., Georgia, southern South Carolina, Florida Gulf Coast). Wave-generated sand transport plays a secondary role in modifying delta shape at these inlets. Because most sand movement is in the onshore-offshore direction, the ebb-tidal delta overlaps a relatively small segment of inlet shoreline. This affects the extent to which this part of the coast undergoes erosional and depositional changes caused by inlet processes.

Wave-dominated inlets tend to be small relative to tide-dominated systems. Their ebb-tidal deltas are driven onshore, close to the inlet mouth by the dominant wave processes. Commonly, the terminal lobe and/or swash bars form a small arc outlining the periphery of the delta (e.g., Nauset Inlet, Massachusetts; Little River, South Carolina; Matanza Inlet, Florida) In many cases the ebb-tidal delta of these inlets is entirely subtidal (e.g., Blind Pass, Florida). In other instances, sand bodies clog the entrance to the inlet, leading to the formation of several major and minor tidal channels.

At mixed-energy tidal inlets, the shape of the delta is the result of tidal and wave processes. These deltas have a well-formed main ebb channel, which is a product of dominant ebb-tidal currents. Their swash platform and sand bodies substantially overlap the inlet shoreline many times the width of the inlet throat due to wave processes and flood-tidal currents (e.g. Northern New England, Southern New Jersey, Southern South Carolina, German Friesian Islands). Asymmetries in ebb delta configuration commonly result from the main ebb channel being deflected along the downdrift barrier shoreline due to the dominant longshore transport direction or the pattern of the dominant back-barrier channels.

12.3 Bedform Distribution

The strength of the tidal currents and abundance of sand at tidal inlets produce a wide distribution of bedforms in tidal channels and on intertidal shoals. Boothroyd and Hubbard (1975) performed extensive field studies of bedforms at inlets in northern Massachusetts using direct underwater measurements and fathometer transects. Since that time, many other investigators have gathered additional information at inlets throughout the world showing similar patterns and migrational behavior of bedforms, including the German Friesian Seegats (Nummedal and Penland 1981), Rangaunu Harbor Entrance, New Zealand (Pickrill 1986), Texel Inlet, The Netherlands (Sha 1989), Martens Plate, German North Sea (Davis and Flemming 1991), New Inlet, Massachusetts (FitzGerald and Montello 1993), Piedras Estuary, Spain (Morales et al. 2001), and the Algarve (Williams et al. 2003). Boothroyd and Hubbard (1975) have demonstrated that tidal inlet channels and shoals have a systematic pattern of bedforms that is governed by three aspects of the current regime: (1) maximum current velocity, (2) velocity asymmetry between the maximum ebb and flood currents, and (3) duration of the velocity initiating bedform formation. Their studies have shown that flood- oriented sandwaves ($L>6$ m; 2-D medium dunes in the scheme of Ashley 1990) occur in the intertidal and shallow subtidal portions of the flood delta, coinciding with the flood ramp and flood channels where there is a pronounced velocity asymmetry and maximum velocities greater than 0.8 m/s. Similar conditions produce ebb-oriented sandwaves flooring the main ebb channel. Where currents exhibit little velocity asymmetry, sandwaves tend to be symmetrical. Boothroyd and Hubbard (1975) also showed that under flow conditions exceeding 0.8 m/s, megaripples become superimposed on the backs of sandwaves. Bedforms are commonly absent at the inlet throat due to the lack of sand or presence of a lag deposit, such as gravel or shell armor, that retards bedform formation. Flood-oriented megaripples ($0.6\ \text{m}<L<6$ m; 2-D small dunes in the scheme of Ashley 1990) occur on the low-intertidal ebb shield areas and in marginal flood channels on the ebb delta. Ebb-oriented megaripples are found in the ebb-spill-over lobes of flood and ebb deltas. The swash bars and bar complexes comprising intertidal portions of the ebb delta are commonly covered by flood-oriented megaripples (FitzGerald 1976).

As demonstrated in Fig. 12.3, the size and orientation of subtidal bedforms at New Inlet, Massachusetts follow closely the patterns reported by Boothroyd and Hubbard (1975) with the exception that mutually evasive channels in the backbarrier exhibit opposing tidal dominance and opposite trending sandwaves (FitzGerald and Montello 1993). A similar pattern of opposing bedform orientations has been documented in the seaward portion of Texel Inlet, (Sha 1989). Nummedal and Penland (1981) describe a system of alternating flood- and ebb- dominant channels on the ebb delta at Friesian tidal inlets having attendant flood- and ebb-oriented bedforms, respectively. The reader is directed to Boothroyd (1985) for additional treatment of bedforms at tidal inlets, including their genesis, migration trends, and sedimentary structures.

12.4 Tidal Inlet Relationships

Tidal inlets throughout the world exhibit several consistent relationships that have allowed coastal engineers and sedimentologists to formulate predictive models: (1) Inlet throat cross-sectional area is closely related to tidal prism, and (2) Ebb-tidal delta volume is a function of the tidal prism.

12.4.1 Inlet Throat Area – Tidal Prism Relationship

The cross section of tidal inlets (A) correlates closely to tidal prism (P) (Eq. 12.1 all units in metric; O'Brien 1931, 1969) and is secondarily affected by the delivery of sand to the inlet channel by wave energy. For example, at jettied inlets, tidal currents can more effectively

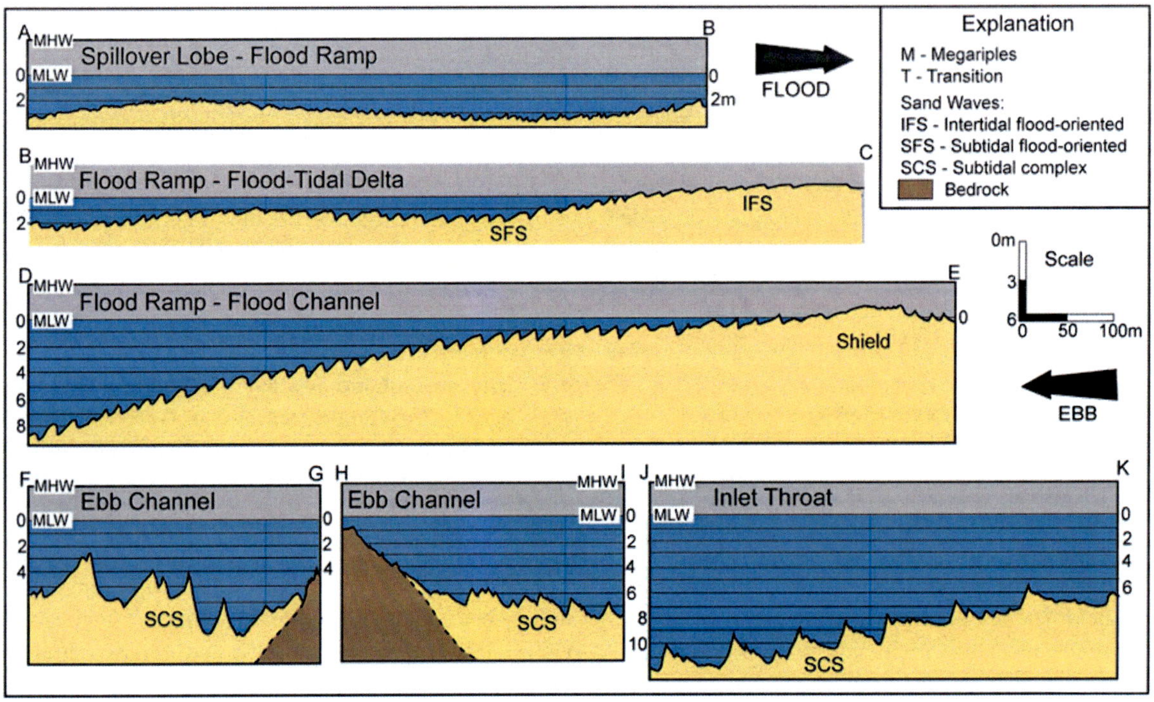

Fig. 12.3 Bedform patterns at New Inlet along the wave-dominated outer coast of Cape Cod, Massachusetts. Note the variety of bedform heights, lengths, and orientations, which is a reflection of flow velocity asymmetry and mutual evasive tidal channels

scour sand from the inlet channel and therefore they maintain a larger throat cross section than would be predicted by the O'Brien Relationship. Thus, Jarrett (1976) has improved the tidal prism – inlet cross-sectional area regression equation for U.S. inlets by taking into account wave energy and separating into three classes the low energy Gulf Coast inlets, moderate energy East Coast inlets, and higher energy West Coast inlets. Even better correlations are achieved when engineered inlets are differentiated from natural systems.

$$A = 3.04 \times 10^{-5} \times P^{1.05} \tag{12.1}$$

12.4.2 Ebb-Tidal Delta Volume – Tidal Prism Relationship

Walton and Adams (1976) showed that the volume of sand contained in ebb-tidal deltas (V) is closely related to tidal prism (P) by the relationship given in Eq. 12.2. This correlation is improved slightly when wave energy is taken into account in a manner similar to Jarrett's end members. Waves are responsible for transporting sand back onshore, thereby reducing the volume of the ebb-tidal delta. Therefore, for a given tidal prism, ebb-tidal deltas in higher wave energy regimes contain less sand than their counterparts along low wave energy coasts. Although the Walton and Adams Relationship works well for inlets all over the world, field studies have shown that the volume of sand comprising ebb-tidal deltas can change through time due to the effects of storms or processes of inlet sediment bypassing, and these effects can change the shoal volume by more than 10% (FitzGerald et al. 1984, Gaudiano and Kana 2001).

$$V = 1.89 \times 10^{-5} \times P^{1.23} \tag{12.2}$$

12.5 Sand Transport Patterns

The movement of sand at a tidal inlet is complex due to reversing tidal currents, effects of storms, and interaction with the longshore transport system. The inlet contains short-term and long-term reservoirs of sand varying from the relatively small sandwaves flooring the inlet channel that migrate meters each tidal cycle, to the large flood-tidal delta shoals where some sand is recirculated but the entire deposit may remain stable for hundreds of years. Sand dispersal at tidal inlets is complicated because, in addition to the onshore-offshore movement of sand produced by tidal and wave-generated currents, there is steady delivery of sand to the inlet and removal by the longshore transport system. In the discussion below, the patterns of sand movement at inlets are described, including how sand bypasses tidal inlets because many of these processes produce the major architectural units and large-scale stratigraphic bounding surfaces comprising ebb-delta deposits.

12.5.1 General Sand Dispersal Trends

Ebb-tidal deltas consist of segregated areas of landward versus seaward sediment transport that are controlled primarily by the way water enters and discharges from the inlet as well as the effects of wave-generated currents. During the ebbing cycle, the tidal flow leaving the backbarrier is constricted at the inlet throat, causing the currents to accelerate in a seaward direction. Once out of the confines of the inlet, the ebb flow expands laterally and the velocity slows. Sediment in the main ebb channel is transported in a net seaward direction and is eventually deposited on the terminal lobe due to this decrease in current velocity.

In the beginning of the flood cycle, the ocean tide rises while water in main ebb channel continues to flow seaward as a result of momentum. Due to this phenomenon, water initially enters the inlet through the marginal flood channels that are the pathways of least resistance. Generally the flood channels are dominated by landward sediment transport and are floored by flood-oriented bedforms. On both sides of the main ebb channel, the swash platform is most affected by landward flow produced by the flood-tidal currents and breaking waves. As waves shoal and break, they generate a landward flow, which augments the flood-tidal currents but retards the ebb-tidal currents. The interaction of these forces acts to transport sediment in a net landward direction across the swash platform. In summary, at many inlets there is a general trend of seaward sand transport in the main ebb channel, which is countered by landward sand transport in the marginal flood channels and across the swash platform. A variation of this pattern occurs along the East Friesian Islands where the updrift swash platform consists of alternating flood- and ebb- dominant channels in which sediment moves onshore-offshore, but in an overall easterly direction coincident with the dominant wave energy flux and net longshore transport direction (Fig. 12.4, Nummedal and Penland 1981).

12.5.2 Inlet Sediment Bypassing

Along most open coasts, particularly in coastal plain settings, angular wave approach causes a net movement of sediment, which along the East Coast of the United States varies from 100,000 to 500,000 m^3/year. The manner whereby sand moves past tidal inlets and is transferred to the downdrift shoreline is called inlet sediment bypassing (Bruun and Gerritsen 1960). The primary mechanisms of sand bypassing natural inlets include: (1) Stable inlet processes, (2) Ebb-tidal delta breaching, and (3) Inlet migration and spit breaching (FitzGerald et al. 2001a). These mechanisms involve channel shifts, and the landward migration and attachment of large bar complexes to the inlet shoreline that produce a distinctive set of tidal inlet facies (Fig. 12.5).

12.5.2.1 Stable Inlet Processes

This mechanism of sediment bypassing occurs at inlets that do not migrate and have main ebb channels that remain approximately in the same position (Fig. 12.5b). Sand entering the inlet via tidal and wave processes is transported to the terminal lobe due to the dominance of ebb-tidal currents in the main channel. Swash bars form in the periphery of the delta (50–150 m long, 50 m wide) and move onshore due to the dominance of landward flow across the swash platform. The coalescence of landward-migrating swash bars forms large bar complexes that may be more than a kilometer in length and up to 3 m in height. The welding of these bar complexes to the landward beach completes the inlet sediment bypassing process. The symmetry of the ebb delta and its overlap along the inlet shoreline control the location of landward bar migrations. As seen at Price Inlet, South Carolina the welding of bar complexes to the beach is responsible for a progradation of the shoreline at specific locations (Fig. 12.6).

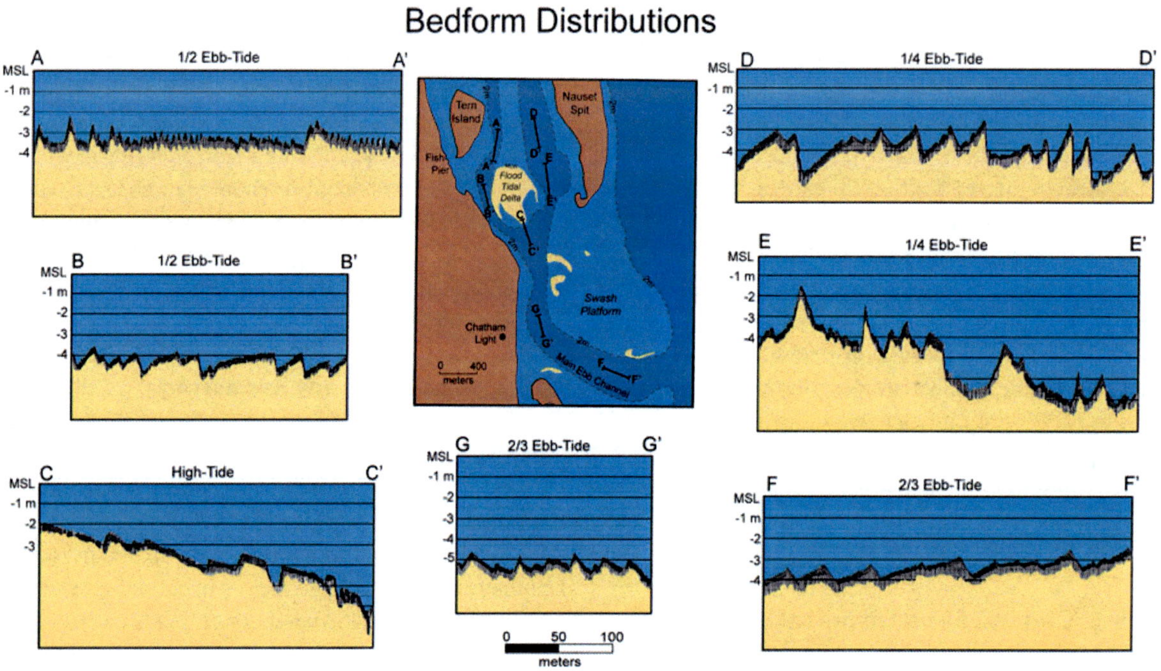

Fig. 12.4 Major pathways of sand transport at Norderneyer Seegat, East Friesian Islands (From Nummedal and Penland 1981). Sand bypasses the inlet from west to east through ebb- and flood- dominant channels

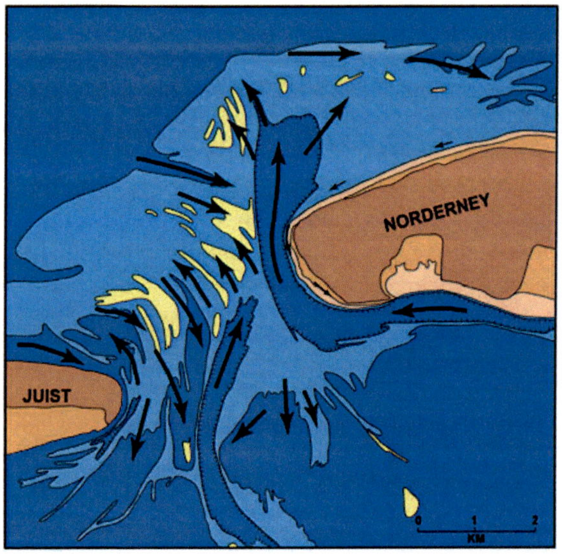

Fig. 12.5 Major pathways of sand transport at Norderneyer Seegat, East Friesian Islands (Modified from Nummedal and Penland 1981). Sand bypasses the inlet from west to east through ebb- and flood- dominant channels

12.5.2.2 Ebb-Tidal Delta Breaching

This mechanism of sediment bypassing occurs at inlets with a stable throat position but with a main ebb channel that migrates through their ebb-tidal deltas like the wag of a dog's tail (Fig. 12.5c). Sand delivered to the inlet is preferentially deposited on the updrift side of the ebb-tidal delta, which causes a deflection of the main ebb channel until it nearly parallels the downdrift inlet shoreline. This circuitous configuration of the main channel results in inefficient tidal flow through the inlet, ultimately leading to breaching, when a new channel cuts through the ebb-tidal delta. The breaching process results in a large packet of sand bypassing the inlet. This process was captured at Capers Inlet, South Carolina between 1917 and 1938 whereby a new main channel breached through a marginal flood channel followed by a 2-km long bar welding to Dewees Island (Fig. 12.7). The truncated beach ridges on Dewees Island and an adjacent deep channel, where the inlet once flowed, are evidence of this process. Truncation of beach ridges along Capers Island attests to this same process occurring at Price Inlet (Fig. 12.7, FitzGerald et al. 1978).

12.5.2.3 Inlet Migration and Spit Breaching

A final mode of inlet sediment bypassing occurs at migrating inlets whereby an abundant sand supply and a dominant longshore transport direction cause spit building and downdrift inlet migration (Fig. 12.5a). Along many coasts, as the inlet is displaced farther

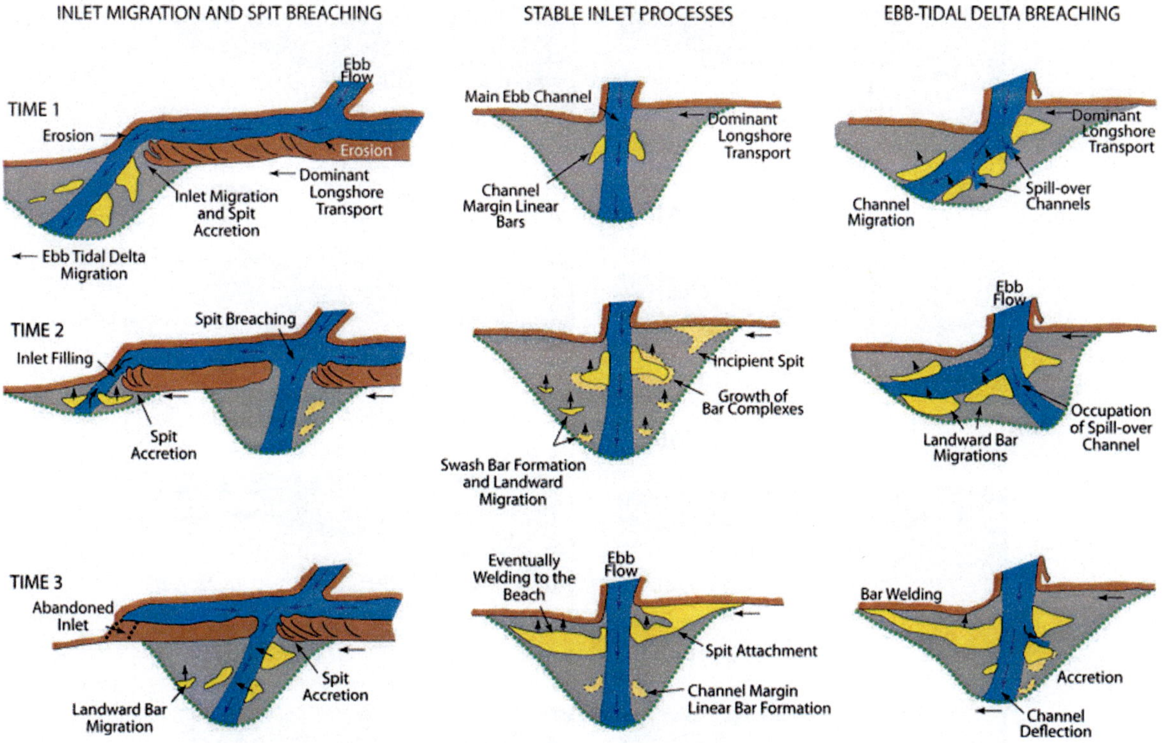

Fig. 12.6 Models of sand bypassing tidal inlets (From FitzGerald et al. 2001a)

along the downdrift shoreline, the channel to the backbarrier lengthens, retarding the exchange of water between the ocean and backbarrier. Ultimately, when the barrier spit is breached and a new inlet is formed in a hydraulically more favorable position, the tidal prism diverts to the new inlet, and the old inlet closes. When this happens, the sand comprising the ebb-tidal delta of the former inlet is transported onshore by wave action, commonly taking the form of a landward migrating bar complex. It should be noted that when the inlet shifts to a new position along the updrift shoreline a large quantity of sand has effectively bypassed the inlet.

12.5.2.4 Bar Complexes

Depending on the size of the inlet, the rate of sand delivery to the inlet, the effects of storms, and other factors, the entire process of bar formation, its landward migration, and its attachment to the downdrift shoreline may take from 6 to 10 years (Gaudiano and Kana 2001).

The volume of sand bypassed can range from 100,000 to more than 1,000,000 m^3. The bulge in the shoreline that is formed by the attachment of a bar complex is gradually eroded and smoothed as sand is dispersed to the downdrift shoreline and transported back toward the inlet. In some instances, a landward-migrating bar complex forms a saltwater pond as the tips of the arcuate bar weld to the beach stabilizing its onshore movement. Although the general shape of the bar and pond may be modified by overwash and dune building activity, the overall shoreline morphology is frequently preserved. Lenticular-shaped coastal ponds or marshy swales become diagnostic of bar migration processes and are common features at many active and relict inlets.

12.6 Stratigraphy and Facies Relationships

Early stratigraphic models of tidal inlets and their related tidal shoals were constructed using surface information consisting largely of bedform and grain size distributions, short cores and box cores, and shallow trenches. Idealized regressive or transgressive tidal inlet sequences were created using a knowledge of inlet morphology and processes and then by stacking characteristic facies from the various inlet environments in accordance with

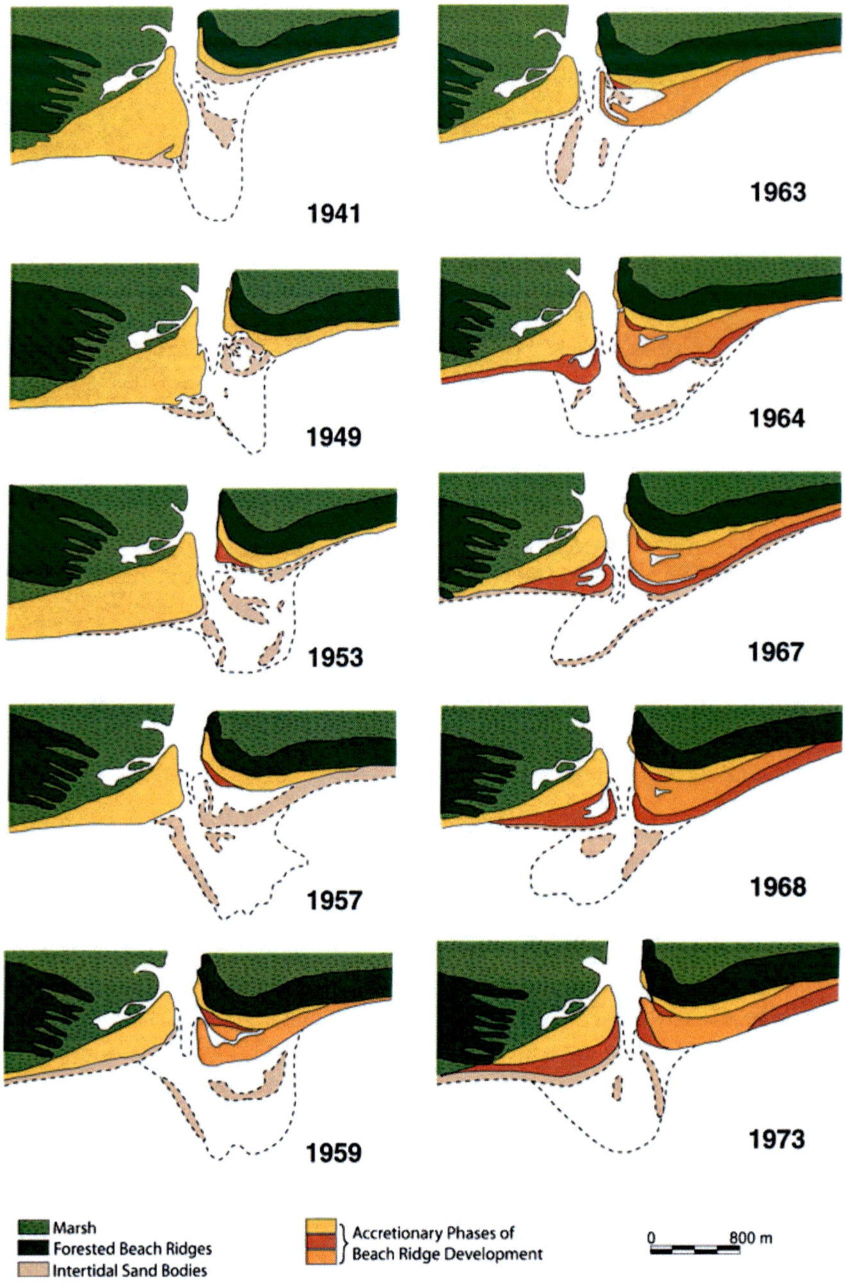

Fig. 12.7 Morphological changes of the ebb-tidal delta at Price Inlet, South Carolina illustrating how inlet shoreline erosional and depositional processes are controlled by the configuration of the ebb delta and sediment bypassing processes (From FitzGerald 1976)

Walther's Law (Kumar and Sanders 1974; Hayes and Kana 1976; Hubbard et al. 1979; Barwis and Hayes 1979; Reinson 1984; Boothroyd 1985). The limited coring studies that existed at that time produced only a rudimentary characterization of tidal-inlet fill deposits indicating a possible presence of coarse layers defining the base of the former channel (Hoyt and Henry 1967; Pierce and Colquhoun 1970). During the past 25 years, a variety of geophysical tools and advanced sediment coring techniques has led to more accurate and detailed tidal inlet sedimentary models. Tidal-inlet fill sequences and ebb- and flood- delta stratigraphy have been imaged using high-resolution shallow-seismic reflection profiling over water and ground-penetrating radar (GPR) on land, the latter offering an order of magnitude finer resolution than seismic-reflection data. The reflectors produced by these systems coincide with large-scale erosional and accretionary surfaces, thereby providing a means of documenting the sedimentation history of tidal-inlet fill sequences and tidal deltas in great detail. Cores taken in conjunction with the geophysical data provide a means of ground-truthing the interpretation of the various reflectors and produce a detail characterization of individual tidal facies. The results of several studies dealing with active tidal inlets and tidal-delta deposits as well as paleo-inlet locations are presented in this section to illustrate the types of facies architecture associated with inlet sequences including their geophysical characterization, when available.

12.6.1 Occurrence of Tidal Inlet Deposits

Tidal-inlet fill sequences are formed at inlets that close or migrate for some distances along shore. At large tidal inlets, they also accumulate where the thalweg shifts laterally within the main inlet channel, such as the inlets along the Friesian Islands (FitzGerald et al. 1984). Complete sections are preserved within regressive deposits, but partial fill deposits are also often preserved during transgressions because inlets are deep, particularly at the throat section (tidal ravinement), and usually erode far below the adjacent barrier lithosome and deeper than the transgressive unconformity produced during shoreface retreat.

As shown by Hayes (1979) and Davis and Hayes (1984), tidal inlets are more numerous and comprise greater stretches of shoreline along mixed-energy coasts and coasts having large bay tidal prisms. Thus, it would be expected that as tidal range and/or bay tidal prism increase, tidal inlet deposits will comprise a greater proportion of the Holocene lithosome. However, as pointed out by several investigators (Moslow and Heron 1978; Heron et al. 1984; Moslow and Tye 1985; Tye and Moslow 1993) wave-dominated barrier coasts can have extensive tidal inlets deposits (30–50% of the barrier length; Moslow and Tye 1985) due to the opening and closing of inlets and channel migration along the coast. A case in point occurs along the central coast of Massachusetts where the small reentrant of Plymouth Bay is fronted by two long spit systems separated by a 2.0 km-wide tidal inlet. The northern spit (13 km long) is pinned to drumlins and its length is riddled with numerous tidal-inlet scars having widths varying from 60 to 285 m. GPR transects revealed at least 18 former inlet channels that have breached the Duxbury barrier, none of which are open today. As depicted in one of these GPR transects, one of the larger paleo-inlets shoaled against a till headland (Fig. 12.8). The reflector geometry and sediment cores taken in this region suggest that inlet filling occurred in pulses whereby high energy events were responsible for transporting pebbly, cobble-rich sand into the channel, forming the strong reflectors. The intervening, more transparent reflectors correspond to periods of lower energy conditions when sand units were deposited (FitzGerald et al. 2001b).

The position of tidal inlets along a coast is commonly stratigraphically-controlled in coastal plain settings and bedrock- or topographically- controlled along glaciated and rocky coasts (i.e., coasts of Oregon, Washington, NSW Australia, and in New Zealand). Many inlets coincide with Pleistocene or younger river valleys as reported along the central East Coast of U.S. (Morton and Donaldson 1973; Halsey 1979; Tye 1984), New England (FitzGerald 1996), and the Friesian Islands (FitzGerald and Penland 1987). Presumably, tidal inlet channels stabilize at these sites due to structural controls or the relative ease in which tidal currents can erode former riverine deposits. Thus, tidal inlet fill sequences may cut through, or be nested within, fluvial sequences; the two types of deposits can be differentiated on the basis of grain size and/or fossil content, or through geophysical imaging. For example, a GPR profile along central Plum Island in northern Massachusetts reveals a former channel cut that is more than 100 m wide, extends from −6 to −13 m in depth, and is overlain by a 6 m-thick tidal inlet fill sequence. This channel aligns perfectly in a landward direction with the Parker River and seaward with a channel system that has been imaged in offshore shallow seismic reflection data (Fig. 12.9, Hein et al. 2007). This deep channel that was subsequently occupied by a tidal inlet has been interpreted to be part of the paleo-drainage formed during the transgression following deglaciation of this region (Hein et al. 2011).

12.6.2 Inlet Fill Sequences

The size, geometry, and facies characteristics of tidal-inlet fills are dependent on a number of factors that

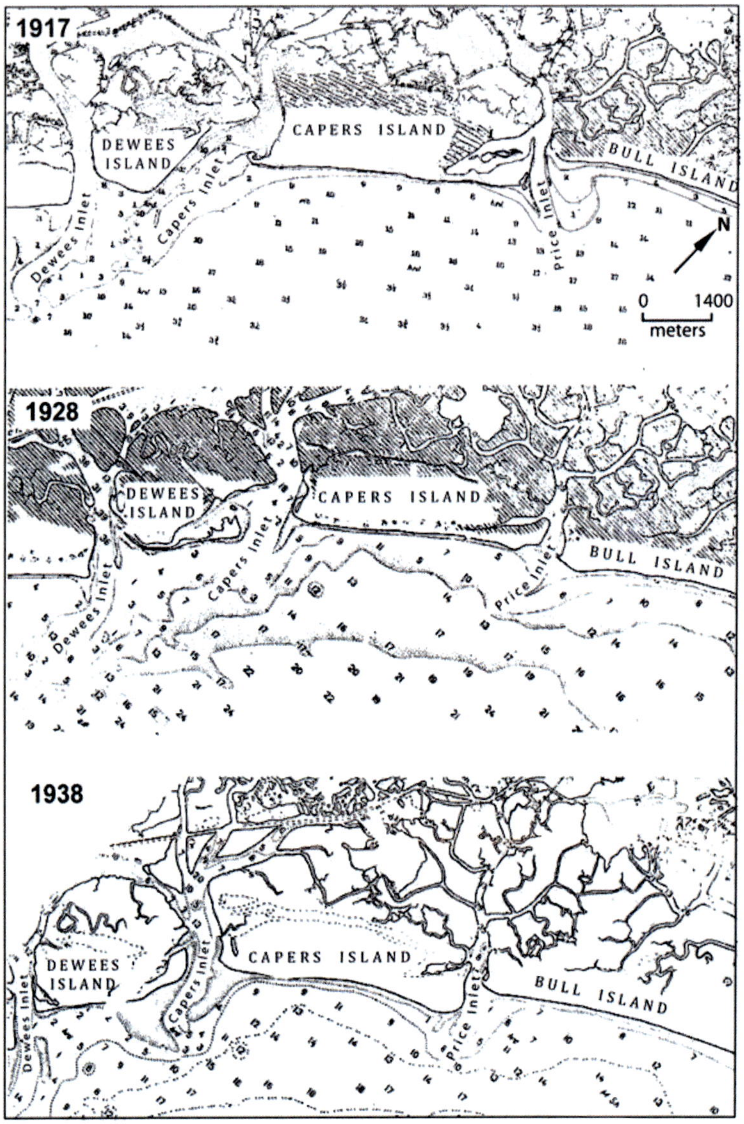

Fig. 12.8 U. S. Coast and Geodetic Survey coastal charts illustrate shoreline changes and processes of ebb-tidal delta breaching at Capers and Price Inlets, SC (From FitzGerald 1988). Note the large bar that attached to Dewees Island (1928) after a new channel was breached through the ebb delta at Capers Inlet (1917). The same breaching process created the closed channel along the northwest end of Capers Island (1917)

define the dimensions of the inlet system, migrational behavior of the inlet channel, and conditions under which the inlet fills with sediment. The size of the inlet channel is controlled by tidal prism that is a function of bay area and bay tidal range. Regions with large to moderate tidal ranges (>2 m) and expansive open water backbarrier areas tend to have large tidal prisms, producing large, deep tidal inlets (d>8 m; i.e. barrier coasts of southern Virginia, South Carolina, Georgia, East and West Friesian Islands, Copper River Delta barriers, some of the Algarve inlets in Portugal), whereas microtidal coasts with small bay areas produce small shallow inlets (<6 m deep; i.e., northern South Carolina; southern North Carolina; Rhode Island; much of the east and west coasts of Florida; Nile River delta). As pointed out by Davis and Hayes (1984), large tidal inlets can also occur along microtidal, wave-dominated coasts, if the inlet connects to a large bay and accesses a large tidal prism (i.e. Barataria Pass, Louisiana, d = 18 m; Pensacola Bay Entrance, Florida, d = 10 m; Beaufort Inlet, North Carolina, d = 10 m, pre-dredging). Likewise, small tidal inlets occur in mesotidal regions where the inlet drains a small tidal basin (i.e. Captain Sam's Inlet, South Carolina; d = 5 m). Thus, the dimensions of inlet deposits along any coast are variable and highly dependent

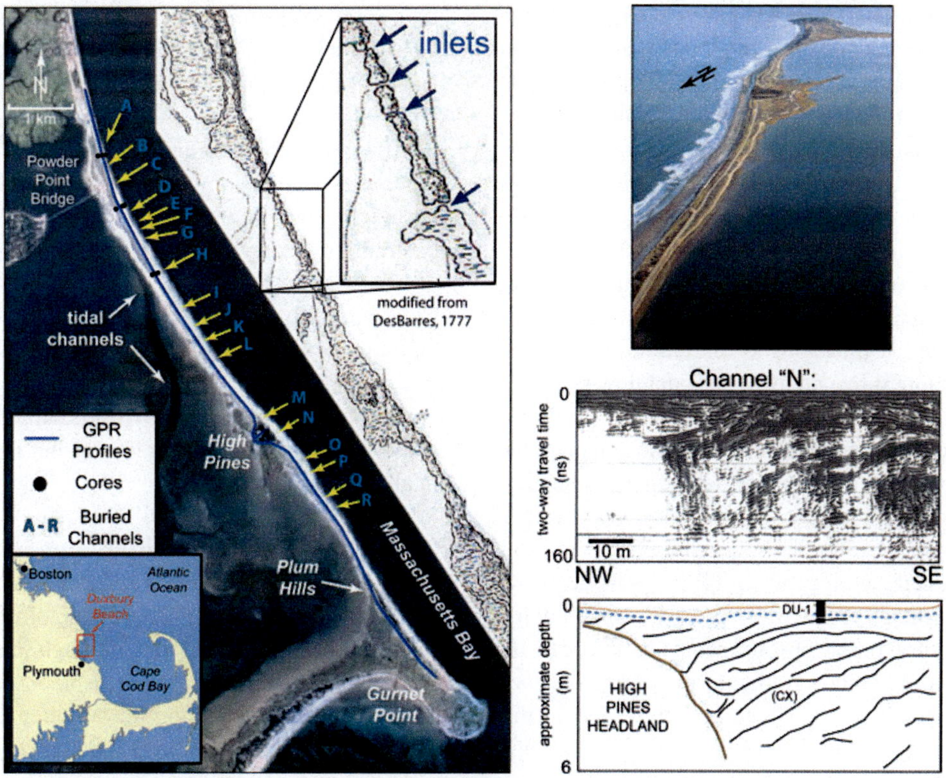

Fig. 12.9 Distribution of former tidal inlets along Duxbury Beach, Massachusetts. A GPR profile illustrates the manner in which an inlet channel gradually filled against a glacial headland (From FitzGerald et al. 2001b)

on the tidal prism and how easily the backbarrier tidal prism is accessed through time.

The geometry and facies characteristics of inlet deposits are also dictated by the dynamics of the inlet channel. Three scenarios of inlet behavior are presented here that build on the pioneering work of many previous authors (Bruun 1966; Kumar and Sanders 1974; Moslow and Heron 1978; FitzGerald et al. 1984, 1988; Tye 1984; Moslow and Tye 1985; Tye and Moslow 1993). These dynamic models provide a framework for developing stratigraphic models for inlet fills and identifying them in the sedimentologic record:

1. Migrating inlets: As the downdrift side of the inlet erodes, the channel fills with sand due to progradation of the updrift spit and thus the strike section of the deposit is dependent on the distance that the inlet migrates along the coast and the depth of the channel. In GPR and shallow-seismic reflection profiles, these deposits are characterized by large-scale, steep- to moderately- dipping, accretionary surfaces oriented in a downdrift direction. In dip sections the deposit is thickest at the inlet throat and thins in both a seaward and landward direction.
2. Breaches: Inlets that are cut during storms (called hurricane passes in the Gulf of Mexico and breaches elsewhere) and later close, produce a fill sequence that is formed through spit accretion from one or both sides of the inlet. Filling may also involve the landward migration of a large bar that closes off the inlet mouth. Subsequent overwash activity fills the channel and/or tidal or wave-induced circulation in the lagoon contributes fine-grained sediment. Thus, the channel fill may consist predominantly of sand, or sand can be mixed or inter-layered with mud. In certain instances, a clay plug may comprise the inner abandoned channel sequence if the influx of sand is prevented due to barrier reconstruction in front of the breach. GPR and shallow-seismic strike sections of these deposits exhibit a variety of channel geometries- including simply conformable, prograded, accretionary, and complex (terminology after Hine et al. 1979; Mitchum et al. 1977).

3. Channel Reorientation: Tidal inlets of this type are characterized by a deflection of the main ebb channel due to the preferential accumulation of sand on the ebb-tidal delta caused the dominant longshore transport system, such that the channel becomes skewed along the downdrift shoreline. The inlet channel can erode deeply into the adjacent barrier as depicted in Fig. 12.7 at Capers and Price Inlets, SC. In this configuration, the low hydraulic gradient of the channel and inefficient tidal exchange eventually cause a breaching of the ebb-tidal delta to effect a more direct pathway for tidal flow (see section on ebb-delta breaching, 12.5.2.2). Initially active tidal deposition characterizes the channel fill until a collapse of ebb-tidal delta and the ensuing landward-migrating bar complex closes the former inlet mouth, resulting in lower energy currents and finer-grained sedimentation (Tye 1984; Moslow and Tye 1985; Tye and Moslow 1993). The inlet fill consists of a fining upward sequence that can be dominated by mud, if the former inlet channel closes completely and there is ample mud in the backbarrier system. This type of succession is commonly capped by marsh deposition (Tye 1984).

In light of the above models and the results of numerous field investigations discussed below, it has been shown that tidal-inlet deposits can be identified within barrier lithosomes, in backbarrier deposits, on continental shelves, and in the rock record by one or more of the following characteristics (Moslow and Heron 1978; Tye 1984; Moslow and Tye 1985; Hine and Snyder 1985; Imperato et al. 1988; Israel et al. 1987; Siringan and Anderson 1993; Tye and Moslow 1993; FitzGerald et al. 2001b; Rieu et al. 2005; Simms et al. 2006):

1. A sharp contact between the base of the inlet channel and underlying strata, which in deep inlets is commonly a scoured Pleistocene surface. In shallow inlets, the base of the inlet is cut into the barrier lithosome and into the shoreface in a seaward direction or into lagoonal deposits in a landward direction.
2. Strike sections of inlet fills imaged in shallow-seismic and GPR profiles exhibit a range of reflector geometries that include various types of single and multiple channel cut-and-fills and/or repetitive sigmoidal-oblique reflectors dipping downdrift that are formed by a migrating inlet and spit system.
3. A general fining-upward sequence that is in contrast to the coarsening-upward grain size trend of most barrier lithosomes (due to their being underlain by either lagoon or nearshore fined-grained sediments). Considerable variability (fining and coarsening trends) can exist within units comprising inlet fills, including layers of mud.
4. The base of the inlet sequence usually consists of a lag deposit and is composed of poorly-sorted shell hash, whole and broken shells, heavy-mineral concentrations, rip-up clasts, coarse sand, and/or fine gravel.
5. Tidal-inlet fill exhibits a wide variety of sedimentary structures, including planar- and trough- cross bedding, graded beds, mud laminations and mud drapes, and shell hash layers. Bioturbation is rare in active sandy channel fill sequences but is common in abandoned inlet fills.

12.6.2.1 Field Studies of Tidal-Inlet Sequences

As presented above, there are several criteria to recognize tidal-inlet sequences in the sedimentologic record, but it should be emphasized that although inlet fills do exhibit some commonality, such as basal sharp contacts, lag deposits, and fining-upward trend, there is also a great deal of variability in the tidal-inlet sequences as will be illustrated in the following discussion.

Texas Coast

A number of stratigraphic studies along the Texas coast shed light on the sedimentological and geophysical facies typifying tidal inlets and tidal deltas along a microtidal sandy barrier island chain having intervening muddy bays and drowned river valleys (Israel et al. 1987; Siringan and Anderson 1993; Simms et al. 2006). Stratigraphic information, including borehole descriptions, sediment cores, grab samples, and high-resolution shallow-seismic data were used to construct stratigraphic sections for San Luis Pass at the southwest end of Galveston Island (Israel et al. 1987, Fig 12.10) and Bolivar Pass at the northeast end of the Galveston Island (Siringan and Anderson 1993, Fig. 12.11). Both of these inlet fills incise into estuarine/bay sediments and are a product of southwesterly migration of the inlet channel, coupled with spit progradation of the updrift barrier. Inlet sequences exhibit sharp basal contacts consisting of shell lags. Inlet fills are composed of fine shelly sand to sandy clay, and generally have graded beds and some degree of bioturbation. Shell hash layers are common and have normal grading. At San Luis Pass, shell and sand content in abandoned inlet fill deposits

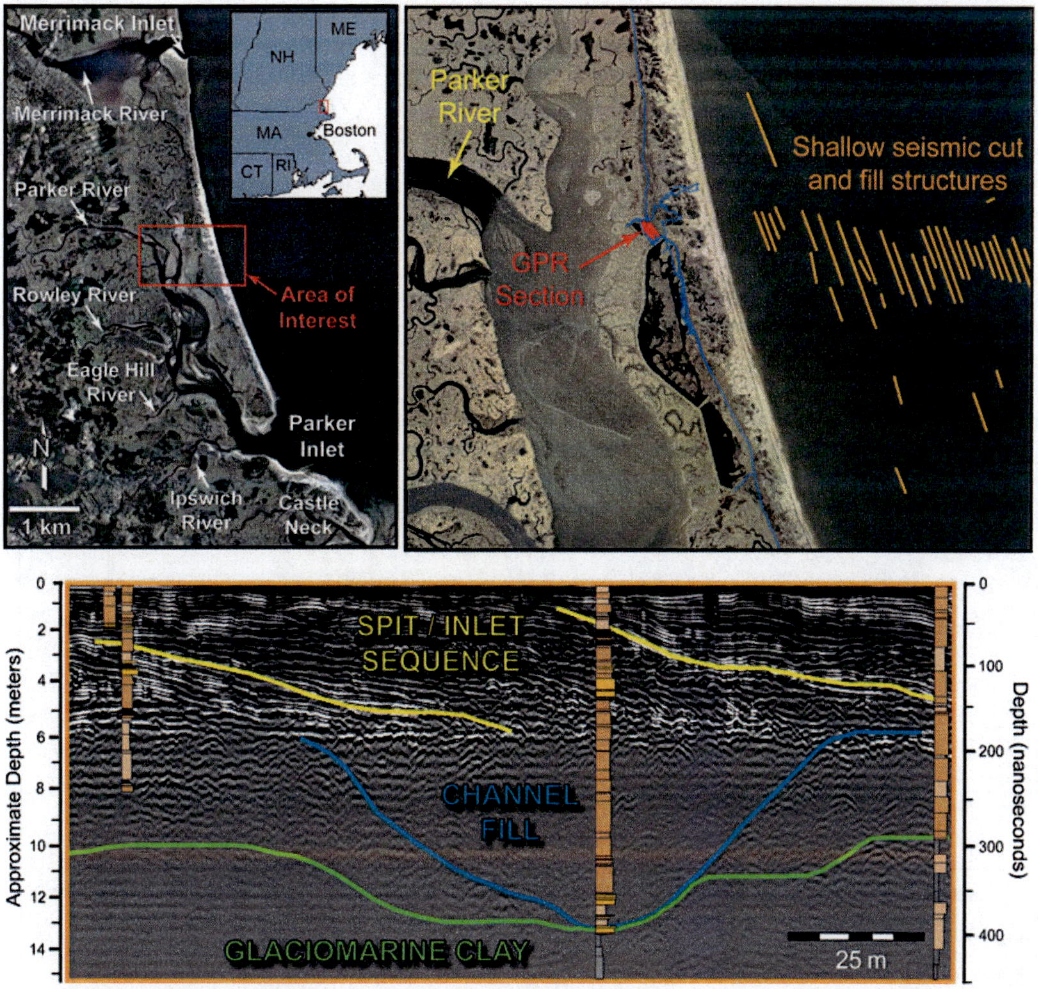

Fig. 12.10 GPR profile showing a Holocene tidal inlet fill sequence at Plum Island, Massachusetts that coincides with late Pleistocene drainage beneath the barrier. This former channel is imaged in offshore shallow-seismic records (From Hein et al. 2011)

decreases upward (Israel et al. 1987). Shallow-seismic profiles of inlet fill at Bolivar Pass reveal channel stacking and cut-and-fill structures as well as pervasive westerly-dipping reflectors (Siringan and Anderson 1993). These clinoforms dip from 0.5 to 4.6 degrees and are shallow dipping along the flanks and more steeply dipping toward the former incised inlet channel. Inlet facies exhibit a high mud content including muddy sand layers, mud lenses, mud clasts, and mud-filled burrows. The muddy character of the inlet fill has been explained by the low wave energy and the contribution of mud to the coastal waters from associated estuaries (Israel et al. 1987; Siringan and Anderson 1993). In contrast, stacked tidal- inlet-fill sequences at Mustang Island (10–15 m thick) consist almost entirely of clean, well-sorted, fine quartz sand. This sequence lacks shell hash, mud lenses, and graded beds commonly associated with inlet deposits, because the sediment was sourced from reworked mature barrier island sand devoid of shells (Simms et al. 2006). These studies of the Texas coast illustrate the importance of sediment source dictating the character of the inlet fill deposits.

North and South Carolinas

The stratigraphy of the Outer Banks of North Carolina is known from numerous coring studies (Moslow and Heron 1978; Susman and Herron 1979; Tye and Moslow 1993; Culver et al. 2006). This body of work provides a basis for characterizing inlet deposits along

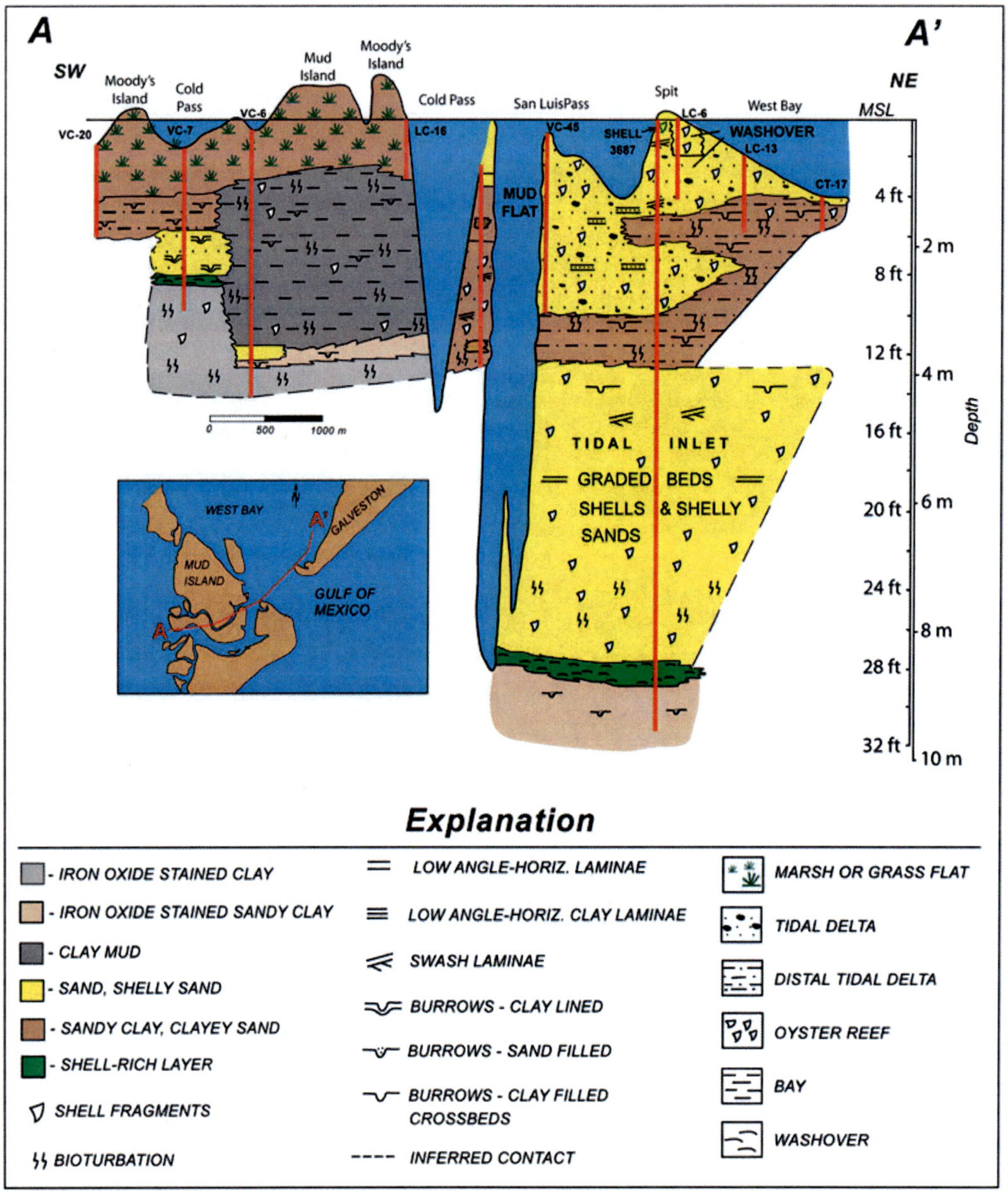

Fig. 12.11 Stratigraphic section for San Luis Pass, Texas based on 10 vibracores. The section shows 8.5 m of inlet fill containing shelly sand and muddy sand (After Israel et al. (1987)

wave-dominated coasts typified by tidal inlets that open during major storms and close during post storm recovery periods, or ones in which the inlet channel migrates several kilometers along the coast. For example, Shackleford Banks, located just west of Cape Lookout, is 14 km long and has tidal inlet sediments that extend to depths of 9–20 m and underlie at least 11 km of the barrier (Susman and Herron 1979). A core through the western end of the island shows a typical tidal-inlet sequence consisting of a coarse, pebbly sand and shell lag sitting atop an erosional contact with the underlying Pleistocene units. The coarse sediment lag is overlain by a 11-m thick, planar to cross-bedded, poorly sorted, fining upward (medium to fine sand) sequence that is topped by spit platform and dune sand (Moslow and Tye 1985, Fig. 12.12).

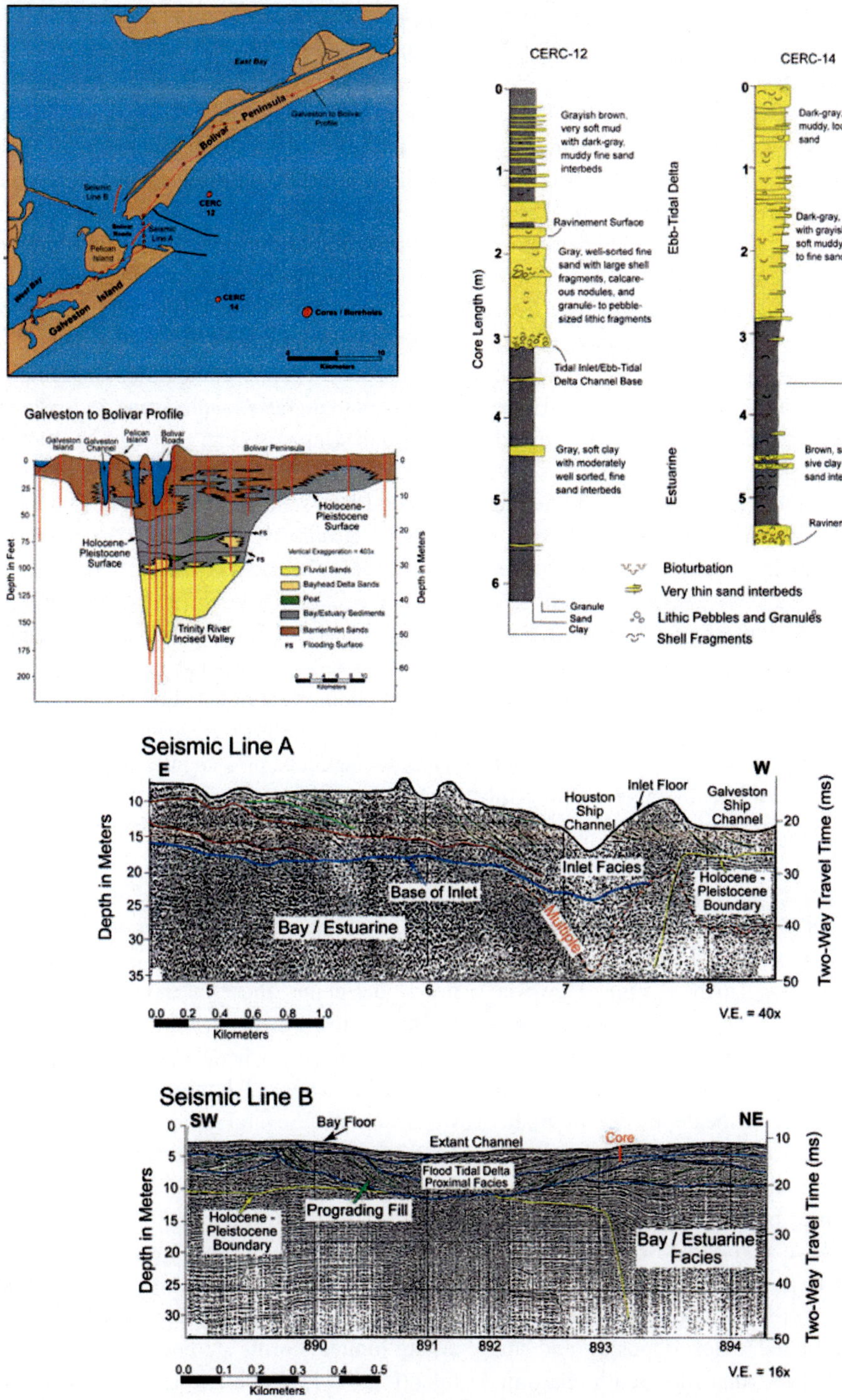

Fig. 12.12 Stratigraphy of tidal-inlet deposits at Bolivar Pass, TX (After Siringan and Anderson 1993). Cores CERC 12 and 14 demonstrate that the ebb delta contains considerable mud in addition to sand. Seismic transect A illustrates the large-scale cut and fill reflectors produced by channel migration

Moslow and Tye (1985) report that wave-dominated barrier island coasts are composed of 30–50% tidal inlet sediments. In strike section, inlet deposits are lenticular to wedge-shaped whereas dip-sections thin in both a landward and seaward direction, interfingering with flood and ebb-tidal delta sediments, respectively. The size and dimension of these inlet deposits are largely a product of tidal prism and geological factors that control channel deepening.

Central South Carolina is a mixed-energy coast having relatively short, stubby barrier islands, numerous tidal inlets, well-developed ebb deltas and a back-barrier consisting of marsh and tidal creeks (Hayes 1975, 1979). Although South Carolina "mixed-energy" inlets tend to be stable, some of them have had histories in which the seaward portion of the inlet channel is deflected downdrift and then breaches back to a straighter, more hydraulically efficient, channel configuration (FitzGerald 1988, Figs. 12.5c and 12.7). Inlet sequences formed by these processes have been studied by Moslow and Tye (1985) and Tye and Moslow (1993) and are presented for Price and Capers Inlets in Fig. 12.13. These types of inlet deposits are more complex than the wave-dominated examples, because, in addition to active channel-fill sand, they contain inactive channel deposits, welded-bar and washover facies, and tidal creek sediments. As depicted in stratigraphic sections in Fig. 12.3, when the inlet channel migrates and erodes the downdrift barrier, active inlet fills are deposited from wave and tidal introduction of sand to the channel. However, after the breaching event occurs and the former inlet mouth is closed by a landward-migrating bar complex (Figs. 12.5c and 12.7), the strong reduction in wave and tidal energy lead to the deposition of muddy sediment (abandoned inlet channel fill). In the case of Price Inlet, this migration and breaching process has occurred twice, leaving behind two active and inactive fill sequences (Tye and Moslow 1993). These units are topped by tidal creek and marsh sediment. The most obvious difference between the North Carolina wave-dominated inlet deposits and the mixed-energy examples from the South Carolina coast is presence of muddy, inactive channel sediment. It should also be recognized that some mixed-energy tidal inlets migrate and produce sedimentary sequences similar to those of North Carolina, such the East Friesian Island inlets (Sindowski 1973; FitzGerald 1996).

Virginia-North Carolina Barrier Coast

Although sedimentological information and, to a lesser extent, shallow-seismic reflection data, have been the traditional means of defining the extent and character of tidal-inlet deposits, GPR is increasingly used on land because subsurface GPR reflections are capable of imaging very subtle lithological transitions, such as slight textural and compositional variations. GPR profiles provide a means of deciphering the mode of channel filling including episodes of channel cut and fill. Historical Old Currituck Inlet, (the original coastal boundary between Virginia and North Carolina) offers an example of a large channel that migrated south during the sixteenth to early eighteenth centuries (Fig. 12.14). The 10-m-deep inlet eventually closed, primarily as a result of a new channel opening to the south and capturing the tidal prism of the Currituck Sound (McBride 1999). The geophysical image of the inlet channel can be used to approximate the channel cross-section and calculate the tidal prism, which in this case exceeds 10 million cubic meters and is comparable to modern large oceanic inlets. Based on successive channel outlines, represented by strong GPR reflections, the tidal prism of the Old Currituck Inlet decreased by an order of magnitude prior to inlet closure.

12.6.2.2 Field Studies of Ebb- and Flood- Tidal Delta Deposits

The size and geometry of ebb-tidal deltas are a function of the inlet tidal prism and wave versus tidal energy of the coast. Although tidal prism also affects the extent of flood-tidal deltas, these interior shoals are primarily a product of the dimensions of the lagoon and frequency of major storms, when large quantities of sand are delivered into the backbarrier due to increased tidal and wave activity. Several field studies provide insights concerning the facies relationships of these deposits.

Central South Carolina

Ebb-tidal deltas in South Carolina are wedged shaped deposits thickening in a landward direction and dominated by sand. Although field investigations of ebb deltas are limited, coring studies of deltas at North Edisto Inlet (Fig. 12.15, Imperato et al. 1988) and Breach Inlet (Fig. 12.16, Nelligan 1983) as well as historical migrational trends of inlet main channels suggest that these deposits range in thickness from 5 to > 15 m. Their stratigraphy largely reflects processes of migration of the main ebb channel and reworking of the delta, formation and abandonment of marginal flood channels, landward

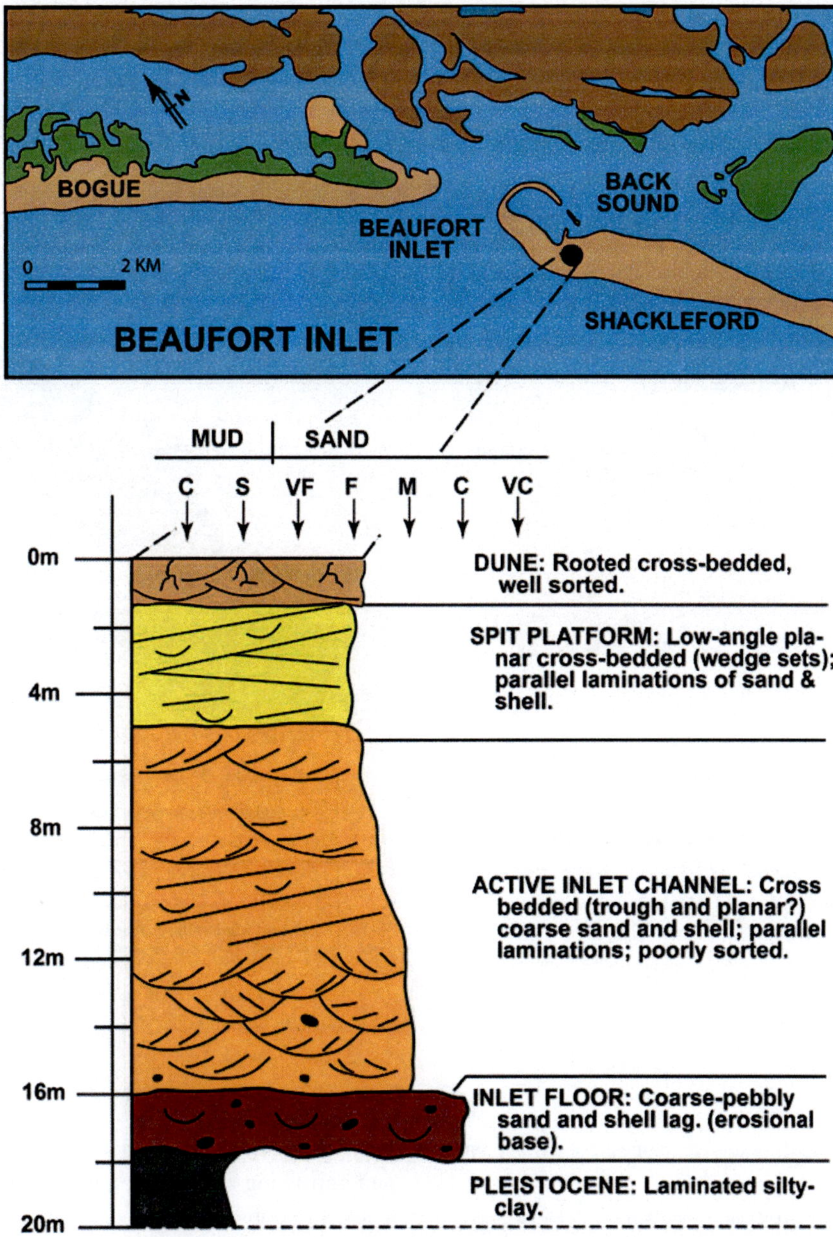

Fig. 12.13 Core log from the western end of Shackleford Banks, NC. Note the coarse lag deposit defining the base of the channel and the overall fining upward inlet sequence (From Moslow and Tye 1985)

migration of individual swash bars and large bar complexes (amalgamated swash bars), and wave shoaling across the swash platform.

North Edisto Inlet is a large inlet, approximately 1.0 km wide, with a well-developed ebb-tidal delta that extends 7 km offshore. Imperato et al. (1988) divided the ebb delta facies of North Edisto into three regions (Fig. 12.15):

1. The delta adjacent to the barriers is dominated by marginal-flood channel deposits (4–5 m thick) consisting of a basal shell-rich coarse sand fining upward into landward-oriented, interbedded, planar to cross-bedded fine to medium sand. These active channel deposits grade upward into a bioturbated, muddy sand with flaser bedding, representing low energy channel fill. This sequence is topped by

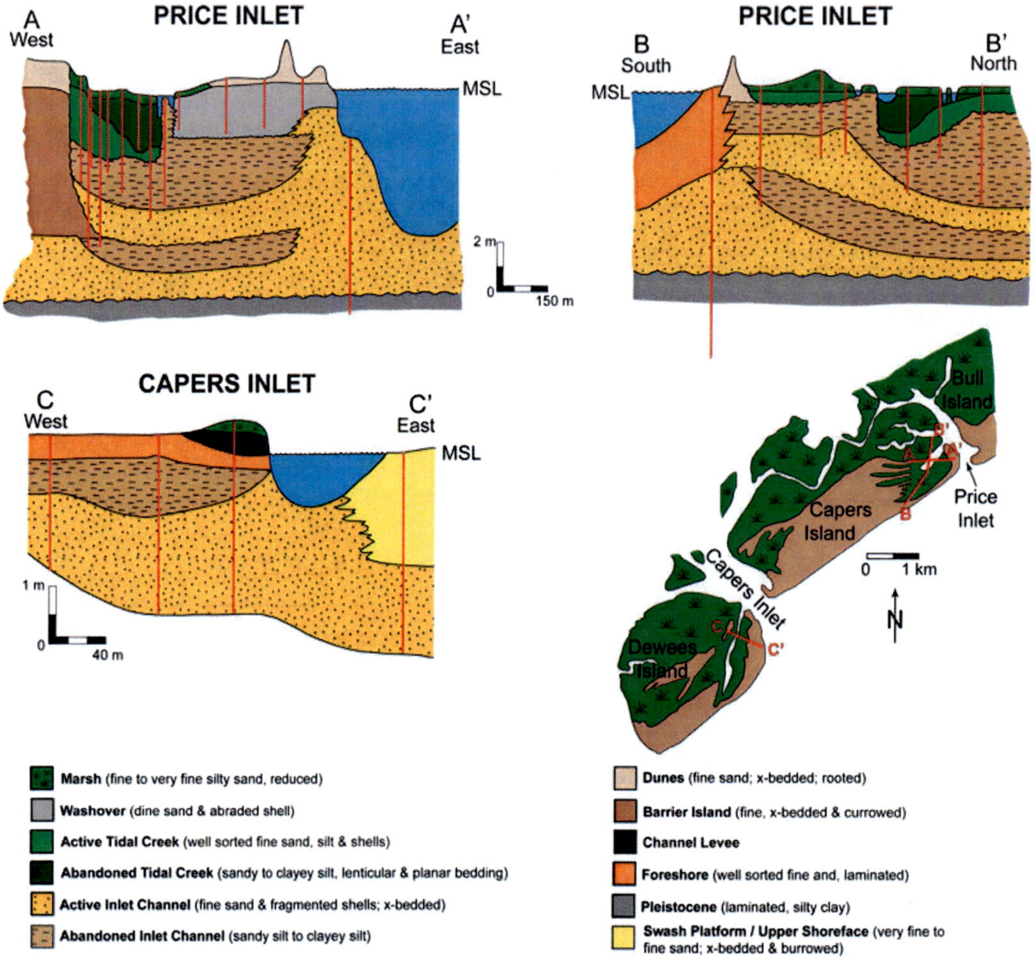

Fig. 12.14 Stratigraphic sections for Price and Capers Inlets, SC (From Tye and Moslow 1993). Note that these inlets did not migrate; rather their main channels were deflected south due to the dominant southerly longshore transport and then breached back to a straight channel course

landward-dipping foresets produced by swash bar migration.

2. A proximal delta region is dominated by a sedimentary sequence produced by the shifting of the main ebb channel. These deposits are up to 20 m thick and have a sharp contact with the underlying Pleistocene sediments. The basal units consist of seaward dipping, cross-bedded, medium to coarse sand with shells grading upward into a well-sorted, planar-bedded fine sand.
3. Distal delta deposits are 1–4 m thick and consist of an overall coarsening upward sequence that interfingers with seaward shoreface sediments. The wave-dominated platform is characterized by planar-bedded to landward-oriented, cross-bedded, very well-sorted, fine sand. Coarse shell hash layers and burrowing are common attributes.

Thus, the ebb-delta deposits exhibit a sharp basal contact with the shoreface and an overall fining-upward sequence except for the distal portion of the delta that coarsens upward. Local coarsening of the sediment may also occur due to the migration of tidal channels or the onshore movement of swash bars.

Nelligan's (1983) study at Breach Inlet, north of Charleston Harbor, SC, recorded the stratigraphy of an ebb delta during a phase of tidal-channel abandonment and encroachment of landward migrating bar complexes (Fig. 12.16, see also Fig. 12.5a). The delta lithosome contains a tidal-channel fill sequence that is similar to that of North Edisto adjacent to the barrier

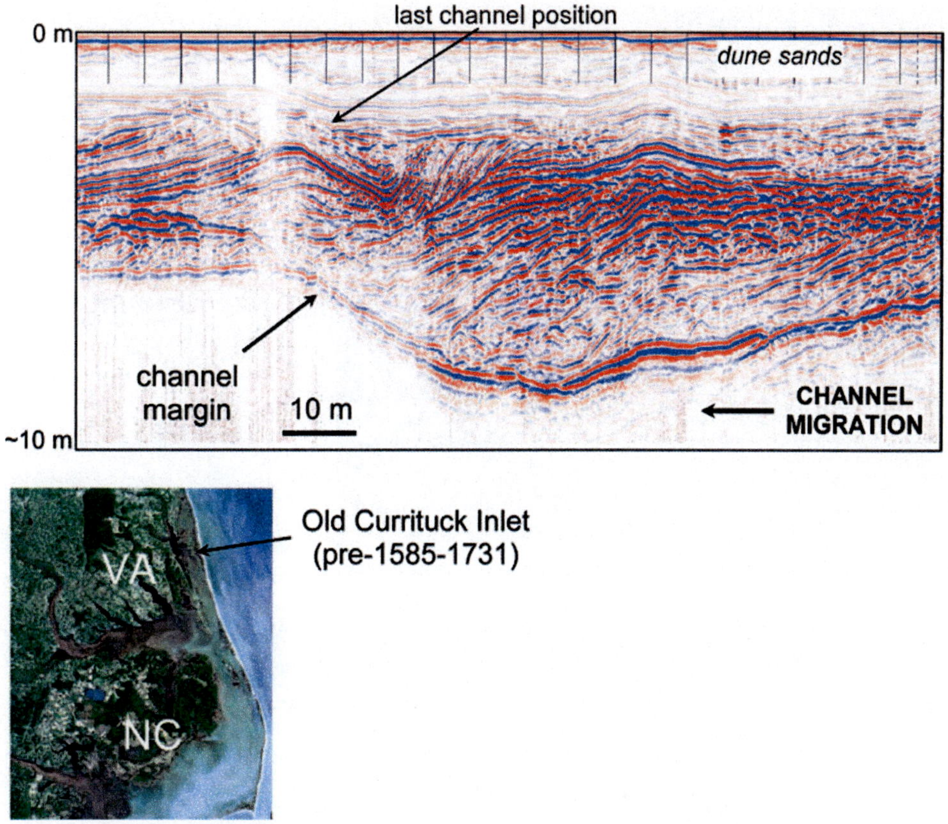

Fig. 12.15 Ground-penetrating radar transect across Old Currituck Inlet, NC. The nested channel sections suggest different periods of excavation and filling (From McBride 1999, McBride et al. 2004)

environment, including a basal coarse sand and shell lag that is 10–15 cm thick, unconformably overlying very fine shoreface sand. Sitting on top of this basal unit are active channel sediments, consisting of clean, horizontal to cross-bedded medium sand with few biogenic structures and inactive channel deposits composed of a highly burrowed, poorly-sorted, medium to fine sand with numerous mud laminations (Nelligan 1983). Overlying the channel fill deposits is a flood platform facies made up of medium to fine sand completely devoid of bedding due to intense bioturbation. As the bar complexes migrate onshore, the entire channel sequence will be capped by a relatively coarse facies consisting of shallow- to steeply- dipping beds of medium to coarse sand having a high shell content.

Baratraia Pass, Louisiana

In contrast to the sand-rich tidal deltas of the U.S. East and Gulf Coasts, and elsewhere in the world (i.e., Copper River Delta, Alaska; Algarve Inlets, Portugal; Friesian Inlets, North Sea), much of the western and central U.S. Gulf Coast deltas contains very muddy deltas. One such system is the ebb-tidal delta of Barataria Pass that is expanding due to backbarrier wetland loss and coincident increasing bay tidal prism (Fig. 12.17). As seen in the longitudinal section of Barataria Pass (Fig. 12.17c), while the inlet throat enlarged between the 1880s and 1980s, the delta prograded seaward approximately 2 km during the same time (List et al. 1994). The sediments comprising the ebb delta coarsen upward and are composed of a proximal facies of up to 25% mud and a distal facies of up to 50% mud (FitzGerald et al. 2004). The proximal facies (1–3 m thick) occurs on both sides of the ebb channel and contains massive to laminated fine sand with mud layers and is highly bioturbated. The distal facies interfingers with landward proximal facies and seaward with shelf sediment. It is relatively thin (0.4–1.2 m thick) and consists of thinly laminated, bioturbated, muddy sand with shell hash layers. The high mud content of this

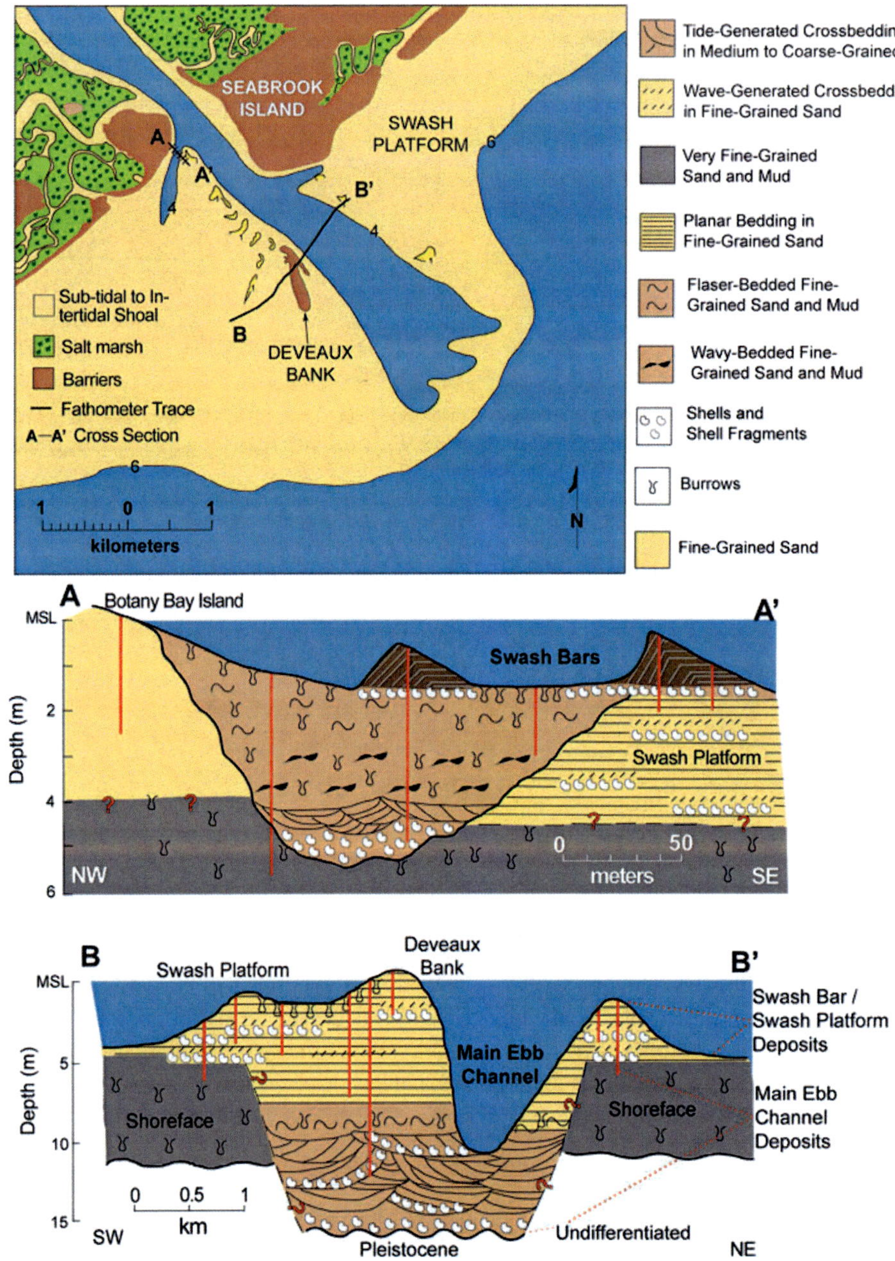

Fig. 12.16 Stratigraphy of North Edisto Inlet, SC based on sediment cores and shallow seismic data (After Imperato et al. 1988). Shelly sand layers define the bottom of channels as well as the base of landward migrating bars

delta is very similar to Bolivar Roads delta (see Fig. 12.11; Siringan and Anderson 1993, Rodriguez et al. 1998). The coarsening upward nature of the Barataria delta sequence is explained by the long-term increase in tidal energy resulting from the enlarging tidal prism (FitzGerald et al. 2004).

The similarities in facies architecture of the Texas and Barataria coast ebb deltas result from similar low-energy hydrographic regimes and the same overall muddy character of the surrounding sedimentary environments. These depositional systems contrast with sand-rich ebb-tidal deltas in which the stratigraphy is

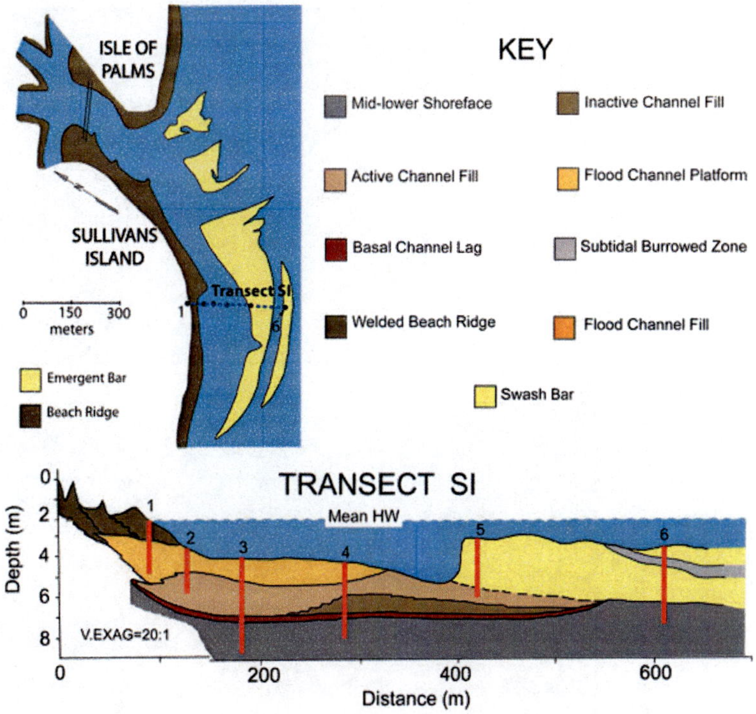

Fig. 12.17 Stratigraphic section across the ebb-tidal delta of Breaches Inlet, SC. The delta sequence is a product of the southerly migration and subsequent abandonment of the main ebb channel (After Nelligan 1983)

dominated by channel cut-and-fills, large-scale landward-dipping foresets produced by onshore swash bar migrations and shallow dipping strata (FitzGerald 1976; FitzGerald and Nummedal 1977; Hubbard et al. 1979; Imperato et al. 1988; Sha 1990a; Sha and de Boer 1991; Smith 1991). Generally, these sandy ebb deltas are subjected to more energetic waves and tides and, therefore, there is little opportunity for mud deposition.

Low Energy Coasts

Flood-tidal deltas are common along microtidal and mesotidal coasts in which there is sufficient open-water area for sand to accumulate landward of a tidal inlet. Intertidal exposure and thickness of flood deltas depend upon sedimentation rates and accommodation space. The Rhode Island coast contains a series of lagoons having multi-lobate deltas that began forming circa 2.5 ka in response to rising sea level and tidal current generation (Boothroyd et al. 1985). Boothroyd et al. (1985) show that Ninigret Pond contains a stacked sequence of landward accreting flood delta lobes separated by silty organic layers, including a sharp basal contact with low-energy lagoon sediments (Fig. 12.18). Delta lobes consist of medium to coarse, horizontal to cross-bedded sand with shell hash layers. Individual lobes are 0.5–1.8 m thick and the entire delta sequence is slightly more than 3 m thick (Boothroyd et al. 1985).

Similar to the Rhode Island examples, along Florida's Gulf Coast flood-tidal deltas are a few meters thick, have a sharp basal contact with muddy lagoon deposits, are often multi-lobate, and are composed of quartz sand with shelly layers, particularly concentrated in channelized regions (Davis et al. 2003). The flood delta associated with Shinnecock Inlet contains a lower unit consisting of a muddy fine sand that grades upward to a proximal delta facies composed of cross-bedded to massive medium sand with numerous shell layers (Hennessy and Zarrillo 1987). The coarsening-upward sequence is a product of increasing tidal energy associated with the opening of Shinnecock during the 1938 Hurricane.

These field investigations show that flood deltas are single- or multi-lobate and can be stacked depending upon the accommodation space. Typically, they are bedded and composed of medium to fine sand with shell layers, having a sharp to gradational contact with muddy lagoonal sediments.

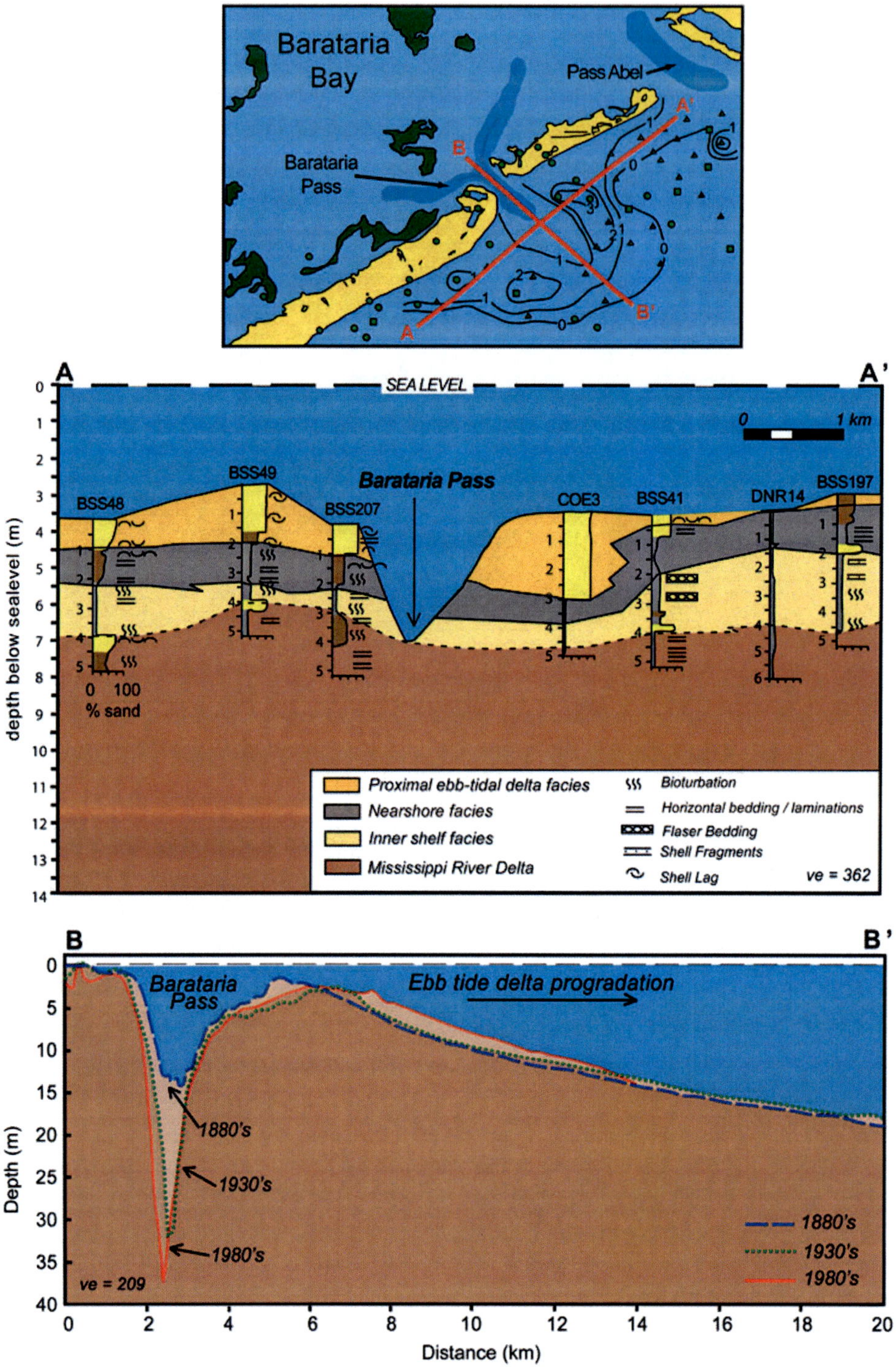

Fig. 12.18 Stratigraphic and historical morphological changes of the Barataria Inlet and ebb delta system along the coast of Louisiana. Section A-A[1] portrays a stratigraphic strike section across the *middle portion* of the delta and B-B[1] demonstrates how the inlet throat and channel and responded to increasing tidal prism (After FitzGerald et al. 2004)

12.7 Preservation Potential

The preservation potential of tidal-inlet fills and tidal-delta deposits is relatively high in regressive sequences, as indicated by the extent of inlet deposits comprising barrier lithosomes and the common occurrence of marsh-covered paleo-deltas behind barrier islands. For example, a series of five stacked flood-tidal deltas was identified in the lagoon behind Mustang Island, Texas (Simms et al. 2006).

In contrast, the relatively thin nature of tidal deltas (commonly < 6 m) and moderate depth of most inlet channels and inlet fills (mostly < 10 m), particularly when compared to the depth of shoreface erosion, indicate that inlet-associated deposits are rarely preserved during a transgression. For instance, a detailed study of Onslow Bay, North Carolina showed that although the onshore barrier lithosomes contain numerous tidal inlet deposits, there is no evidence on the inner shelf of former tidal-inlet deposits, and the channels that do exist are of Pleistocene age (Hine and Snyder 1985). These authors suggested that the landward translation of the shoreface during the Holocene transgression eroded all expression of even the deepest channels. In fact, there are few reported tidal-inlet deposits on continental shelves throughout the world. Possible exceptions occur offshore of Barataria Bay in Louisiana where a channel cut-and-fill is attributed to tidal-inlet migration (Tye and Moslow 1993). Tidal-inlet fills have also been recognized in shallow-seismic transects collected on the inner shelf along the southern Delmarva Peninsula (Foyle and Oertel 1997). These channel fills, which are as much as 25 m deep and extend 25 km offshore from today's coast, are theorized to have developed as the barriers and associated tidal inlets migrated onshore during the Holocene transgression (Foyle and Oertel 1997). The present inlet systems along southern Delmarva are deep (e.g., Wachapreague Inlet > 18 m; Quinby Inlet > 23 m; Great Machipongo Inlet > 20 m) and thus, the paleo-inlets may very well have produced deep inlet scars during the transgression. A similar set of clinoforms in shallow-seismic profiles taken off the West Friesian Islands has also been interpreted to be tidal-inlet and tidal-delta deposits (Sha 1990b). He bases this interpretation on the fact the inlets are deeper (> 30 m) than the depth of shoreface erosion (15 m) during the transgression. However, sediment cores and age-dates to corroborate these seismic interpretations have yet to be taken.

A more detailed study involving a kilometer-size grid of high-resolution shallow-seismic profiles (475 km) along with 80 sediment cores was performed off the western Netherlands coast (Rieu et al. 2005). This study led to the mapping of several tidal-inlet and backbarrier drainage systems and the inferred position of a paleo-barrier island chain (Fig. 12.19). Furthermore, the geometry and migration trends of the partially preserved tidal network suggested that the barrier system and tidal inlets did not migrate landward with the transgression, but rather the barrier system must have been destroyed by rising sea level (Reiu et al. 2005). This study illustrates the type of information that can be gained from the identification and interpretation of former tidal inlet systems. However, as the authors also demonstrated, the preservation of these systems requires deep initial tidal-inlet channels. Using Zoutkamperlaag Inlet as an analogue, it is clear that little of the present channel network is preserved when shoreface erosion removes 6 m of the coastal lithosome (Fig. 12.20). Belknap and Kraft (1981, 1985) showed for the Delaware coast that all but the deepest Holocene systems (>10 m) would be removed by the present rate of sea-level rise and the studies of the Louisiana coast suggest at least a similar magnitude of shoreface reworking (Miner et al. 2009).

12.8 Examples From the Rock Record

Aside from recent inlet systems, several researchers have identified and described inlet-fill sequences and associated tidal deltas from the rock record. For example, Bridges (1976) identified a tidal inlet/ebb-tidal delta complex within Lower Silurian transgressive barrier island facies, southwest Wales. A flood-tidal delta, though not exposed, is proposed have existed in the paleo-lagoon behind the barrier-inlet sequence exposed at Anvil Bay and Maroles Sands (Bridges 1976, Fig. 12.9). In his study of a Carboniferous transgressive succession on a shallow wave-dominated shelf in the South Munster Basin of southern Ireland, MacCarthy (1987) does not refer to inlet facies, but suggests their presence in a paleogeographic interpretation (MacCarthy 1987, Fig. 12.20). It is possible that the inlets were few and their preservation limited on this wave-dominated Carboniferous shelf.

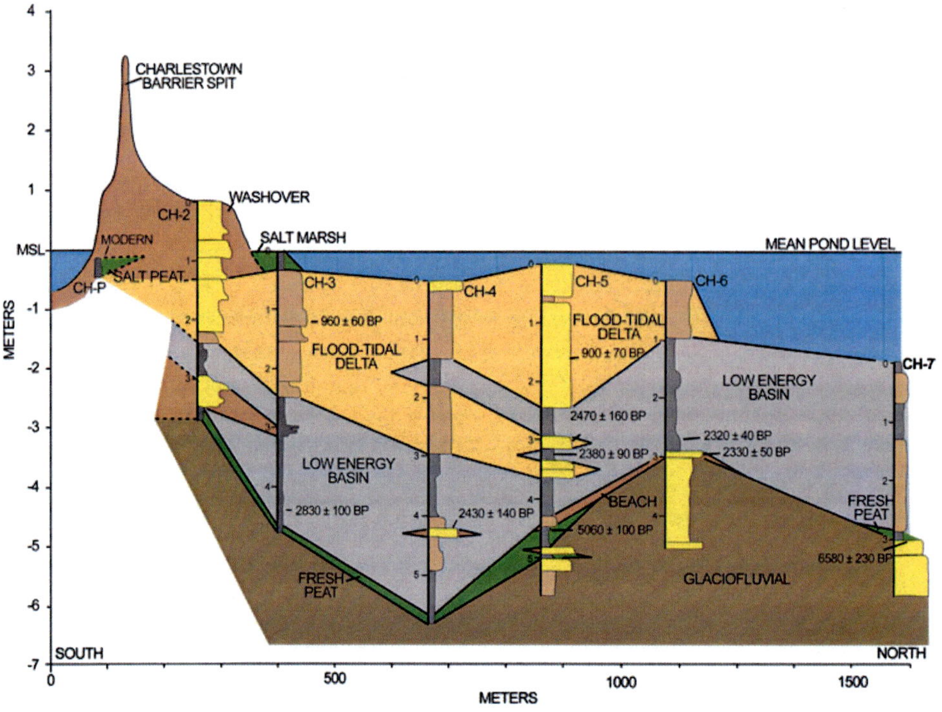

Fig. 12.19 Facies architecture of the Charlestown flood-tidal delta system on the Rhode Island coast (From Boothroyd et al. 1985)

Brownridge, and Moslow (1991) identify a Lower Cretaceous tidally-influenced estuarine and marine facies at Drayton Valley of central Alberta. Cheel and Leckie (1990) describe a tidal-inlet complex in the Upper Cretaceous Milk River Formation of southern Alberta, Canada. A flood-tidal delta complex, with flood- and ebb-dominated facies is clearly identified in the upper Virgelle Member, suggesting optimal preservation during transgression. The upper shale units with ebb-oriented rippled sandstone interbeds are interpreted as low-energy facies (Cheel and Leckie 1990) and are analogous to post-closure mud-plugs in historical inlet sequences described in the previous section. In the contemporary Upper Cretaceous Cliff House sandstone of San Juan Basin, New Mexico, Donselaar (1989) identifies stacked "transgressive" barrier complexes with distinct landward translation ("step-up") phases. In his discussion of shoreline displacement and preservation potential, a scenario of a "deeply scouring and laterally-migrating tidal inlet" is used to demonstrate the differences in the observed sandstone volume (Donselaar 1989, Fig. 12.14b). Such tidal scour (tidal ravinement) and a "step-up" of erosional shoreface ravinement surface are probably responsible for preservation of many tidal-inlet sequences.

As an early Cenozoic example of preserved tidal inlet facies, Ricketts (1991) presents a model of broad, shallow bays connected to the ocean through tidal inlets in the Lower Paleocene of the Canadian Arctic Islands. He used both facies architecture and morphological elements to distinguish these barred estuaries from deep, narrow valleys that had small inlet-spit systems or lacked tidal inlets. An increasing number of studies utilize sequence-stratigraphic principles, combined with sedimentological and ichnological studies, for detailed facies interpretations. In their study of sedimentary sequences in the paleo-Tokyo Bay, Okazaki and Masuda (1995) reconstructed a Pleistocene barrier island, tidal inlet, and tidal delta complex.

Recognition of ancient inlet and tidal delta facies is not only an important element of paleo-environmental reconstruction, but is a key target for petroleum exploration. Substantial amounts of petroleum reserves residing in channel-fill successions and associated tidal delta sequences are well known (Cheel and Leckie 1990. Brownridge and Moslow 1991)

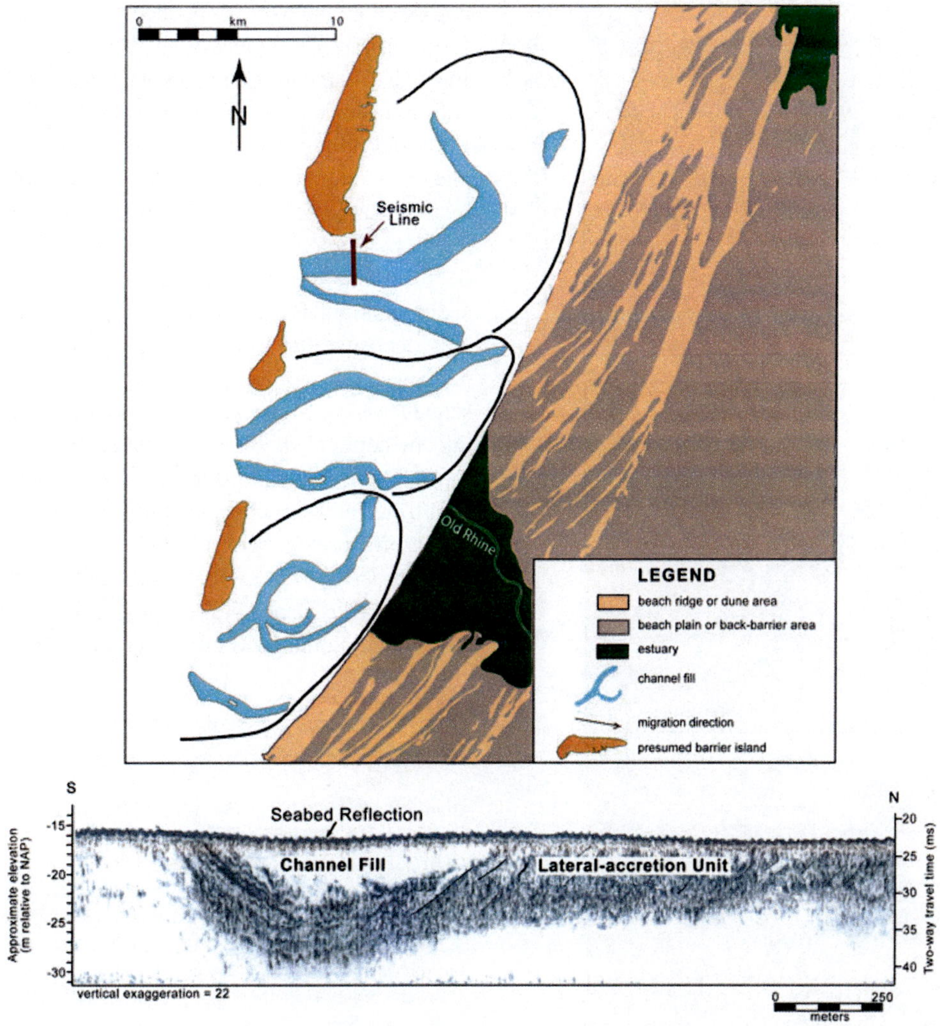

Fig. 12.20 Drainage system inferred from geophysical and sedimentological data collected offshore of the western Netherlands (From Rieu et al. 2005). Note the southerly migration of the tidal channel as seen in the seismic section

and recognition and mapping of tidal inlet complexes will remain an important part of future exploration efforts.

12.9 Summary

Tidal inlets are one of nature's most dynamic coastal systems due to their continuous response to highly variable energy vectors. Their constantly changing morphology reflects short- and long-term adjustments to storm and day-to-day processes. Their morphologic development is facilitated by: (1) the unconsolidated nature of their channel banks, which allows channel migration, and (2) sand delivery to the inlet, which leads to the construction of mobile sedimentary forms including a variety of small bedforms, large swash bars and bar complexes, and tidal deltas. At some inlets, framework geology has stabilized the inlet channel. Distortions of the tidal wave in response to inlet and bay hypsometry produce velocity asymmetries that control the volumes and pathways of sediment transport at the inlet and ultimately, the distribution of sediment to the various sand reservoirs. Sand bypassing the inlet as well as circulated within the inlet creates landward migrating bar complexes on the swash platform that weld to the landward shoreline. Sediment bypassing events and channel

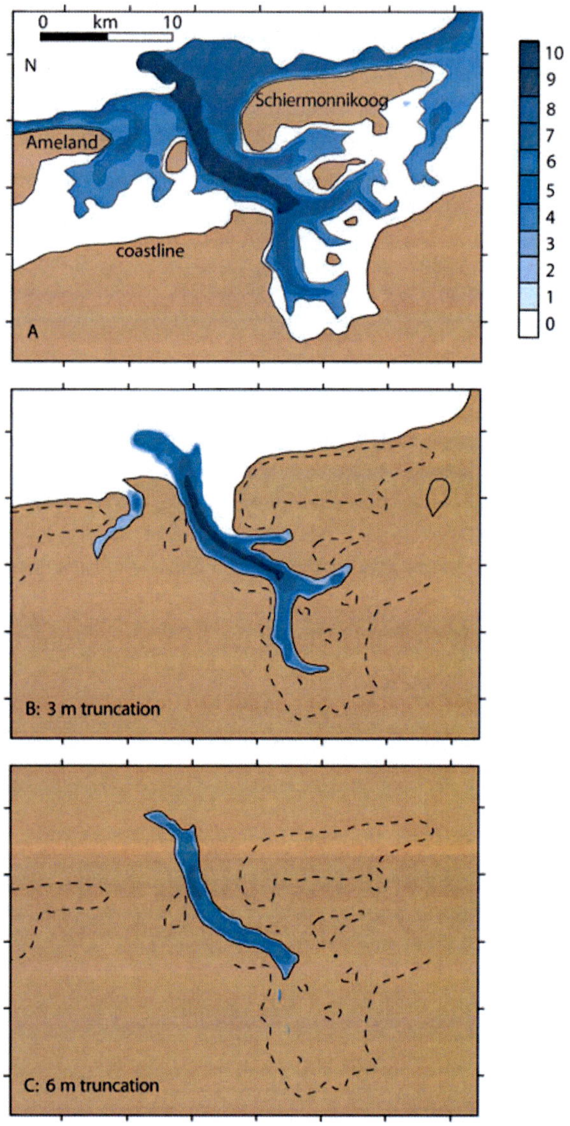

[AU1] **Fig. 12.21** Examples of sediment removal at Zoutkamperlaag Inlet showing that little of the drainage network is preserved under moderated shoreface erosion scenarios (From Reiu et al. 2005)

migration can significantly alter inlet shoreline sedimentation processes.

Tidal-inlet fill comprises substantial portions of barriers in mixed-energy settings and typically a greater fraction of wave-dominated settings due to higher frequency of inlet migration, barrier breachings, and closings. Inlet fills are recognized by the coarse, shell-rich nature of their basal unit forming a sharp contact with underlying sediment. Active inlet fill sequences thin in both a landward and seaward direction and commonly consist of planar to cross-bedded, well-sorted sand that fines upward in contrast to the barrier lithosome. Inactive fills and inlet deposits such as those from Texas and Louisiana coasts contain substantial mud units. Ebb- and flood-tidal deltas are relatively thin deposits and are composed of planar to cross-bedded fine to medium sand. Their structure is controlled by the relative activity of channelization, migration, and closure, bedform and swash bar migration, and wave and tidal sedimentation. Preservation of tidal-inlet and associated deposits, that form a number of important petroleum reservoir sequences, is generally high in regressive sequences, but lower in rapidly transgressing systems due the depth of shoreface erosion, a process that tends to remove all but the deepest parts of inlet channels. Confirmed and suspected inlet fills on the inner shelf all occur offshore of barrier systems having deep inlet channels (>15 m). Recent advances in subsurface imaging and facies analysis are opening new frontiers in understanding the geological legacy of tidal inlets.

References [AU2]

Angulo R, de Lessa GC, Souza MC (2006) A critical review of mid- to late-Holocene sea-level fluctuations on the eastern Brazilian coastline. Q Sci Rev 25:486–506

Ashley GM (1990) Classification of large-scale subaqueous bedforms: a new look at an old problem. J Sed Petrol 60:160–172

Barwis MJH (1978) Recognition of ancient tidal inlet sequences: an example from the Upper Silurian Keyer Limestone in Virginia. Sedi 25:61–82

Barwis JH, Hayes MO (1979) Regional patterns of modern barrier-island and tidal inlet deposits as applied to paleoenvironmental studies, in Ferm JC, and Horne JC, (eds., Carboniferous Depositional Environments in the Appalachian Region: Columbia, South Carolina, Carolina Coal Group, pp 472–508

Belknap DF, Kraft JC (1981) Preservation potential of transgressive coastal lithosomes on the U S Atlantic shelf. Mar Geol 42:429–442

Belknap DF, Kraft JC (1985) Influence of antecedent geology on stratigraphic preservation potential and evolution of Delaware's barrier systems. Mar Geol 63:235–262

Boothroyd JC (1985) Tidal inlet and tidal deltas. In: Davis RA Jr (ed) Coastal sedimentary environments. Springer, New York

Boothroyd JC, Hubbard DK (1975) Genesis of bedforms in mesotidal estuaries, in Cronin JE (ed), Estuarine research, Vol. 2, Geology and engineering: New York, Academic Press, pp 217–234

Boothroyd JC, Friedrich NE, McGinn SR (1985) Geology of microtidal coastal lagoons: Rhode Island. Mar Geol 63:35–76

Bridges PH (1976) Lower Silurian transgressive Barrier islands, southwest Wales. Sedimentology 23:347–362

Bristow CS, Jol HM (2003) Ground penetrating radar in sediments, GSL Special Publication 211. Geological Society, London

Brownridge F, Moslow TF (1991) Tidal estuary and marine facies of the Glauconitic Member, Drayton Valley, central Alberta. In: Smith DG, Reinson GE, Zaitlin BA, Rahmani RA (eds) Clastic tidal sedimentology, Canadian Society of Petroleum Geologists, Memoir 16. Canadian Society of Petroleum Geologists, Calgary

Bruun P (1966) Tidal inlets and littoral Drift, vol 2. Universitelsforlaget, Oslo

Bruun P, Gerritsen F (1959) Natural bypassing of sand at coastal inlets. J Water Harb Div 85:401–412

Bruun P, Gerritsen F (1960) Stability of Coastal Inlets, North Holland Publishing Co., Amsterdam, The Netherlands

Buynevich IV (2003) Subsurface evidence of a pre-1846 breach across Menauhant Barrier, Cape Cod, Massachusetts. Shore Beach 71:3–6

Buynevich IV (2006) Coastal environmental changes revealed in geophysical images of Nantucket Island, Massachusetts, USA. Environ Eng Geosci 12:227–234

Buynevich IV, Donnelly JP (2006) Geological signatures of barrier breaching and overwash, southern Massachusetts, USA. J Coast Res 39:112–116

Buynevich IV, Evans RL, FitzGerald DM (2003) High-resolution geophysical imaging of buried inlet channels. In: Proceedings of the international conference on coastal sediments 2003, World Scientific Publishing Corporation, Corpus Christi

Buynevich IV, Jol HM, FitzGerald DM (2009) Coastal environments. In: Jol HM (ed) GPR: radar theory and applications. Elsevier, Amsterdam

Cheel RJ, Leckie DA (1990) A tidal-inlet complex in the Cretaceous epeiric sea of North America: Virgelle Member, Milk River Formation, southern Alberta, Canada. Sedi 37:67–81

Cleary WJ, Hosier PE, Wells GR (1979) Genesis and significance of marsh islands within southeastern North Carolina lagoons. J Sed Petrol 49:703–709

Culver SJ, Ames DV, Corbett DR, Mallinson DJ, Riggs SR, Smith CG, Vance DJ (2006) Foraminiferal and sedimentary record of late Holocene barrier island evolution, Pea Island, North Carolina: the role of storm overwash, inlet processes, and anthropogenic modification. J Coast Res 22:836–846

Davis RA, Flemming BW (1991) Time-series study of mesoscale tidal bedforms, Martens Plate, Wadden Sea, Germany. In: Smith DG, Reinson GE, Zaitlin BA, Rahmani RA (eds) Clastic tidal sedimentology, Canadian Society of Petroleum Geologists Memoir 16. Canadian Society of Petroleum Geologist, Calgary, p 278

Davis RA, Gibeaut JC (1990) Historical morphodynamics of inlets in Florida: models for coastal zone planning, Sea Grant Technical Paper 55. Florida Sea Grant, Gainesville

Davis RA, Hayes MO (1984) What is a wave-dominated coast? Mar Geol 60:313–329

Davis RA, Cuffe CK, Kowalski KA, Shock EJ (2003) Stratigraphic models for microtidal tidal deltas; examples from the Florida Gulf coast. Mar Geol 200:49–60

Donselaar ME (1989) The Cliff House sandstone, San Juan Basin, New Mexico: model for the stacking of "transgressive" barrier complexes. J Sed Petrol 59:13–27

FitzGerald DM (1976) Ebb-tidal delta of Price Inlet, SC: geomorphology, physical processes, and associated inlet shoreline changes. In: Hayes MO, Kana TW (eds) Terrigenous clastic depositional environments. Coastal Research Division, University of South Carolina, Columbia, pp 158–171

FitzGerald DM (1988) Shoreline erosional-depositional processes associated with tidal inlets. In: Aubrey L, Weishar DG (eds) Hydrodynamics and sediment dynamics of tidal inlets. Springer, New York

FitzGerald DM (1993) Origin and stability if tidal inlets in Massachusetts. In: Aubrey DG, Giese GS (eds) Formation and evolution of multiple tidal inlet systems. AGI, Alexandria

FitzGerald DM (1996) Geomorphic variability and morphologic and sedimentologic controls on tidal inlets. J Coast Res 23:47–71

FitzGerald DM, Montello TM (1993) Backbarrier and inlet sediment response to the breaching of Nauset Spit and formation of New Inlet, Cape Cod, MA. In: Aubrey DG, Giese GS (eds) Formation and evolution of multiple tidal inlet systems. AGI, Alexandria

FitzGerald DM, Nummedal D (1977) Ebb-tidal delta stratification. In: Nummedal D (ed) Beaches and barriers of the central South Carolina coast fieldtrip guidebook for Coastal Zone, Coastal Engineering Conference

FitzGerald DM, Pendleton E (2002) Inlet formation and evolution of the sediment bypassing system: New Inlet, Cape Cod, Massachusetts. J Coast Res, v. SI 36:290–299

FitzGerald DM, Penland S (1987) Backbarrier dynamics of the East Friesian Islands. J Sed Petrol 57:746–754

FitzGerald DM, Hubbard DK, Nummedal D (1978) Shoreline changes associated with tidal inlets along the South Carolina Coast. In: Proceedings of Coastal Zone 1978, American Society of Civil Engineers, New York, pp 1973–1994

FitzGerald DM, Penland S, Nummedal D (1984) Control of barrier island shape by inlet sediment bypassing: East Friesian Islands, West Germany. Mar Geol 60:355–376

FitzGerald DM, Lincoln JM, Fink LK, Caldwell DW (1990) Morphodynamics of tidal inlet systems in Maine. In: Marvenney R (ed) Studies in Maine geology. Maine Geological Survey, Augusta

FitzGerald DM, Buynevich IV, Rosen PS (2001a) Geological evidence of former tidal inlets along a retrograding barrier: Duxbury Beach, Massachusetts, USA. J Coast Res 34: 437–448

FitzGerald DM, Kraus NC, Hands EB (2001b) Natural mechanisms of sediment bypassing at tidal inlets, ERDC/CHL-IV-A, U.S. Army Engineer Research and Development Center, Vicksburg. http://chl.wes.army.mil/library/publications/cetn

FitzGerald DM, Kulp MA, Penland S, Flocks J, Kindinger J (2004) Morphologic and stratigraphic evolution of muddy ebb-tidal deltas along a subsiding coast: Barataria Bay, Mississippi River Delta. Sedimentology 51:1157–1178

Foyle AM, Oertel GF (1997) Transgressive systems tract development and incised- valley fills within a Quaternary Estuary-Shelf System: Virginia Inner Shelf, USA. Mar Geol 137:227–249

Galvin CJ (1994) A conceptual research investigation of tidal inlets. Galvin Engineering Consultants, VA pp 1–7

Gaudiano DJ, Kana TW (2001) Shoal bypassing in mixed-energy inlets: geomorphic variables and empirical predictions for nine inlets. J Coast Res 17:280–291

Halsey SD (1979) Nexus: new model of barrier development. In: Leatherman SP (ed) Barrier Islands: from the Gulf of St. Lawrence to the Gulf of Mexico. Academic, New York

Hayes MO (1975) Morphology of sand accumulations in estuaries: an introduction to the symposium. In: Cronin LE (ed) Estuarine research, vol 2. Academic Press, New York

Hayes MO (1979) Barrier island morphology as a function of tidal and wave regime. In: Leatherman SP (ed) Barrier Islands: from the Gulf of St. Lawrence to the Gulf of Mexico. Academic, New York

Hayes MO, Kana TW (1976) Terrigenous clastic depositional environments, University of South Carolina, Department of Geology. Coast Research Group Technical Report no. 11, pp 81–93

Hein CJ, FitzGerald DM, Barnhardt W (2007) Holocene reworking of a sand sheet in the Merrimack Embayment, Western Gulf of Maine. J Coast Res 50:174–180

Hein CJ, FitzGerald D, Stone BD, Carruthers EA, Gontz AM (2011) The role of backbarrier infilling in the formation of barrier island systems, In: Kraus NC, Rosati JD (eds), Coastal Sediments '11, Proceedings of the 8th International Symposium on Coastal Engineering and Science of Coastal Sediment Processes, 2–6 May 2011, Miami, FL

Hennessy JT, Zarrillo GA (1987) The interrelation and distinction between flood-tidal delta and washover deposits in a transgressive barrier island. Mar Geol 78:35–56

Heron SD, Moslow TF, Berelson WH, Herbert JR, Steele GA, Susman KR (1984) Holocene sedimentation of a wave-dominated barrier island shoreline: Cape Lookout, North Carolina. Mar Geol 60:413–434

Heteren S, FitzGerald DM, McKinlay PA, Buynevich IV (1998) Radar facies of paraglacial barrier systems: coastal New England, USA. Sedi 45:181–200

Hine AC, Snyder SW (1985) Coastal lithosome preservation: evidence from the shoreface and inner continental shelf off Bogue Banks, North Carolina. Mar Geol 63:307–330

Hine AC, Snyder SW, Neumann AC (1979) Coastal plain and inner shelf structure, and geologic history: Bogue Bankks area, North Carolina. Final report to NC Science and Technology Committee, Chapel Hill

Hoyt JH, Henry VJ (1967) Influence of island migration on barrier island sedimentation. Geol Soc Am Bull 78:77–88

Hubbard DK, Oertel G, Nummedal D (1979) The role of waves and tidal currents in the development of tidal inlet sedimentary structures and sand body geometry: examples from North Carolina, South Carolina and Georgia. J Sedi Petrol 49:1073–1092

Imperato DP, Sexton WJ, Hayes MO (1988) Stratigraphy and sediment characteristics of a mesotidal ebb-tidal delta, North Edisto Inlet, South Carolina. J Sedi Petrol 58(6):950–958

Israel AM, Etheridge FG, Estes EL (1987) A sedimentological description of a microtidal, flood-tidal delta, San Luis pass, Texas. J Sed Petrol 57:288–300

Jarrett JT (1976) Tidal prism-inlet area relationships. GITI Report 3 U S Army corps of engineers, Waterways Experiment Station, Vicksburg

Jol HM, Smith DG, Meyers RA (1996) Digital ground penetrating radar (GPR): an improved and very effective geophysical tool for studying modern coastal barriers (examples for the Atlantic, Gulf and Pacific coasts, U.S.A.). J Coast Res 12:960–968

Kumar R, Sanders JE (1974) Inlet sequences: a vertical succession of sedimentary structures and textures created by the lateral migration of tidal inlets. Sedimentology 21:291–323

List JH, Jaffe BE, Sallenger AH, Williams SJ, McBride RA, Penland S (1994) Louisiana Barrier Island erosion study: atlas of seafloor changes from 1878 to 1989, Miscellaneous Investigations Series I-2150-B. US Geological Survey and Louisiana State University, Reston

MacCarthy IAJ (1987) Transgressive facies in the South Munster Basin, Ireland. Sedimentology 34:389–422

McBride RA (1999) Spatial and temporal distribution of historical and active tidal inlets: Delmarva Peninsula and New Jersey, USA. In: Coastal Sediments '99 Proceedings, American Society of Civil Engineers, New York

McBride RA, Buynevich IV, Robinson MM (2004) High-resolution geologic evidence of a former, wave-dominated tidal inlet system: Old Currituck Inlet, VA/NC. In: GSA Northeastern and Southeastern sections abstracts with programs, vol 36, Tysons Corner, Virginia

Miner MD, Kulp MA, FitzGerald DM, Flocks JG, Weathers D (2009) Delta lobe degradation and hurricane impacts governing large-scale coastal behavior; South Central Louisiana. USA Geo-Mar Lett. doi:10.1007/s00367-009-0156-4

Mitchum RM, Vail PR, Sangree JB (1977) Seismic stratigraphy and global changes of sea level, Part 6: Stratigraphic interpretation of seismic reflection patterns in depositional sequences. In: Payton CE (ed) Seismic stratigraphy-applications to hydrocarbon exploration, AAPG Memoir 26. American Association of Petroleum Geologists, Tulsa

Morales JA, Borrego J, Jiminez I, Monterde J, Gil N (2001) Morphostratigraphy of an ebb-tidal delta system associated with a large spit in the Piedras estuary mouth Huelva Coast, Southwestern Spain. Mar Geol 172:225–241

Morton RA, Donaldson AC (1973) Sediment distribution and evolution of tidal deltas along a tide-dominated shoreline, Wachapreague, Virginia: Sedi Geol 10:285–299

Moslow TF, Heron SD (1978) Relict inlets: preservation and occurrence in the Holocene stratigraphy of southern Core Banks, North Carolina. J Sed Petrol 48:1275–1286

Moslow TF, Tye S (1985) Recognition and characterization of Holocene tidal inlet sequences. Mar Geol 63:129–152

Neal A, Richards PK (2003) Sedimentology of coarse-clastic beach-ridge deposits, Essex, southeast England. Sed Geol 162:167–198

Nelligan D (1983) Ebb-tidal delta stratigraphy, (unpublished Masters thesis), Geology Department, University of South Carolina, Columbia

Nummedal D, Fischer I (1978) Process-response models for depositional shorelines: the German and Georgia Bights. In: Proceedings of the 16th coastal engineering conference, Hamburg, West Germany, pp 1215–1231

Nummedal D, Penland S (1981) Sediment dispersal in Norderneyer Seegat, West Germany. Sedimentology 5:187–210

O'Brien MP (1931) Estuary tidal prisms related to entrance areas. Civ Eng 1:738–739

O'Brien MP (1969) Equilibrium flow areas of inlets on sandy coasts. J Water Harb Coast Eng ASCE 95:43–55

Oertel G (1975) Ebb-tidal deltas of Georgia estuaries. In: Cronin LE (ed) Estuarine research, vol 2. Academic, New York, pp 267–276

Okazaki H, Masuda F (1995) Sequence stratigraphy of the late Pleistocene Palaeo-Tokyo Bay: barrier islands and associated tidal delta and inlet. In: Flemming BW, Bartholomä A (eds) Tidal signatures in modern and ancient sediments, International Association of Sediment Special Publication 24. Blackwell Science, Oxford/Cambridge

Pickrill RA (1986) Sediment pathways and transport rates through a tide-dominated entrance, Rangaunu Harbour, New Zealand. Sedimentology 33:887–898

Pierce JW, Colquhoun J (1970) Holocene evolution of a portion of the North Carolina coast. Geol Soc Am Bull 81:3697–3714

Reinson GE (1984) Barrier island and associated strand-plain systems. In: Walker RG (ed) Facies models, Geoscience Canada Reprint Series 1. Geological Association of Canada Publication, Toronto

Ricketts BD (1991) Lower Paleocene drowned valley and barred estuaries, Canadian Arctic Islands: aspects of their geomorphological and sedimentological evolution. In: Smith DG, Reinson GE, Zaitlin BA, Rahmani RA (eds) Clastic tidal sedimentology, Canadian Society of Petroleum Geologists, Memoir 16. Canadian Society of Petroleum Geologists, Calgary

Rieu R, van der Heteren S, Spek AJF, DeBoer PL (2005) Development and preservation of a mid-Holocene tidal-channel network offshore the western Netherlands. J Sed Res 75:409–419

Rodriguez AB, Anderson JB, Bradford J (1998) Holocene tidal deltas of the trinity incised valley: analogs for exploration and production. Gulf Coast Association of Geological Societies Transactions 67:373–380

Sha LP (1989) Sand transport patterns in the ebb-tidal delta off Texel Inlet, Wadden sea, The Netherlands. Mar Geol 86: 137–154

Sha LP (1990a) Surface sediments and sequence models in the ebb-tidal delta of Texel Inlet, Wadden Sea, The Netherlands. In: Sha LP (ed) Sedimentological studies of the ebb-tidal deltas along the West Friesian Islands, The Netherlands, Geological Ultraiectina No. 64. Instituut voor Aardwetenschappen der Rijksuniversiteit te Utrecht, Utrecht

Sha LP (1990b) Preservation potential of ebb-tidal delta and tidal inlets systems in response to sea level rise: examples from the Dutch Wadden Sea. In: Sha LP (ed) Sedimentological studies of the ebb-tidal delta along the West Friesian Islands the Netherlands, Geological Ultraiectina No. 64. Instituut voor Aardwetenschappen der Rijksuniversiteit te Utrecht, Utrecht

Sha LP, de Boer PL (1991) Ebb-tidal delta deposits along the west Friesian Islands (The Netherlands): processes, facies architecture and preservation. In: Smith G, Reinson GE, Zaitlin BA, Rahmani RA (eds) Clastic tidal sedimentology, Canadian Society of Petroleum Geologists, Memoir 16. Canadian Society of Petroleum Geologists, Calgary

Simms AR, Anderson JB, Blum M (2006) Barrier-island aggradation via inlet migration: Mustang Island, Texas. Sed Geol 187:105–125

Sindowski KH (1973) Das ostfriesische Küstengebiet. In: Sammlung Geol. Führer 57, Borntraeger, Berlin, p 162

Siringan FP, Anderson JB (1993) Seismic facies, architecture, and evolution of the Bolivar Roads tidal inlet/delta complex, East Texas Gulf Coast. J Sed Pet 63:794–808

Smith DG, Meyers RA, Jol HM (1999) Sedimentology of an upper mesotidal (3.7 m) Holocene barrier, Willapa Bay, SW Washington, USA. J Sed Res 69:1290–1296

Smith JB (1991) Morphodynamics and stratigraphy of essex river ebb-tidal delta: Massachusetts, Technical Report CERC-91-11, U.S. Army Engineer Waterways Experiment Station, Vicksburg, MS

Susman KR, Herron SD (1979) Evolution of a barrier island-Shackleford Banks, Carteret County, North Carolina. Geol Soc Am Bull 90:205–215

Tye RS (1984) Geomorphic evolution and stratigraphy of Price and Capers Inlets, South Carolina. Sedimentology 31:655–674

Tye RS, Moslow TF (1993) Tidal inlet reservoirs: insights from modern examples. In: Rhodes EG, Moslow TF (eds) Marine clastic reservoirs: examples and analogues. Springer, New York

van der Spek AJF (1995) Reconstruction of tidal inlet and channel dimensions in the Frisian Middelzee, a former tidal basin in the Dutch Wadden Sea. In: Flemming BW, Bartholomä A (eds) Tidal signatures in modern and ancient sediments, International Association of Sediments Special Publication 24. Blackwell Science, Oxford/Cambridge

Walton TL, Adams WD (1976) Capacity of inlet outer bars to store sand. In: Proceedings of 15th coastal engineering conference, ASCE, Honolulu, Hawaii

Williams J, Penland S, Sallenger AH (1992) Atlas of shoreline changes in Louisiana from 1853 to 1989, Miscellaneous Investigation Series I-2150-A. US Geological Survey and Louisiana State University, Reston

Williams JJ, Bell PS, Humphery JD, Hardcastle PJ, Thorne PD (2003) New approach to measurement of sediment processes in a tidal inlet. Continental Shelf Res 23:1239–1254

Zarillo GA, Kraus NC, Hoeke RK (2003) Morphologic analysis of Sebastian Inlet, Florida: Enhancements to the tidal inlet reservoir model, In: Proceedings Coastal Sediments '03. 2003. CD-ROM Published by World Scientific Publishing Corp. and East Meets West Productions, Corpus Christi, Texas, USA

Shallow-Marine Tidal Deposits

13

Jean-Yves Reynaud and Robert W. Dalrymple

Abstract

Shallow-marine tidal deposits form on open shelves, and more specifically in open-mouthed embayments and semi-enclosed epicontinental seas, where the oceanic tide is amplified by resonance. They are also present in straits and seaways where the tidal currents are accelerated by flow constriction. Complex interactions of the tide with the seafloor and coastal topography bring about tidal asymmetry, generating tidal-transport pathways with net, unidirectional transport of sediment over long distances. Tidal currents are commonly capable of resuspending mud in shallow-marine settings, but little is known about the role of tidal currents in the deposition of muddy deposits in the offshore domain. The best-known shelf tidal deposits are sandy and bioclastic transgressive 'lags' that mantle flooding surfaces. These lags are generally thin, but can reach thicknesses of 10–30 m in tidal-current ridges and sand sheets. These deposits are composed of dominantly well-sorted, cross-bedded sands with good reservoir properties. Careful architectural analysis allows the distinction between the deposits of compound dunes, tidal-current ridges and migrating sand sheets. The occurrence of shallow-marine tidal deposits is sensitive to changes in sea level; paleotidal modeling has great potential to help understanding their occurrence in space and time.

13.1 Introduction

Shallow-marine areas, which includes continental shelves and shallow-water seas, are considered here to extend from near the coast to the shelf break, spanning water depths from as shallow as 10–20 m to as much as 150–200 m, and up to 400 m along some formerly glaciated margins. The tides, which generally are created in the open ocean, pass over the shelf on their way to the coast, interacting with the seafloor as they go (Wright et al. 1999; Allen 1997). Geomorphologically, continental shelves and shallow-water seas are diverse. Most continental shelves with significant tidal currents occur on wide passive continental margins, because tidal action typically increases as shelf width becomes greater. Such shelves are commonly straight, with the shoreline essentially parallel to the shelf break for several hundred kilometers. Structural complexities in continental margins create large-scale *embayments*,

J.-Y. Reynaud (✉)
Département Histoire de la Terre – UMR 7193 ISTeP, Muséum National d'Histoire Naturelle, Géologie, CP 48, 43, rue Buffon, F-75005 Paris, France
e-mail: jyr@mnhn.fr

R.W. Dalrymple
Department of Geological Sciences and Geological Engineering, Queens University, Kingston, ON K7L 3N6, Canada
e-mail: dalrymple@geol.queensu.ca

R.A. Davis, Jr. and R.W. Dalrymple (eds.), *Principles of Tidal Sedimentology*,
DOI 10.1007/978-94-007-0123-6_13,

which are here defined as open-mouthed indentations of the coastline such as the North Sea and Yellow Sea. In such epicontinental embayments, the distance from the shoreline to the shelf margin can increase significantly and the flow is partially confined. Broad shallow-marine basins can also occur within continental interiors, far removed from a continental margin, with only a narrow and/or circuitous connection with the open ocean (e.g. Hudson Bay, Canada; the Baltic Sea). Such water bodies are termed semi-enclosed epicontinental seas here. Straits or seaways joining two larger bodies of water commonly exhibit particularly strong tidal water motions (e.g. the Strait of Dover in the English Channel). In all of these offshore settings, currents generated by the tides interact with an array of other processes, including waves, storm/wind-generated currents and geostrophic currents that are part of the global-ocean circulation, giving the potential for the creation of a complex variety of sedimentary deposits.

Because of the large geographic extent and substantial water depth of modern shallow-marine areas, our knowledge of the processes operating there, and of the sedimentary facies generated by these processes, has mostly been obtained by indirect observations. Our understanding of tidal dynamics on shelves has increased markedly over the last few decades, both as a result of improved instrumentation and the application of numerical-modeling approaches. Significant advances have also occurred in our ability to obtain detailed images of the sea floor (e.g. through the use of swath bathymetry), but high-quality 3D seismic imaging of subsurface deposits on modern shelves remains beyond the capability of most academic institutions. In addition, coring techniques have not evolved much over the last several decades; consequently, information on the nature of modern deposits is scanty, although the available database is increasing. As a result, facies models for the deposits of tidal shelves remain poorly developed, as reflected by most textbooks (Stride 1982; de Boer et al. 1988; Suter 2006).

This chapter begins by examining qualitatively some aspects of the dynamics of tides as they progress from the open ocean toward the coast. Then the range of deposits that can occur in tidal settings is considered, namely their composition, surficial morphology (i.e., the bedforms that are present) and internal structure. We focus on the origin and dynamic behavior of *compound dunes* (also called *sand waves*) and *tidal-current ridges* (also called *banks* and *bars*) because they are the largest and most distinctive of the tidally generated bedforms in shallow-marine settings. Reconstructing the Holocene evolution of the offshore ridges in various tidal basins helps to define a model for the transgressive evolution of these large sand bodies, which might have application to the rock record. Finally the potential response of shallow-marine tidal systems to physiographic changes caused by variations in relative sea level is examined briefly, taking examples from both the modern and the rock record. This is coupled with the insights gained from paleotidal modeling, in order to extend our understanding of where tidal deposits are likely to occur in time and space.

13.2 Tidal Processes In shallow Seas

The ability of a basin to develop a large tide depends on the possibility of an amphidromic system to be generated within the basin by the astronomic tide, and on water motions to be amplified by co-oscillation within the basin as the basin borders reflect the tidal wave (co-oscillating tide). The minimum size of a sea (a basin) where the tide is able to generate an amphidromic system is determined by the Rossby radius of deformation of the tidal wave (Pugh 1987), which is the minimum distance required for the Coriolis effect to cause a motion to rotate through 360° (Fig. 13.1a). The Rossby radius decreases as the Coriolis effect increases with increasing latitude. This implies that, for the same basin depth, amphidromic cells are smaller at higher latitude. Wave theory predicts an increase in the celerity of the tidal wave, and hence a larger Rossby radius, as water depth increases: because the perimeter of an amphidromic cell has to be traversed by the tidal wave within one tidal cycle, the larger the amphidromic cell, the higher the celerity the wave must have at its periphery. This is one reason why no shallow, semi-enclosed epeiric sea has significant tides, even though its dimensions are large enough to contain an amphidromic cell. This is, for example, the case in Hudson Bay or the Baltic Sea. Of course, semi-enclosed seas can have large tides even if they are not able to develop tides by themselves, as they can amplify an oceanic tide that enters them through a wide oceanic connection. This is the case in the North Sea and English Channel (Fig. 13.2).

Water motion associated with a rotating Kelvin wave involves two components: the main flow that is perpendicular to the crests and troughs of the wave (i.e., perpendicular to the cotidal lines that show the location of the wave crest as a function of time;

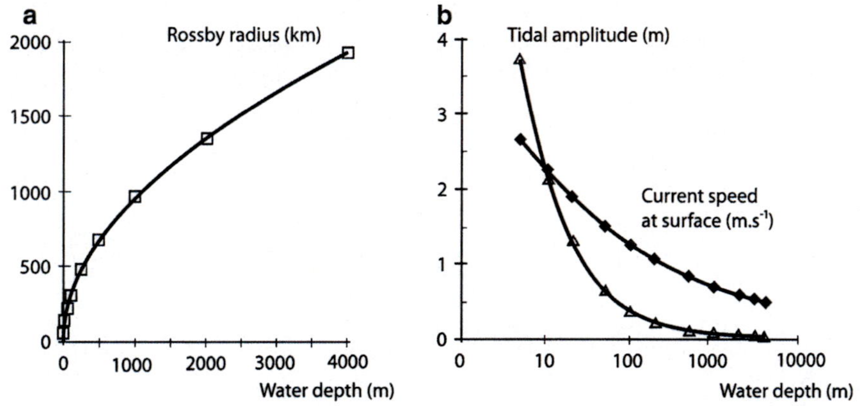

Fig. 13.1 (**a**) Rossby radius (i.e. the dimension of an amphidromic system) as a function of water depth. The Rossby radius is calculated as $R=[(g*d)^{1/2}]/f$ where g is the gravitational acceleration (9.81 m/s^2), d is the water depth and f is the Coriolis parameter (taken at 45° latitude: $10.3*10^{-5}/s$). Amphidromic systems in shallow water have a smaller diameter than those in the deep ocean. (**b**) Amplitude (half the tidal range) and related current velocity at the water surface for a 0.5 m-high, incident tidal wave as it shoals across a continental rise and shelf (see Allen 1997 for details). The relationship between Ad, the amplitude of the tide in deep water, As, its amplitude in shallow water, and Δd, the rate of the decrease in water depth, is expressed by $As=Ad(\Delta d)^{1/4}$. The speed of the surface current, U, is given by $U=A(gd)^{1/2}/d$. In nature, the tidal amplitude and related current velocity do not increase as much as is shown because of the influence of bottom friction (see Fig. 13.3)

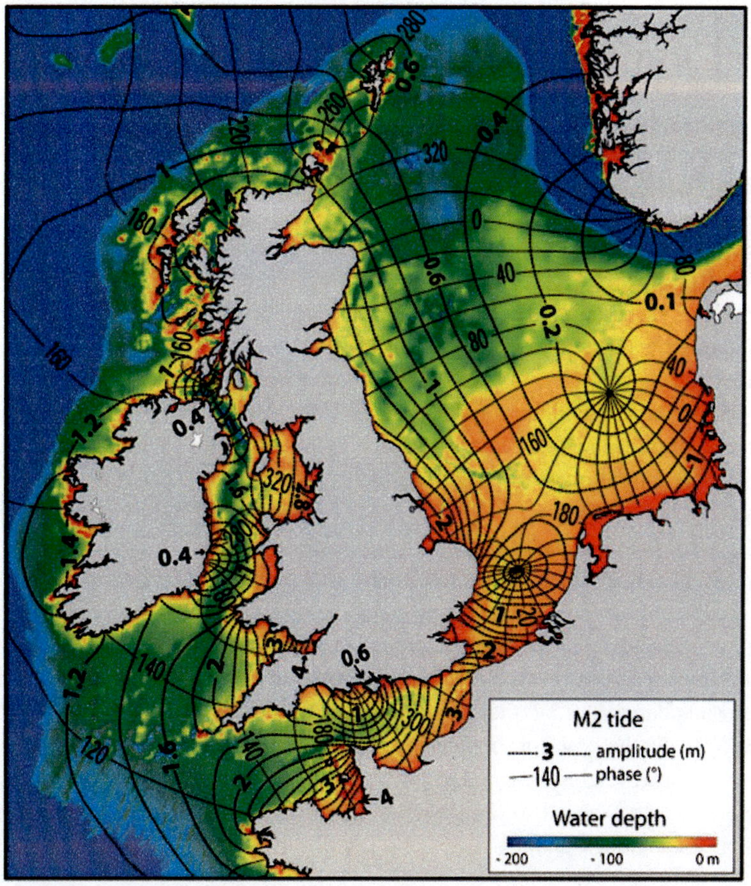

Fig. 13.2 Map showing the amphidromic systems in the seas surrounding the British Isles (After Sinha and Pingree 1997; bathymetry from GEBCO digital atlas, courtesy of Martin Wells). Only the M2 (principal lunar semi-diurnal tide) is considered. Cotidal lines are perpendicular to the coast, which means that the tidal wave travels parallel to the coast, creating tidal currents that are also coast-parallel. Further offshore (e.g. near the Atlantic continental margin), co-tidal (or phase) lines are nearly parallel to the shelf edge, bringing about currents perpendicular to the isobaths. The tidal range increases outward from each amphidromic point, with the highest tidal ranges within embayments such as the German Bight and The Wash, England

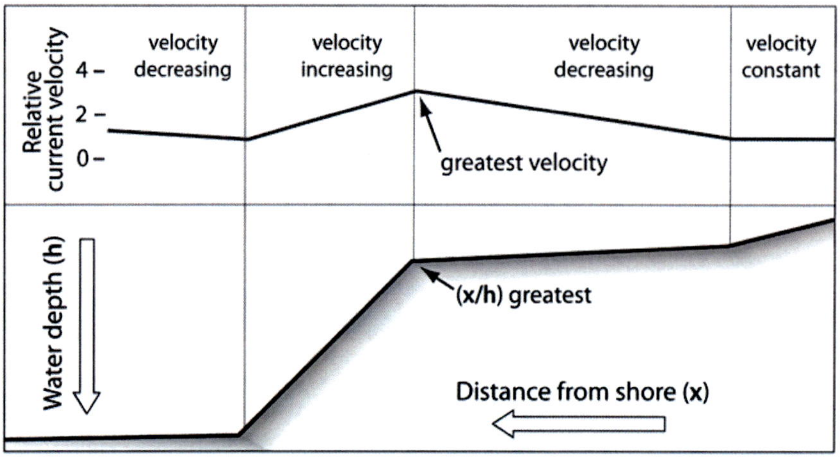

Fig. 13.3 Hypothetical distribution of depth-averaged current speed along a transect perpendicular to a shelf margin. As the tidal wave passes onto the shelf, reduction of the cross-sectional area creates an increase in the current speed. On the shelf, friction in shallow water reduces the tidal-current speed. The result is a zone of maximum current speed near the shelf edge (After Fleming and Revelle 1939)

Fig. 13.2), and a secondary flow that is parallel to the crests and troughs that results from the rotation of the tidal wave. As a consequence, the tidal-current direction at each point in an amphidromic system rotates over a tidal cycle, creating a *tidal ellipse* that traces out the path taken by the tip of successive current vectors. The fastest tidal currents in each tidal cycle (i.e. the major axis of the tidal ellipse) are nearly perpendicular to the co-tidal lines. The propagation direction of the tidal wave and the associated currents are essentially parallel the coast. In river mouths, by comparison, the tidal wave propagates up the river as a standing wave, so that the currents are approximately perpendicular to the nearby coast. On continental shelves, the peak flood and ebb currents are commonly not parallel, because the cotidal lines are not symmetrically distributed within each amphidromic cell, due to the unequal speed of migration of the incoming and outgoing tidal waves (Fig. 13.2).

Where the tide is channelized in a seaway that is much narrower than the radius of the amphidromic cell, the cotidal lines become nearly parallel with each other and are approximately perpendicular to the seaway axis. As a result, the amount of rotation of the currents decreases and they can even become rectilinear. This is the case for the English Channel: the tidal ellipses are greatly elongated and the peak currents are more or less parallel to the direction of travel of the tidal wave and essentially reverse by 180°.

13.2.1 Modification of the Oceanic Tide on the Shelf

The amplitude of the tide in deep oceanic waters is commonly less than 1 m. It increases and, consequently, tidal-current speeds increase, as water depth decreases at the continental margin. This is easily calculated with basic formulae of wave theory (Fig. 13.1b). However, as the tide progresses into shallower water further onto the shelf, frictional dissipation of tidal energy at the sea bed becomes important. Consequently, an area of maximum tidal-current speed is developed near the shelf edge (Fig. 13.3), largely because the tidal prism (i.e., the volume of water passing any point during each half tidal cycle) is greatest near there. The enhancement of tidal currents in this area may also be due to the presence of internal tides that occur along density interfaces in the ocean and break as they impinge on the continental slope (Legg and Adcroft 2003; see Chap. 14). Internal tides are generally important in a zone only a few tens of kilometers wide on the outer shelf and decrease in importance toward the coast. The best studied example is on the outer shelf of the Western Channel Approaches, seaward of the English Channel, where an internal tide is recorded during summer spring tides as brief and pulsed current surges that account for up to 40% of the total current measured near the sea floor. The currents generated by the combined action of the surface and internal tides have created a large, isolated dune field near the shelf edge (Heathershaw et al. 1987).

On the shelf itself, the nature of the tide and tidal currents is strongly controlled by the complex 3D interaction of the tidal wave with the geometry of the shelf and shoreline. On long, straight shelves, the tide is dissipated by friction as it crosses the shelf, such that tidal currents decrease in a landward direction (Fig. 13.3). As the shelf width increases, however, it becomes closer to resonance with the semi-diurnal M2 tide: resonance happens when the tidal wave reflected by the coast is in phase with the incoming wave, which occurs where the shelf width is equal to one-quarter, or 3/4, or 5/4, etc., of the wavelength of the tidal wave, which is a function of the water depth (e.g. Pugh 1987). Due to tidal resonance, the maximum tidal range occurs when the shelf is of the order of 200–400 km wide for typical shelf depths. The influence of changing tidal range on tidal-current speed is direct, but the impact is not uniform over the entire width of the shelf; the greatest change in the strength of the currents occurs near the shelf margin because this is where the change in the tidal prism is greatest.

The situation in embayments and semi-enclosed seas is more complex, with the response of the tidal wave being dependant on the specific configuration of the sea and of its connection with the open ocean. Most open-mouthed embayments accentuate the tide because the cross-sectional area through which the tidal wave passes becomes smaller in a landward direction. Consequently, the tidal range and current speeds are generally higher in embayments than on straight shelves. Examples are given by the English Channel, the North Sea, and the Yellow Sea, and by the Gulf of Bengal, which is a tectonic embayment fully exposed to the ocean. The tidal ellipse is also more elongated and the currents tend toward being rectilinear because of the confinement by the margins of the embayment. Other types of tectonic embayments where the tide is commonly amplified include rifts and foreland basins; in fact, a significant number of the areas with tidal ranges greater than 10 m today are in such settings (Archer and Hubbard 2003). The prediction of resonance in embayments can only be done using numerical modeling, with a full knowledge of the 3D geometry of the shelf and shoreline morphology, as illustrated by studies of the funnel-shaped Gulf of Maine – Bay of Fundy system (Greenberg 1979) and the Western Channel Approaches that might have gone into and out of resonance during the early stages of the last post-glacial transgression (Uehara et al. 2006). By comparison, semi-enclosed seas such as Hudson Bay and the Baltic Sea are more likely to have small tides because the oceanic tidal wave cannot propagate into them effectively, and they are not large enough to have their own tide. Again, the specific response can only be determined by numerical modeling.

Local coastal irregularities such as headlands also perturb the tide. Horizontal flow expansion and constriction on either side of a headland brings about a complex 3D tidal asymmetry, which results in a residual flow that takes the form of time-averaged eddies on either side of the protuberance (e.g. Pingree and Maddock 1979).

Seaways and straits that connect two larger bodies of water are especially prone to pronounced accentuation of the tidal currents because of the constriction. Even a small difference in water elevation at the two ends of a strait can generate strong currents (Pratt 1990). This is the case of the Messina Strait in the modern Mediterranean Sea, despite the fact that the tidal range is less than 10 cm (Androsov et al. 2002), with dunes forming in water depths of more than several hundred meters (Colella 1990).

13.2.2 Residual Tidal Currents

Because each tidal constituent is oscillatory and symmetrical, the net flood and ebb currents should be equal and opposite. However, the examination of measured tidal ellipses show that they are not symmetrical: the peak ebb and flood currents are neither equal in speed, nor are they colinear. This is due to the distortion of the tide and/or to the interplay of more than one tidal constituent. The most important of these is the interaction of the M2 (semidiurnal) tide with its first (M4) harmonic (Pingree and Griffiths 1979; see more below).

Distortion of the tidal wave occurs due to topographic effects. As the tide moves into shallow water, it slows down because of friction, but with the trough decelerating more than the crest because the water depth is less beneath the trough. The consequence is the development of tidal asymmetry, with the front of the tidal wave (i.e., the flood tide) being steeper and of shorter duration than its back (i.e., the ebb tide). This, in turn, brings about an inequality of peak flood and ebb current speeds, creating a tendency for the flood-tidal currents to be faster than the ebb. A similar distortion occurs if the tidal wave enters an embayment, because the progressive, offshore tidal wave cannot continue to propagate freely. Interference of M2 and M4 harmonics of the tide brings about either tidal-phase asymmetry if the M2 and M4 are 90° out of phase or tidal-current inequality if the M2 and M4 are in phase. The tidal motions will be asymmetric in either case: in the first case, the flow in one direction will last longer than in the other, and, in the second, the flow in one direction will be faster although of shorter duration.

Since bedload transport is approximately proportional to the cube of the current speed, any asymmetry in ebb and flood currents will generate inequalities in the sediment transport in the two directions. The result is the creation of a residual sediment transport in one direction (either the ebb or flood). Such inequalities extend over large areas and are referred to as tidal-transport pathways, which are discussed at length later in this chapter.

13.3 Sediment Types on Tidal Shelves

Tidal currents are fast enough on many shelves to transport sand and finer-grained sediment. Much of the existing literature concentrates on sandy deposits, but muddy tidal-shelf deposits are important in areas supplied with large quantities of mud by rivers (e.g. the Amazon and Guyana shelf, the Gulf of Bengal and the Andaman Sea, and the inner portion of the East China Sea). On these shelves, tidal currents contribute significantly to the resuspension of mud (e.g. Viana et al. 1998; Yang and Liu 2007). For example, one of the largest turbid plumes in the world occurs in the Andaman Sea as a result of tidal-current activity (Ramaswamy et al. 2004) with the resulting export of mud to deep water (Rao et al. 2005). On the Amazon shelf, the tidally resuspended mud is advected to the north by wind-driven currents and forms a near-coast nepheloid layer that reduces the bottom friction; consequently, the tide that reaches the coast is larger than would be the case otherwise (Gabioux et al. 2005; Bourret et al. 2008). In the Yellow Sea, tidal resuspension of mud from offshore deposits is responsible for the creation of sandy lags.

Most modern shelves that experience significant tidal-current action are beyond the influence of sediment supplied by rivers. Consequently, older deposits have been reworked by waves and tidal currents that have winnowed away the fine-grained material, leaving behind tidal deposits that are composed predominantly of medium to coarse sand. Such is the case around the British Isles. For the reasons discussed at length by Dalrymple (2010a), the sand becomes finer in the direction of sediment transport, such that coarser sediment, including gravel, can be present at the up-current end of tidal-transport paths, where the currents are fastest, passing down the transport path to fine and very fine sand and even muddy deposits. Tidal currents are an effective sorting agent, and the sorting index is generally high, and increases along the pathway (Gao et al. 1994).

On shelves that are not supplied by large mud-rich rivers, carbonate grains can be an important constituent of the deposits because tidal currents favor the supply and mixing of nutrients coming from the open sea, thereby promoting carbonate production. In cases where there is little or no siliciclastic material, the tidal-shelf deposits can be composed entirely of carbonate grains. In tropical settings, such tidal deposits are commonly composed of ooids, which are believed to be a type of grain formed almost exclusively in tidal settings (e.g. the Bahama Banks; see Chap. 20). In cool- to cold-water settings, herterozoan benthic communities generate abundant bioclastic debris that is particularly prone to reworking by tidal processes (cf. Anastas et al. 1997; James 1997). Tidal-transport pathways exist in carbonate environments (e.g. Harris 1988), but, in such settings, sediment grain size is more strongly controlled by the biota present than by the speed of the tidal currents.

Along a tidal-transport pathway, the nature of the substrate and the strength of the currents control the nature of the benthic biota. Areas scoured by strong currents, where the sea floor consists of exposed bedrock, are dominated by epibenthic, encrusting faunas, whereas depositional tracts with mobile sand are dominated by endobenthic faunas (Wilson 1982); in general, however, the more mobile the substrate, the less diverse the fauna will be. In the modern, relatively little study has been devoted toward linking the fauna with position along a transport pathway, although spatial variations in the composition of small bryozoan particles (Bouysse et al. 1979) or molluscan species (Reynaud et al. 1999c) have been noted. Physical and biogenic destruction of particles occurs during transport. On modern shelves, the intensity of reworking and mixing of grains increases with water depth, as the result of increasing time and decreasing sediment supply through the post-glacial transgression (Wilson 1988). In the Miocene cool-water carbonates of SE France, recurring associations between the fauna and tidal bedforms have been noted (Descote 2010). The largest and coarsest grained dunes contain the highest content of red algae, whereas the small and finer-grained dunes show a larger amount of benthic forams and molluscans. This partitioning is also reflected in the sequence-stratigraphic organization of the deposit. The coarse bioclastic TST deposits are dominated by a bryozoa/echinoderm (Bryonoderm) fauna, which is succeeded by a red algae (Rhodalgal) association, whereas the more muddy HSTs are dominated by a mollusc/benthic foraminifera (Molechfor) association.

13.4 Tidal Dunes

The sandy sediments that are present over large parts of tidal shelves are very commonly molded into a complex array of large bedforms, ranging from flow-transverse dunes of various sizes to nearly flow-parallel tidal-current ridges. Dunes are the most ubiquitous bedforms on continental shelves, occurring both on sand ridges and flat sand sheets, and are responsible for much of the sedimentary record of offshore tidal environments. Therefore, they are discussed at length here.

13.4.1 Morphological Response to Flow

Dunes is the generally accepted term that replaces the older terms *megaripple* and *sandwave* (Ashley 1990). Flume experiments and observations in nature have defined the stability field of dunes as a function of grain size, current speed and water depth (e.g. Rubin and McCulloch 1980; Allen 1982; Southard and Boguchwal 1990). Dunes can be formed in grain sizes between approximately 0.15 mm (i.e. within the range of fine sand) and gravel size (Carling 1999), and by current speeds above about 0.5 m/s. Water depth is not a significant limiting factor on the occurrence of dunes, provided the current speed is sufficient, although an increase in water depth commonly leads to a decrease in current speed and, hence, the disappearance of dunes.

The size and shape of dunes vary widely. Following Ashley (1990) and Dalrymple and Rhodes (1995) we suggest the size distinctions given in Table 13.1. The maximum height of a shelf tidal dune is not well defined, but tidal dunes up to 15 m high are reported on modern shelves (e.g. Berné et al. 1989). The larger the dunes, the lesser their relative relief: in general, the dune wavelength-to-height ratio (= the ripple index; RI) is less than 10 for small dunes but commonly exceeds 30 for large dunes, and may reach 100 for very large ones.

The size and shape of dunes are controlled by water depth, current speed and grain size. Studying dunes in flumes and rivers, Van Rijn (1982, also Southard and Boguchwal 1990) showed that, in the lower part of the dune stability field, increasing current speed brings about an increase of the equilibrium height of dunes. As well, for a given depth and current speed, the dune height increases slightly with grain size (Flemming 1980; Van Rijn 1982). Water depth, which is a proxy for boundary-layer thickness, is generally regarded as being the most important control on dune size, with dune height (H) and wavelength (L) increasing as water depth (h) increases (Ashley 1990). Following Yalin (1964) and based on many examples in nature summarized by Allen (1982), the widely accepted relationships are:

$$L = 6h \tag{13.1}$$

$$H = 0.167h \tag{13.2}$$

These relationships are only applicable in cases where the dunes are fully developed in equilibrium with the flow, and where the sea floor is completely covered by mobile sediment, a condition called fullbedded (Ashley 1990). These relationships do

Table 13.1 Size classes for dunes (From Dalrymple and Rhodes 1995)

	Small	Medium	Large	Very large
Wavelength (L)	0.6–5 m	5–10 m	10–100 m	>100 m
Height (H[a])	0.05–0.25 m	0.25–0.5 m	0.5–3 m	>3 m

[a] Calculated from the Flemming (1988) relationship: $H = 0.0677\ L^{0.8098}$

not apply in deep water where the thickness of the boundary layer, the real control on dune size, is less than the water depth, or to the smaller dunes that are superimposed on the larger dunes in an area to form compound dunes (Ashley 1990). The size of these smaller dunes is generally thought to be related to the presence of an internal boundary layer that is formed on the back of each larger dune (Rubin and McCulloch 1980; Dalrymple 1984).

Because of the widespread occurrence of unidirectional residual sediment transport in tidal-transport pathways, the dunes on tidal shelves are typically strongly asymmetric, with their steeper, lee face inclined in the direction of net sediment movement. In tidal settings, weakly asymmetric dunes, although not common, can be found in areas of weak tidal asymmetry. The speed of dune migration in the direction of residual transport increases as their asymmetry increases, but decreases as their size increases, all else being equal. The average annual distance of migration of small dunes is about 100–300 m, while it is only 25–75 m for large ones and only a few decimeters for very large dunes (e.g. Fenster et al. 1990). The *lag time* of the dunes (i.e. the time needed for them to equilibrate with a changed flow condition) also increases as they become larger; thus, large and very large dunes on the seafloor have the potential to be out of equilibrium with the present-day flow. Also, as the dunes become larger and less active, they have the potential to become more bioturbated.

Although dunes are most commonly oriented with their crest nearly perpendicular to the flow, tidal dunes can be oblique to the peak tidal flow and to the residual transport of sand. Theoretically, this occurs where dominant and subordinate tidal currents are not colinear, with the degree of obliquity depending on the ratio of sediment transport by the dominant and subordinate currents (Rubin and Hunter 1987; Rubin and Ikeda 1990). In shelf settings, however, the dominant and subordinate currents are typically nearly 180° apart, but, despite this, the dunes can be oblique, with the amount of obliquity typically increasing as the dunes become larger. The most widely held suggestion as to why this occurs is that large dunes reflect the impact of infrequent events such as intense storms and wind-driven currents that have higher sediment-transport capacity than the more frequent, but weaker, tidal currents. Also, the obliqueness of large dunes could reflect flow-transverse variations in their migration rate, as a result of inequalities in sediment discharge and bedform height (Dalrymple and Rhodes 1995). Although the smaller dunes may have more variable orientations over large areas, they statistically provide a more reliable indication of the *local* peak tidal-current direction than the large dunes.

13.4.2 Internal Structure of Offshore Tidal Dunes

The available knowledge on the internal structure of tidal dunes comes mainly from studies in modern estuarine settings (Dalrymple 1984), with additional observations from ancient shallow-water successions (Allen and Homewood 1984). Very few observations of the internal structure of shelf dunes exist, and most of those come from seismic records (e.g. Berné et al. 1988, 1989), which do not have sufficient resolution to show the detail seen in outcrops.

The internal structure of simple dunes (i.e. dunes that lack superimposed dunes; Ashley 1990), regardless of size, consists of foreset laminae emplaced by pulses of grain flow on the lee face, which is inclined at an angle close to the angle of repose (ca. 32–35°), and toeset and bottomset laminae that accumulate by the settling of grains from suspension. This produces crossbeds that can extend over hundreds of meters laterally (Fig. 13.4) if formed by a large to very large dune; smaller dunes produce crossbeds of lesser lateral extent. As it is steep, the lee face of a simple dune promotes flow separation and, therefore, possible up-dip migration of ripples on the toesets (Fig. 13.5a). The tidal-bundle successions produced by neap-spring tidal cycles can be present in offshore dunes (e.g. Longhitano and Nemec 2005), but are not likely to be developed extensively because dunes on shelves are typically too large, and move too slowly, to record variations in flow speed and direction over individual tidal cycles. Thus, grain-size segregation in the foreset lamination, which is sometimes referred to as *grain striping*, is generally not related to changes in the speed of the tidal currents, but rather to the effect of pre-sorting of sediment by small superimposed bedforms (Reesink and Bridge 2007, 2009). Similarly, convex-up erosion (reactivation) surfaces within the upper part of these crossbeds (Fig. 13.5a), which are classically attributed to erosion by the subordinate tide (Allen 1980), are more likely to

Fig. 13.4 Succession of vertically stacked carbonate crossbeds formed by simple, large to very large dunes migrating under the influence of unidirectional or highly asymmetric tidal currents, Bonifacio Formation, Corsica (see Brandano et al. 2009; André et al. 2011). The exposure shown is about 25 m high. The vertical stacking of such thick crossbeds in successions up to hundreds of meters thick (250 m in the Bonifacio Formation) is a characteristic feature of the infill of tidal seaways or straits where accommodation is high. The prominent crossbed boundaries correspond to intervals deeply bioturbated by *Thalassinoides*

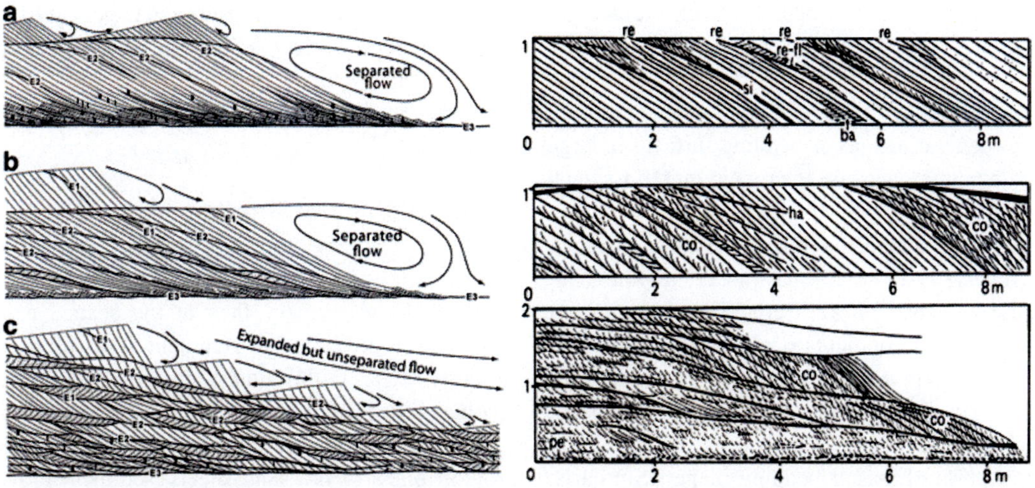

Fig. 13.5 (*Left*) Internal structures formed by compound dunes. Slopes of foresets range from 35° (laminae) to 4° (master beds in **c**). The internal complexity depends on the relative size of the 'master' and superimposed dunes, which controls the amount of erosion on the lee of the large dune and therefore its overall steepness (After Dalrymple 2010b), modified in part from Allen 1980). (*Right*) Outcrop sketches of compound-dune deposits in the Precambrian Lower Sandfjord Formation, Norway (After Levell 1980). *Re* reactivation surfaces, *si* silt drapes, *re-fl* reverse-flow ripples, *ha* hanging set boundaries, *co* convex-up boundaries, *pe*: pebble horizons

be generated by erosion in the troughs of superimposed dunes as they migrate over the brink of the larger bedform (Dalrymple 1984, 2010b; Reesink and Bridge 2009), or to episodic wave action. It is important to note that mud drapes, which are an important signature of tidal sedimentation in estuarine and deltaic settings (Visser 1980; Nio and Yang 1991; Dalrymple 2010b), are rare in offshore tidal deposits because of the presence of rotary tides with no distinct slack-water period, and because suspended-sediment concentrations are generally very low.

Large and very large dunes are typically covered by smaller dunes and have a compound morphology. Such dunes generate *compound crossbedding*, composed of stacked, inclined, planar to trough crossbeds formed by the superimposed smaller dunes (Fig. 13.5b, c). The lee side of compound dunes typically has a much lower slope (commonly <10°) than that of simple dunes. Flow separation does not occur, and the smaller, superimposed dunes migrate continuously down the larger dune's lee side, from the crest to the trough. Flow expansion and flow deceleration bring about deposition on the lee side of the larger dune, so that each superimposed smaller dune leaves behind a crossbed that gets preserved. The continuous accretion of such crossbeds forms the master bedding of the compound dune. The superimposed dunes may themselves be compound, so that 'compound-compound' dunes can occur (Anastas et al. 1997). It is noteworthy that the smaller dunes migrate in essentially the same direction as the larger dune, forming an architecture termed *foreward accretion*. Upslope-climbing ripples or smaller dunes formed by the subordinate current are likely to be preserved, forming herringbone cross-stratification. Tidal-current reversals are generally not capable of generating master-bedding surfaces in large to very large dunes, because the time required for such large dunes to reverse greatly exceeds the duration of a tidal cycle. For example, the time needed for a 4 m-high dune to reverse is about 200 days of continuous bedload transport, based on an average rate of transport typical of tidal environments (Dalrymple and Rhodes 1995). Longer-term flow reversals, such as those associated with seasonal changes in the wind regime or ocean circulation, could, however, cause dune reversal and the creation of master-bedding planes. Similarly, high-energy storms with greatly increased sediment-transport rates could also bring about dune reversal (Houthuys et al. 1994; Le Bot and Trentesaux 2004). The storm-wave activity can also erode the crest of the dune, generating horizontal erosion surfaces (e.g. McCave 1971; Dalrymple 1984; Berné et al. 1991).

The vertical succession of structures produced by a compound dune generally coarsens upward, and the individual cross beds in it commonly become thicker upward (Fig. 13.5). The upward-coarsening trend is caused by the fact that the shear stress exerted by the flow is higher at the crest than in the trough of the compound dune. The downward thinning occurs due to the fact that the superimposed dunes become smaller as they migrate down the larger lee face because they are losing sediment to the cross bed that they leave behind (Rubin 1987). The bottomset region of compound dunes is the area where muddy deposits are more likely to occur; bioturbation is also greater there than elsewhere.

13.5 Offshore Tidal Ridges

Tidal-current ridges are widely developed on tidal shelves and comprise a significant fraction of the total volume of all sandy deposits in the modern. They are the largest bedforms that exist, reaching 200 km in length, 10 km in width and 50 m in height. Offshore tidal ridges generally occur as fields of regularly spaced, parallel *en echelon* ridges (Off 1963) that can cover tens of thousands of square kilometers, as in the Celtic Sea, the North Sea, and the East China and Yellow seas (Fig. 13.6). They may also occur as isolated ridges in the lee of islands and capes (in this case they are called *banner banks*). Offshore tidal ridges are made up by the accretion, at the largest scale, of the crossbeds formed by dunes.

13.5.1 Ridge Morphodynamics

Unlike dunes, ridges are not the expression of the turbulence of the primary flow at the seabed and the related local advection of sediment. The ridges occur in rotary 'cells' within tidal-transport paths where the residual transport retains a higher proportion of the sediment that is in transit. Within these cells, ridges grow in place of flat sand sheets: sediment moves in opposite directions on either side of the ridge crest, with the potential for sand to circulate around bank terminations (McCave and Langhorne 1982; Howarth and Huthnance 1984). Near the coast, these cells can correspond to the eddies that are generated by a headland (Fig. 13.7; Pingree and Maddock 1979). Further from the shore, on the open shelf where linear ridges develop, the rotary circulation is the result of a positive feedback between the reversing tidal flow and the topography of the ridge (Zimmerman 1978; Pan et al. 2007). This rotary residual transport of sand is driven

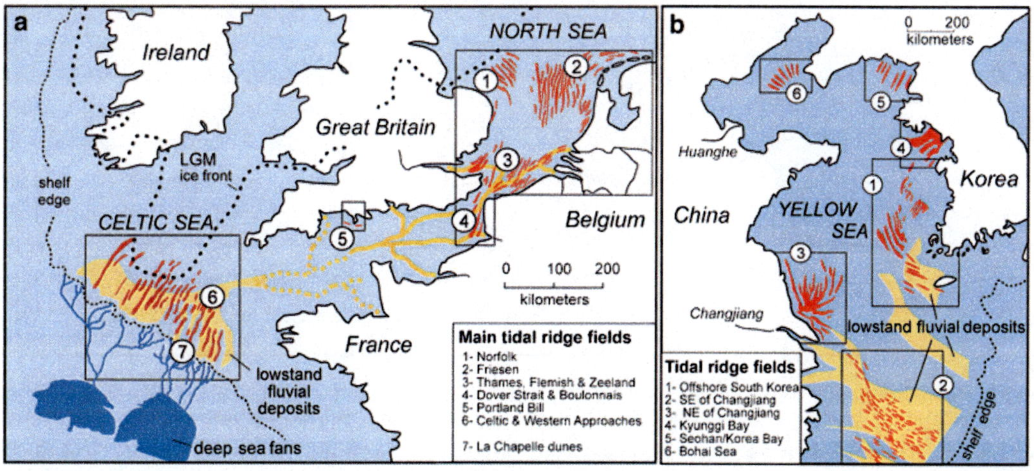

Fig. 13.6 Map showing the distribution of the major tidal ridges (**a**) on the Western European shelves and (**b**) the East China and Yellow seas. All the ridges are linear, offshore 'en echelon' tidal ridges, except those in location 5, which are banner banks associated with a headland. In (**a**), ridge field 1 occurs on a headland-associated shoal-retreat massif (Swift 1975). Ridge field 2 might also occupy a shoal-retreat massif; alternatively, it may represent the redistribution of an east-west-oriented barrier complex that existed earlier in the post-glacial transgression. Ridge fields 3 and 4 may represent an embayment-head type of occurrence (cf. Dyer and Huntley 1999). In (**b**), note the train of tidal ridges along the retreat path of the Changjiang River (ridge field 2). Ridge field 1 may occupy the retreat path of the Huanghe and/or Han River

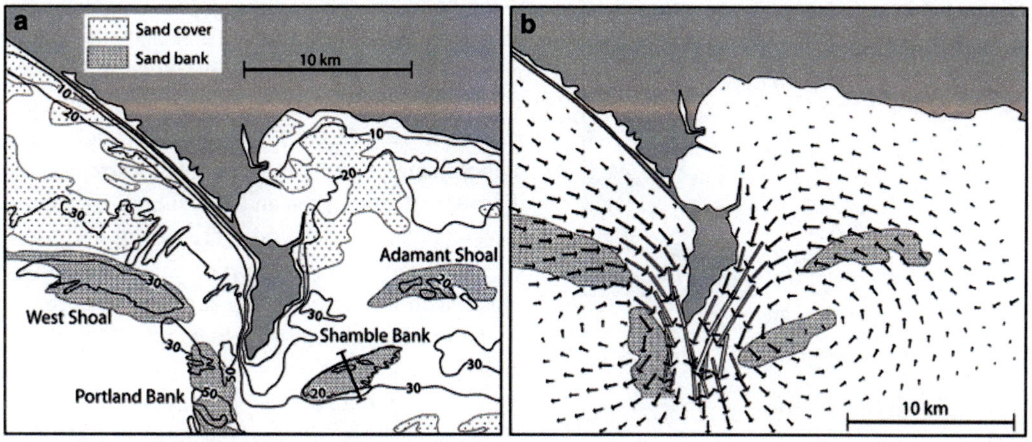

Fig. 13.7 (**a**) Tidal ridges around the Portland Bill headland, English Channel (see location in Fig. 13.6a – ridge field 5). These are typical 'banner banks' that form in the lee of coastal promontories (After Bastos et al. (2003). (**b**) Numerical model of tidal residual circulation (After Pingree and Maddock 1979). Bedform migration directions on Shamble Bank are consistent with the counterclockwise circulation of the modeled eddy. The observed convergence of bedload transport toward the centre of the eddies is explained by a centripetal reduction of bottom shear stress (see detailed explanation in Dyer and Huntley 1999). The process is efficient for small eddies only (ca. 10–20 km in diameter), implying that this process cannot generate longer ridges

by two vorticity forces that increase toward the ridge crest, one due to the increase of bottom friction in the shallower water over the crest, and the other due to an enhanced Coriolis effect (Fig. 13.8). The Coriolis effect either dampens or enhances the friction-driven circulation, depending on whether the ridge is oriented clockwise or counter-clockwise, respectively, to the peak tidal flow. This explains why ridges are mostly skewed in a counter-clockwise sense relative to the peak tidal flow in the northern hemisphere (Fig. 13.9).

The growth of a linear shelf ridge from a small bed perturbation (bump) was first modeled by

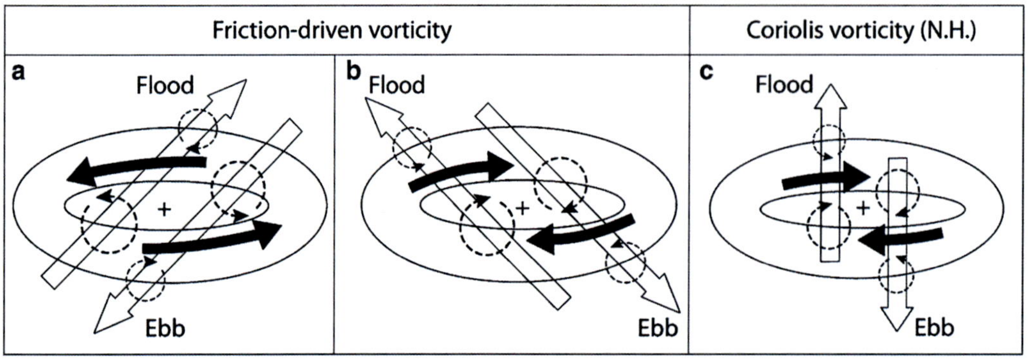

Fig. 13.8 (**a**, **b**) Friction-driven vorticity (*dotted circles*) induced by oblique flow across an elongated bump on the seabed. The open *straight arrows* show the flood and ebb current vectors. The *black arrows* show the resulting residual circulation around the bump (i.e. the tidal ridge). The vorticity increases toward the top of the ridge, acting to move bedload toward the ridge crest. (**c**) Coriolis-driven vorticity induced in the northern hemisphere (*N.H.*) by the flow across the same seabed relief. This results in a clockwise residual circulation (counter-clockwise in the southern hemisphere). This effect enhances or damps the friction-driven residual circulation. In the northern hemisphere, their interplay favors the existence of ridges oriented counter-clockwise to the flow (After Pattiaratchi and Collins 1987)

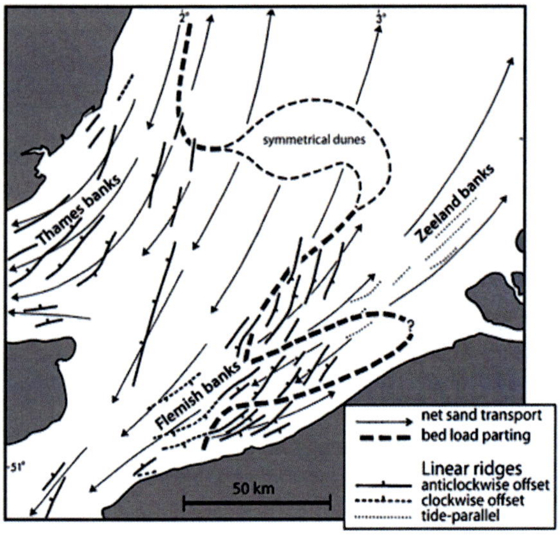

Fig. 13.9 Tidal ridges and tidal-transport pathways in the southern North Sea (ridge field 3 in Fig. 13.6a). Note the alignment of the ridges at a small angle to the net sand-transport paths, which have been determined from the asymmetry of dunes. Most ridges are offset counter-clockwise to the tidal flow, in response to the mechanism described in Fig. 13.8 (After Kenyon et al. 1981)

Huthnance (1973, 1982a, b). The main process involved in the Huthnance model is that the bottom friction associated with the bump delays the upslope current more than it accelerates the downslope one. As a consequence, the transport of sand toward the ridge crest will be higher than the off-ridge transport, thereby causing ridge growth. The transport paths over the ridge reflect the convergence of sediment toward the ridge crest, as noted first by Van Veen (1936). On the side of the ridge facing it, the dominant current is accelerated by flow constriction and the subordinate current is decelerated by flow expansion, so that the residual transport by the regionally dominant current is enhanced (Fig. 13.10). On the ridge side facing the subordinate current, the dominant current is weaker because of sheltering and the subordinate current is accelerated toward the ridge crest by flow constriction, so that the residual transport on this side is commonly dominated by the regionally subordinate current.

Huthnance (1982a, b) calculated that the ridges must become elongated and oblique to the peak flow at an angle of about 20°. Ridge elongation is proportional to the elongation of the tidal ellipse, and the initial bump from which a ridge grows does not have to be elongated or properly oriented itself. Based on tidal ridges in the China and Yellow Seas, Liu et al. (1998) suggest that linear ridges are restricted to areas where tidal M2 ellipticity (i.e., the ratio between the minor and major axes of the tidal ellipse) is < 0.4 (i.e., the tidal ellipses are significantly elongated), whereas sand sheets occur where the currents are more rotary. Hulscher (1996) showed that linear ridges are more prone to develop in deeper water, because the 3D flow structure is more homogenous and there is, therefore, a smaller phase lag between shear stress and the depth-averaged current speed. Theory indicates that ridges

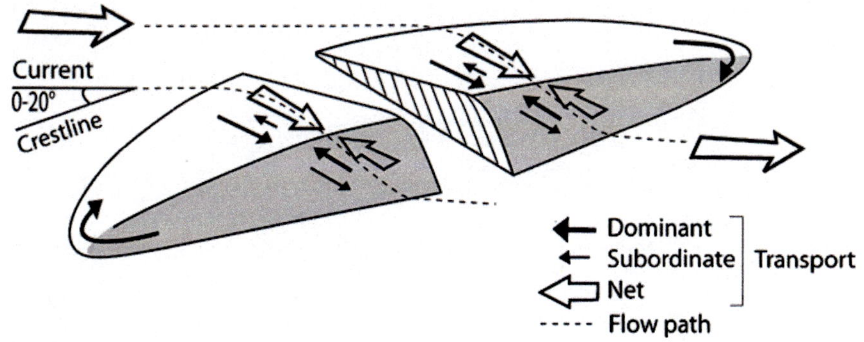

Fig. 13.10 Conceptual model of flow over an offshore tidal ridge that is slightly oblique to the tidal currents. The bending of the flow across the ridge is a consequence of friction at the seabed, which delays the shallower edge of the flow. The crest of the ridge is a convergence zone because of opposed net transport directions on either side of the ridge crest. The circulation of sand around the ends of the ridge has been documented by McCave and Langhorne (1982) and Howarth and Huthance (1984), but it is not a requirement for ridge formation (After Houbolt 1968 and Caston 1981)

must have a spacing that is of the order of 250 times the water depth (i.e. ridge spacing is many kilometers) in order for the residual rotary circulation to be established (Huthnance 1982a). This prediction is supported by observations and may explain why linear tidal ridges are largely restricted to open shelves and seem to be absent in narrow seaways (Harris 1988; Malikides et al. 1988).

13.5.2 Ridge Architecture

As a consequence of the tidal asymmetry, the ridges migrate in the direction of the dominant regional current, but at a rate that is so slow that it cannot be measured with confidence over a few years (Lanckneus et al. 1994). Because of the slight obliqueness of the tidal flow relative to the ridge axis, the ridges migrate laterally, as documented first by Houbolt (1968) (Fig. 13.11), especially in areas with a strong residual transport. In areas of weaker tidal asymmetry, accretion can occur on both sides of the ridge (Davis and Balson 1992). Banner banks, because they are anchored in a coastal eddy, may be aggradational rather than migratory (Fig. 13.11). Lateral migration and aggradation are likely to be disrupted during periods of increased storminess, which can either accelerate deposition or cause significant erosion of the crest, producing flat, wave-planation surfaces, as showed for Sark and Shamble banks in the Western Channel (M'Hammdi et al. 1992; Bastos et al. 2003; Fig. 13.11). Although accretion on the regional down-current flank might be most common, accretion can occur on the side facing the dominant current (McCave and Langhorne 1982; Dalrymple and Rhodes 1995; Reynaud et al. 1999b; Fig. 13.12). This happens when the ridge height is a significant fraction of the water depth, such that bottom friction, which slows the flow toward the ridge crest, is greater than the tendency for acceleration as a result of the flow constriction. In this case, the cross-ridge flow can become accelerated through local low points along the crest, forming oblique channels called *swatchways*. These channels can then lead to splitting of the ridge into two, *en echelon* parts as described by Caston (1972) (Fig. 13.13). By this mechanism, the ridge field can expand throughout the area of the tidal transport pathway, provided there is sufficient sand.

It must be stressed that the available knowledge on the internal structures of offshore tidal ridges is based almost entirely on seismic data and surface morphology. Therefore, it is difficult to propose a generalized facies model that can be used to interpret the rock record. From the facies point of view, it is likely that the deposits within tidal ridges are composed predominantly of crossbedded sand produced by the dunes that mantle the ridges in most situations. However, it is also likely that the deposits record, at least locally, the imprint of storm waves, in the form of gravel lags with large gravel wave ripples, coarse graded storm beds or even HCS if the grain size is too fine to form dunes (cf. Yoshida et al. 2007). Because of the very large sediment volume within a ridge, one or more storms will not destroy the ridge but can be very prominently recorded in its architecture and facies (Houthuys and Gullentops 1988a, b). The deposits are thought to

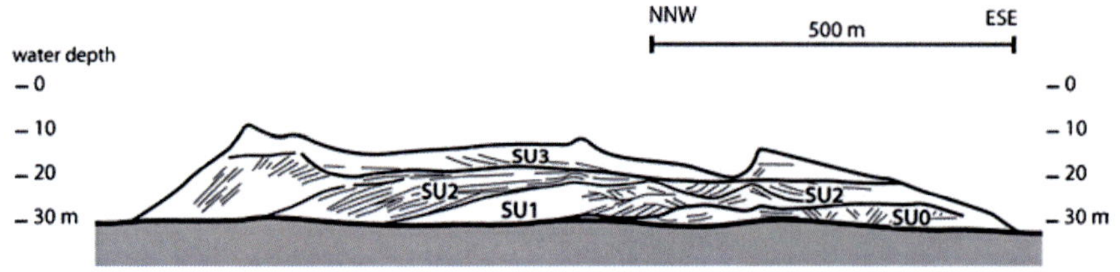

Fig. 13.11 Transverse cross-section of Shambles Bank in the English Channel (see also Fig. 13.7). The flat erosional surfaces that constitute the main master bedding may be the result of erosion by storm waves. The pattern is partly aggradational, because the sand is trapped in the residual eddy that determines the location of the bank. The large superimposed dunes produce compound crossbeds with cosets over 6 m in thickness (After Bastos et al. 2003)

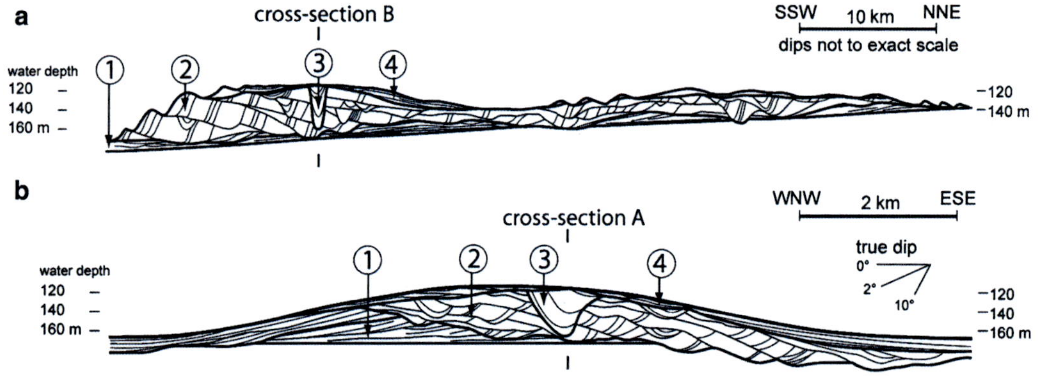

Fig. 13.12 Longitudinal and transverse sections through a deep-shelf tidal ridge in the Celtic Sea. The units are: *1* flank deposits of the early stage of ridge growth; *2* climbing, very large compound dunes of the high-energy phase of ridge growth; *3* swatchway channel cut through the ridge by strong, cross-ridge tidal flows when the ridge crest was shallowest; and *4* abandonment deposits formed by destructive reworking of the ridge crest by wave action after active growth ceased. Dominant current was to the SW. See Fig. 13.6a (area 6) for location (After Reynaud et al. 1999b)

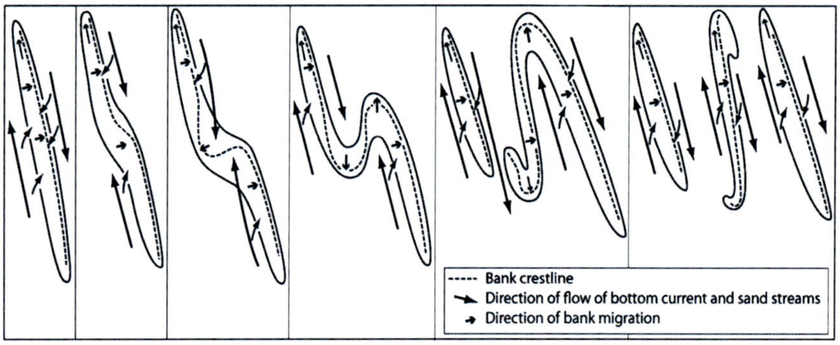

Fig. 13.13 Sequential model of ridge evolution and splitting, based on the Norfolk Ridges (see location in Fig. 13.6a, ridge-field 1). The process of ridge splitting is probably initiated by the development of a cross-ridge swatchway (in steps 2 and 3 from the *left*), which grows in size until the original ridge separates into two parts (After Caston 1972)

coarsen upward, because current speeds and wave action are highest on the ridge crest. If, however, the ridge becomes large enough that it impedes cross-ridge flow, then there may be a tendency for finer sediment to accumulate in the area of weaker currents on the ridge crest. As we will see later, however, the sea-level history during ridge growth and migration can be significant, because ridges are so large that their lag time is likely to be of the same order as the duration of a high-frequency sea-level cycle.

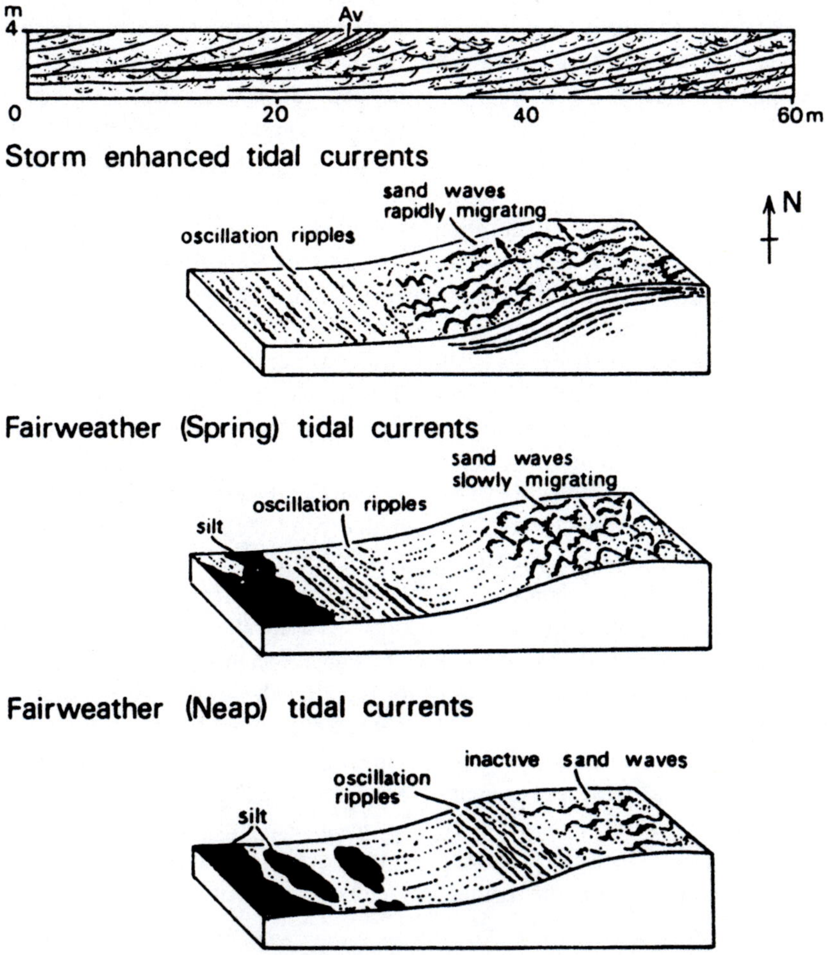

Fig. 13.14 Depositional model for a tidal shelf ridge, based on the Precambrian of north Norway. The ridge is interpreted to be shore-parallel and separated from the coastline by an area of wave erosion. The upper part of the ridge is composed of cross-bedded sandstone created by the along-strike migration of dunes ('sand waves'). Channels cut by bidirectional currents dissect the ridge crest. The ridge growth and migration is related to storm-enhanced tidal activity. It passes gradually offshore to muddy, storm-influenced facies (From Johnson 1977)

13.5.3 Ridges in the Rock Record

There are very few detailed case studies in the ancient that argue convincingly for the existence of tidal-current ridges. The Mutti et al. (1985) model for a 'tidal bar' shows an upward-coarsening succession and the presence of dune cross bedding as predicted from modern tidal ridges, but it shows forward accretion instead of the lateral accretion documented from modern examples. Therefore, the bedforms described by Mutti et al. (1985) are more likely to represent very large compound dunes (cf. Fig. 13.4; Dalrymple 2010b). The Johnson (1977) model for tidal ridges (Fig. 13.14) is more consistent with modern examples because of the presence of lateral accretion and upward coarsening, but the suggestion that migration is always in an offshore direction is inconsistent with many modern examples. The absence of a basal erosion surface is also not what is seen in modern shelf ridges. Erosionally based shelf sand bodies have been documented from the Cretaceous Western Interior Seaway of Northern America (Tillman and Martinsen 1984), but they have been reinterpreted as forced-regressive shorefaces and deltas, detached from the coast by subsequent transgressive ravinement (Walker and Bergman 1993; see Chap. 17).

13.6 Tidal-Transport Pathways

13.6.1 General Characteristics

As has been discussed already, the asymmetry of the tidal currents that results from the complex interaction of the tidal wave with the shelf and shoreline morphology, or with other currents (e.g., storm, wind-induced or geostrophic currents; Flemming 1980; Suter 2006), leads to the existence of *tidal-transport pathways* that extend over large areas. In each pathway, there is unidirectional residual transport of sediment that is reflected in the grain-size of the sediment and the spatial distribution and facing direction of bedforms (e.g. Grochowski et al. 1993; Fig. 13.15). In areas surrounding the British Isles where they are documented in detail, such tidal-transport pathways stretch for tens to hundreds of kilometers (Fig. 13.16). Sand is transported away from erosional areas referred to as *bedload partings* (i.e. locations where the directions of residual transport diverge; Harris et al. 1995), and moves along the transport pathway, much of which is a bypass zone (Stride 1963; Kenyon and Stride 1970; see summary in Johnson et al. 1982). In general, current speed decreases along the length of a transport pathway. Consequently, the sand becomes finer in the residual down-current direction, and is deposited as the transport capacity decreases. The location of the site of maximum deposition, and hence the thickest deposits, depends on the rate of decrease of transport capacity and does not need to be located at the end of the transport pathway. The maximum net deposition might also occur where two pathways meet (i.e. a *bedload convergence* zone), as it is the case to the west of the Strait of Dover (Fig. 13.15).

Smaller scale tidal-transport pathway can also be created by topographic constrictions. For example, straits can create localized areas of scour because of flow acceleration through the narrows. On the other hand, the flow exiting from the constriction experiences flow expansion leading to sediment deposition. Because of the reversing flow, deposition can occur at both ends of the strait, which acts as a localized bedload parting. Several examples of such accumulations have been documented from the rock record. The deposits associated with the straits can take the form of extensive sand sheets (with or without tidal ridges) or more localized delta-like bodies (e.g. Kamp et al. 1988; Reynaud et al. 2006; Fig. 13.17). Because these topographically controlled tidal-transport pathways are spatially more restricted than those occurring on open shelves, crossbedded sands can pass laterally to finer-grained, non-tidal facies over only a few kilometers (Bassant et al. 2005).

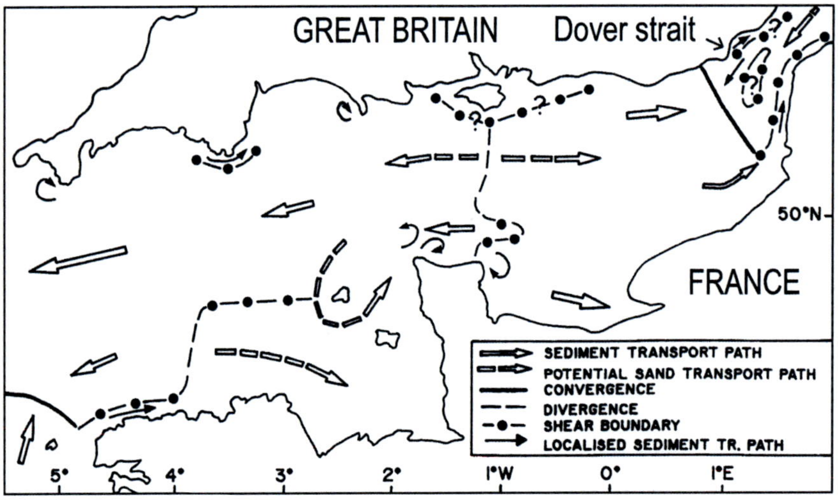

Fig. 13.15 Residual bedload transport directions in the English Channel, based on hydrodynamic modeling of tidal currents. The modeled tidal-transport pathways is in good agreement with the transport directions deduced from active bedforms, except in the Strait of Dover where the convergence zone lies further to the north (After Grochowski et al. 1993)

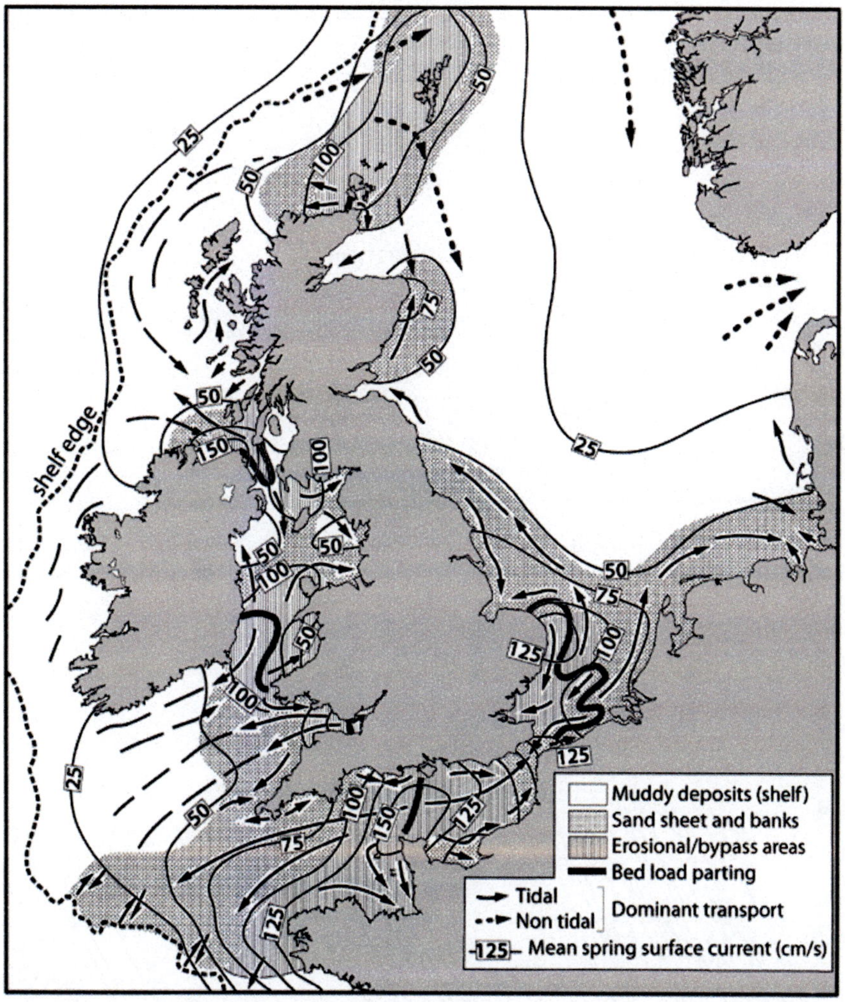

Fig. 13.16 Sea-floor sediment type, surface currents and tidal-transport pathways around the British Isles. Arrows show potential bedload-transport directions that diverge from erosional bedload partings, to convergence areas where deposition occurs (After Howarth 1982 and Johnson et al. 1982)

13.6.2 Bedform Distribution

A predictable progression of bed features occurs along a tidal-transport pathway as a result of changes in the sediment regime (Belderson et al. 1982; Fig. 13.18). In erosional, bedload-parting areas, older deposits are exposed or are covered by a patchy veneer of lag gravel and sand. Erosional features include current-parallel furrows and flute-shaped depressions. Mobile sand in these areas occurs as sand shadows in the lee of bedrock obstacles and as current-parallel sand ribbons. Subaqueous dunes can be present on the sand ribbons. As sand begins to become more abundant down the transport pathway, fields of dunes that migrate in the direction of the residual transport begin to form. In regions of limited sand, the dunes are separated by areas with no sand cover (i.e., they are 'starved') and have a barkhanoïd shape (Fig. 13.18b). If there is a larger amount of sand, the dunes coalesce to produce tidal sand sheets (Fig. 13.18a) and, under proper conditions, tidal sand ridges (Fig. 13.18c). At the extreme end of the transport pathway, sheets and patches of rippled fine and very fine sand occur where the current speed is below the critical velocity for dune stability.

As discussed above, the upper limit on dune size is controlled by the thickness of the boundary layer, which is approximated by water depth in many situations. However, current speed, grain size and sediment

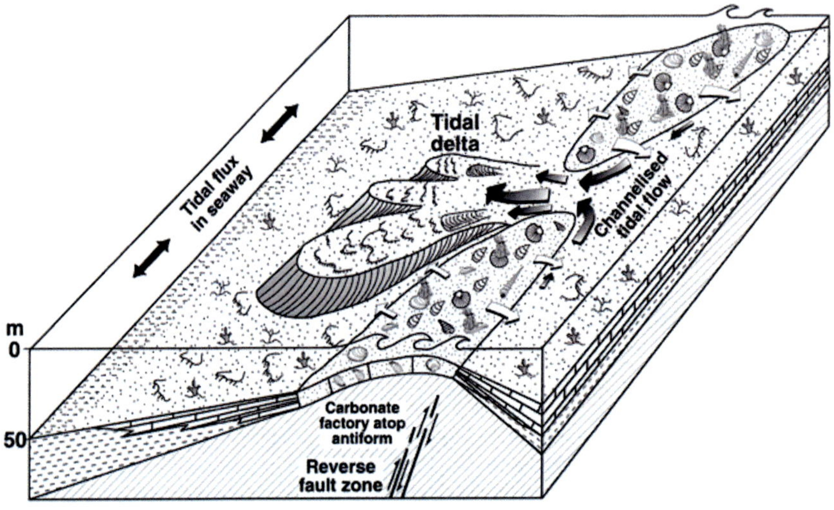

Fig. 13.17 Subaqueous, delta-like body formed of bioclastic carbonates at the 'mouth' of a subaqueous channel that cut through the low part of an anticlinal ridge in the Pliocene forearc basin of the Northern Island of New Zealand. The delta deposits consist in giant tabular crossbeds, with sets 10–40 m thick and foreset dips of 7–36°. The delta pinches out basinward into fine-grained deposits (From Nelson et al. 2003)

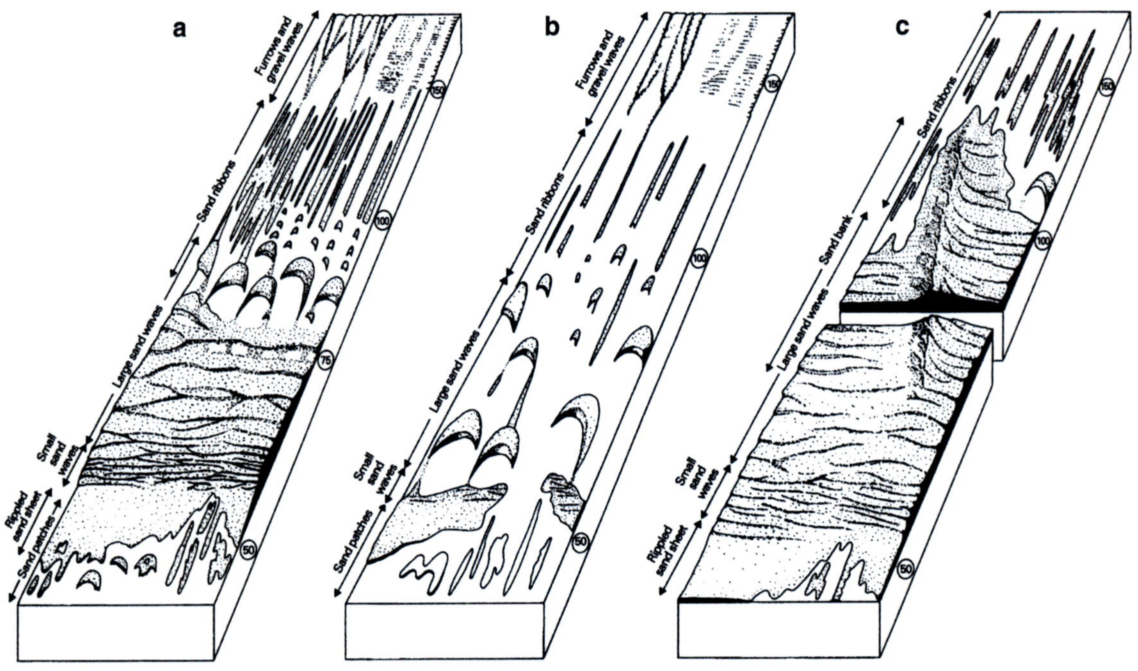

Fig. 13.18 Schematic diagrams showing the spatial distribution of bedform types along tidal-transport pathways (net sediment transport is toward the front of the diagram): (**a**) intermediate case; (**b**) regions with limited sand; and (**c**) areas with abundant sand. The up-current area in all examples experiences net erosion, whereas the downstream portion experiences net deposition because of the down-current decrease in current speed (numbers in circles). In (**c**), tidal sand ridges (tidal sand banks) occur in the upcurrent (higher current speed) portion of the depositional area. Such ridges pass down-current into a tidal sand sheet that is mantled by dunes. All of the features shown are part of the 'transgressive lag' that mantles a flooding surface (From Belderson et al. 1982)

availability also influence the size to which dunes grow. At the scale of an entire tidal-transport pathway, the dunes decrease in size along the pathway due to a decrease in tidal-current speed and grain size (Fig. 13.18a). Thus, the largest dunes are commonly found a short distance down-flow from the point where significant deposition occurs within the pathway, because this is where the current speed, grain size and sediment availability are greatest. At a smaller scale, the dunes in sand sheets are commonly grouped in isolated dune fields that are surrounded by an immobile substrate (e.g. Reynaud et al. 1999a). By contrast to what happens at the larger scale, the dunes in a dune field commonly increase in size in the down-flow direction, because of the increasing amount of available sand. As the dunes get bigger, they also get less mobile, with the larger ones tending to trap sediment brought to them by more rapidly migrating smaller dunes. Sand may not be able to escape from them; consequently, the dune field can terminate abruptly with the largest dunes near or at their leading edge. Within sand-ridge fields, scour in the troughs between the ridges commonly exposes older deposits that can be an internal source of sediment. The coarse-grained, shelly lags that occur in the troughs correlate to, and may be laterally continuous with, the lag facies in the bedload-parting area.

13.6.3 Deposits

Evidence of the former existence of tidal-transport pathways is likely to be preserved on flooding surfaces in ancient successions that contain evidence of tidal action. Over large areas, this evidence will consist of a marine erosion or 'ravinement' surface (see more below) that is mantled by a thin lag. In a down-transport direction, the deposits on this surface will thicken, potentially reaching a few tens of meters in thickness. The most volumetrically significant facies will consist of cross-bedded sands formed by dunes that were part of isolated dune fields, or of more extensive sand sheets and tidal-current ridges. The deposits of tidal-current ridges have been discussed above; here, we examine the deposits of sand sheets (Fig. 13.18a).

Little is known about the organization of these deposits, but we hypothesize that the vertical succession produced by a sand sheet consists of a stacked succession of simple and compound-dune deposits (Fig. 13.5), produced during an episode of increased tidal activity and areal expansion of the bedload parting area because of the transfer of sand to the depositional area (Harris et al. 1995). The lower part of this succession should coarsen upward, with an upward increase in the scale of the crossbeds, as a result of a 'progradation' of up-flow parts of the sand sheet over its more distal portion (Fig. 13.19a). As progressively larger bedforms migrate into the area, the scour surfaces at their base might become more prominent, potentially removing significant a amount of the pre-existing succession. If migration of the sand sheet continues, the succession will be overlain by an erosional surface corresponding to the bypass zone at the upcurrent-end of the transport path (Fig. 13.19a). If, instead, flow speeds decrease because of a change in the tidal regime, then the dunes on the sand sheet should become smaller and finer grained, generating an overall upward fining succession at the top of the sand-sheet deposit.

Full development of such a succession, with a lower upward coarsening/thickening part and an upper upward fining thinning of crossbeds, requires a relatively abundant supply of sediment. Mellere and Steel (1996) and Blackwood et al. (2004) describe successions from a seaway setting that show strong similarities to this model. Perhaps the closest match has been described from a cool-water carbonate environment (Anastas et al. 2006; Fig. 13.19b). In this case, the succession has been interpreted to reflect the migration of an area where the tidal current is above the threshold of dune formation rather than a change of current speed as a result of sea-level rise, but the result of these two processes may be difficult to distinguish. At the fringe of that area, patch reefs are present and there is a tendency for the development of hardgrounds (cf. Fig. 13.19b).

Strong tidal currents are commonly correlated with the supply of nutrients, so that the deposits in tidal sand sheets are likely to be bioturbated and, even in siliciclastic settings, to contain shelly fauna (Wilson 1982). If the dunes in a tidal-transport pathway migrate rapidly, then the intensity of bioturbation will be low, whereas dunes that migrate slowly (i.e. the larger ones, or those toward the distal end of a tidal-transport pathway) can be bioturbated more thoroughly. In general, the bottomsets of the dunes will be more intensely bioturbated than the foresets. The assemblage will be a mixed *Skolithos-Cruziana* Ichnofacies (Seilacher 1967; Ekdale et al. 1984; MacEachern et al. 2005), with vertical burrows subtending from the erosion surfaces that

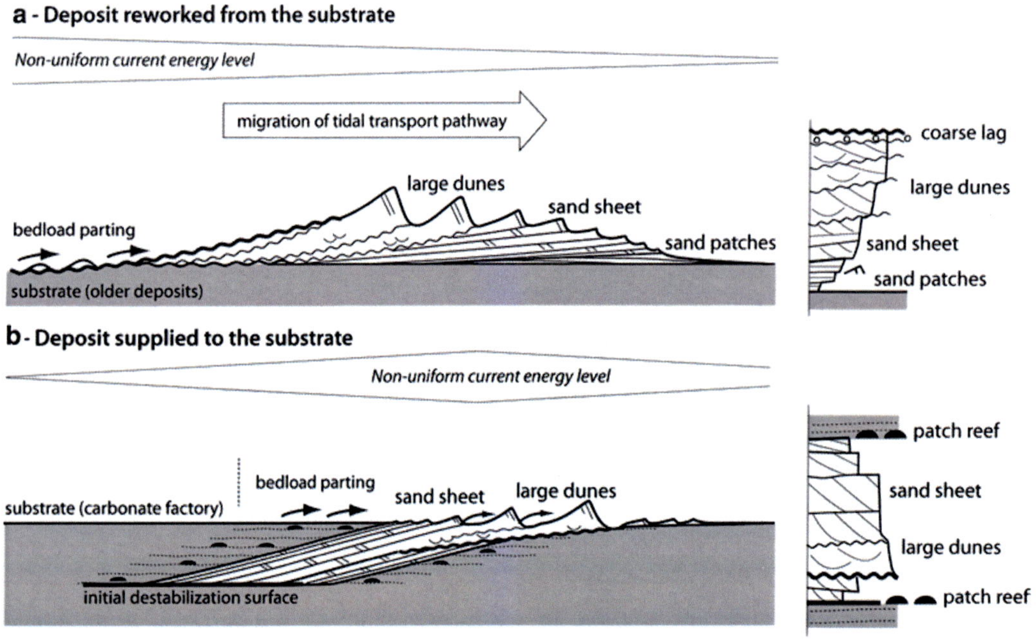

Fig. 13.19 Space-thickness diagrams showing the successions created by a tidal-transport pathway in (**a**) a siliciclastic setting where most sediments are reworked from older deposits and (**b**) a carbonate environment with significant *in situ* production. Note the difference with regard to where the deposit occurs relative to the area of strongest currents. Progressive sediment starvation in the siliciclastic setting and down-flow migration of facies zones take place as the zone of seafloor erosion expands (cf. Harris et al. 1995), producing an upward-coarsening succession topped by an erosional lag. The carbonate system, by contrast, is not sediment starved and records a more symmectrical upward-coarsening/upward-fining succession that forms as a result of migration of the dune field over lower-energy deposits, followed by gradual abandonment as tidal-current speeds decreases while transgression proceeds. Patch reefs are inferred to occur in proximity to the dune field in the carbonate example, because nutrients are supplied by the tidal currents. Cross sections not to scale, but sediment thicknesses can reach 30–50 m in both situations. The horizontal extent is tens of kilometers. (Sketches based on Belderson et al. 1982 and Anastas et al. 2006, respectively)

separate cross beds, and horizontal burrows within the deposits. For additional details on the ichnology of tidal deposits, readers are referred to Chap. 4.

13.7 Transgressive Stratigraphy

As noted already, sandy tidal deposits on modern shelves are transgressive in origin (Fig. 13.20), and such is likely to be the case in the ancient as this is the main time that coarse-grained sediment (coarser than mud) is able to escape from nearshore areas onto the shelf. The siliciclastic sand that forms these deposits is derived primarily from the reworking of older deposits, including such features as drowned valley fills (Reynaud et al. 1999c), lowstand deltas (Posamentier 2002; Berné et al. 2002; Fig. 13.21) and transgressed coastal barriers (Fig. 13.6a). Sand is especially abundant seaward of headlands, because the convergence of longshore drift, coupled with erosional retreat of the shoreline, creates a large headland-attached sand body called a *shoal retreat massif* (Swift 1975). Similar sediment bodies also form seaward of river mouths. As the transgression progresses and the shoreline migrates landward, nearshore ridges become left behind on the shelf where they can either become moribund or continue to be reworked actively. Even where the source deposit is muddy, the tidal currents can winnow away the mud and produce a sandy deposit, as illustrated by some ridges in the East China Sea (Liu et al. 2007).

Erosion of the shoreline and shelf during the transgression is generally termed *transgressive ravinement.* In wave-dominated settings, transgressive ravinement is generally limited to the shoreface, forming a wave ravinement surface. In tidal environments, transgressive ravinement occurs at the coast, both as tidal ravinement at river mouths and as wave ravinement

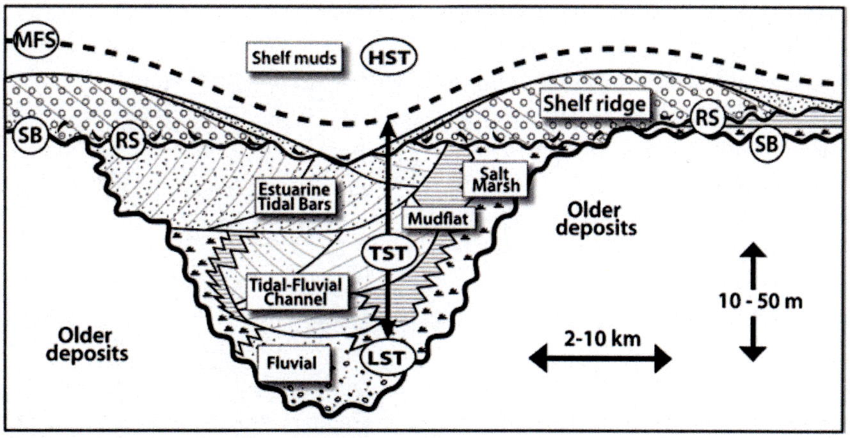

Fig. 13.20 Model of depositional succession on transgressive tidal shelves, supposing that all of the facies tracts are superimposed on each other. The succession begins with a valley-fill deposit that is capped by tidal-current ridges that formed on the shelf. *LST*, *TST*, *HST* lowstand, transgressive, and highstand systems tracts; *SB* sequence boundary; *RS* transgressive ravinement surface; *MFS* maximum flooding surface (After Dalrymple 2010b)

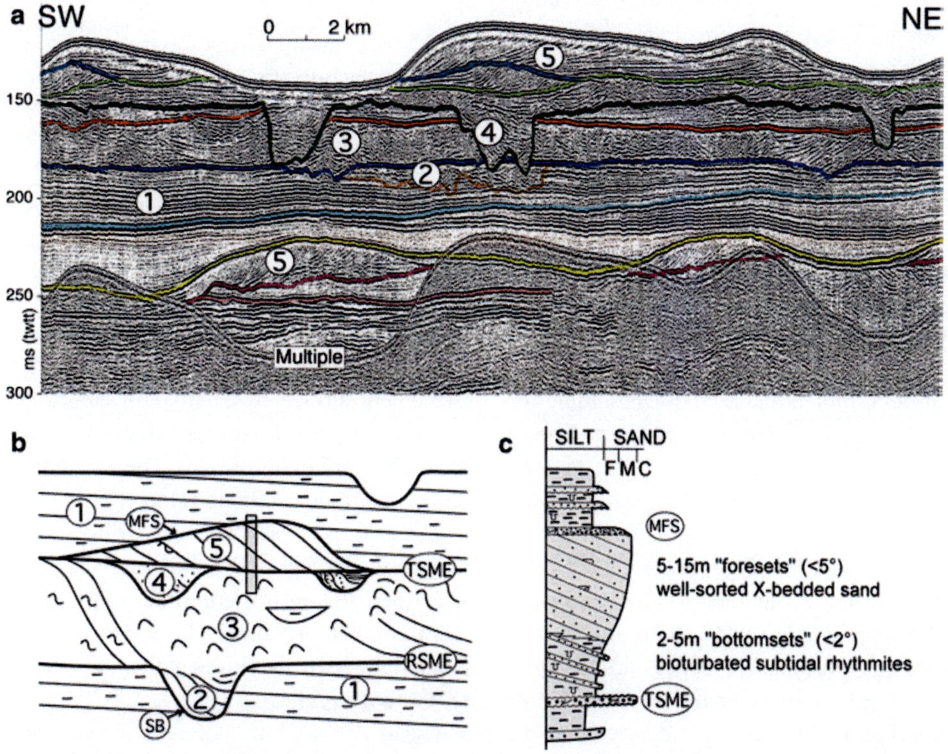

Fig. 13.21 (**a**) Seismic section from the East China Sea, offshore from the Changjiang River (see location in Fig. 13.6b, ridge field 2). Laterally migrating tidal ridges are present on the modern surface; older ridges are present near the bottom of the section, buried beneath prodeltaic mud. (**b**) Interpretation of the succession, as determined from seismic attributes and facies in cores. Depositional environments in (**a**) and (**b**): *1* prodeltaic deposits, *2* deltaic/estuarine channels of the falling-stage and lowstand systems tracts, *3* fluvial deposits of the lowstand systems tract, *4* early transgressive estuarine channels, and *5* transgressive shelf ridges. *RSME* regressive surface of marine erosion, *TSME* transgressive surface of marine erosion, *SB* sequence boundary, *MFS* maximum flooding surface. (**c**) Schematic vertical succession through the ridge (location shown by the *gray rectangle* in **b**). The coarsening-up succession reflects the increase of hydrodynamic energy toward the ridge crest. Note the coarse lag at the bottom of the ridge, which is interpreted as the tidal ravinement surface (From Berné et al. 2002)

along the open-coast shoreface. It continues, however, in the offshore area as a result of tidal scour in widespread, bedload-parting and bypass zones, forming a surface that we term an *offshore tidal ravinement surface* (Reynaud et al. 2003) to distinguish it from ravinement surfaces formed in the coastal zone. The offshore tidal ravinement surface can be continuous with the coastal ravinement surface, or can be stratigraphically distinct, as is the case where the bedload parting area migrates over areas that were formerly depositional (Fig. 13.19a).

If the sand liberated from the seafloor by offshore tidal ravinement is carried away by the residual transport, erosional offshore tidal ridges can form, as is the case in the Yellow Sea and the Korea Strait (Jung et al. 1998; Jin and Chough 2002, Park et al. 2006). The outer shape of the Celtic Banks has also been interpreted by some authors as purely erosional (Berné et al. 1998). The same circulation cells as those shown in Figure 13.8 exist, but in an erosional, rather than depositional, mode. Some of these ridges have been sculpted into muddy estuarine deposits that predate the last glacial-maximum lowstand (Fig. 13.22), and are mantled by a veneer of modern tidal sand. As sea level rises, the erosional area can migrate landward and the ridges can enter a progressively more depositional regime while current speeds decrease. A constructional tidal ridge may then develop above the erosional form that acts as the nucleus for sediment deposition. Numerous examples of offshore tidal ridges that show this erosional-depositional history exist, including the ridges of the Boulonnais along the northern French coast (Lapierre 1975; De Batist et al. 1996) and the Flemish Banks in the southern North Sea (D'Olier 1981; Laban and Schuttenhelm 1981; Berné et al. 1994; Trentesaux et al. 1999; Fig. 13.23).

Thus, extending the evolutionary model proposed by Snedden and Dalrymple (1999), there might be a continuum between erosional (juvenile) and constructional (fully evolved) ridges as conditions change over a transgression (Fig. 13.24). Juvenile ridges contain the initial relief or bump from which they grew. In some instances, this precursor might be composed of sand. This is, for example, the case of one of the Flemish Banks (Trentesaux et al. 1999; Fig. 13.23). There, the 'precursor' consists of a finer-grained, more bioturbated and less well sorted sandbody (U4 in Fig. 13.23) that may correspond to a shoreface-attached ridge,

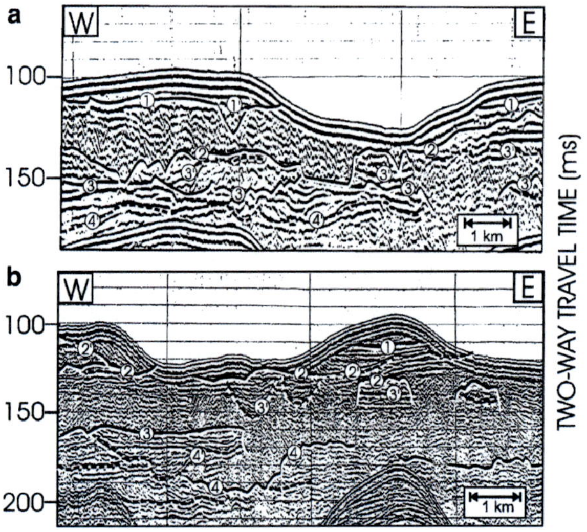

Fig. 13.22 Interpreted seismic sections showing the complex internal architecture of erosional transgressive tidal ridges in the eastern part of the Yellow Sea. The internal bedding is overall aggradational, comprising two seismic facies interpreted as lowstand, probably fluvial, deposits (chaotic reflections) that rest on sequence boundaries (circled numbers without prime), and early transgressive tidal coastal muds (stratified deposits) that rest on tidal ravinement surfaces (circled numbers with prime). The ridge shape is entirely erosional, with the external surface truncating strata within the ridge. Most of the sediment within the ridge is muddy heterolithic deposits that predate the last glacial lowstand (From Jin and Chough 2002)

resting on a pebble lag (the wave ravinement surface). This is overlain, above a prominent offshore tidal ravinement surface, by the modern shelf ridge that consists of well-sorted, coarser, bioclastic-rich crossbedded sand (U6-7 in Fig. 13.23). The longer the ridges are active, the more fully developed they are likely to become. As a consequence, the ridges that originate early in the transgression and are, thus, located on the outer shelf today, are more likely to be fully evolved than those that originated close to the highstand coast, as examplified by the ridges in the English Channel and its Western Approaches (Fig. 13.25).

As the water depth increases during the transgression, the tidal-current speeds can decrease (Fig. 13.26), causing the tidal ridges to become moribund (i.e. they are no longer active). The Celtic Ridges are the best-described examples of moribund ridges (Bouysse et al. 1976; Pantin and Evans 1984; Reynaud et al. 1999a, b). They have a rounded shape, and there is evidence of storm erosion of their crest and the development of

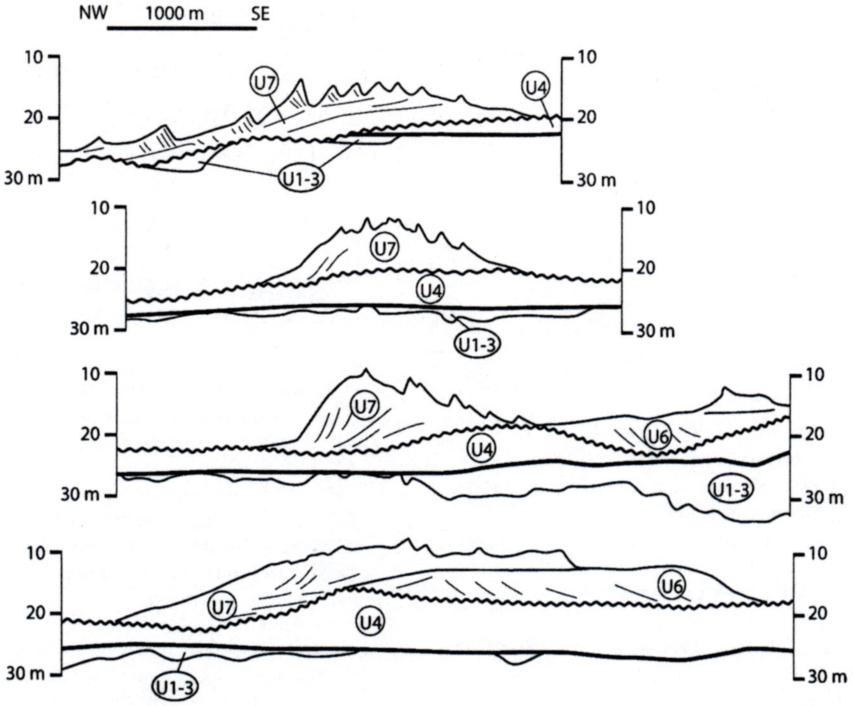

Fig. 13.23 Seismic profiles across Middelkerke Ridge (Flemish Ridges, Southern North Sea, see Fig. 13.6a). *U1–U3* estuarine channels and tidal flats; *U4* coastal barrier, shoreface and ebb-deltas; *U6–U7* active, offshore tidal ridge. The sinusoidal *bold line* is the offshore tidal ravinement surface; the clean, offshore sands of the ridges lie on this surface. The plain thick *black line* is the wave ravinement surface, which is marked by a pebble lag (After Trentesaux et al. 1999)

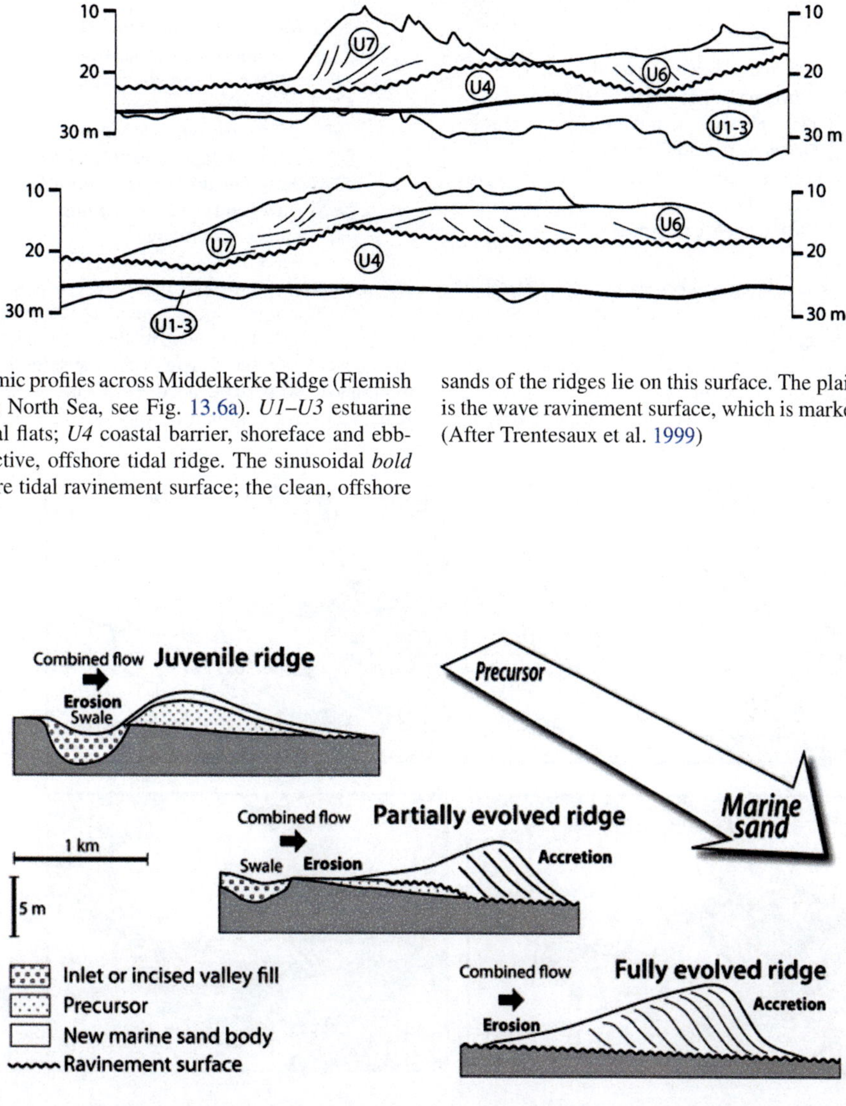

Fig. 13.24 Evolutionary model of offshore ridges. An initial bump, or nucleus, is required to start the Huthnance process. As the ridge migrates in response to the residual current, this initial core ends up being removed by erosion of the upcurrent side of the ridge. Fully developed ridges are exemplified by the deepest Norfolk Ridges in the Southern North Sea (Figs. 13.6a and 13.10). Partially evolved ridges that retain part of the original nucleus are more similar to the Flemish Ridges (Fig. 13.23) that are located closer to the coast and started to form later in the last transgression (From Snedden and Dalrymple 1999)

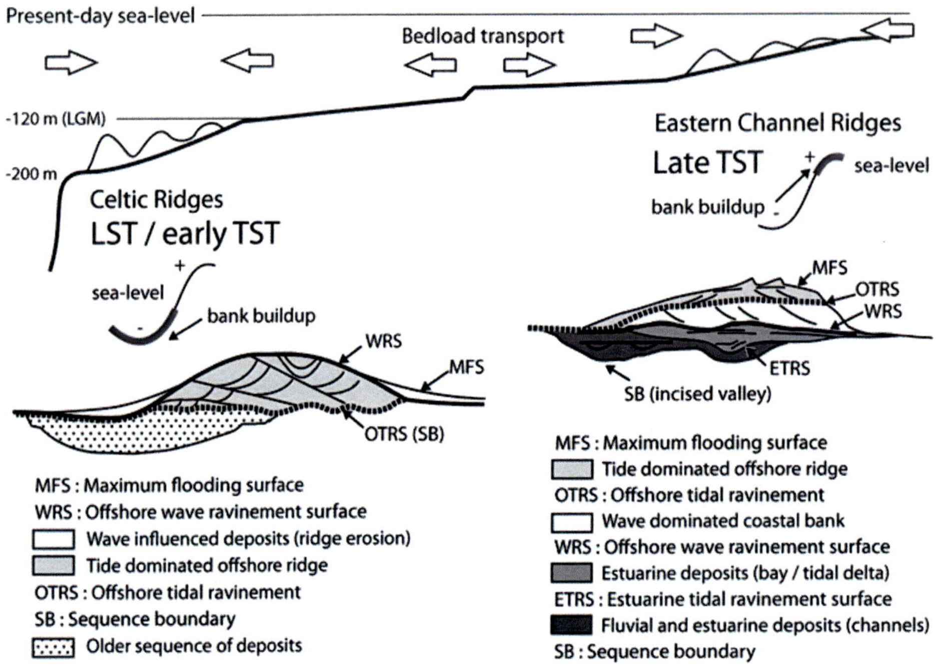

Fig. 13.25 Depositional setting and key sequence-stratigraphic surfaces within tidal sand ridges in the southern Celtic Sea and eastern English Channel (see location in Fig. 13.6a, ridge fields 6 and 4, respectively) that were formed early and late, respectively, in the last post-glacial transgression. The shallower tidal ridges have a core that may preserve coastal sandbodies. By contrast, the offshore tidal ravinement surface has more time to rework the deeper ridges, which are therefore 'fully evolved' (cf. Fig. 13.24). Subsequent to their formation, tidal currents became weaker in the course of the transgression, leaving them "moribund" and partly eroded by storm waves, forming a wave ravinement surface at their crest (After Reynaud et al. 2003)

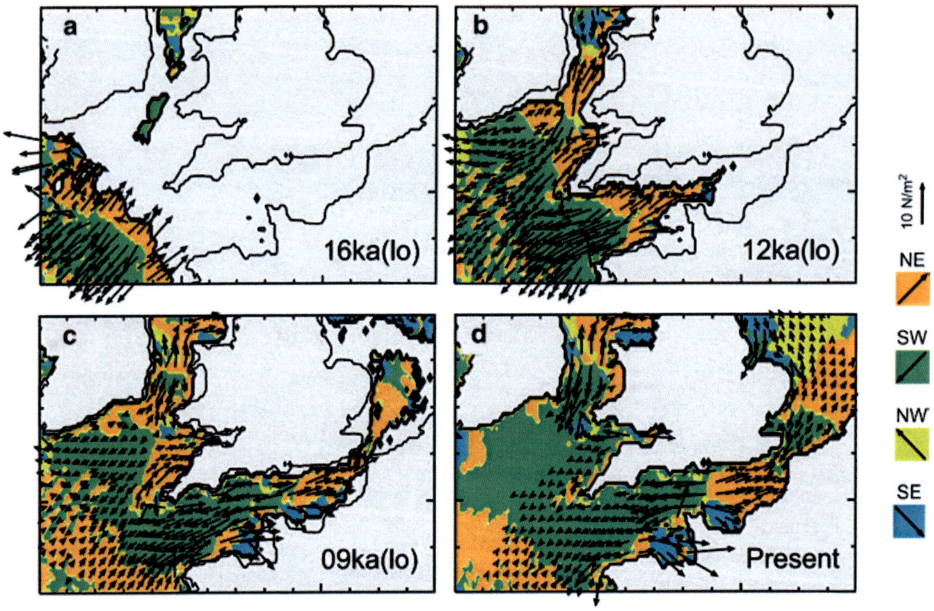

Fig. 13.26 Paleotidal modeling of the English Channel, Irish Sea and southern North Sea for four times during the last transgression. The maps show the peak bed-stress vectors (length is proportional to the current strength) and the directions of potential sand transport, which define the tidal-transport pathways that evolve through time as sea level rises. The peak bed stress near the shelf-margin, in the area occupied by the Celtic Sea sand ridges (Figs. 13.6a and 13.12), is at a maximum around 16 ka, suggesting that these ridges were formed in the early stage of the last transgression. Note that, as transgression proceeds eastward up the English Channel, the shear stress decreases until the Strait of Dover is flooded, at which time conditions are created that allow the formation of a new set of ridges there (cf. Fig. 13.25) (From Uehara et al. 2006)

wedges of sediment that partly bury the ridge flanks (Figs. 13.12 and 13.25). Consequently, their outer slopes are less steeply inclined than those of active ridges, and commonly dip at less than 1° (Stride et al. 1982). Dunes, which typically mantle active ridges, are generally absent from their crest. A decrease in the ability of the tidal currents to bring in more sand leads to an increase in the carbonate content of the sediment, forming a shelly lag on the ridge crest; the constituents present in this bioclastic material should record the increasing water depth (Wilson 1988). The increase in shell content can be either gradual or abrupt (Wilson 1982; Davis et al. 1993). The surface of the ridge can also become pervasively bioturbated. In the ancient, the post-ridge draping deposits might record a transition from tidal to wave dominance, as reported from examples in the Western Interior Cretaceous of the USA (Hein et al. 1991; Mellere and Steel 1995; Yoshida et al. 2007). As would be expected in a transgressive succession (Fig. 13.20), the first deposits above the ridge crest are glauconitic shelf muds that indicate sediment starvation and slow sedimentation (Surlyk and Noe-Nygaard 1991) and the formation of the maximum flooding surface. The dominantly muddy sedimentation that characterizes the overlying highstand systems tract may entirely bury the remnant ridges, as is observed in the Pleistocene of the East China Sea (Berné et al. 2002; Fig. 13.21) and the Miocene of offshore Java (Posamentier 2002).

13.8 Sea-Level and Geomorphic Interactions

Because the oceanic tidal wave interacts strongly with the morphology of the shelf and coastline, changes in the geomorphology of an area as a result of changes in relative sea level can have a profound influence on the occurrence of tidal deposits. As noted above, it is generally believed that the formation of *sandy* tidal-shelf deposits is restricted to transgressive situations. The occurrence of *muddy* tidal-shelf deposits is not well known because they have not been studied as systematically, but they appear to occur mainly in regressive, delta-related settings (e.g. the Amazon River: Gabioux et al. 2005; Bourret et al. 2008; the Irrawaddy River: Viana et al. 1998; Ramaswamy et al. 2004; Rao et al. 2005; the Yellow River: Yang and Liu 2007). Such deposits are discussed elsewhere in this volume (Chap. 7). From a theoretical point of view, however, there is no reason for tidal dominance to occur only during transgressions (Yoshida et al. 2007). The following sections explore the timing of tidal dominance, focusing primarily on the results of paleotidal modeling of both modern and ancient basins.

13.8.1 High-Frequency Changes

Short-term, tectonic- or climate-driven sea-level variations with a period of less than about 100,000 years (fourth-order or higher; Vail et al. 1977) can bring about rapid change in the nature of shelf sedimentation, potentially causing an alternation between tidal and non-tidal deposits (Fig. 13.21). Two contrasting situations can be documented: (i) open shelves on passive margins, and (ii) epicontinental seaways.

On open shelves, the tidal influence is expected to increase with rising sea level, because the increasing shelf width typically brings the system closer to resonance, although the opposite can occur if the width at lowstand is close to one-quarter of the tidal wavelength. If the shelf has embayments, they might be the most sensitive to changes in tidal influence, due to funneling of the flow toward the head of the bay. The increase in tidal influence can be geologically instantaneous in situations where the geomorphology changes rapidly. This was the case in the Gulf of Maine-Bay of Fundy system, which changed from microtidal to extreme macrotidal over a period on only a few thousand years (Greenberg 1979; Dalrymple and Zaitlin 1994; Shaw et al. 2010). One possible ancient analogue is provided by the Woburn Sands in the Cretaceous Greensand Seaway of NW Europe, where the strength of the tides increases upward through the transgressive succession, being stronger in shelf sediments than it is in estuarine deposits at the base (Yoshida et al. 2004), a situation that is the reverse of what might be expected because tidal currents are typically stronger within embayments than on the open shelf. Once tidal resonance has been reached, however, any further increase in sea level will result in the decrease in tidal influence. This might be illustrated by the sudden abandonment and preservation of fossil tidal dune fields beneath Holocene offshore muds in the Southern North Sea (Brew 1996). This could also be the case for the tidal sandbodies of the Devonian Castkill Sea (Ericksen et al. 1990).

It must be noted that different parts of the transgressing sea can become resonant at different times, or

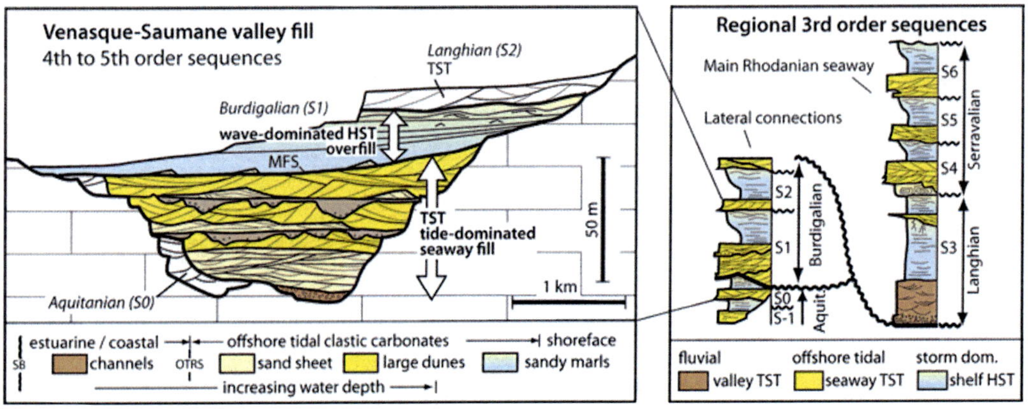

Fig. 13.27 Succession of deposits in the Miocene foreland-basin seaway of SE France, showing the influence of changing sea level and morphology on the development of tidal deposits. (*Right*) Succession of third-order sequences (S1–S6), each of which is composed of a lower, bioclastic-rich transgressive (*TST*) deposit formed by tidal dunes, and an upper marly, wave-influenced highstand (*HST*) deposit. The sequence boundaries are possibly subaerial erosion surfaces, but they were intensely scoured by strong tidal currents during transgression, forming tidal ravinement surfaces. The thickness of the sequence set exceeds 100 m in most places. (*Left*) Detailed architecture of the Saumane-Venasque seaway filled and overtopped by sequence S1. When relative sea level was low and water was confined to the valley, the deposits were tide-dominated, coarse-grained bioclastic material. Once the water level rose above the interfluves, the cross-sectional area increased dramatically, so that tidal-current strength decreased, and the deposits became fine grained and wave-dominated. *OTRS* offshore tidal ravinement surface. This depositional pattern is repeated in each of the six sequences shown in the *right-hand* figure (After Besson 2005 and Besson et al. 2005)

perhaps more than once during a major transgression. This might have been the case for the English Channel during the last sea-level rise. After a first 'tidal climax' early in the post-glacial transgression (i.e., at ca. 16 ka BP), an abrupt loss of tidal resonance was recorded that might correspond to the end of active up-building of the Late Pleistocene Celtic Banks (Reynaud et al. 1999b; Uehara et al. 2006). Much more recently, a second phase of strong tidal action has occurred in the Normandy-Brittany Gulf and Eastern Channel, bringing about the formation of a second set of tidal ridges at a much more landward location (Reynaud et al. 2003; Fig. 13.25).

In the case of broad seaways, the tidal influence is likely to increase with falling sea level, because narrowing of the seaways favors the constriction of tidal flows and restricts the fetch of wind waves (Yoshida et al. 2007). A narrowing of the seaway due to progradation of its margins would have the same effect, even if water depths were increasing (Van der Molen et al. 2004). The Cretaceous Western Interior Seaway of Northern America provides examples of this type of response to sea-level change, with lowstand deposits being tide-dominated, whereas highstand deposits are wave-dominated (see Chap. 17). In a different example, paleotidal modeling by Wells et al. (2007) showed that tidal circulation inside the Late Pennsylvanian Midcontinent Seaway of America would have been diminished during times of maximum flooding, promoting the development of black shales, whereas the lowstand to transgressive intervals, when the seaway was narrower, favored the development of large tides in the eastern embayment of the seaway, as recorded by the coeval fluvial-estuarine transition in Kansas (Lanier et al. 1993). At a smaller scale, the same interpretation is provided by the Miocene incised valleys/seaways of SE France (Besson et al. 2005; Reynaud et al. 2006). During highstands, the valley interfluves are flooded creating broad seaways in which deposition consisted of wave-influenced carbonate-rich mudstones with minimal tidal influence (Fig. 13.27). By contrast, when sea level fell and the interfluves became emergent, flow constriction within the valleys lead to the accumulation of localized tide-dominated deposits. Other examples of water-depth control on the strength of tidal currents in seaways are provided by Anastas et al. (2006; Oligocene of New Zealand) and Surlyk and Noe-Nygaard (1991; Jurassic of Scotland; Fig. 13.28).

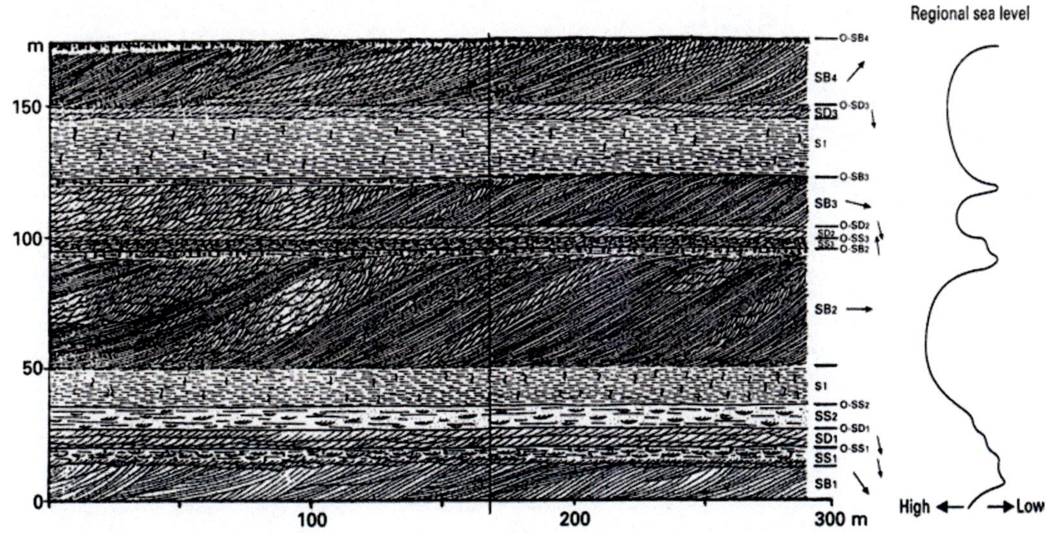

Fig. 13.28 Inferred influence of sea-level change on the architecture of tidal deposits infilling a Jurassic seaway that occupied a North Sea rift basin. The succession consists of an alternation of thick crossbedded deposits formed by large dunes migrating along the axis of the seaway that form a tidal sand ridge during transgressive periods, and finer-grained and more thinly bedded sandstones that accumulated during highstands when the water depth was greater and the currents speeds were less. Each sea-level fall and the start of the subsequent transgression is marked by a discrete pebble lag and bioturbated, glauconitic horizon that underlies the giant crossbeds. Labels in the margin refer to deposit attributes (*SB* sand bank, *SS* sand sheet, *arrows* palaeocurrents) (From Surlyk and Noe-Nygaard 1991)

13.8.2 Long-Term Changes in Basin Morphology

On the much longer time scale of first- to second-order sea-level changes (up to a few hundreds of meters of relative sea-level change, stretching over tens to hundreds of millions of years), two generic end-member situations arise (Fig. 13.29). During overall low sea-level periods (e.g. during the late Cenozoic and present), there is limited flooding of continental interiors. Most shallow-marine sedimentation occurs on narrow shelves at the margins of the continents. Large-scale embayments are restricted primarily to tectonically structured seaways along collisional or transform margin (Kamp et al. 1988; Hoppie 1996). During overall high sea level, such as in the Upper Cretaceous, by contrast, a much larger part of the continents is flooded, creating extensive semi-enclosed seas with a complex topography. Because of their complex paleogeography, these seas experience very complex interactions between friction forces and tide-enhancing processes that cannot be solved without the help of paleotidal modeling.

It is expected that, in the course of an overall first- or second-order transgression, the shelf will gradually grow wider, with the progressive development of a more complex coastline, including tidal embayments that can extend many hundreds of kilometers inland (Houbolt 1982; Houthuys and Gullentops 1988a, b; André et al. 2003). With continued sea-level rise, these embayments can eventually evolve into tidal seaways and straits with a marine connection at both ends (e.g. Anastas et al. 1997; Besson et al. 2005; Longhitano and Nemec 2005). Whereas tidal currents must decrease at the head of an embayment, they can be accelerated through a seaway, making possible the propagation of a progressive tidal wave and the maintenance of strong tidal currents a long distance into the continental interior. This seems to have been the case for the Peri-Alpine, Miocene seaway of southern Europe, which formed a short-lived connection between the Atlantic and the Paratethys during the Burdigalian (Allen et al. 1985; Martel et al. 1994; Bieg 2005; Fig. 13.30). If, however, tidal resonance occurs at the embayment stage, the connection of the head of the embayment to another tidal basin as a

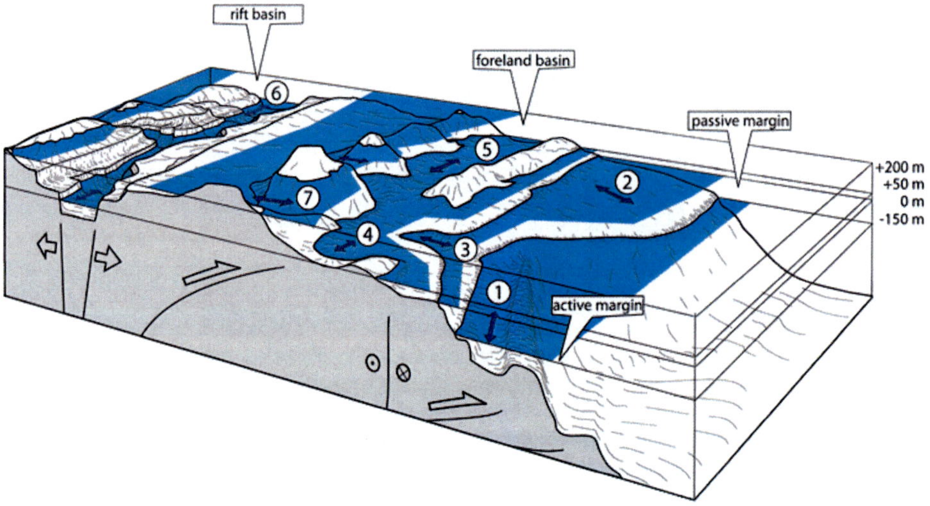

Fig. 13.29 Schematic diagram showing the range of tectonic basin types. In each type, changes in relative sea level cause a change in basin morphology, which in turn controls where and when tidal deposits will form. The *arrows* are oriented parallel to the mean tidal current. The flooding levels displayed correspond to the extremes of long-term (first or second order), high-amplitude sea-level cycles. Drawing not to scale. Active and passive margins are displayed as adjacent to each other only for convenience. Examples: *1*– Te Kuiti Group, New Zealand, Oligocene (Anastas et al. 1997, 2006); Hikurangi fore-arc basin, New Zealand, Pliocene (Kamp et al. 1988; Nelson et al. 2003); Cuyama transform margin, California, Miocene (e.g. Hoppie 1996). *2* Celtic Sea, Late Pleistocene (e.g. Bouysse et al. 1976); South China Sea, Late Pleistocene (e.g. Berné et al. 2002). *3* English Channel, Late Pleistocene to Holocene (e.g. Reynaud et al. 2003). *4* Southern North Sea, Holocene (e.g. Berné et al. 1994; Trentesaux et al. 1999); Yellow Sea, Holocene (e.g. Park et al. 2006); Greensand Seaway, Aptian (Yoshida et al. 2004; Wells et al. 2010). *5* Perialpine Seaway, Burdigalian (e.g. Lesueur et al. 1988; Bieg 2005). *6* North Sea rift basin (Scotland), Jurassic (Blackwood et al. 2004; Mellere and Steel 1996). *7* Bohemian Basin, Turonian (e.g. Mitchell et al. 2010). In *1*, *5*, *6* and *7*, the main control on tidal sedimentation is the constriction of the tidal flow through a narrow passageway. The tidal range does not need to be large. In *2*, *3* and *4*, the main control is the increase in tidal amplitude, related to either an increase in tidal prism (embayments) and/or tidal resonance (shelf)

result of sea-level rise could bring about an abrupt decay of tidal currents, as demonstrated by the post-glacial evolution of Cook Strait, New Zealand (Proctor and Carter 1989). As a result of paleotidal modeling of the Lower Cretaceous Greensands Seaway in Europe, Wells et al. (2010) suggest that, although the southern connection of the seaway to the Neotethys would have been crucial for the development of significant tides, the existence of a coeval connection to the North Sea would have caused tides to decrease (Fig. 13.31).

During times of very high sea level and extensive flooding of the continents, extensive, shallow-marine carbonate sedimentation would occur around emergent archipelagos. Paleotidal modeling shows that such a broad, shallow platform would probably be microtidal, due to the great distance from the open ocean. In such a setting, strong tidal currents would occur only within small constrictions between islands, as shown by modeling of the Bohemian Cretaceous Basin (Mitchell et al. 2010), or of the Lower Jurassic Laurasian Seaway (e.g. Luxembourg sandstones, P. Allison 2010, personal communication).

13.9 Summary

The astronomic tide that exists in the deep ocean basin is amplified as it propagates onto the continental shelf, and is transformed into a co-oscillating tide by complex interactions with the shallow-water topography. Particularly pronounced tidal currents occur where there is tidal resonance in embayments and where there is local enhancement of tidal currents by a flow constriction as in seaways. Tidal currents are efficient in transporting sediments up to medium-grained sand to

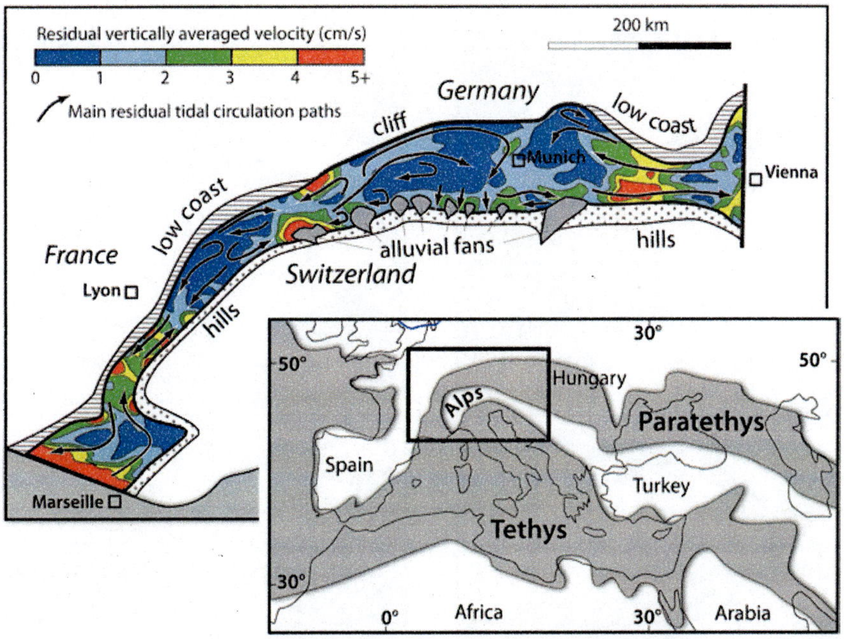

Fig. 13.30 Paleotidal modeling of the Burdigalian Peri-Alpine Seaway. The colors show the velocity of residual tidal circulation and the arrows the tidal-transport pathways. The first paleotidal modeling of this system by Martel et al. (1994) demonstrated that the Swiss Molasse basin, which hosts offshore tidal deposits, would have been meso- to macrotidal only if the tidal wave entered the seaway from both the southwest and eastern ends. More recent work by Bieg (2005) showed that the two tidal waves had to be in phase to get the maximum resonance. The numerically determined current patterns and strengths show a relatively good agreement with the facies distribution and tidal-transport pathways reconstructed from the bedforms in SE France (e.g. Lesueur et al. 1988), Switzerland (e.g. Homewood and Allen 1981), Germany (e.g. Hülsemann 1955) and Hungary (e.g. Sztano and de Boer 1995) (After Bieg 2005). *Inset* map modified from Martel et al. 1994)

depths of approximately 200 m. These currents can deposit large accumulations of well-sorted sand over large areas, and are responsible for the resuspension and redistribution large amounts of mud.

Interactions between various components of the tide, and with the seafloor and coastal morphology, cause the ebb and flood tidal currents to be unequal over large areas. These inequalities generate tidal-transport pathways along which bedload can be transported for distances of tens to hundreds of kilometers, from high-energy erosional and by-pass zones, to lower energy areas of sediment accumulation. The sandy tidal sediments that accumulate in offshore areas are commonly coarser and less muddy than those deposited at equivalent depths on storm-influenced shelves.

The most significant offshore tidal sediment accumulations are sand sheets and ridges. These deposits are made up predominantly of the crossbedded sandstone formed by tidal dunes, which can reach more than 10 m in height. Most of these bedforms are compound dunes, and generate compound crossbedding in which a single the master bedding and smaller cross beds dip in the direction of the residual transport. Herringbone crossbedding can be present in small amounts, but mud drapes, reactivation surfaces and tidal bundles are not likely to be abundant. The ichnology of these deposits reflects the mobile sandy nature of the deposits, and the normal-marine salinity of the water.

Most sandy tidal-shelf deposits are transgressive in origin, with the sand supplied by erosion of the retreating coast or the offshore bedload parting area, and can rest on an offshore tidal ravinement surface that commonly cuts into older coastal deposits. This scour surface can be amalgamated with a sequence boundary, or can be a more prominent distinct surface. The upper boundary of the offshore tidal deposit is expected to be a maximum flooding surface that reflects the decay of tidal currents in the course of a sea-level rise. The large size of some tidal dunes and tidal-current ridges means

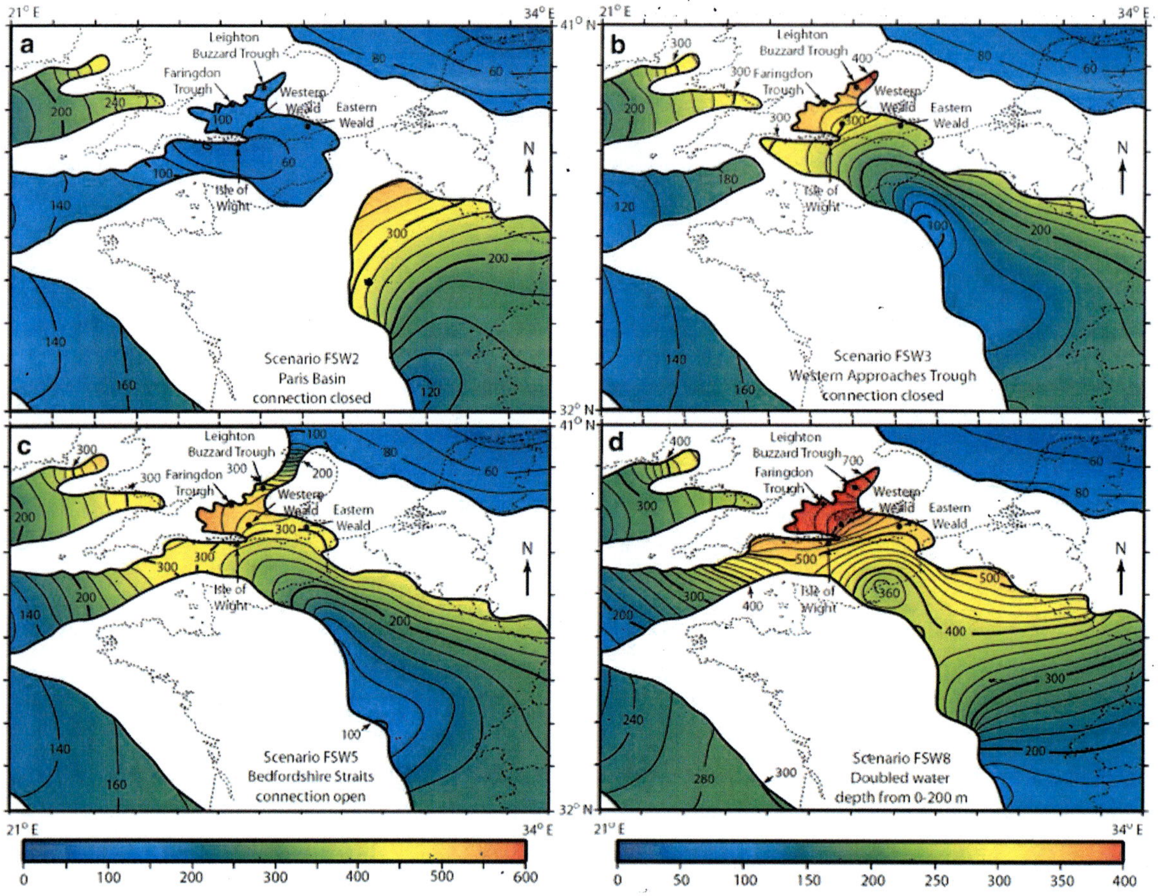

Fig. 13.31 Paleotidal modeling of the Lower Cretaceous seaways of Western Europe. Color bar at the bottom gives the modeled tidal range. Several paleogeographic hypotheses, based on field data of the Greensand Seaway in southern Great Britain and northern France, are compared: (**a**) seaway connected only to the North Atlantic; (**b**) seaway connected only to the Neotethys; (**c**) all three connections open; and (**d**) Neotethys and North Atlantic connections open, and water depth doubled in the areas shallower than 200 m. The results show that (i) only the Neotethys connection was important for the transfer of tidal energy into the seaway (i.e. there is a significant tidal range only when the Neotethys connection is open (**b**, **c** and **d**), and (ii) the tidal range increases when sea level is higher (From Wells et al. 2010)

that original topographic relief may be preserved on the top of offshore tidal deposits.

Shelf tidal sedimentation can, however, occur at any time during a relative sea-level cycle, whenever the sea level interacted with the paleogeography to create embayments and straits that accentuated the tidal currents. Changes in coastal and shelf morphology that result from a change in relative sea level can cause dramatic changes in the strength of the tidal currents on rapid time scales. Paleotidal modeling has great potential to assist in understanding the complex temporal and spacial occurrence of shallow-marine tidal deposits in the rock record.

References

Allen JRL (1980) Sand waves; a model of origin and internal structure. Sediment Geol 26:281–328

Allen JRL (1982) Sedimentary structures: their character and physical basis, Developments in sedimentology 30A and 30B. Elsevier, Amsterdam

Allen PA (1997) Earth surface processes. Wiley-Blackwell, Oxford

Allen PA, Homewood P (1984) Evolution and mechanics of a Miocene tidal sandwave. Sedimentology 31:63–81

Allen PA, Mange-Rajetzky M, Matter A, Homewood P (1985) Dynamic palaeogeography of the open Burdigalian seaway, Swiss Molasse basin. Eclogae Geol Helv 78:351–381

Anastas A, Dalrymple RW, James NP, Nelson CS (1997) Cross-stratified calcarenites from New Zealand; subaqueous dunes in a cool-water, Oligo-Miocene seaway. Sedimentology 44:869–891

Anastas A, Dalrymple RW, James NP, Nelson CS (2006) Lithofacies and dynamics of a cool-water carbonate seaway; mid-Tertiary, Te Kuiti Group, New Zealand. In: Pedley HM, Carannante G (eds) Cool-water carbonates: depositional systems and palaeoenvironmental controls. Geol Soc Lond Spec Publ 255:245–268

André JP, Biagi R, Moguedet G, Buffard R, Clément G, Redois F, Baloge PA (2003) Mixed siliciclastic–cool-water carbonate deposits over a tide-dominated epeiric platform: the Faluns of l'Anjou formation (Miocene, W. France). Ann Paleontol 89:113–123

André JP, Barthet Y, Ferrandini M, Ferrandini J, Reynaud JY, Tessier B (2011) The Bonifacio Formation (Miocene of Corsica): transition from a wave- to tide-dominated coastal system in mixed carbonate-siliciclastic setting. Bull Soc Geol France 182:221–230

Androsov AA, Kagan BA, Romanenkov DA, Voltzinger NE (2002) Numerical modeling of barotropic tidal dynamics in the strait of Messina. Adv Water Res 25:401–415

Archer AW, Hubbard MS (2003) Highest tides of the world. In: Chan MA, Archer AW (eds) Geol Soc Am Spec Paper 370:151–173

Ashley GM (1990) Classification of large-scale subaqueous bedforms: a new look at an old problem. J Sediment Petrol 60:160–172

Bassant P, Van Buchem FHP, Strasser A, Gorur N (2005) The stratigraphic architecture and evolution of the Burdigalian carbonate-siliciclastic sedimentary systems of the Mut Basin, Turkey. Sediment Geol 173:187–232

Bastos AC, Collins M, Kenyon NH (2003) Morphology and internal structure of sand shoals and sandbanks off the Dorset coast, English Channel. Sedimentology 50:1105–1122

Belderson RH, Johnson MA, Kenyon NH (1982) Bedforms. In: Stride AH (ed) Offshore tidal sands: processes and deposits. Chapman & Hall, London

Berné S, Auffret J-P, Walker P (1988) Internal structure of subtidal sandwaves revealed by high-resolution seismic reflection. Sedimentology 35:5–20

Berné S, Bourillet J-F, Durand J, Lericolais G (1989) Les Dunes subtidales geantes de Surtainville (Manche ouest). Bull Cen Rech Expl Elf-Aquit 13:395–415

Berné S, Durand J, Weber O (1991) Architecture of modern subtidal dunes (sand waves), Bay of Bourgneuf, France. Concepts Sedimentol Paleontol 3:245–260

Berné S, Trentesaux A, Stolk A, Missiaen T, De Batist M (1994) Architecture and long term evolution of a tidal sandbank; the Middelkerke Bank (southern North Sea). Mar Geol 121:57–72

Berné S, Lericolais G, Marsset T, Bourillet J-F, de Batist M (1998) Erosional shelf sand ridges and lowstand shorefaces: examples from tide and wave dominated environments of France. J Sediment Res 68:540–555

Berné S, Vagner P, Guichard F, Lericolais G, Liu Z, Trentesaux A, Yin P, Yi HI (2002) Pleistocene forced regressions and tidal sand ridges in the East China Sea. Mar Geol 188:293–315

Besson D (2005) Architecture du bassin rhodano-provençal miocène (Alpes, SE France): relations entre déformation, physiographie et sédimentation dans un bassin molassique d'avant-pays. Ph.D. thesis, Mines Paris Tech

Besson D, Parize O, Rubino J-L, Aguilar J-P, Aubry M-P, Beaudoin B, Berggren WA, Clauzon G, Crumeyrolle P, Dexcoté Y, Fiet N, Iaccarino S, Jimenez-Moreno G, Laporte-Galaa C, Michaux J, von Salis K, Suc J-P, Reynaud J-Y, Wernli R (2005) Un réseau fluviatile d'âge Burdigalien terminal dans le Sud-Est de la France: remplissage, extension, âge, implications. CR Géosci 337:1045–1054

Bieg U (2005) Palaeoceanographic modeling in global and regional scale: an example from the Burdigalian Seaway Upper Marine Molasse (Early Miocene). Ph.D. thesis, University of Tübingen

Blackwood S, Yoshida S, Steel R, Dalrymple R, Martinius A, Gawthorpe R (2004) Comparative studies of modern, Quaternary and ancient seaways; building depositional models for hydrocarbon exploration and production; a review. AAPG Abstr 13:14–15

Bourret A, Devenon J-L, Chevalier C (2008) Tidal influence on the hydrodynamics of the French Guiana continental shelf. Cont Shelf Res 28:951–961

Bouysse P, Horn R, Lapierre H, Le Lann F (1976) Etude des grands bancs de sable du Sud-est de la mer Celtique. Mar Geol 20:251–275

Bouysse P, Le lann F, Scolari G (1979) Les sediments superficiels des Approches occidentales de la Manche. Mar Geol 29:107–135

Brandano M, Jadoul F, Lanfranchi A, Tomassetti L, Berra F, Ferrandini M, Ferrandini J (2009) Stratigraphic architeture of mixed carbonate-siliciclastic system in the Bonifacio Basin (Early-Middle Miocene, South Corsica). Excursion guidebook, 27th IAS meeting of sedimentology, Alghero, 20–24 Sept 2009, pp 299–313

Brew DS (1996) Late Weichselian to early Holocene subaqueous dune formation and burial off the North Sea Northumberland coast. Mar Geol 134:203–211

Carling PA (1999) Subaqueous gravel dunes. J Sediment Res 69:534–545

Caston VND (1972) Linear sand banks in the southern North Sea. Sedimentology 18:63–78

Caston VND (1981) Potential gain and loss of sand by some sand banks in the Southern Bight of the North Sea. Mar Geol 41:239–250

Colella A (1990) Active tidal sand waves at bathyal depths observed from submersible and bathysphere (Messina Strait, southern Italy). In: 13th IAS congress, Abstract, Nottingham, pp98–99

Dalrymple RW (1984) Morphology and internal structure of sand waves in the Bay of Fundy. Sedimentology 31: 365–382

Dalrymple RW (2010a) Introduction to siliciclastic facies models. In: James NP, Dalrymple RW (eds) Facies models 4. Geological Association of Canada, St John's, pp 59–72

Dalrymple RW (2010b) Tidal depositional systems. In: James NP, Dalrymple RW (eds) Facies models 4. Geological Association of Canada, St John's, pp 201–231

Dalrymple RW, Rhodes RN (1995) Estuarine dunes and bars. In: Perillo GME (ed) Geomorphology and sedimentology of estuaries. Elsevier, Amsterdam

Dalrymple RW, Zaitlin BA (1994) High-resolution sequence stratigraphy of a complex, incised valley succession, the Cobequid Bay- Salmon River estuary, Bay of Fundy, Canada. Sedimentology 41:1069–1091

Davis RA, Balson PS (1992) Stratigraphy of a North Sea tidal sand ridge. J Sediment Petrol 62:116–121
Davis RA, Klay J, Jewell P (1993) Sedimentology and stratigraphy of tidal sand ridges, Southwest Florida inner shelf. J Sediment Petrol 63:91–104
De Batist M, Tessier B, Marsset T, Reynaud J-Y, Dimitropoulos D, Llopart X, Proust J-N, Berné S, Chamley H (1996) Analysis by geosonic recordings of the large and small-scale internal structure of the Bassure de Baas sand in the English Channel. In: Heyse I, De Moor G (eds) MAST-II-Starfish Project MAS2-CT92-0029. Final report, Chap. 15
De Boer PL, Van Gelder A, Nio SD (1988) Tide-influenced sedimentary environments and facies. Reidel Publishing, Dordrecht
Descote PY (2010) Relations architecturales, faciologiques et diagénétiques des carbonates bioclastiques du bassin rhodano-provençal (SE France). Ph.D. thesis, Mines Paris-Tech
D'Olier B (1981) Sedimentary events during Flandrian sea-level rise in the south-west corner of the North Sea. In: Nio SD, Schuttenhelm RTE, Van Weering TjCE (eds) Holocene marine sedimentation in the North Sea basin. IAS Spec Publ 5:221–227
Dyer KR, Huntley DA (1999) The origin, classification and modelling of sand banks and ridges. Cont Shelf Res 19: 1285–1330
Ekdale AA, Bromley RG, Pemberton SG (1984) Ichnology: the use of trace fossils in sedimentology and stratigraphy. SEPM Short Course Notes 15
Ericksen MC, Masson DS, Slingerland R, Swetland DW (1990) Numerical simulation of circulation and sediment transport in the late Devonian Catskill Sea. In: Cross TA (ed) Quantitative dynamic stratigraphy. Prentice-Hall, Englewood Cliffs
Fenster MS, Fitzgerald DM, Bohlen WF, Lewis RS, Baldwin CT (1990) Stability of giant sand waves in eastern Long Island Sound, U.S.A. Mar Geol 91:207–225
Fleming RH, Revelle R (1939) Physical processes in the ocean. In: Trask PD (ed) Recent marine sediments. A symposium. Thomas Murby Publishing, London
Flemming BW (1988) Pseudo-tidal sedimentation in a non-tidal shelf environment; Southeast African continental margin. In: De Boer PL, Van Gelder A, Nio SD (eds) Tide-influenced sedimentary environments and facies. Reidel Publishing, Dordrecht
Flemming BW (1980) Sand transport and bedform patterns on the continental shelf between Durban and Port Elizabeth (Southeast African continental margin). Sediment Geol 26:179–205
Gabioux M, Vinzon SB, Palva AM (2005) Tidal propagation over fluid mud layers on the Amazon shelf. Cont Shelf Res 25:113–125
Gao S, Collins MB, Lanckneus J, De Moor G, Van Lancker V (1994) Grain size trends associated with net sediment transport patterns; an example from the Belgian continental shelf. Mar Geol 121:171–185
Greenberg DA (1979) A numerical model investigation of tidal phenomena in the Bay of Fundy and Gulf of Maine. Mar Geol 2:161–187
Grochowski NTL, Collins MB, Boxall SR, Salomon J-C (1993) Sediment transport predictions for the English Channel, using numerical models. J Geol Soc Lond 150:683–695
Harris PT (1988) Sediments, bedforms, and bedload transport pathways on the continental shelf adjacent to Torres Strait, Australia – Papua New Guinea. Cont Shelf Res 8:979–1003
Harris PT, Pattiaratchi CB, Collins MB, Dalrymple RW (1995) What is a bedload parting? In: Flemming BW, Bartoloma A (eds) Tidal signatures in modern and ancient sediments. IAS Spec Publ 24 :2–10
Heathershaw AD, New AL, Edwards PD (1987) Internal tides and sediment transport at the shelf break in the Celtic Sea. Cont Shelf Res 7:485–517
Hein FJ, Robb GA, Wolberg AC, Longstaffe FJ (1991) Facies descriptions and associations in ancient reworked (transgressive) shelf sandstones; Cambrian and Cretaceous examples. Sedimentology 38:405–431
Hoppie BW (1996) The influence of relative sea level on near-shore sedimentary processes in a transform margin basin; the Miocene Cuyama Basin, California. AAPG Absr 5:67
Homewood P, Allen PA (1981) Wave-, tide- and current-controlled sandbodies of Miocene Molasse, western Switzerland. Am Assoc Petrol Geol Bull 65:2534–2545
Houbolt JJHC (1968) Recent sediments in the Southern Bight of the North Sea. Geol Mijn 47:245–273
Houbolt JJHC (1982) A comparison of recent shallow marine tidal sand ridges with Miocene sand ridges in Belgium. In: Scrutton RA, Talwani M (eds) The ocean floor, Bruce Heezen commemorative volume. Wiley, Chichester
Houthuys R, Gullentops F (1988a) The Vlierzele Sands (Eocene, Belgium): a tidal ridge system. In: de Boer PL, van Gelder A, Nio SD (eds) Tide-influenced sedimentary environments and facies. Reidel Publishing, Dordrecht
Houthuys R, Gullentops F (1988b) Tidal transverse bars building up a longitudinal sand body (middle Eocene, Belgium). In: de Boer PL, van Gelder A, Nio SD (eds) Tide-influenced sedimentary environments and facies. Reidel Publishing, Dordrecht
Houthuys R, Trentesaux A, De Wolf P (1994) Storm influences on a tidal sandbank's surface (Middelkerke Bank, southern North Sea). Mar Geol 121:23–41
Howarth MJ (1982) Tidal currents of the continental shelf. In: Stride AH (ed) Offshore tidal sands: processes and deposits. Chapman & Hall, New York
Howarth MJ, Huthnance JM (1984) Tidal and residual currents around a Norfolk sandbank. Estuarine Coast Shelf Sci 19:105–117
Hulscher SJMH (1996) Tidal-induced large-scale regular bedform patterns in a three-dimensional shallow water model. J Geophys Res 101(C9):20727–20744
Hülsemann J (1955) Grossrippeln und Schrägsrichtungengs-Gefüge im Nordsee Watt un in der Molasse. Senck Leth 36:359–388
Huthnance JM (1973) Tidal current asymmetries over the Norfolk sandbanks. Estuarine Coast Mar Sci 1:89–99
Huthnance JM (1982a) On one mechanism forming linear sand banks. Estuarine Coast Mar Sci 14:79–99
Huthnance JM (1982b) On the formation of sand banks of finite extent. Estuarine Coast Mar Sci 15:277–299
James NP (1997) The cool-water carbonate depositional realm. In: James NP, Clarke JAD (eds) Cool-water carbonates. SEPM Spec Publ 56:1–20
Jin JH, Chough SK (2002) Erosional shelf ridges in the mid-eastern Yellow Sea. Geo-Mar Lett 21:219–225

Johnson HD (1977) Shallow marine sand bar sequences; an example from the late Precambrian of North Norway. Sedimentology 24:245–270

Johnson MA, Kenyon NH, Belderson RH, Stride AH (1982) Sand transport. In: Stride AH (ed) Offshore tidal sands, processes and deposits. Chapman & Hall, London

Jung WY, Suk BC, Min GH, Lee YK (1998) Sedimentary structure and origin of a mud-cored pseudo-tidal sand ridge, eastern Yellow Sea, Korea. Mar Geol 151:73–88

Kamp PJJ, Harmsen FJ, Nelson CS, Boyle SF (1988) Barnacle-dominated limestone with giant cross-beds in a non-tropical, tide-swept, Pliocene forearc seaway, Hawke's Bay, New Zealand. Sediment Geol 60:173–195

Kenyon NH, Stride AH (1970) The tide-swept continental shelf sediments between the Shetland isles and France. Sedimentology 14:159–173

Kenyon NH, Belderson RH, Stride AH, Johnson MA (1981) Offshore tidal sand banks as indicators of net sand transport and as potential deposits. In: Nio SD, Schuttenhelm RTE, Van Weering TjCE (eds) Holocene marine sedimentation in the North Sea basin. IAS Spec Publ 5:257–268

Laban C, Schuttenhelm RTE (1981) Some new evidence on the origin of the Zealand Ridges. In: Nio SD, Schuttenhelm RTE, Van Weering TjCE (eds) Holocene marine sedimentation in the North Sea basin. IAS Spec Publ 5:239–245

Lanckneus J, De Moor G, Stolk A (1994) Environmental setting, morphology and volumetric evolution of the Middelkerke Bank (southern North Sea). Mar Geol 121:1–21

Lanier WP, Feldman HR, Archer AW (1993) Tidal sedimentation from a fluvial to estuarine transition, Douglas Group, Missourian-Virgilian, Kansas. J Sediment Petrol 63:860–873

Lapierre F (1975) Contribution a l'etude geologique et sedimentologique de la Manche orientale. Philos Trans R Soc Lond A 1975:177–187

Le Bot S, Trentesaux A (2004) Types of internal structure and external morphology of submarine dunes under the influence of tide and wind-driven processes (Dover Strait, northern France). Mar Geol 211:143–168

Legg S, Adcroft A (2003) Internal wave breaking at concave and convex continental slopes. J Phys Ocean 33:2224–2246

Lesueur J-L, Rubino J-L, Giraudmaillet M (1988) Organisation et structures internes des dépots tidaux du Miocène rhodanien. Bull Soc Geol France 6:49–65

Levell BK (1980) A Late Precambrian tidal shelf deposit, the lower Sandfjord Formation, Finnmark, North Norway. Sedimentology 27:539–557

Liu ZX, Xia DX, Berné S, Yang WK, Marsset T, Tang YX, Bourillet J-F (1998) Tidal depositional systems of China's continental shelf, with special reference to the eastern Bohai Sea. Mar Geol 145:225–268

Liu Z, Berné S, Saito Y, Yu H, Trentesaux A, Uehara K, Yin P, Liu JP, Li C, Hu G, Wang X (2007) Internal architecture and mobility of tidal sand ridges in the East China Sea. Cont Shelf Res 27:1820–1834

Longhitano SG, Nemec W (2005) Statistical analysis of bed-thickness variation in a Tortonian succession of biocalcarenitic tidal dunes, Amantea Basin, Calabria, southern Italy. Sediment Geol 179:195–224

MacEachern JA, Bann KL, Pemberton SG, Gingras MK (2005) The ichnofacies paradigm: high-resolution paleoenvironmental interpretation of the rock record. In: MacEachern JA, Bann KL, Gingras MK, Pemberton SG (eds) Applied ichnology. SEPM Short Course Notes 52:27–64

Malikides M, Harris PT, Jenkins C, Keene J (1988) Carbonate sandwaves in the Bass Strait. Aust J Earth Sci 35:303–311

Martel AT, Allen PA, Slingerland R (1994) Use of tidal-circulation modeling in paleogeographical studies; an example from the Tertiary of the Alpine perimeter. Geology 22:925–928

McCave IN (1971) Sandwaves in the North Sea off the coast of Holland. Mar Geol 10:199–225

McCave IN, Langhorne DN (1982) Sand waves and sediment transport around the end of a tidal sand bank. Sedimentology 29:95–110

Mellere D, Steel R (1995) Facies architecture and sequentiality of nearshore and "shelf" sandbodies; Haystack Mountains Formation, Wyoming, USA. Sedimentology 42:551–574

Mellere D, Steel RJ (1996) Tidal sedimentation in the Inner Hebrides half grabens, Scotland: the Mid-Jurassic Bearreraig Sandstone Formation. In: de Batist M, Jacobs P (eds) Geology of siliciclastic shelf seas. Geol Soc Lond Spec Publ 117:49–79

M'Hammdi N, Berné S, Bourillet J-F, Auffret J-P (1992) Architecture of a tidal sand bank; the Sark Bank (Channel Islands). In: Flemming BW (ed) Modern and ancient clastic tidal deposits. Courier Forschunginstitut Senckenberg 151:59–60

Mitchell AJ, Ulicny D, Hampson GJ, Allison PA, Gorman GJ, Piggott MD, Wells MR, Pain CC (2010) Modeling tidal current-induced bed shear stress and palaeocirculation in an epicontinental seaway: the Bohemian Cretaceous Basin, Central Europe. Sedimentology 57:359–388

Mutti E, Rosell J, Allen GP, Fonnesu F, Sgavetti M (1985) The Eocene Baronia tide dominated delta-shelf system in the Ager Basin. In: Mila MD, Rosell J (eds) Excursion guidebook, 6th IAS European regional meeting, Lleida, pp 579–600

Nio SD, Yang C (1991) Diagnostic attributes of clastic tidal deposits: a review. In: Smith DG, Reinson GE, Zaitlin BA, Rahmani RA (eds) Clastic tidal sedimentology. Can Soc Petrol Geol Mem 16:3–28

Nelson CS, Winefield PR, Hood SD, Caron V, Pallentin A, Kamp PJJ (2003) Pliocene Te Aute limestones, New Zealand; expanding concepts for cool-water shelf carbonates. NZ J Geol Geophys 46:407–424

Off T (1963) Rythmic linear sandbodies caused by tidal currents. AAPG Bull 47:324–341

Pan S, MacDonald N, Williams J, O'Connor BA, Nicholson J, Davies AM (2007) Modelling the hydrodynamics of offshore sandbanks. Cont Shelf Res 27:1264–1286

Pantin HM, Evans CDR (1984) The Quaternary history of the central and southwestern Celtic Sea. Mar Geol 57:259–293

Park SC, Lee BH, Yoo DG, Lee CW (2006) Late Quaternary stratigraphy and development of tidal sand ridges in the eastern Yellow Sea. J Sed Res 76:1093–1105

Pattiaratchi C, Collins M (1987) Mechanisms for linear sandbank formation and maintenance in relation to dynamical oceanographic observations. Prog Oceanograph 19:117–176

Pingree RD, Griffiths DK (1979) Sand transport paths around the British Isles resulting from M (sub 2) and M (sub 4) tidal interactions. J Mar Biol Assoc UK 59:497–513

Pingree RD, Maddock L (1979) The tidal physics of headland flows and offshore tidal bank formation. Mar Geol 32: 269–289

Posamentier HW (2002) Ancient shelf ridges—a potentially significant component of the transgressive systems tract: case study from offshore northwest java. Am Assoc Petrol Geol Bull 86:75–106

Pratt LJ (1990) The physical oceanography of sea straits. Kluwer, Dordrecht

Proctor R, Carter L (1989) Tidal and sedimentary response to the late Quaternary closure and opening of Cook Strait, New Zealand; results from numerical modeling. Paleoceanography 4:167–180

Pugh DT (1987) Tides, surges and mean sea-level; a handbook for engineers and scientists. Wiley, Chichester

Ramaswamy V, Rao PS, Rao KH, Thwin S, Roa NS, Raiker V (2004) Tidal influence on suspended sediment distribution and dispersal in the northern Andaman Sea and Gulf of Martaban. Mar Geol 208:33–42

Rao PS, Ramaswamy V, Thwin S (2005) Sediment texture, distribution and transport on the Ayeyarwady continental shelf, Andaman Sea. Mar Geol 216:239–247

Reesink AJH, Bridge JS (2007) Influence of superimposed bedforms and flow unsteadiness on the formation of cross strata in dunes and unit bars. Sediment Geol 202:281–296

Reesink AJH, Bridge JS (2009) Influence of superimposed bedforms and flow unsteadiness on the formation of cross strata in dunes and unit bars - Part 2, further experiments. Sediment Geol 222:274–300

Reynaud JY, Tessier B, Berné S, Chamley H, De Batist M (1999a) Tide and wave dynamics on a sand bank from the deep shelf of the Western Channel Approaches. Mar Geol 161:339–359

Reynaud JY, Tessier B, Proust JN, Dalrymple RW, Marsset T, De Batist M, Bourillet J-F, Lericolais G (1999b) Eustatic and hydrodynamic controls on the architecture of a deep shelf sand bank (Celtic Sea). Sedimentology 46:703–721

Reynaud JY, Rage A, Tessier B, Néraudeau D, Bracini E, Carriol R-P, Clet-Pellerin M, Moullade M, Lericolais G (1999c) Importations et remaniements de faunes dans les sables de la plate-forme profonde des Approches Occidentales de la Manche. Oceanol Acta 22:381–396

Reynaud JY, Tessier B, Auffret J-P, Berné S, De Batist M, Marsset T, Walker P (2003) The offshore sedimentary cover of the English Channel and its northern and western approaches. J Quart Sci 18:261–282

Reynaud JY, Dalrymple RW, Vennin E, Parize O, Besson D, Rubino J-L (2006) Topographic controls on producing and depositing tidal cool-water carbonates, Uzès basin, SE France. J Sediment Res 76:117–130

Rubin DM (1987) Cross-bedding, bedforms and paleocurrents, vol 1, Concepts in Sedimentology and Paleontology. SEPM, Tulsa

Rubin DM, Hunter RE (1987) Bedform alignment in directionally varying flows. Science 237:276–278

Rubin DM, Ikeda H (1990) Flume experiments on the alignment of transverse, oblique, and longitudinal dunes in directionally varying flows. Sedimentology 37:673–684

Rubin DM, McCulloch DS (1980) Single and superimposed bedforms: a synthesis of San Francisco bay and flume observations. Sediment Geol 26:207–231

Shaw J, Amos CL, Greenberg DA, O'Reilly CT, Parrott DR, Patton E (2010) Catastrophic tidal expansion in the Bay of Fundy, Canada. Can J Earth Sci 47:1079–1091

Seilacher A (1967) Bathymetry of trace fossils. Mar Geol 5:413–428

Sinha B, Pingree RD (1997) The principal lunar semidiurnal tide and its harmonics: baseline solutions for M2 and M4 constituents on the North-West European continental shelf. Cont Shelf Res 17:1321–1365

Snedden JW, Dalrymple RW (1999) Modern shelf sand ridges: from historical perspective to a unified hydrodynamic and evolutionary model. In: Bergman KM, Snedden JW (eds) Isolated shallow marine sand bodies: sequence stratigraphic analysis and sedimentological perspectives. SEPM Spec Publ 64:13–28

Southard JB, Boguchwal LA (1990) Bed configurations in steady unidirectional water flows. Part 2. Synthesis of flume data. J Sediment Petrol 60:649–657

Stride AH (1963) Current-swept sea floors near the southern half of Great Britain. Quart J Geol Soc Lond 119:175–199

Stride AH (ed) (1982) Offshore tidal sands: processes and deposits. Chapman & Hall, London, 222 p

Stride AH, Belderson RH, Kenyon NH, Johnson MA (1982) Offshore tidal deposits: sand sheet and sand bank facies. In: Stride AH (ed) Offshore tidal sands: processes and deposits. Chapman & Hall, London

Surlyk F, Noe-Nygaard N (1991) Sand bank and dune facies architecture of a wide intracratonic seaway; Late Jurassic-Early Cretaceous Raukelv Formation, Jameson Land, East Greenland. In: Miall AD, Tyler N (eds) The three-dimensional facies architecture of terrigenous clastic sediments and its implications for hydrocarbon discovery and recovery. Concepts in Sediment Paleo 3:261–276

Suter JR (2006) Facies models revisited: clastic shelves. In: Posamentier HW, Walker RG (eds) Facies models revisited. SEPM Spec Publ 84:339–397

Swift DJP (1975) Tidal sand ridges and shoal retreat massifs. Mar Geol 18:105–134

Sztano O, De Boer P (1995) Basin dimensions and morphology as controls on amplification of tidal motions (the early Miocene North Hungarian Bay). Sedimentology 42:665–682

Tillman RW, Martinsen RS (1984) The Shannon shelf ridge sandstone complex, Salt Creek Anticline area, Powder River basin, Wyoming. In: Tillman RW, Siemers CT (eds) Siliciclastic shelf sediments. SEPM Spec Publ 34:1–34

Trentesaux A, Stolk A, Berné S (1999) Sedimentology and stratigraphy of a tidal sand bank in the southern North Sea. Mar Geol 159:253–272

Uehara K, Scourse JD, Horsburgh KJ, Lambeck K, Purcell AP (2006) Tidal evolution of the northwest European shelf seas from the last glacial maximum to the present. J Geophys Res 111:C09025

Vail PR, Mitchum RM, Thomson S (1977) Seismic stratigraphy and global changes of sea level; Part 4, Global cycles of relative changes of sea level. In: Payton CE (ed) Seismic stratigraphy; Applications to hydrocarbon exploration. AAPG Memoir 26:83–97

Van der Molen J, Gerrits J, De Swart HE (2004) Modelling the morphodynamics of a tidal shelf sea. Cont Shelf Res 24:483–507

Van Rijn LC (1982) Prediction of bed forms, alluvial roughness and sediment transport. Delft Hydraulics, S 487–11, Delft, The Netherlands

Van Veen J (1936) Onderzoekingen in de Hoofden. Algemene Landsdruckerij's Gravenhage, 252 p

Viana AR, Faugères J-C, Stow DAV (1998) Bottom current controlled sand deposits—a review of modern shallow to deep-water environments. Sediment Geol 115:53–80

Visser MJ (1980) Neap-spring cycles reflected in Holocene subtidal large-scale bedform deposits: a preliminary note. Geology 8:543–546

Walker RG, Bergman KM (1993) Shannon sandstone in Wyoming: a shelf ridge complex reinterpreted as lowstand shoreface deposits. J Sediment Petrol 63:839–851

Wells MR, Allison PA, Piggott MD, Gorman GJ, Hampson GJ, Pain CC, Fang F (2007) Numerical modeling of tides in the late Pennsylvanian Midcontinent seaway of North America with implications for hydrography and sedimentation. J Sediment Res 77:843–865

Wells MR, Allison PA, Piggott MD, Hampson GJ, Pain CC, Gorman GJ (2010) Tidal modeling of an ancient tide-dominated seaway, part 2: the Aptian Lower Greensand seaway of Northwest Europe. J Sediment Res 80: 411–439

Wilson JB (1982) Shelly faunas associated with temperate offshore tidal deposits. In: Stride AH (ed) Offshore tidal sands: processes and deposits. Chapman & Hall, London

Wilson JB (1988) A model for temporal changes in the faunal composition of shell gravels during a transgression on the continental shelf around the British Isles. In: Nelson CS (ed) Non-tropical shelf carbonates; modern and ancient. Sed Geol 60:95–105

Wright J, Colling A, Park D (Open University S330 Course readers) (1999) Waves, tides and shallow-water processes, 2nd edn. Butterworth-Heinemann in association with the Open University, Oxford

Yalin MS (1964) Geometrical properties of sand waves. Proc Am Soc Civil Eng 90:105–119

Yang ZS, Liu JP (2007) A unique Yellow River-derived distal subaqueous delta in the Yellow Sea. Mar Geol 240:169–176

Yoshida S, Johnson HD, Pye K, Dixon RJ (2004) Transgressive changes from tidal estuarine to marine embayment depositional systems: the Lower Cretaceous Woburns Sands of southern England and comparison with Holocene analogs. AAPG Bull 88:1433–1460

Yoshida S, Steel RJ, Dalrymple RW (2007) Changes in depositional processes - an ingredient in the generation of new sequence-stratigraphic models. J Sediment Res 77:447–460

Zimmerman JTF (1978) Topographic generation of residual circulation by oscillatory (tidal) currents. Geophys Astrophys Fluid Dyn 11:35–47

Deep-Water Tidal Sedimentology

14

Mason Dykstra

Abstract

Tides are well-documented in modern deep-water environment, especially around areas of topography on the seafloor such as ocean ridges and continental slopes. Recognition of deep-water tidal deposits in the ancient has lagged far behind, however, with very few examples in the published literature. This paper presents a review of the current state of knowledge about both modern deep-water tidal sediments and ancient deep-water tidal deposits, including new data on tidalites from the Cretaceous Wheeler Gorge channel-levee complex (California), and the Cretaceous Cajiloa submarine canyon (Mexico). In both of these settings detailed analysis of laminae thickness trends revealed cyclicities with frequencies characteristic of tidal deposits.

Recognition criteria for ancient deep-water tidal deposits include statistically significant cyclicities within thin successions (tidalites are unlikely to be very thick) in combination with mud-couplets, mud-bounded ripples, ripples with reactivation surfaces, and, more rarely, bi-directional ripple sets. Typical tidalites successions include cyclically thickening and thinning laminae (5–40 cm thick), and rippled intervals (5–20 cm thick) that exhibit large energy asymmetries (mud drapes) and an overall increase then decrease in ripple size, often arranged in cycles.

Although tides are common in deep-water environments, settings where deposits may be preserved are relatively rare. Such settings must not be subject to erosive turbidity currents yet require a relatively steady sediment supply and local accommodation space. Abandoned meander bends, the backsides of levees, topographic lows on the surface of submarine landslides, and abandoned plunge pools all potentially fit this category. This paper documents a tidalite succession that is preserved within a topographic low above a submarine landslide deposit.

M. Dykstra (✉)
Department of Geology and Geological Engineering,
Colorado School of Mines,
Golden, CO 80401, USA
e-mail: mdykstra@mines.edu

14.1 Introduction

The influence of tides in deep-water settings was first documented in the 1950s and 1960s when researchers noticed that the measured periods of some internal

R.A. Davis, Jr. and R.W. Dalrymple (eds.), *Principles of Tidal Sedimentology*,
DOI 10.1007/978-94-007-0123-6_14, © Springer Science+Business Media B.V. 2012

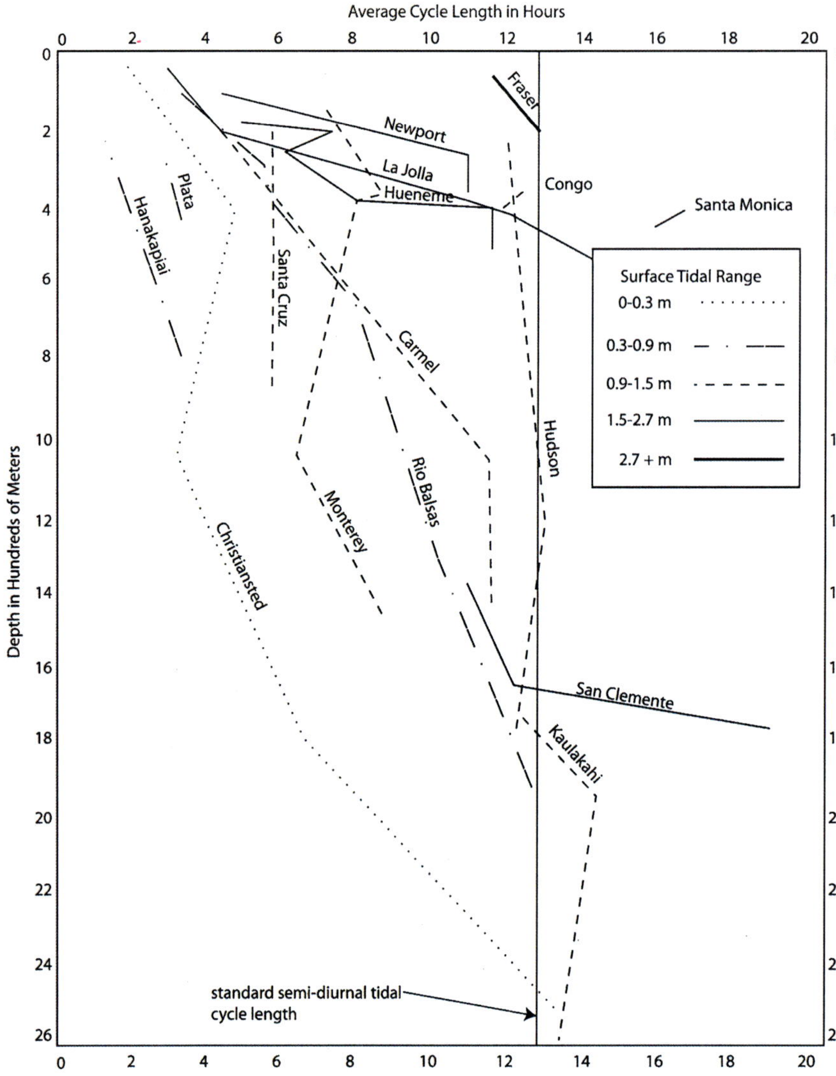

Fig. 14.1 Depth versus cycle length plot for a number of submarine canyons, showing the increasing cycle length vs depth trend typical of the canyons. With increasing tidal range, the cycle length more closely approximates the semidiurnal tidal cycle at shallower depths (Modified from Shepard et al. 1979)

waves coincided with those of the diurnal or semidiurnal tides (LaFond 1962). This work was followed up in the 1960s and 1970s by groups at Scripps Institution of Oceanography (Lonsdale et al. 1972; Shepard 1976; Shepard et al. 1979) who demonstrated through many measurements in a number of different submarine canyons and on seamounts that internal tides do occur, but are generally only present in water depths exceeding about 250 m. In addition, they found an inverse correlation between surface tidal range, and the minimum depth below the surface at which internal tides match frequency with surface tides (Fig. 14.1). Above this minimum depth, internal wave periods exhibit a higher frequency than the surface tides, presumably due to interference from other external factors (e.g. wind currents, river plumes, longshore currents, etc.). These measurements showed that the velocities of tidal currents in submarine canyons for outlier events can exceed 250 cm/s down-canyon, and 110 cm/s up-canyon, although on a daily basis are typically in the range of 10–50 cm/s, both up and down-canyon (Shepard and Marshall 1973; Zhenzhong et al. 1998). Even within

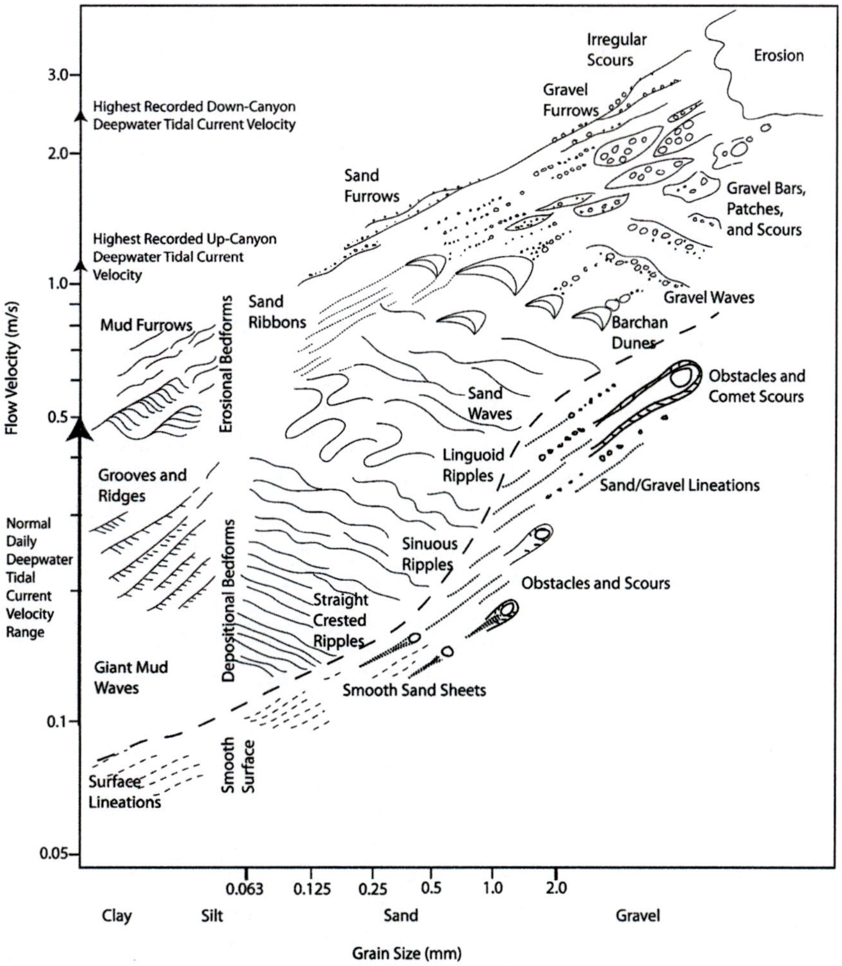

Fig. 14.2 Simple bedform velocity matrix for deep-water sediment movement, showing the range of common and unusual current velocities in deep-water tides, and the potential corresponding bedforms. Note that this figure does not take into account either the duration of a current or sediment availability, both of which are important controls on the development of bedforms (Modified from Stow et al. 2009)

the normal daily velocity range of 10–50 cm/s shown by these measurements, sediment up to fine gravel can be moved, and sandy bedforms including ripples and dunes can form, depending on sediment availability (Fig. 14.2) (Stow et al. 2009). Recognition of tidal deposits in ancient deep marine settings, however, has not kept up with studies of the modern, and currently there are only a handful of papers that deal with convincing deep-water tidalites (Shanmugam 2003; Zhenzhong and Eriksson 1991). The purpose of this paper is to present background from modern deep-water tidal currents and seafloor sediments, from outcrops interpreted as deep-water tidal in origin, and to discuss recognition criteria and the significance of tidalites in deep-marine rocks.

14.2 Deep-Water Processes

The major current-driven processes in operation on the seafloor that tend to move sediment include turbidity currents, contour currents, wind-driven currents, internal waves, and tidal currents. In a process sense, these all share some similarities and differences, and there may even be significant crossover in some settings. Turbidity currents are currents driven by gravity and buoyancy forces acting on a density contrast between the sediment-water mixture in the current and that in the surrounding water. Contour currents consist of flows that normally follow bathymetric contours, are produced by differences in the density of water masses, and can be greatly influenced by the geostrophic gyre (Heezen et al. 1966), although they are also modified and influenced by wind-driven currents, boundary currents, and tidal currents. Wind-driven currents only occur when strong winds blow in the same direction for an extended period of time, and generally do not affect really deep waters, although some have been known to persist down to at least 1,500 m depth (Vidal et al. 1992) with velocities up to 25 cm/s at 500 m depth (Shanmugam 2008), a velocity high enough to form sinuous coarse-sand ripples and to move pea-sized gravel (Fig. 14.2). Because they only last while the wind blows (usually seasonally), and the bottom current portion commonly joins the geostrophic flow (Vidal et al. 1992), their effects are often hard to differentiate from contour currents, especially when dealing with sediments or rocks. Internal waves are waves that occur along density boundaries or gradients within water bodies. These only contact the seafloor when the density boundary or gradient is at the seafloor, such as against a seamount or continental slope. They probably have many different formation and modification mechanisms including wind, gravity settling of dense fluid, and convection. Internal waves include deep-water tidal currents (baroclinic tides), which are internal waves with tidal periodicities. In deep-water settings, tidal currents consist of internal waves that form along density boundaries in the ocean by conversion of surface tides (barotropic tides – really the 'whole ocean' tide, as they affect more than just the water surface where they can most easily be measured) to internal waves along topographically rough surfaces such as continental slopes, continental shelves, submarine landslides, seamounts, or any other rough bit of seafloor (Garrett and Kunze 2007).

None of these processes operate in isolation, however. A contour current flowing through a constriction, for example, may resuspend enough sediment to generate a turbidity current (Mulder et al. 2006; Özsoy et al. 2001). Because the contour current will continue along at more or less the same depth, and the turbidity current will flow down-slope, the two currents should split up relatively soon after generation of the turbidity current. Likewise it has long been suspected that tidal currents may be forcing mechanisms for the generation of turbidity currents, especially near the head of submarine canyons, and a coincidence of strong down-canyon flows with high surface tides reinforces this contention (Shepard et al. 1979). In some settings internal tidal currents and contour currents may interfere such that an internal tidal current can augment the velocity of a contour current part of the time, and negate it part of the time (McCave et al. 1980). The largest unknown is what the relative contribution of each process is on both the modern seafloor, and to preserved sediments in the rock record. An understanding of the former can only come from modern observations, while an understanding of the latter must come from detailed outcrop studies, keeping in mind the recognition of turbidites, tidalites, and contourites is sometimes enigmatic and certainly problematic.

14.2.1 Internal Waves and Internal Tides

Density boundaries in water bodies along which internal waves can form can be due to the variable salinity of the water, particulate matter (including organic matter, sediment from river plumes, and eolian dust), and temperature stratification. Internal waves vary greatly in amplitude, from a few centimeters to hundreds of meters. Likewise their wavelengths can vary from a few centimeters or meters to thousands of kilometers (Zhenzhong et al. 1998). The period of internal waves can vary from a few minutes to days or possibly longer (seasonal). LaFond (1962) showed that some internal waves impart oscillatory shear at the fluid boundary, indicating that they may sometimes behave like surface waves; this type of wave may therefore also generate symmetrical ripples or hummocky cross-strata (Fig. 14.3) (Heezen and Rawson 1977; Kneller et al. 1997).

Internal tides are internal waves that approximate the diurnal or semi-diurnal tidal cycle in period, although they may be significantly out of phase with the surface tides. Internal tides are thought to be a significant factor in helping to mix the deep oceans (Legg 2004; Munk and Wunsch 1998; St. Laurent and Garrett 2002), and in providing energy to deep marine areas that otherwise might not see much energy flux from currents, such as mid-ocean ridges or abyssal plains (Egbert and Ray 2000). Internal tides are probably

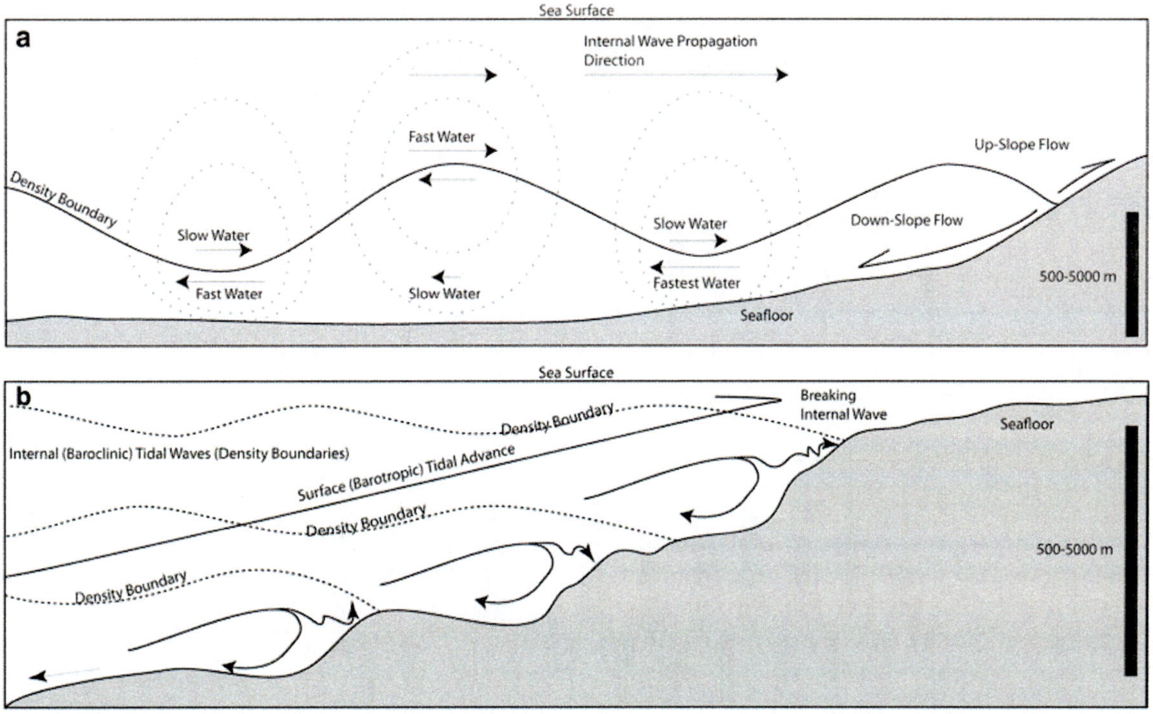

Fig. 14.3 Dynamics of internal waves, showing the velocity relationships of water above and below the internal wave, and how that is impacted when the internal wave touches ground along a slope. (**a**) Water below the internal wave boundary moves opposite to the propagation direction of the wave as a whole, whereas water above the boundary moves in the same direction as the wave (Modified after LaFond 1962). (**b**) If more than one wave is present in the ocean at different depths, which can occur in well-stratified water with significant seafloor topography (e.g. Robertson 2005), current directions along the seafloor can become quite complicated

primarily generated by the interaction of surface tides with seafloor topography (Cacchione et al. 1988; Huthnance 1989), although in quite a complicated fashion (Garrett and Kunze 2007; Sherwin et al. 2002). Once generated, internal tides may behave differently than surface tides, as the internal tidal waves can radiate out from the generation location in different directions (Garrett 2003), and in fact a number of separate internal tides may occur at differing depths, sometimes slightly out of phase with each other (Fig. 14.3) (Mulder et al. 2009; Robertson 2005; Shepard et al. 1979). One implication of this is that the dominant flow direction at the seafloor can differ depending on the depth of the seafloor, and which internal tidal wave is involved. Above the density boundary of an internal tidal wave, the flow will typically be up-slope, whereas below this boundary, flow is down-slope (Fig. 14.3). If a series of internal tidal boundaries are present (Robertson 2005), current flow related to the same tidal cycle can be reversed at different locations, creating a confusion of sediment transport directions along and down the continental slope (Fig. 14.3b). Current meter evidence for this exists from a number of localities, where the current direction during the same time period at the seafloor is reversed at different localities and depths (Mulder et al. 2009; Shepard et al. 1979; Xu et al. 2008). The time-velocity curve of internal tidal currents is apparently quite variable; some currents start abruptly and strongly and then taper off, whereas others start slowly, grow in strength, and end rather abruptly, and still others appear to have a more normal velocity distribution, with a gradual increase in velocity, a peak, and a gradual decrease in velocity (Fig. 14.4) (Nash et al. 2006; Shepard and Emery 1973; Shepard and Marshall 1973; Shepard et al. 1979; Xu et al. 2008). Why different time-velocity curves exist at the same locations during succeeding cycles is unclear at present, but may have to do with harmonic convergence

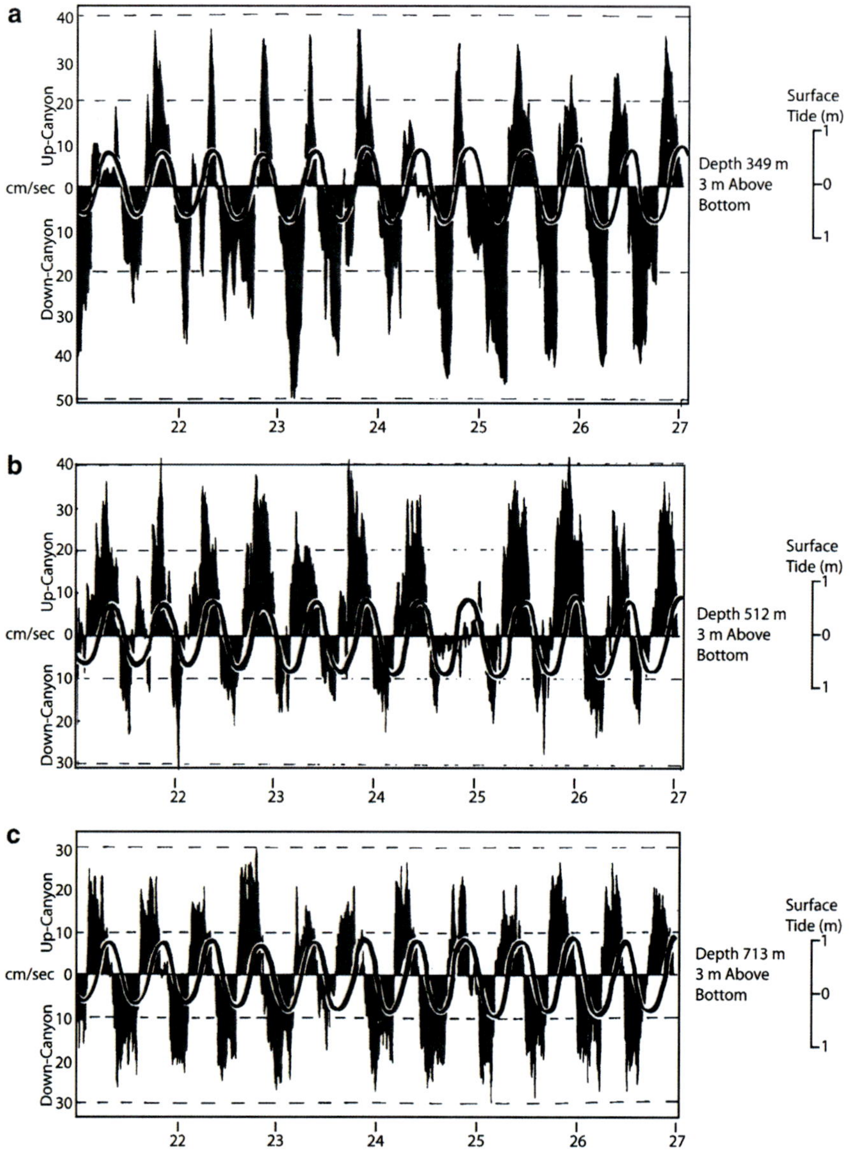

Fig. 14.4 Current velocity vs time records for Hydrographer Canyon, New England, US, at different depths. The surface tidal predictions are shown by the smooth curve. Notice that although the currents in the canyon match the frequency of the surface tide, they are all slightly out of phase with both the surface tide, and with each other, which may have to do with either how the internal tide is advancing (up or down the canyon, or along the slope), or how the internal tide is being generated – i.e. at the same or different depths (Fig. 14.3) (Modified from Shepard et al. 1979)

of tidal constituents, just as in surface tides (Kvale 2006). It is also possible there is an effect of sampling bias due to the position of current meters, generally suspended at least a few meters off the seafloor, such that if the height of the tidal wave varies during a tidal cycle, and between cycles, then the part of the tidal wave impacting the current meter may differ from cycle to cycle (and even within cycles), and, depending on the geometry of the internal tidal waves, this will affect the local velocity of the current (see Fig. 14.3). The current direction of any given internal tide may also change in orientation during a tidal cycle, commonly progressing consistently either clockwise or anticlockwise during the cycle (Shepard et al. 1979), probably as a function of the orientation of the seafloor slope relative to the propagation direction of

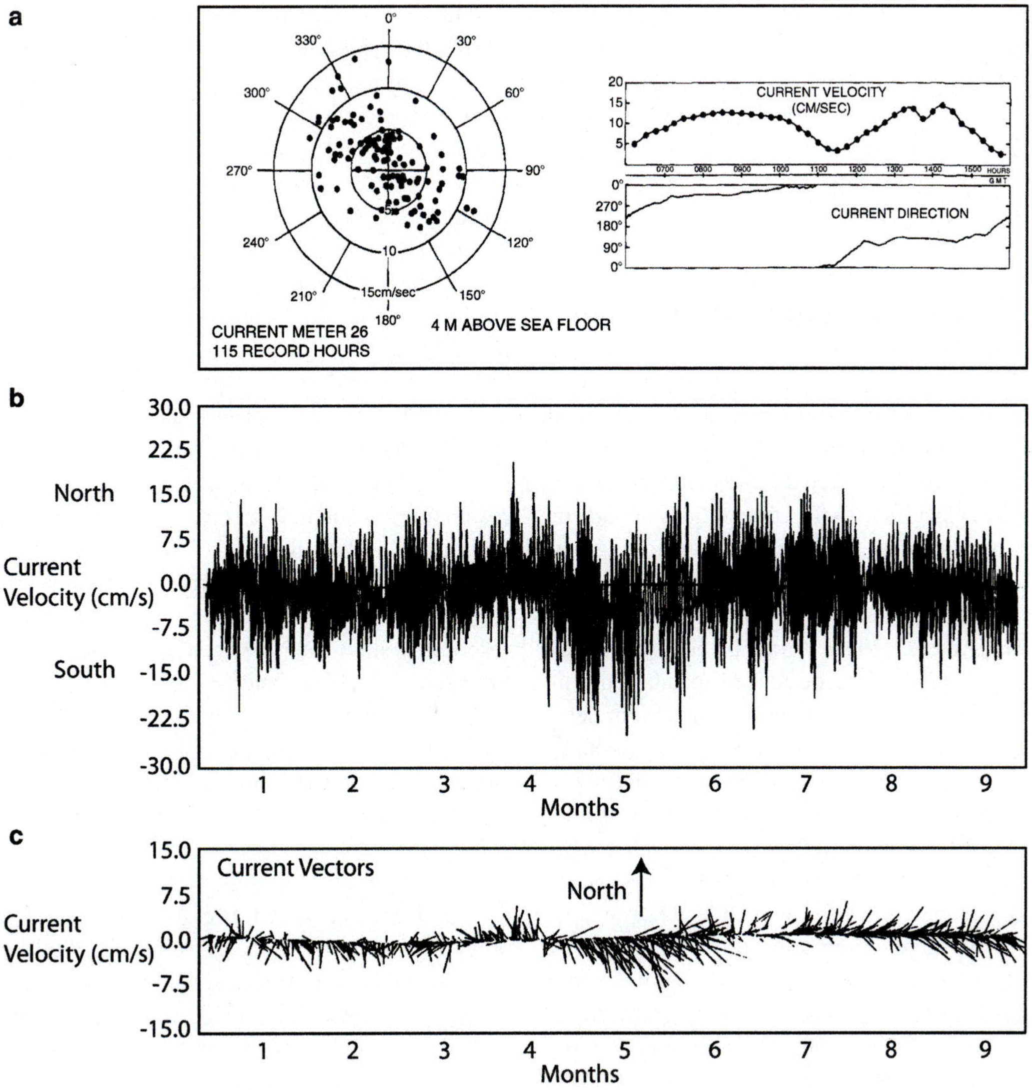

Fig. 14.5 (**a**) Current velocities and orientations from Horizon Guyot, Pacific Ocean, in 1,700 m water depth. Notice that the orientation of the current shifts progressively during the change in the tidal current. This progressive change seems to be typical of unconfined internal tidal settings, as opposed to confined settings (e.g. within submarine canyons) where the change is often abrupt (Modified from Lonsdale et al. 1972). (**b**) Current velocities showing a well-defined semi-diurnal reversal of direction for Horizon Guyot, Pacific Ocean (Modified from Cacchione et al. 1988). (**c**) Residual vectors for (**b**), showing consistent changes in the orientation of the total current direction over long time scales

the internal tide, modified by the Coriolis effect, similar to amphidromic systems in surface tides (Fig. 14.5a) (Kvale 2006). Changes in the orientation of internal-tide related currents has also been shown to occur on a longer time-scale in the open ocean, although the cause is again not well understood (Fig. 14.5b, c) (Cacchione et al. 1988; Noble et al. 1988). A few current measurements within deeply incised submarine canyons also show occasional cross-canyon current directions, however, these do not generally correlate to either a tidal period or the tidal cycle, suggesting they are either contour currents or internal waves spun-off from internal tidal waves (Rudnick et al. 2003), or are generated by another mechanism altogether, such as surface wind shear (Shepard et al. 1979).

Currently an active area of research, much of the state of knowledge on internal waves and internal tides comes from a lot of numerical modeling and a geographic scattering of in-situ measurements at arbitrary depths, most commonly at least a few meters above the sea floor, which although numerous are far from a comprehensive look at the deep oceans.

14.3 Modern Examples of Deep-Water Tidal Deposits

Excellent but fairly short-term records of currents in deep-water settings are plentiful, and clear and convincing correlation of many deep-water currents with tidal periods exist (Fig. 14.4). Observations include the direction and strength of currents, and resulting bedforms (Figs. 14.5 and 14.6) (Cacchione et al. 1988; Noble et al. 1988; Shanmugam et al. 2009; Shanmugam et al. 1993; Xu et al. 2008). Common bedforms on the modern seafloor that can be related to tidal currents include plane beds, furrows and scours, asymmetric and symmetric ripples, and small to large dunes. A major problem in modern deep-water settings is direct assignment of formative currents to bedforms, as most of the photographs of bedforms are taken when the currents are not strong or are absent. However, the common presence of bedforms such as these in areas where internal tidal currents appear to dominate over contour currents strongly suggests internal tides are a major factor in their formation (Heezen and Rawson 1977; Okada and Ohta 1993). Heezen and Rawson (1977) reported observations of internal tide and other bottom-current driven erosion and deposition on the Cocos Ridge in 650–2,000 m water depth. The erosional channels they reported range from tens of meters to several kilometers wide, and hundreds of meters deep. They interpreted the common occurrence of symmetrical ripples in foraminiferal sand along the seamounts they examined as a result of internal-tide reworking. Other areas of major erosion attributed to tidal currents include canyons 100–300 m deep and several km wide on the Ecuadoran slope, and local scours hundreds of meters wide and deep in calcareous sediment on ridges in the Panama Basin, the latter of which are apparently carved by dissolution aided by tidal currents (within the scours tidal currents are consistently fast, >15 cm/s) (Lonsdale 1976). Xu et al. (2008) documented large-scale dunes with 1–2.3 m amplitude and 20–75 m wavelength in the Monterey submarine canyon which they interpreted as the products of internal tidal currents, although whether these were generated by internal tidal currents or simply modified by them is unclear. Cacchione et al. (1988) showed that strong internal tides are present near Horizon Guyot in the Pacific Ocean at depths of 1,100 m, where large-wavelength (30 m) sand dunes with rippled stoss sides had previously been observed (Lonsdale et al. 1972). These dunes appear to migrate up-slope on both sides of the Guyot (Lonsdale et al. 1972), possibly due to transport by different phases of the internal tides, equivalent to flood and ebb barotropic tides in nearshore environments.

In modern fjord settings, water depths can easily reach several hundred meters, and not uncommonly can exceed 500 m (Benn and Evans 1998; Eyles et al. 1990). The influence of internal waves generated by the interaction of the surface tides with topography (especially sills) has long been recognized (Allen and Simpson 1998; Hein and Syvitski 1992; Stigebrandt 1976; Stigebrandt 1979), and appears to be a major factor in water exchange (Vlasenko et al. 2002). Generation and propagation of internal tides in fjords is greatly enhanced by strong vertical density gradients due to temperature, sediment concentration, and salinity changes. Bottom current velocities due to internal tides in fjords can exceed several tens of centimeter per second (Inall et al. 2004; Stashchuk et al. 2007), sufficient not only to cause vertical mixing, but to create and move significant bedforms (Fig. 14.2). Direct observation of internal tidal bedforms or deposits in modern fjords are not well documented, however, and most tidal rhythmites in fjords are attributed to the action of the surface and not internal tides (Cowan et al. 1997; Cowan et al. 1998). Like many other deep-water phenomena, however, observation is often the most difficult part.

14.4 Ancient Examples

Ancient deposits reported to be of deep-water tidal origin include the Ordovician Bays Formation, Virginia, U.S.A. (Zhenzhong and Eriksson 1991), the Ordovician of Tonglu and the Tarim Basin, the Lower Cambrian in Hunan, the Devonian to Triassic in Western Qingling, and the Mesoproterozoic in Xiushui, China (He et al. 2008), the Devonian Greenland Group, New Zealand

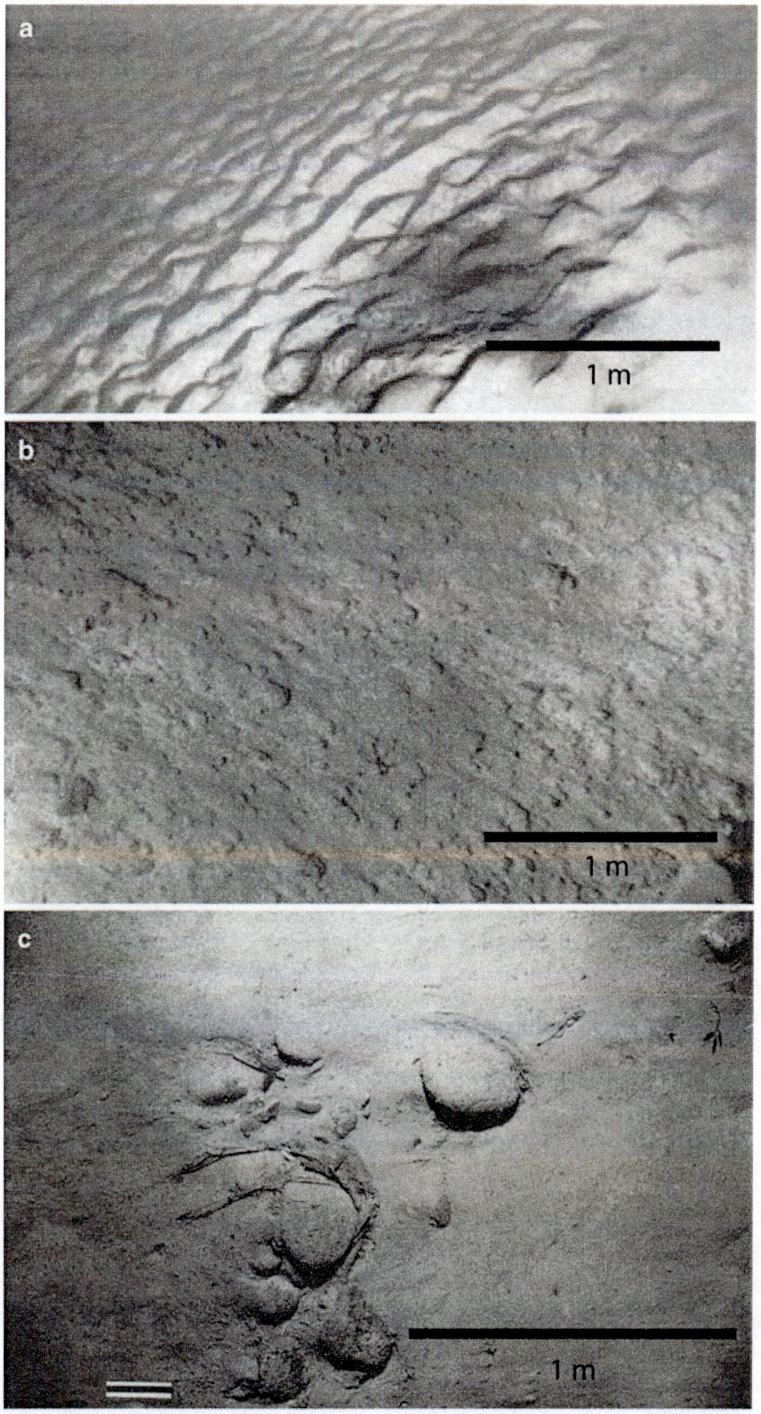

Fig. 14.6 Bedforms possibly produced or modified by internal tidal currents. (**a**) Unidirectional linguoid ripples from the Scotia Sea southeast of Tierra del Fuego in 4,010 m water depth (Modified from Heezen and Hollister 1964). (**b**) Current lineations from a flat-floored trough near Reunion Island, Indian Ocean in 4,909 m water depth (Modified from Heezen and Hollister 1964). (**c**) Scours around pebbles and cobbles in 1,304 m water depth, Suruga Bay, Japan (Modified from Okada and Ohta 1993)

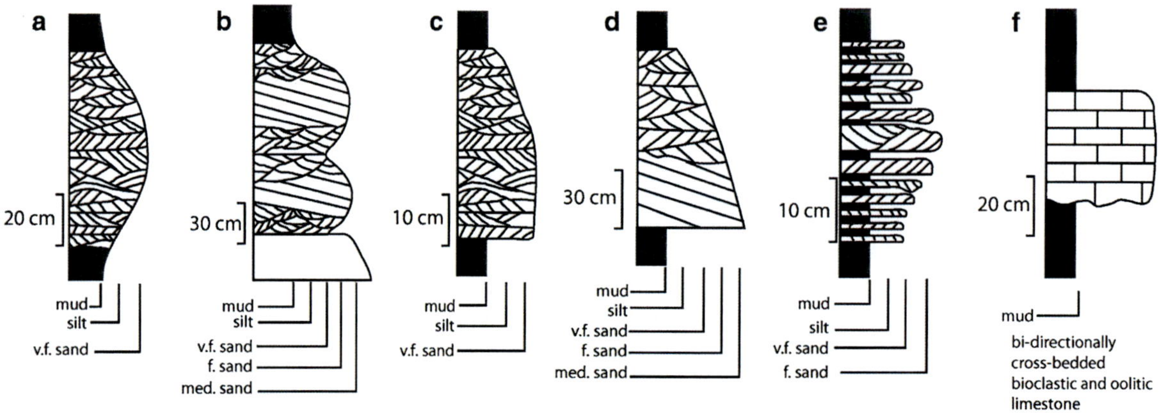

Fig. 14.7 Bedding patterns of purported internal tide deposits, compiled by He et al. (2008). (**a**) inverse to normally graded sandstone/siltstone exhibiting bi-directional cross-stratification. (**b**) inverse to normally-graded sandstone with bi-directional cross-stratification. (**c**) Normally graded sandstone with bi-directional cross-stratification. (**d**) Same as (**c**) but with large-scale cross-stratification (unidirectional) at the base. (**e**) Thickening then thinning upward succession of sand-mud couplets. The sand exhibits bi-directional cross-stratification. (**f**) Bioclastic or oolitic limestone, commonly with bi-directional cross-stratification

(Laird 1972), the Cretaceous of the Ontong-Java Plateau (Klein 1975), the Cretaceous Rosario Formation, Mexico (this study), the Cretaceous Wheeler Gorge conglomerate, California, U.S.A. (this study), the Eocene-Oligocene Annot Formation, SW France (Shanmugam 2003), the Eocene Torrey submarine canyon, California, U.S.A. (May et al. 1983), the Miocene Salir Formation, SW Turkey (Hayward 1984), and the Pliocene of the deep-water Krishna-Godavari Basin, India (Shanmugam et al. 2009). A few of these deposits are described in more detail below.

Zhenzhong and Eriksson (1991) presented a relatively shallow (<200 m water deep) submarine canyon-confined part of the Ordovician Bays Formation, where they interpreted sedimentary structures consistent with deposition from bi-directional currents as the deposits of internal tides. This interpretation is based on several 40–75 cm thick intervals of well-sorted, very-fine grained, cross-laminated sandstone with bidirectional paleocurrent orientations (foresets inclined both up and down-canyon), two of which exhibit inverse to normal grading, and a thin (10–13 cm), normally-graded interval that contains unidirectional ripple cross-lamination which dip up-canyon. These 'internal tide deposits' are interbedded with dark shales and normally-graded, poorly sorted beds interpreted to be of turbidite origin. Deposited under less than 200 m of water, however, this setting could have experienced a major contribution from the surface tides, and therefore the relative contribution of surface to internal tidal currents is unresolved.

Zhenzhong et al. (1998) interpreted a 30 m thick section of thinly interbedded fine-grained sandstone and mudstone of the Upper Ordovician Yankou Formation (China) as tidal rhythmites deposited in an unchannelized upper continental-slope setting, although they present neither a clear understanding of the paleobathymetry nor any rigorous test of tidal origin (e.g. harmonic analysis). These deposits are arranged into 20–100 cm thick intervals of thickening then thinning upward couplets of very-fine to fine-grained sandstone and mudstone (Fig. 14.7e). The couplets vary in thickness from 1 to 2 cm at the base and top of a succession to 3–7 cm thick in the middle. The sandstones are coarsest in the middle of a succession. The sandstones exhibit bidirectional paleocurrent directions, foresets dipping up and down the slope, and contain abundant lenticular, wavy, and flaser bedding. Subsequently quite a lot of work on internal tide deposits in outcrop and well-bores has been done in China, primarily by Chinese scientists at Yangtze University (Chengxin et al. 2005; Gao et al. 1997; He and Gao 1999; He et al. 2008; He et al. 2007). This group of researchers has recognized four main types of sedimentary facies associations they think are characteristic of internal tidal deposits: (1) inverse to normally-graded successions with bidirectional cross-lamination (up and down-channel dipping) (Fig. 14.7a, b), (2) normally graded successions with bidirectional cross-lamination (up and down-channel dipping) (Fig. 14.7c, d), (3) thickening-then-thinning upward successions of sandstone-mudstone couplets with bidirectional cross-lamination (up and down-slope oriented) (Fig. 14.7e), and (4) resedimented bioclastic (coral and shelly debris from the shelf or foraminiferal sand) and oolitic limestone-mudstone successions with bidirectional cross-lamination (Fig. 14.7f). The water depths in which these successions are reported to have been deposited are unclear, although the authors mainly claim the upper slope as the depositional environment. Alternative hypotheses for the origin of many of the features described in these papers could be contour currents or hyperpycnal currents (see Recognition Criteria, below) (Mulder et al. 2002; Mulder et al. 2001), although the abundance of bi-directional paleo-current indicators does point to a frequently changing current regime. Additionally, because all of the deposits they report are quite thick (Fig. 14.7), if they are indeed tidal in origin, their preservation would have necessitated a very high sediment supply combined with local accommodation space. Although this issue is not addressed in the papers reviewed above, nor indeed in any of the papers presenting internal tidal deposits, I do review some of these controls in more detail in the section on Preservation Potential, below.

Current reworked foraminiferal and volcaniclastic sand has also been recognized in Cretaceous cores on the Ontong-Java Plateau in what were interpreted as very deep-water settings (2,200–3,000 m water depth) (Klein 1975). Klein (1975) interpreted wavy, lenticular, and flaser bedding with at least two paleocurrent orientations, and an apparent cyclicity, as probably of tidal origin. Similar sedimentary structures have been found in the modern in deep-water foraminiferal sand (see Modern Examples of Deep-water Tidal Deposits, above) (Heezen and Hollister 1964; Heezen and Rawson 1977; Lonsdale and Malfait 1974).

Shanmugam (2003) reexamined outcrops from the Peira Cava outlier of the Eocene Annot Formation, from which the first vertical turbidite facies models were produced (Bouma 1962). His interpretation was that the tops of many turbidite beds had been reworked by internal tides, generating sedimentary structures such as planar laminae, ripple and sigmoidal (dune-scale) cross-stratification, and mud couplets. Planar laminae, ripples, and dunes, however, are commonly also formed by turbidity currents, a fact that has been well documented in flume experiments (Fedele and García 2009; Kneller 1995). True mud couplets (*sensu* Visser 1980), are more difficult to explain with turbidity currents, however, as they do imply slack-water conditions between strong currents, in this case of opposed direction (Fig. 14.8). While reflection of turbidity currents, especially within enclosed basins (such as the Peira Cava outlier), may be one possibility to account for this (Kneller et al. 1991), the conditions to set up flows that can reflect and yet develop slack-water conditions between passage of the primary flow and its reflection seem less likely to occur than do deep-water tidal currents. Therefore, it is not unreasonable to think there may have been an influence of internal tides on at least part of the Annot Formation.

Shanmugam et al. (2009) interpreted some parts of a cored interval within a submarine canyon in the Pliocene of the deep-water Krishna-Godavari Basin (689–920 m, modern water depths (Shanmugam et al. 2009)) as internal-tidal in origin. They distinguish between sandy tidalites and muddy tidalites. Their interpretation of sandy tidalites is based on the presence of very rhythmic lamination, mud couplets, lenticular and wavy bedding, parallel and ripple laminae, mud-draped ripples, and thick-thin bundles that they interpret as spring-neap cycles, respectively. Their interpretation of muddy tidalites is based mainly on rhythmic bedding and mud couplets. From the core photographs it is unclear how rhythmic these deposits are, or if any characteristic cyclicity is evident in the deposits. Other workers interpret similar features as the product of hyperpycnal currents (river-flood generated turbidity currents) (Mas et al. 2010; Nakajima 2006).

May et al. (1983) interpreted the final fill of the upper-slope, Eocene Torrey submarine canyon, exposed along Black's Beach in La Jolla, California, as exhibiting a tidal influence due to multiple (mainly up and down canyon) directions of ripple foresets.

The Cretaceous Wheeler Gorge Conglomerate has been interpreted as the deposit of a channel-levee complex, with axial facies represented by the conglomerate, and levee or interchannel facies represented by thinly interlaminated very fine-grained sandstones and siltstones, termed zebra-striped intervals, which both underlie and overlie the conglomeratic interval (Nelson et al. 1977; Walker 1985) (Fig. 14.9). The author re-examined the lower zebra-striped interval in detail, carefully logging and measuring the sandstone and siltstone intervals and plotting the resulting data

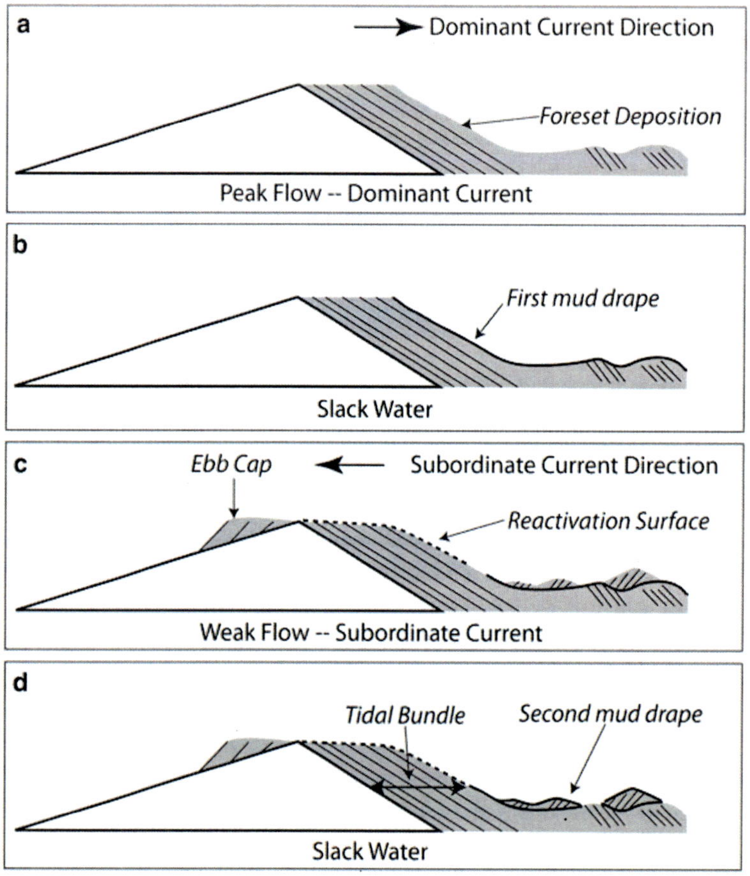

Fig. 14.8 Developmental model for ripple or dune-related mud couplets. (**a**) During peak flow of the dominant current, cross-strata build (in this case dune-scale, but they could be ripple-scale or even planar lamination). (**b**) During the subsequent slack-water, mud drapes the final depositional surface of (**a**). (**c**) The subordinate current (flowing in a different direction, but not necessarily directly opposite to the dominant current) reworks the previous cross-strata, eroding the upper part but leaving the lower mud-drape intact, now with some coarser rippled sediment deposited above it (this could also be planar laminated and finer-grained than that of the dominant current). An ebb cap of rippled sand also develops on the stoss side of the dune (lee side for the subordinate current), composed of sediment reworked from the crest of the dune. (**d**) The succeeding slack-water sees deposition of a final mud drape, making the mud couplet complete (Modified from Visser 1980 and Dalrymple 2010)

(Fig. 14.9). A clear, albeit noisy, trend of regular thick-thin alternations and of longer-term symmetrical thickening and thinning cycles are apparent in the data (Fig. 14.9c). Fourier analysis of the thickness data demonstrates that thick-thin intervals with a period of 2 are very common (Fig. 14.10). This suggests not only that the succession is tidal in origin, but that it is a semi-diurnal tide. Additional peaks at various periods suggest the tidal regime was influenced by a number of different tidal constituents; possible interpretations of these peaks are shown (Fig. 14.10).

The Cretaceous Rosario Formation, Baja California, Mexico, preserves a large volume of upper to middle continental slope sediments representing submarine canyon, channel, levee, and open continental slope depositional settings (Dykstra and Kneller 2007; Dykstra and Kneller 2009; Kane et al. 2009; Kane et al. 2007; Morris 1992). The Upper Cretaceous Cajiloa submarine canyon is incised over 150 m into the Upper Cretaceous, fluvial-alluvial el Gallo Formation (Morris and Busby-Spera 1988). Paleo-water depths in the submarine canyon were bathyal, and it was probably situated on the upper to mid-continental slope (Morris and Busby-Spera 1988). The submarine canyon fill consists of conglomeratic and sandstone-dominated channel axis deposits flanked and overlain

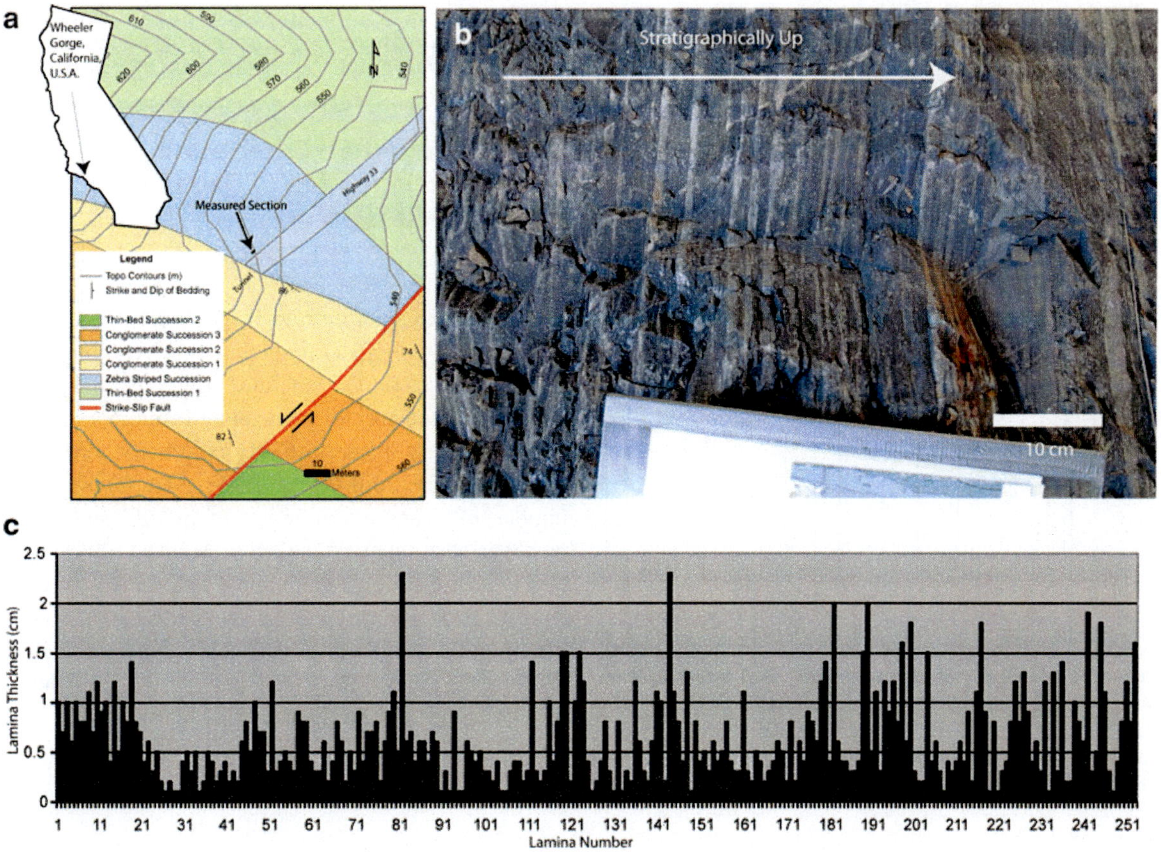

Fig. 14.9 Interpreted tidalites in the deep-water Cretaceous Wheeler Gorge, California. (**a**) Location and geologic map, showing the location of the measured section in the thin-bedded 'Zebra-Striped' succession. (**b**) Outcrop photograph of the measured section. Despite the color difference, there is not much difference in mean grain-size between the *dark* layers and the *light* layers, but the *light* layers are much cleaner with more evident sedimentary structures, primarily ripple cross-strata. (**c**) Graph of lamina thicknesses versus lamina number counted. The laminae are generally arranged in thick-thin couplets, and there is a cyclicity that approximates 20–30 laminae

by a relatively thin-bedded succession dominated by both in-place and remobilized (slumped) turbidite sandstones and siltstones interpreted to be of channel overbank origin (Fig. 14.11). The remobilized sections alternate laterally and vertically with in-place beds (Fig. 14.12), and consist of slumped thin-beds intercalated with and often injected by pebble-cobble conglomerates similar to those of the channel axis. The remobilized intervals usually have a topographically complex upper surface that formed highs and lows on the paleo-seafloor (Fig. 14.12). The in-place sediments are commonly partially 'ponded' or laterally confined by this seafloor topography. Interbedded with the in-place turbidites are numerous intervals 5–40 cm thick, that I have interpreted as deep-water tidalites; these are present only in areas overlying mass-failure generated seafloor topography. Common sedimentary structures in these tidalite beds include plane-parallel and wavy lamination, flaser lamination, and ripple cross-lamination which sometimes is bi-directional, and is commonly encased in mud drapes (Fig. 14.13). Detailed analysis of one of these intervals (32 cm thick) shows that groups of laminae preserved in this short succession are arranged into thick-thin couplets and into thickening and thinning successions exhibiting inverse to normal grading (Figs. 14.13 and 14.14a). A power-spectral analysis of the laminae thickness data (Fig. 14.14b) shows a very clear signal with a frequency of 2, equivalent to a semi-diurnal frequency, as well as frequency peaks around 17, 18, and 34, which may constitute tropical, synodic, and anomalistic half-months. Detailed grain-size analysis of some of the tidalite and turbidite intervals showed that tidalite sands tend to be much better sorted than turbidite sands of the equivalent thickness and modal grain size.

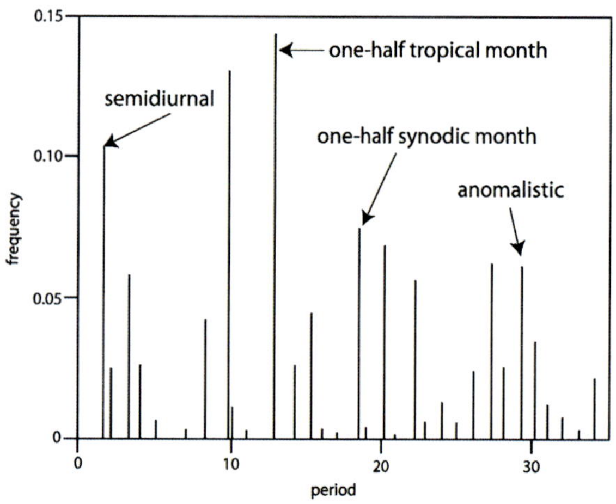

Fig. 14.10 Fourier transform (power spectrum) of the Wheeler Gorge data (see Fig. 14.9c), showing clustering of frequencies at the semi-diurnal period, and several other periods, possibly the tropical, synodic, and anamolistic periods. The frequency cluster around period 28 is particularly well-defined

14.5 Recognition Criteria and Discrimination of Deep-Water Tidalites

Tidally-dominated deposits (tidalites) may be confused on the modern seafloor and in the rock record with the deposits of both contour currents (contourites), and both river-flood generated (hyperpycnal) and surge-generated turbidity currents (turbidites), especially relatively dilute turbidity currents where bed-load transport dominates. Table 14.1 presents a number of criteria that may aid in differentiating between these types of deposits. Archer (1996) suggested the most robust method for distinguishing tidal deposits from other deposits is with spectral analysis of lamina thickness patterns to tease out any characteristic cyclicities. However, spectral analysis alone cannot distinguish between trends forced by the Earth-Moon-Sun system versus other external trends such as seasonal variability. In addition, where

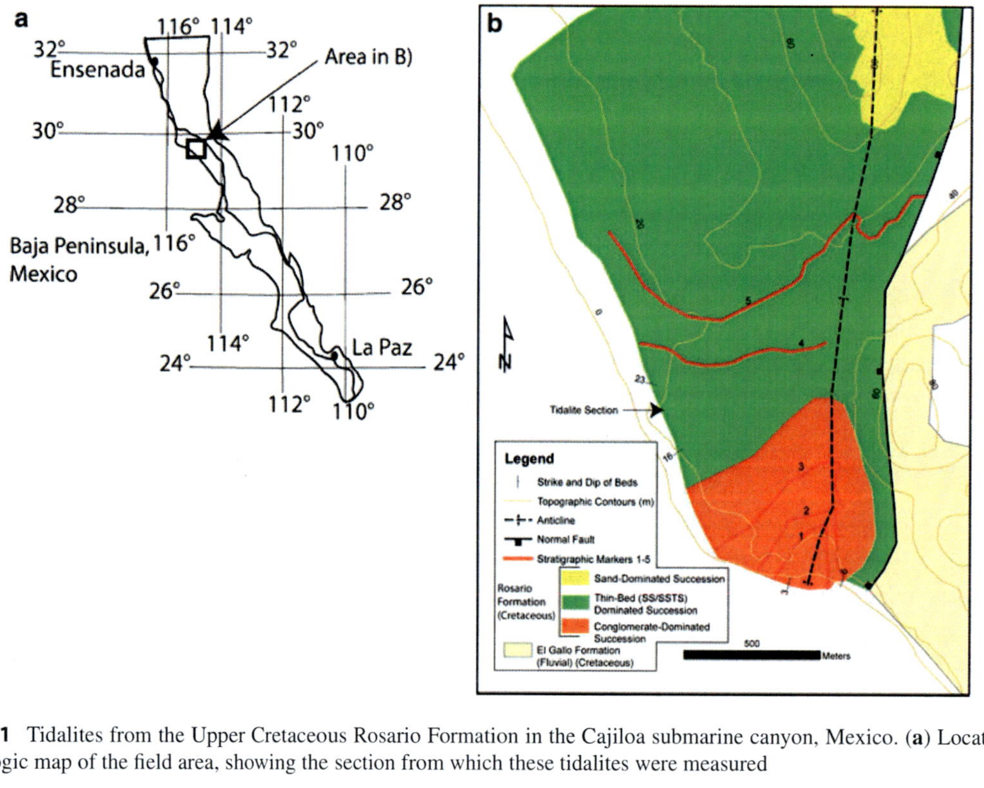

Fig. 14.11 Tidalites from the Upper Cretaceous Rosario Formation in the Cajiloa submarine canyon, Mexico. (**a**) Location map. (**b**) Geologic map of the field area, showing the section from which these tidalites were measured

Fig. 14.12 (**a**) Photomosaic and (**b**) interpretation of the tidalite section in the Cajiloa submarine canyon. The section is underlain by a slump (MTD 1) which created a local topographic low on the seafloor. This low was a critical factor in the deposition and preservation of the tidalites, as it created accommodation space for the tidalites to fill, and a location sheltered from the strongest turbidity currents. The slump is overlain by a 4–8 m thick interval of interbedded turbidites and tidalites, and is overlain by another slump (MTD 2). The locations of the detailed tidalites sections in Fig. 14.13 are shown

incomplete records are laid down or preserved (e.g. neap cycles too weak to move sediment or spring tides strong enough to erode, removing laminae), spectral analysis will not necessarily show a good correlation to any typical tidal cycle. Cowan et al. (1998), for example, showed that semi-diurnally driven sedimentation in a macrotidal deep-water estuary in Alaska left a record of thick-thin couplets that varied in number 3–18 per tropical month (the lunar orbital period determined by the moon crossing over the Earth's equator), with an average of only 7.4 couplets per cycle, showing that even in a very well constrained setting variability in deposition and preservation can be extreme (Dalrymple et al. 2003).

Due to the paucity of study of internal tide deposits, it is difficult to nail down distinct recognition criteria. Zhenzhong et al. (1998) suggest that in channelized settings, bidirectional current indicators oriented both down-channel and up-channel may provide good evidence of internal tidal deposits. In certain cases, however, turbidity currents are known to be able to produce reflections off of obstacles (channel-bends, slumps, fault-scarps) which may send a bore up-channel and can move, erode, rework, and deposit sediment in that direction (Morris and Alexander 2003); in field observations bi-directional current indicators in deep-water deposits have therefore most commonly been attributed to turbidity current reflections or deflections. Many of these interpretations are based on a divergence between sole structures in a bed and ripples higher in the same, apparently normally-graded bed (Kneller and McCaffrey 1999). In such a case where the bed appears continuously graded, and the difference in current directions is via indicators formed during very different energy conditions (e.g. sole structures = erosional; ripples = depositional), it stands to reason the same current is likely responsible for development of both current indicators, suggesting reflection. If, however, bi-directional current indicators consist of sedimentary structures formed under similar energy conditions, or they appear to be part of different event-beds, especially if separated by a drape of mud or hydraulically equivalent grains, then it stands to reason they were formed by different

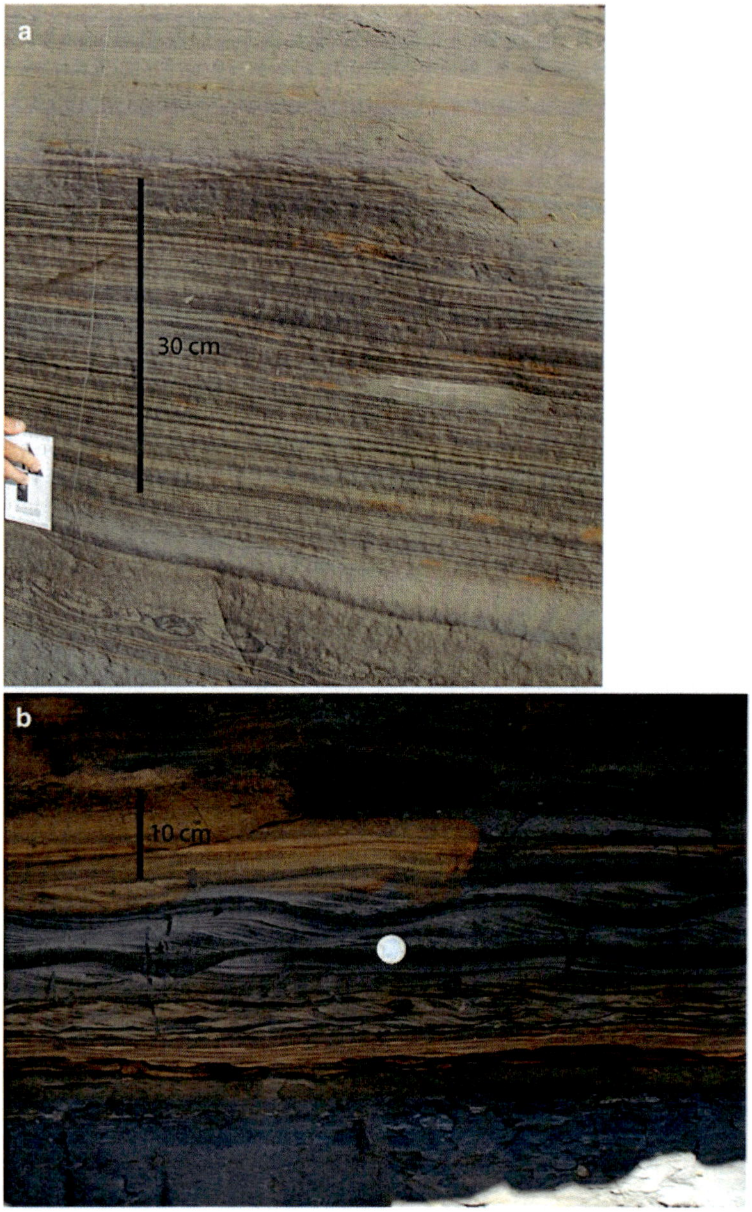

Fig. 14.13 (**a**) The tidalites section measured in detail, showing the tidal cycles, which are evident in Fig. 14.14. (**b**) A typical ripple-dominated tidalite interval, exhibiting multiple mud-draped ripples with clear and systematic lateral changes in foreset thicknesses indicative of changes in flow competence. Some reactivation surfaces can be picked out to the *left* of the coin

currents, which may suggest the action of opposing internal tides.

Shanmugam (2003) proposed that many of the criteria for the recognition of tidal deposits in shallow marine settings may be applicable in deep-water, including: (1) Heterolithic facies, (2) rhythmites of sandstone-shale, (3) thick-thin bundles, (4) alternation of parallel and cross-laminations, (5) climbing ripples, (6) double mud layers (mud couplets), (7) cross-beds with mud-drapes, (8) superposed bidirectional cross-bedding, (9) sigmoidal cross-bedding with mud drapes, (10) reactivation surfaces, (11) crinkled laminae, (12) elongate mudstone clasts, (13) flaser beds, (14) wavy bedding, and (15) lenticular bedding. Many of these

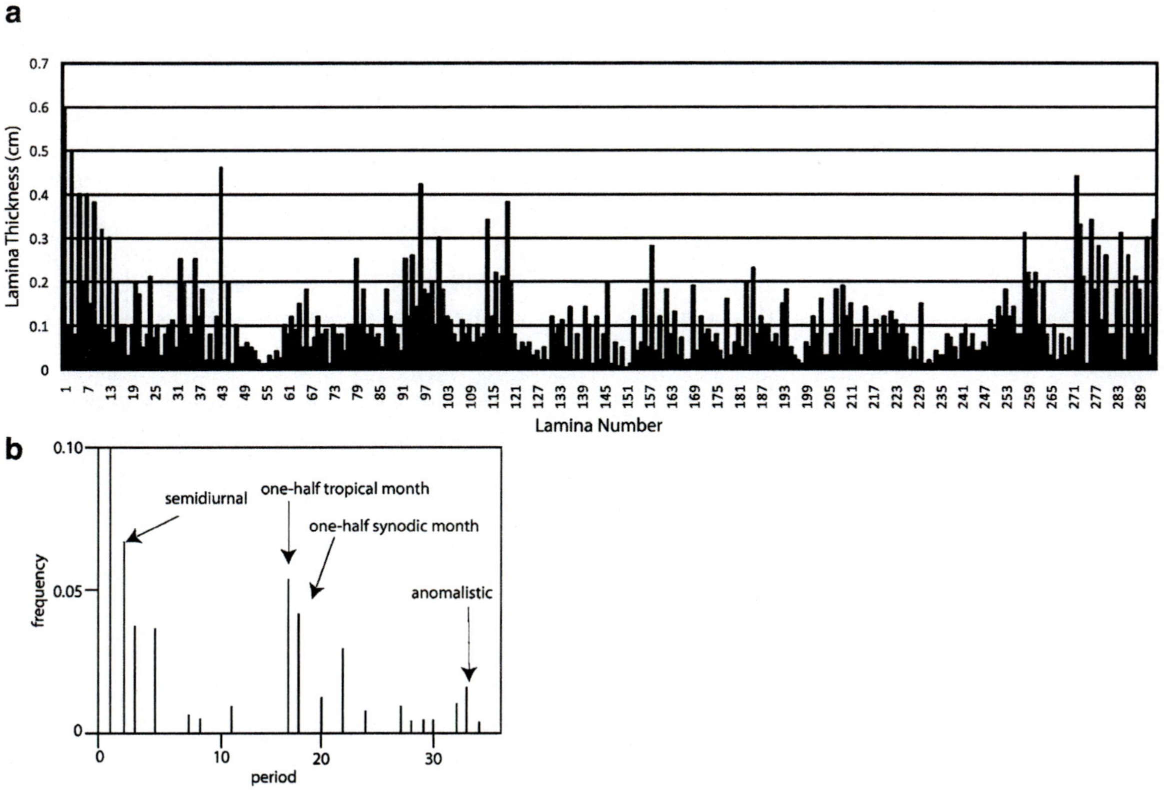

Fig. 14.14 (**a**) Lamina thickness versus lamina number for the Cajiloa submarine canyon (see Fig. 14.11). Notice the inequality of the laminae thicknesses is quite pronounced, especially in the beginning and end of the graph. (**b**) Fourier transform (power spectrum) of the data from (**a**), with the semidiurnal, synodic half-month, tropical half-month, and anomalistic periods highlighted. The presence of the other peaks may be indicative of other currents or other tidal constituents

criteria, however, can be produced by turbidity currents or contour currents (Kneller 1995; Rebesco et al. 2008), with the possible exception of a statistically-significant occurrence of thick-thin bundles (De Boer et al. 1989) and mud couplets *sensu* Visser (1980) (Fig. 14.8).

He et al. (2008) suggested that bidirectional cross-lamintions or cross bedding, flaser, wavy, and lenticular beds are normally observed in internal tide deposits. Since all of these sedimentary structures can be produced in turbidity currents, however, they are not good independent diagnostic criteria (Kneller 1995). He et al. (2008) also suggested that there are characteristic vertical successions in internal tide deposits, the most recognizable being inverse to normal grading of intervals (Fig. 14.7a, b, e). While inverse to normal grading undoubtedly occurs in tidalites, it can also be a product of flood-generated turbidity currents (hyperpycnal flows) or contour currents (Mulder et al. 2002; Mulder et al. 2001). If we look at what is genetically unique to internal tidal currents, what stands out is the cyclic nature of internal tides. Therefore, the most convincing diagnostic criterion should reflect this cyclicity, either in statistically meaningful numbers of thick-thin couplets or in repeated thickening-thinning cycles, or both (Figs. 14.7, 14.9–14.14). These cycles may be coupled with other criteria that aid in the interpretation such as mud drapes or couplets, reactivation surfaces, and bi- or multi-directional cross-bedding, but without the evidence of cyclicity, the other structures can as easily be explained by continuous or seasonal contour currents or by episodic hyperpycnal or surge-type turbidity currents, which after all can vary radically both spatially and temporally in energy and thus in erosion and deposition (Kneller and McCaffrey 2003).

Table 14.1 Important features of turbidites, hyperpycnites, contourites, and tidalites are compared (Modified from Mulder et al. 2002)

Bed type	Turbidite sequence (Bouma-like)	Hyperpycnal turbidite sequence (hyperpycnite)	Contourite sequence	Tidalite sequence
Flow type	Turbulent surge	Turbidity current	Contour current	Internal tides
Flow behaviour	Unsteady, mainly waning; unidirectional	Mainly steady, waxing then waning; unidirectional	Almost completely steady, waxing then waning; unidirectional	Waxing, waning, reversing, repeat; sometimes abrupt changes in direction, sometimes gradual
Flow regime	Turbulent	Turbulent	Turbulent	Laminar to turbulent
Flow duration and time for deposition	Minutes to days	Hours to months	1,000s to 10,000s years	Hours in each direction
Base contact	Erosive to sharp	Gradational	Gradational	Gradational to erosive
Top contact	Gradational	Gradational	Gradational	Gradational
Intrabed contact	Occurs sometimes between facies	Frequent, erosive to sharp	None	Gradational to erosive, depending on when formed during tidal cycle (spring or neap)
Grading	Clear, normal	Clear, normal then inverse	Crude, normal then inverse	Clear to crude; normal or inverse to normal
Bioturbation	Absent to intense	Absent to intense	Thorough and intense	Absent to slight
Ichnofacies	Few	Few	Many	Few
Structures	Well-developed parallel and cross-bedding, convolutes	Well-developed parallel and cross-bedding, climbing ripples frequent	Crude and sparse parallel and cross-bedding, frequent mottles and lenses	Parallel and wavy lamination, ripple-scale cross-bedding, flaser bedding, mud drapes and mud couplets common
Fauna/flora	Allochthonous, mainly marine	Allochthonous, mainly continental, frequent plant and wood fragments	Mainly autochthonous	Unknown – probably mainly autochthonous

14.6 Stratigraphic Successions

Zhenzhong et al. (1998) point out that due to the neap-spring-neap cycles inverse-to-normal grading is to be expected in vertical successions. This symmetrical grading may also apply to relative thicknesses such that couplets of sandstone-mudstone for example may be relatively thicker during the spring part of the cycle (Fig. 14.7). They also point out that due to energy concentration in channelized environments (Hotchkiss and Wunsch 1982), the neap-spring part of the cycle may frequently be absent, preserving only asymmetrically graded cycles from the spring to the neap tides. Any such cycle will also naturally be modulated by sediment availability; as in any system, if only mud is available, the succession will look different than if only sand and no mud is available, or if only carbonate grains are available. In non-channelized settings, vertical successions are more likely to preserve the neap to the spring tide part of the cycle, and thus be inverse to normally graded (Zhenzhong et al. 1998). In non-channelized successions, however, the flood and ebb internal tides (adopting terminology from surface tides) may not follow the same paths such that unidirectional currents may dominate, although in the case of the Yankou Formation in an unchannelized open slope setting bidirectional current indicators were apparently common (Zhenzhong et al. 1998).

In the Cajiloa submarine canyon succession (see Ancient Examples, above), the successions vary, some consist of a series of stacked thickening and thinning intervals, 5–40 cm thick, comprised of plane-parallel to wavy laminae (Fig. 14.13a), while others consist of laterally accreted bundles of ripple cross-laminae (Fig. 14.13b). The latter tend to be organized into successions 5–20 cm thick, with smaller-scale ripples at the bottom and top of a succession, and larger-scale ripples in the middle.

14.7 Preservation Potential

While it is obvious from current velocity records in modern submarine canyons that internal tidal currents are strong enough in some cases to move fairly coarse sediment in traction or even suspension on diurnal or semi-diurnal intervals, preservation of this sediment as deposits in the rock record is less likely, due to a high probability of sediment reworking by later, more powerful turbidity currents in the canyon. Deep-water tidalites are therefore much more likely to be preserved where current strength is a bit lower, and the power of intervening turbidity currents is reduced, such as in abandoned meander bends (Damuth et al. 1988), on the back-sides of levees (both master-bounding and internal, *sensu* (Kane et al. 2007)), within topographic lows such as those created by structural deformation on the surface of submarine landslides (Figs. 14.12 and 14.15), including submarine landslides on the inside of levees (Dykstra 2005), and within plunge-pools along submarine canyons (Gamberi and Marani 2007). Additionally, preservation potential probably increases during the waning phase of any given energy cycle that might be helping to externally control turbidity currents entering the deep-water (e.g. sea-level rise, decrease in sediment flux due to climate change, a decrease in tectonic activity onshore, etc.) (Fig. 14.16).

Both Zhenzhong and Eriksson (1991) and May et al. (1983) reported on tidalites preserved in the upper part of submarine canyon fills, stratigraphically well above the coarse-grained turbidites that comprised the lower fill. In the Ordovician Bays Formation, the tidalites are overlain by a thick slump consisting of outer shelf material, which in this case may have plugged the submarine canyon and aided in final preservation (Zhenzhong and Eriksson 1991). They interpreted the internal tidal interval as that of a highstand of sea-level, which in this case appears to have cut off the energetic turbidity-current system which in the lower part of the canyon fill overwhelmed any potential internal tidal signal. In other cases, however, internal tidal deposits can be below (Wheeler Gorge, see Ancient Examples, above) or interbedded with coarse channel-fill facies (Shanmugam 2003; Shanmugam et al. 2009), yet were apparently sheltered from erosional removal. This could occur if the subsequent turbidity currents were over-capacity when they arrived such that deposition began immediately, precluding erosion of the bed (Kneller and Branney 1995). In the Cajiloa submarine canyon, Cretaceous Rosario Formation (see Ancient Examples, above), the tidalites were in a position both lateral to the main channel axis (Fig. 14.11), and were sheltered by seafloor topography from an underlying slump, and were subsequently buried by an overlying slump (Figs. 14.12 and 14.15). In the Salir Formation, Turkey, Hayward (1984) found an inverse relationship between proximity to the active channel and the abundance of bottom-current reworked chalk beds (currents which he interpreted as tidal in origin), also suggesting that being lateral to the active channel can aid preservation.

Zhenzhong et al. (1998) found internal tidal deposits in an open-slope setting, where the potential for erosional removal by turbidity currents is lower, and if the slope was aggradational (Dykstra and Kneller 2007), then preservation potential might have been quite high. This example from the Yankou Formation (see Ancient Examples, above) is in fact the thickest deposit (30 m) yet documented. On some continental slopes mass-failure is the rule rather than the exception, however, and many tidal records may thus be removed by slumping (Posamentier and Walker 2006). Additionally, some very energetic internal tidal systems may actually cause net erosion rather than leave any depositional record (see discussion of Heezen and Rawson (1977) in Modern Examples, above, and discussion of Cacchione et al. (2002) in Morphological Impact of Tides in Deep-Water Settings, below).

Another major factor in the preservation potential of tidalites must be sedimentation rates. If sediment availability is low, then the internal tides may rework sediment a bit, but not much can accumulate, and preservation of anything that does accumulate becomes less likely. If sediment availability is high, then the converse may be true. The case of good sediment availability is exemplified by the Cajiloa submarine canyon (see Ancient Examples, above). In that case, a slump created a local seafloor low (Figs. 14.12 and 14.15). Into this (and presumably on the seafloor high surrounding the low), a turbidite was deposited which mantled the base of the seafloor low but did not fill it (Fig. 14.15b). Above that turbidite is the first appearance of tidalites in the succession (Figs. 14.12 and 14.15c). Therefore, what probably happened was that tidal currents running through the submarine canyon reworked the turbidite from the seafloor above the

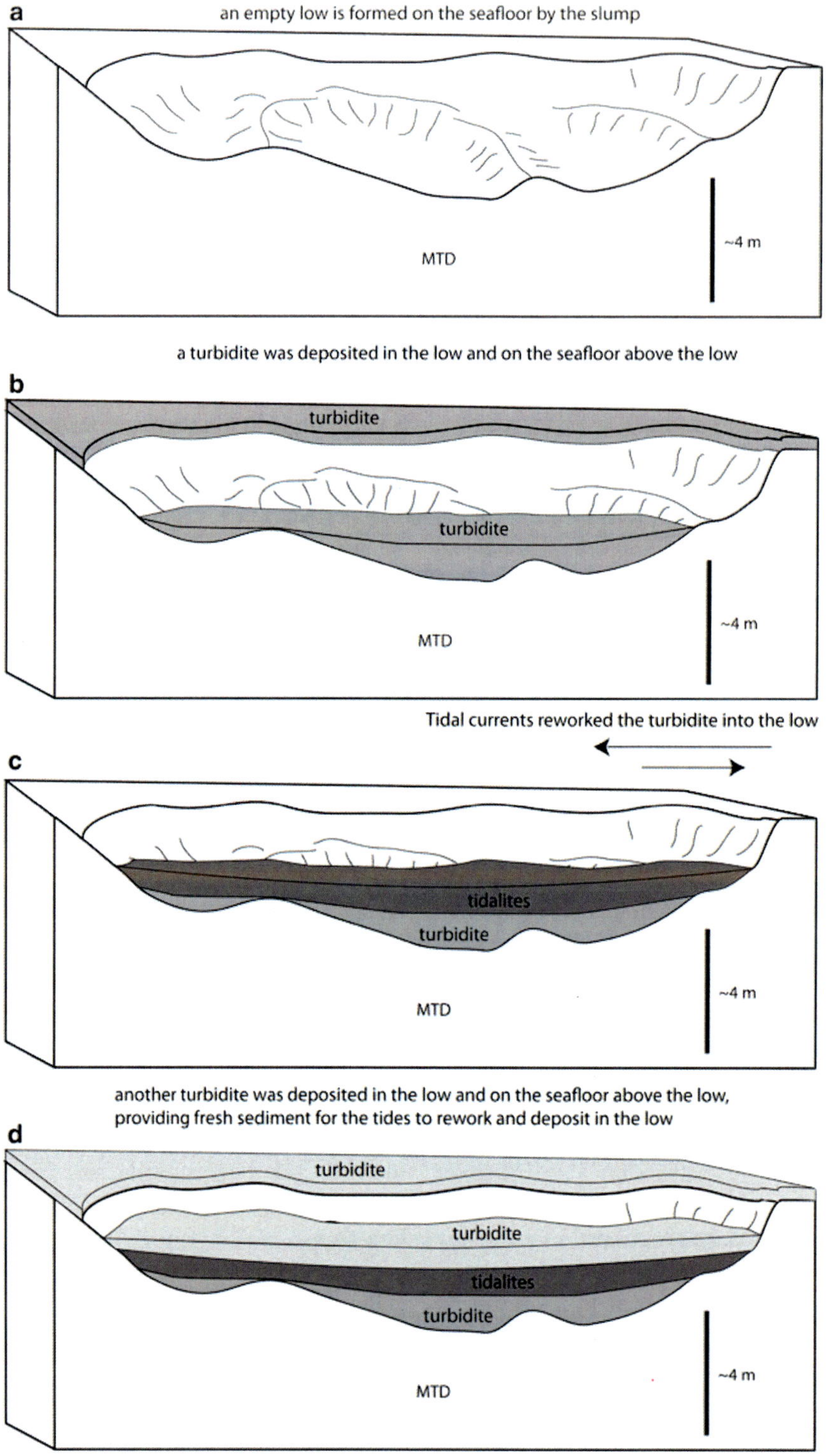

Fig. 14.15 Cartoons of the evolution of the Cajiloa tidalites succession, showing (**a**) the empty seafloor low created by emplacement of the slump (MTD 1 in Fig. 14.12); (**b**) the first large turbidite bed which mantled the seafloor low and partially smoothed out the seafloor roughness; (**c**) the turbidite from (**b**) above the seafloor low was reworked by tidal currents and redeposited into the seafloor low. After a hiatus during which the interval above the lower tidalites became extensively bioturbated; (**d**) another turbidite was deposited and reworked into more tidalites in the seafloor low. This sequence was repeated a number of times until the seafloor low was filled and subsequently overlain by another slump. Other tidalite successions in the Cajiloa system exhibit a very similar evolution

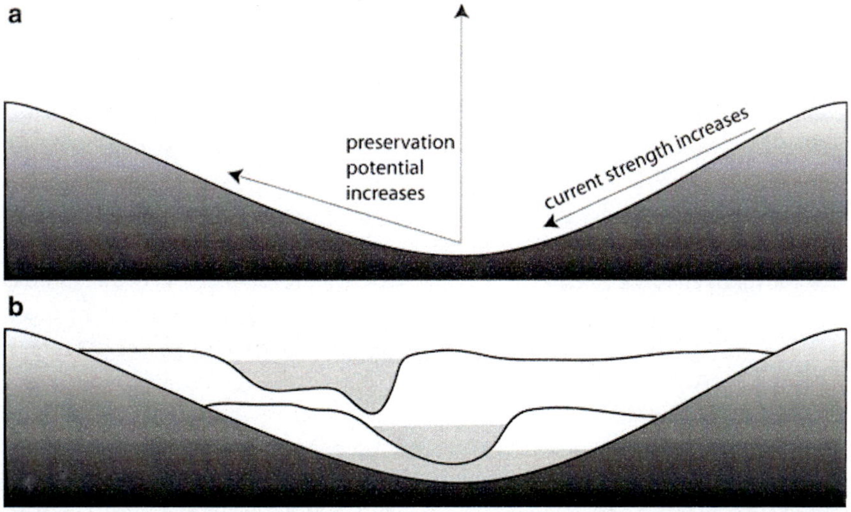

Fig. 14.16 Cartoons illustrating preservation potential within a submarine canyon. (**a**) Preservation potential increases both towards the margins and upward within the fill of a submarine canyon. (**b**) Cartoon showing axial coarse clastics (*dark gray*), and finer-grained overbank (*white*). Tidalites are much more likely to be preserved in the upper part of the submarine canyon fill because the energy in the system is lower (the fill is generally moved under lower energy conditions than whatever currents cut the canyon), and because the currents have further to spread out laterally, decreasing the amount of energy available per unit volume even further than lower down in the submarine canyon

topographic low, and deposited the reworked sediment into that low. Subsequently another turbidite was deposited, both within the low and on the surrounding seafloor, which then was also reworked and deposited by tidal currents into the low (Fig. 14.15d). This was repeated episodically until the low was filled. In between periods of sediment availability tidal currents undoubtedly continued to pump up and down the submarine canyon, but without sediment to rework the only record of that are intervals of greatly increased bioturbation (see the top of the tidalite intervals in Fig. 14.13 as an example).

14.8 Sequence Stratigraphy

Zhenzhong and Eriksson (1991) developed a sequence stratigraphic model based on the Ordovician Bays Formation, Virginia. They suggested that internal tidal deposits are best preserved during the late transgression or early highstand part of a sea-level cycle. May et al. (1983) also showed that tidal current indicators were more common in the upper part of a submarine canyon fill, although they did not put this into a strict sequence stratigraphic framework. Shanmugam et al. (2009) interpreted internal tidal deposits to have been present in a submarine canyon at the same time as powerful turbidity currents and debris flows. Additionally, Shanmugam (2003) suggested that the tops of many beds throughout the Peira Cava outlier of the Annot Formation (Amy et al. 2007) were reworked by tidal currents. Although I disagree that all or even most of the beds in the Peira Cava area with rippled tops are reworked by tidal currents, the apparent unsystematic occurrence of potential tidalites within any given sequence suggests there may not be a strong external control on their preservation. Perhaps only local conditions such as accommodation space, are important.

The strength of surface tides during times of lowered sea-level with the same plate configuration has been calculated to increase, potentially by 50% or more (Egbert et al. 2004; Hall and Davies 2004; Neill et al. 2010; Uehara et al. 2006). In addition, the dissipation of tidal energies in the deep oceans has been shown to increase during lowered sea-level, with 1.25–3 times more energetic deep ocean internal tides globally during the Last Glacial Maximum than at present (Egbert et al. 2004), when sea-level was between 120 and 140 m lower than today (Adkins et al. 2002; Lambeck and Chappell 2001). Although based on modeling, this result suggests that a eustatic cycle should have a significant impact on the action of deep-water tides. The strongest tides in the deep oceans may therefore generally occur during sea-level lowstands (greatly modified, no doubt, by any given basin's bathymetry and geometry), when preservation may be more of an issue, especially within submarine canyons. For the present, however, we have a very poor understanding of deep marine tidal deposits and their relationship with sea-level or energy cycles.

14.9 Morphological Impact of Tides in Deep-Water Setting

Altimetry analysis of the oceans have demonstrated that a huge amount of tidal energy is dissipated in the deep oceans (about 10^{12} W (1 TW) – one-quarter to one-third of the total tidal energy) (Egbert and Ray 2000). This energy is dissipated in areas of rough seafloor topography such as seamounts, continental slopes, and ridges, and may be a major formative agent in the morphology of these features, and a major factor in abyssal circulation (Cacchione et al. 2002; Mitchell and Huthnance 2008; Munk and Wunsch 1998). Cacchione et al. (2002) calculated that reflected semi-diurnal internal tides may provide enough energy to continental slopes to inhibit deposition of very fine-grained sediment over a characteristic depth range dependant on the physiography of the slope, oceanographic conditions, and the orientation and strength of incident internal tidal waves. This inhibition of deposition, they suggest, helps to create the morphology of the continental slopes on the large-scale by creating a zone of nondeposition or resuspension on the slope. While their formulation suggested that a particular angle of incidence of the internal wave to the angle of the slope is required to cause shear coupling with the substrate, which is in agreement with other analytical solutions (Cacchione et al. 2002; Garrett and Kunze 2007; Thorpe 1992), nonlinear analyses suggest that in fact the angle of incidence relative to the slope plays little role in energy transfer (Legg and Adcroft 2003), and therefore shear should be generated at any boundary acted upon by internal waves. Observational data from offshore Oregon show that turbulent mixing can occur quite strongly due to internal tidal forcing, and appears to be aided significantly by seafloor topography, such as rugosity above a submarine landslide (Moum et al. 2002). Hotchkiss and Wunsch (1982) demonstrated that internal tidal energy can become concentrated in submarine canyons, which may aid in enlargement of the canyons by erosion of the walls.

Much of the current surface tidal energy is dissipated on continental shelves and in shallow seas (Egbert and Ray 2003). Many of these were smaller or absent of water during the Last Glacial Maximum, and therefore the excess tidal energy, both of the surface and the internal tides, had to be directed elsewhere. The first effect of this is that the conversion of surface to internal tidal energy would have been greater, meaning there was significantly more energy available in the deep-water via internal tides during periods of lowered sea-level (Egbert et al. 2004). One would expect, therefore, that any effect of erosion or non-deposition such as that proposed by Cacchione et al. (2002) should be more of a factor during times of lowered sea-level, and that internal tides would have more of or a slightly different morphological impact on continental slopes during those times. While the morphological impact of wave, surface, and internal tides has been examined by trying to forward model seismic profiles of some upper continental slopes using the gravity effect (Mitchell and Huthnance 2008), the same has not been done looking at really deep-water settings, thus the sensitivity of seafloor morphology to internal tidal forcing is very poorly understood.

14.10 Summary

Internal tidal currents have been well documented in the modern oceans, from the upper slope into deep ocean basins. The strength of internal tides is commonly enough to remobilize coarse sediment and form bedforms in sand and gravel, and may be a significant factor in sediment movement over long time periods in some deep-water settings (Fig. 14.2). Recognition criteria for deep marine tidal deposits include evidence of cyclicity, including short-term (semi-diurnal, diurnal) and long-term cyclicities (bi-monthly, monthly, and longer), commonly reflected in tidal couplets or in inverse-to-normally graded intervals (Fig. 14.7). Additionally, the extremely asymmetrical current energies involved in tidal regimes often leave evidence in the way of mud-draped ripples or cross-strata with evidence of reactivation surfaces, sometimes in different or even opposing directions. Although some work has been done on internal tidal deposits in outcrop, compared with the extensive literature on shelf, shallow-marine, and estuarine tidal deposits, deep marine tidal deposits have been and remain largely untouched. Whether this is worker bias or a true reflection of their preservation in the rock record remains to be seen.

References

Adkins F, McIntyre K, Schrag DP (2002) The salinity, temperature, and delta 18O of the glacial deep ocean. Science 298:1769–1773

Allen GL, Simpson JH (1998) Reflection of the internal tide in Upper Loch Linnhe, a Scottish Fjord. Estuarine Coast Shelf Sci 46:683–701

Amy LA, Kneller BC, McCaffrey WD (2007) Facies architecture of the Gres de Peira Cava, SE France: landward stacking patterns in ponded turbiditic basins. J Geol Soc 164:143–162

Archer AW (1996) Reliability of lunar orbital periods extracted from ancient cyclic tidal rhythmites: Earth and Planetary Science Letters 141:1–10

Benn DI, Evans DJ (1998) Glaciers & glaciation. Arnold, London, 734 p

Bouma AH (1962) Sedimentology of some flysch deposits: a graphic approach to facies interpretation. Elsevier, Amsterdam, 168 p

Cacchione D, Schwab W, Noble M, Tate G (1988) Internal tides and sediment movement on Horizon Guyot, Mid-Pacific mountains. Geo-Mar Lett 8:11–17

Cacchione DA, Pratson LF, Ogstona AS (2002) The shaping of continental slopes by internal tides. Science 296:724–727

Chengxin L, Zhenzhong G, Youliang J, Xiaolong W (2005) Ordovician deep-water traction current deposits on the southwestern margin of the Ordos Basin Haiyang Dizhi yu Disiji Dizhi. Mar Geol Quat Geol 25:31–36

Cowan EA, Cai J, Powell RD, Clark JD, Pitcher JN (1997) Temperate glacimarine varves: an example from Disenchantment Bay, southern Alaska. J Sediment Res 67:536–549

Cowan EA, Cai J, Powell RD, Seramur KC, Spurgeon VL (1998) Modern tidal rhythmites deposited in a deep-water estuary. Geo-Mar Lett 18:40–48

Dalrymple RW, Baker EK, Harris PT, Hughes MG (2003) Sedimentology and stratigraphy of a tide-dominated, foreland-basin delta (Fly River, Papua New Guinea). In: Tropical deltas of Southeast Asia. SEPM Spec Publ 76:147–173

Dalrymple RW (2010) Tidal depositional systems. In: James NP, Dalryple RW (eds) Facies Models 4:201–231

Damuth J, Flood R, Kowsmann R, Belderson R, Gorini M (1988) Anatomy and growth pattern of Amazon deep-sea fan as revealed by long-range side-scan sonar (GLORIA) and high-resolution seismic studies. Am Assoc Petrol Geol Bull 72:885

De Boer PL, Oost AP, Visser MJ (1989) The diurnal inequality of the tide as a parameter for recognizing tidal influences. J Sediment Res 59:912–921

Dykstra M (2005) Dynamics of sediment mass-transport from the shelf to the deep sea. Ph.D dissertation, University of California, Santa Barbara, 152 p

Dykstra M, Kneller B (2007) Canyon San Fernando, Baja California, Mexico: a deep-marine channel-levee complex that evolved from submarine canyon confinement to unconfined deposition. In: Nilsen TH, Shew HD, Steffens GS, Studlick JRJ (eds) Atlas of deep-water outcrops; *CD ROM*. American Association of Petroleum Geologists Studies in Geology 56, 14 p

Dykstra M, Kneller B (2009) Lateral accretion in a deep-marine channel complex: implications for channelized flow processes in turbidity currents. Sedimentology 56:1411–1432

Egbert GD, Ray RD (2000) Significant dissipation of tidal energy in the deep ocean inferred from satellite altimeter data. Nature 405:775–778

Egbert GD, Ray RD (2003) Semi-diurnal and diurnal tidal dissipation from TOPEX/Poseidon altimetry. Geophys Res Lett 30:1907–1910

Egbert GD, Ray RD, Bills BG (2004) Numerical modeling of the global semidiurnal tide in the present day and in the last glacial maximum. J Geophys Res Ocean 109:1–15

Eyles N, Mullins HT, Hine AC (1990) Thick and fast: sedimentation in a Pleistocene fiord lake of British Columbia, Canada. Geology 18:1153–1157

Fedele J, García MH (2009) Laboratory experiments on the formation of subaqueous depositional gullies by turbidity currents. Mar Geol 258:48–59

Gamberi F, Marani M (2007) Downstream evolution of the Stromboli slope valley (southeastern Tyrrhenian Sea). Mar Geol 243:180–199

Gao Z, He Y, Li J, Li W, Luo S, Wang Z (1997) The first internal-tide deposits found in China. Chin Sci Bull 42:1113–1116

Garrett C (2003) Ocean Science: enhanced: internal tides and ocean mixing. Science 301:1858–1859

Garrett C, Kunze E (2007) Internal tide generation in the deep ocean. Ann Rev Fluid Mech 39:57–87

Hall P, Davies AM (2004) Modelling tidally induced sediment-transport paths over the northwest European shelf: the influence of sea-level reduction. Ocean Dyn 54:126–141

Hayward AB (1984) Hemipelagic chalks in a clastic submarine fan sequence: Miocene SW Turkey. Geol Soci Lond Spec Publ 15:453–467

He Y, Gao Z (1999) The characteristics and recognition of internal-tide and internal-wave deposits. Chin Sci Bull 44:582

He Y, Gao Z, Luo S, Peng D, Wang H, Luo J (2007) Discovery of internal-tide deposits from the third member of Pingliang Formation in Longxian area, Shaanxi Province Shiyou Tianranqi Xuebao. J Oil Gas Technol 29:28–33

He Y, Gao Z, Luo J, Luo S, Liu X (2008) Characteristics of internal-wave and internal-tide deposits and their hydrocarbon potential. Petrol Sci 5:37–44

Heezen BC, Hollister C (1964) Deep-sea current evidence from abyssal sediments. Mar Geol 1:141–174

Heezen BC, Rawson M (1977) Visual observations of contemporary current erosion and tectonic deformation on the Cocos Ridge crest. Mar Geol 23:173–196

Heezen BC, Hollister CD, Ruddiman WF (1966) Shaping of the continental rise by deep geostrophic contour currents. Science 152:502–508

Hein FJ, Syvitski JPM (1992) Sedimentary environments and facies in an arctic basin, Itirbilung Fiord, Baffin Island, Canada. Sediment Geol 81:17–45

Hotchkiss FS, Wunsch C (1982) Internal waves in Hudson Canyon with possible geological implications. Deep Sea Res 29:415–442
Huthnance JM (1989) Internal tides and waves near the continental shelf edge. Geophys Astrophys Fluid Dyn 48:81–106
Inall M, Cottier F, Griffiths C, Rippeth T (2004) Sill dynamics and energy transformation in a jet fjord. Ocean Dyn 54:307–314
Kane IA, Kneller BC, Dykstra M, Kassem A, McCaffrey WD (2007) Anatomy of a submarine channel–levee: an example from Upper Cretaceous slope sediments, Rosario Formation, Baja California, Mexico. Mar Petrol Geol (online)
Kane IA, Dykstra ML, Kneller BC, Tremblay S, McCaffrey W (2009) Architecture of a coarse-grained channel-levee system: the Rosario Formation, Baja California, Mexico. Sedimentology 56:2207–2234
Klein GD (1975) Resedimented pelagic carbonate and volcaniclastic sediments and sedimentary structures in Leg 30 DSDP cores from the western Equatorial Pacific. Geology 3:39–42
Kneller BC (1995) Beyond the turbidite paradigm: physical models for deposition of turbidites and their implications for reservoir prediction. In: Marley AJ, Prosser DJ (eds) Characterisation of deep marine clastic systems. Geol Soc Lond Spec Paper 94:29–46
Kneller BC, Branney MJ (1995) Sustained high-density turbidity currents and the deposition of thick massive sands. Sedimentology 42:607–616
Kneller BC, McCaffrey WD (1999) Depositional effects of flow non-uniformity and stratification within turbidity currents approaching a bounding slope: deflection, reflection, and facies variation. J Sediment Res 69:980–991
Kneller BC, McCaffrey WD (2003) The interpretation of vertical sequences in turbidite beds: the influence of longitudinal flow structure. J Sediment Res 73:706–713
Kneller BC, Edwards D, McCaffrey W, Moore R (1991) Oblique reflection of turbidity currents. Geology 19:250–252
Kneller BC, Bennett SJ, McCaffrey WD (1997) Velocity and turbulence structure of density currents and internal solitary waves: potential sediment transport and the formation of wave ripples in deep water. Sediment Geol 112:235–250
Kvale EP (2006) The origin of neap–spring tidal cycles. Mar Geol 235:5–18
LaFond EC (1962) Internal waves: part I. In: Hill MN (ed) The sea: ideas and observation on progress in the study of the seas, vol 1. Wiley, New York, 864 p
Laird MG (1972) Sedimentology of the Greenland Group of the Paparoa Range, west coast, South Island, NewZealand. N Z J Geol Geophys 15:372–393
Lambeck K, Chappell J (2001) Sea level change through the last glacial cycle. Science 292:679–686
Legg S (2004) Internal tides generated on a corrugated continental slope. Part I: cross-slope barotropic forcing. J Phys Oceanogr 34:156–173
Legg S, Adcroft A (2003) Internal wave breaking at concave and convex continental slopes. J Phys Oceanogr 33:2224–2246
Lonsdale P (1976) Abyssal circulation of the southeastern Pacific and some geological implications. J Geophys Res 81:1163–1176
Lonsdale P, Malfait B (1974) Abyssal dunes of foraminiferal sand on the Carnegie Ridge. Geol Soc Am Bull 85:1697–1712
Lonsdale P, Normark WR, Newman WA (1972) Sedimentation and erosion on Horizon Guyot. Geol Soc Am Bull 83:289–316
Mas V, Mulder T, Dennielou B, Schmidt S, Khripounoff A, Savoye B (2010) Multiscale spatio-temporal variability of sedimentary deposits in the Var turbidite system (North-Western Mediterranean Sea). Mar Geol 275:37–52
May JA, Warme JE, Slater RA (1983) Role of submarine canyons on shelfbreak erosion and sedimentation: modern and ancient examples. In: The shelfbreak: critical interface on continental margins. SEPM Spec Publ 33:315–332
McCave IN, Lonsdale PF, Hollister CD, Gardner WD (1980) Sediment transport over the Hatton and Gardar contourite drifts. J Sediment Petrol 50:1049–1062
Mitchell NC, Huthnance JM (2008) Oceanographic currents and the convexity of the uppermost continental slope. J Sediment Res 78:29–44
Morris WR (1992) The depositional framework, paleogeography and tectonic development of the Late Cretaceous through Paleocene Peninsular Range forearc basin in the Rosario Embayment, Baja California, Mexico. Ph.D. dissertation, University of California, Santa Barbara, 240 p
Morris SA, Alexander J (2003) Changes in flow direction at a point caused by obstacles during passage of a density current. J Sediment Res 73:621–629
Morris WR, Busby-Spera CJ (1988) Sedimentologic evolution of a submarine canyon in a forearc basin, Late Cretaceous Rosario Formation, San Carlos, Mexico. Bull Am Assoc Petrol Geol 72:717–737
Moum JN, Caldwell DR, Nash JD, Gunderson GD (2002) Observations of boundary mixing over the continental slope. J Phys Oceanogr 32:2113–2130
Mulder T, Migeon S, Savoye B, Faugères JC (2001) Inversely graded turbidite sequences in the deep Mediterranean: a record of deposits from flood-generated turbidity currents? Geo-Mar Lett 21:86–93
Mulder T, Migeon S, Savoye B, Faugères JC (2002) Reply to discussion by Shanmugam on Mulder et al. (2001, Geo-Mar Lett 21:86–93) Inversely graded turbidite sequences in the deep Mediterranean. A record of deposits from flood-generated turbidity currents? Geo-Mar Lett 22:112–120
Mulder T, Lecroart P, Hanquiez V, Marches E, Gonthier E, Guedes JC, Thiébot E, Jaaidi B, Kenyon N, Voisset M, Perez C, Sayago M, Fuchey Y, Bujan S (2006) The western part of the Gulf of Cadiz: contour currents and turbidity currents interactions. Geo-Mar Lett 26:31–34
Mulder T, Zaragosi S, Garlan T, Mavel J, Cremer M, Sottolichio A (2009) Deep-sea currents related to internal tides in Canyons of the Bay of Biscay, implications on the behaviour of submarine Canyons. In: Pascucci V, Andreucci S (eds) 27th IAS meeting, Alghero, Italy, pp 616
Munk W, Wunsch C (1998) Abyssal recipes II: energetics of tidal and wind mixing. Deep Sea Res Part I: Oceanogr Res Pap 45:1977–2010
Nakajima T (2006) Hyperpycnites deposited 700 km away from river mouths in the Central Japan Sea. J Sediment Res 76:60–73

Nash JD, Kunze E, Lee CM, Sanford TB (2006) Structure of the baroclinic tide generated at Kaena Ridge, Hawaii. J Phys Oceanogr 36:1123–1135

Neill S, Scourse J, Uehara K (2010) Evolution of bed shear stress distribution over the northwest European shelf seas during the last 12,000 years. Ocean Dyn 60:1139–1156

Nelson C, Mutti E, Ricci Lucchi F (1977) Upper Cretaceous resedimented conglomerates at Wheeler Gorge, California: description and field guide discussion. J Sediment Petrol 47:926–934

Noble M, Cacchione DA, Schwab WC (1988) Observations of strong Mid-Pacific internal tides above Horizon Guyot. J Phys Oceanogr 18:1300–1306

Okada H, Ohta S (1993) Photographic evidence of variable bottom-current activity in the Suruga and Sagami Bays, central Japan. Sediment Geol 82:221–237

Özsoy E, Di Iorio D, Gregg MC, Backhaus JO (2001) Mixing in the Bosphorus Strait and the Black Sea continental shelf: observations and a model of the dense water outflow. J Mar Sys 31:99–135

Posamentier HW, Walker RG (2006) Deep-water turbidites and submarine fans. In: Posamentier HW, Walker RG (eds) Facies models revisited. SEPM Spec Publ 84:399–520

Rebesco M, Camerlenghi A, Van Loon AJ (2008) Contourite research: a field in full development (Chapter 1). In: Rebesco M, Camerlenghi A (eds) Developments in sedimentology, vol 60. Elsevier, Amsterdam/Oxford, pp 3–10

Robertson R (2005) Baroclinic and barotropic tides in the Weddell Sea. Antarct Sci 17:461–474

Rudnick DL, Boyd TJ, Brainard RE, Carter GS, Egbert GD, Gregg MC, Holloway PE, Klymak JM, Kunze E, Lee CM, Levine MD, Luther DS, Martin JP, Merrifield MA, Moum JN, Nash JD, Pinkel R, Rainville L, Sanford TB (2003) From tides to mixing along the Hawaiian Ridge. Science 301:355–357

Shanmugam G (2003) Deep-marine tidal bottom currents and their reworked sands in modern and ancient submarine canyons. Mar Petrol Geol 20:471–491

Shanmugam G (2008) Deep-water bottom currents and their deposits (Chap. 5). In: Rebesco M, Camerlenghi A (eds) Developments in sedimentology, vol 60. Elsevier, Amsterdam/Oxford, pp 59–81

Shanmugam G, Spalding TD, Rofheart DH (1993) Traction structures in deep-marine, bottom-current-reworked sands in the Pliocene and Pleistocene, Gulf of Mexico. Geology 21:929–932

Shanmugam G, Shrivastava SK, Das B (2009) Sandy debrites and tidalites of Pliocene reservoir sands in upper-slope canyon environments, Offshore Krishna-Godavari Basin (India): implications. J Sediment Res 79:736–756

Shepard FP (1976) Tidal components of currents in submarine canyons. J Geol 84:343–350

Shepard FP, Emery KO (1973) Congo submarine canyon and Fan Valley. Am Assoc Petrol Geol Bull 57:1679–1691

Shepard FP, Marshall NF (1973) Currents along floors of submarine canyons. Am Assoc Petrol Geol Bul 57:244–264

Shepard FP, Marshall NF, McLoughlin PA, Sullivan GG (1979) Currents in submarine canyons and other sea valleys. American Association of Petroleum Geologists, Tulsa, 173 p

Sherwin TJ, Vlasenko VI, Stashchuk N, Jeans DRG, Jones B (2002) Along-slope generation as an explanation for some unusually large internal tides. Deep Sea Res 49:1787–1799

St. Laurent L, Garrett C (2002) The role of internal tides in mixing the deep ocean. J Phys Oceanogr 32:2882–2899

Stashchuk N, Inall M, Vlasenko V (2007) Analysis of supercritical stratified tidal flow in a Scottish Fjord. J Phy Oceanogr 37:1793–1810

Stigebrandt A (1976) Vertical diffusion driven by internal waves in a Sill Fjord. J Phys Oceanogr 6:486–495

Stigebrandt A (1979) Observational evidence for vertical diffusion driven by internal waves of tidal origin in the Oslofjord. J Phys Oceanogr 9:435–444

Stow DAV, Hernandez-Molina FJ, Llave E, Sayago-Gil M, Diaz del Rio V, Branson A (2009) Bedform-velocity matrix: the estimation of bottom current velocity from bedform observations. Geology 37:327–333

Thorpe SA (1992) Thermal fronts caused by internal gravity waves reflecting from a slope. J Phy Oceanogr 22:105–108

Uehara K, Scourse JD, Horsburgh KJ, Lambeck K, Purcell AP (2006) Tidal evolution of the northwest European shelf seas from the Last Glacial Maximum to the present. J Geophys Res 111:C0902

Vidal VM, Vidal FV, Pérez-Molero JM (1992) Collision of a loop current anticyclonic ring against the continental shelf slope of the western Gulf of Mexico. J Geophys Res 97:2155–2172

Visser MJ (1980) Neap-spring cycles reflected in Holocene subtidal large-scale bedform deposits: A preliminary note: Geology 8:543–546

Vlasenko V, Stashchuk N, Hutter K (2002) Water exchange in fjords induced by tidally generated internal lee waves. Dyn Atmos Ocean 35:63–89

Walker RG (1985) Mudstones and thin-bedded turbidites associated with the Upper Cretaceous Wheeler Gorge conglomerates, California: a possible channel-levee complex. J Sediment Petrol 55:279–290

Xu JP, Wong FL, Kvitek R, Smith DP, Paull CK (2008) Sandwave migration in Monterey Submarine Canyon, Central California. Mar Geol 248:193–212

Zhenzhong G, Eriksson KA (1991) Internal tide deposits in an Ordovician submarine channel: previously unrecognized facies? Geology 19:734–737

Zhenzhong G, Eriksson KA, Youbin H, Shunshe L, Jianhua G (1998) Deep-water traction current deposits. Science Press/VSP, New York/Utrecht, 128 p

Precambrian Tidal Facies

15

Kenneth A. Eriksson and Edward Simpson

Abstract

The Precambrian stratigraphic record dating back to 3.2 billion years is replete with examples of interpreted tidal facies. This chapter discusses relevant qualitative as well as quantitative criteria that support tidal interpretations. Qualitative criteria include herringbone cross bedding, bimodal-bipolar paleocurrent patterns, tidal bedding and modified ripples. Quantitative criteria in the form of tidal rhythmites which display semidiurnal, fortnightly and monthly hierarchical bundling patterns provide the best evidence for tidal processes during the Precambrian Era. Banded iron-formations (BIF's), which are unique to the Precambrian rock record, may record evidence of tidal modulation in the form of Earth-tidal rather than ocean-tidal rhythms. Preservation of tidal and particularly tidal-flat facies in the Precambrian was enhanced by sediment stabilization as recorded in microbially induced sedimentary structures (MISS). Tidal facies in the Precambrian record are preserved in both transgressive and highstand systems tracts, the latter as progradational delta front-prodelta deposits. Data from the Precambrian record reveal that despite a closer Earth-Moon distance, at the least in the Archean Era, bedforms were of comparable scale to those existing today and tidal ranges were probably mostly macrotidal but not extreme.

15.1 Introduction

In the absence of body fossils, coupled with the dearth of trace fossils, physical sedimentary structures of tidal origin provide the best evidence for marine as opposed to lacustrine conditions in Precambrian basins. Recognition of marine paleoenvironments is critical to sequence stratigraphic interpretations of the Precambrian rock record including construction of relative sea level curves. Individual qualitative criteria that have been used in the past in support of a tidal interpretation are equivocal but in combination may favor such an interpretation. In contrast, quantitative data provide better support for tidal processes on the early Earth and also can be used to constrain Earth-Moon orbital dynamics. Thus, the purpose of this chapter is to reevaluate qualitative criteria that have been proposed in favor of a tidal origin and to summarize those studies based on quantitative data

K.A. Eriksson (✉)
Department of Geosciences, Virginia Tech,
Blacksburg, VA 24061, USA
e-mail: kaeson@vt.edu

E. Simpson
Department of Physical Sciences, Kutztown University,
Kutztown, PA 19530, USA
e-mail: simpson@kutztown.edu

R.A. Davis, Jr. and R.W. Dalrymple (eds.), *Principles of Tidal Sedimentology*,
DOI 10.1007/978-94-007-0123-6_15, © Springer Science+Business Media B.V. 2012

that provide more robust evidence for a tidal origin for stratigraphic units ranging in age from 3.25 to 0.6 billion years. Specifically, examples are from the Moodies Group (~3.25 Ga) and the Witwatersrand Group (~3.0–2.8 Ga) in South Africa; the upper Mount Guide Group (~1.8 Ga) and the Elatina Group (0.6 Ga) in Australia. Rhythmic bedding preserved in banded iron formations (BIF's) of the Weeli Wolli Formation (2.45 Ga) in the Hamersley Basin of Western Australia is evaluated in terms of whether or not these record a tidal signal.

15.2 Tidal Processes

The gravitational attraction of the Moon (and Sun) raises a tidal bulge in the solid Earth and in oceans. But, because of tidal friction, there is a delay in Earth's response causing the tidal bulge to lead the Earth-Moon axis by a small angle (Lambeck 1980). The Moon exerts a torque on the tidal bulge that results in a slowing down of Earth's rotation and an increase in the length of the day. The torque that the Earth's tidal bulge exerts on the Moon leads to an acceleration of the Moon's orbital motion that causes the Moon to retreat from Earth. The retreat curve is dependent on the equation for the secular growth of the lunar orbit (Lambeck 1980). This equation requires that the Earth-Moon distance was closer in the past than today but less clear is the effect of a closer Earth-Moon distance on tidal amplitudes and thus on tidal current velocities. Based on the data presented for the five Precambrian examples, consideration is given to whether tidal currents were stronger than today and whether macrotidal conditions predominated in the Precambrian.

Interpretations of tidal processes in the rock record are based on evidence from inferred comparative Holocene environments of: (1) opposing (bimodal-bipolar) and landward-directed paleocurrents; (2) unidirectional ebb or flood tidal currents characterized by semi-diurnal inequality in which successive currents are alternately stronger and weaker (de Boer et al. 1989); (3) periodic acceleration and deceleration of tidal currents in response to variations in gravitational attraction of the Sun and the Moon on the Earth. Monthly tidal signals are synodic, anomalistic or tropical (see Kvale et al. 1999); (4) alternating bedload and slack-water suspension sedimentation (Dalrymple et al. 1991); and (5) ebb runoff and exposure of tidal flats (Klein 1977).

15.3 Facies and Facies Associations

Facies in Precambrian stratigraphic units that have been used to infer a tidal origin include: (1) herringbone cross bedding and bimodal-bipolar paleocurrent patterns that record opposing tidal currents; (2) cross-bed foreset bundles (product of lateral accretion) and horizontal laminae (a product of vertical accretion) that display thick-thin pairs, reflecting diurnal inequality of tidal current velocities, and rhythmic thickening and thinning of laminae, reflecting fortnightly neap-spring-neap cyclicity; (3) alternating thicker and thinner neap-spring-neap bundles that record perigee and apogee effects; and (4) mudstone drapes on cross-bed and ripple-bedded foresets and between horizontally bedded sandstone laminae that record slack-water suspension sedimentation; (5) flaser, wavy and lenticular bedding (tidal bedding) that reflect alternating bedload and suspension sedimentation; and (6) modified ripples, often developed and preserved because of the development of microbial mats (microbially induced sedimentary structures – MISS), include ladder-backed, flat-topped, washed-out and superimposed forms that record ebb runoff and exposure of tidal flats. These criteria are evaluated with respect to four Precambrian siliciclastic stratigraphic units before discussing possible tidal signals preserved in banded iron formations.

15.3.1 Elatina Formation, Australia

The best preserved record of tidal sedimentation in the Precambrian Era is from the Neoproterozoic (~620 Ma) Elatina Formation in the Flinders Ranges of South Australia (Fig. 15.1). The Reynella Siltstone, a stratigraphic member of the Elatina Formation near Adelaide (Fig. 15.1; Preiss 1987), also contains rhythmites of tidal origin (Williams 1989, 2000).

The Reynella Siltstone contains graded laminae of fine-grained sandstone and siltstone up to 2 cm thick that commonly have a thin mudstone cap. The Elatina rhythmites comprise graded laminae 0.2–3.0 mm thick of very fine-grained sandstone and siltstone. Rhythmites of the Reynella Siltstone and parts of the Elatina

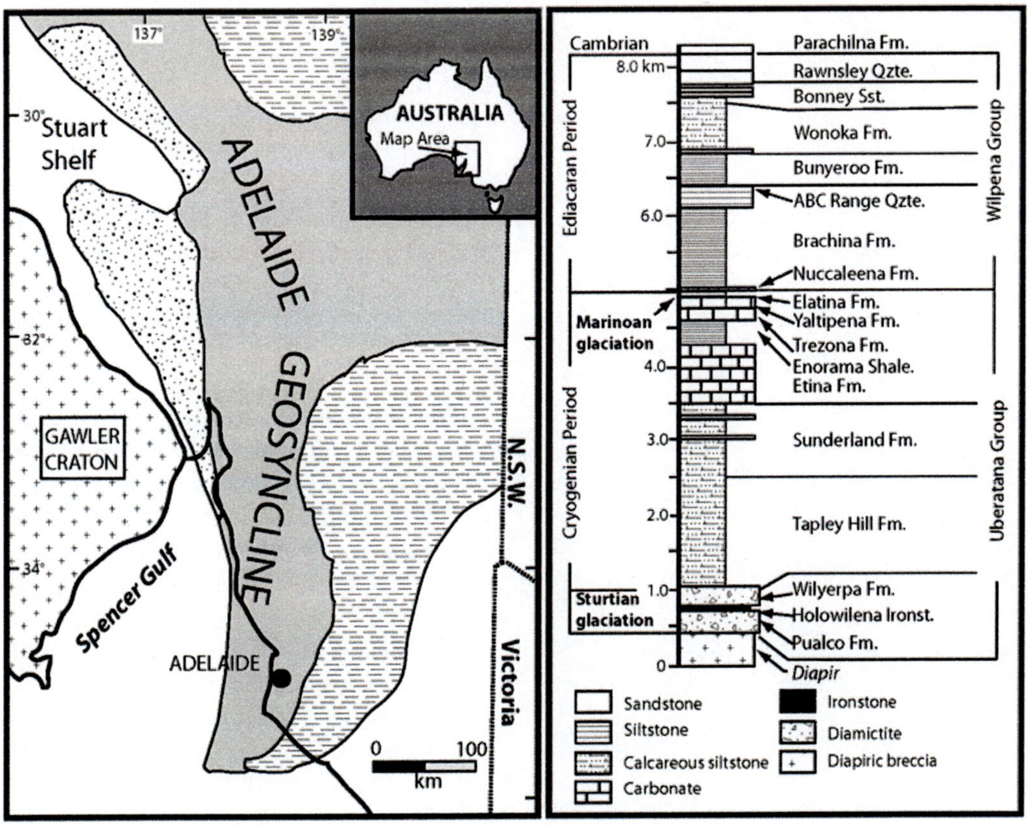

Fig. 15.1 Location map of the Adelaide Geosyncline (From Schmidt and Williams 1995) and generalized stratigraphic column of the Umberatana and Wilpena groups of the central Flinders ranges, South Australia (From Knoll et al. 2004)

Formation commonly display laminae that are arranged in thick-thin pairs (Williams 1989, 1991). Such systematic alternation of relatively thick and thin laminae is observed in modern tidal deposits and uniquely records the diurnal inequality of the tides through its influence on the strength of successive semidiurnal tidal currents (e.g. de Boer et al. 1989; Dalrymple et al. 1991; Kvale and Archer 1991). Laminae in the Reynella Siltstone are grouped in cycles from 0.5 to >6 cm thick in which the laminae thicken and thin. Alternating relatively thick and thin cycles occur in many places. Mudstone drapes commonly bound the cycles. However, mudstone drapes are thinner and less conspicuous for some of the thicker cycles which commonly contain 14–15 laminae. Thinner cycles are up to 2 cm thick and contain 8–16 laminae. Conspicuous thickening and thinning cycles in the Elatina Formation are up to 2 cm thick, contain 8–16 laminae and are bounded by mudstone drapes (Fig. 15.2). The cycles are comparable to neap-spring-neap cycles that occur in modern tidal deposits (Dalrymple et al. 1991; Tessier 1993) and record the fortnightly tidal cycle. The tidal interpretation of the Reynella and Elatina rhythmites is reinforced by the similarity of the rhythmite patterns to modern tidal records (Williams 2000).

15.3.2 Upper Mount Guide Quartzite, Australia

The Upper Mount Guide Quartzite in the Mount Isa region of Australia (Fig. 15.3) overlies rift-related bimodal volcanics, conglomerates and feldspathic sandstones of Bottletree Formation and Upper Mount Guide Quartzite (Fig. 15.4). The Upper Mount Guide Quartzite consists entirely of supermature quartz arenite and is interpreted as a response to intracratonic thermal contraction (Eriksson et al. 1994). Felsic volcanic rocks of the Bottletree Formation and cross cutting

Fig. 15.2 Tidal rhythmites from the Elatina Formation, South Australia, showing five complete thickening and thinning (neap-spring-neap) cycles consisting of between 10 and 14 graded sandy to silty laminae. Cycles are bounded by thin mudstone partings that developed during the neap phases of the tidal cycle (Published with permission of G.E. Williams)

felsic dykes constrain the age of the Upper Mount Guide Quartzite to between 1,800 and 1,740 Ma (Page 1983a, b).

Facies in the Upper Mount Guide Quartzite are arranged in parasequences (cycles) between 0.5 and 12 m thick and consist of cross-bedded arenites capped by thin-bedded arenites (Fig. 15.5). The parasequences are interpreted to record shoaling from subtidal-sand-wave to tidal-flat conditions (Eriksson and Simpson 1990; Simpson and Eriksson 1991). Qualitative evidence for tidal processes is recorded in both facies. Three types of sand-wave deposits are recognized: (1) tabular cross-bed sets and cosets (0.5–0.3 m thick) consisting of planar, tangential, and trough cross strata (Fig. 15.6); (2) compound cross-bed cosets (up to 10.0 m thick) characterized by three hierarchical orders of bounding surfaces (E1, E2, E3); and (3) large-scale trough cross-bed cosets (up to 5.0 m thick). Medium sand is the predominant grain size.

Internal structures in the first two sand-wave types consist of ripple stratification and grain flows (sand flows; Hunter 1977) that typically are associated in acceleration-deceleration flow cycles. Ripple stratification predominates in the large-scale trough cross-beds. Paleocurrent data display a bimodal-bipolar pattern with a prevailing mode to the southwest and a subordinate mode to the northeast. Tabular cross-bed sets and cosets are interpreted to be the product of migration of Type I and Type II megaripples of Dalrymple et al. (1978), whereas large-scale trough cross-bed cosets reflect migration of sinuous-crested sand waves. Migration of megaripples on sand waves or sand ridges produced the compound cross-bedded cosets (cf. Dalrymple et al. 1978). Allen (1980) interpreted the hierarchy of E surfaces to represent erosion surfaces generated by the movement of superimposed bedforms (E1 and E2) and a change in flow dynamics within a tidal regime (E3). Comparable sigmoidal reactivation surfaces bounding acceleration and deceleration flow cycles to those developed in this facies have been identified from tidal sand-wave deposits and have been related to fluctuating tidal current velocities (Boersma and Terwindt 1981; Kreisa and Moiola 1986). The above criteria, together with a lack of exposure features and the dominant westerly paleocurrent mode indicate that this facies was deposited in a subtidal setting dominated by tidal flow to the southwest. The thin-bedded arenite facies contains a variety of structures: (1) asymmetric, slightly asymmetric, and symmetric ripples and megaripples; (2) modified ripples including ladder-back, round-crested, flat-topped, washed-out forms (Figs. 15.7 and 15.8); (3) inversely graded stratification and adhesion ripples and warts (cf. Kocurek and Fielder 1982); and (4) desiccation cracks. The variety and types of preserved sedimentary structures within the thin-bedded arenite facies indicate that the depositional interface frequently was emergent to intertidal and possibly supratidal conditions. Comparable parasequences to those developed in the Upper Mount Guide Quartzite have also been described from the Quilalar Formation higher up in the Haslingden Group (Fig. 15.4; Jackson et al. 1990).

15.3.3 Witwatersrand Supergroup, South Africa

The gold-bearing Witwatersrand Supergroup (Fig. 15.9) is upward of 7 km thick (Tankard et al. 1982) and

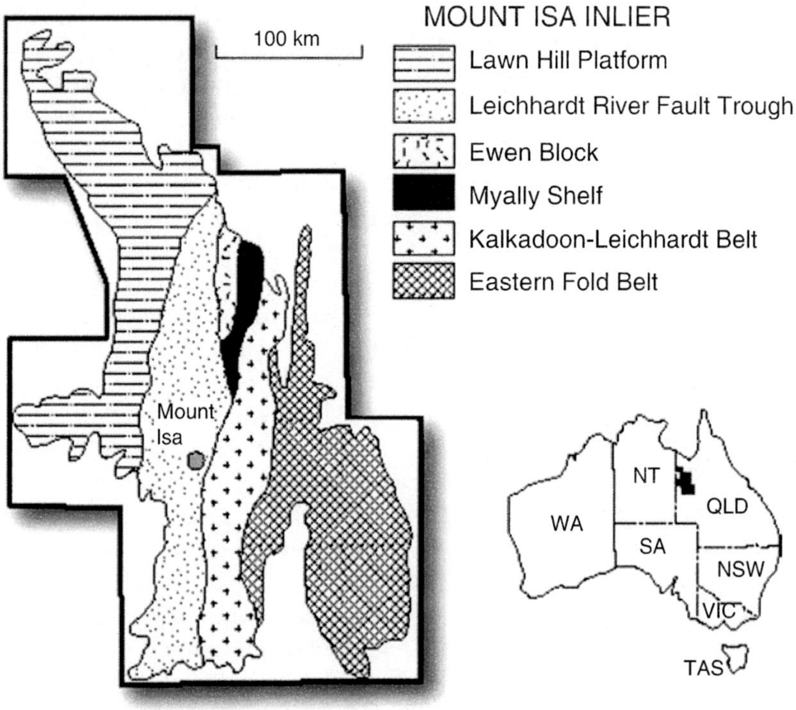

Fig. 15.3 Location map showing the tectonic framework of the Mount Isa Inlier, Queensland, Australia (Based on Blake 1987)

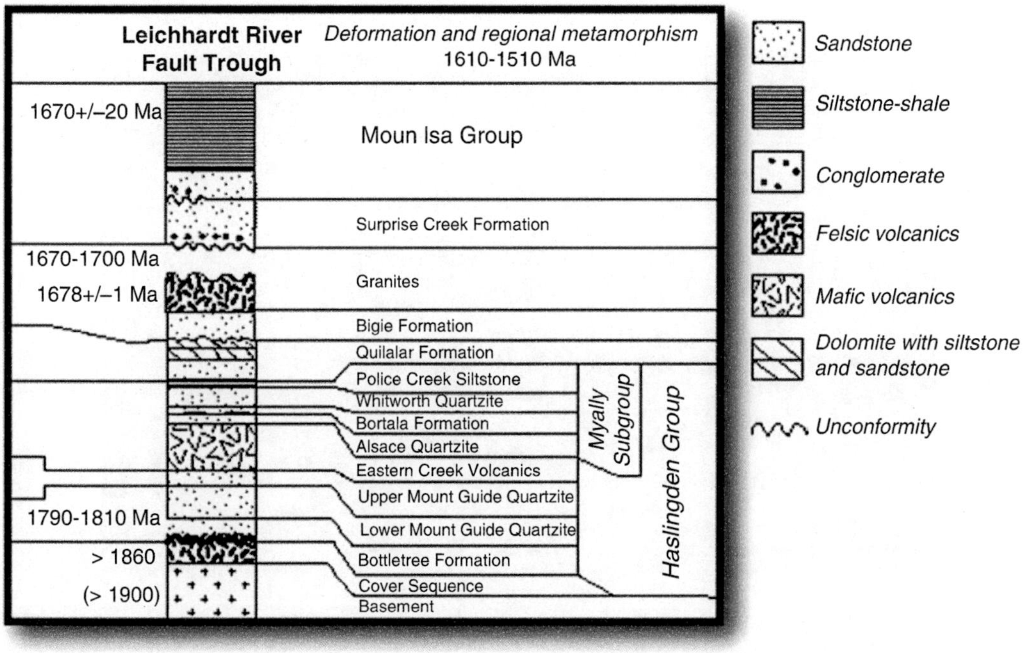

Fig. 15.4 Generalized lithostratigraphic column of cover rocks in the Leichhardt River Fault Trough of the Mount Isa Inlier (Based on Blake 1987). Note the position of the Upper Mount Guide Quartzite in the Haslingden Group

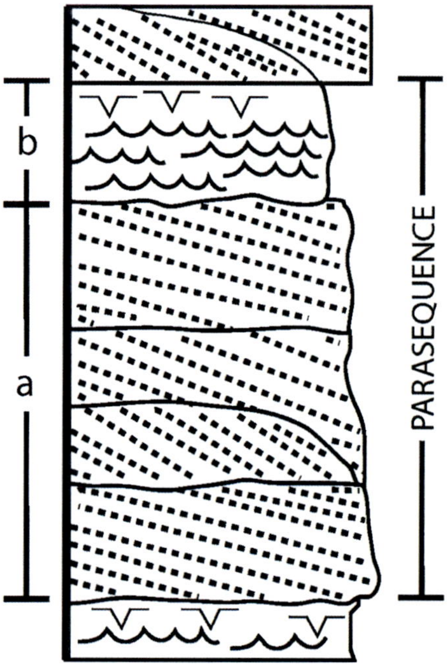

Fig. 15.5 Stacking patterns of facies in parasequences in the Upper Mount Guide Quartzite of the Mount Isa Inlier (Based on Simpson and Eriksson 1991). Parasequences range in thickness from 0.5 to 12 m and record shoaling from subtidal sandwaves (**a**) to tidal flats (**b**)

consists primarily of conglomerate and sandstone with subordinate mudstone and rare volcanic horizons (Fig. 15.10). Based on U-Pb SHRIMP dating of lavas in the older Dominion Group, in the Jeppestown Subgroup and in the overlying Venterdorp Supergroup (Fig. 15.10; Armstrong et al. 1991), the age of the Witwatersrand is constrained to ca. 3.0–2.7 billion years. Quantitative evidence for tidal processes in the form of herringbone cross bedding (Fig. 15.11), bimodal-bipolar paleocurrent patterns and three-dimensional modified ripples (Fig. 15.12) have been reported from the Hospital Hill and Johannesburg Subgroups (Eriksson et al. 1981). Rhythmically interbedded sandstones and mudstone developed in the Coronation Formation, and below the Livingstone Reef (placer) at the base of the Luipaardsvlei Formation (Fig. 15.10) may provide quantitative evidence in support of tides.

The Coronation Formation is an unconformity-bounded sequence underlain and overlain by coarse-grained sedimentary rocks of mostly braided-alluvial origin. Internally, the Coronation Formation is a coarse- to fine-grained siliciclastic unit that consists of an upward-fining interval overlain by an upward-coarsening interval. Facies stacking patterns are considered to represent a transition from braided alluvial

Fig. 15.6 Medium-scale tabular-tangential cross-bed set in Upper Mount Guide Quartzite showing variation in dip angle. This facies represents the deposit of a simple subtidal sandwave that developed under conditions of accelerating and decelerating tidal current flow velocities

Fig. 15.7 Flat-topped ripples from tidal-flat caps to parasequences in the Upper Mount Guide Quartzite

Fig. 15.8 Washed-out ripples from tidal-flat caps to parasequences in the Upper Mount Guide Quartzite

to shallow-marine followed by progradational deltaic environments (Tankard et al. 1982; Winter and Brink 1991). Intercalated diamictites are interpreted as glaciogenic deposits (Tankard et al. 1982; Crowell 1999). Rhythmically interbedded facies (Fig. 15.13) sampled in core from the base of the progradational component of the formation are interpreted as hemipelagic, bottomset deposits of a prograding delta the proximal equivalents of which were eroded along the upper sequence boundary (Winter and Brink 1991). The rhythmically bedded facies consist of vertically accreted alternating fine-grained sandstone/siltstone and mudstone couplets. Alternating, thick-thin sandstone pairs record semi-diurnal, dominant and subordinate currents. Cyclic variations in laminae thickness permit identification of neap-spring-neap cycles. Bar graphs of sandstone laminae thickness reveal thickening and thinning trends that are interpreted as neap-spring-neap cycles (Fig. 15.14; cf. Kvale et al. 1999).

Rhythmically bedded facies in the Central Rand Group (Fig. 15.15) are preserved in outcrop at the top of the Randfontein Formation (Fig. 15.10). Most laminae are normal graded and range in thickness from 1 to 8 mm. Sandstone laminae are separated by siltstone/mudstone partings. Bar graphs of sandstone laminae thickness (Fig. 15.16) reveal a hierarchy of laminations that are interpreted as semi-diurnal (thick-thin pairs) and possible neap-spring-neap tidal periodicities

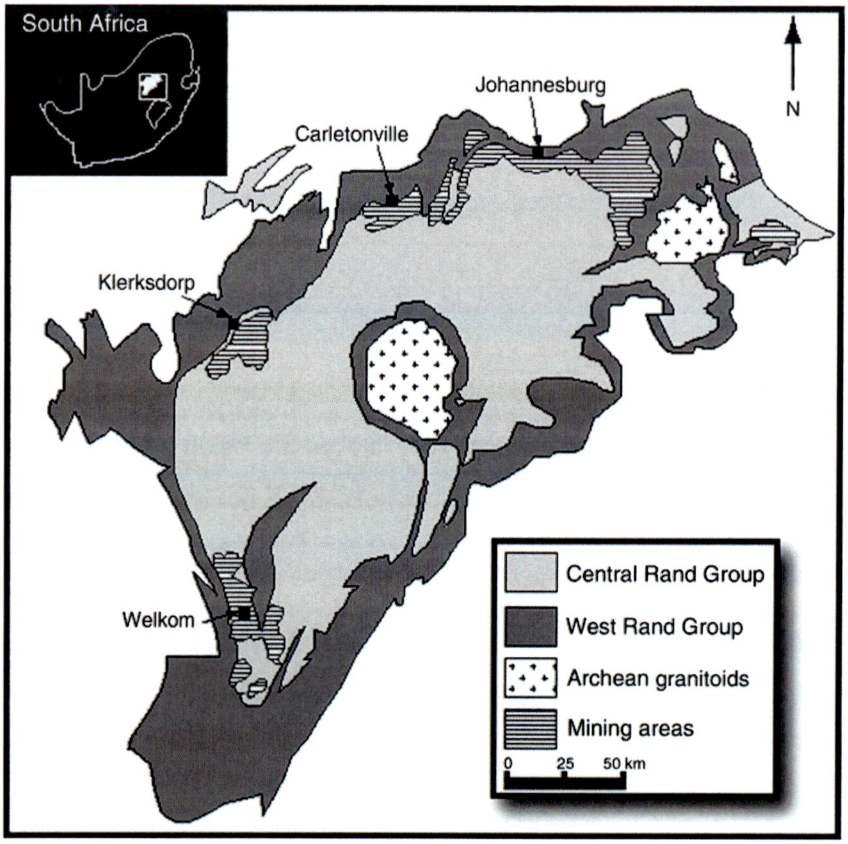

Fig. 15.9 Location map of the Witwatersrand structural basin, South Africa showing the distribution of the West Rand and Central Rand groups (Modified from Catuneanu and Biddulph 2001)

(thickening followed by thinning of laminae) (cf. Kvale et al. 1999). Spectral analysis using a Fast Fourier Transform program on the complete data set reveals a peak at 2.3 and on the data set from which inferred subordinate (thinner) laminae had been removed (Fig. 15.16) reveals peaks at 8.6 and 10.6.

15.3.4 Moodies Group, Barberton Greenstone Belt, South Africa

The Moodies Group in the Barberton Greenstone Belt (Fig. 15.17) is the uppermost of three stratigraphic intervals that compose the Swaziland Group. The age of the Moodies Group is well constrained at around 3.25 billion years (Kamo and Davis 1994; Heubeck et al. 1993). Sedimentary structures of inferred tidal origin were first described from the Moodies Group by Eriksson (1977) who reported herringbone cross bedding, bimodal-bipolar paleocurrent patterns, mudstone drapes, and flaser, wavy and lenticular bedding. More recently, Eriksson and Simpson (2000) and Eriksson et al. (2006) have documented rhythmically interlaminated sandstones and mudstones (tidal rhythmites) in which different orders of tidal cyclicity can be recognized. Tidal rhythmites in the Moodies Group are preserved as flat-laminated rhythmites, as described above from the Elatina-Reynella succession in South Australia, but more commonly as bundles of sandstone foresets separated by mudstone drapes.

In Dycedale Syncline, three facies associations are recognized: (1) structureless conglomerate; (2) cosets of trough and tabular cross-bedded sandstone; and (3) interlaminated sandstone, siltstone and mudstone (Eriksson et al. 2006). These facies are arranged in 45–140 cm-thick, fining-upward packages in which the proportion of interlaminated sandstone, siltstone and mudstone increases upwards. Cross-bedded sandstone ranges in grain size from very coarse to fine sand. Locally, pebble stringers define set boundaries.

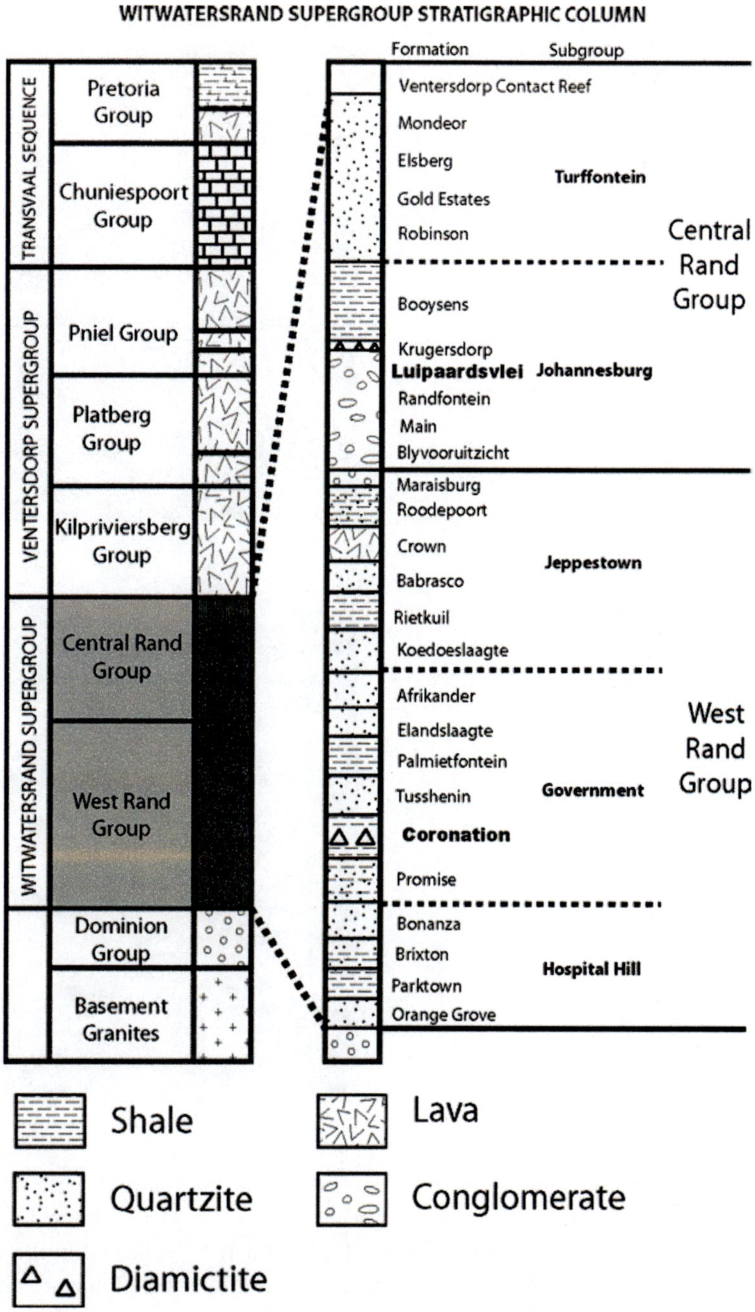

Fig. 15.10 Generalized lithostratigraphical column of the Witwatersrand Supergroup and its relationship to underlying and overlying stratigraphic units (Modified from Catuneanu and Biddulph 2001). Note the positions of the Coronation and Luipaardsvlei formations

Cosets vary from 20 to 210 cm thick. In several cases, laminated sandstone, siltstone and mudstone, and wave- and combined-flow ripple bedforms are preserved below coset boundaries. Within sets, foresets are tangential, planar or sigmoidal in shape and, toward the top of upward-fining packages, commonly are draped with mudstone and equate to tidal bundles (Fig. 15.18). In general, thin foresets have continuous mudstone drapes whereas thicker foresets have discontinuous drapes or are separated by mudstone chips. In bedding plane views, these chips display polygonal desiccation cracks. Reactivation surfaces are present throughout the section. Laterally within sets a systematic thickening and thinning of foresets occurs with a corresponding increase in development of mudstone drapes associated with thinner foresets. Some foresets

Fig. 15.11 Herringbone cross bedding from the Johannesburg Subgroup, Witwatersrand Supergroup. Note pen for scale

Fig. 15.12 Large-scale wave ripples with flattened crests from the Hospital Hill Subgroup, Witwatersrand Supergroup

contain internal ripple cross laminations directed up the foresets. These ripple cross laminations show a complex pattern of mudstone drapes. The laminated sandstone, siltstone and mudstone facies association attains a maximum thickness of 25 cm but commonly is absent at the top of fining-upward packages as a result of erosion. Vertically within the facies association, thick-thin pairs and systematic thickening and thinning of laminations are developed. Desiccation cracks are ubiquitous but are best preserved at the top of upward-fining packages. Where laminations are absent, mudstones are black and desiccated.

The vertical succession of strata within upward-fining packages records the increased influence of tidal currents with time at the expense of fluvial processes. The switch to a dominance of tidal processes is reflected in the upward decrease in the proportion of conglomerate, the increase in abundance of mudstone drapes on foresets, the presence of cyclic foresets, and the occurrence of interlaminated sandstone, siltstone and mudstone at the top of upward-fining packages. Conglomerates reflect channel processes whereas cosets of trough and tabular cross-bedded sandstone and the laminated sandstone, siltstone and mudstone were generated by flows modified by various tidal beats. Cosets of trough and tabular cross-bedded sandstone with or without mudstone drapes reflect lateral accretion of sediment, whereas interlaminated sandstone,

Fig. 15.13 Thin section photomicrograph of rhythmically bedded siltstone/mudstone in the Coronation Formation, Government Subgroup, Witwatersrand Supergroup. *Red arrow* indicates correlative laminae across a break in the core

siltstone and mudstone records vertical accretion. In both facies associations, mudstone developed during slack water phases whereas sand and/or silt transport took place during the ebb and flood stages. Within both laterally and vertically accreting facies, alternating thin-thick laminations reflect diurnal twice-daily tides. Thinner groupings of foresets and thinner intervals of vertically stacked sandstone/siltstone/mudstone laminations formed during neap tides whereas thicker groupings of foresets and laminations developed during spring tides. Desiccated mudstone drapes on foresets indicate that bedforms rarely were exposed during some portion of the tidal cycle.

In the Eureka Syncline, tidal facies are represented by mudstone-draped cross-bed foresets (Fig. 15.19) that reflect intermittent migration of a sand wave (Eriksson and Simpson 2000). Foreset-bundle thicknesses, when plotted on a histogram of foreset bundle thickness versus foreset bundle number (Fig. 15.20), reveal a hierarchy of diurnal, semi-monthly, and monthly tidal periodicities. Thick-thin pairs of foreset bundles (Fig. 15.20a) are considered to reflect deposition from semidiurnal dominant and subordinate flood-tidal currents, respectively. Similar thick-thin diurnal pairs are widely developed in Holocene tidal sediments. Cyclic variations in foreset bundle thicknesses record longer period changes in strength of the dominant semidiurnal tidal currents consistent with semi-monthly anomalistic, perigean-apogean tidal signatures. Fast Fourier Transform analysis on the data set reveals strong peaks at 13.11, 9.83 and 2.18 (Eriksson and Simpson 2000). The last two peaks are consistent with the interpretation of diurnal and neap-spring cyclicity discussed above whereas the 13.11 peak is considered to record neap-spring cycles in which both dominant and subordinate semi-diurnal, subordinate-tide foreset bundles had been removed (Fig. 15.20b) and reveal only one well-developed peak at 9.33 that is interpreted as a strong semi-monthly signature. Close inspection of Fig. 15.20b reveals that monthly perigean-apogean cycles in the Moodies sandwave deposit have a maximum of 20 foreset bundles. This is a record of the minimum number of days in the synodic month during the middle Archean because of missing neap-tide foreset bundles especially within the apogean component of the monthly cycle when tidal current velocities are less than during perigee.

15.3.5 Other Precambrian Siliciclastic Examples

The Precambrian rock record is replete with other examples of inferred tidal facies: these are shown in Table 15.1 in comparison to the examples discussed above. Sedimentary structures of tidal origin are extensively developed in the Big Cottonwood Formation in Utah and include heterolithic tidal rhythmites that record four tidally forced cycles, sigmoidal cross-bed bundles with reactivation surfaces, tidal bedding, current ripples with rounded crests (Chan et al. 1994; Sonett et al. 1996; Ehlers and Chan 1999). The Ortega Quartzite and Uncompahgre Formation in New Mexico

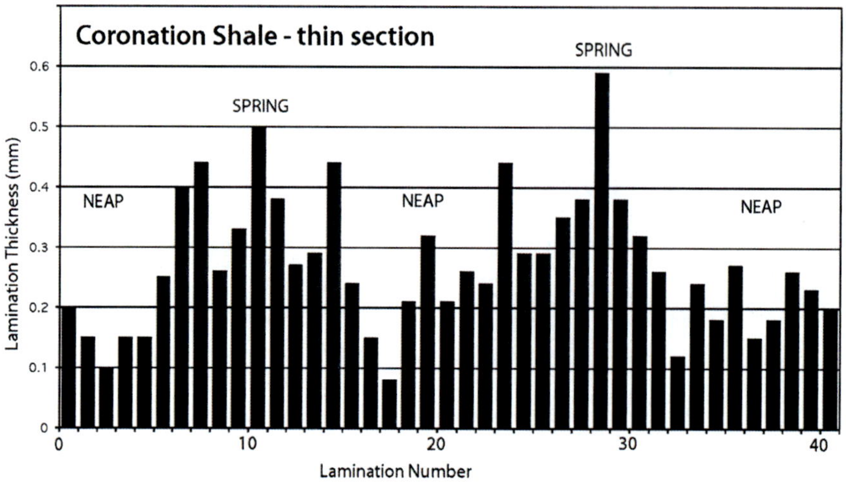

Fig. 15.14 Bar graphs of siltstone laminae thicknesses for rhythmically bedded siltstone/mudstone in the Coronation Formation. Note rhythmic thickening and thinning of laminae interpreted as neap-spring-neap cycles

Fig. 15.15 Cut slab of rhythmically bedded sandstone/mudstone at the top of the Randfontein Formation, Johannesburg Subgroup, Witwatersrand Supergroup. Slab is 7.5 cm wide

and Colorado, respectively, have been interpreted as tidal shelf deposits (Soegaard and Eriksson 1985; Harris and Eriksson 1990) on the basis of qualitative criteria. Both formations contain thinly interlaminated siltstone and mudstone but rhythmic patterns have not been identified. The Waterberg Group in South Africa is a dominantly braided alluvial succession but thin intervals of flaser, wavy and lenticular bedding have been interpreted as a product of tidal reworking in embayments between braid deltas (Vos and Eriksson 1977; Eriksson and Vos 1979). Paleocurrent data for the inferred tidal facies are based on ripple lee-face azimuths and a wave (lacustrine) origin for this facies remains a viable alternative. The Pokegama Quartzite and equivalents in the Lake Superior region have long been interpreted as tidal deposits based on bimodal-bipolar paleocurrent patterns, herringbone cross bedding and tidal bedding (Ojakangas 1983). More recently, Ojakangas (1996) has identified an interval of alternating thicker and thinner laminae in the lower-most Pokegama that provide evidence for semidiurnal inequality of tidal currents. Runzel marks (wrinkle marks), mudcracks and raindrop impressions in finer-grained facies of the Pokegama Quartzite are considered to provide evidence for periodic exposure possibly on tidal flats (Ojakangas 1983).

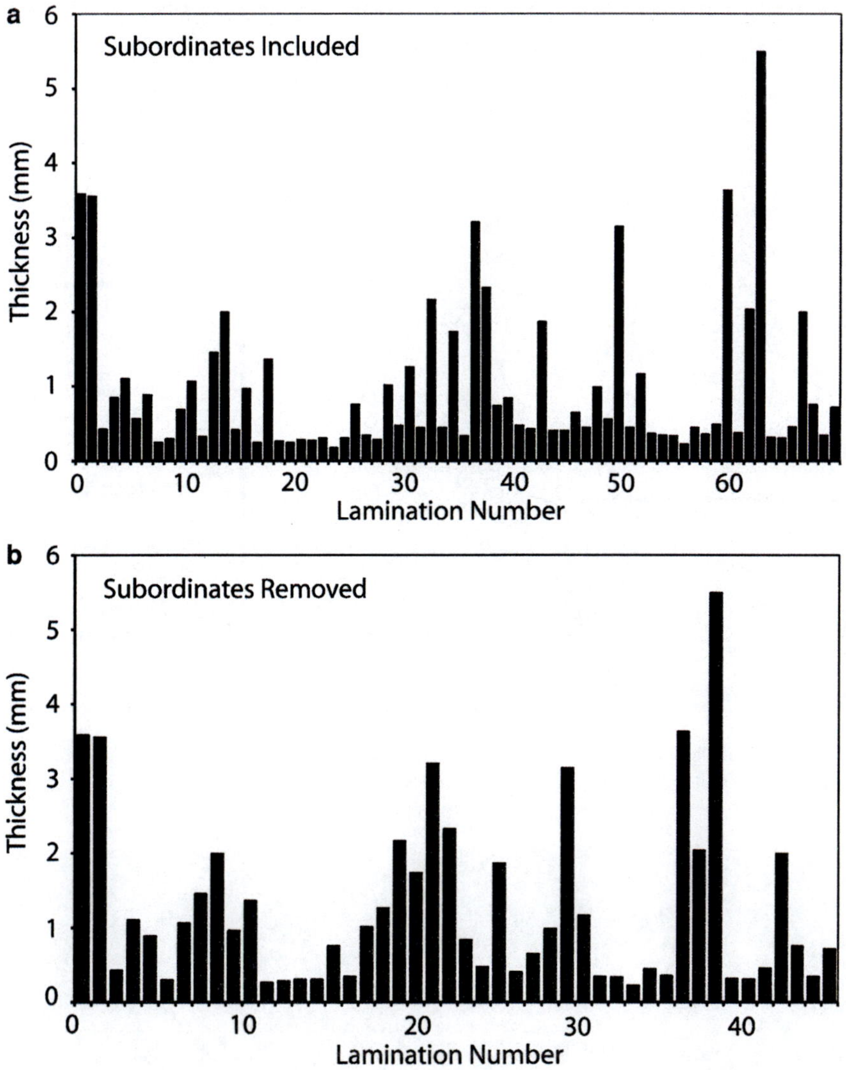

Fig. 15.16 Bar graphs of siltstone laminae thicknesses for rhythmically bedded siltstone/mudstone at the top of the Randfontein Formation, Johannesburg Subgroup, Witwatersrand Supergroup; (**a**) complete data set; (**b**) data set with inferred diurnal subordinate laminae removed. Note rhythmic thickening and thinning of laminae best expressed after removal of subordinates and interpreted as possible neap-spring-neap cycles

15.3.6 Weeli Wolli Iron Formation, Australia

The Weeli Wolli Formation in the Hamersley Basin of Western Australia is one of a number of iron-formation-dominated stratigraphic intervals of the Hamersley Group (Fig. 15.21). The age of the Weeli Wolli Formation is well-constrained at ca. 2.5 billion years (Trendall et al. 1990; Pidgeon and Horwitz 1991). Cyclicity in the iron-formation is expressed as regular variations in thickness of chert-rich and hematite-rich components, giving the facies a characteristic striped appearance. So-called microbands (Trendall 1973) are typically 0.05 mm or less thick, and only the cyclic stripes are readily discernible (Fig. 15.22a). Locally, silicification has prevented diagenetic compaction allowing mineral couplets to be identified (Fig. 15.22b). Counts carried out by Williams (1989) on thin sections of silicified pods indicate as many as 28–30 couplets per microband. Cycles containing fewer couplets usually show evidence of amalgamation of hematite laminae; counts for such cycles probably underestimate the true cycle period (Williams 2000).

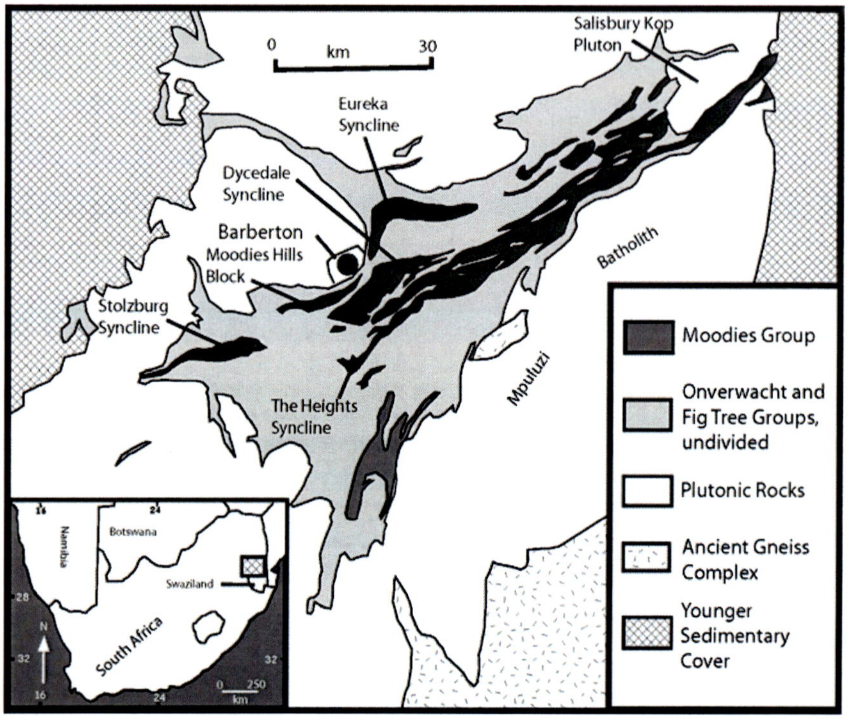

Fig. 15.17 Geological map of the Barberton Greenstone Belt and surroundings, South Africa. The examples of tidal facies discussed in the chapter are from the Eureka and Dycedale synclines

Fig. 15.18 Tabular-tangential cross bed with mudstone-draped foresets, Dycedale Syncline. Note that the number of mudstone partings decrease then increase from *right* to *left*. This pattern is interpreted as a neap-spring-neap cycle. Note hand lens for scale

Cyclic rhythmites in banded iron-formations of the Weeli Wolli Formation are considered to provide a record of tidal modulation although earth-tidal rhythms may be recorded rather than ocean-tidal rhythms (Williams 1989, 2000). Submarine, volcanic-associated hydrothermal or fumarolic activity has been proposed for the origin of numerous iron-formations (e.g. Simonson 1985; Fralick 1987). Such an origin for the Weeli Wolli

Fig. 15.19 Close-up view of mudstone-draped foresets within a large-scale sandwave deposit, Moodies Group, Eureka Syncline. Note the cyclic thickening and thinning of mudstone-draped foreset laminae. Scale is in centimeters

BIF is supported by the local presence of intercalated volcanic ash beds (Trendall and Blockley 1970). Because some geyser activity today is modulated by earth tides (Rinehart 1972a, b), the question is raised by Williams (2000) as to whether the Weeli Wolli cyclicity records earth-tidal rhythms that modulated the discharge of silica- and (or) iron-bearing fumarolic waters.

Two possible tidal interpretations of the Weeli Wolli cyclicity are suggested by Williams (2000): (a) the mineral couplets are semidiurnal increments grouped in lunar fortnightly cycles. By this interpretation, there were about 28–30 lunar days per lunar month at about 2,500 Ma; (b) the mineral couplets are lunar fortnightly increments that are arranged in annual cycles related to seasonal influences on sedimentation. This would indicate about 28–30 lunar fortnights, or about 14–15 lunar months, per year at about 2,500 Ma. Williams (2000) favors the latter interpretation because geothermal areas usually are so sluggish mechanically that the semidiurnal and diurnal components are filtered out, whereas the activity of geysers may be influenced by the fortnightly tidal component (Rinehart 1974). Furthermore, an annual origin for the Weeli Wolli cyclic stripes gives sedimentation rates for the compacted facies that are comparable to presumed rates for other iron-formations in the Hamersley Group in which microbanding is regarded as annual (see Trendall and Blockley 1970; Trendall 1983). Such an origin for the cyclicity also finds support in the presence of between 15 and 27 laminae (depending on the observer) in a thick microband, presumed to represent 1 year of accretion, from the Brockman Iron-formation (Fig. 15.21; Ewers and Morris 1981).

15.4 Microbially Induced Sedimentary Structures (MISS)

Microbial structures in carbonates and cherts are formed by mineral precipitation whereas equivalent structures in siliciclastic lithologies originate by the physical interaction of benthic microbiota with erosion and deposition of sediment (Noffke et al. 2003a). Microbial mats respond to erosion by biostabilization or react to deposition of sediment by baffling, trapping and binding (Noffke and Krumbein 1999; Noffke et al. 2003a). This distinctive biotic-physical interaction creates a variety of characteristic sedimentary structures that, due to their unique mode of formation, have been categorized as their own group termed 'microbially induced sedimentary structures – MISS' (Noffke et al. 2003a). MISS have been described from a number of Neoproterozoic tidal flat to shallow shelf

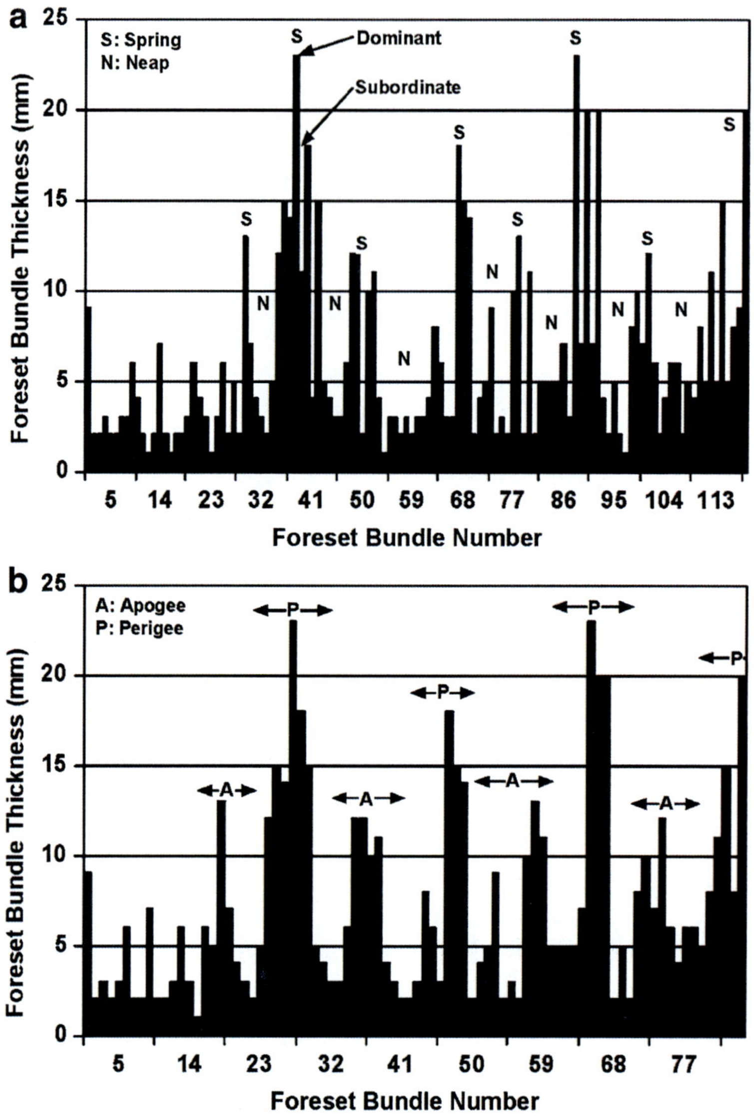

Fig. 15.20 Bar graphs of foreset sandstone laminae thicknesses in a large-scale sandwave deposit, Moodies Group, Eureka Syncline; (**a**) complete data set; (**b**) data set with inferred diurnal subordinate laminae removed. Note rhythmic thickening and thinning of laminae best expressed after removal of subordinates and interpreted as possible neap-spring-neap cycles. Neap-spring-neap cycles are alternately thicker and thinner and are interpreted as perigee and apogee cycles, respectively

successions (e.g. Hagadorn and Bottjer 1999; Gehling 2000; Noffke et al. 2002) and from Archean sedimentary intervals in South Africa (Noffke et al. 2003b, 2006, 2008; Heubeck 2009). Wrinkle structures that resemble runzel marks are a common MISS and may imply that all such structures described from the rock record are microbial in origin.

MISS in the Moodies Group are developed in tidal channel and tidal flat facies in the Dycedale and Saddleback synclines (Fig. 15.17; Noffke et al. 2006; Heubeck 2009). In the Saddleback Syncline, wrinkle structures are preserved on bedding planes of fine-grained sandstone. The wrinkles are 5 mm in wavelength and about 3 mm in height, and in one example define a 5–10 cm-wide sinuous belt. Wrinkle structures record crinkling and dewatering of a microbial mat during burial by freshly deposited sand (Gehling 2000; Noffke et al. 2002), possibly implying a syneresis origin. Wrinkle structures in the Moodies Group are covered by a pattern of cracks, which indicate that the loose grains of the ancient sandy surface must have been bound together by a cohesive medium before cracking occurred. Desiccated mudstone drapes are common in the facies containing the wrinkles supporting a tidal flat setting (Eriksson 1977). Microbial binding of sediment is indicated by roll-up mudstone flakes preserved in a sandstone matrix (Fig. 15.23). A roll-over structure is preserved in a 2–3 cm-thick, fine-grained sandstone

Table 15.1 Tidal sedimentary structures

	Bim-Bip Palcurr.	H-bone. X-bed.	Tidal bedding	Modified ripples	Rhythmic bedding	Foreset bundles
Elatina (0.6 Ga)	Yes	Yes	Yes	No	Yes	Yes
Big Cwood (0.9 Ga)	No	No	Yes	Yes	Yes	Yes
Uncomp. (1.7 Ga)	No	No	Yes	No	No	Yes
Ortega (1.7 Ga)	Yes	Yes	Yes	No	No	No
U Mt Guide (1.8 Ga)	No	No	No	Yes	No	Yes
Waterberg (1.8 Ga)	Yes	No	Yes	No	No	No
Pokegama (~1.9 Ga)	Yes	Yes	Yes	No	Yes	No
Witsrand (~2.9 Ga)	Yes	Yes	Yes	Yes	Yes	Yes
Moodies (3.25 Ga)	Yes	Yes?	Yes	No	Yes	Yes

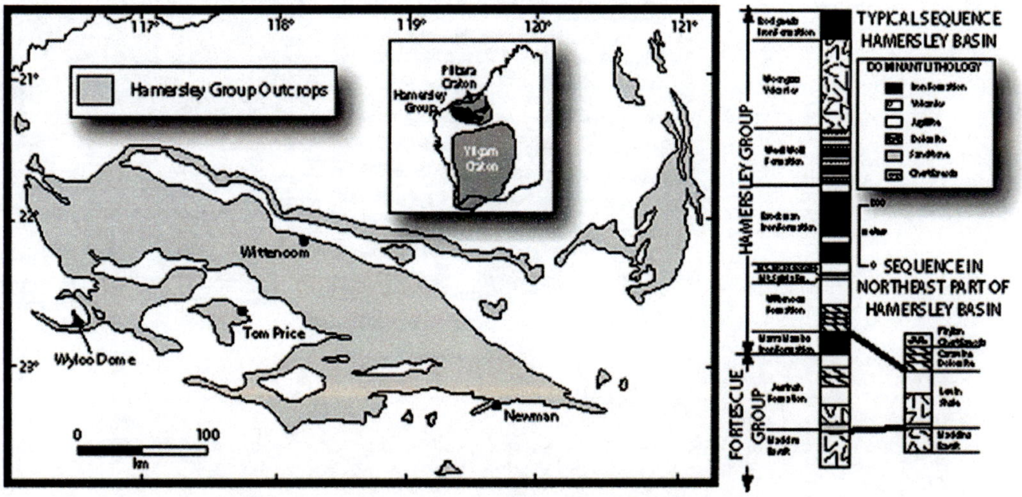

Fig. 15.21 Geological map showing the distribution of the Hamersley Group and generalized stratigraphic column of the Fortescue and Hamersley groups, Pilbara Craton, Western Australia (Adapted from Trendall 1983)

bed near the top of a tidal channel cycle in the Dycedale Syncline. This structure is lens-shaped with dimensions of about 3 cm by 2.5 cm and is composed of alternating mm-thick sandstone, and sub-mm-thick carbon-rich laminae. Such roll-over microbial mats are not uncommon on modern tidal flats (Noffke et al. 2001) and are produced by bottom currents that overfold desiccated microbial mats. In thin-section, the MISS samples from the Moodies Group reveal a wavy-crinkly pattern of dark, opaque laminae characteristic of ancient microbial mats in sandstones (Noffke et al. 2006). The dark, opaque laminae are between 50 and 500 μm thick and alternate with mm-thick quartz sand laminae. This laminated pattern is also very characteristic of modern, unlithified tidal sand deposits that include microbial mat layers.

15.5 Preservation Potential and Sequence Stratigraphy

An analysis of the stratigraphic location and sequence stratigraphic interpretation of the Precambrian tidal facies reviewed in this chapter permits conclusions to be drawn concerning factors that promoted their preservation in the rock record. Tidal-shelf and tidal-flat facies in the Moodies Group of the Eureka Syncline are developed above braided-alluvial deposits and are capped by a banded iron-formation (Fig. 15.24). BIF's developed in predominantly siliciclastic successions have been compared with Holocene pelagic deposits that are concentrated in environments not diluted with siliciclastic sediment (Eriksson 1983). Thus, the BIF in the Moodies Group is interpreted as a

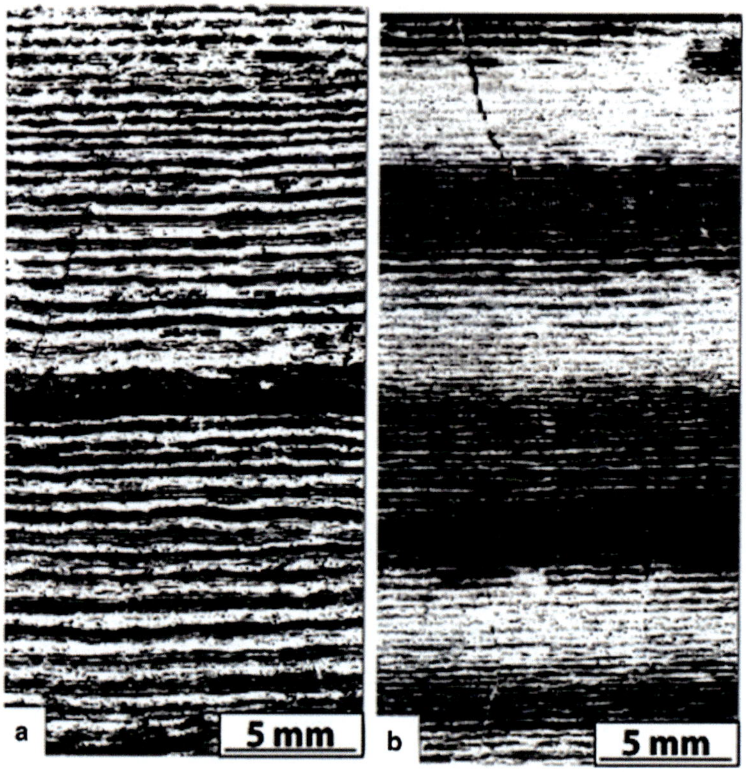

Fig. 15.22 Thin section photomicrographs of iron-formation from the Weelie Wolli Formation, Hamersley Basin, Western Australia; (**a**) microbands of alternating hematite and chert in compacted iron-formation; (**b**) silicified pods containing couplets of hematite and chert arranged in cycles that are alternately richer and poorer in hematite. See text for details

Fig. 15.23 Rolled-up mudstone flakes (shown by *arrows*) in a matrix of coarse sand, Moodies Group Saddleback Syncline, Barberton Greenstone Belt, South Africa. Preservation of rolled-up mudstone flakes is attributed to biostabilization. Scale in centimeters

drowned-shelf deposit. The vertical transition of facies in the Eureka Syncline records progressive deepening of the depositional interface (Fig. 15.24) with the BIF representing the equivalent of a condensed section (maximum flooding surfaces). The vertical succession of facies in the Moodies Group of the Dycedale Syncline records a gradual increase in tidal influence at the expense of braided-fluvial processes and thus an overall upward-deepening of the depositional interface. Thus, tidal facies in the Moodies Group are interpreted

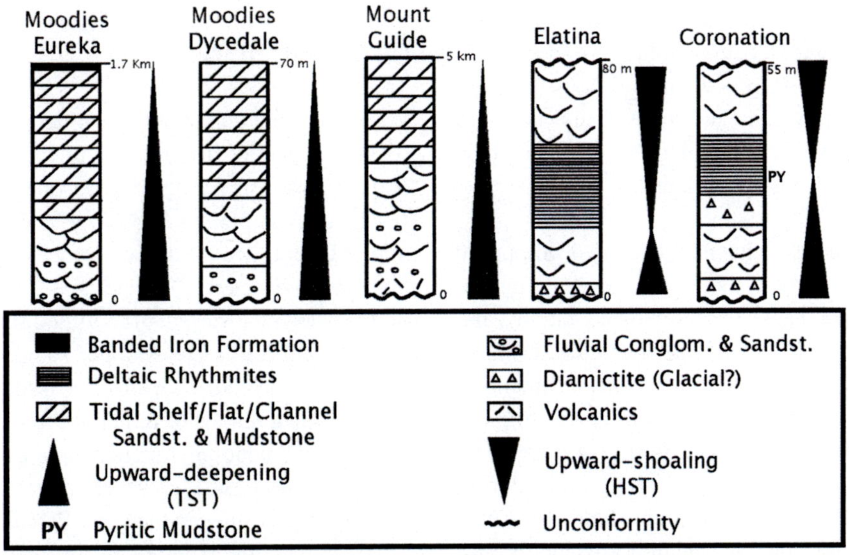

Fig. 15.24 Generalized vertical sections showing stratigraphic positions of tidal facies (tidal shelf, tidal flat, tidal channel and rhythmites) in relation to underlying and overlying facies. Also shown are inferred upward-deepening and upward-shoaling trends that are interpreted to represent transgressive and highstand systems tracts, respectively (Data for Moodies Group of Eureka Syncline adapted from Eriksson 1977; Moodies of Dycedale Syncline from Eriksson et al. 2006; Mount Guide Quartzite from Simpson and Eriksson 1991; Elatina Formation adapted from Lemon and Gostin 1990); Coronation Formation based on core observations. See text for details on sedimentology of tidal facies

as transgressive systems tract deposits and accommodation leading to their preservation likely resulted from a combination of sea-level rise and subsidence. The Upper Mount Guide Quartzite likewise overlies braided alluvial facies of the Lower Mount Guide Quartzite and records upward-deepening. In common with the Moodies Group in the Dycedale Syncline, no condensed-section deposit is preserved. Notwithstanding, the vertical succession of facies in the Mount Guide Quartzite is compatible with a transgressive systems tract (Fig. 15.24). The maturity of the Upper Mount Guide Quartzite and the repetitive nature of the similar shallow-subtidal to tidal-flat parasequences reflect a balance between sediment supply and long-term subsidence both of which are consistent with a basin that was undergoing thermotectonic subsidence (Eriksson et al. 1994). Stacked parasequences in the Upper Mount Guide Quartzite are considered to record low-amplitude/high-frequency sea-level fluctuations rather than jerky subsidence (Eriksson and Simpson 1990). Fischer plots of parasequence thickness versus time reveal a longer-term sea level change on the order of 1.5 Ma (Eriksson and Simpson 1990).

Rhythmites in the Elatina Formation in Australia and the Coronation Formation in South Africa are developed within upward-shoaling components of unconformity-bounded sequences (Fig. 15.24). Tidal rhythmites in the Elatina Formation are interpreted as progradational deltaic facies (Williams 2000) in which preservation resulted from aggradation in a subtidal setting during highstand of sea level. Paleocurrent data and paleogeographic reconstruction for the Elatina Formation (Preiss 1987) indicate that the Reynella and Elatina rhythmites were deposited near the margin of a marine gulf in distal ebb tidal delta setting (Williams 1991). Rhythmites in the Coronation Formation are similarly interpreted as the deposits of a prograding delta at highstand of sea level. The sequence stratigraphic setting of the rhythmites beneath the Livingstone Reef (placer) is not known nor is that of the Weeli Wolli iron-formation but, in both cases, the absence of wave- and current-produced structures implies sufficient accommodation to maintain the depositional interface below storm wave base.

15.6 Stratigraphic Successsions and Modern Analogs

Holocene analogues are widely developed for the Precambrian tidal facies highlighted in the preceding section. A modern counterpart for the Moodies Group cross-bedded facies in the Eureka Syncline is subtidal sand shoals in the Oosterschelde estuary in the Netherlands that similarly contain mud-draped bundles of foresets typically arranged in semi-diurnal thick-thin pairs (Boersma and Terwindt 1981; de Boer et al. 1989) and in which neap-spring-neap cycles have been identified (Visser 1980).

Analogues for tidal channel deposits developed in the Moodies Group in the Dycedale Syncline are developed in the inner river-dominated but marine-influenced zones of Holocene tide-dominated estuaries (Dalrymple et al. 1992). Channel bank sediments in modern tidal channels consist of interlaminated sand and mud comparable to those in the upper parts of fining-upward packages in the Moodies Group (Eriksson et al. 2006). An important difference between Holocene tidal channel deposits and those in the Moodies Group is the coarseness of the sediment particularly at the base of the packages where pebbles of extrabasinal origin are ubiquitous. Eriksson et al. (2006) inferred a proximal source area in a tectonically active basin to explain this distinction.

Cross-bedded sandstones that comprise the major portion of parasequences in the Upper Mount Guide Quartzite have analogues in the form of subtidal sand waves in many Holocene settings including the Bay of Fundy (Dalrymple 1984). Comparable modified ripple types to those present in the thin-bedded facies of the upper Mount Guide Quartzite are present on Holocene tidal flats including the North Sea, northwestern Australia, the Bay of Fundy and The Wash, and develop as a result of ebb runoff and emergence (Klein 1977). Eolian modification of tidal flats is reflected in the preservation of adhesion warts and ripples (Kocurek and Fielder 1982) and inversely graded wind-ripple stratification (Hunter 1977).

In the absence of direct evidence for the existence of a barrier island in the Elatina Formation, an alternative depositional setting to an ebb tidal delta is a tide-dominated delta such as the Fly River, Yangtze and Amazon deltas. Millimeter- to decimeter-scale, sand-mud alternations are present in the delta front/prodelta settings of all three deltas (Jaeger and Nittrouer 1995; Dalrymple et al. 2003; Hori et al. 2002; Harris et al. 2004). In the case of the Amazon Delta, a neap-spring signal is discernable in the tidal laminites (Jaeger and Nittrouer 1995). Similar Holocene tide-dominated delta analogues are inferred for rhythmites of the Coronation Formation.

15.7 Summary

Some of the singular qualitative criteria used previously to support a tidal origin are equivocal but repetitive associations of structures may warrant, a tidal interpretation. For example, stacked meter-scale parasequences in the Upper Mount Guide succession containing acceleration-deceleration cycle capped by thinly bedded sandstones with a variety of modified ripples and other exposure indicators are strongly suggestive of a tidal origin. In the Moodies Group, the association in some stratigraphic intervals of bimodal-bipolar paleocurrent patterns, tidal bedding, rare herringbone cross bedding and mudstone-draped foreset bundles strongly support the existence of tides in the early Precambrian Era.

The most convincing evidence for tidal forcing in the Precambrian is provided by rhythmites that display semi-diurnal, fortnightly (neap-spring-neap) and, in some cases, monthly (perigee-apogee) hierarchical bundling patterns. Data of these types presented earlier strongly indicate the existence of tides during deposition of the Elatina-Reynella and Moodies successions. Data on rhythmites from the Witwatersrand succession are more "noisy" but are suggestive of tidal forcing.

The land-ocean interface in the Precambrian was likely much different to most coastlines that exist today. In the absence of rooted land plants, point sources of sediment supply to the ocean were unlikely. Instead, the land-ocean interface was probably in the form of braid deltas with tidal modification taking place in river channels (Moodies Group), on tidal flats within embayments between delta lobes (Moodies Group), in delta front and prodeltaic settings (Elatina and Coronation), and on the shallow shelf (Moodies and Upper Mount Guide). Data from cross beds in the Moodies Group and the Upper Mount Guide reveal that bedforms were of comparable scale to those existing in Holocene estuaries supporting the conclusion that tidal current velocities were similar to those existing today in spite of a closer Earth-Moon distance at least in the Archean Era. Similarly there is no evidence for tidal ranges on the order of tens of meters as inferred previously by Von Brunn and Hobday (1976) for the 2.9 billion year old Mozaan Group in South Africa based on thicknesses of inferred progradational tidal-flat cycles. The lack of barrier-beach facies in association with examples discussed may indicate that coastlines were tide-dominated but there is no evidence for tidal ranges greater than the maximum of 13 m on Earth today (Archer and Hubbard 2003).

References

Allen JRL (1980) Sandwaves: a model of origin and internal structure. Mar Geol 26:281–328

Archer AW, Hubbard MS (2003) Highest tides in the World. In: Extreme depositional environments: mega end members in geologic time. Geol Soc Am Spec Publ 370:151–174

Armstrong RA, Compston W et al (1991) Zircon ion microprobe studies bearing on the age and evolution of the Witwatersrand triad. Precamb Res 53:243–266

Blake DH (1987) Geology of the Mount Isa Inlier and environs, Queensland and Northern Territory. Bur Min Res Bull 225, 83 p

Boersma JR, Terwindt JHJ (1981) Neap-spring sequences in intertidal shoal deposits in a mesotidal estuary. Sedimentology 28:51–170

Catuneanu O, Biddulph MN (2001) Sequence stratigraphy of the Vaal Reef facies associations in the Witwatersrand foredeep, South Africa. Sediment Geol 141–142:113–130

Chan MA, Kvale EP et al (1994) Oldest direct evidence of lunar-solar tidal forcing encoded in sedimentary rhythmites, Proterozoic Big Cottonwood Formation, central Utah. Geology 22:791–794

Crowell JC (1999) Pre-Mesozoic ice ages: their bearing on understanding the climate system. Geol Soc Am Mem 192:71–73

Dalrymple RW (1984) Morphology and internal structure of sand waves in the Bay of Fundy. Sedimentology 31:365–382

Dalrymple RW, Knight RJ, Lambiase JJ (1978) Bedforms and their hydraulic stability relationships in a tidal environment, Bay of Fundy, Canada. Nature 275-A:100–104

Dalrymple RW, Makino Y, Zaitlin BA (1991) Temporal and spatial patterns of rhythmite deposition on mudflats in the macrotidal, Cobequid Bay-Salmon River estuary, Bay of Fundy, Canada, Clastic tidal sedimentology. Can Soc Petrol Geol Mem 16:137–160

Dalrymple RW, Zaitlin BA, Boyd R (1992) Estuarine facies models: conceptual basis and stratigraphic implications. J Sediment Petrol 62:1130–1146

Dalrymple RW, Baker EK, Harris PT, Hughes MG (2003) Sedimentology and stratigraphy of a tide-dominated, foreland-basin delta, Fly River, Papua New Guinea. In: Tropical deltas of Southeast Asia and vicinity-sedimentology, stratigraphy, and petroleum geology. SEPM Spec Publ 76:147–173

de Boer PL et al (1989) The diurnal inequality of the tide as a parameter for recognizing tidal influences. J Sediment Petrol 59:912–921

de la Winter HR, Brink MC (1991) Chronostratigraphic subdivision of the Witwatersrand Basin based on a Western Transvaal composite column. South Afr J Geol 94:191–203

Ehlers TA, Chan MA (1999) Tidal sedimentology and estuarine deposition of the Proterozoic Big Cottonwood Formation, Utah. J Sediment Res 69:1169–1180

Eriksson KA (1977) Tidal deposits from the Archaean Moodies Group, Barberton Mountain Land, South Africa. Sediment Geol 18:257–281

Eriksson KA (1983) Archean iron-formations: environments of deposition and controls on formation. J Geol Soc Austr 30:473–482

Eriksson KA, Simpson EL (1990) Recognition of high-frequency sea-level changes in Proterozoic siliciclastic tidal deposits, Mount Isa, Australia. Geology 18:474–477

Eriksson KA, Simpson EL (2000) Quantifying the oldest tidal record: the 3.2 Ga Moodies Group, Barberton greenstone Belt, South Africa. Geology 28:831–834

Eriksson KA, Vos RG (1979) A fluvial fan depositional model for middle Proterozoic red beds from the Waterberg Group, South Africa. Precamb Res 9:169–188

Eriksson KA, Turner BR, Vos RG (1981) Evidence of tidal processes from the lower part of the Witwatersrand Supergroup. Sediment Geol 29:309–325

Eriksson KA, Simpson EL, Jackson MJ (1994) Stratigraphic evolution of a Proterozoic rift to thermal-relaxation basin, Mount Isa Inlier, Australia: constraints on nature of lithospheric extension. Int Assoc Sediment Spec Publ 20:203–221

Eriksson KA, Simpson EL, Mueller W (2006) Depositional and geodynamic setting of fluvio-tidal facies in the 3.2 Ga Moodies Group, South Africa. Sediment Geol 190:13–24

Ewers WE, Morris RC (1981) Studies of the Dales Gorge Member of the Brockman Iron Formation, Western Australia. Econ Geol 76:1929–1953

Fralick P (1987) Depositional environment of Archean iron formation: inferences from layering in sediment and volcanic hosted end members, Precambrian iron-formations. Theophrastus Publications, Athens, pp 251–266

Gehling JG (2000) Environmental interpretation and a sequence stratigraphic framework for the terminal Proterozoic Ediacara member within the Rawnsley Quarzite, South Australia. Precamb Res 100:65–95

Hagadorn JW, Bottjer DJ (1999) Restriction of a Late Neoproterozoic biotape: suspect-microbial structures and trace fossils at the Vendian-Cambrian transition. Unexplored microbial worlds. Palaios 14:73–85

Harris CW, Eriksson KA (1990) Allogenic controls on the evolution of storm to tidal shelf sequences in the early Proterozoic Uncompahgre Group, southwest Colorado, U.S.A. Sedimentology 37:189–213

Harris PT, Hughes MG et al (2004) Sediment transport in distributary channels and its export to the pro-deltaic environment in a tidally dominated delta: Fly River, Papua New Guinea. Cont Shelf Res 24:2431–2454

Heubeck C (2009) An early ecosystem of Archean tidal microbial mats (Moodies Group, South Africa, 3.2 Ga). Geology 37:931–934

Heubeck C, Wendt JI et al (1993) Timing of deformation of the Archean Greenstone Belt, South Africa: constraints from zircon dating of the Salisbury Kop Pluton. South Afr J Geol 96:1–8

Hori K, Saito Y et al (2002) Architecture and evolution of the tide-dominated Changjiang (Yangtze) River delta, China. Sediment Geol 146:249–264

Hunter RE (1977) Basic types of stratification in small eolian dunes. Sedimentology 24:361–388

Jackson JM, Simpson EL, Eriksson KA (1990) Facies and sequence stratigraphic analysis in an intracratonic, thermal-relaxation basin: the middle Proterozoic, lower Quilalar Formation, Mount Isa Orogen, Australia. Sedimentology 37:1053–1078

Jaeger JM, Nittrouer CA (1995) Tidal controls on the formation of fine-scale sedimentary strata near the Amazon river mouth. Mar Geol 125:259–281

Kamo SL, Davis DW (1994) Reassessment of Archean crustal development in the Barberton Mountain Land, South Africa, based on U-Pb dating. Tectonics 13:167–192

Klein GD (1977) Clastic tidal facies. Continuing Education Publication Company, Champaign IL, 149 p

Knoll AH, Walter M et al. (2004) The Ediacaran Period: a new addition to the Geologic Time Scale. Terminal Proterozoic Subcommission of the International Commission on Stratigraphy

Kocurek G, Fielder G (1982) Adhesion structures. J Sediment Petrol 52:1229–1241

Kreisa R, Moiola RJ (1986) Sigmoidal tidal bundles and other tide-generated sedimentary structures of the Curtis Formation, Utah. Geol Soc Am Bull 97:381–387

Kvale EP, Archer AW (1991) Characteristics of two, Pennsylvanian-age, semi-diurnal tidal deposits in the Illinois Basin, USA. Can Soc Petrol Geol Mem 16:179–188

Kvale EP, Johnson HW et al (1999) Calculating lunar retreat rates using tidal rhythmites. J Sediment Res 69:1154–1168

Lambeck K (1980) The earth's variable rotation: geophysical causes and consequences. Cambridge University Press, Cambridge, 449 p

Lemon NM, Gostin VA (1990) Glacigenic sediments of the late Proterozoic Elatina Formation and equivalents, Adeliade Geosyncline, South Australia. In: Evolution of a Late Precambrian-Early Palaeozoic Rift complex: the Adelaide Geosyncline. Geol Soc Aust Spec Publ 16:149–163

Noffke N, Krumbein WE (1999) A quantitative approach to sedimentary surface structures contoured by the interplay of microbial colonization and physical dynamics. Sedimentology 46:417–426

Noffke N, Gerdes G et al (2001) Microbially induced sedimentary structures indicating climatological, hydrological and depositional conditions within recent and Pleistocene coastal facies zones, southern Tunisia. Facies 44:23–30

Noffke N, Knoll AH, Grotzinger JP (2002) Sedimentary controls on the formation and preservation of microbial mats in siliciclastic deposits: a case study from the upper Neoproterozoic Nama Group, Namibia. Palaios 17:1–14

Noffke N, Gerdes G, Klenke Th (2003a) Benthic cyanobacteria and their influence on the sedimentary dynamics of peritidal depositional systems (siliciclastic, evaporate salty and evaporitic carbonatic). Earth Sci Rev 12:1–14

Noffke N, Hazen R, Nhleko N (2003b) Earth's earliest microbial mats in a siliciclastic marine environment (Mozaan Group, 2.9 Ga, South Africa). Geology 31:673–676

Noffke N, Eriksson KA, Hazen RE, Simpson EL (2006) A new window into early life: microbial mats in Earth's oldest siliciclastic tidal flats (3.2 Ga Moodies Group, South Africa). Geology 34:253–256

Noffke N, Beukes N et al (2008) An actualistic perspective into Archean worlds – (cyano) bacterially induced sedimentary structures in the siliciclastic Nhlazatse Section, 2.9 Ga Pongola Supergroup, South Africa. Geobiology 6:5–20

Ojakangas RW (1983) Tidal deposits in the early Proterozoic basin of the Lake Superior region-the Palms and the Pokegema Formations: evidence for subtidal-shelf deposition of Superior-type banded iron-formations. Early Proterozoic Geology of the Great Lakes Region. Geol Soc Am Mem 160:49–66

Ojakangas GW (1996) Cyclic tidal laminations in the Early Proterozoic Pokegama Formation: digital image analysis and computer modeling (abstrat). In: 42nd Institute of Lake Superior geology, pp 44–45

Page RW (1983a) Timing of superposed volcanism in the Proterozoic Mount Isa Inlier, Australia. Precamb Res 21:223–245

Page RW (1983b) Chronology of magmatism, skarn formation and uranium mineralization, Mary Kathleen, Queensland, Australia. Econ Geol 85:838–853

Pidgeon RT, Horwitz RC (1991) The origin of olistoliths in Proterozoic rocks of the Ashburton Trough, Western Australia, using zircon U-Pb isotopic characteristics. Austr J Earth Sci 38:55–63

Preiss WV (1987) The Adelaide Geosyncline. South Aust Dept Mines Energy Bull 53, 438 p

Rinehart JS (1972a) Fluctuations in geyser activity caused by variations in earth tidal forces, barometric pressure, and tectonic stresses. J Geophys Res 77:342–350

Rinehart JS (1972b) 18.6-year earth tide regulates geyser activity. Science 177:346–347

Rinehart JS (1974) Geysers. Am Geophys Union Trans 56:1052–1062

Schmidt PW, Williams GE (1995) The Neoproterozoic climatic paradox: equatorial palaeolatitude for Marinoan glaciation near sea level in South Australia. Earth Planet Sci Lett 134:107–124

Simonson BM (1985) Sedimentological constraints on the origins of Precambrian iron-formations. Geol Soc Am Bull 96:244–252

Simpson EL, Eriksson KA (1991) Depositional facies and controls on parasequence development in siliciclastic tidal deposits from the early Proterozoic, upper Mount Guide Quartzite, Mount Isa Inlier, Australia. Can Soc Petrol Geol Mem 16:371–387

Soegaard K, Eriksson KA (1985) Evidence for tidal, wave and storm interaction on a Precambrian shelf: the 1.7 Ga Ortega Group, New Mexico. J Sediment Petrol 55:672–684

Sonett CP, Kvale EP et al (1996) Late proterozoic and Paleozoic tides, retreat of the moon and rotation of the earth. Science 273:100–104

Tankard AJ, Jackson MPA et al (1982) Crustal evolution of Southern Africa. Springer, New York, 523 p

Tessier B (1993) Upper intertidal rhythmites in the Mont-Saint-Michel Bay (NW France): perspectives for paleoreconstruction. Mar Geol 110:355–367

Trendall AF (1973) Varve cycles in the Weeli Wolli Formation of the Precambrian Hamersley Group, Western Australia. Econ Geol 68:1089–1097

Trendall AF (1983) The Hamersley Basin. Iron-formation: facts and problems. Elsevier, Amsterdam, pp 69–129

Trendall AF, Blockley JG (1970) The iron formations of the Precambrian Hamersley Group, Western Australia. Geol Surv West Aust Bull 119:366p

Trendall AF et al (1990) Percise zircon U-Pb chronological comparison of the volcano-sedimentary sequences of the Kaapvaal and Pilbara Cratons between about 3.1 and 2.4 Ga. In: 3rd International Archean Symposium, Perth, pp 81–83

Visser MJ (1980) Neap-spring cycles reflected in Holocene subtidal large-scale bedform deposits: a preliminary note. Geology 8:543–546

Von Brunn V, Hobday DK (1976) Early Precambrian tidal sedimentation in the Pongola Supergroup of South Africa. J Sed Petrol 46:670–679

Vos RG, Eriksson KA (1977) An embayment model for tidal deposits occurring within a fluvially-dominated middle Proterozoic sequence in South Africa. Sediment Geol 18:161–173

Williams GE (1989) Late Precambrian tidal rhythmites in South Australia and the history of the Earth's rotation. J Geol Soc Lond 146:97–111

Williams GE (1991) Upper Proterozoic tidal rhythmites, South Australia: sedimentary features, deposition, and implications for the earth's paleorotation. Clastic tidal sedimentology. Can Soc Petrol Geol Mem 16:161–177

Williams GE (2000) Geological constraints on the Precambrian history of Earth's rotation and the Moon's orbit. Rev Geophys 38:37–59

Hypertidal Facies from the Pennsylvanian Period: Eastern and Western Interior Coal Basins, USA

16

Allen W. Archer and Stephen F. Greb

Abstract

Siliciclastic tidal facies have been recognized in Pennsylvanian coal measures of the Eastern Interior (Illinois) and Western Interior (Forest City) basins. In particular, rhythmic tidal laminations or tidal bundles are recorded in shale-rich, heterolithic estuarine and coastal paleofacies, as well as within tidal and fluvio-estuarine channels. The tidal facies are recurring and range from the upper Morrowan (Early Pennsylvanian) through at least the Desmoinesian (late Middle Pennsylvanian). Laminae-thickness series within tidal facies in both basins exhibit a variety of well-developed, tidal cycles that include semidiurnal, diurnal, neap-spring, apogean-perigean, and seasonal to annual periodicities.

Study of modern analogs, predominantly from hypertidal settings, provides evidence to suggest the presence of elevated paleotidal ranges in the Pennsylvanian seaways in both basins. The tidal facies are best developed within transgressive systems, particularly within incised valley-fill sequences. During sediment accumulation, the extreme tidal dynamics resulted in widespread deposition of rhythmites. Cyclic rhythmites that contain high-resolution records of daily to yearly periodicities are much more locally restricted. Preservation of tidal rhythmites was likely aided by (1) rapid, high-magnitude changes in global paleosealevel, (2) strongly resonant depositional embayments, (3) formation of large tropical Pangean rivers during lowstand that were converted to estuaries during subsequent periods of glacial melting and the resultant sea-level rise, and (4) a strongly resonant, extensive global paleo-ocean. Preservation of cyclic tidal rhythmites that contain high-resolution records were likely controlled by the generation of local accommodation space via (1) peat compaction, (2) faulting, and (3) tidal and fluvial channel avulsion.

A.W. Archer (✉)
Department of Geology, Kansas State University, Manhattan, KS 66506, USA
e-mail: aarcher@ksu.edu

S.F. Greb
Kentucky Geological Survey, University of Kentucky, Lexington, KY 40506, USA
e-mail: greb@uky.edu

16.1 Introduction

16.1.1 Geographic and Geologic Setting

A variety of tidally influenced facies occur within Pennsylvanian strata in the interior coal basins of the eastern half of the United States. In general, the bedrock geology of the basins consists of flat-lying, cratonic

R.A. Davis, Jr. and R.W. Dalrymple (eds.), *Principles of Tidal Sedimentology*,
DOI 10.1007/978-94-007-0123-6_16, © Springer Science+Business Media B.V. 2012

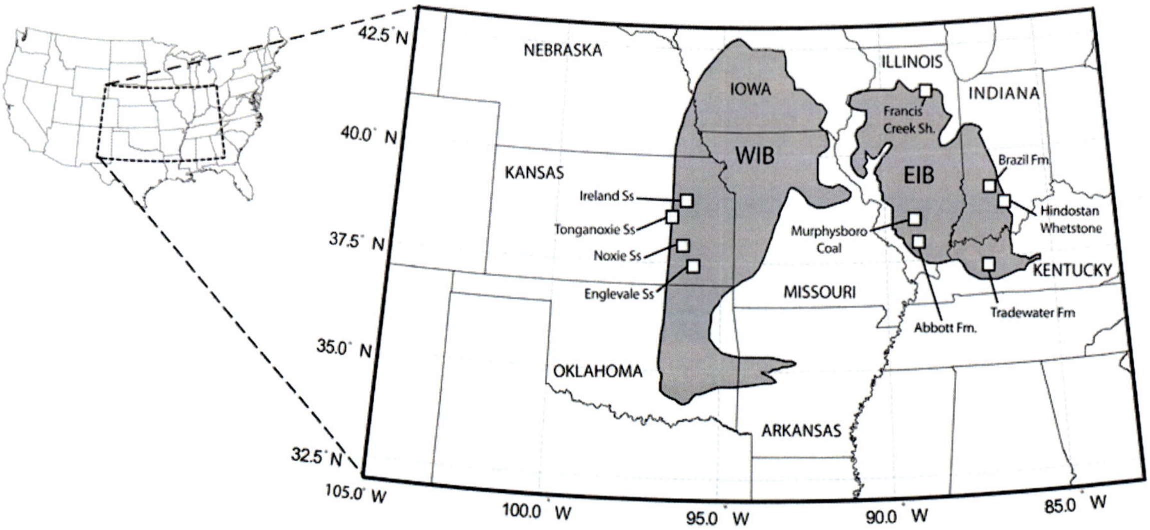

Fig. 16.1 Location of study sites and stratigraphic units in the Eastern Interior Basin (*EIB*) and Western Interior Basin (*WIB*) in the central U.S. The EIB is located mostly in the state of Illinois, but also include large parts of Indiana and western Kentucky. The WIB contains large areas of the states of Iowa, Missouri, Kansas, and Oklahoma. Small areas in southeastern Nebraska and west-central Arkansas are also included in the WIB

rocks and natural outcrops are rare. Exposures are mostly limited to quarries, open-pit and underground coal mines, relatively rare roadcuts, and valleys of the larger rivers. Shallow coal-exploration cores have also provided much useful data.

The Eastern Interior Basin (EIB) is regionally referred to as the Illinois Basin and includes parts of the states of Illinois, Indiana, and western Kentucky as well as small parts of southeastern Iowa (Fig. 16.1). The EIB began as an aulocogen in western Kentucky during the Cambrian and evolved into a broader, intracratonic basin throughout the Paleozoic (Soderberg and Keller 1981; Heidlauf et al. 1986). During the Pennsylvanian Period, there were essentially two major depocenters, one above the old aulocogen in the Rough Creek Graben of western Kentucky, and a second above a semi-circular structural depression sometimes referred to as the Fairfield Basin in southeastern Illinois (e.g. Wanless 1975). Pennsylvanian strata thicken above these depocenters and thin laterally toward the basin margins. Many of the known exposures of rhythmites occur along the basin margins.

During Pennsylvanian deposition the EIB was essentially a broad structural embayment that was open to the south and connected to the Ouachita Trough on the southern margin of the craton. Similarly, there were times during the Pennsylvanian, when deposition was continuous across the Midcontinent between the EIB and WIB (Wanless and Wright1978; Greb et al. 2003). Uplift of the Pascola Arch on the southern margin of the basin after the late Pennsylvanian resulted in a closure of the EIB. Thus, age-equivalent rocks in the EIB, WIB, and northeastern Arkansas are now geomorphically separated.

The Western Interior Coal Basin (WIB) is also named the Forest City Basin (Fig. 16.1). The basin includes parts of southwestern Iowa, southeastern Nebraska, eastern Kansas and central Missouri. The southern extent of the Paleozoic depositional basin also includes parts of Arkansas and Oklahoma. The WIB has a complex Paleozoic history and initial movements began in the Ordovician (Lee 1943). Subsidence in northern Kansas formed the ancestral basin, which was subsequently bisected by uplifts along the western margin (Nemaha Anticline). During the late Mississippian and early Pennsylvanian, widespread erosion produced a surface with more than 70 m of relief.

Renewed uplift of the Nemaha Anticline created the western margin and downwarping to the east created the WIB. Pennsylvanian strata, ranging from the Atokan to Virgilian, attain thicknesses of approximately 600 m in the center of the basin (Anderson and Wells 1968). Unlike the EIB, no lower Pennsylvanian (Morrowan) rocks have been preserved.

16.1.2 Lithostratigraphy

Throughout the study area Pennsylvanian stratigraphic successions exhibit repetitions of lithologies. This repetition is most notable in the marine, limestone-rich parts of the Middle and Upper Pennsylvanian section. In the EIB, the oscillations of nonmarine (sandstones, nonfossiliferous shale and coals) to marine (fossiliferous shale and limestone) gave rise to the formerly widespread concept of "cyclothems" (Weller 1930, 1931; Wanless and Weller 1932). An "ideal cyclothem" model was developed in Illinois and was, for a brief period, utilized as a formalized lithostratigraphic unit. The concept was subsequently applied to the Western Interior Basin (Moore 1935, 1964; Moore et al. 1951), and attempts were also made to use cyclothems as principal components within formal lithostratigraphic nomenclature. Application of these lithostratigraphic models provoked widespread debate regarding the origin of cyclothems (e.g. Heckel 1977, 1986). Much of the discussion focused upon the origins of widespread baselevel (eustatic) oscillations. Some workers advocated regional basinal subsidence (e.g. Sloss 1963) whereas other workers invoked sea level changes, particularly those related to Gondwanan paleoglacial cycles and the resultant glacio-eustatic variations (e.g. Heckel 1994).

Archer (2008) presented a critique of cyclothem models noting that they oversimplify the lateral variability that is characteristic of the clastic components. For parts of the Pennsylvanian in the EIB, the stratigraphic section does not offer a simple fit into a standard cyclothem model. Lower Pennsylvanian (Morrowan) strata are dominated by thick, laterally discontinuous and channel-filling sandstone units rather than cyclic successions of strata (Fig. 16.2). These sandstone units can directly overlie the Mississippian-Pennsylvanian regional unconformity. Thicker sands occur where there was greater erosional incision and relief. The lower Middle Pennsylvanian (Atokan) section exhibits significant lateral variability in facies and thickness. The upper Middle Pennsylvanian (Desmoinesian) coal-bearing parts of the section exhibits more lateral continuity and is where the concept of cyclothems originated and was most applied. Even within the Desmoinesian, the gray shale parts of coal-bearing cycles exhibit at least some degree of regional variability. Locally, outcrops contain thin, discontinuous sandstone beds and laminae (Archer and Kvale 1993). In such exposures, mm- to cm-scale lamina can be laterally traced for considerable distances before being truncated by low-angle reactivations (Fig. 16.3).

In the WIB, a similar stratigraphic succession can be delineated (Fig. 16.2). Thick sandstone units were deposited over a regional unconformity. Locally, these sandstone units exhibit 10-m thicknesses in surface sections and 30-m thicknesses in the subsurface. The remaining parts of the sequence are dominated by laterally variable gray shale that contains persistent m-scale limestone and cm-scale coal seams. Coals are generally thinner and limestones are better developed than in the WIB as compared to the EIB

16.1.3 Stratigraphy and Common Lithofacies

Bundled rhythmites and tidal bundles in crossbedded facies have been documented in several lithofacies in the EIB and WIB (Fig. 16.2). A variety of depositional models (i.e., "ideal cyclothems") have been used to describe parts of the stratigraphic and sedimentological successions in both basins. The marine parts are characterized by laterally persistent, meter-scale limestone units that contain a diverse suite of marine fossils, and dark gray to black shales. Organic-rich, black-shale beds, which are commonly only a meter or less in thickness, are regionally widespread. Various types of gray shale or heterolithic strata comprise the most volumetrically dominant lithofacies relative to rhythmite preservation. Marine sandstone units are rare. In general, mostly marine facies are best developed in the upper Middle to Upper Pennsylvanian (Desmoinesian and younger) strata, and are poorly developed or laterally restricted in the lower Middle (Atokan) and Lower Pennsylvanian (Morrowan) (Fig. 16.2) (Greb et al. 1992; Greb et al. 2002).

The dominantly terrestrial parts of the Pennsylvanian stratigraphic succession include persistent coals and the paleosols beneath them. These tend to be laterally restricted in the Lower Pennsylvanian in both basins, but become widespread and can be readily correlated directly between the interior basins in upper Middle Pennsylvanian (Desmoinesian) and younger strata (Wanless and Weller 1932; Heckel 1986, 1994). In contrast, large-scale, trough-crossbedded, fluvial sandstones are more locally restricted in both basins. The thicker sandstone units (10 m+) commonly exhibit

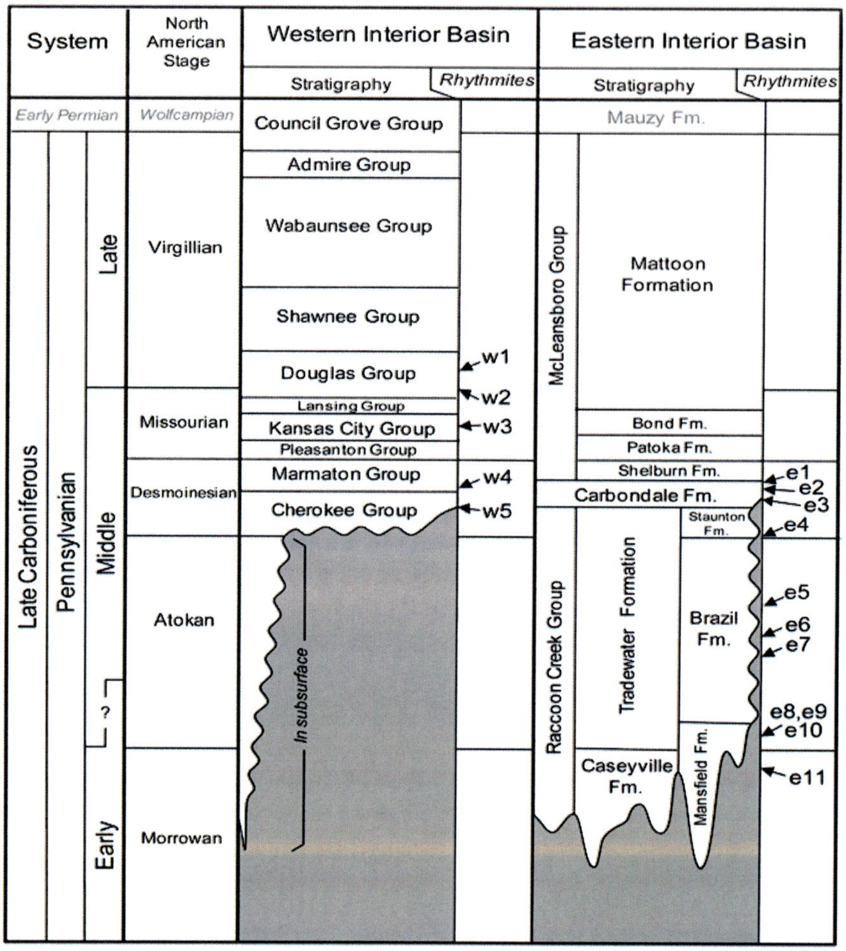

Fig. 16.2 Stratigraphic column of the Pennsylvanian (Late Carboniferous) Period in the EIB and WIB. The Morrowan and Atokan Stages are largely absent from the WIB. Stratigraphic intervals studied in the WIB include: *w1*: Ireland Sandstone, *w2*: Tonganoxie Sandstone, *w3*: Noxie Sandstone, *w4*: Englevale Sandstones, and *w5*: Cherokee Group. In the EIB, the stratigraphic intervals include: *e1*: Herrin Coal/Energy Shale, *e2*: Springfield Coal/Dykersburg Shale, *e3*: Colchester coal/Francis Creek Shale, *e4*: Murphysboro Coal, *e5*: Western Kentucky No. 4 coal, *e6*: Elm Lick coal, *e7*: Abbott Formation, *e8*: above Lower Block coal, *e9*: Hindostan whetstones, *e10*: above Caseyville incised valley, *e11*: upper Caseyville incised valley-fill sequence

an unconformable base and fill incised paleovalleys (IVFs), which are cut down from the more regionally extensive paleosols (Feldman et al. 1995). In the lower Pennsylvanian, the valley-filling sandstones of the EIB may be more than 60 m thick (Potter and Desborough 1965). Valley fills are complex but conglomeratic sandstones, with extrabasinal quartz pebbles are common (Sedimentation Seminar 1978; Greb et al. 1992; Archer et al. 1994; Archer and Greb 1995). Younger paleochannel sandstones in both basins may also contain extrabasinal quartz clasts, but to a lesser extent than in the lower Pennsylvanian. Lithoclasts in Middle and Upper Pennsylvanian incised sandstones commonly contain intrabasinal lithologies, with limestone, shale, and sideritic clasts being the most abundant. Quartzarenites dominate the Lower Pennsylvanian, while litharenites and sublitharenites dominate the Middle and Upper Pennsylvanian. Carbonized plant material is locally abundant and ranges in size from large, fossil-tree trunks down to sand-sized material ("coffee grounds"). Sandstone bodies can exhibit an elongate trend, mostly south to southwest, and have been historically described as "shoestring sands" (Bass 1934, 1936; Potter 1962). Sandstone units typically fine upward into heterolithic strata or gray shale that lack fossils of marine macroinvertebrates. The shale

Fig. 16.3 Outcrop of heterolithic rhythmites from western Kentucky (see Greb and Archer 1995, 1998). (**a**) Roadcut exhibiting lateral continuity of mm- to cm-scale rhythmites and low angle, large-scale reactivations. Exposure is 5 m thick. (**b**) Closeup of planar to rippled lamina that exhibit prominent dm-scale bundling. The more sand-rich zones are less deeply weathered than the intervening shale-rich zones. Scale is 10-cm long

units can contain abundant, well-preserved, carbonized plant fossils. Many of the shale units are heterolithic and exhibit a variety of lenticular, wavy, and flaser bedding (Kvale and Archer 1990). Economic coals are also commonly overlain by heterolithic facies.

16.2 Facies Containing Tidal Rhythmtes

16.2.1 Range of Associated Lithofacies

Many of the incised valley-filling sandstones are oriented downdip, which is generally south to southwest, and are interpreted as fluvial in origin (Potter 1962; Wanless and Wright 1978; Archer et al. 1995a). Cross bed foresets are sometimes delineated by thin drapes of coalified and fragmentary plant materials ("coffee grounds") (Fig. 16.4a). Conglomeratic lags are common within the lower parts of IVF sandstones. For the EIB, the conglomerates can include extrabasinal vein-quartz pebbles. Conversely, in the WIB, the clasts consist of intrabasinal facies, such as limestone, sandstone, and sideritic clay chips (Fig. 16.4b).

In some cases, however, the upper parts of fluvial sandstones may contain tidal features indicative of estuarine influences, or the sandstones may be overlain by heterolithic facies of coastal-estuarine origin. In the EIB and WIB, fluvial facies may be overlain by heterolithic tidal channel or heterolithic tidal flat facies. In some cases, apparently fluvial channels are bordered by gray shale wedge facies, which have tidal indicators.

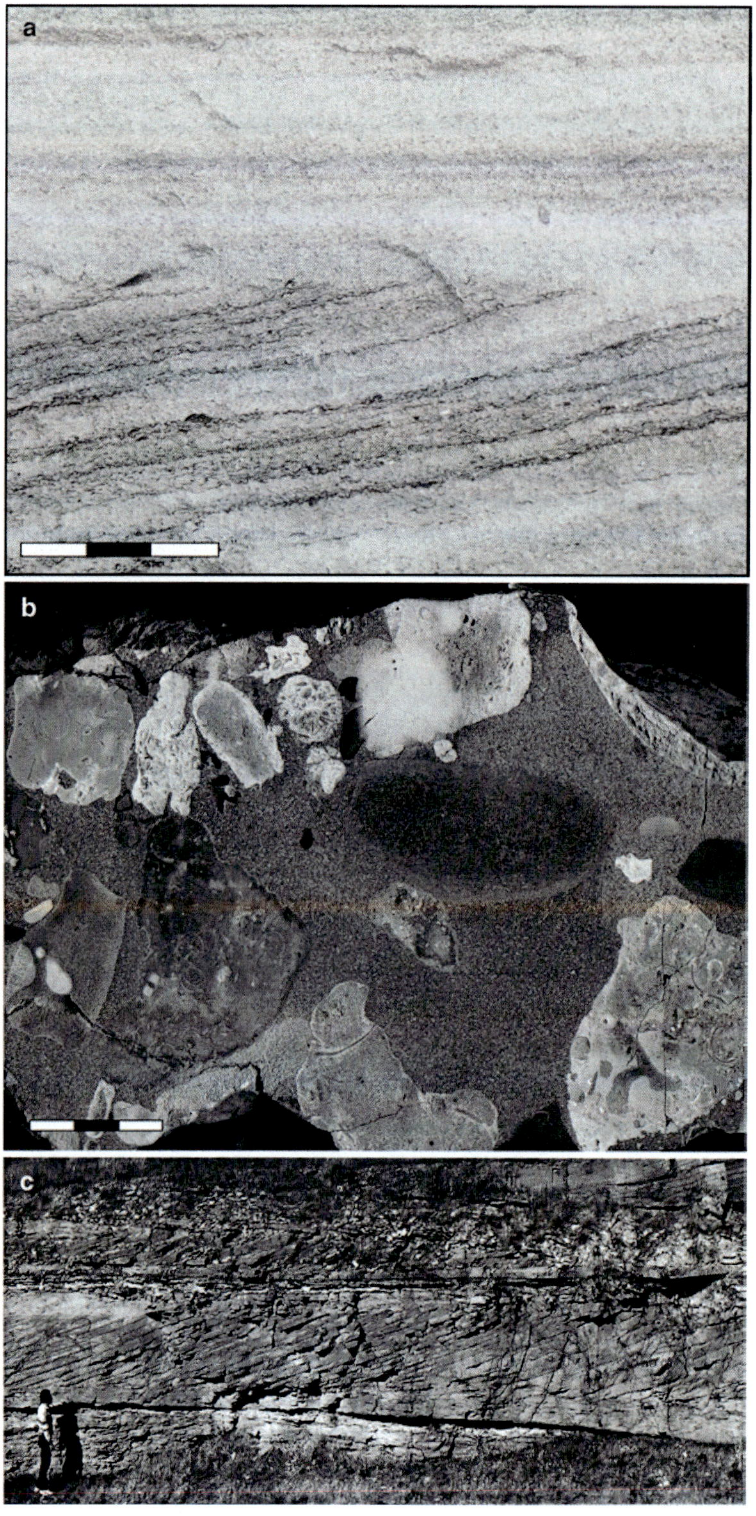

Fig. 16.4 Fluvial to estuarine facies in rhythmites from southern Illinois. (**a**) Low angle planar forsets in medium-grained sandstone. Foresets are delineated by sand- and silt-sized accumulations of coalified plant debris. Sample from immediately above Murphysboro Coal in southwestern Illinois. (**b**) conglomerate within lower part of IVF sequence consisted of locally derived clasts of limestone (*light colored*), poorly cemented sandstone, sideretic chips. From lower IVF in Tonganoxie Sandstone, east-central Kansas. (**c**) Outcrop in southern Illinois of Abbott Formation exhibiting mostly planar forests delineated by iron-cemented zone (originally mud drapes). Cyclicity within this outcrop has been described by Kvale and Archer (1991) and Archer (1996a)

16.2.2 Heterolithic Tidal Channel Facies

Tidal rhythmites within channel-form scours represent fluvial channels that have been converted to tidal-estuarine channels, tidal channels, or abandoned channels (tidal or fluvial) that filled with tidal flat facies. Modern tidal channels may contain bedforms that exhibit unimodal to bimodal bedding, mud-draped foresets, or bundled foresets. Pennsylvanian heterolithic tidal channel facies in the EIB and WIB, commonly exhibit mud-draped planar foresets. Successive thicknesses of foresets can exhibit systematic thickening and thinning or bundling (Fig. 16.4c).

In western Kentucky a lower Middle Pennsylvanian channel described by Greb and Archer (1995), exhibited heterolithic laminae arranged in thickening and thinning bundles on low-angle bedding surfaces, rather than true foresets (Fig. 16.5). Thickening and thinning laminae bundles amalgamated vertically within the channel fill into ripple-laminated and ripple-bedded sandstone. At first glance, stacked successions of thickening and thinning laminae bundles are generally similar to other documented examples of Pennsylvanian rhythmites in which the bundling was interpreted to represent neap-spring cycles. More detailed analyses, however, suggested that these cm- to dm-scale bundles could also be interpreted as annual bundles (Fig. 16.5).

16.2.3 Heterolithic Tidal-Flat Facies

Flat-lying heterolithic facies with tidal bedding or lamination are common in the Pennsylvanian of both the EIB and WIB. These facies are similar to the deposits of modern tidal flats. Well-preserved successions of tidal rhythmites are not common on all modern tidal flats. The best-preserved rhythmites in modern tidal-flat settings are documented in hypertidal systems (Dalrymple and Makino 1989; Tessier 1993; Archer 2004). Heterolithic tidal-flat facies in the Pennsylvanian of the EIB and WIB consist of mixed sandstone, siltstone and shale. Extrabasinal clasts are lacking. Sandstones within the units are commonly well sorted. Mud-chip conglomerates can be locally evident. Within sandstones, ripple-scale features are very common and a great variety of ripple marks and fine-scale, exposure-related features have been reported in both basins (Kvale and Archer 1991; Lanier et al. 1993; Greb and Archer 1995, 1998).

Some rhythmites in the heterolithic tidal flat facies are predominantly silt rather than alternations of mud and sand. Laminated siltstone facies are volumetrically rare and are known from only a few outcrops in southern Indiana and Kansas. Individual lamina range in thickness from a few millimeters to as much as several centimeters. Thus these rocks contain finely interspersed thick laminae and thin beds. Each individual lamina or bed exhibits a relatively abrupt lower boundary and is capped by a finer grained lamina drape. Rhythmic lamination contains gradationally thickening and thinning clay-draped laminae. In the EIB, rhythmites commonly contain alternating thicker- and thinner laminae pairs within laminae bundles (Fig. 16.6a, e.g. Kvale et al. 1989). Similar, but geologically younger facies in the WIB do not commonly exhibit prominent laminae pairing (Lanier et al. 1993).

Tidal rhythmites in the heterolithic tidal flat facies may contain mm- and cm-scale cyclicities, which are laterally continuous at the scale of an outcrop (Kuecher et al. 1990; Kvale and Archer 1991). At other localities, however, erosional features and small-scale reactivations are common in the laminated siltstones, and rhythmites are less laterally continuous. Relatively discontinuous rhythmites in heterolithic tidal flat facies have been reported above coal beds (Fig. 16.6b). At these locations, low-angle reactivations and soft-sediment deformation are evident. In some cases, a single lamination has been overturned and truncated by overlying, planar laminae (Fig. 16.6c). These are similar to overturned foresets that can occur in cross-bedded sandstone. Disrupted or discontinuous rhythmites in the tidal-flat facies in these situations, likely result from syndepositional compaction of the underlying peat during rapid loading of tidal sediments. Generally there is very little internal bioturbation (vertical burrowing) of tidal rhythmites in the heterolithic tidal-flat facies. Heterolithic tidal-flat facies do occur with well-developed bioturbation in both basins, but in these cases (as in the modern), tidal lamination is disrupted, so that bedding is no longer rhythmic. Where tidal rhythmites are well-developed in modern tidal settings, burrowing is absent or limited (Archer 2004). This is not to say that tidal rhythmites contain no biogenic structures. Where clay drapes are well developed in rhythmites, the rocks can be readily split along this surface, and surficial (horizontal) biogenic structures are locally common. Because of the well-sorted, fine-grained sediment, many types of biogenic structures (horizontal trace fossils) are unusually well

Fig. 16.5 Ripple-dominated rhythmites from western Kentucky (Tradewater Formation) showing prominent cm-scale bundling, which is interpreted as annual cyclicity. (**a**) Outcrop view of non-weathered rhythmites. Thick mud-rich zones contain mm-thick streaks of sandstone. (**b**) Thick mud-rich zones separated by ripples with rounded crests

preserved (Archer and Maples 1984; Maples and Archer 1987). Also, a variety of intricate and well-preserved erosional and depositional sedimentary structures, such as foam casts, have been described by Lanier et al. (1993) from laminated rhythmites in Kansas. A diverse suite of similar features has been described from modern hypertidal settings and directly compared to the strata deposited during the Pennsylvanian Period (Tessier et al. 1995; Archer 2004).

Most occurrences of the heterolithic tidal-flat facies in the Pennsylvanian of the EIB and WIB directly overlie coal seams and are in turn, overlain by gray shale (marine) or shales that coarsen upward into heterolithic strata or sandstone. In some cases, upright lycopod trees (2–3 m high) are encased in tidal rhythmites above coal seams (Kvale et al. 1989; Archer 2004). Because the tidal-flat facies overlies a coal bed, which originated as a terrestrial, freshwater peat, the facies occupies a transgressive position and this facies is perhaps most common in transgressive (and possible highstand) tracts. The facies, however, is not always marine (see discussions).

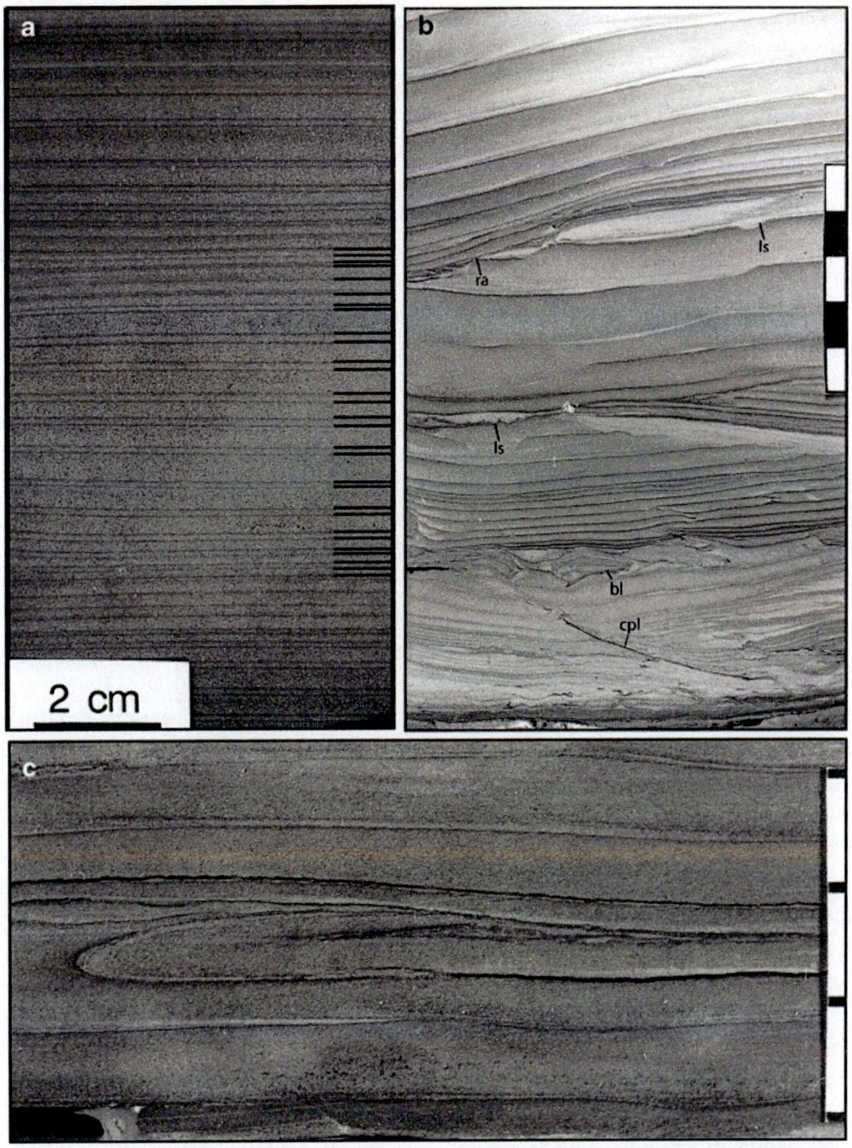

Fig. 16.6 Silty rhythmites from the EIB and WIB. (**a**) Section of core from the Francis Creek Shale, northeastern Illinois. Rhythmites within this unit have been described by Kuecher et al. (1990); Archer (1996). *Short black lines* delineate thick-thin pairing of individual lamina as well as the extent of a larger scale (neap-spring) cycle. (**b**) Cut of polished slab of rhythmites from the Tonganoxie Sandstone, east-central Kansas. Sedimentary features within this unit have been described in detail (Lanier et al. 1993). Comparison to modern analogs is discussed in Archer (2004). Note the mm- to cm-scale lamina ranging upward into thin beds of siltstone. Reactivations (*ra*), and small-scale loading structures (*ls*) are common. Base of this sample immediately overlies a coal seam. Very thin, *inclined*, *black-colored*, linear feature in lowest bundle is a single, coalified plant leaf. Note the differences in lamina on each side of the leaf. Upright leaves coated by mm-scale mud lamina (*bl*). (**c**) Overturned and truncated lamina, same locality as Fig. 16.6b. The deformation is largely constrained to a single, cm-thick lamination. Laminae above and below deformed lamina are planar indicating that the soft-sediment deformation was restricted to a single (tidal?) event

16.2.4 Gray-Shale Wedge Facies

Within the EIB, this facies was first uniquely defined as gray shale wedges along inferred fluvial channels above Desmoinesian coal beds (Wanless 1964; Gluskoter and Hopkins 1970). Shale wedges are thick above and alongside paleochannels and thin laterally away from the channels. Shale wedges exhibit a high degree of lateral variability in thickness and bedding. The unique juxtaposition along paleochannels and lateral variability within the wedges is quite different from the typical "cyclothemic" successions common within parts of the Desmoinesian and younger strata. Away from the paleochannels and gray-shale wedges, coals are overlain by a more common succession of limestones or coarsening-upward, gray shale and siltstone capped by sandstone or another coal bed. Gray shale wedges were widely mapped in the EIB because the sulfur content of the underlying coal seemed to bear a more-or-less direct relationship to the relative thickness of overlying gray-shale wedge. Lower sulfur contents were found in coals beneath the thicker parts of the shale wedges, and more typical higher sulfur contents were found toward the thinning margins of the wedges (e.g. Gluskoter and Hopkins 1970).

Gray shale wedges in the Desmoinesian of the EIB were formerly interpreted as (1) levee deposits of the adjacent fluvial channels, (2) crevasse-splay deposition into floodplain lakes, or (3) lacustrine varves (Archer and Maples 1984). Various types of fine-scale rhythmites (Fig. 16.7), however, are common in the shale wedges, suggesting tidal, rather than fluvial influences. Cyclic rhythmites in the gray shale wedges were first noted by Kvale and Archer (1990, 1991). In this facies, there is a complex continuum from thin-bedded, rippled sandstone near the channels, to mud-draped sandstone beds, and ultimately to mudstone-dominated, heterolithic bedding, which includes flaser, wavy, and lenticular bedding (Reineck and Wunderlich 1968). In addition, mm-thick, planar sand streaks (Fig. 16.7a) are frequent and have been termed "pinstripe bedding" (Kvale and Archer 1990). A pinch-and-swell texture, created by small-scale truncations and reactivations are locally abundant (Fig. 16.7b). Very similar gray-shale dominated facies that are adjacent to paleochannels occur in the WIB and, in may cases, directly overlie coal seams.

As in rhythmites of the heterolithic-tidal flat facies, there is little to no bioturbation in rhythmites of the gray-shale wedge facies. Fine-scale depositional fabric is commonly very well preserved (Fig. 16.7d). In zones where bioturbation (horizontal traces) does occur, however, the biological activity indicates a low-diversity infauna of burrow-making organisms (Archer and Maples 1984). In modern settings, such organisms are termed "opportunistic" because they commonly exhibit a high density of one or more individual species, but have low overall biotic diversity.

16.3 Discussion

Prior to the late 1980s, facies now interpreted as tidal were commonly considered to have formed in nonmarine, fluvial-deltaic settings. Later, sedimentological research focused on the laminae and laminae bundles in these facies. Various types of cyclicity and related features were used to reinterpret the depositional setting as tidally influenced (Kvale et al. 1989). Then, increasing detail was focused on the types of cyclicity that could be extracted from long, laminae-thickness series and the remarkable apparent completeness of tidal records preserved in some ancient tidal rhythmites in the basins. In southern Illinois, a long and continuous series of foreset thicknesses from the Abbott Formation exhibited what appears to be one of the most complete paleotidal records from the Pennsylvanian (see Kvale and Archer 1991; Archer 1996a). Rhythmites from both basins were found to preserve a variety of short- and longer-term tidal cycles.

Fig. 16.7 Heterolithic rhythmites from open-pit coal mines in Indiana and Illinois. (**a**) Polished slab from Brazil Formation (Kvale and Archer 1991) exhibiting prominent bundling of mostly planar to slightly rippled sandstone streaks. Note the cm-thick dark, mud-rich zones near the *bottom* and *top*. Originally described as neap-spring tidal cycles, the bundled could also reflect seasonal (yearly) periodicities. (**b**) Small-scale ripples and truncations within heterolithic rhythmites from Murphysboro Coal in southern Illinois. Note more-organized zone at top, which is similar to Fig. 16.7a. (**c**) Unusual rhythmites with several scales of periodicities. Neap-spring cycles range from about 1-cm thick at bottom to as much as 3-cm thick at top. Note the alternation of thicker neap-spring cycles overlain by thinner neap-spring cycles that occurs throughout the sample. This pattern can be interpreted as related to perigean-spring tides (see Archer 1996a). Thicker neap-spring cycles would be produced during lunar perigee and the thinner cycles would have been deposited during lunar apogee. Preservation of such detailed records of paleotides is remarkable. (**d**) Polished slab of infaunally bioturbated heterolithic rhythmites from the EIA. Along the right side, the only bioturbation consists of a few, tubular, 5-mm diameter burrows. Along the *left side*, particularly in the *lower left*, extensive bioturbation has selectively and nearly completely destroyed the rhythmite fabric. Rectangle on right delineates the highest-order cyclicity (*dark lines*), intermediate-order cyclicity (*thin, solid lines*), and lowest-order cyclicity (*thin, dashed lines*)

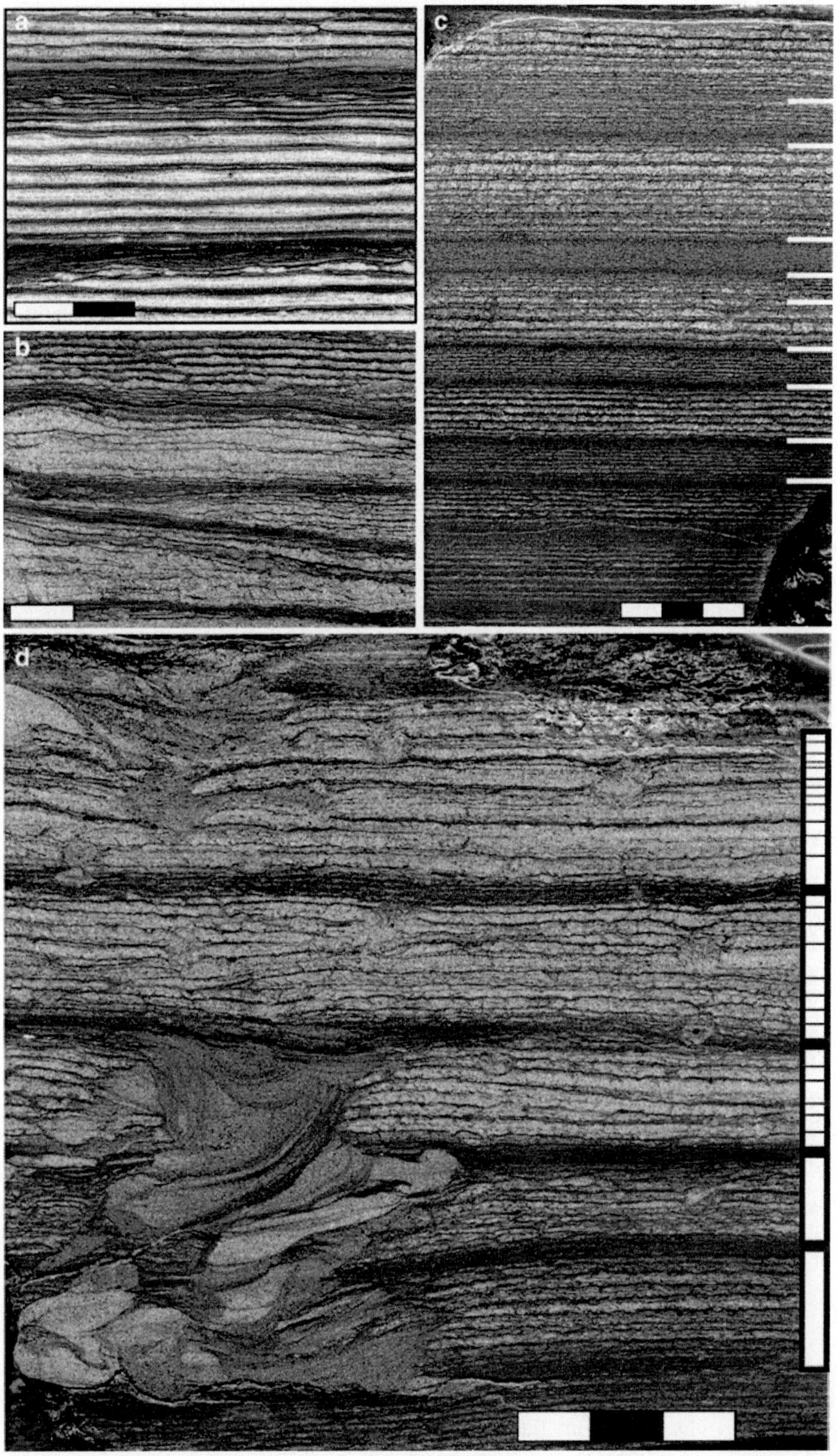
a
b
c
d

16.3.1 Shorter-Term Tidal Cycles

Within the rhythmites and texturally-banded facies discussed herein, a variety of tidal cycles have been described. At the finest scale, banded facies exhibit well-defined lamina or beds that were formed during the subdaily to daily (semidiurnal to diurnal) rise and fall of lunar tides. Pairing of a thicker lamina with an overlying thinner lamina is common (Fig. 16.6a) in the EIB (e.g. Kvale et al. 1989).

Tidal lamina pairing, termed "doublets" or "couplets" (Kvale et al. 1989; Kvale and Archer 1991). Occurrence of couplets will be best developed within a tidal system that exhibits a mixed, predominantly semidiurnal regime. This type of tidal system has a well-developed diurnal inequality such that a higher-high and lower-high tide occur each tidal day. The higher-high tide produces a thicker lamina than the lower-high tide, resulting in a laminae couplet. Predominantly semidiurnal systems are generally not able to produce couplets because each high tide is essentially of the same height. Preservation of successive couplets indicates that the original sediments were deposited within a setting that had a strong asymmetry between flood- and ebb-tidal velocities, as occurs in many modern tidal settings. Statistical techniques have been proposed that can determine if the couplets are statistically significant (De Boer et al. 1989; Tessier 1993) and thus support a tidal-depositional interpretation. Detailed analyses of ripple-scale features in Pennsylvanian rhythmites from such units indicate that any depositional effects of the subordinate tide, which could be either the flood or the ebb tide, are much reduced (Kvale and Archer 1991).

In addition to simple two-part couplets, rhythmites have been described that are actually composed of pairs of couplets. Such complex rhythmites, termed four-part rhythmites (Archer et al. 1995a), preserve both ebb- and flood deposition in a twice daily (semidiurnal) tidal system.

Perhaps the most common and prominent periodicity in texturally banded, cyclic rhythmites is the neap-spring cycle (Fig. 16.6a, 16.7c, d). This cycle relates to changes in lunar phase as observed from the earth. During new or full moon (syzygy), a linear alignment within the earth-moon-sun system results in higher, or spring, tides. The synodic month is the duration of one lunar orbit around the earth. There are two periods of spring tides separated by periods of lower, neap tides. The moon also undergoes changes in declination relative to the earth and this is the "tropical" period (see Kvale and Archer 1991; Archer 1996a). Extraction and delineation of tropical periodicities require a rhythmite that has essentially continuous preservation of all tidal events, which are relatively uncommon. A number of such near-continuous cycles, however, has been documented in Pennsylvanian rhythmites of the EIB and WIB (Kvale et al. 1989; Archer 1996a).

Another prominent shorter-term tidal cycle includes changes in lunar distance from the earth as related to the varying eccentricity of the lunar orbit. During lunar apogee, the moon is farther from the earth. During perigee, the moon is significantly closer to the earth. When lunar perigee closely corresponds to new or full moon (syzygy), tidal ranges can be considerable amplified. Conversely, during the preceding or following lunar apogee the neap-spring tides are significantly reduced. Apogean-perigean periods can be very distinctive in vertically accreted tidal facies, particularly when they are in phase with neap-spring periods. Gray shale facies in the EIB can exhibit this combined effect (Fig. 16.7c).

Many examples of Proterozoic and Phanerozoic rhythmites also exhibit these combined periodicities (e.g. Archer 1996a). It is probably not unusual that such combined cycles occur within many laminae-thickness series. In an analysis of modern tides, Wood (1986) pointed out that the co-occurrence of spring tides and perigee would result in unusually high tides, which he termed "perigean-spring tides." Because of higher tidal velocities, rhythmites deposited during perigean-spring tides could be thicker and more rapidly accreted. A combination of these factors could greatly increase the preservational potential of rhythmites. Although, apogee-perigee cycles have been interpreted in many ancient rhythmites (Archer 1996a), there are few modern examples. One example has recently been documented from tidal flats in Turnagain Arm, Alaska, a hypertidal estuary (Greb and Archer 2006).

16.3.2 Longer-Term Cycles in Similar-Appearing Rhythmites

A sample of rhythmites can appear to contain neap-spring cycles, but simple similarity doesn't preclude that the observed cycles might represent an entirely different magnitude of periodicity. As an example, within the cycles described in the gray-shale facies described above, some of the rhythmites appear to contain several "magnitudes" of periods (Fig. 16.7c). First of all, there appears to be a series of well-developed neap-spring cycles. Further examination indicates that the thickness of successive neap-spring cycles varies. A thicker neap-spring cycle is overlain by a thinner neap-spring cycle. The presence of these two co-existing periods, a result of the differential alignment of synodic and anomalistic months, represents an apogee-perigee cycle.

Not all thickening and thinning tidal laminae bundles represent "neap-spring" periods. Care should always be taken to test apparent neap-spring bundling with alternative hypotheses in mind. Figure 16.6a illustrates "pinstripe bedding" produced by very thin sand streaks separated by mud drapes. As the sand streaks thin, the drapes begin to merge and a condensed zone of amalgamated drapes, or "dark band" occurs. If the mud deposition was sequentially getting thinner, a thinner interval of amalgamated drapes would be expected. In this case, the total thickness of the dark band appears anomalously thick, or "over developed." The dark bands could represent a period significantly longer than a week-long period of neap tides. Also, some of the thicker sand streaks exhibit smaller-scale sub-laminae. Thus, even the sand-rich zones show evidence that the sand streaks might be related to multiple, rather than single, events. If more than 7 days events are recorded in the inferred spring- or neap-part of a tidal laminae bundle, it might not represent a neap-spring periodicity. It may still be tidal, but represent longer-term depositional cycles.

Figure 16.7d exhibits a number of apparent "neap-spring" bundles from a mine in southwestern Illinois. There is extensive and localized bioturbation, apparently by a burrowing sea anemone (the trace fossil *Conostichnus*). If an individual lamina is interpreted as the product of daily tidal events, then the cm-scale cycles would represent "neap-spring" cycles. It seems unlikely that a burrowing organism, such as a sea anemone, might have been able to tolerate such rapid rates of deposition. In fact, most Pennsylvanian cyclic rhythmites exhibit very little to no bioturbation. An alternative interpretation, based upon bioturbation, would be that the apparent "neap-spring" cycles are actually a longer-duration periodicity. The prominent "dark band" shows some similarities with similar features in other rhythmites (Fig. 16.7a). This band might suggest that the "neap-spring" cycles within these pinstripe bedding might be yearly cycles, related to seasonal precipitation (Kvale et al. 1994) or some other type of regularly recurring variability. In addition, seasonal variations in fluvial input of freshwater can dramatically affect the location of the inland tidal limit. Thus, a particular depositional setting might oscillate between a fluvial to an estuarine system during the course of a year, and produce a repeated succession of tidal laminae that perhaps mimicked neap-spring cyclicity, but actually represented stacked spring tidal bundles that were separated by longer hiatus.

Detailed viewing of potentially "yearly bands" in the example indicates that some of the sand streaks appear to contain several, very thin mud drapes. If these streaks were the product of a semidiurnal or diurnal tidal event, a single, simple drape would have been preserved. Similarly, (Greb and Archer 1995, 1998) noted these apparent "neap-spring" cycles. Multiple drapes are difficult to explain and would seemingly require multiple sedimentary events followed by stillstands. The sand streaks may represent an entire, but condensed neap-spring cycle (or longer duration). In this case, the causality of the interior drapes becomes easier to understand. Bundling in a heterolithic channel facies sometimes was documenting longer annual or seasonal sedimentation periods. In two Pennsylvanian examples from Kentucky, complex internal draping within the sand lamina of "spring-like" bundles suggested more than 7 days of spring-tide sedimentation. Also, in the channel fill shown in Fig. 16.5, shaly intervals between the sandy "spring" bundle were relatively thick or "over-developed." Such rhythmites can easily be misinterpreted as neap-spring cycles if internal details of lamination and bedding are not tested to ensure that inferred clay-draped laminae or laminae couplets likely represent 1-day's tidal events, rather than the amalgamation of multiple very thin events into a single bed or lamina.

16.3.3 Salinity of Late Paleozoic Tidal Systems

A potential conceptual problem exists with the term tidal when interpreting rhythmite or texturally banded facies. To many, it suggests marine conditions or at least a setting predominantly influenced by marine energies and elevated salinities. In some modern settings, however, tidal influences can extend well into freshwater settings. Low salinities can greatly influence biological activity and patterns of sedimentation.

In hypertidal systems, tidal energies can propagate for considerable distances into embayments and up into fluvial settings. Many tidal bores consist mostly of freshwater that is being pushed back upstream during the initial flood tide. The high-energy conditions related to bore passage can result in significant amounts of erosion as well as inland transport and deposition of suspended sediment. The notion of the potential scale and importance of freshwater tidal systems can be understood by ongoing research on modern analogs in the lower reaches of the Amazon River system (Archer 2004, 2005). The titanic outflow from the Amazon mouth results in a freshwater cap that extends for approximately 200 km out onto the Atlantic Ocean continental shelf. No saline-water intrusion has ever been documented in Amazon River waters and saline-freshwater mixing occurs at considerable distance from the coastline. On the upstream end, tidal influences can be measured more than 1,000 km inland from the coast. The combined result is a 1,200-km long and 300-km wide, fresh-water tidal system (Archer 2005). Obviously, the Amazon is a mega-end-member depositional system and has the highest freshwater flux in the world. For the Pennsylvanian basins discussed herein, however, it may be a useful modern analog both in terms of scale and in terms of the potential for the lateral, inland extent of a vast freshwater tidal zone (Archer and Greb 1995).

At several localities in the EIB and WIB, tidal rhythmites encase in situ lycopod trunks. These trees rooted in the upper part of the peat and represent non-peat accumulating forested wetlands that succeeded the underlying peat-forming wetlands, which formed the coal as the water table rose. Lycopods were apparently restricted to freshwater settings (Habib and Groth 1967; Phillips and DiMichele 1992). That these lycopods were encased in rhythmites indicates either the transgression of a heterolithic tidal flat with brackish to marine water that killed the trees, or if the trees remained living for some time during initial burial, freshwater tidal conditions. Based upon geochemical, petrographic, and sedimentologic evidence, Kvale and Mastalerz (1998) determined that rhythmites above some EIB coals were formed within freshwater settings.

16.3.4 Influences on Late Paleozoic Tidal Ranges

During the Pennsylvanian Period the earth had a number of unique features that could have influenced tidal modulation in the EIB and WIB. The basins under discussion were close to the paleoequator and would have had tropical climates (Heckel 1986). Widespread coals, especially in the upper Middle Pennsylvanian (Desmoinesian), indicate the periodic establishment of vast tropical rainforests within, and sometimes extending between the EIB and WIB (e.g. Greb et al. 2003). The tremendous size of the landmass of Pangea may have resulted in the common occurrence of mega paleorivers (Potter 1978; Archer and Greb 1995). The relatively flat topography of the interior Pennsylvanian basins meant that large areas could have been influenced by water-level changes, both from the seaward and the landward end. Also, a global paleo-ocean spanned most of the planet during the Paleozoic Era and unusual tidal resonances could have occurred (Archer 1996b).

16.3.5 Late Paleozoic Glaciations

Another factor that could have exerted major controls on rhythmite deposition and preservation was glacio-eustacy. Vast areas of Gondwanaland were affected by continental glaciation and deglaciation, which was manifested by high-frequency and high-magnitude glacio-eustatic cycles (Heckel 1986, 1994). Successive alternation of Pennsylvanian facies have long been interpreted as resulting from glacial-eustacy (e.g., Wanless and Weller 1932). Perhaps the resulting oscillations in global sealevels created a potential for strong, basinal resonances during specific periods of time. Extensive paleovalleys were incised during periods of sea level lowstand, and inferred glaciation. During transgression, these would have been backfilled and converted to estuaries. Funnel-shaped estuaries within low-relief basins with strong, basinal resonances would favor tidal conditions and local hypertidal systems, resulting in a variety of tidal facies with locally well-developed tidal rhythmites in which multiple orders of Pennsylvanian tidal cyclicity were preserved.

16.4 Summary

Vertically accreted rhythmites, in which small-scale tidal cycles are preserved, are common in the Pennsylvanian Period coal basins of the central U.S. Repetitive cycles of textural banding or cyclic rhythmites are preserved in tidal and estuarine channel facies, heterolithic tidal-flat facies, and gray-shale wedge facies along major paleochannels. Laminae bundling within these facies preserves several orders or frequencies of tidal periodicity, including semi- and diurnal flood and ebb of the tides, diurnal inequality of the tidal system, synodic tidal periods, and apogean-perigean effects. At a somewhat large scale, annual cycles also appear to be common.

A unique combination of concurrent processes may have resulted in the widespread deposition and preservation of tidal facies. These factors include: (1) rapid, high-magnitude changes in global paleosealevel, (2), occurrence of strongly resonant depositional embayments within the sedimentary basins (3) conversion of large tropical Pangean rivers into landward-funneling estuaries during glacial meltdown and subsequent sealevel rise.

On a more local scale, rhythmite preservation could be related to rapid generation of accommodation space by: (1) peat compaction, (2) basinal faulting and subsidence, and (3) avulsion of tidal and fluvial channels. These were likely common in both basins during the Pennsylvanian.

References

Anderson KH, Wells JS (1968) Forest city Basin of Missouri, Kansas, Nebraska, and Iowa. Am Assoc Pet Geol Bull 52:264–281

Archer AW (1996a) Reliability of lunar orbital periods extracted from ancient cyclic tidal rhythmites. Earth Planet Sci Lett 141:1–10

Archer AW (1996b) Panthalassa: paleotidal resonance and a global paleocean seiche. Paleoceanography 11:625–632

Archer AW (2004) Recurring assemblages of biogenic and physical sedimentary structures in modern and ancient extreme macrotidal estuaries. J Coast Res 43:4–22

Archer AW (2005) Review of Amazonian depositional systems. Spec Publ Int Assoc Sediment 35:17–39

Archer AW (2008) Cyclic sedimentation (cyclothem). In: Gornitz V (ed) Encyclopedia of paleoclimatology and ancient environments. Springer, Dordrecht, pp 226–228

Archer HR, Feldman AW, Archer AW, Feldman HR (1995a) Incised valleys and estuarine facies of the Douglas Group (Virgilian): implications for similar Pennsylvanian sequences in the U.S. Mid-Continent. In: Hyne N (ed) Sequence stratigraphy of the mid-Continent. Tulsa Geological Society, Tulsa, pp 119–140

Archer AW, Greb SF (1995) An Amazon-scale drainage in the early Pennsylvanian of Central North America. J Geol 103:611–628

Archer AW, Kvale EP (1993) Origin of gray-shale lithofacies ("clastic wedges") in U.S. midcontinental coal measures (Pennsylvanian): an alternative explanation. In: Cobb JC, Cecil B (eds) Modern and ancient coal-forming environments, Geological Society of America, special paper 286. Geological Society of America, Boulder, pp 181–192

Archer AW, Maples CG (1984) Pennsylvanian nonmarine trace-fossil assemblages: southwestern Indiana. J Paleontol 58:448–466

Archer AW, Lanier WP, Feldman HR (1994) Stratigraphy and depositional history within incised-paleovalley and related facies, Douglas Group (Missourian/Virgilian; Upper Carboniferous) of Kansas, U.S.A. In: Dalrymple RW et al (eds) Incised-valley systems: origin and sedimentary sequences, SEPM Special Publication 51. SEPM, Tulsa, pp 176–190

Archer AW, Kuecher GJ, Kvale EP (1995b) The role of tidal-velocity asymmetries in the deposition of silty tidal rhythmites (Carboniferous, Eastern Interior Coal Basin). J Sediment Res A65:408–416

Bass NW (1934) Origin of Bartlesville shoestring sands, Greenwood and Butler Counties, Kansas. AAPG Bull 18:1313–1345

Bass NW (1936) Origin of the shoestring sands of Greenwood and Butler Counties, Kansas. State Geol Surv Kansas Univ Bull 23:135 p

Dalrymple RW, Makino Y (1989) Description and genesis of tidal bedding in the Cobequid Bay-Salmon River estuary, Bay of Fundy, Canada. In: Taira A, Masuda F (eds) Sedimentary facies of the active plate margin. Terra Publishing, Tokyo, pp 151–177

De Boer PL, Oost AP, Visser JJ (1989) The diurnal inequality of the tide as a parameter for recognizing tidal influences. J Sediment Petrol 59:912–921

Feldman HR, Gibling MR, Archer AW, Wightman WG, Lanier WP (1995) Stratigraphic architecture of the Tonganoxie paleovalley fill (Lower Virgilian) in northeastern Kansas. AAPG Bull 79:1019–1043

Gluskoter HJ, Hopkins ME (1970) Distribution of sulfur in Illinois coals. In: Smith WH et al (eds) Depositional environments in parts of the Carbondale formation – Western and northern Illinois, Illinois Geological Survey Guidebook Series 8. Illinois State Geological Survey, Urbana, pp 89–95

Greb SF, Archer AW (1995) Rhythmic sedimentation in a mixed tide and wave deposit, eastern Kentucky, U.S.A. J Sediment Res B65:96–106

Greb SF, Archer AW (1998) Annual sedimentation cycles in rhythmites of Carboniferous tidal channels. In: Tidalities: processes and products, SEPM Special Publication 61. SEPM, Tulsa, pp 75–83

Greb SF, Archer AW (2006) Apogee-perigee cycles preserved in the fluvio-estuarine transition in Turnagain Arm, Alaska; implications for ancient tidal rhythmites (abs). Geol Soc Am Abs Progr 38(7):185
Greb SF, Williams DA, Williamson AD (1992) Geology and stratigraphy of the Western Kentucky coal field. Kentucky Geol Surv Ser 11 Bull 2:77 p
Greb SF, Eble CF, Chesnut DR Jr (2002) Comparison of the Eastern and Western Kentucky Coal Fields, U.S.A—Why are coal distribution patterns and sulfur contents so different in these coal fields? Int J Coal Geol 50:89–118
Greb SF, Andrews WM, Eble CF, DiMichele W, Cecil CB, Hower JC (2003) Desmoinesian coal beds of the Eastern interior and surrounding basins: the largest tropical peat mires in earth history. In: Chan MA, Archer AW (eds) Extreme depositional environments: mega-end members in Geologic time, Geological Society of America, special publication 370. Geological Society of America, Boulder, pp 127–150
Habib D, Groth PKH (1967) Paleoecology of migrating carboniferous peat environments. Palaeogeogr Palaeoclimatol Palaeoecol 3:185–195
Heckel PH (1977) Origin of phosphatic black shale facies in Pennsylvanian cyclothems of mid-continent North America. AAPG Bull 61:1045–1068
Heckel PH (1986) Sea-level curve for Pennsylvanian eustatic marine transgressive-regressive depositional cycles along the Midcontinent outcrop belt, North America. Geology 14:330–334
Heckel PH (1994) Evaluation of evidence for glacio-eustatic control over marine Pennsylvanian cyclothems in North America and consideration of possible tectonic effects. SEPM Concept Sedimentol Paleontol 4:65–87
Heidlauf DT, Hsui AT, Klein GD (1986) Tectonic subsidence analysis of the Illinois Basin. J Geol 94:779–794
Kuecher GM, Woodland BG, Broadhurst FM (1990) Evidence of deposition from individual tides and of tidal cycles from the Francis Creek Shale (host rock to the Mazon Creek Biota, Westphalian D (Pennsylvanian), northeastern Illinois). Sediment Geol 68:211–221
Kvale EP, Archer AW (1990) Tidal deposits associated with low-sulfur coals, Brazil Fm. (Lower Pennsylvanian), Indiana. J Sediment Petrol 60:563–574
Kvale EP, Mastalerz M (1998) Evidence of ancient freshwater tidal deposits. In: Alexander C et al (eds) Tidalities: processes and products, Society of Sedimentary Geology (SEPM) Special Publication, 61. SEPM, Tulsa, pp 95–107
Kvale EP, Archer AW, Johnson HR (1989) Daily, monthly, and yearly tidal cycles within laminated siltstones of the Mansfield Formation (Pennsylvanian) of Indiana. Geology 17:365–368
Kvale EP, Archer AW (1991) Characteristics of two Pennsylvanian-age semidiurnal tidal deposits in the Illinois Basin, U.S.A. In: Smith DG (ed) Clastic tidal sedimentology, Mem. Canadian Society of Petroleum Geologists 16. Canadian Society of Petroleum Geologists, Calgary, pp 179–188
Kvale EP, Fraser GS, Archer AW, Zawistoski A, Kemp N, McGough P (1994) Evidence of seasonal precipitation in Pennsylvanian sediments of the Illinois Basin. Geology 22:331–334
Lanier WP, Feldman HR, Archer AW (1993) Tidally modulated sedimentation in a fluvial to estuarine transition, Douglas Group, Missourian-Virgilian, Kansas. J Sediment Petrol 63:860–873
Lee W (1943) The stratigraphy and structural development of the Forest City Basin in Kansas. Kansas Geol Soc Bull 51:142 p
Maples CG, Archer AW (1987) Trace-fossil holotypes from the freshwater Hindostan Whetstone Beds (Late Carboniferous) of Indiana. J Paleontol 61:890–897
Moore RC (1935) Stratigraphic classification of the Pennsylvanian rocks of Kansas. Kansas Geol Surv Bull 22:256 p
Moore RC (1964) Paleoecological aspects of Kansas Pennsylvanian and Permian cyclothems. In: Merriam DF (ed) Symposium on cyclic sedimentation. Kansas Geol Surv Bull 169(1):287–380
Moore RC, Frye JC, Jewett JM, Wallace L, O'Connor HG (1951) The Kansas rock column. Kansas Geol Surv Bull 89:132 p
Phillips TL, DiMichele WA (1992) Comparative ecology and life-history biology of arborescent lycopsides in Late carboniferous swamps of Euramerica. Ann Mo Bot Gard 79:560–588
Potter PE (1962) Regional distributions patterns of the Pennsylvanian sandstones in the Illinois Basin. Am Assoc Pet Geol 46:1890–1911
Potter PE (1978) Significance and origin of big rivers. J Geol 86:13–33
Potter PE, Desborough GA (1965) Pre-Pennsylvanian Evansville paleovalley and Caseyville (Pennsylvanian) sedimentation in the Illinois basin. Illinois State Geol Surv Circ 384:16 p
Reineck H-E, Wunderlich F (1968) Classification and origin of flaser and lenticular bedding. Sedimentology 11:99–104
Sedimentation Seminar (1978) Sedimentology of the Kyrock Sandstone (Pennsylvanian) in the Brownsville paleovalley, Edmonson and Hart Counties, Kentucky. Kentucky Geol Surv Ser X Rep Investigat 21
Sloss LL (1963) Sequences in the cratonic interior of North America. Geol Soc Am Bull 74:94–114
Soderberg RK, Keller GR (1981) Geophysical evidence for deep basin in western Kentucky. Am Assoc Pet Geol 65:226–234
Tessier B (1993) Upper intertidal rhythmites in the Mont-Saint-Michel Bay (NW France): perspectives for paleoreconstruction. Mar Geol 110:355–367
Tessier B, Archer AW, Lanier WP, Feldman HR (1995) Comparison of ancient tidal rhythmites (Carboniferous of Kansas, U.S.A.) with modern analogs (the Bay of Fundy and Mont-Saint-Michel, N.W. France). In: Proceedings of the 1992 Tidal Deposition Conference, Special Publications of International Association of Sedimentologist 24, pp 259–271
Wanless HR (1964) Local regional factors in Pennsylvanian cyclic sedimentation. In: Merriam DF(ed) Symposium on cyclic sedimentation. Kansas Geol Surv Bull 169(2):593–606
Wanless HR (1975) The Appalachian region. In: McKee ED, Crosby EJ (eds) Paleotectonic investigations of the Pennsylvanian System in the United States. U.S. Geological Survey Professional Paper 853-C, 62 p
Wanless HR, Wright CR (1978) Paleoenvironmental maps of Pennsylvanian rocks, Illinois basin and northern midcontinent region. Geol Soc Am MC-23, 32 p
Wanless HR, Weller JM (1932) Correlation and extent of Pennsylvanian cyclothems. GSA Bull 43:1177–1206
Weller JM (1930) Cyclical sedimentation of the Pennsylvanian period and its significance. J Geol 38:97–135
Weller JM (1931) The conception of cyclical sedimentation during the Pennsylvanian period. Ill State Geol Surv Bull 60:163–177
Wood FJ (1986) Tidal dynamics. Reidel, Dordrecht

Tidal Deposits of the Campanian Western Interior Seaway, Wyoming, Utah and Colorado, USA

17

Ronald J. Steel, Piret Plink-Bjorklund, and Jennifer Aschoff

Abstract

The large-scale effects of tidal waves entering the Cretaceous Western Interior Seaway from the Gulf of Mexico have previously been modeled, but the field evidence for tides in the Campanian succession has never been assembled. Tidal deposits in deltaic, estuarine and barrier-lagoon successions along the southwestern margin of the seaway, in Utah, Colorado and Wyoming are documented. Tidal currents dominated the distal, subaqueous segments of many regressive deltaic transects (setting 1), and tidal influence was strong during the transgressive backstepping (setting 2) of shorelines. Marked tidal influence in setting 2 was likely due to increased tidal constriction and coastline irregularity after valley incision as well as possible tidal resonance with the increase of shelf width accompanying sea-level rise. In the regressive deltaic setting the common basinward cross-shelf trend from wave- to tide-dominated probably resulted from tidal amplification as sea level fell (albeit few tens of meters). The seaway narrowed and possibly became restricted to the north during lowstand periods, enhancing the counter-clockwise, Coriolis-driven current gyre in the southern half of the basin. In addition, there is notable increase in tidal influence along all of the 77.5–75 Ma shorelines, irrespective of sea-level stand. These more embayed shorelines (contrasting with straight wave shorelines before and after) are likely due to irregular widespread shallowing around embryonic, subaqueous basement-involved topography, as the seascape adjusted to a slight basinward tilt (as opposed to the earlier backtilt of the foreland basin) and a much more irregular, shallow bathymetry during the Sevier-Laramide transition.

17.1 Background

17.1.1 Historical Recognition

There has been a significant change of view regarding the influence of tides in the Late Cretaceous Western Interior Seaway (WIS) (Fig. 17.1) of North America since Shaw (1964) advocated 'tideless' epeiric seas. Thanks to improved recognition criteria, new field data

R.J. Steel (✉)
Department of Geological Sciences,
University of Texas – Austin, Austin, TX 78712, USA
e-mail: rsteel@mail.utexas.edu

P. Plink-Bjorklund • J. Aschoff
Department of Geology and Geologic Engineering,
Colorado School of Mines, Golden, CO, USA
e-mail: pplink@mines.edu; jaschoff@mines.edu

R.A. Davis, Jr. and R.W. Dalrymple (eds.), *Principles of Tidal Sedimentology*,
DOI 10.1007/978-94-007-0123-6_17,

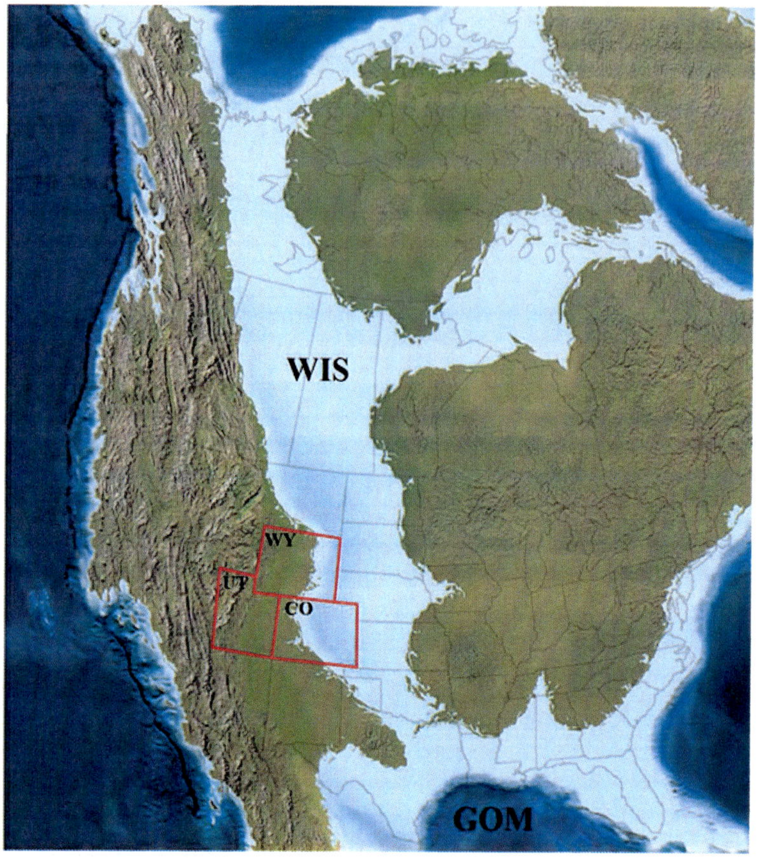

Fig. 17.1 The Western Interior Seaway at 75 Ma as portrayed by Blakey, NAU Geology. The study areas of Utah, Colorado and Wyoming are highlighted. The map here would be typical of the wide seaway at sea-level highstand. During the Campanian there were repeated, high-frequency (several 100 ky) changes in the width (100s of km) of the basin, with the western margin lowstand shorelines positioned as far east as the eastern borders of Wyoming and Colorado. Note the narrow seaway opening to the south, imposing general microtidal conditions in the basin

have demonstrated ample tidal influence or even tidal dominance along some Campanian coastlines of this seaway. As early as mid-1960s tidal signals were well documented within Campanian strata of S Wyoming and N Colorado (Weimer 1966; Masters 1966). By late 1970s there was a more general awareness of the importance of tides in shelf seas (Klein and Ryer 1978). By late 1980s-early 1990s tidal estuarine deposits (e.g., Rahmani 1988), and strong tidal influence in transgressive strata (e.g., Cross 1988) were well known. In a key paper Devine (1991) argued for a reinterpretation of the uppermost parts (commonly interpreted as upper shoreface or distributary channel deposits) of some Campanian regressive units. He argued that they are commonly transgressive estuarine deposits, because of the common occurrence of capping units of muddy lagoonal deposits overlain by thick (up to 15 m) channelized deposits showing evidence of transport by flood-oriented tidal currents. Through the 1990s, tidal deposits in the region of the Book Cliffs of Utah had been well documented (Van Wagoner 1991; Kirschbaum and Hettinger 1998). At this time there was some reluctance to accept in principle that regressive deltas could be tide-dominated (e.g., Walker 1992; Bhattacharya and Walker 1992) though later such designation became accepted (Bhattacharya and Willis 2001). The study region had become important for more intensive research on tidal deposits at this time, notably on three themes: (1) the recognition of significant tidal influence on units such as Shannon Sandstone (Suter and Clifton 1999, Bergman and Sneddon 1999) or Sego Sandstone (Van Wagoner 1991) and the ensuing debate about whether tidal units such as Sego Sandstone occupied mainly estuarine incised valleys (e.g., Van Wagoner 1991; Wood 2004), or represented deltas with large-scale tidal scouring, not necessarily related to base-level changes (Willis and Gabel 2001, 2003), (2) the importance of tidal deposits (within short-lived marine incursions) for correlation of marine into coeval nonmarine strata (e.g., the classic study of Shanley et al. 1992, as well as that of McLaurin and Steel 2000; but see also Yoshida et al. 2001), and (3) the proposal that the tide-dominated deltas of the lower Haystack Mountains Fm. of S. Wyoming were the falling-stage to lowstand basinal equivalents (albeit southerly deflected) of wave-dominated highstand shorelines of the WIS (Mellere and Steel 1995a, b, 2000; Hampson 2010).

17.1.2 Tectonic Setting

The Western Interior Seaway occupied a Cretaceous retroarc foreland basin that extended at times from Gulf of Mexico to the Arctic Ocean, on the eastern flank of the Sevier fold-and-thrust belt. The fill of the basin is characteristically asymmetric, a westward-thickening Cretaceous sediment wedge. However, from the point of view of the landscape and its tidal influence, there was a key change in tectonic setting around 77.5 Ma. At about this time flexural loading by the thin-skinned Sevier thrust-belt was being irregularly replaced in places by basement-involved, steep Laramide faulting and block uplifts, thought to be related to sub-lithosphericic loading and cooling induced by a shallowing of subducted oceanic (Farallon) plate (Liu and Nummedal 2004). Cumulative westward backtilting and thickening of strata in the basin was being replaced by an irregular eastward sloping landscape, but with a marked shallowing of water locally above Laramide uplifts. As discussed below, this tectonic shift that varied slightly in its timing across the region, caused a major change in the morphology of the WIS western shorelines, from an abundance of straight, wave-dominated coasts to more highly embayed, tide-influenced coasts, at least within the interval 77.5 through 75 Ma. (Aschoff and Steel in press).

17.1.3 Modelling of WIS Tides

Early opinion about the role of tides in the WIS was that co-oscillating tides entering the epeiric WIS platform from the southern oceanic area (Fig. 17.1) could not propagate large distances within the seaway because of rapid attenuation of tidal wave energy (Keulegan and Krumbein 1949; Shaw 1964; Irwin 1965). However, as empirical evidence accumulated from facies data, an opposing viewpoint emerged. This new viewpoint was that ancient, shallow epeiric seas and shelves, where there is often a correlation between tidal amplitudes, tidal current velocities and shelf width (Redfield 1958), should have been dominated by tidal action and were therefore an ideal setting for tides and preservation of tidal deposits (Klein and Ryer 1978). Today there is general agreement that tides can become significant even in large epicontinental seas due to localized funneling of tidal currents, local or sub-regional shoaling, tidal resonance and the effects of Coriolis acceleration and amphidromic circulation patterns in large seaways (Dalrymple 2010).

The present study area covers only the southwestern reaches of the Campanian Western Interior Seaway, a large epicontinental sea which covered much of the foreland basin east of the Cordilleran fold-and-thrust belt (Fig. 17.1). Of main relevance for the study area would have been the southerly incoming tide from the Gulf of Mexico. The shallowness of the Campanian seaway as well as its restricted oceanic opening to the south (Fig. 17.1) would predictably have dampened tides and reduced the tidal range in the southern reaches of the seaway (see also Wells et al. 2010). A modeling of storm and tidal conditions within the entire seaway was made by Ericksen and Slingerland (1990), on the basis of paleogeography and paleobathymetry (provided largely by Kauffman 1984), as well as tides and storm-wind stresses. Later modeling by Slingerland et al. (1996) and by Slingerland and Keen (1999) demonstrated that the normal surface circulation pattern in the seaway was likely a counterclockwise gyre, and that added storm conditions enhanced this to produce dominantly southward-directed currents along the western side of WIS. These models produced important results, suggesting (1) that the seaway would have been dominated by winter storm (passing west to east) and hurricane (running northwards) conditions that caused southerly longshore drift of sediment along the studied west coast of the seaway, and southerly geostrophic currents on the shelf, and (2) that the overall tidal regime is likely to have been microtidal (0–2 m tidal range) on the southwestern seaway coastlines, though on the southeastern side of the seaway there is likely to have been a meso- to macrotidal regime.

17.1.4 Modelling Results and Field Data

Although we have no evidence to suggest a regional tidal range greater than a microtidal one, it is clear that local shoreline morphology varied greatly (not least due to syn-sedimentary faulting, as emphasized by Martinsen 2003b, c), with ample evidence of tidal currents locally strong enough to transport medium-grained sand in simple dunes, compound dunes and bars.

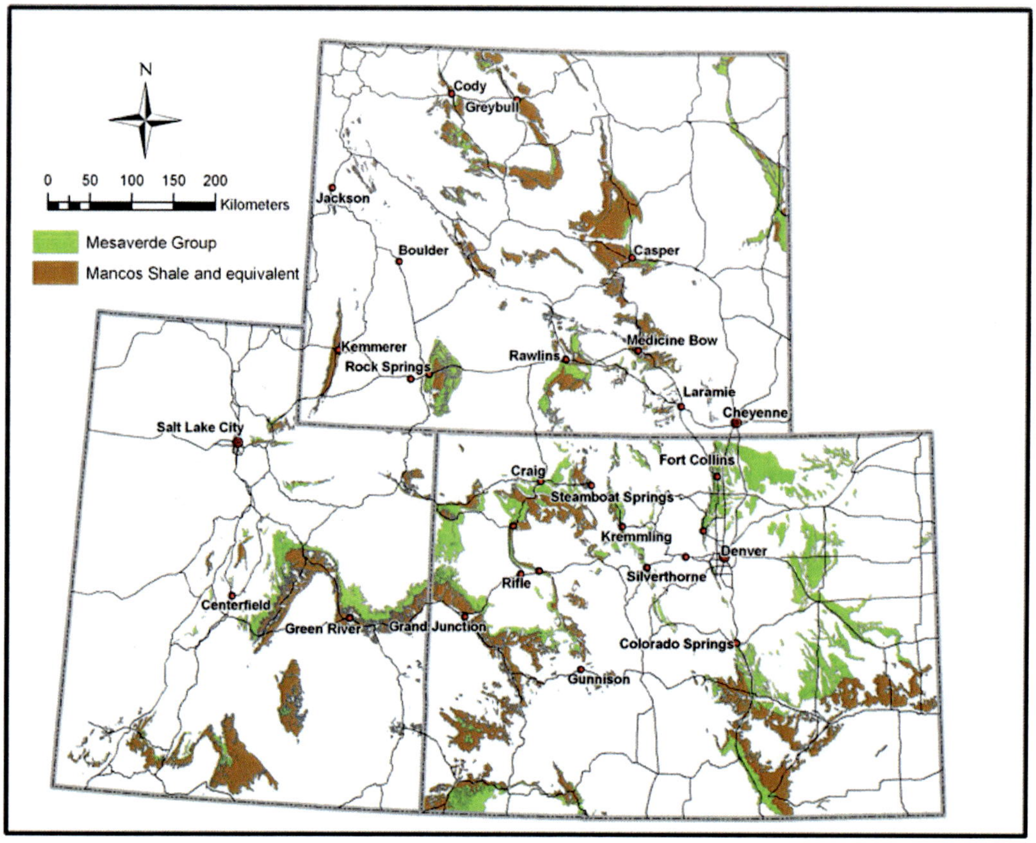

Fig. 17.2 Map showing outcrops of Mesaverde Group and associated Mancos Shale (Compiled after Green (1992) (Colorado), Green and Drouillard (1994) (Wyoming), and Hintze and others (2000) (Utah)). Map was assembled by J. Leva-Lopez.

At the same time, field evidence of storm-wave conditions is clear and prevalent through much of the southwestern WIS stratigraphy (Hampson 2010), as predicted by the modeling, though it is important to note that such conditions applied particularly to the western, highstand shorelines, rather than the lowstand ones. The WIS modeling described above was generalized for only a single Campanian scenario. The dominant Campanian stratigraphic theme outlined in this paper was one characterized by a rapidly changing regressive and transgressive paleogeography, landscape and bathymetry. The frequency (few 100 ky) and extent (repeated cross-shelf shoreline migrations of 100s of kms) of these changes caused the seaway to change its shape significantly and relatively rapidly. This involved equally frequent changes in shoreline morphology from regressive deltas and strandplains to transgressive estuaries and barrier-lagoon systems. In addition came basinwide coastline morphology changes that accompanied the Sevier-Laramide tectonic transition. The scenarios emerging from modeling studies account well for basin-scale processes, for the highstand, storm-wave dominated shorelines, and for the now well-documented southerly deflection of sediment dispersal that appears to be preferential at the tips of clastic wedges and sub-wedges. However, the models were less able to reproduce the significant environmental changes over the short time scales of the regressive-transgressive cycles, and in particular the tidal amplification that accompanied the frequently changing width of the seaway during these cycles, or that resulting from the changing Sevier-to-Laramide tectonic setting of the basin.

17.1.5 Study Area and Objectives

Figures 17.2 and 17.3 show the outcropping Campanian Mesaverde Group in Wyoming, Colorado and Utah, and the stratigraphy of the successions being considered.

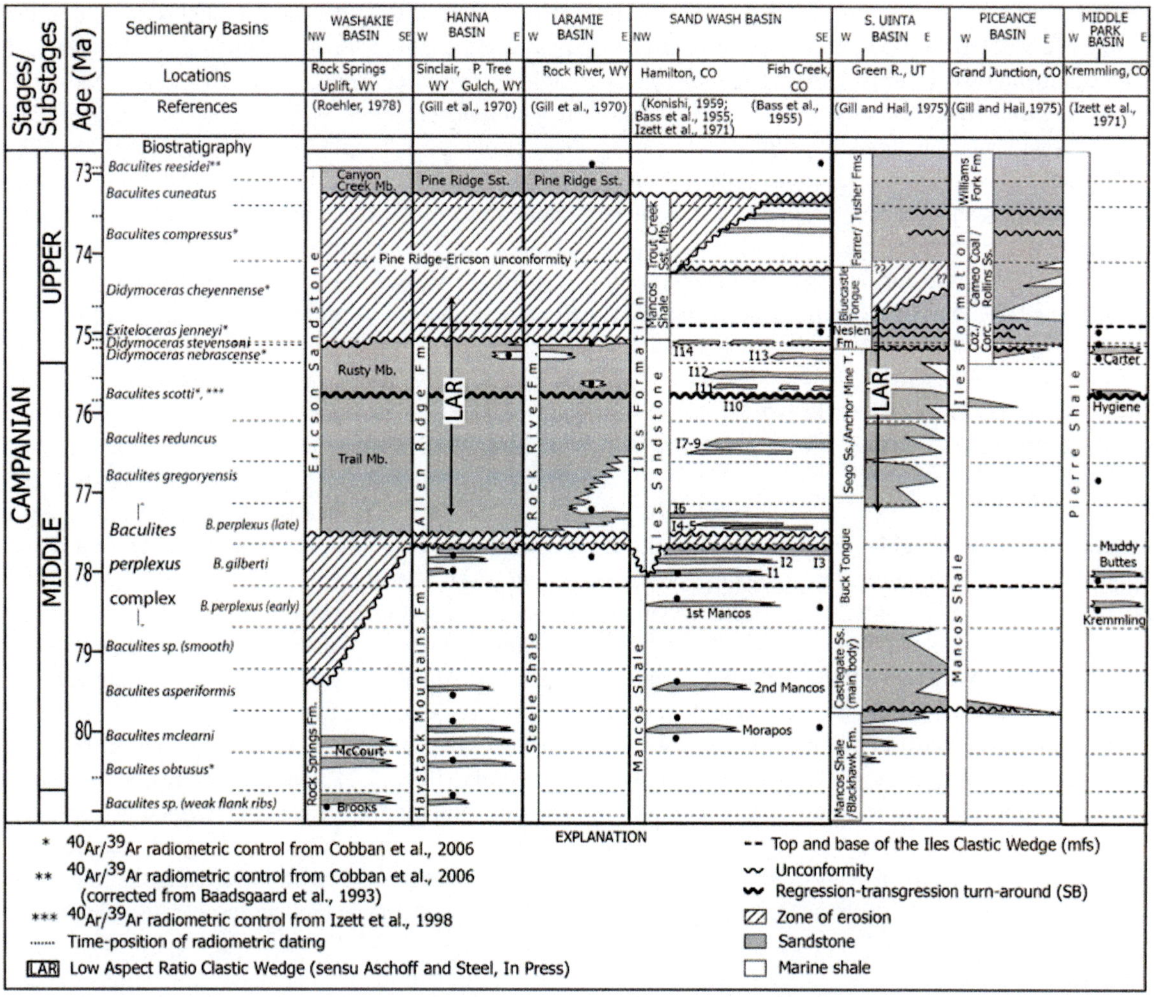

Fig. 17.3 Stratigraphic data across N. Utah, S. Wyoming and N. Colorado showing clastic wedges prograding progressively basinwards through Lower to Upper Campanian. The settings discussed in this paper where there are well-developed tidal deposits (see also Fig. 17.18) are (1) many of the most basinal, regressive sands, on the distal fringes of clastic wedges (e.g., Haystack Mountains tongues in Hanna Basin; Morapos and Mancos sandstones in Sand Wash Basin; Kremmling, Muddy Buttes, Hygiene and Carter sandstones in Middle Park Basin), (2) the transgressive systems tracts capping almost all the marine shoreline sands. In addition, there is an unusual degree of tidal influence throughout the 77.5–75 Ma low-accommodation interval (see LAR interval marked on chart). The Iles sandstones associate with both marine shales and with coaly coastal plain deposits in the Sand Wash Basin column (Diagram modified from Gomez-Veroiza and Steel 2010). *LAR* (low-aspect ratio) is terminology of Aschoff and Steel (in press)

The treatment of tidal records is not exhaustive, but is rather problem-oriented, emphasizing that there are two main categories of tidal deposits, regressive (mainly in a lowstand position in the basin) and transgressive, and there is a generally strong tidal influence (all sea level stands) within the time interval 77.5–75 Ma. We organize the discussion of tidal deposits as follows:

1. Tidal sandbodies of *Setting 1*, representing *regressive*, distally situated, tide-dominated, subaqueous marine deltas, or deltas variably reworked by tidal currents in the seaway. These regressive occurrences occur at or near the progradational limits of many clastic wedges that built from the Sevier fold-and-thrust belt into the Western Interior Seaway (Fig. 17.3). Because of their spatial position, far into the basin at the extremities of 80–100 km shoreline transits, they are likely to be falling-stage and lowstand sandbodies, occurring near the regressive-transgressive turnaround of the wedges or sub-wedges.

2. Tidal deposits of *Setting 2*, occurring frequently in time and space within the entire *transgressive* parts of Campanian sequences. They occur most commonly as thin (m-scale) transgressive caps to regressive strata, but sometimes accumulate to greater thicknesses, especially where incised valleys or other high-accommodation areas developed and became infilled. An increase in the thickness and extent of preserved tidal deposits during transgression is a well-established phenomenon (Cattaneo and Steel 2003).

17.2 Criteria for Recognizing Tidal Deposits in the Western Interior Seaway

General criteria for recognizing tidal signals in recent and ancient strata are relatively well known (De Raaf and Boersma 1971; Nio and Yang 1991; Dalrymple 1992, 2010; Fenies et al. 1999), and have also been well used in the WIS (e.g., Rahmani 1988; Shanley et al. 1992). However, we choose here to look at tidal criteria within the context of sub-environments within coastal depositional systems. Following the tidal environmental subdivisions of Dalrymple and Choi (2007) we summarize criteria from the key environments (Fig. 17.4). Tidal signals in regressive and transgressive strata are not greatly different, though tidal bars and compound dunes tend to be somewhat more thickly developed in estuaries, and some facies successions (e.g., compound dunes fining upwards into tidal flat and supratidal muds) are more common in estuaries, whereas in deltas compound dunes tend to cap upward-thickening parasequences.

17.2.1 Tidal Criteria: Campanian Delta-Front and Distributary-Mouth Bars

Many of the most sand-rich Campanian tidal deposits are interpreted to occur within the subaqueous distributary mouth bars and delta-front reaches (Fig. 17.4b) of tide-influenced or tide-dominated deltas, e.g., in the Morapos Sandstone of N. Colorado (Hampson et al. 2008a), Blair Sandstone (Devlin et al. 1993; Martinsen 2003c), O'Brien Springs, Seminoe and Hatfield 1 sandstones of S. Wyoming (Mellere and Steel 1995b; Martinsen 2003c) and Sego Sandstone in N. Utah (Willis and Gabel 2001). The grain size and sand-rich nature of these tidal deposits implies significant current energy, but there are also thin mudstones and organic-matter concentrations. The key criterion is the presence of thick (>5 m), stacked, well-ordered sets of planar or trough cross stratification (Figs. 17.5 and 17.6) (see also Willis 2005; Dalrymple 2010), commonly in 'blocky' or upward-coarsening and thickening regressive successions. Individual sets of cross strata that are sited seaward of the distributary mouth commonly have a sigmoidal geometry, with evidence of landward-directed (flood tide) currents (Fig. 17.5) in addition to seaward-directed paleocurrents. Willis and Gabel (2003) documented upward-coarsening, delta-front bedsets in Lower Sego Sandstone that are relatively steeply inclined (5–15°) and 6–12 m thick. The commomly sharp erosional base of the Sego bedsets was interpreted in terms of tidal current scour within a tidal-channel mouth-bar system (Willis and Gabel 2001, 2003).

Thin muddy or organic drapes are common in most tidal cross-stratal foresets, because of the frequency of slack-water periods in tidal settings (Fig. 17.6). Other reliable criteria are double mud drapes on foresets and bottomsets (Fig. 17.7) indicative of two slack-water periods bounding the weaker of the flood or ebb tidal current intervals (Fenies et al. 1999) and thick-to-thin foreset bundling along the transport length of the dune set (Fig. 17.8), suggesting spring-neap tidal bundling (Dalrymple and Choi 2007).

In addition to the tidal signals within single dunes, it is a characteristic feature of WIS upper delta-front sandbodies that individual dunes combine to form compound dunes (see also Dalrymple 2010), within which both ebb and flood tidal currents can be recorded (Dalrymple and Rhodes 1995) (Fig. 17.9). Although compound dunes occur also in some fluvial systems (Collinson 1970), the tidal cases tend to show evidence of bi-directional paleocurrents, as well as being associated with tide-influenced mouth bars and transgressive estuarine channels and bars (as in the Chimney Rock Sandstone of N. Utah; Plink-Bjorklund 2008).

ManyWIS tide-dominated, delta-front sandbodies additionally show a characteristic occurrence of more extensive (>100 m) but very thin (mm to cm) mudstone beds, that occur between the bedsets and sets of cross-stratified and ripple-laminated sandstone, (see also Willis 2005), seen well in the O'Brien Springs Member of Hackstack Mountains Formation (Fig. 17.10). Such extensive mud drapes and layers may sometimes

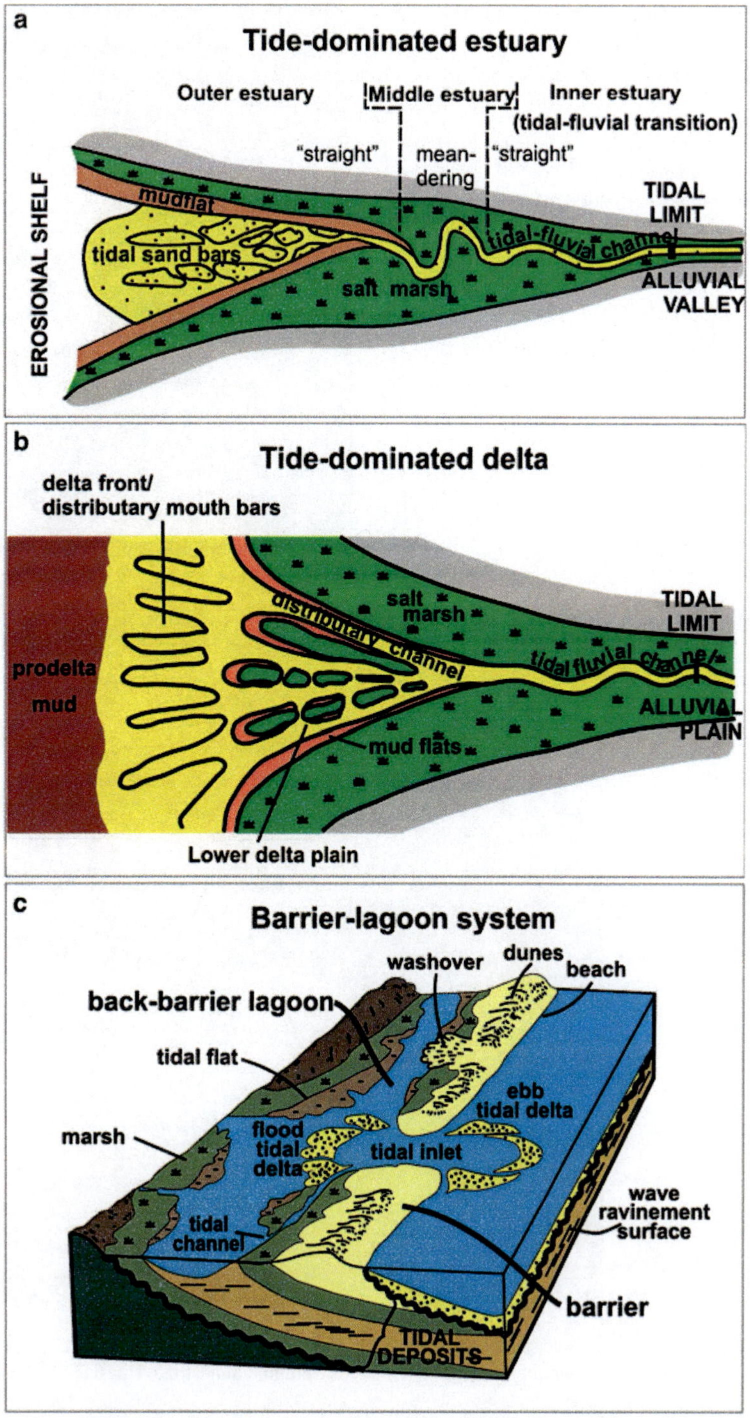

Fig. 17.4 Reference diagrams for tide-dominated and tide-influenced environments and sub-environments used throughout the text section 'Criteria for recognizing tidal deposits in the Western Interior succession' (Compiled from Dalrymple and Choi 2007, and from Reinson 1992)

originate from slack water during a single tidal cycle, but more commonly from intervals of spring tides, or spring tides enhanced by river floods, when stronger currents carry much mud in suspension, or from longer-term seasonal causes. In cases where the mudstone beds on the tide-dominated delta front are unlaminated and unbioturbated, and are more than a cm thick, they are likely to represent rapidly deposited fluid muds

Fig. 17.5 Thickly stacked (11 m) sets of mainly planar cross-strata within a coarsening and thickening-upward, tide-dominated, delta-front unit in Seminoe-1 Sandstone, Haystack Mountains Formation, near Sinclair, Wyoming. Many of the dunes were west and northwest migrating (landward) (Photo courtesy S. Ahmed)

Fig. 17.6 Very thin mudstone or organic drapes on tangential cross-stratal foresets are a common tidal signal. There are also thin ripple-laminated sets within the larger cross-strata. Seminoe Sandstone, Haystack Mountains Formation, south of Rawlins, Wyoming

brought out from the tidal distributaries (Ichaso and Dalrymple 2009). Such mudstone layers, in alternation with medium-grained sandstone beds, create a characteristic grain-size bi-modality within the delta-front facies succession (Fig. 17.10), contrasting greatly with the more homogeneous grain-size character of storm wave-dominated delta fronts, that are much more common in WIS shoreline successions. Another feature of the prodelta and lower delta front deposits of such tide-dominated deltas is a somewhat stressed ichnofacies

Fig. 17.7 Repeated double mud drapes on 40 cm-high, cross-stratal foresets record semi-diurnal tidal cycles. Ripped-up mud clasts are also very common. Note the rippled cap of the large set, as well as ripple-lamination within the large set. The latter are frequently reverse-flow tidal ripples and not flow separation-eddy ripples that would be confined to a lowermost foreset position. Sego Sandstone, Jim Canyon, Colorado

Fig. 17.8 Alternations of thicker and thinner foresets (possible spring-neap tide bundling) within a 40 cm high, sigmoidal cross-set recording migration of 2-D dune, Sego Sandstone, Jim Canyon, Colorado

assemblage, typical of areas near a river mouth where salinity is lowered and sedimentation rates relatively high. In the lower delta-front to prodelta reaches of WIS cases in S Wyoming, bioturbated very fine sandstones and siltstones with *Teichichnus, Ophiomorpha, Cylindrichnus* and *Rosselia* are recorded (Mellere and Steel 2000). In the more distal reaches of the same deltas in N Colorado, Hampson et al. (2008a) additionally recorded *Schaubcylindrichnus, Paleophycus and Conichnus* traces.

Fig. 17.9 Thickly stacked sets of cross strata, some with markedly bi-directional paleocurrents (eastward and westward), in an upper delta-front unit in upper O'Brien Springs Sandstone, Haystack Mountains Formation, S. Wyoming (scale is 15 cm). The inclined accretional arrangement of the sets of cross strata suggest a compound dune. As suggested by Dalrymple and Choi (2007), such occurrences of well-developed herring-bone cross-strata are likely to occur mainly in compound dunes, or on a crestal location in an elongate tidal bar

Fig. 17.10 Thick (55 m) upward-coarsening and thickening succession of tide-dominated, delta-front deposits in O'Brien Springs Sandstone, Haystack Mountains Formation, near Rawlins, Wyoming. The tidal signals include some of the thin (1–2 cm) but pervasive mudstone layers separating fine to medium-grained sandstone sets, as well as sigmoidal sets in the upper third of the succession

17.2.2 Tidal Criteria: Campanian Backbarrier Lagoonal Deposits

Transgressive back-barrier successions are dominated by lagoonal, tidal channel, flood-tidal delta and tidal-inlet deposits (Fig. 17.4c). Lagoonal deposits are recognized as coaly mudstones and siltstones with thin coal and sandstone beds. *Planolites*, *Paleophycus*, root traces, and more occasionally *Arenicolites*, *Teichichnus* and *Ophiomorpha* occur in the lagoonal deposits (e.g. Bullimore et al. 2008, in the Trout Creek Sandstone of N Colorado). The lagoonal mudstones and siltstones together with the flood-tidal delta and tidal inlet deposits occur as coarsening-upward, landward-accreting strata (see also Kamola and Van Wagoner 1995). The landward-accreting flood-tidal delta deposits have a strike extent of 50–100 m (Kamola and Van Wagoner 1995) with a downdip length of 100s to 1,000m. Flood-tidal deposits in Chimney Rock Sandstone of N Utah (Plink-Bjorklund 2008) consist of dominantly

Fig. 17.11 Middle Castlegate fluvial-tidal transition succession along Willow Creek, Utah. The lower third of the 35 m succession shows two fluvial-tidal channel belts with multiple heterolithic point bars containing turtle fragments and *Teredolites*-bored logs. The recessive interval just below center contains a thick, muddy intertidal-flat succession (wavy and lenticular beds) passing up to organic-rich, supratidal salt-marsh deposits (likely to be estuarine due to association with bayhead delta deposits off photo to right). Short cliffs above the center contain marine trace fossils. This succession is mapped to connect some 70 km basinward to the Sego Sandstone and the transgressive Anchor Mine Tongue of the Mancos Shale. The Bluecastle Tongue forms the upper cliffs

landward-directed, ripple- and climbing-ripple laminated, trough cross-stratified, plane-parallel laminated sandstones with mud drapes and double mud drapes. They also show flaser to wavy bedding, and some cross-strata with bidirectional paleocurrent directions. Tidal inlet deposits in the same Utah succession occur as intervals of erosionally-based bidirectional cross-stratified, low-angle planar cross-stratified, plane-parallel laminated and current-ripple laminated sandstones with occasional mud drapes and multiple erosion surfaces lined by shell debris and clay clasts. In places bed tops are wave reworked and bioturbated by *Planolites*, *Ophiomorpha*, *Diplicraterion* or *Rhizocorallium*.

17.2.3 Tidal Criteria: Campanian Fluvial Tidal Transition (Inner to Mid Estuary) Zones

Inner estuary deposits, particularly from tide-dominated estuaries, are dominated by channel fills (Fig. 17.4a) that tend to change from being relatively sandy at the down-dip margin of the inner estuary to markedly heterolithic and sinuous in the high suspended-sediment concentration zone, before becoming sandy again into the landward river system where few, if any, tidal structures occur (Dalrymple and Choi 2007). Some of the best known Campanian successions that contain deposits of this most proximal estuarine zone include the strata seaward of the fluvial parts of Ericson Formation, and landward of the Twenty-Mile, Trout Creek and Iles marine shorelines in SW Wyoming and N. Colorado (Fig. 17.3) (Gomez-Veroiza and Steel 2010), the middle Castlegate Sandstone and Neslen Formation that lie landward of the Sego, Corcoran and Cozzette shoreline deposits in Utah and western Colorado (Fig. 17.3) (Aschoff and Steel in press), and the Chimney Rock Sandstone near the Utah-Wyoming border (Plink-Bjorklund 2008).

The most easily recognizable tidal signals in these deposits occur at the seaward end of the fluvial-tidal transition area, where point bars and inclined heterolithic strata (Thomas et al. 1987; Rahmani 1988) develop along the sinuous reaches of distributary and tidal channels. Good examples of such inclined heterolithic point bars containing brackish-water trace fossils and pervasive, muddy, slack-water drapes occur in Middle Castlegate (Fig. 17.11) and Neslen (Fig. 17.12)

Fig. 17.12 Examples of tidally influenced fluvial channels in the Neslen Formation, Floy Basin, Utah. The inclined heterolithic strata that fill the channels (4 m high in **a**, 10 m high in **b**) contain brackish-water trace fossils in places, and abundant tidally deposited mud drapes and layers

strata of N. Utah and W. Colorado. The tidal signals are often most clear in the upper parts of some of these sinuous channel units as thick, intertidal flaser, wavy and lenticular strata, with their capping supratidal mudstones and coaly salt marsh deposits (Figs. 17.11 and 17.12). In the more proximal reaches of the fluvial-tidal transition, coarser fluvial deposits with the wood-borer *Teredolites* can be seen to interdigitate with more heterolithic point bar strata that contain small brackish-water traces of *Teichichnus*.

The importance of recognizing tidal effects within mainly fluvial strata, in locations far landward of the open shoreline, was emphasized in the classic work of Shanley and others (1992), and has become a key method of correlating Campanian shoreline successions, via tidally-influenced estuarine incursions into Campanian coastal and alluvial plains, 10s to 100 km behind the pinch-out of coeval shoreline bodies (McLaurin and Steel 2000; Aschoff 2008; Gomez-Veroiza and Steel 2010).

17.2.4 Tidal Criteria: Campanian Middle-to-Outer Estuary Zones

Typical sites in *middle to outer estuarine zones* where tidal signals can be identified are on the upper flow-regime sand flats and associated channels of tide-dominated

Fig. 17.13 Photo of upward-coarsening bayhead delta deposits in the Neslen Formation, Coal Canyon, Utah. Tidal indicators occur primarily in the lower delta front and lagoonal deposits. Note the fluvial cap of the prograding delta succession. Height of foreground cliff is 19 m

estuaries (Fig. 17.4a), on bayhead deltas of wave-dominated estuaries, and a combination of bayhead deltas and tidal channels in mixed-energy estuaries (Dalrymple et al. 1992) (Fig. 17.4). Bayhead deltas have a scale and geometry superficially like inclined hererolithic point bars, but they coarsen upwards, often have a concave-up slope profile as opposed to slightly convex-up for point bars, build onto areas of muddy tidal channels and are capped by distributary channels. An example of a bayhead delta deposit from the lower Neslen Formation near Coal Canyon in Utah, landward of a Sego shoreline, is shown in Fig. 17.13. In this example the tidal indicators are developed in the lower bayhead-delta front, where some sigmoidal fine sandstone sets, as well as flaser and wavy-laminated heterolithics interfinger with organic-rich mudstones. The upper delta front shows very fine-grained, climbing-ripple and planar cross stratified sandstones alternating with mudstones. Sedimentary structures generally record higher-energy flow conditions near the top of the unit and lower energy conditions near the toe (Fig. 17.13).

Middle to outer estuary zones of tide-dominated estuaries typically contain upper flow-regime sandflats and channel-bar systems, with flanking mudflats and supratidal marshes (Fig. 17.4a) (Dalrymple and Choi 2007). One of the best Campanian examples of an inner-middle estuary that developed from an abandoned delta-plain during regional transgression is the Rusty Member of the Ericson Formation in the Rock Springs area of S. Wyoming (Martinsen et al. 1999; Gomez-Veroiza 2009). In this example the estuary system that replaced the underlying brackish-water, lower delta plain is characterized by channelized, clean white sands (Fig. 17.14) that show bi-directional paleocurrents and paired, organic-rich drapes on cross-strata foresets. The tidal signals are similar to those shown in Figs. 17.5–17.8, but tidal compound dunes are also prominent features in these deposits. Martinsen et al. (1999) argued and provided evidence that the broadly lenticular (km scale) white sandstone units in Fig. 17.14 occupy valleys that incised the abandoned delta plain.

Middle estuary zones of mixed-energy estuaries typically contain tidal bars in association with bayhead

Fig. 17.14 The light and rusty-colored sandstones and mudstones of the Rusty Member, Ericson Sandstone occupy the lower half of the photo. The white sandstones (third way up photo), about 15 m thick, are interpreted as inner estuary dune systems (compound dunes with bi-directional paleocurrents, Martinsen et al. 1999) occupying a broad channel or valley (white sand pinches out beyond left margin of photo). These sandstones grade up into supratidal mudstones and multiple paleosol horizons, and then to a second tidal unit of upper flow-regime sandflats. The upper third of the photo shows the unconformity-based Canyon Creek Member fluvial sandstones

deltas in the middle-estuarine zone (e.g. Dalrymple et al. 1992; Allen and Posamentier 1994; Vis 2009). These tidal bars form because tidal currents are stronger in mixed-energy estuaries, compared to wave-dominated estuaries, and thus able to rework the river-derived sediments in bayhead deltas into tidal bars.

Note that the above tidal bars are distinctly different from the outer-estuarine tidal bars described below, in that they consist of river-derived sands, muds and organic matter, in contrast with the marine sediment supplied to the outer-estuarine tidal bars. A Campanian example of a mixed-energy estuary that developed in an incised valley is the middle portion of the Chimney Rock Sandstone of the Rock Springs Formation, exposed on the Utah/Wyoming border (Plink-Bjorklund 2008). The inner-to-middle estuarine tidal bars in this interval characteristically occur as elongate sandbodies (2–7 m thick and 1,000–3,500 m long) that contain lateral accretion sets, with bidirectional cross-stratified and compound cross-stratified sandstones that contain single and double mud drapes (Fig. 17.15). These tidal bars occur landward of the central basin of the estuary and are associated with bayhead delta deposits (Fig. 17.15).

Outer estuary zones of tide-dominated estuaries are dominated by elongate tidal bars (Fig. 17.4a) of greater length and lateral extent than seen further landwards within the narrower reaches of the estuaries (Dalrymple and Choi 2007). Tidal sands in an outer-estuary setting have been identified in the Hatfield Sandstone of

Fig. 17.15 An example of tidal bars in a middle-estuary zone of a mixed-energy estuary. Note the large-scale accretion (**a** and **b**) of the tidal bars. Close-ups (**c** and **d**) show typical cross-stratified sandstones with ubiquitous mud drapes within the bars. Chimney Rock Sandstone (Modified from Plink-Bjorklund 2008)

S. Wyoming within the transgressive or back-stepping estuarine units that cap shallow-water, lowstand deltas (Mellere and Steel 2000). Compared to middle estuary tidal bars, the sandstones here are slightly coarser (medium- rather than fine-grained), the sets of cross strata are generally larger (commonly >40 cm), channels are abundant and landward-directed paleocurrents are prominent.

Another Campanian example is the upper portion of the Chimney Rock Sandstone, where the outer-estuarine tidal bars occur as erosionally based bodies, 4–7 m thick and 4–10 km long, comprising compositionally and texturally well sorted sandstones (Fig. 17.16; Plink-Bjorklund 2008). These sandstones occur as obliquely landward-accreting sets with bidirectional cross strata and sigmoidal cross strata. In contrast with inner-middle estuarine tidal deposits, these outer-estuarine tidal bars are better sorted, contain fewer mud drapes, are associated with upper-flow-regime plane-parallel laminated sandstones, and contain a few interbedded hummocky-cross stratified sandstones (Fig. 17.16).

17.3 Tidal Deposits of Setting 1: Regressive Tide-Dominated, Subaqueous Deltas Developed Preferentially in the Basinward Reaches of Campanian Clastic Wedges

Some of the best known tide-influenced and tide-dominated sandbodies in the WIS are restricted to the most 'basinal' reaches of regressive clastic-wedges as noted first by Mellere and Steel (1995b, 2000) and by Hampson (2010) (Figs. 17.17 and 17.18). Because all large-scale clastic wedges consist internally of a series of thinner tongues, and the shoreline progradations of these tongues extend progressively farther into the basin through time, the tidal sandbodies being discussed here occur relatively far west in the Early-Middle Campanian and up to hundreds of km farther east (basinwards) in the Middle-Late Campanian (Fig. 17.3). Their key characteristic is that they tend to occupy the most distal 10–50 km of the high-frequency regressive tongues, the most basinward increment of

Fig. 17.16 Example of outer-estuarine tidal bars in a tide-dominated estuarine system, Chimney Rock Sandstone. (**a**, **b**) White, clean sandstones of the outer-estuarine tidal bars are cut above by central-estuarine tidal channels and inner-estuarine tide-influenced fluvial channels. (**c**, **d**) The outer estuarine tidal bars typically consist of inclined sets of bidirectional cross strata (Modified from Plink-Bjorklund 2008)

Fig. 17.17 O'Brien Springs and overlying Seminoe sandstones, interpreted as tide-dominated deltas, Rawlins, Wyoming. These sandbodies typically occupy a near-maximum regressive position in clastic wedges, contain no capping delta-plain deposits and are overlain and underlain by shelf mudstones (photo courtesy R. Martinsen)

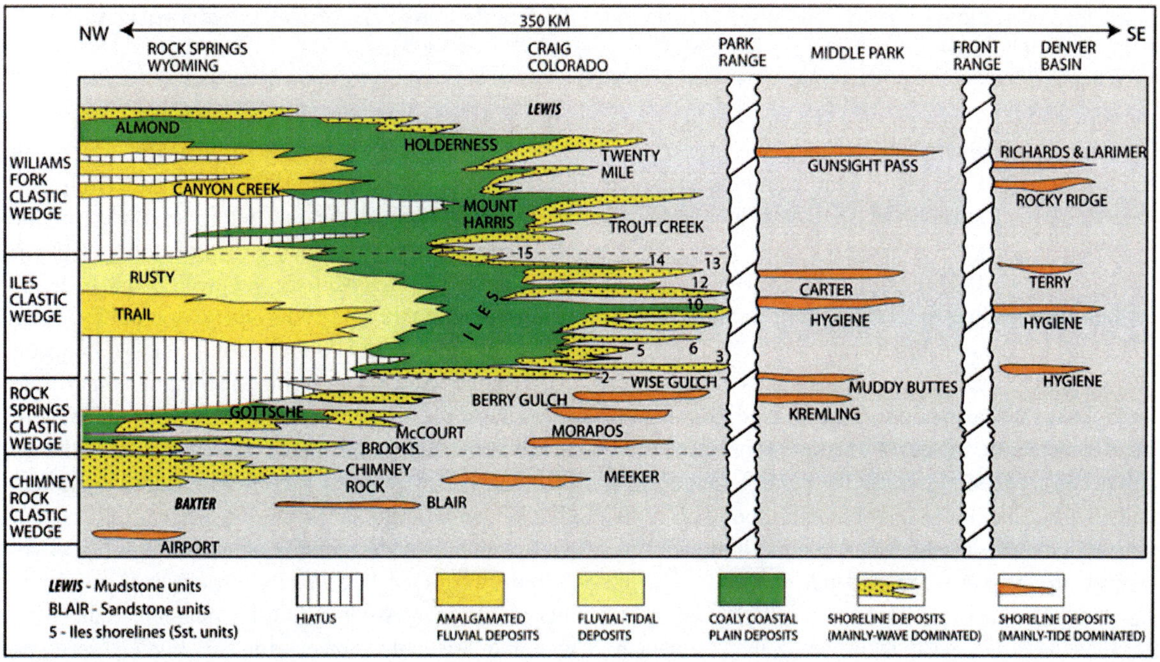

Fig. 17.18 A Campanian transect from S Wyoming through N Colorado showing a series of four clastic wedges reaching from Rock Springs Uplift to the Denver Basin (Wedges modified from Crabaugh 2001). The sandstones of Setting 1 (*in red*) include the most basinal sands, on the distal fringes of basinward-stepping clastic wedges. Note that these sands migrated eastwards through time as the main clastic wedges built basinward

regression and associated sand deposition. For this reason, the tidal deposits of Setting 1 have been referred to as falling-stage or lowstand shorelines on the shelf (Mellere and Steel 2000). This is not to say that all lowstand sands in any sequence were strongly tidally influenced; in fact, there were coeval and along-strike shoreline sands that were wave-dominated (Mellere and Steel 1995a; Hampson et al. 2008a). It is argued that relative sea level is likely to have been lowered at these times because it is difficult for comparatively small rivers and deltas to maintain regression across a slightly seaward-sloping, pre-existing shelf for such great distances (up to 200 km) without the forcing aid of falling relative sea level (Muto and Steel 1997). At some stratigraphic levels the sandstones of Setting 1 extend many tens of km basinward into the muddy seaway (e.g., into the Denver Basin). In these cases some of the deltaic sandbodies have been partly or extensively reworked (e.g., Krystinik 1995) by south-directed currents in the seaway (Martinsen 2003b). Another aspect of these sandstones is that most of them lack distributary channels so they appear to have been dominantly subaqueous in character. The Blair Sandstone is one of the few that does have a distributary channel.

Within the study region, the tidal sandstones of Setting 1 occur in the following groups:

- Fishtooth, Sussex and Shannon sandstones of the Bighorn-Powder River basins in northern Wyoming (Tillman and Martinsen 1987; Walker and Bergman 1993; Fitzsimmons 1994, 1999; Sullivan et al. 1997; Bergman 1999; Fitzsimmons and Johnson 2000)
- Tapers Ranch, O'Brien Springs, Seminoe 1–3, and Hatfield 1 sandstones in Hanna Basin of south-central Wyoming (Smith et al. 1965; Gill et al. 1970; Tillman and Martinsen 1985; McClurg 1990; Davies 1990; Roehler 1990; Mellere and Steel 1995a, 2000; Uroza 2008; Ahmed 2008)
- Hygiene, Terry, Rocky Ridge, Larimer and Richards sandstones of the Denver Basin, Colorado (Kiteley and Field 1984)
- Kremmling, Muddy Buttes, Carter, Hygiene and Gunsight Pass sandstones of the Middle Park Basin, Colorado (Izett et al. 1971; Krystinik 1995; A. Petter, 2007, personal communication; J. Crabaugh, 2007, personal communication)

- Meeker, Morapos, Berry Gulch, Wise Gulch, and Duffy Mountain sandstones of the Sand Wash Basin (Boyles and Scott 1982; Boyles 1983; Hampson et al. 2008a)
- Airport and Blair sandstones in the area of the Rock Springs Uplift, S Wyoming and Utah (Devlin et al. 1993; Martinsen et al. 1998; Roehler 1989, 1990).
- Mancos Sandstones of Uinta Basin, Utah (Pattison 2005; Hampson et al. 2008a; Hampson 2010).

The sandbodies of this list are interpreted here (on the basis of the criteria given above) as dominantly but not exclusively tidal in origin. Some of them additionally show evidence of waves, storms and other processes, and these vary along strike. In the light of the attention paid to the Shannon Sandstone, it should be noted that this type of sandstone unit occurs throughout the Campanian, from the Airport Sandstone just below the Campanian-Santonian boundary in SW Wyoming, to the Gunsight Pass Sandstone just below the Campanian-Maastrichtian boundary in NE Colorado.

17.3.1 Relationship of Setting-1 Tidal Deposits to the Campanian Clastic Wedges

The preferential distal position of the tidal sandbodies of Setting 1 are illustrated in a 350 km long, NW-SE transect through southern Wyoming and northern Colorado (Fig. 17.18). In Fig. 17.18 we propose that the Carter, Hygiene and Terry sandstones form the distal reaches of a late-middle Campanian wedge (named the Iles Clastic Wedge by Crabaugh 2001), the Gunsight Pass, Richards, Larimer and Rocky Ridge sandstones relate to an Upper Campanian wedge named here Williams Fork Clastic Wedge, and the Haystack Mountains tongues (not shown in Fig. 17.18), Muddy Buttes, Kremmling, Duffy Mountain, Wise Gulch, Berry Gulch and Morapos sandstones relate to an early-middle Campanian wedge named the Rock Springs Clastic Wedge. The lower Campanian-Santonian Blair and Airport sandstones of southernmost Wyoming and N. Utah relate to a yet older Chimney Rock Clastic Wedge. Liu and others (Liu and Nummedal 2004; Liu et al. 2005) included both Chimney Rock and Rock Springs clastic wedges in their Megasequence 3, whereas the Iles and Williams Fork clastic wedges are included in their Megasequences 4 and 5 respectively.

17.3.2 Variability of the Tidal Sandstones in Setting 1

The most detailed descriptions of these tidal sandstones are of the Shannon (Tillman and Martinsen 1987; Walker and Bergman 1993) and Sussex (Bergman 1999) sandstones in the Powder River Basin, of the O'Brien, Seminoe and Hatfield sandstones (Tillman and Martinsen 1985; Mellere and Steel 1995a, b, 2000) in the Hanna Basin, and of the Duffy Mountain (Boyles and Scott 1982) and Mancos sandstones (Hampson 2010) in the Sand Wash Basin. The Shannon Sandstone is perhaps the most infamous of this group, having been interpreted and re-interpreted as a shelf sandbody (Tillman and Martinsen 1987), a lowstand shoreface (Walker and Bergman 1993), an incised valley infill (Sullivan et al. 1997) and as estuary mouth tidal bars (Elliot 1997) (see Suter and Clifton 1999 for a summary of these interpretations). The Shannon Sandstone was an unfortunate choice as the best-known representative of these tidal sandstones as it outcrops at fairly distal locations in the system where the tidal characteristics are not always clear.

It should be emphasized, and it is clear from their debated interpretation (see Suter and Clifton 1999), that the bodies referred to here as tidal sandstones are quite variable in their character and some of them are clearly of mixed-energy (tides and waves) origin. For this reason some researchers have simply referred to them as 'isolated shallow-marine sandstones' (Hampson et al. 2008a), though these latter authors also make it clear that the dominant sandbodies are likely to be large tide-dominated deltas (as did Mellere and Steel 1995a, b, 2000). The key aspect of the bodies, justifying the term 'tidal sandstones', is the abundance of thickly stacked tidal cross strata (see above) in their proximal reaches. Not generally appreciated, is the fact that when their proximal reaches are exposed, the up-dip attachment to WIS highstand or falling-stage shorelines can occasionally be mapped (Mellere and Steel 1995a, b, 2000), so the term 'isolated' can be incorrect, an impression given when viewing distal outcrops or 2-D seismic data. The outcrops, for example, of the O'Brien Springs, Morapos and Blair sandstones are representative of middle to proximal reaches of these sandbodies, and best illustrate their tide-dominated character, though even in these examples there are also wave-generated facies evident along strike. Only the Blair Sandstone contains downcutting distributary channels, a main feature named by Suter and Clifton (1999) as missing in the Shannon Sandstone to confirm a tide-influenced estuary or delta interpretation. Because of the common lack of distributaries we interpret the middle and distal reaches of these sandbodies to be subaqueous. The outcrops of the Shannon, Kremmling and Airport sands have a more 'distal' muddy character and they can contain mixed-energy facies, though they also show limited thicknesses of tidal cross strata in their uppermost parts. The best overall appreciation of these sandbodies can be gained by viewing the distal Berry Gulch and Wise Gulch sands of N. Colorado (Hampson et al. 2008a), together with their proximal equivalents in the Haystack Mountains of S Wyoming (Mellere and Steel 2000). Hampson et al. (2008a, their Fig. 19A) proposed a large-scale reconstruction suggesting southward-deflected currents as the deltas built out into the seaway. Figure 17.19, based on facies as well as paleocurrent patterns (Uroza 2008) document this deflection, with sediment dispersal shift from eastwards to southwards for the O'Brien Springs Sandstone. Martinsen (2003c) noted that syn-sedimentary tectonics also had significant influence on coastline physiography and bathymetry through much of the Upper Cretaceous and that deltas may have been confined and protected from waves at times by structurally-generated embayments. The role played by the counter-clockwise basin circulation and southward-sweeping currents versus the effects of syn-sedimentary structure at lowstand (Martinsen 2003b) in creating the southward-elongated character of many of the sandbodies is still unclear.

17.3.3 Interpretation of the Tidal Sandbodies in Setting 1

Following the long-lasting WIS shelf-sands debate (see Suter and Clifton 1999 for summary) the sandbodies of Setting 1, when seen in their proximal as well as distal expression, have most of the attributes of falling-stage to lowstand deltaic shorelines. The bodies have muddy to rippled prodelta and lower delta-front reaches with a restricted ichnofauna, and upper delta-front dunefields that display both landward and seaward paleocurrents (Fig. 17.19), though usually dominated by the latter. The persistent basinal location of the sandstones (with respect to their westerly highstand-equivalent shorelines) is also consistent with this sequence stratigraphic interpretation. Further, the most recent researchers who brought new data to this debate (Mellere and Steel 2000; Uroza 2008; Hampson et al. 2008a; Hampson 2010) tend to agree that the tidal signals in the sandbodies, the rare storm-wave signals (though there are some associated wave-generated sandbodies) and the common medium-grained character of the sands all suggest that the sandbodies are derived from rivers in the west and northwest, and that the strong currents deflecting the sandbodies to the south (Fig. 17.19) are likely to have been tidal currents, rather than purely wave-driven longshore currents or other seaway currents. Modeling by Slingerland et al. (1996) suggested that the steady-state surface circulation pattern of the WIS was a counter-clockwise gyre, with inflowing river currents from the west deflecting to the south and in the east deflecting to the north, due to Coriolis acceleration. Slingerland and Keen (1999) further developed this model by simulating an 8-day storm superimposed on the steady-state gyre, and computed the net sediment transport on the western margin of the seaway to be dominantly southerly. However, the modeling emphasized mainly storm and wind-driven effects on the seaway, probably most relevant during sea-level highstands and wide shelf conditions. It is likely that waves would have been dampened and tidal currents accentuated in the counter-clockwise gyre during sea-level lowstand periods, as discussed below. Their regressive character indicates that most of these tidal sandbodies were deltaic, though estuaries are likely in the transgressive phases. Those farthest out in the seaway may have become sufficiently reworked so as to become transgressive shelf sand ridges (Snedden and Dalrymple 1999).

Figure 17.20 shows the southward-skewed Yangtze tide-dominated deltas, proposed by Hampson and others (2008a) as the best recent analog for the distal lowstand deltas on the western coast of the WIS. However, note that the scale of the deflected system is quite different in Figs. 17.19 and 17.20, and the cause of the coast-parallel currents need not be the same in both cases

17.3.4 Causes of Tidal Amplification in Setting 1: Lowstand Narrowing and Northward Constriction of the Seaway

Individual tidal sandbodies of Setting 1 occupy a maximum regressive position in their host sequence, i.e., they occupy the outermost 100+ km of regressive shoreline transits that were 100s of km wide. In addition they tend to occur progressively farther basinwards, as

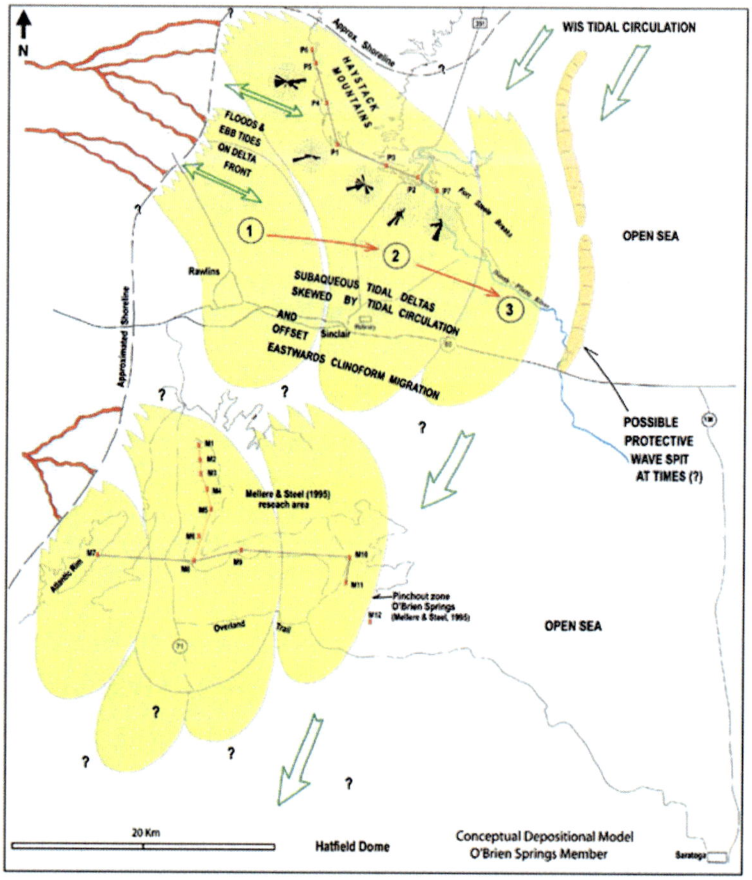

Fig. 17.19 An interpretation of the O'Brien Springs Sandstone tide-dominated deltas of Setting 1 in the Hanna Basin area of southern Wyoming (From Uroza 2008). Paleocurrent patterns (see northerly set of lobes) suggest that the shorelines initially prograded east and southeastwards and were of mixed wave-tidal energy, but were then deflected southwards by southerly directed seaway tidal currents. Lettered locations on southern lobes are from Mellere and Steel (1995b)

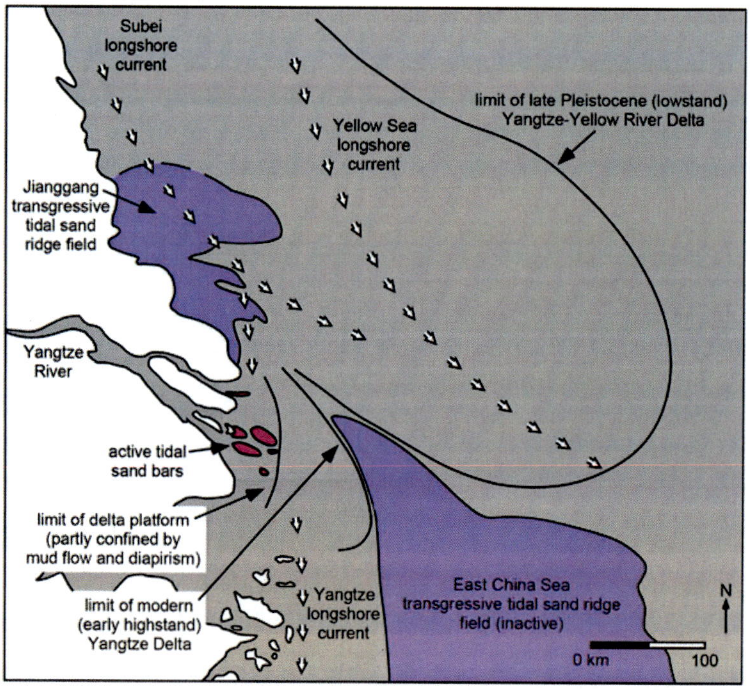

Fig. 17.20 Late Pleistocene (*lowstand*) Yangtze-Yellow River delta and modern Yangtze delta, both of which are tide dominated (Hori et al. 2002) and have an orientation reflecting southward-directed, wave-driven longshore currents and Coriolis-deflected tidal currents (Chen et al. 2000) (Diagram compiled and suggested as good analog for the WIS distal shorelines by Hampson et al. 2008a)

the larger clastic wedge on which they are perched built out irregularly into the Western Interior Seaway (Fig. 17.18). From the oldest (Blair) to the youngest (Gunsight Pass) of these Campanian sandbodies, the clastic wedges had cumulatively prograded some 600 km into the WIS during 10 My. Because these tidal sandbodies systematically changed their location in the basin (migrated southeastwards) through time, their occurrence is less likely to be related to tectonic lineaments, sediment supply pulses, subsidence rate changes or even to tidal resonance related to particular shelf widths. Their changing location but constant occurrence near the tip of high-frequency sandstone tongues rather suggests that their lowstand position in the seaway was critical. Sea-level falls of only a few tens of meters (in an already very shallow seaway), as likely in late Cretaceous Greenhouse times (see Miller et al. 2004) during the observed regressive shelf transits, would likely have caused some narrowing and northward constriction of the WIS at lowstand intervals. This scenario somewhat resembles the southern North Sea setting where there is a narrowing towards the English Channel-Straits of Dover. In this case the tide comes in as a Kelvin wave between Scotland and Norway, and Coriolis acceleration forces it to bend to the right in the direction of propagation, so that it `bunches-up' on the southeast UK coast, before turning all the way around the N Sea Basin.

17.4 Tidal Setting 2: Tidal Deposits in Transgressive Settings

Tide-influenced transgressive strata are the best known tidal accumulation in the WIS and they occur as three types of succession. *Thin transgressive tidal deposits* are widespread and well documented throughout Campanian statigraphy (Table 17.1), normally forming the linkage between successive regressive shoreline units. There has been some tendency to misinterpret these coarse-grained, transgressive estuarine or barrier-remnant sandstones on the tops of regressive shorelines as regressive distributary channels or upper shoreface units, as was early pointed out by Land (1972), Devine (1991), and Cross (1998). The latter two authors emphasized the importance of preserved tidal signals in the transgressive systems tract of sequences. *The second type of transgressive occurrence* involves the accumulation of thick, estuarine deposits within incised valleys, or at least occurring above basal fluvial erosion surfaces (Table 17.2). Estuarine incised valley fills in the WIS Campanian strata were first recognized by Van Wagoner (1991) in the Lower Sego Sandstone and by Taylor and Lowell (1991) in the Kenilworth Member of the Blackhawk Formation, as conceptual models for incised valleys were being developed. A *third type of transgressive occurrence* and the most unusual in the Campanian WIS, is a thick, vertically stacked estuarine accumulation that is not constrained to a valley, recognized only in Lower Campanian Chimney Rock clastic wedge of the Utah-Wyoming border area (Plink-Bjorklund 2008).

17.4.1 Thin, Widespread Transgressive Tidal Strata

These thin tidal accumulations cap most regressive clastic tongues and wedges throughout the Campanian succession (Fig. 17.21). There are especially many examples from the Lower Campanian succession in Utah and New Mexico, and from Upper Campanian strata of NW Colorado. The main occurrences are listed in Table 17.2. Most examples of these transgressive deposits are <5 m thick, but occasionally exceed 10 m. (Table 17.1). Figure 17.21 shows the early example where Devine (1991) clearly pointed to transgressive tidal deposits occupying a relatively thick capping to a larger-scale regressive unit, all of which could be interpreted as regressive without facies information. All of the transgressive accumulations of this type have been interpreted in terms of a back-barrier setting as lagoonal or non valley-confined estuarine accumulations with tidally influenced fluvial channel sandstones, bayhead delta and lagoonal sediments, as well as flood-tidal delta and tidal-inlet deposits.

17.4.2 Tidal Incised Valley Fills

Incised valley-fill tidal accumulations are less well known in Upper Cretaceous strata than in the classic Lower Cretaceous Dakota Formation valleys in the Front Range and Denver Basin area (Weimer 1983) and the Muddy Sandstone valleys of NE Wyoming (Weimer 1984; Martinsen 2003b). Nevertheless, Campanian incised valleys are fairly common (Table 17.2, Fig. 17.22). In most cases incision depths are 10–30 m deep and they are filled with fluvial to estuarine sediments. The valley fills are based by regional

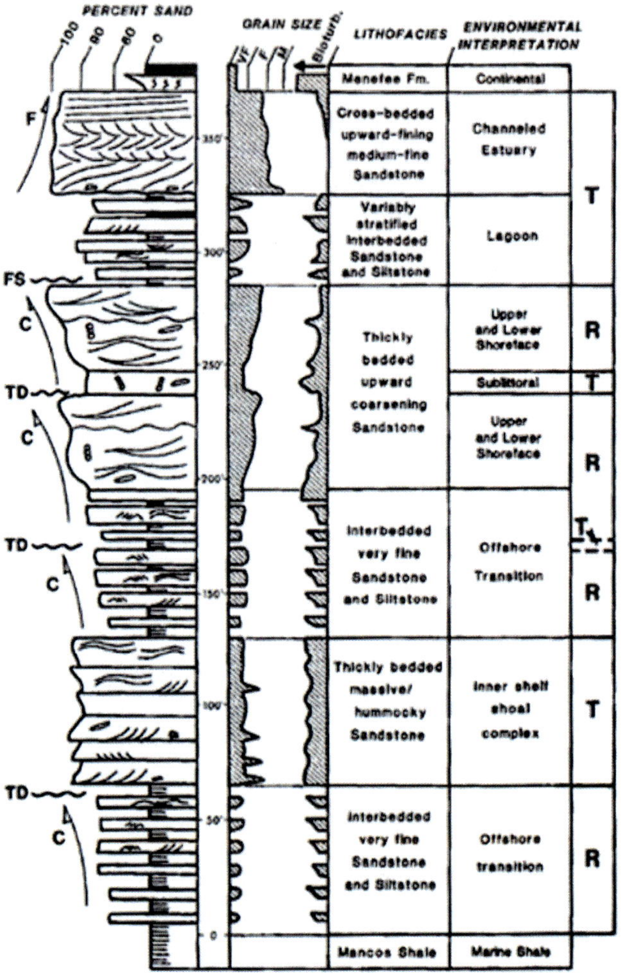

Fig. 17.21 A tendency to misinterpret transgressive tidal deposits on the tops of regressive shorelines as distributary channels or upper shoreface units in the Cretaceous Western Interior Seaway was highlighted by Devine (1991). He emphasized the common regressive (R)-transgressive (T) nature of sequences, the tide-influenced or tide-dominated nature of the transgressive deposits, and the presence of transgressive ravinement surfaces (TD)at the regressive-transgressive turnarounds. (Modified from Devine 1991)

erosion surfaces (e.g. Van Wagoner et al. 1990; Van Wagoner 1991; Plink-Bjorklund 2008) (Fig. 17.24), reflect an initial significant seaward shift of facies, where fluvial deposits occur above the valley base (e.g. Hampson et al. 2008a, b), have adjacent interfluve surfaces (e.g., O'Byrne and Flint 1995) and show onlap of infill strata onto the valley walls (e.g. Plink-Bjorklund 2008; Charvin et al. 2010). The valleys were incised during times of sea-level fall and lowstand, and are interpreted to have been filled mainly during lowstand and transgression (eg., Kenilworth Member of Blackhawk Fm and Chimney Rock Sandstone), or primarily during transgression (Desert, Grassy and Sunnyside Members of Blackhawk Fm). Valley incisions were particularly common during the 77.5–75 Myear interval as discussed below, and an example of the detail within a 20 m-thick valley-fill from Iles 3 sandstone tongue (Crabaugh 2001) is shown in Fig. 17.24.

The Campanian WIS valley fills occur as wave-dominated, mixed-energy or tide-dominated estuary deposits (*sensu* Dalrymple et al. 1992). The wave-dominated estuarine fills are similar to those in thin, non-valley-confined tidal accumulations and consist of tide-influenced fluvial, bayhead delta, central basin (lagoonal) and outer-estuarine wave-dominated barrier, tidal inlet, flood-tidal delta deposits (e.g. Yoshida 2000). The only mixed-energy estuary fill is that reported from the Chimney Rock Sandstone, which is similar to a wave-dominated estuary, except for the occurrence of inner-estuarine tidal bars in association with bayhead deltas (Plink-Bjorklund 2008; Fig. 17.23). The tide-dominated estuarine accumulations consist of tide-influenced fluvial, tidal channel, outer-estuarine tidal bar and marginal tidal flat deposits (e.g. Fitzsimmons 1994).

Two of the tidal accumulations included here with incised valley fills are associated with basal fluvial

Table 17.1 Thin, widespread transgressive tidal accumulations occurring outside incised valleys

Example	Location	Setting	Thick	Depositional environment	References
Haystack Mountains Fm.	SE Wyoming	Transgressive parts of regressive-transgressive fourth-order sequences	2–3 m	Lagoonal-estuarine deposits	Mellere and Steel (1995a, 2000)
Iles Fm. sandstones	NW Colorado	Transgressive parts of regressive-transgressive fourth-order sequences	5–7 m	Back barrier	Crabaugh (2001)
Twentymile Sandstone	NW Colorado	Transgressive parts of regressive-transgressive fourth-order sequences	3–5 m	Back barrier tidal channel	Seidler and Steel (2001)
Trout Creek Sandstone, Mount Harris Mbr. of Iles Fm.	NW Colorado	Transgressive parts of regressive-transgressive fourth-order sequences	10–15 m	Lagoonal-estuarine deposits with tidal inlet, flood-tidal delta, lagoonal and tidally-influenced fluvial deposits	Bullimore et al. (2008)
Holderness Mbr, Williams Fork Fm.	NW Colorado	Transgressive parts of regressive-transgressive, fourth-order sequences	ca 5 m	Back-barrier lagoonal with tidal channel fills	Benda (2000), Benda and Steel (2000), Seidler and Steel (2001)
Grassy Mbr. of Blackhawk Fm.	Book Cliffs, Utah	Transgressive parts of parasequences	5–7 m	Lagoonal	O'Byrne and Flint (1995)
Spring Canyon Mbr. of Blackhawk Fm.	Book Cliffs, Utah	Transgressive parts of parasequences	5–7 m	Back-barrier lagoonal with tidally influenced channel sandstones, bayhead deltas, lagoonal sediments, flood-tidal and tidal- inlet deposits	Kamola and Van Wagoner (1995)
Point Lookout Sandstone of the Menefee Fm.	San Juan basin, New Mexico	Transgressive parts of regressive-transgressive couplets	9–14 m	Lagoonal-estuarine deposits with tidal inlet, flood-tidal delta, lagoonal and tidally-influenced fluvial deposits	Devine (1991)
Panther Tongue	Book Cliffs Utah	Transgressive Cap	1 m	Lag on ravinement	Hwang and Heller (2002)

Table 17.2 Tidal accumulations overlying fluvial incision surfaces, usually within incised valleys

Example	Location	Setting	Thickness	Depositional environment	References
Holderness Mbr., Williams Fork Fm.	NW Colorado	Incised valley fill in the LST/TST of a fourth-order regressive-transgressive wedge	15–20 m	Wave-dominated estuary with flood-tidal delta, tidal inlet, bayhead delta, crevasse splay and central basin deposits	Benda (2000), Benda and Steel (2000), Seidler and Steel (2001)
Iles Fm. sandstones	NW Colorado	Incised valley fill in the LST/TST of a fourth-order sequences	10–25 m	Estuarine	Crabaugh (2001)
Lower Sego Sandstone	Book Cliffs, Utah	LST incised valley fills	10–25 m	Tide-dominated and mixed energy (wave-tide) estuaries	Van Wagoner (1990, 1991), Willis and Gabel (2001, 2003), Wood (2004)
Lower Castlegate Sandstone	Book Cliffs, Utah	TST incised valley fill	30–40 m	Wave-dominated estuary with flood-tidal delta, tidal inlet, bayhead delta, crevasse splay and central basin deposits	Yoshida (2000)
Desert Mbr., Blackhawk Fm.	Book Cliffs, Utah	TST incised valley fill	15–25 m	Wave-dominated estuary with flood-tidal delta, tidal inlet, bayhead delta, crevasse splay and central basin deposits	Yoshida (2000)
Grassy Mbr., Blackhawk Fm.	Book Cliffs, Utah	TST incised valley fill	5–15 m	Wave-dominated estuary with flood-tidal delta, tidal inlet, bayhead delta, crevasse splay and central basin deposits	O'Bryne and Flint (1995)
Sunnyside Mbr., Blackhawk Fm.	Book Cliffs, Utah	TST incised valley fill	20–25 m	Heterolithic to sandy estuarine deposits	Howell et al. (1997)
Kenilworth Mbr., Blackhawk Fm.	Book Cliffs, Utah	LST-TST incised valley fill	Up to 15–20 m	Fluvial to estuarine	Taylor and Lovell (1991, 1995), Ainsworth and Pattison (1994), Pattison (1995)
Aberdeen Mbr., Blackhawk Fm.	Book Cliffs, Utah	LST incised valley fill	15–25 m	Fluvial to estuarine	Kamola (1992), Charvin et al. (2010)
Chimney Rock Sandstone, Rock Springs Fm.	NE Utah/SW Wyoming	Late LST to TST valley fill	15–30 m	Mixed-energy estuary with bayhead delta, inner-estuarine tidal bar, central basin, flood-tidal delta and inlet deposits	Plink-Björklund (2008)
Fishtooth and Virgelle Mbrs., Eagle Fm	Bighorn Basin, Wyoming	TST incised valley fills		Heterolithic, partially wave-reworked distal estuarine, and proximal, sandy tide-dominated estuarine respectively	Fitzsimmons (1994)
Masuk Fm	Henry Mnts., SE Utah	Transgressive parts of regressive-transgressive couplets, above incised fluvial basal surface	5–25 m	Tide-influenced fluvial, lagoonal and tidal channel deposits	Birgenheier et al. (2009)
Muley Canyon Sandstone	Henry Mnts., SE Utah	Transgressive parts of regressive-transgressive fourth-order sequences, above incised fluvial basal surface	5–15 m	Tidally-influenced fluvial channel deposits and tidal channel deposits	Birgenheier et al. (2009)

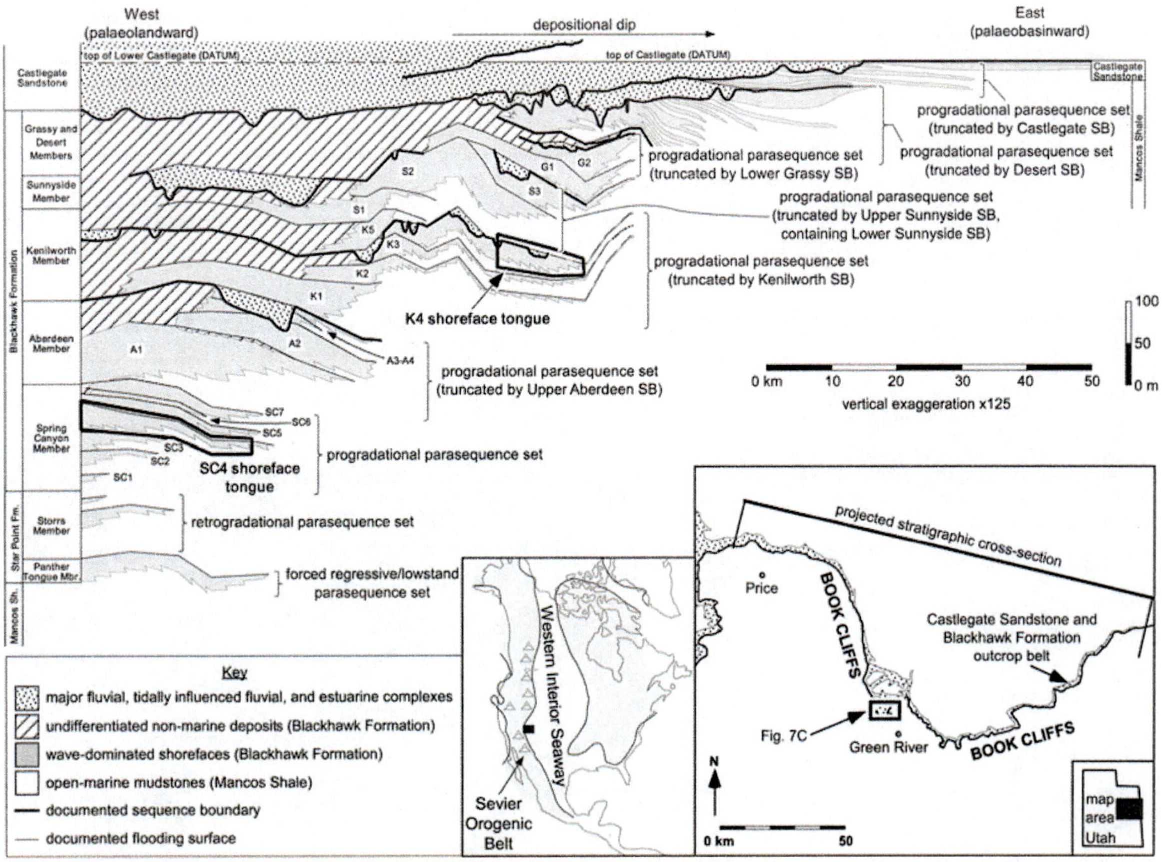

Fig. 17.22 Transgressive tidal accumulations in incised valleys are relatively common in the Western Interior Seaway succession, e.g. through the Campanian Blackhawk Formation, Book Cliffs, Utah, USA (From Hampson et al. 2008b, after Balsley 1980; Hampson and Howell 2005). Note that the valley abundance and depth increases upwards, from higher to lower accommodation settings

incisions, but the incisions have not been interpreted as an incised valley base (Birgenheier et al. 2009). These two accumulations are in the Masuk Formation and Muley Canyon Sandstone in SE Utah. These accumulations are 5–25 and 5–15 m thick respectively, and interpreted as transgressive tide-influenced fluvial, lagoonal and tidal channel deposits that overlie fluvial incisions associated with a seaward shift of facies (Birgenheier et al. 2009).

17.4.3 Thick Non-valley-confined Transgressive Tidal Accumulations

This third, transgressive tidal setting is the most uncommon, and is documented in the uppermost part of the Chimney Rock Sandstone of the Rock Springs Fm in SW Wyoming and NE Utah, where it occurs as a 60 m thick stack of estuarine deposits (Plink-Bjorklund 2008; Fig. 17.23, Table 17.3). The lowermost of the four vertically stacked tide-dominated estuarine units (units overlying TR2-5 in Fig. 17.23) is an erosional remnant and only 10 m thick, whereas the subsequent three units are 17, 24 and 22 m thick. Each of these units has a tidal ravinement surface at its base, and is overlain by a landward-stepping lower part and by an aggradational and partially (in inner- and central-estuarine zone) seaward-stepping upper part. This vertically stacked tide-dominated estuarine succession is not confined by a valley, nor is it based by a fluvial incision surface. Instead the four stacked units reflect repeated tidal ravinement, reshaping of the river mouth, and successive *in situ* infilling (Plink-Bjorklund 2008). The four vertically aggrading units consist of tide-dominated or

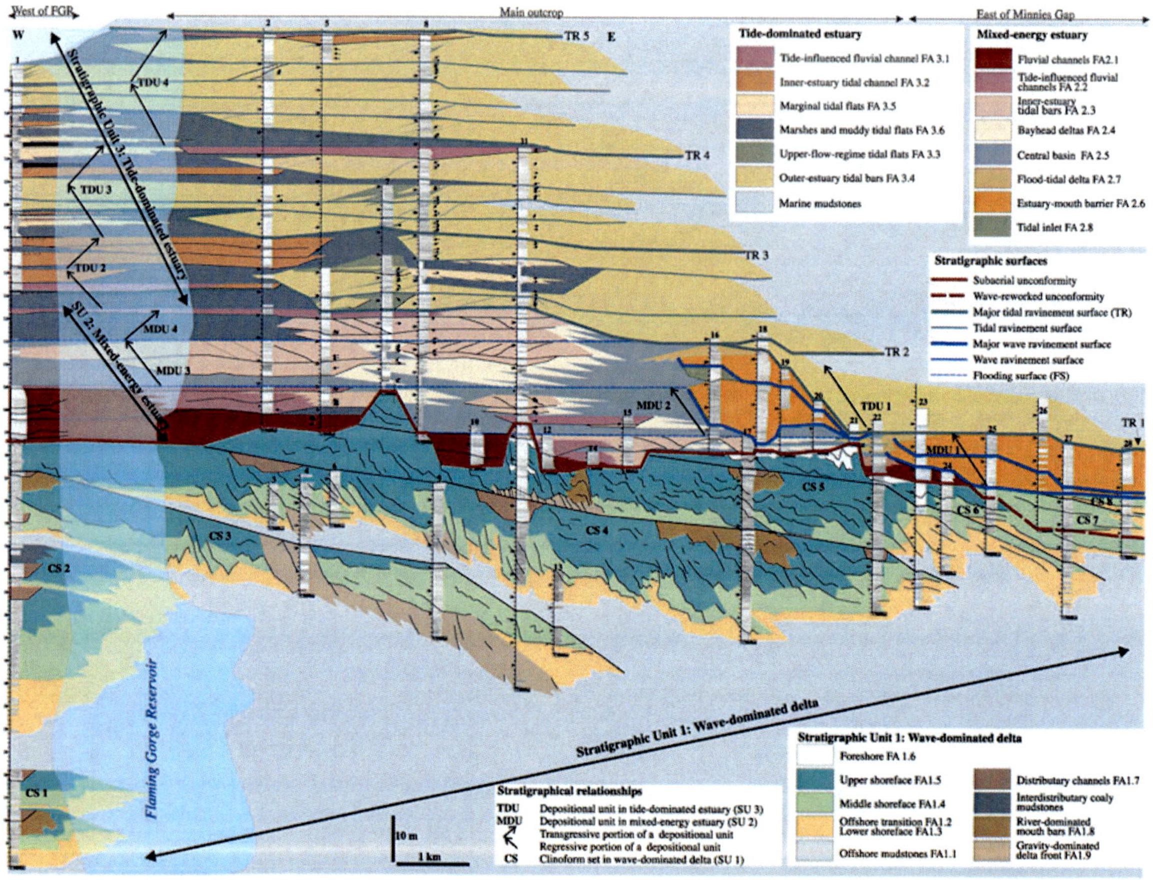

Fig. 17.23 Correlation panel across Chimney Rock Sandstone near Minnies Gap, Wyoming-Utah border. Note the great thickness of the backstepping tide-dominated estuarine sandstones stacked on four tidal ravinement surfaces (uppermost Chimney Rock Sandstone), accumulated above an incised valley fill with a mixed-energy estuary fill (middle part of the Chimney Rock Sandstone), incised into a set of regressive, river- and wave-dominated deltaic units (lowermost Chimney Rock Sandstone) (From Plink-Björklund 2008)

open-mouth estuarine deposits (*sensu* Dalrymple et al. 1992) with tide-influenced fluvial, tidal channel, upper-flow-regime tidal flat, outer-estuarine tidal bar and marginal tidal flat deposits.

17.4.4 Causes of Increased Tidal Influence in Transgressive Settings

Thin tidal accumulations indicate drowning and cannibalization of strandplains and formation of back-barrier lagoons, or drowning of delta plains and river mouths and formation of wave-dominated estuaries or back-barrier lagoons. These transgressive accumulations often display a greater degree of tidal influence than their regressive counterparts. The tendency for the increased abundance of tidal deposits in transgressive tracts may be simply due to the shelf coming into tidal resonance as it widens during transgression. The thinness of most of these transgressive intervals (typically <5 m) probably reflects a lack of significant relief during rapid transgressive drowning or a low sediment supply, whereas thicker transgressive accumulations (9–15 m thick) reflect a higher sediment supply and a steeper transgressive trajectory.

In contrast, the transgressive tidal accumulations in incised valleys commonly reach significant thicknesses (up to 40 m). This happens preferentially within low-accommodation settings where valley abundance and depth of incision are preferentially increased. Valley

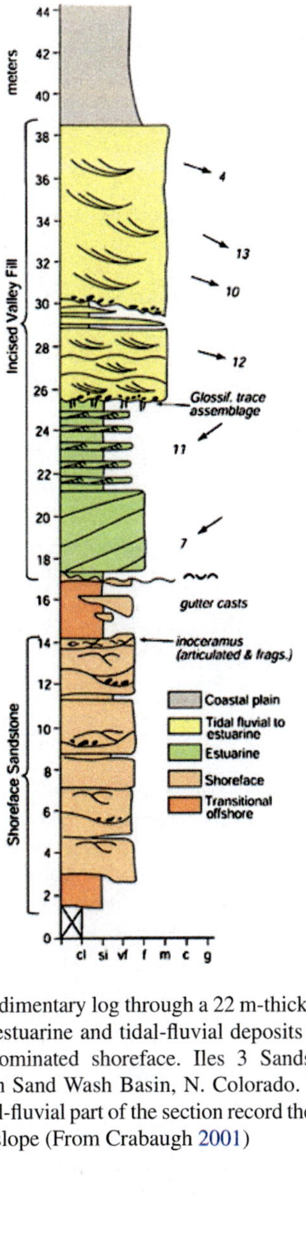

Fig. 17.24 Sedimentary log through a 22 m-thick incised valley succession of estuarine and tidal-fluvial deposits that cut down into a wave-dominated shoreface. Iles 3 Sandstone Tongue, Horse Gulch in Sand Wash Basin, N. Colorado. The paleocurrents in the tidal-fluvial part of the section record the southeasterly regional paleoslope (From Crabaugh 2001)

incision increases coastline irregularity, locally dissipating wave energy and enhancing tidal currents as the valleys are drowned during subsequent transgression. Thus, in contrast to the thin tidal accumulations, the valley-fill tidal accumulations are estuarine, some wave-dominated with a wave-built barrier mouth, others tide-dominated.

The thick non-valley accumulation in the uppermost Chimney Rock Sandstone, resulting from vertical stacking of four estuarine units, is attributed to locally high subsidence rates, caused by an embryonic Laramide Uinta uplift, compensated by a high sediment supply (Plink-Bjorklund 2008). Because the middle portion of the Chimney Rock Sandstone was a mixed-energy estuary, and was then overlain by vertically stacked tide-dominated estuarine units, it is likely that the system entered a window of tidal amplification during transgressive widening of the shelf (see also e.g. Dalrymple and Zaitlin 1994; Yoshida et al. 2007).

17.5 Widespread Tidal Deposits in the Interval Bounded by Ammonite Zones *B. perplexus* (late) Through *D. stevensoni* (ca 77.5–75 Ma)

There is an important third occurrence of WIS tidal deposits. Although this occurrence contains the same types of regressive and transgressive tidal deposits as described above, the unusual aspect is that the tidal deposits are ubiquitous and developed throughout highstand, lowstand and transgressive tracts of multiple cycles within a particular short time interval from Central Utah eastwards to Colorado and southern Wyoming. Aschoff and Steel (2010 and in press) referred to this as an 'anomalous clastic wedge' because of its architectural contrast with other Cretaceous clastic wedges in the Utah-Colorado segment of the seaway. The anomalous wedge is bounded by ammonite zones *D.stevensoni* and *Baculites perplexus* (late), though in Utah it begins earlier. The interval represents about a 2.5 My period (ammonite zones and radiometric calibration of Cobban et al. 2006) and is seen to be anomalously amalgamated and highly progradational on the regional well-log cross section of Kirschbaum and Hettinger (2004) (Fig. 17.25). It was studied in detail in outcrop and subsurface by Aschoff (2008), who extended the correlation 100 km westward to the thrust-belt and illustrated the high-frequency sequence architecture of this wedge.

17.5.1 Characteristics of the Widespread Low-Accommodation Stratigraphic Interval

The 400 km-long transect, referred to above, through Campanian strata from the Wasatch Plateau of Utah to western Colorado includes alluvial and estuarine deposits of the Middle Castlegate sandstones and Neslen

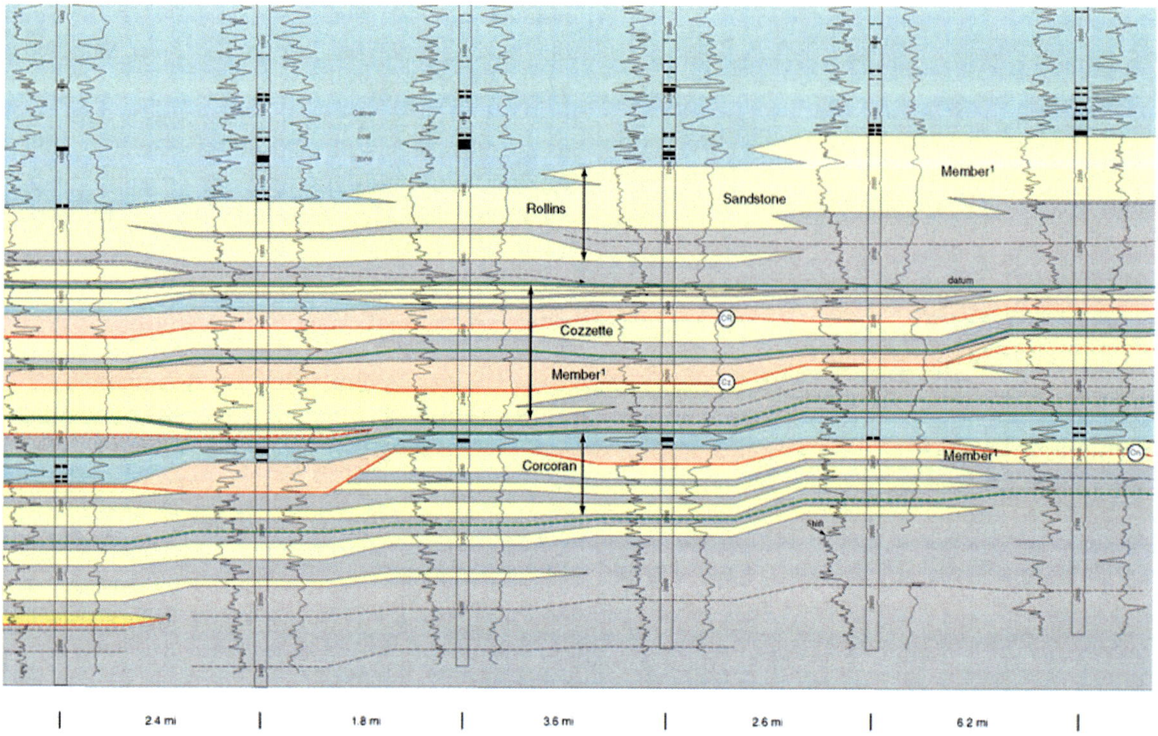

Fig. 17.25 A ca 17 km NW-SE transect along part of the Book Cliffs of Western Colorado from Kirschbaum and Hettinger 2004. Notice how the architecture of the 'anomalous' clastic wedge (Aschoff 2008) (complex lower half of diagram) contrasts with that of the wedge above (containing Rollins Sandstone shorelines). This anomalous wedge has stacked, thin, extensively progradational shorelines (*yellow*) that are frequently incised by channels and valleys containing estuarine (*pink*) and coastal plain (*green*) deposits. This architectural complexity, probably reflecting irregular coastline morphology compared to the straight shorelines of the overlying Rollins Sandstone, also contains an unusual degree of tidal influence

Table 17.3 Thick, transgressive non-valleyed tidal accumulations

Example	Location	Setting	Thickness	Depositional environment	References
Uppermost Chimney Rock Sandstone of Rock Springs Fm	NE Utah/SW Wyoming	Uppermost transgressive part of a regressive-transgressive fourth-order sequence	60 m	Four vertically stacked tide-dominated estuarine accumulations	Plink-Björklund (2008)

Formation that correlate basinwards to multiple marine shorelines of the Sego, Corcoran and Cozzette Sandstones and their coeval Mancos Shale basinal equivalents. To the north, along southern Wyoming and northern Colorado, the time interval includes the alluvial-estuarine Trail and Rusty Members of the Ericson Formation that correlate basinwards to about 12 marine shoreline tongues of the Iles Formation and their coeval Mancos Shale equivalents. The characteristic features of this anomalous stratigraphic interval that contrast with the underlying (Blackhawk Formation in Utah; Rock Springs Formation in Wyoming) and overlying (Rollins Sandstone, Trout Creek and Twenty-Mile Sandstones in Colorado) successions are:

1. A relatively slow, average sediment accumulation rate of 42 m/My – less than half the rate seen in the underlying Rock Springs Formation in S. Wyoming, and in the Blackhawk Formation in Utah, as well as a fast rate of overall shoreline progradation (~208 km/My) relative to underlying and overlying

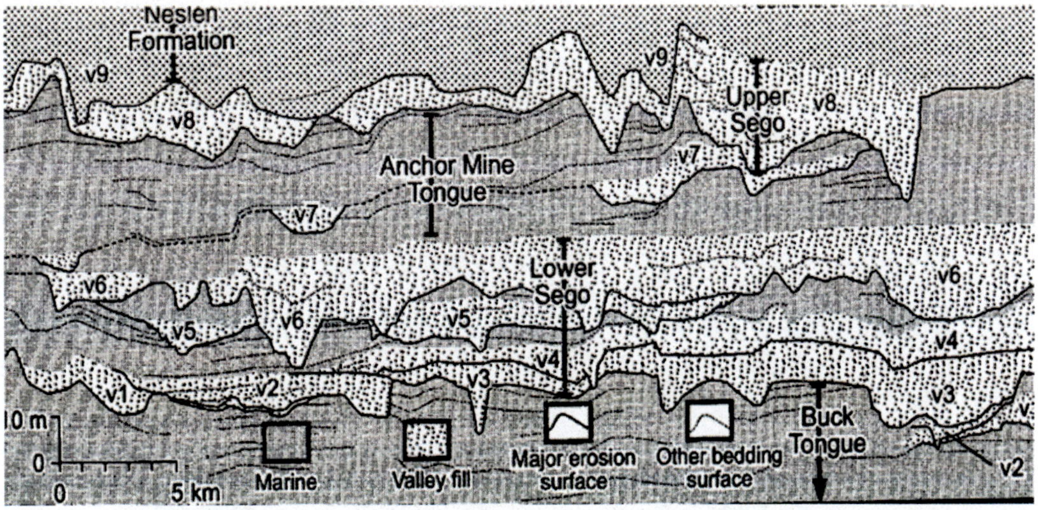

Fig. 17.26 Strike section through the Sego Sandstone showing multiple deep channels and valleys (the V-units). Willis and Gabel (2001, 2003) interpret many of these incisions as deep tidal scours formed during the forced regression of tide-dominated deltas, rather than as incised-valley fills (Figure is from Willis and Gabel 2001)

clastic wedges that built basinwards at rates of 50–80 km/My (Aschoff and Steel in press). A sense of the great basinward extent of individual shoreline progradations can be seen in the lower half of the Fig. 17.25 succession, in great geometric contrast to the more limited progradational lengths of individual shorelines in the overlying Rollins Sandstone.

2. A highly amalgamated and incised sequence architecture (Fig. 17.25) with well-developed, forced regressive sequence sets seen in the Sego/Neslen/Corcoran/Cozzette interval in E. Utah and W. Colorado (Aschoff 2008), and in some of the Iles shoreline complexes in N. Colorado (Crabaugh 2001). The frequent Sego Sandstone incisions (Fig. 17.26), some of which are interpreted as deep tidal scours during regressive delta building (Willis and Gabel 2001), are among the best-known examples of incised and amalgamated architectures.
3. During regressions, the high-frequency shorelines have the character of mixed tidal-wave shoreline deposits, but with a significant percentage of tide-dominated segments. Regressive units are frequently erosively based, as noted above.
4. During transgressions, the high-frequency shorelines of both Iles and Sego/Corcoran/Cozzette intervals deposited thick tide- and wave-influenced estuarine deposits (orange colored units in succession of Fig. 17.25) (Van Wagoner 1991; Crabaugh 2001; Kirschbaum and Hettinger 2004; Aschoff 2008; Gomez-Veroiza and Steel 2010) making probably the highest stratigraphic concentration of estuarine strata within the entire Cretaceous succession of WIS.

17.5.2 Characteristics of Tide-Influenced Strata Within the Low-Accommodation Interval

There are no special or unusual sedimentary processes for the deltas, estuaries and coastal plain systems that developed during the 77.5–75 Ma stratigraphic interval, but the particular geometric and stacking characteristics referred to above, imparted by the low accommodation setting, is emphasized. In addition to the features already listed above, there is a notable amalgamation of fluvial and tidal-fluvial channels in lower alluvial plain deposits (Fig. 17.27), a marked clustering of bayhead delta deposits preserved at the head of estuaries, great distances (80–100 km) of delta regression with very thin (<10 m) prodelta deposits, and a complex juxtaposition of upward-fining with upward-coarsening units due to deep channel incisions cutting through regressive shoreline tongues (Figs. 17.25 and 17.26).

Fig. 17.27 A series of five stacked and amalgamated fluvial to fluvial-tidal facies successions (each 20–25 m thick). Each succession has a sharply erosive base (*black lines*) overlain by medium-grained, trough cross-stratified sandstones, then draped by one or more sets of inclined heterolithic strata. The trough cross-stratified intervals (apparently structureless in the photo) are interpreted as fluvial deposits, whereas the intervening heterolithic intervals contain brackish traces (small *Teichichnus*) suggesting they represent transgressively deposited lower coastal plain or inner estuary point bars. Trail Member, Ericson Formation on Hwy 430, Rock Springs Uplift, Wyoming

17.5.3 Causes of Tidal Amplification in the 77.5–75 My Interval

The low average sediment accumulation rate (see above) strongly indicates that deposition during this particular stratigraphic interval was within a low accommodation setting. The very extensive marine transgressions of the Iles and Sego/Corcoran/Cozzette shorelines and the even lengthier brackish/tidal incursions back onto the coastal plains (Crabaugh 2001; Kirschbaum and Hettinger 2004; Aschoff 2008; Gomez-Veroiza 2009) also suggest that the coastal and alluvial plains at this time had an extremely low gradient. Liu and Nummedal (2004) and Liu and others (2005) identified the base of our low-accommodation interval (base of their Megasequence 4) as a time of tectonic change from relatively rapidly subsiding foreland basin conditions to isostatic fold-belt uplift and erosion, following earlier suggestions by DeCelles and Mitra (1995). However, the tectonic 'quiescence' of the thrust activity (Liu et al. 2005) at this boundary appears to have been challenged by Horton et al. (2004) who show evidence of thrust movements related to Castlegate Sandstone at this time. Recent work by Aschoff (2008) also suggests that thrust activity did not necessarily cease at this time, but that foreland basin subsidence was counteracted by early Laramide movements and a change in slab subduction style (Gurnis 1992). Irrespective of the precise tectonic causes, this new tectonic influence converted the earlier back-tilted foredeep of the Western Interior Basin to a flat or slightly seaward-tilted, low-gradient landscape/seascape that was hundreds of kilometers wide. The embayed shorelines of this landscape in Utah-Colorado (Aschoff and Steel 2010) and across S Wyoming-N. Colorado (Crabaugh 2001; Gomez-Veroiza and Steel 2010) repeatedly prograded for great distances, with significant tidal influence. The tectonic effects alone do not 'cause' the tidal influence in the seaway, but they had an important controlling influence on landscape slope and bathymetry as well as basin width and coastline shape, that in turn was optimal for amplifying tides during this interval in the epicontinental seaway. Subsequently, major flooding with a tongue of widespread Mancos Shale in the ammomite zone E. *jenneyi* (Fig. 17.3) led to a new dominance of storm-wave generated shorelines (Rollins Sandstone in Colorado, Trout Creek Sandstone in S Wyoming), with significant tidal influence only in lowstand and transgressive shoreline units.

17.6 Summary

The field evidence for tidal deposits and processes in the 13My Campanian interval of the Western Interior Seaway of Wyoming, Utah and Colorado shows that tidal signals were recorded predictably along many of the open shorelines as well as in back-barrier and estuarine environments. Less predictable is (1) the preferential occurrence of tidal influence or dominance along lowstand shorelines that developed 100s of km basinwards of their storm-wave dominated highstand positions, and (2) the occurrence of a ca. 2.5 My anomalous stratigraphic interval (mainly in uppermost Middle Campanian) when regressive and transgressive shorelines at all sea-level stands were significantly tidally influenced. The former occurrence is suggested to result from stronger tidal currents and dampened waves in the central reaches of the basin when sea level was low. This scenario is in contrast to passive margin basins where an increase in storm-wave conditions is commonly documented in the distal parts of sequences, on the outer reaches of the shelf and at the shelf edge (Porebski and Steel 2006; Yoshida et al. 2007). The latter occurrence is likely due to initial (77.5–75 Ma) Laramide tectonic movements in the region that began to change the configuration of the basin from a back-tilted foreland basin to a slightly forward-tilted, irregularly shallowed seascape, with a corresponding change in shoreline morphology.

References

Ahmed S (2008) Tide-influenced deltas in the Haystack Mountains Formation, SW Wyoming. Senior Honors thesis, University of Texas, Austin

Ainsworth BR, Pattison SAJ (1994) Where have all the lowstands gone? Evidence for attached lowstand systems tracts in the Western Interior of North America. Geology 22:415–418

Allen GP, Posamentier HW (1994) Transgressive facies and sequence architecture in mixed tide- and wave-dominated incised valleys; example from the Gironde Estuary, France. In: Dalrymple RW, Boyd R, Zaitlin BA (eds) Incised-valley systems; origin and sedimentary sequences, SEPM Special Publication 51. SEPM, Tulsa, pp 225–240

Aschoff J (2008) Controls on the development of clastic wedges and growth strata in foreland basins: examples from Cretaceous Cordilleran Foreland Basin strata, USA. Ph.D. thesis, University of Texas, Austin

Aschoff J, Steel RJ (2010) Anatomy and development of a low-accommodation clastic wedge,Upper Cretaceous, Cordilleran Foreland Basin, USA. Sedimentary Geology

Aschoff J, Steel RJ (in press) Anomalous clastic wedge development during the Sevier-Laramide transition, North American Cordilleran Foreland basin, USA. Geol Soc Am Bull

Balsley MD (1980) Cretaceous wave-dominated delta systems, Book Cliffs, East-Central Utah. In: AAPG continuing education course field guide. American Association of Petroleum Geologists, San Francisco

Benda TL (2000) Facies, architecture and sequence stratigraphy of a fourth-order clastic wedge: Upper Cretaceous Twentymile Sandstone (Mesaverde Group) NW Colorado. MS thesis, University of Wyoming, Laramie

Benda TL, Steel RJ (2000) Architecture, facies, and sequence stratigraphy of a fourth-order clastic wedge: Twentymile Sandstone, Mesaverde Group, NW Colorado, AAPG Annual Meetings and Abstracts. AAPG, San Antonio, p A12

Bergman KM (1999) Shannon Sandstone in Hartzog Draw-Heldt Draw Fields (Cretaceous, Wyoming, USA) reinterpreted as lowstand shoreface deposits. J Sediment Res B64:184–201

Bergman KM, Snedden JW (1999) Isolated shallow marine sandbodies: sequence stratigraphic analysis and sedimentological perspectives, SEPM special publications 64. SPEM, Tulsa

Bhattacharya J, Walker RG (1992) Deltas. In: Walker RG, James NP (eds) Facies models: response to sea level change. Geol Association of Canada, St. Johns

Bhattacharya J, Willis B (2001) Lowstand deltas in the Frontier Formation, Wyoming: implications for sequence stratigraphic models. Am Assoc Petrol Geol Bull 85:261–294

Birgenheier LP, Fielding CR, Corbett MJ, Kesler C (2009) Sequence stratigraphic assessment of the Muley Canyon Sandstone and Masuk formation, Henry Mountains syncline: implications for understanding the Muley Canyon Cole Zone, UGS report, 106 p

Boyles JM (1983) Depositional history and sedimentology of Upper Cretaceous Mancos Shale and Lower Mesaverde Group, northwestern Colorado: migrating shelf-bar and wave-dominated shoreline deposits. Ph.D. thesis, University of Texas, Austin

Boyles JM, Scott AJ (1982) A model for migrating shelf-bar sandstones in upper Mancos Shale (Campanian), northwestern Colorado. Am Assoc Petrol Geol Bull 66:491–508

Bullimore SA, Helland-Hansen W, Henriksen S, Steel RJ (2008) Shoreline trajectory and its impact on coastal depositional environments; an example from the Upper Cretaceous Mesaverde Group, northwestern Colorado, U.S.A. In: Hampson et al (eds) Recent advances in models of siliciclastic shallow-marine stratigraphy, SEPM special publications 90. SPEM, Tulsa, pp 209–236

Cattaneo A, Steel RJ (2003) Transgressive deposits: a review of their variability. Earth-Sci Rev 62:187–228

Charvin K, Hampson GJ, Gallagher KL, Labourdette R (2010) Intra-parasequence architecture of an interpreted asymmetrical wave-dominated delta. Sedimentology 57:760–785

Chen Z, Song B, Wang Z, Cai Y (2000) Late Quaternary evolution of the subaqueous Yangtse Delta, China: sedimentation, stratigraphy, palynology and deformation. Mar Geol 162: 423–441

Cobban WA, Obradovich JD, Walaszcyk I, McKinney KC (2006) A USGS zonal table for the Upper Cretaceous middle Cenomanian-Maastrichtian of the Western Interior of the United States based on ammonites, Inoceramids, and radiometric ages. USGS open-file report 2006-1250

Collinson JD (1970) Bedforms of the Tana River, Norway. Geogr Ann 52A:31–56

Crabaugh JP (2001) Nature and growth of nonmarine-to-marine clastic wedges: examples from the Upper Cretaceous Iles Formation, Western Interior (Colorado) and the lower Paleogene Wilcox Group of the Gulf of Mexico Basin (Texas). Ph.D. thesis, University of Wyoming, Laramie

Cross TA (1988) Controls on coal distribution in transgressive-regressive cycles, Upper Cretaceous, Western Interior, U.S.A. In: Wilgus CK, Hastings BS, Ross CA, Posamentier H, Van Wagoner JC, Kendall C (eds) Sea-level changes; an integrated approach, SEPM special publications 42. SPEM, Tulsa, pp 371–380

Cross TA (1998) The shoreface/tidal couplet; a mechanism for tapping into coarse sediment, Program with Abstracts. AAPG Annual Meeting, Salt Lake City

Dalrymple RW (1992) Tidal depositional systems. In: Walker RG, James NP (eds) Facies models: response to sea level change. Geological Association of Canada, St. John's, pp 195–218

Dalrymple RW (2010) Tidal depositional systems. In: James RW, Dalrymple NP (eds) Facies models 4, GEOText6. Geological Association of Canada, St. John's, pp 199–208

Dalrymple RW, Choi K (2007) Morphologic and facies trends through the fluvial-marine transition in tide-dominated depositional systems: a schematic framework for environmental and sequence-stratigraphic interpretation. Earth Sci Rev 81:135–174

Dalrymple RW, Rhodes RN (1995) Estuarine dunes and bars. In: Perillo GME (ed) Geomorphology and sedimentology of Estuaries. Elsevier, Amsterdam, pp 359–422

Dalrymple RW, Zaitlin BA (1994) High-resolution sequence stratigraphy of a complex, incised valley succession, Cobequid Bay-Salmon River estuary, Bay of Fundy, Canada. Sedimentology 41:1069–1091

Dalrymple RW, Zaitlin BA, Boyd R (1992) Estuarine facies models; conceptual basis and stratigraphic implications. J Sediment Res 62:1130–1146

Davies H (1990) Quantitative field studies of Cretaceous shelf ridge sandstones of the western Interior Basin, USA. In: Abstract, 13th international sedimentological congress, England, pp 119–120

De Raaf JFM, Boersma JR (1971) Tidal deposits and their sedimentary structures. Geol Mijn 50:479–503

DeCelles PG, Mitra G (1995) History of the Sevier orogenic wedge in terms of critical taper models, northeast Utah and southwest Wyoming. Geol Soc Am Bull 107:454–462

Devine PE (1991) Transgressive origin of channeled estuarine deposits in the Point Lookout Sandstone, NW New Mexico: a model for Upper Cretaceous cyclic regressive parasequences of the US Western Interior. Am Assoc Petrol Geol Bull 75:1039–1063

Devlin WJ, Rudolph KW, Shaw CA, Ehman KD (1993) The effect of tectonic and eustatic cycles on accommodation and sequence-stratigraphic framework in the Upper Cretaceous foreland basin of southwestern Wyoming. In: Posamentier WH, Summerhayes CP, Haq BU, Allen CP (eds) Sequence stratigraphy and facies associations, International Association of Sedimentology special publications 18. Blackwell Scientific, Boston, pp 501–520

Elliot T (1997) Physical processes, depositional settings and stratigraphic context of the Shannon sandstone, Wyoming, U.S.A. In: Swift DJP, Snedden J, Plint AG (conveners) Tongues, ridges and wedges: highstand versus lowstand architecture in shallow marine basins. SEPM research conference program, Powder River and Bighorn basins, Wyoming, 24– 29 June 1995

Ericksen MC, Slingerland RL (1990) Numerical simulations of tidal and wind-driven circulation in the Cretaceous interior seaway of North America. Geol Soc Am Bull 102:1499–1516

Fenies H, de Ressequier A, Tastet JP (1999) Intertidal clay-drape couplets (Gironde Estuary, France). Sedimentology 46:1–15

Fitzsimmons RJ (1994) Field-based testing of high resolution sequence stratigraphic concepts; the Campanian Eagle Formation, Mesaverde Group, Bighorn Basin, Northeastern Wyoming, USA. Ph.D. thesis, University of Liverpool, Liverpool

Fitzsimmons R (1999) A tale of two valley fills; tidal depositional systems of the Campanian Eagle Formation of the Bighorn Basin, Program with abstracts. AAPG Annual Meeting, San Antonio, p A42

Fitzsimmons RJ, Johnson SD (2000) Forced regressions: recognition, architecture and genesis in the Campanian of the Bighorn Basin, Wyoming. In: Hunt D, Gawthorpe RL (eds) Sedimentary responses to forced regressions, Geological Society London special Publications 172. Geological Society, London, pp 113–139

Gill JR, Merewether EA, Cobban, WA.(1970) Stratigraphy and nomenclature of some Upper Cretaceous and Lower Tertiary rocks in south-central Wyoming. U.S. Geological Survey Professional Paper 667

Gomez-Veroiza CA (2009) Clastic wedge development and sediment budget in a source-to-sink transect (Late Campanian Western Interior Basin, SW Wyoming and N Colorado). Ph.D. thesis, University of Texas, Austin

Gomez-Veroiza CA, Steel RJ (2010) Iles Clastic Wedge development and sediment partitioning in a 300 km, fluvial to marine Campanian transect (3my), Western Interior Seaway, SW Wyoming and N Colorado. Am Assoc Petrol Geol Bull 94:1349–1377

Green GN (1992) The digital geologic map of Colorado in ARC/INFO format, U.S. Geological Survey, open-file report OF-92-507, scale 1:500000

Green GN, Drouillard PH (1994) The digital geologic map of Wyoming in ARC/INFO format, U.S. Geological Survey, open-file report OF-94-425, scale 1:500000

Gurnis M (1992) Rapid continental subsidence following the initiation and evolution of subduction. Science 255:29–32

Hampson GJ (2010) Sediment dispersal and quantitative stratigraphic architecture across an ancient shelf. Sedimentology 57:96–141

Hampson GJ, Howell JA (2005) Sedimentologic and geomorphic characterization of ancient wave-dominated shorelines: examples from the late Cretaceous Blackhawk Formation, Book Cliffs, Utah. In: Bhattacharya JP, Giosan L (eds) Deltas old and new, SEPM special publication 83. SPEM, Tulsa, pp 133–154

Hampson GJ, Procter EJ, Kelly C (2008a) Controls on isolated shallow-marine sandstone deposition and shelf construction: late Cretaceous western interior seaway, northern Utah and Colorado, U.S.A. In: Hampson GJ, Steel RJ, Burgess PM,

Dalrymple RW (eds) Recent advances in models of Siliciclastic shallow-marine stratigraphy, SEPM special publications 90. SPEM, Tulsa

Hampson GJ, Rodriguez AB, Storms JEA, Johnson HD, Meyer CT (2008b) Geomorphology and high-resolution stratigraphy of progradational wave-dominated shoreline deposits; impact on reservoir-scale facies architecture. In: Hampson GJ, Steel RJ, Burgess PM, Dalrymple RW (eds) Recent advances in models of siliciclastic shallow-marine stratigraphy, SEPM special publications 90., pp 117–142

Hintze LF, Willis GC, Laes DYM, Sprinkel DA, Brown KD (2000) Digital geologic map of Utah. Utah Geological Survey, Map 179DM, scale 1:500000

Hori K, SaitoY ZQ, Wang P (2002) Architecture and evolution of the tide-dominated Changjiang (Yangtze) River delta, China. Sediment Geol 146:249–264

Horton BK, Constenius KN, DeCelles PG (2004) Tectonic control on coarse-grained foreland-basin sequences: an example from the Cordilleran foreland basin, Utah. Geology 32:637–640

Howell JA, Kamola DL, Flint SS (1997) The anatomy of incised valley fill deposits within the Sunnyside member, Blackhawk formation, Book Cliffs, Utah (abstract). AAPG annual convention, Dallas, A 53

Hwang I-G, Heller PL (2002) Anatomy of a transgressive lag: Panther Tongue Sandstone, Star Point Formation, central Utah. Sedimentology 49:977–999

Ichaso AA, Dalrymple RW (2009) Tidal and wave-generated fluid mud deposits in the Tilje Formation (Jurassic), offshore Norway. Geology 37:539–542

Irwin ML (1965) General theory of epeiric clear water sedimentation. Am Assoc Petrol Geol Bull 49:445–459

Izett GA, Cobban WA, Gill JR (1971) The Pierre Shale near Kremmling, Colorado, and its correlation to the east and the west. U.S. Geological Survey Professional Paper 684-A

Kamola DL (1992) Sequence boundary variations within the Aberdeen member, Cretaceous Blackhawk formation, Utah (abstract). AAPG annual convention, A61

Kamola DL, Van Wagoner JC (1995) Stratigraphy and facies architecture of parasequences with examples from the Spring Canyon Member, Blackhawk Formation, Utah. In: Van Wagoner JC, Bertram GT (eds) Sequence stratigraphy of foreland basin deposits – outcrops and subsurface examples from the Cretaceous of North America, AAPG Memoir 64. AAPG, San Antonio, pp 27–54

Kauffman EG (1984) Paleobiogeography and evolutionary response dynamic in the Cretaceous Western Interior Seaway of North America. In: Westermann GEG (ed) Jurassic-Cretaceous biochronology and biogeography of North America, Geological Association of Canada special paper 27. Geological Association of Canada, St. John's, pp 273–306

Keulegan GH, Krumbein WC (1949) Stable configuration of bottom slope in a shallow sea, and its bearing on geological processes. Am Geophys Union Trans 30:855–861

Kirschbaum MA, Hettinger RD (1998) Stratigraphy and depositional environments of the Late Campanian coal-bearing Neslen/Mt. Garfield Formations, eastern Book Cliffs, Utah and Colorado. US Geological Survey, open file report, pp 98–43

Kirschbaum, MA, Hettinger RD (2004) Facies analysis and sequence stratigraphic framework of Upper Campanian strata (Neslen and Mount Garfield Formations, Bluecastle Tongue of the Castlegate Sandstone, and Mancos Shale), Eastern Book Cliffs, Colorado and Utah. USGS digital data report DDS-69-G

Kiteley LW, Field WE (1984) Shallow-marine depositional environments in the Upper Cretaceous of northern Colorado. In: Tilman RW, Siemers CT (eds) Siliciclastic shelf sediments, SEPM special publications 34. SPEM, Tulsa, pp 179–204

Klein GD, Ryer TA (1978) Tidal circulation patterns in PreCambrian, Paleozoic and Cretaceous epeiric and mioclinal shelf seas. Geol Soc Am Bull 89:1050–1058

Krystinik LF (1995) Lateral facies relationships in the Kremmling and Muddy Buttes Sandstones of northern Colorado (abstract). In: Swift DJP, Snedden JW, Plint AG (eds) Tongues, ridges and wedges: highstand versus lowstand in marine basins, SEPM research conference. SEPM, Tulsa

Land CB (1972) Stratigraphy of Fox Hills Sandstone and associated formations, Rock Springs Uplift and Wamsutter Arch Area, Sweetwater County, Wyoming; a shoreline-estuary sandstone model for the late Cretaceous, Quarterly of the Colorado School of Mines 1906, 67

Liu S, Nummedal D (2004) Late Cretaceous subsidence in Wyoming: quantifying the dynamic component. Geology 32:397–400

Liu SF, Nummedal D, Yin PG, Luo HJ (2005) Linkage of Sevier thrusting episodes and Late Cretaceous foreland basin megasequences across southern Wyoming (USA). Basin Res 17:487–506

Martinsen RS (2003a) Depositional remnants, part 1: common components of the stratigraphic record with important implications for hydrocarbon exploration and production. Am Assoc Petrol Geol Bull 87:1869–1882

Martinsen RS (2003b) Depositional remnants, part 2: examples from the Western interior Cretaceous basin of North America. Am Assoc Petrol Geol Bull 87:1883–1909

Martinsen RS (2003c) Syn-sedimentary tectonic and sediment supply influences on basin physiography and depositional regime of the western interior Cretaceous Seaway, Wyoming. AAPG Search and Discovery Article #90013, AAPG annual meeting, Salt Lake City, 2003

Martinsen OJ, Steel R, Martinsen R, Krystinik L (1998) The Mesaverde Group, Rock Springs Uplift, Wyoming. In: The Mesaverde Group, Rock Springs Uplift, Wyoming, AAPG annual convention, field guide. AAPG, San Antonio

Martinsen OJ, Ryseth A, Helland-Hansen W, Flesche H, Torkildsen G, Idil S (1999) Stratigraphic base level and fluvial architecture: Ericson Sandstone (Campanian), Rock Springs Uplift, SW Wyoming, USA. Sedimentology 46:235–259

Masters C (1966) Sedimentology of the Mesaverde group of the upper part of the Mancos Formation, northwestern Colorado. Ph.D. thesis, Yale University, New Haven

McClurg TA (1990) Lithosome relationships and depositional environments of the O'Brien spring Sandstone (Late Cretaceous) south-central Wyoming. MS thesis, University of Wyoming, Laramie

McLaurin BT, Steel RJ (2000) Fourth-order nonmarine to marine sequences, middle Castlegate Formation, Book Cliffs, Utah. Geology 28:359–362

Mellere D, Steel RJ (1995a) Facies architecture and sequentiality of nearshore and 'shelf' sandbodies: Haystack Mountains Formations, Wyoming, USA. Sedimentology 42:547–551

Mellere D, Steel RJ (1995b) Variability of lowstand wedges and their distinction from forced regressive wedges in the Mesaverde Group, SE Wyoming. Geology 23:803–806

Mellere D, Steel RJ (2000) Style contrast between forced regressive and lowstand/transgressive wedges in the Campanian of south-central Wyoming. Geol Soc Lond Spec Publ 172:51–75

Miller KG, Sugarman PJ, Browning JV, Kominz MA, Olsson RK, FeigensonMD HJC (2004) Upper Cretaceous sequences and sea-level history, New Jersey Coastal Plain. Geol Soc Am Bull 116:368–393

Muto T, Steel RJ (1997) Principles of regression and transgression; the nature of the interplay between accommodation and sediment supply. J Sediment Res 67:994–1000

Nio SD, Yang CS (1991) Diagnostic attributes of clastic tidal deposits: a review. In: Smith DG, Reinson GE et al (eds) Clastic tidal sedimentology, Canadian Society of Petroleum Geologists 16. Canadian Society of Petroleum Geologists, Calgary, pp 3–28

O'Byrne CJ, Flint SS (1995) Sequence, parasequence and intra-parasequence architecture of the Grassy member, Blakhawk Formation, Book Cliffs, Utah. In: Van Wagoner JC, Bertram GT (eds) Sequence stratigraphy of foreland basin deposits – outcrops and subsurface examples from the Cretaceous of North America, AAPG Memoir 64. AAPG, San Antonio, pp 225–255

Pattison SAJ (1995) Sequence stratigraphic significance of sharp-based lowstand shoreface deposits, Kenilworth Member, Book Cliffs, Utah. Am Assoc Petrol Geol Bull 79:444–462

Pattison SAJ (2005) Storm-influenced prodelta turbidite complex in the Lower Kenilworth Member at Hatch Mesa, Book Cliffs, Utah, USA: implications for shallow-marine facies models. J Sediment Res 75:420–439

Plink-Bjorklund P (2008) Wave-to-tide facies change in a Campanian shoreline complex, Chimney Rock Tongue, Wyoming-Utah, USA. In: Hampson GJ, Steel RJ et al (eds) Recent advances in models of Siliciclastic shallow-marine stratigraphy, SEPM special publications 90. SPEM, Tulsa, pp 265–291

Porebski SJ, Steel RJ (2006) Deltas and sea-level change. J Sediment Res 76:1–14

Rahmani RA (1988) Estuarine tidal channel and nearshore sedimentation of a Late Cretaceous epicontinental sea, Drumheller, Alberta, Canada. In: de Boer PL, van Gelder A, Nio SD (eds) Tide-influenced sedimentary environments and facies. Reidel Publishing, Dordrecht, pp 433–481

Redfield AC (1958) The influence of the continental shelf on the tides of the Atlantic coast of the United States. J Mar Res 17:432–448

Reinson GE (1992) Transgressive barrier island and estuarine systems. In: Walker RG, James NP (eds) Facies models: response to sea level change. Geological Associations of Canada, St. John's, pp 179–194

Roehler HW (1989) Surface-subsurface correlations of the Mesaverde Group and associated Upper Cretaceous formations, Rock Springs to Atlantic Rim, Southwest Wyoming, U.S. Geological Survey Miscellaneous Field Studies Map MF-2078

Roehler HW (1990) Stratigraphy of the Mesaverde Group in the central and eastern greater Green River basin, Wyoming, Colorado, and Utah, U.S. Geological Survey Professional Paper 1508

Seidler L, Steel R (2001) Pinch-out style and position of tidally influenced strata in a regressive-transgressive wave-dominated deltaic sandbody, Twentymile Sandstone, Mesaverde Group, NW Colorado. Sedimentology 48:399–414

Shanley KW, McCabe PJ, Hettinger RD (1992) Tidal influence in Cretaceous fluvial strata from Utah, USA: a key to sequence stratigraphic interpretation. Sedimentology 39:905–930

Shaw DP (1964) Time in stratigraphy. McGraw-Hill, New York

Slingerland R, Kump LR, Arthur MA, Fawcett PJ, Sageman BB, Barron EJ (1996) Estuarine circulation in the Turonian Western Interior Seaway of North America. Geol Soc Am Bull 108:941–952

Slingerland R, Keen TR (1999) Sediment transport in the Western Interior Seaway of North America: predictions from a climate-ocean-sediment model. In: Bergman KM, Snedden JW (eds) Isolated shallow marine sand bodies; sequence stratigraphic analysis and sedimentologic interpretation, SEPM special publications 64. SPEM, Tulsa, pp 179–190

Smith JW, Gras VB, Fabini EM (1965) Southeast Hatfield Dome (measured section). In: DeVoto RH, Bitter RK (eds) Sedimentation of Late Cretaceous and Tertiary Outcrops, Rock Springs Uplift. Wyoming Geological Association guidebook, 19th Field Conference

Sullivan MD, Van Wagoner JC, Jennette DC, Foster ME, Stuart RM, Lovell RW, Pemberton SG (1997) High-resolution sequence stratigraphy and architecture of the Shannon Sandstone, Hartzog Draw Field, Wyoming: implications for reservoir management (abstract). GCSSEPM foundation 18th annual conference, Shallow marine and non-marine reservoirs

Suter JR, Clifton HE (1999) The Shannon Sandstone and isolated linear sand bodies: interpretation and realizations. In: Bergman KC, Snedden J (eds) Isolated shallow-marine sandbodies: sequence stratigraphic analysis and sedimentologic perspectives, SEPM special publication 64. SPEM, Tulsa, pp 321–356

Taylor DR, Lowell RWW (1991) Recognition of high-frequency sequences in the Kenilworth Member of the Blackhawk Formation, Book Cliffs, Utah. In: Van Wagoner JC et al (eds) Sequence stratigraphy applications to shelf sandstone reservoirs – Outcrop to subsurface examples, AAPG field conference guidebook. American Association of Petroleum Geologists, Tulsa

Taylor DR, Lowell RWW (1995) High-frequency sequence stratigraphy and paleogeography of the Kenilworth member, Blackhawk Formation, Book Cliffs, Utah, USA. In: Van Wagoner JC, Bertram GT (eds) Sequence stratigraphy of foreland basin deposits – outcrops and subsurface Examples from the Cretaceous of North America, AAPG Memoir 64. American Association of Petroleum Geologists, Tulsa, pp 257–275

Thomas RG, Smith DG, Wood JM, Visser J, Calverly-Range GA, Koster EA (1987) Inclined heterolithic stratification – terminology, description, interpretation and significance. Sediment Geol 53:123–179

Tillman RW, Martinsen RS (1985) Upper Cretaceous and Haystack Mountains field trip, Wyoming: guidebook for Rocky Mountain section field trip no. 11. SEPM mid-year meeting, Golden Colorado

Tillman RW, Martinsen RS (1987) Sedimentologic characteristics and production model of Hartzog Draw Field, Wyoming, Shannon shelf-ridge sandstone. In: Tillman RW, Weber KJ (eds) Reservoir sedimentology, SEPM special publications 40. SPEM, Tulsa, pp 15–112

Uroza CA (2008) Processes and architecture of deltas in shelf break and ramp platforms: examples from the Eocene of West Spitsbergen, the Pliocene paleo-Orinoco Delta (SE Trinidad), and the Cretaceous Western Interior Seaway (S Wyoming and NE Utah). Ph.D. thesis, University of Texas, Austin

Van Wagoner JC (1991) High frequency sequence-stratigraphy and facies architecture of the Sego Sandstone in the Book Cliffs of western Colorado and eastern Utah. In: Van Wagoner JC, Clive R, Taylor DR, Nummedal D, Jennette DC, Riley GW (eds) Sequence stratigraphy applications to shelf sandstone reservoirs; outcrop to subsurface examples, American Association of Petroleum Geologists Field Conference. American Association of Petroleum Geologists, Tulsa

Van Wagoner JC, Mitchum RM, Campion KM, Rahmanian VD (1990) Siliciclastic sequence stratigraphy in well logs, cores, and outcrops; concepts for high-resolution correlation of time and facies, AAPG methods in exploration series 7. American Association of Petroleum Geologists, Tulsa

Vis G-J (2009) Fluvial and marine sedimentation at a passive continental margin: the late Quaternary Tagus depositional system. Ph.D. thesis, Vrije Universiteit, Amsterdam

Walker RG (1992) Facies, facies models and modern stratigraphic concepts. In: Walker RG, J+James NP (eds) Facies models: response to sea level change. Geological Association of Canada, St. John's

Walker RG, Bergman KM (1993) Shannon Sandstone in Wyoming: a shelf ridge complex reinterpreted as lowstand shoreface deposits. J Sediment Petrol 63:839–851

Weimer RJ (1966) Time stratigraphic analysis and petroleum accumulations, Patrick Draw Field, Sweetwater County, Wyoming. Am Assoc Petrol Geol Bull 50:2150–2175

Weimer RJ (1983) Relation of unconformities, tectonics, and sea level changes, Cretaceous of the Denver Basin and adjacent areas. In: Reynolds MW, Dolly ED (eds) Mesozoic Paleogeography of West-Central United States, Rocky Mountain Section. SEPM, Tulsa, pp 359–376

Weimer RJ (1984) Relation of unconformities, tectonics, and sea-level changes, Cretaceous of Western Interior, U.S.A. In: Schlee R (ed) Interregional unconformities and hydrocarbon accumulation, American Association of Petroleum Geologists Memoir 36. American Association of Petroleum Geologists, Tulsa, pp 7–35

Wells MR, Allison PA, Piggot MD, Hampson GJ, Pain CC, Gorman GJ (2010) Tidal modeling of an ancient tide-dominated seaway, Part 1: model variation and application to global Early Cretaceous (Aptian) tides. J Sediment Res 80:1–16

Willis BJ (2005) Deposits of tide-influenced river deltas. In: Giosan L, Bhattacharya JP (eds) River deltas – concepts, models and examples, SEPM special publications 83. SPEM, Tulsa, pp 87–129

Willis BJ, Gabel SL (2001) Sharp-based tide-dominated deltas of the Sego Sandstone, Book Cliffs, Utah. Sedimentology 48:479–506

Willis BJ, Gabel SL (2003) Formation of deep incisions into tide-dominated river deltas: Implications for the stratigraphy of the Sego Sandstone, Book Cliffs, Utah, U.S.A. J Sediment Res 73:246–263

Wood LJ (2004) Predicting tidal sand reservoir architecture using data from modern and ancient depositional systems, integration of outcrop and modern analogs in reservoir modeling. AAPG Mem 80:45–66

Yoshida S (2000) sequence and facies architecture of the upper Blackhawk Formation and the Lowr Castlegate Sandstone (Upper Cretaceous), Book Cliffs, Utah, USA. Sediment Geol 136:239–276

Yoshida S, Miall AD, Willis A (2001) Fourth-order non-marine to marine sequences, Middle Castlegate, Book Cliffs Utah, Comment. Geology 29:187–188

Yoshida S, Steel RJ, Dalrymple RW (2007) Depositional process changes: an ingredient in a new generation of sequence stratigraphic models. J Sediment Res 77:447–460

Contrasting Styles of Siliciclastic Tidal Deposits in a Developing Thrust-Sheet-Top Basins – The Lower Eocene of the Central Pyrenees (Spain)

18

A.W. Martinius

Abstract

Lower Eocene tidal deposits in the Tremp-Graus-Ager Basin in the southern Pyrenees (Spain) are well-developed and include typical examples of tidal bars, compound tidal dunes, tidal bundles and tide-dominated back-barrier lagoons as well as tidally-influenced fluvial systems. They occurred in a relatively narrow (up to 60 km) and long (up to 200 km in total) semi-enclosed sea which had an open connection to the Atlantic ocean in the west. Two groups of tidal deposits are recognised related to two stages of the obliquely migrating thrust-sheet-top basin which affected the position and relative dimensions of the foredeep and shelf sections.

Compound tidal dune fields and large tidal bars developed mainly in the initially underfilled foredeep in relatively deep water (at least up to 40 m) during the Early Ypresian. Favourable conditions existed for basin scale tidal current circulation patterns, locally significantly amplified and modified by pronounced bathymetric variations (related to developing blind thrust related folds and blind ramps) and a variable, and probably distinct, structurally controlled, coastline morphology. Small shoal-water fan deltas and larger Gilbert-type delta(s) and associated tidal bars developed along the basin margins near, often long-lived, sediment entry points.

During the Late Ypresian to Late Lutetian the basin shelf area filled-up by a rapidly developing axial east to west prograding alluvial to deltaic system. This reduced tidal amplification in the basin and shallow-water tidalites developed only in a narrow (approximately 10 km) zone, located above an oblique lateral ramp system, including the in-shore parts of the delta distributaries and the subaqueous part of the, partly barred, delta top.

18.1 Introduction

The central Pyrenean Eocene thrust-sheet-top basins in Spain have been well known for their siliciclastic tidal deposits since the mid 1960s when first the Dutch (Mey et al. 1968; van Eden 1970), soon followed by the Spanish, Italian and French workers, began to study the well-exposed outcrops in detail. A major step forward was the understanding of tidal bundle successions based on studies of excavations in the modern and historically well-documented Oosterschelde estuary (The Netherlands) during the late 1970s and early 1980s (Nio et al. 1980; Visser 1980; van den Berg 1981, 1982) as well as sedimentological studies in the Wadden Sea (Sha 1990; Oost 1995) that

A.W. Martinius (✉)
Statoil Research and Development, Arkitekt Ebbels vei 10, N-7005 Trondheim, Norway
e-mail: awma@Statoil.com

R.A. Davis, Jr. and R.W. Dalrymple (eds.), *Principles of Tidal Sedimentology*,
DOI 10.1007/978-94-007-0123-6_18, © Springer Science+Business Media B.V. 2012

were applied to interpret analogous deposits in the Spanish Pyrenees (Donselaar and Nio 1982; Yang and Nio 1985, 1989). The tidal nature of the deposits was recognized early on, but the larger environmental setting remains the subject of discussion and re-evaluation.

The tide-influenced and tide–dominated successions discussed in this chapter form part of a number of formations of Ypresian and Lutetian age covering a continuum of almost 16 Ma. Tidal deposits of Maastrichtian age in the Aren Formation (Tremp Group; Nagtegaal et al. 1983; De Boer 1985, and a wealth of recent but unpublished data collected by Spanish workers) and of Upper Lutetian age in the Sobrarbe Formation (Campodarbe Group; Hall 1997; Dreyer et al. 1999) are also present but are not included in this chapter.

This chapter discusses examples of two different styles of tidal amplification that existed during the Ypresian and Lutetian in the central Pyrenean thrust-sheet-top Tremp-Graus-Ager Basin. The first style was controlled by the underfilled foredeep stage of basin development, the second style by the overfilled shelf stage of basin development. The two styles were closely related to significant differences in basin configuration characteristics and controlled by thrust-sheet-top basin (cf. DeCelles and Giles 1996) development and associated changes in basin paleogeography, morphology and dimensions. Therefore, usage of the terms foredeep and shelf in this chapter is strictly related to the particular development of the thrust-sheet-top basin as outlined hereafter. In addition, eustatic sea-level fluctuations influenced the degree of tidal amplification. A summary of the characteristics of the two styles is given in Table 18.1 and discussed hereafter.

18.2 Geological Framework

18.2.1 Development of Lower Eocene Thrust-Sheet-Top Basins

The Pyrenean Range (Capote et al. 2002) consists of the Aragonese-Catalan Pyrenees in the east and the Basque-Cantabrian Pyrenees in the west. The Aragonese-Catalan Pyrenees are subdivided into the Eastern, Central and West-Central Pyrenees (Fig. 18.1). The focus in this chapter will be exclusively on the thrust-sheet-top basins of the Spanish Central Pyrenees (the south Pyrenean central unit of Séguret 1972). For details about other aspects of structural evolution of the Pyrenees, one is referred to Van der Voo (1969), Puigdefàbregas and Souquet (1986), Galdeano et al. (1989), Malot (1989), Choukroune et al. (1990), Malod and Mauffret (1990), Srivastava et al. (1990), Roest and Srivastava (1991), Olivet (1996), Meigs and Burbank (1997), and Capote et al. (2002).

An important phase in the development of the Pyrenees started around 84 Ma (Late Santonian) when the Iberian Plate and the African Plate collided, and subduction along the northern plate margin was initiated (Guimerá 1984, 1996). This caused conversion from a foregoing extension phase to compression with a near N-S shortening direction, and produced inversion of Mesozoic extensional rift basins and the rise of the Pyrenees (Beaumont et al. 2000; Capote et al. 2002).

Two main foreland basins, the Aquitanian basin in the north and the Pyrenean Foreland basin to the south, developed from the Late Santonian (84 Ma) to the Miocene in conjunction with the convergence from extension to compression (Cámara and Klimowitz 1985; Puigdefàbregas et al. 1992; Teixell and Muñoz 2000; Beaumont et al. 2000). Four main compressional stages are recognized in the Pyrenean Foreland Basin (Puigdefàbregas et al. 1992) of which Stage III is the focus of this chapter.

Three main thrusts developed during Stage III (Early and Middle Eocene): from north to south the Bóixols, Montsec and Sierras Marginales. They were initiated successively in time but their displacement periods overlapped; they are particularly clear in the ECORS seismic profile (Cámara and Klimowitz 1985; Choukroune et al. 1989, 1990; Figs. 18.1 and 18.2). In the central Pyrenees, the size and shape of the thrust sheets were determined by the inverted Mesozoic fault pattern which controlled the location of oblique (with respect to the thrust motion) and lateral ramps that are defined as dividing two different segments of a thrust belt and consequently the distribution of accommodation space and thrust-sheet-top basin facies (Puigdefàbregas et al. 1992; Vergés 2007).

Initiation of the Montsec Thrust occurred at the end of the Paleocene (Cámara and Klimowitz 1985; Puigdefàbregas et al. 1986, 1992; Nijman 1998; Mascle and Puigdefàbregas 1998). The thrust underlies more

Table 18.1 Summary of the characteristics of the underfilled foredeep and overfilled shelf stage of the Tremp-Graus-Ager Basin. See text for discussion

	Underfilled foredeep	Overfilled shelf
Chronostratigraphic period	Early Ypresian (55.5 Ma to 51.5 Ma)	Late Ypresian to Lutetian (51.5 Ma to 43 Ma)
Basin shape	Narrow and elongate, closed in the E, open to the Atlantic Ocean in W	Narrow and elongate, closed in the E, open to the Atlantic Ocean in W
Basin width (total marine section)	50 km	40 km
Basin length (total marine section)	200 km	150 km
Estimated water depth in present-day Tremp-Graus and Ager Basin area	Up to 60 m on average	0 to 20 m on average
Sea-floor morphology	Shelf floor typified by gentle ridges and swells above blind thrust-related ramps and faults. Indented northern basin margin. Significant shelf area present. Slope located in present-day Ainsa Basin area.	Almost no shelf present (delta and alluvial plain). Present-day Ainsa Basin locus of structurally fixed, narrow and steep shelf platform-to-slope transition, and slope.
Length of (marine) shelf	Approximately 50 km	Approximately 20 km decreasing to 5 km
Nature of main basin filling processes on shelf	Dominantly shallow marine siliciclastic deposition along basin margin in inner parts, with gradual transition to central basin pelagic deposition and basinmargin fringing carbonate margins in present-day Jaca Basin area	Extensive alluvial and lower delta plain with relatively narrow delta front. Major shelf slope collapse and basin floor turbidites in present-day Ainsa and Jaca Basins with carbonate platform along margin.
Position of tidalites in basin	1) Narrow sea; 2) on platform shelf in front of (detached or attached to) deltas; 3) estuaries and embayments	Subaquous delta top along indented shoreline in inshore part of channels and embayments
Type of tidal deposits	1) Tidal dunes in embayment in front of Gilbert-type delta lobes; 2) offshore compound tidal dune fields on narrow shelf; 3) detached or attached delta-front tidal bars	1) Tide-influenced mouth bars and point bars in meandering channels; 2) interdistributary bay deposits; 3) tidal inlets and back-barrier lagoons
Examples	Alveolina Lst (Serraduy Fm), Roda Fm, Baronia Fm and Ametlla Fm	Castigaleu and Montllobar Fms (Lower Montanyana Group) and Capella and Pano Fms (Upper Montanyana Group)

than 3 km of displaced, mainly Upper Cretaceous limestones and Maastrichtian and Paleocene deposits of the Tremp Group. These were deposited in the authochtonous South Pyrenean Foreland Basin when it was not broken-up and partly displaced. The Montsec Thrust carries a Lower Eocene allochtonous thrust-sheet-top basin (or piggyback basin cf. Ori and Friend 1984, see also Ricci Lucchi 1986) named the (present-day Tremp-Graus Basin). The remaining autochtonous part of the Pyrenean Foreland Basin is the present-day Ebro Basin (Fig. 18.1).

As a consequence, the western part of the southward-moving Montsec Thrust developed as a northward curved oblique blind thrust with associated faults (cf. Dunne and Ferrill 1988) and acted as a lateral ramp separating shelf and slope deposition during Ypresian sedimentation. The formation of secondary blind thrusts, associated with the Montsec Thrust, is expressed as near-surface subtle, low-amplitude, gentle folding in intrabasinal areas (de Boer et al. 1991; López-Blanco et al. 2003; Clevis et al. 2004).

The southward moving Montsec Thrust sheet, including the present-day Tremp-Graus Basin, became incorporated into the cover of the next developing thrust, the Sierras Marginales, when this was initiated in the early Lutetian and overthrusted to form several smaller imbricated thrust units and thrust-sheet-top basins; the connected Tremp-Graus,

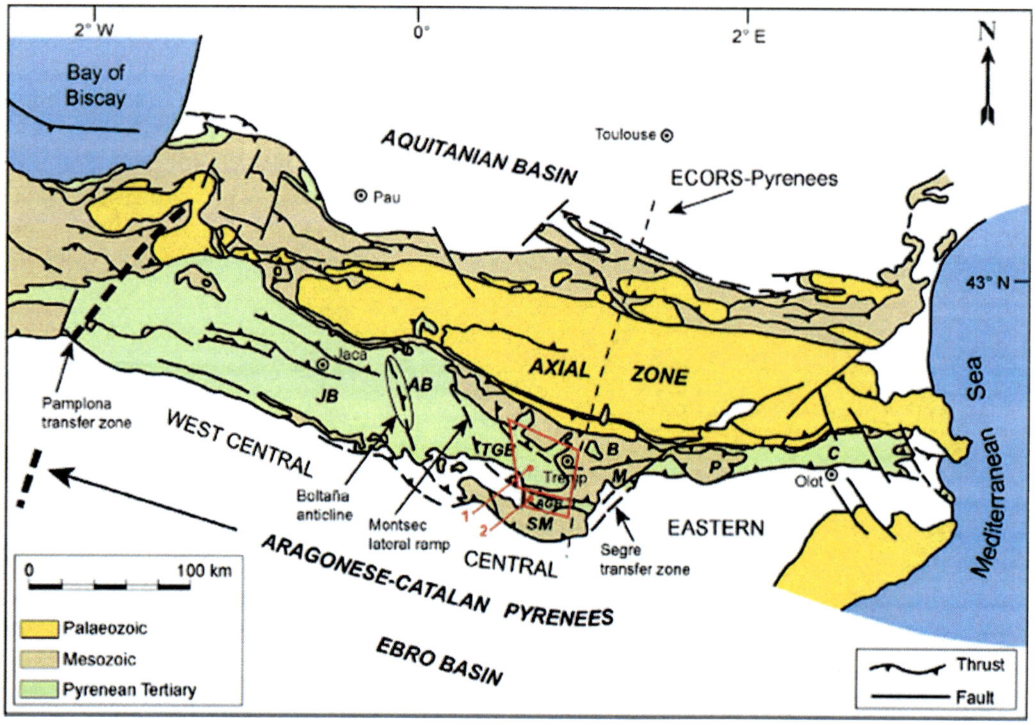

Fig. 18.1 Structural map of the Aragonese-Catalan Pyrenees (modified after Capote et al. 2002). The present-day Tremp-Graus and Ager Basins are located south of the axial zone in the central and eastern Pyrenees and south of a structural divergence axis in the west-central Pyrenees. The eastern boundary of the Central Pyrenees is formed by the NE-SW oriented oblique Segre transfer zone and the western boundary is formed by the NW-SE oriented oblique Boltaña anticline. The southern margin is formed by the E-W oriented frontal thrust of the Sierras Marginales (SM). B Bóixols thrust sheet, C Cadi thrust sheet, M Montsec thrust sheet, P Pedraforca thrust sheet, SM Sierras Marginales thrust sheet, AB present-day Ainsa Basin, AgB present-day Ager Basin, JB Jaca Basin, TGB present-day Tremp-Graus Basin. Inset 1: geological map of Fig. 18.10; inset 2: geological map of Fig. 18.11b

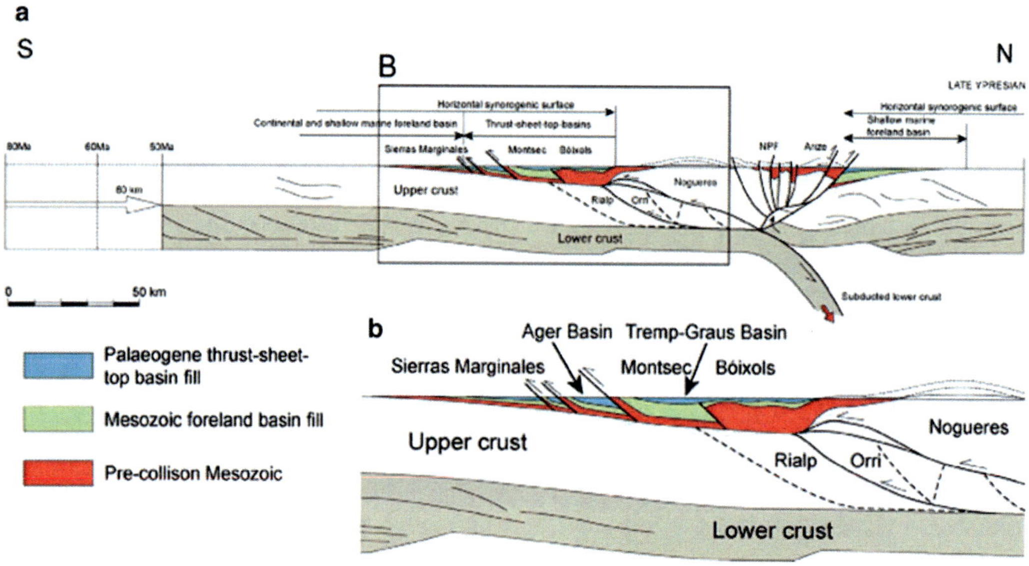

Fig. 18.2 (**a**) Partially restored cross-section of the crust through the central Pyrenees at the location of the ECORS profile (Choukroune et al. 1989, 1990) illustrating the tectonic style and the three main imbricated thrust sheets (Bóixols, Montsec and Sierras Marginales). Reference frame holds the European Plate and *arrows* indicate estimated total convergence. *NPF* North Pyrenean Fault (Modified after Beaumont et al. (2000)). (**b**) Enlargement of the *boxed area* in **a** illustrating the position of the present-day Ager and Tremp-Graus Basins

Ager and Ainsa Basins (Mascle and Puigdefàbregas 1998). The rate of thrusting was highest from the end of the Paleocene to the Lutetian (Vergés et al. 1995; Millán et al. 1995) producing a relatively deep and wide basin and the broadest expansion of marine deposits (Puigdefàbregas et al. 1992; Burbank et al. 1992). Syntectonic contemporaneous sequences are present in neighboring basins (Pocoví 1978; Martínez-Peña and Pocoví 1988; Vergés et al. 1995; Mascle and Puigdefàbregas 1998). Paleomagnetic data from the present-day Ainsa Basin, which rotated 30° clockwise during Montsec Thrust displacement (Poblet et al. 1998), indicates the rotation of the entire thrust-sheet-top basin (Nijman 1989; Poblet et al. 1998). This complex structural development had a profound effect on basin morphology and sedimentation patterns and has influenced basin-scale tidal current patterns.

18.2.2 Lower Eocene Paleoclimate, Eustatic Sea Level, and Stratigraphy

During the upper lower Ypresian, the upper Ypresian and the Lutetian, the study area was located at a latitude of approximately 35°N and was characterized by a warm and stable, sub-tropical to tropical climate (Early Eocene Climatic Optimum; Zachos et al. 2001) and generally warm and arid to semiarid conditions in the Pyrenees (Haseldonckx 1972; Schmitz and Pujalte 2003; Pearson et al. 2007).

The Ypresian and Lutetian are periods of pronounced long-term tectonic development (see above) controlling local relative sea-level changes, that are likely to have varied spatially and temporally in the Pyrenean foreland Basin. Egger et al. (2009), in contrast, concluded that the effects of Paleogene changes in eustatic sea-level exceeded the effects of regional tectonic activity in the shelves of the European and Adriatic Plates as a result of the much shorter magnitude and time scales over which these processes operate. This is illustrated by the occurrence of a widespread and relative rapid marine transgression during the lowermost Ypresian (Pujalte et al. 2009). However, in general and in comparison Alpine tectonic activity in the Pyreneen orogenic zone was much stronger than elsewhere (Vergés et al. 1995; Meigs and Burbank 1997; Capote et al. 2002).

18.3 Tidalites and Paleogeography, Bathymetry and Fill – Underfilled Foredeep Versus Overfilled Shelf

18.3.1 Underfilled Foredeep

The underfilled foredeep stage (Table 18.1, Fig. 18.3b) is stratigraphically represented by deposits of the Ager Group (Figs. 18.4 and 18.5; see next section) and occurred during the early stage of thrust-sheet-top basin development (Early Ypresian, 55.5 Ma to 51.5 Ma). As discussed hereafter, the associated and characteristic configuration of basin morphology parameters indicate that the basin was within the tidal amplification window (cf. Sztanó and de Boer 1995; Fig. 18.6a, c).

The present-day Tremp-Graus and Ager Basins, now separated by the Montsec Thrust, formed one entity, Tremp-Graus-Ager (T-G-A) Basin. The T-G-A Basin was elongated along an E-W line and connected to the Atlantic Ocean via the Jaca Basin (Fig. 18.7a). A subaerial topographic high, formed by a thrusted anticlinal system, bounded the T-G-A Basin in the south (Figs. 18.2 and 18.3b). In the area east and south-east of Tremp, the T-G-A Basin was most likely closed and is referred to as the Gulf of Ager by some workers (Fonnesu 1984; Eichenseer 1988; Eichenseer and Luterbacher 1992; Mutti et al. 1994; Barberà et al. 1997; Waehry 1999). It included a southern limb that likely extended farther to the east than the present-day closure (Maestro-Maideu et al. 1991; Dreyer and Fält 1993; Dercourt et al. 2000; Vincent 1993; Olariu et al. 2008a; Fig. 18.3b). Any connection with the Mediterranean is uncertain and unlikely as it would generate strong tidal currents for which no support is found in the sedimentary record. A number of NW-SE oriented anticlines developed along the central northern basin margin controlled by the long-lived inverted Mesozoic fault structures. They developed either as (i) lateral or oblique ramps (or their associated hanging-wall anticlines) to S-directed upper cover-thrust sheets (cf. Muñoz 1992) or, alternatively, (ii) linked to the Bóixols frontal thrust (Cámara and Klimowitz 1985). It is suggested here that the periodic re-activation of these structures was expressed by an indented coastline along the northern basin margin locally forming estuaries and/or embayments causing tidal amplification and ebb-flood cyclicity.

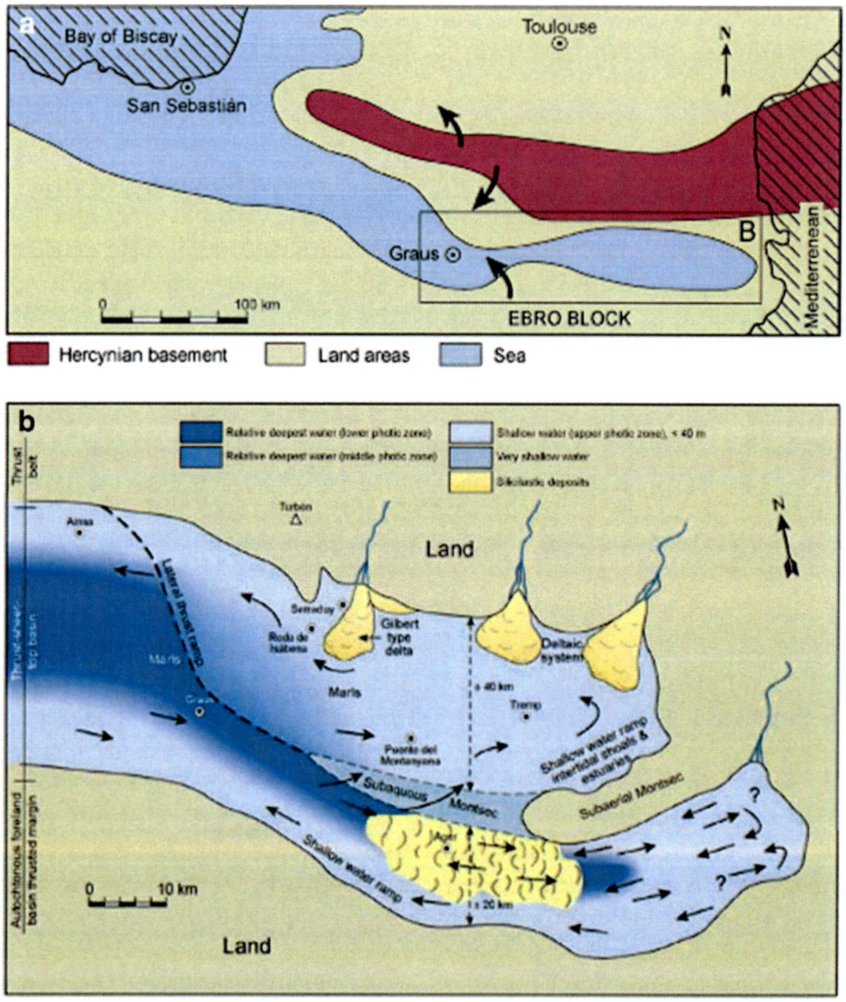

Fig. 18.3 (**a**) Sketch of the estimated maximum extent of the allochtonous south Pyrenean Foreland Basin illustrating the narrow, elongated geomorphology of the semi-enclosed sea encompassing the Jaca and T-G-A Basins (modified after Plaziat 1981). (**b**) Sketch of the inferred paleogeography and water depths of the T-G-A Basin sea during the underfilled foredeep stage of the Late Ilerdian (approximately 53 Ma) based on the work of the authors referred to in the text. The main siliciclastic depositional systems are indicated

The basin had an asymmetrical transverse section with the strongest subsiding parts along the northern basin margin and along the south-side of the developing Montsec Thrust. Highest Early Eocene subsidence rates occurred during the Ypresian. Sedimentation rates, however, were also highest along the northern basin margin, such that a transverse section through the basin may have been approximately symmetrical or even asymmetrical with a depositional basin axis further south. The basin had pronounced sea-floor topography due to the effects of the southward moving thrust sheets. The E-W oriented Montsec Thrust probably formed a shallow-water anticlinal topographic high on the sea floor across which exchange (intermittent?) of marine waters could occur between the areas of the present-day Ager Basin and Tremp-Graus Basin. This situation was sustained by the relatively rapid eustatic sea-level rise at the start of the Early Ypresian (Pujalte et al. 2009).

As a result, the total basin length was approximately 200 km (Mutti et al. 1985a) with an approximately 50 km long and shallower (approximately 40–60 m) T-G-A Basin section east of the Montsec oblique blind thrust (Fig. 18.3). The paleo-width of the T-G-A Basin

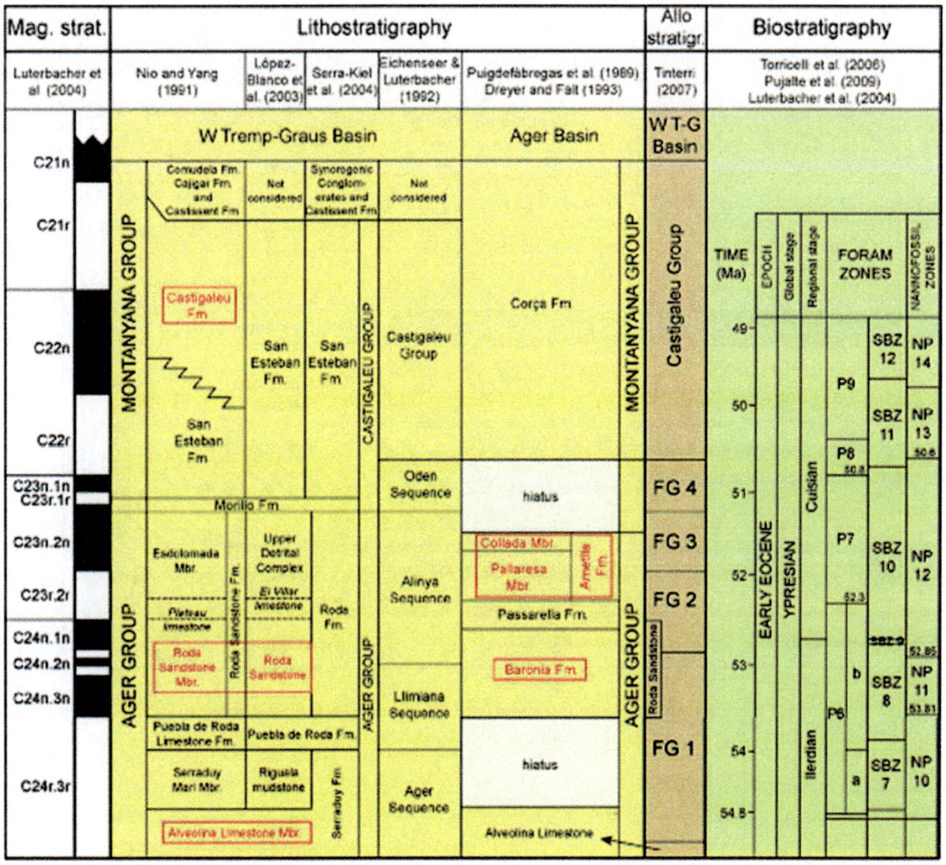

Fig. 18.4 Summary correlation diagram of Ypresian stratigraphy of the present-day Tremp-Graus and Ager Basins with a focus on the Ager Group. The base of the Ypresian is estimated at 55.8 Ma and the top at 48.6 Ma (Luterbacher et al. 2004). The magnetic polarity change from chron C24 to chron C23 occurs at 52.6 Ma, and the change from chron C23 to chron C22 occurs at 50.8 Ma (Luterbacher et al. 2004). The former change coincides with a dating of 52.6 Ma for the Plateau Limestone (López-Blanco et al. 2003). Unpublished data of the top of a *Turritella*-dominated mudstone interval in the Esdolomada Mbr (which is on average 180 m thick) occurring 29 m above the Plateau Limestone at the stratigraphic level of the El Villar Limestone gives an age of 52.4 Ma which coincides with top P6 and a maximum flooding of the Roda Fm in the Isabéna valley (confirmed by Torricelli et al. 2006). Following this data, the Roda Sst Mbr covers approximately 0.9 Ma and the Esdolomada Mbr covers approximately 1.6 Ma. Note, however, that López-Blanco et al. (2003), using magnetostratigraphic data obtained from the Roda Fm (Bentham and Burbank 1996) and the magnetic polarity scale of Cande and Kent (1995), estimated approximately 400 Ka for the Roda Sst Mbr and approximately 600 Ka for the Esdolomada Mbr. The allostratigraphic subdivision of the Figols Group advocated by Mutti and his co-workers, based on the philosophy that comparable facies throughout the basin are combined into groups or depositional systems (for example, the Figols Group for terrigeneous facies of mainly deltaic character, the Campodarbe-Montañana Group for continental facies and the Hecho Group for basin floor facies), is not incorporated. Note that tidal deposits of Maastrichtian age in the Aren Formation (Tremp Group; Nagtegaal et al. 1983; De Boer 1985 and a wealth of recent but unpublished data collected by Spanish workers) and of Upper Lutetian age in the Sobrarbe Formation (Campodarbe Group; Hall 1997; Dreyer et al. 1999) are also present but not included in this review

is uncertain but is estimated at up to 60 km (larger than it is at present after post-Ypresian compression, uplift and erosion).

In conclusion, the combined Jaca and T-G-A Basin was within the tidal amplification window (cf. Sztanó and de Boer 1995) with favourable average depth and length for resonant amplification (Figs. 18.3 and 18.6a, c), favorable paleobathymetric parameters, but unfavourable depth-width configuration for development of amphidromic point(s) as the basin was too narrow.

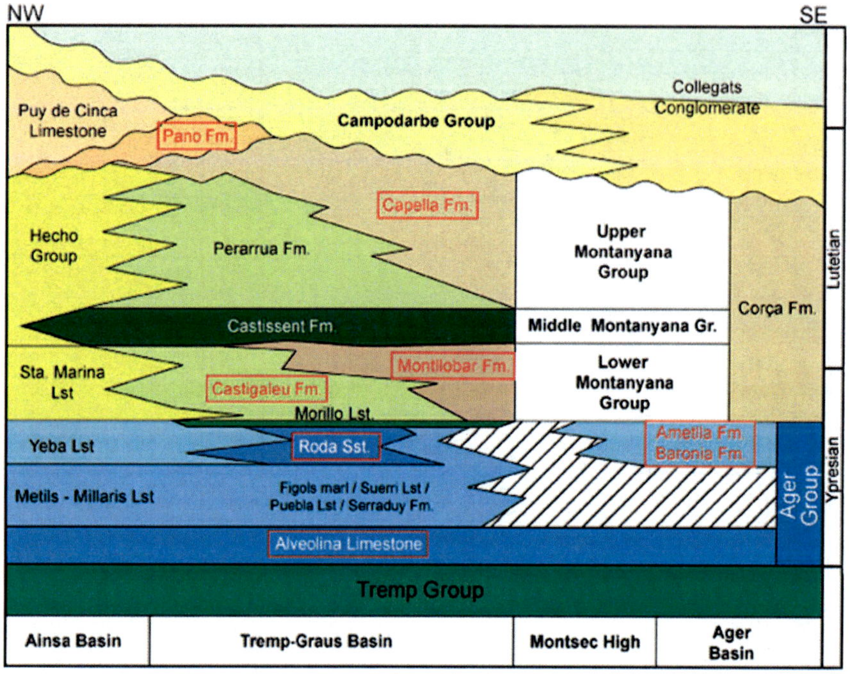

Fig. 18.5 Summary correlation diagram of Ypresian and Lutetian stratigraphy of the present-day Tremp-Graus and Ager Basins with a focus on the Montanyana Group (modified after Nijman 1998)

18.3.2 Overfilled Shelf

Tidalites of the overfilled shelf stage (Table 18.1, Fig. 18.7) are represented by facies of the Middle to Late Ypresian and Lutetian (51.5 Ma to 43 Ma) Montanyana Group present across the entire T-G-A Basin (Figs. 18.4 and 18.5). The T-G-A Basin most probably moved to the margin of the tidal amplification window (cf. Sztanó and de Boer 1995; Fig. 18.6a, c). As discussed hereafter, this was caused by the development of increasingly more pronounced basin-floor topographic features formed as the result of continuing thrust movement. These had a progressively more important control on basin morphology, depositional environments and tidal resonance.

The overfilled shelf stage resulted from the conversion of the (segmented) underfilled foredeep of the T-G-A Basin to a well-defined shelf and slope configuration as a result of ongoing thrusting and associated southward thrust-sheet-top basin translation along the Montsec Thrust as well as shortening along a N-S direction. The western part of the southward moving Montsec Thrust further developed into a pronounced northward curved oblique (NW-SE) blind thrust (Nijman and Nio 1975; Dunne and Ferrill 1988; Cuevas Gozalo 1989; Donselaar 1996a; Nijman 1998; Poblet et al. 1998; Clevis et al. 2004; Figs. 18.1, 18.2, 18.7 and 18.8) and acted as a lateral ramp separating shelf from slope and basin floor deposition during Late Ypresian and Lutetian sedimentation (52 Ma to 43 Ma; Cámara and Klimowitz 1985; Puigdefàbregas et al. 1992).

Associated smaller blind thrusts and associated ramps developed contemporaneously. An example of such an additional thrust is the Lascuarre reverse fault system (E of Graus; Fig. 18.9), with a NNE-SSW orientation, which acted as the most important sea-floor topographic expression (Cámara and Klimowitz 1985; Puigdefàbregas et al. 1992) east of the main oblique lateral ramp of the Montsec Thrust from 55.8 Ma to 48.6 Ma. It formed the transition from upper delta plain environments of the Montanyana Group in the T-G-A Basin on the hanging wall of the lateral ramp of the Montsec Thrust to contemporaneous lower delta plain deposition on the shelf margin and mass-flow deposition on the slope (the latter now making up a significant part of the present-day Ainsa Basin; Nijman and Nio 1975; Cuevas Gozalo 1989; Cuevas Gozalo and de Boer 1991; Donselaar 1996a; Nijman

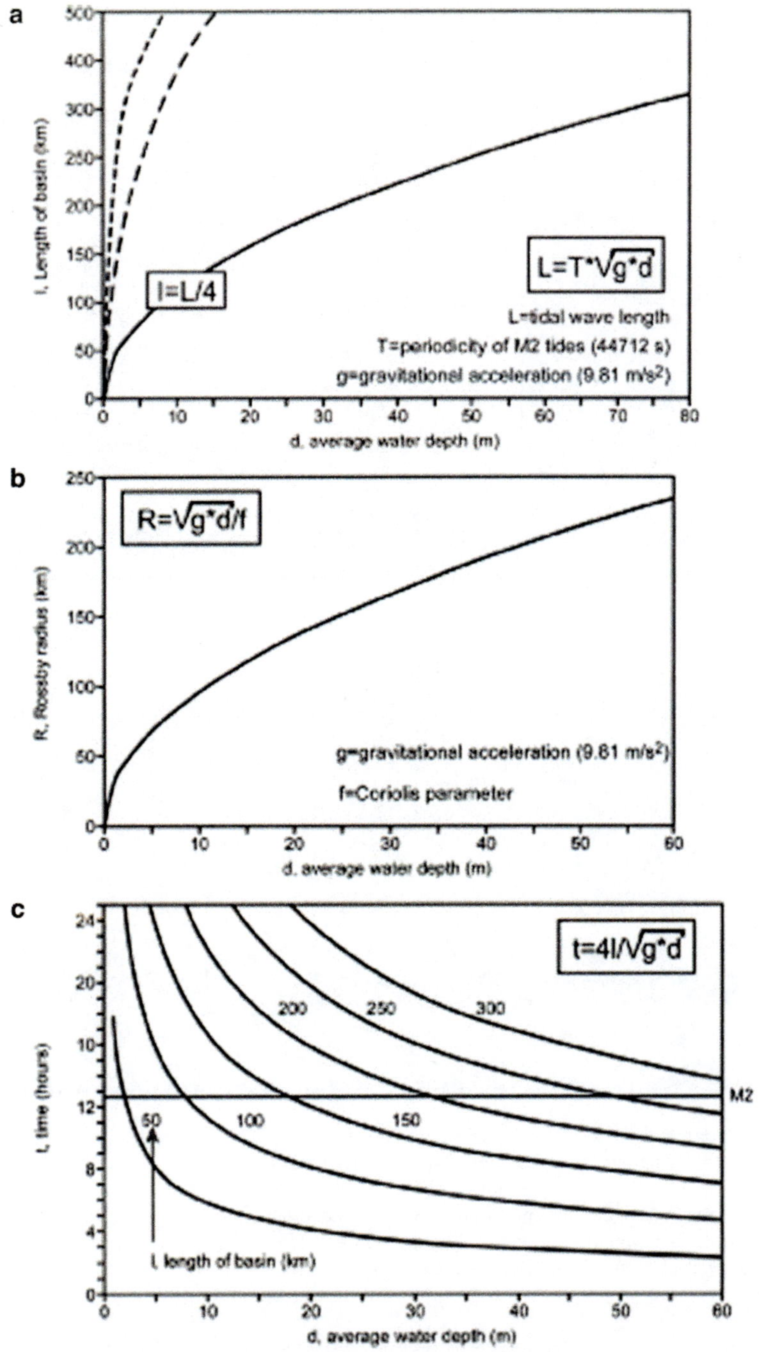

Fig. 18.6 The "tidal amplification window" (Sztanó and de Boer 1995) includes a number of variables (but does not necessarily need to simultaneously honor all of these) such as favourable depth and length for resonant amplification, and depth and width for the development of (an) amphidromic point(s) (Pugh 1996; see also Chap. 13), and funnelling (cf. Bay of Fundy). Basin-scale tidal amplification thus may happen during specific phases of (i) overall basin development as well as (ii) relative sea-level cycles during one phase of basin development during which the necessary requirements for resonance are met. (**a**) The average depth of a basin determines the celerity and the length of the propagating tidal wave. For resonant amplification the critical basin length should be an odd multiple of the quarter of the tidal wave length. The *solid line* indicates the relation between the average water depth and the length of the basin for one times the quarter of the tidal wave length. The two successively steeper *dashed lines* indicate the same relationship for three and five times (the two following odd multiples) the quarter of the tidal wave length respectively. (**b**) The Rossby deformation radius of the Kelvin wave describes how wide a basin should be for the development of an amphidromic system. (**c**) The natural period of oscillation of basins of different length depends on water depth. Cross-points of the line of the M2 tide (12.42 h) with the hyperbolical curved oscillation period time line for different basin lengths indicate basin length and depth required for conditions near resonance of the semidiurnal tide (modified after Sztanó and de Boer (1995) and based on Pugh (1987, 1996))

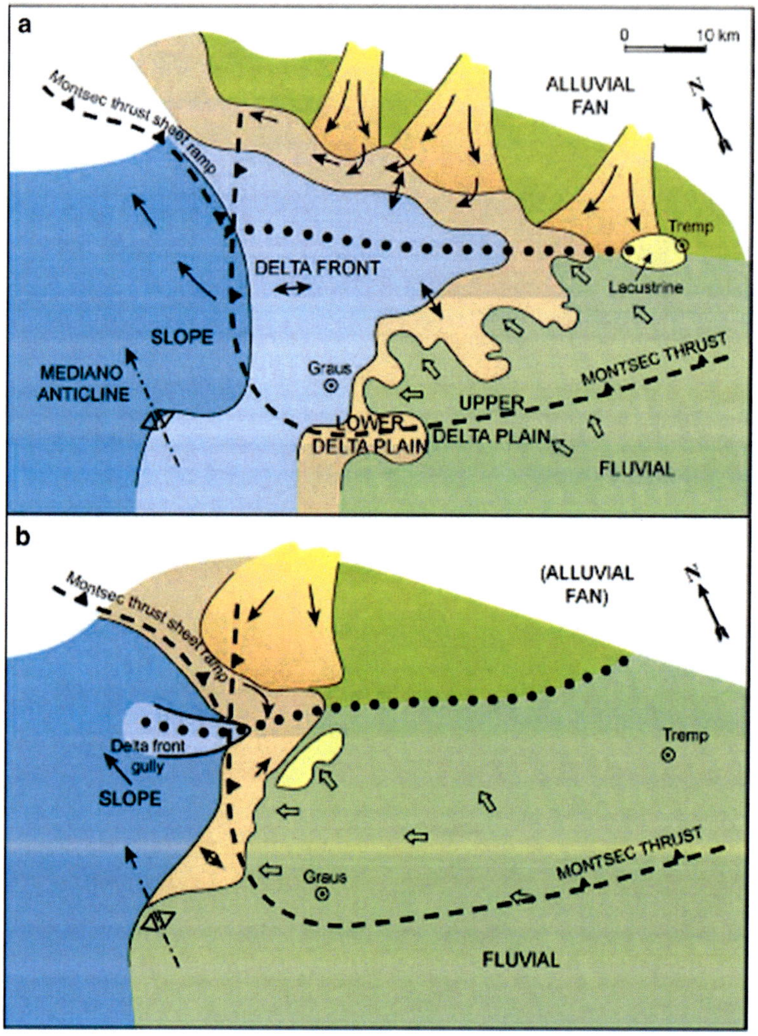

Fig. 18.7 Sketch of the inferred paleogeography and water depths of the T-G-A basin sea during the overfilled shelf stage of the Lutetian. The *dotted line* indicates the position of the basin axis. (**a**) The Montanyana system at initial progradation. Alluvial fans along the northern basin margin and a fluvial system in the eastern and central parts of the T-G-A Basin. A relatively large delta front area existed with interdistributary bays and mouth bars. (**b**) The Montanyana system at advance progradation. River systems dominated and extended out almost to the shelf break filling-up the shelf area. Only a narrow delta front area remained (modified after Nijman 1998)

1998; Poblet et al. 1998; Clevis et al. 2004; Figs. 18.7 and 18.8). It also prevented the formation of well-developed clinoforms (Puigdefàbregas et al. 1989). Later, from 48 Ma to 43.5 Ma (Early Lutetian), the Mediano anticline (Figs. 18.7 and 18.9) developed over the propagating Montsec oblique blind thrust, approximately 20 km to the west of the Lascuarre fault system, causing further steepening of the slope, over a time span of 5.5 Ma (Garrido-Megías 1973; Nijman and Nio 1975; Poblet et al. 1998). In between, the Isábena Depression developed (Cuevas Gozalo 1989; Fig. 18.9).

During Middle to Late Ypresian and Lutetian, the eastern and central parts of the T-G-A basin had begun to be progressively filled with alluvial and deltaic sediments from the north and east towards the west.

As a consequence, water depths in the early stages of infill are estimated to have been approximately 20 m

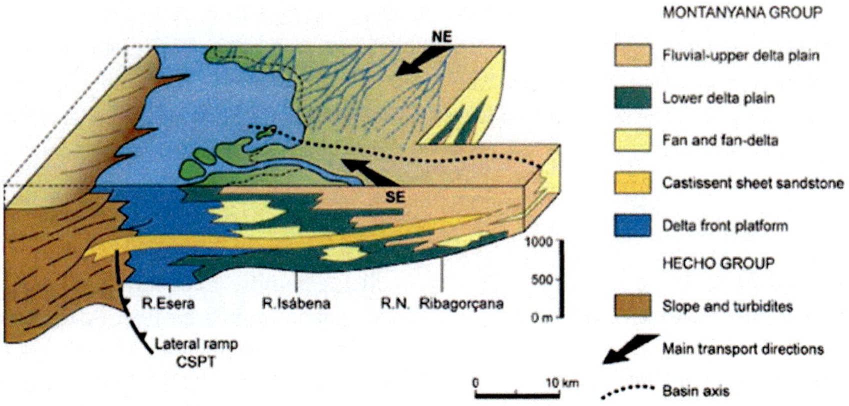

Fig. 18.8 Block diagram of the Montanyana delta showing alluvial fan and fluvial feeder systems combining in one delta front with break-in-slope above the lateral ramp of the underlying Montsec thrust sheet. *CSPT* Central South Pyrenean Thrust system including the Montsec Thrust and its lateral ramps (modified after Marzo et al. 1988)

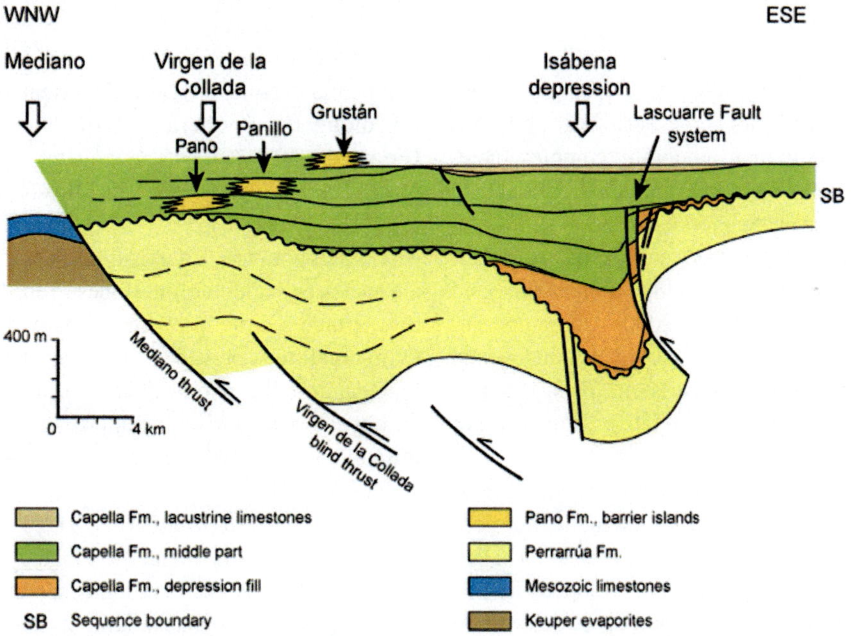

Fig. 18.9 Correlation profile of the Capella and Pano Fms in the Isábena depression and Virgen de la Collada ramp areas (modified after Cuevas Gozalo 1989, and Donselaar 1996a)

in front of the Montanyana shoreline in the west (Nijman 1998) but quickly decreased during westward progradation of the system to water depths significantly less than 20 m (and up to 0). This reduction of the shelf area reduced tidal amplification. It is estimated, however, that the T-G-A Basin width and length were largely unchanged (approximately 40 km wide and 50 km long). Sediment pathways continued into contemporaneous extensive basin floor fan systems (Hecho Group; Mutti et al. 1972, 1973, 1975, 1985a) in the trough of the strongly subsiding E-W oriented Jaca Basin on the footwall. The open connection to the Atlantic Ocean still existed; no major structural barrier was present (Mutti et al. 1972, 1985a, 1988; Nijman and Nio 1975).

In conclusion, it is envisaged that the combined Jaca and T-G-A Basin was located in the margin of the tidal amplification window (cf. Sztanó and de Boer 1995). Despite the fact that average basin length was favourable for resonant amplification (Figs. 18.3 and

18.6a, c), the paleobathymetry became unfavourable as water depths on the (very) narrow shelf were (very) low and increased over (very) short (slope) distances to relatively deep. In addition, the depth-width configuration was unfavourable for development of amphidromic point(s) as the basin was too narrow.

18.4 Underfilled Foredeep Tidalites – The Ager Group

18.4.1 Ager Group Stratigraphy and Depositional Environments (Early Ypresian – 55.5 Ma to 51.5 Ma)

The Lower Ilerdian Alveolina Limestone is part of the Serraduy Fm (sensu Cuevas Gozalo et al. 1985; Figs. 18.4, 18.5 and 18.10) which, in general, comprises a low-relief carbonate ramp platform facies association developed along the basin margins. It is typified by a spatially and temporally complex facies architecture comprising, for example, intertidal and supratidal flats, carbonate shoreface facies, sublittoral sand bars, as well as outer ramp and open marine shelf environments (Eichenseer 1988; Eichenseer and Luterbacher 1992; Payros et al. 2000). Contemporaneous deposition of bioclastic carbonate shales and nodular limestones (Metils-Milaris Fm; Figs. 18.4 and 18.5) took place in the area of relative deepest water (Fig. 18.3b).

Tidal deposits of the Roda Fm (sensu Cuevas Gozalo et al. 1985; Fig. 18.4) occur along the northern margin of the Tremp-Graus Basin in the Isábena valley (Fig. 18.10) and form part of a well-developed NE-SW prograding, tide-influenced Gilbert-type delta (Nio and Siegenthaler 1978; Cuevas Gozalo et al. 1985; Yang and Nio 1985, 1989; Jimenez 1987; Eichenseer 1988; Tosquella 1988; Nio and Yang 1991; Serra-Kiel et al. 1994; Molenaar and Martinius 1990, 1996; Martinius and Molenaar 1991; Joseph et al. 1993; López-Blanco 1996a, b; López-Blanco et al. 2003; Torricelli et al. 2006; Tinterri 2007; Olariu et al. 2011; Leren et al. 2010; Michaud 2011).

In the Isábena valley, the Roda Formation is divided into the lower Roda Sandstone Mbr (approximately 120 m thick) and the upper Esdolomada Mbr (approximately 180 m thick, Nio and Yang 1991; López-Blanco et al. 2003; Fig. 18.4). The Roda Sandstone Member comprises at least six lobate-shaped sandstone wedges of a Gilbert-etype delta (López-Blanco et al. 2003; Leren et al. 2010) that displays an overall progradational (the Roda Sandstone Member) to retrogradational (lower part of the Esdolomada Mbr) pattern. The retrogradational part of the Roda Gilbert-type delta is overlain by the El Villar Limestone (Fig. 18.4). Each lobe has been subdivided into a number of smaller subunits. Each of these shows a generally lobate shape formed by large-scale (up to 15 m high) foresets with a dip angle of up to 32° (Yang and Nio 1989; Joseph et al. 1993; López-Blanco 1996a; López-Blanco et al. 2003; Tinterri 2007).

Approximately age-equivalent Roda Fm outcrops in the closed eastern part of the Tremp-Graus Basin, between the Noguera Ribagorçana and Noguera Pallaresa Rivers (Fig. 18.10), are less well described. Tidally dominated channel fills and bars formed in an embayment setting suggest NW-oriented tidal reworking along the northern and southern basin margin (Fonnesu 1984; Cuevas Gozalo et al. 1985; Mutti et al. 1994; Waehry 1999, his allostratigraphic units Figs. 18.4 and 18.5).

Contemporaneous deposition of bioclastic carbonate shales and nodular limestones (Yeba Fm; Fig. 18.5) took place in the area of relative deepest water. Oxic conditions prevailed, in places close to the lower boundary of the photic zone (Torricelli et al. 2006) with water depths reaching approximately 80 m. Near the Roda Gilbert-type delta, water depths decreased to about 40 m (Jimenez 1987).

The Baronia Fm (Mutti et al. 1972, 1973; Figs. 18.4, 18.5 and 18.11) is a tide-dominated succession of interbedded sandstones and sandy siltstones with generally large-scale tabular bedding and no recognizable channels. Trace- and body-fossil assemblages throughout the formation indicate fully marine conditions (Mutti et al. 1985b; Rubino et al. 1985; Wonham 1993; Olariu et al. 2008a). The dominant paleoflow direction was towards the east, with a subordinate component towards the west. However, in some parts of the depositional system, westward oriented transport directions parallel to the orientation of the subaqueous Montsec Thrust prevailed (Rubino et al. 1985; Olariu et al. 2008a). Sediment must have been derived from the basin margin in the south and/or east and from reworking of older deposits (note the hiatus below the Baronia Fm; Figs. 18.4 and 18.5).

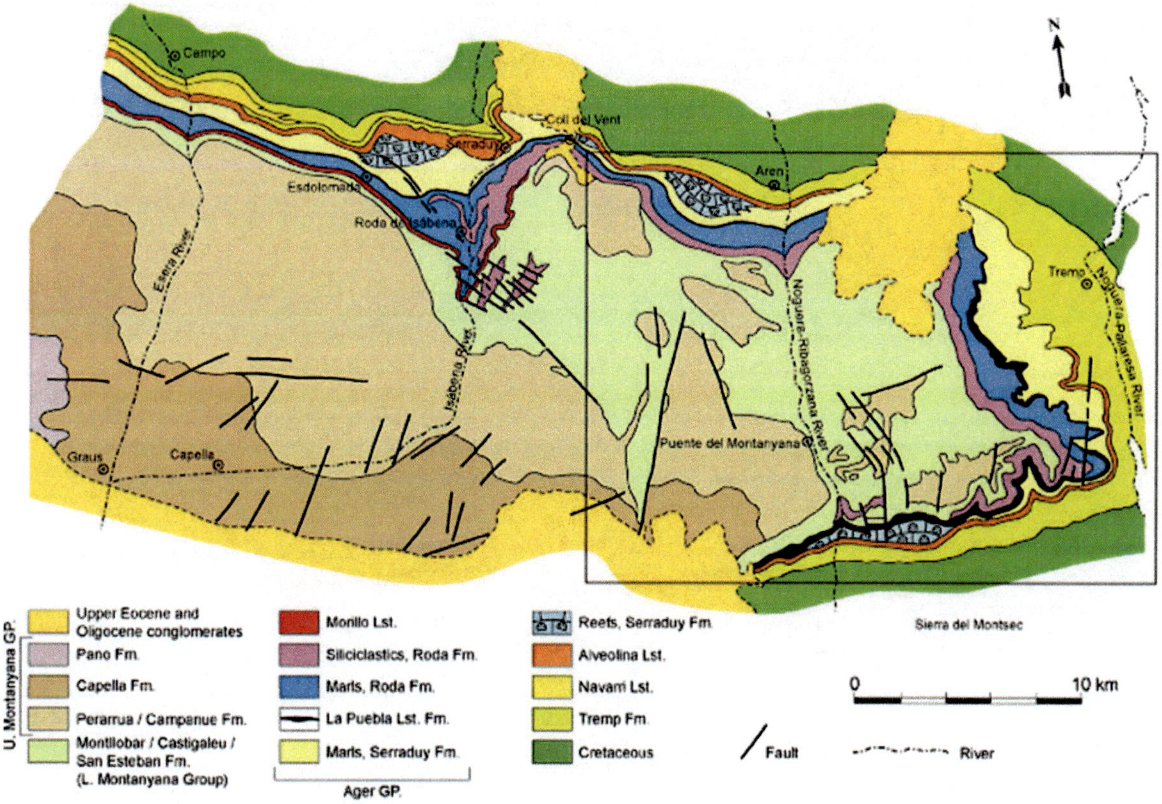

Fig. 18.10 Simplified geological map of the present-day Tremp-Graus Basin showing the location of the outcropping Alveolina Limestone and the Roda, Capella and Pano Fms (modified after Serra-Kiel et al. 1994, who used data from Fonnesu 1984, Samsó 1988 and Tosquella 1988, and Cuevas Gozalo 1989). The *box* indicates the area in part covered by Fig. 19.11a

Mutti et al. (1972, 1973, 1975) divided the formation into three units (lower, middle and upper). The lower and upper units consist of a series of tabular sandstone bodies characterised by an upward-coarsening grain-size profile and a sigmoidal geometry of accretion surfaces (Mutti et al. 1985b; Olariu et al. 2008a). The upper unit of the Baronia Fm is interpreted to have been deposited in shallower water, subject to stronger tidal currents than the lower unit (Mutti et al. 1985b).

Wonham (1993) divided the Baronia Fm into two low-order sequences. The lower sequence is composed of a transgressive estuary succession with barriers at the mouth of the estuary developed above an erosional unconformity formed by lowstand fluvial incision. A tripartite estuarine facies distribution was recognized with ebb-tidal delta deposits in the distal western part of the basin and bay-head delta bars in the proximal astern part of the basin. Higher-order relative sea-level changes resulted in repeated basinward shifts of facies and a successive broadening of the estuary in time. The estuary deposits are overlain by a relatively thin (up to 12 m) succession of transgressive offshore and shelf deposits. The upper sequence is also formed by a tide-dominated estuary overlying an unconformity formed by lowstand fluvial incision (Wonham 1993). The wide estuary had no barriers at its mouth and was filled with compound cross-stratified beds formed by ebb- and flood-directed cosets containing tidal bundles; water depths were interpreted to decrease from about 20 m at the base to 5–10 m at the top. The transgressive estuary fill is overlain by a retrogradational set of lower shoreface and offshore deposits (Wonham 1993).

The Ametlla Fm (Mutti et al. 1972, 1973; Figs. 18.4, 18.5 and 18.11) rests conformably on the offshore siltstones of the Passarella Fm. It is informally

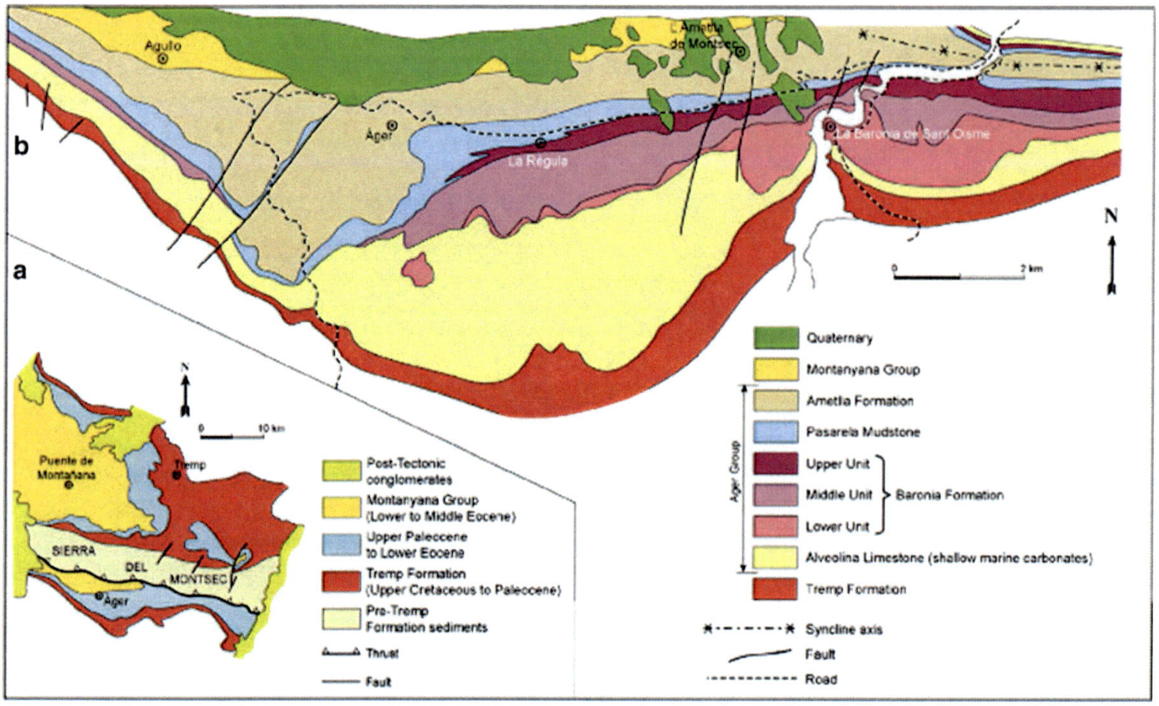

Fig. 18.11 (**a**) Strongly simplified geological sketch map of the eastern sector of the present-day Tremp-Graus Basin and the Ager Basin. (**b**) Geological sketch map of the present-day Ager Basin (modified after Mutti et al. 1985b)

subdivided into two members (the Pallaresa and Collada members; Dreyer and Fält 1993; Fig. 18.4) and tidal deposits occur in both members (Ghibaudo 1975). The Pallaresa member is comprised of offshore siltstone intervals which may be up to 45 m thick, and six laterally continuous tide-influenced and – dominated sandstone bodies deposited in deltaic, estuarine and tidal shelf settings which may be up to 32 m thick (Fig. 18.12). The Collada member is interpreted to have accumulated in a coastal plain to tidal flat environment (Dreyer and Fält 1993; Dreyer 1994). Sediments were shed from an extensive topographic high to the south (the developing Sierras the Marginales thrust system) and to the east. The formation was deposited during active pulsating thrust-sheet development, resulting in a number of high-frequency relative sea-level changes (Mutti et al. 1988). The highest relative subsidence occurred in the eastern part of the basin, where the Ametlla depocentre was located. Water depths, however, increased westward (Dreyer and Fält 1993). During deposition of the Ametlla Fm, the eastern closure of the T-G-A basin was (significantly) nearer to its present-day position (Dreyer and Fält 1993).

18.4.2 Tidal Bars of the Alveolina Limestone

The combined effect of the relatively rapid eustatic sea level rise at the start of the Early Ypresian (Pujalte et al. 2009) and southward displacement of the Bóixols Thrust and associated lateral or oblique ramps resulted in NW-SE oriented anticlines and synclines along the central-northern margin of the T-G-A Basin. Two anticlines developed near Serraduy (Fig. 18.10; the Roda and Coll de Vent anticlines; Eichenseer 1988; Vincent 2001; López-Blanco 1996a, b) with a syncline in between (the Serraduy-Sis syncline; López-Blanco et al. 2003). This area became the locus for an approximately NE-SW oriented warm-water and southward opening narrow coastal embayment ('Serraduy Bay' of Unit 2 in the upper part of sequence V of Eichenseer 1988), typified by strong tidal currents, during the relative sea-level rise. Deposits formed in this embayment are part of the Alveolina Limestone Mbr of the Serraduy Fm (Fig. 18.4). Shallow-water carbonate banks and reefs developed particularly on the flank of the Roda anticline (Pool 1983; Eichenseer 1988). The embayment was dominated by WSW directed

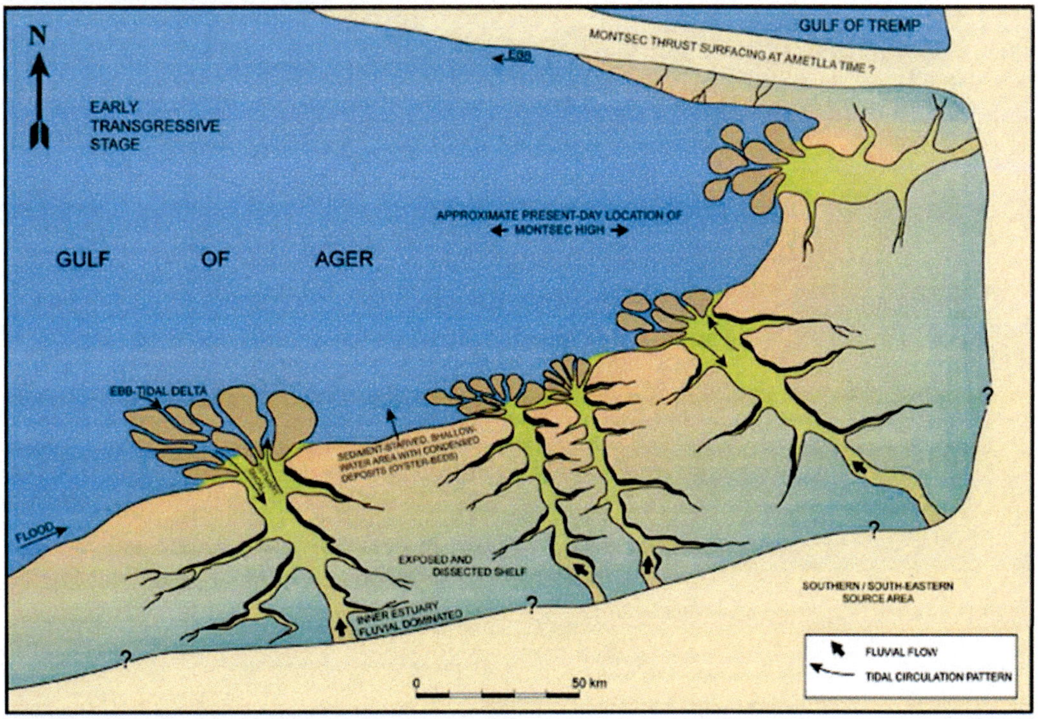

Fig. 18.12 Schematic illustration of the early transgressive stage in the inferred paleogeographical development of the T-G-A Basin during deposition of the Ametlla Fm (modified after Dreyer and Fält 1993)

ebb-oriented dunes of up to 50 cm height; oppositely directed cross strata (flood currents) were subordinate. The dunes are part of thick (up to 12 m) bioclastic (grainstone) tidal bedforms, interpreted as bars (cf. Pool 1983; Eichenseer 1988; Fig. 18.13) based on the observation that cross-bedded sets show paleocurrent directions oriented obliquely to the accretion surfaces. The bars formed in tidal channels with a deeply scoured base and often covered by a lag of coralgal breccias derived from adjacent patch reefs. The bars contain several accretionary units with superposed dunes separated by discontinuity planes. Southward at short distances (200 m), small ebb tidal delta lobes with dominant WSW current directions developed, which, in places, contain double mud drapes; the existence of barrier islands, however, is not reported but inferred. The lateral equivalent of these tidal channel deposits east of Coll del Vent are formed by lagoonal to shallow-water bay deposits with occasional patch reefs. The succession is overlain by a coralgal reef and subsequently wide-spread shallow water nodular limestones of the Serraduy Fm.

18.4.3 Tidal Bars of the Esdolomada Member

Tidal bars, in this case transgressive shore parallel sandstone bodies, were developed in front of retrogradational fan-delta lobes and mouth bars of the lower part of the Esdolomada Mbr and are either attached to or detached (offshore) from the sandy delta lobe front. Underneath the El Villar Limestone (Fig. 18.4), a well-developed detached example (Fig. 18.14) is exposed near Roda de Isábena. This is formed by slightly inclined (1.6–4.6°) master bedding surfaces and contains stacked sets of high-angle (average dip angle 21°) cross-stratification up to 70 cm thick (Olariu et al. 2011). The crest of this bar is oriented sub-parallel to the tidal paleocurrent and to the nearby paleo-shoreline; the bar was built by oblique accretion, migrating transverse to the tidal currents towards the SW; it has a width to length ratio of approximately 1:10 (Olariu et al. 2011). Other examples, higher in the stratigraphy, also lack wave-generated structures and generally migrated obliquely towards the W driven by ebb-tidal

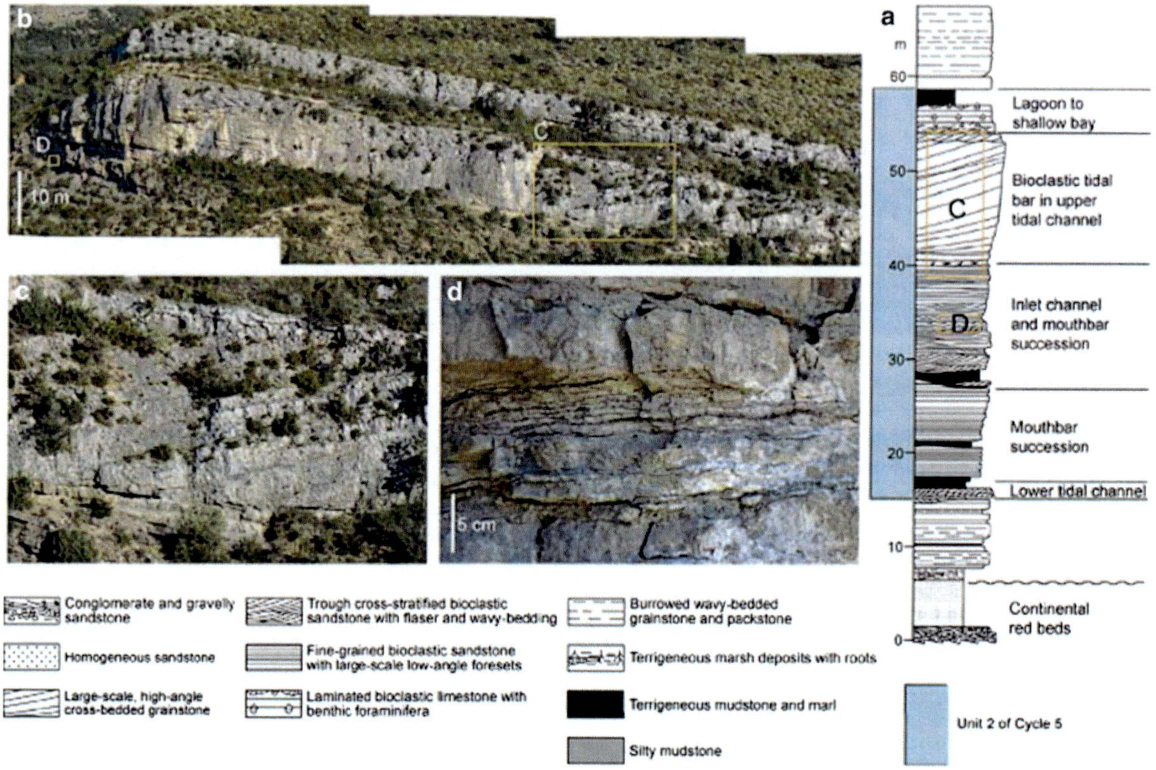

Fig. 18.13 (**a**) Sedimentary section through Unit 2 of Cycle 5 of Eichenseer (1988) of the Alevolina Limestone in the village of Serraduy (Fig. 18.10). (**b**) Outcrop image of bioclastic tidal bar deposits in the upper tidal channel indicated in (**a**). (**c**) Detail of the tidal bar deposits in the upper tidal channel showing large-scale inclined stratification (master bedding surfaces) and the erosive tidal channel base. (**d**) Bottomsets of cross-stratified sets with mud-draped laminae formed in an inlet channel and mouthbar succession (cf. Eichenseer 1988) underneath the upper tidal channel

currents flowing towards the NW. Tidal currents were deflected by the Gilbert-type delta lobe front along the NW-SE oriented paleoshoreline, reworking the lobe front.

The tidal bars appear to have developed during the transgressive phase of sedimentary cycles (López-Blanco 1996a; Olariu et al. 2011; Michaud 2011) in response to delta lobe abandonment after sediment delivery to the delta ceased. The ensuing high-frequency (10s of millenia) relative sea-level rise initiated favourable conditions for a period of reworking of the Gilbert-type delta lobe front by tidal currents and the formation of tidal bars similar to headland tidal banks (cf. Michaud 2010) that subsequently drowned or, in some cases, became moribund. In the latter case, they are capped by mudstone. In the former case, when the bars became stationary, carbonate buildups developed on top (Michaud 2010; several examples in the Esdolomada Member).

18.4.4 Tidal Bars of the Ametlla Formation

The lowermost three sandstone units of the Pallaresa mbr are dominantly composed of tide-influenced, delta-front mouth bars (Dreyer and Fält 1993). Sandstone unit 4 was studied in more detail (Dreyer 1994) and is composed of three parts. The lower part was deposited in a tide-dominated estuarine environment during flooding (early transgressive stage) of a previously created incised valley with a pronounced unconformity at the base (Dreyer 1994). Units formed by stacked cross-stratified sets are up to 4 m thick and separated by fine-grained sandstone (Dreyer and Fält 1993) resemble the compound dunes of the Baronia Fm. In the data presented, however, no specific information is provided enabling the assessment of the progradation direction of the master bedding surfaces in relation to paleoflow directions of superposed cross-stratified sets. It is therefore unclear whether the

Fig. 18.14 (a) A tidal bar at the base of the Esdolomada Mbr on the E side of the Isábena River SE of Roda de Isábena. Note the sheet-like geometry and well-developed slightly inclined (1.6–4.6° towards the SW) master bedding surfaces. This sandstone body migrated laterally (i.e., transverse to the tidal currents) towards the SW. (b) Detail of (a) showing stacked sets of high-angle (average 21°) cross stratification formed by dunes that migrated in a NW direction, that is approximately coast-parallel

sandstone bodies represent tidal bars (sensu Mutti et al. 1985b) or compound tidal dunes (sensu Olariu et al. 2008a).

This estuarine valley fill is overlain by mouth bar deposits of a bay-head delta (middle part of unit 4) which prograded towards the NW. The delta was initially fluvially dominated and influenced by tidal processes but transformed into a tide-dominated delta that prograded out into offshore inner shelf sediments during the later stages of the transgression (upper part of unit 4). No barrier further to the west is reported that created a back-barrier lagoon into which the bay-head delta prograded. The described characteristics illustrate that a certain degree of uncertainty is associated with the published interpretations and that the tide-influenced and tide-dominated depositional setting was characterized by a high degree of spatial and temporal variability of facies and sedimentary processes (Dreyer 1994).

The overlying sandstone units 5 (Fig. 18.15a, b) and 6 were formed by tide-dominated, near-shore to inner shelf 'sandbar complexes' during the late transgressive stage with common large-scale sigmoidal cross-stratified sets of up to 5 m thick. Average foreset dip angle is 23° to the west (Dreyer and Fält 1993) which is comparable to the dip angle observed in the tidal bar at the base of the Esdolomada Member (Olariu et al. 2011). Compound sets, up to 30 m thick and up to 3 km long, most probably have their long axis oriented parallel to the regional tidal flow direction (Dreyer and Fält 1993). These characteristics tend to classify these bodies as tidal bars, although a lack of relevant observational data prevents the distinction between tidal bars and compound tidal dunes.

18.4.5 Tidal Bars Versus Compound Tidal Dunes in the Baronia Formation

Tidal sandstone bodies of the lower part of the Baronia Fm in the T-G-A Basin (Mutti et al. 1985b; Fig. 18.16a–d) have long served as a well-documented and classical example of shelf tidal bars in front of a delta (Dalrymple 1992; Mellere and Steel 1996; Willis 2005). Recent work (Olariu et al. 2008a) on the lower unit of the Baronia Fm, however, led to the conclusion that the tidal sandstone bodies were formed by compound tidal dunes deposited in a narrow, approximately 10 km wide, sea that extended farther eastward than the eastern closure of the present-day Ager Basin. The tabular sandstone bodies are generally 4–6 m (but up to 10 m; Fig. 18.16a–c) thick and display an alignment transverse to paleoflow. They continue over 100s of

Fig. 18.15 Outcrop images of Sandstone Unit 5 of the Pallaresa mbr of the Ametlla Fm. east of the old railroad station of the town of Ametlla. (**a**) Overview of Sandstone Unit 5 behind the railroad station. (**b**) Overview of a part of the stratigraphy with Sandstone Units 4, 5 and 6 looking westward towards the railroad station. (**c**) Tabular cross-stratified set with mud-draped bottomsets in the outcrop of (**a**). (**d**) Tidal bundle succession at the base of Sandstone Unit 5 filling up a scour in the channel floor (2 km east of the railroad station). Note the reactivation surfaces and the neap-spring cycles. (**e**, **f**) – Details of (**d**) showing mud-draped foresets, reactivation surfaces and neap-spring cyclicity

m to a km, both along depositional dip and depositional strike. They commonly have a gradational base and are formed by stacked siliciclastic and bioclastic cross-stratified, planar- and trough-bedded sets, ripple-laminated sandstone and highly bioturbated sandstone. No oscillatory wave produced sedimentary structures have been reported. Cross-stratified sets show unidirectional or bi-directional paleocurrent directions and have occasional mud drapes on the foresets. The rippled sandstone is finer grained and contains thicker mud drapes (Mutti et al. 1985b; Olariu et al. 2008a, b). The bedforms shingled by migrating one over the other and offlaping (Fig. 18.17d).

Importantly, single dunes in the stacked sets (compound dunes) are inclined in the same direction (eastward) as the compound-dune master surfaces, that is, the surfaces on which the larger compound dune migrated by forward accretion (Olariu et al. 2008a, b; Fig. 18.17). This observation classifies the sandstone bodies as tidal compound dunes with their crest oriented normal to the tidal currents and internal accretion surfaces that dip in the same direction as the tidal currents. This interpretation stands in contrast to the interpretation as tidal bars proposed by Mutti et al. (1985b) for sandstone bodies of the lower unit which have their long axis parallel with the tidal currents and internal accretion surfaces that migrate laterally (at a high angle to the tidal currents). Note that Wonham (1993) interpreted bedforms in the overlying upper unit (up to 6–8 m) as compound tidal dunes on the same grounds. In contrast, however, to the typical orientation of compound dunes (sensu Olariu et al. 2008a, b) inferred to have been aligned parallel to paleoflow.

At basin scale, migration directions were primarily controlled by seafloor topography (dunes migrated to fill adjacent deeper parts of the basin) and dominant tidal current directions. Additionally, migration directions of the largest compound dunes were controlled by relative sea-level changes because dunes respond to

Fig. 18.16 (**a–c**) Outcrop images of compound tidal dunes of the lower unit of the Baronia Fm. east of La Baronia (Fig. 18.11b); *arrows* in (**a**) indicate the base and top. Cross-stratified sets show unidirectional or bi-directional paleocurrent directions and have occasional mud drapes on the foresets. The bedforms shingled by migrating one over the other and offlaping. Single dunes in the stacked sets (compound dunes) are inclined in the same direction as the compound-dune master surfaces, that is, the surfaces on which the larger compound dune migrated by forward accretion (Olariu et al. 2008a, b). (**d**) Heterolithic fine-grained facies interpreted to have been formed in the distal part of compound tidal dunes and areas in between compound tidal dunes. (**e**) Tidal bundle succession of approximately 10 m long forming part of a compound dune in an erosional depression on the seafloor either associated with an estuarine channel (cf. Mutti et al. 1985b) or a large tidal scour (cf. Olariu et al. 2008a, b) in the upper unit of the Baronia Fm at the village of La Règula (see Fig. 18.8); the latter interpretation is considered more likely. Dune foresets and master bedding surfaces dip towards the WNW and are draped with mud (Olariu et al. 2008a, b). Successively increasing and decreasing bundle thicknesses are interpreted to represent successive neap and spring periods (cf. Mutti et al. 1985b)

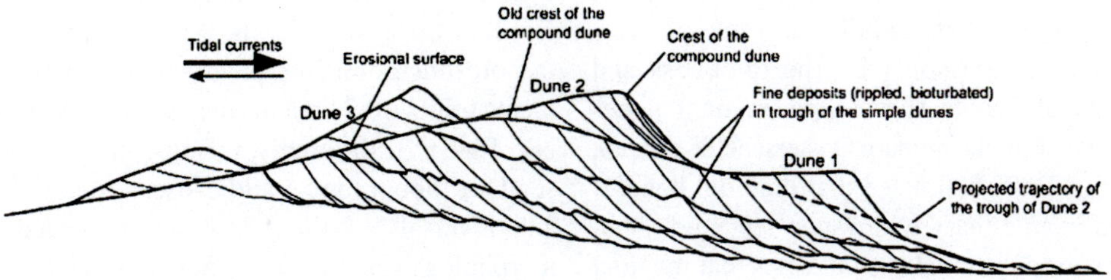

Fig. 18.17 Compound tidal dune model for the Baronia Fm showing the formation of a compound dune. Note that the inferred trajectory of successive dune troughs (*dashed line*) caused truncation of the previous cross-strata (modified with permission after unpublished data of Cornel Olariu, University of Texas at Austin)

changes in water depth. Olariu et al. (2008a) suggested water depths between 25 and 36 m at a minimum average for the lower unit. The sandstone bodies are intercalated with strongly bioturbated muddy sandstones up to tens of meters thick that represent low-energy fringes of amalgamated dune fields or periods of drowning.

Wonham (1993) and Olariu et al. (2008a) compared the depositional setting of the Baronia Fm with the outer part of the San Francisco Bay where the sea floor is covered by a tidal dune field (Rubin and Hunter 1982; Barnard et al. 2006). Berné et al. (1988) used the Baronia bedforms as an ancient analogue for the modern compound dunes of Surtainville in the English Channel that are formed by strong tidal currents. It is at present, however, unclear what the geography and morphology of the south-eastern section of the T-G-A Basin was at the time of deposition of the Baronia Fm. If the basin was closed towards the southeast this closure must have been located at least 10 or more kilometres away from the location of the (preserved) Baronia bedforms.

18.4.6 Tidal Bundles

The distally located toesets of almost all progradational Gilbert-type delta lobes of the Roda Sandstone Member in the Isábena valley were modified by tidal currents during periods between fluvially-derived sediment influxes and are represented by tidal dunes. These are small (up to 20 cm) in the lowermost three lobes and increase in size in the upper three lobes (preserved thickness up to 1 m). All have mud-draped foresets (Figs. 18.18 and 18.19). Bottomsets of the Gilbert-type delta lobe that reached farthest into the basin overlie an approximately 100 m wide belt formed by an at least 5 m thick preserved succession of tidal dunes with distinct neap-spring-neap cyclicity (Yang and Nio 1985, 1989; Nio and Yang 1991). This is well displayed in an outcrop along the Isábena River close to Roda de Isábena.

Based on a comparison with the thickness and characteristics of bundle successions in tidal dunes formed in channels of the modern Oosterschelde estuary, Yang and Nio (1985) estimated that the tidal bundles were formed in an estuary with water depths of about 15 m. Tidal periodicity analysis indicates that the tidal bundles were formed in a meso- to macrotidal semi-diurnal regime (M2 dominant) with an estimated mean tidal range of 3.6 m (Fig. 18.19). Large irregular deviations from the expected equality in the bundle sequence were interpreted to reflect incidental storm influence.

A universally applicable dependency between tidal current velocity and the tidal range was used to estimate the tidal range. However, only a linear relationship between the volume of water flowing through a tidal channel during the dominant tidal period and the wet cross-sectional surface of a tidal channel below mean water level has been proven (O'Brien 1931; van de Kreeke and Haring 1979; Van den Berg 1986). Although maximum current velocities vary between spring tide and neap tide, and deeper channels are associated with somewhat higher local current velocities than shallow channels, theoretically equal tidal current velocities for all channel depths can be expected if tidal current velocity is replaced with shear velocity (J.H. van den Berg, personal communication 2010). Thus, no relationship exists between tidal shear velocity and tidal range and, hence, the estimated tidal range derived from the Roda bundle succession is questionable.

López-Blanco et al. (2003, his Fig. 6) suggested that the tidal dunes migrated over the lower part of the sandy delta front as part of attached tidal bars and that they were driven generally towards the NW by tidal currents. Low-amplitude, gentle NW-SE oriented folds resulted in a seafloor topography that caused funnelling of tidal currents in a NW-SE direction; low water depth would have contributed to the effectiveness of this process (López-Blanco et al. 2003). However, from observations in the Rhine-Meuse delta, including the Oosterschelde (Siegenthaler 1982, his Fig. 1), it appears that offshore tidal currents close to the coastline follow a rotary path in contrast to inshore estuarine tidal currents that show distinct reversals of current direction approximately along a linear flow path. The latter situation requires a funnel-shaped land constriction and, consequently, it is concluded that the Roda tidal bundles, exposed along the Isábena River close to Roda de Isábena, formed in a NW-SE oriented, restricted, inshore, tide-dominated environment such as an embayment (following Nio and Siegenthaler 1978; Yang and Nio 1985). Additionally, it is doubtful whether gentle seafloor topography could cause sufficient funnelling and reversals of tidal flow. The indented coastline morphology resulted from movements along the same pre-existing NW-SE oriented blind thrust and associated faults mapped by López-Blanco et al. (2003) and which also controlled the location of the 'Serraduy

Fig. 18.18 (**a**) Overview of the tidal bundle outcrop locality along the Isábena River. The tidal bundles are located in the lowermost 5 m of the cliff section as indicated by the box. (**b–e**) Details of the tidal bundles showing the mud draped foreset and bottomset laminae. Paleoflow from *left* to *right* (SE to NW)

Bay' (Eichenseer 1988). The Roda Sandstone Gilbert-type delta lobes debouched into this embayment.

The uncommon occurrence of a tidal bundle succession of approximately 10 m long (Fig. 18.16e) in the Baronia Fm near the village of La Règula (Fig. 18.11) in an erosional depression on the seafloor is interpreted to be either associated with an estuarine channel and shoal (cf. Mutti et al. 1985b) or a large tidal scour filled by a forward accreting compound tidal dune (cf. Olariu et al. 2008a, b). Given the paleogeographic setting of the Baronia Fm, the latter interpretation is considered more likely.

Sigmoidal and bidirectional cross-stratified beds with double mud-draped toesets and tidal bundles as well as herringbone cross-bedding and reactivation surfaces occur in sandstone bodes 4 and 5 of the Pallaresa member of the Ametlla Fm (Dreyer 1994). Large-scale sigmoidal cross-stratified sets of up to 5 m thick contain tidal bundles with well-developed double mud drapes, mostly in the toesets, and reactivation surfaces (Fig. 18.15c–f). Average foreset dip angle is 23° to the NW interpreted to have been formed by the ebb-dominant currents (Dreyer and Fält 1993).

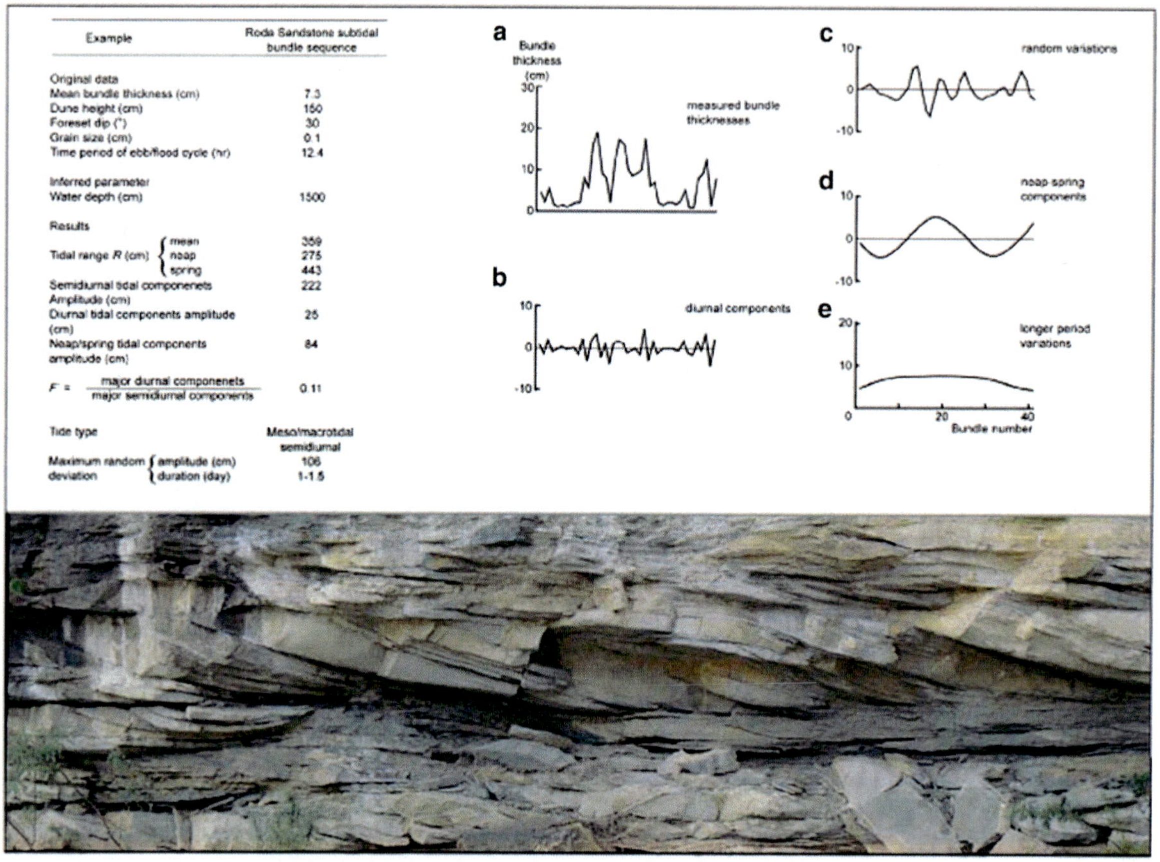

Example	Roda Sandstone subtidal bundle sequence
Original data	
Mean bundle thickness (cm)	7.3
Dune height (cm)	150
Foreset dip (°)	30
Grain size (cm)	0.1
Time period of ebb/flood cycle (hr)	12.4
Inferred parameter	
Water depth (cm)	1500
Results	
Tidal range R (cm) mean	359
Tidal range R (cm) neap	275
Tidal range R (cm) spring	443
Semidiurnal tidal componenets Amplitude (cm)	222
Diurnal tidal components amplitude (cm)	25
Neap/spring tidal components amplitude (cm)	84
$F = \frac{\text{major diurnal componenets}}{\text{major semidiurnal components}}$	0.11
Tide type	Meso/macrotidal semidiurnal
Maximum random deviation amplitude (cm)	106
Maximum random deviation duration (day)	1-1.5

Fig. 18.19 Summary of the estimated paleotidal ranges, components and maximum random deviations of the tidal bundle deposits of the Roda Fm (*upper left*). Paleotidal components (**b**), random variations (**c**) and longer-period variations (**d**, **e**) derived from filtering analysis (*upper right*) (Modified after Yang and Nio 1985). The photograph shows the tidal bundles along the Isábena River, same locality as Fig. 18.18

18.5 Overfilled Shelf Tidalites – The Montanyana Group

18.5.1 Montanyana Group Stratigraphy and Depositional Environments (Late Ypresian to Late Lutetian – 51.5 Ma to43 Ma)

The Montanyana Group (Figs. 18.5, 18.8, 18.10 and 18.20) unconformably overlies shallow marine sediments of the Ager Group and had a dispersal pattern in which the main sediment transfer zones moved southward across the basin through time maintaining an E-W orientation. This was caused by the interplay between the progressive uplift in the inner parts of the thrust system along the northern basin margin and synchronous growth of transverse alluvial systems (Nijman and Nio 1975; Nijman 1998). The Montanyana system has been divided into a Lower, Middle and Upper part (van Eden 1970; Nijman and Nio 1975; Nijman 1998; Fig. 18.5). Lower delta plain facies of the Lower Montanyana Group are mapped as the brackish water facies of the Castigaleu Fm. These are contemporaneous with and interfinger with upper delta plain and alluvial facies of the Montllobat Fm. Equivalent facies of the Upper Montanyana Group are mapped as the Perrarua and Capella Fms respectively (Garrido-Megías 1968; Nijman and Nio 1975; Van der Meulen 1989; Puigdefàbregas et al. 1989; Figs. 18.5 and 18.20). The Middle Montanyana is an incised fluvial sheet sandstone that prograded rapidly westwards across the basin (the Castissent Fm; Figs. 18.5 and 18.8). The Montanyana system was fed by alluvial fans along the northern and eastern basin margin (Nijman 1998).

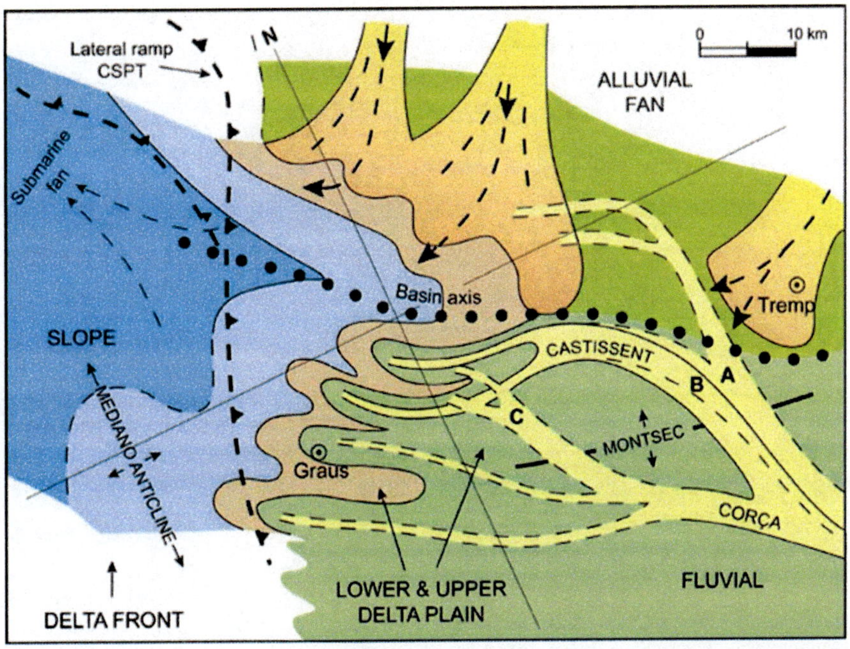

Fig. 18.20 Paleodrainage pattern of the Castissent Sandstone and related Corca Fm. across the underlying Castigaleu Fm. (lower and upper delta plain) and time equivalent Montllobat Fm. (fluvial); see also Fig. 18.4 (modified after Nijman 1998). Units A to C correspond to successive fluvial depositional cycles. *CSPT* Central South Pyrenean Thrust system including the Montsec Thrust and its lateral ramps

The Castigaleu Fm is on average 400 m thick and is formed by a number of up to 12 m thick sandstone units intercalated with thick fine-grained (mud to very fine sand) intervals (Nijman and Nio 1975; van der Meulen 1989; Puigdefàbregas et al. 1989). Sandstone units, dominantly formed by fluvial channel fills and bars, are interpreted to generally have a meandering planform style. Commonly, sheet river-flood deposits are conglomeratic in nature. These channel fills intercalate with distributary mouth bars and interdistributary muddy, brackish bay deposits (Fig. 18.21a, e).

Fluvial sandstone units in the distal part of the Montllobat Fm are up to 6 m thick and occur as (1) ribbons, (2) tabular bodies with lateral accretion and (3) lenticularly-bedded sheet bodies. Wide (up to 3 km) incised alluvial valley fills are formed by amalgamated sandstone complexes (up to 20 m thick) deposited in trunk rivers on the lower alluvial plain (Fig. 18.21a). In places, they are influenced by brack water and tidal processes as a consequence of having been formed in the fluvial to tidal transition zone (cf. Ghosh et al. 2005; Van den Berg et al. 2007; Martinius and Gowland 2010; Fig. 18.22).

Alluvial upper and lower delta plain fluvial facies of the Capella Fm reach a thickness of approximately 1,000 m. Sediments were deposited in the depression formed by subsidence on the footwall of the Lascuarre Fault system and the lateral ramp of the Montsec Thrust (Fig. 18.9). Due to the relative high subsidence rate, a significant volume of fluvial sediment was stored in the Isábena Depression (Fig. 18.9) largely preventing progradation of the system. Phases of source area rejuvenation were characterized by an initially low-relief alluvial profile allowing tidal processes to increase their effect on the dominantly muddy lower delta plain environments (Cuevas Gozalo 1989) despite the short shelf.

The Pano Fm (uppermost Montayana Group; Fig. 18.5) was deposited as a transgressive coastal sandstone wedge forming the shallow-marine and time equivalent continuation of the uppermost part of Capella Fm (Nijman and Nio 1975; Nio and Donselaar 1978; Cuevas Gozalo 1989; Donselaar 1996a; Fig. 18.9). The Virgen de la Collada ramp, located between the Mediano anticline and the Lascuarre reverse fault system (Donselaar 1996a; Fig. 18.9), controlled sedimentation patterns and the position of

Fig. 18.21 (**a**) Outcrop impression of the alluvial valley deposits of the Gargalluda sandstone complex, a 2 km wide tidally-influenced trunk river alluvial valley exposed 2 km south of Puente de Montanyana (Fig. 18.10). (**b**) Lower delta plain embayment deposits into which the alluvial valley incised at this location. (1) Well-bioturbated brack-water marls and very-fine grained thin sandstone layers with oysters. (2) Non-bioturbated planar bedded and laminated very fine grained sandstone and siltstone layers showing rhythmic deposition inferred to have been controlled by tides and possibly indicating neap-spring cyclicity. (3) Medium-grained fluvial sandstone bed. (**c, d**) Medium-grained sandstone beds showing (i) increasing to decreasing bottomset thickness, (ii) increasing to decreasing foreset dip and shape (from convex to concave), (iii) in places increasing-decreasing organic particle concentrations, and (iv) occasional reactivation surfaces. These features interpreted as tidally-influenced fluvial accretionary bedforms resulting from successive fluctuations in flow regime conditions from lower (during flood tidal retardation) to higher (during ebb tidal drawdown) current velocities (cf. Martinius and Gowland 2010). (**e**) Outcrop image of a mouth bar complex associated with a lower delta-plain distributary channel in the Perarrua Fm (Fig. 18.5) directly east of the town of Salanova (8 km south of La Puebla de Roda; Fig. 18.10). (**f**) Mud-draped ripple-laminated (form)set forming part of the bottomset of a tabular cross-stratified bed in the upper part of the mouth bar shown in (**e**) and interpreted to have been formed during bank-full stage and reversal after flood combined with low current velocities in the channel (cf. Cuevas Gozalo and de Boer 1991)

internal Pano Fm facies boundaries. Only a short (approximately 10 km) shelf was present and a number of tectonically induced relative sea-level changes controlled sedimentation. These were related to a relatively high rate of subsidence alongside the growing Mediano anticline. The Pano Fm is divided in two third order sequences and each of these is further divided in a number of fourth order sequences (Cuevas Gozalo 1989; Donselaar 1996a, Fig. 1.-23).

The Ypresian and Lutetian T-G-A Basin fill is overlain by upper Eocene and Oligocene fluvial and alluvial fan deposits mainly sourced from the north (Puigdefàbregas et al. 1989).

18.5.2 Tidally-Influenced Fluvial Point-Bars and Mouth Bars

Tidal influence is encountered in fluvial sandstone bodies of the Castigaleu, Montllobar and Capella Fms and two tidally-influenced fluvial point-bar models with current reversals were proposed (Cuevas Gozalo

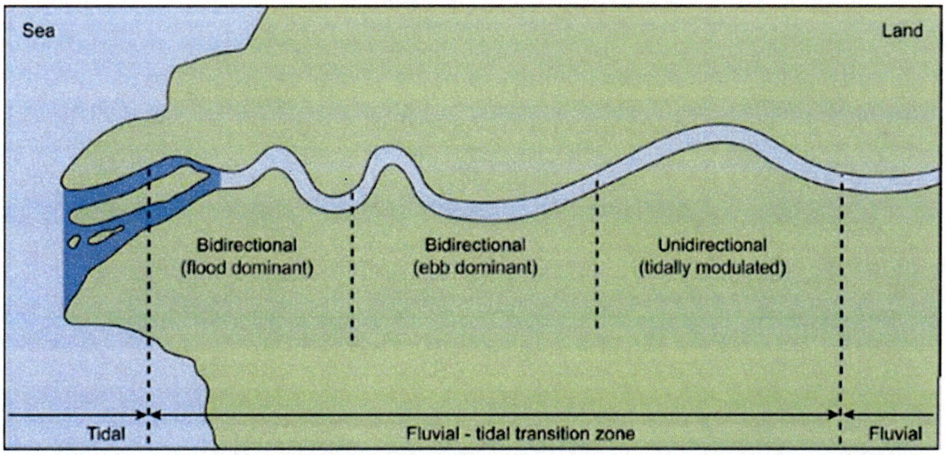

Fig. 18.22 Schematic illustration of the fluvial to tidal transition zone showing zones of variable influence of the tides (after Martinius and Gowland 2010)

and de Boer 1991; Fig. 18.23), highlighting the segregation of flood- and ebb-current generated structures around the meander bend.

Tidally-influenced deposits of the Castigaleu Fm are formed in meandering channels and mouth bars associated with distributary channels (Cuevas Gozalo and de Boer 1991, their stop 4). These are typified by tabular and trough cross-stratified well-sorted sandstone beds with common reactivation surfaces and occasional herringbone structures; ripple-laminated sets occur in the top and mud drapes occur particularly in mouth bars entering brackish bays (Nijman and Nio 1975; Marzo et al. 1988; Hoornweg 1988; Fig. 18.21e, f). Some examples contain abundant brack-water to normal marine ichnofacies and oyster beds, and the top is commonly mottled. The thick fine-grained intervals between the sandstone units are deposited in inshore brack-water lagoons or shallow-water embayments based on body fossil content.

Some isolated meandering channels of the delta plain of the Montllobat Fm in the Noguera Ribagorçana River valley, originally described by Puigdefàbregas and van Vliet (1978) and Van der Meulen (1982), as well as the Gargalluda sandstone complex, a 2 km wide trunk river alluvial valley stratigraphically 30 m higher (Marzo et al. 1988), show features interpreted as tidal influence on fluvial accretionary bedforms during bankfull stage and reversal after flood combined with low current velocities in the channel (Cuevas Gozalo and de Boer 1991, their stop 9 and 11; Fig. 1.21a–d). These occur in a few stratigraphic positions indicating propagation of stronger tidal currents up into the meandering and distributary channels during certain phases of delta development while the shelf area was not completely filled yet.

The Capella Fm contains tidally-influenced fluvial channels throughout its stratigraphy (Cuevas Gozalo 1985a, b, 1989; Cuevas Gozalo and de Boer 1991). Typically, trough cross-stratified sets in the deepest part of the fluvial channels that are tidally-influenced have foresets showing mud drapes, bundling, reactivation surfaces not formed by dune overtaking and bipolar (but unevenly distributed) current directions while fluvially generated cross-stratified sets are clean, coarse and unidirectional. Tidally-influenced point bars contain large-scale avalanching foresets, undermined banks and bioturbated tops (de Boer 1998). Additionally, Cuevas Gozalo (1985a, 1989, her page 80/81) and Cuevas Gozalo and de Boer (1991) defined ebb-oriented, tidally-influenced fluvial spill-over lobes that formed in areas of fluvial flow expansion associated with fluvial channel bifurcation on the intertidal plain (Fig. 18.24). They are single sedimentary bedforms; the feeder channel shallows towards the lobe flat (Cuevas Gozalo 1985a, 1989). Furthermore, in some cases an upstream transition from a sand-dominated part of the point bar to sandy-muddy part of the point bar is observed. Cuevas Gozalo and de Boer (1991) suggest that upstream fines deposited in the upper part of the point bar were protected by flood dominated swales and/or inner parts of the channel.

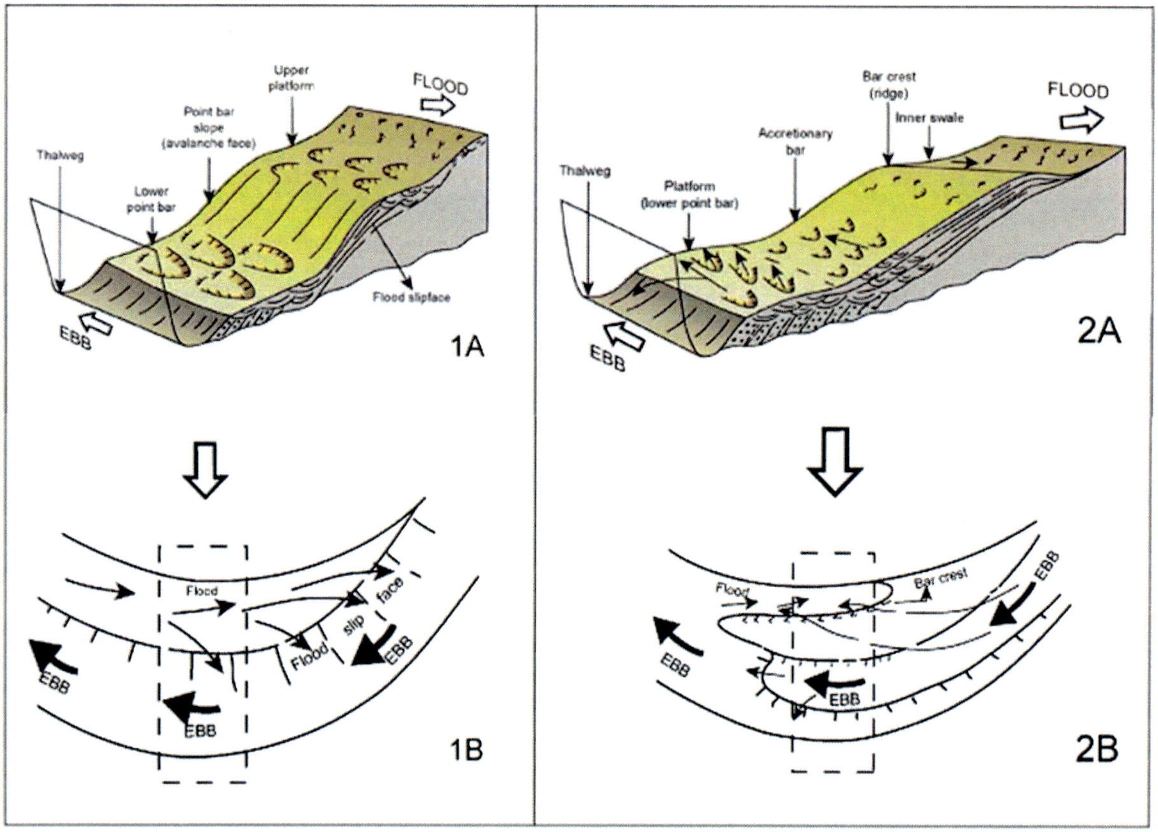

Fig. 18.23 Two models (*1* model 1, and *2* model 2) for tide-influenced fluvial point-bars with current reversals developed for the Castigaleu Fm and Capella Fm highlighting the segregation of flood- and ebb-current generated structures around the meander bend. (**a**) perspective view, and (**b**) plan view (modified after Cuevas Gozalo and de Boer 1991)

18.5.3 Heterolithic Tidal Lagoon Deposits

The tidally-influenced and tidally-dominated environments in third order sequence 1 of the Pano Fm (Fig. 18.25) are formed as part of a retrogradational fourth order sequence set of a N-S to NE-SW oriented barrier and back-barrier complex succession with a shoestring geometry (Pano, Panillo and Grustán barrier chains respectively; Figs. 18.26 and 18.27; Donselaar 1996a). Each of these is associated with tidal inlets and back-barrier lagoons including tidal channels. Their formation was attributed to eustatic sea-level rise and coeval (punctuated) basin floor subsidence. In particular, sedimentation of the Pano coastal barrier complex was strongly influenced by tidal action as witnessed by the tidal channels, flood-tidal delta deposits and overall bimodal currents directions. Cross-stratified sets contain mud drapes on foresets, sigmoidal laminae shapes and convex-up reactivation surfaces. The Panillo barrier complex is wave dominated but the overlying Grustán barrier complex is a mixed-energy system with tidal (inlet) channels (Donselaar 1996a).

The tidally-influenced environments of third order sequence 2 (Fig. 18.25) are formed by a retrogradational succession (parasequence set) formed by a tidally-influenced embayment fill at the base and ensuing transgressive, up to 35 m thick flood-tidal delta deposits overlain by highstand carbonate deposits (Donselaar 1996a). Time-equivalent barrier and/or inlet deposits are not reported and are assumed absent in the studied area or not preserved. The lowermost parasequence is located in the tidally-influenced (restricted) embayment and is formed by heterolithic (mud and very fine sand) deposits with double mud drapes at the base and neap-spring cyclicity (fourth order sequence 2.1; Fig. 18.28) forming an aggradational cyclic succession (Donselaar 1996a, b). The top part of the heterolithic succession forms the transition into the toe of the

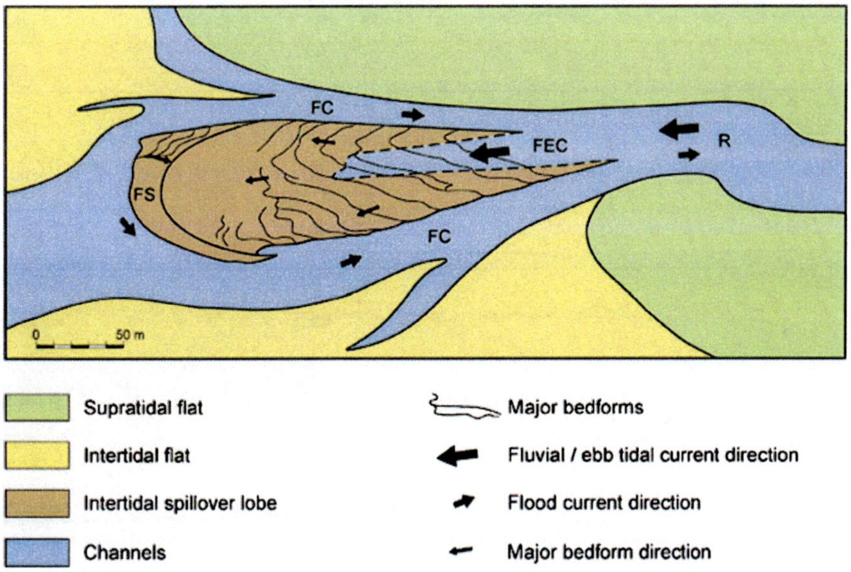

Fig. 18.24 Sedimentary model for the Capella Fm for spill-over lobes in an intertidal plain. *R* tidally-influenced river, *FEC* fluvial-ebb tidal channel, *FC* marginal flood channel, *FS* flood shield (modified after Cuevas Gozalo 1985a, 1989)

Fig. 18.25 Chrono- litho- and sequence stratigraphy of the Pano Fm in the area NW of Graus (modified after Donselaar 1996a)

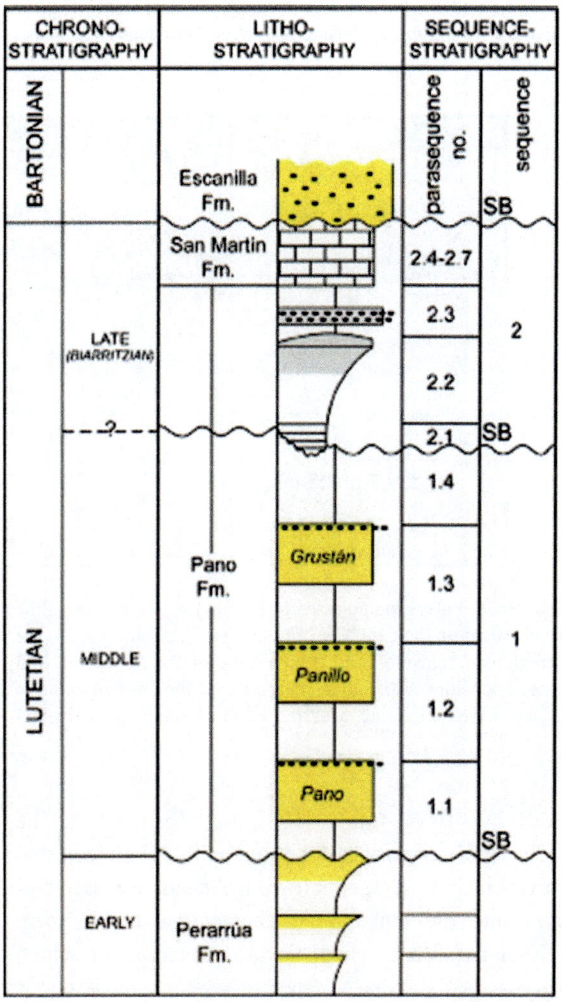

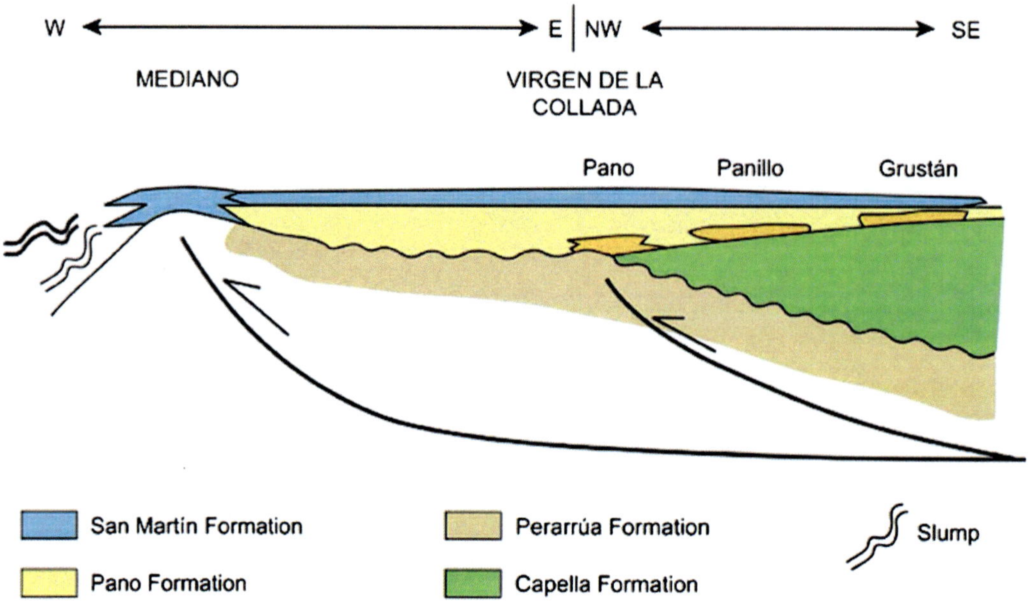

Fig. 18.26 Schematic west-southeast cross-section through the Mediano and Virgin de la Collada lateral ramps showing the SE-ward stratigraphic displacement of barrier islands, inlets and tidally-dominated back-barrier environments. (Not to scale, see also Fig. 18.9. Modified after Donselaar 1996a)

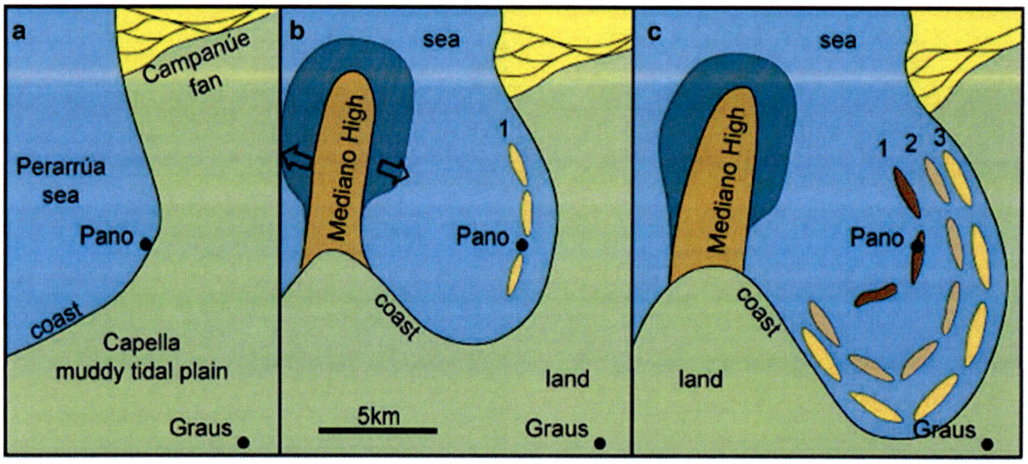

Fig. 18.27 Paleogeographic reconstruction of Lutetian coastline development in the area NW of Graus. (**a**) Situation prior to the initiation of transgression. (**b**) Doming of the Mediano High caused subsidence of adjacent areas and the start of relative sea-level rise. Flooding occurred of the Capella coastal plain and the Pano barrier chain was formed. (**c**) Phase of maximum flooding and development of the Grustán barrier chain (3) and tide-dominated back-barrier area; the Pano (1) and Panillo (2) barrier chains are abandoned and drowned (modified after Donselaar 1996a)

overlying flood-tidal delta that formed in the embayment supposedly behind an inlet between two barrier islands. Large inclined avalanche foresets dipping into the embayment characterise the deposit. Associated feeder channels are preserved in which cross-laminated sets have an E-W bimodal flood-dominated foreset dip distribution. Small-scale planar cross-stratified sets with opposite paleocurrent directions on top of convex-up reactivation surfaces illustrate the tidal signature of the deposit (Donselaar 1996a).

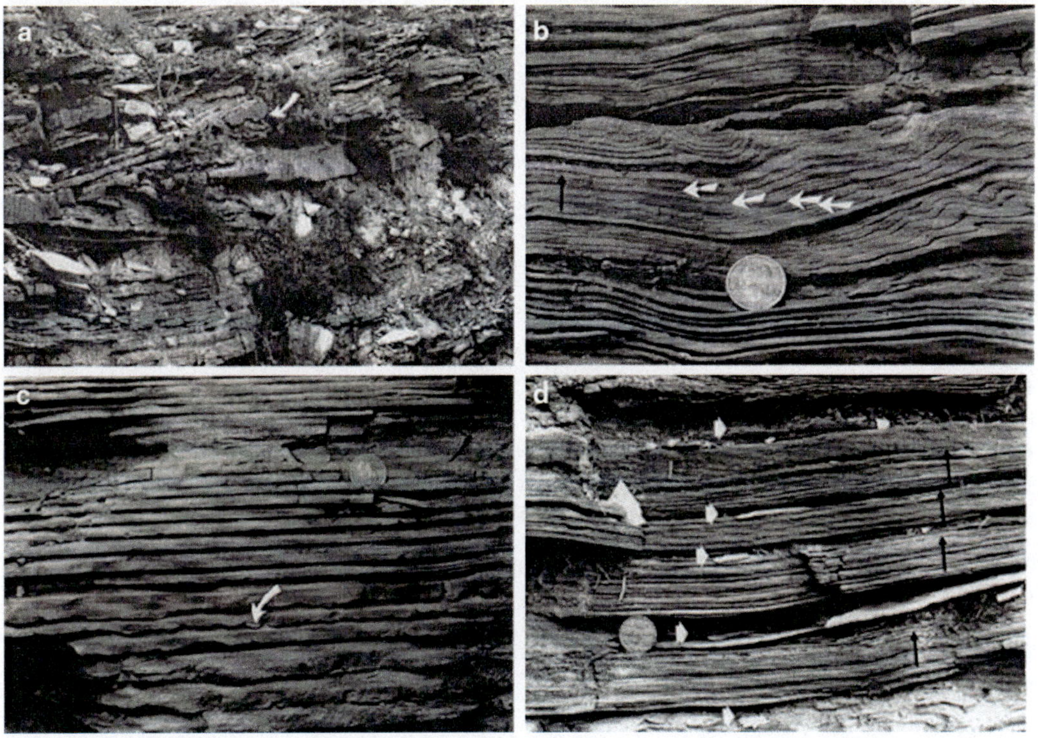

Fig. 18.28 (**a–d**) Planar bedded and laminated inclined heterolithic facies of the lowermost parasequence in sequence 2.1 (Fig. 18.25) of the Pano Fm developed above a major unconformity. Erosive surfaces separate subsequent units of inclined heterolithic strata (*white arrow* in **a**). The deposits are interpreted to have been formed in a tidally-dominated part of an (restricted) embayment (cf. Donselaar 1996a, b) and are formed by heterolithic (mud and very fine sand) deposits with double mud drapes (*white arrows* in **b**) at the base and neap-spring cyclicity (*black arrows* in **d**). Regular vertical decrease and subsequent increase of bed thickness is interpreted to indicate spring to neap tide cycles with the thinnest beds formed during neap (*black arrow* in **b**; *white arrows* in **d**). Some layers have a deformed base (*white arrow* in **c**). (After Donselaar 1996b. With permission)

18.6 Summary

Lower Eocene tidalites in the T-G-A Basin in the southern Pyrenees (Spain) were deposited in response to developing thrust related folds and blind ramps which determined the position of facies belts and focussed and enhanced tidal currents. Two distinct stages of basin configuration can be recognised which share a general basin outline typified by a relatively narrow (up to 60 km) and long (up to 200 km in total) semi-enclosed sea which had an open connection to the Atlantic ocean in the west. They differ, however, significantly in water depth distribution, basin floor topography and coastal morphology. The two stages are directly related to two different configurations of basin dimensions favourable for resonant amplification and dominantly controlled by thrust sheet development.

The underfilled foredeep stage occurred during the Early Ypresian and was a period favourable for the formation of compound tidal dune fields and large tidal bars in the foredeep in relatively deep water (at least up to 40 m). Conditions were favourable for circulating and amplified outward flowing tidal currents. No major axially draining fluvial systems existed, but instead locally fed relatively small shoal-water fan deltas and larger (a) Gilbert-type delta(s) developed along the northern basin margin, dominantly located in structurally controlled areas of the coastline. Additionally, a large offshore compound tidal dune field was present in the south-eastern part of the basin as a result of sufficiently strong and confined tidal currents flowing E-W in a narrow sea.

The overfilled shelf stage became manifest during the Late Ypresian to Late Lutetian during which modest tidal amplification occurred in shallow water (up to 10 m), in-shore parts of delta distributaries and the subaqueous part of the delta top. A shelf formed behind the developing oblique lateral ramp of the Montsec Thrust, and a distinct shelf break was located above the lateral ramp. The shallow shelf sea was relatively rapidly filled by the axial east to west prograding Montanyana alluvial to deltaic system, which restricted tidal amplification. Alluvial fans fringed the basin margin. During Lower Montanyana times the depositional shoreline to shelf break distance was approximately 15 km but only a few km's remained during Upper Montanyana times when the fluvial system almost reached the shelf margin. A relatively short and steep slope existed westward of the shelf break (the area of the present-day Ainsa Basin) and the basin floor (the present-day Jaca Basin) was relatively deep and narrow. These areas were characterised by turbidite deposition.

References

Barberà X, Marzo M, Reguant S, Samsó JM, Serra-Kiel J, Tosquella J (1997) Estratigrafía del Grupo Fígols (Paleógeno, Cuenca de Graus-Tremp, NE de España). Rept Soc Geol España 10:67–81

Barnard P, Hanes DM, Rubi DM, Kvitek RG (2006) Giant sand waves at the mouth of San Francisco Bay. EOS Trans AGU 87(29):285–289

Beaumont C, Muñoz JA, Hamilton J, Fullsack P (2000) Factors controlling the Alpine evolution of the Central Pyrenees inferred from a comparison of observations and geodynamic models. J Geophys Res 105:8121–8145

Bentham P, Burbank DW (1996) Chronology of Eocene foreland basin evolution along the western oblique margin of the south-central Pyrenees. In: Friend PF, Dabrio C (eds) Tertiary basins of Spain. The stratigraphic record of crustal kinematics. Cambridge University Press, Cambridge, pp 144–152

Berné S, Auffret JP, Walker P (1988) Internal structure of subtidal sandwaves revealed by high-resolution seismic reflection. Sedimentology 35:5–20

Burbank DW, Puigdefàbregas C, Muñoz JA (1992) The chronology of the Eocene tectonic and stratigraphic development of the eastern Pyrenean foreland basin, northeast Spain. Geol Soc Am Bull 104:1101–1120

Cámara P, Klimowitz J (1985) Interpretación geodinámica de la vertiente centr-occidental surpirenaica (cuencas de Jaca y Tremp). Estudios Geológicos 41:391–404

Cande SC, Kent DV (1995) Revised calibration of the geomagnetic polarity time scale for the Late Cretaceous and Cenozoic. J Geophys Res 100:6093–6095

Capote R, Muñoz JA, Simón JL, Liesa CL, Arlegui LE (2002) Alpine tectonics I: the Alpine system north of the Betic Cordillera. In: Gibbons W, Moreno T (eds) The geology of Spain. Geological Society, London

Choukroune P, ECORS Team (1989) The ECORS Pyrenean deep seismic profile reflection data and the overall structure of an orogenic belt. Tectonics 8:23–39

Choukroune P, Roure F, Pinet B, ECORS Team (1990) Main results of the ECORS Pyrenees profile. Tectonophysics 173:411–423

Clevis Q, de Jager G, Nijman W, de Boer PL (2004) Stratigraphic signatures of translation of thrust-sheet top basins over low-angle detachment faults. Basin Res 16:145–163

Cuevas Gozalo M (1985a) Sedimentary lobes in a tidally influenced alluvial area, Capella Formation, Tremp-Graus Basin, southern Pyrenees, Spain. Geol Mijnb 64:145–157

Cuevas Gozalo M (1985b) Geometry and lithofacies of sediment bodies in a tidally influenced alluvial area, Middle Eocene, southern Pyrenees, Spain. Geol Mijnb 64:221–231

Cuevas Gozalo M (1989) Sedimentary facies and sequential architecture of tide-influenced alluvial deposits: an example from the middle Eocene Capella formation. Ph.D. thesis, University of Utrecht, Geologica Ultraiectina 61, 152 p

Cuevas Gozalo M, de Boer PL (1991) Tide-influenced fluvial deposits; examples from Eocene of the southern Pyrenees. In: Marzo M, Puigdefàbregas C (eds) Guidebook to the 4th international conference on fluvial sedimentology. Publicacions del Servei Geològic de Catalunya, 92 p

Cuevas Gozalo M, Donselaar ME, Nio S-J (1985) Eocene clastic tidal deposits in the Tremp-Graus Basin (Provs. of Lérida and Huesca). In: Milá MD, Rosell J (eds) 6th International Association of sediment. European Regional Meeting, Lérida, Institut d'Estudis Ilerdencs, Excursion guidebook, pp 215–266

Dalrymple RW (1992) Tidal depositional systems. In: Walker RG, James NP (eds) Facies models, response to sea level change. Geological Association of Canada, St. John's, Canada, pp 195–218

De Boer PL (1985) Paleozoic/Mesozoic sedimentary development. In: Donselaar MP, Geel CR (eds) Guide to the sedimentology of the Tremp-Graus Basin. Open file report, University of Utrecht, pp 19–65

De Boer PL (1998) Intertidal sediments: composition and structure. In: Eisma D (ed) Intertidal deposits: river mouths, tidal flats, and coastal lagoons. CRC Press/LLC, Boca Raton, pp 345–361

De Boer PL, Pragt JSJ, Oost AP (1991) Vertically persistent sedimentary facies boundaries along growth anticlines and climate-controlled sedimentation in the thrust- sheet-top south Pyreneean Tremp-Graus Foreland Basin. Basin Res 3:63–78

DeCelles PG, Giles KA (1996) Foreland basin systems. Basin Res 8:105–123

Dercourt J, Gaetani M, Vrielinck B, Barrier E, Biju-Duval B, Brunet MF, Cadet JP, Crasquin S, Sandulescu M (2000) Atlas of Peri-Tethys, Palaeogeographical Maps. Commission de la Carte Géologique de Monde (CCGM/CGMW), Gauthier-Villars, Paris, 269 p, 13 maps, 1 pl

Donselaar ME (1996a) Barrier island coasts and relative sea level rise: preservation potential, facies architecture and sequence analysis. Ph.D. thesis, University of Delft, NUGI 816, 223 p

Donselaar ME (1996b) Inshore heterolithic deposits. In: Johnson HD, Wonham JP, Gupta R, Donselaar ME, van de Weerd AA, Mutterlose J, Stadler A, Ruffell AH (eds) Geological characterisation of shallow marine sands for reservoir modelling and high resolution stratigraphic analysis. JOULE II final report, Section 1, pp 1–19

Donselaar ME, Nio SD (1982) An Eocene tidal inlet/washover type barrier island complex in the south Pyrenean marginal basin, Spain. Geol Mijnb 61:343–353

Dreyer T (1994) Architecture of an unconformity-based tidal sandstone unit in the Ametlla Formation, Spanish Pyrenees. Sediment Geol 94:21–48

Dreyer T, Fält L-M (1993) Facies analysis and high-resolution sequence stratigraphy of the Lower Eocene shallow marine Ametlla Formation, Spanish Pyrenees. Sedimentology 40:667–697

Dreyer T, Corregidor J, Arbues P, Puigdefàbregas C (1999) Architecture of the tectonically influenced Sobrarbe deltaic complex in the Ainsa Basin, northern Spain. Sediment Geol 127:127–169

Dunne M, Ferrill DA (1988) Blind thrust systems. Geology 16:33–36

Egger H, Heilmann-Clausen C, Schmitz B (2009) From shelf to abyss: record of the Paleocene/Eocene-boundary in the Eastern Alps (Austria). Geol Acta 7:215–227

Eichenseer H (1988) Facies geology of late Maestrichtian to early Eocene coastal and shallow marine sediments (Tremp-Graus Basin, northeastern Spain). Unpublished Ph.D. thesis, University of Tübingen, 237 p

Eichenseer H, Luterbacher H (1992) The marine Paleogene of the Tremp region (northeast Spain) – depositional sequences, facies history, biostratigraphy and controlling factors. Facies 27:119–152

Fonnesu F (1984) Estratigrafía física y análisis de facies de las secuencias de Fígols entre el río Noguera Pallaresa e Iscles (provs. De Lérida y Huesca). Unpublished Ph.D. thesis, Universitat de Barcelona, 317 p

Galdeano A, Moreau MG, Pozzi JP, Brethou PY, Malod JA (1989) New paleomagnetic results from Cretaceous sediments near Lisboa (Portugal) and implications for the rotation of Iberia. Earth Planet Sci Lett 92:95–106

Garrido-Megías A (1968) Sobre la estratigrafía de los conglomerados de Campanúe (Santa Lietsra) y formaciones superiores del Eocene (extremo occidental de la cuenca Tremp-Graus, Pirineo central, provincia de Huesca). Acta Geol Hispanica 3:39–43

Garrido-Megías A (1973) Estudio geológico y relacíon entre tectónica y sedimentación del Secundario y Terciario de la vertiente meridional pirenaica en su zona central (prov. Huesca y Lerida). Unpublished Ph.D. thesis, University of Granada, 395 p

Ghibaudo G (1975) Depositi di barra di foce nel Paleogene della valle di Ager (Provincia di Lerida, Spagna). Boll Soc Geol Ital 94:2131–2154

Ghosh SK, Chakraborty C, Chakraborty T (2005) Influence of fluvial-tidal interactions on the nature of cross-stratified packages in a deltaic setting: examples from the Barakar Coal Measure (Permian), Satpura Gondwana Basin, central India. Geol J 40:65–81

Guimerá J (1984) Paleogene evolution of deformation in the north-eastern Iberian Peninsula. Geol Mag 121:413–420

Guimerá J (1996) Cenozoic evolution of eastern Iberia: structural data and dynamic model. Acta Geol Hispánica 29:57–66

Hall MT (1997) Sequence stratigraphy and early diagenesis: the Sobrarbe Formation, Ainsa Basin, Spain. Unpublished Ph.D. thesis, University of Manchester, 136 p

Haseldonckx P (1972) The presence of Nypa palms in Europe; a solved problem. Geol Mijnb 51:645–650

Hoornweg A (1988) Interaction between a fan delta and a fluvial system in the Eocene Castigaleu and La Rocca Formations, Isábena valley, Spanish southern Pyrenees. Unpublished M.Sc. thesis, University of Utrecht, 150 p

Jimenez C (1987) Paleoecologie et Valeur Chronostratigraphique des Foraminiferes Benthiques dans des Systemes Sedimentaires Littoraux et ltaiques. Applicationaux Series Ilerdiennes de Roda (Versant Sud-Pyreneen) et Coustouge (Corbieres). Unpublished Ph.D. thesis, University Paul-Sabatier de Toulouse, 302 p

Joseph P, Hu LY, Dubrule O, Claude D, Crumeyrolle P, Lesueur JL, Soudet HJ (1993) The Roda deltaic complex (Spain): From sedimentology to reservoir stochastic modelling. In: Eschard R, Doligez B (eds) Subsurface Reservoir characterization from Outcrop Observations. Éditions Technip, Paris, pp 97–110

Leren BL, Howell JA, Enge HD, Martinius AW (2010) Controls on stratigraphic architecture in contemporaneous delta systems from the Eocene Roda Sandstone, Tremp-Graus Basin, northern Spain. Sediment Geol 229:9–40

López-Blanco M (1996a) Estratigrafia Secuencial de Sistemas Deltaicos en Cuencas de Antepais: Ejemplos de Sant Lorenc del Mont, Montserrat y Roda (Paleogeno, Cuenca de Antepais Surpirenaica). Unpublished Ph.D. thesis, Universitat de Barcelona, 238 p

López-Blanco M (1996b) Estratigrafía secuencial de sistemas deltaicos en cuencas de antepais: Ejemplos de Sant Llorenc del Munt, Montserrat y Roda (Paleogeno, Cuenca de Antepais Surpirenaica). Acta Geol Hispánica 31:91–95

López-Blanco M, Marzo M, Muñoz JA (2003) Low-amplitude, synsedimentary folding of a deltaic complex: Roda Sandstone (lower Eocene), south-Pyrenean Foreland Basin. Basin Res 15:73–95

Luterbacher HP, Ali JR, Brinkhuis H, Gradstein FM, Hooker JJ, Monechi S, Ogg JG, Powell J, Röhl U, Sanfilippo A, Schmitz B (2004) The Paleogene period. In: Gradstein FM,t Ogg JG, Smith AG (eds) A Geologic Time Scale 2004. Cambridge University Press, Cambridge, pp 384–408

Maestro-Maideu E, Betzler C, van den Hurk AM, Serra-Rolg J (1991) El Ilerdiense de la Serra D'Aubens. Correlacion con la Vall d'Ager. Geogaceta 10:58–61

Malod JA, Mauffret A (1990) Iberian plate motions during the Mesozoic. Tectonophysics 184:261–278

Malot JA (1989) Ibérides et plaque ibérigue. Bull Soc Géol France 5:927–934

Martínez-Peña MB, Pocoví A (1988) El amortiguamiento frontal de la estructura de la cobertera surpirenaica y su relación con el anticlinal de Barbastro-Balaguer. Acta Geol Hispánica 2:81–94

Martinius AW, Gowland S (2011) Tide-influenced fluvial bedforms and tidal bore deposits (Late Jurassic Lourinhã Formation, Luisitanian Basin, Western Portugal). Sedimentology 58:285–324

Martinius AW, Molenaar N (1991) A coral-mollusc (*Goniaraea – Crassatella*) dominated hardground community in a siliciclastic – carbonate sandstone (the Lower Eocene Roda Formation, southern Pyrenees, Spain). Palaios 6:142–155

Marzo M, Nijman W, Puigdefabregas C (1988) Architecture of the Castissent fluvial sheet sandstones, Eocene, south Pyrenees, Spain. Sedimentology 35:719–738

Mascle A, Puigdefàbregas C (1998) Tectonics and sedimentation in foreland basins: results from the Integrated Basins Studies project. In: Mascle A, Puigdefàbregas C, Luterbacher HP, Fernàndez M (eds) Cenozoic Foreland Basins of Western Europe. Geol Soc Spec Publ 134:1–28

Meigs AJ, Burbank DW (1997) Growth of the south Pyrenean orogenic wedge. Tectonics 16:239–258

Mellere D, Steel RJ (1996) Tidal sedimentation in inner hebrides half-grabens, Scotland: the mid-Jurassic Bearreraig sandstone formation. In: De Batist M, Jacobs P (eds) Geology of siliciclastic shelf seas. Geol Soc Lond Spec Publ 117:49–79

Mey PHW, Nagtegaal PJC, Roberti KJ, Hartevelt JJA (1968) Lithostratigraphic subdivision of post-hercynian deposits in the south-central Pyrenees, Spain. Leidse Geologische Meded 41:221–228

Michaud K (2011) Facies architecture and stratigraphy of tidal ridges in the Eocene Roda Formation, northern Spain. M.Sc. Thesis, Queen's University, Kingston, 128 p

Millán H, Den Bezemer T, Vergés J, Zoetemeier R, Cloetingh S, Marzo M, Muñoz JA, Puigdefàbregas C, Roca A, Cerés (1995) Palaeo-elevation and effective elastic thickness evolution at mountain ranges: inferences from flexural modeling in the Eastern Pyrenees and Ebro Basin. Mar Petrol Geol 12:917–928

Molenaar N, Martinius AW (1990) Origin of nodules in mixed siliciclastic-carbonate sandstones, the Lower Eocene Roda Sandstone Member, southern Pyrenees, Spain. Sediment Geol 66:277–293

Molenaar N, Martinius AW (1996) Fossiliferous intervals and sequence boundaries in shallow marine, siliciclastic deposits (Early Eocene, fan-deltaic deposits in the southern Pyrenees, Spain). Palaeogeogr Palaeoclim Palaeoecol 121:147–168

Muñoz JA (1992) Evolution of a continental collision belt: ECORS – Pyrenees crustal balanced cross-section. In: McClay KR (ed) Thrust tectonics. Chapman and Hall, London, pp 235–246

Mutti E, Luterbacher HP, Ferrer J, Rosell J (1972) Schema stratigrafico e lineamenti di facies del Paleogene marino nella zone central sud-pirenaica tra Tremp (Catalogna) e Pamplona (Navarra). Mem Soc Geol Ital 11:391–416

Mutti E, Obrador A, Rosell J (1973) Sedimenti deltizii di piana dim area nel Paleogene della valle de Ager (Provincia de Lérida, Spagna). Bull Soc Geol Ital 92:517–528

Mutti E, Rosell J, Ghibaudo G, Obrador A (1975) The Paleogene of the Ager Basin. In: 9th international sedimentological congress, International Association of Sedimentologists, Nice, Excursion guidebook part B 1, pp 1–6

Mutti E, Remacha E, Sgavetti M, Rosell J, Valloni R, Zamorrano M (1985a) Stratigraphy and facies characteristics of the Eocene Hecho Group turbidite systems, south-central Pyrenees. In: Mila MD, Rosell J (eds) Excursion guidebook: VI European Regional Meeting, IAS, Lerida, Excursion 1, pp 521–576

Mutti E, Rosell J, Allen GP, Fonnesu F, Sgavetti M (1985b) The Eocene Baronia tide dominated delta shelf system in the Ager Basin. In: Mila MD, Rosell J (eds) Excursion guidebook: VI European Regional Meeting, IAS, Lerida, Excursion 13, pp 579–600

Mutti E, Séguret M, Sgavetti M (1988) Sedimentation and deformation in the tertiary sequences of the southern Pyrenees. AAPG Mediterranean Basins Conference Field Trip 7, Special Publication, Institute of Geology, University of Parma, 153 p

Mutti E, Sgavetti M, Waehry A, Carminatti M, Davoli G, Ghielmi M, Figoni M, Mora S (1994) Regional stratigraphy and sequence-stratigraphic aspects of the Figols Group. In: Mutti E, Davoli G, Mora S and Sgavetti M (eds) The eastern sector of the south-central folded Pyrenean Foreland: criteria for stratigraphic analysis and excursion notes. Second high-resolution sequence stratigraphic conference, Tremp, pp 37–41

Nagtegaal PJC, van Vliet A, Brouwer J (1983) Syntectonic coastal offlap and concurrent turbidite deposition: the Upper Cretaceous Aren Sandstone in the south-central Pyrenees, Spain. Sediment Geol 34:185–218

Nijman W (1989) Thrust sheet rotation ? – the south Pyrenean Tertiary basin configuration reconsidered. Geodin Acta 3:17–42

Nijman W (1998) Cyclicity and basin axis shift in a piggyback basin: towards modelling of the Eocene Tremp-Ager Basin, south Pyrenees, Spain. In: Mascle A, Puigdefàbregas C, Luterbache CPr, Fernàndez M (eds) Cenozoic Foreland Basins of Western Europe. Spec Publ Geol Soc 134:135–162

Nijman W, Nio S-D (1975) The Eocene Montañana delta (Tremp-Graus Basin, provinces of Lérida and Huesca, southern Pyrenees, N. Spain). In: 9th international Sedimentological Congress, IAS, Nice, Excursion guidebook, 19, part B, 1–20 and Appendix (36 p)

Nio S-D, Donselaar ME (1978) Field guide to transgressive siliciclastic complexes in the southern Pyrenean Basin, Spain: part Two. An Eocene transgressive barrier complex near Graus. Open file report no. 18, University of Utrecht, pp 45–58

Nio S-D, Siegenthaler C (1978) Field guide to transgressive siliciclastic complexes in the southern Pyrenean Basin, Spain: part One. A Lower Eocene estuarine-shelf complex in the Isabena valley. Open file report no. 18, University of Utrecht, pp 1–44

Nio S-D, Yang CS (1991) Sea-level fluctuations and the geometric variability of tide-dominated sandbodies. Sediment Geol 70:161–193

Nio S-D, van den Berg JH, Goesten M, Smulders F (1980) Dynamics and sequential analysis of a mesotidal shoal and intershoal cannel complex in the Eastern Scheldt (southwestern Netherlands). Sediment Geol 26:263–279

O'Brien MP (1931) Estuary tidal prisms related to entrance areas. Trans ASCE 1:738–739

Olariu C, Steel RJ, Dalrymple RW, Gingras MK (2008a) Tidal deposits of Baronia sandstone, Lower Eocene, Ager Basin, Spain: compound tidal dunes and architecture of Lower Baronia. Report for BITE Joint Industry Consortium Phase 1, 46 p

Olariu C, Steel RJ, Dalrymple RW, Gingras MK and Rubino J-L (2008b) Tidal dunes of the Eocene Baronia Sandstone, Ager Basin, Spain: distinguishing tidal dunes from tidal bars; Why bother? AAPG Annual Convention, San Antonio, TX

Olariu MI, Olariu C, Steel RJ, Dalrymple RW, Martinius AW (2011) Anatomy of a laterally migrating tidal bar in front of a delta system: Esdolomada Member, Roda Formation, Tremp-Graus Basin, Spain. Sedimentology 58 (in press)

Olivet JL (1996) La cinématique de la plaque ibérique. Bull Cent Rech Explor 20:131–195

Oost AP (1995) Dynamics and sedimentary development of the Dutch Wadden Sea with emphasis on the Frisian Inlet. Ph.D. thesis, University of Utrecht, 454 p

Ori GG, Friend PF (1984) Sedimentary basins formed and carried piggyback on active thrust sheets. Geology 12:475–478

Payros A, Pujalte V, Baceta JI, Bernaola G, Orue-Etxebarria X, Apellaniz E, Caballero F, Ferrandez (2000) Lithostratigraphy and sequence stratigraphy of the Upper Thanetian to Middle Ilerdian strata of the Campo section (southern Pyrenees, Spain): revision and new data, Rev Soc Geol España 13, pp 213–226

Pearson PN, van Dongen BE, Nicholas CJ, Pancost RD, Schouten S, Singano JM, Wade BS (2007) Stable warm tropical climate through the Eocene epoch. Geology 35:211–214

Plaziat JC (1981) Late Cretaceous to Late Eocene paleogeographic evolution of southwest Europe. Palaeogeogr Palaeoclimatol Palaeoecol 36:263–320

Poblet J, Muñoz JA, Travé A, Serra-Kiel J (1998) Quantifying the kinematics of detachment folds using three-dimensional geometry: application to the Mediano anticline (Pyrenees, Spain). Geol Soc Am Bull 110:111–125

Pocoví A (1978) Estudio geológica de las Sierras marginales catalanas. Acta Geol Hispanica 13(73):70

Pool W (1983) Laterale en verticale faciesveranderingen in de Onder-Eocene Serraduy-, Roda Marl- en Sandstone Formation (Ager Group) tussen Serraduy en La Puebla de Roda, provincie Huesca, Spanje. Unpublished M.Sc. thesis, Open file report, University of Utrecht, 100 p

Pugh DT (1987, 1996) Tides, surges and mean sea-level. Wiley, Chichester

Puigdefàbregas C, Souquet P (1986) Tectonosedimentary cycles and depositional sequences of the Mesozoic and tertiary from the Pyrenees. Tectonophysics 129:173–203

Puigdefàbregas C, van Vliet A (1978) Meandering stream deposits from the tertiary of the southern Pyrenees. In: Miall AD (ed) Fluvial sedimentology. Can Soc Petrol Geol Mem 5:469–485

Puigdefàbregas C, Muñoz JA, Marzo M (1986) Thrust belt development in the Eastern Pyrenees and related depositional sequences in the southern foreland basin. In: Allen PA, Homewood P (eds) Foreland basins. Int Assoc Sediment Spec Publ 8:229–246

Puigdefàbregas C, Nijman W, Muñoz JA (1989) Alluvial deposits of the successive Foreland basin stages and their relation to the Pyrenean thrust sequences. In: Marzo M, Puigdefàbregas C (eds) Guidebook to the 4th international conference on fluvial sedimentology. Publicacions del Servei Geològic de Catalunya, 176 p

Puigdefàbregas C, Muñoz JA, Vergés J (1992) Thrusting and foreland basin evolution in the southern Pyrenees. In: McClay K (ed) Thrust tectonics. Chapman and Hall, London, pp 247–254

Pujalte V, Baceta JI, Schmitz B, Orue-Etxebarria X, Payros A, Bernaola G, Apellaniz E, Caballero F, Robador A, Serra_kiel J, Tosquella J (2009) Redefinition of the Ilerdian stage (early Eocene). Geol Acta 7:177–194

Ricci Lucchi F (1986) The Oligocene to recent foreland basins of the northern Apennines. Int Assoc Sedimentol Spec Publ 8:105–139

Roest WR, Srivastava SP (1991) Kinematics of the plate boundaries between Eurasia, Iberia and Africa in the North Atlantic from the Late Cretaceous to the present. Geology 19:613–616

Rubin DM, Hunter RE (1982) Bedform climbing in theory and nature. Sedimentology 29:121–138

Rubino JL, Leo M, Fonnesu F(1985) Detailed stratigraphy of a tidal bar complex in the Eocene Baronia sandstones, Ager Basin, south-central Pyrenees. IAS 6th European Regional Meeting, Lerida, abstract book, pp 657–660

Samsó JM (1988) Estudi sedimentològic i biostratigràfic de la Formació St. Esteve del Mall (Eocé, conca Tremp-Graus). Unpublished M.Sc. thesis, University of Barcelona, Barcelona

Schmitz B, Pujalte V (2003) Sea-level, humidity, and land-erosion records across the initial Eocene thermal maximum from a continental-marine transect in northern Spain. Geology 31:689–692

Séguret M (1972) Étude tectonique des nappes et séries décollées de la partie centrale du versant sud des Pyrénées. Publications de l'Université de Sciences et techniques de Languedoc, série Geologie Structurale no. 2, Montpellier, 155 p

Serra-Kiel J, Canudo JI, Dinarés J, Molina E, Ortoz N, Pascual JO, Samsó JM, Tosquella J (1994) Cronoestratigrafía de los sedimentos marinos del Terciario inferior de la Cuenca de Graus-Tremp (Zona Central Surpirenaica). Rev Soc Geol España 7:273–297

Sha LP (1990) Sedimentological studies of the Ebb-Tidal Deltas along the West-Frisian Islands, the Netherlands. Ph.D. thesis, University of Utrecht, Utrecht, 159 p

Siegenthaler C (1982) Tidal cross-strata and the sediment transport rate problem: a geologists approach. Mar Geol 45:227–240

Srivastava SP, Roest WR, Kovacs LC, Oakey LC, Lévesque S, Verhoef J, Macnab R (1990) Motion of Iberia since the Late Jurassic: results from detailed aeromagnetic measurements in the Newfoundland Basin. Tectonophysics 18:229–260

Sztanó O, de Boer PL (1995) Basin dimensions and morphology as controls on amplification of tidal motions (the Early Miocene North Hungary Bay). Sedimentology 42:665–682

Teixell A, Muñoz JA (2000) Evolucíon tectono-sedimentaria del Pirineo meridional durante el terciario: una síntesis basada en la transversal del río Noguera Ribagorçana. Rev Soc Geol España 13:251–264

Tinterri R (2007) The Lower Eocene Roda Sandstone (south-central Pyrenees): an example of a flood-dominated river-delta system in a tectonically controlled basin. Riv Ital Paleontol Strat 113:223–255

Torricelli S, Knezaurek G, Biffi U (2006) Sequence biostratigraphy and paleoenvironmental reconstruction in the Early Eocene Figols Group of the Tremp-Graus Basin (south-central Pyrenees, Spain). Palaeogeog Palaeoclim Palaeoecol 232:1–35

Tosquella J (1988) Estudi Sedimentològic I Biostratigràfic de la Formació Gresos de Roda (Eocè, Conca Tremp-Graus). Unpublished M.Sc. thesis, University of Barcelona, Barcelona

Van de Kreeke J, Haring J (1979) Equilibrium flow areas in the Rhine-Meuse Delta. Coast Engr 3:97–111

Van den Berg JH (1981) Rhythmic seasonal layering in a mesotidal channel fill sequence, Oosterschelde Mouth, the Netherlands. In: Nio S-D, Schüttenhelm RTE, van Weering

TjCE (eds) Holocene marine sedimentation in the North Sea Basin. Spec Publ Int Assoc Sedimentol 5:147–159
Van den Berg JH (1982) Migration of large-scale bedforms and preservation of crossbedded sets in highly accretional parts of tidal channels in the Oosterschelde, SW Netherlands. Geol Mijnb 61:253–263
Van den Berg JH (1986) Aspects of sediment- and morphodynamics of subtidal deposits of the Oosterschelde (the Netherlands). Ph.D. thesis, University of Utrecht, Utrecht, 126 p
Van den Berg JH, Boersma JR, van Gelder A (2007) Diagnostic sedimentary structures of the fluvial-tidal transition zone. Evidence from deposits of the Rhine and Meuse. Geol Mijnb 86:287–306
Van der Meulen S (1982) The sedimentary facies and setting of Eocene point bar deposits, Montllobat Formation, Southern Pyrenees, Spain. Geol Mijnb 61:217–227
Van der Meulen S (1989) The distribution of Pyrenean erosional material, deposited by Eocene sheetflood systems and associated fan-deltas: A fossil record in the Montllobat and adjacent Castigaleu Formations, in the drainage area of the present Rio Noguerra Ribagorzana, provinces of Huesca and Lérida, Spain. Geologica Ultraiectina, 59, 125 p
Van der Voo R (1969) Paleomagnetic evidence for the rotation of the Iberian Peninsula. Tectonophysics 7:5–56
Van Eden JG (1970) A reconnaissance of deltaic environments in the Middle Eocene of the south-central Pyrenees, Spain. Geol Mijnb 49:145–157
Vergés J (2007) Drainage responses to oblique and lateral thrust ramps: a review. In: Nichols G, Williams EA, Paola C (eds) Sedimentary processes, environments and basins. A tribute to Peter Friend. Int Assoc Sedimentol Spec Publ 38:29–47
Vergés J, Millán H, Roca E, Muñoz JA, Marzo M, Cirés J, den Bezemer T, Zoetemeijer R, Cloetingh S (1995) Eastern Pyrenees and related foreland basins: pre-, syn- and post-collisional crustal-scale cross-sections. Mar Petrol Geol 12:893–915
Vincent SJ (1993) Fluvial paleovalleys in mountain belts: an example from the south-central Pyrenees. Unpublished Ph.D. thesis, University of Liverpool, Liverpool, 407 p
Vincent SJ (2001) The Sis palaeovalley: a record of proximal fluvial sedimentation and drainage basin development in response to Pyrenean mountain building. Sedimentology 48:1235–1276
Visser MJ (1980) Neap-spring cycles reflected in Holocene subtidal large-scale bedform deposits: a preliminary note. Geology 8:543–546
Waehry A (1999) Facies analysis and physical stratigraphy of the Ilerdian in the eastern Tremp-Graus Basin (south-central Pyrenees, Spain). Ph.D. thesis, University of Genève, Terre et Environment, 15, XII + 191 p
Willis BJ (2005) Deposits of tide-influenced river deltas. In: Giosan L, Bhattacharya JP (eds) River deltas – concepts, models and examples. SEPM Spec Publ 83:87–129
Wonham JP (1993) Sedimentology and sequence stratigraphy of tidal sandstone bodies: implications for reservoir characterisation. Chapter 4 – controls on the facies architecture of estuarine incised valley fill sandstone bodies from the Eocene Figols Group, Ager Basin, Spain. Unpublished Ph.D. thesis, University of Liverpool, Liverpool, pp 154–226
Yang CS, Nio S-D (1985) The estimation of palaeohydrodynamic processes from subtidal deposits using time series analysis methods. Sedimentology 32:41–57
Yang CS, Nio S-D (1989) An ebb-tide delta depositional model – a comparison between the modern Eastern Scheldt tidal basin (southwest Netherlands) and the Lower Eocene Roda Sandstone in the southern Pyrenees (Spain). Sediment Geol 64:175–196
Zachos J, Pagani M, Sloan L, Thomas E, Billups K (2001) Trends, rhythms, and aberrations in the global climate 65 Ma to present. Science 292:686–693

Holocene Carbonate Tidal Flats

19

Eugene C. Rankey and Andrew Berkeley

Abstract

Carbonate tidal flats of the Bahamian archipelago and the Arabian Gulf have served as important analogs for interpreting and understanding ancient tidal flat systems. Geomorphic associations include well-zoned subtidal, intertidal, and supratidal environments and their deposits, each with distinctive associations of biota and biologic and physical sedimentary structures. Although they include broadly similar facies associations in each environment within and between tidal flats, the occurrence and distribution of specific facies across landscapes differs markedly between tidal flats. Depending on the details of climate, tidal amplitude, regional setting and energy level, Holocene carbonate tidal flats include systems penetrated by numerous sinuous channels with adjacent levees and ponds, areas with broad, flat progradational intertidal and supratidal plains, and regions with shorelines that appear to have abruptly stepped oceanward or eroded. Stratigraphically, each different type of tidal flat includes a shallowing-upward facies succession, although in many areas, a basal transgressive unit is present.

19.1 Introduction

Shallowing-upward peritidal facies successions, passing from shallow subtidal up through intertidal and supratidal deposits, are a hallmark of many carbonate successions throughout the stratigraphic record (Wilson 1975; Hardie 1986; Pratt and James 1986; Grotzinger 1986; Lehrmann and Goldhammer 1999). Some stacked peritidal deposits are hundreds of meters thick and occur across areas of several 1,000 km^2. In seeking to understand these successions, the geologic history that they record, and the hydrocarbon, mineral, and water resources that they contain, sedimentologists have studied several modern carbonate tidal flats.

The purpose of this chapter is to outline general aspects of sedimentology, geomorphic character, and the Holocene stratigraphic record of some of these tidal flats. The focus is on the tidal flats of the Bahamian Archipelago, but for completeness, the chapter includes brief discussions of patterns and processes on the more arid tidal flats of the Arabian Gulf.

E.C. Rankey (✉)
Department of Geology, University of Kansas,
1475 Jayhawk Blvd., 120 Lindley Hall, Lawrence,
KS 66045, USA
e-mail: grankey@ku.edu

A. Berkeley
Department of Environmental & Geographical Sciences,
Manchester Metropolitan University,
John Dalton Extension Building, Chester Street, Manchester,
M1 5GD, UK

R.A. Davis, Jr. and R.W. Dalrymple (eds.), *Principles of Tidal Sedimentology*,
DOI 10.1007/978-94-007-0123-6_19, © Springer Science+Business Media B.V. 2012

Detailed documentation of many of these aspects or areas is well beyond the scope of this chapter; interested readers are referred to the primary literature on the topics as cited. Similarly, this discussion omits other carbonate peritidal systems with marked differences with the Bahamas, such as Shark Bay, Western Australia (Logan et al. 1970, 1974), or Florida Bay (Enos and Perkins 1979; Enos 1989) and does not consider sandy carbonate flats such as those of Joulter Cays or Shroud Cay in the Bahamas (Harris 1977, see Chap. 20, this volume).

19.2 Distribution of Carbonate Tidal Flats

Although Holocene carbonate tidal flats are much less spatially expansive than their ancient epeiric counterparts, several important and interlinked factors that collectively influence the distribution, geomorphology, and sedimentologic character are probably similar. At a fundamental level, tidal flats are broad, nearly planar areas near mean sea level, alternately flooded and exposed by tides, and formed by accumulation of generally muddy sediment transported and deposited in the absence of appreciable wave energy. A primary condition for tidal flat development is a shallow-water setting with low energy. Within this context, several factors influence the nature of tidal flats, including geologic influences (bedrock configuration, coastline orientation), tidal amplitude, winds (daily, related to the passage of cold fronts, or tropical depressions), climatic setting, and sediment source. Of course, sea-level change also markedly influences these systems; this aspect is discussed in some detail in a later section, in the context of the evolution and stratigraphic record of Holocene tidal flats.

Geologic framework and history controls the orientation of the shoreline, the topography from which tidal flats nucleate and evolve, and the characteristics of the nearshore shallow-marine system. The orientation and geographic setting of the coastline, for example, are among the most marked broad-scale controls on tidal-flat distribution and character because they influence waves and tides, the physical mechanisms that drive energy and sediment fluxes impacting the coast. Accordingly, tidal flats in the Bahamian Archipelago are most extensive on the western or southern platformward flanks of Pleistocene bedrock highs and islands – areas typically sheltered from prevailing wind and wave energy. At a more local scale, irregular bedrock topography, especially bedrock highs that extend above sea level (such as on Crooked-Acklins and southwest Andros; see below) within a tidal-flat complex, can exert a pronounced influence on local geomorphic and facies patterns.

Tidal range is an important control on carbonate tidal flats, regulating the volume of water and, indirectly, sediment passed onto and off the tidal flats, and exerting an important influence on tidal hydrodynamics, geomorphology, and the distribution of flora and fauna. Most modern carbonate tidal flats occur in microtidal settings, with tidal range less than 2 m. For example, in the Bahamas near the open ocean, the semi-diurnal tides have spring tidal amplitudes of <1.2 m, but range in the protected platform interiors and the tidal flats on the west side of Andros Island are ~46 cm at the mouth of creeks, decreasing to 29 cm in the inner tidal-flat ponds (Hardie 1977). Likewise, open-ocean spring-tidal range in the southern Arabian Gulf can exceed 2 m, but decreases into the lagoons and onto the sabkha complex (Purser 1973). Despite its potential significance, the influence of tidal range on the character of carbonate tidal flats has not received systematic study, and objective criteria for recognition of tidal range in the geologic record have not been developed.

Sediments from terrestrial sources are negligible in most humid carbonate tidal flats, and unlike many carbonate systems, sediment is not produced *in situ* to any great extent. As a result, these areas are intimately tied to their nearshore subtidal sediment sources. There are few detailed studies of the influence of the nearshore subtidal realm on tidal flats, although in a notable exception, Gebelein (1977) estimated that up to 94% of the sediment on the Cape Sable tidal-flat complex of southwestern Florida was generated from offshore sources. Comparable qualitative statements are made by Shinn et al. (1969) and Hardie (1977) concerning the Andros Island tidal flats, and the trend is probably similar for Caicos and Crooked-Acklins tidal flats as well. In contrast, arid tidal flats such as those in the Arabian Gulf may have markedly different sediment dynamics. In these areas, eolian quartz sand from terrestrial sources can add considerable sediment to the tidal flat system (Shinn 1973a, b). Nonetheless, the lack of supply from terrestrial or onshore sources is distinct from many siliciclastic systems.

Wind related to tropical depressions, cold fronts (in the Bahamian archipelago), or shamals (in the Arabian Gulf) can also impact tidal-flat sedimentation by influencing both suspended-sediment concentrations, and patterns of water movement across the flats. For example, tropical depressions commonly stir up and suspend muddy offshore sediments, and storm surges can flood the tidal flat with sediment-laden water, which then settles as the storm passes (Shinn et al. 1969; Hardie 1977; Wanless et al. 1988a, b). Numerous workers have considered these to be the most important physical process impacting tidal-flat sedimentation (e.g. Hardie 1977; Shinn 1986; Gebelein 1975). These sedimentologic interpretations notwithstanding, relatively few direct observations of the impacts of storms on carbonate tidal flats have been published (Ball et al. 1967; Perkins and Enos 1968; Wanless et al. 1988a; Rankey et al. 2004), but the limited impact observed from some tropical depressions suggest that the tide stage and forward velocity of storms at landfall strongly influence the impacts that these events have on the tidal flats.

Persistent onshore cold-front wind, although less destructive, can result in the flooding of much of the Bahamian tidal flats for days. Conversely, strong offshore wind can cause continuous tidal-flat drainage (e.g. Hardie and Garrett 1977). Sustained wind so strongly modulate the tide in shallow, more restricted parts of the Bahamas, that these flats might be considered "wind tidal flats" rather than tidal flats.

Wind also influences the Arabian Gulf systems. Here, strong, northerly, shamal wind drives southward longshore transport on the Qatari shoreline (Shinn 1973a, b). This wind also pushes water onshore in southern Gulf (UAE), flooding the sabkha with a meter or more of water (Schneider 1975).

Although it does not directly influence where they occur, the climatic setting of tidal flats influences biota, salinity, early fluid flow, and diagenetic features. Climate is therefore an important factor influencing the types of sediments, their susceptibility to erosion, and early diagenesis on tidal flats. For example, on the tidal flats of Andros Island with mean annual rainfall >120 cm, tidal ponds and algal marshes are abundant and evaporites are ephemeral to absent. In the Arabian Gulf, with annual rainfall < 10 cm, tidal ponds and creeks are essentially absent, algal marshes are very limited in extent, gypsum and anhydrite are common depositional and diagenetic features, bioturbation of intertidal sediment is less intense than in the Bahamas, and eolian transport provides ample sediment. Although less marked than these contrasts, subtle facies changes from the northern Bahamas to the southern Bahamas and Caicos islands (discussed below) have been interpreted to reflect the southward increase in aridity.

19.3 Humid Tidal Flats of the Bahamian Archipelago

Most understanding of Holocene humid carbonate tidal flats has come from the study of those in the Bahamian Archipelago, reaching from Little Bahama Bank in the north to Caicos platform in the south (Fig. 19.1). The most expansive tidal flats here fall on the platformward side of larger islands (Great Abaco, Andros, Grand Bahama, Crooked, Acklins, North, Middle and East Caicos), although narrower or less areally expansive tidal flats occur on parts of other islands as well (e.g. Long Island, San Salvador). Precipitation in this region decreases from ~150 cm/year in the northern Bahamas to ~75 cm/year on Caicos, and a majority falls in the wet season between May and October (northern Bahamas) and September and December (Caicos).

19.3.1 Zonation and Subenvironments

Carbonate tidal flats of the Bahamas and Caicos include spatially complex patchworks of subtidal, intertidal, and supratidal zones (Shinn et al. 1969; Hardie 1977; Shinn 1986; Wanless et al. 1989; Rankey 2002), their characteristics being generally related to elevation relative to mean tidal level. In a classic study of the Three Creeks area of Andros Island, Ginsburg et al. (1977) developed the "exposure index," a quantitative measure of the percentage of time that given elevations are exposed, which is closely related to the ecologic and sedimentologic characteristics across the tidal flat (Fig. 19.2). Although the details of the exposure index – subenvironment relationship varies somewhat with climate, sediment supply, and shoreline orientation along the archipelago (Wanless et al. 1989), broadly similar subenvironments, each with generally distinct associations of biota and sedimentary structures (Fig. 19.2), occur in tidal flats across the region; as such, they are discussed together here.

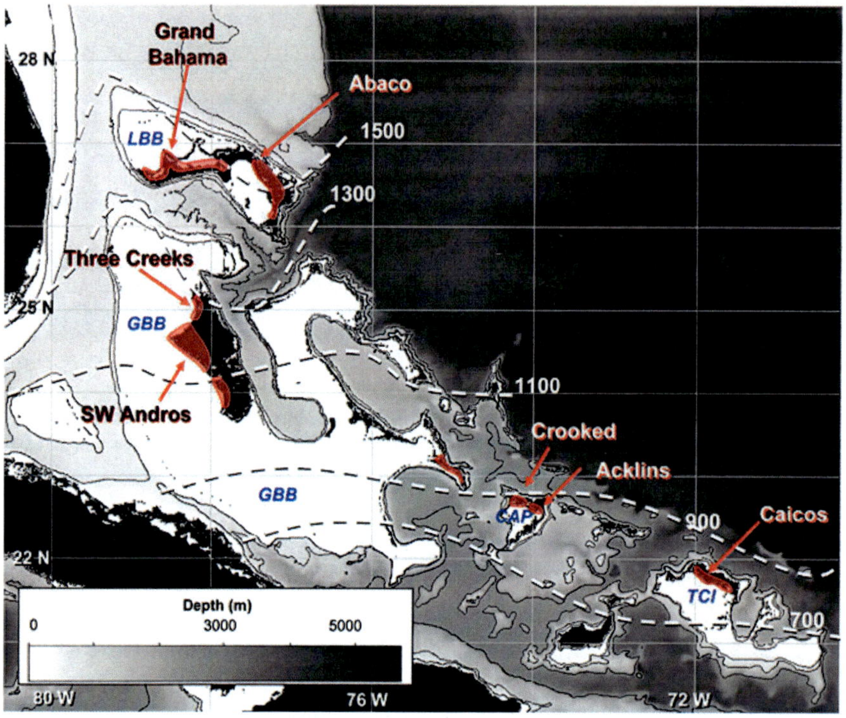

Fig. 19.1 General overview of the Bahamian Archipelago. The locations of some of the largest tidal flats are schematically highlighted by the *red* areas, and areas mentioned in the text are labeled. *Dashed white lines* represent lines of equal annual rainfall (labels in mm) (Data from Pierson 1982). Bathymetric data from General Bathymetric Chart of the Oceans (GEBCO) of the Intergovernmental Oceanographic Commission (IOC) and the International Hydrographic Commission (IHC) Contours are illustrated as *solid lines* for 0, 500, 1,000, and 1,500 m; other depths are shaded

Subtidal regions are those which are more-or-less permanently submerged, and in the Bahamas include nearshore marine regions, ponds, and tidal channels (Fig. 19.3). The nearshore marine regions may be sparsely covered with seagrasses (*Thalassia* and *Syringodium*) and calcareous algae (Fig. 19.3a, b), and include grey peloidal lime mud with scattered skeletal fragments, including miliolid and soritid foraminifera and mollusks. These sediments are commonly extensively bioturbated by the ghost shrimp *Callianassa* (Fig. 19.3a). On the Bahamian platforms and Caicos platform, these shallow marine settings bounding the tidal flat complexes extend kilometers offshore. Within the Andros tidal-flat complexes, ponds are ubiquitous (Fig. 19.3c), bounded by creek levees, beach ridges, palm hammocks, or intertidal flats. They are much less common on Caicos and Crooked-Acklins tidal flats further south. On Andros, the ponds are completely drained only during periods of prolonged offshore winds, thus they are essentially subtidal. Pond sediments are yellow-brown peloidal mud with scattered foraminifera and gastropods (Fig. 19.3c) and, being extensively burrowed, form deposits similar to those offshore from the tidal flat, but with lower biodiversity. In contrast, deposits of subtidal tidal channels (Fig. 19.3d, e) can include a coarse basal lag of cemented crust and fragments of Pleistocene bedrock (generally only the lowest few cm), gastropods and foraminifera, and intraclasts of indurated laminated levee deposits. In more outboard areas with wider tidal creeks, the creeks can exceed 2 m depth, and expose the top-Pleistocene bedrock. Creeks generally shallow and narrow landward, however, and inland creeks can simply be subtle, muddy depressions extending into ponds (Fig. 19.3f).

Intertidal areas lie between mean low tide and mean high tide, and are flooded and exposed twice daily, in the absence of persistent onshore or offshore winds. In the Bahamas, intertidal regions support mangrove and microbial marshes ("algal marshes" of older literature) between the subtidal open ponds and the supratidal levees flanking the creeks. In general, low intertidal

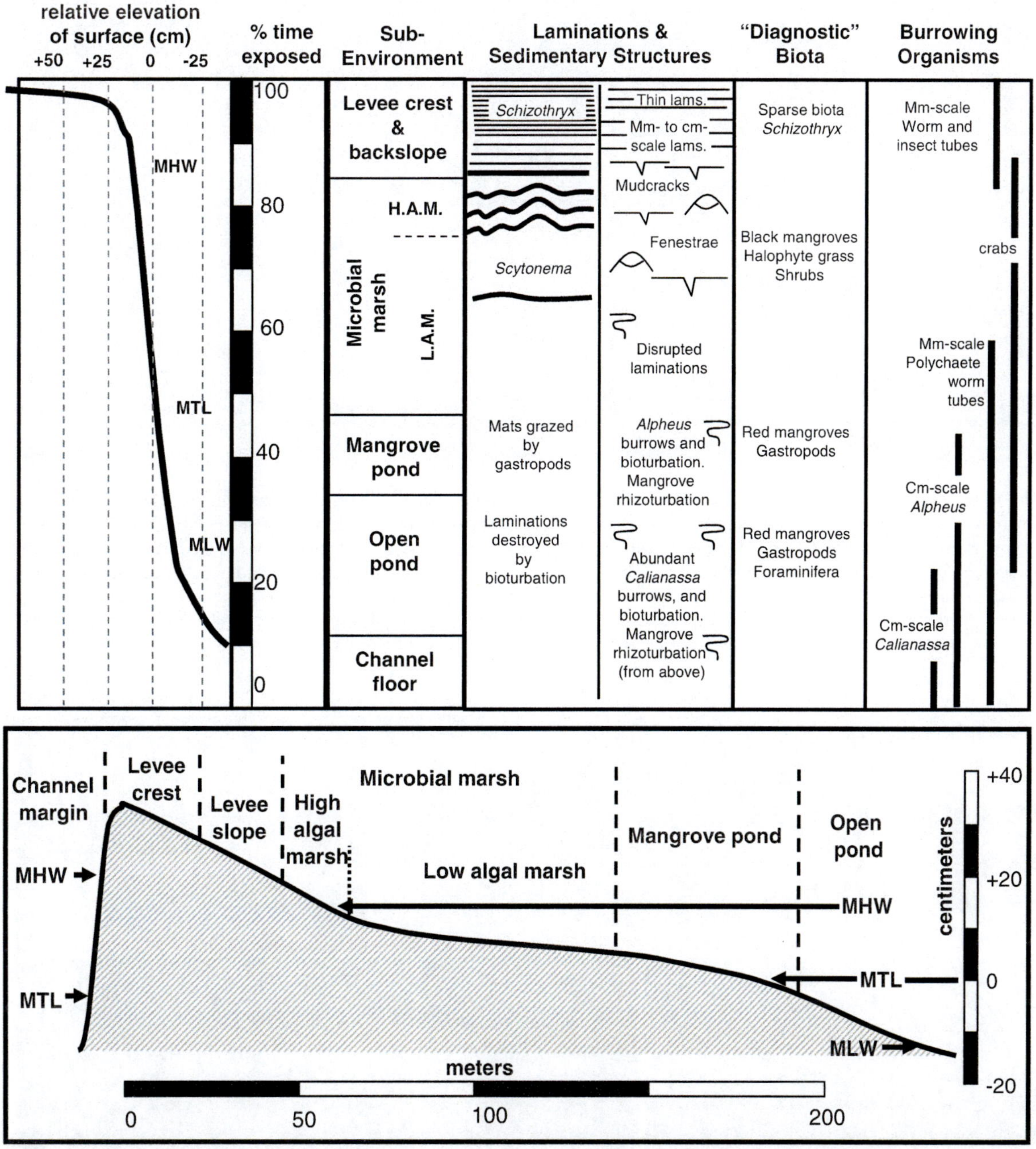

Fig. 19.2 Relations among environments, topography, and relative amount of time exposed to the atmosphere (Modified from Ginsburg et al. 1977), illustrating close relations between ecologic-depositional environments and the relative frequency of exposure ("Exposure Index" of Ginsburg 1977) and relation to relative elevation and subenvironment, (*upper*). In this area, each elevation range has a unique assemblage of biota and sedimentary features. Geomorphic display (*bottom*) of distribution of subenvironments on a representative channel-to-pond topographic transect; note the dampening of the tidal range into the pond

areas include red mangroves (*Rhizophora mangle*) (Fig. 19.4a), but sediments have a less diverse biota than adjacent pond deposits, and only cerithid gastropods and *Batophora* (a non-calcifying green alga) are common (Fig. 19.4b). Low-mid intertidal deposits are grey, reduced carbonate mud (Fig. 19.4c), with a distinct H_2S odor. Laminations here are destroyed by bioturbation (polychaete worm and

Fig. 19.3 Field photos of representative subtidal and associated environments, Bahamian Archipelago. (**a**) Shallow subtidal region, nearshore Crooked Island. The patchy pattern in the foreground is formed by burrow mounds and adjacent depressions. (**b**) Underwater photograph of shallow subtidal, nearshore marine environment, nearshore from the same area as illustrated in part A. Note seagrass and calcareous green algae (*arrows*), and the irregular topography caused by burrowing. (**c**) Shallow subtidal to lower intertidal area in a pond, Andros Island. Note sparse mangroves and the burrows in the foreground. (**d**) Large tidal channel and upper intertidal levee, Three Creeks, Andros Island. (**e**) Large tidal channel bordered by mangrove swamp, Caicos. (**f**) Termination of small tidal creek into pond, and "delta" like form (outlined by *dashes*), Andros Island

Fig. 19.4 Field photos of representative intertidal environments and sediments, Bahamian Archipelago. (**a**) Mangrove pond, with widespread dwarf red mangroves, Andros Island. (**b**) Sediments from mangrove pond, including abundant high-spired gastropods and Batophora, a non-calcifying green alga, Andros Island. Pencil for scale. (**c**) Short push core, illustrating light *grey* (oxidizing) color and thin laminations in lower intertidal zone, Andros Island. This sample illustrates the dark reduced band below the brownish algal mat at the surface. (**d**) Scytonema pincushions from the microbial marsh ("low algal marsh" of Hardie 1977), Andros Island. Bootie for scale. (**e**) Mudcracked, leathery microbial mat penetrated by black mangrove pneumatophores from the upper part of the microbial marsh ("high algal marsh" of Hardie 1977), Andros Island. Pencil for scale. (**f**) Lithified crusts associated with microbial mats, lower part of high algal marsh, Andros Island. Pencil (*circled*) for scale

Alpheus shrimp burrows, red mangrove rooting). These deposits can be indistinguishable from underlying subtidal deposits, especially with gradational contacts due to aggradation from subtidal up into lower intertidal elevations, and downward penetration of mangrove roots.

Marked changes in depositional conditions occur between lower intertidal and upper intertidal environments, causing pronounced differences in biota and sediment facies. In particular, halophyte grasses (especially *Spartina*), shrubs, and black mangroves (*Avicennia sp. nitida?*) appear in the upper intertidal zone, red mangroves are less abundant, and the sediment surface is covered by microbial mats (*Scytonema* pincushions at lower elevations Fig.19.4d; mixed with *Schizothryx* mats at higher levels, Fig.19.4e). Short cores from upper intertidal zone reveal that, due to the decreased density of vegetation and burrowing, sedimentary structures are more commonly preserved. Sediments typically are a light buff color, and include fenestrae (multigranular-roofed pores; Shinn 1983a), desiccation cracks and resultant intraclasts. These occur in lenticular layers with mm- to cm-scale laminations. Gastropods, foraminifera, and polychaete worms are quite uncommon relative to their high abundance in the mangrove and open ponds. Cemented crusts are locally abundant in the intertidal zone, associated with *Scytonema* (Fig. 19.4f).

Supratidal regions lie above mean high tide, and are flooded only during spring tides, storms, or prolonged onshore winds. Supratidal areas from across the archipelago include creek-adjacent levees, beach ridges, and supratidal plains. Some tidal flats in the Bahamas (parts of Andros, Crooked-Acklins), include hammocks, supratidal areas which support cabbage palm trees and grasses, and are essentially terrestrial.

Supratidal deposits include the most diagnostic suite of sedimentary structures on the tidal flat. On parts of the Andros tidal flat, supratidal levees flank the larger creeks near their outlet to the open ocean, and are bounded by an abrupt channel margin with abundant *Uca* crab burrows (Fig. 19.5a). These levees typically have a sparse vegetation cover (black mangrove, scattered grasses and shrubs) (Fig. 19.5b), and are covered with a distinct continuous, firm mat of *Schizothryx* (Fig. 19.5c–e). Although "air domes" (incipient fenestrae; Fig. 19.5d) and small, shallow mud cracks occur on the levee backslope, on the levee crest, this mat is tightly bound and mud cracks are absent. Short cores reveal that the sediments include mm-scale discontinuous normally graded laminations of peloidal silt and fine sand (Fig. 19.5e).

Beach ridges with grainy deposits occur on the shorelines of some tidal flats, most notably those of the northwest-facing parts of Andros tidal flats and the east-facing appendages of Crooked-Acklins and Caicos tidal flats. These ridges commonly have an erosional oceanward face with small, scalloped embayments, irregular at a decameter scale (Fig. 19.5f). This face cuts into fine, laminated peloid-skeletal sand (Fig. 19.5g, h), and locally is covered by thin coarse skeletal debris in the shoreline recesses (Fig. 19.5h). In numerous locations on the northwest Andros tidal flats near Three Creeks, thin, platy cemented crusts, similar to those found in the marsh on the beach-ridge backslope, occur within the erosional faces. Beach ridges parallel the shoreline and are generally less than 100 m wide, but reach up to ~250 m in a few exceptional cases; in scattered locations, discontinuous wedges of storm-derived coarse skeletal sand and gravel extend landward from the shoreline (Fig. 19.5i). The beach ridges pass landward and downdip into marshes and ponds either abruptly (in the presence of sandy skeletal wedges) or gradationally (mimicking the lateral changes from levees down to ponds).

In contrast to the beaches on the northwest-facing tidal flats of Andros Island, the shorelines of the southwest-facing Andros tidal flats are oblique to both normal trade winds and the brisk winds from southeastward-moving cold fronts. These areas include supratidal beach ridges with a different character. Along much of this coast, the shoreface is essentially a low scarp with up to 1 m relief (Fig. 19.6a) cutting into gastropod-bearing muddy carbonates. The scarp is capped with grasses, bushes, and small trees, indicating that it is rarely flooded, and gravels are uncommon.

Inboard supratidal plains form some of the most areally extensive regions on the tidal flats, perhaps best developed on parts of Andros complexes and on the Crooked Island system. These features are extremely flat; gentle gradients of 10 cm/km (1:10,000) (Hardie 1977), with only subtle undulations, are not uncommon (Fig. 19.6b). These areas are covered by a vast expanse of dark *Scytonema* pincushions (Andros) or scattered pincushions with a hardened crust (Crooked Island, Fig. 19.6c), with only rare stunted black mangroves and scattered grass. Flooding is limited to major storms.

Fig. 19.5 Field photos of representative supratidal environments and sediments, Bahamian Archipelago. (**a**) Margin of large tidal creek (*left*) with adjacent levee (*right*), Andros Island. The flanks of the levee are extensively burrowed by the fiddler crab *Uca*. Mangrove in foreground is ~50 cm tall. (**b**) Supratidal levee crest, illustrating the relatively sparse vegetation, including only sparse low shrubs and grasses. Backpacks for scale. Three Creeks, Andros. (**c**) Close-up of supratidal levee, illustrating the absence of mudcracks and sparse vegetation, Andros. Field notebook long axis ~22 cm. Note rill marks formed as flooding tide tops levee. (**d**) Air-filled blisters on supratidal levee that could become fenestrae, Andros Island. Pencil for scale. (**e**) Upper part of small trench through supratidal levee sediments, Andros Island, illustrating thinly laminated appearance. Pencil for scale. Apparent waves in lamination are artifact of irregularities in trench wall. (**f**) Irregular shoreline, southwest Andros Island. To the *left* of the observer, the shoreline is sandy, whereas to the *right*, it is made of eroding thinly laminated sediments. (**g**) Finely laminated sediments eroding at the shoreline, Acklins Island, Bahamas. (**h**) Close up from (**g**), illustrating the thin laminations. Pencil for scale. (**i**) Landward-dipping, partly stabilized, sandy deposits on the tidal flat. These deposits probably resulted from large (storm) waves on the adjacent beach. Observer and backpack (to *left*) for scale

Short cores from this area reveal that the sediments include generally muddy sediment with distinct cm-scale laminations, thicker than elsewhere on the tidal flat (Fig. 19.6d), likely reflecting mud settling out from surges driven by the infrequent, but pronounced, flooding associated with hurricanes (Shinn et al. 1969).

On parts of southwest Andros tidal flats and on the tidal flats south of Crooked Island, bedrock highs protruding above the sediment form another supratidal environment. Hardie (1977) illustrates as much as 3 m relief on the top of bedrock around these features, and tidal flats of Crooked Island include similar irregularity on the bedrock surface (Berkeley and Rankey, in review). These features, termed hammocks, have patchy soil capped with a diverse terrestrial biota, including grasses, palmettos, and palm trees, as well as *Cerion*, a tree snail (Fig. 19.6e). These bedrock highs are ringed by fringes of laminated mud, locally mudcracked, with sparse vegetation (Fig. 19.6f). Many areas include cemented crusts and locally continuous pavements. These pavements on Andros are locally known as "sidewalks" or "runways," because some are

Fig. 19.6 Field photos of representative supratidal environments and sediments, Bahamian Archipelago. (**a**) Eroding supratidal shoreline of southwest Andros Island. The shoreline here is stabilized by grasses, which pass (northeast, to the *right*) into cabbage palm hammocks. (**b**) Supdatidal plain of blackened crusts, thin microbial mats, and small, stunted mangroves, Crooked Island. (**c**) Close-up of the continuous cemented crust. Crooked Island. The slightly darker areas are covered with a thin, soft microbial mat; the lighter areas are cemented crust. (**d**) Thick laminations, common in the broad supratidal inland marsh, Andros Island. (**e**) Edge of cemented pavement, and the adjacent cabbage palm hammock, southwest Andros Island. (**f**) Coarse rubble of cemented clasts from a cemented pavement southwest Andros Island

so extensive that they were used as landing strips for small planes in more rough-and-tumble days.

19.3.2 Contrasts in Tidal Flat Geomorphology – Bahamian Archipelago

The tidal flats from the Three Creeks area are among the best studied carbonate flats in the world and represent what is commonly portrayed as the 'humid channeled belt' morphotype (e.g., Wright 1984; Shinn 1986). Although informative, they are not necessarily representative; there are several other geomorphically distinct types of tidal flats in this region.

19.3.2.1 Three Creeks Area, Andros Island, Bahamas

The 'classic' Andros Island tidal flats (Black 1933; Shinn et al. 1969; Gebelein 1974; Hardie 1977, 1986; Shinn 1986) occur in the Three Creeks area of northwest Andros Island (Fig. 19.1), where they form an onlapping wedge of muddy subtidal, intertidal and supratidal sediments. Shinn et al. (1969) recognized three general geomorphic zones: inland marsh, channeled belt, and adjacent marine. The inland marsh (supratidal plain) comprises a more-or-less continuous microbial mat up to 6 km wide and 25 km long that fills an arcuate embayment in Pleistocene outcrops (Fig. 19.7a). This mat is penetrated only by a few very shallow (<10–20 cm) and narrow (meter scale) creeks and is extremely flat. Some creeks represent continuations of depressions in bedrock from the island to the east, and water in these can be fresh to brackish.

The channeled belt, the 'heart' of the tidal flat, lies between the inland algal marsh and the offshore marine zones (Fig. 19.7b). This area, up to 5 km wide, includes a complex network of tidal creeks, levees, ponds, and algal marshes, mostly dominated by muddy sediments. Near their emergence on the coast, tidal creeks are flanked by elevated levees (up to ~200 m wide, though most are much less) that gradually slope down and outward through an algal marsh zone into mangroves and open ponds (Fig. 19.7b). As the creeks reach headward, levees are less well developed until the creeks gradually become narrow, shallow gullies within the mangroves and are impenetrable by boat. Although studies of modern (Shinn et al. 1969) and ancient (Cloyd et al. 1990) tidal flats suggest the importance of tidal channel migration, little evidence for pronounced

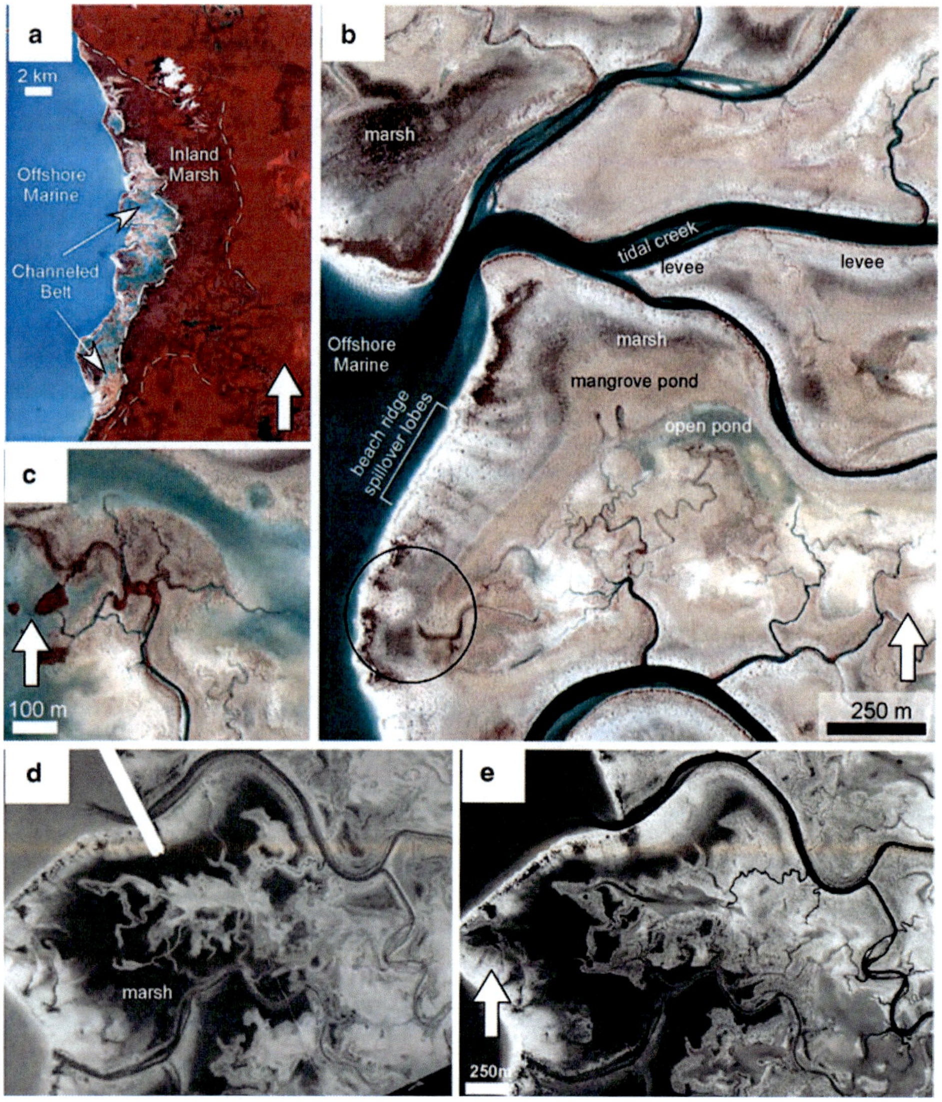

Fig. 19.7 Remote-sensing and aerial images of the Three Creeks area, Andros Island. (**a**) Image illustrating the nearshore marine, channeled belt, and inland marsh located, within a broad embayment in the Andros coastline. (**b**) Ikonos image of part of the channeled belt. Note the different geomorphic elements, including well-developed creeks and flanking levees, marshes and ponds. One spillover lobe (cf. Fig. 19.5i) is circled. (**c**) A small delta, extending into an open pond (to the north). (**d–e**) Paired historical (1943) and recent (2001) remote sensing image from the southern part of the Three Creeks area, illustrating some changes. Note the contraction of the marshes (*dark areas*) and the expansion of ponds (Remote sensing images (**b**), (**c**), and (**e**) copyright GeoEye)

migration is evident in historical changes (Rankey and Morgan 2002) or cores (e.g., Shinn 1986 showed only ~75 m migration) (cf. Wright 1984).

The most influential processes impacting the geomorphology of the channeled belt are driven by water and sediment transport through the creeks and into adjacent environments. Most sediment on the tidal flats is transported from the nearshore marine areas via these conduits, probably associated with meteorological events that suspend and transport offshore mud into the channeled belt (Shinn et al. 1969; Ginsburg and Hardie 1975; Hardie 1977, cf. Rankey et al. 2004). Because sediment is transported from the ocean landward through creeks, Shinn et al. (1969) described the tidal flat system as a delta "turned inside out." The same is true more literally at a finer scale, as illustrated by distributary systems (small birdfoot deltas) within ponds at the end of some creeks (Fig. 19.7c) illustrating appreciable landward sediment delivery. Similarly, in the southern Three Creeks area where there are no active creeks, ponds are much more extensive. Since no sediment is being delivered into these areas and sea level continues to rise (Rankey and Morgan 2002), the ponds here areas are expanding (Fig. 19.7d, e).

The channeled belt is flanked on the oceanward side by grainy beach ridges at the shoreline (Fig. 19.7b). The beach ridges and spillovers are generally less than 100 m wide, but reach up to ~250 m inboard in a few exceptional cases, and include either fine, laminated peloid-skeletal sand or coarse skeletal debris. Locally, intraclasts of crusts are abundant.

The Three Creeks area is changing on historical time scales. For example, Rankey and Morgan (2002) documented shoreline erosion at rates of just under 1 m/year with varied geomorphic signatures: (1) pond and levee backslope sediments are exposed in the intertidal zone seaward of the beach ridge in many areas, (2) stumps of *Casurina,* a freshwater conifer living only on beach ridges, are also present in the intertidal zone; and (3) beach-ridge spillover lobes cover or block creeks (Fig. 19.7b, circle), illustrating landward encroachment. The cause of the change from aggradational or progradational (representing the accumulation of the tidal–flat record) to the present erosional state is not well understood. Wright (1984) suggested changes in sediment supply could have influenced this change, but this hypothesis has remained untested, as has the suggestion that it may be related to a relative rise in sea level (Rankey and Morgan 2002). At a longer time scale, the tidal flats have been stepping landward; nearshore marsh deposits are found beneath channeled belt sediments and, locally, several 100 meters offshore (see below; Shinn et al. 1969; Shinn 1986). Obviously, at some point the tidal flat built seaward or upward, only to undergo erosion again.

19.3.2.2 West Andros Island, Bahamas

Relative to the tidal flats of the Three Creeks area, those of west Andros Island are less well studied. This broad area forms the westernmost tip of Andros Island, extending up to 25 km outboard from patchily-exposed Pleistocene bedrock. This area rims western Andros, from ~25 km northeast of Williams Island (an erosional vestage of the tidal flat; Gebelein 1974), around the tip, and >45 km along shore from near Williams Island to the Bights (Fig. 19.8), and beyond. This area includes both erosional (northwest-facing) and progradational (southwest-facing) shorelines, with the transition between each corresponding with a marked change in shoreline orientation. Because these tidal flats have such distinct geomorphic characteristics, they are described separately.

Southwest-facing margin. Offshore of the southwest-facing Andros tidal flats is a broad expanse of shallow subtidal, muddy sediments that are <1 m deep at low tide and exhibit an extremely low bathymetric gradient (Queen 1976). The shoreline is straight along much of its length along this margin, and is penetrated by only a few broad, shallow tidal creeks. The present shoreline has no grainy beach, but at least in the western area, instead consists of a low scarp (Fig. 19.6a), suggesting that it presently is erosional.

Tidal channels extending into the tidal flat are relatively rare, with only five channels along the 45 km coast (Fig. 19.8a). These channels, although much less common than in the Three Creeks area, are much larger, reaching nearly 3 km wide, but are less than a few metres deep. Most do not branch for much of their extent, but instead are wide, straight, and shallow. Creeks in the southeasternmost part of the tidal-flat complex (Fig. 19.8b) pass from a classic estuarine-like V-shape to narrower creeks with branching distributary systems. and include stabilized and vegetated levees (Fig. 19.8b), unlike those in the Three Creeks region.

Along much of the coast, landward of the present shoreline, a broad supratidal plain gently passes northeasterly into mangrove ponds and open freshwater ponds (Fig. 19.8a). The supratidal plain here is wide and flat, and covered by a thin blackened crust, scattered microbial mats, and mangroves. It is thus broadly similar to the inland algal marsh in the Three Creeks area, but with fewer *Scytonema* pincushions and more expansive crusts. Further inboard, however, ponds become more common, and are the dominant landscape element. These ponds are circular to highly irregular, can form pond networks ranging in size up to 6 km across, and are mostly shallow, but can be up to several meters deep (Fig. 19.8a, c). Networks of ponds may collectively form larger bodies of water several kilometers across, broken only where they abut hammocks. The arcuate margins of some ponds appear to be influenced by spit accretion, possibly controlled by winds.

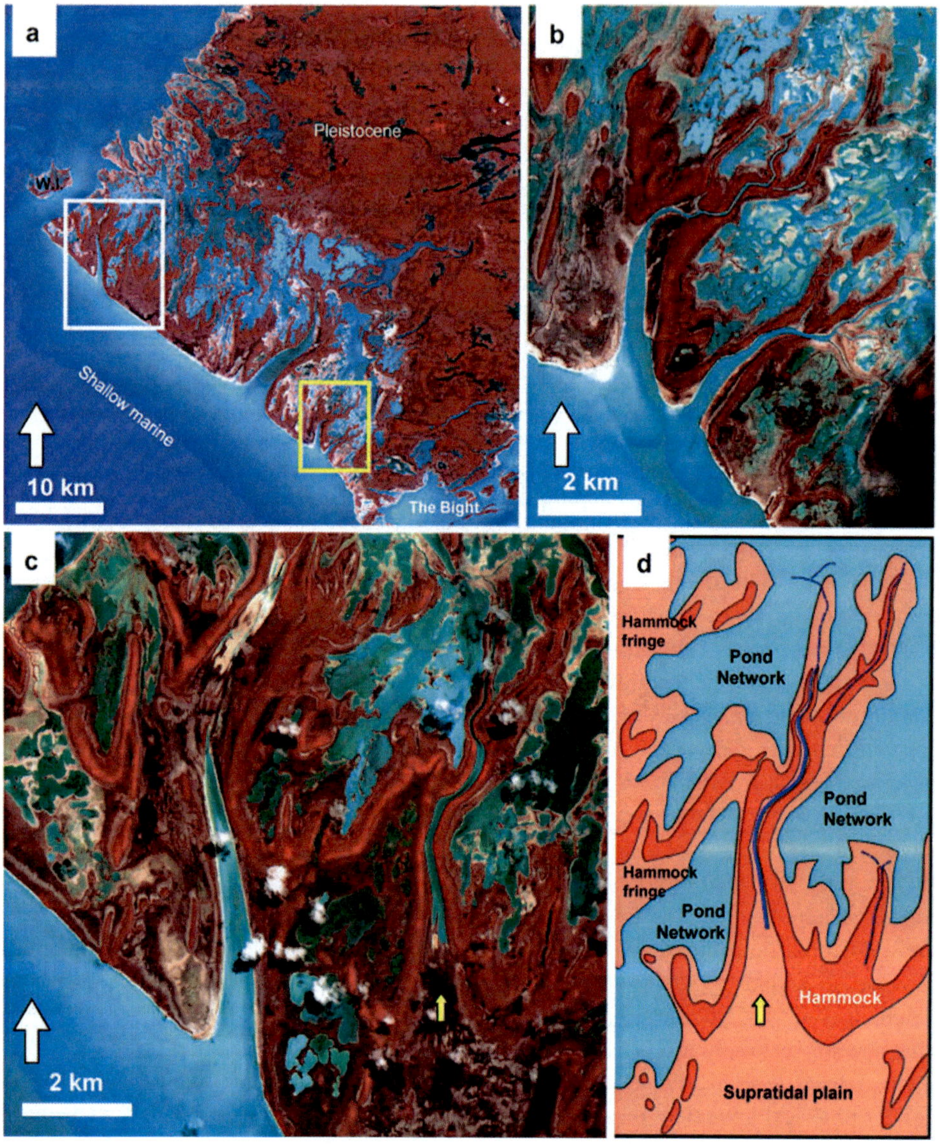

Fig. 19.8 Geomorphic characteristics of the tidal flats of parts of southwest-facing Andros Island. (**a–c**) False-color (NIR-G-B) remote-sensing images. In these images, *redder* areas indicate denser vegetation. (**a**) Overview image. Note the broad tidal-flat complex, with an irregular northwest-facing coast and a straight southwest-facing coast with relatively few tidal creeks. "W.I." indicates the location of Williams Island, "The Bight" refers to the narrow passage through central Andros. (**b**) Remote sensing image of part of the southeastern expanse of the tidal flat (*yellow box* in **a**). In this area, the shallow creeks are flanked by levees with dense vegetation, and the more inland areas include shallow ponds and mudbanks. (**c–d**) Remote sensing image (**c**) and simplified interpretive sketch (**d**) of an area southeast of Williams Island (*white box* in **a**). Here, regions near the coast include a broad supratidal plain and palm hammocks. These pass landward (northeast) into a complex of ponds and linear hammocks with morphology akin to the levee flanks on creeks further southeast (see area in **b**). *Yellow arrow* points to the same closed creek mouth in both parts (Remote sensing images (**b**) and (**c**) copyright GeoEye.com)

Within the broad tidal flat of southwest Andros Island, several elongate to V-shaped ridges are present (Fig. 19.8c, d). These ridges presently support cabbage palms, grasses, and several are flanked by dolomitic crusts ("runways" Shinn et al. 1965) (Fig. 19.6e, f). These ridges are aligned normal to the coast, with long 'tails' that extend landward. In several places, two or more parallel ridge lines appear (Fig. 19.8c, d). By

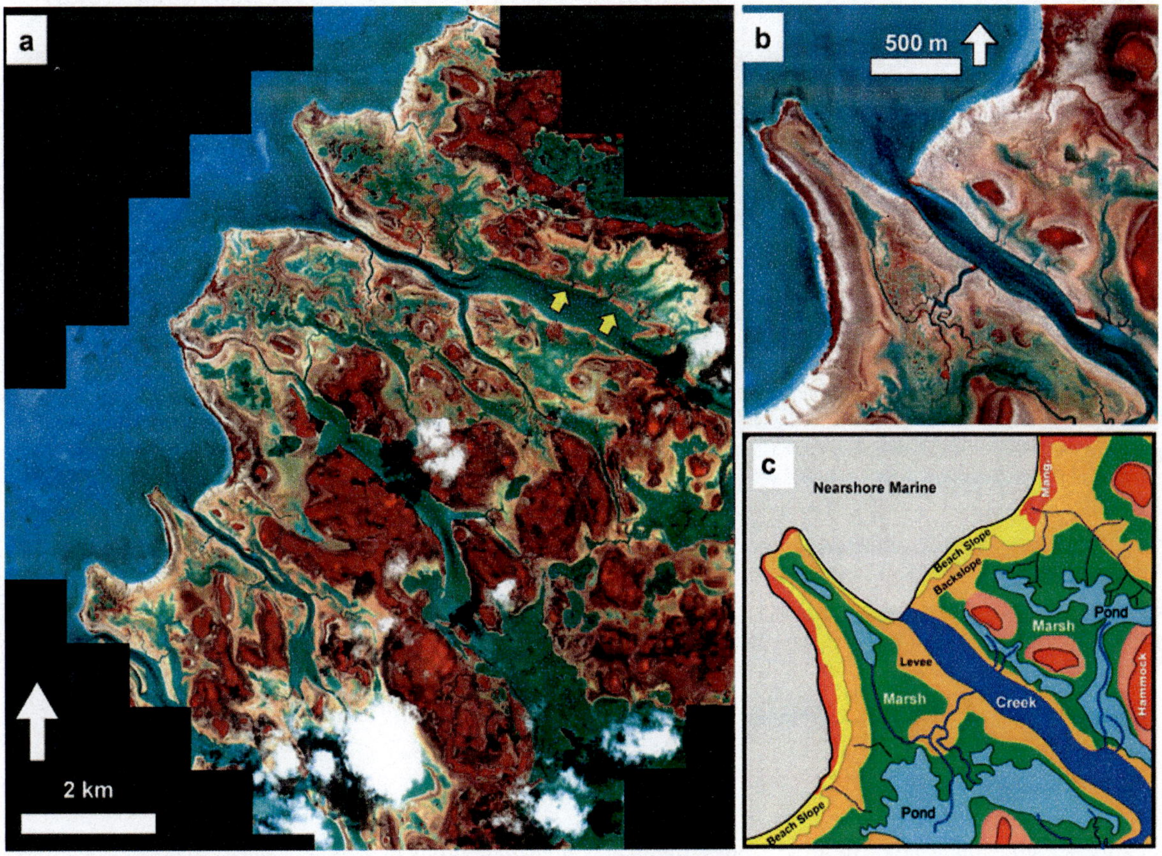

Fig. 19.9 Geomorphic characteristics of the tidal flats of west Andros Island southeast of Williams Island. (**a–b**) False-color (NIR-G-B) remote sensing images from tidal flats of southwestern Andros Island. (**a**) Overview of part of the coastline, illustrating the irregular shoreline morphology (cf. Fig. 19.5f), abundant creeks, ponds, and hammocks. (**b–c**) Remote-sensing image (**b**) and interpretive sketch (**c**) of an illustrative area (Remote sensing images copyright GeoEye.com)

analogy with the characteristics of parts of the present coast (see Fig. 19.8b), these features can be reasonably interpreted to represent the positions of relict shorelines and channel levees. Gebelein (1974) interpreted a succession of these features to represent a succession of progradational shorelines. The distinct hammocks that preserve relict shorelines suggest that the shoreline prograded in a series of pulses, rather than through progressive accretion at the shoreface, although the stratigraphic succession of this area is not known in detail.

Northwest-facing margin. The northwest-facing margin of the western Andros system includes some of the most complex subfacies patterns on the tidal flats of Andros Island. As in the Three Creeks area, the shoreline is irregular (and likely erosional) at several scales, including several km-scale promontories and recesses (Figs. 19.6f and 19.9). Many promontories include bedrock exposed at or near the low-tide level, whereas bedrock outcrops are absent in most of the recesses, illustrating that bedrock highs protect shorelines from erosion.

Unlike the Three Creeks area, which has generally similar depths (~2–3 m) to bedrock across most of the channeled belt, or the southwest-facing margin, which has deeper bedrock (>5 m; Gebelein 1974), the facies patterns in this area are strongly influenced by the irregular bedrock topography. Many of the larger hammocks (red in Fig. 19.9a) include bedrock highs, some of which reach a meter or so above high tide. Similarly, even many smaller hammocks (e.g. Fig. 19.9b, c) have bedrock highs near the surface, whereas subtidal channels between these features reach 3 m depth. Many of these channels, and trends in the hammocks, are

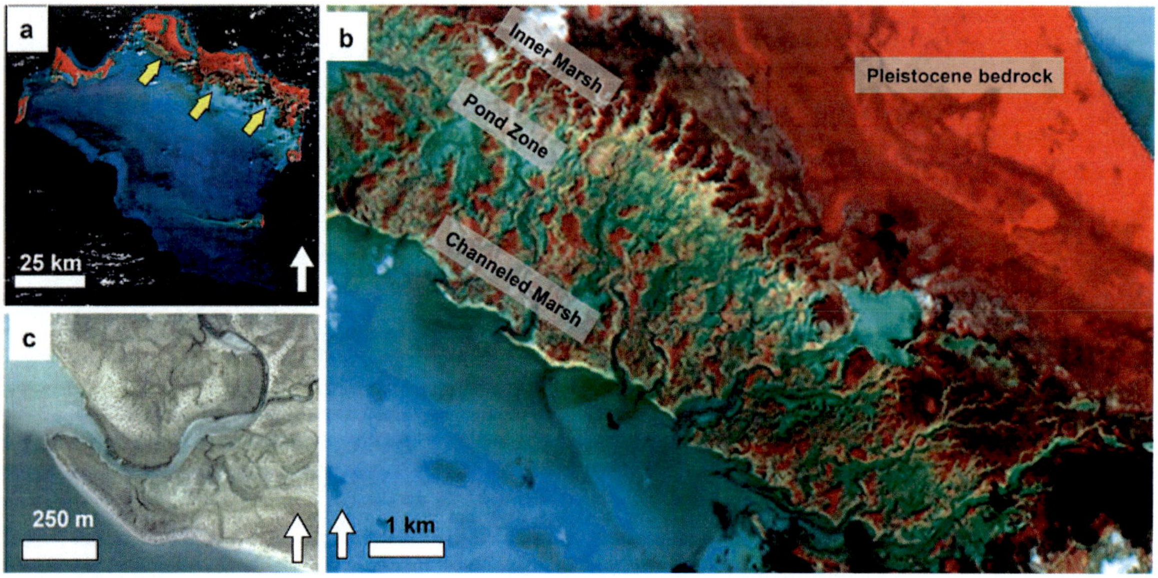

Fig. 19.10 Geomorphic characteristics and setting of the tidal flats of the Caicos platform. (**a**) Image (NIR-G-B) illustrating the entire platform, with the location of the most expansive tidal flats on the southern flanks of the islands (*arrows*). (**b**) Remote sensing image illustrating general zonation of part of the tidal-flat complex. (**c**) Remote sensing image illustrating the westward bend in a creek near its termination into the platform interior, related to longshore currents from easterly winds (Remote sensing image (**c**) copyright DigitalGlobe)

roughly perpendicular to the coastline, suggesting that pre-existing bedrock topography strongly influences facies patterns here.

Because of the numerous bedrock highs, this region lacks a well-defined channeled belt like that in the Three Creeks region. Instead, landward of the shoreline are a series of subtidal open ponds (Fig. 19.9a), lower intertidal mangrove ponds, and scattered algal marshes. Small supratidal levees and upper intertidal high algal marshes flank some of the larger channels and most of the hammocks, and slope down to the ponds. Some of the ponds are large and deep enough to have sufficient fetch to develop flanking grainy beaches (see yellow arrows in Fig. 19.9a).

19.3.2.3 Tidal Flats, Caicos Platform

The best-developed tidal flats of the Caicos platform occur on the low-energy south to southwestern flanks of Middle and North Caicos islands. Facies patterns in this area include subtle variations from those in the Three Creeks area of Andros Island. As on Andros, the nearshore is a shallow, highly burrowed subtidal zone dominated by mud and soft peloidal sand that slopes gently away from the coast. Tidal flats here include three general geomorphic zones: channeled flats, a pond zone, and an inner marsh (Fig. 19.10b) (Wanless et al. 1989) zone. The shoreline is a low (30–40 cm), erosional scarp. Relative to the northwest Andros, this shoreline levee is lower and broader (up to ~1 km wide) and has less area covered with thinly laminated *Schizothryx* mats. Like the Andros tidal-flat shoreline, however, it too includes evidence for basal transgressive deposits. For example, Wanless et al. (1988a, b) documented that some offshore marine sediments were underlain by tidal-flat pond sediments. They also highlighted the influence of bioturbation in re-working the sedimentary record of transgression.

Along the shoreline, the mouths of several tidal channels are offset to the west-northwest due to growth of spits, presumably driven by the prevailing easterly wind (Fig. 19.10c). Sinuous tidal channels extend up to 5 km into the tidal flat, but are flanked by lower, broader, and less well-developed levees than in the *Schizothix*-dominated levees of the Three Creeks area. Most creeks extend landward and empty into the pond zone, a region of shallow ponds continuous along strike. The ponds are flanked landward by an inner microbial marsh 1–2 km wide. This marsh is a broad zone dominated by

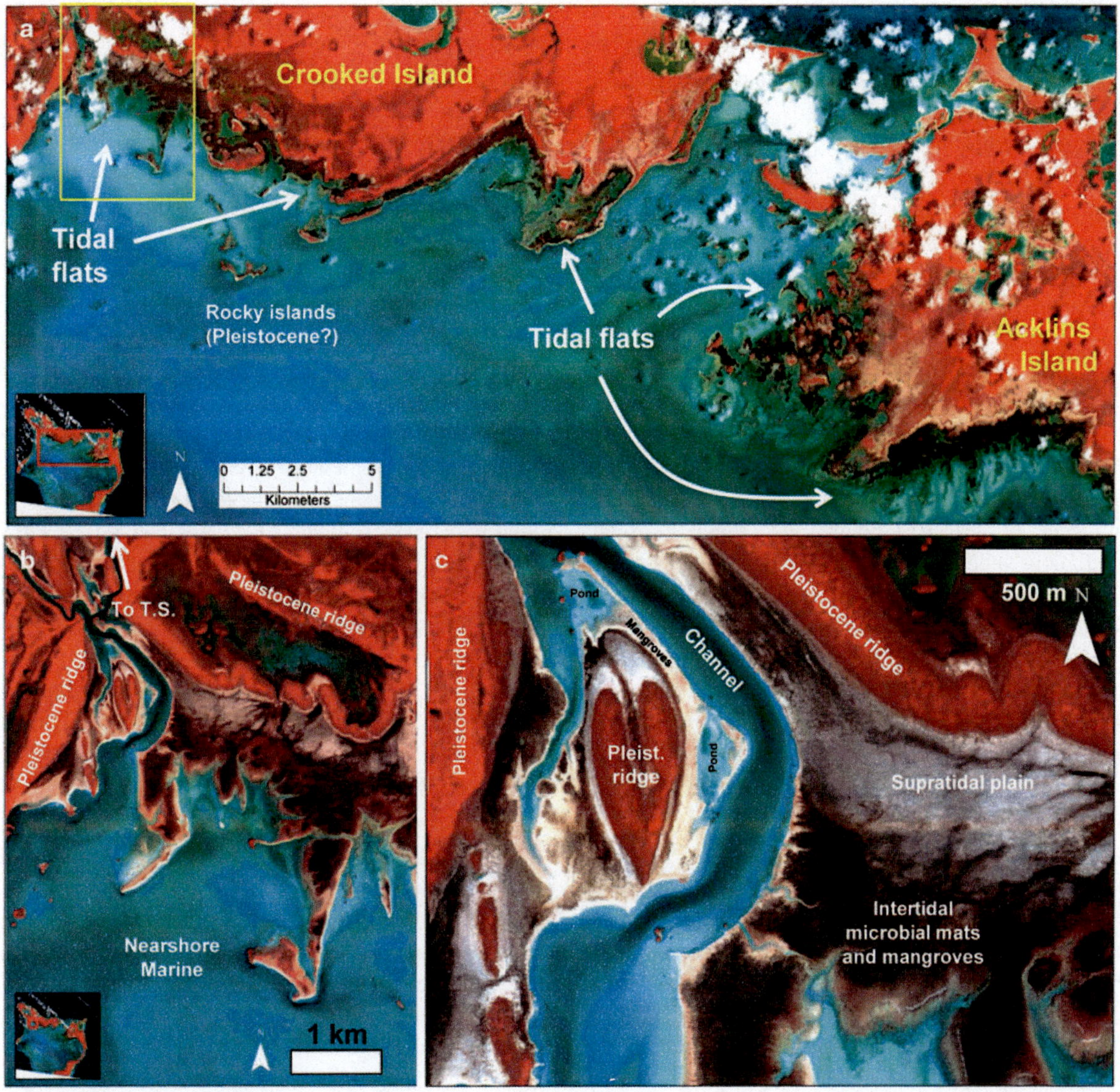

Fig. 19.11 Geomorphic characteristics and setting of the tidal flats of part of the Crooked-Acklins Platform, illustrated in false-color (NIR-G-B) remote sensing images. (**a** and **b**) General character of tidal flats flanking the southern margin of Crooked Island, including a general zonation that includes Pleistocene outcrops, a supratidal pavement, microbial marshes, and mangoves, with relatively few creeks. (**c**) Detail of one area near the creek, illustrating considerable variability (Remote sensing images copyright DigitalGlobe)

soft *Scytonema* mats and scattered cemented crusts; it grades landward into terrestrial environments.

19.3.2.4 Tidal Flats, Crooked-Acklins Platform, Bahamas

Tidal flats form a continuous band 1–2 km wide and ~18 km long on the southern (platformward) flank of Crooked Island, and exhibit geomorphic patterns distinct from those on Andros Island or the Caicos platform (Fig. 19.11). The tidal flats are bound to the north and west by Pleistocene ridges, 3–5 m high, and platformward by a broad, shallow, subtidal nearshore region (Fig. 19.11) (Rankey and Reeder 2010). A conspicuous feature of this system is the paucity of well-developed tidal channels. A single, deep (>2 m) tidal channel occurs near the western margin of the tidal flats, forming the only marine conduit through Pleistocene outcrops to an otherwise landlocked shallow embayment (locally known as Turtle Sound) (Fig. 19.11b). The broad, irregular, supratidal plain of the tidal flat is bordered to the south by a gently dipping intertidal shoreface. The thickness of Holocene sediments is typically ~2 m beneath this tidal flat (thinning landward) but exceeds 3 m close to the channel, suggesting that the prevailing geomorphic pattern is strongly influenced by the antecedent topography (Berkeley and Rankey, in review).

In the near-absence of channels and associated geomorphic features such as levees and ponds, most of this system is characterized by a supratidal plain dipping towards the nearshore marine environments (Fig. 19.11b, right side). Supratidal areas include extremely subtle topographic relief (centimeters across the kilometer-wide tidal flats), and are sparsely vegetated with scattered grasses and black mangroves. The infrequent flooding imparts strong evaporative and freshwater influences on surface sediments, and a thin (~1 cm) but continuous indurated surface with scattered intraclast debris is present in many areas. Downdip (i.e. southward) from this supratidal zone in the upper intertidal zone, the sediment surface is colonized by thin (~1 cm), smooth *Scytonema* mats, which pass downdip into thicker (5–10 cm) 'pincushions' upper intertidal zone with black mangroves. Further towards the coast, around mean tide level, red mangroves colonize a narrow belt, and low intertidal to shallowest subtidal areas are non-vegetated except for localized occurrences of the non-calcifying alga *Batophora* and scattered seagrass.

The shoreline delineating the tidal flats is notably irregular, and includes numerous intertidal to supratidal extensions into the nearshore shallow subtidal areas (Fig. 19.11). The absence of beach ridges suggests that shoreward-directed wave and tide energy is not appreciable. The irregular shoreline and subtle (~10 cm) relief may reflect locally variable rates of sediment accumulation, perhaps related to patterns of mangrove colonization.

Adjacent to the single well-developed tidal channel, a number of intensely burrowed lower intertidal to subtidal ponds have become partially enclosed by the development of mangrove-colonized intertidal bars (Fig. 19.11c). These features occur on a broad meander bend where subtle levees with only a few centimeters of relief have accreted to mean high water. The outer channel bank is steeper, and the presence of slabs of thin, indurated crust suggests erosion and exhumation of previously buried lithified horizons, like those present in adjacent supratidal areas. Unambiguous evidence for marked channel migration is absent in core, however (Berkeley and Rankey, in review).

19.4 Arid Tidal Flats of the Arabian Gulf

Sabkhas (arid tidal flats) can be found around much of the western and southern Arabian Gulf (e.g., Purser 1973). Some of the best studied and most illustrative examples occur in the United Arab Emirates (UAE; the "Trucial Coast" of older literature) and on the eastern flank of the Qatar peninsula (Fig. 19.12) (Evans et al. 1969; Kendall and Skipwith 1969; Purser and Evans 1973; Shinn 1973a, b, 1983a, 2010 in press, Alsharhan and Kendall 2003). Due to the low precipitation (several cm/year), these sabkhas include important sedimentologic contrasts with the humid Bahamian examples. Like the Bahamas, subtidal, intertidal and supratidal zones include pronounced sedimentologic differentiation; nonetheless, the nature and distribution of sediments and structures are quite distinct from those in comparable zones in the Bahamas, and vary even in different parts of the Gulf.

19.4.1 Eastern Qatar – Zonation, Subenvironments, and Geomorphology

The nation of Qatar occupies a club-shaped, north-south trending peninsula that protrudes into the Arabian Gulf. On the peninsula's eastern coastline, several areas include illustrative peritidal geomorphic patterns (Figs. 19.12 and 19.13).

On the northeast coast of Qatar near Khor (Fig. 19.13a, b; Shinn 1973a, b; 2010), a carbonate-dominated shoreline has prograded up to 5 km from low-lying, highly irregular outcrops of Tertiary dolomite.

Geomorphic processes here are strongly influenced by the shoreline orientation, which is roughly parallel to the northwesterly shamal winds. This setting creates a strong longshore drift from north to south, resulting in beaches with hook-shaped spits at their southern end (Figs. 19.13c, d and 19.14). These beach-spit complexes are up to 12 km long and 50–100 m wide and are comprised of cross-bedded, coarse, bioclastic grainstone (Fig. 19.14e, f) with admixed quartz sand supplied by eolian dunes.

Subtidal sediments include soft, muddy, grey, reduced, peloidal sediments, which occur immediately

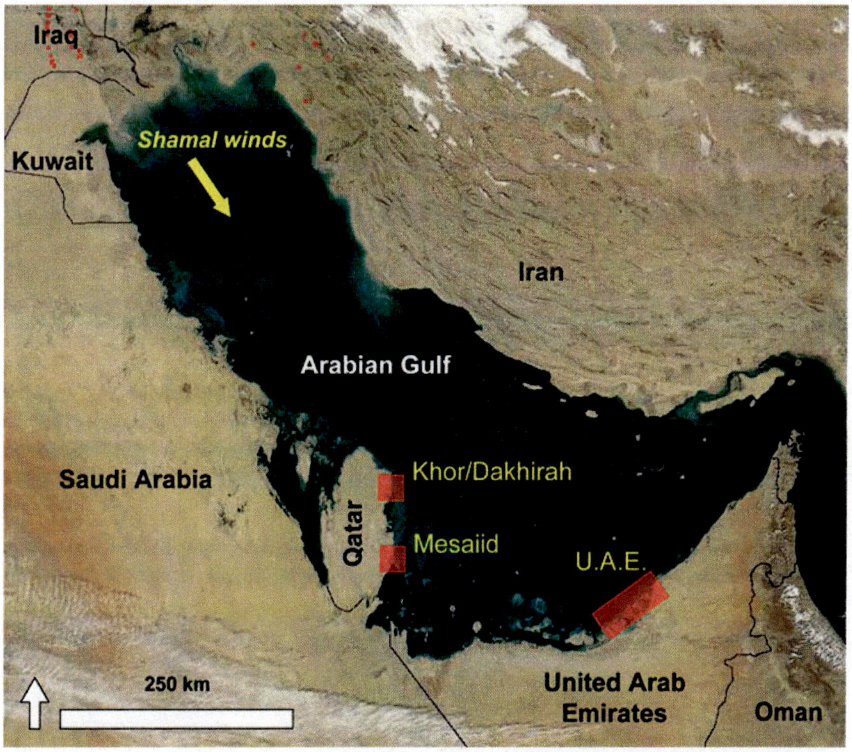

Fig. 19.12 Overview image of the Arabian Gulf, illustrating the location of the well studied sabkhas discussed in the text

offshore (Shinn 1973a, b). These sediments contain up to 30% dolomite, which can be penecontemporaneous (de Groot 1973), or blown in from nearby eroding Tertiary dolomites. Well-sorted, cross-laminated sand occurs in the shoreface, and poorly sorted muddy sand is found in some of the channels that extend inland. Probing and coring by (Shinn 1973a, b) suggest that bedrock lies more than 8 m below some of the lagoons and present-day beach ridges, indicating considerable Holocene accumulation. He did not describe any direct bedrock influences on sedimentary patterns.

Although the entire area is underlain by muddy subtidal sediments, suggesting long-term progradation, (Shinn 1973a, b) noted that some beach deposits directly overlie intertidal and supratidal deposits. Similarly, some beach ridges and spillover lobes cut back and partially fill tidal channels (Fig. 19.14c, d). These geomorphic observations suggest that the beaches are migrating landward, at least locally (Shinn 2010 in press).

As the beaches and spits migrate to the south (at rates up to 40 m/year; Shinn 1973a, b), they form a local energy barrier that leads to development of protected lagoons and tidal flats, deposited in a shore-parallel zone up to several km wide behind the beach ridges (Figs. 19.13 and 19.14). Wide tidal flats include predominantly muddy sediments and abundant meandering channels (Fig. 19.14b, c); narrow flats are grainy and lack channels.

Landward of the intertidal zone lies a broad supratidal sabkha up to 5 km wide. The sabkha in this area is flooded only during the summer months, when winds blow from the east. It includes predominantly muddy sediments with thin storm lags of skeletal-rich sediment. Unlike the sabkhas further south, the sabkhas of eastern Qatar have only thin evaporites (2–3 cm thick crusts) (Shinn 1973a, b). Where not removed by deflation, microbial laminations and mudcracks can be found, and fenestrae are more widespread than in the UAE tidal flats.

Further south on the Qatar peninsula, in the area of Messaid (Fig. 19.12), southward-migrating accreting beach ridges are evident as well. This area contrasts with the Khor region, however, in that well-developed tidal creek networks do not occur between older beach ridges.

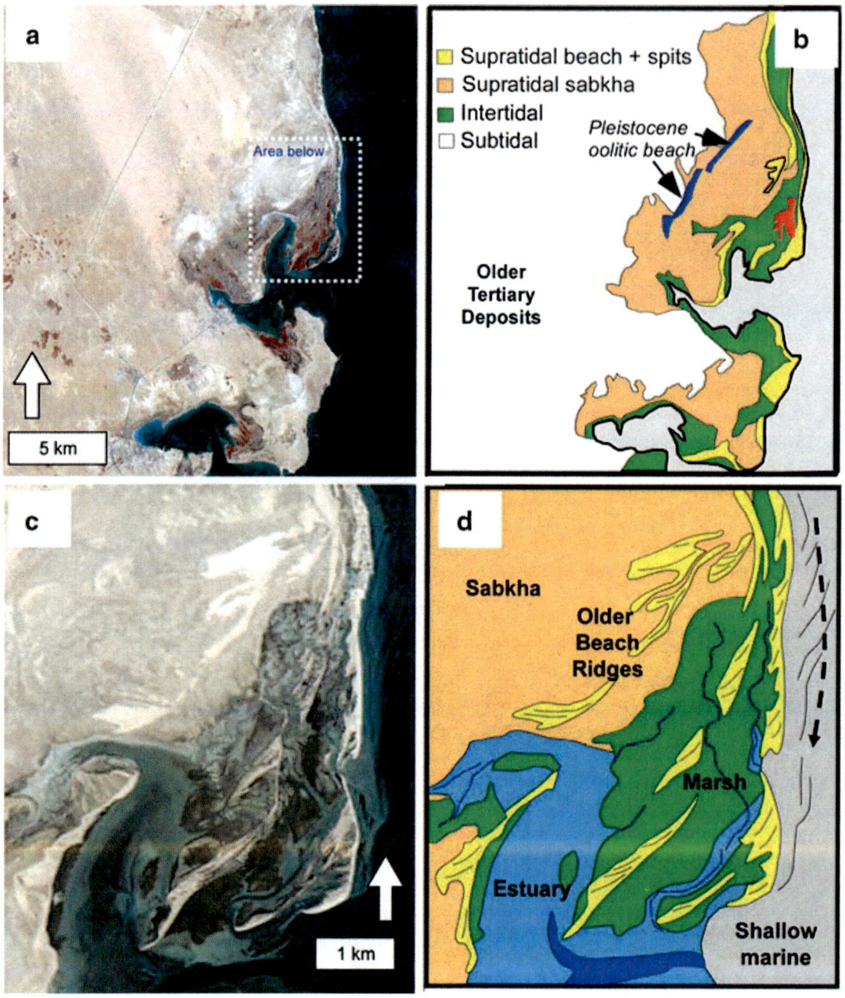

Fig. 19.13 Geomorphic characteristics of the sabkha, Khor region, Qatar (Fig. 19.12). (**a**, **b**) Remote sensing image (**a**) and interpretive facies patterns (**b**). (**c**, **d**) Remote sensing image (**c**) and interpretive facies patterns (**d**) of the area highlighted in (**a**), illustrating the progradational and offlapping pattern associated with southward longshore transport (*arrow*) (Modified from original in Shinn (1973a, b))

Instead, many inter-ridge lows appear to have aggraded to supratidal levels, although a subtle depression running normal to the shore includes marshy algal mats and seasonal evaporites (white areas in Fig. 19.15).

19.4.2 United Arab Emirates – Zonation, Subenvironments, and Geomorphology

At the southern end of the Arabian Gulf, carbonate coastal sediments of the United Arab Emirates (Figs. 19.12 and 19.16) overlie Miocene and Pleistocene bedrock that crops out along an arcuate, low, NE-SW oriented ridge. This coastline includes an irregular complex of subtidal shoals, reefs, tidal deltas and passes, protected intertidal lagoons, and supratidal sabkhas that extends several 100 km along strike and is up to 15 km wide, generally narrowing to the northeast. The shamal blows onshore in this region (see above), creating waves oriented roughly perpendicular to the coast. The power of the waves varies along the coast, however, as a result of the shielding effect of the Qatar peninsula, which creates a more protected setting in its lee, to the west along this coast (Fig. 19.16). More exposed areas (generally east of the Zubaiya peninsula, Kendall and Skipwith 1969, see also Purser and Evans 1973) include a narrow lagoon separated from

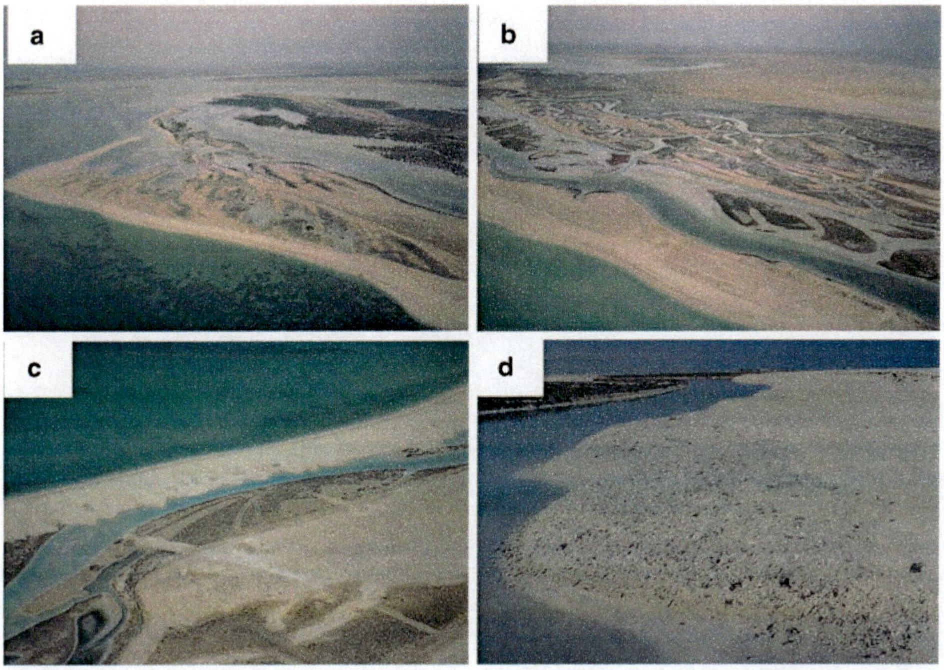

Fig. 19.14 Photos from the Khor area, courtesy of Peter Scholle. (**a**) Aerial photo (view to *south*) of southward prograding shorelines. (**b**) Detail of back-barrier shoreline area (view to *southwest*), with numerous tidal creeks. (**c**) Close-up of part of the shoreline (view to *east*), which appears to be spilling landward (*bottom*) over a creek. (**d**) Coarse sand and gravel that make up the shoreline spillover deposits

Fig. 19.15 Remote sensing (**a**) and interpretive (**b**) image of area near Mesaiid, Qatar Peninsula (Fig. 19.12). As further north, this region includes a series of southward stepping progradational beach ridges, bordered landward by sabkha deposits (Remote sensing image copyright DigitalGlobe)

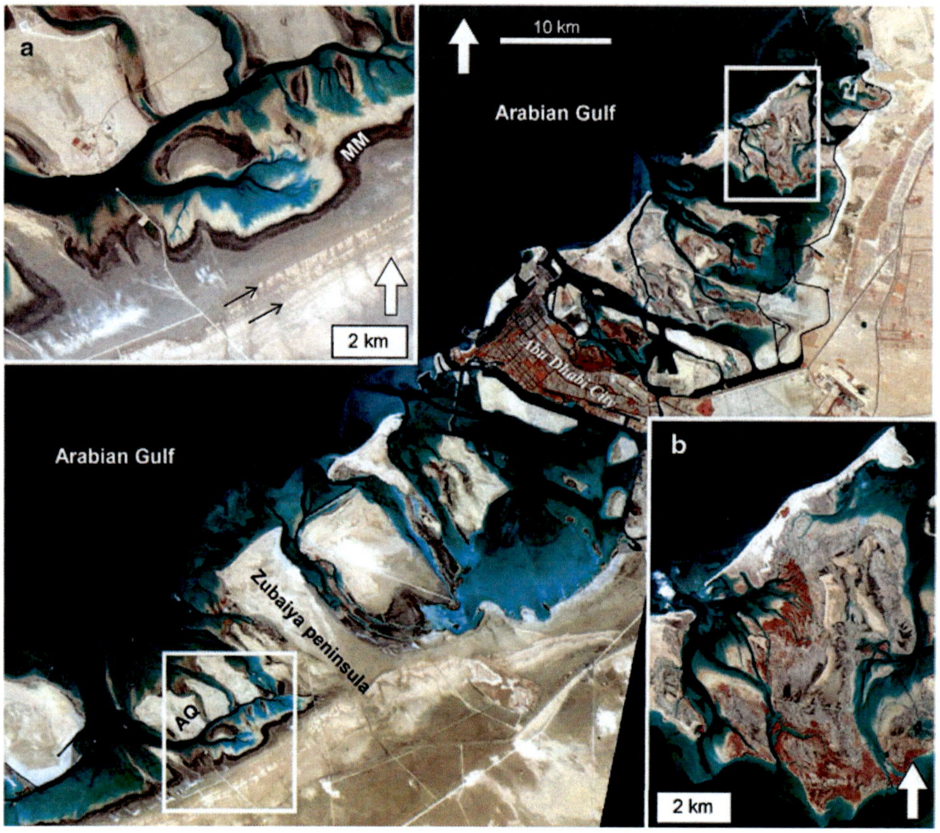

Fig. 19.16 Character of the coastal system, Abu Dhabi, United Arab Emirates, southern Arabian Gulf (Fig. 19.12). In this area, the shoreline faces into the dominant winds. (**a**) Khor al Bazam area, south of Al Qanatir Island (labeled "AQ"). In this region, a series of older beach ridges (*black arrows*; Kirkham 1997) are bordered oceanward by a series of microbial-mat facies belts ("MM") that generally parallel the shoreline (Kendall and Skipwith 1968). (**b**) Area to the east. A broad intertidal to supratidal mangrove flat (*reddish* color), penetrated by tidal creeks

the open ocean by closely spaced "T"-shaped islands (Fig. 19.16). The islands have one long axis parallel to the NE-SW oriented shoreline, with an elongate perpendicular stem extending leeward (NW-SE). Beaches on the exposed, seaward sides of these islands can have beach ridge dunes up to 5 m high, reflecting the high-energy conditions. These supratidal ridges are composed primarily of medium to coarse skeletal sand, including abundant gastropods, and unlike beach sand of Bahamian tidal flats, these sediments are commonly reworked into eolian dunes. The type of sand (skeletal, oolitic) is closely related to the nearshore sediment.

In the subtidal realm, large oolitic tidal deltas are found seaward of tidal inlets between islands in this area (Fig. 19.16); reefs also may be present in areas in front of the islands protected from lagoonal waters extruded through tidal passes (Alsharhan and Kendall 2003). Extensive early cementation, creating hardgrounds up to 30 cm thick, occurs in areas oceanward of the islands. In many places, the elongate leeward (landward) stems of islands are flanked by broad intertidal flats which gradually slope into the protected lagoon (Fig. 19.16b). On these flats the lower intertidal mangrove swamp includes abundant black mangrove, *Avicennia sp.*, which grows in areas penetrated by small tidal channels and flooded daily. These mangroves may facilitate the formation of small levees, and these areas tend to accumulate oxidized muddy sediments that are extensively burrowed by crabs. Microbial mats are not as expansive here as in other supratidal areas; instead, the lateral transition from the lagoon to the sabkha is relatively abrupt.

Areas west of the Zubaiya peninsula are downwind of the Qatar peninsula, and are not exposed to the full wave fetch of the Arabian Gulf. The lower wave exposure leads to a different geomorphic character. Unlike east of here, subtidal areas do not have oolitic deltas, but instead include well-developed reefs offshore, perhaps reflecting the more limited flux of toxic lagoonal waters. Supratidal islands landward of the Gulf shoreline are larger, more equant (not "T" shaped) (Fig. 19.16 left side), and pass into the lagoon without a high-energy beach.

In the restricted, higher salinity, subtidal areas behind the protection of the barrier islands all along this coast, deposits are dominantly grey, reduced, muddy skeletal sand with gastropods and foraminifera, sediments broadly analogous to those found offshore in many areas of the Bahamas. In exposed shallow subtidal to lowermost intertidal regions, winnowing creates thin deposits of muddy sand that are locally cemented to crusts less than 10 cm thick, which can subsequently be bored and fractured (Shinn 1986, 1969).

On the landward side of the lagoon are well-defined, shore-parallel facies zones, from muddy subtidal lagoonal sediments, to intertidal cemented gastropod-rich sediments, to upper intertidal dark-colored laminated cyanobacterial mats, intertidal to supratidal evaporites and the sabkha (Fig. 19.16a). As in the Bahamas, intertidal deposits of this area have a distinct change in character from the lower intertidal to upper intertidal zone. Due to the larger tidal amplitude (up to 2 m in some areas, less in more restricted lagoonal environments), however, intertidal deposits are thicker here, with some accumulations reaching 2 m thick.

Lower intertidal deposits here are similar in many ways to the underlying shallow subtidal deposits, and include predominantly burrowed peloidal carbonate mud with scattered skeletal fragments or thin sandy layers. With the exception of laminar or irregular fenestrae, sedimentary structures are generally not preserved due to burrowing, and this lower intertidal zone has few microbial mats, in large part due to the activity of grazing cerithid gastropods.

Upper intertidal deposits differ considerably from shallow subtidal and lower intertidal sediments (Shinn 1986; Alsharhan and Kendall 2003). In much of the area, upper intertidal deposits include a distinct zone of well-developed algal mats up to 30 cm thick. These mats preserve thin, mm-scale, organic-rich laminations of wind-blown eolian silt or flooding-related silt and sand. The mats include a distinct vertical zonation (Kendall and Skipwith 1968), ranging from: (1) a warty black surface (topographically lowest); (2) a mudcracked zone, with cracks forming polygons that can exceed 1 m across; (3) a crinkle zone with scattered gypsum; and (4) a zone with a flat, smooth algal surface (topographically highest). These mats may not extend into the supratidal zone due to the intense evaporation and deflation.

Supratidal sediments of sabkha in the United Arab Emirates have a character transitional with upper intertidal sediments. In many areas, the uppermost intertidal zone is characterized by precipitation of a gypsum mush interbedded with storm-deposited carbonates. These deposits are flanked by extremely flat (slopes of 1:1,000) supratidal sabkhas characterized by anhydrite and halite, mixed with eolian quartz-rich sand. Trenches through this area reveal that the thin evaporite layers commonly are deformed, buckled and bent (ptygmatic-like folds). Butler (1970) suggested that the evaporites and the deformation were the result of evaporation of waters brought to the sabkha during extreme flooding events. Between flooding events, deflation is pronounced and actively removes mudcracks and truncates folds in the evaporites.

As on Andros Island, the surficial environments in this area have been changing in recent years (Fig. 19.17). Beyond the pronounced human imprint (e.g., the canal just onshore, dredging and creation of artificial islands – highlighted by white arrows in some areas), several islands and mangrove swamps (reddish areas, and red boxes in Fig. 19.17) are expanding. The aggradation reflects a gradual infilling of the lagoon by progradation in the lee of the islands into the shallow lagoon, rather than by building out from the mainland shoreline (see below).

19.5 Facies Successions of Holocene Carbonate Tidal Flats

The relatively few Holocene carbonate tidal flat successions that have been cored and described in detail collectively illustrate several consistent trends in sedimentologic and stratigraphic features among locations. To a large extent, these trends reflect the ubiquitous shallowing-upwards that occurs as a result of aggradation and progradation outpacing rising sea levels, and the close linkages between elevation and depositional facies which occurs across tidal flat systems in general. Nonetheless, bioturbation and rooting by mangroves in tidal flat successions can modify or destroy original depositional textures (Boudreau

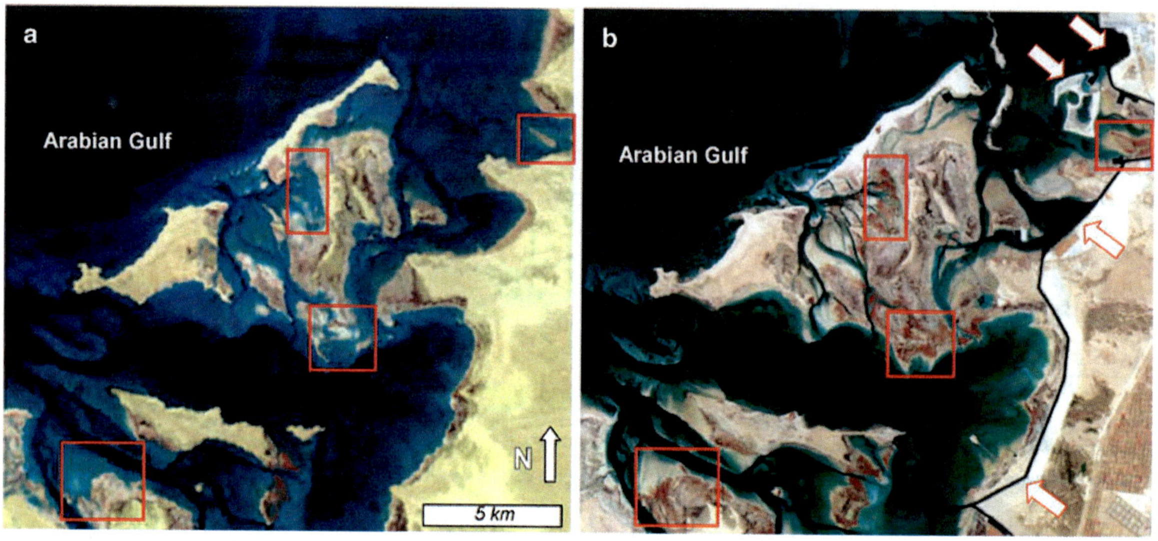

Fig. 19.17 Paired false-color (NIR-G-B) remote sensing images from area just northeast of Abu Dhabi City (Fig. 19.16, illustrating changes between 1972 (**a**) and 2001 (**b**). Although there are numerous anthropogenic/cultural changes on or near the terrestrial, mainland shoreline (e.g. the canal highlighted by the *white arrows* with *red* borders), other more offshore areas have changed as well. The most obvious changes (highlighted by the *red boxes*) include sediment aggradation and consequent expansion of mangrove swamps

1994; Tedesco and Wanless 1991; Tedesco and Aller 1997; Berkeley and Rankey in review). As with previous sections, our goal is to provide a general flavor of the spectrum of types of vertical facies successions that can occur in carbonate tidal flats. Readers interested in more stratigraphic details for each area should refer to the primary literature, reviews that emphasize stratigraphic aspects (e.g., James 1979; Shinn 1983a, b; Wright 1984; Hardie 1986; Pratt et al. 1992), or the AAPG movies "Arid Carbonate Coastlines" (Scholle et al. 2005) or "Stratigraphic Traps: The Tidal Flat Model" (Shinn 1981) which graphically illustrate many of these features.

19.5.1 Three Creeks Area, Andros Island

Approximately 2 m of Holocene sediments overlie Pleistocene bedrock in the Three Creeks area (Shinn et al. 1969). The basal component in many areas consists of thinly-bedded freshwater marsh sediments (<40 cm thick), which has been identified from cores from up to 1 km offshore from the contemporary tidal flat, indicating a substantial Holocene transgression (Fig. 19.18) (Shinn et al. 1969; Hardie and Ginsburg 1977). Nonetheless, these freshwater marsh deposits occur only sporadically beneath the overlying bioturbated subtidal and intertidal mud (Fig. 19.18), and the sharp discontinuity separating the freshwater marsh and overlying subtidal mud may partly reflect erosion of the underlying deposits (Hardie and Ginsburg 1977). Where accretion has reached upper intertidal and supratidal elevations (e.g. beach ridge and levee, crest and back slopes), the thick (up to 2 m) subtidal and intertidal mud is overlain by a finely laminated supratidal cap up to 30 cm in thickness. Notably, the Three Creeks succession shows no clear indication of systematic seaward progradation or migration of tidal channels (although some are abandoned), but rather includes a dominantly aggradational signal. Tidal-flat sedimentation appears to have initiated following the development of an initial shoreline with low energy conditions in its lee. Sediment accretion has continued, evolving into the complex of channels, ponds, levees and mangrove belts observed today (Hardie and Ginsburg 1977), driven by tidal creeks which act essentially as distributary networks.

As discussed above, because there is limited evidence for tidal-channel migration in the Three Creeks area, and widespread evidence for aggradation and infilling within the low-energy ponds, the most likely stratigraphic product of a "complete" succession would

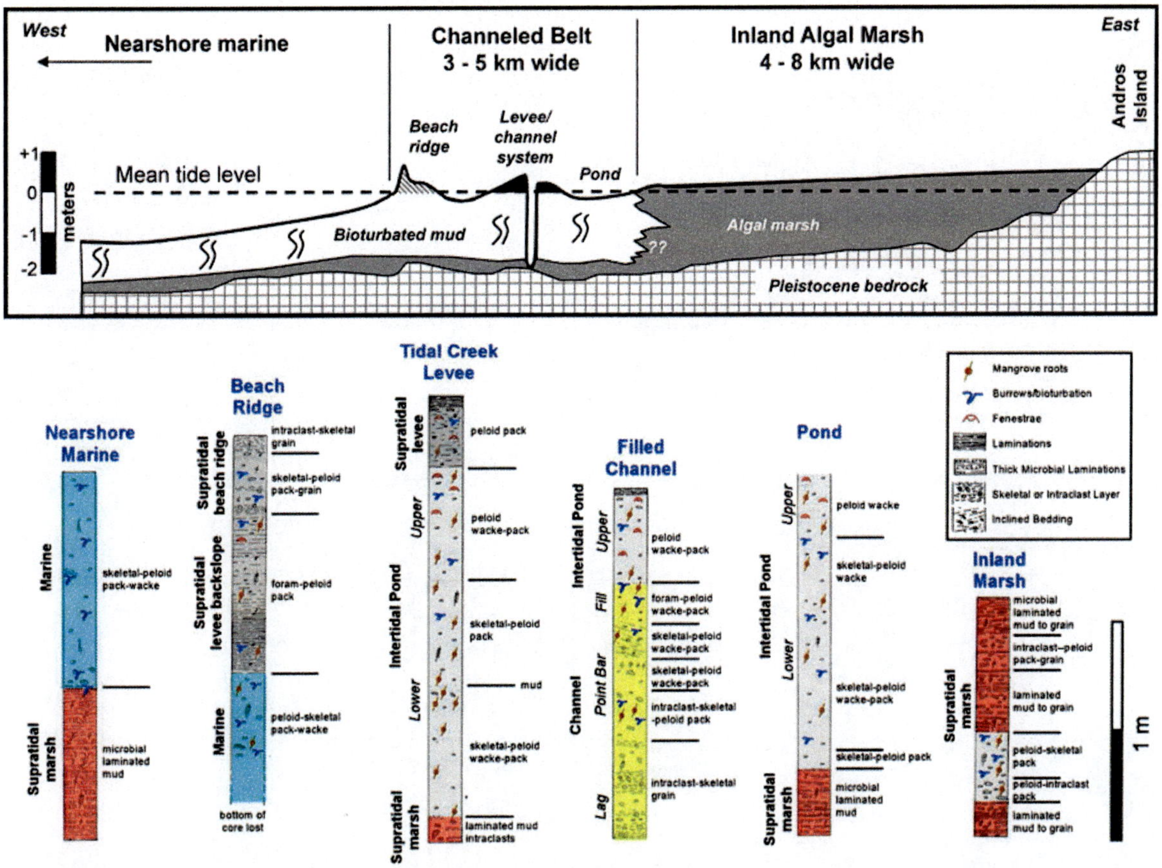

Fig. 19.18 (*Upper*) General stratigraphic patterns in the tidal flats of Three Creeks area, Andros Island, Bahamas. In this region, the tidal flat complex (including part of the nearshore marine region) is underlain by a transgressive algal marsh deposits, suggesting a long-term onlapping pattern (Modified from Hardie 1977). (*Lower*) Generalized core descriptions and interpretations from a suite of subenvironments, Andros tidal flats. To the right of each core is a "Dunham" texture related to amount and support of mud versus grains, but without a "-stone" suffix. These are generally arranged from offshore (*left*) to onshore (*right*), but come from several areas along the shore, so no horizontal scale is implied. Each of the cores penetrated to bedrock except the beach ridge core (Redrafted from core descriptions graciously provided by Paul Enos, from original descriptions in 1965)

include scattered (transgressive) marsh deposits at the base. These sediments would be overlain by subtidal bioturbated peloid mud deposits, with gastropods and foraminifera. Areas with continued shallowing could include marsh or levee deposits, if near a creek (Fig. 19.18). Nonetheless, at a larger scale, if the creeks and levees are not migrating markedly, aggradation and pond filling (Fig. 19.7c) may instead facilitate progradation of the inland marsh over the channeled belt.

19.5.2 Crooked Island

The Holocene tidal-flat package on Crooked Island includes a platformward-thickening wedge of fine-grained sediments that can exceed 2 m thickness (Fig. 19.19). Across the area, ~20–30 cm of organic-rich mud with conspicuous mangrove root fragments and gravel- to pebble-sized intraclasts and lithoclasts lie directly atop the underlying Pleistocene bedrock, interpreted to represent transgressive reworking and initial mangrove colonization. Soft, white mud with foraminifera and rare *Halimeda* plates overlie this basal unit, suggesting continued relative rise in sea level (Fig. 19.19b). The homogeneity of the subtidal mud-rich unit may partly reflect deep burrowing or mangrove rhizoturbation (Fig. 19.19c), which may obscure a clear turnaround. In more landward areas, this subtidal unit is only 20–30 cm thick, but it thickens platformward, and can exceed 2 m in the present-day

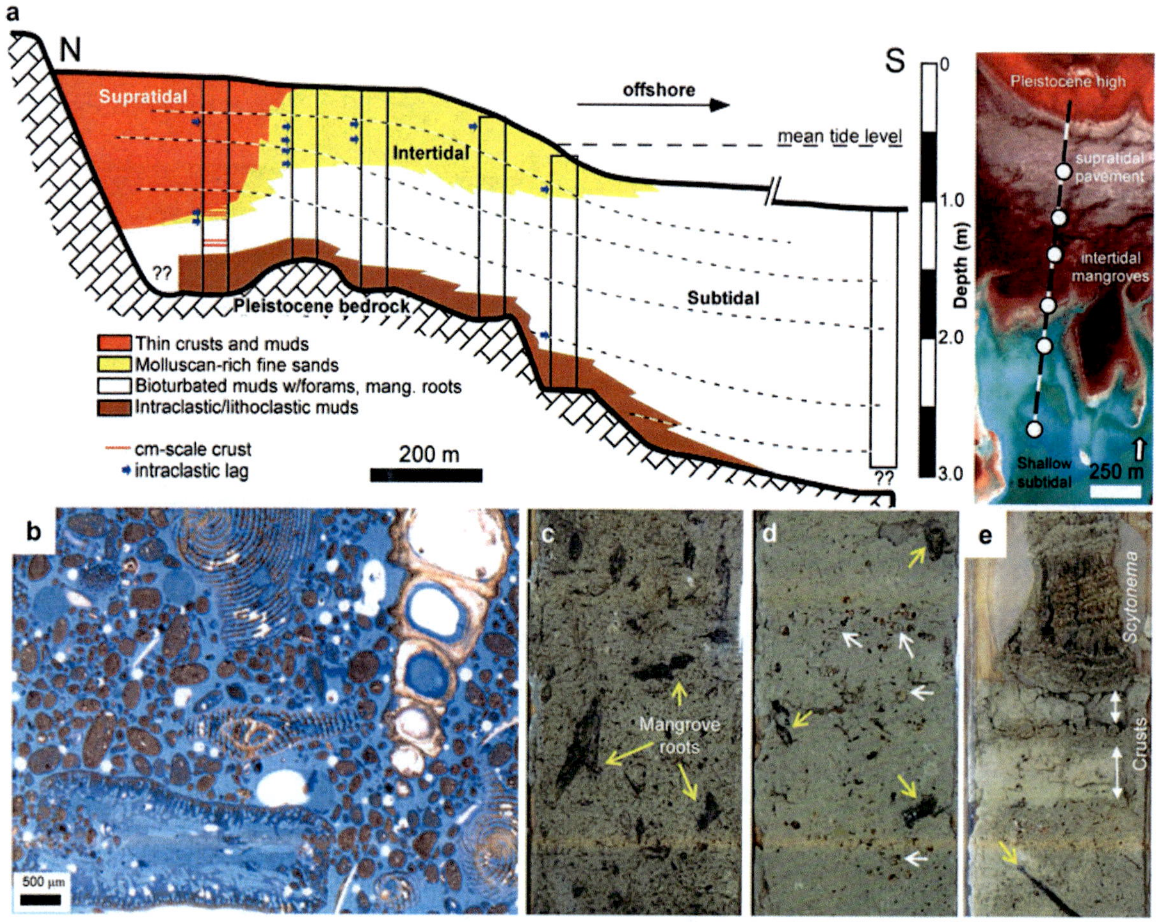

Fig. 19.19 General stratigraphic patterns in the tidal flats south of Crooked Island, Bahamas. (**a**) Stratigraphic cross-section. Here, a lower layer of organic-rich intraclastic and lithoclastic mud (transgressive deposits) overlies bedrock and is in turn overlain by an offlapping, muddy succession. (**b**) Subtidal deposits, including the most diverse sediments, such as *Halimeda*, foraminifera, and gastropods. (**c**) Bioturbated subtidal deposits, with common mangrove roots. (**d**) Intertidal deposits. Note the *pale yellow/buff* color, and the abundant gastropods (more *orangish* color, see *white arrows*). Mangrove roots are highlighted by the *yellow arrows*. (**e**) Intertidal-supratidal deposits, illustrating upper *Scytonema* pincushion (warped due to dessication during core preparation), two thin crusts (slightly *lighter* colored, noted by the *white arrows*). A sole mangrove root is highlighted by the *yellow arrow*. Modified from Berkeley and Rankey (submitted)

subtidal realm. Towards the platform interior, the basal transgressive lag and thick subtidal peloid-skeletal mud pass up into fine sand with conspicuous molluscs (especially cerithid gastropods) (Fig. 19.19d), collectively interpreted as intertidal deposits. Above the subtidal unit in landward areas is a ~60-cm-thick succession of interbedded thin (~1 cm) indurated horizons (Fig. 19.19e) with angular, gravel-sized intraclasts and ~40 cm of fine sand, reflecting shallowing from upper intertidal to supratidal deposition. Generalized facies patterns (Fig. 19.19) suggest that this system has included marked shoreface progradation, a notion consistent with measured accumulation rates greater in the intertidal (~0.28 cm year^{-1}) than supratidal (~0.04 cm year^{-1}) environments (Berkeley and Rankey, in review).

19.5.3 Abu Dhabi

The stratigraphy of the Abu Dhabi sabkhas indicate extensive shoreline progradation (schematically indicated in Fig. 19.20), although the details of the facies patterns vary (Alsharhan and Kendall 2003). For example, in

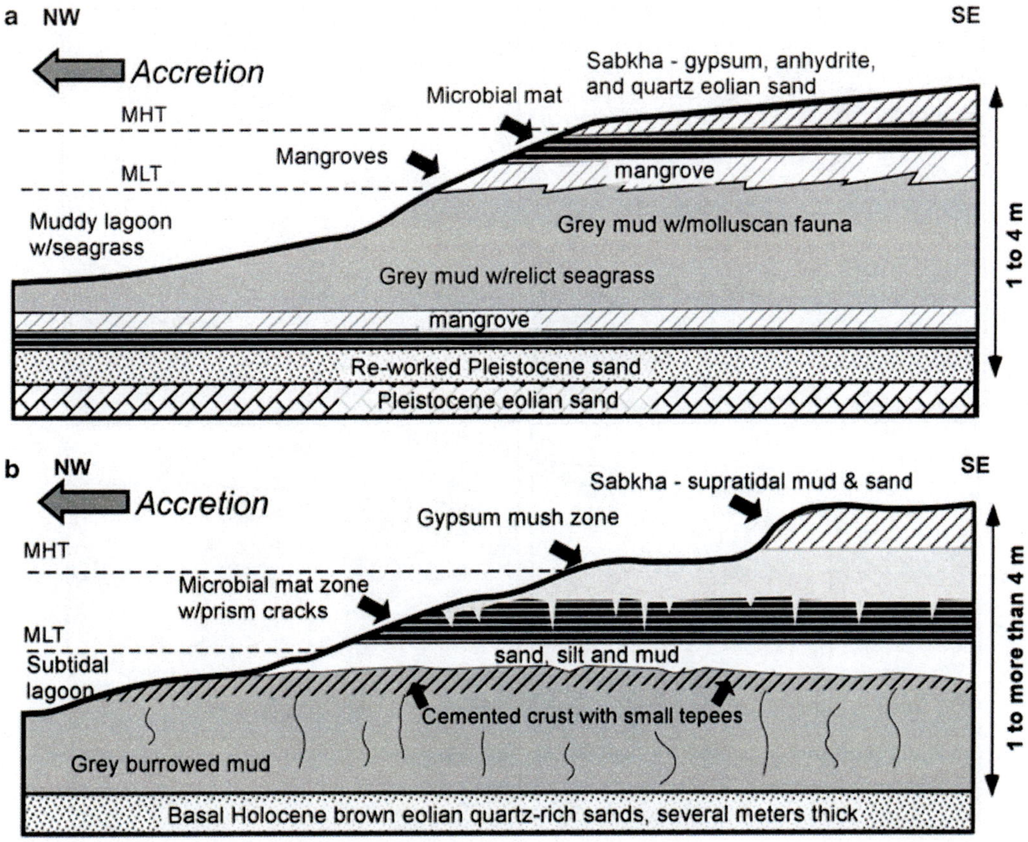

Fig. 19.20 Schematic conceptual models of shoreline progradation and resultant facies patterns from different areas in the United Arab Emirates. See text for discussion (*upper* – Modified from Kenig et al. 1990; the *lower* – Modified from Shinn 1986)

some areas (Kenig et al. 1990) at the base of the Holocene succession, reworked bioclastic and oolitic deposits include clasts of Pleistocene eolianite in a unit up to 1 m thick (Fig. 19.20a). This unit is overlain by transgressive intertidal microbial mat and mangrove deposits up to 35 cm in thickness. The lowermost 15 cm of these contain well-preserved sedimentary structures such as polygonal desiccation cracks and mangrove root casts, but the upper parts are bioturbated, resulting in the loss of primary sedimentary structures and nearly half of the original organic content. Above these microbial and mangrove deposits, 1–2 m of subtidal bioturbated mud contains seagrass roots and leaves. Above these lagoonal deposits, a 40 cm thick interval of sand, silt and mud originally deposited in non-vegetated, lower intertidal environments. This interval is overlain by 25 cm of organic-rich, mangrove deposits with dense accumulations of root and leaf material; roots deeply penetrate and disrupt the underlying sand, silt and mud. The intertidal package is capped by microbial mat deposits, containing fining-upwards laminae. This unit is typically ~35 cm thick, although the upper ~15 cm contain gypsum precipitates which cause the destruction of fine laminations and the organic matter. In contrast with the more humid examples of the Bahamian Archipelago, the regressive mangrove and microbial mat facies show no evidence of post-depositional reworking through bioturbation. Up to 1 m of supratidal eolianite and evaporite deposits ('chicken-wire' anhydrite, gypsum), typically containing negligible organic matter, cap the succession.

In contrast, other regions that lack mangroves exhibit a slightly different facies pattern (e.g. Shinn 1986; Warren 1991; Alsharhan and Kendall 2003), with a general shoaling upward trend (Fig. 19.20b). In these areas, basal Holocene brown eolian quartzose carbonate sand (which can exceed 6 m in thickness) is

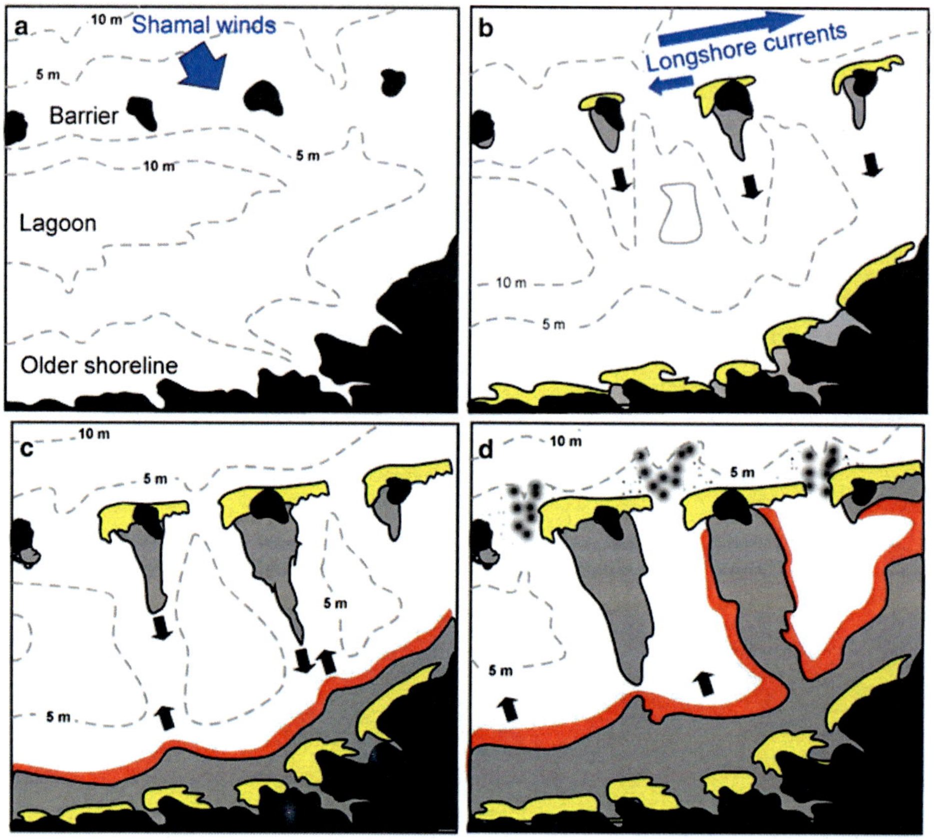

Fig. 19.21 Schematic of evolution of sabkhas on Trucial Coast (Modified from Purser and Evans 1973). (**a**) Offshore Pleistocene islands (*black*), separated from the mainland (also *black*) by a trough, the Khor al Bazm lagoon. (**b**) Formation of beaches (*yellow*) of bioclastic sand on windward side of islands, tidal flats (*gray*) on leeward side. Sand beaches form on the mainland shoreline (see Fig. 19.16a). (**c**) Beaches expand laterally due to longshore currents, tidal flats (*grey*) fronted by microbial mats (*red*) nucleate on mainland once wave energy is sufficiently restricted. (**d**) Sufficient restriction occurs between islands leads to development of oolitic tidal deltas (*fuzzy dots*). Coral reefs grow oceanward of islands, protected from toxic lagoon waters. Tails of islands continue to accrete landward and lagoons gradually infill

overlain by grey, subtidal burrowed, sparsely fossiliferous mu that in turn grades upwards into slightly winnowed muddy skeletal sand, silt and mud (Evans et al. 1969; Shinn 1986). In the absence of the disrupting influence of mangroves, the upper parts of these sandy, shallow subtidal deposits can be well cemented, with syndepositional polygonal fractures (teepees) and borings. This succession is overlain by 1–2 m of peloidal mud and organic-rich, laminated mud with prism cracks and fenestrae, corresponding to the well-developed microbial mat zones (e.g. Kendall and Skipwith 1968). Although many surface mats include mudcracks, Shinn (1986) noted that mudcracks are rare in core from this area, likely due to erosion by wind. This succession is in turn overlain by a complex succession of evaporite and carbonate that can include nodules, "chicken-wire" texture, or contorted beds ("gyspsum mush zone," Fig. 19.20) (Warren 1991), locally truncated by deflation (e.g., Kirkham 1997). Cross-bedded eolian quartz or carbonate sand or supratidal mud, the sabkha deposits, locally caps the succession.

At a larger scale, the geologic history of this coastline is quite complex (Fig. 19.21) (Kirkham 1998; Walkden and Williams 1998). Purser and Evans (1973) suggested that the general geomorphic configuration was established by a pre-Holocene, westerly plunging structural depression (presently manifest as the Khor al Bazm lagoon to the west of the area of Fig. 19.16), flanked to the south by the mainland, and to the north by a subtle offshore high (Fig. 19.21a). These subtle highs provided enough relief for establishment of patch reefs, which then produced abundant sediment, and led to expansion of the islands by wind and current action (Fig. 19.21b). With continued leeward expansion, tombolo tails eventually reached the mainland (Fig. 19.21c, d). Longshore transport and lateral island accretion focused tidal flow in the inlets and led to development of oolitic deltas, as well as isolation of lagoons, leading to lower-energy conditions and muddy deposits. Within the lagoons, subenvironments along the coast are prograding and infilling the lagoons (Fig. 19.21c; cf. Fig. 19.17) at rates averaging 0.75 m/year (Lokier and Steuber 2008).

19.6 Summary

The Holocene carbonate tidal flats of the Bahamas and Turks and Caicos have provided important conceptual models that have been widely applied to understand ancient successions. The most expansive of these tidal flats occur on leeward or protected flanks of bedrock islands, in low-energy settings in which muddy sediments accumulate as a result of sediment transport onto the intertidal and supratidal areas. Tidal flats commonly have only subtle relief, but due to the stressful environment, even cm-scale changes in topography lead to marked shifts in flora, fauna, and sedimentary structures. These changes result in accumulations that are very strongly zoned with the topography, although the spatial patterns of different subenvironments is quite variable within and among systems. Carbonate tidal flats in the Bahamian Archipelago can include a range of geomorphic forms, and include channel networks, broad supratidal plains, and offlapping or erosional shorelines. The stratigraphy of several tidal flats record basal transgressive deposits, overlain by a generally shallowing-upward succession.

The arid tidal flats and sabkha of the Arabian Gulf illustrate several contrasts to the more humid systems of the Bahamas and Caicos. There, significant quantities of eolian sand can be supplied to the shoreline, where they can be re-worked by longshore currents to form chenier complexes. Likewise, due to the more arid setting, fewer organisms (such as mangroves) thrive, and evaporites are common. These systems are commonly interpreted to form broadly offlapping successions.

References

Alsharhan AS, Kendall CGStC (2003) Holocene coastal carbonates and evaporites of the southern Arabian Gulf and their ancient analogue. Earth Sci Rev 61:191–243

Ball MM, Shinn EA, Stockman KW (1967) The geologic effects of Hurricane Donna in South Florida. J Geol 75:583–597

Berkeley A, Rankey EC (in review) Carbonate tidal flats of Crooked Island, southern Bahamas Sedimentology.

Black M (1933) The algal sediments of Andros Island, Bahamas. R Soc Lond Philos Trans Ser B 222:165–192

Boudreau BP (1994) Is burial velocity a master parameter for bioturbation? Geochem Cosmoch Acta 58:1243–1249

Butler GP (1970) Holocene gypsum and anhydrite of the Abu Dhabi, Trucial Coast, Persian Gulf: An alternative explanation of origin. Proc 3rd Int Salt Symp, Northern Ohio Geol Soc, Cleveland, pp 120–152

Cloyd KC, Demicco RV, Spencer RJ (1990) Tidal channel levee and crevasse-splay deposits from a Cambrian tidal channel system: a new mechanism to produce shallowing upward sequences. J Sediment Petrol 60:73–83

De Groot K (1973) Geochemistry of tidal flat brines at Umm Said, SE Qatar, Persian Gulf. In: Purser BH (ed) The Persian Gulf. Springer, Berlin/New York, pp 377–394

Enos P (1989) Islands in the bay – a key habitat of Florida Bay. Bull Mar Sci 44:365–386

Enos P, Perkins RD (1979) Evolution of Florida Bay from island stratigraphy. Bull Geol Soc Am 90:59–83

Evans G, Schmidt V, Bush P, Nelson H (1969) Stratigraphy and geologic history of the sabkha, Abu Dhabi, Persian Gulf. Sedimentology 12:145–159

Gebelein CD (1974) Guidebook for Modern Bahamian Platform Environments. Geological Society of America annual meeting field trip guide, 93 p

Gebelein CD (1975) Holocene sedimentation and stratigraphy southwest Andros Island, Bahamas. 9th International sedimentological congress, Nice, Theme 5, 1, pp 193–198

Gebelein CD (1977) Dynamics of recent carbonate sedimentation and ecology. Cape Sable Florida. In: EJ Brill, Leiden 120 p

Ginsburg RN, Hardie LA (1975) Tidal and storm deposits northwestern Andros Island Bahamas. In: Ginsburg RN (ed) Tidal deposits: a casebook of recent examples and fossil counterparts. Springer, Heidelberg

Ginsburg RN, Hardie LA, Bricker OP, Garrett P, Wanless H (1977) Exposure index: a quantitative approach to defining position within the tidal zone. In: Hardie LA (ed) Sedimentation on the modern carbonate tidal flats of northwest Andros Island, bahamas, The John Hopkins University Press, Baltimore, pp 7–12

Grotzinger JP (1986) Cyclicity and palaeoenvironmental dynamics Rocknest platform northwest Canada. Bull Geol Soc Am 97:1208–1231

Hardie LA (1977) Sedimentation on the modern carbonate tidal flats of northwest Andros Island Bahamas, The Johns Hopkins University studies in geology no. 22. The Johns Hopkins University Press, Baltimore, 202 p

Hardie LA (1986) Stratigraphic models for carbonate tidal-flat deposition. Colo Sch Min Quart 81:59–74

Hardie LP, Garrett P (1977) General environmental setting. In: Hardie LA (ed) Sedimentation on the modern carbonate tidal flats of northwest Andros Island Bahamas, The Johns Hopkins University studies in geology. The Johns Hopkins University Press, Baltimore, pp 12-49

Hardie LA, Ginsburg RN (1977) Layering: the origin and environmental significance of lamination and thin bedding. In: Hardie LA (ed) Sedimentation on the modern carbonate tidal flats of northwest Andros Island Bahamas, The Johns

Hopkins University studies in geology. The Johns Hopkins University Press, Baltimore, pp 50–123
Harris PM (1977) Sedimentology of the Joulters Cays Ooid Sand Shoal, Great Bahama Bank. Unpublished PhD dissertation, University of Miami, Coral Gables, 452 p
James NP (1979) Shallowing-upward sequences in carbonates. In: Walker R (ed) Facies models. Geoscience Canada reprint series 1, pp 109–132
Kendall CGStC, Skipwith PAd'E (1968) Recent algal mats of a Persian Gulf lagoon. J Sediment Petrol 38:1040–1058
Kendall CGStC, Skipwith PAd'E (1969) Holocene shallow-water carbonate and evaporite sediments of Khor al Bazam, Abu Dhabi, southwest Persian Gulf. Am Assoc Petrol Geol Bull 53:841–869
Kenig F, Huc AY, Purser BH, Oudin J-L (1990) Sedimentation distribution and diagenesis of organic matter in a recent carbonate environment, Abu Dhabi, UAE. Adv Org Geochem 16:735–747
Kirkham A (1997) Shoreline evolution aeolian deflation and anhydrite distribution of the Holocene, Abu Dhabi. GeoArabia 2:403–416
Kirkham A (1998) A quaternary proximal foreland ramp and its continental fringe, Arabian Gulf, UAE. In: Wright VP, Burchette TP (eds) Carbonate ramps. Geological Society of London Special Publication 149, pp 15–41
Lehrmann DJ, Goldhammer RK (1999) Secular variation in facies and parasequence stacking patterns of platform carbonates: a guide to application of the stacking patterns technique in strata of diverse ages and settings. In: Harris PM Saller AH, Simo JA (eds) Recent advances in carbonate sequence stratigraphy; applications to reservoirs outcrops and models. SEPM Special Publication 63, pp 187–226
Logan BW, Read JF, Davies GR (1970) History of carbonate sedimentation, Quaternary Epoch, Shark Bay, Western Australia. In: Logan BW, Davies GR, Read JF and Cebulski DE (eds) Carbonate sedimentation and environments, Shark Bay, Western Australia. Am Assoc Petrol Geol Mem 13:38–84
Logan BW, Read JF, Hagan GM, Hoffman P, Brown RG, Woods PJ, Gebelein CD (eds) (1974) Evolution and diagenesis of quaternary carbonate sequences, Shark Bay, Western Australia. Am Assoc Petrol Geol Mem 22, 358 p
Lokier S, Steuber T (2008) Large scale intertidal polygonal features of the Abu Dhabi coastline. J Sediment Res 78:423–431
Perkins RD, Enos P (1968) Hurricane Betsy in the Florida-Bahama area; geologic effects and comparison with Hurricane Donna. J Geol 76:710–717
Pierson BJ (1982) Cyclic sedimentation, limestone diagenesis and dolomitization in upper Cenozoic carbonates of the southeastern Bahamas. Unpublished PhD dissertation, University of Miami, Coral Gables, 286 p
Pratt BR, James NP (1986) The St. George Group (Lower Ordovician) of western Newfoundland: tidal flat island model for carbonate sedimentation in shallow epeiric seas. Sedimentology 33:313–343
Pratt BR, James NP, Cowan CA (1992) Peritidal carbonates. In: Walker RG, James NP (eds) Facies models response to sea-level change, 2nd edn. Geological Association, Canada, pp 303–322
Purser BH (ed) (1973) The Persian Gulf: Holocene carbonate sedimentation and diagenesis in a shallow epicontinental sea. Springer, New York, 471 p
Purser BH, Evans G (1973) Regional sedimentation along the trucial coast SE Persian Gulf. In: Purser BH (ed) The Persian Gulf. Springer, Berlin/New York, pp 211–232
Queen JM (1976) Hydrology, sedimentology and ecology of pelleted carbonate muds, west of Andros Island, Bahamas. PhD dissertation, State University, New York, Stony Brook, 366 p
Rankey EC (2002) Spatial patterns of sediment accumulation on a Holocene carbonate tidal flat Northwest Andros Island, Bahamas. J Sediment Res 72:591–601
Rankey EC, Morgan JJ (2002) Quantified rates of geomorphic change on a modern carbonate tidal flat, Bahamas. Geology 30:583–586
Rankey EC, Reeder SL (2010) Controls on platform-scale patterns of surface sediments, shallow Holocene platforms, Bahamas. Sedimentology 57:1545–1565
Rankey EC, Enos P, Steffen K, Druke D (2004) Lack of impact of Hurricane Michelle on tidal flats Andros Island, Bahamas: integrated remote sensing and field observations. J Sediment Res 74:654–661
Schneider J (1975) Recent tidal deposits Abu Dhabi, U.A.E., Arabian Gulf. In: Ginsburg RN (ed) Tidal deposits. Springer, New York, pp 209–214
Scholle PA, Shinn EA, Halley RB, Harris PM (2005) "Arid Carbonate Coastlines" AAPG Video, Tulsa, OK
Shinn EA (1973a) Carbonate coastal accretion in an area of longshore transport, NE Qatar, Persian Gulf. In: Purser BH (ed) The Persian Gulf – Holocene carbonate sedimentation and diagenesis in a shallow epicontinental sea. Springer, Berlin, pp 179–191
Shinn EA (1973b) Sedimentary accretion along the leeward SE coast of Qatar Peninsula, Persian Gulf. In: Purser BH (ed) The Persian Gulf – Holocene carbonate sedimentation and diagenesis in a shallow epicontinental sea. Springer, Berlin, pp 199–209
Shinn EA (1981) Stratigraphic traps: the tidal flat model. AAPG Video, Tulsa, OK, 17 min
Shinn EA (1983a) Birdseyes fenestrae, shrinkage pores, and loferites: a reevaluation. J Sediment Petrol 53:619–628
Shinn EA (1983b) Tidal flat environment. Am Assoc Petrol Geol Mem 33:172–210
Shinn EA (1986) Modern carbonate tidal flats: their diagnostic features. Colo Sch Min Quart 81:7–35
Shinn EA (2010) The interplay between quaternary sedimentation and diagenesis and implications for hydrocarbon exploitation: return to sabkha of Ras Umm Said, Qatar
Shinn EA, Ginsburg RN, Lloyd RM (1965) Recent supratidal dolomites from Andros Island Bahamas. In: Pray LC, Murray RC (eds) Symposium on dolomitization and limestone diagenesis. SEPM Special Publication 13:112–123
Shinn EA, Lloyd RM, Ginsburg RN (1969) Anatomy of a modern carbonate tidal flat. J Sediment Petrol 53:1202–1228
Tedesco LP, Aller RC (1997) 210Pb chronology of sequences affected by burrow excavation and infilling: examples from shallow marine carbonate sediment sequences: Holocene South Florida and Caicos Platform, British West Indies. J Sediment Res 67:36–46
Tedesco LP, Wanless HR (1991) Generation of sedimentary fabrics and facies by repetitive excavation and storm infilling of burrow networks: Holocene of South Florida and Caicos Platform, BWI. Palaios 6:326–343
Walkden GM, Williams A (1998) Carbonate ramps and the Pleistocene – recent depositional systems of the Arabian

Gulf. In: Wright VP, Burchette T (eds) Carbonate ramps. Geological Society of London Special Publication 149, pp 45–53

Wanless HR, Tyrell KM, Tedesco LP, Dravis JJ (1988a) Tidal-flat sedimentation from Hurricane Kate Caicos Platform British West Indies. J Sediment Petrol 58: 724–738

Wanless HR, Tedesco LP, Tyrrell KM (1988b) Production of subtidal tubular and surficial tempestites by Hurricane Kate, Caicos Platform, British West Indies. J Sediment Petrol 58:739–750

Wanless HR, Tedesco LP, Rossinsky V, Dravis JJ (1989) Carbonate environments and sequences of caicos platform with an introductory evaluation of South Florida. American Geophysical Union 28th international geological congress field trip guidebook T374, 75 p

Warren JK (1991) Tepees modern and ancient (Permian) – a comparison. Sediment Geol 34:1–19

Wilson JL (1975) Carbonate facies in geologic history. Springer, New York, 471 p

Wright VP (1984) Peritidal carbonate facies models: a review. Geol J 19:309–325

Tidal Sands of the Bahamian Archipelago

20

Eugene C. Rankey and Stacy Lynn Reeder

Abstract

Tidal sands consisting entirely of carbonate sediments are ubiquitous in the Bahamian archipelago. These sands include a diversity of sediment types, including ooids, peloids, and skeletal fragments. Sands transported by tides, waves, and currents create barforms in tidal sand complexes with a range of shapes and sizes. These features are shaped by, and in turn modify, tidal currents that move on and off the shallow platforms; waves and wave-driven currents play a subordinate but locally important role in their genesis and architecture. Collectively, barforms make up shallow shoal complexes. These shoal complexes are focused in areas with elevated tidal currents (locally in excess of 200 cm/s) near platform margins, and can exceed 10 km in width. The diversity of barforms and shoal morphology evident in Holocene examples is reflected in the stratigraphic record of numerous ancient tidal sand shoals, with preservation favored by the early cementation ubiquitous in these carbonate systems.

20.1 Introduction

Tidal sands of the Bahamian Archipelago have been the focus of study for over 100 years (Agassiz 1896; Rich 1948; Illing 1954; Newell et al. 1960; Purdy 1961; Ball 1967; Hine 1977; Gonzalez and Eberli 1997). These sands have attracted the attention of geologists not only because they occur in tropical waters, but also because they form modern analogs for carbonate successions throughout the geologic record from Archean onward. Carbonate strata form prolific reservoirs that collectively host more than 50% of the world's hydrocarbons and much of its groundwater; many of the largest carbonate reservoirs are in tidal sands, including oolitic units (e.g., Harris and Weber 2006).

The objective for this chapter is to provide an overview summary of aspects of carbonate tidal sand shoals in the Bahamas. The purpose is not to explore any one shoal in detail, but rather it aims to provide a broad overview of sediments, bars, shoals, their geologic history, and aspects of their dynamics. As such, it is intended to be a synthesis; for details of data or methods, refer to the primary literature cited throughout.

E.C. Rankey (✉)
Department of Geology, University of Kansas, 1475 Jayhawk Blvd., 120 Lindley Hall, Lawrence, KS 66045, USA
e-mail: grankey@ku.edu

S.L. Reeder
Schlumberger-Doll Research, One Hampshire Street, Cambridge, MA 02139, USA
e-mail: sreeder@slb.com

R.A. Davis, Jr. and R.W. Dalrymple (eds.), *Principles of Tidal Sedimentology*,
DOI 10.1007/978-94-007-0123-6_20, © Springer Science+Business Media B.V. 2012

20.2 Boundary Conditions: Tides, Waves, and Saturation State

The Bahamian Archipelago, located in the tropical to sub-tropical Caribbean (~20–28°N latitude, ~69–79°W longitude), includes a series of flat-topped, isolated carbonate platforms separated by deeper water passages (Fig. 20.1). It is bounded by the Atlantic Ocean to the north and east, the Old Bahama Channel to the south, and the Straits of Florida to the west. Water depths on the tops of the larger platforms (Little Bahama Bank, Great Bahama Bank, Crooked-Acklins Platform, Caicos Platform) are less than 20 m, and many areas are less than 10 m. At their margins, these shallow-topped platforms slope precipitously to depths of 500 m (Straits of Florida) or more (Tongue of the Ocean, Northwest Providence Channel, Exuma Sound, Crooked Island Passage).

This area is affected largely by the easterly trade winds, which affect the latitudes between 4° and 30°. These winds are strong throughout the year, but the speeds peak during the winter where they can blow steadily at speeds greater than 10 m/s. Several cold fronts affect this area annually (Roberts et al. 1982, 1992). These fronts pass over the archipelago from the northwest to the southeast, bringing winds that shift from south to west (frequently the strongest) to north or northeast before returning to the dominant easterly direction. Tropical cyclones are common in this part of the Caribbean from June through November, so hurricane strength winds can be expected during these months. Neumann et al. (1978) estimated that, on average, hurricanes struck the southern Bahamas/Caicos once every 5 years, and Florida and the northern Bahamas once every 7 years.

Despite the generally high wind speeds of the Bahamian Archipelago, the waves have little effect on most platform interiors. Large, open water swells decrease in size due to shoaling caused by the steep platform slopes and shallower waters of the margins and platform interiors or breaking either along the reef-rimmed margins

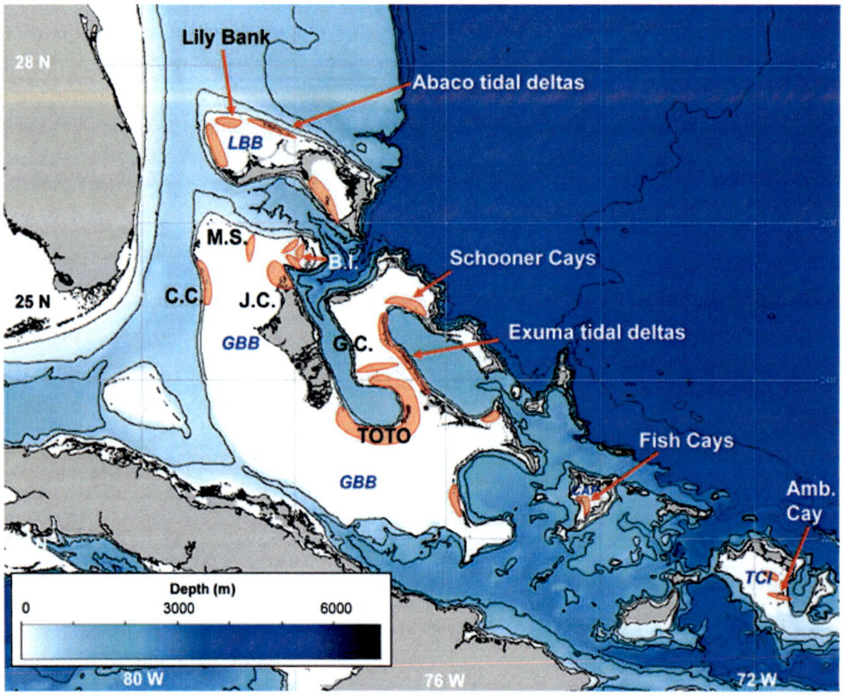

Fig. 20.1 Locations of study areas. Regional bathymetric setting in the eastern Caribbean, illustrating location of major shoal complexes (*red blobs*) including those discussed in the text and labelled: Mackie Shoal (*M.S.*), Cat Cays (*C.C.*), Joulter Cays (*J.C.*), Green Cay (*G.C.*), Tongue of the Ocean (*TOTO*), Ambergris Cay (*Amb. Cay*). Platforms include: Great Bahama Bank (*GBB*), Little Bahama Bank (*LBB*), Crooked-Acklins Platform (*CAP*), and Caicos Platform (*TCI*); Bathymetric data from General Bathymetric Chart of the Oceans (GEBCO) of the Intergovernmental Oceanographic Commission (IOC) and the International Hydrographic Organization (IHO)

or on the beaches of the islands that occur on the eastern, windward margins of the platforms. As a result of these factors, waves in the platform interiors are small, with most significant wave heights less than 1 m (Reeder and Rankey 2009a; Rankey and Reeder 2010).

Tidal currents, however, can locally influence the sediment distribution on these flat-top platforms. Although the platforms are microtidal (spring tidal amplitude of approximately 1 m near shelf margins throughout the area), the tidal exchange floods and drains large platform areas, and in many locations currents are focused through localized channels. Whereas current speeds in sandy siliciclastic channels reach a dynamic equilibrium point at 1 m/s, within these channels and surrounding shoals, the current speeds can exceed 1 m/s or even 2 m/s, especially where bedrock islands occur (Gonzalez and Eberli 1997; Reeder and Rankey 2009b). On the platform interior, however, the tidal currents are much lower, rarely exceeding 0.50 m/s (Smith 1995; Rankey and Reeder 2010).

Many tidal sand shoals of the Bahamas are dominated by ooids, so much of this chapter focuses on these grains and their oolitic shoals. Because oolitic laminations are mainly physiochemical precipitates, the occurrence of oolitic tidal sand shoals in the Bahamas is influenced by the geochemical characteristics of Bahamian seawater. Recent global compilations (Royal Society 2005; Lee et al. 2006) illustrate an elevated carbonate saturation state and pH of waters in this part of the Atlantic, factors that would favor abiotic precipitation of aragonitic ooids, given the right hydrodynamic conditions (Rankey and Reeder 2009, 2010; discussed below).

20.3 Tidal Sands of the Bahamas and Caicos

20.3.1 Setting and Distribution

The modern sedimentologic patterns on the tops of the shallow platforms are the net result of complex interactions among physical, biological, and chemical processes, all acting during the late Holocene sea-level rise within the geographic framework established by Pleistocene bedrock (Newell and Rigby 1957; Purdy 1963; Enos 1974; Hine 1977; Hine and Neumann 1977; Wanless et al. 1989). In general terms, highs of Pleistocene reefs, marine sands, and eolianites on the eastern and northern flanks of the platforms form the largest islands (Abaco, Andros, Eleuthera, Long, Crooked, Acklins, Caicos chain) (Beach and Ginsburg 1980; Hearty and Kindler 1993; Aurell et al. 1995). Many windward islands are bordered oceanward by reefs and reef-derived sands. Tidal flats flank the leeward sides of many islands, towards the platform interior (Rankey and Berkeley, this volume), and pass laterally into shallow subtidal settings.

Across much of the archipelago, the best developed and thickest tidal sands occur at or near bank margins, where there is an open exchange between the platform interior and the surrounding deep basins (Hine et al. 1981). Areas with broad expanses of tidal sands (Fig. 20.1, red patches) can be found on windward margins (Joulter Cays, Exuma tidal deltas), at the end of deep-water embayments into the platform (Tongue of the Ocean, Schooner Cays), leeward margins (Cat Cay, Fish Cays, Berry Islands), and on margins oblique to the dominant easterlies (Lily Bank, Abaco tidal deltas) (e.g., Hine et al. 1981). In general terms, shoals are most probable in concavities extending into the platform, and the distribution of tidal sands suggests that margin orientation relative to the dominant winds is not the sole factor controlling where tidal sands occur. Instead, although the region is microtidal, the strong reversing tidal currents near the margins facilitate the development of expansive tidal sand bodies (Ball 1967; Reeder and Rankey 2008). Shoal areas generally lack well-developed reefs outboard, due to inimical waters of elevated salinity or temperature flowing off the banks during ebb tide (Newell et al. 1959; Neumann and Macintyre 1985; Ginsburg and Shinn 1993), and many sediments are oolitic (Illing 1954; Newell et al. 1960), although there can be a pronounced skeletal or peloidal component (Hine et al. 1981) (see next section).

Although why they occur where they do is generally well known, the controls on the considerable variability in morphology of ooid shoals are less understood. For example, the complicated geometries present in shoals have led to summative statements such as "the form of the oolite shoals is less subject to generalization than the distribution of these deposits" (Purdy 1961, p. 54–55), and "each marginal trend is unique" (Hine et al. 1981, p. 286). Some have even gone so far as to suggest that the variability reflects a "marginal-facies mosaic" characterized by "facies complexities" (Hine and Neumann 1977 p. 376–377).

The scheme most frequently used to describe the morphology of carbonate sand accumulations in the Bahamas utilizes several different classes: marine sand belts, tidal bar belts, platform interior sand blankets, and eolian ridges (Ball 1967). This scheme has been amplified and refined in several reviews (e.g., Halley et al. 1983; Handford 1988; Wanless and Tedesco 1993; Tucker and Wright 1996), resulting in addition of several new classes, including tidal deltas (expressed by Halley et al. 1983), shorefaces and "ooid sediment production restricted through shoaling" (Wanless and Tedesco 1993). [Some of the classes (eolian ridges, shorefaces) are not dominated by tidal processes, and so are not discussed further in this chapter.]

In these summative schemes, however, there can be considerable ambiguity. For example, the Joulter Cays sand shoal complex has been variously described as a marine sand belt (Halley et al. 1983, although they also suggested that parts formed a tidal bar belt – p. 472), an area with "ooid sediment production restricted through shoaling" (Wanless and Tedesco 1993), and a "marine sand belt to sand flat … with some spit-like tidal bars" (Tucker and Wright 1996). Similarly, the archetypal tidal bar belt, the Schooner Cays shoal complex (Ball 1967), only has well-developed tidal bars oriented roughly normal to flow in its eastern third; the western part of the system is dominated by bars with a markedly different morphology.

These ambiguities suggest that lumping any particular shoal complex into one class may result in either misleading oversimplification or misrepresentation. As such, this summary takes a slightly different approach to characterizing shoals, building up in scale from sediments, to bed forms, to bar forms found in the shoals, focusing on the influences of tides on their genesis. A few examples of the spatially variable geometries present in several shoal complexes further illustrate some of the geomorphic forms and depositional processes in these systems.

20.3.2 Sediment Composition and Early Diagenesis

Tidal shoals of the Bahamian Archipelago are composed exclusively of carbonate sands, with no siliciclastics. Grain types include ooids, composite grains, peloids, and skeletal fragments present in varying amounts within and among the shoals (Fig. 20.2); intraclasts and lithoclasts may occur locally. Unlike siliciclastic tidal sands, which must be reworked from earlier deposits or transported to the depositional system from elsewhere, these sedimentary particles can be (and many are) produced *in situ*, with relatively little net transport.

Skeletal grains are the remains of any of a diverse group of flora and fauna (Fig. 20.2a, b). On sand shoals in the Bahamas and Caicos, the most abundant grain types include whole skeletons or fragments of bivalves, gastropods, miliolid and peneropolid foraminifera, and, less abundantly, fragments of coral, red algae, or *Halimeda* (a green alga) or other minor skeletal components. Many skeletal fragments on tidal sand shoals are broken or abraided, and may have been transported to the area by the strong currents or they may come from organisms that resided in the area. Skeletal grains may serve as nuclei for ooids.

Peloids are non-discript aggregates of cryptocrystalline carbonate less than 2 mm in diameter (Fig. 20.2c). Peloids can be of various origins; they may represent fecal pellets, aggregates from erosion and transport of semi-cohesive muds, or extensively bored or micritized skeletal or oolitic grains in which all primary textures have been obliterated (Bathurst 1975). On the bank top and in many tidal shoals, peloids occur in every subenvironment and water depth. Hardened peloids may form a substantial percentage of grains in some tidal shoals, and peloids can serve as nuclei for ooids.

Ooids are sand-sized ovoid to circular carbonate particles less than 2 mm in diameter that include one or more concentric laminae (collectively forming a cortex) around a nucleus (Fig. 20.2d–g). These non-skeletal grains can be found in strata of almost any age, from Archean to recent (e.g., Opdyke and Wilkinson 1990; Sumner and Grotzinger, 1993), and are found today on most platforms in the Bahamas and Caicos. Nuclei from these areas are either peloids or skeletal fragments, and oolitic laminations are made of tangentially arranged aragonite needles that can make up more than 80% of the volume of some grains. The number and type of laminations varies considerably, from classic *concentric ooids* coated with up to 90 or more concentric laminations (Fig. 20.2e, f; Newell et al. 1960), to *superficial ooids* (Fig. 20.2g; Illing 1954; Bathurst 1967) with only a few, thin (1–5 μm) laminations, and *irregular ooids* (Fig. 20.2d; Illing 1954; Wanless and Tedesco 1993) that illustrate grain degradation such as

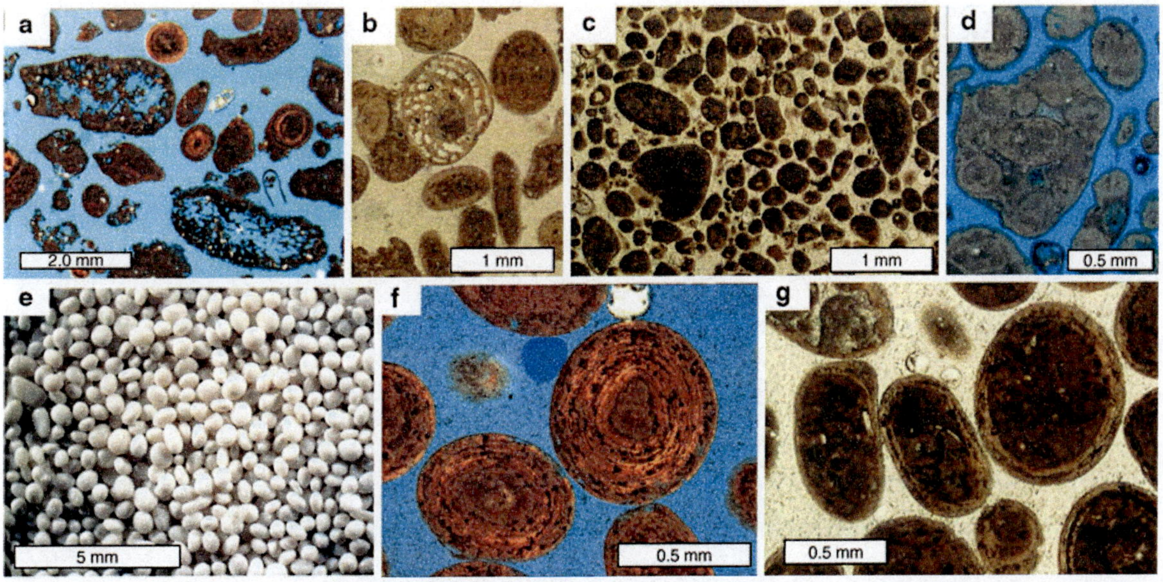

Fig. 20.2 Petrographic character of several types of sand present in Bahamian tidal sand shoals. (**a**) Thin section photomicrograph of *Halimeda* fragments, such as those that can occur with oolitic sand, Schooner Cays. (**b**) Thin section photomicrograph of a peneropolid foraminifera, Lily Bank. (**c**) Thin section photomicrograph of poorly sorted peloidal sand, Lily Bank. (**d**) Thin section photomicrograph of a composite grain made of numerous ooids, Fish Cays. (**e**) Photograph of loose ooids with a pearly luster from Lily Bank. (**f**) Thin section photomicrograph of ooids with numerous laminations, Lily Bank. (**g**) Thin section photomicrograph of superficial ooids with only one or two laminae, Abaco tidal deltas

borings and binding from early cementation by fibrous aragonite or encrusting by foraminifera or microbes that collectively form aggregates with clumpy shapes. The latter are essentially a sub-type of *composite grains* (Newell et al. 1960; "grapestone" of Illing 1954), which can be formed of aggregations of peloids or skeletal fragments as well as ooids, and which are found in abundance in some shoals. Ooids commonly are interpreted to represent direct physiochemical precipitates from seawater (Newell et al. 1960; Bathurst 1975; Deelman 1978; Duguid et al. 2010; compare with Folk and Lynch 2001), whose formation is favored by: (a) water supersaturated with respect to calcium carbonate; (b) a source of nuclei; and (c) a means of agitation (e.g., Cayeux 1935; Deelman 1978; Davies et al. 1978; Sumner and Grotzinger 1993; Reeder and Rankey 2008; Rankey and Reeder 2009). Ooids on Bahamian platforms and Caicos are most abundant in agitated waters less than 2 m deep (Newell et al. 1960), and most likely are formed there, although they can be found in less abundance in deeper waters across parts of some platforms.

In summary, many of the grains that form the tidal sand shoals have a proximal source – they are formed on or near the sand shoals themselves. Besides sediment production on the sand shoals, another characteristic of carbonate tidal sand shoals that is distinct from siliciclastic tidal sands is the presence of early marine or meteoric cementation.

Marine cementation can be ubiquitous on sand shoals, forming expansive *hardgrounds*, or early lithified surfaces (Fig. 20.3a; Dravis 1979). Hardgrounds are formed by fibrous acicular aragonite cements (Fig. 20.3b–d) or by filamentous algae that partly fill intergranular pores and bind sediments (Hillgärtner et al. 2001). These marine cements can occur rapidly, in periods of less than a year in some cases (Grammer et al. 1999). Many hardgrounds have sharp upper surfaces that are extensively bored or encrusted, but cementation intensity decreases downward, and they grade to unconsolidated sands. Hardgrounds have several important sedimentologic influences. First, because they resist erosion more than unconsolidated sands, they can armor surfaces from erosion. Second, they form a substrate for colonization by organisms that otherwise are not well suited to live on the shifting sands of the tidal sand shoals, such as corals or sponges. Finally, if eroded, these lithified sediments can form *intraclasts* which can subsequently be transported as sedimentary particles again.

Fig. 20.3 Characteristics of several types of sand present in Bahamian tidal sand shoals. (**a**) Underwater photo of an intraclast, formed as part of a cemented crust. Clast is ~40 cm across. Schooner Cays. (**b**) Thin section photomicrograph of rim of fibrous aragonite that cements ooid sand, forming clasts such as that in (**a**), Fish Cays. (**c**) SEM image of an ooid (*right*, note laminations) and fibrous aragonite (*left*) from the same sample as in (**b**). (**d**) SEM close-up of the ooid-cement boundary. (**e**) Low-angle aerial photo of a parabolic barform with an oolitic sand island, Schooner Cays. Island is ~160 m across at its widest. (**f**) Cemented Holocene beach–dune succession; this is common on the flanks of islands such as that illustrated in (**e**) (Budd 1984), Schooner Cays. (**g**) Outcrop photo of fining-up laminations in oolitic beachrock sand, Fish Cays. Approximately 3 cm of the width of a hand-held GPS unit is in the lower right. (**h** and **i**) Thin section photomicrograph of clear equant calcite cement with meniscus morphology, indicative of vadose-zone freshwater diagenesis, from outcrop illustrated in (**g**)

Meteoric cementation occurs in areas where sand shoals have aggraded above sea level and are exposed to the effects of freshwater (Fig. 20.3e–g). These slightly acidic waters dissolve unstable aragonite (e.g., in ooids) in near-surface layers, and re-precipitate much of the calcium carbonate as low-magnesium calcite lower in the column, near grain contacts or as pendant cements (Fig. 20.3h, i; Halley and Harris 1979; Budd 1984). As with marine cements, meteoric cementation can be quite rapid, as indicated by observations of bottles, cans, and other trash cemented within beach rock, and entire islands can be cemented within a 1,000 years (Halley and Harris 1979).

20.3.3 Bedforms

Although carbonate sediments are "born, not made" (James 1983), they can be suspended by waves and moved by currents (Braithwaite 1973; Wanless et al. 1981; Kench and McLean 1996; Prager et al. 1996). Tidal sand shoals of the Bahamian Archipelago show ample evidence of transport by reversing tidal currents, or less frequently and importantly, by tidal surges or waves related to the passage of fronts or storms.

As sands are transported, the processes create bedforms of various sizes, from cm-scale ripples to cm- to m-scale subaqueous dunes to even larger barforms.

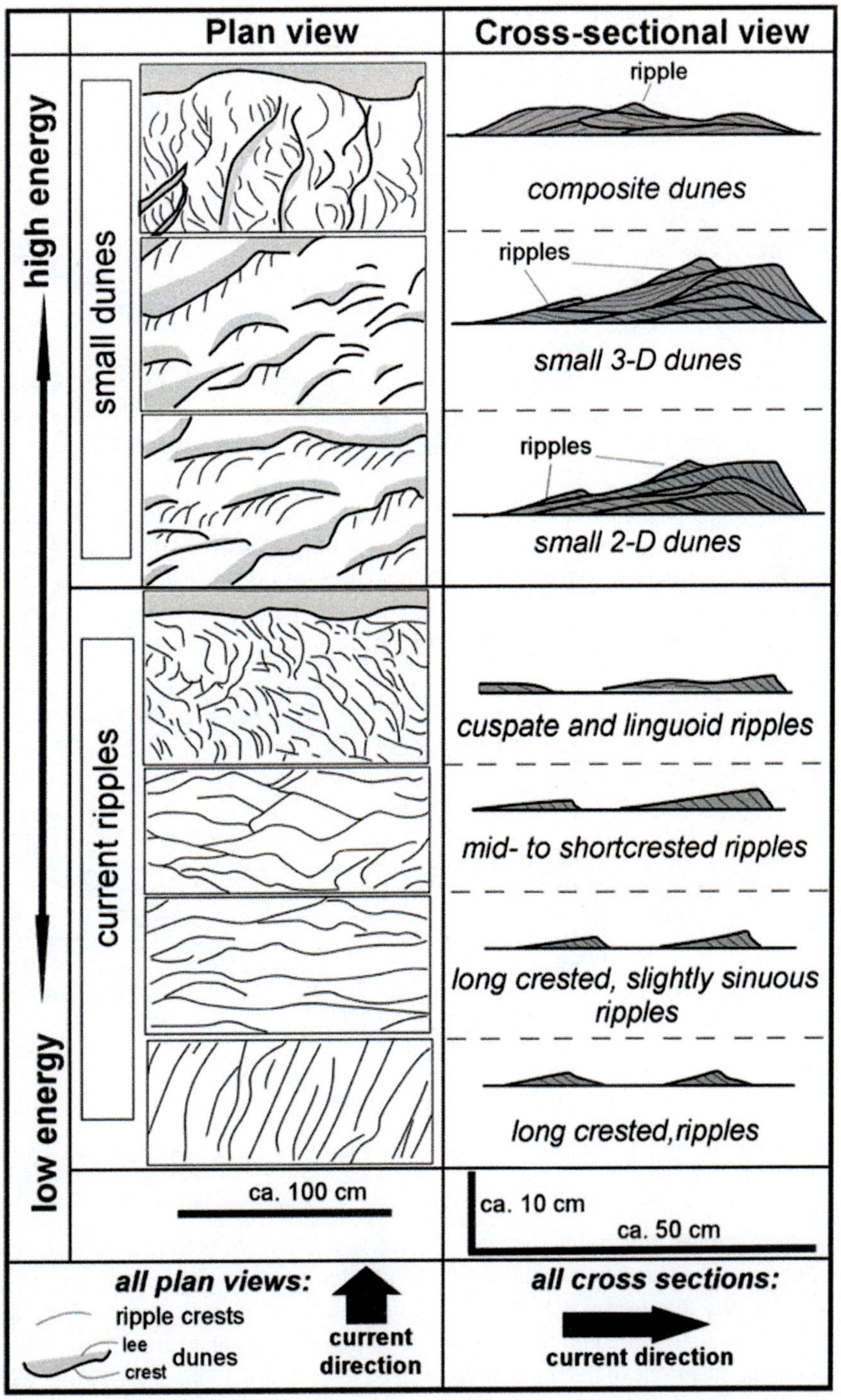

Fig. 20.4 Types of ripples and small dune forms found on larger subaqueous dunes (Modified from Gonzalez and Eberli (1997))

In spite of their importance in generation and maintenance of carbonate sand shoals, the details of the generation and dynamics of physical bedforms in these systems have not been systematically studied in the Bahamas, with the exceptions of Imbrie and Buchanan (1965) and Gonzalez and Eberli (1997). Here, we follow the terminology of the latter study (Figs. 20.4 and 20.5), which focused on patterns from an inlet system in the Exuma island chain.

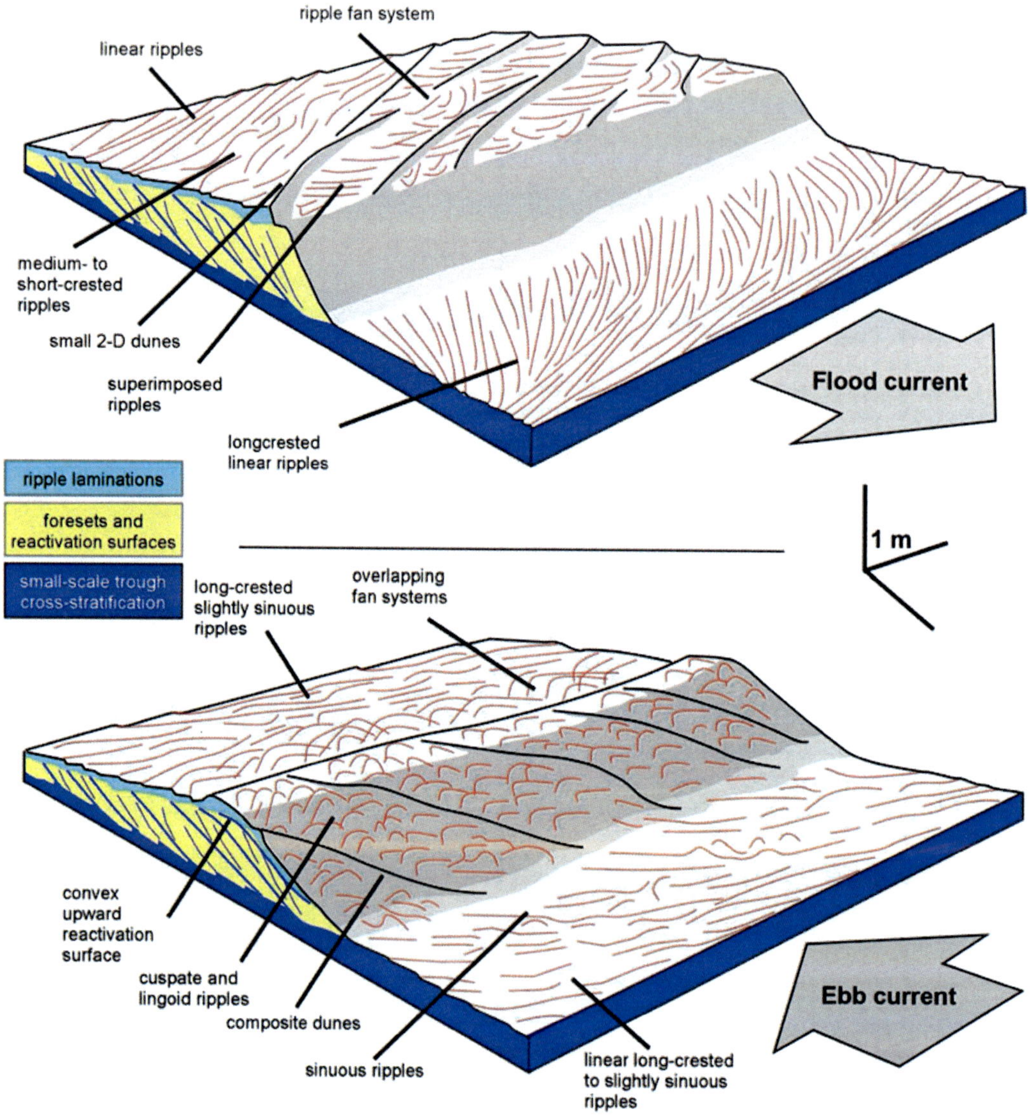

Fig. 20.5 Schematic summary of changes in ripple attributes with changing tides, and internal structure of a compound dune (Modified from Gonzalez and Eberli 1997)

Current ripples are the most ubiquitous bedform found on all shoal complexes. These have heights of several cm and wavelengths that can exceed 0.3 m. Asymmetric ripples are associated with uni-directional bottom velocities, with the steep side of the ripple indicating the direction in which the current is flowing. Based on crest geometry, Gonzalez and Eberli (1997) recognized several different types of these tidal-current related ripples, which they also related to current velocity (Fig. 20.4). These types include: (1) *long-crested linear ripples* that have very little sinuosity and are oriented parallel to subparallel to the dominant direction of the currents; (2) *slightly sinuous ripples*, which have amplitudes of 1–2 cm, with crests ~1+m long that taper off at their ends (Fig. 20.6a, b). These are oriented normal to flow, have variable sinuosity, and grade into (3) *sinuous ripples*, with comparable amplitude, but which branch every meter or two. With increasing currents, these pass into (4) *cuspate and lingoid ripples*, characterized by irregular crests and dip directions (Fig. 20.6c). The distribution of different types of ripples is quite variable in both time and space on carbonate tidal sand shoals, because of the abrupt and repeated changes in bathymetry and

Fig. 20.6 Field photos of illustrative bedforms. (**a**) Small linear ripples from ~1.5 m water depth, Abacos. (**b**) Straight-crested ripples, with a smaller secondary set superimposed, forming ladder-back ripples, Abaco tidal deltas. Handle on *yellow* and *black* tape measure is 18 cm across. (**c**) Cuspate to lingoid C is ~15 cm across. (**d**) Symmetric ripples and scattered finger coral fragments. Ripples have spacing of 30–40 cm. Fish Cays, ~1.5 m water depth. (**e**) Simple dunes, with height of 20–30 cm, and superimposed ripples. Schooner Cays, ~2 m water depth. (**f**) Crest of an asymmetric medium to large subaqueous dune. Height of dune is ~50 cm. (**g**) Surface photo of the crest of a subaqueous dune from Lily Bank. From this perspective, these features appear as "stripes" in the water. (**h**) Crest of shoal and subaqueous dunes exposed at low tide. (**i**) Partly stabilized bottom, off the crest of an active shoal, Abaco tidal deltas, ~3 m water depth. Handle on *yellow* and *black* tape measure is 18 cm across

current direction and magnitude, and the ease with which current ripples can be modified because of their small size.

Beyond these current-generated ripples, wave-generated *symmetric ripples,* frequently straight-crested and around 2–10 cm in height, can be found locally associated with tidal shoals (Fig. 20.6d). These occur either on the slightly deeper flanks of shoal complexes or, more ephemerally, on the shallowest intertidal crests of some of the bar forms where small waves break. The combination of waves breaking on shoal-marginal shallow bar crests and the strong daily tidal currents results in a paucity of symmetric wave ripples across most parts of shoal complexes.

At a larger scale, several distinct types of subaqueous dunes are found on carbonate tidal sand shoals (Gonzalez and Eberli 1997), including: (1) *small 2-D dunes* with crests that are straight to sinuous (with amplitudes between 0.2 and 1.0 m; note that Ashley (1990) would not refer to these as "small" dunes) and spaced between 1 and 3 m apart, with superimposed ripples (Fig. 20.6e); (2) *small 3-D dunes*, with heights of decimeters, and spacing on the order of 1 m, with crest bifurcations every few m; and (3) *complex composite dunes* with heights up to 0.1 m and which migrate normal to the current. The largest scale of these bedforms include *medium to large 2-D dunes* and *compound dunes* of Ashley (1990). These features

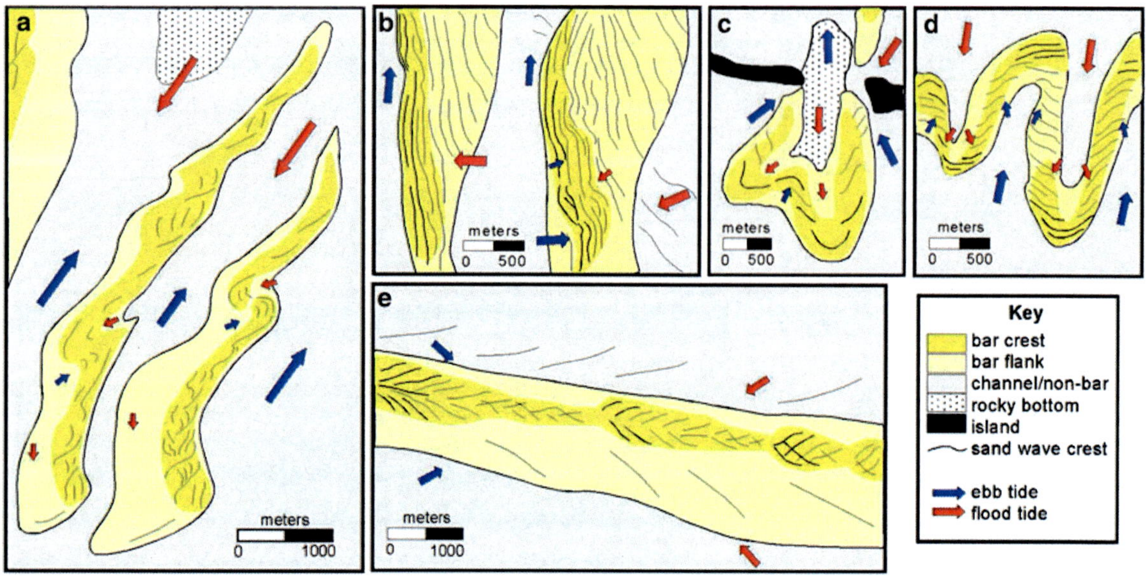

Fig. 20.7 Schematic of patterns of bars and flood (*red*) and ebb (*blue*) tidal flows in several end-member morphodynamic classes. *Arrows* illustrate general trends in direction of flow, not residual sediment transport directions. (**a**) Longitudinal tidal sand ridges; (**b**) Transverse shoulder bars; (**c**) Tidal deltas; (**d**) Parabolic bars; (**e**) Isolated longitudinal sand bars. See text for detailed discussion

can range from markedly asymmetrical (Fig. 20.6f) to practically symmetrical, have amplitudes that may exceed 0.5 m, and be continuous for 1 km or more. They are ubiquitous on the crests of many tidal bars in the Bahamas (Fig. 20.6g), may be exposed at low tide (Fig. 20.6h), and are the smallest bedforms discernable on high resolution (QuickBird or IKONOS, ≤4 m^2 pixels) remote sensing data.

Not all bedforms on tidal sand shoals are presently active. On and around most shoals, subtidal bedforms stabilized by seagrass, calcareous or non-calcareous algae, non-calcifying microbial mats, corals, and/or sponges, occur in waters more than a few meters deep, or on the platform adjacent to the shoals (Fig. 20.6i). These features have been variously interpreted to represent relict deposits (Hine et al. 1981) or deposits active only during extreme events (Wanless and Tedesco 1993), and they can include complex geometries, perhaps caused by several generations of activity.

20.3.4 Barform Morphology

Just as the transport of individual grains constructs bedforms, the movement of individual bedforms constructs barforms, which in turn generate tidal sand shoal complexes. Within the tidal sand shoals of the Bahamian Archipelago, sand bars exhibit a wide range of sizes and shapes. These bars can be grouped into five end-member classes, each related to distinct morphodynamic processes (Fig. 20.7). Gradations between bar-form classes are evident (e.g., shoulder bars with progressively increasing sinuosity and asymmetry pass into parabolic bars with linked crests, Rankey et al. 2006), and these different barforms can occur spatially juxtaposed (e.g., the flank of a parabolic barform can form part of an adjacent shoulder barform, Rankey et al. 2006) or, in some cases, superimposed (e.g., small parabolic barforms on the crest of tidal sand ridges).

This empirical classification captures the essential architectural building blocks of shallow carbonate tidal sand shoals in the Bahamas. Although it has broad parallels with, and includes aspects of, previous descriptions from Holocene carbonate shoals (e.g., Ball 1967; Hine 1977; Harris 1979; Rankey et al. 2006; Reeder and Rankey 2008), it utilizes distinct barform geometries as the basis for description. This classification shares similarities with aspects of classifications proposed for siliciclastic systems (e.g., Off 1963; Dalrymple and Rhodes 1995; Dyer and Huntey 1999; Wood 2004), but it differs at a fundamental level because, relative to siliciclastic counterparts, these carbonate systems all

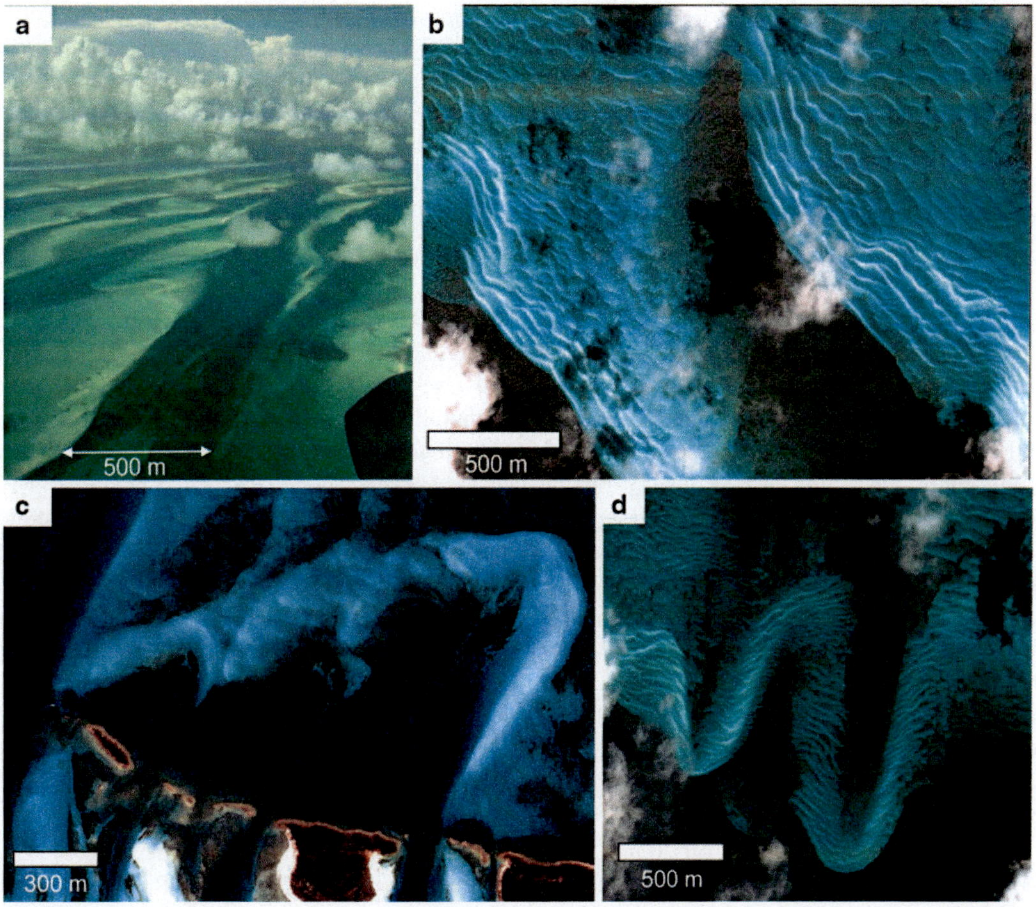

Fig. 20.8 Aerial and remote sensing images of barforms. (**a**) Longitudinal tidal sand ridges, Schooner Cays. Photo from 2/2007. (**b**) Transverse shoulder barforms, Lily Bank. Image acquired 8/19/2002. (**c**) Ebb tidal deltas, western Abacos. Image acquired 3/16/2005. (**d**) Parabolic bars, Lily Bank. Image acquired 8/19/2002 (Images in (**b**) and (**c**) copyright GeoEye, image in (**d**) copyright DigitalGlobe.com)

occur in shallow (<10 m) water and have very different geographic boundary configurations (for example, there are no estuary or "river mouth" carbonate sand examples; see Off 1963; Wood 2004).

Longitudinal tidal sand ridges are generally straight (Fig. 20.8a), although they can have slightly sinuous crests and bends or even recumbent hooks at their platformward terminations. The long axis of tidal sand ridges is oriented at a small oblique angle to peak tidal flow, which in many cases is almost normal to the shelf margin, although sets of tidal sand ridges locally create a radiating pattern because of the divergent flow pattern. For example, in the shoals at the southern margin of Tongue of the Ocean, tidal sand ridge crests are oriented between 50° and 90° oblique to orientation of the shelf margin.

Tidal sand ridges can be laterally extensive, reaching almost 25 km long with 3.5 km spacing. Tidal sand ridges in the southern TOTO area are the largest in the Bahamas, with mean length of 10.4 km and spacing of 1.6 km; those from Schooner Cays average 7.9 km long, and (now stabilized, earlier Holocene) examples from Lily Bank have mean length of just over 1 km and spacing of ~250 m. In general terms, many tidal sand ridges broaden on-platform, and can exceed 1 km in width at their platformward end. Tidal sand ridges can reach up to ~8 m tall, but most are on the order of 3–6 m. Although Ball (1967; echoed by Halley et al. 1983) suggested that these features reflect current speeds exceeding 1 m/s, recent measurements suggest that velocities amongst tidal sand ridges in Schooner Cays shoal complex do not exceed 0.9 m/s (Rankey and Reeder 2011).

The shallowest parts of the crests of these ridges can be exposed at low tide. Many ridge crests are ornamented with small simple dunes up to 1 m tall, and include the best sorted and most oolitic sediments on these bars. Crests can have early marine cementation, which can facilitate the growth of corals or other organisms that require a hard substrate to grow. Ridges have steep flanks, locally at the angle of repose, although a systematic asymmetry is absent in many tidal sand ridge fields, unlike that which occurs in some siliciclastic examples (e.g. Houbolt 1968).

Transverse shoulder bars are asymmetric, straight to slightly sinuous crested bars, with crests oriented roughly normal to the dominant tidal flow, which in many cases is roughly normal to the shelf margin (Fig. 20.8b). These bars can reach up to several meters in height (generally less than tidal sand ridges, however), and have a distinct steep side (up to the angle of repose) and a more gently sloping flank (<2°). Nonetheless, because they are normal to the shelf margin, shoulder bars generally are at high angles to either ebb or flood tidal currents, and one tide will be dominant across much of the barform (Hine 1977; Rankey et al. 2006), enhancing their asymmetry. These features can be quite large; on Lily Bank, individual shoulder bars extend along strike for 5 km and reach up to 2 km in breadth, and in the western parts of the Schooner Cays shoal complex, shoulder bars can exceed 30 km long and 5 km in breadth. These features can have up to 5 m relief.

The crests of many shoulder barforms are not exposed at low tide. Elevated tidal flow velocities lead to active sediment transport over the crest. As with tidal sand ridges, the crests of shoulder bars are ornamented with symmetrical small sandy dunes up to 1 m tall, and include some of the best-sorted oolitic sands in the system. These dunes can be oriented at a range of orientations, from normal to flow (Rankey et al. 2006), to oblique to flow, to sinuous, depending on the details of the patterns of ebb and flood tides across the crest. Although scattered cemented zones occur there, the crests of shoulder bars of Lily Bank shoal complex do not have widespread hardgrounds, probably also related to the active sand transport. Sedimentologically, shoulder bar crests have the best-sorted and most oolitic sediments, and they pass laterally into channels with less-well sorted sands and silts that can include up to 20% mud (Rankey et al. 2006).

Parabolic bars are complex, lobate features with variable crest orientations that change systematically by more than 90°. Broadly speaking, and as used here, parabolic bars include both tidal deltas (Fig. 20.8c), frequently associated with bedrock highs in the Bahamas, and open-marine forms not associated with any local *a priori* restriction, because the morphology and sedimentology of both are strongly influenced by mutually exclusive flood- and ebb-dominated conduits (Fig. 20.8d; van Veen 1936; Rankey et al. 2006; Reeder and Rankey 2008). Although perhaps broadly analogous, we avoid the use of "spillover lobes" because of the implicit (or explicit) implication that these are created by storms (Ball 1967; Hine 1977; Wanless and Tedesco 1993) (see discussion below).

With the exception of tidal deltas anchored by bedrock highs, parabolic bars do not occur as isolated features; in many cases, the flank of flood parabolic bars form the margin of adjacent ebb parabolic bars and vice versa, forming a sinuous-crested barform (Fig. 20.8d; cf. Caston 1972). Parabolic bars have crests with distance from flood-oriented apex to ebb-oriented apex (amplitude, in map-view) ranging from 0.3 to 3.0 km and aperture (width) between 0.5 and 3.0 km; mean amplitude: aperture ratios for sets of parabolic bars range from 1:2 (in tidal deltas of the Abacos) to almost 1:1 (in the southern TOTO area) (Rankey et al. 2006).

Parabolic bars can include up to ~5 m bathymetric relief, although most are between 2 and 4 m. In most cases, bar crests remain submerged by 0.5 m or more at low tide (e.g., Lily Bank; Abaco tidal deltas, Ocean Cay tidal deltas), but in some, bars have aggraded and now form stabilized islands (e.g., Schooner Cays). Linked parabolic bars pair flood-dominant bars with ebb-dominated bars; commonly, this pair is ornamented by subaqueous dunes up to 1 m tall with systematically variable orientations that reflect the reversing ebb- and flood-tides that flow over their crests (Rankey et al. 2006).

As with the other barforms, the best-sorted and most ooid-rich sediments occur on the crests of parabolic barforms. In many open-marine parabolic bars (such as those on Lily Bank or Schooner Cays), the crests pass laterally into poorly sorted muddy, skeletal-peloidal sands in grass-stabilized lows between the bars. Although bar crests pass laterally into less well-sorted, less oolitic sediments in Bahamian tidal deltas, they differ markedly in that the deepest parts of the inlets at the center of the parabolic tidal deltas have hard

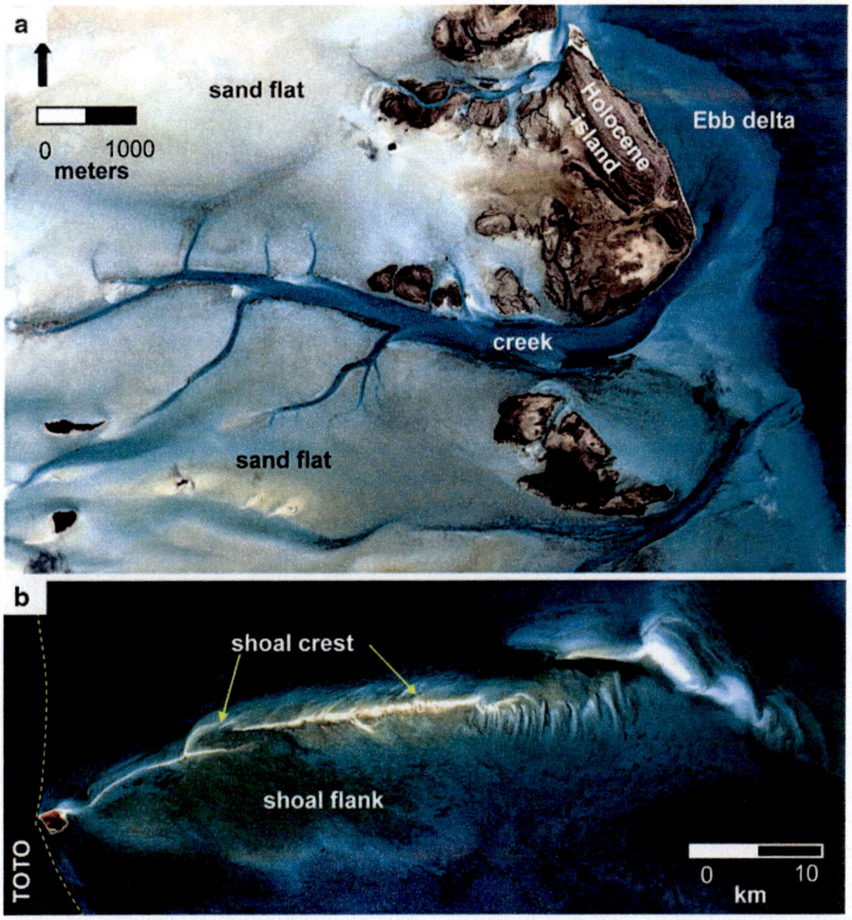

Fig. 20.9 Remote sensing images of barforms. (**a**) Joulter sand flat. Note the tidal channels extending onto the platform (to the *left* in this image), the Holocene islands, and the ebb tidal delta. Image acquired 12/27/2003, and is copyright DigitalGlobe.com. (**b**) Isolated longitudinal sand bars, on Green Cay shoal. See text for detailed discussion

bottoms. These rocky (Pleistocene or coral-covered) floors, along with the bedrock islands, help focus the tidal currents in these inlets, allowing the tidal velocities to exceed the equilibrium velocities (1 m/s) observed in siliciclastic analogs, which can lead to larger tidal deltas in carbonate settings.

Sand flats are expansive, flat-topped, sandy, intertidal to shallowest subtidal accumulations penetrated by tidal channels that are oriented roughly normal to the platform margin (Fig. 20.9a). Sand flats (Joulter Cays, Fish Cays) have discontinuous islands near the platform margin that are flanked platformward by broad intertidal to shallow subtidal flats that gradually slope down to the subtidal platform interior without a well-defined break in slope. Islands associated with most sand flats are of Holocene age, are associated with longshore or on-bank transport, and are extensively cemented by meteoric processes (Harris 1979; Halley and Harris 1979, see above). Sand flats can be quite extensive; the Joulter Cays sand flat reaches up to 20 km onto the platform, and the Fish Cays sand flat is up to 7.5 km wide. In both examples, however, the broad shallow flat is broken along strike by tidal channels several 100 m wide and 3–5 m deep that either taper and branch or broaden into diffuse, shallow depressions as they extend onto the platform. The large areas of the crests of sand flats are intertidal, and can be stabilized by seagrass and calcareous green algae (Joulter Cays), and may also include pastures of red algae, seagrass, and sponges (Fish Cays). Constituent sediments reflect this assemblage, and these intertidal flats include fewer ooids than higher-energy channels or ocean-facing margins of the shoal complexes (Harris 1979; Rankey and Reeder 2010).

Isolated longitudinal sand bars are individual elongate, narrow bars oriented normal to the shelf margin (Fig. 20.9b) that occur apparently "in the middle of nowhere." Several of these occur in the Bahamian Archipelago, and include Ambergris, Mackie, and Green Cay shoals. These features are the least well-understood barforms, in large part due to the fact that they occur away from easily-accessible islands. These elongate barforms can reach up to 30 km long, have current- or wave-agitated crests <1 km wide (narrower in many places), and can have several meters total relief. Unlike longitudinal tidal sand ridges, these isolated forms have no flanking, flow focusing channels, and for much of their extent have straight crests with no superimposed parabolic forms or switchbacks. The crests are ornamented with low relief subaqueous dunes of various orientations, and slope gradually to waters 4–5 m deep. Of these bars, the sedimentology has been studied in detail only on Ambergris shoal on the Caicos platform (Rankey et al. 2008). There, the bar crest includes well- to very-well-sorted, coarse ooid sands; the flanks include less well-sorted and less ooid-rich sands, with discontinuous hardgrounds.

20.3.5 Associated Geomorphic Elements

The occurrence and distribution of *channels* between bars is implicit in the preceding discussion, but these merit some discussion of their attributes as well. Channels can be of variable width, depth and orientation, and some reach as deep as the Pleistocene bedrock, which forms a hard substrate resistant to erosion. The sedimentology of the channel floors is dictated by the speeds of the currents passing through them. Many channels passing through inlets between rocky islands (either of Pleistocene bedrock or lithified Holocene sands) have rocky (Pleistocene) floors with only patchy, rubbly sediments and local patch reefs that pass into bare sands with ripples or small dunes away from the restricted opening. Where channel floors are mostly seagrass-covered, bioturbated muddy sands, the current velocities rarely exceed 0.5 m/s, and the margins are commonly composed of erodible sediments instead of the rocky inlet margins. In channels where currents are sufficient for active sediment transport (e.g., in the oceanward parts of Joulter Cays tidal channels; amongst parabolic bars of eastern Tongue of the Ocean), however, the channel floors consist of sandy substrate in the form of subaqueous dunes and little to no seagrass. In a few channels in the Exuma Island chain, subtidal columnar stromatolites are found in tidal inlets (Dill et al. 1986).

Holocene islands, another associated type of deposit, form on the crests of several sand shoals. Many of these islands have well-developed sandy, oolitic beaches, include a succession of beach ridges and eolian dunes, and are lithified by early meteoric cementation (Halley and Harris 1979; Budd 1984). In some cases, shoreline spit-like geometries suggest their genesis may initiate longshore transport, producing a feedback mechanism allowing the islands to grow.

Platform interior sediments occur bankward of many shoals, in lower energy environments. These regions generally have a flat bottom, with no active physical sedimentary structures (subaqueous dunes or bars). In some areas, such as on Little Bahama Bank platformward of Lily Bank and on Great Bahama Bank platformward of Joulter Cays, sediments are mud dominated. In contrast, grain-dominated fine to coarse sands occur in the interior of the Berry Islands area, on Crooked-Acklins Platform, platformward of Fish Cays, and on the platform west of the Exuma Islands. Nonetheless, although the character of these sediments is highly variable, these areas are almost always burrowed, and often (although not always) are partly stabilized by seagrass or calcareous algae; cementation can occur locally (Taft et al. 1968). These attributes, along with infrequent ripples, illustrate that these regions are less frequently agitated and that the rate of biological reworking is greater than the rate of physical reworking.

20.3.6 Examples of Shoal Complexes

Although they all occur in shallow, agitated environments, the tidal sand shoals of the Bahamas and Caicos include a wide variety of sizes and shapes, collectively formed by complex associations of individual bedforms and bars. The variability is related to historical or spatial contingencies, such as bedrock topography and the presence of ridges or islands, age of the shoal, orientation relative to predominant winds and waves, and shelf-break curvature. This section explores how different bedforms and bars combine to create vastly different shoal systems by discussing the details of a few examples of tidal sand shoals from the Bahamian Archipelago.

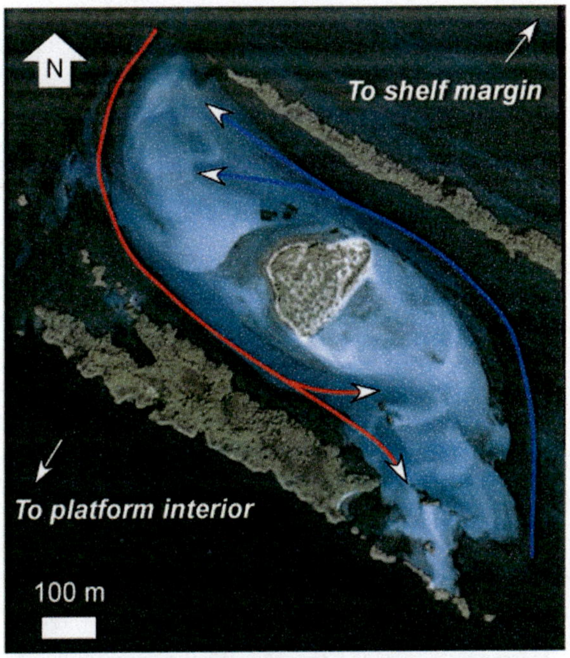

Fig. 20.10 Remote sensing image of Double Breasted Cays, Abacos, Bahamas, and schematic patterns of dominant path of ebb (*blue*) and flood (*red*) tides (Modified from Reeder and Rankey (2008))

In some carbonate tidal systems, pre-existing topography shaped by Pleistocene bedrock highs determines the location of shoals and markedly influences their sedimentology and hydrodynamics. For example, the *Double Breasted Cays sand shoal*, in the Abaco island chain, occurs between two parallel bedrock highs. In this area, two channels flanking the Pleistocene island ridges set up a mutually evasive flow pathway whereby the flood tidal currents are stronger in the southern channel while the ebb tidal currents are focused in the northern channel. This tidal channel configuration sets up a net circular flow pattern around the central oolitic shoal, a dynamic termed the 'spin cycle' by Reeder and Rankey (2008) (Fig. 20.10). In this carbonate system, the flow pattern establishes a situation in which the sediment can be agitated and transported to permit the growth of oolitic cortices, without the particles being removed from the geomorphic system. Reeder and Rankey (2008) suggested that this general process was an important element for the generation of ooids in the tidal-dominated settings of the Bahamas.

The Double Breasted Cays oolitic shoal is one extreme, where the shoal is situated between bedrock islands, which highly restrict the possible flow patterns. However, tidal sand shoals can also form in less-restricted settings, such as the *tidal deltas of the Abaco Islands*. In this area of ~50 km^2 along the northwestern Abaco island chain, ten inlets with associated flood and ebb tidal deltas of various sizes occur (Fig. 20.11) (Reeder and Rankey 2009b). Here, the sandy deltas are almost exclusively parabolic bars forming around inlets in the island chain. The parabolic bars terminate at the edges of the islands and are convex outwards from the main channels which extend perpendicularly to the inlet opening. In some cases, a multi-inlet system creates complex shoal geometries by combining multiple parabolic bars into one.

The crests of the deltas, some of which may be exposed at the lowest tides, are largely bare oolitic sands (>75%) covered with ripples and subaqueous dunes. The inner lobes of the deltas are located between the islands and the sandy crests; they are stabilized by dense seagrass and may contain patch reefs (ebb deltas). The sediments of the inner lobes are bioturbated peloidal-skeletal sand and mud with few ooids (<1%). The main channels passing through the inlets consist of a hard bottom including some sponges and corals, but no loose sediments.

The sedimentology of the tidal deltas is closely linked to the hydrodynamics of the region (Reeder and Rankey 2009b). The main channels experience tidal velocities >1.5 m/s, explaining the lack of loose sediments. They also form part of a mutually evasive flow pattern, whereby the main channel of the flood lobe experiences high velocities during flood tide, but the ebb tides are focused down marginal channels flanking the lobe (broadly comparable to the model of Hayes 1975). The presence of the shoal also increases velocities over the shoal, allowing adequate agitation on the bar. This sets up a 'spin cycle' on the shoal, leading to the agitation, cortex formation, and transport of ooids on the shoal itself, but a lack of ooids off the shoal. As the velocities pass through the inlets or over the shoal, the flow expands outward, so the velocities in the deeper, more peloidal, seagrass stabilized regions of the inner lobes and outside of the tidal deltas are much lower (~0.25 m/s).

The size of the deltas varies systematically with the size of the inlet opening. Each inlet experiences a different volume of water exchange between low and high tides (tidal prism), and in each delta, the distance to the delta apex is closely related to the tidal prism

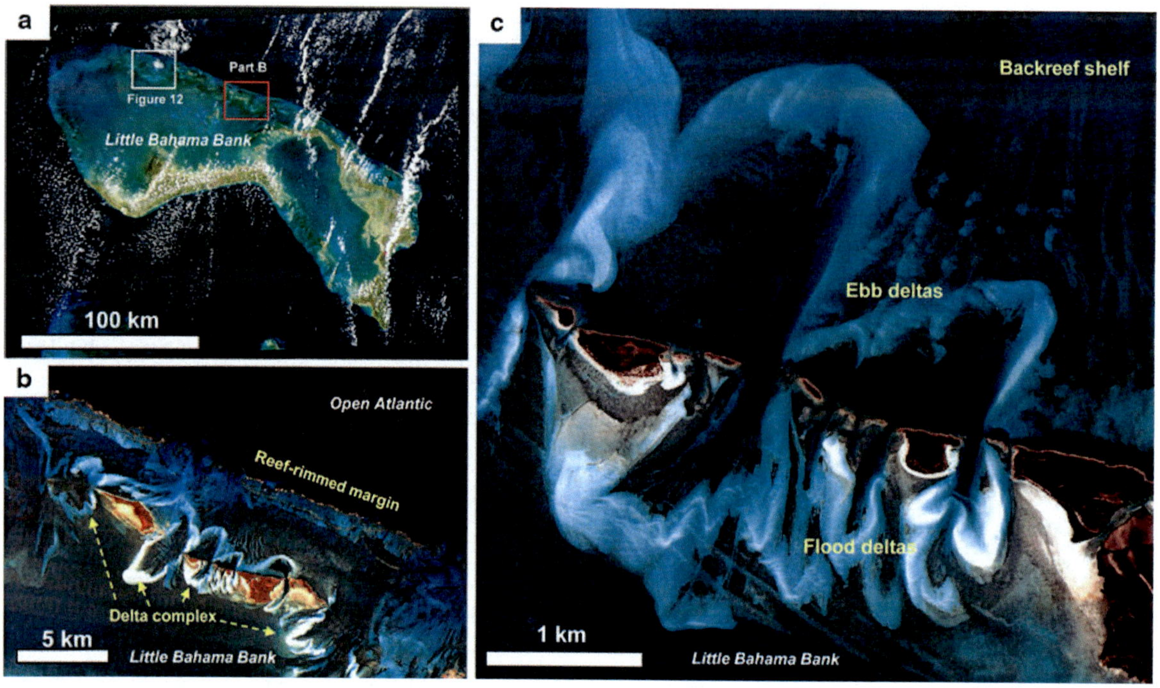

Fig. 20.11 Setting and character of tidal deltas, western Abaco Islands, Little Bahama Bank. (**a**) Image illustrating location of tidal delta complex (*red box*) and Lily Bank (*white box*) on northern margin of Little Bahama Bank. (**b**) Image of the northern platform margin. Note the reef-rimmed northern margin, the discontinuous island chain, and the ebb- and flood-tidal deltas. (**c**) Image illustrating details of the ebb- and flood-tidal deltas. Note the discontinuous islands, which restrict flow, and the associated tidal deltas. Rocky ridges form the WNW-ESE "stripes" in the southern part of this image. Image acquired 3/16/2005 (Image copyright DigitalGlobe.com)

(Reeder and Rankey 2009b). This process-response relationship shows a strong geomorphic similarity with siliciclastic analogs.

Lily Bank shoal complex also occurs on the northeastern margin of Little Bahama Bank (Fig. 20.11a) and covers ~270 km^2, but, unlike Double Breasted Cays or the Abaco Tidal Deltas, it is not associated with any bedrock islands (Hine 1977; Rankey et al. 2006). The shoal complex falls along a part of the margin that has no Pleistocene bedrock exposed above sea level and is generally unrimmed, although three discontinuous reef patches along the margin appear to provide sufficient lateral flow restriction to focus currents enough to establish an ooid shoal. The complex includes two downdip systems of radiating tidal sand ridges (up to 2 m tall, in waters several meters deep) that radiate outward (Fig. 20.12, top). These areas are now largely stabilized by seagrass, and were interpreted by Hine (1977) to represent earlier Holocene deposits. As sea-level continued to rise, the active oolitic shoals of Lily Bank backstepped further onto the platform, where they form a 4 km-wide complex that extends up to 15 km onto the platform. The active shoal complex includes a mix of shoulder bars and parabolic bars extensively ornamented with sand waves, formed by the complex patterns of ebb- and flood-tidal currents (Fig. 20.12, bottom). Shallow bar crests are ~1 m deep at low tide, and include well-sorted medium sands with abundant (>90%) ooids; in contrast, 3–5 m deep channels are composed of poorly sorted, more peloidal and skeletal sediments, and include up to 16% mud. The shoals end abruptly, dipping platformward into waters 5–8 m deep covered by dense seagrass, with sediments broadly similar to those in the channels.

At the largest scale, because it backsteps through the Holocene rather than aggrades (in contrast to Joulter Cays, see below), this complex illustrates one means by which shoals may evolve in response to changes in sediment supply and rising sea level. Similarly, Lily Bank barforms show that parabolic bars in carbonate systems can be formed as a result of feedbacks among tidal flow, sediment transport, and geomorphology, and do not require bedrock highs or storms for their formation

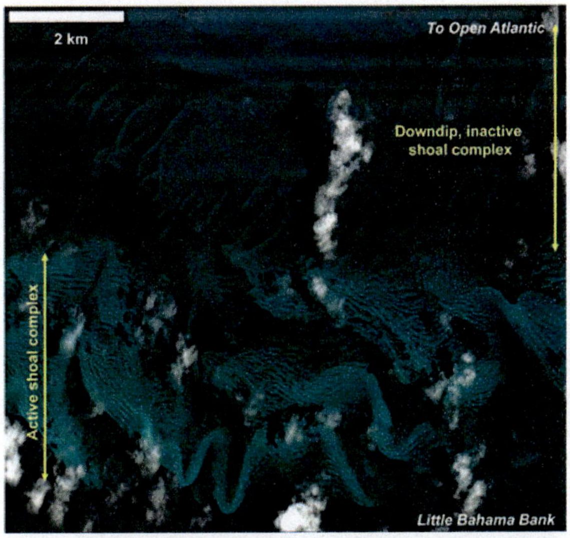

Fig. 20.12 Image of part of Lily Bank, westernmost Abaco Island chain, Little Bahama Bank, from the area illustrated in Fig. 20.11a. In this area, Lily Bank includes a downdip (*top* of the image) inactive shoal complex of generally flow-parallel tidal sand ridges and an updip complex of active ooid sands with abundant parabolic bars and subaqueous dunes. Image acquired 8/19/2002. Image copyright GeoEye.com. See text for discussion

(Rankey et al. 2006; cf. Caston 1972; Hine 1977). Finally, the system illustrates barform geometries analogous to those found in some siliciclastic middle estuary systems (Dalrymple and Rhodes 1995; Wood 2004), highlighting some geomorphic similarities between carbonates and siliciclastics.

Tongue of the Ocean shoal complex is the largest oolitic shoal system in the world, but is also the least well studied system in the Bahamas due to its isolation (Palmer 1979). It lies on the shallow shelf of the Great Bahama Bank bordering the broad deep-water embayment of Tongue of the Ocean (TOTO). As it wraps around the TOTO for more than 130 km (Fig. 20.13), the shoal complex includes various barform geometries. To the east, parabolic bars are quite extensive, and some appear to have sand waves in the intervening channels (unlike those on Lily Bank, which have seagrass-stabilized bottoms) (e.g., Fig. 20.13b, c). TOTO is best known for its spectacular longitudinal tidal sand ridges (Fig. 20.13d), west of the parabolic bars. These tidal sand ridges average over 10 km long, have widths of several 100 m, and are separated by deeper channels that average over 1.2 km wide and can be up to 8 m deep. The isolated longitudinal sand barform of Green Cay (Fig. 20.9b) extends east-west just north of the ooid shoal complexes.

Joulter Cays shoal complex occurs on the northern flank of the Pleistocene high of Andros Island, in an embayment in the platform bordered by the deep-water Tongue of the Ocean and an unrimmed, skeletal-sand rich shelf (Harris 1979) (Fig. 20.14). The shoal complex includes a broad sand flat (~400 km^2), bordered on its eastern margin by an active mobile sand belt with small ebb tidal deltas and on its northern margin by a number of longitudinal sand ridges. The shoal includes a number of Holocene islands across its extent, but the highest islands (up to 6 m above sea-level) are on the eastern margin, due to active longshore transport, accretion and cementation on beaches, as well as eolian processes. The sand flat is penetrated from the east by a number of tidal channels, and gradually slopes down to the platform interior to the west. Much of the vast expanse of the sand flat is exposed at spring low tide and stabilized by seagrass and algae. Surface sediments are dominantly non-skeletal. Clean, well-sorted oolitic sand is most abundant on the mobile sand belt on the eastern and northern flanks of the shoal (average abundance of ooids = 83%; Harris 1979). Much of the sand flat is burrowed, and includes fine micritized ooids and peloids.

As it represents an aggraded shoal, Joulter Cays shoal complex highlights the importance of prolific carbonate sediment production on carbonate shoals. Aggradation may have progressively filled accommodation space on the shoal crests, decreasing the tidal prism, which in turn decreased currents, leading to further aggradation. Indeed, the sedimentology of Joulter Cays shoals illustrates decreased energy in the interior of the sand belt (Harris 1979). Similarly, Joulter Cays shoal occurs on a windward margin. The dispersion and refraction of impending waves sets up a northward longshore transport, which is most likely responsible for the formation of islands. Similar combinations of tidal and wave-driven hydrodynamics on the outskirts of the shoal facilitate the prolific generation of oolitic sands in this belt (Carney and Boardman 1993). Aggradation and island formation, and associated feedbacks, may have facilitated the widespread aggradation. Given that many ancient ooid shoal complexes include evidence for aggradation and stabilization, the shoal may represent the "ultimate fate" of many ooid shoals (but – note the contrast with Lily Bank, above).

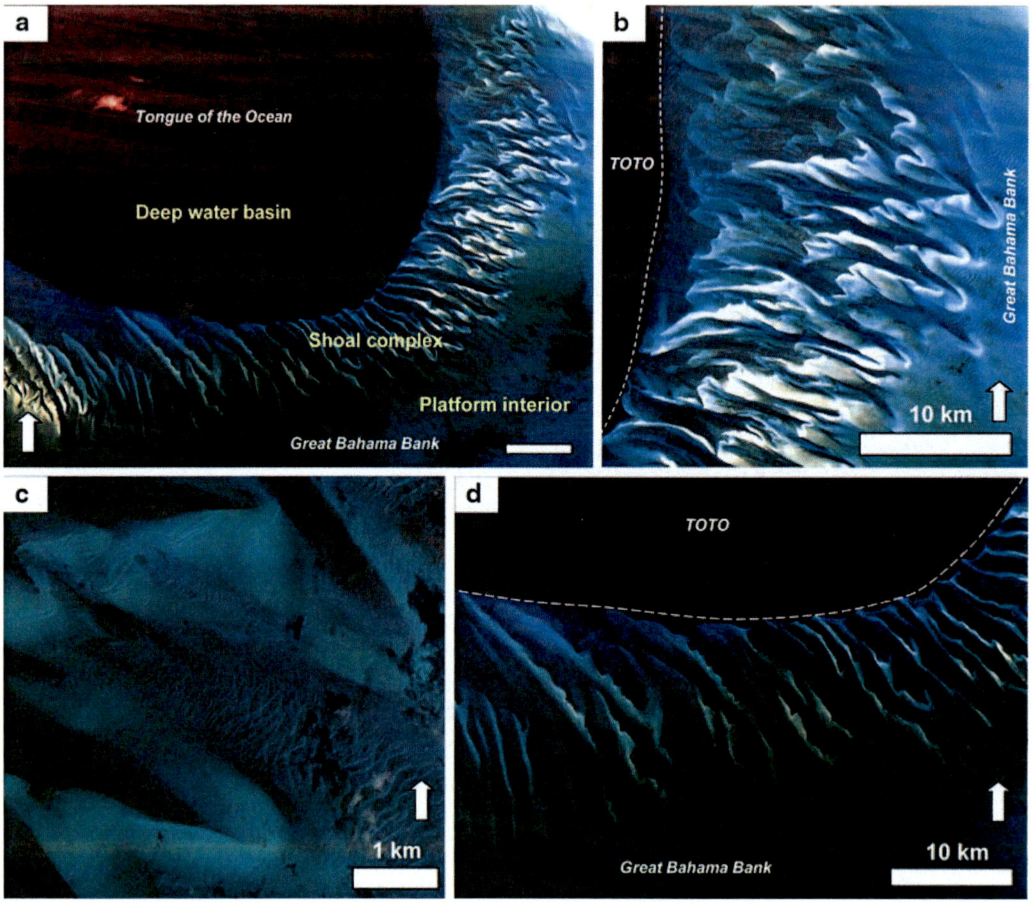

Fig. 20.13 Remote sensing images of Tongue of the Ocean shoal complex, Great Bahama Bank. (**a**) Overview of the arcuate trend ringing the deeper water Tongue of the Ocean. Image acquired 1/29/1985. (**b**) Image from the eastern part of the complex, with well-developed parabolic bars. (**c**) Close up of part of the eastern part, illustrating the complex patterns of parabolic bars and subaqueous dunes in this area. Image acquired 10/21/02. Image copyright GeoEye.com. (**d**) Close up of the southern margin, with numerous longitudinal tidal sand ridges

The possible role of winds on the generation of ooid shoals has been highlighted in studies of the *Ambergris shoal* on Caicos Platform south of the Bahamas (Wanless et al. 1989; Wanless and Tedesco 1993; Rankey et al. 2008). This 20 km long isolated longitudinal sand barform is oriented roughly parallel to the predominant trade winds. The shoal includes coarse sand-sized ooids and low amplitude (generally <0.5 m) subaqueous dunes on its crest, which can be exposed at low tide. The crest is immediately flanked by a rocky hardground up to 300 m wide with a slightly deeper (3 m) wave-rippled bottom to the north and a 2–4 m deep flat, intraclast-rich bottom to the south, before passing into burrowed deeper (>4 m) platform interior sands. Wanless et al. (1989) and Wanless and Tedesco (1993) emphasized the role of "agitation by wind-generated waves and the cross-platform current" on the morphology and sedimentology of the Ambergris shoal complex, but Rankey et al. (2008) pointed out that many of the geometries of superimposed barforms suggested a "more complicated situation," influenced in part by tides. Similarly, in a more general sense, the geometry of other isolated longitudinal sand bars, including Mackie Shoal (long axis normal to predominant winds) and Green Cay shoal (shoal in the stoss of an island), suggest that wind-driven waves may not be a fundamental control on the geometry of these systems. Yet, to date, there have been no measurements of waves or tides to test these alternative concepts, and these represent the least well-understood of all shoal complexes.

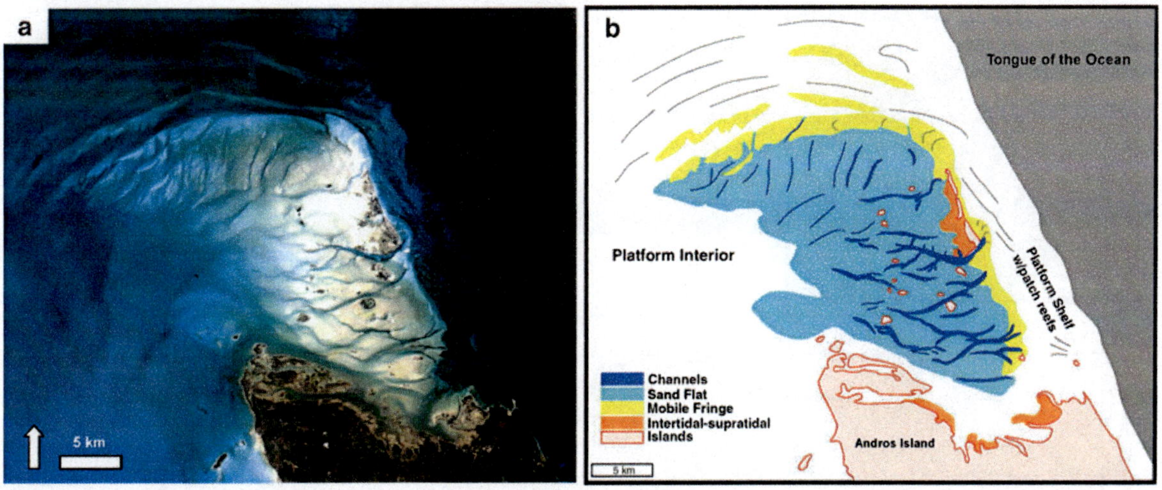

Fig. 20.14 Morphology of Joulter Cays shoal complex, Great Bahama Bank. (**a**) Entire shoal, bounded by deep water to the east, and platform to the west. Remote sensing image from 4/29/1986. (**b**) Interpretive diagram illustrating general facies patterns (Modified from Harris (1979))

20.4 Possible Stratigraphic Record and Geologic Examples

20.4.1 Holocene Shoals

Although the sedimentology and morphology of Holocene ooid shoals in the Bahamas and Caicos have been studied for over a 100 years, there are strikingly few extensive published analyses of their potential application for interpreting the stratigraphic record. The two areas that have been most extensively studied are the Cat Cay shoal complex (Ball 1967; Cruz 2008) and the Joulter Cays shoal complex (Harris 1979).

The Cat Cay shoal complex is on the western, leeward margin of Great Bahama Bank, flanking the Straits of Florida. This ooid-rich shoal complex extends ~14 km along strike, and is ~2 km wide. In this area, a bedrock high runs roughly parallel to the margin, west of the shoal complex, passing eastward to a flat top-Pleistocene surface ~6 m underneath the shoal complex (Fig. 20.15; Cruz 2008, compare with Purdy 1961). Within the shoal complex, a core transect revealed a basal burrowed skeletal-peloidal fine sand that is sharply to gradationally overlain by a ~4 m thick succession of cross-bed sets that thin upwards (Ball 1967). The gradational lower contact and the admixed ooids in the basal peloidal sands suggest a burrowed contact (Ball 1967). More recent high-resolution seismic data suggest that, at least locally, some cross-bed sets lie directly on bedrock (Cruz 2008; Neal et al. 2008) and illustrate that the cross-bed sets can dip either bankward (on the east side) or basinward (on the west side).

Joulter Cays shoal complex illustrates spatial and temporal variability in depositional facies that developed in response to the Holocene relative rise in sea level (Figs. 20.16 and 20.17). The shoal complex was deposited just north of present-day Andros Island, a high of Pleistocene bedrock, and probe data suggests that a nose of Pleistocene high plunges northward (Fig. 20.16a) beneath the present shoal. Cores from this area (Harris 1979) illustrate a typical vertical succession of a basal interval of lithoclast packstone and peloid wackestone, a medial fine-grained peloid packstone, and an upper ooid packstone. This succession includes an upwards-coarsening succession with an increase in the abundance of ooids. Within the sand flat, the thickness of ooid packstone thins bankward, as the fine peloid packstone thickens. In this shoal, ooid grainstone is abundant only on the eastern, windward margin, where it locally overlies bedrock and interfingers with the packstones onto the platform.

The shoal complex is interpreted to reflect three general stages of growth (Harris 1979): (1) an early stage during initial flooding, in which the lithoclast packstone and wackestone accumulated in subtle lows on the Pleistocene bedrock surface; (2) a shoaling stage of generation and accumulation of ooids, initiated by bedrock highs that focused tidal currents; and

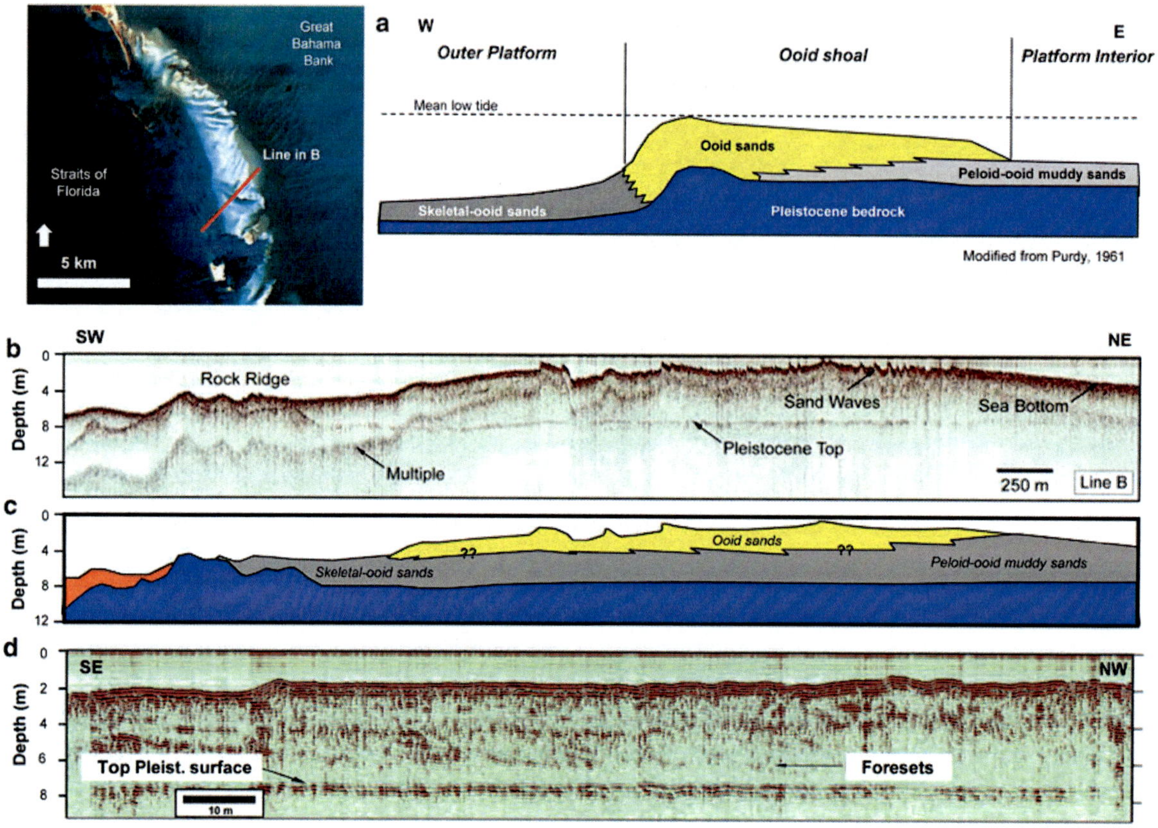

Fig. 20.15 Character of bedrock, Cat Cay ooid shoals, Great Bahama Bank. (**a**) Schematic figure of interpretations of a bedrock high under the shoal complex. Modified from Purdy (1961). (**b**) and (**c**) Representative shallow Chirp subbottom profile (**b**) and interpretation (**b**) of cross section across the shoal complex. This line, and others in the area, include no evidence for a bedrock high underneath this shoal complex. (**d**) Detail of part of a shallow Chirp subbottom profile, illustrating stacked cross-bedded units above the Pleistocene surface in some areas (Figure modified from Cruz (2008))

(3) a maturation stage in which the production and distribution of sands led to aggradation and shoaling to intertidal levels, along with expansion of the shoal complex. As the shoal aggraded to shallow depths, active bioturbation led to mixing of oolitic and peloidal sands and muds, leading to the ooid packstone of the sand flat interior. In this final phase, the clean oolitic sands form a 2 km wide and 2–3 m thick accumulation on the eastern and northern peripheries of the shoal, where wave-driven or tidal currents agitate and actively transport the sediments (Fig. 20.17).

20.4.2 Ancient Analogs

Holocene tidal sands of the Bahamas have several well-studied ancient analogs, preservation of which may be favored by their early lithification. One relatively young ancient analog occurs in the Upper Pleistocene Miami Oolite, exposed near Miami, Florida (Fig. 20.18) (Hoffmeister et al. 1967; Halley et al. 1977; Halley and Evans 1983; Evans 1987). Here, an ooid-rich late Pleistocene shoal complex is preserved as a wedge-shaped low ridge over 60 km long. The shoal morphology is broadly similar to that of the Holocene system of Joulter Cays (Fig. 20.14). The shoal complex has been dated at 130 ka (marine isotope stage 5e), and was deposited when sea level was ~7 m higher than present.

To the west, the lower-energy bryozoan-rich peloidal sand sediments (Hoffmeister et al. 1967) of the platform interior west of the shoal complex gently dip up (average gradients of 0.02 m/km, up to the east) to a broad shoal complex. The shoal system is broken into broad flat areas ~5 km across, separated by sinuous depressions several 100s of meters wide and several

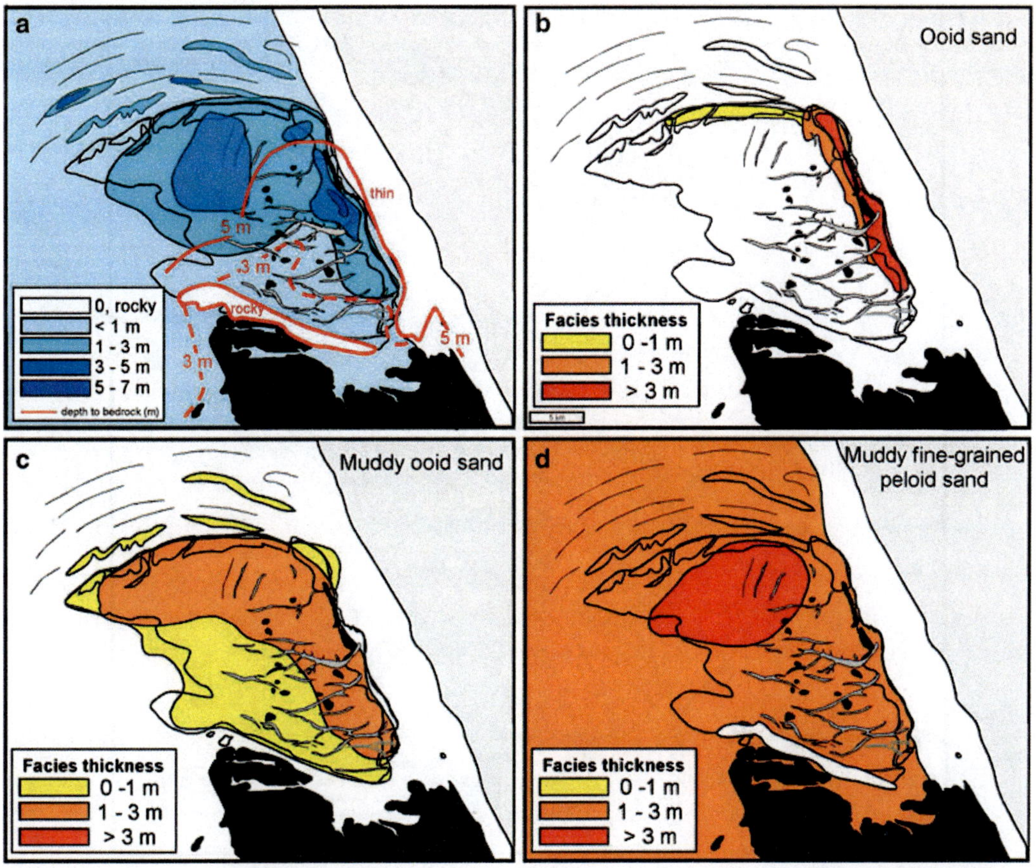

Fig. 20.16 Character and interpreted geologic history of Joulter Cays ooid shoal complex, Great Bahama Bank, modified from Harris (1979). (**a**) Thickness (*colors*) and depth to bedrock (*red lines*) in the area. Note that the shoal lies above a bedrock high that plunges to the north. (**b–d**) Thicknesses of various facies (**b** – ooid sand, **c** – muddy ooid sand, **d** – muddy fine peloidal sand) in the shoal complex. See text for discussion

kilometers long, oriented roughly normal to the trend of the complex. Cores from this sand flat and channel system suggest it averages between 5 and 8 m thick and is dominated by burrowed oolitic sand (Fig. 20.18d), which makes up an average of 60% of the succession. Cross-bedded ooid sand makes up the remainder, but are laterally discontinuous, occurring preferentially on the margins of the channels (Evans 1987).

In contrast, flanking part of the eastern margin of the sand flat and channel belt, a 32 km × 0.8 km barrier bar system developed, apparently post-dating much of the system to the west. This system is up to 11 m thick and attains higher elevations than the sand flat-channel system to the west. In the barrier bar system, cross-bedded facies are common, making up 60% of the succession, and the less abundant burrowed facies are discontinuous. The cross-bedded deposits have complex internal geometries, related to migration of inlets (along strike) and subaqueous dunes (across strike, with the ebb and flood tides); individual outcrops reveal decimeter- to meter-scale tabular- or sigmoidal-cross bedded oolitic grainstone (Fig. 20.18e) with dips that can exceed 20°. The eastern margin of this complex slopes towards the paleo-open ocean (topographic gradients of 0.3–3.0 m/km to the east), and this ooid-rich succession passes abruptly (across a few 100 m; Halley et al. 1977) to subtidal skeletal sands to the east. The barrier bar system probably reflects a combination of amalgamation of ebb tidal deltas and longshore transport (Halley et al. 1977; Evans 1987; Grasmueck and Weger 2002; Neal et al. 2008).

Joulter Cays is not the only Holocene example to have ancient analogs; trends similar to those recognized in tidal sand ridges from Schooner Cays or Tongue of the Ocean have been recognized in several

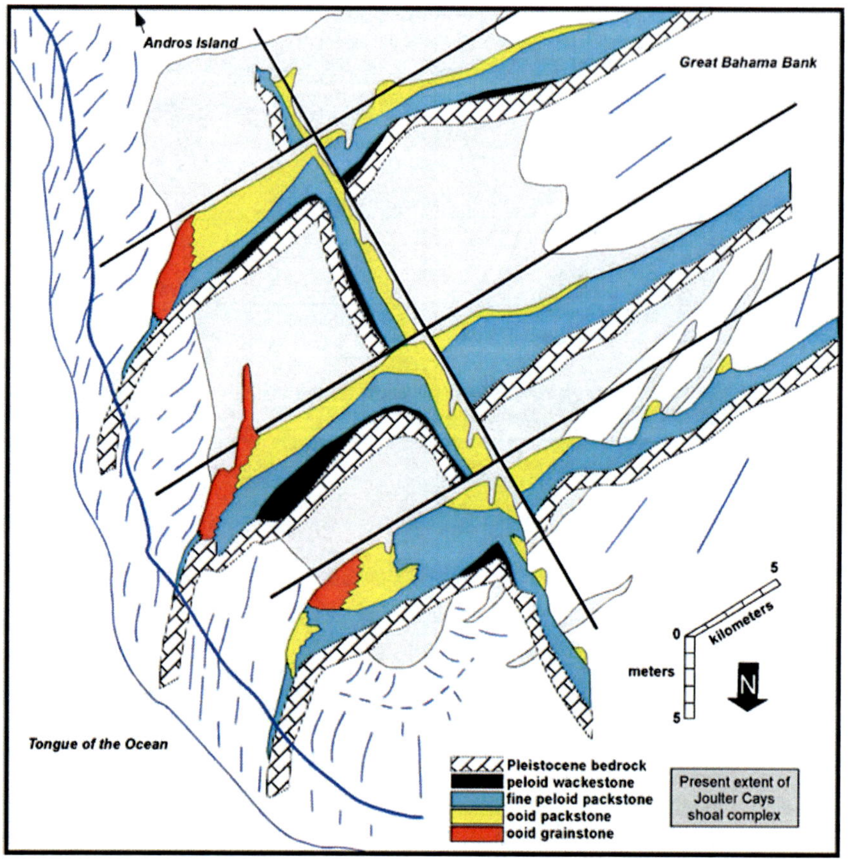

Fig. 20.17 Fence diagram of facies patterns, Joulter Cays ooid shoal complex, Great Bahama Bank, modified from Harris (1979). Note that ooid sand is dominant only on the eastern, windward fringes of the shoal complex. North is toward the *bottom* of the diagram. See text for discussion

ancient ooid shoal complexes. The best documented ancient analogs occur in Carboniferous strata (e.g., see summary in Keith and Zuppann 1993). In one Lower Carboniferous example from West Virginia, in a series of oolitic shoals, "...each shoal consists of a central cross-bedded oolite-grainstone facies surrounded by a transitional grainstone/packstone facies. Burrowed packstone occupies the adjacent tidal channels..." (Cavallo and Smosna 1997) (Fig. 20.19a). These facies, directly analogous to Holocene examples, are closely correlated with changes in reservoir quality in these systems. In another study, Carr (1973) documented facies patterns in Lower Carboniferous ooid shoals of the Illinois Basin. There, elongate ooid shoal bars, interpreted to represent flow-parallel tidal bars, include porosity zones up to 4 m thick which are directly correlated to the occurrence of clean oolitic grainstone (Fig. 20.19b).

A study of outcrop and shallow cores in an Upper Carboniferous (Pennsylvanian, Missourian) shoal complex in the Bethany Falls and Mound Valley limestones in southeast Kansas (Fig. 20.20) (French and Watney 1993) illustrates several important concepts. In a broad embayment in the area, underlying these oolitic units, a shaly interval in the Pleasanton Formation thins to the southeast, creating a prominent break in depositional slope. In this embayment, on top of this break in slope, tidal currents were focused, and a series of at least five elongated, longitudinal tidal sand ridges between 1.6 and 4.8 km wide and at least 16 km long were deposited. In this succession, the Bethany Falls Limestone Member (Swope Formation) includes a lower skeletal wackestone with echinoderms, brachopoids, bryozoans, and phylloid algae, capped by a cryptic, but regionally extensive, subaerial exposure surface. This lower unit is overlain by a

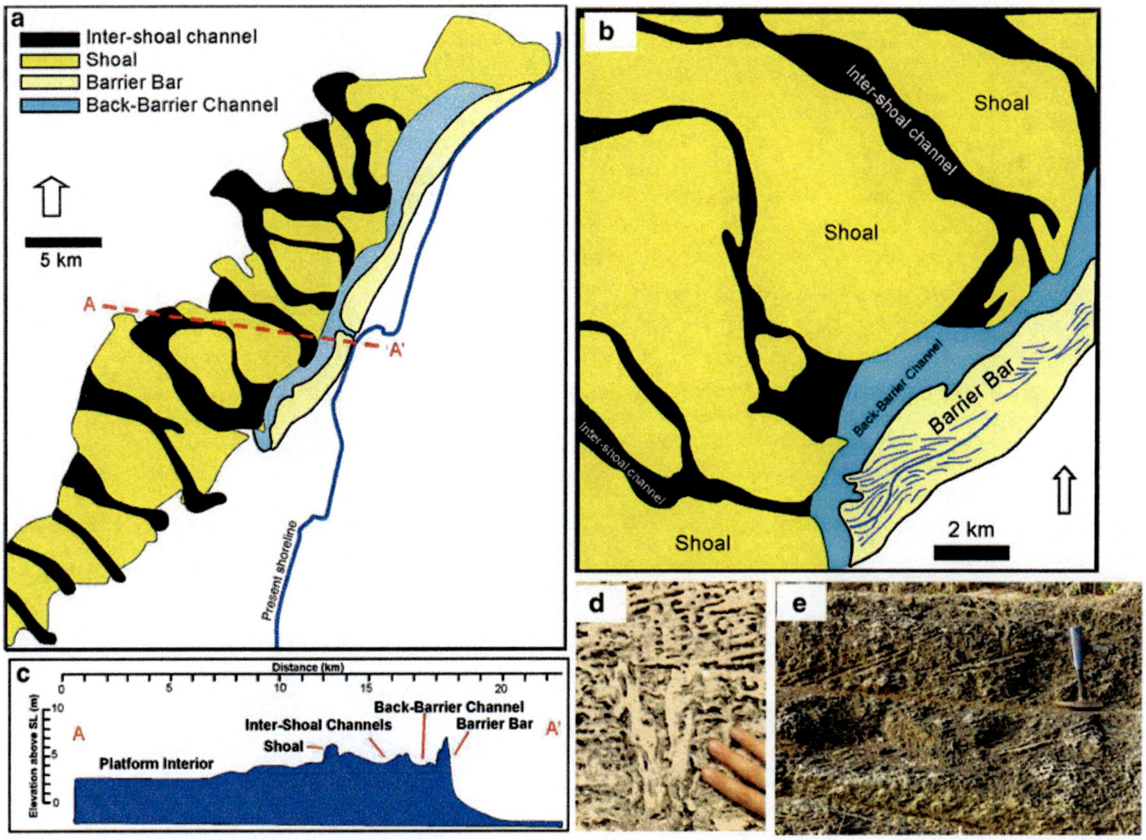

Fig. 20.18 Geomorphic patterns in an ancient ooid shoal complex, Pleistocene Miami Oolite, Florida, modified from Halley et al. (1977). (**a**) General facies patterns and dimensions of part of the coastal ridge in the Miami area, and the paleogeomorphic interpretations. (**b**) Detail of one area, illustrating the earlier shoal and inter-shoal channels, and the later barrier bar. (**c**) Representative topographic profile across the bar system. In terms of lateral and vertical scale, this system is broadly comparable with the Joulter Cays system (cf. Fig. 20.17). (**d**) Downward-oriented, chevron-like anemone burrow cutting through cross-bedding. (**e**) Preserved cross-bed sets, 30–40 cm tall. Both sets are oriented in the same direction, as in many systems, probably related to flood- or ebb-dominance in a specific location

porous, cross-laminated ooid-skeletal grainstone, and eventually with another subaerial exposure surface and paleosol (Ladore Shale). Overlying the paleosol is another oolitic grainstone deposit (Mound Valley Limestone) which has variable thickness, ranging from 3 m to absent, but is capped with another paleosol (the Galesburg Shale). Comparable to Holocene examples, this Upper Carboniferous example illustrates how breaks in slope and subtle embayments that focus tidal currents can be important factors influencing where these bars occur. Similarly, the Mound Valley ooid shoals stacked on the Bethany Falls high illustrate the important role of subtle paleotopographic highs that can focus tidal flows and favor formation of ooids.

Unlike Holocene examples, however, many ancient systems have been extensively modified by diagenesis. For example, a well-sorted oolitic sand of a Holocene accumulation may include abundant interparticle pore space (Fig. 20.2e–g). In many cases, the pore space is partly filled by calcite or aragonite cement, of marine (Fig. 20.3b, c) or meteoric (Fig. 20.3h, i) origin. The aragonite that makes up the ooids is diagenetically unstable, however, and a change in conditions (e.g., subaerial exposure and flow of meteoric fluids through the sand) may lead to partial or complete dissolution of ooids, leaving behind molds of the former grains (Fig. 20.21). With progressive diagenesis, more grains and even cements may be recrystallized (Fig. 20.21c, d). Clearly, the heartbreak of diagenesis can lead to many difficulties in trying to discern the details of the geologic history or distribution of porosity and permeability in ancient oolitic successions.

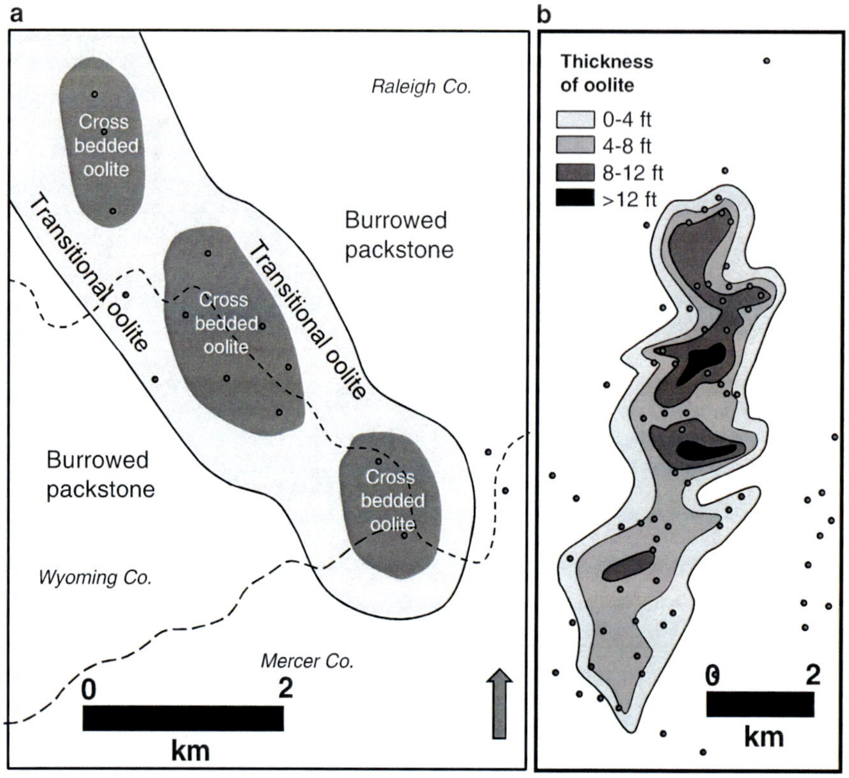

Fig. 20.19 Patterns of thickness and facies in ancient analogs. See text for discussion. (**a**) Mississippian example from West Virginia, modified from Cavallo and Smosna (1997). (**b**) Mississippian example from the Illinois Basin, modified from Carr (1973). *Dots* indicate well locations

Although many tidal sands of the Bahamas include abundant ooids, tidal sands are not always oolitic. One well-studied example that includes classic tidal sedimentary structures is the Upper Carboniferous (Mississippian) Salem Limestone of the United States Midcontinent (Sedimentology Seminar 1966; Brown et al. 1990). Although it is oolitic in some areas, in the St. Louis, Missouri, area, this unit consists largely of skeletal (foraminifera, crinoid, bryozoan) grains, peloids, and other skeletal debris. Although ooids are not abundant, evidence for tidal currents is found in the presence of ubiquitous current ripples with variable directions (Fig. 20.22a), muddy toes (Fig. 20.22b) or topsets in cross-bed sets, and cross-bedded units with tidal bundles (Fig. 20.22c e.g., Brown et al. 1990).

20.5 Summary and Perspectives

In the clear, shallow, blue-green waters of the Bahamas, tidal sands represent more than a snorkeler's or sailor's paradise. These sands are strongly influenced by tides that form the most spectacular seascapes in the tropics, and they are made exclusively of carbonate grains, with many shoals consisting largely of ooids. These tidal sands are locally derived (and in many cases, formed on the shoals themselves) and, reflecting the dominant physical influences, include a hierarchy of bedforms and barforms. Barforms in these systems include a wide array of geometries and sizes, both within and among shoals, but most shoal complexes include close relations between sediments, hydrodynamics, and morphology due to feedbacks. Ancient tidal sand systems can preserve overall morphology comparable to Holocene analogs, although clearly not all do. Preservation of geomorphic forms is probably enhanced by the early cementation of these systems, but diagenesis can also lead to dissolution of grains and complicated patterns of the distribution of porosity and permeability.

As illustrated above, the sedimentology and geomorphology of many carbonate tidal sand systems in the Bahamas have been extensively documented. In spite of the wealth of study, important gaps remain in

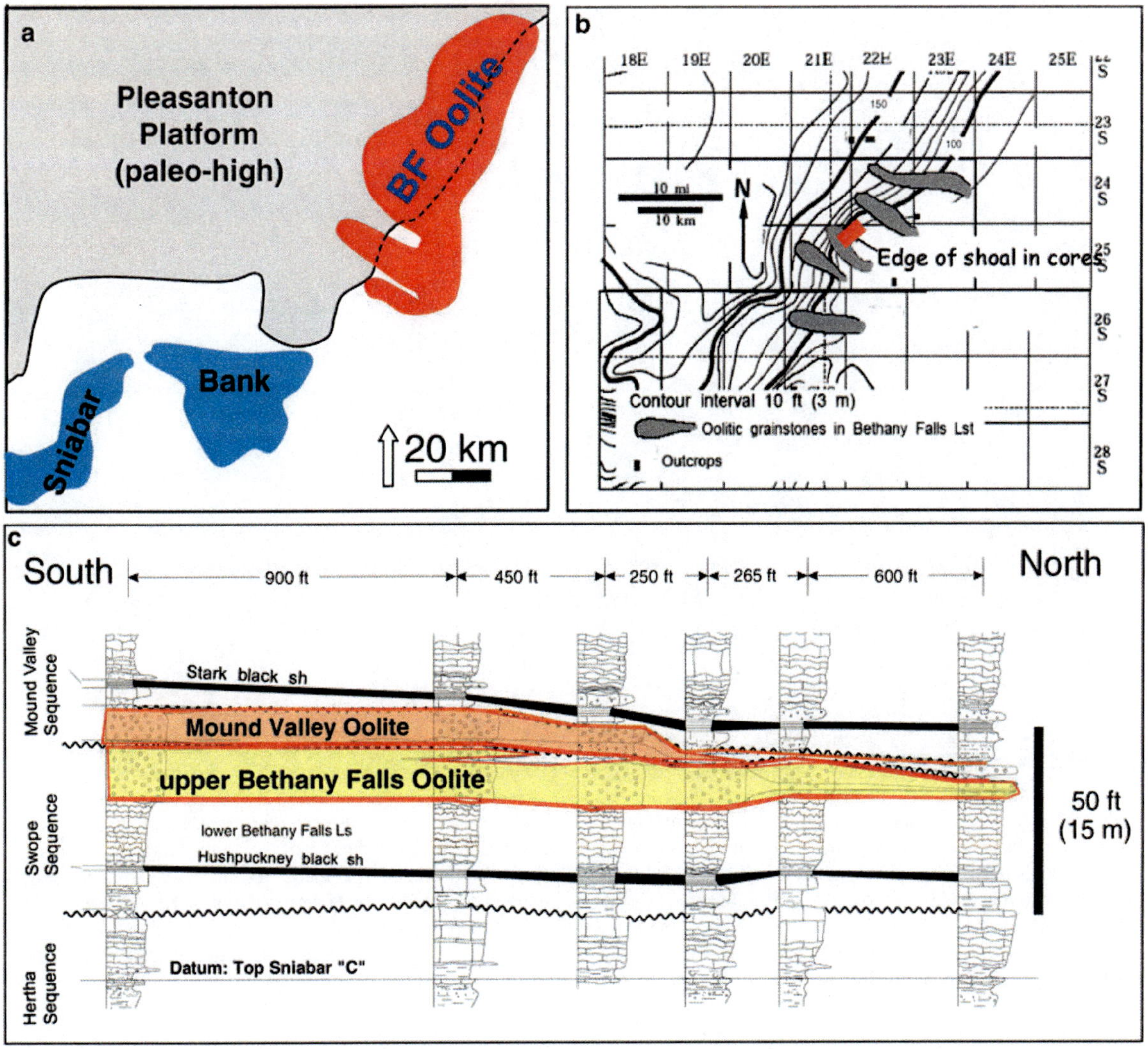

Fig. 20.20 Geomorphic and facies patterns in Pennsylvanian tidal shoal complex, Kansas (modified from French and Watney 1993). (**a**) General setting. The shoal complex (*red*) occurs on the flank of a shallow embayment of the Pleasanton Platform (*blue*). (**b**) Detailed coring and outcrop analysis suggests that the ooids occur in flow-parallel bodies (*grey color*). Line of section in (**c**) is highlighted in the *red box*. (**c**) Stratigraphic cross-section across the shoal complex. The system includes two stacked ooid shoal systems (the upper Bethany Falls Oolite and The Mound Valley Oolite) separated by a subaerial exposure surface

understanding the aspects of carbonate tidal sand systems. Important questions include:

1. What is the internal geometry and heterogeneity present in Holocene ooid shoals? Although data illustrate relationships among geomorphology, sedimentology, and hydrodynamics that are associated with the shoals today, they provide only speculative insights into stratigraphic architecture, facies variability, and depositional geometries. To develop more accurate facies models for ancient successions, the stratigraphic character of a suite of shoals could be examined. Do shoals with distinct geometries have unique stratigraphic signatures, such that a certain succession of sedimentary structures, facies, or granulometric changes could be used to predict the barform geometry or dimensions in ancient analogs?
2. How have oolitic systems evolved through the Holocene? Observations on internal geometries should provide insights into how these systems have

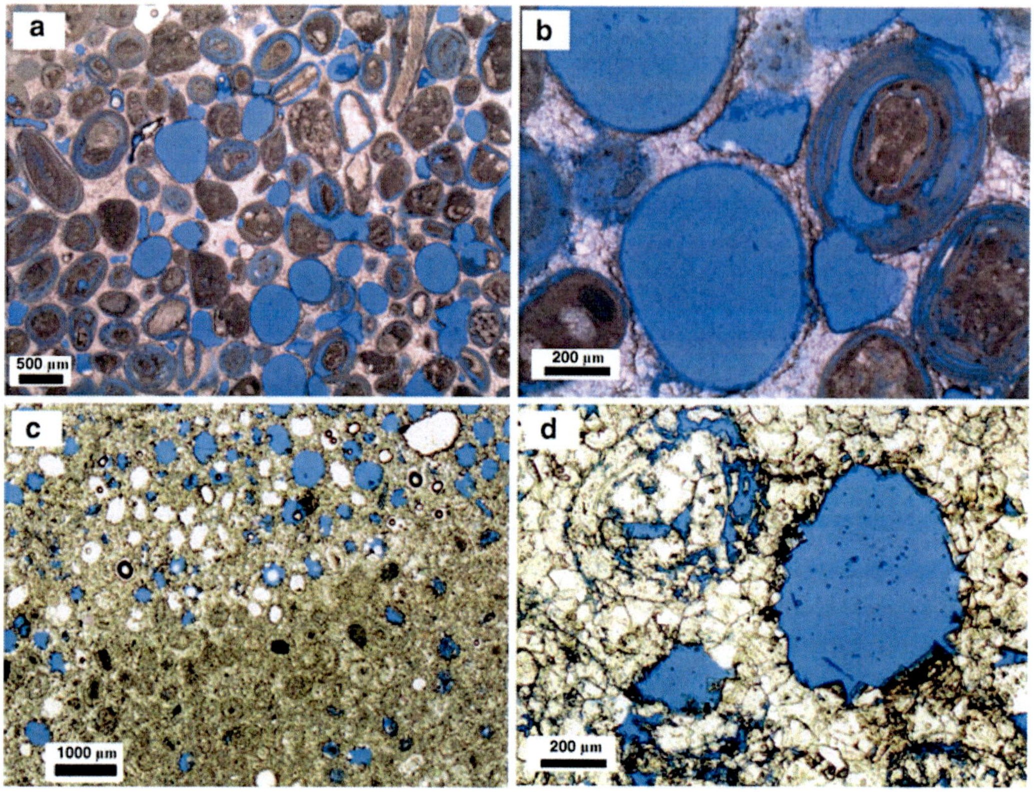

Fig. 20.21 Thin section photomicrographs of ancient oolitic sand. (**a** and **b**) Pleistocene oolite, from Crooked Island, Bahamas. Note how ooids may be completely or partly dissolved, but their form is preserved as molds by the early cement. (**c** and **d**) Oolite from a Pennsylvanian oolite in the Bethany Falls Limestone, Kansas. As in the Pleistocene example, some ooids are completely dissolved, others are only partly dissolved. Unlike the Pleistocene sample, this entire rock has been recrystallized. In this rock, most porosity is in molds, and much of the original interparticle porosity has been occluded by cement. In some examples such as this, although porosity may be high, if the molds are not connected or are connected by only small pore throats, permeabilities may be low (note the white voids in C – these are voids that remained unfilled during vacuum impregnation of the sample)

changed through the Holocene, in response to both local and global change. When did shoals originate? Are all oolitic shoals of the same age? Why have some complexes backstepped onto the platform (Lily Bank), whereas others have aggraded and filled much of the available accommodation (Joulter Cays, Fish Cays), and still others may not have changed during this rise?

3. How does the granulometry of Holocene shoals compare with ancient shoals? Because the detailed stratigraphic records of most Holocene ooid shoals are not well documented, Holocene analogs are most commonly applied in general, qualitative terms. Likewise, the detailed granulometry of ancient analogs are only rarely quantitatively documented. Are relationships between morphology and facies comparable to those documented herein preserved and discernable in the rock record?
4. At a larger scale, all of the examples documented in this chapter are from the tidally-dominated, flat-topped isolated platforms of the Bahamas, with distinct (and geologically unique?) wave conditions, tides, and geochemical settings. How analogous are platform-top Bahamian Holocene examples to many ancient systems, including ramp systems? Is it possible that Bahamian Holocene oolitic systems provide a limited or non-unique suite of actualistic models or incomplete sampling of the possible range of variability?
5. The Bahamas include some of the most expansive accumulations of oolitic sand today. Recent studies (Rankey and Reeder 2009, 2010) have suggested that this is related to the elevated supersaturation

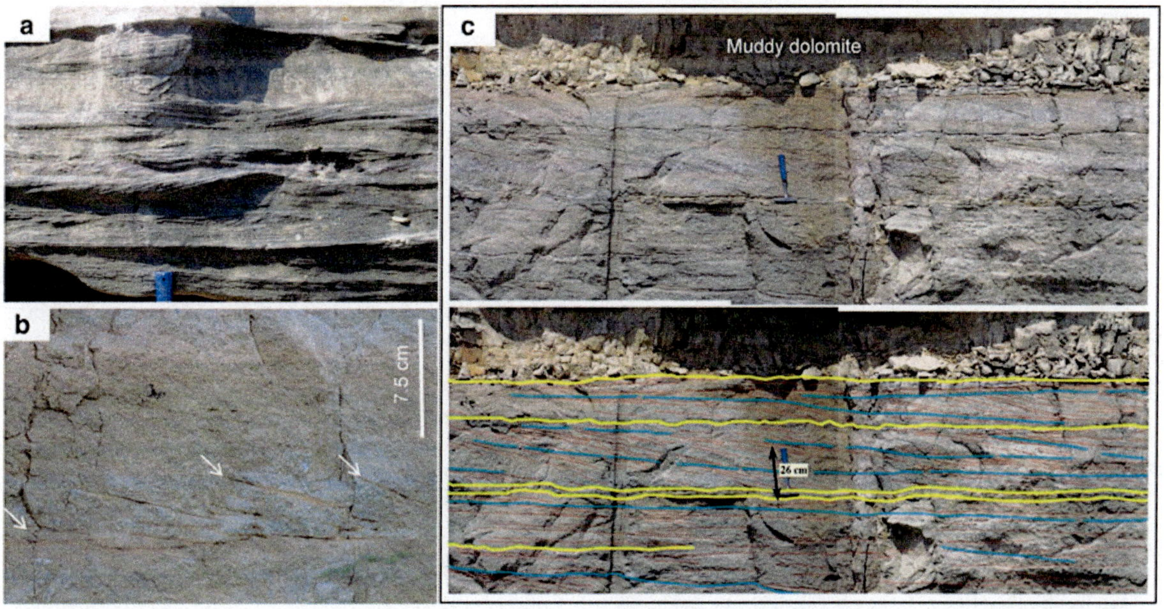

Fig. 20.22 Representative outcrop photos of a non-oolitic tidal sand unit, the Mississippian Salem Limestone, near St. Louis, Missouri. (**a**) Current ripples in a foraminiferal grainstone. Hammer handle in bottom center for scale. (**b**) Muddy bottomsets on small trough cross-bed set. Some muddy layers are highlighted by *arrows*. (**c**) Paired uninterpreted (*upper*) and annotated (*lower*) photos of a cross-bedded interval overlain by a dolomitic muddy unit. The cross-bedded unit includes a set of laminae (some highlighted by *red lines*) stacked into bundles (*blue lines*), some of which may include muddy toesets. Sets of bundles are separated by thin (<4 cm) muddy layers (*yellow lines*)

and the vigorous tides impacting these platforms. Many ancient systems include what appear to be tidal sands but are not oolitic. How do the sedimentology, geomorphology, and diagenetic character of these systems differ from their oolitic relatives?

Clearly, many important and interesting questions regarding tidal sand shoals and their geologic record await future study.

References

Agassiz A (1896) The elevated reef of Florida. Bull Mus Comp Zool 28:1–62

Ashley GM (1990) Classification of large-scale sub-aqueous bedforms: a new look at an old problem. J Sediment Petrol 60:160–172

Aurell M, McNeill DF, Guyomard T, Kindler P (1995) Pleistocene shallowing-upward sequences in New Providence, Bahamas: signature of high-frequency sea-level fluctuations in shallow carbonate platforms. J Sediment Res B65:170–182

Ball MM (1967) Carbonate sand bodies of Florida and the Bahamas. J Sediment Petrol 37:556–591

Bathurst RGC (1967) Oolitic films on low energy carbonate sand grains, Bimini Lagoon, Bahamas. Mar Geol 5:89-109.

Bathurst RGC (1975) Carbonate sediments and their diagenesis, vol 12, Developments in sedimentology. Elsevier, Amsterdam, 658 pp

Beach DK, Ginsburg RN (1980) Facies succession of Pliocene-Pleistocene carbonates, northwestern Great Bahama Bank. Am Assoc Petrol Geol Bull 64:1634–1642

Braithwaite CJR (1973) Settling behaviour related to sieve analysis of skeletal sands. Sedimentology 20:251–263

Brown MA, Archer AA, Kvale EP (1990) Neap-spring tidal cyclicity in laminated carbonate channel-fill deposits and its implications; Salem Limestone (Mississippian), south-central Indiana, U.S.A. J Sediment Res 60:152–159

Budd DA (1984) Freshwater diagenesis of Holocene ooid sands, Schooner Cays, Bahamas. Unpublished Ph.D. dissertation, University of Texas, Austin, 491 p

Carney C, Boardman MR (1993) Trends in sedimentary microfabrics of ooid tidal channels and deltas. In: Rezak R, Lavoie DL (eds) Frontiers in sedimentary geology: carbonate microfabrics. Springer, New York, pp 29–39

Carr DD (1973) Geometry and origin of oolite bodies in the Ste. Genevieve Limestone (Mississippian) in the Illinois basin. Indiana Geol Surv Bulletin 48, 81 p

Caston VND (1972) Linear sand banks in the southern North Sea. Sedimentology 18:63–78

Cavallo LJ, Smosna R (1997) Predicting porosity distribution within oolitic bars in J.A. Kupecz, J. Gluyas, and S. Block (eds.), reservoir quality prediction in sandstones and carbonates. Am Assoc Petrol Geol Mem 69:211–229

Cayeux L (1935) Les Roches Sedimentares de France. Roches Carbonatees Masson, Paris, 463 p

Cruz FE (2008) Processes, patterns and petrophysical heterogeneity of grainstone shoals at Ocean Cay, Western Great Bahama Bank. Unpublished PhD dissertation, University of Miami

Dalrymple RW, Rhodes RN (1995) Estuarine dunes and barforms. In: Perillo GM (ed) Geomorphology and sedimentology of Estuaries, vol 53, Developments in sedimentology. Elsevier, Amsterdam, pp 359–422

Davies PJ, Bubela B, Ferguson J (1978) The formation of ooids. Sedimentology 25:703–729

Deelman JC (1978) Experimental ooids and grapestones: carbonate aggregates and their origin. J Sediment Petrol 48:503–512

Dill RF, Shinn EA, Jones AT, Kelly K, Steinen RP (1986) Giant subtidal stromatolites forming in normal salinity waters. Nature 324:55–58

Dravis J (1979) Rapid and widespread generation of recent oolitic hardgrounds on a high energy Bahamian platform, Eleuthera Bank, Bahamas. J Sediment Petrol 49:195–208

Duguid SMA, Kyser TK, James NP, Rankey EC (2010) Microbes and ooids. J Sediment Res 80:236–251

Dyer K, Huntey DA (1999) The origin, classification, and modeling of sand banks and ridges. Cont Shelf Res 19:1285–1330

Enos P (1974) Surface sediment facies map of the Florida-Bahamas Plateau. Geol Soc Am Map Series MC-5, Boulder, CO, 5 p

Evans CC (1987) The relationship between the topography and internal structure of an ooid shoal sand complex: the upper Pleistocene Miami Limestone, p 18–41. In: Maurasse FJ-MR (ed) Symposium on South Florida geology, Miami Geological Society, Miami, 233 pp

Folk RL, Lynch FL (2001) Organic matter, putative nannobacteria, and the formation of ooids and hardgrounds. Sedimentology 48:215–229

French JA, Watney WL (1993) Stratigraphy and depositional setting of the lower Missourian (Pennsylvanian) Bethany Falls and Mound Valley limestones, analogues for age-equivalent ooid grainstone reservoirs. Kansas Geological Survey Bulletin, Kansas, pp 27–39

Ginsburg RN, Shinn EA (1993) Preferential distribution of reefs in the Florida reef tract: the past is the key to the present, Global Aspects of Coral Reefs: Health, Hazards, and History. University of Miami Press, Miami, pp H21–H26

Gonzalez R, Eberli GP (1997) Sediment transport and sedimentary structures in a carbonate tidal inlet; Lee Stocking Island, Exuma Islands, Bahamas. Sedimentology 44:1015–1030

Grammer M, Crescini CM, McNeill DF, Taylor LH (1999) Quantifying rates of syndepositional marine cementation in deeper platform environments – new insights into a fundamental process. J Sediment Res 69:202–207

Grasmueck M, Weger RJ (2002) 3D GPR reveals complex internal structure of Pleistocene oolitic sandbar. Lead Edge 21:634–639

Halley RB, Evans CC (1983) The Miami Limestone: a guide to selected outcrops and their interpretation. Miami Geological Society, Coral Gables, 67 pp

Halley RB, Harris PM (1979) Fresh-water cementation of a 1,000-year old oolite. J Sediment Petrol 49:969–987

Halley RB, Shinn EA, Hudson JH, Lidz BH (1977) Pleistocene barrier bar seaward of ooid shoal complex near Miami, Florida. Am Assoc Petrol Geol Bull 61:519–526

Halley RB, Harris PM, Hine AC (1983) Bank margin environments. In: Scholle PA, Bebout DG, Moore CH (eds) Carbonate depositional environments. Am Assoc Petrol Geol Mem 33:463–506

Handford CR (1988) Review of carbonate sand-belt deposition of ooid grainstones and application to Mississippian reservoir, Damme Field, southwestern Kansas. Am Assoc Petrol Geol Bull 72:1184–1199

Harris PM (1979) Facies anatomy and diagenesis of a Bahamian ooid shoal, vol VII, Sedimenta. University of Miami, Miami, 163 pp

Harris PM, Weber LJ (eds) (2006) Giant hydrocarbon reservoirs of the world: from rocks to reservoir characterization and modeling. Am Assoc Petrol Geol Memoir 88, 469 p

Hayes MO (1975) Morphology of sand accumulation in estuaries and introduction to the symposium. In: Cronin LE (ed) Estuarine research. Academic, New York, pp 3–22

Hearty PJ, Kindler P (1993) New perspectives on Bahamian geology. J Coast Res 9:577–594

Hillgärtner H, Dupraz C, Hug W (2001) Microbially induced cementation of carbonate sands: are micritic meniscus cements good indicators of vadose diagenesis? Sedimentology 48:117–131

Hine AC (1977) Lily Bank, Bahamas: history of an active oolite sand shoal. J Sediment Petrol 47:1554–1581

Hine AC, Neumann AC (1977) Shallow carbonate-bank-margin growth and structure, Little Bahama Bank, Bahamas. Am Assoc Petrol Geol Bull 61:376–406

Hine AC, Wilber RJ, Neumann AC (1981) Carbonate sand bodies along contrasting shallow bank margins facing open seaways in northern Bahamas. Am Assoc Petrol Geol Bull 65:261–290

Hoffmeister JE, Stockman KW, Multer HG (1967) Miami Limestone of Florida and its recent Bahamian counterpart. Geol Soc Am Bull 78:75–190

Houbolt JJHC (1968) Recent sediments in the Southern Bight of the North Sea. Geol Mijnb 47:245–273

Illing LV (1954) Bahamian calcareous sands. Am Assoc Petrol Geol Bull 38:1–95

Imbrie J, Buchanan H (1965) Sedimentary structures in modern carbonate sands of the Bahamas. In: Middleton GV (ed) Primary sedimentary structures and their hydrodynamic interpretations. Soc Econ Paleontol Mineral Spec Publ 12:149–172

James NP (1983) Reef environment (Chap. 8). In: Scholle PA, Bebout DG, Moore CH (eds) Carbonate depositional environments. Am Assoc Petrol Geol Memoir 33:346–462

Keith BD, Zuppann CW (eds) (1993) Mississippian oolites and modern analogs. American Association of Petroleum Geologists Studies in Geology, Tulsa, 35

Kench PS, McLean RF (1996) Hydraulic characteristics of bioclastic deposits: new possibilities for environmental interpretation using settling velocity fractions. Sedimentology 43:561–570

Lee K, Tong LT, Millero FJ, Sabine CL, Dickson AG, Goyet C, Park G-H, Wanninkhof R, Feely RA, Key RM (2006) Global relationships of total alkalinity with salinity and temperature in surface waters of the world's oceans. Geophys Res Lett 33:L19605. doi:10.1029/2006GL027207

Neal A, Grasmueck M, McNeill DF, Viggiano DA, Eberli GP (2008) Full-resolution 3D radar stratigraphy of complex

oolitic sedimentary architecture: Miami Limestone, Florida, U.S.A. J Sediment Res 78:638–653
Neumann AC, Macintyre I (1985) Reef response to sea level rise: keep-up, catch-up or give-up. In: Proceedings of the 5th international coral reef congress, Tahiti, 27 May–1 June 1985, vol 3, pp 105–110
Neumann CJ, Cry GW, Caso EE, Jarvinen BR (1978) Tropical cyclones of the North Atlantic Ocean, 1871–1977. National Climatic Center, Asheville, 170 p
Newell ND, Rigby JK (1957) Geological studies on the Great Bahama Bank. In: Le Blanc RJ, Breeding JG (eds) Regional aspects of carbonate deposition. Soc Econ Paleontol Miner Spec Publ 5:15–72
Newell ND, Imbrie J, Purdy EG, Thurber DL (1959) Organism communities and bottom facies, Great Bahama Bank. Bull Am Mus Natl Hist 117:177–228
Newell ND, Purdy EG, Imbrie J (1960) Bahamian oolitic sand. J Geol 68:481–497
Off T (1963) Rhythmic linear sand bodies caused by tidal currents. Am Assoc Petrol Geol Bull 47:324–341
Opdyke BN, Wilkinson BH (1990) Paleolatitude distribution of Phanerozoic marine ooids and cements. Palaeogeogr Palaeoclim Palaeoecol 78:1–14
Palmer MS (1979) Holocene facies geometry of the leeward bank margin, Tongue of the Ocean, Bahamas. MS thesis, University of Miami, Miami, FL, 199 p
Prager EJ, Southard JB, Vivoni-Gallart ER (1996) Experiments on the entrainment threshold of well-sorted and poorly sorted carbonate sands. Sedimentology 43:33–40
Purdy EG (1961) Bahamian oolite shoals. In: Peterson JA, Osmond JC (eds) Geometry of sandstone bodies. American Association of Petroleum Geologists Studies in Geology, Tulsa, pp 53–63
Purdy EG (1963) Recent calcium carbonate facies of the Great Bahama Bank. 2. Sedimentary facies. J Geol 71:472–497
Rankey EC, Reeder SL (2009) Holocene ooids of Aitutaki Atoll, Cook Islands, South Pacific. Geology 37:971–974
Rankey EC, Reeder SL (2010) Controls on platform-scale patterns of surface sediments, shallow Holocene platforms, Bahamas. Sedimentology 57:1545–1565
Rankey EC, Reeder SL (2011) Holocene oolitic marine sand complexes of the Bahamas. J Sediment Res 81:97-117
Rankey EC, Riegl B, Steffen K (2006) Form and function in a tidally dominated ooid shoal, Bahamas. Sedimentology 53:1191–1210
Rankey EC, Reeder SL, Correa TBS (2008) Geomorphology and sedimentology of Ambergris ooid shoal, Caicos Platform. In: Morgan WA, Harris PM (eds) Developing models and analogs for isolated carbonate platforms – Holocene and Pleistocene carbonates of Caicos Platform, British West Indies SEPM (Society for Sedimentary Geology) Core workshop, vol 22, pp 127–132
Reeder SL, Rankey EC (2008) Interactions between tidal flows and ooid shoals, northern Bahamas. J Sediment Res 78:175–186
Reeder SL, Rankey EC (2009a) An integrated field, remote sensing, and modeling study examining the impact of Hurricanes Frances and Jeanne on carbonate systems, Bahamas. In: Swart PK, Eberli GP, McKenzie JA (eds) Perspectives in carbonate geology: a tribute to the career of Robert Nathan Ginsburg. IAS special publication 41. Wiley-Blackwell, Oxford, pp 75–90
Reeder SL, Rankey EC (2009b) Controls on morphology and sedimentology of carbonate tidal deltas, Abacos, Bahamas. Mar Geol. http://dx.doi.org/10.1016/j.margeo.2009.09.010
Rich JL (1948) Submarine sedimentary features on Bahama Banks and their bearing on distribution patterns of lenticular oil sands. Am Assoc Petrol Geol Bull 32:767–779
Roberts HH, Rouse LJ Jr, Walker ND, Hudson JH (1982) Cold-water stress in Florida Bay and northern Bahamas – a product of winter cold-air outbreaks. J Sediment Petrol 52:145–155
Roberts HH, Wilson PA, Lugo-Fernandez A (1992) Biologic and geologic responses to physical processes: examples from modern reef systems of the Caribbean-Atlantic region. Cont Shelf Res 12:809–834
Sedimentology Seminar (1966) Cross-bedding in the Salem Limestone of central Indiana. Sedimentology 6:95–114
Smith NP (1995) On long-term net flow over Great Bahama Bank. J Phys Oceanogr 25:679–684
Society R (2005) Ocean acidification due to increasing atmospheric carbon dioxide. The Royal Society, London, pp 1–55, Policy Document 12/05
Sumner DY, Grotzinger JP (1993) Numerical modeling of ooid size and the problem of Neoproterozoic giant ooids. J Sediment Petrol 63:974–982
Taft WH, Arrington F, Haimovitz A, MacDonald C, Woolheater C (1968) Lithification of modern marine carbonate sediments at Yellow Bank, Bahamas. Bull Mar Sci 18:762–828
Tucker ME, Wright VP (1996) Carbonate sedimentology. Blackwell Science, Oxford
Van Veen J (1936) Onderzoekingen in de Hoofden in verband met de gesteldheid previous termvannext term de Nederlandse kust. (Measurements in the Straits of Dover, and their relation to the Netherlands coast). Algemeene Landsdrukkerij, The Hague, 252 pp
Wanless HR, Tedesco LP (1993) Comparison of oolitic sand bodies generated by tidal vs. wind-wave agitation. In: Keith BD, Zuppan ZW (eds) Mississippian oolites and modern analogs. Am Assoc Petrol Geol Stud Geol 35:199–225
Wanless HR, Burton EA, Dravis JJ (1981) Hydrodynamics of carbonate fecal pellets. J Sediment Petrol 51:27–36
Wanless HR, Tedesco LP, Rossinsky V, Dravis JJ (1989) Carbonate environments and sequences of Caicos Platform with an introductory evaluation of South Florida. American Geophysical Union, 28th international Geological congress field trip guidebook T374, 75 pp
Wood LJ (2004) Predicting tidal sand reservoir architecture using data from modern and ancient depositional systems. In: Grammer MG, Harris PM, Eberli GP (eds) Integration of outdrop and modern analogs in reservoir modeling. Am Assoc Petrol Geol Mem 80:45–66

Ancient Carbonate Tidalites

21

Yaghoob Lasemi, Davood Jahani,
Hadi Amin-Rasouli, and Zakaria Lasemi

Abstract

Carbonate tidalites are sediments deposited in supratidal, intertidal and the adjacent shallow subtidal environments by tidal, biogenic, chemical and diagenetic processes and are among the most common deposits in ancient carbonate platform successions. This chapter illustrates sedimentary facies, environments of deposition, and stratigraphy of carbonate tidalites and describes a few analogs from ancient deposits that commonly are encountered in the geological record. Ancient carbonate tidalites consist of a variety of constituents and diagnostic features formed during deposition and early diagenesis in different environments of a tidal system. Peritidal facies are arranged into meter-scale, commonly shallowing-upward succession of subtidal- to tidal flat facies known as parasequence, and may constitute the bulk of the transgressive and highstand packages of a depositional sequence. The geological record of ancient carbonate tidalites indicates deposition in the proximal areas of a tropical sea, particularly during global relative sea level highstands, in carbonate platforms and environments that have recurred many times since the Paleoproterozoic.

21.1 Introduction

Tidalites are sediments deposited by tidal currents and are characterized by a distinct combination of sedimentary structures, textures, lithologies, and vertical successions reflecting various phases of tidal sediment transport in carbonate and siliciclastic tidal flat and shallow subtidal environments (Klein 1971, 1998). The term tidalites is somewhat synonymous with peritidal sediments, sediments formed near the tidal zone (Wright 1984; Flügel 2010; Pratt 2010), which

Y. Lasemi (✉) • Z. Lasemi
Illinois State Geological Survey, Prairie Research Institute, University of Illinois at Urbana-Champaign, Champaign, IL 61820, USA
e-mail: ylasemi@illinois.edu; zlasemi@illinois.edu

D. Jahani
Department of Geology, Faculty of Basic Sciences, North Tehran Branch, Islamic Azad University, Tehran, Iran
e-mail: d_jahani@iau-tnb.ac.ir

H. Amin-Rasouli
Department of Geosciences, University of Kurdistan, Sanandaj, Iran
e-mail: H.Aminrasouli@uok.ac.ir

R.A. Davis, Jr. and R.W. Dalrymple (eds.), *Principles of Tidal Sedimentology*,
DOI 10.1007/978-94-007-0123-6_21, © Springer Science+Business Media B.V. 2012

designate deposits of a tidal flat system comprising supratidal, intertidal, and the adjacent shallow subtidal environments. Carbonate tidalites, formed as a result of tidal, biogenic, chemical and diagenetic processes, are among the most common deposits in ancient tropical carbonate platform successions. They have been deposited under arid or humid conditions in carbonate platforms (Read 1985; Pomar 2001) associated with a variety of sedimentary basins. This chapter first summarizes the diagnostic features and characterizes sedimentary facies, environments of deposition, and stratigraphy of ancient carbonate tidalites. This will be followed by a few illustrative examples of ancient carbonate tidalites related to passive margins, intracratonic, failed rift, and foreland basins.

Our understanding of the deposition and early diagenesis of carbonate tidal deposits grew significantly during the 1960s and 1970s as a result of comprehensive studies of many modern shallow and marginal marine carbonate environments (reviews in Bathurst 1975; Tucker and Wright 1990; Flügel 2010) that included studies in the Persian Gulf (e.g. Purser 1973), the Bahamas (e.g. Hardie 1977), south Florida (e.g. Enos and Perkins 1977) and western Australia (e.g. Logan et al. 1970). The application of these results and observations obtained from modern siliciclastic tidal flat environments (e.g. Reineck 1972) to ancient carbonate deposits (e.g. Ginsburg 1975; Hardie and Shinn 1986; Carozzi 1989), using Walther's Law (Middleton 1973) and the comparative sedimentology approach of Ginsburg (1974), led to accurate interpretations of facies, depositional environments and sequences of carbonate tidal deposits in the sedimentary record. Carbonate tidalites encompass a wide variety of characteristic depositional and diagenetic features analogous to their modern counterparts (e.g., Grotzinger 1989; Demicco and Hardie 1994, Flügel 2010). These features record important information on water level and tidal range, depth and energy level, salinity, climate and sea level history during deposition.

21.2 Tidal Processes

Tide is a periodic fluctuation in water level in a marine realm that is created by the gravitational pull of the moon and sun (Davis 1983; Dalrymple 1992) with the moon, being closer to earth, exerting the most gravitational force. In a tidal cycle, rising water generates the landward flood tidal current, but the fall generates the seaward and normally weaker ebb tidal current. A tidal cycle normally occurs twice daily (semidiurnal), but once a day (diurnal) or mixed diurnal and semidiurnal cycles may occur depending on tidal regime or local conditions (Davis 1983). The height of the water column between normal high tide and low tide levels is known as tidal range. During new and full moon, alignment of the moon, sun and earth generates a greater than normal tidal range (spring tide); conversely, during first and third quarters of a lunar cycle, a minimal tidal range (neap tide) occurs (Davis 1983). Based on variation of tidal range, shorelines are classified into macrotidal (>4 m), mesotidal (2–4 m) and microtidal (<2 m). In macrotidal areas, tide dominates over other processes and most mesotidal and microtidal areas are wave and storm dominated, but tide domination may even occur in a protected microtidal coast where wave action is limited (Dalrymple 1992). In contrast to their siliciclastic counterparts, most modern peritidal carbonate environments are microtidal (Wright 1984; Pratt et al. 1992).

21.3 Depositional Environments

The daily fluctuation of water level subdivides the tidal system into three bathymetric belts including subtidal (below mean low tide level), intertidal and supratidal environments (Fig. 21.1a) that are characterized by a set of biological, physical and chemical processes. The subtidal environment may extend for hundreds of kilometers offshore depending on tectonic and geographic settings. It is the major source of sediment for the adjacent tidal flat setting and comprises low energy back barrier lagoons of various salinity and areal extent, platform margin (carbonate barriers/sand shoals) and open marine environments (Fig. 21.1). The intertidal environment lies between mean low tide and mean high tide levels (Fig. 21.1a) and is exposed once or twice daily during each tidal cycle. This zone in part includes isolated ponds and meandering tidal channels that are essentially subtidal environments occurring within the intertidal belt (Fig. 21.2). The supratidal environment lies above the mean high tide level (Fig. 21.1a) and is flooded only during spring tides (twice each month) and less frequent storm tides. It is widespread in mainland coasts, but narrow supratidal environment develops on channel levees and beach ridges within the intertidal belt (Fig. 21.2b). In an arid climate, the supratidal environment is evaporitic and is

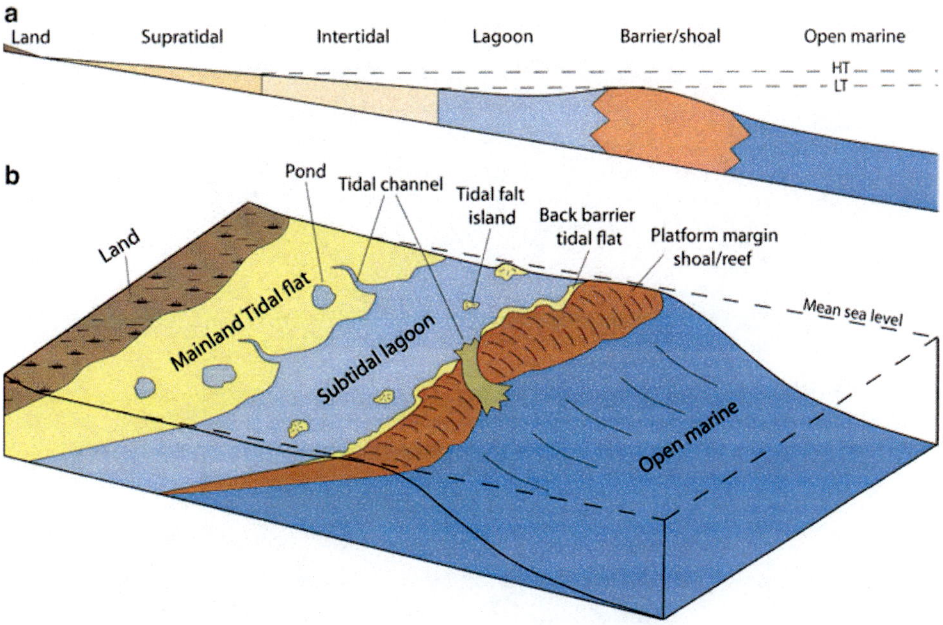

Fig. 21.1 Carbonate platform paleogeography and various depositional environments of a tidal system. (**a**) A gently sloping carbonate ramp profile showing depositional environments and their relationship to sea level (**b**) Block diagram of a carbonate shelf showing various areas of tidal flat development. Note that platform margin barrier/shoal and tidal flat environments are dissected by tidal channels. Abbreviations: *HT* high tide, *LT* low tide

known as "sabkha" (Fig. 21.2a) named after the evaporitic supratidal flats of the southern part of the Persian Gulf. In humid climate, supratidal flat is characterized by an extensive freshwater marsh (Fig. 21.2b).

Tidal flats normally develop in shorelines protected from waves and fluvial-deltaic influence and are the most extensive in mainland coasts, but narrower tidal flats occur in the back of islands or carbonate barriers/shoals at the platform margin (Fig. 21.1b). In ancient carbonate platforms, the back barrier tidal flats could have been quite extensive (see the Precambrian tidalites in Sect. 21.7.1). Ancient carbonate tidal flats could have also developed on low-relief supratidal islands and intertidal banks surrounded by subtidal environment (Pratt and James 1986; Pratt 2010). On a windward-facing tidal flat, beach ridges at the seaward edge of intertidal flat (Fig. 21.2b) or distinct barrier islands separated from the intertidal flats by a subtidal lagoon of variable width (Fig. 21.2a), analogous to siliciclastic barrier islands, may develop (Shinn 1986). In unprotected coasts exposed to high energy waves, such as the eastern and western parts of the Persian Gulf Abu Dhabi Embayment and the windward northern side of the Persian Gulf barrier islands (Purser and Evans 1973), tidal flats are not well developed. In these coasts, the shoreline is covered by high energy beach environment. In cool-water settings, too, tidal flats are scarce and the shoreline deposits are characterized by high energy sand- to gravel-size carbonate beach facies commonly backed by carbonate aeolianites (James 1997).

21.4 Processes of Sedimentation

Carbonate sediments form *in situ*, mainly by carbonate secreting organisms in the subtidal environments (e.g. Wilson 1975; Flügel 2010); part of the subtidal carbonate sediment is transported landward by storms and tidal currents and deposited in the intertidal and supratidal environments of the tidal flat system. In the subtidal lagoon, current energy and grain size normally decreases seaward toward the deeper outer shelf-lagoon. In this setting, quiet water condition results in accumulation of lime mud of biological (e.g. Robbins and Blackwelder 1992; Pratt 2001) and possibly chemical origin (Shinn et al. 1989), which may be stabilized by sessile organisms, such as seagrass and calcareous algae. In the subtidal and adjacent intertidal settings, bioturbation by burrowing organisms (except for extreme conditions, such as hypersalinity

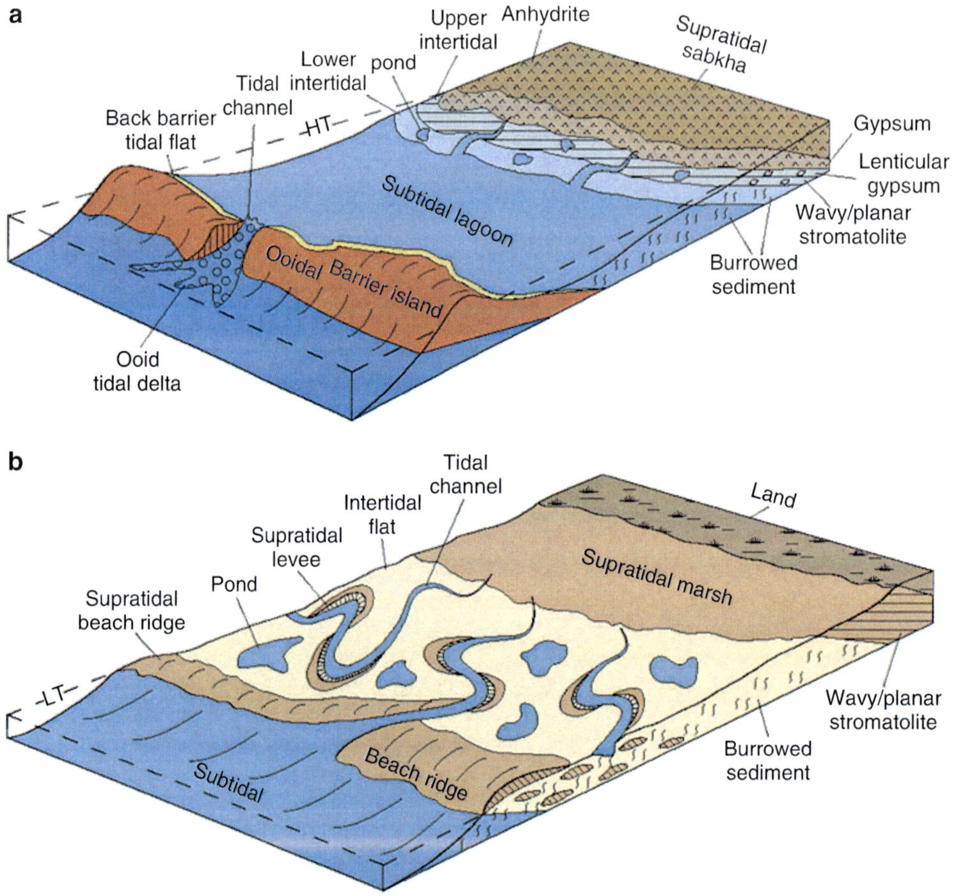

Fig. 21.2 Diagrams showing major depositional settings of modern arid (Persian Gulf) and humid (Bahamas) tidal systems (Modified from Shinn 1983a). (**a**) Major subtidal and tidal flat facies and sedimentary environments of the southern Persian Gulf inner ramp setting. An ooidal barrier island separates the open sea from the quiet back barrier subtidal lagoon. Note the development of ebb ooid tidal delta and a narrow tidal flat in the back of the barrier island. Note also the presence of bioturbated sediment in the lower intertidal, lenticular gypsum bearing planar/wavy stromatolite in the upper intertidal and gypsum/anhydrite deposits in the supratidal zone. (**b**) Major facies and environments of the Andros Island tidal flat system. Supratidal zones are shown in *light brown*. Note the burrowed deposits in the intertidal and planar/wavy stromatolite in the supratidal freshwater marsh environments, respectively. Note also the presence of beach ridge at the seaward edge of intertidal zone, and the supratidal areas on the beach ridge and tidal channel levees. Abbreviations: *HT* high tide, *LT* low tide

or strong current energy) may destroy primary fabrics and structures of sediments; intensity of bioturbation and skeletal diversity decrease with increasing salinity. In various environments of a tidal flat system, current energy and grain size decreases in a landward direction and various processes, such as desiccation, cementation and dolomitization operate. In various environments of a tidal system, binding and trapping of sediment and carbonate precipitation by bacteria and cyanobacteria lead to microbial deposits (see the following section). In the tidal flat setting, evaporation results in higher salinity leading to cementation and primary micritic dolomite formation. Early cementation prevents compaction of soft and muddy sediments, leading to preservation of peloids, intraclasts and primary sedimentary fabrics and structures.

The high energy conditions of the platform margin result in the development of barrier reefs and/or carbonate sand shoals, which separate the often restricted back-barrier environments from the open sea. As a result of tidal current activity, the margin is normally dissected by tidal channels that connect the open-ocean water with that of the quiet back-barrier subtidal environment. Depending on tidal regime, the mouths of these channels may develop tidal deltas. In the Abu Dhabi region of southern Persian Gulf, for example, ebb ooid tidal deltas are developed seaward of tidal channels (Fig. 21.2a) that cut through the wave-formed oolitic barrier island complex (Purser and Evans 1973).

21.5 Facies Belts and Their Diagnostic Features

Periodic changes in the direction and speed of tidal currents, differences in environmental conditions within the inner shelf, biogenic activity of organisms, and intermittent exposure of the proximal areas of a tropical carbonate platform result in various facies in different environments of a tidal system. These facies are characterized by a variety of constituents and diagnostic features formed during deposition and early diagenesis. Tidal flat deposits commonly consist of thin bedded and laminated lime mudstones/dolomudstones or microbial laminae known as laminites (Fig. 21.3a–c). They are tan to light brownish gray in the field as opposed to gray-colored subtidal facies formed under a relatively reducing condition (Fig. 21.3d). Some tidal flat deposits consist of mixed carbonate and clay- to sand-sized siliciclastics indicating proximity to aeolian or fluvial/deltaic systems or intermittent advance of siliciclastics from a nearby upland area during floods (see the Middle Cambrian tidalites in Sect. 21.7.2). Tidal flat and subtidal facies are interlayered with various erosive-based, graded intraclastic storm deposits ("tempestites") in some stratigraphic intervals (e.g., Wignall and Twitchett 1999, Y. Lasemi et al. 2008) recording deposition in a storm dominated platform (see the Middle Cambrian and the Lower Triassic tidalites in Sects. 21.7.2 and 21.7.5.1). Carbonate storm beds commonly consist of intraclasts (edgewise conglomerate) of various facies (Fig. 21.4a, b) and/or ooids, peloids and bioclasts of mixed fauna and normally display hummocky cross-stratification, lamination, and gutter casts (Fig. 21.4c, d). Repeated storm events may partially or completely remove the previously deposited tidal sediments leading to stacked storm deposits (Fig. 21.4 c, d).

21.5.1 Subtidal Belt

The quiet water, back-barrier lagoon facies generally are muddy and consist of characteristic gray and bioturbated mudstone to packstone texture (Figs. 21.3d and 21.5a, b). These sediments generally contain microbial pisoids known as oncoids (see below), peloids, low diversity skeletal components including gastropods, ostracods, green algae, and benthic foraminifera or abundant numbers of a certain individual biota depending on climate and salinity (Fig. 21.5c, d). Open marine and restricted subtidal lagoon sediments may comprise microbial structures known as microbialites.

Microbialites are organosedimentary deposits formed by interactions between biological, environmental and diagenetic processes as a result of benthic microbial organisms that trap and bind sediment and/or form the locus of calcium carbonate precipitation (Burne and Moore 1987). Trapping and binding mechanisms and early diagenetic calcification and/or precipitation by bacteria appear to be the major processes for the formation and preservation of microbialites in modern marine subtidal and tidal flat environments (e.g., Riding 2000; Reid et al. 2000; Dupraz et al. 2009). Marine microbialites include the non-laminated structures with macroscopic, dark-colored clotted fabric (Fig. 21.6a) referred to as thrombolites and laminated forms (Fig. 21.6b, f) known as stromatolites (e.g., Pratt and James 1982; Riding 1999, 2000) commonly constructed by filamentous calcified (Fig. 21.6c) and non-calcified (Fig. 21.6d, f) bacteria. Microbialites represent arid upper intertidal (Fig. 21.2a), humid supratidal marsh (Fig. 21.2b), hypersaline subtidal- to intertidal and normal marine platform margin environments.

Stromatolite pisoidal (Figs. 21.5d and 21.6c, d) and columnar microbialite structures (Fig. 21.6b, e) forms represent both modern and ancient open marine and restricted subtidal sediments as old as 3.5 Ga (e.g. Grotzinger 1989; Riding 2000), but thrombolites are normally found in Phanerozoic subtidal deposits. Sediment of a high energy restricted lagoon may contain columnar stromatolite and ooids/skeletal grains derived from the nearby platform margin (see The Precambrian and Middle Cambrian tidalites in Sects. 21.7.1 and 21.7.2). Laterally extensive interlayered gypsum/anhydrite-bearing dolomudstone/limestone and layered nodular anhydrite with no evidence of subaerial exposure (Fig. 21.7) may record deposition in a subtidal lagoon environment where the water became increasingly saline due to poor circulation.

The upper subtidal sediment in high energy exposed coasts typically consists of beach facies characterized by flat-laminated to cross-bedded grainstone facies. These strata may contain skeletal grains and/or ooids,

Fig. 21.3 (**a–c**) Laminated tidal flat deposits and (**d**) intertidal and lagoonal deposits: (**a**) Thin section photograph showing interlamination of mud and fine sand- to silt-size carbonate. Note cement-filled desiccation crack in the *lower left*. Thickness variation of the laminae may reflect daily variation of tidal range (neap-spring cycles); Cave Hill Member of the Mississippian Kinkaid Formation in the Buncombe Quarry, southern Illinois. (**b**) Mud-cracked laminated tidal flat facies in the Middle Devonian Vernon Fork Member of the Jeffersonville Limestone, southwest Indiana (Photo courtesy of Dr. B.D. Keith, Indiana geological Survey). (**c**) Thin section photograph of interlaminated planar- to wavy stromatolite (*darker laminae*) and dolomudstone. The graded intraclastic upper lamina was formed by a storm tide; Cave Hill Member of the Mississippian Kinkaid Formation in the Kinkaid Creek section, southern Illinois. (**d**) Field photograph of a succession composed of *bluish gray* subtidal limestone overlain by *light grayish brown* to tan intertidal dolomudstone. The contact between the subtidal and intertidal facies appears to be sharp; Lower Triassic lower member of the Elika Formation, Alborz Mountains, northern Iran

peloids and intraclasts (Fig. 21.8a) and may grade landward into a narrow belt of lower energy intertidal (foreshore) facies. The tidal channel deposits at the platform margin may consist of laminated horizontal strata and/or cross-bedded (sometimes bidirectional) gravel- and sand-size intraclastic ooid/bioclast grainstones (Fig. 21.8b) and/or columnar microbialites.

21.5.2 Intertidal Belt

The intertidal belt is flooded and exposed once or twice daily and consists of various facies with diagnostic features. The arid upper intertidal sediment commonly consists of mud-cracked laminated facies deposited by tidal currents (Fig. 21.3a, b) and/or planar to wavy

Fig. 21.4 Tidal flat storm deposits: (**a**) Photomicrograph of a thin section showing graded tidal flat storm bed composed of intraclasts (mainly mud-cracked stromatolite fragments), peloids and mixed fauna (the majority of the light grains are echinoderm debris); Cave Hill Member of the Mississippian Kinkaid Formation, Buncombe Quarry, southern Illinois. (**b**) Field photograph of a tidal flat succession intercalated by an erosive-based intraclast grainstone storm bed. The storm bed consists of yellow tidal flat dolomudstone; Lower Triassic lower member of the Elika Formation, central Alborz Mountains, northern Iran (magic marker is 14.5 cm long). (**c**) Stacked fining-upward erosive-based storm beds (*black arrows* at the basal erosional contacts) with intraclasts of subtidal/intertidal facies and hummocky cross-stratification (*white arrow*) (*coin* diameter is 2.5 cm). (**d**) *Reddish brown* very thin-bedded and laminated intertidal deposit intercalated by several erosive based intraclastic storm layers. Note the large gutter cast (*arrow*) that has cut through the underlying intertidal facies (above the hammer in the *lower right*) (Photo courtesy of Dr. M. Ghomashi, Sistan-Baluchistan University, Zahedan, Iran); Lower Triassic lower member of the Elika Formation, Bibishahrbano Mountain, northern Iran

stromatolite (Figs. 21.3c, 21.6f and 21.9d) (see the Triassic tidalites in Sect. 21.7.5 for an example). The lower intertidal environment adjacent to a hypersaline subtidal setting may comprise domal stromatolite structure (Fig. 21.9a–c) changing to wavy and planar stromatolite in a landward direction (see the Precambrian and Middle Cambrian tidalites in Sects. 21.7.1 and 21.7.2, respectively). Other features common in arid

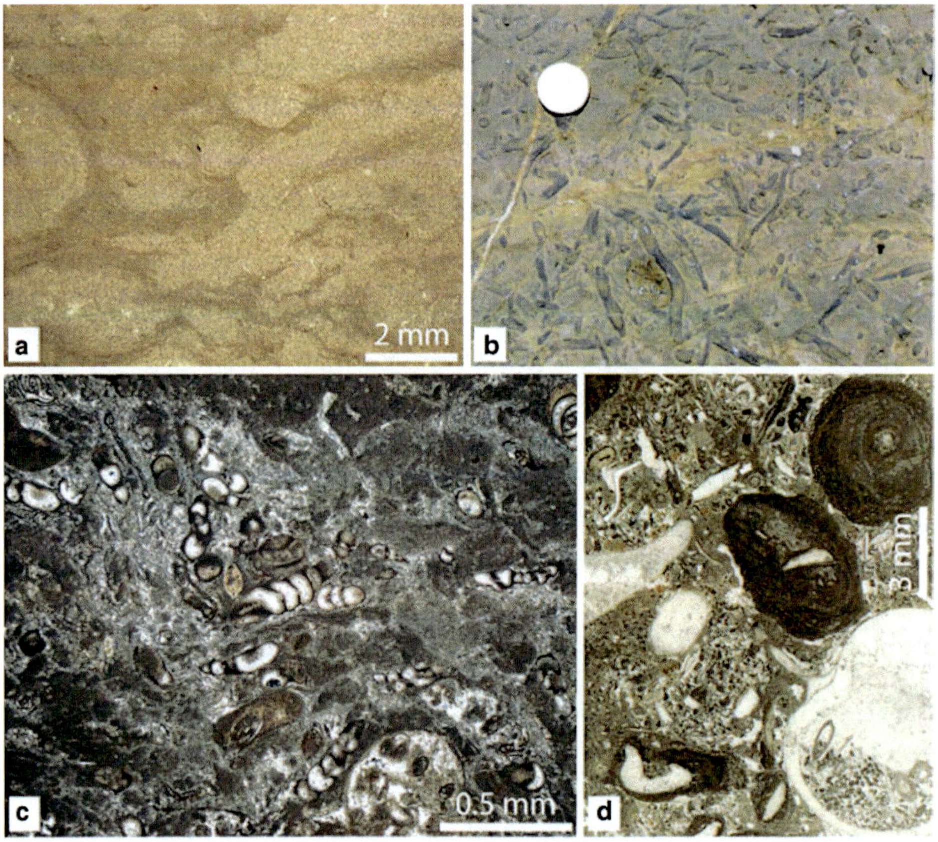

Fig. 21.5 Subtidal lagoon facies: (**a–b**) Heavily bioturbated lime mudstone subtidal facies in the Lower Triassic lower member of the Elika Formation of the Alborz Mountains, northern Iran. (**a**) Photomicrograph of a vertically oriented thin section showing abundant horizontal burrows in a lime mudstone (**b**) Upper surface of a heavily bioturbated subtidal lime mudstone bed (*coin* diameter is 2.5 cm). (**c**) Photomicrograph of a subtidal lagoon facies composed of peloids, intraclasts and restricted fauna (mainly dentritinid forams). Note a miliolid foram in the *upper right*; Lower Miocene Asmari Formation, Zagros Mountains, southwest Iran. (**d**) Thin section photograph of a peloidal bioclast oncoid packstone. Note the nearly concentric oncoids (pisoidal form of microbialites) in the *upper right*. Note also the geopetally filled gastropod shell mold in the *lower right* of the photograph; Negli Creek Member of the Mississippian Kinkaid Formation, southern Illinois

intertidal sediments include tepee structures (Fig. 21.10a–c) and gypsum crystals or their pseudomorphs (Figs. 21.9d and 21.10d, e).

Tepees as defined by Adams and Frenzel (1950) are structures having an inverted V-shaped profile similar to the American Indians tents. However, this form of tepee is only occasionally present and it normally appears as irregular and low ridges (Pratt 2002). Tepee structures (Fig. 21.10a–c) are common to peritidal deposits and form as a result of desiccation, cementation and crystal growth, thermal expansion, and contraction of partially lithified sediment in arid tidal flat or high energy shallow subtidal sediments (Kendall and Warren 1987). They reflect the polygonal antiform ridges arched along polygonal cracks in a cemented surface crust as seen in plan view (Demicco and Hardie 1994). In high energy subtidal settings, cementation of carbonate grainstone layers can lead to expansion and development of centimeter-scale to giant polygonal cracks that may be folded or thrusted at the margins forming tepee structures (e.g. Kendall and Warren 1987; Lokier and Steuber 2008). Tepees may also form as a result of brecciation of lithified sediment, regardless of their depositional setting, by syndepositional fault movement and subsequent cementation (Pratt 2002).

The lower intertidal sediment in high energy tide-dominated coasts, commonly consists of planar laminated and/or cross-bedded bioclast/peloid intraclast grainstone facies that grade landward to heterolithic

Fig. 21.6 Subtidal (**a** through **e**) and intertidal microbialites: (**a**) Shallow subtidal dome-shaped thrombolite structures with dark microbial clots and dolomite-field fenestral voids; Middle Cambrian member 1 of the Mila Formation, east central Alborz Mountains in northern Iran. (**b**) A branching columnar stromatolite from the Silurian deposits of the southern Tabas Basin, east central Iran. (**c**) Photomicrograph under normal light of a part of pisoidal form of stromatolite (*oncoid*) formed by calcified tubular cyanobacteria (*Girvanella*); Negli Creek Member, Mississippian Kinkaid Formation, east of Princeton, Kentucky. (**d**) Thin section photograph of a part of an oncoid in a peloidal bioclast oncoid packstone. The oncoid appears to have been formed by non-calcified cyanobacteria; Negli Creek Member, Mississippian Kinkaid Formation, southern Illinois. (**e**) A compound subtidal columnar stromatolite complex from the Middle Cambrian member 2 of the Mila Formation in central Alborz Mountains, northern Iran. (**f**) Photomicrograph of flat- to wavy-laminated stromatolite with cement-filled planar birdseyes (see Sect. 21.5.2) and molds of filamentous cyanobacteria (*arrows*); Middle Triassic middle member of the Elika Formation in the Alborz Mountains, northern Iran

layers of sand- to coarse silt-sized and mud-size sediments showing flaser to wavy and lenticular bedding. This millimeter- to centimeter-scale interlayering, referred to as rhythmites (Reineck and Singh 1980) and tidal bedding or heterolithic stratification (Demicco and Hardie 1994) form due to declining tidal current energy and the resulting change in sand to mud ratio in a landward direction, and represent deposition in the

Fig. 21.7 (**a**) Cycles consisting of gypsum-anhydrite-bearing limestone (*gray*) and anhydrite (*white*) with chicken wire fabric. (**b–c**) Photomicrographs of the gray limestone portion in (**a**) under normal light (**b**) and under polarized light (**c**), showing scattered crystals and lenticular forms of gypsum and anhydrite in a lime mudstone groundmass. Absence of subaerial exposure features indicates deposition in a subtidal lagoon or in an intertidal pond settings; Upper Jurassic Mozduran Formation in the Kopet Dagh Basin, northeast Iran

lower to middle intertidal settings. Such stratification is common in ancient carbonate tidal deposits (Figs. 21.11 and 21.12a–c) where the sand-sized layers may consist of quartz sandstone (e.g. Y. Lasemi 1986, Y. Lasemi et al. 2008; Ghomashi 2008) or grainstone (e.g., Demicco 1983; Amin-Rasouli 1999) indicating deposition by high energy ebb or flood tidal currents. The muddy dolomudstone or lime mudstone layers represent deposition by the waning current during the slack water period of a tidal cycle. It may be mudcracked (Fig. 21.12a) or bioturbated by organisms leaving vertical to sub-horizontal traces (Fig. 21.12c). Periodic pumping of water by tidal currents through lowermost intertidal carbonate sands and subsequent evaporation, results in cementation and formation of a lithified crust known as beachrock (Scoffin and Stoddard 1983) (Fig. 21.13a, b).

A feature diagnostic to tidal flat environments is birdseyes (fenestral fabric), which are millimeter-size irregular voids and occur in stromatolite structures and carbonates ranging from grainstone to mudstone in texture (Figs. 21.6f, 21.14 and 21.15). They are commonly filled with cement (Figs. 21.6f and 21.15) or geopetal internal sediment (Fig. 21.14a). Birdseyes commonly form as a result of air or gas bubble formation, desiccation shrinkage, wrinkles in the laminated bacterial deposits or development of trapped air bubbles in the pore spaces during flood tide and subsequent rapid cementation (Shinn 1983b, 1986).

The lower intertidal/beach ridge facies of mainland coasts with high salinity conditions are normally characterized by packstone-grainstone facies containing peloids, intraclasts and/or a restricted range of bioclasts, commonly small gastropods (Fig. 21.13b).

Fig. 21.8 High energy beach and platform margin tidal channel facies: (**a**) Photomicrograph of a bioclastic ooid intraclast grainstone under polarized light; Mississippian upper Salem Limestone (lower St. Louis Limestone equivalent), southwestern Illinois. (**b**) A fining-upward ramp margin tidal channel deposit (*lower arrow* at the basal erosional contact) within a predominantly ooid grainstone sequence. Note that facies of this channel deposit include flat-bedded and laminated gravel- to sand-sized ooid intraclast grainstone overlain by herringbone cross-bedded grainstone, which in turn grades to laminated ooid grainstone. The channel deposit is overlain, with a sharp contact (*upper arrow*), by rocks of a transgressive open marine mudstone facies; Mississippian Ste. Genevieve Limestone, Alton Bluff section, Madison County, southwest Illinois

These sediments, similar to a beach facies (Inden and Moore 1983), commonly are characterized by fenestral grainstone facies (e.g. Y. Lasemi 1995) (Figs. 21.14 and 21.15c). Prolonged exposure in arid to semiarid climate leads to partial dissolution of carbonate grains by meteoric water, pore enlargement, and formation of iron oxide stained coated grains (Fig. 21.14c).

In protected modern and ancient Phanerozoic Humid intertidal deposits, lamination and other diagnostic features are absent due to intense bioturbation by metazoans (see the Ordovician example in Sect. 21.7.3). In arid climate similar to the southern part of the Persian Gulf (Fig. 21.2a), however, due to high salinity and evaporite formation, burrowers and browsers are practically absent in the upper intertidal deposits, whereas, the lower intertidal facies are intensely bioturbated. These deposits, except for their lower fossil diversity and lighter color due to more oxidizing condition, are similar to the adjacent subtidal deposits (Shinn 1983a).

The low energy and isolated intertidal pond facies is thin bedded and laterally discontinuous. In a humid condition, it is characterized by thin beds of bioturbated lime mudstone to wackestone containing restricted-bioclast. In an arid climate, the pond facies may consist of planar stromatolite (Fig. 21.16a) and/or evaporitic dolomudstone, depending on geographical location and salinity. Intertidal channel point bar deposits are characterized by erosive based fining-upward gravel- and sand-size sediment (may be laminated and/or herringbone cross-bedded) skeletal intraclast/peloid grainstone to packstone facies. They are laterally discontinuous and normally are capped by intertidal deposits (Fig. 21.16b, c).

21.5.3 Supratidal Belt

Sediment of the supratidal environment is transported from the subtidal carbonate factory during spring and storm tides. Because most of the sediment is carried by storms that transport large quantities of sediment, carbonate facies of the supratidal belt are intraclastic and commonly are characterized by thick laminae and thin to very thin beds (Figs. 21.17a, b, d and 21.18c). Periodic exposure of muddy tidal flat deposits particularly in the upper intertidal and supratidal settings results in desiccation and the formation of mud cracks and mud polygons (Figs. 21.3a, b, 21.9d, 21.12a, and 21.17). These cracks are strikingly different than diastasis cracks (Cowan and James 1992) that result from differential mechanical behavior of stiff muddy sediment, under stress, in any subtidal environment. Desiccation cracks may show a variety of sizes depending on exposure time, layer thickness, and the presence or absence of microbial mats (Shinn 1986). Mud cracks in carbonates may develop a variety of cross-sectional shapes ranging from classical "v" to wide and parallel-walled or deep

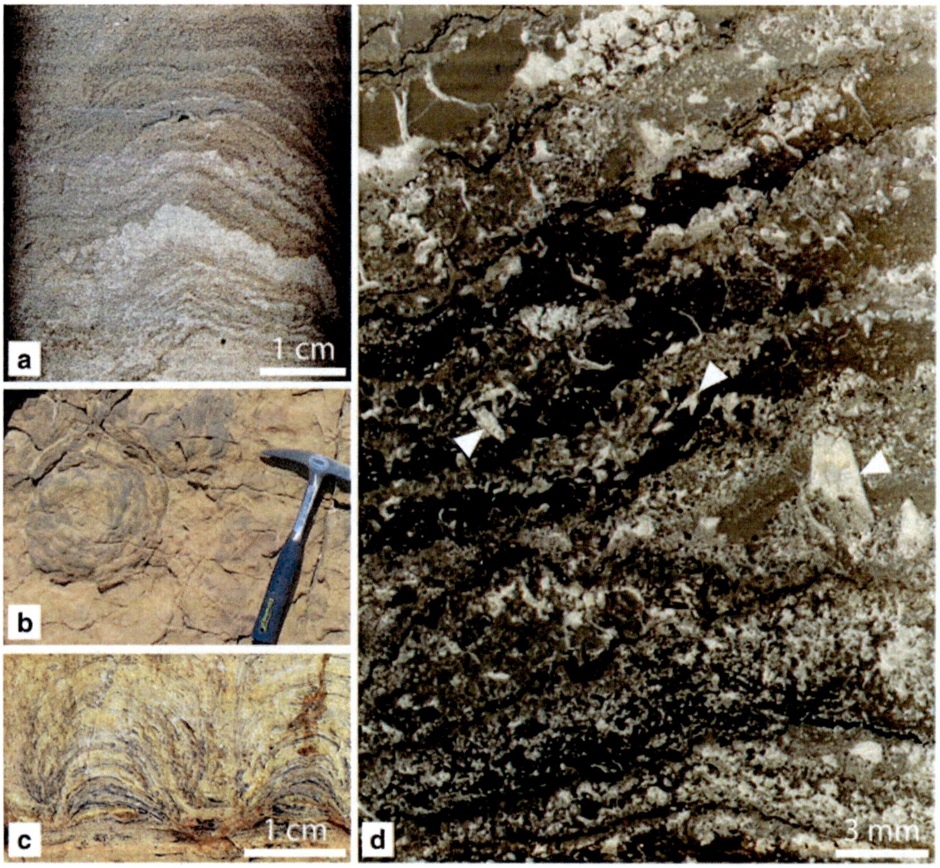

Fig. 21.9 Intertidal stromatolites: (**a**) Core photograph from a dolomitized wavy to domal intertidal stromatolite; Middle Devonian Grand Tower Limestone, Tuscola Quarry, Douglas County, east-central Illinois. (**b–c**) Photographs from the upper bedding plane (**b**) and vertical section (**c**) of a domal lower intertidal stromatolite; Middle Cambrian member 2 of the Mila Formation in central Alborz Mountains, northern Iran. (**d**) Thin section photograph showing mud-cracked wavy stromatolite with calcite pseudomorphs after gypsum crystals (*arrows*). Note the disrupted laminae throughout the sample as a result of desiccation and crystal growth. Note also the geopetal vadose sediment inside a gypsum crystal mold in the middle right; Cave Hill Member of the Mississippian Kinkaid Formation, western Kentucky

and narrow shapes. These cracks typically are filled with sediment or carbonate cement (Demicco and Hardie 1994) and their walls are not generally smooth (Figs. 21.3a, b, 21.9d, 21.12a and 21.17a, b, d). In the supratidal facies, as a result of prolong exposure, desiccation related polygonal mud cracks (Figs. 21.3b and 21.17) are larger than those found in the arid upper intertidal belt and on supratidal beach ridges and levees.

Modern and ancient arid supratidal deposits generally comprise dolomudstone which may be interbedded with gypsum/anhydrite and collapse breccias (see the Mississippian, Middle Triassic and Miocne examples in Sects. 21.7.4, 21.7.5.2 and 21.7.6). The dolomudstone layers are commonly fenestral and may contain lenses, nodules or rosettes of gypsum/anhydrite or their pseudomorphs (Figs. 21.15a, b and 21.18). Halite crystals cast (Fig. 21.19c) may be present, but are not always indicative of supratidal conditions (see the Middle Cambrian tidalites in Sect. 21.7.2). In arid supratidal sediment, microbial boundstone is normally absent or occurs as thin crinkly and discontinuous laminae (Fig. 21.17b) due to dry climate, prolonged exposure and wind energy as opposed to humid supratidal setting. The supratidal freshwater marsh facies in humid climates, such as the rainy tidal flats of the Bahamas (Fig. 21.2b), on the other hand, is characterized by interlayered planar to wavy laminated

Fig. 21.10 Tepee structures and calcite pseudomorphs after lenticular gypsum: (**a**) Tepee structure in thin-bedded and laminated rocks of the tidal flat facies; Middle Triassic member 2, Elika Formation, eastern Alborz Mountains, northern Iran (scale is 14 cm long). (**b**) A tepee structure (center of the photograph) in a tidal flat succession; lower part of the Mississippian lower St. Louis Limestone at Bussen Quarry, St. Louis County, Missouri. The height of the teepee is about 50 cm. (**c**) Photograph of a thin section from an arid upper intertidal planar stromatolite showing a small tepee structure; Cave Hill Member, Mississippian Kinkaid Formation, Buncombe Quarry, southern Illinois. Note that all the light crystals are calcite pseudomorphs after lenticular gypsum as seen in (**d**). Note also the disruption of the laminae as a result of desiccation and gypsum formation. (**d**) Photomicrograph of a part of sample (**c**) under normal light showing calcite pseudomorphs after lenticular gypsum, microbial laminae and microcrystalline dolomite. Circular objects in the dark microbial laminae are calcispheres believed to be green algal sporangium. (**e**) Photomicrograph under normal light of a microbial lamina containing calcite pseudomorphs after gypsum covered by a sediment-rich lamina composed of intraclastic dolomudstone; Cave Hill Member, Mississippian Kinkaid Formation, Kinkaid Creek section, southern Illinois

stromatolities that typically display fenestral fabric, root casts, desiccation cracks and storm-generated deposits (e.g. Shinn 1986). A well preserved humid tidal flat and the associated coastal marsh facies has been described by Mitchell (1985) from the Middle Ordovician St. Paul Group in central Appalachians (see Sect. 21.7.3). Deposition of calcium carbonate and/or gypsum in tidal flat environments leads to the formation of early diagenetic microcrystalline dolomite by either direct precipitation or replacement of the previously deposited calcium carbonate (e.g. Hardie 1987, Z. Lasemi et al. 1989).

Fig. 21.11 Photographs of hand specimens showing millimeter- to centimeter-scale interlayering of quartz sand (*gray*) and mud-sized carbonate (*tan*) (heterolithic stratification) in the lower part of the Mississippian Bayport Formation, Bayport, Michigan. (**a**) Lenticular and wavy bedding of quartz sand in carbonate mud laminae in the lower part, changing upward to finer scale lamination. Note rain drop impressions in the carbonate lamina on the top of the specimen (*coin* diameter is 2.5 cm). (**b**) Interlayered planar- to wavy-bedded and rippled quartz sandstone (*dark gray*) and carbonate mud laminae (*light gray* to tan). The ripple in the *lower part* of the photograph and its internal trough cross-lamination indicate tidal current direction (possibly flood tide) to the right. Note also that the smaller current ripples in the upper right indicate tidal current reversal (ebb tide). In both ripples carbonate mud fills the ripple troughs forming flaser bedding

Other features common in supratidal setting include very thin lenticular bedding (Fig. 21.17b), root casts (Fig. 21.15b), rain drop impressions (Fig. 21.11a) and tepee structures (see the previous section). The supratidal facies may be capped by wind-blown bimodal and super-mature quartz sandstone (Fig. 21.19d) during seaward progradation (see the Lower Triassic tidalites in Sect. 21.7.5.1). The supratidal facies of beach ridges and channel levees (Fig. 21.2b) are laterally discontinuous and have a very low preservation potential. They consist of very thin bedded, interlaminated mud cracked fenestral mudstone laminae with a thin microbial coating and crinkly stromatolite boundstone (Shinn 1986).

21.6 Peritidal Cycles and Sequence Stratigraphy

The peritidal facies commonly are arranged into meter scale, shallowing-upward successions of subtidal- to tidal flat facies (e.g. Wilson 1975; James 1984; Hardie and Shinn 1986; Grotzinger 1986b, 1989; Pratt et al. 1992; Pratt 2010). In a single shallowing-upward cycle, facies boundaries are transitional in accord with Walther's Law (for an example see figure 21.21a), and the contact between the shallow-water and deeper facies of the overlying cycle is erosional or sharp due to non-deposition or prolong exposure (Fig. 21.21a, b).

Fig. 21.12 Planar, wavy and ripple bedding in mixed carbonate-siliciclastic and pure carbonate intertidal deposits: (**a**) Millimeter- to centimeter-thick tidal lamination in interlaminated lime mudstone and quartz sandstone (*gray*). Note disrupted laminae, intraclasts, smooth-walled sand-filled desiccation cracks and lenticular bedding in the lower part; Mississippian Bayport Formation, Bayport, Michigan. (**b–c**) Intertidal heterolithic interlayering of dolomudstone and grainstone in the Middle Cambrian member 2 of the Mila Formation from the Alborz Mountains, northern Iran: (**b**) Interlayered wavy and rippled grainstone and dolomudstone containing lenticular bedding. (**c**) Wavy, ripple and lenticular bedding. Note the vertical and sub-vertical burrows in the middle of the photograph that was filled during the deposition of the overlying current ripple. The uppermost layer is a bioclastic intraclast grainstone transgressive lag deposit

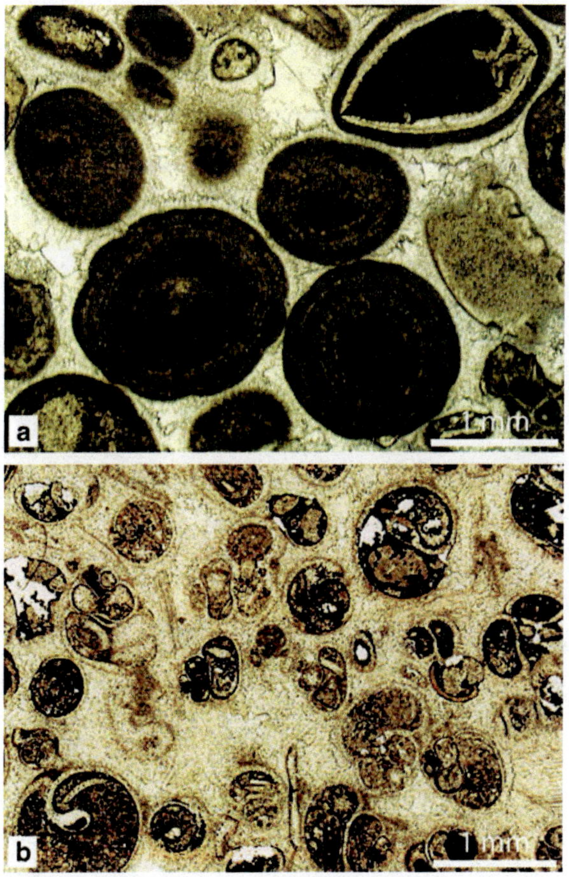

Fig. 21.13 (**a**) Fibrous and irregular sparry calcite cement fringes (beachrock cement) in a bioclastic ooid grainstone interpreted as a barrier beach facies; Upper Mississippian Negli Creek Member of the Kinkaid Formation, southern Illinois. (**b**) An example of a beach ridge grainstone composed solely of small gastropod shells with void-filling hematite and calcite cements. Note the irregular radial fibrous beachrock cement fringe on the grains; Lower Triassic lower member of the Elika Formation in the central Alborz Mountains, northern Iran

Even though shallowing-upward cycles dominate the peritidal successions, deepening-upward and deepening- to shallowing-upward trends or aggradational cycles consisting of a single facies are also present (see Fig. 21.3d for an example) (e.g. Burgess 2006; Spence and Tucker 2007; Bosence et al. 2009; Zecchin 2010) depending on rates of deposition and sea level fluctuations during the course of their development. Some ancient successions consist of stacked cycles characterized by a subtidal facies capped by karstic or calichie

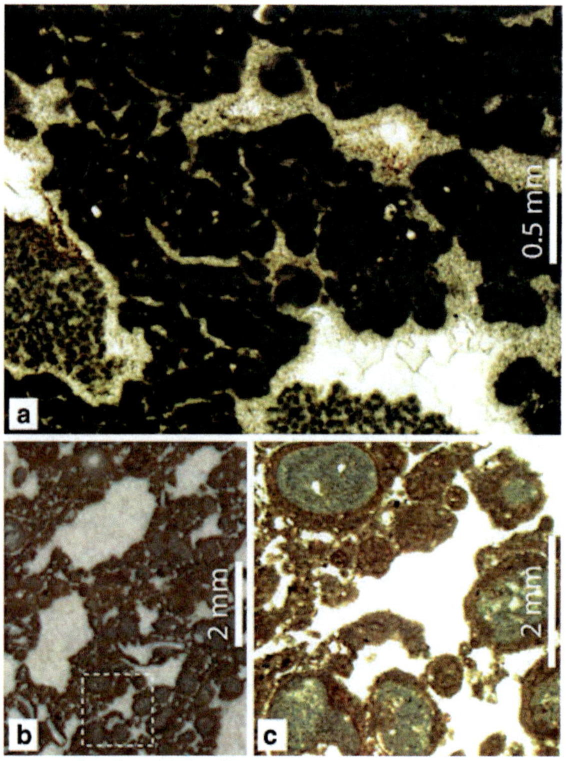

Fig. 21.14 (**a**, **b**, **c**) Subaerial exposure features in intertidal fenestral grainstone from the Upper Jurassic Mozduran Formation in the Kopet Dagh back arc basin, northeast Iran: (**a**) Photomicrograph under normal light of a fenestral peloidal intraclast grainstone. Note that the geopetal sediment (silt-sized peloids) overlies earlier *light brown* vadose cement and is overlain by late stage phreatic drusy mosaic cement. (**b–c**) Thin section photograph of fenestral grainstone facies showing subaerial exposure features. Photograph (**c**) is the enlarged portion of the *lower part* of (**b**) showing partial dissolution of the grains, enlarged pores and formation of iron oxide stained coated grains

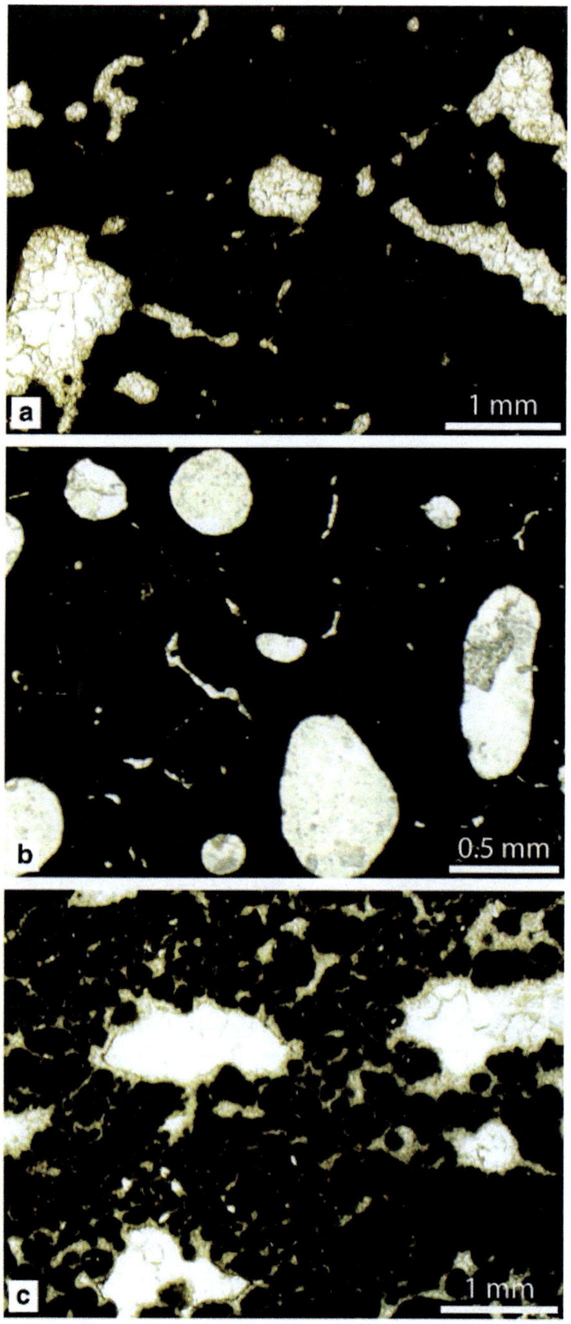

Fig. 21.15 (**a**, **b**) Photomicrographs of calcite cemented fenestral fabric (birdseyes) in lime mudstone interpreted as supratidal facies. Note the calcite-cemented root casts and small birdseyes in b; Cave Hill Member of the Mississippian Kinkaid Formation, southern Illinois. (**c**) Photomicrograph of the lowermost intertidal calcite-cemented fenestral peloid intraclast grainstone under plane-polarized light showing early vadose cement (*light yellowish brown crystals*) that lines the larger fenestrae and fills the smaller voids. Note that the vadose cement is overlain by later phreatic cement indicating subaerial exposure; Upper Jurassic Mozduran Formation in the Kopet Dagh back arc basin, northeast Iran

soil horizon without any intervening tidal flat facies as a result of rapid sea level fall and prolonged exposure (Hardie 1986; Preto et al. 2004). Examples include the karstic soil-capped cycles of the Plio-Pleistocene deposits of the Bahamas (Beach and Ginsburg 1980) and the calichie-capped tepee dominated diagenetic cycles of the Middle Triassic Latemar Limestone of northern Italy (Hardie et al. 1986) formed under humid and arid- to semiarid conditions, respectively. The meter-scale peritidal cycles are known as parasequences (e.g., Van Wagoner et al. 1988; Lehrmann and Goldhammer 1999; Burgess 2006; Spence and Tucker

Fig. 21.16 (**a**) Photomicrograph of a bioturbated ostracod dolomudstone with microbial laminae from an intertidal flat sequence interpreted as the intertidal pond facies; Mississippian Bayport Formation, Bayport, Michigan. (**b, c**) Intertidal channel deposits from the Middle Triassic middle member of the Elika Formation in east central Alborz Mountains, northern Iran. (**b**) Erosive-based fining-upward intertidal channel sequence (*arrow* at the erosional contact) composed of a grainstone bed with gravel-size intraclasts grading upward into a trough cross-bedded sand-size grainstone that in turn grades to herringbone cross-bedded grainstone (*above the scale*) covered by an intertidal deposit (pen for scale is 14 cm long). (**c**) A tidal channel deposit (pen on the left side of the channel facies is 14.5 cm long) within a laminated and thin-bedded tidal flat succession. Note that the channel facies thins to the *right* of the photograph and pinches out completely toward the *left*

2007; Catuneanu et al. 2009), the fundamental unit of stratigraphic sequences.

Many ancient peritidal successions, at least in part, lack any stratal order and comprise random vertical facies transition and thickness patterns (e.g. Wilkinson et al. 1997). The random stratigraphic patterns have been interpreted to be the result of deposition in a set of randomly distributed environments (e.g. Wilkinson et al. 1996, 1999). Recent studies (e.g. Lehrmann and Goldhammer 1999; Lehrmann and Rankey 1999; Spence and Tucker 2007) and investigation on the Holocene carbonates of the Bahamas (Rankey 2002), on the other hand, indicate highly ordered facies transitions in peritidal facies tracts and the resulting vertical successions. The random facies patterns in ancient successions may have been the consequence of incomplete stratigraphic record and extrabasinal forcing mechanisms (Rankey 2002; Lehrmann and Goldhammer 1999). Random vertical successions normally develop during icehouse conditions, due to unfilled accommodation space created by higher-frequency, higher-amplitude sea level changes (Lehrmann and Goldhammer 1999; Burgess 2006).

21.6.1 Origin of Meter-Scale Cyclicity

Shallowing-upward cycles form when tidal flats aggrade to sea level and prograde into the adjacent subtidal setting. This is the consequence of high sedimentation rate in tidal flat areas that normally exceeds the available accommodation space created by high-frequency 5th- to 4th-order sea level rise. Progradation leads to a seaward thickening wedge of tidal flat sediments and a succession of subtidal- to intertidal- to supratidal facies (e.g. Hardie 1986). Repeated deepening and filling of accommodation space in response to relative sea level changes (the sum of eustatic sea level change, subsidence and sediment supply), lead to

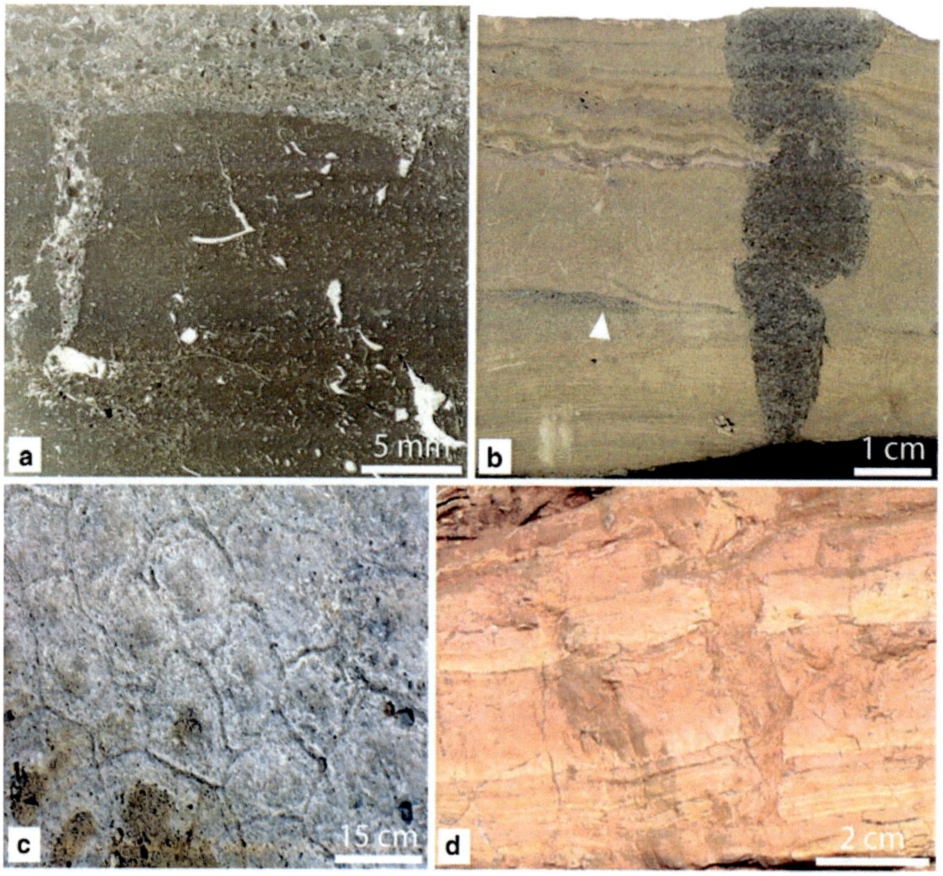

Fig. 21.17 Desiccation cracks in laminated supratidal sediment: (**a**) Sediment filled mud crack in a peloid wackestone lamina covered by an intraclast grainstone transgressive lag deposit; Cave Hill Member of the Mississippian Kinkaid Formation, southern Illinois. (**b**) Quartz sand-filled 'V' shaped mud crack in a mixed carbonate-siliciclastic tidal flat sequence. Note the irregular wall of the crack and a small quartz sand-filled lenticular bed (*dark gray*) in thickly-laminated dolomudstone toward the *middle* of the *left side* of the photograph. Fine-scale laminations in the *upper part* comprise dark crinkly stromatolite and dolomudstone laminae. (**c**) Plan view of mud cracks in the upper surface of a supratidal deposit; Mississippian lower St. Louis Limestone, Columbia Quarry, St. Clair County, southwest Illinois. (**d**) Dolomudstone-filled mud crack in tan to *pink laminated* deposits of the supratidal facies (Photo courtesy of Dr. M. Ghomashi, Sistan-Baluchistan University, Zahedan, Iran); Lower Triassic Sorkh Shale Formation, Tabas failed rift basin, east central Iran

stacked complete and incomplete meter-scale successions (see the illustrative examples). Progradation to develop stacked peritidal meter-scale cycles can be generated by intrabasinal autocyclic processes and extrabasinal allocyclic mechanisms including eustatic sea level fluctuation and tectonic subsidence (e.g., Hardie 1986; Pratt et al. 1992; Pratt 2010).

21.6.1.1 Autocyclicity

Autocycles form in response to processes operating within the environment of deposition and include tidal flat shoreline and island progradation and lateral migration of tidal channels. Tidal flat progradation is the dominate process during greenhouse periods (small polar ice volume) due to lower-amplitude high-frequency sea level changes (Lehrmann and Goldhammer 1999; Burgess 2006). Tidal flat shoreline progradation to generate stacked shallowing-upward peritidal cycles (Ginsburg 1971; Hardie 1986) assumes gradual subsidence, slow sea level rise or stillstand and changes in sedimentation rate during deposition. High sedimentation rate in tidal flat areas results in progradation

Fig. 21.18 Supratidal facies: (**a–b**) Rosettes and laths of calcite pseudomorphs after anhydrite within a fenestral lime mudstone from the Cave Hill Member of the Mississippian Kinkaid Formation, southern Illinois. Note that in a (thin section photograph) birdseyes are almost totally filled with vadose silt. Note also that in (**b**) (photomicrograph of the enlarged portion of the *upper left* of (**a**) under normal light) the rosettes are filled with coarsely crystalline calcite cement. (**c**) Field photograph of a thickly laminated fenestral dolomudstone containing lenticular gypsum casts (the silver end of the pen in the upper left is 3 cm long); Middle Triassic middle unit of the Elika Formation in the Ghoznavi section, eastern Alborz Mountains, northern Iran

and rapid infilling of the subtidal carbonate factory. Subsidence and subsequent sea level rise result in the resumption of carbonate deposition (after a lag period) and formation of a new asymmetric shallowing-upward cycle. An alternative autocyclic mechanism, the tidal flat island model, has been proposed for peritidal shallowing-upward cycles that are laterally discontinuous (Pratt and James 1986; Pratt 2010). In this model, deposition would take place on small low relief islands and intertidal banks separated by subtidal source areas. Progradation and lateral growth of the islands may generate shallowing-upward peritidal cycles of limited areal extent with random stratigraphic distribution and variable thicknesses (Pratt et al. 1992; Pratt 2010).

Lateral migration of tidal channels is another autocyclic process forming discontinuous small-scale cycles. Migration of intertidal channel leads to point bar deposits which are laterally discontinuous, erosive-based, fining-upward successions of gravel- and sand-size sediments capped by a laminated tidal flat facies (Fig. 21.16b, c). Tidal channel migration in platform margin shoals generates an erosive-based fining-upward succession (Fig. 21.8b), capped by offshore, tidal flat, lagoonal or coarsening-upward beach facies (Inden and Moore 1983).

Fig. 21.19 (**a**) Interlayered *light green-* to *reddish brown* laminated argillaceous dolomudstone/dolomitic shale and *reddish brown* and fine- to very fine-grained lithic sandstone from the Middle Cambrian member 2 of the Mila Formation, eastern Alborz Mountains, northern Iran. (**b**) Close up view of a part of (**a**) showing the laminated shale/dolomite and the capping sandstone facies (Lens cap diameter is 5.5 cm). The laminated shale-carbonate facies contain hopper halite casts and has no subaerial exposure features interpreted as a lowstand coastal salina pond deposit. (**c**) Cast of hopper halite crystals mentioned in (**b**). (**d**) A laminated supper mature quartz arenite bed overlies and underlies, with abrupt contacts, a fenestral dolomudstone (supratidal facies) forming a pure carbonate and pure siliciclastic double cycle (Y. Lasemi et al. 2008); base of the transgressive systems tract in the Lower Triassic Sorkh Shale Formation, south of the Tabas failed rift basin, east central Iran

21.6.1.2 Allocyclicity

In many ancient carbonate tidalites, cyclicity has been interpreted based on Milankovitch-band periodic climate changes (e.g. Goldhammer et al. 1987; Koerschner and Read 1989; Goldhammer et al. 1993; Strasser et al. 1999; Preto et al. 2004). In the Milankovitch orbital forcing model, three parameters including precession, obliquity and eccentricity could generate high frequency eustatic sea level cycles with approximate durations of 20,000, 41,000 and 100,000 years, respectively. Evidence for Milankovitch-band periodic eustatic sea level changes in the stratigraphic record include lateral continuity of cycles on a regional and interregional scale, 5:1 grouping of 5th-order small-scale cycles to form larger 4th-order cycles of 100,000 years duration, and high frequency subtidal cycles with karstic or calichie soil caps (e.g. Goldhammer et al. 1987; Koerschner and Read 1989; Preto et al. 2004). Allocyclicity is strong during icehouse periods due to higher-amplitude relative sea level changes (Lehrmann and Goldhammer 1999; Burgess 2006).

Subsidence due to high-frequency extensional fault movements ("yo-yo" and "yo" tectonics) can also create accommodation space for the formation of stacked peritidal cycles (Hardie 1986; Hardie et al. 1991; De Benedictis et al. 2007; Bosence et al. 2009). According to Bosence et al. (2009), the cycles form due to filling of accommodation created by episodic rapid downwarping followed by slow subsidence in the hanging wall and graben sites ("yo" tectonics) and concurrent rapid uplift and slow subsidence in the footwall sites ("yo-yo" tectonics). The most common cycle type is asymmetric shallowing-upward, but symmetric deepening then shallowing-upward, asymmetric

Fig. 21.20 Panorama of conspicuously bedded peritidal carbonate successions along the Azadshahr Highway in eastern Alborz Mountains: (**a**) Highstand lagoonal shallowing-upward cycles of the Upper Permian Ruteh Formation (*right*) capped by a laterite horizon, which is in turn overlain by the peritidal cycles of the Lower Triassic lower member of the Elika Formation. (**b**) The Lower and Middle Triassic peritidal deposits of the lower and middle members of the Elika Formation comprising two depositional sequences. The lower sequence is bounded by the lowstand laterite horizon at the Permian-Triassic boundary (*right*) and a quartz sandstone bed at the Lower-Middle Triassic boundary in the middle of the photograph, and consists of lagoonal- to intertidal cycles intercalated by numerous storm beds. The Middle Triassic middle member (*upper left* of the photograph) is a depositional sequence composed almost entirely of intertidal- to supratidal shallowing-upward cycles

with subaerial exposure surface or karstic/calichie soil cap, and asymmetric deepening-upward can develop depending on tectonic setting. The cycles have the same thickness and frequency as eustatic or autocyclic processes, however, lateral variation of cycle types and cycle stacking and irregular non-bundled stacking of the cycles support an overriding tectonic control (De Benedictis et al. 2007; Bosence et al. 2009).

21.6.1.3 Sequence Stratigraphy

The meter-scale cycles are superimposed on an underlying lower-frequency 3rd-order relative sea level cycle (e.g. Goldhammer et al. 1993; Kerans and Tinker 1997; Spence and Tucker 2007) (see the Middle Triassic tidalites in Sect. 21.7.5.2 for an example). The 3rd-order cycle has a duration of 1–3 my (Haq et al. 1987) or 1–10 my, (Kerans and Tinker 1997; Lehrmann

Fig. 21.21 (**a**) Brownish gray lagoonal deposits (*right*) grading into *light brown* intertidal and supratidal deposits. Note the abrupt upper contact (*dashed line*) with the overlying lagoonal deposit (the *encircled tree* is about 3 m tall); Lower Miocene middle member of the Asmari Formation in the Mish Mountain, Zagros Mountain range, southwest Iran. (**b**) *Dark gray* bioturbated lime mudstone (subtidal lagoon facies) overlies, with an erosional contact, a *yellow* dolomudstone (tidal flat facies) indicating transgression over a tidal flat cycle cap; Lower Triassic lower member of the Elika Formation in the central Alborz Mountains, northern Iran

and Goldhammer 1999) and represent a depositional sequence. A depositional sequence as defined by Mitchum et al. (1977) is a stratigraphic unit composed of a relatively conformable succession of genetically related strata bounded by unconformities (Fig. 21.20) or their correlative conformities. Catuneanu et al. (2009) recommended using the term 'stratigraphic sequence', a sedimentary succession deposited during a full cycle of change in accommodation or sediment supply (Fig. 21.22), which is compatible with various existing sequence stratigraphic models (For a detailed review of sequence stratigraphic principles and models see Miall 1997; Schlager 2005; Catuneanu 2006; Catuneanu et al. 2009).

A depositional sequence can be subdivided into systems tracts (linkage of contemporaneous depositional systems) as defined by Brown and Fisher (1977), which are defined on the basis of parasequence stacking patterns, position within the sequence and types of bounding surfaces (Van Wagoner et al. 1988, 1990). A depositional sequence may consist of up to four types of systems tracts depending on the shape of base-level curve, type of depositional system, basinal setting, and post depositional erosion at the sequence boundary (e.g. Catuneanu 2006; Catuneanu et al. 2009). These packages include lowstand, transgressive, highstand and falling stage systems tracts (Fig. 21.22).

The lowstand systems tract (LST) overlies the sequence boundary and comprises normal regressive sediments deposits after the onset of relative sea level rise. The transgressive systems tract (TST) lies between the transgressive surface (ts) above the LST or it overlies the sequence boundary and is capped by maximum flooding surface (mfs). It forms when the rate of rise exceeds the rate of deposition, displays a deepening-upward facies trend and a characteristic retrogradational parasequence stacking pattern. The highstand systems tract (HST) forms during the late stage of base-level rise, when the rate of deposition exceeds the rate of accommodation being created (shoreline normal regression). HST deposits underlie the FSST or the sequence boundary and are characterized by shallowing-upward facies trend displaying aggradational to progradational parasequence stacking pattern. The falling stage systems tract (FSST) includes the strata deposited during base-level fall (forced regression) and underlies the sequence boundary.

Carbonate platforms produce sediment during periods of base-level rise mainly during transgressive and highstand stages of a base-level curve. During falling stage and lowstand sea level rise, the entire platform or the platform interior is exposed to karstification and calichie (calcrete) development in humid and dry climates, respectively, forming a pronounced subaerial unconformity. Therefore, FSST-LST packages are normally absent and the sequence boundary commonly is capped by transgressive systems tract of the overlying sequence. In a carbonate ramp setting of a passive

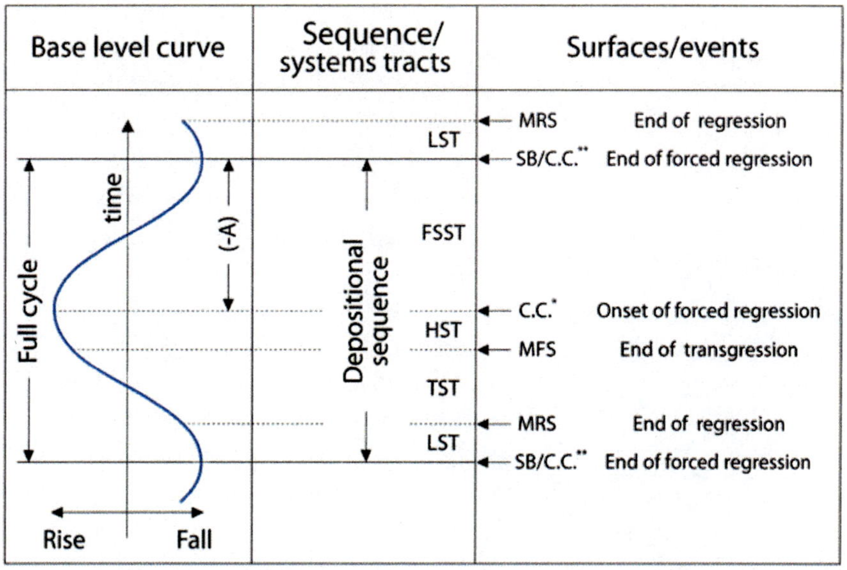

Fig. 21.22 *Base-level curve*, depositional sequence ('stratigraphic sequence'), systems tracts, and sequence stratigraphic surfaces defined in relation to base-level curve (Modified from Catuneanu 2006; Catuneanu et al. 2009). Abbreviations: (**a**), accommodation; *C.C**, correlative conformity *sensu* Posamentier and Allen 1999; *C.C.***, *sensu* Hunt and Tucker 1992; *MFS* maximum flooding surface, *MRS* maximum regressive surface, SB sequence boundary, *LST* lowstand systems tract, *TST* transgressive systems tract, *HST* highstand systems tract, *FSST* falling stage systems tract

margin succession, however, lowstand peritidal deposits may form during lowstand slow base-level rise consisting of siliciclastics, evaporites or mixed carbonate-siliciclastic peritidal deposits (see the Middle Cambrian tidalites in Sect. 21.7.2 and the Mississippian St. Louis Formation in Sect. 21.7.4).

In mature passive margins, peritidal deposits may constitute the bulk of the transgressive and highstand packages (see the Middle Cambrian and Middle Triassic tidalites in Sects. 21.7.2 and 21.7.5.2). Platform sequences in proximal areas of a basin may consist almost exclusively of tidal flat deposits as in the Lower Triassic Sorkh Shale Formation (see Sect. 21.7.5.1). Sequences in the distal areas of a basin, on the other hand, consist dominantly of subtidal facies with tidal flat facies occurring in the upper part of the highstand systems tracts (see the Lower Triassic Elika Formation in Sect. 21.7.5.1). Absence of peritidal deposits in the transgressive tracts may be the result of rapid base level rise. In foreland basin successions, tidal flat deposits commonly comprise the upper part of the highstand systems tract (see the Miocene tidalites in Sect. 21.7.6); rapid subsidence in response to overthrust loading may prevent tidal flat deposition in the transgressive systems tract.

21.7 Ancient Examples of Carbonate Tidalites

Numerous examples of ancient carbonate tidal facies, practically comparable to their modern counterparts, have been reported from the Precambrian and Phanerozoic successions (e.g. Ginsburg 1975; James 1984; Hardie and Shinn 1986; Tucker and Wright 1990; Demicco and Hardie 1994; Flügel 2010). Ancient carbonate tidalites were deposited on huge carbonate platforms that developed along the Atlantic type passive margins and in smaller platforms associated with failed rifts, intracratonic, foreland, back-arc/fore-arc and pull-apart basins (Read 1985; Grotzinger 1989; Demicco and Hardie 1994). The thickest and most extensive tidal deposits, however, were laid down during Proterozoic through Late Mesozoic times on the vast carbonate platforms that developed along the passive (Atlantic type) continental margins, such as the Iapetus (proto-Atlantic), Paleo-Tethys and Neo-Tethys ocean margins and associated intracratonic basins.

Proterozoic carbonate tidalites consist of microbialites (both thrombolites and stromatolites), ooids, intraclasts and micritic carbonates reflecting carbonate production by chemical and microbial processes.

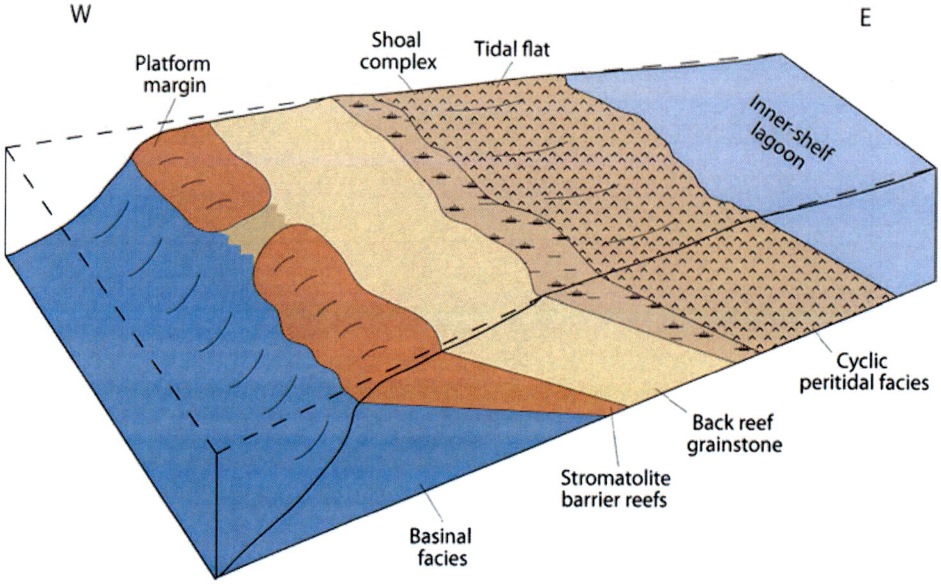

Fig. 21.23 Platform paleogeography during the deposition of the Rocknest Formation (Modified from Grotzinger 1986a). Cyclic peritidal deposits were formed in an extensive back barrier ('*shoal complex*') tidal flat facies setting adjacent to a shallow subtidal lagoon to the east of the platform

These deposits commonly are characterized by the paucity of grainstone facies and the dominance of stromatolite reefs and fine-grained carbonate facies (Grotzinger 1989). During the Precambrian, stromatolites were able to flourish in any setting from deep marine to supratidal and normal to hypersaline waters (e.g. Hoffman 1976; Grotzinger 1989). Although skeletal metazoans were absent prior to Cambrian time, the Precambrian carbonates were deposited in platforms and environments surprisingly similar to those of their Phanerozoic counterparts (e.g. Hardie and Shinn 198, Grotzinger 1989; Flügel 2010).

In the Phanerozoic deposits, in addition to microbialites, micritic carbonates, ooids, oncoids and intraclasts that appeared in the Precambrian platform deposits, various new components, including skeletal grains and fecal pellets emerged. A number of metazoans and calcareous algae constructed patch, fringing and barrier reefs and by late Phanerozoic time seagrass probably contributed in stabilizing mud size sediments in the subtidal setting. During the Phanerozoic, burrowing activity by metazoans and increased supply of sand- to silt-sized bioclasts and peloids restricted microbialites mainly to areas of low sediment influx (Pratt 1982) and to more ecologically stressed environments. The following section of this chapter summarizes facies, paleo-environment, and cycle/sequence stratigraphic analyses of a few specific examples of ancient carbonate tidal deposits representing the Proterozoic through Tertiary successions. The examples represent arid and humid carbonate and mixed carbonate-siliciclastic tidal deposits that occur in various systems tracts of depositional sequences related to passive and active continental margins.

21.7.1 Tidalites of the Rocknest Formation of Northwest Canada

The Lower Proterozoic (older than 1,800 Ma) Rocknest Formation is a carbonate platform succession up to 1,200 m thick that was deposited in a westward-facing passive margin in Northwest Territories, Canada (Grotzinger 1989). This summary describes a back barrier mixed carbonate-siliciclastic peritidal succession (Hoffman 1975; Grotzinger 1985, 1986a, b). In the Rocknest platform, an extensive belt of tidal flat environment developed behind a barrier ('shoal-complex') that separated a wave-dominated windward platform margin to the west from an extensive low energy inner-shelf lagoon to the east (Fig. 21.23). The proximal area of the 'shoal-complex' was seldom

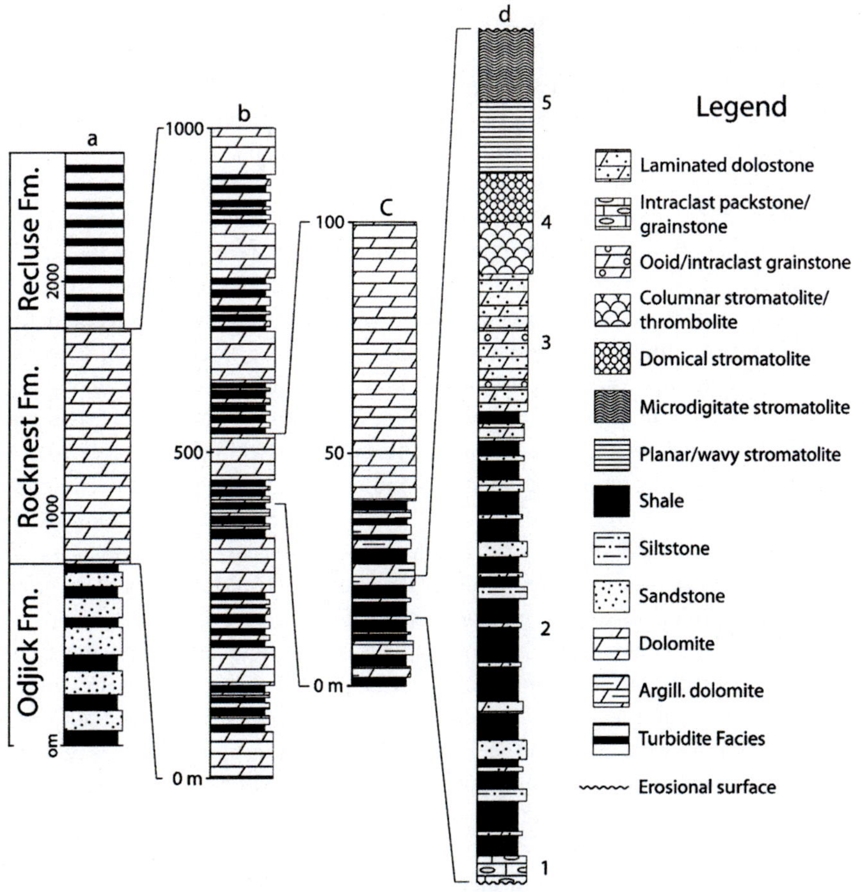

Fig. 21.24 Stratigraphic columns (**a–c**) and the peritidal shale-based cycle of the inner shelf cyclic facies (**d**) of the Rocknest Formation in the Lower Proterozoic carbonate platform succession. Numbers 1 through 5 denote lagoonal and tidal flat facies or facies groups within the cycle (Modified from Hoffman 1975 and Grotzinger 1986b). *1*: transgressive lag deposit; *2*: distal inner-shelf lagoon facies; *3*: proximal inner-shelf lagoon facies; *4*: upper subtidal-lower intertidal facies; *5*: upper intertidal-supratidal facies

submerged, preventing communication between the open sea and the low energy inner shelf lagoon. The tidal flat and lagoonal deposits of the Rocknest Formation are characterized by mixed carbonate and siliciclastic facies that grade eastward to siliciclastic deposits.

The Rocknest peritidal facies are arranged into asymmetric shallowing-upward cycles (Fig. 21.24d), which display various basal lithofacies reflecting their paleogeographic location on the platform. Consequently, the shale-based cycles of the distal inner-shelf lagoon pass westward into laminated dolostone-based cycles and stromatolite-based cycles of the 'shoal-complex'. The cycles were initiated by rapid transgression and flooding of tidal flats followed by eastward progradation of the tidal flat facies over the lagoonal facies of the inner shelf during sea level fall. Shale-based cycles up to 15 m thick contain the most diverse facies, which from base to top include (Fig. 21.24d): (1) intraclast packstone to grainstone, 5–30 cm thick that covers erosional tops of cycles and commonly consists of rounded stromatolitic fragments derived from the upper part of the underlying cycle (transgressive lag deposit); (2) mixed carbonate-siliciclastic facies consisting of below- to above-wave-base interstratified argillaceous dolomudstone and shale containing intercalations (up to 30 cm thick) of massive, planar laminated or hummocky cross-stratified sheets and lenses of siltstone, sandstone and sand-size dolostone (distal inner-shelf lagoon); (3) laminated dolostone that consists of very thin-bedded laminated or wave-rippled silt-size dolostone with common intercalations of dolomitized oolitic and intraclastic grainstone layers, suggesting deposition close to fair weather wave base (proximal inner-shelf lagoon); (4) stromatolitic and thrombolitic dolostone (10–150 cm thick) containing isolated to laterally-linked columnar and domal stromatolite and/or thrombolite that show an upward decrease in relief (upper subtidal to lower intertidal); (5) stromatolites (up to 1–2 m thick) as planar tan- to buff-colored crinkly to wavy laminites overlain by gray to black planar or discrete to partially-linked microdigitate forms less than 1 cm wide and less than 10 cm high. This facies contains desiccation cracks, irregular to planar birdseyes, tepee structures, abundant halite casts and rare gypsum and/or anhydrite pseudomorphs (arid upper intertidal to supratidal).

The erosional surface at the base of asymmetric depositional cycles become more pronounced towards the shelf margin, suggesting prolonged exposure of the proximal 'shoal-complex'. The 'shoal complex' peritidal deposits overlie the back reef grainstone and stromatolite reef facies of the platform margin as a result of basinward progradation forming the highstand packages of long-term progradational sequences. Abundant halite casts and tepee structures and rare gypsum and anhydrite pseudomorphs in the shelf facies (Grotzinger 1985, 1986a, b) suggest that, contrary to Hoffman (1975), arid conditions prevailed during deposition of the Rocknest Formation.

21.7.2 Middle Cambrian Tidalites in Northern Iran

The Middle and Upper Cambrian was a time of extensive shallow marine carbonate ramp deposition in Iran. The ramp covered the length of the northern Alborz Mountains in northern Iran and the Persian Gulf in southwest Iran (Fig. 21.25) in the north-northeast facing Proto-Paleotethys passive margin of northern Gondwana (Y. Lasemi 2001), a width of several thousands of kilometers at the time of deposition. In the Alborz Mountains, the Middle- to Upper Cambrian deposits includes member 1 carbonates, member 2 mixed carbonate-siliciclastics and member 3 carbonates of the Mila Formation (Stocklin et al. 1964). In this summary, the tidal deposits of the lower part of member 2 in the Tuyeh section (location 1 in Fig. 21.25) of the eastern Alborz Mountains (Amin-Rasouli 1999) will be described. In the Tuyeh section, member 2 is 100 m thick and its lower part comprises two depositional sequences built mainly by peritidal facies. The lower sequence comprises lowstand, transgressive and highstand systems tracts built by meter-scale shallowing-upward cycles (Fig. 21.26).

The lowstand systems tract consists of interlayered light green- to redish brown, laminated argillaceous dolomudstone/dolomitic shale capped by reddish brown, fine- to very fine-grained lithic sandstone (Figs. 21.19a, b and 21.26a). The laminated shale-carbonate facies contain hopper halite casts (Fig. 21.19c) and has no subaerial exposure features. This facies association is interpreted as a coastal salina pond deposit. It resembles the carbonate-shale cycles described by Spencer and Demicco (1993) from the Middle Cambrian passive margin deposits of the Canadian Rocky Mountains. The brown lithic sandstone layers were probably deposited during the advance of siliciclastics into the coastal plain area during periods of continental flooding. The lower part of the transgressive systems tract consists of small-scale shallowing-upward cycles in which interlayered carbonate and siliciclastic facies just described above caps wavy- to planar stromatolite facies (Fig. 21.26b), which in turn change upward to planar- to wavy- to domal stromatolite capped by columnar stromatolite facies (Fig. 21.6e) recording the most transgressive facies. Columnar stromatolite bioherms are characterized by trapped peloids, ooids and intraclasts and absence of bioclasts recording deposition in a hypersaline and high energy subtidal lagoon to lower intertidal environments similar to Hamelin Pool in Shark Bay, Western Australia (e.g. Logan, et al. 1970). The highstand systems tract consists of meter-scale peritidal cycles. A typical and complete cycle that occurs in the lower part of the highstand tract (Fig. 21.26c) consists of (1) basal individual and compound columnar stromatolite bioherms subtidal lagoonal facies (Fig. 21.6e) grading upward into (2) laterally linked domal stromatolites lower intertidal facies (Fig. 21.9b, c), which in turn grades into (3) wavy to planar stromatolite upper intertidal facies containing desiccation cracks, planar fenestrae and calcite pseudomorphs after gypsum/anhydrite capped by (4) supratidal facies comprising fenestral and laminated dolomudstone with gypsum/anhydrite casts. The cycle is interpreted to have formed by progradation of an extensive tidal flat over a high energy lagoon covered

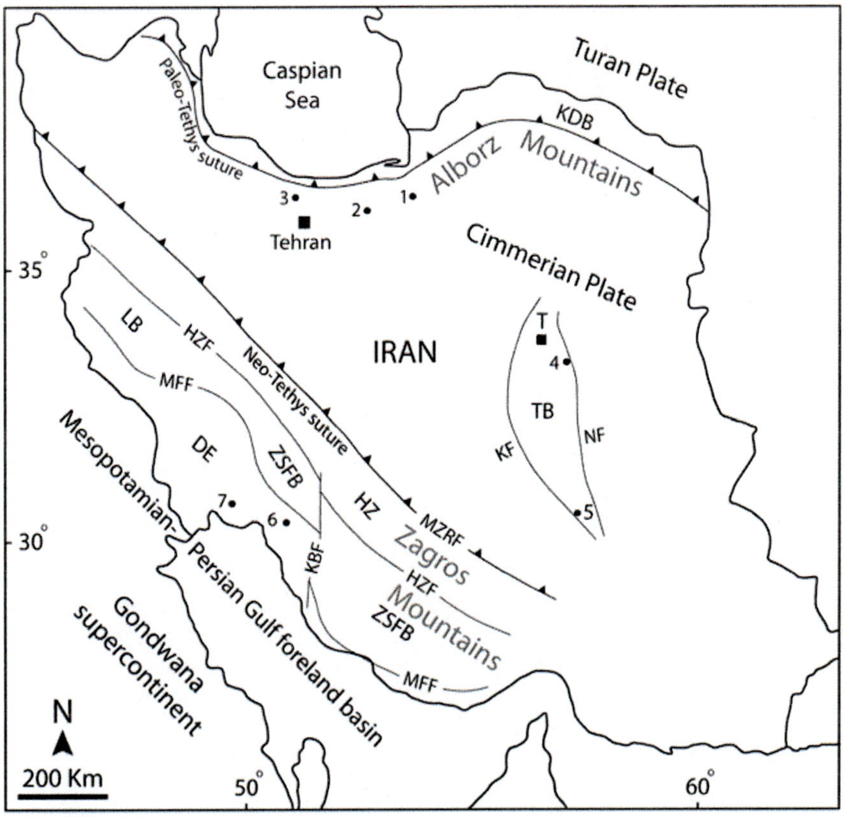

Fig. 21.25 Location map of Iran showing the structural features, plate boundaries and the basins/sub-basins mentioned in the text. The Cimmerian Plate between the Paleo-Tethys and Neo-Tethys sutures includes central Iran and the Alborz Mountains of northern Iran. The Zagros Mountains cover the area between the Neo-Tethys suture and Mountain Front Fault (*MFF*) in southwest Iran (fault traces are according to Berberian 1995 and Alavi et al. 1997). Numbered localities: (*1*) Tuyeh section; (*2*) Veresk section; (*3*) Elika section (*4*) Godare Sorkh section; (*5*) Eslamabad section; (*6*) Mish Mountain section; (*7*) Agha-Jari section. Abbreviations: *DE* Dezful Embayment, *HZ* High Zagros, *HZF* High Zagros Fault, *KBF* Kazrun-Borazjan Fault, *KDB* Kopet Dagh Basin, *KF* Kalmard/Kuhbanan Fault, *LB* Lurestan Sub-basin, *MZRF* Main Zagros Reverse Fault, *MFF* Mountain Front Fault, *NF* Nayband Fault, *T* Tabas (city), *TB* Tabas Basin, *ZSFB* Zagros Simply Folded Belt

with stromatolite reefs. Similar cycles are found in many other Cambrian successions (e.g. Demicco 1985; Spencer and Demicco 1993).

The overlying sequence (Fig. 21.26) includes transgressive and highstand systems tracts built mainly by small-scale shallowing-upward peritidal cycles consisting of (1) bioturbated bioclast/peloid wackestone-packstone (subtidal lagoon); (2) thin-bedded, interlayered dolomudstone and peloid bioclast grainstone (Fig. 21.12b, c) with ripple- wavy- lenticular bedding and vertical burrows (lower intertidal); (3) wavy to flat laminated stromatolite (upper intertidal); and (4) fenestral, mud-cracked dolomudstone with common lamination (disrupted in places forming collapse breccias) and calcite pseudomorphs after gypsum/anhydrite crystals or nodules (supratidal).

Tidal deposits of the Middle Cambrian member 2 of the Mila Formation were deposited in an arid homoclinal ramp, a vast epeiric sea that bordered the Proto-Paleotethys Ocean. Peritidal facies (predominantly tidal flat) constitute the bulk of the transgressive and highstand systems tracts and are arranged into high frequency 4th to 5th-order shallowing-upward cycles (Fig. 21.26b–f). They are interlayered with erosive-based fining-upward storm deposits consisting of intraclasts, peloids, ooids and bioclasts of mixed fauna, which suggest intermittent storm conditions during deposition (Y. Lasemi and Amin-Rasouli 2002).

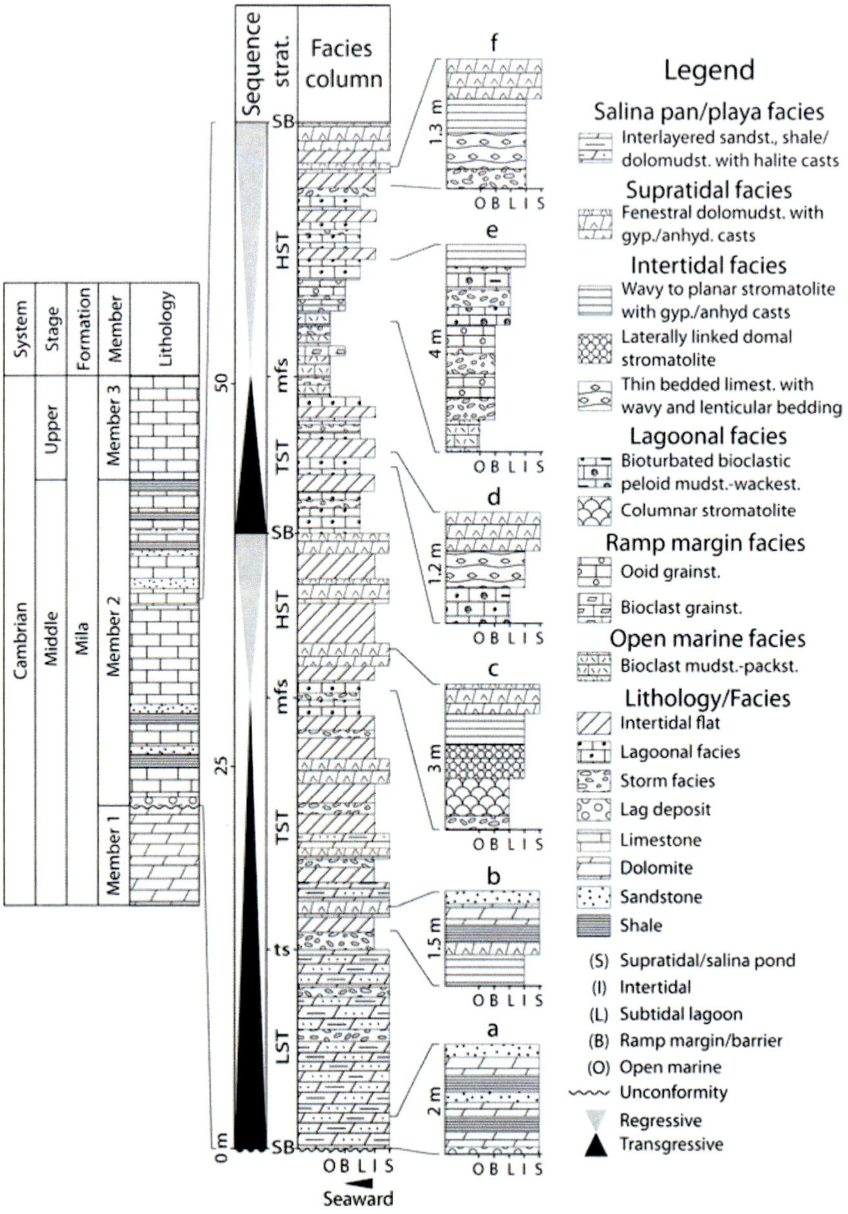

Fig. 21.26 Stratigraphic nomenclature and facies stratigraphy of the Middle Cambrian member 2 carbonates of the Mila Formation in the Tuyeh section (locality 1 in Fig. 21.25). Sequences (3rd-order cycles) are shown to the *left* of the facies column. Detail of a few small-scale shallowing-upward cycles is shown to the right. Note that the transgressive and highstand systems tracts of the lower sequence and the main part of the transgressive and highstand systems tracts of the upper sequence are composed of peritidal facies. These facies are arranged into small-scale shallowing-upward cycles composed of lagoonal- to intertidal- to supratidal or intertidal- to supratidal facies. Note also that the cycle stacking pattern is aggradational to progradational in the lowstand and highstand systems tracts; the pattern is aggradational to retrogradational in the transgressive packages. Black and gray triangles represent transgression and regression, respectively. Abbreviations: *HST* highstand systems tract, *LST* lowstand systems tract, *mfs* maximum flooding surface, *SB* sequence boundary, *ts* transgressive surface, *TST* transgressive systems tract

21.7.3 Tidalites of the Middle Ordovician St. Paul Group in Central Appalachians

The St. Paul Group (150–200 m thick) is a part of the Cambro-Ordovician platform carbonate succession in the central Appalachians, USA. In Maryland and Pennsylvania, the St. Paul succession consists mainly of lagoonal, tidal flat and freshwater coastal marsh facies (Mitchell 1985; Hardie 1986; Demicco and Hardie 1994) including: (1) thin-bedded interlayered LLH (Logan et al. 1964) stromatolite, unfossiliferous mudstone containing molds of cyanobacteria and intraclast packstone facies (coastal freshwater lake/supratidal marsh); (2) laminite facies (0.2–1 m thick) characterized by interlaminated planar- to wavy peloidal mudstone and crinkly stromatolite boundtone with birdseyes and desiccation cracks (supratidal levee); (3) bioturbated mudstone-wackestone facies (0.1–1.5 m thick) with low diversity fauna (mainly ostracods) and no internal layering (intertidal pond); (4) coarse peloidal sand- to pebble-sized micritic intraclast grainstone-packstone up to 0.3 m thick (meandering point bar deposit); (5) thick-bedded bioturbated lime mudstone facies with low to moderate diversity fauna (restricted to semi-restricted inner shelf lagoon to intertidal); (6) thick-bedded bioturbated wackestone/packstone and grainstones with diverse fauna including tabulate corals and bryozoans (open outer shelf lagoon).

In the St. Paul tidal deposits, bioturbated mudstone capped by mud cracked laminite facies dominate the lower part of tidal flat facies succession and are progressively replaced upward by thin-bedded and laminite facies that become more abundant upward near the boundary with the overlying coastal freshwater lake facies. According to Mitchell (1985), the closest modern analog of the St. Paul facies is the tidal flat system of the Bahamas including the Great Bahama Bank, Andros Island tidal flat and the inland freshwater marsh described by (Shinn 1983a, 1986). The absence of reefs and ooid grainstone, dominance of bioturbated peloidal mudstone, low diversity fauna, abundance of stromatolites and the absence of any trace of evaporites in the supratidal facies, all suggest the existence of a low energy, rainy tidal flat during the deposition of the St. Paul Group in central Appalachians (Mitchell 1985, Demicco and Hardie 1994).

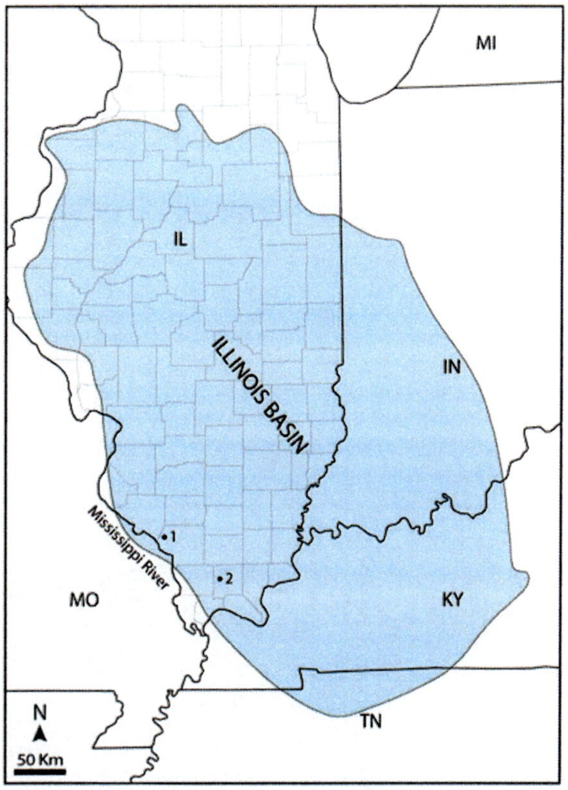

Fig. 21.27 Map of Illinois and neighboring states showing an outline of the Illinois Basin (From Buschbach and Kolata 1991) and the locations of sections used to prepare the composite section shown in Fig. 21.28. Numbered localities: (*1*) Kinkaid Creek section, Jackson County, Illinois; (*2*) Buncombe quarry section, Johnson County, Illinois

21.7.4 Mississippian Tidalites in the Illinois Basin

The intracratonic Illinois Basin (Kolata and Nelson 1991) covers parts of the states of Illinois, Indiana, Kentucky and Tennessee (Fig. 21.27) in the mid-continent of the United States (Buschbach and Kolata 1991). During the Mississippian (Early Carboniferous), several hundred meters of shallow marine carbonates were deposited on carbonate ramp platforms that opened into the deep Ouachita Trough to the south. Extensive peritidal facies have been described from the Cave Hill Member of the Upper Mississippian Kinkaid Formation (Y. Lasemi 1980, Y. Lasemi and Carozzi 1981). The Cave Hill Member (Swann 1963) is bounded by two limestone units namely the Negli

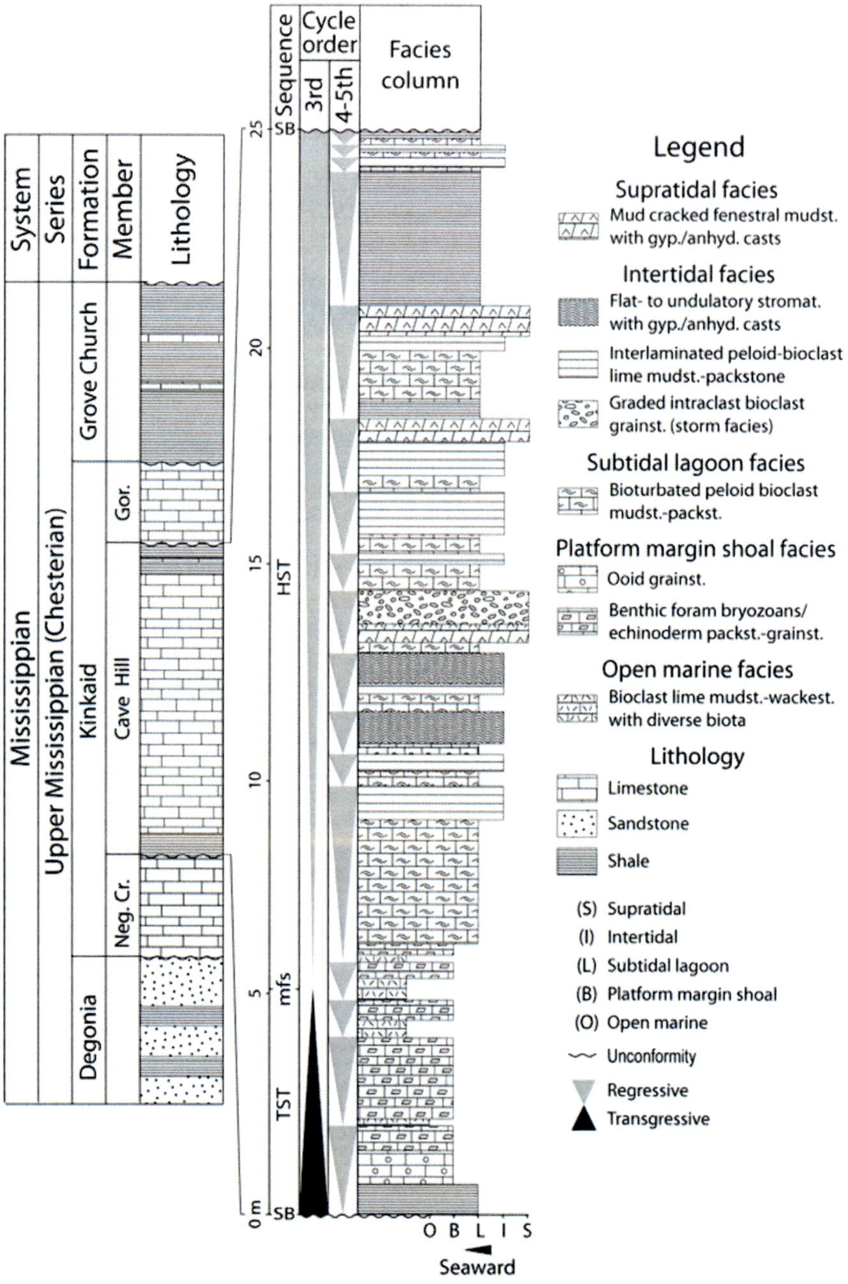

Fig. 21.28 Stratigraphic nomenclature and facies stratigraphy of the Cave Hill Member of the Kinkaid Formation in southwestern Illinois: The Cave Hill is an unconformity bounded sequence (3rd-order cycles) on which, several 4th to 5th-order shallowing-upward cycles are superimposed. The section is composed of the lower 6.5 m of the lower Cave Hill Member from locality 1 and the upper 18.5 m of the Cave Hill Member from locality 2 (see Fig. 21.27 for the location of sections). Note that small-scale peritidal cycles display progradational and aggradational stacking patterns and comprise the bulk of the highstand systems tract. *Black* and *gray triangles* represent transgression and regression, respectively. Abbreviations: *HST* highstand systems tract; *mfs* maximum flooding surface, *SB* sequence boundary, *TST* transgressive systems tract

Creek and the Goreville Members (Fig. 21.28). This summary concentrates on a composite section (Fig. 21.27, localities 1 and 2) of the Cave Hill Member in southwestern Illinois.

The Cave Hill Member mainly consists of cyclic peritidal deposits (Fig. 21.28) related to an arid and gently sloping homoclinal carbonate ramp similar to the modern Persian Gulf. Peritidal facies of the Cave

Hill include: (1) fenestral lime mudstone with calcite pseudomorphs after anhydrite and/or thickly interlaminated peloidal grainstone to mudstone with irregular to tabular fenestrae, root casts, and desiccation cracks (supratidal) (Figs. 21.15a, b, 21.17a and 21.18a, b); (2) millimeter-thick interlamination of micrite-rich and fine sand- to silt-sized bioclast/peloid-rich carbonates or flat- to wavy stromatolites commonly containing dolomudstone, molds of filamentous cyanobacteria, desiccation cracks and calcite pseudomorphs after lenticular gypsum (upper intertidal) (Figs. 21.3a, c, 21.9d, and 21.10c–e); and (3) bioturbated lime mudstone to packstone commonly containing peloids, oncoids and bioclasts (Fig. 21.5d) of a restricted fauna that commonly includes gastropods, ostracods, and benthic foraminifera (lower intertidal/subtidal lagoon). The Cave Hill Member represents an unconformity bounded depositional sequence (3rd-order cycle) on which are superimposed numerous small-scale fourth- to fifth-order cycles (Fig. 21.28). In the Cave Hill sequence, peritidal small-scale cycles comprise the bulk of the highstand systems tract and display aggradational and progradational stacking patterns (Fig. 21.28). The absence of peritidal deposits in the transgressive tract may be the consequence of a rather fast sea level rise during the onset of deposition of the Cave Hill Member.

A similar carbonate tidal deposit occurs in the lower part of the Middle Mississippian St. Louis Limestone of the Illinois Basin and its equivalent in the Michigan Basin. In southwestern Illinois, a peritidal succession with a basal unconformable boundary is present in the lower part of the lower St. Louis Limestone (Z. Lasemi and Norby 1999). This interval includes cyclic bioturbated lime mudstone to intraclast/bioclast peloid and/or oncoid wackestone-grainstone containing a low-diversity fauna (subtidal) capped by wavy- to planar stromatolites or laminated peloid mudstone-grainstone (lower intertidal) overlain by mud-cracked peloidal dolomudstone/lime mudstone (Fig. 21.17c) with fenestral fabric, calcite pseudomorphs after gypsum or dissolution collapse breccia (supratidal). The breccia beds change laterally to gypsum and anhydrite beds in the subsurface, thus, they are related to collapse of the overlying limestone layers after dissolution of gypsum and anhydrite beds (Saxby and Lamar 1957). The peritidal facies grades upward to deeper marine facies and is interpreted here as the lowstand systems tract of a depositional sequence. This interval is correlated with the lower part of the Bayport Formation in the Michigan Basin, which includes: (1) millimeter to centimeter thick beds of interlayered quartz sandstone and dolomudstone/lime mudstone with calcite pseudomorphs after gypsum, lamination, desiccation cracks, rain drop impressions, birdseyes, microbial lamination and heterolithic stratification including wavy, flaser and lenticular bedding (Figs. 21.11 and 21.12a) recording deposition in an arid tidal flat adjacent to an aeolian sand flat; (2) dark gray bioturbated ostracod mudstone and/or microbial laminites (Fig. 21.16a) interpreted as an intertidal pond facies; and (3) sandy peloid, restricted-fauna bioclast mudstone to packstone lagoonal facies (Y. Lasemi 1986). In the St. Louis and Bayport Formations, the peritidal deposits occur in the lower part of the sequence and are here interpreted to have been deposited during lowstand and transgressive sea level rise.

21.7.5 Triassic Tidalites of Northern and Central Iran

During the Triassic, thick shallow marine carbonates were deposited under arid conditions on carbonate ramps that covered the northern Cimmerian Plate of central and northern Iran (Paleo-Tethys margin) and the northeast Gondwanan continent (Neo-Tethys margin) of southwest Iran (Fig. 21.25). The Elika Formation (up to 1000 m thick) in northern Iran is the uppermost unit of a thick platform carbonate succession related to the north-facing Paleo-Tethys passive margin that existed from Devonian through early Late Triassic times (Y. Lasemi 2001). This summary discusses the lower and middle Elika (Lower-Middle Triassic) in the central Alborz Mountains of northern Iran (Paleotethys passive margin) and the Sorkh Shale Formation (the lower Elika equivalent) in the Tabas failed rift basin of east central Iran (Fig. 21.25).

21.7.5.1 Lower Elika Member

The unconformity bounded lower Elika is up to 200 m thick and consists of thin to thick-bedded limestone with thin shale intercalations. This summary concentrates on the lower Elika of the Veresk section and the middle Elika of the type locality (Fig. 21.25, localities 2 and 3, respectively) and is adopted from Jahani (2000). The lower Elika member consists of subtidal open marine and grainstone shoal facies in the lower part changing upward to lagoonal and intertidal facies (Fig. 21.29). Peritidal facies comprise the middle and upper parts and include (1) heavily bioturbated gray bioclastic peloid lime mudstone-packstone with a restricted fauna (subtidal lagoon) (Figs. 21.3d, and 21.5a, b) that may be overlain by (2) an iron oxide-stained, 20–70 cm thick, bioclast (mainly small gastropods) grainstone (Fig. 21.13b) that may contain fenestral fabric and exposure features (beach ridge) and/or (3) very thin- to medium-bedded, laminated tan to light brown lime mudstone/dolomudstone with desiccation cracks (intertidal) or graded grainstone to lime mudstone containing bioclast/peloid and intraclast grains (intertidal channel). These facies are interlayered with numerous erosive-based, commonly graded intraclastic storm facies of various thicknesses (Fig. 21.4b–d) recording deposition under a storm dominated platform. Peritidal facies are arranged into 4th to 5th-order shallowing-upward cycles and comprise the main part of the highstand systems tract exhibiting progradational stacking pattern (Fig. 21.29).

In the northern Tabas failed rift basin (Fig. 21.25, locality 4), the lower Elika equivalent Sorkh Shale Formation consists almost entirely of tidal flat facies (Fig. 21.30) intercalated with numerous storm beds and comprise the transgressive and highstand systems tracts of a depositional sequence (Ghomashi 2008, Y. Lasemi et al. 2008). Southward, the carbonate-rich peritidal deposits grade into mixed-carbonate and quartz sandstone containing bimodal, spherical, and well-rounded sand grains with polished (frosted) surface, tabular bedding with internal laminations, herringbone cross-laminations/cross bedding, flaser and lenticular bedding. In the Eslamabad section (locality 5 in Fig. 21.25), sandstone beds are overlain and underlain, with sharp contacts, by shallowing-upward carbonate tidal flat cycles forming pure siliciclastic-pure carbonate double cycles (Fig. 21.19d). Sedimentary structures, super mature rounding and vertical association with carbonate tidal flat facies suggest deposition in tidal flat setting. The frosting and bimodal size distributions of the quartz grains suggest proximity to a desert environment (e.g. Klein 1977). The sand grains are interpreted as coming from aeolian dune sands which were transported over the carbonate tidal flat (at the time of emergence) and were subsequently reworked in the intertidal environment during the next platform flooding (Y. Lasemi et al. 2008).

21.7.5.2 Middle Elika Member

The middle member (up to 700 m thick) consists of very thin- to thick-bedded dolomite and dolomitic limestone and overlies the lower unit with a distinct interregional unconformity. In the type locality (Fig. 21.25), it is 200 m thick and consists almost exclusively of peritidal facies (Fig. 21.31) including: (1) laminated fenestral dolomudstone with desiccation cracks, tepee structures and calcite pseudomorphs after gypsum/anhydrite (supratidal facies) (Figs. 21.10a and 21.18c); (2) wavy to flat-laminated stromatolite with desiccation cracks (upper intertidal facies) (Fig. 21.6f). In some localities of the central Alborz Mountains, domal stromatolite/thrombolite bioherms related to upper subtidal/intertidal depositional settings have been recognized near the base of the member; (3) laminated peloid bioclast packstone to mudstone/dolomudstone with desiccation cracks that may display heterolithic stratification (lower intertidal facies); (4) fenestral bioclast peloid/intraclast grainstone (beach ridge facies); (5) erosive-based and laterally discontinuous layers of bioclast/peloid intraclast grainstone that grade upward to mudstone with tabular bedding and herringbone cross-bedding (intertidal channel facies) (Fig. 21.16b, c); and (6) gray peloid bioclast mudstone to packstone (lagoonal facies).

Tidal flat deposits constitute the bulk of the middle Elika and occur in the transgressive and highstand systems tract (Fig. 21.31). They are characterized by their light tan to cream color, thin to very thin bedding and common presence of micritic dolomite, desiccation cracks, tepee structures, laminations, birdseyes, gypsum/anhydrite casts/molds and collapse breccias. Short-term seaward progradation resulted in numerous shallowing-upward high frequency cycles that are superimposed on long-term third-order cycles (Fig. 21.31). In sharp contrast to lower Elika, the Middle Triassic middle Elika was deposited primarily under a fair weather condition.

21.7.6 Tidalites of the Upper Miocene Asmari Formation

The Upper Oligocene to Lower Miocene Asmari Formation (James and Wynd 1965) was deposited in the elongate Persian Gulf-Mesopotamian foreland

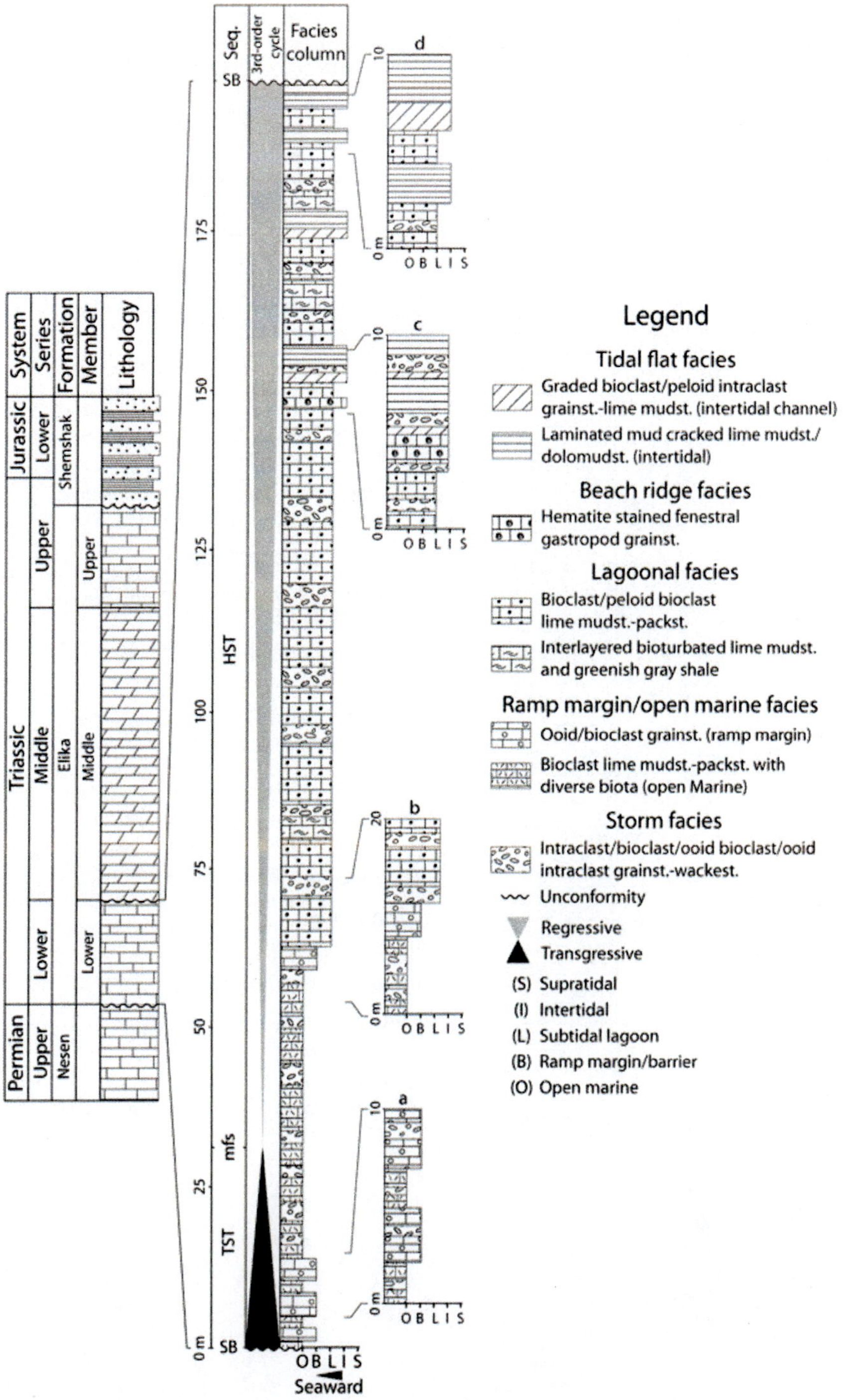

Fig. 21.29 Stratigraphic nomenclature and facies stratigraphy of the Lower Triassic lower member of the Elika Formation in the Veresk section of the Alborz Mountains (locality 2 in Fig. 21.25). Typical small-scale shallowing-upward cycles are shown to the right of facies column. Note that intertidal facies caps the small-scale cycles in the upper part of the highstand systems tract. Abundant storm facies (see the shallowing-upward cycles a through (**c**) indicate frequent storm conditions during the deposition of the lower Elika depositional sequence. Black and gray triangles represent transgression and regression, respectively. Abbreviations: *HST* highstand systems tract; *mfs* maximum flooding surface, *SB* sequence boundary, *TST* transgressive systems tract

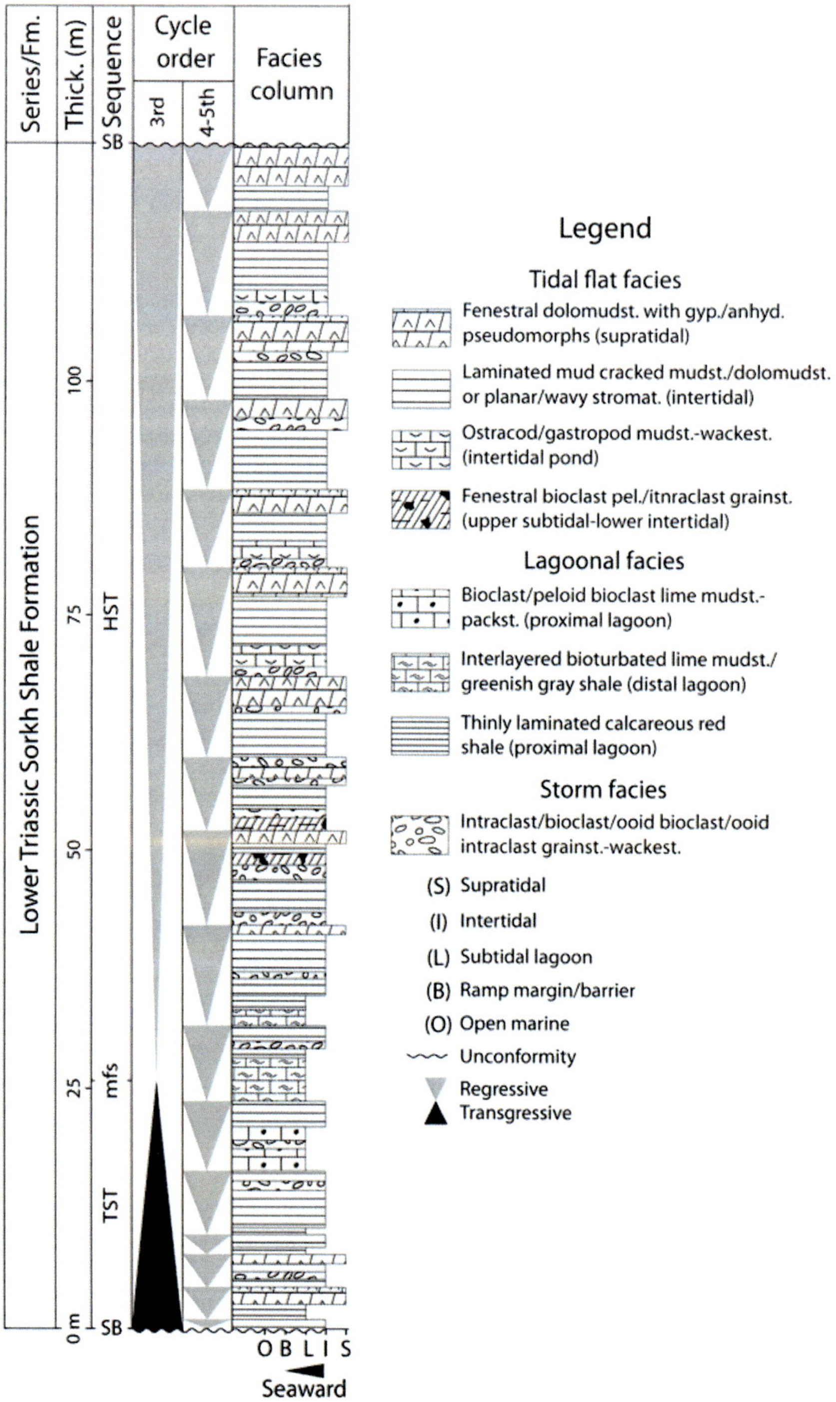

Fig. 21.30 Facies stratigraphy of the Sorkh Shale Formation in the Tabas rift basin, east central Iran (locality 4 in Fig. 21.25). Note that, except for the upper part of the transgressive systems tract, the sequence is entirely composed of tidal flat facies intercalated with storm deposits containing intraclasts, ooids and bioclasts derived from distal areas of the basin (not all storm layers are shown). Note also that the small-scale shallowing-upward cycle stacking pattern is retrogradational in the transgressive and progradational to aggradational in the highstand systems tracts, respectively (for comparison with the distally located Veresk section see Fig. 21.29). Black and gray triangles represent transgression and regression, respectively. Abbreviations: *HST* highstand systems tract, *mfs* maximum flooding surface, *SB* sequence boundary, *TST* transgressive systems tract

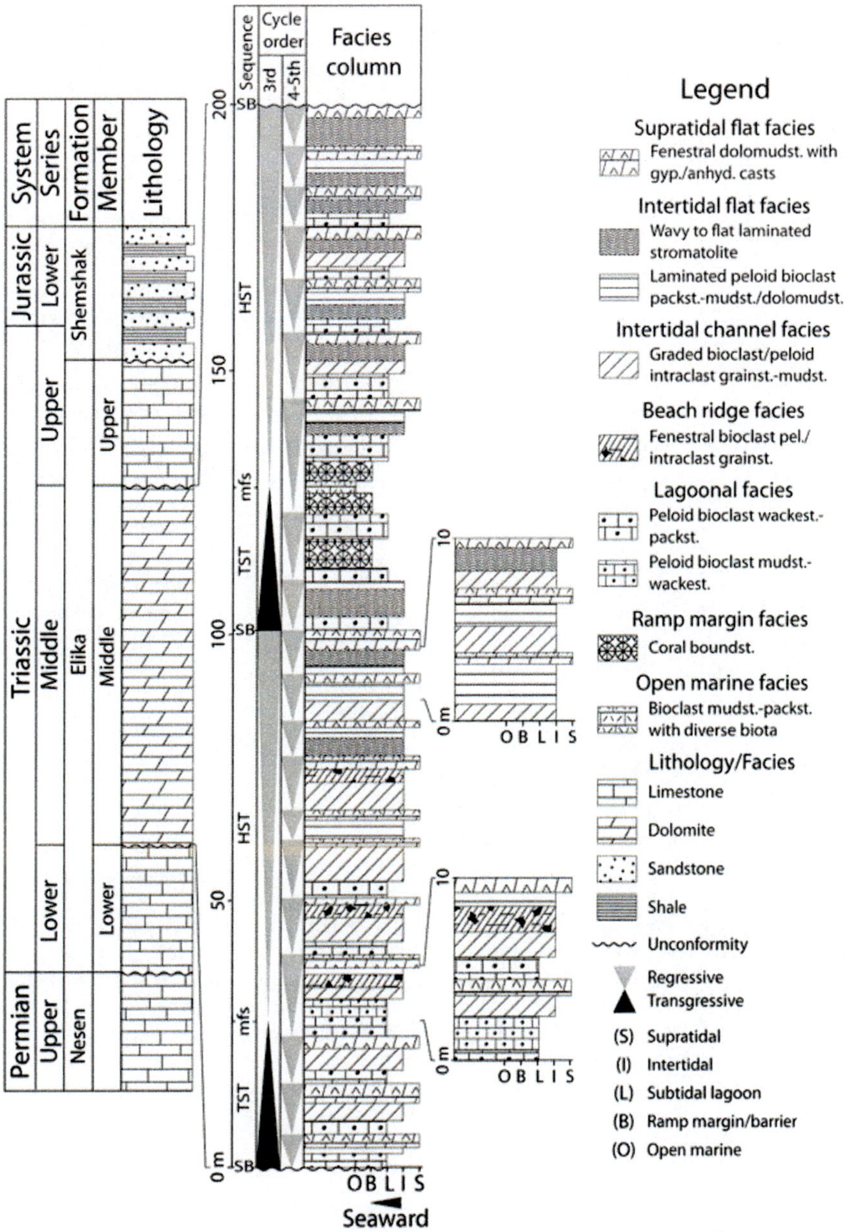

Fig. 21.31 Stratigraphic nomenclature and facies stratigraphy of the Middle Triassic middle Elika member in the type locality (locality 3 in Fig. 21.25). Note that the lower sequence and the highstand systems tract of the upper sequence are solely composed of peritidal facies that are arranged into small-scale shallowing-upward cycles composed of lagoonal- to intertidal- to supratidal or intertidal- to supratidal facies. Note also the progradational- to aggradational stacking pattern of the small scale cycles in the highstand systems tracts. Black and gray triangles represent transgression and regression, respectively. Abbreviations: HST highstand systems tract, *mfs* maximum flooding surface, *SB* sequence boundary, *TST* transgressive systems tract

basin in southwest Iran (Fig. 21.25). It is up to 500 m thick and consists mainly of shallow marine carbonates. This summary is adopted from Amin-Rasouli (2007) and describes the tidal deposits and sequences of the Lower Miocene upper member of the Asmari Formation in the Dezful Embayment, the Mish outcrop section and the Agha-Jari Well No. 61 (Fig. 21.25, localities 6 and 7, respectively).

The carbonate tidalites of the upper Asmari comprise (Fig. 21.32): (1) interlayered anhydrite and mud-

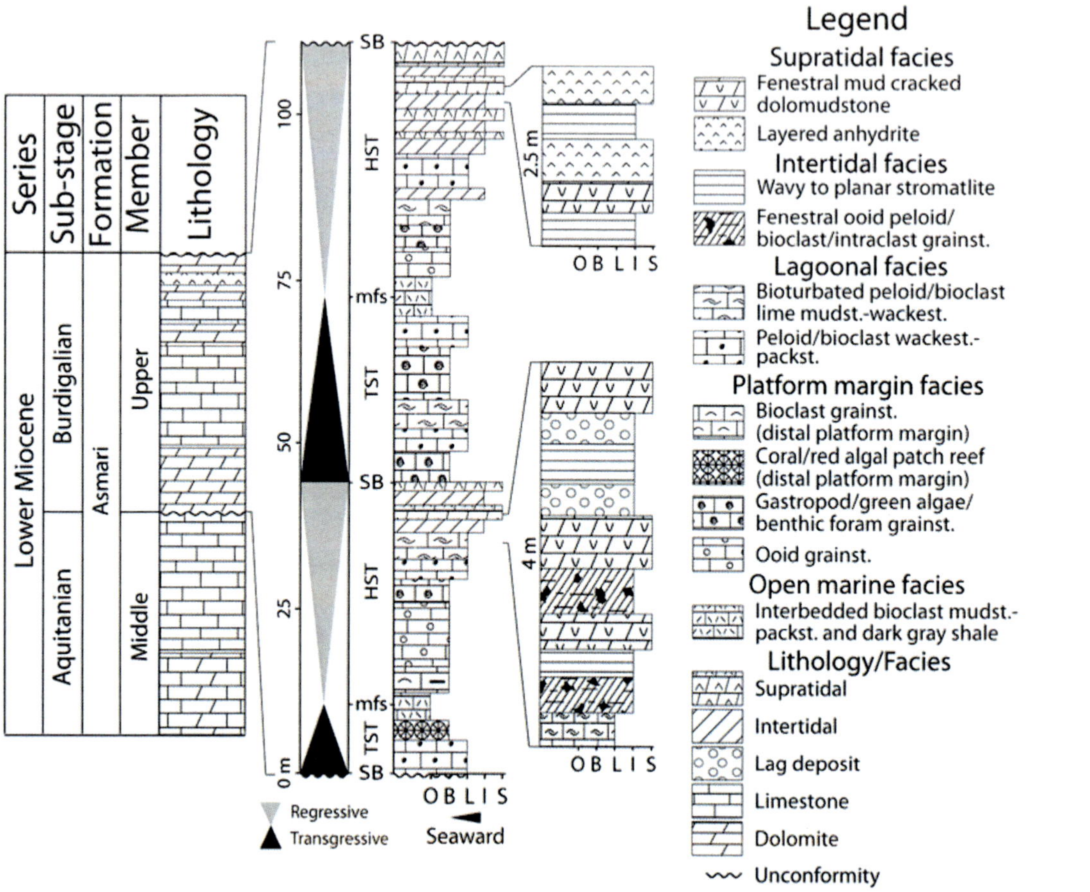

Fig. 21.32 Facies column and sequence stratigraphy of the Lower Miocene upper member of the Asmari Formation in southwest Iran (localities 6 and 7 in Fig. 21.25). The upper Asmari facies are arranged into numerous meter-scale shallowing-upward cycles superimposed on two depositional sequences consisting of transgressive and highstand systems tracts. Two typical small-scale shallowing-upward cycles are shown to the right of the facies column. Note that tidal flat deposits occur in the upper part of the highstand packages. Black and gray triangles represent transgression and regression, respectively. Abbreviations: *HST* highstand systems tract, *mfs* maximum flooding surface, *SB* sequence boundary, *TST* transgressive systems tract. Letters at the base of columns denote facies including: supratidal (*S*), intertidal (*I*), subtidal lagoon (*L*), ramp margin barrier/shoal (*B*), and open marine (*O*)

cracked dolomudstone containing fenestral fabric and/ or crystals/nodules or molds of gypsum/anhydrite (supratidal facies); (2) dolomitized wavy and flat-laminated stromatolite boundstone with desiccation cracks, birdseyes, small tepee structures and gypsum and/ or anhydrite crystals or pseudomorphs intercalated with erosive-based intraclast grainstone-packstone laminae (upper intertidal facies); (3) dolomitized fenestral ooid peloid/bioclast intraclast or fenestral bioclast grainstone with skeletal fragments of a restricted fauna (mainly miliolids) (beach ridge/lower intertidal facies); and (4) bioturbated peloid/bioclast lime mudstone to wackestone or packstone; bioclasts include ostracods, benthic foraminifera (mainly dentritinids and miliolids), gastropods and green algae (lagoonal facies) (Fig. 21.5c). These facies are arranged into meter-scale shallowing-upward cycles and display retrogradational and progradatiol stacking pattern in the transgressive and highstand systems tracts, respectively (Fig. 21.32). Tidal flat deposits are as much as 20 m thick and occur in the upper part of the highstand packages. Carbonate tidal sediments of the Asmari Formation were deposited in the ancestral Persian Gulf Foreland Basin in environments that are quite similar to the southern portion of the present day arid Persian Gulf (e.g. Purser 1973).

A quite different tidal deposit of Late Miocene age (Messinian) has been reported from the Mediterranean Basin of southeast Spain (Riding et al. 1991). The Messinian carbonate tidalites are characterized by closely packed thrombolite and stromatolite domes forming a 12 m thick bioherm surrounded by ooids. Individual domes (up to 1.5 m high and 4 m across) are typically composite and consist of distinct juxtaposed stromatolite and thrombolite parts containing fine-medium grained ooid and peloid sand. Corals, red algae, vermitid gastropods and encrusting foraminifera are present in the stromatolite, suggesting a normal marine environment during deposition. Thrombolite is more abundant than stromatolite and consists of distinct clotted fabric and irregular fenestrae. These microbialites are closely comparable with modern "giant" subtidal columnar stromatolites observed at the Exuma chain of islands on the Bahamian platform (Dill 1991; Planavsky and Ginsburg 2009).

21.8 Concluding Remarks

The geological record of ancient carbonate peritidal deposits indicates deposition in the proximal areas of a tropical sea, particularly during global relative sea-level highstands, in carbonate platforms and environments that have recurred many times since the Early Proterozoic. As shown in the illustrative examples, ancient carbonate tidalites commonly consist of stacked high-frequency 4th- to 5th-order shallowing-upward cycles and may constitutes the bulk of the transgressive and highstand systems tract of a depositional sequence. The peritidal cycles in the geological record generally are characterized by evidence of a dry climate in the capping supratidal facies. Although supratidal marshes are widespread in the Bahamas and the neighboring modern humid platforms, ancient humid tidal flat and coastal marsh deposits like the Middle Ordovician St. Paul Group (see Sect. 21.7.3) are scarce. Rapid sea level fall and prolonged exposure of the tidal flat in a humid climate could lead to karstification and removal of the highstand packages of a depositional sequence, which commonly consists of peritidal cycle as in the existing Plio-Pleistocene record of the Bahamas. Comparison of the features of modern deposits with those in ancient carbonates suggests that sedimentary processes in carbonate platforms and environments have occurred repeatedly throughout geologic time.

References

Adams JE, Frenzel HN (1950) Capitan barrier reef, Texas and New Mexico. J Geol 58:289–312

Alavi M, Vaziri H, Seyed-Emami K, Lasemi Y (1997) The Triassic and associated rocks of the Nakhlak and Aghdarbad area in central and northeastern Iran as remnants of the southern Turanian active continental margin. Geol Soc Am Bull 109:1563–1575

Amin-Rasouli H (1999) Microfacies, depositional environments and sequence stratigrapohy of Shale and Top Quarzite units of the Lalum Formation and members 1 and 2 of the Mila Formation, Eastern Alborz (in Persian with English abstract), MS thesis, Tarbiat Moallem University, Tehran, Iran

Amin-Rasouli H (2007) Sequences stratigraphy of the Asmari Formation in Folded Zagros, Southwest Iran (in Persian with English abstract). Ph.D. thesis, Tarbiat Moallem University, Tehran, Iran

Bathurst RGC (1975) Carbonate sediments and their diagenesis. Elsevier, Amsterdam, 658 p

Beach DK, Ginsburg RN (1980) Facies succession of Pliocene-Pleistocene carbonates, northwestern Great Bahama Bank. Am Assoc Petrol Geol Bull 64:1634–1642

Berberian M (1995) Master "blind" thrust faults hidden under the Zagros folds: active basement tectonics and surface morphotectonics. Tectonophysics 241:193–224

Bosence D, Procter E, Aurell M, Kahla AB, Boudagher-Fadel M, Casaglia F, Cirilli S, Mehdie M, Nieto L, Rey J, Scherreiks R, Soussi M, Waltham D (2009) A dominant tectonic signal in high-frequency, peritidal carbonate cycles? A regional analysis of Liassic platforms from western Tethys. J Sed Res 79:389–415

Brown LF Jr, Fisher WL (1977) Seismic stratigraphic interpretation of depositional systems: examples from Brazilian rift and pull apart basins. In: Payton CE (ed) Seismic stratigraphy—applications to hydrocarbon exploration, American association ofpetroleum geological Memoir 26. American Association of Petroleum Geologists, Tulsa, pp 213–248

Burgess PM (2006) The signal and the noise: forward modelling of allocyclic and autocyclic processes influencing peritidal stacking patterns. J Sed Res 76:962–977

Burne RV, Moore LS (1987) Microbialites: organosedimentary deposits of benthic microbial communities. Palaios 2:241–254

Buschbach TC, Kolata DR (1991) Regional setting of Illinois Basin. In: Leighton MW, Kolata DR, Oltz DF, Eidel JJ (eds) Interior cratonic basins, American association petroleum geological Memoir 51. American Association Petroleum Geologists, Tulsa, pp 29–55

Carozzi AV (1989) Carbonate rock depositional models. A microfacies approach. Prentice Hall, Englewood Cliffs, 604 p

Catuneanu O (2006) Principles of sequence stratigraphy. Elsevier, Amsterdam, 375 p

Catuneanu O, Abreu V, Bhattacharya JP, Blum MD, Dalrymple RW, Eriksson PG, Fielding CR, Fisher WL, Galloway WE, Gibling MR, Giles KA, Holbrook JM, Jordan R, Kendall CGStC, Macurda B, Martinsen OJ, Miall AD, Neal JE, Nummedal D, Pomar L, Posamentier HW, Pratt BR, Sarg JF, Shanley KW, Steel RJ, Strasser A, Tucker ME, Winker C (2009) Towards the standardization of sequence stratigraphy. Earth Sci Rev 92:1–33

Cowan CA, James NP (1992) Diastasis cracks; mechanically generated synaeresis-like cracks in Upper Cambrian shallow water oolite and ribbon carbonates. Sedimentology 39:1101–1118

Dalrymple RW (1992) Tidal depositional systems. In: Walker RG, James NP (eds) Facies Models response to sea level change. Geological Association of Canada, St John's, pp 195–218

Davis RA (1983) Depositional systems. Prentice Hall, New Jersey, 669 p

De Benedictis D, Bosence DWJ, Waltham DA (2007) Tectonic control of peritidal carbonate parasequence formation: an investigation using forward tectono-stratigraphic modeling. Sedimentology 54:587–605

Demicco RV (1983) Wavy and lenticular bedded carbonate ribbon rocks of the Upper Cambrian Conococheague Limestone, Central Appalachians. J Sed Petrol 52:1121

Demicco RV (1985) Platform and off-platform carbonates of the Upper Cambrian of western Maryland, U.S.A. Sedimentology 32:1–22

Demicco RV, Hardie LA (1994) Sedimentary structures and early diagenetic features of shallow marine carbonates, Atlas Series 1, 265p

Dill RF (1991) Subtidal stromatolites, ooids, and crusted-lime muds at the Great Bahama Bank margin. In: Osborne RH (ed) From shoreline to abyss. SEPM Special Publication 46:147–171

Dupraz C, Reid RP, Braissant O, Decho AW, Norman RS, Visscher PT (2009) Processes of carbonate precipitation in modern microbial mats. Earth Sci Rev 96:141–162

Enos P, Perkins RD (1977) Quaternary sedimentation in South Florida, GSA Memoir 147. Geological Society of America, Boulder, 198 p

Flügel E (2010) Microfacies of carbonate rocks, analysis, interpretation and application. Springer, Berlin, 984 p

Ghomashi M (2008) Depositional environments and sequence stratigraphy of the Sorkh Shale and Shotori Formations (Lower-Middle Triassic) in the Tabas Block (in Persian with English abstract). Ph.D. thesis, Tarbiat Moallem University, Tehran, Iran

Ginsburg RN (1971) Landward movement of carbonate mud: new model for regressive cycles in carbonates (abstract): Am Assoc Petrol Geol Bull 55: 340

Ginsburg RN (1974) Introduction to comparative sedimentology of carbonates. Am Assoc Petrol Geol Bull 58:781–786

Ginsburg RN (1975) Tidal deposits, a casebook of recent examples and fossil counterparts. Springer, New York, 428 p

Goldhammer RK, Dunn PA, Hardie LA (1987) High frequency glacio-eustatic sea level oscillations with Milankovitch characteristics recorded in Middle Triassic platform carbonates in Northern Italy. Am J Sci 287:853–892

Goldhammer RK, Lehmann PJ, Dunn PA (1993) The origin of high-frequency platform carbonate cycles and third-order sequences (Lower Ordovician El Paso Gp, West Texas): constraints from outcrop data and stratigraphic modeling. J Sed Petrol 63:318–359

Grotzinger JP (1985) Evolution of early Proterozoic passive-margin carbonate platform, Rocknest Formation, Wopmay orogen, N. W. T., Canada. Ph.D. thesis, Virginia Polytechnic Institute, Blacksburg, Virginia, USA

Grotzinger JP (1986a) Evolution of early Proterozoic passive-margin carbonate platform, Rocknest Formation, Wopmay orogen, Northwest Territories, Canada. J Sed Petrol 56:831–847

Grotzinger JP (1986b) Cyclicity and paleoenvironmental dynamics, Rocknest platform, northwest Canada. Geol Soc Am Bull 97:1208–1231

Grotzinger JP (1989) Facies and evolution of Precambrian carbonate depositional systems: emergence of the modern platform archetype. In: Crevello PD, Wilson JL, Sarg JF, Read JF (eds) Controls on carbonate platform and basin development, SEPM Special Publication 44. SEPM, Tulsa, pp 79–106

Haq BU, Hardenbol J, Vail PR (1987) Chronology of fluctuating sea levels since the Triassic (250 million years ago to present). Science 235:1156–1166

Hardie LA (1977) Sedimentation on the modern carbonate tidal flats of northwest Andros Island, Bahamas, Johns Hopkins studies in geology 22. Johns Hopkins University Press, Baltimore, 202 p

Hardie LA (1986) Carbonate tidal-flat deposition: ten basic elements. Colo Sch Min Quart 81:3–6

Hardie LA (1987) Dolomitisation: a critical view of some current views. J Sed Petrol 57:166–183

Hardie LA, Shinn EA (1986) Carbonate depositional environments, modern and ancient. Part 3, Tidal flats. Colo Sch Min Quart 81, 74 p

Hardie LA, Bosellini A, Ginsburg RN (1986) Repeated subaerial exposure of subtidal carbonate platforms, Triassic, northern Italy: evidence for high frequency sea level oscillations on a 104 year scale. Paleoceanography 1:447–457

Hardie LA, Dunn PA, Goldhammer RK (1991) Field and modelling studies of Cambrian carbonate cycles, Virginia Appalachian—Discussion. J Sed Petrol 61:636–646

Hoffman P (1975) Shoaling-upward shale-to-dolomite cycles in the Rocknest Formation (Lower Proterozoic), Northwest Territories, Canada. In: Ginsburg RM (ed) Tidal deposits. Springer, Berlin, pp 257–265

Hoffman P (1976) Stromatolite morphogenesis in Shark Bay, Western Australia. In: Walter MR (ed) Stromatolites. Elsevier, Amsterdam, pp 261–271

Hunt D, Tucker ME (1992) Stranded parasequences and the forced regressive wedge systems tract: deposition during base-level fall. Sed Geol 81:1–9

Inden RF, Moore CH (1983) Beach environment. In: Scholle PA, Bebout DG, Moore CH (eds) Carbonate depositional environments, American association petroleum geologists memoir 33. American Association Petroleum Geologists, Tulsa, pp 211–265

Jahani D (2000) Sedimentary basin analysis of the Elika Formation in central and eastern Alborz (in Persian with English abstract). Ph.D. thesis, Islamic Azad University Science and Research Branch, Tehran, Iran

James NP (1984) Shallowing-upward sequences in carbonates. In: Walker KG (ed) Facies models, 2nd edn, Geoscience

Canada Reprint Series. Geological Association of Canada, Toronto, pp 213–228

James NP (1997) The cool-water carbonate depositional realm. In: James NP, Clarke JAD (eds) Cool-water Carbonates, SEPM Special Publication 56. SEPM (Society for Sedimentary Geology), Tulsa, pp 1–20

James GA, Wynd JG (1965) Stratigraphic nomenclature of Iranian oil consortium agreement area. Am Assoc Petrol Geol Bull 49:2182–2245

Kendall CGStC, Warren J (1987) A review of the origin and setting of tepees and their associated fabrics. Sedimentology 34:1007–1027

Kerans C, Tinker SW (1997) Sequence stratigraphy and characterization of carbonate reservoirs: SEPM short course notes 40, 130 p

Klein GdeV (1971) A sedimentary model for determining paleotidal range. Geol Soc Assoc Bull 82:2585–259

Klein GdeV (1977) Clastic tidal facies. Continuing Education Publishing Co, Champaign, 149 p

Klein GdeV (1998) Clastic tidalites_A partial retrospective view. In: Alexander CR, Davis RA, Henry VJ (eds) Tidalites: processes and products, SEPM Special Publication 61. SEPM (Society for Sedimentary Geology), Tulsa, pp 5–14

Koerschner WF, Read JF (1989) Field and modeling studies of Cambrian carbonate cycles, Virginia Appalachians. J Sed Petrol 59:654–687

Kolata DR, Nelson WJ (1991) Tectonic history of the Illinois Basin. In: Leighton MW, Kolata DR, Olts DF, Eidel JJ (eds) Interior cratonic Basins, American Association of Petroleum Geological Memoir 51. American Association Petroleum Geologists, Tulsa, pp 263–85

Lasemi Y (1980) Carbonate microfacies analysis and depositional environments of the Kinkaid Formation (Upper Mississippian) of the Illinois Basin. Ph.D. thesis, University of Illinois at Urbana-Champaign, Urbana, Illinois, USA

Lasemi Y (1986) Deepening-upward tidal-flat sequence within Bayport Formation, Upper Mississippian, Michigan Basin (abs) Am Assoc Petrol Geol Bull 70: 1068

Lasemi Y (1995) Platform Carbonates of the upper Jurassic Mozduran Formation in The Kopet Dagh Basin, NE Iran-facies, palaeoenvironments and sequences. Sed Geol 99: 151–164

Lasemi Y (2001) Facies analysis, depositional environments and sequence stratigraphy of the Upper Precambrian and Paleozoic rocks of Iran (in Persian). Geological Survey of Iran Publication, Tehran, 180 p

Lasemi Y, Amin-Rasouli H (2002) Microfacies and depositional environments of the storm deposits in the lower part of Member 2 of the Mila Formation in Tuyeh-Darvar area (in Persian with English abstract). Univ Tehran J Sci 28:33–52

Lasemi Y, Carozzi AV (1981) Carbonate microfacies and depositional environments of the Kinkaid Formation (Upper Mississsippian) of the Illinois Basin, U.S.A. In: Proceedings of 8th Congress Geol Argentino, San Luis. Actas 2:357–384

Lasemi Y, Ghomashi M, Amin-Rasouli H, Kheradmand A (2008) The Lower Triassic Sorkh Shale Formation of the Tabas Block, East Central Iran: succession of a failed-rift basin at the Paleo-Tethys margin. Carbonates and Evaporites 23:21–38

Lasemi Z, Norby RD (1999) Stratigraphy, paleoenvironments, and sequence stratigraphic implications of the Middle Mississippian carbonates in western Illinois. In: Lasemi Z, Norby RD, Devera JA, Fouke BW, Leetaru HE, Denny FB (eds) Middle Mississippian Carbonates and siliciclastics in Western Illinois, Illinois State Geological Survey Guidebook 31. Illinois State Geological Survey, Champaign, pp 1–18

Lasemi Z, Boardman MR, Sandberg PA (1989) Cement origin of supratidal dolomite, Andros Island, Bahamas. J Sed Petrol 59:249–257

Lehrmann DJ, Goldhammer RK (1999) Secular variations in parasequence and facies stacking patterns of platform carbonates a guide to application of stacking patterns analysis in strata of diverse ages and settings. In: Harris PM, Saller AH, Simo JA (eds) Advances in carbonate sequence stratigraphy: application to reservoirs, outcrops, and models, SEPM, Special Publication 63. SEPM, Tulsa, pp 187–225

Lehrmann DJ, Rankey EC (1999) Do meter-scale cycles exist? A statistical evaluation from vertical (1-D) and lateral (2-D) patterns in shallow-marine carbonate–siliciclastics of the "fall in" strata of the Capitan reef, Seven Rivers Formation, Slaughter Canyon, New Mexico. In: Saller AH, Harris PM, Kirkland BL, Mazzullo SJ (eds) Geological framework of the Capitan Reef, SEPM, Special Publication 65. SEPM (Society for Sedimentary Geology), Tulsa, pp 51–62

Logan BW, Rezak R, Ginsburg RN (1964) Classification and environmental significance of algal stromatolites. J Geol 72:68–83

Logan BW, Davies GR, Read JF, Cebulski DE (1970) Carbonate sedimentation and environments, Shark Bay, Western Australia. AAPG Mem 13:223p

Lokier S, Steuber T (2008) Large-scale intertidal polygonal features of the Abu Dhabi coastline. Sedimentology 56: 609–621

Miall AD (1997) The geology of stratigraphic sequences. Springer, Berlin, 433 p

Middleton GV (1973) Johannes Walther's law of correlation of facies. Geol Soc Am Bull 84:979–988

Mitchell RW (1985) Comparative sedimentology of shelf carbonates of the Middle Ordovician St. Paul Group, Central Appalachians. Sed Geol 43:1–41

Mitchum RM Jr, Vail PR, Thompson S III (1977) Seismic stratigraphy and global changes of sea-level, part 2: the depositional sequence as a basic unit for stratigraphic analysis. In: Payton CE (ed) Seismic stratigraphy—applications to hydrocarbon exploration, American Association of Petroleum Geologists Memoir 26. American Association Petroleum Geologists, Tulsa, pp 53–62

Planavsky N, Ginsburg RN (2009) Taphonomy of modern marine Bahamian microbialites. Palaios 24:5–17

Pomar L (2001) Types of carbonate platforms: a genetic approach. Basin Res 13:313–334

Posamentier HW, Allen GP (1999) Siliciclastic sequence stratigraphy: concepts and applications. SEPM Concepts in Sedimentology and Paleontology 7, 210 p

Pratt BR (1982) Stromatolite decline – a reconsideration. Geology 10:512–515

Pratt BR (2001) Calcification of cyanobacterial filaments: *Girvanella* and the origin of lower Paleozoic lime mud. Geology 29:763–766

Pratt BR (2002) Tepees in peritidal carbonates: origin via earthquake-induced deformation, with example from the Middle Cambrian of western Canada. Sed Geol 153:57–64

Pratt BR (2010) Peritidal carbonates. In: James NP, Dalrymple RG (eds) Facies models (3rd edn) Geological Association of Canada, St. John's, (in press)

Pratt BR, James NP (1982) Cryptalgal-metazoan bioherms of early Ordovician age in the St. Georges Group, western Newfoundland. Sedimentology 29:543–569

Pratt BR, James NP (1986) The St. George Group (Lower Ordovician) of western Newfoundland: tidal flat island model for carbonate sedimentation in shallow epeiric seas. Sedimentology 33:313–343

Pratt BR, James NP, Cowan CA (1992) Peritidal carbonates. In: Walker RG, James NP (eds) Facies Models response to sea level change. Geological Association of Canada Publication, St. Johns, pp 303–322

Preto N, Hinnov LA, DE Zanche V, Mietto P, Hardie LA (2004) The Milankovitch interpretation of the Latemar Platform cycles (Dolomies Itlay): implications for geochronology, biostratigraphy and Middle Triassic carbonate accumulation. In: D'Argenio B, Fischer AG, Silva IP, Weisert H, Ferreri V (eds) Cyclostratigraphy: approaches and case histories, SEPM, Special Publication 81. SEPM (Society for Sedimentary Geology), Tulsa, pp 167–182

Purser BH (1973) The Persian Gulf- Holocene carbonate sedimentation and diagenesis in a shallow epicontinental sea. Springer, New York, 471p

Purser BH, Evans G (1973) Regional sedimentation along the Trucial Coast, SE Persian Gulf. In: Purser BH (ed) The Persian Gulf- Holocene carbonate sedimentation and diagenesis in a shallow epicontinental sea. Springer, New York, pp 211–231

Rankey EC (2002) Spatial patterns of sediment accumulation on a Holocene carbonate tidal flat, Northwest Andros Island, Bahamas. J Sed Res 72:591–601

Read JF (1985) Carbonate platform facies models. Am Assoc Petrol Geol Bull 66:860–879

Reid RP, Visscher PT, Decho AW, Stolz JF, Bebout BM, Dupraz C, MacIntyre LG, Paerl HW, Pinckney JL, Prufert-Bebout L, Steppe TF, Des Marais DJ (2000) The role of microbes in accretion, lamination and early lithification of modern marine stromatolites. Nature 406:989–992

Reineck HE (1972) Tidal flats. In: Rigby JK, Hamblin WK (eds) Recognition of ancient sedimentary environments, SEPM Special Publication 16. Society of Economic Paleontologists and Mineralogists, Tulsa, pp 146–159

Reineck HE, Singh IB (1980) Depositional sedimentary environments, 2nd edn. Springer, New York, 551p

Riding R (1999) The term stromatolite: towards an essential definition. Lethaia 32:321–330

Riding R (2000) Microbial carbonates: the geological record of calcified bacterial–algal mats and biofilms. Sedimentology 47:179–214

Riding R, Braga JC, Martin JM (1991) Oolite stromatolites and thrombolites, Miocene, Spain: analogues of recent giant Bahamian examples. Sed Geol 71:121–127

Robbins LL, Blackwelder PL (1992) Biochemical and ultrastructural evidence for the origin of whitings: a biologically induced calcium carbonate precipitation mechanism. Geology 20:464–468

Saxby DB, Lamar JE (1957) Gypsum and anhydrite in Illinois: Illinois State Geological Survey Circular 226, 26p

Schlager W (2005) Carbonate Sedimentology and Sequence Stratigraphy. SEPM Concepts in Sedimentology and Paleontology #8, 200p

Scoffin TP, Stoddard DR (1983) Beachrock and intertidal cements. In: Goudie AS, Pye K (eds) Chemical sediments and geomorphology. Academic Press, London, pp 401–425

Shinn EA (1983a) Tidal flat environment. In: Scholle PA, Bebout D, Moore CH (eds) Carbonate depositional environments, American Association Petroleum Geologists Memoir 33. American Association Petroleum Geologists, Tulsa, pp 172–210

Shinn EA (1983b) Birdseyes, fenestrae, shrinkage pores and loferites: a re-evaluation. J Sed Petrol 53:619–628

Shinn EA (1986) Modern carbonate tidal flats: their diagnostic features. Colo Sch Min Quart 81:7–35

Shinn EA, Steinen RP, Lidz BH, Swart PK (1989) Whitings, a sedimentological dilemma. J Sed Petrol 59:147–161

Spence GH, Tucker ME (2007) A proposed integrated multi-signature model for peritidal cycles in carbonates. J Sed Res 77:797–808

Spencer RJ, Demicco RV (1993) Depositional environments of the Middle Cambrian Arctomys Formation, southern Canadian Rocky Mountains. Bull Can Petrol Geol 41:373–388

Stocklin J, Ruttner A, Nabavi M (1964) New data on the lower Paleozoic and Precambrian of north Iran. Geological Survey of Iran, report 1, 29 p

Strasser A, Pittet B, Hillgartner H, Pasquier JB (1999) Depositional sequences in shallow carbonate dominated sedimentary systems: concepts for a high-resolution analysis. Sed Geol 128:201–221

Swann DH (1963) Classification of the Genevievan and Chesterian (Late Mississippian) rocks of Illinois. Illinois State Geological Survey RI 216, 91 p

Tucker ME, Wright VP (1990) Carbonate sedimentology. Blackwell, Oxford, 482p

Van Wagoner JC, Posamentier HW, Mitchum RM, Vail PR, Sarg JF, Loutit TS, Hardenbol J (1988) An overview of the fundamentals of sequence stratigraphy and key definitions. In: Wilgus CK, Hastings BS, Kendall CG St C, Posamentier HW, Ross CA, Van Wagoner JC (eds) Sea-level changes, an integrated approach, SEPM Special Publication 42. Society of Economic Paleontologists, Mineralogists, Tulsa, pp 39–45

Van Wagoner JC, Mitchum RM Jr, Campion KM, Rahmanian VD (1990) Siliciclastic sequence stratigraphy in well logs, core, and outcrops: concepts for high-resolution correlation of time and facies, AAPG Methods in Exploration Series 7. American Association of Petroleum Geologists, Tulsa, 55 p

Wignall PH, Twitchett RJ (1999) Unusual intraclastic limestones in the Lower Triassic carbonates and their bearing on the aftermath of the end-Permian mass extinction. Sedimentology 46:303–316

Wilkinson BH, Diedrich NW, Drummond CN (1996) Facies succession in peritidal carbonate sequences. J Sed Res 66:1065–1078

Wilkinson BH, Drummond CN, Rothman ED, Diedrich NW (1997) Stratal order in peritidal carbonate sequences. J Sed Res 67:1068–1082
Wilkinson BH, Drummond CN, Diedrich NW, Rothman ED (1999) Poisson processes of carbonate accumulation on Paleozoic and Holocene platforms. J Sed Res 69:338–350
Wilson JL (1975) Carbonate acies in geologic history. Springer, Berlin, 471 p
Wright VP (1984) Peritidal carbonate facies models: a review. Geol J 19:309–325
Zecchin M (2010) Discussion – pomar Towards the standardization of sequence stratigraphy: is the parasequence concept to be redefined or abandoned? Earth Sci Rev 102:117–119

Index

R.A. Davis, Jr. and R.W. Dalrymple (eds.), *Principles of Tidal Sedimentology*,
DOI 10.1007/978-94-007-0123-6, © Springer Science+Business Media B.V. 2012

Q

R

S

T